W9-CHK-387

# GENETICS
## Second Edition

Charlotte J. Avers
*Rutgers University*

Willard Grant Press Boston

**PWS PUBLISHERS**

Prindle, Weber & Schmidt • Willard Grant Press • Duxbury Press •
Statler Office Building • 20 Park Plaza • Boston, Massachusetts 02116 • 617-482-2344



Cover Photo: © Gopal Murti / Science Source / Photo Researchers, Inc.

*Genetics, 2nd edition* was prepared for publication by the following people:
Production Editor: Robine Storm van Leeuwen
Cover Designer: Trisha Hanlon
Typesetting by Jonathan Peck, Typographers, Ltd.; covers printed by Lehigh Press Lithographers; text printed and bound by Halliday Lithograph

Library of Congress Cataloging in Publication Data

Avers, Charlotte J.
Genetics.
Bibliography: p.
Includes index.
1. Genetics. I. Title.
QH430.A89 1984 575.1 83-20842

ISBN 0-87150-779-X

Printed in the United States of America

88 87 86 85 84 - 1 2 3 4 5 6 7 8 9 10

to Maurice

*a friend in need is a friend indeed*

# PREFACE

In this second edition as in the first, I have tried to provide a balanced, comprehensive review of genetics in a book of reasonable length. The increasingly rapid pace of progress in genetics research makes the subject more exciting with each passing year, but it also presents a greater challenge to authors, instructors, and students to assimilate new information that unabatedly pours forth in journals, symposia, and monographs. The use of the historical perspective throughout *Genetics, second edition* acknowledges our debt to the pioneering geneticists and biochemists who paved the way in this field with their new experimental methods for probing gene structure, function, and regulation.

Many of the chapters in this edition were substantially rewritten to clarify and update information. Some topics were rearranged to comply with the suggestions of users of the earlier edition—for example, an introduction to probability and statistics is now offered in Chapter 2—and to enhance the logical flow of topics in accordance with newer areas of genetics research.

An expanded set of problems and questions (with answers provided in the back of the book) concludes each chapter and offers students ample opportunity to practice and test the concepts presented in that chapter. A list of pertinent readings and references follows each set of questions and problems. New to this edition is a glossary of almost six hundred terms, which provides definitions for virtually every boldfaced term in the text.

I wish to extend my appreciation to the reviewers of the first edition for their helpful suggestions and comments: Dave Axelrod, who helped me to sharpen my focus time and again as the project took shape; Sally Allen, University of Michigan at Ann Arbor; William Birky, Ohio State University; Philip Hedrick, University of Kansas; William Moore, Wayne State University; Simon Silver, Washington University; Herbert Wiesmeyer, Vanderbilt University.

I also very much appreciate the assistance of the reviewers of the second edition: Audrey Barnett, University of Maryland; William Birky, Ohio State University; Joseph Chimici, Virginia Commonwealth University; Robert Fowler, San Jose State University; Carl Heuther, University of Cincinnati; James Wild, Texas A & M University; and the many professors who kindly responded to a prerevision survey. In particular, I wish to thank Maurice Rosenstraus and David Fox for their patience and their meticulous attention to every part of the manuscript, by which it was substantially improved. I am grateful to the literary executor of the late Sir Ronald A. Fisher, F.R.S., to Dr. Frank Yates, F.R.S., and to Longman Group, Ltd., London, for their permission to reprint Table II from Statistical Tables for Biological Agricultural and Medical Research by Fisher and Yates.

**Charlotte J. Avers**

# CONTENTS

# CHAPTER 1

## Mendelian Inheritance

The science of genetics is young, relative to physics or chemistry. In 1900 three European botanists independently discovered and cited the earlier studies of Gregor Mendel to help explain their own studies of inheritance. For more than eight years, Mendel, an Austrian monk at the monastery of St. Thomas in Brünn, Czechoslovakia, conducted hybridization experiments using the garden pea (Box 1.1). In 1865 Mendel reported his experimental results and conclusions to the Brünn Natural History Society, and the report was published in the journal of the Society in 1866. Until 1900 Mendel's publication was not fully appreciated because his proposed rules of inheritance could not be fit into the existing biological framework. In 1900 other discoveries facilitated a widespread understanding of Mendel's work and the confusing nature of the inheritance of biological characteristics began to reveal its mysteries to twentieth-century geneticists. Genetic studies have continued without interruption since 1900, and Mendelian rules of inheritance still provide the basis for interpreting heredity in sexually reproducing species.

## Mendelian Rules of Inheritance

Before Mendel's time scientists had a general recognition and appreciation of inheritance based on hundreds of years of hybridization studies. These studies provided the basis for selective breeding programs for crops and ornamental plants and for domestic animals such as sheep and cattle. Indeed, the phenomenon of **inheritance,** that is, the transmission of characteristics from generation to generation in a family, has been recognized for thousands of years. For example, the Jewish Talmud has laws negating the requirement of circumcision for sons of women whose family members had disorders that lead to uncontrolled bleeding. In this and other cases, a practical approach to some particular problem or need often existed but little *generalization* was made to allow an understanding of inherited traits or to provide a means of predicting their occurrence and pattern of transmission. Each trait was studied and treated independently of other inherited characteristics, either in the same species or in different species of plants and animals. As a result of his experiments, Mendel was the first to formulate rules of inheritance that were generally applicable to any trait or species in question. His conclusions have been shown to apply to all organisms, including human beings.

Mendel's methods and interpretations can be appreciated even more today in view of our current knowledge of the cell nucleus, DNA, and chromosomes that was unknown to Mendel and his contemporaries. His methods of analysis are as useful and applicable today as they were over 100 years ago. Mendel chose his experimental organism carefully and used a systematic approach to design and conduct his experiments. He was an innovator in these regards, and his approach was fundamental to the success of his experiments. In particular, Mendel

***1.*** concentrated on one inherited characteristic at a time in an experiment;

***2.*** selected single traits that showed clearly different, alternative forms;

***3.*** kept accurate records for each experimental plant;

## BOX 1.1 Gregor Johann Mendel

Gregor Johann Mendel, 1822–1884.

Johann Mendel was born in the Silesian town of Heinzendorf, which was part of Austria before 1918 and is now part of Czechoslovakia. He was the only son of a poor farmer and was interested in science from boyhood on. When he became a monk of the Augustinian order, he took the name of Gregor. Gregor Mendel, who taught high school classes for boys in his earlier years and later rose to the rank of prelate and abbot of his monastery in Brünn, conducted experiments in plant hybridization for many years in a small garden on the grounds of the monastery.

Mendel's reports on his experiments to the Brünn Society of Natural Science (which he helped to establish) in February and March of 1865 went entirely unappreciated. According to the recollections of some who attended those meetings, no discussion took place and no questions were asked after the presentations. Scientists celebrate 1900, the year Mendel was discovered, as the beginning of the science of genetics, rather than 1865, the year Mendel reported his discoveries on the mechanism of heredity. Mendel died long before his monumental contributions were recognized and appreciated by the scientific community.

His contemporaries in church and government considered Mendel a nonconformist because he was actively associated with liberal political and human causes. From all we know about him, he appeared to be dearly loved by his students, parishioners, and friends as a kind and generous man who was always willing to help others.

Like Charles Darwin, whose ideas and work he admired and accepted, Mendel proposed a scientific theory in place of mystical forces then believed to direct the natural world and its inhabitants. The mechanism proposed by Mendel to explain heredity and variation was later shown to fit beautifully into the concept of evolution by natural selection proposed by Darwin in 1869. Both Mendel and Darwin profoundly changed our ideas about, our natural place in, and our relationship to the world. In place of mysteries governed by magical forces, these two giants of nineteenth century science provided the theoretical foundations and explanations for natural mechanisms that govern change in living systems.

***4.*** counted the different kinds of individuals produced in each experimental cross and could therefore quantitate information precisely; and

***5.*** kept track of the pedigree, or history of transmission, of each trait in each set of plants over a number of generations.

Systematic attention to details and overall planning of sets of experiments enabled Mendel to formulate rules of inheritance. In view of the historical importance of Mendel's studies and their applicability to modern studies of inheritance, an examination of them in some detail will be instructive.

## BOX 1.2 Garden Pea Flower Structure and Hybridization Method

The garden pea is normally self-fertilizing, that is, pollen and eggs from the same flower engage in fertilization that leads to a new generation of plants. The pollen-bearing **stamen** (male reproductive structure) and the ovule-bearing **ovary** (female reproductive structure) are enclosed within a floral part called the **keel.** Pollen from another flower cannot reach the receptive female structures because of the keel enclosure, but pollen from the same flower has complete access to the ovary of that flower. When left undisturbed, self-fertilization takes place in each flower of the plant.

In order to cross a plant of one type with that of another, the keel must be removed, the stamens must be cut off, but the ovary left untouched. Pollen from another flower can then be applied to the female component of the experimental flower. The procedure is performed reciprocally for reciprocal crosses, as the diagram indicates. After fertilization, the ovary enlarges as the ovules, which contain the fertilized eggs, enlarge and develop. The ovary develops into the fruit, or pod, and its ovules develop into seeds (peas, in this case). Each seed contains an embryo that, upon germination and growth, develops into the plant of the next generation. The plants of this next generation can then be analyzed to determine their phenotypic features.

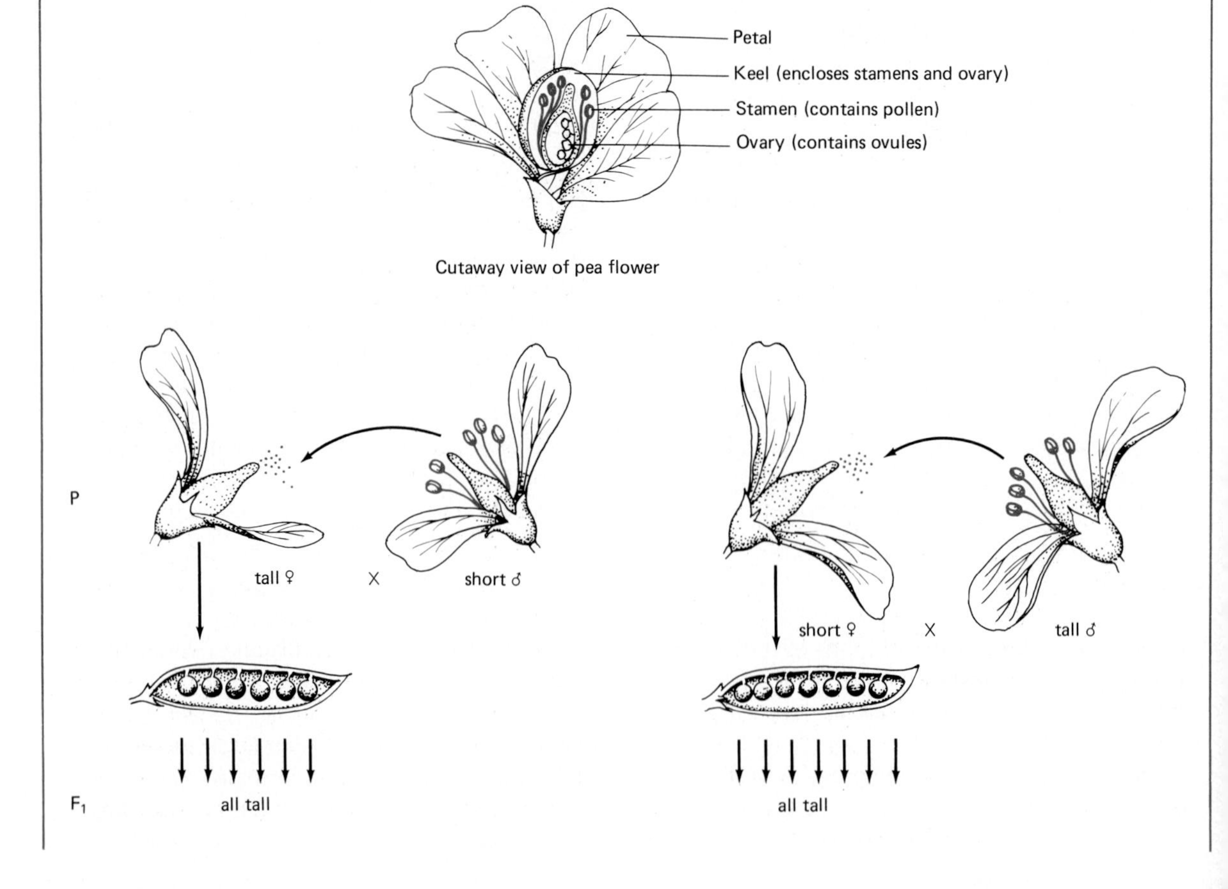

In a typical experiment, plants of different strains or phenotypes are cross-fertilized, and the seeds are collected when ripe. The seeds grow into the $F_1$ generation of plants. In order to proceed to the $F_2$ generation, $F_1$ plants are allowed to self-fertilize by natural means, that is, the flowers are left intact. The seeds that are produced in pods of the $F_1$ plants are then planted and grow into the $F_2$ individuals, which, in turn, are analyzed.

## 1.1 Systematic Approach

Mendel collected 14 varieties of the garden pea (*Pisum sativum*) that differed in seven distinct traits, or physical characteristics. Each trait was represented by paired contrasting varieties, and each member of a pair showed one of the two possible contrasting forms of the selected characteristic. For example, a *tall* variety and a *short* variety represented contrasting forms of the characteristic of *height.* Hence each pair of varieties in the final collection was part of a carefully selected experimental group, with easily distinguished characteristics occurring in unambiguously contrasting forms. For example, there was no difficulty in distinguishing tall plants that were 6 to 7 feet high from short plants less than 2 feet in height, or green seeds from yellow seeds.

Using peas as the experimental organism had several advantages. Pea plants, which are easily raised in cultivation, are normally *self-fertilizing* (or inbred); that is, before the flower opens each set of female (♀) sex cells, or eggs, is fertilized by the male (♂) sex cells, or pollen, from the same flower. Many other species are normally *cross-fertilizing;* that is, the eggs in one flower are fertilized by pollen from the flowers of another plant.

Before initiating the main set of experiments, Mendel spent two years testing the purity of each of the 14 varieties to verify that the desired characteristic appeared in each of the plants and in each generation of breeding and was therefore an inherited trait. Determining whether each variety was pure-breeding simply required planting seeds and examining the features of the seeds and plants of the subsequent generation in each of the two years.

Mendel initiated the hybridization experiments by performing artificial cross-fertilization, removing the pollen-bearing anthers of the flower and dusting the female floral structures with pollen taken from a different plant (Box 1.2). He then covered each experimental flower or group of flowers to prevent accidental contamination by foreign pollen. This practice is still common in modern hybridization procedures for many plants.

Since Mendel was studying seven pairs of contrasting inherited characteristics, he performed seven separate experiments in the first year. In each experiment he made *reciprocal crosses* between the contrasting strains of the paired varieties. For example, he used pollen from the short variety to fertilize the eggs of the tall variety (symbolized as tall ♀ × short ♂) and pollen from tall plants to fertilize the eggs of the short plants (short ♀ × tall ♂). Since the original parents of the experiment are involved, these crosses can be designated as the parental or P crosses, or as the **P generation** (Table 1.1, see next page).

The progeny resulting from P crosses comprise the first filial generation, or **$F_1$ generation.** In peas we may examine the seeds of P plants and score their characteristics, or we may plant these seeds and inspect the resulting adult plants and their flowers for the characteristic under study. Both the seeds and the plants grown from these seeds are members of the $F_1$ generation. Since Mendel obtained similar results for all seven sets of reciprocal crosses, the results of crosses between tall and short plants

***Table 1.1*** Seven traits studied in the garden pea by Mendel and results of reciprocal crosses between strains with contrasting expressions for each trait.

| **trait** | **crosses between plants of alternative characteristics** | **appearance of $F_1$ progeny plants** |
|---|---|---|
| height | tall × short | tall |
| seed shape | round × wrinkled | round |
| cotyledon color | yellow × green | yellow |
| seed coat color | gray × white | gray |
| pod shape | inflated × constricted | inflated |
| pod color | green × yellow | green |
| position of flowers | axial × terminal | axial |

will serve to illustrate this general observation. All $F_1$ plants were tall regardless of whether the cross was tall ♀ × short ♂ or short ♀ × tall ♂. There were no short plants among the numerous $F_1$ progeny; the characteristic had seemingly disappeared.

In order to analyze the nature of the tall $F_1$ plants, Mendel permitted them to self-fertilize and produce the second filial or **$F_2$ generation.** He then collected seeds of each self-fertilized $F_1$ plant and planted them to obtain adult plants of the $F_2$ generation. When these were inspected, Mendel counted 787 tall and 277 short, a ratio of approximately 3 tall:1 short in the total group of 1064 individual plants. Each self-fertilized $F_1$ plant had produced $F_2$ plants that exhibited a ratio of 3 tall plants for every short plant, regardless of the absolute number of individuals counted. The tall and short $F_2$ plants were indistinguishable from their tall and short grandparents (P generation), $F_2$ talls being about 6 to 7 feet high and shorts about 2 feet high. The short alternative of the characteristic of plant height had reappeared unchanged from the original parental generation.

## 1.2 Analysis of Monohybrid Crosses

Each of Mendel's seven sets of experiments focused on a single inherited characteristic. Each was therefore a **monohybrid cross** involving transmission of a pair of contrasting expressions of one characteristic. The progeny ($F_1$) of each parental cross resembled only one of the two parental forms. When the $F_1$ plants were allowed to self-fertilize, both parental varietal forms were represented among their progeny ($F_2$). The parental variety that appeared in the $F_1$ was represented in ¾ of the $F_2$ plants. The other variety was represented in ¼ of the $F_2$ plants. The two parental varieties were expressed in the $F_2$ generation in a ratio of 3:1 (Table 1.2).

Based on observations of monohybrid ratios, Mendel hypothesized that the $F_1$ plants contained hereditary factors from both parents, even though the plants resembled only one of the original parents. In the case of tall and short alternatives for height, the tall $F_1$ plants must have possessed a hereditary factor for tall since ¾ of the $F_2$ offspring were tall. In addition, each $F_1$ plant must also have carried a hereditary factor for short because ¼ of their $F_2$ offspring were short. The characteristic that was expressed in the $F_1$ generation Mendel called **dominant,** and the characteristic that was hidden, or masked, in the $F_1$ he called **recessive.** We say tall is dominant over short. We call an individual a dominant or a recessive.

It was also apparent to Mendel that all the tall plants were not genetically alike. In two years of preliminary tests, self-fertilized tall P plants had produced only tall progeny, thereby showing that the tall variety was pure-breeding. Self-fertilized tall hybrid $F_1$ plants, on the other hand, produced both tall and short $F_2$ progeny. We must, therefore, make a distinction between the appearance of an individual, or its **phenotype,** and its underlying genetic constitution, or **genotype.**

Mendel tested his interpretations by conducting further experiments in which $F_2$ plants were allowed to self-fertilize, thereby producing an $F_3$ generation. Repeating such

***Table 1.2*** Data from some of Mendel's monohybrid crosses showing numbers and ratios of $F_2$ progeny types after self-fertilizations of $F_1$ plants.

| **appearance of $F_1$ plants** | **number of $F_2$ plants observed** | **$F_2$ ratios calculated** |
|---|---|---|
| round seeds | 5474 round : 1850 wrinkled | 2.96 : 1 |
| yellow cotyledons | 6022 yellow : 2001 green | 3.01 : 1 |
| gray seed coats | 705 gray : 224 white | 3.15 : 1 |
| inflated pods | 882 inflated : 299 constricted | 2.95 : 1 |
| green pods | 428 green : 152 yellow | 2.85 : 1 |
| axial flowers | 651 axial : 207 terminal | 3.14 : 1 |
| tall plants | 787 tall : 277 short | 2.84 : 1 |

self-fertilizations through six generations of progeny, he found tall and short plants in each generation of progeny; both tall and short plants were typically parental in phenotype.

When Mendel observed the breeding behavior of *each* individual plant, however, it was clear that $F_2$, $F_3$, $F_4$, and $F_5$ progeny differed from the $F_1$ in the transmission pattern of the inherited characteristic (Fig. 1.1, p. 8). All the hybrid $F_1$ plants were tall, and each self-fertilized $F_1$ plant produced both tall and short $F_2$ descendants in a ratio of 3:1. The $F_2$ generation, therefore, included both parental phenotypes, tall and short, and not just the dominant tall phenotype expressed in the $F_1$ generation. The recessive short $F_2$ plants produced only short $F_3$ progeny by self-fertilization and were, therefore, true-breeding like the original short P plants. In contrast, self-fertilized tall $F_2$ plants either were true-breeding or produced both short and tall plants. From the actual counts of the two kinds of self-fertilized tall $F_2$ plants, Mendel found that ⅓ were true-breeding for the dominant phenotype and ⅔ *segregated* 3 tall : 1 short $F_3$ descendants. He found the same pattern in each of the subsequent self-fertilized generations, namely, all the recessives bred true, ⅓ of the dominant type bred true, and ⅔ of the dominant phenotype segregated 3 dominant : 1 recessive in the next generation. The $F_2$ genotypic ratio, therefore, was ¼ pure-breeding dominant : ½ segregating dominant : ¼ pure-breeding recessive.

These experimental results verified the dominance and recessiveness of the hereditary factors. Since ⅔ of the tall plants continued to segregate both tall and short progeny, they must have carried the masked recessive factor as well as the expressed dominant factor. The plants true-breeding for the dominant or recessive phenotype must have carried only the dominant or only the recessive factor and, therefore, produced progeny that were identical to each other and to the true-breeding parental plants. Only by examining the breeding behavior of an individual, can we determine whether the individual carries one kind or both kinds of alternative hereditary factors of a pair.

It is easy to see that the segregating tall individuals must have carried two hereditary factors for height—one for tall and one for short—but how can we argue for the presence of two factors in an individual true-breeding for the dominant or recessive phenotype? The answer lies in comparing the contributions of hereditary factors that were made by parents to gametes in a reciprocal cross of the P generation. No matter which parent contributed each factor, all $F_1$ hybrids received one of each factor so there could not have been more factors contributed by one parent than by the other. Since

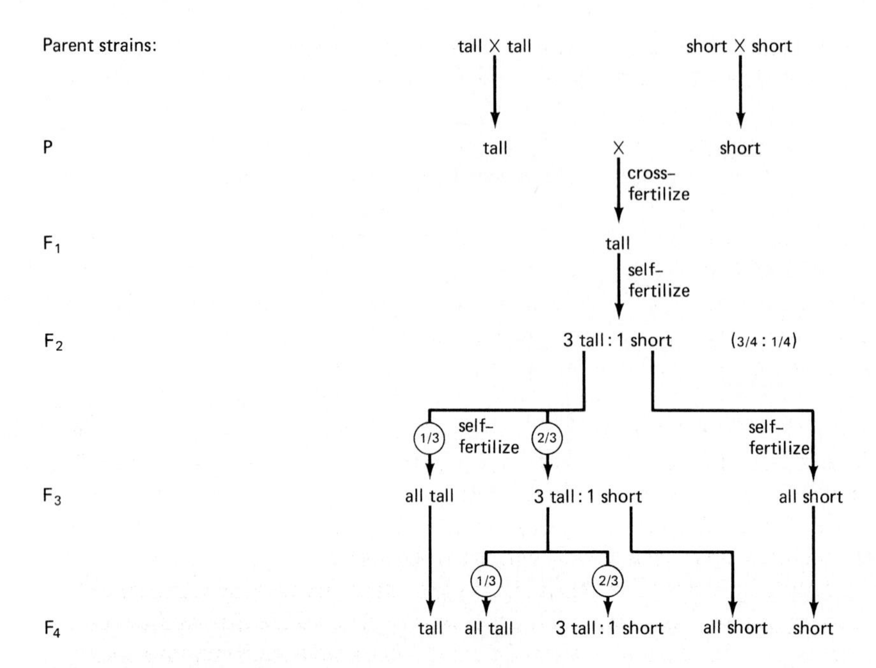


***Figure 1.1*** Breeding analysis of tall and short plants reveals dominance and recessiveness of the pair of alternative traits and distinguishes true-breeding from segregating individuals. By allowing self-fertilization of each plant, Mendel found that each $F_1$ tall plant segregated 3 tall:1 short in the $F_2$ progeny. In contrast, self-fertilized tall plants of the $F_2$ and later generations were either true-breeding (⅓ of the individuals) or segregating (⅔ of the individuals). All the short plants, however, were true-breeding. Segregating tall plants must therefore carry both the masked recessive hereditary factor and the expressed dominant hereditary factor for height. True-breeding talls or shorts carried only the dominant or recessive hereditary factor, respectively.

each $F_1$ hybrid contained two hereditary factors for height, one derived from each parent, it is likely that any individual arising from the union of two gametes must possess two hereditary factors for a particular characteristic. It follows that since the tall parent produced only tall offspring by self-fertilization, it must have had two factors for tall. Similarly, the short parent, which produced only short offspring by selfing, must have possessed two hereditary units for shortness. If each true-breeding tall or short parent carried two factors for the same alternative expression of height, then it follows that any true-breeding tall or short plant will also have a pair of identical hereditary factors for height. This is true for any true-breeding trait and for any filial generation.

At this stage we should introduce some more specific terms for genetic analysis. Mendel's hereditary factors, or units of inheritance, are **genes; alleles** are the alternative forms of a gene. The terms *dominant* and *recessive* can refer to these allelic forms or to the phenotypes that these alleles produce. The alternative forms of the gene for height are the dominant allele for tall and the recessive allele for short stature. In our example, tall is the dominant phenotype for height, and short is its recessive alternative.

Individuals having two identical alleles of a gene are **homozygous** for the allele under consideration, and such individuals are called **homozygotes.** The tall parent was homozygous for the dominant allele for height, and the short parent was homozygous for the recessive allele for height. We can describe true-breeding tall plants as homozygous dominants and true-breeding short plants as homozygous recessives. They are homozygous for the dominant and homozygous for the recessive allele, respectively. It is helpful as well as conventional to use italicized alphabetical symbols to represent alleles of a gene for any inherited characteristic. For peas, the symbols of the gene controlling height are *T* to represent the dominant allele and *t* for its recessive alternative. The dominant allele is indicated by a capital letter; and the recessive, by the same letter in lower case. We can now refer to homozygous tall plants as having the genotype *TT,* and the homozygous recessive plants as being genotypically *tt.*

Hybrid individuals, such as tall $F_1$ plants, are said to be **heterozygous** for the alleles in question, or to be **heterozygotes.** Their genotype is *Tt* because they possess two different alleles for the height gene. Tall heterozygotes, whether from the $F_1$ or any other filial generation, are phenotypically identical to any of the true-breeding, or homozygous, tall plants from the P or any other generation. We can distinguish homozygous tall plants (*TT*) from heterozygous tall plants (*Tt*) only by conducting breeding tests or additional crosses. Such crosses can reveal the genotype of a tall individual according to the phenotype of its progeny. By self-fertilization, a tall plant of the *TT* genotype will produce only tall offspring, that is, it will be true-breeding. Tall plants of the *Tt* genotype, however, will produce progeny that segregate into tall and short plants.

We can illustrate these features of inheritance by comparing the phenotypes that Mendel observed in breeding experiments involving tall and short plants with the genetic interpretations of these phenotypes (Fig. 1.2). Mendel's interpretations, or his deductions

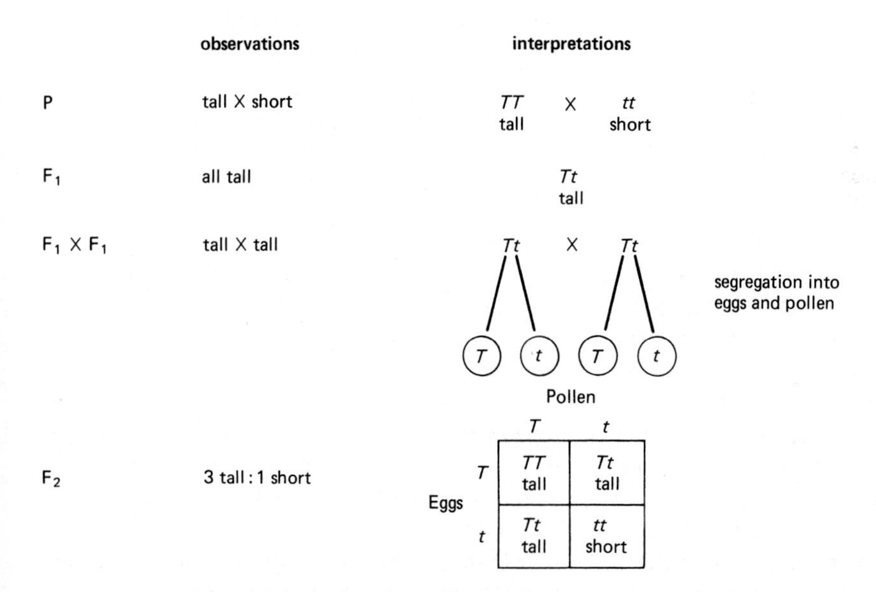


***Figure 1.2*** Observations of parental and progeny phenotypes and the ratio of phenotypes produced are interpreted in genotypic terms. $F_1$ tall plants must be heterozygous since their $F_2$ progeny segregate into talls and shorts. By self-fertilization each $F_2$ plant can be assigned a specific genotype: *TT* homozygous dominants will breed true for tall, *Tt* heterozygotes will segregate 3 tall : 1 short in a progeny, and homozygous recessive *tt* will breed true for short height.

based on observations, included a number of conceptually important points:

***1.*** Inheritance of a particular characteristic is based on unchanging unit factors transmitted from parents to progeny and not on a "blending" of parental characteristics since the phenotypes of the offspring are identical to and not intermediate between the two parental phenotypes. We now call these factors alleles.

***2.*** Each individual possesses a pair of alleles governing a particular characteristic, such as height, seed color, or some other physical feature. The alleles may be identical or different.

***3.*** Each member of a pair of alleles is sorted out into an individual gamete during sexual reproduction.

***4.*** Fertilization results in new combinations of the pair of alleles in the offspring.

***5.*** Each member of a pair of alleles does not change or vanish during its transmission from one generation to the next.

Mendel also pointed out that the phenotypic and deduced genotypic ratios in monohybrid inheritance required the production of equal proportions of gametes of each allelic type by each parent. A heterozygous tall plant would produce sex cells carrying either the *T* allele or its *t* alternative, and on the average it would produce 50% *T* gametes and 50% *t* gametes. Stated in terms of probability, we would say that the chance for *T* segregating in a gamete is ½, and the chance for its alternative, *t*, is also ½. In a large enough sample of gametes, therefore, 50% would carry one of the alleles and 50% would carry the alternative member of the *Tt* pair.

If the chance is ½ (or 50%) that a gamete from a *Tt* heterozygote will carry the *T* allele, then the chance is also ½ that any gamete taken at random from the population will be *T*. The probability that one gamete carrying *T* will be fertilized by another gamete also carrying *T* is ½ × ½, or ¼. The production of each gamete in sexual reproduction is an independent event that is not influenced by other gametes produced by the same or by a different individual. Once we know the probability of one event occurring by chance alone, we can predict the probability of two such events occurring at the same time. The probability of two independent events occurring together *at random* is the *product of their separate probabilities.* Any combination of a pair of alleles can be analyzed as the probability of two independent events coming together. In the case of a *Tt* heterozygote, we know that the chance of one gamete carrying *T* is ½. The probability of two *T* gametes producing an individual of genotype *TT* is the product of ½ × ½, or ¼. In the case of one pair of alleles, the $F_2$ phenotypes will be produced in a ratio that is the outcome of chance encounters between pairs of gametes; these gametes are of two allelic types and they are produced in equal proportions during sexual reproduction.

For the cross between *Tt* × *Tt*, or in the self-fertilization of *Tt* individuals, we explain the 1:2:1 genotypic ratio as follows:

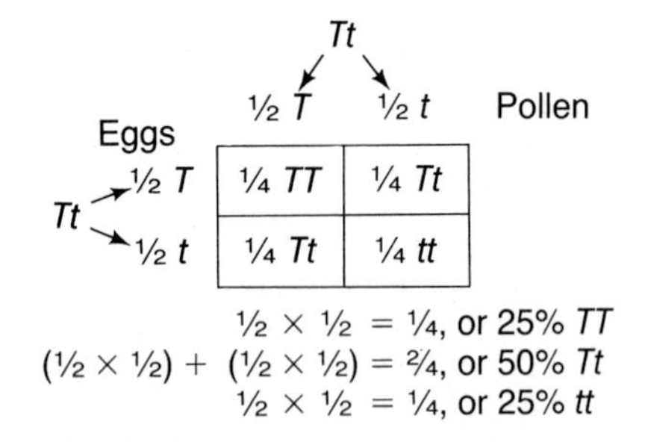


The three allelic combinations (*TT*, *Tt*, *tt*) can be viewed in another way to calculate the chance that any $F_2$ individual is carrying at least one *T* allele (disregarding the second allele for the moment). The answer is ¼ + 2/4 = ¾, or 75% probability. Stated in terms of the ratio of $F_2$ phenotypes, a series of crosses between *Tt* and *Tt* (or *Tt* self-fertilization) is expected to produce progeny in a ratio of 3 *T*–:1*tt* (where *T*– stands for either *TT* or *Tt* or the sum of both), or 3 tall:1 short.

The predictability of these genotypic and phenotypic ratios is due to the same laws of probability that state that we have a 50% chance to toss a coin heads up and a 25% chance to toss two coins together and have them both land heads up ($0.5 \times 0.5 = 0.25$, or $\frac{1}{2} \times \frac{1}{2} = \frac{1}{4}$). If we toss one coin often enough, we expect an average of ½ heads : ½ tails. If we have enough trials with two coins, we expect ¼ both heads : 2/4 heads + tails : ¼ both tails. These results are expected on the basis of chance alone, and the law applies equally well to independent events that involve a pair of alternative sides of a coin or a pair of alleles.

Mendel's experimental results and the simple mathematical interpretations of the observed ratios led him to propose the concept that is now known as **Mendel's First Law of Inheritance:** there is a *segregation of alleles* during sexual reproduction.

## 1.3 Dihybrid Crosses

A monohybrid cross involves parents that differ in one pair of alleles; a **dihybrid cross** involves parents that differ in two pairs of alleles. Mendel studied the inheritance of the seven characteristics of peas in combinations of twos and threes, as well as singly. In every case he found that the dominant and recessive alternatives behaved the same whether studied singly or in combination with other genes. His analysis of dihybrid (two-gene) $F_2$ and $F_3$ data revealed that each of the observed phenotypic ratios could be explained using the same simple rules of probability as in monohybrid crosses. By continuing to the $F_3$ generation, Mendel could again identify the genotpes of the $F_2$ plants as being true-breeding or heterozygous on the basis of $F_3$ progeny phenotypic classes (Table 1.3).

When studied in monohybrid crosses, tall was dominant over short and round seed shape was dominant over wrinkled. In combination, tall and round were still dominant to their recessive alternatives. Whether the parental cross was tall, round × short, wrinkled, or tall, wrinkled × short, round, the $F_2$ ratio of phenotypes was

9/16 tall, round
3/16 tall, wrinkled
3/16 short, round
1/16 short, wrinkled

The observed 9:3:3:1 ratio of phenotypes is explained most simply if each characteristic is inherited independently of the other. We expect ¾ tall : ¼ short and ¾ round : ¼ wrinkled in the $F_2$ of monohybrid crosses. If each characteristic is inherited independently of the other (height and seed shape), then the product of

***Table 1.3*** Predicted phenotypes of $F_2$ progeny resulting from self-fertilizations of $F_2$ progeny individuals segregating for seed color and seed shape.

| $F_2$ phenotypes | proportion of $F_2$ phenotypes | postulated $F_2$ genotypes | | predicted phenotypes of $F_2$ progeny from self-fertilized $F_2$ individuals |
|---|---|---|---|---|
| yellow, round | 9/16 | 1/16 *YYRR* | → | all yellow, round |
| | | 2/16 *YYRr* | → | 3 yellow, round : 1 yellow, wrinkled |
| | | 2/16 *YyRR* | → | 3 yellow, round : 1 green, round |
| | | 4/16 *YyRr* | → | 9:3:3:1 ratio, as in $F_2$ generation |
| yellow, wrinkled | 3/16 | 1/16 *YYrr* | → | all yellow, wrinkled |
| | | 2/16 *Yyrr* | → | 3 yellow, wrinkled : 1 green, wrinkled |
| green, round | 3/16 | 1/16 *yyRR* | → | all green, round |
| | | 2/16 *yyRr* | → | 3 green, round : 1 green, wrinkled |
| green, wrinkled | 1/16 | 1/16 *yyrr* | → | all green, wrinkled |

their separate probabilities is equal to the probability of the two characters occurring together in the $F_2$:

¾ tall × ¾ round = 9⁄16 tall, round
¾ tall × ¼ wrinkled = 3⁄16 tall, wrinkled
¼ short × ¾ round = 3⁄16 short, round
¼ short × ¼ wrinkled = 1⁄16 short, wrinkled

When Mendel raised $F_3$ plants from self-fertilized $F_2$ individuals, he deduced the occurrence of 9 different genotypes among the 4 $F_2$ phenotypic classes (Fig. 1.3).

Crosses involving pea plants differing in three characteristics (**trihybrid crosses**), confirmed the general principle that inheritance is based on discrete units existing in alternative forms. Analyses of the $F_3$ generations revealed 3 genotypes in a monohybrid cross, 9 genotypes in a dihybrid cross, and 27 genotypes in a trihybrid cross. These results are explained by the simple relationship of **$3^n$ genotypic classes,** where $n$ is the number of different pairs of alleles involved in a cross between heterozygous parents: $3^1 = 3$ genotypes for 1 pair of alleles of one gene, $3^2 = 9$ genotypes for 2 pairs of alleles, and $3^3 = 27$ genotypes for 3 different pairs of alleles. He interpreted this simple, consistent relationship to mean that inheritance was particulate and was not the result of blending or mixing of responsible factors.

The ratios of $F_2$ phenotypic classes observed in Mendel's monohybrid, dihybrid, and trihybrid crosses were 3:1, 9:3:3:1, and 27:9:9:9:3:3:3:1, respectively. This relationship can be expressed as **$2^n$ phenotypic classes,** where $n$ is the number of different genes involved: $2^1 = 2$, $2^2 = 4$, and $2^3 = 8$ $F_2$ phenotypic classes in this series (Table 1.4). These results led to what is called **Mendel's Second Law of Inheritance:** members of different pairs of alleles undergo *independent assortment* during sexual reproduction.

***Table 1.4*** Relationship between the number of pairs of alleles and the number of kinds of gametes, gamete combinations at fertilization, genotypic classes, and phenotypic classes that can be produced by heterozygotes.

| number of allele pairs (genes) | number of kinds of gametes formed by each sex | number of gamete combinations produced by random fertilization | number of genotypic classes in progeny | number of phenotypic classes in progeny |
|---|---|---|---|---|
| 1 | 2 | 4 | 3 | 2 |
| 2 | 4 | 16 | 9 | 4 |
| 3 | 8 | 64 | 27 | 8 |
| 4 | 16 | 256 | 81 | 16 |
| 5 | 32 | 1,024 | 243 | 32 |
| 6 | 64 | 4,096 | 729 | 64 |
| 7 | 128 | 16,384 | 2,187 | 128 |
| 8 | 256 | 65,536 | 6,561 | 256 |
| 9 | 512 | 262,144 | 19,683 | 512 |
| 10 | 1,024 | 1,048,576 | 59,043 | 1,024 |
| $n$ | $2^n$ | $4^n$ | $3^n$ | $2^n$ |

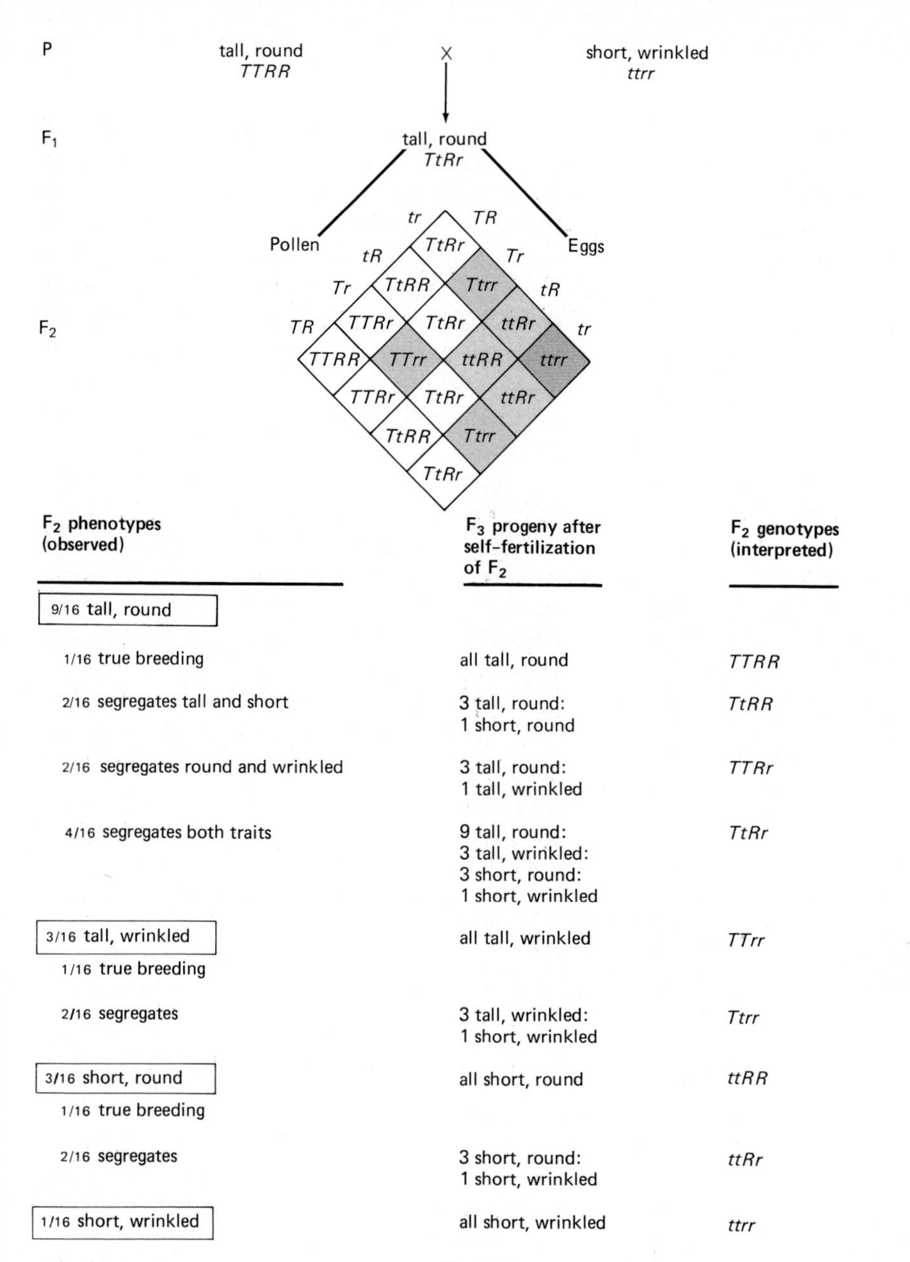


***Figure 1.3*** Mendelian ratios in $F_2$ progeny of a dihybrid cross between homozygous parents differing in plant height and seed shape. Upon self-fertilization, the various $F_2$ types would produce $F_3$ offspring in the ratios shown. These $F_3$ ratios provide the basis for interpreting specific genotypes of $F_2$ individuals, even when genotypically different individuals have the same phenotype.

## 1.4 Confirming a Prediction by Testcrosses

One important assumption in Mendel's interpretations leading to the two laws of inheritance was that egg cells and pollen cells of different genetic types were produced in approximately equal percentages. When calculating the $F_2$ ratios of phenotypes and genotypes, we *assume* equal proportions of each allele represented among the reproductive cells of each parent. These numerical values are not shown in the usual diagrammatic presentations because we expect them to be equal. For the two hypothetical pairs of alleles $A$ and $a$ and $B$ and $b$, the expected proportions are the following:

Monohybrid crosses $F_1 \times F_1$:

*Aa* × *Aa*

| | ½ *A* | ½ *a* |
|---|---|---|
| ½ *A* | ¼ *AA* | ¼ *Aa* |
| ½ *a* | ¼ *Aa* | ¼ *aa* |

*Bb* × *Bb*

| | ½ *B* | ½ *b* |
|---|---|---|
| ½ *B* | ½ *BB* | ¼ *Bb* |
| ½ *b* | ½ *Bb* | ¼ *bb* |

Dihybrid cross $F_1 \times F_1$:

*AaBb* × *AaBb*

| | ¼ *AB* | ¼ *Ab* | ¼ *aB* | ¼ *ab* |
|---|---|---|---|---|
| ¼ *AB* | 1/16 *AABB* | 1/16 *AABb* | 1/16 *AaBB* | 1/16 *AaBb* |
| ¼ *Ab* | 1/16 *AABb* | 1/16 *AAbb* | 1/16 *AaBb* | 1/16 *Aabb* |
| ¼ *aB* | 1/16 *AaBB* | 1/16 *AaBb* | 1/16 *aaBB* | 1/16 *aaBb* |
| ¼ *ab* | 1/16 *AaBb* | 1/16 *Aabb* | 1/16 *aaBb* | 1/16 *aabb* |

Mendel tested the assumption of equal proportions of gametes of different genetic types by making reciprocal crosses between doubly heterozygous $F_1$ plants, such as *AaBb*, with their homozygous *AABB* and *aabb* parental types. These are called **backcrosses** (of progeny to parent or to parental type). The most informative kind of backcross is the **testcross,** which refers specifically to breeding progeny with the recessive parent or parental type, as in the dihybrid testcross between a doubly heterozygous offspring and a doubly recessive parent: *AaBb* × *aabb*. The testcross is a basic design in genetic analysis.

Mendel predicted all of the results of a series of backcrosses involving various characteristics. We can look at the dihybrid backcrosses involving the pair of alleles for height, $T$ and $t$, and the pair of alleles for seed shape, $R$ (round) and $r$ (wrinkled). The particularly significant crosses were made *reciprocally* between doubly heterozygous tall, round (*TtRr*) and doubly recessive short, wrinkled (*ttrr*) plants. If the hereditary units existed in pairs and segregated into the reproductive cells in equal proportions, then four different phenotypic classes of equal size would be produced in the testcross progeny. Using the *branching notation,* we have the following:

Dihybrid testcross:

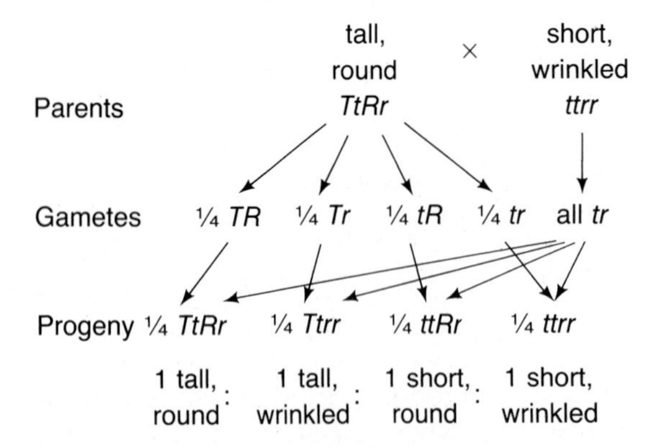


Testcross ratio: 1:1:1:1

Segregation and independent assortment of alleles in reproduction

Monohybrid testcross:

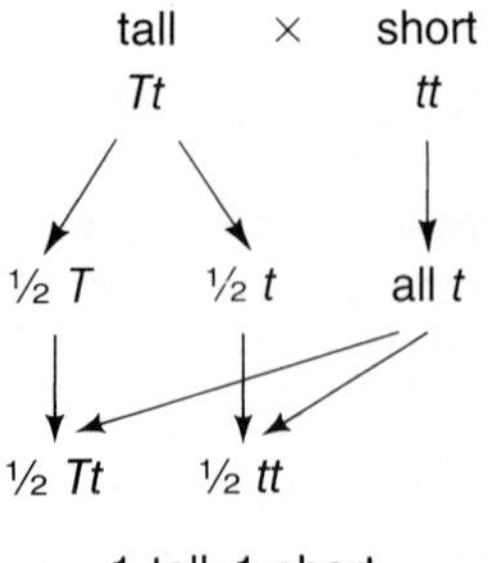


Testcross ratio: 1:1

Similarly, in a trihybrid testcross involving the alleles for height, seed shape, and seed color ($Y$, dominant allele for yellow; $y$, recessive allele for green), Mendel predicted eight phenotypic classes occurring in equal proportions. Using a *checkerboard notation,* we have the following:

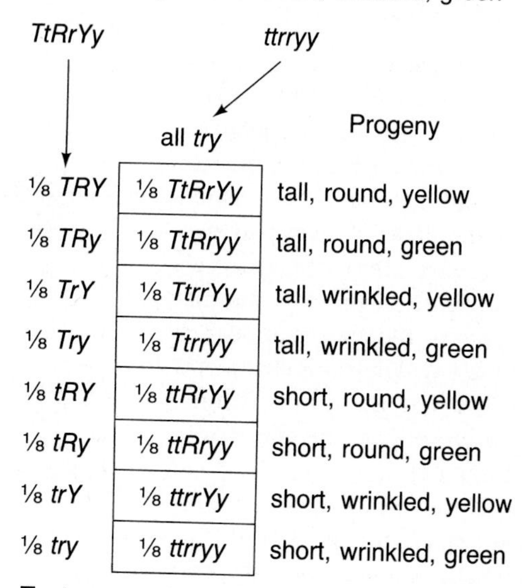


In order to verify the genotypes of testcross progeny, Mendel allowed individuals to self-fertilize. The ratio of phenotypic classes was exactly as he had predicted. For example, self-fertilized, tall, round, yellow testcross offspring gave rise to progeny consisting of eight phenotypic classes in the ratio of 27:9:9:9:3:3:3:1. The enormous value of the testcross can be seen in the two examples above: *each phenotypic class is represented by only one genotype.* Simple inspection of testcross progeny can therefore show the genotypes present, without the trouble of proceeding to another generation as Mendel had to do when he was establishing the principles. Whenever a phenotype can be uniquely interpreted on a genotypic basis, time and effort are saved. The convenience of the testcross and its accurate interpretation have made it a basic method in the breeding analysis of many organisms. Testcross phenotypic ratios of 1:1, 1:1:1:1, and 1:1:1:1:1:1:1:1, for one, two, and three pairs of alleles, respectively, are evidence for segregation of alleles and for their independent assortment during sexual reproduction.

## Variations in Mendelian Phenotypic Ratios

Very soon after 1900 investigators reported inheritance patterns that gave phenotypic ratios different from those established by Mendel. In every case, once the apparent "exceptions" were understood, the results strengthened and broadened the basis of Mendelian inheritance. The observed variations included dominance and recessiveness relationships between alleles of a pair and the number of alleles of a gene. A few examples will illustrate these variations in inheritance patterns and how Mendelian principles explained each of these different phenotypic ratios.

## 1.5 Dominance Relationships

Mendel chose seven characters that provided unambiguous pairs of contrasting alternatives. The progeny were not intermediate between parental types; in fact, on the basis of the phenotype, the progeny could not be distinguished from their parents. One allele of each pair was dominant over its recessive alternative since all the $F_1$ resembled only one of the two parents and since only two phenotypic classes appeared in monohybrid $F_2$ progeny. All seven characters studied by Mendel yielded the same results, namely, *complete dominance* of one allele over the other allele of the pair.

In early studies of the four-o'clock and certain other flowering plants, members of the $F_1$ generation were intermediate between the two parental phenotypes for some traits. The heterozygous $F_1$ plants did not resemble either parent. When four-o'clocks with red flowers were crossed with others having white flowers, all the $F_1$ progeny had pink flowers. Since complete dominance was not observed, did this mean that the alleles for red and white flower color were not transmitted to the $F_1$ in accordance with Mendelian predictions?

The situation was clarified when pink $F_1$ plants were interbred to produce the $F_2$ generation. The $F_2$ progeny consisted of three phenotypic classes: ¼ red:2/4 pink:¼ white (Fig. 1.4). This inheritance pattern could be interpreted as showing that the alleles for red and white flower color were present in the $F_1$ plants, did segregate from each other into gametes, and were passed on, unchanged, to the $F_2$ progeny. To test whether this was indeed the case, plants having pink flowers were interbred over a number of generations. These pink × pink crosses invariably produced progeny showing the same ratio of 1 red:2 pink:1 white, regardless of which generation of pinks were used. If inheritance were not based on segregation of members of a pair of alleles, then flower color should have become paler in successive generations. The consistency of the expected 1:2:1 ratio also strengthened the interpretation that only one pair of alleles was segregating since ¾ of the progeny were colored and ¼ were not colored (white).

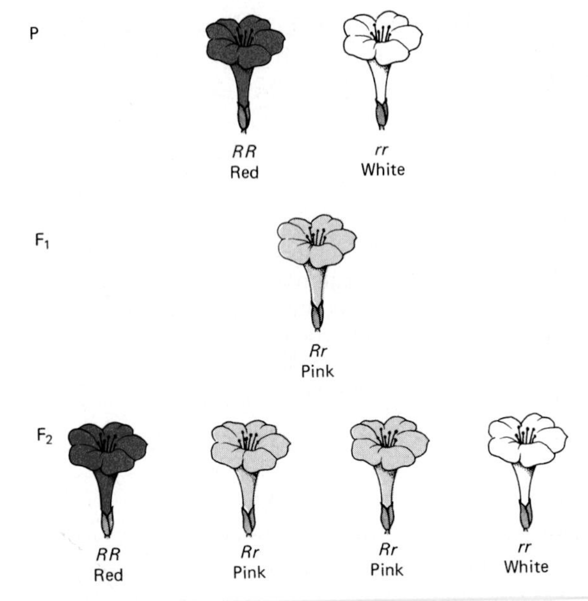


***Figure 1.4*** Incomplete dominance in four-o'clocks is evident from the pink-flowered $F_1$ progeny and from the 1:2:1 ratio of genotypes and phenotypes in the $F_2$.

To further test the hypothesis that only one pair of alleles of one gene governed the three flower color types, testcrosses were made between pink- and white-flowered plants. This kind of breeding test would produce a 1:1 ratio of pinks and whites if pinks were heterozygous for one pair of alleles and whites were recessive. If this were not the case, then flower color would become paler as more generations of testcrosses were obtained by the continued interbreeding of pinks and whites in each of a series of testcross generations. In testcrosses between pinks and whites, the results invariably showed 1 pink:1 white phenotypic ratio. The two kinds of breeding tests, therefore, confirmed the hypothesis that the red and white alleles of the flower-color gene adhered to the Mendelian rules for segregation and assortment of alleles in sexual reproduction.

The differences between the $F_1$ and $F_2$ phenotypes and ratios observed for four-o'clocks and those observed by Mendel can easily be explained by *incomplete dominance* of the red allele over the white allele. Homozygous red- and homozygous white-flowered four-o'clocks produce a new phenotype when crossed. The pink heterozygotes carried one allele of each kind, $R$ and $r$, and the pair of alleles together in an individual cannot direct the production of either of the two homozygous parental phenotypes. Other cases of incomplete dominance are recognized by these same results, namely, the heterozygous phenotype is different from the parental phenotypes. The $F_2$ phenotypic ratio of 1:2:1 is characteristic of incomplete dominance of a single pair of alleles. In effect, each genotype produces a unique phenotype. The phenotypic and genotypic ratios are identical in crosses involving a pair of alleles that shows incomplete dominance. The crucial supporting evidence for such an interpretation is provided by testcrosses between heterozygotes and recessives. Testcross progeny will consist of only the heterozygous and recessive types in the expected 1:1 ratio.

Another kind of variation involving a pair of alleles of a single gene is the phenomonon of **codominance.** Codominant alleles in a heterozygote are expressed equally and completely in the phenotype so that both allelic expressions can be identified in the single heterozygous individual. There is no blending of characters or diminished expression of either one of the alternative forms, as in incomplete dominance. A good example of codominance is provided by the antigens on the surface of red blood cells, phenotypically symbolized as M and N, and governed by a pair of alleles, $L^M$ and $L^N$. Antigens are proteins that can be identified by a clumping, or *agglutination reaction* when mixed with the appropriate antibodies. Antibodies are produced in the serum of an individual sensitized to the specific antigen under consideration. This sensitized serum is called the antiserum. If blood from an individual is mixed with antiserum from another individual, cells may clump together or remain freely suspended in the mixture. Agglutination takes place when the antiserum contains the specific antibodies that can react with the surface antigen of the red blood cells. A person with M antigen can be identified by the agglutination of a sample of his or her blood when mixed with serum containing anti-M antibodies. Blood cells having N antigen will remain suspended and will not undergo clumping when mixed with anti-M serum, but these cells will agglutinate in a mixture with anti-N serum. If both M and N antigens are present in the blood sample, agglutination will take place when the red blood cells are mixed with anti-M or with anti-N serum. The agglutination test, therefore, reveals the antigenic phenotype of the person in question.

When tests are conducted on a single blood sample using both anti-M and anti-N serum separately, some people's red blood cells show the agglutination reaction only with anti-M serum, some with only anti-N serum, and some with both kinds of antiserum. Persons who are homozygous for the $L^M$ allele produce only M antigen; and those who are homozygous for $L^N$ make only the N antigen. Heterozygotes have the genotype $L^ML^N$ and possess M and N antigens on their red blood cells (Table 1.5). The $L^M$ and $L^N$ alleles are codominant since each makes an equal and full contribution to the phenotype.

***Table 1.5*** The MN blood group in humans.

| alleles | genotypes | antigens produced | phenotypes (blood group) |
|---|---|---|---|
| | $L^ML^M$ | M | M |
| $L^M$, $L^N$ | $L^ML^N$ | M, N | MN |
| | $L^NL^N$ | N | N |

Another example of codominant alleles is provided by a globin gene governing hemoglobin synthesis. Different genes direct the production of each of the globin components that make up hemoglobin molecules in red blood cells. One globin gene, $Hb_\beta$, is responsible for the manufacture of beta-globin ($\beta$-globin), which forms part of adult hemoglobin molecules. In human populations the most common allele of the $\beta$-globin gene is $Hb^A$, which is responsible for synthesis of globin that forms part of *normal adult hemoglobin*. Another allele of the $Hb_\beta$ gene, $Hb^S$, is responsible for aberrant $\beta$-globin that forms part of *sickle-cell hemoglobin*. The $\beta$-globin alleles $Hb^A$ and $Hb^S$ are codominant. Individuals homozygous for the $Hb^A$ allele make only normal adult hemoglobin, homozygotes for the $Hb^S$ allele make only sickle-cell hemoglobin, and $Hb^AHb^S$ heterozygotes make both kinds of molecules. Each kind of hemoglobin molecule can be recognized by its behavior when subjected to electrical forces in an electrophoresis apparatus (Fig. 1.5). To test for molecule type, a sample of hemoglobin is allowed to migrate in a gel substance under the influence of an electrical field. Each kind of molecule comes to occupy a particular place in the gel and can be identified in that place after suitable treatment. Both normal adult hemoglobin and sickle-cell hemoglobin are present in blood samples of a heterozygous person; only one kind of hemoglobin is found in a homozygous individual. In heterozygous persons, sickle-cell hemoglobin causes some red blood cells to sickle. In homozygotes, the sickling is chronic and sickle-cell anemia results (Fig 1.6). Deformed red blood cells tend to clump, and these clumps cause the very narrow blood capillaries to clog. Blood flow is impeded, and various body parts may be damaged as a result of clogging episodes in the blood system of an afflicted person. These episodes, or crises, are painful as well as debilitating during the individual's relatively shortened lifetime.

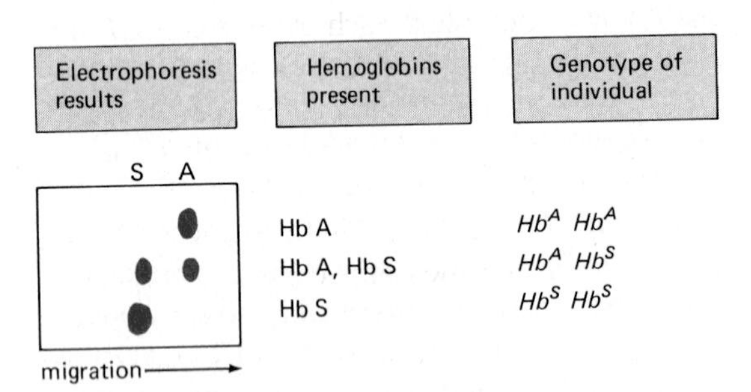


***Figure 1.5*** When hemoglobin molecules migrate in a gel under the influence of an electrical field, normal (Hb A) and sickle-cell (Hb S) hemoglobins come to rest in different but specific places in the gel. By observing the kinds and relative amounts of hemoglobin molecules from a blood sample, we can determine the genotype of an individual. Heterozygotes have equal amounts of Hb A and Hb S molecules, thereby indicating codominance (equal expression) of the $Hb^A$ and $Hb^S$ alleles of the $Hb_\beta$ gene specifying the $\beta$-globin component of hemoglobin protein.

This example of codominant alleles of a gene also provides an illustration of the *relative nature of dominance and recessiveness*. The $Hb^S$ and $Hb^A$ alleles are codominant at the molecular level of the phenotype, but not at the level of the whole organism. People enjoy the same good health whether they have two normal alleles or one normal and one $Hb^S$ allele. At the phenotypic level of the whole organism, therefore, the normal allele behaves as a dominant over the sickle-cell allele. Dominance and recessiveness are *relative behaviors of alleles* and are not due to some special chemical or physical characteristic of the alleles themselves. This general principle applies to all genes and their alleles.

The ABO major human blood groups furnish another example of codominant alleles of a single gene. The alleles $I^A$ and $I^B$ are expressed equally and fully in a person having blood type AB; such an individual is heterozygous for this pair of alleles. Someone of blood type A has one or two $I^A$ alleles; a person of blood type B carries one or two $I^B$ alleles; only someone with blood type AB will have both $I^A$ and $I^B$ alleles of the blood-group gene. Identification for blood type AB is made by the same kind of agglutination test as is used for the M and N antigens. If a sample of red blood cells clumps when mixed with anti-A serum, those cells carry the A antigen and the

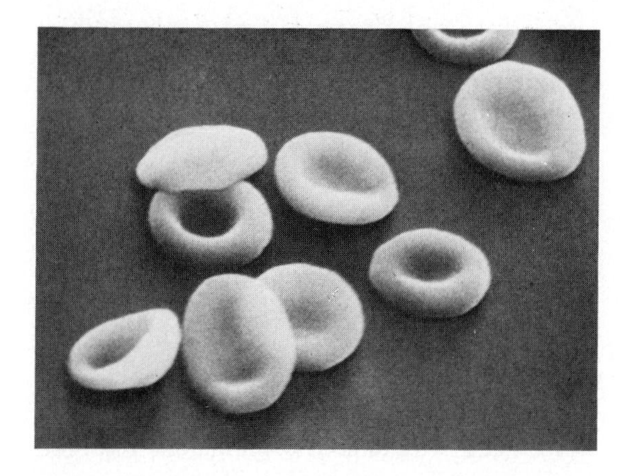

***Figure 1.6*** Red blood cells from individuals having normal hemoglobin or sickle-cell hemoglobin: (left) cells showing typical disk-shaped normal red blood cells, and (right) deformed red blood cells characteristic of individuals homozygous for the sickle-cell hemoglobin allele. (Courtesy of Dr. Marion Barnhart, Wayne State University School of Medicine.)

person must possess the $I^A$ allele. If red blood cells clump in anti-B serum, the cells carry the B antigen on their surface. Agglutination of a sample of red blood cells in either anti-A or anti-B serum indicates that A antigens and B antigens are present and thus the person is genotypically $I^AI^B$. The codominant alleles are fully expressed in an individual of this heterozygous type.

In addition to the blood groups A, B, and AB, a fourth group, O, exists. The type O phenotype is produced in an individual who carries neither the $I^A$ nor the $I^B$ allele but who is homozygous for a third allele of the ABO blood-group gene. This allele, which is symbolized as $i^O$, behaves as a recessive to both of the other alleles of the blood-group gene since it is expressed only in homozygous $i^Oi^O$ individuals. Because the $i^O$ allele is recessive, we cannot distinguish between individuals of blood group A who have the homozygous $I^AI^A$ genotype or the heterozygous $I^Ai^O$ genotype. We are also unable to make this distinction for people who are blood type B; they may be either $I^BI^B$ or $I^Bi^O$ to be phenotypically blood type B. Someone of the $I^AI^B$ genotype cannot carry an $i^O$ allele because in a single individual each gene is represented by only two alleles (Table 1.6).

Although these examples of codominance involve blood proteins, the phenomenon is not restricted to such molecules. There are numerous examples of codominant alleles of genes that govern enzymes and other kinds of molecules. In many cases expression of codominant alleles can be determined unambiguously by tests that identify the presence of one or both

***Table 1.6*** The ABO blood group in humans.

| alleles | genotypes | antigens produced | phenotypes (blood type) | transfusions accepted from |
|---|---|---|---|---|
| $I^A$, $I^B$, $i^O$ | $I^AI^A$, $I^Ai^O$ | A | A | type A, type O |
| | $I^BI^B$, $I^Bi^O$ | B | B | type B, type O |
| | $I^AI^B$ | A, B | AB | any donor |
| | $i^Oi^O$ | none | O | type O |

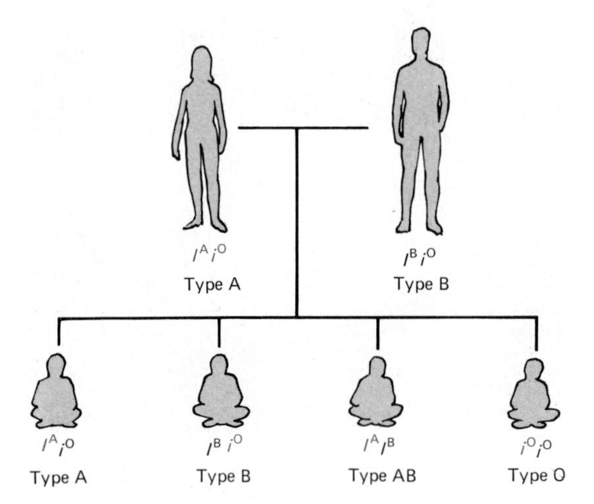


***Figure 1.7*** Inheritance of the ABO blood groups in humans is based on multiple alleles of one gene, with alleles $I^A$ and $I^B$ codominant and the $i^O$ allele recessive to both $I^A$ and $I^B$. Blood group genotypes and phenotypes in a family will vary in proportion, depending on the particular genotypes of the parents. A heterozygous type A mother and heterozygous type B father can have children with any of four possible genotypes and phenotypes. Each parent contributes one allele of the pair to each child, through the gametes. Alleles of the mother are in color, those of the father are in black.

kinds of molecules in an individual. For any characteristic the pattern of transmission of alleles can provide essential information regarding the number of genes involved and the number of different alleles of a gene that influence the phenotype of a particular trait (Fig. 1.7).

## 1.6 Multiple Alleles

Discussion of the ABO blood-group gene introduced another variation on Mendelian inheritance involving a single gene, namely, the occurrence of more than two alleles of a gene in a population. The ABO blood-group gene exists in **multiple alleles,** of which we have described only three. Other allelic forms of this gene are known to influence varieties of the A and B antigens produced in an individual. Multiple allelic forms have been identified for many genes, and they influence a variety of traits. Contrary to Mendel's observations of two allelic forms for each gene, we know today that any gene may exist in many alternative allelic forms. In any one diploid individual, only two alleles of any gene will be present. The whole set of allelic alternatives for the gene can be found in populations. The larger the sample, the higher the probability of discovering even the rarest of alleles for a given gene.

How can we determine whether different phenotypes represent alleles of one gene or of more than one gene? The simple test relies on basic Mendelian principles of segregation of members of a pair of alleles and of assortment of members of different pairs of alleles during reproduction. We expect different phenotypic and genotypic ratios for monohybrid inheritance versus dihybrid inheritance. In addition, for genes whose alleles exhibit complete dominance and recessiveness, we usually expect to find only two phenotypes in monohybrid crosses and four $F_2$ phenotypic classes in dihybrid crosses. Even for alleles showing incomplete dominance or codominance, we can predict the numbers of phenotypic and genotypic classes for monohybrid and dihybrid crosses.

In *Drosophila melanogaster,* the common fruit fly and a favorite experimental organism of geneticists past and present, a number of genes govern eye color. Red is the most common, or **wild-type,** eye color and results from a mixture of bright red and brown pigments that are produced in separate biochemical pathways during development. A number of genes control the various steps during pigment production in these biochemical pathways. Consider three homozygous strains of *Drosophila:* one (the wild type) with red eyes, one with brown eyes, and one with bright scarlet (an orange-red color) eyes. In order to determine whether these eye colors are governed by different genes or by multiple alleles of a single gene, we can perform crosses of all possible combinations of the three true-breeding (and, therefore, homozygous) strains.

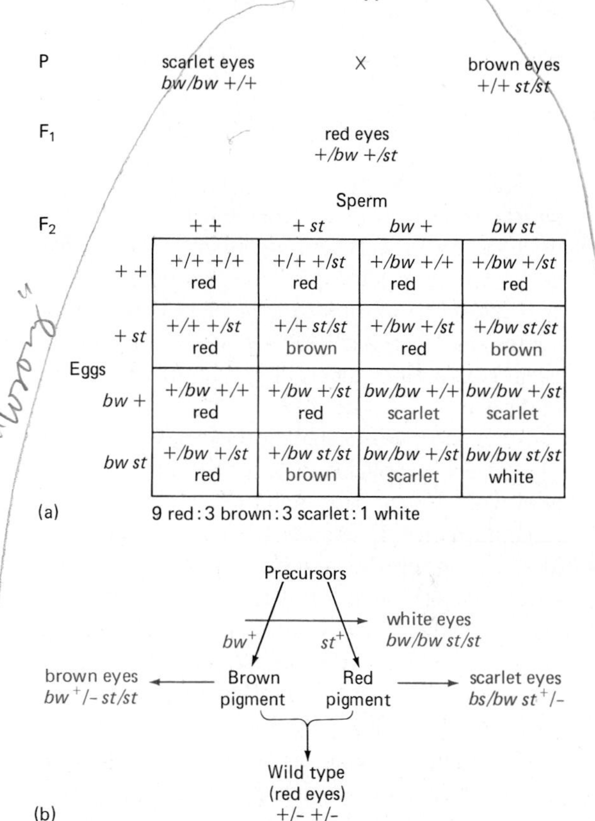


***Figure 1.8*** Inheritance of eye color in *Drosophila melanogaster.* (a) Breeding analysis reveals a dihybrid inheritance pattern of two independently assorting genes. The significant clues in this pattern include production of only red-eyed (dominant phenotype) $F_1$ progeny in mutant × mutant crosses, and a typical 9:3:3:1 dihybrid phenotypic ratio in the $F_2$. (b) Pigment synthesis in wild-type flies involves production of brown and scarlet pigments from precursors in two different biochemical pathways, each governed by a different gene. If either one of these two genes occurs in homozygous recessive form, only one of the pigments is synthesized. If the genotype is doubly recessive, neither pigment is made and flies will have white eyes.

When red-eyed flies are crossed with brown-eyed flies, all the $F_1$ progeny have red eyes. Interbreeding between $F_1$ heterozygotes leads to $F_2$ progeny consisting of 3 red:1 brown. These results show that one pair of alleles is involved and that red is dominant over brown. The red-eyed flies can be symbolized as $bw^+/bw^+$, or as +/+ in the usual notations used in *Drosophila* genetics. The recessive allele of the gene is *bw*, and brown-eyed flies are genotypically *bw/bw*. The genes in *Drosophila* are identified by the mutant (nonwild) phenotype, and their wild-type alleles are indicated by + (Box 1.3).

Crosses between parental stocks with red eyes and those with scarlet eyes produce $F_1$ progeny with red eyes and $F_2$ progeny consisting of 3 red:1 scarlet. In this case, too, it is clear that the alternative phenotypes are governed by one pair of alleles and that red is dominant over scarlet.

The wild type can now be studied in relation to the scarlet-eyed type and the brown-eyed type by the interbreeding of scarlet and brown recessives. If *st* and *bw* are alleles of the same gene, then we would expect an $F_2$ progeny ratio of 3:1 or 1:2:1, according to the relationship of complete dominance, incomplete dominance, or codominance of the alleles. We find instead that the $F_2$ progeny include four phenotypic classes in the ratio of 9/16 red:3/16 brown:3/16 scarlet: 1/16 white eyes. This is the expected 9:3:3:1 phenotypic ratio for a dihybrid cross involving two different and independently assorting pairs of alleles (Fig. 1.8). Thus brown and scarlet eye colors are produced by recessive alleles of two different genes, the brown gene (*bw*) and the scarlet gene (*st*). Brown and scarlet are nonallelic.

The 9/16 wild-type progeny in the $F_2$ result from the presence in the genotype of dominant alleles of both the brown and the scarlet gene. The 3/16 brown-eyed progeny have at least one dominant allele for the brown gene but are homozygous for the recessive *st* allele of the scarlet gene. The 3/16 scarlet-eyed progeny have at least one dominant allele for the scarlet gene but are homozygous for the recessive *bw* allele of the brown gene. The novel phenotypic class of white-eyed flies has the genotype of exclusively recessive alleles for both genes. Neither red pigment nor brown pigment is produced in these flies, so their eyes are white. The brown-eyed flies make only brown pigment and the

## BOX 1.3 Notation of Genes and Their Allelic Alternatives in Genetics

Since the inception of genetics in 1900, little effort has been made to provide rules for genetic notation among different organisms. Symbols for genes, for their alleles, and for genotypes as well are different for diploid (two sets of chromosomes) and haploid (one set of chromosomes) organisms and for different species of diploids in particular.

One accepted designation of the two alleles of a gene is a capital italic letter for the dominant allele and a lowercase italic letter for its recessive alternative. Dominant allele *A* and recessive allele *a* signify the *A*,*a* pair of one gene. The letter chosen for a known gene will often be representative of the dominant trait, such as *T*,*t* for dominant tall and recessive short alternatives for height. The diploid genotypes possible for such an allelic pair would be *TT*, *Tt*, and *tt*.

The convention used with *Drosophila* employs information about dominance and recessiveness in this diploid species as well as about the relative frequency of the various alleles in populations. The gene is designated by a lowercase letter representative of the mutant trait, such as *bw* for the brown eye gene. The most common allele of the brown gene in populations is $bw^+$, the **wild-type** alternative. The superscript + indicates the wild-type allele of a gene, and it is used whether the mutant trait is dominant or recessive. The bar eye locus *B* has a dominant mutant allele, *B* (bar), and a recessive wild-type allele $B^+$ (nonbar). The symbols thus indicate whether the mutant form is dominant or recessive and which allele is the wild-type alternative.

Multiple alleles of a gene are distinguished by superscripts, such as the alleles at the white eye locus in *Drosophila, w*. The wild-type allele $w^+$ is dominant. Other alleles of this gene include $w^e$ (eosin), $w^a$ (apricot), $w^{ch}$ (cherry), and several more. The wild-type allele for genes in *Drosophila* and other organisms is usually abbreviated to +, and the gene itself is identified by the mutant allele present, such as *w* or *B*. Wild-type flies would have homozygous (++) or heterozygous (+*w*) genotypes if the wild-type allele is dominant or homozygous only (++) if the wild-type allele is recessive to its mutant alternative(s).

The diploid genotype is often indicated as +/+, +/*w*, or some similar notation. The slash mark indicates that the two alleles of a gene are present on the pair of chromosomes that have the same set of genes (homologous chromosomes). This notation is particularly helpful in distinguishing genes on the same chromosome from genes on different chromosomes, as we will discuss in Chapter 4. Using the hypothetical genes *a* and *b*, the genotype notations *ab*/++ or *a*+/+*b* indicate that genes *a* and *b* occur on the same chromosome. If these two genes are on different chromosomes, the genotypes would be written *a*/*a* *b*/*b*, *a*/+ *b*/+, or some other combination of the two pairs of alleles of a particular individual. The slash marks show immediately whether or not the genes are on the same chromosome. If the location of the genes is unknown, the genotypes may be written *aabb*, *a*+*b*+, *aab*+, and so on for a diploid species, since their genes and chromosomes occur in pairs.

Codominant alleles are distinguished by superscripts. In humans, genes are symbolized in various ways. Alleles of the beta-globin gene ($Hb_\beta$) specifying one of the globins in hemoglobin include two codominants, $Hb^A$ and $Hb^S$, among others. The alleles of the ABO blood group locus are the two codominant alleles $I^A$ and $I^B$, plus the recessive third allele $i^O$ (sometimes shown only as *i*). The gene governing synthesis of the enzyme thymidine kinase is represented by the dominant allele *Tk* (enzyme is synthesized) and its recessive alternative *tk* (enzyme missing or defective).

Different notations are used in describing bacteria and several eukaryotic microorganisms.

In these cases allelic superscripts + and − indicate presence or absence of a molecule or a function. For example, cells that grow independently of arginine in the nutrient medium are $arg^+$; those requiring arginine for growth are designated $arg^-$. These superscript notations do not imply dominance or recessiveness. We will describe these and other allelic notations in succeeding chapters.

scarlet-eyed flies make only red pigment; the 9/16 wild-type $F_2$ progeny, however, produce both pigments and have the wild-type dark red eye color. This is an example of the interaction of genes in development of a single phenotypic character. The *bw* and *st* alleles are not alternative forms of the same gene but are instead allelic alternatives of different genes. These conclusions are based on unambiguous dihybrid phenotypic ratios.

Breeding analysis of another eye-color gene in *Drosophila* provides an example of a series of multiple alleles. The gene produces the wild-type red eye color and several mutant, or nonwild, colors, including apricot, eosin, and cherry, in addition to white. The particular gene responsible for this series of colors is the white gene, symbolized by *w*. Crosses between the wild type and any nonwilds of this series invariably produce $F_1$ progeny with red eyes. All the alleles of the white gene are therefore recessive to the wild type allele, $w^+$, which we designate as +. When $F_1$ heterozygotes interbreed to produce the $F_2$ generation, the phenotypic ratio is 3 red:1 white or eosin or apricot, depending on the parental strain used (Fig 1.9). The critical tests involve crosses between homozygous mutant strains. In these crosses, such as eosin × apricot, eosin × white, apricot × white, the $F_2$ progeny always consist of flies with eye colors like the two parents. No wild-type flies appear in the $F_2$. Since each cross produces a monohybrid phenotypic ratio, all the alleles must be alternative forms of a single gene. Allelic forms of the white gene include *w* (white), + (red), $w^e$ (eosin), $w^a$ (apricot), $w^{ch}$ (cherry), and others.

In principle, therefore, we can determine whether a phenotypic character is governed by pairs of alleles of different genes or by multiple alleles of the same gene. On the basis of breeding analyses, we expect two different pairs of alleles to show dihybrid ratios and multiple alleles of the same gene to produce $F_2$ ratios characteristic of monohybrid inheritance.

| | Phenotypes |
|---|---|
| P | red × apricot |
| $F_1$ | red |
| $F_2$ | 3 red: 1 apricot |
| P | red × white |
| $F_1$ | red |
| $F_2$ | 3 red: 1 white |
| P | apricot × white |
| $F_1$ | apricot |
| $F_2$ | 3 apricot: 1 white |

***Figure 1.9*** Monohybrid inheritance pattern for multiple alleles of the white gene in *Drosophila melanogaster*. The occurrence of a 3:1 $F_2$ phenotypic ratio in both the wild type × mutant and the mutant × mutant crosses indicates that alleles of the same gene govern inheritance of these eye colors. If two or more genes were involved, dihybrid or more complex phenotypic ratios in $F_2$ progeny would occur (see Fig. 1.8).

## 1.7 Nonallelic Gene Interactions

Numerous examples exist of different (nonallelic) genes that contribute to the same trait. We have already discussed an example in which actions of both the brown gene and the scarlet gene are needed for development of wild-type red eye color in *Drosophila*. This case supports Mendelian inheritance, although it was exceptional in two ways: (1) the $F_1$ had red eyes and, therefore, did not resemble either parent (brown and scarlet) and (2) the $F_2$ progeny included a novel phenotypic class (white eyes). In all other ways, the breeding results followed Mendelian rules of segregation and assortment of members of pairs of alleles. The $F_2$ showed both the genotypic ratio and the 9:3:3:1 phenotypic ratio expected for a dihybrid cross.

In other cases of nonallelic gene interaction in production of a trait, the $F_2$ phenotypic ratio may be modified. Analysis of the genotypic ratio, however, usually shows the typical proportions and distribution of the expected classes. The 9:3:3:1 $F_2$ phenotypic ratio is predicted on the basis of Mendel's First and Second Laws and depends on the assumption that alleles of different genes do not interact to produce a phenotype. If this assumption is invalidated, then Mendel's Laws lead to predictions of modified ratios.

A modified phenotypic ratio is sometimes the result of a phenomenon called **epistasis,** in which the expression of one gene prevents or interferes with the expression of another, different gene. The gene that masks or interferes with the expression of another is said to be *epistatic* to it. The gene that is so influenced is said to be *hypostatic*. Cyanide production in white clover plants provides an example of an epistatic interaction.

Some clover strains have high cyanide (HCN, hydrocyanic acid) levels; others have lower levels of cyanide in their leaves. Cattle do not seem to be affected by the cyanide they consume while eating these plants. The usual result of crossing high-cyanide and low-cyanide strains of clover is a uniform $F_1$ with high cyanide and an $F_2$ that segregates 3 high:1 low for cyanide content. Such results point to a single pair of alleles of one gene, with low-cyanide recessive to its alternative. We would therefore expect crosses between homozygous recessive low-cyanide strains to produce progeny consisting only of low-cyanide phenotypes. In some crosses, however, both the $F_1$ and $F_2$ progeny are different from expected:

| | |
|---|---|
| Parents | low strain-1 × low strain-2 |
| $F_1$ | high-cyanide |
| $F_2$ | 9 high-cyanide:7 low-cyanide |

Looking carefully at this, we can see that the $F_2$ ratio must be modified from the usual dihybrid 9:3:3:1 phenotypic ratio such that $\frac{9}{16}$ are high cyanide and $\frac{7}{16}$ ($\frac{3}{16} + \frac{3}{16} + \frac{1}{16}$) are low cyanide. The fact that the $F_1$ are all high cyanide further indicates that this expression must be dominant over low cyanide. If we look for a common feature among the $\frac{7}{16}$ that are low cyanide—one that does not occur in the $\frac{9}{16}$ high cyanide or in the $F_1$—we can put the story together more easily. The common feature among $\frac{7}{16}$ of an $F_2$ progeny is the presence of at least one pair of recessive alleles; $\frac{9}{16}$ of a dihybrid $F_2$ progeny have at least one dominant allele for each of the two genes involved:

$\frac{9}{16}$ are *A–B–* (– stands for either allele)
$\frac{3}{16}$ are *A–bb*
$\frac{3}{16}$ are *aaB–*
$\frac{1}{16}$ are *aabb*

If we construct the breeding scheme so that two genes govern the characteristic of cyanide content and both genes must be present in the dominant form for high cyanide content to develop, we have the following:

*A–* = high  *B–* = high
*aa* = low  *bb* = low

P  *AAbb* × *aaBB*
  low  low

$F_1$  *AaBb*
  high

$F_2$  9 *A–B–* : 3 *A–bb* : 3 *aaB–* : 1 *aabb*
  9 high : 7 low

In this particular case, two different chemical compounds must be produced in order to get high cyanide levels. Each compound is made only when each gene is represented by at least one dominant allele *A–B–*. The *A–bb* low-cyanide strain can make only one of these compounds, and the *aaB–* can make only the other compound; *aabb* cannot make either compound. In all three genotypes producing the low-cyanide phenotype, therefore, the chemical reactions leading to high cyanide content do not take place.

In mice several genes interact to produce coat color. The wild-type color pattern is called *agouti,* and it is the result of subtle black and yellow alternate banding on individual hairs. The total effect is that of a gray color, which is also very common in other rodents. Among the known variations in coat color in mice is the familiar *albino,* which is white due to the absence of pigments. Mice that do not make the yellow pigment have *black* coat color. Both albino and black varieties breed true, and both are recessive to the wild-type agouti character. Recessiveness is evident in two crosses: agouti × albino, producing only agouti $F_1$ progeny, and 3 agouti : 1 albino in the $F_2$; and agouti × black, producing only agouti $F_1$, and 3 agouti : 1 black in the $F_2$. When black mice are crossed with albinos, the $F_1$ progeny are all agouti. Interbreeding of the $F_1$ mice produces $F_2$ progeny consisting of 9/16 agouti, 3/16 black, and 4/16 albino mice (Fig. 1.10). The $F_2$ ratio indicates dihybrid inheritance, but the last two terms of the usual 9 : 3 : 3 : 1 ratio have been combined, thereby indicating that two of the usually different phenotypic classes cannot be distinguished in this $F_2$ progeny.

The explanation of this 9:3:4 modified dihybrid ratio is that two genes exist: one gene whose dominant allele, *C,* is necessary for development of color and another gene whose dominant allele, *A,* is necessary for the blending of black and yellow pigments to produce the agouti pattern. Agouti mice are genotypically *C–A–*; black mice are genotypically *C–aa* and develop black coat color but not the agouti pattern; and albinos are homozygous for the recessive allele of the color gene, *cc,* so that no

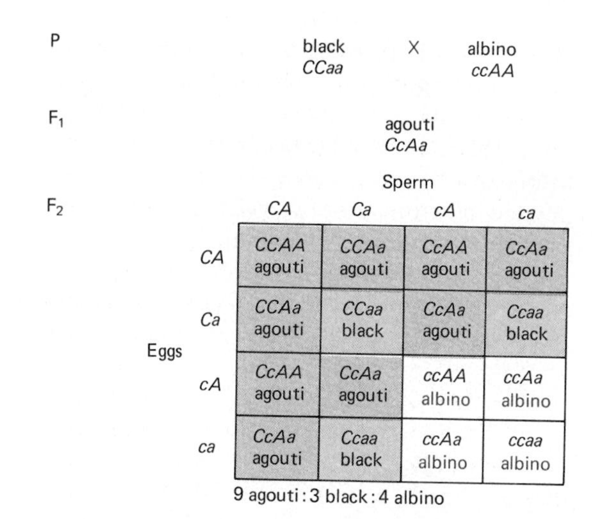


***Figure 1.10*** Epistatic gene interaction in the inheritance of coat color in mice. Crosses between black and albino mutants produce wild-type agouti $F_1$ progeny that have wild-type alleles for both the color gene (*C*) and the agouti pattern gene (*A*). $F_2$ progeny segregate into 9/16 agouti, with at least one wild-type allele for each of the two genes; 3/16 black, which are recessive for the agouti gene but have at least one wild-type allele for the color gene; and 4/16 albino. The albino phenotype is expressed in mice recessive for the color gene, regardless of the nature of the alleles for the agouti gene in the same genotype. The recessive genotype of the color gene is epistatic to the agouti gene, since *cc* will mask the expression of the *A* allele of the agouti gene in a genotype. Epistasis accounts for phenotypic identity of all 4/16 of the $F_2$ progeny carrying *cc* in the genotype since all 4/16 fail to produce pigments and are phenotypically albino.

color develops, whether the dominant allele for agouti pattern is present or absent. The presence of *cc* in the genotype, therefore, masks the expression of the *A* allele of the agouti gene, and the 1/16 doubly recessive class (*ccaa*) looks the same as the 3/16 of the $F_2$ progeny that is *ccA–* in genotype. The recessive genotype of the color gene (*cc*) is, therefore, epistatic to the dominant allele of the agouti gene.

In the common bread wheat (*Triticum vulgare*), a number of color genes contribute to red coloration of the wheat kernels. In crosses between certain true-breeding red varieties, the red-kerneled $F_1$ progeny produce an $F_2$ generation consisting of plants with red or with white kernels in the phenotypic ratio of 15:1 (Fig. 1.11). This modified dihybrid ratio is due to the action of two color genes that undergo independent assortment of two pairs of alleles. Red pigment is produced when either gene or both genes are represented by at least one dominant allele in the genotype. The only class that has no red color is the double recessive. Upon closer inspection of the red coloration, we observe a gradation from very dark red through intermediate to light red tones. Each dominant allele contributes to pigment production, so that the $R_1R_1R_2R_2$ genotype produces the deepest red color, and paler shades are the result of the presence of three or two or one of the four possible dominant alleles of the two genes. In the complete absence of dominant alleles of these two genes, no color develops. On the basis of red or white, the $F_2$ phenotypic ratio is 15 red:1 white. On the basis of dominant alleles interacting in pigment production, we detect 1/16 darkest red (four dominants present):4/16 dark red (three dominants present):6/16 red (two dominants present):4/16 pale red (one dominant present):1/16 white (no dominants present).

In this example, as in the case of brown × scarlet eye color in *Drosophila*, the nonallelic interactions do not involve epistasis. They do, however, show that conventional Mendelian phenotypic ratios can be modified when two or more genes contribute to the development of the same phenotypic character. Epistasis involves a special kind of gene interaction in that one gene masks the presence of another gene in

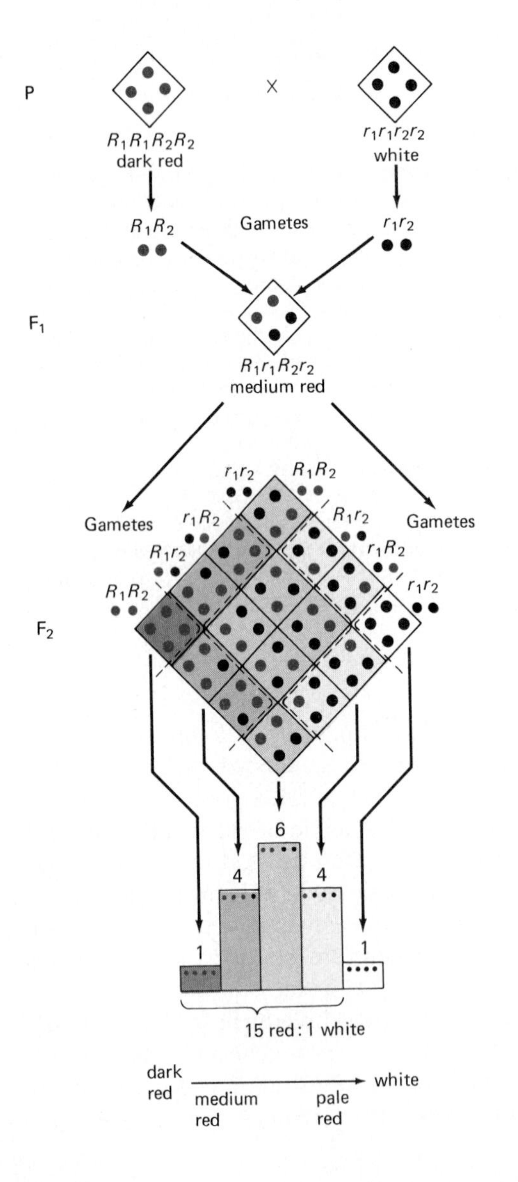


***Figure 1.11*** Nonallelic gene interaction leading to a modified dihybrid ratio in the inheritance of kernel color in wheat. Crosses between dark red and white plants produce $F_1$ progeny with medium red kernels. When self-fertilized, the $F_1$ gives rise to $F_2$ progeny consisting of 15/16 red:1/16 white phenotypic classes. This modified dihybrid ratio may also be described as a ratio of 1/16 darkest red:4/16 dark red:6/16 medium red:4/16 pale red:1/16 white. Each of these five phenotypic classes differs in the number of dominant alleles of the two color genes. In the diagram, each dominant allele is represented by a colored dot, and the summation of genotypes is shown below as a bar diagram. Whether the $F_2$ phenotypic ratio is considered to be 15:1 or 1:4:6:4:1, the explanation is that a modified Mendelian dihybrid ratio results from the equal and additive effects of dominant alleles of two different and independently assorting color genes. (Adapted with permission of Macmillan Publishing Company from *Genetics,* 2nd ed. by M. Strickberger. Copyright © 1976 by Monroe W. Strickberger.)

the same genotype. Whether or not one gene is epistatic to another, interactions between different genes can be analyzed according to conventional Mendelian principles and by genotypic and phenotypic ratios can be shown to follow the Mendelian Laws of Inheritance as expected. By appropriate breeding tests, we can distinguish among different genotypes, even when these produce the same phenotype.

Through Mendelian methods, it has repeatedly been shown that segregation and independent assortment of different pairs of alleles are responsible for inheritance of many traits in plants and animals. None of the observed variations in $F_2$ phenotypic ratios are due to changes in behavior of pairs of alleles in reproduction. Changes from expected phenotypic ratios are due to interactions between different genes in phenotype development. Ratios such as 15:1, 12:3:1, 13:3, 12:3:4, 9:7, and 9:3:4 have all been shown to be modifications of the basic 9:3:3:1 dihybrid ratio of $F_2$ phenotypic classes (Box 1.4, next page).

## 1.8 Environmental Influences on Phenotypic Development

Phenotypic expression usually depends on both the genotype of the organism and its environment. For example, tall plants will develop if the genotype is *TT* or *Tt* and if conditions are suitable for growth. A genotypically tall plant may develop into a short, stunted plant, despite its alleles, if it grows in unfavorable conditions of lighting, watering, and soil quality.

One of the axioms of scientific analysis is the need to establish and maintain experimental conditions that are *standard,* or consistent, and *optimum* for the system being studied. By reducing the variability of the system and its surroundings, we can place more confidence in the differences we may observe under controlled conditions of the experiment. Mendel followed these simple rules by pretesting pea varieties and selecting strains that showed consistent development of characteristics to be studied and by growing the plants under the best possible conditions for their development. Any variations he observed in the experiments could then be related to inherited variation and not to random effects caused by unknown features of the plants or changes in their surroundings.

This general approach can also help to show whether an inherited feature behaves differently in different environmental conditions, once it has been established that the feature is inherited and its pattern of inheritance is known. In the high- and low-cyanide strains of white clover, the expression of one gene could be influenced by another gene in the individual and is an example of variation in phenotypic expression in relation to the genotype or the genetic environment. Numerous examples exist of variation in phenotypic expression in relation to the physical environment in which the individual develops, such as a stunted plant with genes for *potential* development of tallness. Phenotypic development is the outcome of cooperative interaction between genes and nongenic environmental factors. To put this into a familiar axiom or slogan, development is the result of nature *plus* nurture, not nature *or* nurture.

Phenylketonuria (PKU) is a condition in humans that is inherited as a simple recessive trait and can lead to extreme mental retardation. The primary effect in recessives is failure to convert the amino acid phenylalanine into tyrosine during protein processing and degradation:

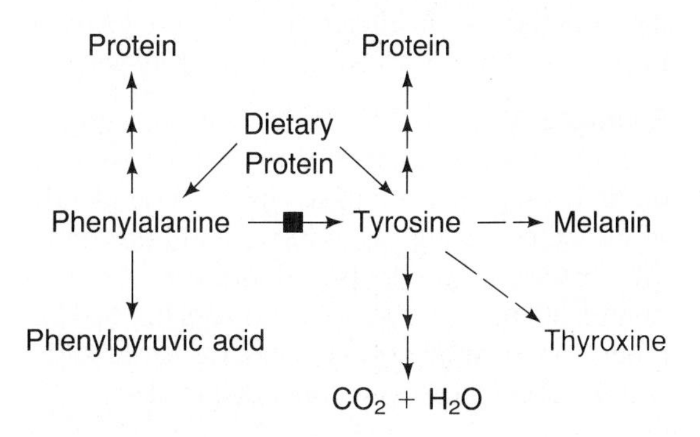


If a phenylketonuric child is given a suitable low-protein diet during the first six years or so of

## BOX 1.4 Mendelian Dihybrid Ratios

In crosses between $F_1$ heterozygotes *AaBb* × *AaBb*, Mendel found the phenotypic $F_2$ ratio to be 9 : 3 : 3 : 1 in all cases involving pairs of the seven traits he analyzed in garden peas. After 1900, geneticists found variations in the standard Mendelian $F_2$ dihybrid ratio; these variations, however, followed the laws of Mendelian inheritance despite modified $F_2$ ratios. In a typical Mendelian dihybrid cross, the two traits were inherited independently of each other. Since these traits concerned different phenotypic features, it was easy to see that their inheritance was the outcome of noninteracting genes governing different developmental pathways. Modified $F_2$ ratios, however, were usually produced in crosses involving two nonallelic genes that governed a single phenotypic character. In the first nine examples shown below, one allele at each locus is completely dominant over its recessive alternative allele. The differences in $F_2$ phenotypic ratios are a consequence of nonallelic gene interaction in the development of a single phenotypic trait in examples 2 through 10; in example 1, the standard or typical Mendelian $F_2$ progeny have segregated for two different phenotypic traits (seed color and seed shape). In example 11, the two loci govern different sets of blood proteins; the phenotypic ratio is the same as the genotypic ratio because of the codominance of the pair of alleles at each gene locus. A brief explanation for each example follows.

**Example 1.** Yellow is dominant over green, and round is dominant over wrinkled. Each pair of alleles segregates and assorts independently of the other during reproduction and produces the expected 9/16 double dominant, 3/16 one dominant and one recessive expression, 3/16 the other dominant and other recessive expression, and 1/16 doubly recessive phenotypic class.

**Example 2.** Each gene governs a step in a pathway leading to pigment synthesis, but the pathways are independent of each other. The dominant wild-type eye color is the result of a doubly dominant genotype (9/16), brown-eyed flies have the dominant brown allele and recessive *st/st* alleles and can make only brown pigment (3/16 *A–bb*), scarlet-eyed flies have the dominant scarlet allele but the *bw/bw* alleles and can make only scarlet pigment (*aaB–*), and the double recessive has white eyes since it cannot make either of the two pigments (1/16 *aabb*).

**Example 3.** The rose allele is dominant over the nonrose, and the pea allele is dominant over the nonpea. When dominant alleles are present at both gene loci, genic interaction leads to the novel phenotype of walnut comb shape. The double recessive also produces a novel phenotype of single comb shape. The situation of nonallelic gene interaction involving independent pathways leading to development of a particular phenotypic trait is the same in this example as in example 2.

**Example 4.** One gene determines agouti pattern and a different gene determines synthesis of black pigment. The doubly dominant animal is agouti, 3/16 of the progeny are black nonagouti, 3/16 are albino because the homozygous recessive alleles at the color locus mask the presence of the agouti genotype, and 1/16 are albino double recessives. The color gene is epistatic to the agouti gene.

**Example 5.** Dominant alleles must be present at both gene loci if purple color is to develop. One gene governs development of color versus no color, and the other gene governs purple versus white flowers. When either gene locus is represented by homozygous recessive alleles, flowers are white (not colored), as is also the case for the double recessive. Either recessive homozygote is epistatic to the effects of the other gene.

**Example 6.** All genotypes that have the dominant allele for white (*A*) produce white fruits. The *aa* genotypes produce colored fruit, either yellow (*B–*) or recessive green (*bb*). The dominant allele for the white gene masks the presence of the color gene, that is, the dominant white allele is epistatic to the alleles of the color gene.

**Example 7.** Inhibition of feather color development is dominant over color; but in the absence of the dominant inhibitor allele of one gene, feathers will be colored if dominant alleles of the color gene are present. The double recessive is white since it lacks a dominant color allele, even though no inhibitor is produced. The dominant allele of the inhibitor gene is epistatic to the color allele of the second gene; its presence is, therefore, masked and white feather color results.

**Example 8.** At least one dominant allele at either gene locus is required for development of leg feathers. Only the doubly recessive homozygote has no feathers on its legs.

**Example 9.** The dominant allele of either gene governs development of spherical fruit when the second gene is homozygous recessive. When dominant alleles of both genes are present, the novel disk phenotype develops as an exaggerated spherical form. The long fruit shape is produced only in doubly recessive homozygotes.

**Example 10.** Multiple genes with additive effects of dominant alleles of each gene produce graduated shades of redness of the wheat kernel. White kernels are produced only in the absence of dominant alleles of these genes. Additive effects are apparent since progressively lighter shades of red are produced with progressively fewer dominant alleles present (see Fig. 1.17).

**Example 11.** The presence in a genotype of codominant alleles of each of two different genes affecting two different sets of blood proteins leads to a phenotypic ratio that is identical to the genotypic ratio. Each genotype is expressed as a unique phenotype since each allele contributes to phenotypic development. In this hypothetical situation the $F_1$ heterozygotes producing nine $F_2$ phenotypic classes would have been $I^AI^B/L^ML^N \times I^AI^B/L^ML^N$.

| | | | genotype and proportion | | | | | | | | |
|---|---|---|---|---|---|---|---|---|---|---|---|
| organism | character | ratio | *AABB* 1/16 | *AaBB* 2/16 | *AABb* 2/16 | *AaBb* 4/16 | *Aabb* 2/16 | *AAbb* 1/16 | *aaBB* 1/16 | *aaBb* 2/16 | *aabb* 1/16 |
| 1. peas | seed color and shape | 9:3:3:1 | yellow, round | | | | yellow, wrinkled | | green, round | | green, wrinkled |
| 2. *Drosophila* | eye color | 9:3:3:1 | red | | | | brown | | scarlet | | white |
| 3. chickens | comb shape | 9:3:3:1 | walnut | | | | rose | | pea | | single |
| 4. mice | coat color | 9:3:4 | agouti | | | | black | | albino | | |
| 5. peas | flower color | 9:7 | purple | | | | white | | | | |
| 6. squash | fruit color | 12:3:1 | white | | | | | | yellow | | green |
| 7. chickens | feather color | 13:3 | white | | | | | | colored | | white |
| 8. chickens | leg feathers | 15:1 | present | | | | | | | | absent |
| 9. squash | fruit shape | 9:6:1 | disk | | | | sphere | | | | long |
| 10. wheat | kernel color | 1:4:6:4:1 | dark red | medium dark red | | medium red | light red | medium red | | light red | white |
| 11. humans | ABO and MN blood types | 1:2:2:4:2:2:1:1:1 | type A type M | type AB type M | type A type MN | type AB type MN | type AB type N | type A type N | type B type M | type B type MN | type B type N |

its life, relatively little phenylalanine is metabolized, and the devastating effects of mental retardation can be avoided. Past six years of age, there appears to be relatively little danger of mental retardation developing in youngsters with the recessive genotype. In a large part of the United States, newborn babies are tested routinely to identify PKU positives, who are immediately placed on a low-protein diet. PKU-positives excrete large amounts of phenylalanine and phenylpyruvic acid in the urine. Individuals who are homozygous or heterozygous for the dominant allele can convert phenylalanine into tyrosine and have normal levels of these amino acids in their urine. By modifying the environment through diet, recessives can avoid the phenotypic expression of mental retardation. Regardless of their diet, however, PKU-positives retain the recessive genotype and can transmit recessive alleles to their own children. Modification of the environment does not alter the gene or the genotype, but it can alter the phenotype.

In experimental studies we find similar situations in which one genotype can give rise to different phenotypes in different environments. The common colon bacillus, *Escherichia coli,* can process the milk sugar lactose and related compounds into simpler metabolites. Normal *E. coli* has the inherited ability to synthesize the enzyme β-galactosidase for lactose metabolism. When *E. coli* cells are incubated in nutrient media containing the sugar glucose, very little enzyme activity can be detected (Fig. 1.12). When such cells are incubated in lactose-containing media, in the absence of glucose, β-galactosidase activity increases sharply and rapidly as enzyme protein is synthesized in the cells. Synthesis of this enzyme must be under gene control because in some mutant strains the enzyme is not synthesized under any conditions. Normal *E. coli* may or may not synthesize the enzyme depending on its environmental surroundings (Table 1.7). Such gene-environment interactions during cell metabolism are a common feature in all species studied. These observed interactions form the basis of the important concept that ultimate phenotypic expression is the outcome of many prior steps involving interactions between genes and between genes and environment during development and differentiation of the organism.

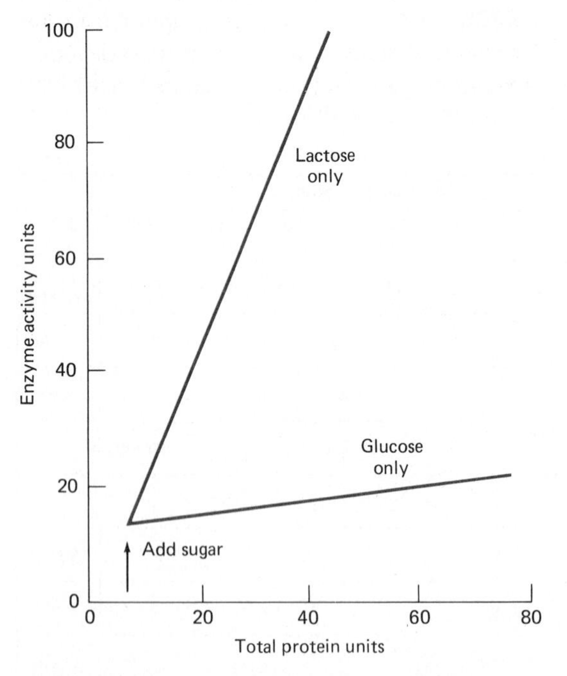


***Figure 1.12*** *E. coli* synthesizes little detectable β-galactosidase when only glucose is present in the growth medium but makes substantial amounts of the enzyme if lactose is present instead. Different phenotypes (amount of enzyme made) may therefore arise among genotypically identical cells in different environments.

***Table 1.7*** Phenotypes, observed as presence or absence of the enzyme β-galactosidase, in genetically different strains of *E. coli* grown in the presence of the sugars glucose or lactose in their environment.

| | **genotype** | |
|---|---|---|
| **sugar** | + | − |
| glucose | no enzyme | no enzyme |
| lactose | enzyme | no enzyme |

## 1.9 Pleiotropy, Penetrance, and Expressivity

So far we have discussed genes that produce all-or-none effects in phenotypic development and that influence a single characteristic. Some genes exhibit more complex effects on phenotypic development. Examples of three particular phenomena can illustrate the nature of these complex effects.

If a gene causes phenotypic development of two or more characteristics that are not obviously related, the gene is called **pleiotropic,** or is said to have multiple effects on the phenotype. The primary effect of the sickle-cell allele of the hemoglobin gene is on synthesis of abnormal hemoglobin molecules. However, the abnormal hemoglobin manufactured in a homozygous recessive individual leads to the disorder of sickle-cell anemia as well as other effects in the afflicted person (Fig. 1.13). When these secondary effects are examined individually, many appear to be unrelated; however, by tracing these symptoms back to the primary molecular defect and the abnormal behavior of red blood cells containing sickle-cell hemoglobin, we can understand how they contribute to organ damage in the individual.

If we examine other characteristics carefully, we may recognize them to be pleiotropic effects, or multiple consequences of the primary genic effect on the phenotype. We usually focus attention entirely on the most striking change in the phenotype and often ignore multiple secondary effects of the gene. If we examined individuals very closely and in greater detail, we would find that many, or perhaps most, genes are pleiotropic. Genetic analysis of inherited characters may be difficult, however, if the multiple effects are not related to the primary consequence of the gene. Simple monohybrid inheritance may underwrite the basic genetic pattern, but the complexities of pleiotropy can at first obscure the simple pattern. Just as we can show that different genes may contribute to the development of a single phenotypic character, we can

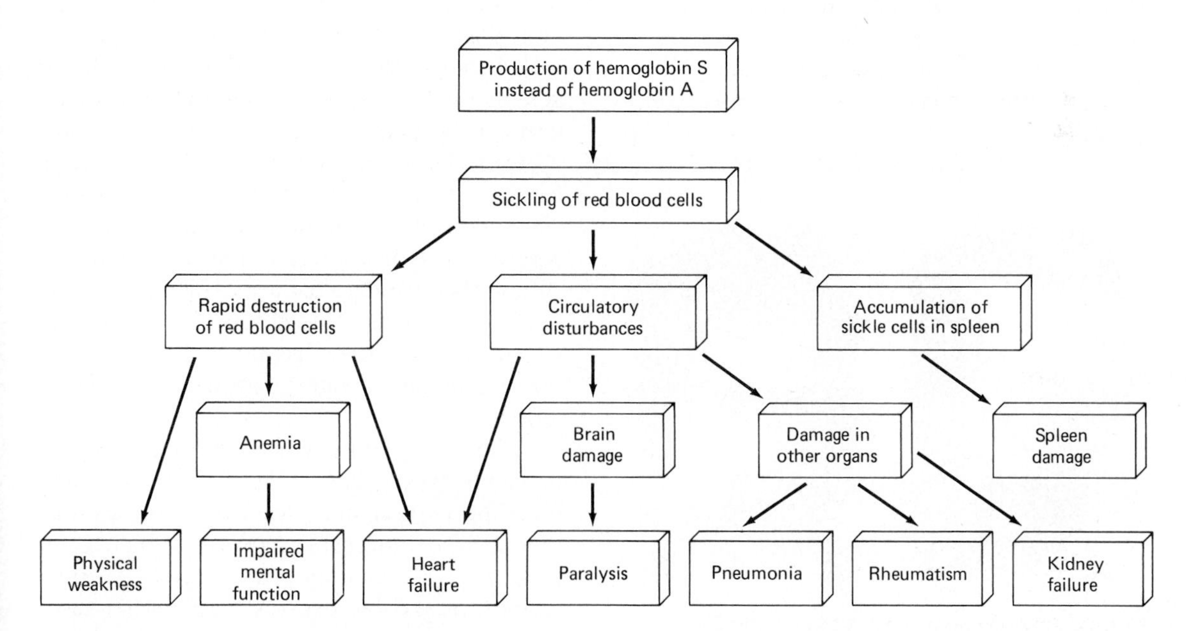


***Figure 1.13*** Pleiotropic effects of the sickle-cell hemoglobin allele in homozygous recessive individuals. Synthesis of sickle-cell hemoglobin is under direct genetic control, but the cascade of clinical effects is due to the abnormal behavior of red blood cells containing hemoglobin S molecules.

also show that a single gene may contribute to modifications in many traits even when these traits may seem to be unrelated at first glance. Breeding analysis is a powerful tool that can be used to unravel apparently complicated inheritance patterns.

A gene that produces the same effect in every individual of the proper genotype is said to have complete **penetrance.** Every *Drosophila* fly that is homozygous for the white eye allele will develop white eyes; the white gene is, therefore, completely penetrant. In some situations, however, individuals with the same genotype may not develop the same phenotype. In these cases the gene is usually *incompletely penetrant,* that is, the gene is expressed only in some individuals of the proper genotype. The degree of penetrance varies from a maximum of 100% for complete penetrance through a range of lower percentages for genes showing reduced penetrance. Incompletely penetrant genes are obviously more difficult both to analyze and to identify in relation to an inheritance pattern.

One example of reduced penetrance is seen in humans in the developmental anomaly called *cleft lip,* in which the upper lip is divided, or cleft. In identical twins, who are produced from the same fertilized egg and are, therefore, genotypically identical, one twin may have the cleft lip condition and the other not (Fig. 1.14).

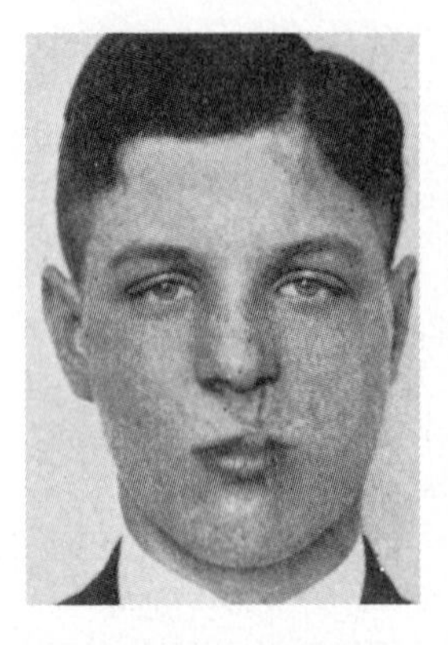

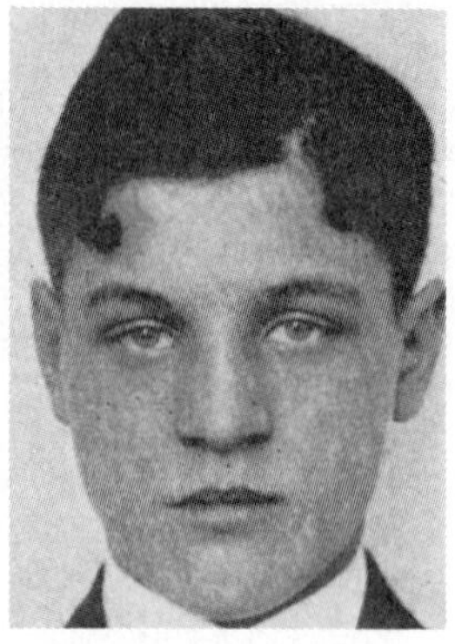

***Figure 1.14*** Demonstration of incomplete penetrance in identical twins: *left,* cleft lip present; *right,* cleft lip absent. (From F. Clausen. 1939. *Z. Abstgs. Vererb.* **76**:30.)

Twin studies are very useful in genetic analysis, particularly when controlled breeding analysis is not possible. Because identical twins have exactly the same genotype, twin studies can identify incompletely penetrant genes that govern a known inherited trait. Where breeding analysis can be conducted, as with mice, flies, and other organisms, inheritance can be demonstrated and the degree of reduced penetrance can be determined on the basis of percentage of appearance of the expected phenotype in a genotypically identical group of progeny.

Genes that show variable **expressivity** produce a range of phenotypes in genotypically identical individuals. Variably expressive genes may show either complete or reduced penetrance since the two phenomena are unrelated. In the case of *polydactyly* in humans, individuals carrying a dominant allele of the gene may have more than five fingers or more than five toes on one or both hands and feet (Fig. 1.15). Polydactylous persons having five fingers on each hand may have six toes on each foot, or those with six fingers may have five toes, or some may have six digits on both hands and feet. Some individuals have a different number of fingers on each hand or toes on each foot. Although the gene is variably expressive, the inheritance pattern can be determined on the basis of transmission of the trait in a family. Dominance is indicated by the appearance of the trait in each generation of the family; by its appearance in approximately half the offspring when only one parent has the trait (a backcross ratio of $Pp \times pp \rightarrow 1\ Pp : 1\ pp$), and its absence in families in which neither parent has the trait ($pp \times pp$). These phenotypic ratios indicate monohybrid inheritance.

Whether the inheritance pattern is analyzed in family (pedigree) studies or in controlled breeding experiments, we can often see that monohybrid inheritance governs certain complex traits showing pleiotropy, reduced penetrance, or variable expressivity. In other cases two or more genes may govern the trait. Whatever the number of genes involved, inheritance patterns can reveal the explanation for a par-

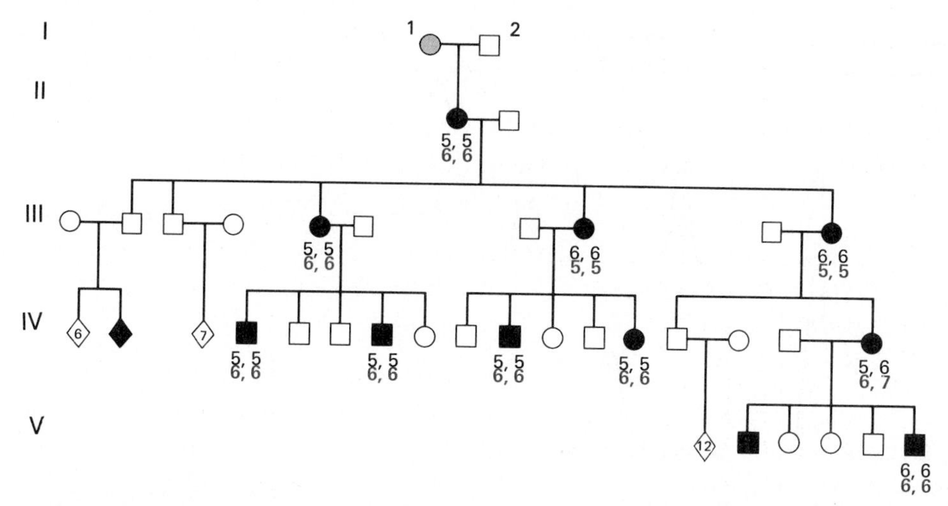


***Figure 1.15*** Pedigree of polydactyly showing variable expressivity of this dominant trait. The phenotype of individual I-1 is uncertain but was probably polydactylous. Affected individuals are shown by black symbols, below which is shown the number of digits on each hand (black) and on each foot (color). Females are shown by circles, males by squares, and combinations of males and females by a diamond with the number of individuals represented. Five generations (I–V) are included in this family history.

ticular situation. Variations in gene expression during phenotype development will not alter the pattern of gene transmission. Segregation of members of pairs of alleles and their assortment into new combinations will still take place according to Mendelian principles.

## Quantitative Inheritance

Mendel and many other geneticists early in this century studied characteristics that could easily be assigned to discrete phenotypic classes, such that plants were either tall or short and flowers were either red, pink, or white. This kind of variation is called **qualitative,** or **discontinuous,** that is, the phenotypic classes are distinct, different, and not overlapping. Many other traits show a range of **quantitative,** or **continuous variation** from one extreme of a range to another. When this is the case it is difficult or impossible to separate individuals into separate and unique phenotype classes. For example, traits such as stature usually vary from one extreme of shortness to the other extreme of tallness (Fig. 1.16). Mendel's tall and short peas were somewhat exceptional. In most cases we cannot easily separate increments of height differences into anything other than arbitrary categories.

Many important features, such as crop yield, milk production, height, weight, and intelligence exhibit continuous variation. Early geneticists, therefore, found it important to determine whether Mendelian inheritance was responsible for continuous variation as it had been shown to be for discontinuous inherited variation. Plant geneticists such as H. Nilsson-Ehle in Sweden and R.A. Emerson and E. East in the United States provided the first genetic analyses and explanations in Mendelian terms for the inheritance of continuously variable traits. Their studies still stand as the major experimental basis for understanding *quantitative inheritance,* the transmission of traits showing continuous variation in phenotypic expression.

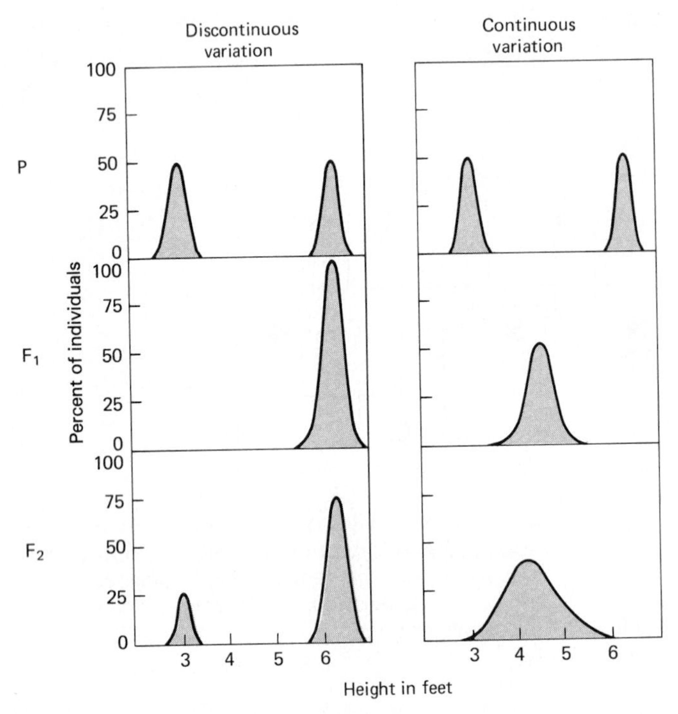


***Figure 1.16*** Discrete phenotypic classes are characteristic of discontinuous variation; most continuously variable traits, such as height, are distributed across a continuous gradation between one phenotypic extreme and the other.

## 1.10 Multiple Gene Hypothesis

Nilsson-Ehle, Emerson, East, and others proposed that quantitative traits are expressions due to the *additive* influence of three or more genes that govern the same trait. They postulated that each gene contributes a portion of the total inheritance, that is, **multiple genes,** or **polygenes** as they are called today, are responsible for inheritance of quantitative traits. In Nilsson-Ehle's experiments, quantitative expression of kernel color in wheat was genetically analyzed. We can use his information to see how the principles of quantitative inheritance were derived.

Crosses between different strains of wheat (*Triticum vulgare*) produced $F_2$ progenies with 3:1, 15:1, or 63:1 ratios of red:white kernels. These ratios reflect monohybrid, dihybrid, and trihybrid phenotypic ratios, respectively, and indicate that red kernels are produced by plants having dominant alleles of one or more color genes. If the dihybrid $F_2$ progeny plants were examined more carefully, five phenotypic classes could be identified instead of just the two classes of red and white. In addition to kernels that were as dark red or as white as the two parental types, there were three intermediate classes showing gradually lighter red–colored kernels, in a phenotypic ratio of 1:4:6:4:1 (Fig. 1.17). Similarly, the trihybrid ratio of 63 red:1 white could be characterized instead by 1:6:15:20:15:6:1, according to the different shades of red that were observed.

With dominant alleles of three genes interacting additively to produce a phenotype, it is still possible to see slight discontinuities between classes. We can interpret these discontinuities in Mendelian terms according to segregation and independent assortment of members of pairs of alleles during reproduction (Fig. 1.18). If more than three genes contribute to the phenotypic expression of a trait, the discontinuities become smaller and smaller, until we cannot resolve the separate phenotypic classes. Variation in the trait will appear to range continuously across a gradation of expressions from one extreme to the other in a progeny or population. By extension of the multiple gene hypothesis derived from studies of one, two, or three genes, we can theoretically accommodate situations in which more than three genes influence a single quantitative trait. When we obtain a measure of some quantitative trait for each individual in a population sample and then group these according to the frequency of individuals showing different measurements, the distribution approximates a bell-shaped curve of

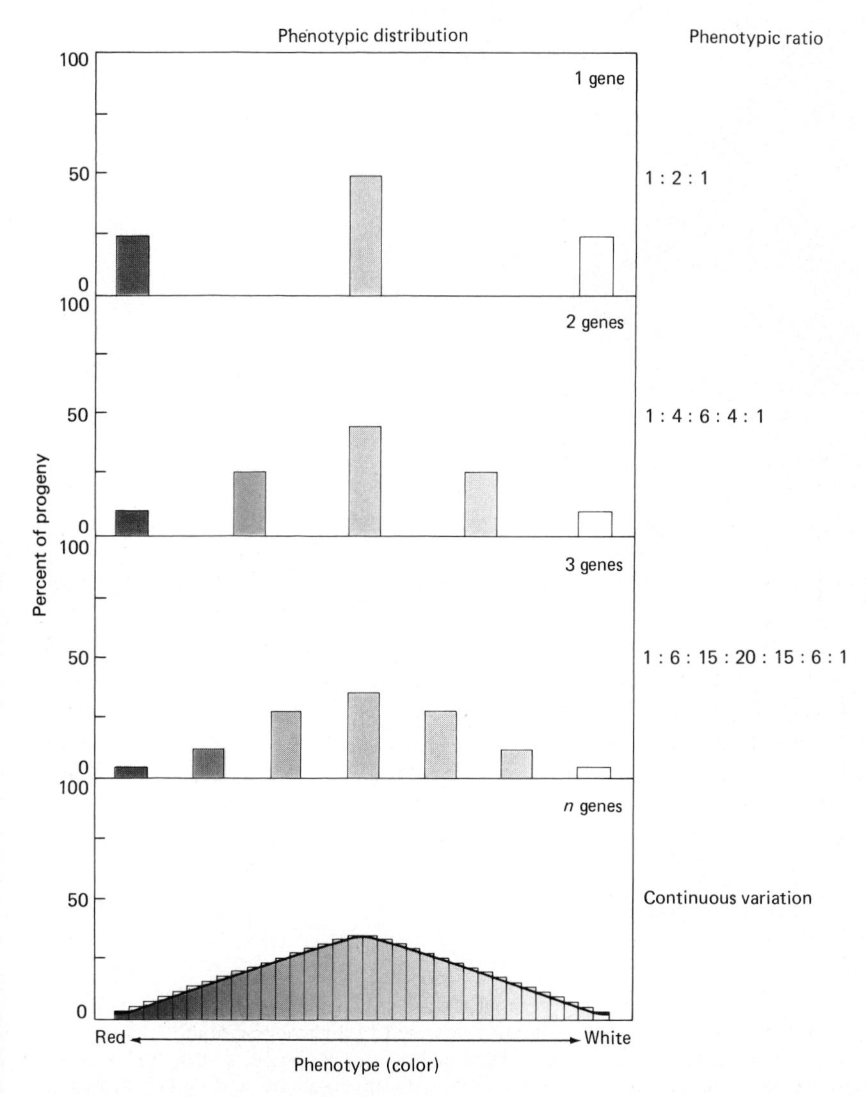


***Figure 1.17*** Distribution of phenotypes across a spectrum of color differences from the extreme of darkest color to the other extreme of white, according to the number of genes involved in the phenotypic expression. As the number of governing genes increases so does the number of phenotypes. With multiple genes (polygenes) the numerous phenotypic classes merge into a spectrum showing continuous (graduated) variation in color from dark to white.

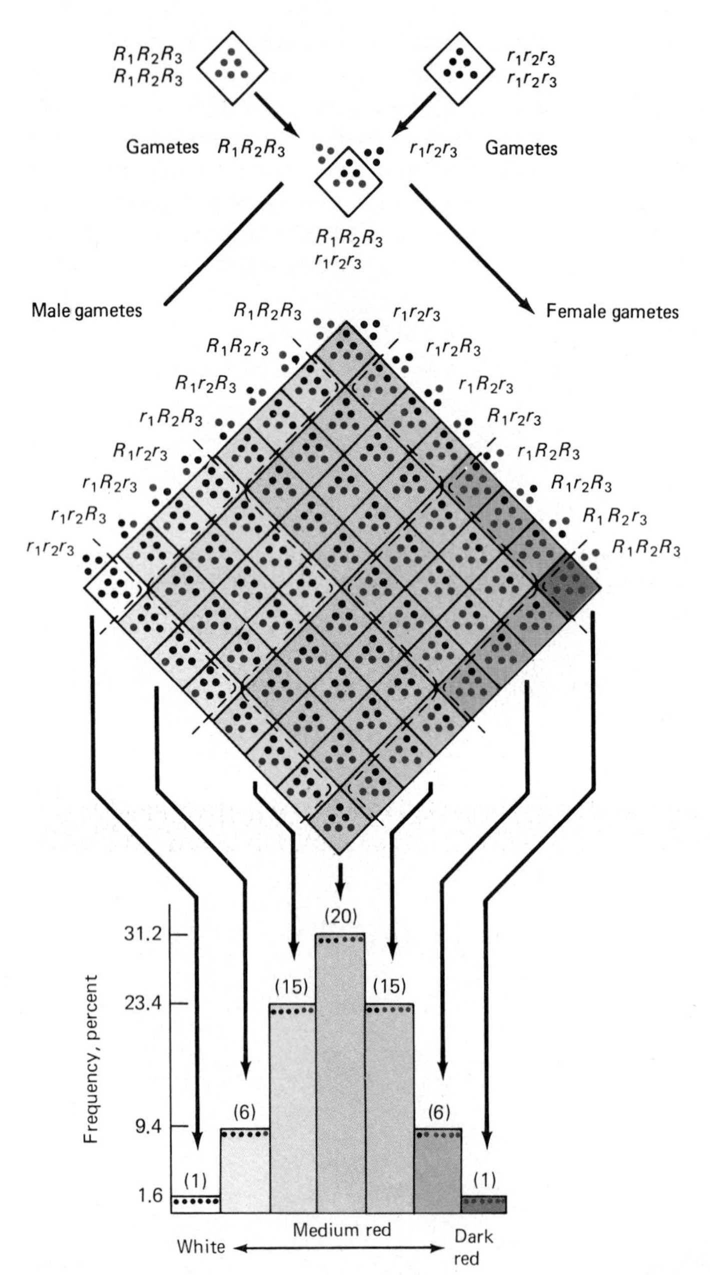


*Figure 1.18* Interaction of three genes in expression of kernel color in wheat. Crosses between red plants that are homozygous dominant for all three genes and triply recessive white plants produce medium red $F_1$ progeny. Self-fertilized $F_1$ individuals give rise to $F_2$ progeny consisting of $63/64$ red : $1/64$ white. This trihybrid phenotypic ratio can be further refined according to the particular shade of red in the colored plants, as shown in the summarizing histogram below. Each shade is the phenotypic result of a particular number of dominant alleles in a genotype, with lightest shades due to fewer dominant alleles and darker shades due to more of the six possible dominant alleles being present. (Adapted with permission of Macmillan Publishing Company from *Genetics,* 2nd ed. by M. Strickberger. Copyright © 1976 by Monroe W. Strickberger.)

continuous variation. Such a bell-shaped curve is typical of a *normal distribution* of continuous variation (Fig 1.19). We will discuss this further in Section 2.5.

We can sometimes estimate the number of genes involved in determining a quantitative trait. We saw that the proportions of the two extreme classes were progressively reduced as the number of genes increased from one to three for red kernel color in wheat. With one pair of alleles the phenotypic extremes were $\frac{1}{4}$ red and $\frac{1}{4}$ white in the $F_2$, with two pairs of alleles they were $\frac{1}{16}$ red and $\frac{1}{16}$ white, and with three pairs of alleles they were $\frac{1}{64}$ red and $\frac{1}{64}$ white. In other words, when $F_1$ parents are heterozygous for $n$ genes governing a single phenotypic trait, $(\frac{1}{4})^n$ of the $F_2$ progeny will resemble one of the original homozygous parents. As $n$ increases, the value of $(\frac{1}{4})^n$ decreases progressively and the number of extreme phenotypes may be determined only from large populations or from adequate samples of such populations. Since the number of phenotypic classes within the range of variation also increases, it becomes virtually impossible to distinguish separate classes. Variation is therefore observed to be essentially continuous, since clear-cut discontinuities cannot be resolved except on an arbitrary basis.

In human beings, a number of different genes are believed to be responsible for skin color variations between blacks and whites. This estimate is based on pedigree analyses, one of the earliest of which was reported by C.B. Davenport in 1913. Davenport recognized five grades of pigmentation in different individuals. He then investigated a number of marriages between blacks and whites and found that the $F_1$ offspring were exactly intermediate in skin color to both parents. He interpreted these observations as showing that pigmentation differences arose from additive effects of different genes governing this phenotypic trait (Fig. 1.20). From distributions of skin color measurements among 32 offspring of $F_1 \times F_1$ matings, the frequencies

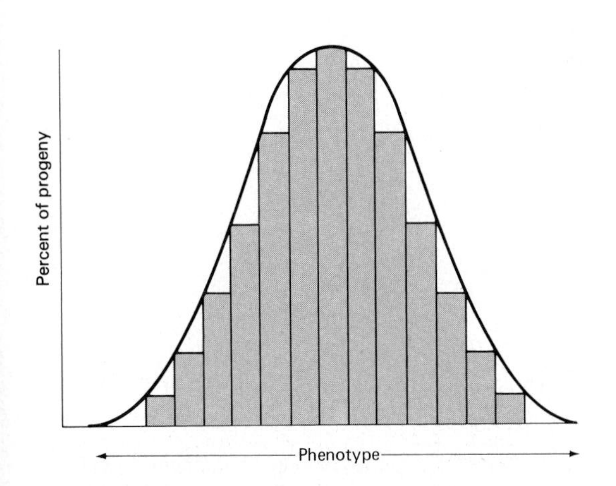


***Figure 1.19*** Bell-shaped curve typical of a normal distribution of frequencies of a continuously variable phenotypic trait. Such a curve is typical of polygenically inherited quantitative traits.

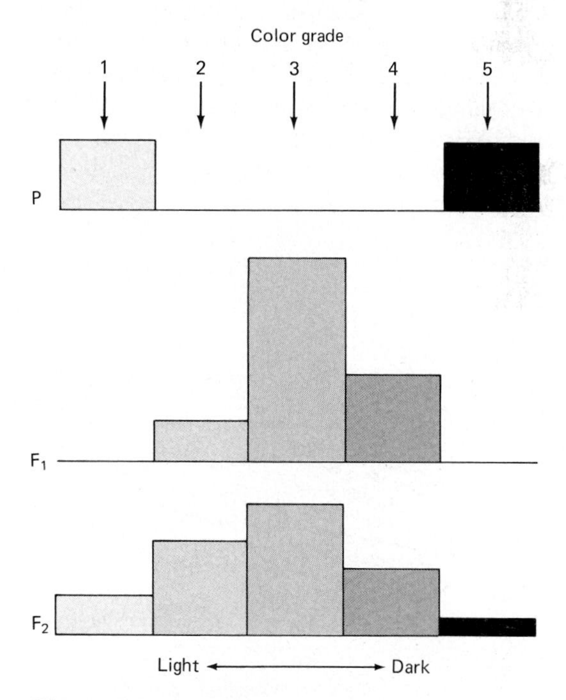


***Figure 1.20*** C.B. Davenport postulated the occurrence of five phenotypic classes of skin color, with the distribution shown. Such a distribution of $F_1$ and $F_2$ phenotypic classes led to the interpretation that skin pigmentation in humans is due to the equal and additive effects of two different pairs of alleles. (See Fig. 1.17 for general reference.)

of all five color grades were interpreted to be due to two genes governing skin color since there was approximately a 1:4:6:4:1 phenotype ratio in the $F_2$.

Davenport's classification of five color grades was rather arbitrary. More recent information on skin color, according to reflectance measured photometrically, has shown that there is a continuous gradation of skin color from light to dark rather than a stepwise distribution (Fig. 1.21). From wider samplings of American populations, more than two genes appear to be involved in pigmentation. The best estimate at present is that four or five different genes contribute to skin color variations from dark to light.

Pigmentation is clearly a quantitative trait based on polygenic inheritance. The genes have at least some additive effect in producing the final phenotype. Two individuals, each having some alleles for dark skin, can produce children as dark or somewhat lighter or darker than themselves. It is genetically unlikely that two light-skinned parents will produce a very dark-skinned child, however, since alleles for dark skin are not known to be recessive to alleles for white skin color.

## 1.11 Hybrid Vigor and Gene Interactions

When two different inbred lines of corn are crossed, the hybrid progeny are almost always considerably more vigorous than their parents. This phenomenon, called **hybrid vigor,** or **heterosis,** is not restricted to corn. Many expressions of quantitative traits in plants and animals show striking differences between parents and progeny. Although the $F_1$ hybrids are much more vigorous than either parent, vigor and uniformity generally decline in the $F_2$ and later generations. This observation should immediately suggest that genetic segregation and recombination take place in the $F_2$ and later generations and are responsible for an increasing spread of variability when compared with the $F_1$.

There is no consensus of opinion on the genetic explanation for hybrid vigor; several different explanations have been proposed. According to some geneticists, heterozygosity itself is required for heterosis. In effect, they have proposed that the heterozygous combination of alleles is superior to any homozygous com-

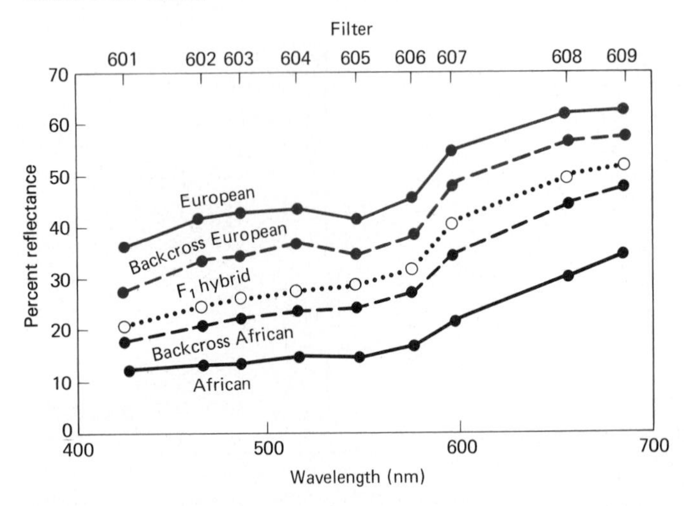


*Figure 1.21* Mean reflectance from the skin of Europeans, Africans, and various hybrid groups as measured with a reflectance spectrophotometer using filters of selected wavelengths (in nm) of visible light. The percent reflectance at each selected wavelength, according to the filter used, is shown for each phenotypic class. (Reprinted by permission of the publisher, Cambridge University Press, from G. Harrison and J. Owen. 1964. Studies on the inheritance of human skin color. *Ann. Human Genet.* **28**:27.)

bination of alleles. The underlying implication is that the different alleles do different things, and the sum of their interaction yields a more vigorous individual than when only a single kind of allele is present. This hypothesis is known as *overdominance*.

Another hypothesis favored by many geneticists is that heterosis results from ordinary dominance and recessiveness relationships between alleles and not from overdominance. It is simply due to dominance of alleles that are favorable to vigor and recessiveness of alleles that are unfavorable to vigor. Since inbreeding of normally cross-fertilizing corn will lead to homozygosity of a number of genes, some genes will be homozygous for recessive alleles and other genes will be homozygous for dominant alleles. If the different recessive alleles are even mildly unfavorable to vigor, inbreeding of strains will result in decline in vigor. When two different inbred lines are crossed, however, their hybrid may have favorable dominant alleles for more different genes than either parent had. For example, if we consider only six different pairs of alleles in a hypothetical cross, we might have the following situation:

| **inbred 1** | | **inbred 2** |
|---|---|---|
| *aabbCCDDeeFF* | × | *AABBccddEEFF* |

$F_1$ *AaBbCcDdEeFF*

In this hypothesis of vigor, heterozygosity is incidental to the expression of hybrid vigor and is not a requirement. Theoretically, we should expect plants of the genotype *AABBCCDDEEFF* to be just as vigorous as those having genotype *AaBbCcDdEeFf* or any other genotype in which either one or two dominant alleles are present for each gene concerned with vigor. Only the presence of a dominant allele is important, not whether it occurs in homozygous or heterozygous combination.

If hybrid vigor is due to a group of favorable dominant genes, we should be able to obtain lines of corn that breed true for vigor. If, on the other hand, heterosis is due to allelic interactions and overdominance, then heterozygosity is a required condition and we would not expect to find stable, homozygous lines that breed true for vigor. Although it seems a simple matter to test the validity of one hypothesis or the other by performing the appropriate crosses, it is probably impossible on a practical level. The number of different genotypic classes segregating in the progeny of a cross between two heterozygotes for $n$ genes is $3^n$. There are 729 genotypic classes ($3^6$) for 6 heterozygous pairs of alleles and tens of thousands or even millions of genotypic classes that are possible for 10 or more heterozygous pairs of alleles. The land area needed to grow an adequate progeny population could easily exceed the total land area of the Earth, and the chance of finding even one homozygous genotype is virtually zero. Attempts to isolate lines of corn that are true-breeding for vigor have not been successful, as we might predict, because of the technical problems of obtaining a sufficiently large progeny population.

In addition to the practical problems in relation to population size, other difficulties in detecting some particular genotype among the progeny exist. Quantitative traits generally are influenced by environmental factors, and these would have to be sorted out from genetic effects. At the present time, there is no clear evidence in favor of any proposed explanation for hybrid vigor nor any way of choosing among the several alternative and compromise explanations.

Although the theoretical basis for hybrid vigor remains in doubt, its practical use is a long-standing agricultural approach. Once a grower obtains suitable inbred lines that can serve as parents for the commercial crop, the business of producing hybrid seeds for the farmer can be established. In the case of corn, hybrid seeds for the farmer are usually produced through double crosses between different $F_1$ hybrid lines (Fig. 1.22). In this procedure, $F_1$ hybrids are obtained from crosses between inbred lines A and B and between inbreds C and D. The $F_1$ AB and the $F_1$ CD hybrids are then interbred to yield double-cross hybrid seeds that are genetically (AB)(CD).

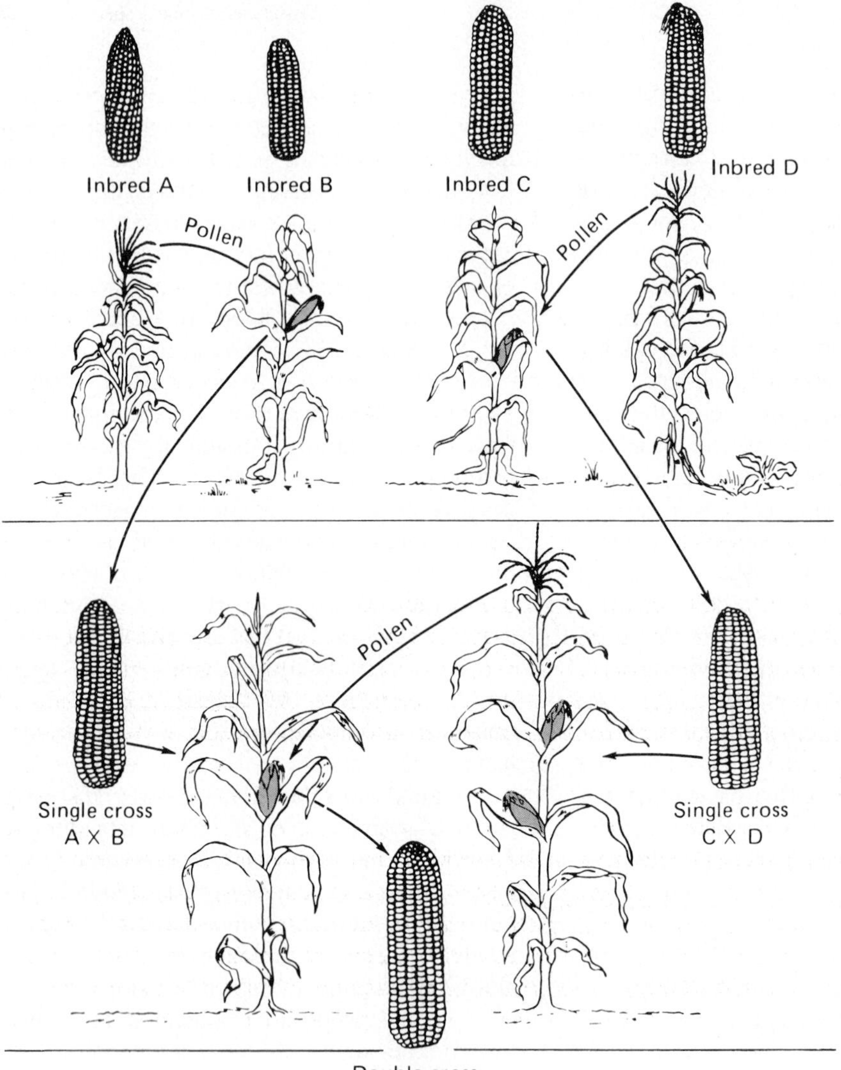


***Figure 1.22*** The double cross method used to obtain (A × B) × (C × D) hybrid corn seed for planting. (After Dobzhansky.)

The reason for the double-cross procedure is that inbred lines of corn are generally small plants with little vigor, and the maternal parent of the $F_1$ AB or $F_1$ CD hybrids produces small ears with a low yield of $F_1$ seeds. These $F_1$ seeds develop, of course, in the ears produced on the maternal inbred plant. If these $F_1$ AB and CD seeds are collected and planted, they produce vigorous $F_1$ AB or CD hybrid plants. These $F_1$ single-cross hybrids produce large and uniform ears when pollinated by other $F_1$ single-cross hybrids used as male parents. The double-cross hybrid seeds harvested from the single-cross $F_1$ maternal parent plants are then sold to the farmer. These double-cross seeds develop into vigorous, high-yielding, uniform $F_1$ double-cross plants.

A new batch of $F_1$ double-cross seeds must be purchased each year from the hybrid seed growers. If the farmer used seeds from the $F_1$

double-cross plants for the next year's crop, the plants grown from these seeds would be $F_2$ progeny. Considerable variation rather than uniformity would occur in such an $F_2$ population, and many of the segregants would be less vigorous than the original $F_1$ double-cross plants. To profit from the advantages of hybrid vigor, the farmer must obtain a new supply of $F_1$ hybrid corn seeds each year.

In corn and in other species that show hybrid vigor in $F_1$ progeny produced by inbred parent lines, many genes contribute to the expression of the quantitative trait. These polygenic systems apparently do not operate on the cumulative effects of additive genes. Instead, they appear to represent gene systems in which dominance interactions and nonallelic interactions contribute to quantitative inheritance. Vigor describes an increase in quantitatively measured characteristics, such as height, yield, and weight. The $F_1$ progeny greatly exceed the parental types in these measured traits, rather than being intermediate in phenotypic expression. Two short inbred corn lines that produce small ears with few seeds can give rise to $F_1$ progeny of considerably greater height that produce large ears with many seeds. Whether the $F_1$ progeny is intermediate or considerably different from the parental types depends on the nature of gene action and interaction in phenotypic expression. Quantitative inheritance underlies those traits that express continuous variation and are not separable into discrete phenotypic classes. The genetic evidence in all cases of quantitative inheritance reveals a polygenic basis. The gradation of continuous variation of expression from one extreme to another extreme is due in all of these cases to segregation and reassortment of alleles of many genes influencing the same trait.

In quantitative inheritance we can clearly see that the behavior of alleles during reproduction follows Mendelian principles; these alleles, however, contribute *additively* to the expression of the phenotype. A detailed analysis of quantitative inheritance requires statistical methods. We will discuss these methods as well as others used to analyze Mendelian phenotypic ratios in Chapter 2.

Throughout this chapter we have seen that Mendelian rules of inheritance can be applied to complex as well as to simple genetic interactions. In every case members of pairs of alleles segregate and undergo independent assortment during reproduction. Dominance relationships, multiple alleles of a single gene, nonallelic gene interactions, variable expression of phenotypic characters, and additive effects of alleles of different genes can be interpreted according to the basic Mendelian laws of allelic segregation and independent assortment. These basic principles, which remain valid today, have provided the solid foundation for genetic analysis in this century.

## Questions and Problems

***1.1*** For one of the genes governing coat color in mice, a dominant allele *B* produces black and its recessive alternative allele *b* produces white.

***a.*** Six types of mating are possible for the three different genotypes, *BB*, *Bb*, and *bb*. For each mating shown below, indicate the genotypes and phenotypes of the $F_1$ progeny and the ratios in which these are expected to occur. (1) *BB* × *BB*, (2) *BB* × *Bb*, (3) *BB* × *bb*, (4) *Bb* × *Bb*, (5) *Bb* × *bb*, (6) *bb* × *bb*.

***b.*** A black male of unknown genotype was crossed with a white female, and the progeny consisted of approximately equal numbers of black and white offspring. What is the genotype of the male parent?

***1.2*** Tomatoes can be either yellow or red. Plants of these two phenotypes were crossed as follows:

| parents | progeny |
|---|---|
| red × red | 61 red |
| red × red | 47 red, 16 yellow |
| red × yellow | 58 red |
| yellow × yellow | 64 yellow |
| red × yellow | 33 red, 36 yellow |

***a.*** What phenotype is dominant?
***b.*** What are the genotypes of the parents and progeny in each cross?
***1.3*** In *Drosophila,* scarlet eye color is produced in homozygous recessive flies; normal flies have red eyes.
***a.*** If homozygous red-eyed females are crossed with scarlet-eyed males, what will be the eye color of the offspring?
***b.*** If males and females of the $F_1$ progeny are interbred, what phenotypes can we expect in the $F_2$ and in what proportions?
***c.*** If an $F_1$ female is crossed to a scarlet-eyed male, what kinds of progeny will be produced and in what proportions?
***1.4*** A black guinea pig was crossed to a white guinea pig, and the litter consisted entirely of black offspring. The $F_1$ offspring were used in the following crosses:

| parents | progeny |
|---|---|
| $F_1$ black × $P_1$ black | 10 black |
| $F_1$ black × $F_1$ black | 8 black, 3 white |
| $F_1$ black × $P_1$ white | 5 black, 4 white |

***a.*** Which is the dominant and which is the recessive phenotype?
***b.*** If *B* represents the dominant allele and *b* the recessive allele, what are the genotypes of the original parents?
***c.*** What are the genotypes of the $F_1$ black guinea pigs?
***d.*** What are the genotypes of the black progeny in the third cross?
***e.*** If $F_1$ black was crossed to $F_2$ white, what kinds of offspring would be produced and in what proportions?
***1.5*** If you have a single black guinea pig of unknown parentage, how would you determine whether the animal is homozygous or heterozygous for coat color?

***1.6*** Four generations of guinea pigs are shown in the pedigree below, from crosses between parents I-1 and I-2 and brother-sister matings between II-3, II-4, and III-2, III-3. Black is dominant over white (use *B*, *b* as allelic symbols), ○ = ♀ white, □ = ♂ white, ● = ♀ black, ■ = ♂ black.

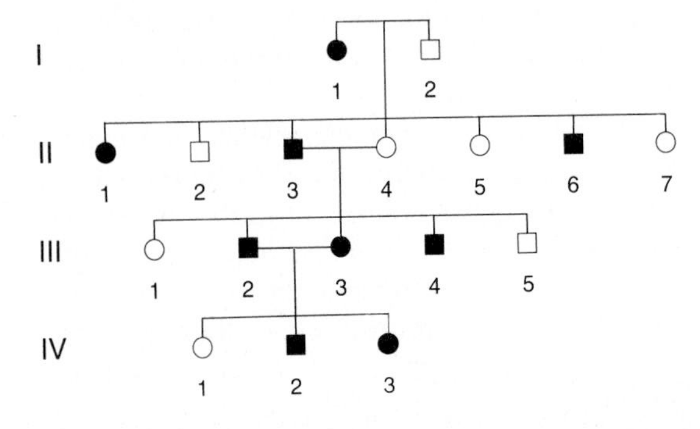


***a.*** What are the genotypes of parents I-1 and I-2?
***b.*** Which of the black guinea pigs are definitely heterozygous and which are possibly homozygous for the dominant allele?
***c.*** What would be the color of offspring produced by brother-sister mating between III-1 and III-5?
***1.7*** In corn, yellow seedlings (*g*) is recessive to green seedlings (*G*), and waxy endosperm (*wx*) is recessive to starchy endosperm (*Wx*). The two genes assort independently. A cross between a homozygous green, waxy plant and a homozygous yellow, starchy plant produced $F_1$ progeny that were phenotypically green, starchy.
***a.*** If $F_1$ individuals are interbred, what phenotypes would be found in the $F_2$ generation and in what proportions?
***b.*** If some $F_1$ plants are crossed to plants from the green, waxy variety, what genotypes and phenotypes would be found in their progeny and in what ratios?
***c.*** If particular $F_2$ plants are self-fertilized in an experiment, such $F_2$ plants could produce $F_3$ progeny as follows:

| $F_2$ phenotype | $F_3$ progeny |
|---|---|
| green, starchy | 33 green, starchy; 10 green, waxy |
| green, starchy | 51 green, starchy |
| green, starchy | 101 green, starchy; 34 green, waxy; 35 yellow, starchy; and 12 yellow, waxy |
| green, starchy | 44 green, starchy; 15 yellow, starchy |

What is the genotype of each of the four self-fertilizing $F_2$ plants listed?

***d.*** What cross would you perform to determine the genotype of a green, starchy plant of unknown parentage, if self-fertilization could not be carried out?

***1.8*** In *Drosophila,* black body color is due to a recessive allele *b* and wild-type (gray) body color, to its dominant allele $b^+$; and sepia eye color is due to a recessive allele *s* and wild-type (red) eye color, to its dominant allele $s^+$.

***a.*** What would be the genotypes and phenotypes of $F_1$ and $F_2$ progeny produced in crosses between homozygous wild-type females and males with black body color and sepia eyes, and in what ratios would these be expected to occur?

***b.*** What would be the genotypes and phenotypes of $F_1$ and $F_2$ progeny produced in crosses between homozygous females with black body color and red eyes and homozygous males with gray body color and sepia eyes, and in what ratios would these be expected to occur?

***c.*** Wild-type males of unknown genotype are backcrossed to females with black body color and sepia eyes. What would be the ratios of genotypes and phenotypes in the progeny if the wild-type males are (1) homozygous for both traits, (2) heterozygous for both traits, or (3) homozygous for one trait and heterozygous for the other?

***1.9*** A cross between two snapdragon plants produced 83 plants with pink flowers, 35 reds, and 36 whites.

***a.*** What are the genotypes and phenotypes of the two parent plants?

***b.*** What phenotypes would you expect and in what ratios, among the progeny of the following crosses: (1) pink × pink, (2) red × red, (3) red × white, and (4) pink × white?

***1.10*** Coat colors of the Shorthorn breed of cattle are governed by a pair of codominant alleles, $C^R$ and $C^W$. Red cattle have the genotype $C^RC^R$, white cattle are genotypically $C^WC^W$, and $C^RC^W$ heterozygotes are roan (a mixture of red and white).

***a.*** What would be the ratios of $F_1$ genotypes and phenotypes in the following crosses: (1) red × white, (2) red × roan, (3) roan × roan, and (4) roan × white?

***b.*** What percentage of the $F_1$ progeny will be true-breeding in each of the crosses listed in (a)?

***1.11*** In *Drosophila,* the Tetraptera allele (*T*), which causes development of four wings, is dominant over the wild-type allele ($T^+$), which governs development of normal flies with two wings. The dominant trait is expressed only in heterozygotes since *T/T* is a lethal condition. What phenotypic and genotypic ratios would be expected in progeny of the following crosses: (1) wild-type × tetraptera, and (2) tetraptera × tetraptera?

***1.12*** Multiple alleles of the ABO blood-group gene govern the four major blood types in humans (types A, B, AB, and O). Alleles $I^A$ and $I^B$ are codominant and allele $i^O$ is recessive to both codominant alleles. A woman of blood type O is suing a man of blood type AB in a paternity suit concerning her child, who is blood type A.

***a.*** Can this man be the father of her child? Explain.

***b.*** What blood types would definitely rule out this man as the father of this woman's child? Explain.

***c.*** Suppose the woman and her child were both of blood type AB. What possible blood types would you expect to characterize the father of this child? Explain.

***1.13*** Four babies, two of whom are fraternal twins (produced from two different fertilized eggs), were born in a hospital during the same night. Their blood types were A, B, AB, and O. The three sets of parents had blood types: B and O, A and B, and AB and O.

***a.*** If the fraternal twins were known to have blood type A and blood type B, and the other two babies were blood types AB and O, which babies belong to which parents?

***b.*** Having assigned each baby to its parents, list the genotypes of each baby and its parents.

***1.14*** In chickens, two different genes govern comb shape, such that genotype *R–P–* produces walnut comb, *R–pp* rose comb, *rrP–* pea comb, and *rrpp* single comb.

***a.*** How many different genotypes will produce a walnut comb phenotype? What are these genotypes?

***b.*** A cross between walnut and rose parents produces progeny consisting of 31 walnut, 28 rose, 10 pea, and 9 single. What are the genotypes of the parents?

***c.*** If chickens with walnut, rose, and pea combs are readily available in all possible genotypic combinations, what would be the genotypes and phenotypes of the animals you would select to produce the highest percentage of single-combed chickens in a single progeny?

***1.15*** Crosses between a true-breeding white onion strain and a pure yellow strain produce $F_2$ progeny exhibiting a ratio of 12 white:3 red:1 yellow. A different pure white strain crossed with a pure red strain produced $F_2$ progeny exhibiting a ratio of 9 red:3 yellow:4 white.

***a.*** What is the inheritance pattern governing color in each case?

***b.*** If the two different pure white strains were crossed, what would be the ratios of genotypes and phenotypes of $F_1$ and $F_2$ progeny? (*Hint*: one strain is homozygous dominant and the other strain is recessive for all genes involved.)

***1.16*** Crosses between pure-breeding red and pure-breeding white swine produce red $F_1$ progeny. Upon interbreeding red $F_1$ animals, however, sandy-colored offspring are produced along with reds and whites. From a number of crosses it was found that the three phenotypes occurred in the $F_2$ progeny in a ratio of $\frac{9}{16}$ red:$\frac{6}{16}$ sandy:$\frac{1}{16}$ white. What is the nature of the inheritance pattern?

***1.17*** Crosses between pure-breeding white and pure-breeding green squash plants produce white $F_1$ progeny, but the $F_2$ segregates $\frac{12}{16}$ white:$\frac{3}{16}$ yellow:$\frac{1}{16}$ green. In the $F_3$ offspring produced by self-fertilization, all the green plants were found to be true-breeding, but about $\frac{1}{3}$ of the $F_2$ yellow plants bred true while about $\frac{2}{3}$ segregated 3 yellow:1 green.

***a.*** What is the nature of the inheritance pattern?

***b.*** What ratios of genotypes and phenotypes would be produced in backcrosses between individual $F_2$ white plants and green plants?

***c.*** What crosses would you perform to identify homozygous and heterozygous yellow plants if self-fertilization was not possible?

***d.*** How many genotypically different varieties of white squash could you isolate as pure-breeding types? What are their genotypes?

***1.18*** Two pure strains of corn with white kernels were crossed. All the $F_1$ progeny had red kernels; the $F_2$ progeny consisted of 91 red and 69 white. What is the nature of the inheritance pattern for kernel color in this case?

***1.19*** The presence of extra fingers and/or toes (polydactyly) in humans is governed by a dominant allele *P*, which is incompletely penetrant since some individuals of genotype *Pp* are not polydactylous. Four generations of a family history are shown in the pedigree at the top of the next column, ○ = ♀ normal, □ = ♂ normal, ● = ♀ polydactylous, ■ = ♂ polydactylous:

***a.*** What are the probable genotypes of members I-1 and I-2? Explain.

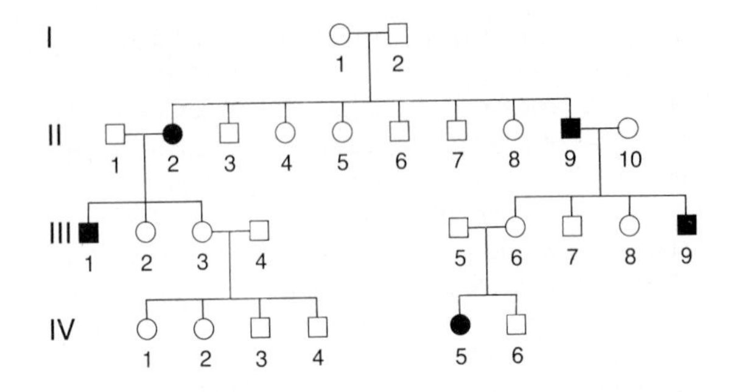


***b.*** How would you explain the polydactylous condition of IV-5?

***c.*** Which parent of IV-5 is more likely to have the genotype *Pp*? Why?

***d.*** What information in this pedigree indicates incomplete penetrance of the polydactyly allele?

***e.*** What information in this pedigree indicates dominant inheritance of the rare trait of polydactyly?

***f.*** What additional information should we have to determine whether or not this rare trait shows variable expressivity?

***1.20*** Kernel color in wheat is governed by a number of independently assorting genes whose dominant alleles contribute additively to color development. Plants produced from crosses between multiply heterozygous parents included 10 with white kernels and more than 2000 with kernels showing various shades of red.

***a.*** How many different color genes were probably involved in this case?

***b.*** If 10 different color genes segregated in the progeny of a cross between multiply heterozygous parents, what would be the minimum number of progeny needed to produce a single homozygous recessive plant with white kernels?

***1.21*** In crosses between large and small true-breeding gerbils, all $F_1$ progeny are intermediate in size. Among 452 $F_2$ progeny, 106 are large and 108 are small. Assuming the simplest case of inheritance, how many pairs of alleles contribute to gerbil size?

***1.22*** From repeated matings between two mice, each having a tail length of 12.5 cm, the following progeny were found to have tail lengths as indicated:

| Tail length, cm | 2.5 | 7.5 | 12.5 | 17.5 | 22.5 |
|---|---|---|---|---|---|
| Number of progeny | 9 | 37 | 57 | 34 | 12 |

***a.*** How many pairs of alleles regulate this characteristic?

***b.*** Give the expected ratio of each class of tail length and the proportion of alleles in each genotype that would produce the above results.
***c.*** What offspring phenotypes would be expected from a mating between progeny having tail lengths of 2.5 cm and 17.5 cm?

***1.23*** Two inbred lines of guinea pigs differ in weight. You know that the difference between strains is due to three pairs of alleles, where each dominant allele adds 40 g to the weight and each "neutral" allele adds only 20 g.
***a.*** If the average weight of strain A is 120 g and that of strain B is 240 g, what are the genotypes of these two strains?
***b.*** What would be the average weight of the $F_1$ progeny of strain A × strain B animals?
***c.*** With respect to weight, what classes of progeny would occur in the $F_2$?
***d.*** How many $F_2$ guinea pigs must you raise to recover four individuals similar in weight to their strain B grandparents?

***1.24*** A cross between two inbred plants that had seeds weighing 30 g and 50 g, respectively, produced an $F_1$ progeny with seeds that uniformly weighed 40 g. The $F_2$ offspring consisted of 4000 plants; 4 had seeds weighing 30 g, 4 had seeds weighing 50 g, and the remaining 3992 plants produced seeds with weights varying between these two extremes. Calculate the probable number of allele pairs involved in the determination of this trait.

***1.25*** In the laboratory albino rat, five pairs of alleles contribute with equal and additive effect to total body weight. Two homozygous inbred strains, one with very high and other with very low body weight, are crossed and their progeny show weights that vary between the two parental body weight extremes. Suppose that the average cost of raising a rat to maturity is $2. What will be the cost to recover in the $F_2$ a rat that shows the phenotype of the "high" body weight parent?

***1.26*** A mouse breeder notes that when pure-breeding varieties with long tails (100 mm) and short tails (50 mm) are crossed, their progeny uniformly have intermediate length tails (75 mm).
***a.*** What is the simplest genetic explanation for this result?
***b.*** When $F_1$ mice having 75 mm-long tails are interbred, however, results such as those shown below are obtained repeatedly in the $F_2$ progeny:

| tail length, mm | 50 | 62 | 75 | 88 | 100 |
|---|---|---|---|---|---|
| number of offspring | 5 | 20 | 30 | 20 | 5 |

How must your answer in (a) be modified?
***c.*** What are the genotypes for each of the five tail-length phenotypes in the $F_2$ progeny?

***1.27*** A farmer produced a bumper crop of hybrid corn from seed purchased from a hybrid corn seed company. He plans to save some of the harvested corn to use for planting the following season. What is your advice to the farmer? Explain.

## References

Correns, C. 1900. G. Mendels Regel über das Verhalten der Nachkommenschaft der Rassenbastarde. *Ber. deutsch. bot. Ges.* **18**:158–168. (English translation in *Genetics* **35,** Suppl. to No. 5, Part 2, pp. 33–41, 1950.)

Davenport, C.B. 1913. Heredity of skin color in Negro-white crosses. *Carnegie Inst. Wash. Publ. No. 554,* Washington, D.C.

Dunn, L.C. 1965. *A Short History of Genetics.* New York: McGraw-Hill.

East, E.M. 1910. A Mendelian interpretation of variation that is apparently continuous. *Amer. Nat.* **44**:65.

Harrison, G.A. and J.J.T. Owen. 1964. Studies on the inheritance of human skin color. *Ann. Hum. Genet.* **28**:27.

Iltis, H. 1951. Gregor Mendel's life and heritage. In L.C. Dunn, ed., *Genetics in the 20th Century.* New York: Macmillan.

Mendel, G. 1866. Versuche über Pflanzen Hybriden. *Verh. naturf. Ver. in Brünn, Abhandlungen, iv.* (English translation in J.A. Peters, ed. 1959. *Classic Papers in Genetics.* Englewood Cliffs, N.J.: Prentice-Hall.)

Nilsson-Ehle, H. 1909. Kreuzungsuntersuchungen an Hafer und Weizen. *Lunds Univ. Aarskr, N.F. Afd.,* Ser. 2, **5**(2):1–122.

Ravin, A.W. 1965. *The Evolution of Genetics.* New York: Academic Press.

Stent, G. Dec. 1972. Prematurity and uniqueness in scientific discovery. *Sci. Amer.* **227**:84.

# CHAPTER 2

# Probability and Statistics

Genetic analysis of Mendelian inheritance patterns usually relies on interpretations of numerical values obtained directly as observations of progeny generations. These numerical data are summarized as ratios of phenotypic classes or of some equivalent as in the case of quantitative inheritance. By applying various statistical methods to the genetic analysis of a characteristic, the geneticist is able to: (1) judge whether or not the observed numerical data fit the expected ratio for the inheritance pattern of the characteristic, (2) calculate the probability of occurrence of an individual or an event, once the inheritance pattern is established, and (3) statistically judge whether or not there is a genetic basis for transmission of the trait from parents to progeny. In all these cases, statistical methods provide an objective means by which some hypothesis can be evaluated according to the numerical data obtained directly as observations of progeny populations in breeding analysis or as samples of individuals drawn from populations at large. Various statistical methods have been adapted to deal with different situations, such as discontinuous versus continuous variation and observations from breeding analysis versus those from a more general sampling of individuals.

# Predicting Outcome of Crosses

In Chapter 1, genotypes resulting from a cross were diagrammed with a checkerboard or a branching system of notation. In more complex breeding tests these methods are somewhat tedious and are subject to inadvertant errors. For most genetic analyses, some shortcut is necessary to reduce the time required to analyze data as well as to provide a check on the conclusions made. Probability formulations offer convenient shortcuts for predicting the proportions and ratios of genotypic and phenotypic classes produced in a cross involving one or more genes. In addition, the chi-square ($\chi^2$) test provides a means for making an objective judgment of the fit between numbers of individuals actually produced in each progeny class in a cross and the numbers expected on the basis of the ratio expected for a particular inheritance pattern.

## 2.1 Probability and Its Application

The familiar genotypic ratio of 1 *AA*:2 *Aa*:1 *aa* is the outcome of a cross between monohybrid heterozygous parents, *Aa* × *Aa*, in which each parent produces *A* gametes and *a* gametes in approximately equal proportions. The probability of the three allelic combinations occurring in the ratio of ¼:½:¼ is based on the random fusions between two gametes of either allelic type because the probability of two independent events occurring simultaneously is the product of their separate probabilities:

egg *A* fertilized by sperm *A*:

$$\tfrac{1}{2} \times \tfrac{1}{2} = \tfrac{1}{4}\,AA$$

egg *A* fertilized by sperm *a*:

$$\tfrac{1}{2} \times \tfrac{1}{2} = \tfrac{1}{4}\,Aa$$

egg *a* fertilized by sperm *A*:

$$\tfrac{1}{2} \times \tfrac{1}{2} = \tfrac{1}{4}\,Aa$$

$$\tfrac{1}{4} + \tfrac{1}{4} = \tfrac{1}{2}\,Aa$$

egg *a* fertilized by sperm *a*:

$$\tfrac{1}{2} \times \tfrac{1}{2} = \tfrac{1}{4}\,aa$$

If *A* is completely dominant to *a*, then we predict the ratio of phenotypic classes to be

3 dominant:1 recessive. If *A* is incompletely dominant to *a*, then we predict 1 parental *AA*:2 intermediate *Aa*:1 parental *aa* ratio of phenotypes. Codominant alleles would also lead to a phenotypic ratio of 1:2:1, with both of the alleles contributing to the phenotypic expression of the *Aa* heterozygous class.

By applying the same probability formulations, we can predict the outcome of a dihybrid cross in which each parent is heterozygous for two pairs of independently assorting alleles. In the cross *AaBb* × *AaBb*, where alleles *A* and *B* are both completely dominant to their recessive alternatives, we predict that each pair of alleles will segregate and reassort at random to produce the phenotypic ratio of 9 doubly dominant:3 dominant only for *A*:3 dominant only for *B*:1 doubly recessive. This ratio is predicted on the basis of the simple rule that the probability of two independent events occurring simultaneously is equal to the product of their separate probabilities. We can predict the types and proportions of each genotypic class by the same rule. We therefore predict that progeny from the cross *AaBb* × *AaBb* will be produced in the 9:3:3:1 ratio of phenotypes, as follows:

both dominant traits expressed: $\frac{3}{4} \times \frac{3}{4} = \frac{9}{16}$

dominant *A* trait and recessive *bb*: $\frac{3}{4} \times \frac{1}{4} = \frac{3}{16}$

recessive *aa* and dominant *B* trait: $\frac{1}{4} \times \frac{3}{4} = \frac{3}{16}$

both recessive traits expressed: $\frac{1}{4} \times \frac{1}{4} = \frac{1}{16}$

The proportion of progeny expected to be of a particular genotype can be predicted by the same rule. For example, the proportion expected of genotype *Aabb* is $\frac{1}{2} \times \frac{1}{4} = \frac{1}{8}$; the proportion of genotype *AaBb* is $\frac{1}{2} \times \frac{1}{2} = \frac{1}{4}$; the proportion of genotype *aabb* is $\frac{1}{4} \times \frac{1}{4} = \frac{1}{16}$.

Probability formulations can be used to predict the outcome of virtually any type of cross, as long as the situation conforms to two simple requirements: (1) the genetic events under consideration are independent of one another and (2) the simultaneous occurrence of two or more such events is due to chance alone. We can predict the proportion of genotypes and phenotypes expected in crosses involving three or four or more genes just as easily as cases of monohybrid or dihybrid inheritance.

Consider the cross *AabbCcDd* × *AaBBccDd*, which involves four genes. In probability calculations for several genes, it is essential to work with each gene separately. For the alleles *A* and *a*, the cross is between heterozygotes, and we can expect a genotypic ratio of $\frac{1}{4}$ *AA*:$\frac{1}{2}$ *Aa*:$\frac{1}{4}$ *aa* and a 3:1 phenotypic ratio in the progeny. For the alleles *B* and *b*, the cross is between a homozygous recessive and a homozygous dominant, and all the progeny will be heterozygous for this pair of alleles and identical genotypically and phenotypically. For alleles *C* and *c*, the cross is a testcross, which will produce a 1:1 genotypic and phenotypic ratio. For alleles *D* and *d*, the cross involves two heterozygotes, and we expect a 1:2:1 genotypic ratio and a 3:1 phenotypic ratio in the progeny.

To predict the proportion of progeny that will show the dominant phenotype for all four traits, our calculation is $\frac{3}{4} \times 1 \times \frac{1}{2} \times \frac{3}{4}$, or $\frac{9}{64}$. The proportion that will show the dominant phenotype for *A* and *B* and the recessive phenotype for *c* and *d* will be $\frac{3}{4} \times 1 \times \frac{1}{2} \times \frac{1}{4}$, or $\frac{3}{32}$. We can also calculate the proportion of the progeny that will show a particular genotype. We expect the genotype *aaBbccDd* to be produced with the frequency of $\frac{1}{4} \times 1 \times \frac{1}{2} \times \frac{1}{2}$, or $\frac{1}{16}$, and the genotype *AaBbCcDd* to be produced with the frequency of $\frac{1}{2} \times 1 \times \frac{1}{2} \times \frac{1}{2}$, or $\frac{1}{8}$. We can use probability calculations to predict the frequency of occurrence of each of the possible genotypes expected from this cross.

Using simple probability formulations as a shortcut, we can predict the outcome of any genetic cross that conforms to the two basic premises mentioned above. When different pairs of alleles assort independently of one another during reproduction and their simultaneous occurrence is due to chance alone, the probability that different allelic combinations will occur randomly together is the product of their separate probabilities.

The same probability formulations apply to any sets of separate events occurring simultaneously, not just to genetic crosses. We have already mentioned that since the chance of tossing a coin heads up is ½ and the chance of tossing it tails up is ½, the probability that two coins will land heads up is ½ × ½, or ¼. The mathematical principles that underlie predictions of outcome of simultaneously occurring independent events apply equally well to nonbiological and biological systems.

## 2.2 Binomials and Factorials

The **binomial expansion** provides another convenient shortcut for predicting the frequencies of genotypes and phenotypes involved in various breeding analyses and in pedigree analyses of human family histories. The expansion is useful for situations in which either one of two alternative outcomes can occur by chance for each independent event, such as heads or tails in numerous coin tosses. In the general binomial expression $(a + b)^n$, the letters $a$ and $b$ represent the probability of one outcome, $a$, or its alternative, $b$, for each of $n$ independent events or individuals.

If we deal with two independent events, then $n = 2$ and the probability for all possible combinations is $(a + b)^2$; for three events it is $(a + b)^3$; for four events it is $(a + b)^4$. The number of individuals, or independent events, therefore, determines the power to which the binomial is raised (Fig. 2.1). After expanding the binomial, we can select the appropriate term to calculate a numerical value for the predicted frequency of a particular combination of outcomes. Suppose we toss three coins simultaneously and we wish to predict the frequency with which two coins will land heads up and the third coin will land tails up. The probability of heads is ½ for each coin, and the probability for tails is ½ for each coin. The probability of heads, or $a$, is ½; the probability for tails, or $b$, is ½. There are three coins (individuals, or independent events), so the appropriate binomial is $(a + b)^3$. The expansion of the binomial is $a^3 + 3a^2b + 3ab^2 + b^3$. The term that fits our question of the predicted frequency for two heads and one tails is $3a^2b$, and we can now substitute the numerical probabilities of $a = ½$ and $b = ½$. The expansion term is therefore $3(½)^2(½)$, which is equal to ⅜. The predicted frequency in percentage is 37.5%. If we had asked for the frequency of one heads + two tails, we would select the term $3ab^2$; the correct term for three heads would be $a^3$, and for three tails it would be $b^3$. These four terms handle all possible combinations of outcomes, or heads and tails, for three coins.

The binomial expansion works quite well when relatively few events are involved. It becomes awkward, however, to find the proper term when a larger number of events is in

| Binomial | Power of binomial ($n$) | No. terms in expansion ($n + 1$) | Number of combinations | Expanded binomial |
|---|---|---|---|---|
| $(a + b)$ | 1 | 2 | 2 | $a + b$ |
| $(a + b)^2$ | 2 | 3 | 4 | $a^2 + 2ab + b^2$ |
| $(a + b)^3$ | 3 | 4 | 8 | $a^3 + 3a^2b + 3ab^2 + b^3$ |
| $(a + b)^4$ | 4 | 5 | 16 | $a^4 + 4a^3b + 6a^2b^2 + 4ab^3 + b^4$ |
| $(a + b)^5$ | 5 | 6 | 32 | $a^5 + 5a^4b + 10a^3b^2 + 10a^2b^3 + 5ab^4 + b^5$ |
| $(a + b)^6$ | 6 | 7 | 64 | $a^6 + 6a^5b + 15a^4b^2 + 20a^3b^3 + 15a^2b^4 + 6ab^5 + b^6$ |

***Figure 2.1*** Expansion of the binomial $(a + b)^n$, where $n$ = the power of the binomial, $n + 1$ = the number of terms in the corresponding expansion, and $a = b = ½$. The coefficient of the first and last term is always 1, and the coefficient of the other terms is the sum of the two coefficients above the term in question. The values of the coefficients form a symmetrical distribution, and the number of possible combinations doubles with each successive increase in the power of the binomial.

question. Whether we are dealing with a small or a large number of individuals, or events, rather than expanding the binomial to find the proper term, we can apply the factorial formula to determine probabilities for a particular question. The *factorial formula* for determining probability is

$$\frac{n!}{x!(n-x)!} \times (a)^x(b)^{n-x}$$

This formula includes the total number of events or individuals ($n$), the number of individuals, or events, with probability $a$ ($x$), and the number of individuals or events with probability $b$ ($n - x$). The symbol $n!$, or $n$ **factorial**, is the product of all the positive integers to the specified term, $n$. For example, if $n = 5$, then $5! = 1 \times 2 \times 3 \times 4 \times 5$; if $x = 3$, then $3! = 1 \times 2 \times 3$. We should note that $0! = 1$. The symbols $a$ and $b$ stand for the probabilities of the two alternative outcomes, just as they did in the binomial expansion. The first part of the factorial formula, $n!/x!(n-x)!$, gives the number of possible combinations of outcomes and the second part of the formula, $(a)^x(b)^{n-x}$, gives the probability of a particular set of possible outcomes occurring in all possible orders.

Suppose a couple wanted to know their chances for having 3 boys and 2 girls in 5 births. Each birth represents an independent event, and the probability for the birth of a boy is ½, for the birth of a girl it is ½. The probability for a boy can be represented by $a$ and the probability for the birth of a girl is $b$. The total number of events (children) is 5, of which 3 are boys and 2 are girls. We can substitute these numbers for $n$ (5), $x$ (3), and $n - x$ (2). With $n$, $x$, $n - x$, $a$, and $b$ stipulated, we can proceed with the calculation as follows:

$\frac{5!}{3!2!} = \frac{120}{12} = 10$ combinations (possible orders of the 5 births that produce 3 boys and 2 girls).

$10(1/2)^3(1/2)^2 = 10(1/8)(1/4) = \frac{10}{32}$, the probability of 3 boys and 2 girls in any order of their birth.

The factorial formula permits us to calculate the particular term of the binomial expansion applicable to the question. The term of $(a + b)^5$ needed to solve for the probability of 3 boys and 2 girls is $10a^3b^2$, precisely what we calculated using the factorial formula.

Both the binomial theorem and the factorial formula can be useful in predicting risk as well as other features of inheritance and are particularly useful in many kinds of studies in population genetics.

## 2.3 Predicting Risk in Human Inheritance

The knowledge of Mendelian ratios and Mendelian patterns of inheritance can be applied to human beings as well as to other species. One particular instance in which this knowledge is of great practical importance is in predicting risk of occurrence of some inherited affliction in a family. In essence, we can predict the outcome of matings in terms of the risk, or probability, that two parents will produce a child who is normal or who has inherited some affliction that is known to run in the family, according to its past history. Such predictions are of importance to the families of afflicted individuals as well as to these individuals themselves.

Suppose two normal parents have produced a child with the inherited disease *cystic fibrosis*. The condition is inherited as a simple recessive, and the phenotype is characterized by defects in mucus secretion, subsequent respiratory problems, and susceptibility to various diseases. People who have cystic fibrosis have greatly shortened life expectancies but with improved medical care many individuals survive into their early twenties. The parents would want to know whether or not they may expect another child to have the same affliction.

Mendelian ratios can be used to predict the risk that children will exhibit genetic defects inherited from their parents. In our example, both parents are normal, and therefore each must have a dominant normal allele of the gene. But since their child has the disease, each parent must have contributed one recessive allele to produce a child with the recessive phenotype. Each parent must therefore be heterozygous for the alleles of this gene.

The risk of the couple's having another child with cystic fibrosis can be calculated on the basis of *Cc* × *Cc*, which represents the parental genotypes. On the average, ¾ of the progeny of many such pairs of heterozygous parents would be *C*–, and ¼ would be *cc*. The same 3:1 phenotypic ratio can be used to predict risk, or the chance or probability, of the birth of a normal versus an afflicted child. The crucial point is that *each* birth is an independent event, and the probability in *each* birth is 75% (¾) for *C*– and 25% (¼) for *cc*.

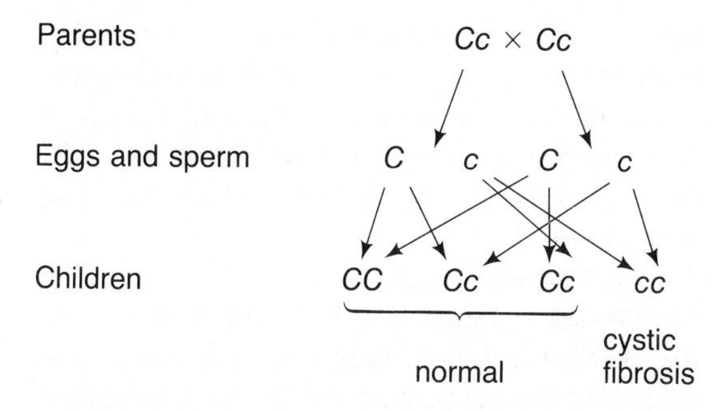


It makes no difference whether the couple has previously had a normal child or one with cystic fibrosis. Each conception has a statistical probability for an egg carrying the *C* or *c* allele to be fertilized by a sperm carrying *C* or *c*. On the basis of four possible allelic combinations, produced randomly, there is a 25% probability (risk, in this case) for conception to lead to the *cc* genotype and to cystic fibrosis in the child.

We will discuss this topic further in Chapter 4, but this example illustrates how a knowledge of basic Mendelian principles can be put to practical use in human societies. The methods for establishing a genetic basis for some characteristic and for determining its pattern of inheritance rely on the principles first established by Mendel. It is rare to be able to set such a firm date and identify one individual responsible for a field of study; genetics, which is one such field, can be traced specifically to Gregor Mendel's studies in 1865. It is all the more remarkable that the same principles he established are applied to new genetic situations, to the present day. Mendel's scientific contribution stands as an intellectual milestone in human history.

## 2.4 Chi-Square Test

The **chi-square test** ($\chi^2$) is a simple statistical test used to determine whether or not a given set of data fits a particular ratio. In genetic analysis we expect particular ratios of genotypes and of phenotypes to be produced among the progeny, according to a particular hypothesis concerning the distribution of alleles from parents to progeny. For example, if our hypothesis is that height in peas is governed by a single gene and that the allele for tall is completely dominant over the allele for short, we expect a phenotypic ratio in the $F_2$ generation of 3 tall:1 short. This ratio is based on actual counts of the different kinds of offspring produced in any cross, and enables us to deduce the genotypes of the progeny. In Mendel's crosses there were 787 tall:277 short plants in the $F_2$ generation, yielding a phenotypic ratio of 2.84:1. Is this calculated ratio close enough to the expected ratio of 3:1, based on the hypothesis of monohybrid inheritance in which tall is completely dominant over short? The answer to this question can be determined objectively rather than intuitively by use of the chi-square statistical test of the hypothesis for this cross.

The formula for chi-square is

$$\chi^2 = \Sigma(d^2/e)$$

where $\Sigma$ stands for "the sum of," $d$ stands for deviation, and $e$ is the expected value. To use the chi-square test, we must first establish the hypothesis to be tested. Our hypothesis is that tall and short appear in the ratio of 3:1, respectively. If the total progeny conform exactly to the expected 3:1 ratio, we expect the $F_2$ of 1064 individual plants to consist of 798 (¾ · 1064) tall and 266 (¼ · 1064) short plants. The expected values differ somewhat from Mendel's observed values of 787 tall and 277 short plants. This difference between what was observed and what was expected for each phenotypic class is the deviation ($d$) for each class. Sometimes the deviation ($d$) is expressed more specifically as observed ($o$) minus expected ($e$), or ($o - e$). These values are used in the chi-square test as follows:

| | **$\chi^2$ test for progeny of 1064 $F_2$ plants** | | **total** |
|---|---|---|---|
| hypothesis | 3 tall | 1 short | |
| observed ($o$) | 787 | 277 | 1024 |
| expected ($e$) | 798 (= 1064 × ¾) | 266 (= 1064 × ¼) | 1024 |
| deviation ($d$) or ($o - e$) | −11 | 11 | 0 |
| $d^2$ or $(o - e)^2$ | 121 | 121 | |
| $d^2/e$ | $121/798 = 0.152$ | $121/266 = 0.455$ | |

$\chi^2 = 0.152 + 0.455 = 0.607$

After the $\chi^2$ value has been calculated, we must determine one other value, the **degrees of freedom** ($df$), before we can use the $\chi^2$ table to interpret our data in relation to the hypothesis of a 3:1 phenotypic ratio in the $F_2$. The number of degrees of freedom is always one less than the number of classes in the ratio because we calculate the expected number ($e$) for all but one class, which must therefore contain all remaining progeny. For the case of only two classes, as in our example, we may assign the data to either class. But once that class is designated, the other class is automatically determined. For example, once we assign an expected value of 798 for the tall phenotypic class, the other class must include the remaining 266 individuals in the progeny of 1064 plants. In our present example, therefore, one degree of freedom is available.

Once we have calculated the values for $\chi^2$ and degrees of freedom, we can turn to the $\chi^2$ table and determine the **probability** ($P$) that our data fit the expected ratio and that the deviations observed can be attributed to chance (Table 2.1). We compare the calculated $\chi^2$ value with those given in the row of the table for one degree of freedom. After finding the closest match between the calculated $\chi^2$ value and a tabulated value, we consult the heading at the top of that column for the corresponding probability value. In the present case, our chi-square value of 0.607 corresponds to a probability of between 0.5 and 0.3. We would therefore expect the probability of finding a deviation as great as that observed to occur by chance in 30% to 50% of the same kind of crossing experiments, or in 30 to 50 out of 100 such crosses.

The criterion for interpreting the meaning of the $P$ value is that when $P$ is greater than 0.05, the deviation is not statistically significant and such a deviation can be expected on the basis of chance alone. If $P$ is 0.05 or less, the deviation between the expected and observed ratios is statistically significant and must be due to some factor other than chance. In other words, if the probability of obtaining the observed ratio is equal to or less than 5 in 100 ($P \leqq 0.05$), some factor other than chance may be involved. A significant deviation may indicate that the hypothesis is incorrect, that individuals in one of the phenotypic classes are less likely to survive, that the environment favors one class more than another, or that some other nonchance factor is operating.

In general, most statisticians agree that a deviation by chance alone in 5% or less of the trials or experiments is statistically significant. Many, however, consider $P = 0.05$ to be borderline, and these investigators prefer to establish 1% probability ($P = 0.01$) as the cutoff point between statistically significant and nonsignifi-

***Table 2.1*** Table of $\chi^2$ (chi-square)

| df | **P = 0.95** | **0.90** | **0.70** | **0.50** | **0.30** | **0.20** | **0.10** | **0.05** | **0.01** |
|---|---|---|---|---|---|---|---|---|---|
| 1 | 0.004 | 0.016 | 0.148 | 0.455 | 1.074 | 1.642 | 2.706 | 3.841 | 6.635 |
| 2 | 0.103 | 0.211 | 0.713 | 1.386 | 2.408 | 3.219 | 4.605 | 5.991 | 9.210 |
| 3 | 0.352 | 0.584 | 1.424 | 2.366 | 3.665 | 4.642 | 6.251 | 7.816 | 11.345 |
| 4 | 0.711 | 1.064 | 2.195 | 3.357 | 4.878 | 5.989 | 7.779 | 9.488 | 13.277 |
| 5 | 1.145 | 1.610 | 3.000 | 4.351 | 6.064 | 7.289 | 9.236 | 11.070 | 15.086 |
| 6 | 1.635 | 2.204 | 3.828 | 5.348 | 7.231 | 8.558 | 10.645 | 12.592 | 16.812 |
| 7 | 2.167 | 2.833 | 4.671 | 6.346 | 8.383 | 9.803 | 12.017 | 14.067 | 18.475 |
| 8 | 2.733 | 3.490 | 5.527 | 7.344 | 9.524 | 11.030 | 13.362 | 15.507 | 20.090 |
| 9 | 3.325 | 4.168 | 6.393 | 8.343 | 10.656 | 12.242 | 14.684 | 16.919 | 21.666 |
| 10 | 3.940 | 4.865 | 7.267 | 9.342 | 11.781 | 13.442 | 15.987 | 18.307 | 23.209 |

Abridged from Table II of Fisher and Yates, *Statistical Tables for Biological, Agricultural and Medical Research* (1953), published by Longman; with permission of the authors and publisher.

cant deviation from the expected. Everyone will agree that if there is only one chance in a hundred that a deviation could be due to chance, then such a deviation is significant. Not everyone will agree that $P = 0.05$ is the cutoff point for significance, but they will look very carefully at the data and at the hypothesis nevertheless.

The chi-square test can be used for ratios containing more than two terms. For example, in a dihybrid cross in which the hypothesis predicts a phenotypic ratio of 9:3:3:1 in the $F_2$ generation, we can calculate how closely the data fit the predictions when three degrees of freedom are available. We can use data from one of Mendel's experiments (see the table below) to illustrate the situation. There is more than a 90% probability that the observed deviation is due to chance, and we can conclude that Mendel's data fit the hypothesis very closely.

Chi-square, therefore, provides a valuable objective basis by which data may be tested for **goodness of fit** to a proposed hypothesis. The progeny or population samples that are tested must be sufficiently large and representative if the statistical test is to be meaningful. In general, the chi-square test can only be used in cases where the smallest class contains five or more individuals. It is therefore most useful in genetic analysis when controlled matings are possible and when progeny size can be planned in ad-

**$\chi^2$ test for progeny of 556 $F_2$ plants**

| hypothesis | 9 round, yellow | 3 round, green | 3 wrinkled, yellow | 1 wrinkled, green | | **total** |
|---|---|---|---|---|---|---|
| observed (*o*) | 315 | 101 | 108 | 32 | = | 556 |
| expected (*e*) | 313 | 104 | 104 | 35 | = | 556 |
| deviation (*d*) | 2 | −3 | 4 | −3 | = | 0 |
| $d^2$ | 4 | 9 | 16 | 9 | | |
| $d^2/e$ | 0.013 | 0.086 | 0.154 | 0.257 | | |

$\chi^2 = 0.013 + 0.086 + 0.154 + 0.257 = 0.510$

$df = 3$

$P > 0.9$

vance to include an adequate number of individuals representative of all the expected phenotypic classes.

The chi-square test for goodness of fit between observed and expected values can only provide one item of evidence for or against a certain hypothesis. We should always seek to obtain independent evidence in support of the hypothesis. When two or more different kinds of experimental results show the same support for the hypothesis, then we can be more confident of the truth of the conclusions drawn from the analysis. For $F_2$ data indicating monohybrid inheritance for one pair of alleles with complete dominance, goodness of fit is a good indication of the truth of the hypothesis. The hypothesis should not be accepted on one basis, however, but should be studied further. In this case we could make testcrosses to see if the expected 1 : 1 ratios of genotypes and phenotypes are obtained. We can also judge the testcross data by using the chi-square test. Breeding $F_2$ progeny and obtaining the expected genotypic and phenotypic ratios in the $F_3$ would provide us with another useful check on the validity of the working hypothesis. Once we have obtained independent lines of evidence that all point to the hypothesis as correct, we can accept the hypothesis. Similarly, if chi-square analysis indicates that a hypothesis is incorrect, we can usually conduct other experiments to see whether or not this conclusion is indeed the case. The stronger the experimental support for a hypothesis, the greater the confidence in the conclusions drawn from statistical analysis of the data.

## Statistical Analysis of Quantitative Traits

Analysis of quantitative traits, those showing continuous variation, requires a different set of statistical tools from ones used to analyze discontinuous variation, in which phenotypic classes are nonoverlapping and unambiguous. Mendelian inheritance of tall and short height is evaluated according to the ratio of tall to short individuals in progenies consisting of these two discrete phenotypic classes. In cases in which a range of heights from tallest to shortest with no distinct separation into different phenotypic classes exists, there is no apparent ratio. Statistical analysis of individual measurements must be carried out in order to characterize a quantitative trait. We will examine some of the central concepts and some of the more widely used statistics that are routinely determined in many kinds of studies of quantitative traits.

### 2.5 Mean and Variation around the Mean

The first step in analyzing quantitative data is to calculate the **arithmetic mean,** symbolized as $\bar{x}$, by simply summing up the observed values and then dividing by the number of values (Table 2.2). For greater convenience of calculation, we usually group the individual values into *classes* and indicate the *frequency* of individual values in each designated class. In the example of the measured height of a sample of 100 American men, the data can be plotted in the form of a bar graph, or histogram, or in the form of a curve

***Table 2.2*** Frequency distribution of height measurements for a sample of 100 American men.

| class value, $v$ (inches) | 60 | 61 | 62 | 63 | 64 | 65 | 66 | 67 | 68 | 69 | 70 | 71 | 72 | 73 | 74 | 75 | 76 | 77 | 78 |
|---|---|---|---|---|---|---|---|---|---|---|---|---|---|---|---|---|---|---|---|
| frequency, $f$ | 1 | 0 | 1 | 1 | 3 | 4 | 6 | 9 | 14 | 19 | 16 | 11 | 6 | 4 | 2 | 1 | 1 | 0 | 1 |
| $fv$ | 60 | 0 | 62 | 63 | 192 | 260 | 396 | 603 | 952 | 1311 | 1120 | 781 | 432 | 292 | 148 | 75 | 76 | 0 | 78 |

$$\text{Mean} = \frac{\Sigma fv}{N} = \frac{6905}{100} = 69.0 \text{ inches}$$

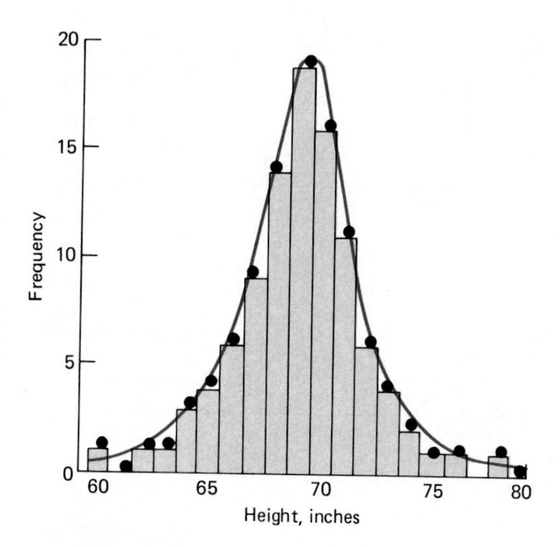


***Figure 2.2*** Height measurements (black dots) can be plotted in the form of a histogram or as a curve implying a continuous distribution of heights.

constructed by connecting the plotted values (Fig. 2.2). In either representation of these data, a *symmetrical distribution* of values is apparent on either side of the mean height of 69.0 inches. The familiar bell-shaped curve indicates that the spread of variation follows a **normal distribution** in which the mean value occurs at the apex of a balanced distribution of frequencies, and these frequencies diminish regularly and symmetrically in both directions away from the mean value (Box 2.1, next page).

The mean value is the most useful term for further statistical analysis. The *central tendency* of a sample can be described in two other ways as well: the median value and the mode. The **median** value defines the center of a set of data such that there are as many values, or numbers of measurements, on one side of the mean as on the other side. In our sample of height measurements, the median value is 68.9 inches. Half the measurements were more than 68.9 inches (69.0 inches or more) and half were less than 68.9 inches. The **mode** is defined as the most frequent value in a set of data. In our sample, the mode is 69.0 inches. We can now say that the average or mean height of American men in our

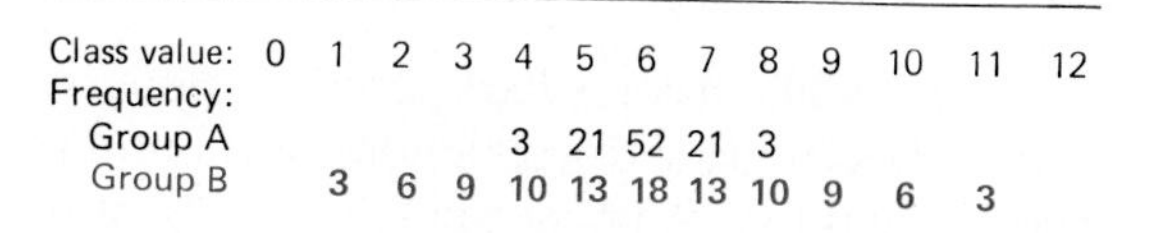

| Class value: | 0 | 1 | 2 | 3 | 4 | 5 | 6 | 7 | 8 | 9 | 10 | 11 | 12 |
|---|---|---|---|---|---|---|---|---|---|---|---|---|---|
| Frequency: | | | | | | | | | | | | | |
| Group A | | | | | 3 | 21 | 52 | 21 | 3 | | | | |
| Group B | | 3 | 6 | 9 | 10 | 13 | 18 | 13 | 10 | 9 | 6 | 3 | |

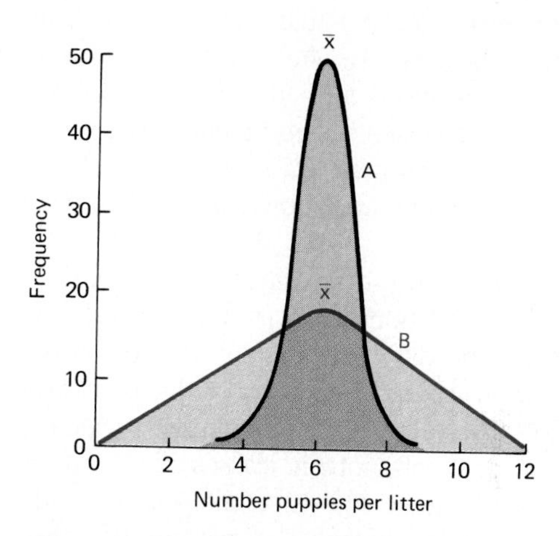


***Figure 2.3*** Populations having the same mean value for a quantitative feature may show very different spreads of variability around the mean. Population A is much less variable (3 to 9 puppies per litter) than population B (0 to 12 per litter), but the mean value (6) is the same in each population of 100 animals.

sample is 69.0 inches, that this is the most common measurement obtained, and that half the men are taller than the average and half the men are shorter than the average.

The mean value describes one aspect of the sample or population, but different *spreads of variation* may produce the same arithmetic mean (Fig. 2.3). We can provide a more complete picture of a population by measuring the variation around the mean. Two particular statistics can be calculated for this purpose: the **variance** ($V$ or $\sigma^2$), which is the average squared difference from the mean, and the square root of the variance, called the **standard deviation** of the mean ($\sigma$). The formulae for these statistics and their computations for our measurements of height are shown in Table 2.3 (next page). The statistic of standard deviation has a broader use than the variance for many studies of quantitative characteristics.

## BOX 2.1 Normal Distribution

Genes with major effects tend to produce phenotypes showing discontinuous distribution. The distribution of phenotypes in the $F_2$ of a monohybrid cross can be plotted as tall and short, for example, or actual measurements can be made showing the influence of environmental modifications such that the phenotypic measurements cluster around some mean value for each major progeny class of tall and short. These distributions are shown in the first two plots.

If there is no dominance, the expected distribution of phenotypes in a population of individuals of different heights can be computed according to expansion of the binomial $(a + b)^n$. The third illustration shows expected distributions of phenotypic classes for $F_2$ ratios involving 1–4 gene loci (2–8 alleles) with additive and equal effects. The symmetrical distribution known as Pascal's triangle shows the coefficients for terms in the expanded binomials raised to different powers; for example, $(a + b)^2 = a^2 + 2ab + b^2$, and $(a + b)^6 = a^6 + 6a^5b + 15a^4b^2 + 20a^3b^3 + 15a^2b^4 + 6ab^5 + b^6$, where $a = b = ½$. When the relative frequencies (shown as coefficients) of these classes are plotted, the bell-shaped curve indicative of the normal frequency distribution is approached, as shown in the fourth illustration. A population with less variation (dashed line) may have the same mean as a population with greater variance (histogram plot).

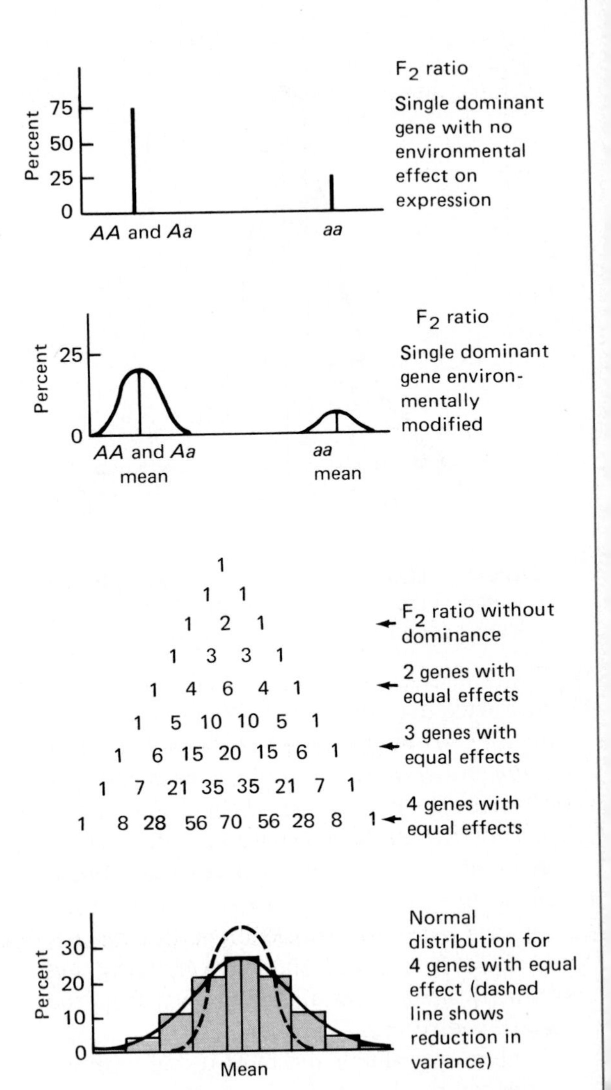


With the mean and the standard deviation of the mean, the set of 100 measurements has been reduced to two precise values by which the sample can be described. The average or mean height is 69.0 inches, but there is a spread of variation around this mean. In our sample, 81% of the individuals fall within the range between 3.6 inches above and 3.6 inches below the mean (within one standard deviation), 97% range between 61.8 and 76.2 inches in height, and 100% of this sample are between 58.2 and 79.8 inches tall (Fig. 2.4).

These results vary somewhat from the theoretical. In a perfect distribution, one standard deviation to either side of the mean would include 68.26% of the area under the curve; two standard deviations would include 95.44% of the area; and three standard deviations would include 99.74% of the area under the curve (Fig. 2.5). These relationships hold for any normal distribution, regardless of the actual spread of variation around the mean. Whether the spread is narrow, producing a steeply sided curve, or broader, producing a flattened bell-shaped

***Table 2.3*** Statistical analysis of height measurements in sample of 100 American men ranging between 60 and 78 inches tall.

| class | frequency ($f$) | deviation ($d$) | $d^2$ | $fd^2$ |
|---|---|---|---|---|
| 60 | 1 | −9 | 81 | 81 |
| 61 | 0 | −8 | 64 | 0 |
| 62 | 1 | −7 | 49 | 49 |
| 63 | 1 | −6 | 36 | 36 |
| 64 | 3 | −5 | 25 | 75 |
| 65 | 4 | −4 | 16 | 64 |
| 66 | 6 | −3 | 9 | 54 |
| 67 | 9 | −2 | 4 | 36 |
| 68 | 14 | −1 | 1 | 14 |
| 69 | 19 | 0 | 0 | 0 |
| 70 | 16 | 1 | 1 | 16 |
| 71 | 11 | 2 | 4 | 44 |
| 72 | 6 | 3 | 9 | 54 |
| 73 | 4 | 4 | 16 | 64 |
| 74 | 2 | 5 | 25 | 50 |
| 75 | 1 | 6 | 36 | 36 |
| 76 | 1 | 7 | 49 | 49 |
| 77 | 0 | 8 | 64 | 0 |
| 78 | 1 | 9 | 81 | 81 |
| | $N = 100$ | | | $\Sigma = 1293$ |

$$V = \frac{\Sigma fd^2}{N-1} = \frac{1293}{99} = 13.06$$

$$\sigma = \sqrt{\frac{\Sigma fd^2}{N-1}} = \sqrt{13.06} = 3.6$$

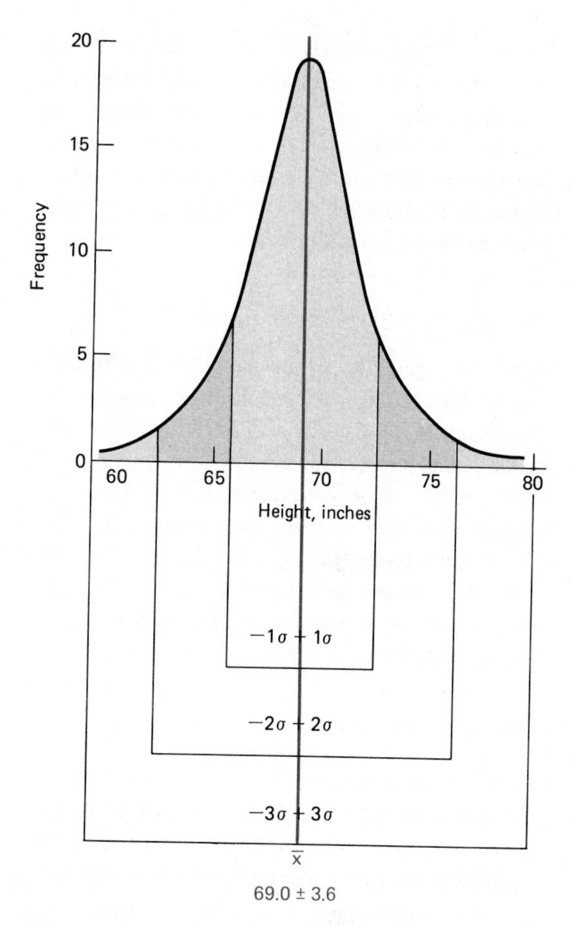


***Figure 2.4*** (Above) Normal distribution of measurements, showing the proportion of the distribution included between $\pm 1\sigma$, $\pm 2\sigma$, and $\pm 3\sigma$, with reference to the mean of 69.0 inches and a standard deviation ($\sigma$) of 3.6.

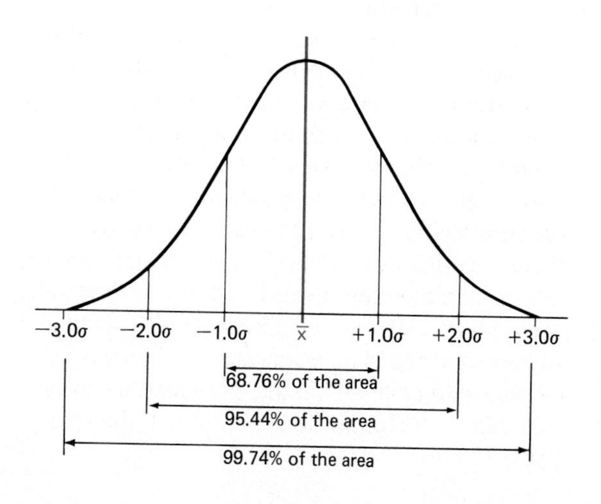


***Figure 2.5*** (Right) In a perfect normal distribution, regardless of the actual spread of variation around the mean, specific proportions of the area under the curve are characteristic for $\pm 1$, 2, and 3 standard deviations of the mean.

curve, one, two, or three standard deviations on either side of the mean will represent the normal distribution of the particular data. Our sample of 100 men comes close to the expected distribution, although a higher percentage than the expected falls within the range of one standard deviation from the mean.

## 2.6 Standard Error of the Mean

How accurately does our sample of 100 men reflect stature in the whole population? We could sample repeatedly and find the means and standard deviations for many samples or even for the entire American adult male population, but this would be tedious and unnecessary. We can take a statistical shortcut by determining the **standard error** of the mean ($S.E._{\bar{x}}$), which can be calculated from the standard deviation of any sample, as follows:

$$S.E._{\bar{x}} = \frac{\sigma}{\sqrt{N}}$$

where $\sigma$ is the standard deviation and $N$ is the size of the sample. In our sample of 100 individuals, with a standard deviation of 3.6, the standard error of the mean value of 69.0 inches is

$$S.E._{\bar{x}} = \frac{3.6}{\sqrt{100}} = \frac{3.6}{10} = 0.36, \text{ or } 0.4.$$

We may now state the mean height in the sample as 69.0 ± 0.4 inches.

The standard error can be considered as the standard deviation of many means. It is therefore a standard deviation of the theoretical population mean. The standard error tells us the *reliability* of the calculated mean value. There is a 68% chance that the true mean lies within one standard error of the calculated mean, or between 68.6 and 69.4 inches. There is about a 95% chance that the true mean lies within two standard errors, or between 68.2 and 69.8 inches, and about a 99% chance that the true mean is 1.2 inches more or less than the calculated mean of 69.0 inches. It should be apparent that the standard error is used in much the same way that one, two, or three standard deviations is used to indicate our *confidence* in the calculated mean value. The estimates of the range within which the population mean should lie are called confidence intervals. We may therefore say that we are 95% confident or 99% confident that the true mean lies somewhere between 68.2 and 69.8 inches or between 67.8 and 70.2 inches, respectively.

When characterizing the mean value by its standard deviation, we are making a statement about the observed spread of variation around the particular mean in a particular sample. On the other hand, characterizing the mean value in terms of its standard error leads to a statement about the mean value in terms of an infinite population or, in the usual notation, a description of the mean plus or minus the standard error of the mean. It is then a simple matter to add and subtract two or three standard errors from the mean value in order to see at a glance how the calculated mean value differs from the expected value for the entire population, at the 95% or 99% level of confidence.

## 2.7 Standard Error of the Difference between Means

The means and standard errors of two different samples or populations can be compared to determine if the populations are the same or different. Suppose stature in a population of European adult men is measured, and it is found that the mean height in a sample of 100 individuals is 67.0 inches. The American and European population samples can be crudely compared by inspection of the two sets of statistics, as shown in Table 2.4. It is apparent at a glance that in this case the two samples do not overlap at either the 95% or the 99% confidence levels, and we may conclude that there is a significant statistical difference between them. In many cases, however, the values may overlap. In such cases a more precise comparison is required in order to determine whether the difference is significant or merely due to chance.

***Table 2.4*** Statistical information on height measurements for two different populations of adult men.

| statistic | Americans | Europeans |
|---|---|---|
| mean height in inches ($\bar{x}$) | 69.0 | 67.0 |
| standard deviation ($\sigma$) | 3.6 | 2.0 |
| standard error of the mean ($S.E._{\bar{x}}$) | 0.4 | 0.2 |
| range of variation: $\bar{x} \pm 1$ $S.E._{\bar{x}}$ | 68.6–69.4 | 66.8–67.2 |
| $\bar{x} \pm 2$ $S.E._{\bar{x}}$ | 68.2–69.8 | 66.6–67.4 |
| $\bar{x} \pm 3$ $S.E._{\bar{x}}$ | 67.8–70.2 | 66.4–67.6 |

To determine whether or not overlapping values are statistically significant, we compare the difference between the means of the two samples with the standard error of this difference. We have already calculated means and standard errors of these means and have found that the mean height in the sample of American men is 69.0 ± 0.4, and in the sample of European men it is 67.0 ± 0.2 inches. We then obtain the standard error of the difference in means ($S.E._D$) using the equation

$$S.E._D = \sqrt{(S.E._{\bar{x}_1})^2 + (S.E._{\bar{x}_2})^2}$$

where $S.E._{\bar{x}_1}$ is the standard error of the mean of sample 1 and $S.E._{\bar{x}_2}$ is the standard error of the mean of sample 2. Substituting the calculated values, we determine that

$$S.E._D = \sqrt{(0.4)^2 + (0.2)^2}$$
$$= \sqrt{0.16 + 0.04}$$
$$= \sqrt{0.20}$$
$$= 0.4$$

The standard error of the difference between the two means is therefore ±0.4. Two such standard errors would be ±0.8, and three standard errors would be ±1.2. The expected difference between the means of the samples is 0 if the two populations are identical. The actual difference between the means is 2.0. When we compare these values it is obvious that the actual difference between means is much greater than even three standard errors of the difference in means (2.0 versus 1.2), that is, 2.0 is much more than three standard errors away from 0. We can conclude, therefore, that there is a statistically significant difference in mean height between these two samples and that they indeed represent different populations. We can be more than 99% confident that this difference in means is significant, or we may say that there is less than 1% probability ($P < 0.01$) that the two populations differ because of chance alone. We do not know the factors that are responsible for the observed difference, but we do know it is not simply a reflection of chance variation. We might then proceed to analyze the basis for the significant difference in mean height between the two populations. From the statistics alone we do not know whether a genetic or an environmental basis underlies the difference. Other tests are required to make such assessments, as we shall see shortly.

## Influences on Expressions of Quantitative Traits

Populations of organisms contain a large amount of variability in both discontinuous and continuous traits. Some part of this expressed variability is the result of environmental influences, but a large part is genetic. Quantitative traits are generally affected by both genetic and environmental factors, and it is important to determine the relative proportions of these two sets of factors in natural populations and in programs of plant and animal breeding. There are several ways to estimate the proportion of phenotypic expression that is due to the genotype and the proportion that varies according to environmental influences.

## 2.8 Effectiveness of Selection

Since the expression of quantitative traits varies across a range of values, we might intuitively expect that a vigorous program of selection for the most desirable variant type could lead to an increase in the proportion of such a desirable type in later generations. This would be true, of course, only if there were at least some genetic basis for the observed variation. The effectiveness of a program of selection depends on the relative influences of genotype and environment in phenotypic expression. The Danish botanist Wilhelm Johannsen first demonstrated this fundamental principle convincingly in the early years of this century.

Johannsen studied the possibilities of selecting for seed weight in experiments carried out with the Princess variety of kidney bean (*Phaseolus vulgaris*). The species is self-fertilizing, so it was only a matter of obtaining seeds and allowing these to develop into mature plants, which in turn produced their own seeds. Starting with a mixture of seeds obtained from many different plants, Johannsen found that progeny derived from heavier-weight seeds produced heavier seeds than progeny obtained from lighter-weight seeds. This finding showed that selection had been effective.

In further experiments, Johannsen selected 19 different plants and collected one seed from each. The 19 seeds grew into progeny plants that produced their own sets of seeds. Johannsen kept each set of seeds separated from all other sets. Within each of the 19 lots of seeds, weights varied from one seed to another, and the total display of variation in seed weight within each lot followed an approximately normal distribution (Fig. 2.6). For a number of generations, Johannsen continued to select the heaviest and the lightest seeds from each of the 19 lines and propagated each line every year for the duration of the experiments. He found that the average weight of seeds *within a line* was remarkably constant, year after year, regardless of whether the line was reproduced by its heaviest or by its lightest seeds. As shown in Table 2.5, in pure line

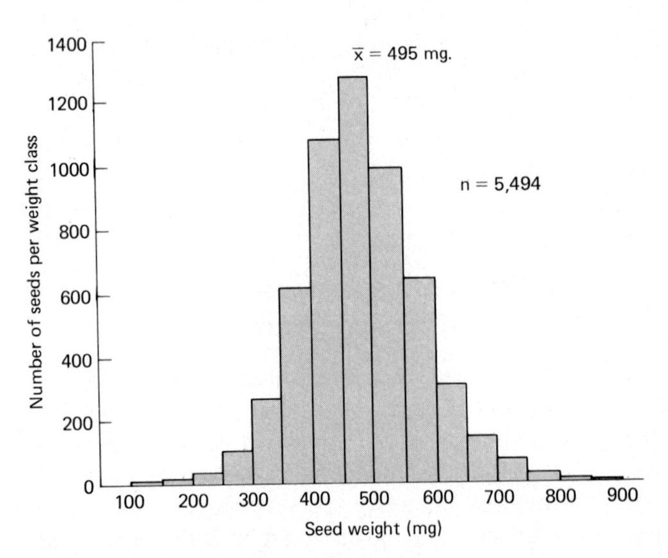


***Figure 2.6*** Distribution of seed weights of 5494 kidney beans showing a normal probability distribution, taken from records kept by Wilhelm Johannsen.

***Table 2.5*** Seed weight in pure line No. 19 of the Princess variety of kidney bean over a period of six years of selection.

| | **average weight of selected parent seeds, mg** | | **average weight of progeny seeds, mg** | |
|---|---|---|---|---|
| **harvest year** | **lighter seeds** | **heavier seeds** | **lighter seeds** | **heavier seeds** |
| 1902 | 300 | 400 | 360 | 350 |
| 1903 | 250 | 420 | 400 | 410 |
| 1904 | 310 | 430 | 310 | 330 |
| 1905 | 270 | 390 | 380 | 390 |
| 1906 | 300 | 460 | 380 | 400 |
| 1907 | 240 | 470 | 370 | 370 |

Source: Johannsen, W., *Elemente der Exacten Erblichkeitslehre.* Jena: Gustav Fischer, 1926.

number 19, the heaviest parent seeds produced a progeny with the same average seed weight as the progeny that was produced by the lightest parent seeds. By the sixth generation of selection, the average weight of progeny seeds was 370 milligrams for both the heaviest and the lightest parent seeds. Clearly, selection had not been effective in this series of experiments.

We can easily understand why selection was effective in the first experiments using a mixture of seeds from different plants, and why it was not effective in the experiments using 19 separate lines. The parent seeds with which Johannsen started his second set of experiments were all homozygous, having arisen by self-fertilizations from inbred plants. Each parent seed was the basis for initiating a different **pure line,** all of which would produce genetically homozygous progeny after each round of self-fertilizations in each generation. We would not expect homozygous plants to show genetic segregation, no matter how many generations of inbreeding were studied. The variation in seed weight that was expressed among the progeny of any one pure line was due to environmental influences only. These influences have no effect on the inherited properties of the individuals, but they may cause some seeds to be heavier and others to be lighter at maturity than their parent seeds (Fig. 2.7).

The program of selection was effective in Johannsen's first set of experiments because the population from which he obtained seeds was a mixture of pure lines and was, therefore, genotypically as well as phenotypically variable. When he selected heavy- and light-weight seeds from the mixed population of plants, he was able to obtain plants producing heavier or lighter seeds because he had separated out different genotypes having different inherent properties relative to seed weight.

Johannsen's experiments were very important in the early years of Mendelian genetics and

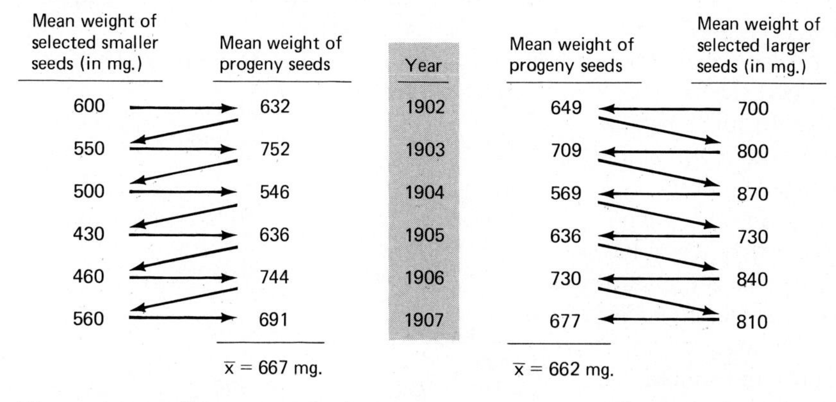


***Figure 2.7*** Ineffectiveness of selection in a pure line. There was no significant difference in mean weight of kidney bean seeds during six years of selection using selected smaller or selected larger seeds. Based on experiments by Wilhelm Johannsen.

contributed to our understanding of the genetic basis for evolution in several ways:

***1.*** We can distinguish between heritable and nonheritable variation for quantitative traits, even where the actual genotypes are not known.

***2.*** We can predict that selection within a genetically diverse group can lead to changes in characteristics of subsequent populations. As a corollary, we can predict that selection within a homozygous population will not produce changes in subsequent populations. Selection per se does not create variation.

***3.*** We can see that inbreeding leads to genetic homozygosity and that no amount of selection will alter the homogeneous character of the inbred population.

Johannsen's experiments and those of other investigators showed very clearly that quantitative traits are remarkably sensitive to environmental influence. These studies also showed that any meaningful program of agricultural improvement or analysis of quantitative variation in natural populations absolutely requires information by which heredity and environment can be distinguished and, hence, can be subjected to some sort of control.

## 2.9 Heritability and Twin Studies

The value called **heritability** is used to express the degree to which the phenotypic expression of a trait is influenced by genetic factors. A heritability of 1.0 indicates that the phenotype in question was produced only by the action of the genotype and was not influenced at all by environment. The ABO major blood-group phenotypes are an example of a trait with a heritability of 1.0. A heritability of 0 means that the phenotype is due entirely to environmental influences, such as the accidental loss of a finger or a tail. Values between 0 and 1 represent estimates of the relative contribution of heredity in the expression of a trait. A heritability of 0.65 is an estimate that 65% of the expression of the trait is due to genotype. The remainder of 35% would then be the proportionate influence estimated to be due to environmental influence on phenotypic expression. The percentage heritability, that is, *the proportion of phenotypic expression that is due to genetic factors,* varies for different traits and even for the same trait in different species. Calculated values for some particular traits of agricultural importance are given in Table 2.6.

In plant and animal breeding programs, heritability estimates are useful in predicting how closely the progeny will resemble the parents chosen as breeding stock. Heritability estimates are only approximations, however, since all of the contributions of the genotype are not considered in the calculations. Dominant and epistatic gene effects are usually minimized or even ignored, and the performance of genotypes in different environments is rarely determined. For these reasons, we consider heritabilities only as crude indications of the relative input of the genotype in phenotypic expression.

Calculations of heritability for quantitative traits in human beings are made on the basis of differences between pairs of monozygotic twins and pairs of dizygotic twins. **Monozygotic twins,** or identical twins, arise from the division of a single fertilized egg early in development. Monozygotic twins are, therefore, genetically identical and, of course, always of the same sex. **Dizygotic twins,** also called fraternal twins, develop from two different fertilized eggs that

***Table 2.6*** Heritabilities for some traits of economic importance.

| characteristic | heritability |
|---|---|
| Cattle | |
| conception rate | 0.05 |
| milk production | 0.30 |
| slaughter weight | 0.85 |
| Corn | |
| ear length | 0.17 |
| yield | 0.25 |
| plant height | 0.70 |
| Poultry | |
| egg production | 0.20 |
| egg weight | 0.60 |

happen by chance to have been released at the same time. Other than being of the same age, dizygotic twins are no more genetically alike than any brothers or sisters (siblings) in the same family. Since each egg is fertilized by a different sperm, dizygotic twins may be of the same sex or of different sexes, depending on whether X-carrying sperm or Y-carrying sperm are involved in both fertilizations or whether one of each type of sperm is involved in the dual events. The observed frequencies of 25% twin boys, 25% twin girls, and 50% twin boy and girl show that fertilizations are chance events.

Estimates of heritability for a particular trait are not always comparable between different twin studies. The difficulty is due in part to the different ways in which estimates can be calculated and to various interpretations of the statistic of heritability. The usefulness of the statistic is further limited by two assumptions that are implicit in any heritability estimate: (1) that the environments of both kinds of twins are entirely equivalent and (2) that no genotype-environment interaction is involved in the expression of the particular trait. Despite all these reservations, interpretations of heritability estimates still offer the simplest and most convenient way to summarize twin data on quantitative characteristics.

The similarities between monozygotic twins due to their having similar environments can be determined, to some extent, by comparing monozygotic twins reared apart with those reared together. Since the frequency of monozygotic twin births is only about 0.4%, and fewer than one pair in a thousand pairs is separated early and reared apart, there are practical problems in assembling an adequate number of individuals for data analysis. In a comprehensive study reported by J. Shields in 1962, 44 pairs of separately reared monozygotic twins were compared for a number of quantitative characteristics, along with monozygotic twins raised together who served as control pairs. Comparisons for dizygotic twins were also included. The differences for height, weight, and IQ measurements between members of these three kinds of twin pairs are shown in Fig. 2.8. For the

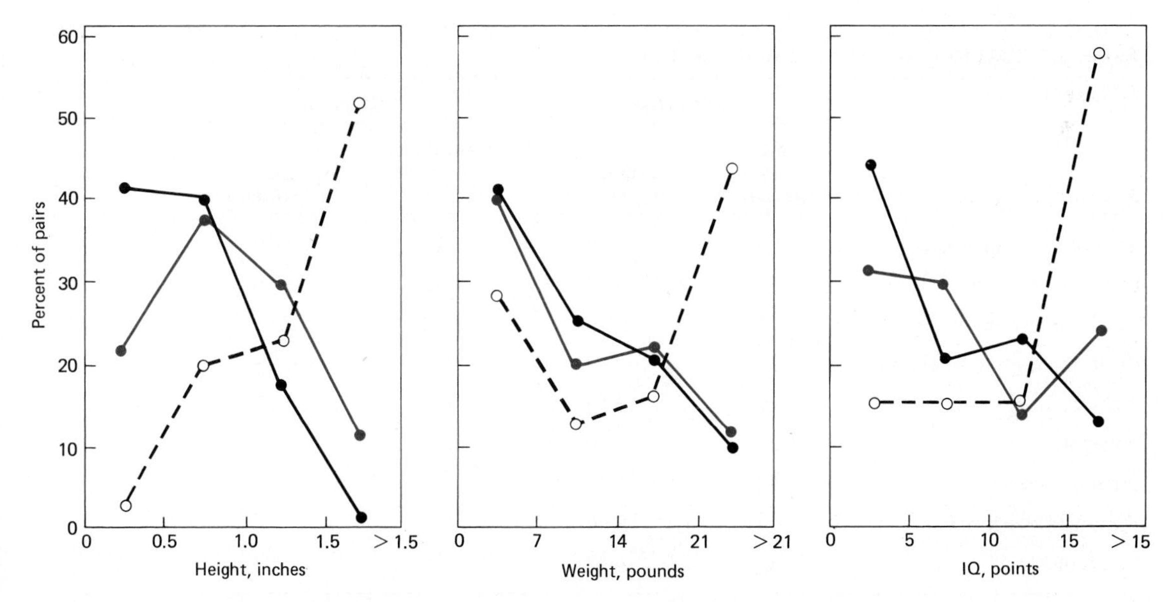


***Figure 2.8*** Distribution of height, weight, and IQ differences between monozygotic twins brought up together (solid, black line) or separately (colored line), and dizygotic twins (dashed line), demonstrating the relative input of genes and environment in determining quantitative traits. (Data from J. Shields. 1962. *Monozygotic Twins Brought Up Apart and Brought Up Together.* London: Oxford University Press.)

three characteristics, the similarity between monozygotic twins appeared to be only slightly decreased by their early separation, relative to the differences between dizygotic twins. This finding supports the existence of a strong genetic component in the determination of size and IQ, and similar data have been reported by others in equivalent studies. In the case of IQ, heritability values of 40% to 80% have been reported. Regardless of the absolute differences in this value in different studies, all the available twin data suggest that a fairly high heritability for IQ exists. Additional support for this conclusion has been obtained in more broadly based studies in which pairs of individuals of varying genetic relationship have been compared (Box 2.2). The closer the genetic relationship between members of a pair, the higher the correlation observed in their phenotypic expression of IQ scores. This kind of analysis also shows that a strong genetic component is involved in IQ.

Very little information is available concerning the influences on IQ of environmental variables such as educational opportunities, cultural differences, and living conditions. In addition, we really have no firm understanding of the relationship between IQ and overall intelligence. In fact, there is no universally accepted description or definition of intelligence. Measurements of intelligence, as defined by IQ, are decidedly different from measurements of height, weight, or hip width. The individual being measured for IQ must be an active participant in the test. Performance may vary according to motivation, interaction with the person administering the test, and similar variables. Measurements of height, for example, are not subject to these influences at the time the measurement is made. In general, we cannot say how IQ scores can be translated into measurements of overall human intelligence, much less into a genetic statement about mental abilities.

## 2.10 Concordance and Discordance

Another measure of the genetic component involved in quantitative character expression can be applied for traits that are either present or

***Table 2.7*** Twin concordance for various diseases.

| | MZ twins | | DZ twins | |
|---|---|---|---|---|
| **disease** | **number of pairs studied** | **percent concordance** | **number of pairs studied** | **percent concordance** |
| arterial hypertension | 80 | 25.0 | 212 | 6.6 |
| bronchial asthma | 64 | 47.0 | 192 | 24.0 |
| cancer at any site | 207 | 15.9 | 212 | 12.9 |
| death from acute infection | 127 | 7.9 | 454 | 8.8 |
| diabetes mellitus | 76 | 47.0 | 238 | 9.7 |
| epilepsy | 27 | 37.0 | 100 | 10.0 |
| mental retardation | 18 | 67.0 | 49 | 0 |
| rheumatoid arthritis | 47 | 34.0 | 141 | 7.1 |
| tuberculosis | 135 | 37.2 | 513 | 15.3 |

Source: B. Harvald and M. Hauge, in *Genetics and the Epidemiology of Chronic Diseases* (eds. Neel J.V., M.W. Shaw, and W.J. Schull), U.S. Public Health Service Publication No. 1163 (1965). Washington, D.C.: U.S. Department of Health, Education and Welfare.

## BOX 2.2 Correlations and Relationships

Although there is considerable disagreement about the meaning of IQ, there appears to be a strong genetic component underlying whatever it is that IQ tests measure. Evidence in support of this genetic component comes from various lines of study, including studies showing that increasing genetic relationship is increasingly correlated with "intelligence" measurements between individuals. The illustration shows a summary of correlation coefficients compiled from various sources by L. Erlenmeyer-Kimling and L.F. Jarvik. Each horizontal line shows the range of correlation coefficients (dots) between individuals identified by the left-hand columns. The mean value is shown by a vertical line for each group of values. There is some environmental influence, as seen in the greater variability for one-egg twins reared apart versus those reared together, and for two-egg twins of like sex compared with all one-egg twins.

| Genetic and nongenetic relationships studied | | | Genetic correlation | Range of correlations (0.00 0.10 0.20 0.30 0.40 0.50 0.60 0.70 0.80 0.90) | No. of studies |
|---|---|---|---|---|---|
| Unrelated persons | | Reared apart | 0.00 | | 4 |
| | | Reared together | 0.00 | | 5 |
| Foster-parent-child | | | 0.00 | | 3 |
| Parent-child | | | 0.50 | | 12 |
| Siblings | | Reared apart | 0.50 | | 2 |
| | | Reared together | 0.50 | | 35 |
| Twins | Two-egg | Opposite sex | 0.50 | | 9 |
| | | Like sex | 0.50 | | 11 |
| | One-egg | Reared apart | 1.00 | | 4 |
| | | Reared together | 1.00 | | 14 |

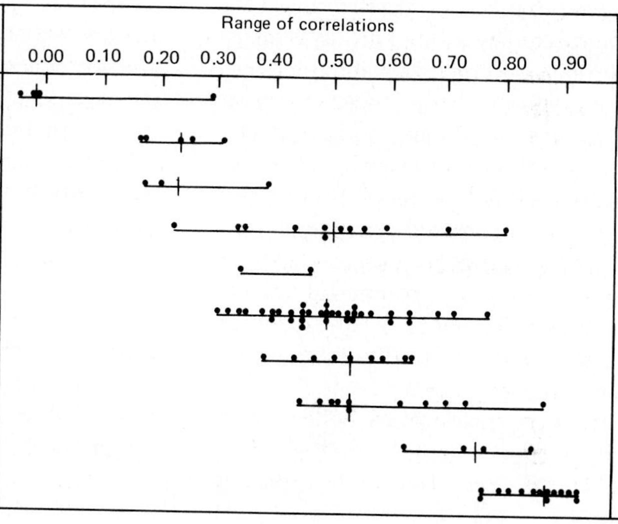

absent ("all or none") in one or both members of a twin pair. A twin pair is *concordant* with respect to a character if both twins have it or both do not. The pair is *discordant* for the character if one twin has it and the other does not. The **concordance** frequency, which is calculated *separately* for monozygotic and dizygotic twins, is the proportion of concordant twin pairs to the total number of pairs studied.

$$\% \text{ concordance} = \frac{\text{number of concordant twin pairs}}{\text{total number of twin pairs studied}}$$

If the concordance frequency for a quantitative trait is significantly higher for monozygotic twins than for dizygotic twins, it is considered to be evidence for a significant genetic component in determination of the trait (Table 2.7).

Complications in genetic analysis may be introduced by single genes that show incomplete penetrance, that is, where only a percentage of the individuals who have the gene will express it phenotypically (see Fig. 1.14). In cases in which the genetic basis for character expression is unknown or uncertain, we may find it difficult to determine whether the trait is polygenic in inheritance or whether the observed variability is due to one or more genes with reduced penetrance. In either situation, concordance frequency will be reduced somewhat since both members of a monozygotic twin pair may not show the same phenotype. This inconsistency has led to different notions about the basis for inheritance of schizophrenia, a psychotic disorder prevalent in human populations.

Twin data reveal a range of concordance frequencies for schizophrenia, but it is clear that there is a significantly higher frequency for monozygotic than for dizygotic twins (Table 2.8). Investigators who believe there is a genetic basis for schizophrenia have interpreted the data in different ways. According to one group, one or two dominant genes that show incomplete penetrance may lead to the disorder; the same data have been interpreted by another group to mean that schizophrenia is polygenically inherited. According to the polygenic theory, there would be some point in the scale of many additive genes that would be the *threshold* above which the trait would be expressed. Below this threshold, the trait would not be expressed. Phenotypic expression in appropriate genotypes may also be elicited by one or more environmental factors. Phenotypic variance in genotypically identical individuals may therefore be due to environmental variance alone.

A third group of investigators believes that schizophrenia is not inherited. In this view, schizophrenia results entirely from disturbances in interpersonal relations because of early social environment and upbringing. A fourth viewpoint is that the disorder is caused primarily by heredity, but that social situations may trigger the appearance of schizophrenic symptoms through some undetermined developmental pathway.

Overall, there is substantial evidence in support of a genetic basis for schizophrenia. At the present time, however, the precise nature of this genetic component is uncertain.

Twin studies are an important part of the analysis of quantitative inheritance, but such studies *do not provide information on the nature of the inheritance pattern.* Once we have determined the nature of the pattern from family studies, we can calculate the genetic input into phenotypic expression of quantitative traits and the degree of penetrance or range of expressivity of single genes.

In summary, the application of an appropriate statistical method can provide a more objective basis on which to analyze hypotheses and data concerning inheritance patterns. Once the inheritance pattern is known, other statistical methods can also be useful in making predictions of the probability of some event or events occurring. Genetic interpretations can be made with greater confidence when supported by statistical analysis.

***Table 2.8*** Percent concordance for occurrence of schizophrenia in monozygotic (MZ) and dizygotic (DZ) twins.

| | | **MZ twins** | | **DZ twins** | |
|---|---|---|---|---|---|
| **year of study** | **country** | **number of pairs studied** | **percent concordance** | **number of pairs studied** | **percent concordance** |
| 1928 | Germany | 19 | 58 | 13 | 0 |
| 1946 | U.S.A. | 174 | 69 | 296 | 11 |
| 1953 | Great Britain | 26 | 65 | 35 | 11 |
| 1961 | Japan | 55 | 60 | 11 | 18 |
| 1964 | Norway | 8 | 25 | 12 | 17 |
| 1965 | Denmark | 7 | 29 | 31 | 6 |

Data from various sources.

## Questions and Problems

***2.1*** Heterozygous black mice (*Bb*) are interbred to obtain black and white offspring.
***a.*** What is the probability that the first three offspring wll be (1) white–white–black, in that order; (2) black–white–black, in that order; (3) all black?
***b.*** What is the probability that a litter of four offspring will be (1) two black and two white, in that order; (2) two black and two white, in any order?

***2.2*** Two normal parents have a child with the recessive inherited disorder of cystic fibrosis. Determine the probability of each of the following events:
***a.*** The next child will have cystic fibrosis.
***b.*** These parents have two more children both of whom will have cystic fibrosis.
***c.*** These parents will have three children, the first and the third will be normal but the middle child will have cystic fibrosis.

***2.3*** In a family of five children, what is the probability of the following events:
***a.*** All five children are sons.
***b.*** All five children are of the same sex.
***c.*** Their first child will be male and the next four children will be female.

***2.4*** Crosses are made between parents whose genotypes are *AABbCcddee* and *aabbCcDdEe*. What proportion of the progeny will be the following:
***a.*** Heterozygous for all five pairs of alleles.
***b.*** Recessive for traits governed by *b*, *c*, *d*, and *e*.

***2.5*** If a coin is tossed 7 times, what is the probability of obtaining the following:
***a.*** 6 heads, 1 tails.
***b.*** 3 heads, 4 tails.
***c.*** Heads on the first three tosses and tails on the next four tosses.

***2.6*** In families of 8 children, what is the theoretical percentage expected to have the following:
***a.*** 6 girls and 2 boys.
***b.*** 6 children of one sex and 2 of the other.

***2.7*** Five separate experiments were performed to obtain $F_2$ progeny by inbreeding $F_1$ fruit flies heterozygous for the recessive traits black body (+/*b*) and rough eyes (+/*ro*). The data obtained are shown at the top of the next column.
***a.*** Calculate the $\chi^2$ for each experiment. Which of the experimental results fit a 9:3:3:1 distribution?
***b.*** Does a significant deviation appear in any of these experimental results?

| **experiment number** | **normal color, normal eyes** | **normal color, rough eyes** | **black color, normal eyes** | **black color, rough eyes** |
|---|---|---|---|---|
| 1 | 190 | 58 | 34 | 6 |
| 2 | 276 | 94 | 64 | 26 |
| 3 | 226 | 66 | 74 | 18 |
| 4 | 304 | 120 | 110 | 42 |
| 5 | 124 | 60 | 42 | 30 |

***2.8*** A coin is tossed 10 times and lands heads up 6 times and tails up 4 times.
***a.*** Do these results fit the predicted 1:1 ratio?
***b.*** If the same coin is tossed 100 times and lands heads up 60 times and tails up 40 times, can the hypothesis be accepted despite the observed deviation from the expected 1:1 ratio?

***2.9*** A survey was made of 208 families, each having four children, and the following results were obtained:

| number of boys per family: | 4 | 3 | 2 | 1 | 0 |
|---|---|---|---|---|---|
| number of girls per family: | 0 | 1 | 2 | 3 | 4 |
| number of families: | 10 | 51 | 80 | 54 | 13 |

***a.*** Does this distribution fit the prediction that the probability of having a boy is the same as the probability of having a girl?
***b.*** If the survey had included five times as many families, what would be the theoretical distribution of boys and girls in families with four children each?

***2.10*** Crosses between two pure strains of flowering plants produced white $F_1$ progeny, but $F_2$ offspring from crosses between $F_1$ plants gave the following results in three separate experiments:

| **Experiment number** | **flower color: white** | **flower color: purple** |
|---|---|---|
| 1 | 150 | 42 |
| 2 | 176 | 48 |
| 3 | 1290 | 310 |

***a.*** Calculate $\chi^2$ for each experiment, assuming a 3:1 ratio and a 13:3 ratio.
***b.*** Is flower color due to one gene or to two interacting genes showing epistasis?

***2.11*** Heterozygous tomato plants with purple, hairy stems are backcrossed to pure-breeding tomato plants with green, hairless stems. The progeny consists of 28 purple, hairy:38 purple, hairless:33 green, hairy:41 green, hairless.

***a.*** Using the $\chi^2$ test, determine whether each pair of alleles has segregated in the expected ratio in the backcross progeny.

***b.*** Does the $\chi^2$ test support the hypothesis that the two pairs of alleles segregate independently?

***2.12*** Leg length (in mm) was measured for 20 insects of the same species, with the following results: 2.4, 1.9, 1.8, 1.7, 2.2, 2.1, 2.6, 2.5, 2.3, 2.9, 1.4, 1.5, 2.5, 1.8, 1.9, 1.7, 1.5, 2.7, 2.3, 1.4. Calculate the mean, the standard deviation, and standard error of the mean.

***2.13*** A sample of 2000 male college students was measured for height, with the results shown at right.

***a.*** Draw a histogram of these data. Do these values plot a normal curve? Do these data fit a normal distribution?

***b.*** Calculate the mean, the standard deviation, and the standard error of the mean.

***c.*** What percentage of this sample is included within the limits of $1\sigma$, $2\sigma$, and $3\sigma$?

***d.*** Based on the standard error, is this a representative sample of male college students?

| **height (cm) midclass value** | **frequency** |
|---|---|
| 154 | 4 |
| 156.5 | 4 |
| 159 | 40 |
| 161.5 | 96 |
| 164 | 150 |
| 166.5 | 234 |
| 169 | 268 |
| 171.5 | 314 |
| 174 | 280 |
| 176.5 | 242 |
| 179 | 160 |
| 181.5 | 114 |
| 184 | 52 |
| 186.5 | 26 |
| 189 | 10 |
| 191.5 | 4 |
| 194 | 2 |
| | 2000 |

***2.14*** In experiments on inheritance of ear length in corn, R.A. Emerson and E.M. East obtained the following results:

| | **ear length, cm** | | | | | | | | | | | | | | | | |
|---|---|---|---|---|---|---|---|---|---|---|---|---|---|---|---|---|---|
| | **5** | **6** | **7** | **8** | **9** | **10** | **11** | **12** | **13** | **14** | **15** | **16** | **17** | **18** | **19** | **20** | **21** |
| parent 60 (Tom Thumb) | 4 | 21 | 24 | 8 | | | | | | | | | | | | | |
| parent 54 (Black Mexican) | | | | | | | | | 3 | 11 | 12 | 15 | 26 | 15 | 10 | 7 | 2 |
| $F_1$ (60 × 54) | | | | | 1 | 12 | 12 | 14 | 17 | 9 | 4 | | | | | | |
| $F_2$ | | | 1 | 10 | 19 | 26 | 47 | 73 | 68 | 68 | 39 | 25 | 15 | 9 | 1 | | |

***a.*** Calculate the mean, the standard deviation, and the standard error of the mean for each parent strain and for the $F_1$ and $F_2$ progeny.

***b.*** Calculate the standard error of the difference between means for the two parent strains, the $F_1$ and $F_2$ progeny, and parent 54 (Black Mexican corn) and the $F_2$. Indicate whether the difference is statistically significant for each pair of strains.

***c.*** Construct histograms from plotted data to show the differences graphically between the parental types and their $F_1$ and $F_2$ progeny. Indicate the mean for each group with an arrow.

***2.15*** If we assume that the parental Tom Thumb and Black Mexican strains of corn are genotypically similar except for three different genes governing ear length, the cross can be shown as *AABBCC* × *aabbcc*. We may further assume that their similar genotypic background leads to ears of 6 cm mean minimum length in both strains and that *each* dominant allele of the three genes governs an additional increment of 2 cm length and *each* recessive allele of the three genes has no effect on ear length. Making these assumptions, answer the following questions:

***a.*** What is the expected average (mean) ear length

for each parental strain?
***b.*** What is the expected average (mean) ear length for $F_1$ offspring?
***c.*** How would you explain the greater variability in $F_2$ than in $F_1$ progeny?
***2.16*** Twin concordance for five traits showed the following results:

| | MZ twins | | DZ twins | |
|---|---|---|---|---|
| **trait** | **number of pairs studied** | **percent concordance** | **number of pairs studied** | **percent concordance** |
| 1 | 25 | 30.1 | 102 | 27.3 |
| 2 | 110 | 100.0 | 232 | 24.7 |
| 3 | 36 | 82.8 | 74 | 13.2 |
| 4 | 56 | 46.4 | 105 | 44.1 |
| 5 | 88 | 63.0 | 122 | 10.2 |

***a.*** Which traits appear to have a substantial genetic basis to expression?
***b.*** Which trait appears to be entirely genetically determined?
***2.17*** In Table 2.7, which shows twin concordance for various diseases, there is a significant difference between monozygotic and dizygotic twin concordances for tuberculosis infection, thereby indicating the influence of genetic factors in development of the disease. How can you explain this result in view of the fact that tuberculosis is an infectious disease caused by a bacterium and is not an inherited disease?

## References

Bodmer, W.F. and L.L. Cavalli-Sforza. Oct. 1970. Intelligence and race. *Sci. Amer.* **223**:19.

Brewbaker, J.L. 1964. *Agricultural Genetics.* Englewood Cliffs, N.J.: Prentice-Hall.

Cavalli-Sforza, L.L., and W.F. Bodmer. 1971. *The Genetics of Human Populations.* San Francisco: Freeman.

East, E.M. 1916. Studies on size inheritance in *Nicotiana. Genetics* **1**:164.

Feldman, M.W., and R.C. Lewontin. 1975. The heritability hang-up. *Science* **190**:1163.

Lerner, I.M. 1968. *Heredity, Evolution, and Society.* San Francisco: Freeman.

Srb, A.M., R.D. Owen, and R.S. Edgar. 1965. *General Genetics.* San Francisco: Freeman.

Thompson, J.N. 1975. Quantitative variation and gene number. *Nature* **258**:665.

# CHAPTER 3

## Cellular Reproduction and Inheritance

Mendel was able to formulate basic rules of inheritance even though he knew nothing about the nature of genes, their location, or the mechanisms by which genes are transmitted from generation to generation. It is a tribute to his incisive analytical skills that he provided the foundations for genetic analysis despite the meager amount of biological information available to him. Modern genetics is grounded in a firm knowledge of cellular and chromosomal reproductive systems. In fact, the ready acceptance of Mendelian principles in 1900, and not in 1866, was due in large measure to the ease with which Mendel's abstract inheritance factors could be fit into the biological framework discovered shortly before and after Mendel's death in 1884.

Our knowledge of cells, chromosomes, organisms, and the chemistry of heredity will be useful as we study and understand how genes produce phenotypes. We can proceed beyond the stage of predicting outcomes of crosses and can analyze the biological bases for genetic phenomena. A solid biological framework is the foundation for a deeper and more comprehensive understanding of the genetic systems in living organisms. In this chapter and the next one, we will discuss some of the important features of biology that help us understand inheritance patterns in terms of cellular parts and processes.

## Cellular Distribution of Genes: Mitosis and Meiosis

All cellular organisms can be classified as **prokaryotes** or **eukaryotes,** according to their cellular organization. Prokaryotes include all the bacteria and blue-green algae; all other cellular organisms are eukaryotes (Table 3.1). In prokaryotes, genes are localized in a region of the cell called the nucleoid, which is surrounded by cytoplasm encased in a plasma membrane and bounded by a cell wall (Fig 3.1). There is no permanent membrane other than the plasma membrane, and the nucleoid is not separated from the cytoplasm by a nuclear membrane. A membrane-bounded nucleus characterizes eukaryotic cells. The basic distinction between the two groups of cellular organisms is the *presence or absence of a membrane-bounded nucleus.*

***Table 3.1*** Kingdom classification systems.

| "traditional" | Dodson, 1971 | Whittaker, 1969 |
|---|---|---|
| Plantae | Monera | Monera |
| bacteria | bacteria | bacteria |
| blue-green algae | blue-green algae | blue-green algae |
| chrysophytes | Plantae | Protista |
| green algae | chrysophytes | chrysophytes |
| red algae | green algae | protozoa |
| brown algae | red algae | Fungi |
| slime molds | brown algae | slime molds |
| true fungi | slime molds | true fungi |
| bryophytes | true fungi | Plantae |
| tracheophytes | bryophytes | green algae |
| Animalia | tracheophytes | red algae |
| protozoa | Animalia | brown algae |
| metazoa | protozoa | bryophytes |
| | metazoa | tracheophytes |
| | | Animalia |
| | | metazoa |

Data from E.O. Dodson, 1971. The kingdoms of organisms. *Systematic Zoology* **20**:265–281; and R.H. Whittaker, 1969. New concepts of the kingdoms of organisms. *Science* **163**:150–160.

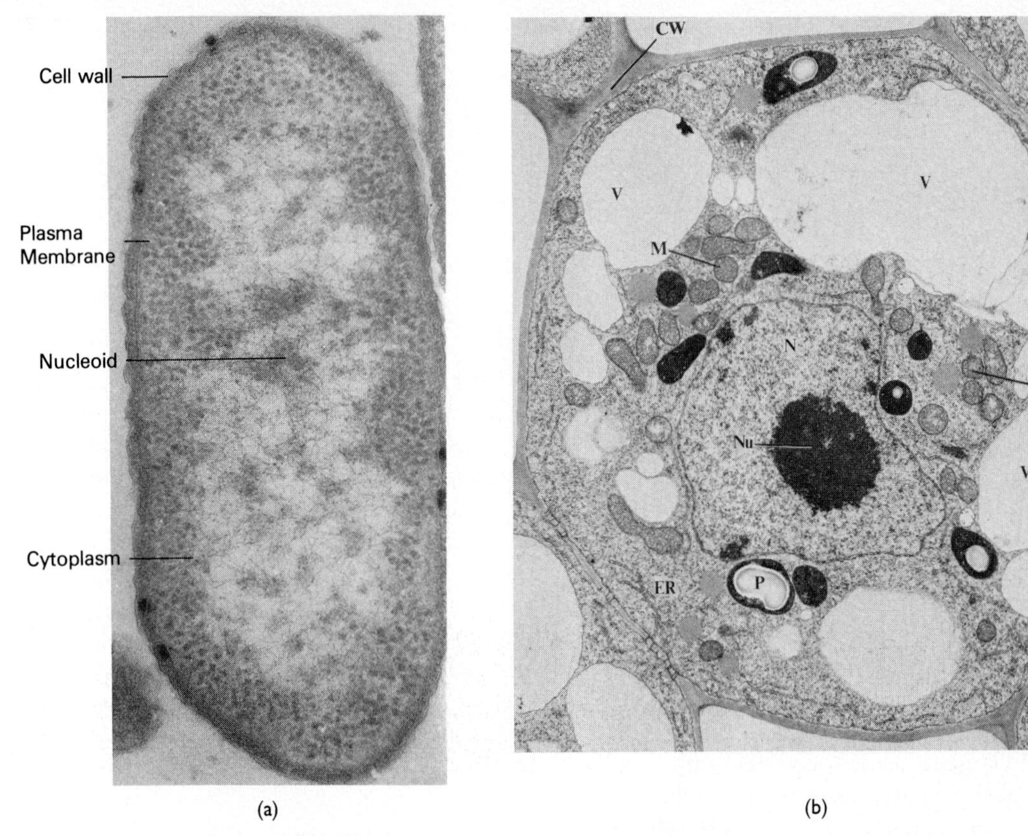


***Figure 3.1*** Electron micrographs of thin sections of a prokaryotic and a eukaryotic cell type: (a) Rod-shaped bacterium (*Pseudomonas aeruginosa*), showing typical prokaryotic cellular plan of a nucleoid region surrounded by cytoplasm, all enclosed within the plasma membrane; the living protoplast is surrounded by a cell wall. × 60,000 (Courtesy of H.P. Hoffmann). (b) Mature root cell from the eukaryotic species *Potamogeton natans,* a flowering plant, showing the nucleus containing a nucleolus and enclosed by a nuclear membrane system. The cytoplasm surrounding the nucleus is filled with various differentiated components, including mitochondria, plastids containing starch, endoplasmic reticulum, and prominent vacuoles. The plasma membrane is surrounded by a cell wall. × 7,500 (Courtesy of M.C. Ledbetter).

Mitosis, meiosis, and fertilization are the processes that transmit genes from one generation to the next. They occur exclusively in eukaryotes. Mitosis is one phase in a cycle of cellular events that leads to a new generation of identical cells or organisms, which arise by **asexual reproduction.** In sexual species, meiosis and fertilization take place in addition to mitotic activities. During **sexual reproduction,** gametes are produced, and when gametes fuse at fertilization the new sexual generation is initiated. *Genetic constancy* characterizes asexual reproduction, and *genetic variety* characterizes sexual reproduction.

## 3.1 Mitosis and the Cell Cycle

Each new generation of cells or individuals is the result of reproduction. Since progeny resemble their parents, mechanisms must exist that ensure faithful *increase* and *transmission* of genetic information. Increase is essential because more copies of the **genome** (one set of genes) must be made if progeny are to receive all the information necessary to grow up and produce their own offspring in turn. Once multiplied, the genes must be delivered from parents to progeny with great accuracy. Unless the processes of gene increase and gene transfer are accomplished with considerable fidelity, progeny will not resemble their parents, as they in fact do.

Our understanding of the mechanics of gene and chromosome replication came from a healthy infusion of biochemistry into biological study, but we can easily see the chromosomes themselves by microscopy. From combined studies using biochemistry, genetics, and cytology (the study of cells by microscopy), we know that chromosome replication is accomplished during an interval in the **cell cycle** preceding the delivery of replicated chromosomes by mitosis (Fig. 3.2).

The whole interval between one mitosis and the next is subdivided into three separate phases called $G_1$, $S$, and $G_2$. During $G_1$, preparations get under way for replication of the chromosomes. During $S$, the new chromosomes are made through processes of replication; and during $G_2$, preparations begin for delivery of the parental and progeny chromosomes to two daughter cells. The actual process of delivery is mitosis.

As we will see in Chapter 4, genes are found on chromosomes. Chromosomes are duplicated and then distributed to new cells; thus, genes are passed on to new cells. In each cell cycle, we see chromosomal events, but we relate these events directly to gene increase and gene transfer to the progeny. Mitosis is an asexual process by which daughter cells are virtually assured of receiving equal and identical copies of all the genes from the parent cell. Mitosis is the only mechanism available in asexually reproducing eukaryotes for the transfer of genetic information to successive generations. In sexual species of multicellular eukaryotes, mitosis is responsible for accurate delivery of the genes in each somatic (body cell) nuclear division event. The trillions of cells in one human being arise by countless mitotic divisions of the original fertilized egg during development. Replacement of old cells by new cells during a lifetime also takes place by mitotic divisions. The unique combinations of alleles in each person are accurately copied and accurately delivered in every cell cycle in each part of the individual.

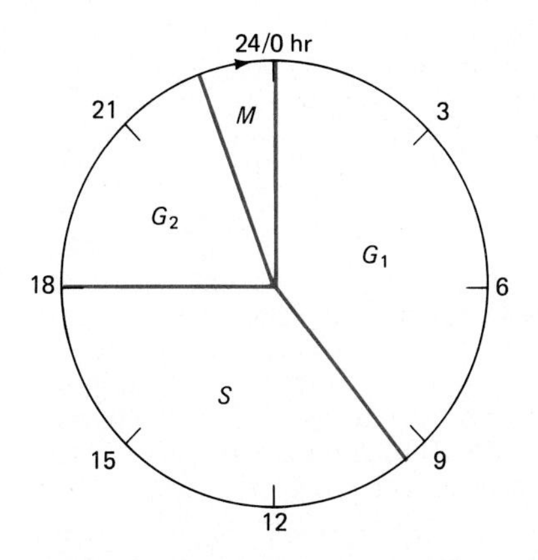


***Figure 3.2*** In a cell cycle, which may be 24 hours long, the interval of mitosis (*M*) is relatively brief. During interphase, between successive mitotic divisions, $G_1$ is variable in duration but it is the time when preparations get under way for chromosome replication; $S$ is the phase of chromosome replication; and $G_2$ is the post-replication phase when preparations are made to deliver sets of chromosomes to daughter nuclei by mitosis.

## 3.2 Stages of Mitosis

The continuous sequence of events in mitosis is conventionally divided into arbitrary stages (Fig. 3.3): **prophase** (*pro,* before), **metaphase** (*meta,* between), **anaphase** (*ana,* back), and **telophase** (*telo,* end). The entire interval between mitotic divisions is called **interphase,** and it includes the $G_1$, $S$, and $G_2$ phases of a somatic cell cycle.

As prophase begins, the stringy tangle of chromosomes in the interphase nucleus begins to shorten and thicken. By about mid-prophase individual chromosomes can be seen and recognized as double (duplicated) structures. As mitosis proceeds, the chromosomes continue to condense until they line up at the equatorial plane of the **spindle,** and metaphase begins. Each metaphase chromosome can clearly be seen as a replicated structure made up of two identical **chromatids** (half-chromosomes) held together at the common centromere region (Fig.

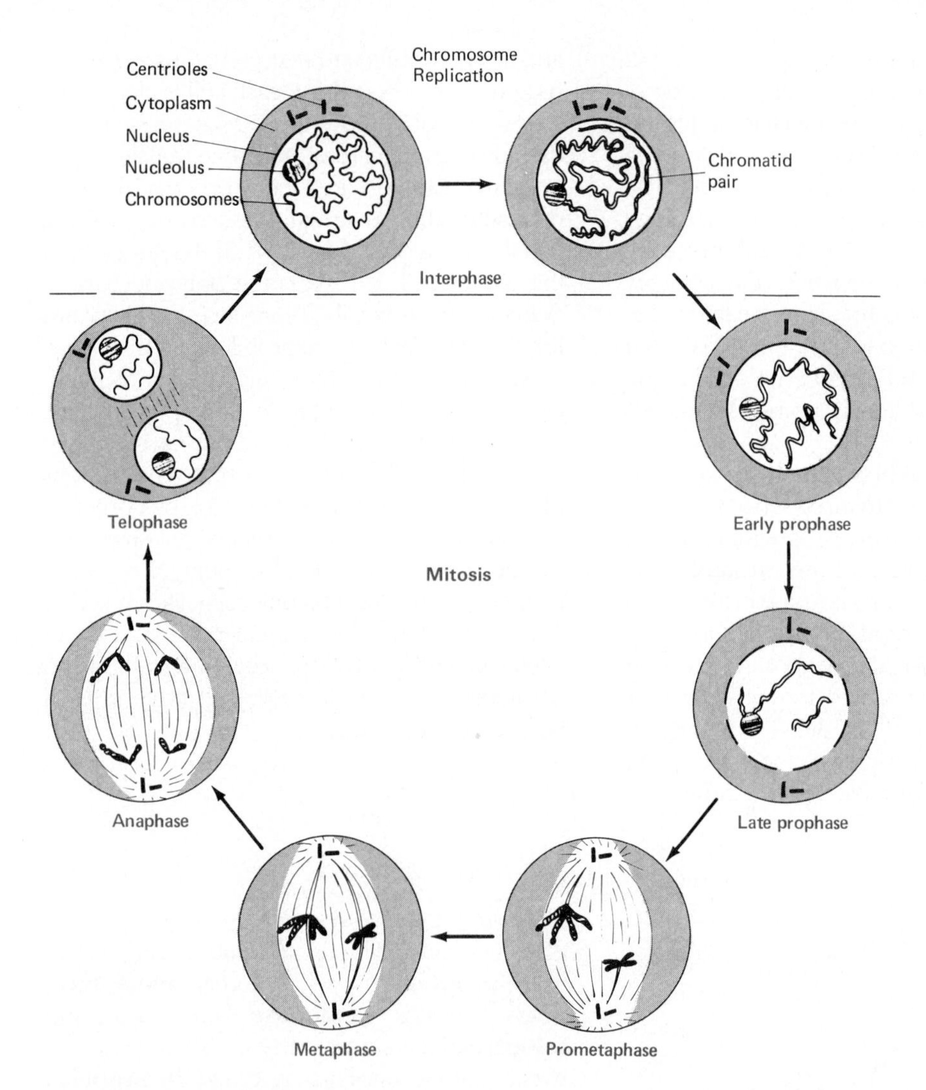


***Figure 3.3*** Mitosis. Chromosomes replicate during interphase and begin to condense in prophase. The nuclear membrane usually disappears, as do nucleoli, by the end of prophase. The spindle forms and during prometaphase the chromosomes move toward the equatorial plane of the spindle and become aligned there in metaphase. Sister chromatids of each replicated chromosome separate and move toward opposite poles in anaphase, and nuclei are reorganized in telophase. Afterward, each nucleus enters the interphase state, and may or may not undergo other mitotic divisions.

3.4). Each sister chromatid of a chromosome has its own **centromere,** or **kinetochore,** and the two centromeres face opposite poles of the cell. When anaphase begins, sister chromatids separate and move toward opposite poles because of their initial metaphase orientation on the spindle, aided by spindle fibers extending from the centromere to one pole of the cell. When chromatids separate from each other, each becomes a full-fledged and independent chromosome, no longer acting together with its sister. During the final stage of telophase, the condensed chromosomes in each daughter nucleus unfold and gradually assume their interphase appearance. Nuclear reorganization takes place, and the sequence ends with two new

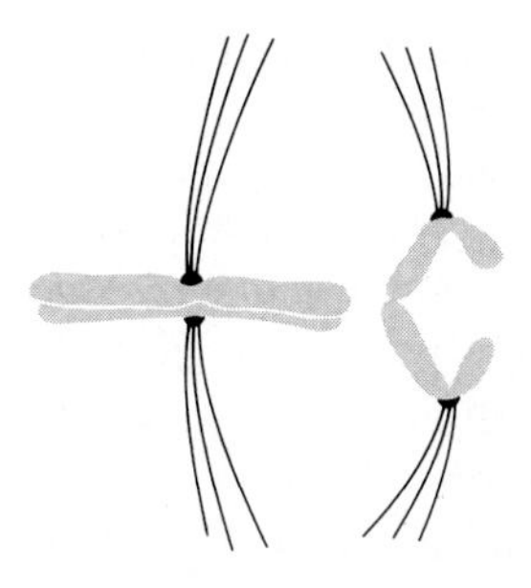

***Figure 3.4*** Sister chromatids of a replicated chromosome are more likely to separate and move toward opposite poles since their individual centromeres are oriented toward opposite poles of the cell during metaphase. The sister centromeres are held together in a common centromere region until anaphase separation begins.

interphase nuclei produced from the original parent nucleus (Fig. 3.5).

Each daughter nucleus contains a set of chromosomes and genes identical to that of the parent nucleus and to that of the other daughter nucleus because the replication, which took place during the preceding interphase, produced two identical sets of chromosomes that were distributed accurately during mitosis itself. The *same distribution mechanism* operates for all eukaryotic cells, regardless of their chromosome number, the number of chromosome sets, or aberrancies in number (Table 3.2). The fidelity of distribution is responsible for ensuring identical chromosome complements and, therefore, for *genetic constancy* in mitotic generations. Variations can arise by mutation or by other random modifications of chromosomes,

***Table 3.2*** Chromosome numbers found in various organisms.

| **organism** | **diploid number** |
|---|---|
| human (*Homo sapiens*) | 46 |
| chimpanzee (*Pan troglodytes*) | 48 |
| rhesus monkey (*Macaca mulatta*) | 42 |
| dog (*Canis familiaris*) | 78 |
| cat (*Felis domestica*) | 38 |
| horse (*Equus caballus*) | 64 |
| toad (*Xenopus laevis*) | 36 |
| housefly (*Musca domestica*) | 12 |
| mosquito (*Culex pipiens*) | 6 |
| nematode (*Caenorhabditis elegans*) | 11♂,12♀ |
| tobacco (*Nicotiana tabacum*) | 48 |
| cotton (*Gossypium hirsutum*) | 52 |
| kidney bean (*Phaseolus vulgaris*) | 22 |
| broad bean (*Vicia faba*) | 12 |
| onion (*Allium cepa*) | 16 |
| potato (*Solanum tuberosum*) | 48 |
| tomato (*Lycopersicon esculentum*) | 24 |
| bread wheat (*Triticum aestivum*) | 42 |
| rice (*Oryza sativa*) | 24 |
| baker's yeast (*Saccharomy cescerevisiae*) | 34 |

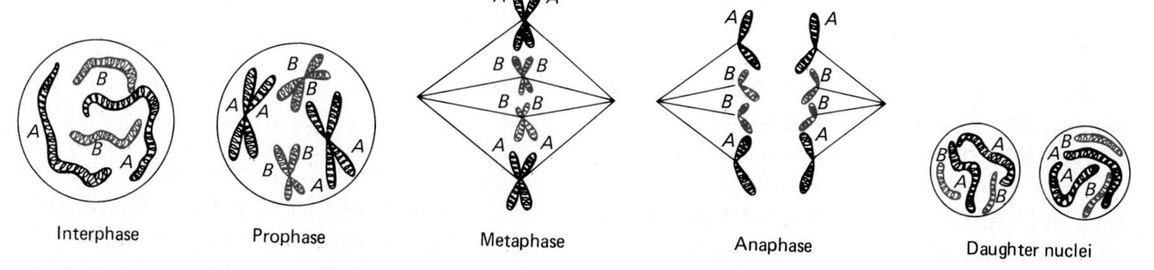


***Figure 3.5*** Each daughter nucleus has an identical set of chromosomes and genes, unchanged from the original parental nucleus that underwent chromosome replication and subsequent mitosis.

but these variations in turn will be replicated and transmitted faithfully to all descendants of the modified cell or individual. Asexually reproducing species are, therefore, genetically rather uniform.

Each body (somatic) cell in a multicellular organism is essentially identical in genetic content. Differences in cell appearance, function, and activity arise during development and differentiation through regulation of gene action, a topic we will cover in Chapters 10 and 14 and elsewhere in this book. For now, the main point is to realize that mitosis delivers the genes on chromosomes to progeny nuclei during reproduction, and other processes then direct the expression of genetic potential into the variety of cells, tissues, and organs of the individual. The genetic consequence of mitosis is that the two new cells contain identical copies of the same genes and that these genes are the same as the ones that were present in the cell from which they came.

## 3.3 Overview of Meiosis

Only sexually reproducing species have cells that can switch from mitotic divisions to meiotic divisions of the nucleus at specified times in a life cycle. Unicellular eukaryotes switch from a phase of asexual, or vegetative, multiplication by mitosis to meiosis during a sexual phase. The prokaryotic bacteria and blue-green algae have different gene delivery systems, and the acellular viruses reproduce by methods that are unique to them. We will discuss examples of these kinds of organisms later in the chapter in order to underscore the direct relationship between systems for gene transmission and reproduction in a life cycle.

Two consecutive divisions make up the total process of nuclear division by meiosis. Cells that are **haploid** (having one set of chromosomes) normally do not undergo meiosis; cells with two sets (**diploid**) or more than two sets (**polyploid**) of chromosomes have meiotic potential. In a typical diploid species, such as garden peas or human beings, meiotic divisions reduce the diploid number of chromosomes *by one-half* to the haploid number (Fig. 3.6).

If reduction in chromosome number is the only criterion and chromosome number is reduced by one-half at the end of the first meiotic division, we may wonder why meiosis involves a second division. The answer lies in a consideration of the number of copies of the genome when meiosis begins and when it is concluded.

Before meiosis begins, chromosomes replicate during interphase and the number of genome copies is doubled in the mother cell nucleus. When the mother cell is diploid, there are 2 copies of the genome before replication and 4 copies afterward. The chromosome number remains diploid, however, since sister chromatids remain associated in the replicated chromosomes. When meiosis begins, therefore, the diploid nucleus contains 2 replicated chromosomes of every kind so that 4 genomes of every kind are distributed among the pairs of chromatids (Fig. 3.7). When anaphase of the first meiotic division takes place, partner chromosomes separate and move to opposite poles, and the chromosome number is reduced by one-half. Each chromosome consists of two chromatids, however, so that each anaphase nucleus has two copies of each set of genes. This number is the same as in the original mother cell before replication, so *gene number* has not yet been reduced by one-half (to the haploid level). In the second meiotic division, sister chromatids and the alleles they carry separate into different nuclei. Each product of the second phase of meiosis, thus, ends up with one set of chromosomes carrying one copy of the genome. Reduction from the genetic diploid to the genetic haploid state, therefore, requires two consecutive divisions for one complete meiotic cycle.

The four nuclear products of a meiotic cell, or **meiocyte,** are usually genetically different from one another (see Fig. 3.7). This provides the biological basis for Mendel's deductions from breeding analysis that all possible kinds of germ cells are produced in equal numbers, on

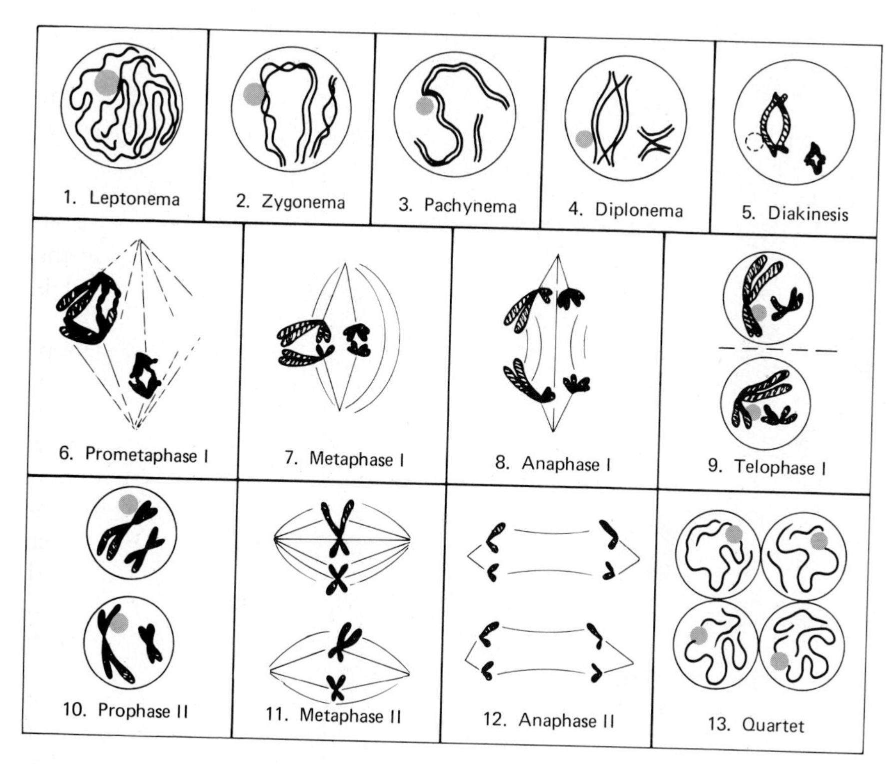


***Figure 3.6*** Meiosis. Chromosomes proceed through an extended prophase in division I (1–5), followed by movement to the spindle equator (6), alignment at the equator (7), and separation (8) and possible reorganization of the two nuclei (9) when Meiosis I ends. During Meiosis II, conventional prophase (10), metaphase (11), and anaphase (12) activities eventually lead to a quartet of nuclear products of the original meiotic nucleus. Significant events include chromosome pairing at zygonema (2), reduction of chromosome number by one-half at anaphase I (8), and reduction of gene numbers or DNA content in anaphase II (12) to one-half the level that was present in the premeiotic nucleus. In this example the chromosome number has been reduced from $2n = 4$ to $n = 2$.

the average, by each parent in a cross. If we follow one pair of alleles alone, for example *A*/*a*, we will find equal proportions of the two alleles of the gene in the meiotic products. If we follow two pairs of alleles, we will find *AB*, *Ab*, *aB*, and *ab* in equal proportions. With three pairs of alleles in triply heterozygous meiotic cells, we will find eight different genotypes among the meiotic products of a random sampling of meiocytes (Fig. 3.8). Each pair of chromatids can be oriented in either one of two different ways on the basis of purely random alignment on the metaphase plate in the first meiotic division.

The genetic consequence of meiosis is segregation and random assortment of members of pairs of alleles into the germ cells, thereby leading to various combinations of alleles in the gametes. When gamete fusions take place at random in the fertilization phase of a sexual cycle, a variety of genotypes can arise among the progeny.

## 3.4 First Meiotic Division

The meiotic nucleus proceeds through a sequence of prophase, metaphase, anaphase, and telophase stages in a continuing series of events. These designations are made for convenience in analysis and discussion and not because there are interruptions in the division process. Because the prophase of **Meiosis I** (the first division) is the most complex, protracted, and

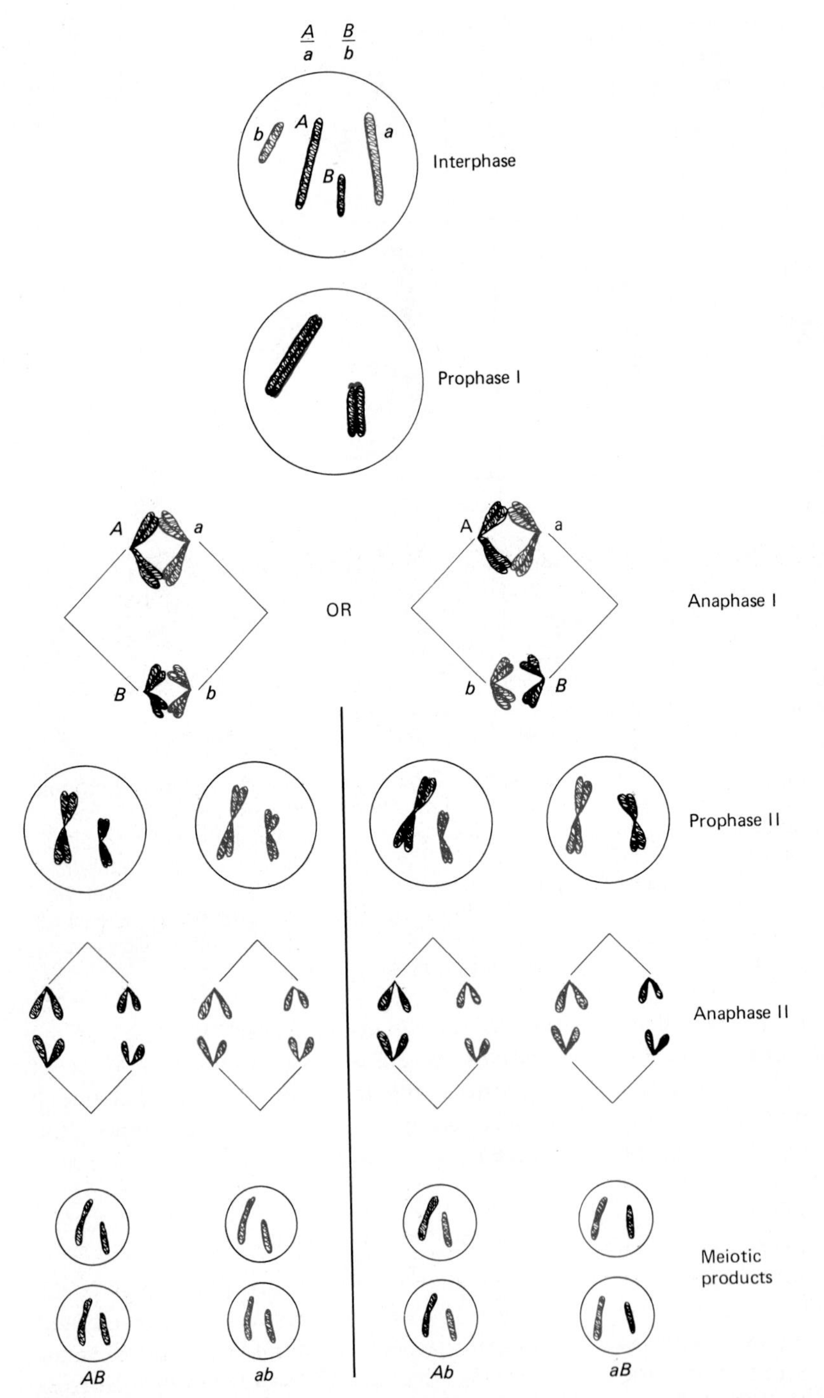


***Figure 3.7*** During meiosis, the chromosome number in the nuclear products is reduced to one-half the number present in the parent meiocyte nucleus. Reduction of chromosome number and of gene number require two consecutive divisions. At the end of meiosis, genetically different progeny nuclei may arise from a heterozygous meiocyte.

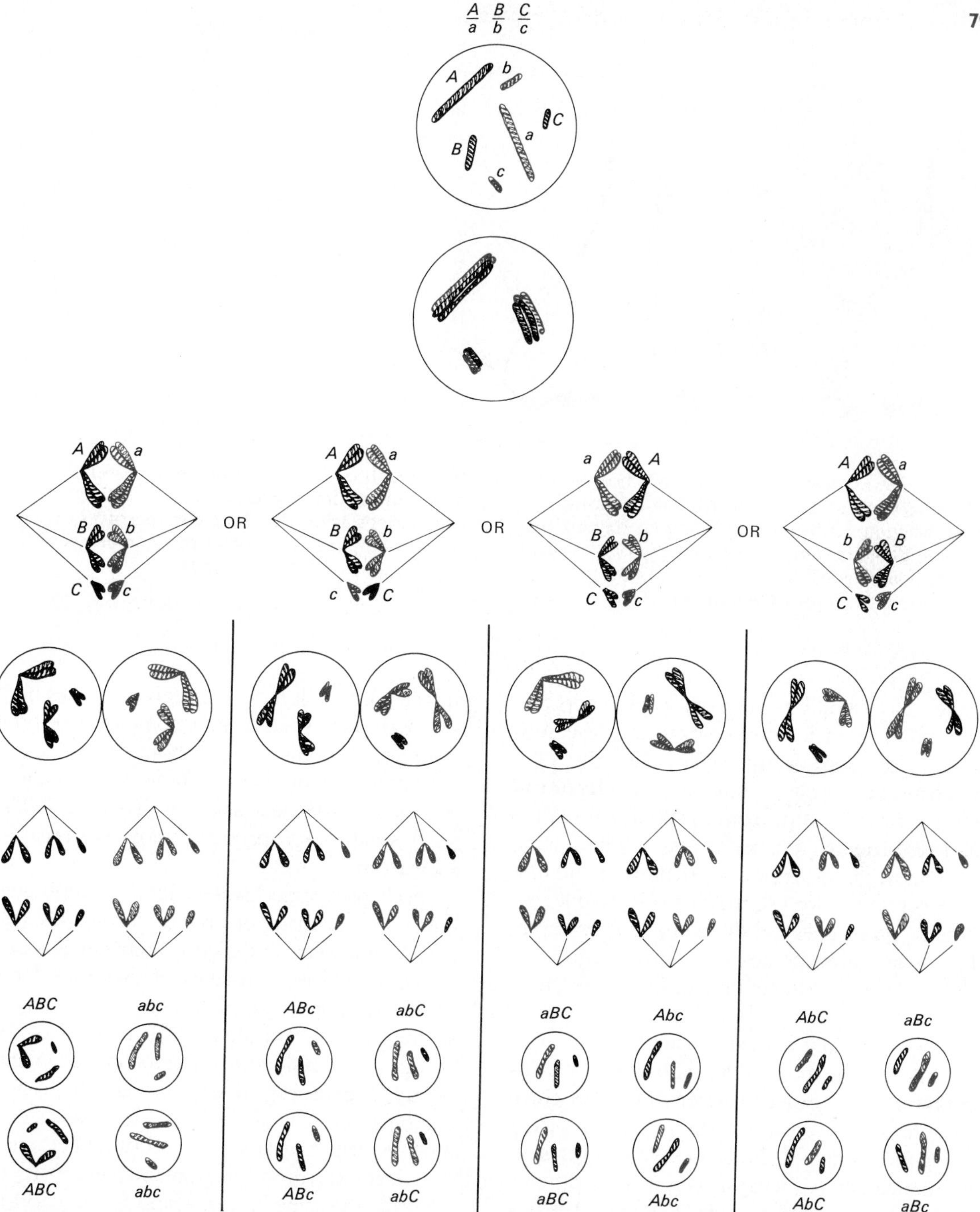


***Figure 3.8*** Equal proportions of eight possible genotypic combinations may be found in a random sampling of meiotic products arising from triply heterozygous meiocytes. Segregation of members of a pair of alleles and independent assortment of different pairs of alleles, which are the Mendelian Laws of Inheritance, are the genetic consequences of meiosis. Distributions of chromosomes carrying *A/a*, *B/b*, and *C/c* are independent of one another and they assort at random into all possible combinations of genotypes which, therefore, are produced in approximately equal proportions among the meiotic products.

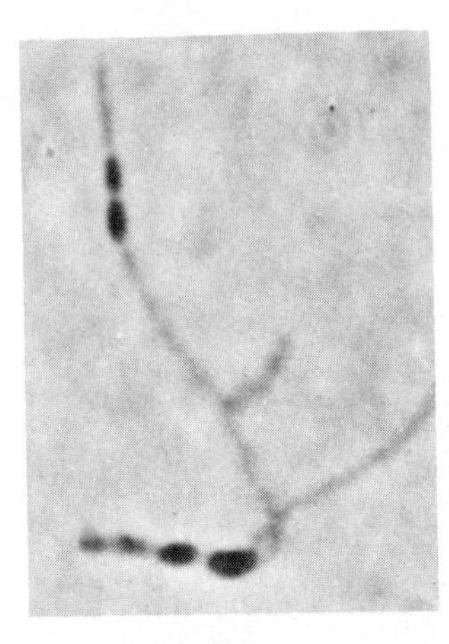

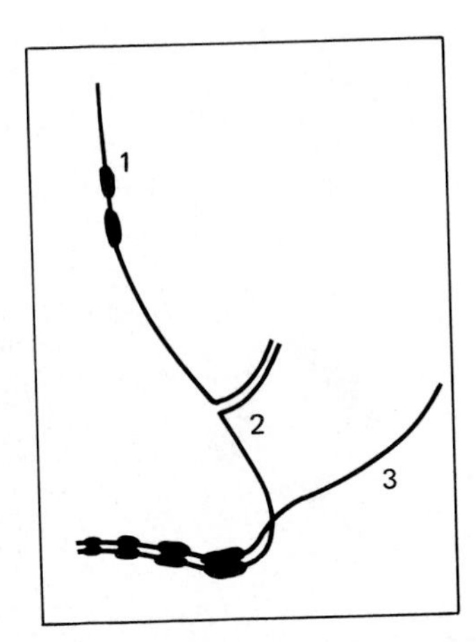


***Figure 3.9*** Synapsis involves only two homologous chromosome segments at any particular pairing site, even when more than two homologous regions are present in the nucleus. In this example of a trivalent (three homologous, synapsed chromosomes) in castor bean (*Ricinus communis*) meiosis, the photograph of trivalent chromosome 9 is interpreted in the drawing as having only two-by-two pairings, with one of the chromosomes (2) paired with parts of the other two homologues (1, 3). (Courtesy of G. Jelenkovic.)

genetically significant interval, it has been further subdivided for convenience into the substages of **leptonema** ("slender thread"), **zygonema** ("yoked thread"), **pachynema** ("thick thread"), **diplonema** ("double thread"), and **diakinesis** ("divided across"). When using an adjective to describe some feature of the first four substages, we use the terms leptotene, zygotene, pachytene, and diplotene. For example, leptotene chromosomes pair during zygonema to produce pachytene bivalents of four chromatids each.

Meiosis begins in the replicated nucleus with leptonema, the first of the substages of **prophase I.** The beaded, slender threads of chromosomes soon begin to pair at various places along their length in the process of **synapsis,** as zygonema begins. Synapsis is a specific process that involves pairing only between **homologous chromosomes** or parts of such genetically matched chromosomes (Fig. 3.9). A structure called the **synaptonemal complex** begins to form between paired chromosome regions. It runs completely down the length of each chromosome pair when zygonema ends and all parts of homologous chromosomes have synapsed. The synaptonemal complex is responsible for keeping paired homologous chromosomes together and in register throughout pachynema, the next stage of prophase I (Fig. 3.10). A pair of synapsed chromosomes is called a **bivalent**.

Each pachytene bivalent looks unreplicated but is actually made up of four chromatids, two per chromosome in the pair. Pachynema is one of the significant substages of prophase I because exchanges take place between chromatid segments in the bivalents. The process of exchange of homologous chromosome segments is called **crossing over,** during which new combinations of alleles on the *same* chromosome pair may arise. The process of crossing over leads to recombinations between pairs of alleles in the same chromosome and is a critical feature in recombination analysis and in gene mapping, as we will see in Chapters 5 and 6. The presence of the synaptonemal complex is essential for crossing over to take place, since homologous chromosomes may separate before crossing over occurs unless they remain in reg-

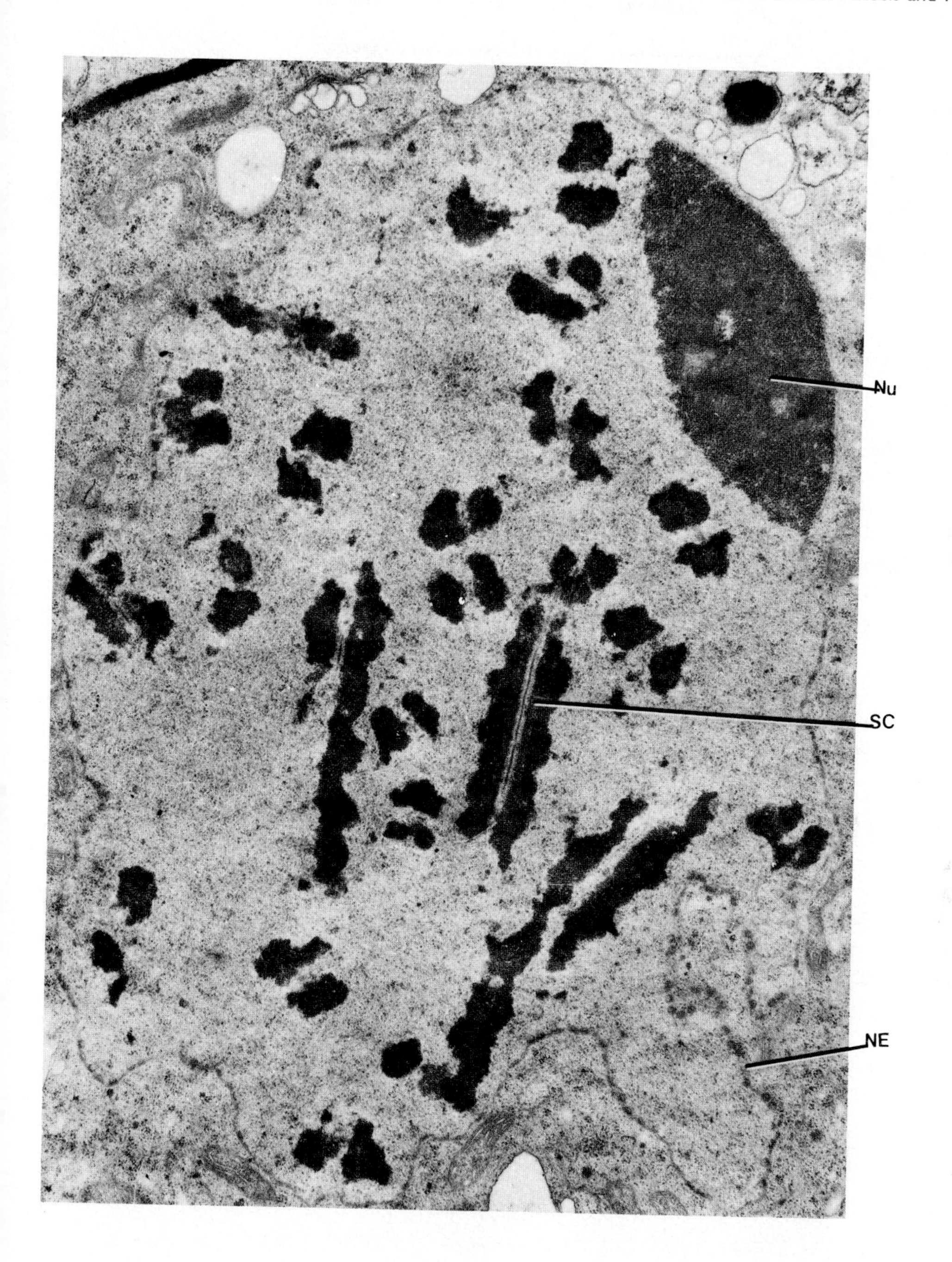


***Figure 3.10*** Electron micrograph of a thin section through a meiotic cell of the fungus *Neottiella,* a member of the Ascomycetes. One of the pachytene bivalents has been sectioned favorably and displays the synaptonemal complex (SC) in the space between the paired homologous chromosomes. The nuclear envelope (NE) is visible, as well as a prominent nucleolus (Nu). × 16,000. (Courtesy of D. von Wettstein, from M. Westergaard and D. von Wettstein. 1970. *Compt. rend. Lab. Carlsberg* **37**:239, Fig. 1.)

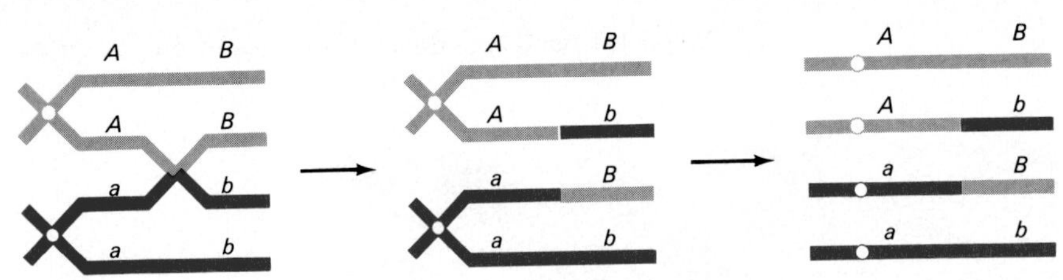


***Figure 3.11*** Crossing over in a pachytene bivalent heterozygous for two pairs of alleles. The homologous chromosomes are held in register and are closely paired as a result, which increases the probability that one or more exchanges may take place between homologous chromosomes. New genotypic combinations may arise in this way (*Ab*, *aB*), as well as parental combinations of alleles (*AB*, *ab*).

ister and closely paired during pachynema (Fig. 3.11).

The beginning of diplonema is signaled by the opening out of the bivalents at various places along their length. The homologues remain associated only at places called chiasmata (sing., **chiasma**), which are believed to be the first visible evidence of previous crossover events within the bivalents. From genetic analysis (to be described later), it is quite clear that *each* crossover event involves only two of the four chromatids in a bivalent. In favorable cytological materials such as insect spermatocytes, it can clearly be seen that two chromatids participate in each exchange but that all four chromatids can be involved in multiple exchange events in a bivalent (Fig. 3.12).

During diplonema, the synaptonemal complex is shed everywhere along the bivalent except at chiasmata. Even these fragments of the

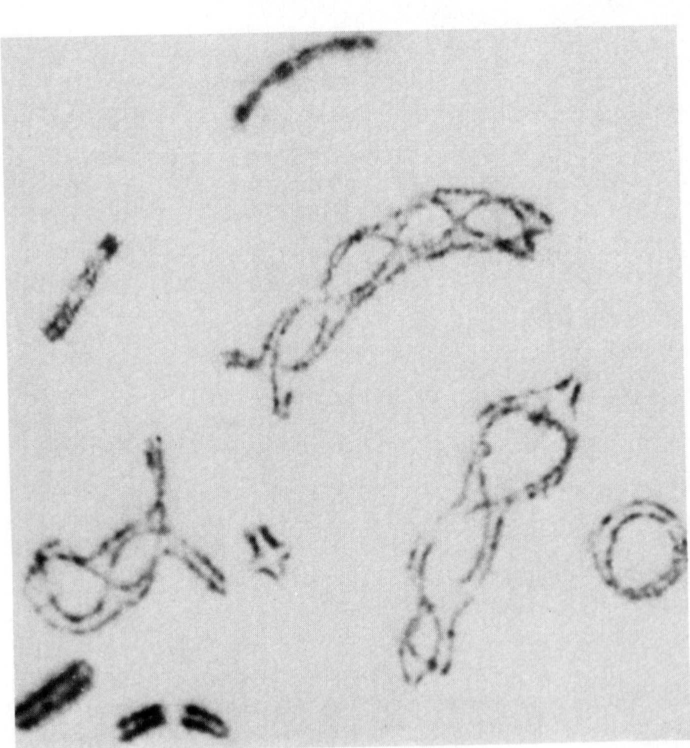

***Figure 3.12*** Diplotene bivalents in grasshopper spermatocyte. Each chiasma (site of a previous crossover event) involves two of the four chromatids in a bivalent, but all four chromatids may be involved in multiple exchange events in a bivalent. (courtesy of B. John.)

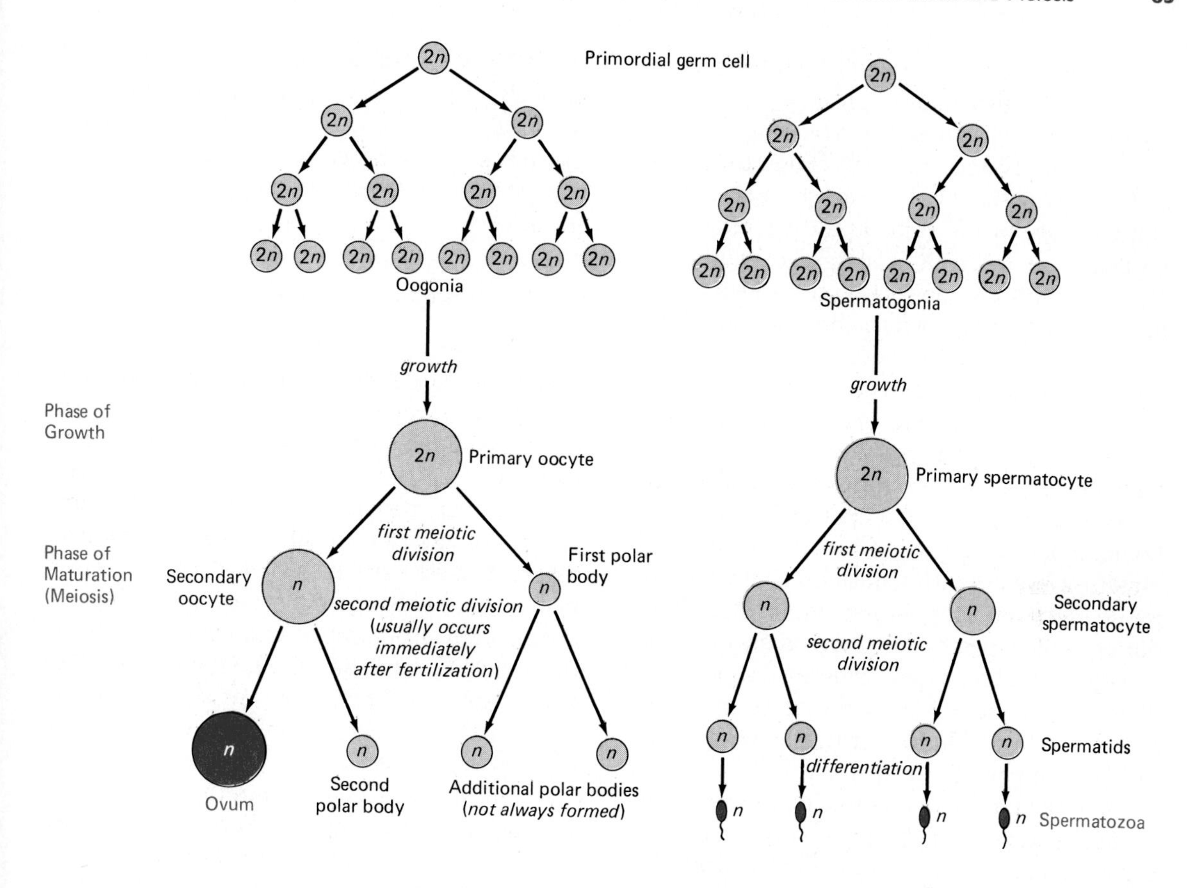


***Figure 3.13*** In humans and other animals oogenesis, or formation of eggs, leads to one functional egg per oocyte, spermatogenesis, or formation of sperm, results in four functional sperm per spermatocyte at the conclusion of meiosis.

complex are usually shed before diplonema is over. Diplonema may last for weeks, months, or years in some species, and length may vary between sexes. In the human female embryo, all the oocytes that lead to egg production are produced for a lifetime, but they remain suspended in diplonema until the girl reaches the age of puberty. During each menstrual cycle between puberty and menopause, when a woman is biologically capable of reproduction, an oocyte resumes meiosis and completes the first division (Fig. 3.13). It is in this stage that the oocyte (usually but inaccurately called the egg or ovum at this stage) is released from the ovary during ovulation, midway through a menstrual cycle. The second meiotic division will not take place unless the oocyte is penetrated by a sperm. Once penetration occurs, the oocyte nucleus completes the second meiotic division. The resulting single egg nucleus fuses with the sperm nucleus to form the first nucleus of the new individual, which begins as a fertilized egg cell.

In human males, on the other hand, sperm production in the testes does not begin until the boy reaches puberty. Sperm are then produced in the hundreds of millions every day until relative old age. If sperm are not released or eggs are not fertilized, eggs and sperm disintegrate within a relatively short time and are resorbed into the tissues, just like any other kind of worn-out cell. In general, gametes do not undergo further development unless they unite at fertilization. They are highly specialized reproductive cells with no other known function.

Diplotene chromosomes continue to condense until they reach their most contracted state during diakinesis. Bivalents are short and thick, chiasmata are plainly evident, and the individual bivalents are relatively well spaced within the nucleus. Chromosome counts are easily made at this stage in many species, and aberrant chromosomes or chromosome rearrangements are seen most readily here.

The meiotic bivalents come into alignment on the spindle equator at metaphase I. At this time the ends of the chromosome arms are positioned at the equator and centromeres of homologous chromosomes are as far apart as physically possible. This situation is just the opposite of the situation during mitotic metaphase during which centromeres are aligned at the equator and chromosome arms wave about in all directions on either side of this zone (compare Figs. 3.3 and 3.6). This orientation at metaphase I makes it highly likely that homologous centromeres will move to opposite poles of the cell.

Each pair of sister chromatids making up one chromosome of a bivalent is called a **dyad.** We can, therefore, say that separation of one dyad from its sister dyad is almost guaranteed by the simple mechanism of centromere positioning at the equatorial plane of the metaphase spindle. Mistakes do occur on occasion; for example, both dyads of the bivalent may move to the same pole. We will discuss such errors later on.

Anaphase begins when homologous chromosomes (dyads or replicated chromosomes) separate and move to opposite poles of the cell. Each chromosome is a double structure containing two copies of each gene of that chromosome. On the basis of chromosome number, however, reduction to one-half has been achieved.

Telophase is a stage of nuclear reorganization, just as it is in mitosis. Considerable variability among species occurs in the events between anaphase I and metaphase of the second meiotic division. In some cases nuclei in anaphase I enter almost directly into metaphase II. At the other extreme some species proceed through anaphase I, telophase I, interphase between divisions, and prophase II, before entering metaphase II. Every variation between these two extremes has been observed in one or more species.

## 3.5 Second Meiotic Division

Nothing is particularly striking or unusual about Meiosis II. Nuclei proceed through stages not unlike the stages in mitosis; however, it is not mitosis at all but, rather, the concluding phase of meiosis.

The main result of Meiosis II is the separation of the chromatids of each dyad into different nuclei, at which time each chromatid becomes a full-fledged chromosome acting independently from other chromosomes. Since every nucleus now has one genome, reduction to one-half the content of the chromosmes, DNA, and genes has been accomplished. Reduction division is complete only at this point.

In most cases, separation of the four nuclear products of meiosis into four separate cells takes place at the end of Meiosis II. In other cases a cell division may occur after Meiosis I to produce two cells, each of which will divide again at the end of Meiosis II. In both situations there are four products of meiosis, and the fate of these four cells varies among different groups of eukaryotes. In the more complex plants and animals, among the four products produced by a meiocyte in the female, only one is functional and the other three are aborted; however, all four products of a meiocyte are functional in the male (see Fig. 3.13).

## 3.6 Genetic Consequences of Meiosis

Meiosis leads to segregation of members of pairs of alleles into the gametes that sooner or later develop from the meiotic products. Subsequent fertilization allows new and different combinations of alleles in every sexual generation. Almost every individual in a sexual species has a unique genotype even though all the members of a species have the same genes; the differences are due to the variety of combinations of alleles for these genes.

Mendel's First Law is explained by meiosis: the two members of each pair of alleles are segregated into different gametes when homologous chromosomes are segregated into separate haploid nuclei. Mendel's Second Law is also explained by meiosis: different pairs of alleles are assorted independently from other allele pairs since nonhomologous chromosomes align independently and are segregated independently into haploid gametes.

If $3^n$ is the number of different genotypes theoretically possible when $n$ = the number of chromosome pairs, each carrying one pair of heterozygous alleles, then peas may produce $3^7$ different genotypes when differing in only 7 pairs of alleles on their 7 pairs of chromosomes in diploids. In human beings with 23 pairs of chromosomes, $3^{23}$ different genotypes are possible when only one pair of heterozygous alleles is present on each pair of chromosomes. Since there are thousands of genes in the human genome, and since many of these exist in the heterozygous state, it is statistically almost impossible for identical genotypes to arise by chance. Except for identical twins and other identical siblings, no two people have, ever have had, or are ever likely to have the same genotype by chance alone. From such considerations we can see that sexual reproduction is a system that essentially guarantees *genetic variety* in every generation.

Asexually reproducing species are relatively uniform within a population since meiosis and fertilization do not occur. Mitosis in eukaryotes is dedicated to accurate delivery of genome copies, and progeny resemble their parent in such species. Variation arises primarily by **mutation,** leading to some differences between populations of asexual species. Mutation is a random event and different mutations may arise in different populations at different times. Mutations also occur in sexual species, thereby enhancing the degree of variability and providing the new genetic information that is tested during evolution. According to many biologists, the greater the genetic variability, the better the chances that some inherited features will prove advantageous and will thus provide the foundation for change that leads toward adaptation during evolutionary time. According to this idea, the explosive increase in life forms over the past billion years was due in large measure to the appearance of sexual reproduction in certain ancestral populations. For the first three billion years or so, life was present in limited variety, according to the fossil record.

## Life Cycles of Some Sexually Reproducing Species

In animals and in a number of other kinds of organisms, the haploid cells resulting from meiosis function directly as sex cells and are gametes. The first cell formed by gamete fusion is the diploid **zygote,** which develops into the new individual in diploid species through development and differentiation of mitotic-cell lineages. In diploid land plants, haploid spores are produced by meiosis in the **sporophyte phase,** an intermediate stage in the life cycle. These spores subsequently develop into structures or systems called **gametophytes,** which later produce gametes by mitotic divisions of haploid cells (Fig. 3.14, see next page). Gamete fusion restores the diploid state and the zygote develops into the diploid plant through many mitotic divisions. In all of these cases, gametes arise only from cells that are meiocytes or that can be traced back directly to meiocyte divisions by meiosis.

In haploid eukaryotic organisms, the diploid phase of the life cycle may consist only of the zygote nucleus or cell, which immediately undergoes meiosis and is restored to the haploid state. The products of such meiotic divisions are not gametes; they are ordinary vegetative cells or spores with the haploid chromosome number. But two such haploid nuclei or their mitotic descendants may become sexually active under suitable conditions and unite to produce

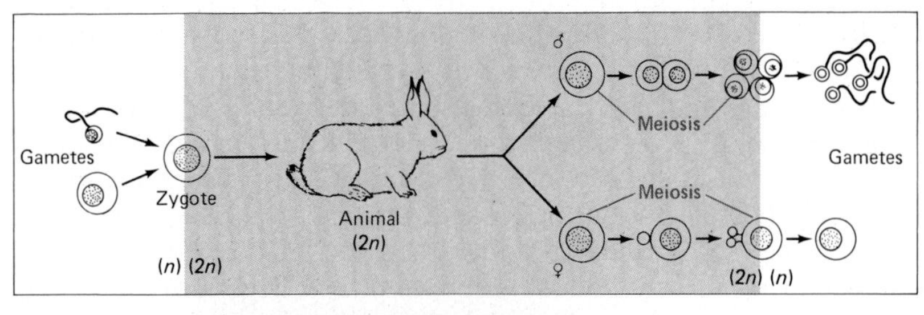


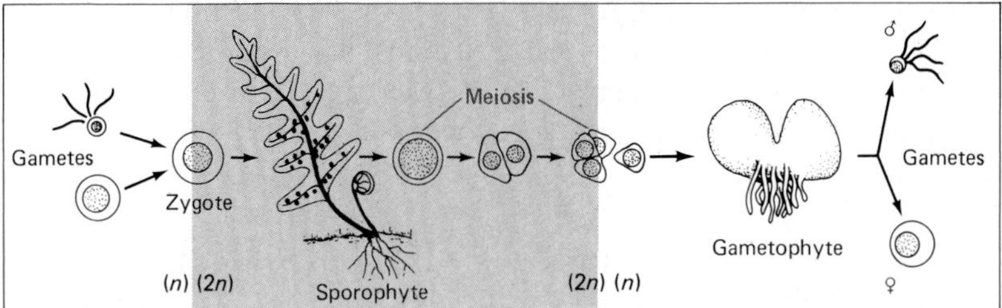


***Figure 3.14*** Comparison of animal and plant life cycles, showing differences in the time of meiosis in sexual reproduction. The events shown for animals are typical of almost all animal species, but some differences in detail characterize various plant groups.

another zygote and another generation of the species by sexual processes.

In multicellular sexually reproducing organisms, therefore, the sexual cycle of gamete fusion–fertilization–gamete fusion occupies only a part of the life cycle. The mitotic divisions giving rise to new cells and the development and differentiation of the organism very often take up the major part of a life cycle, that is, the time between the start of one generation and the initiation of the next generation.

The sampling of organisms we will describe represent a few of the species that have an important place in genetic research and species that are fairly typical of their group.

## 3.7 *Drosophila*

Both the vertebrate and invertebrate animals share the common feature of spending almost all of their life cycle in the diploid state (Fig. 3.15). The only haploid cells are the gametes themselves, which restore the diploid state when they unite at fertilization and initiate the new generation. Many invertebrates, including insects like *Drosophila* and most of the amphibians, undergo relatively little embryonic development before the egg hatches into an immature, or juvenile, stage. The juvenile tadpole or larval form undergoes additional mitotic divisions and a series of developmental changes during metamorphosis and finally emerges in the adult form of the species.

One of the great advantages in genetic studies of *Drosophila* is that specimens of each developmental stage can easily be collected for detailed analysis of chromosomal activities and for determination of the time when the effect of mutant alleles occurs. The giant, banded chromosomes in nuclei of the larval salivary glands and of some other organs can be studied at a level of detail that is almost impossible with other species (Fig. 3.16). These huge chromosomes are important in gene mapping studies, in studies of the molecular organization of chromosome structure, and in relating the activity of a particular gene with its phenotypic expression at a particular time in larval development.

*Drosophila melanogaster* was the experimental organism used in the laboratory of Thomas Hunt Morgan, the first geneticist to receive the Nobel prize for his pioneering studies during the early decades of this century.

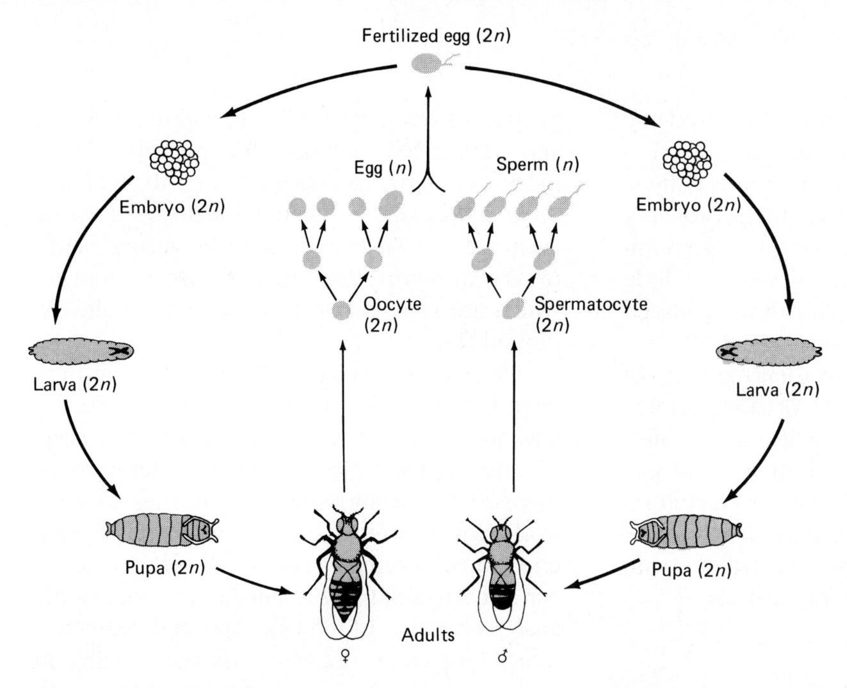


***Figure 3.15*** Life cycle of *Drosophila melanogaster,* showing the succession of developmental stages leading from fertilized egg to embryo, larva, and pupa, before emergence of the adult fly, or imago. Females have five dark stripes across the dorsal surface of the abdomen, males have only three dark stripes. Except for the gametes, all other cells are diploid (2*n*).

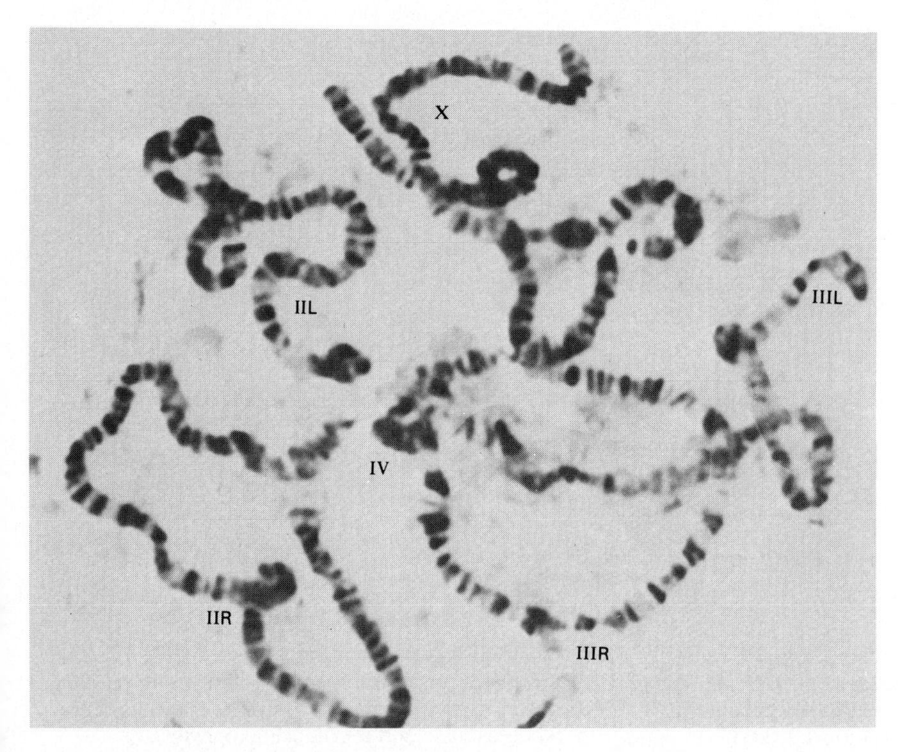


***Figure 3.16*** Light micrograph of the salivary gland chromosomes of *Drosophila melanogaster,* stained to show the band patterns in high contrast. The individual chromosomes or chromosome arms are identified (chromosomes X, II–IV; Right and Left arms [R, L] where appropriate). × 1,000

Morgan and his colleagues had been studying the embryology of *Drosophila* and continued to analyze the well-known system from a genetic standpoint. The generation time for *Drosophila,* about two weeks, is relatively short; large numbers of flies can be raised in relatively little space, and numerous mutants can be analyzed by breeding tests and chromosome study. Females are easily recognized by their markings of five bands on the upper surface of the abdomen, in contrast to three bands in males. Since, after fertilization, females store sperm in an organ called a spermotheca, it is essential to use virgin females that are recently hatched from pupae and kept isolated until they are put together with appropriate males for breeding analysis.

## 3.8 Corn (Maize)

The agricultural importance of *Zea mays* made this species a favored experimental organism for geneticists associated with agricultural schools and experiment stations. We probably know more about the genetics of corn than of any other seed plant. The general features of its life cycle are fairly typical of other flowering plants, or *angiosperms,* like peas, tobacco, tomato, which are species that are often genetically exploited (Fig. 3.17).

Despite its relatively long life cycle and the need for acres of space and a warm climate, the advantages of studying corn are that: (1) large numbers of its known mutant characteristics are mapped on chromosomes, (2) matings are relatively easy to control, (3) large numbers of progeny can be obtained from the seeds on a single ear of corn, and (4) it is suitable for cytogenetic analysis because of the large size and distinctive morphology of the ten chromosomes making up the genome.

The prominent corn plant is a sporophyte in which female flowers are borne in the *ears* along

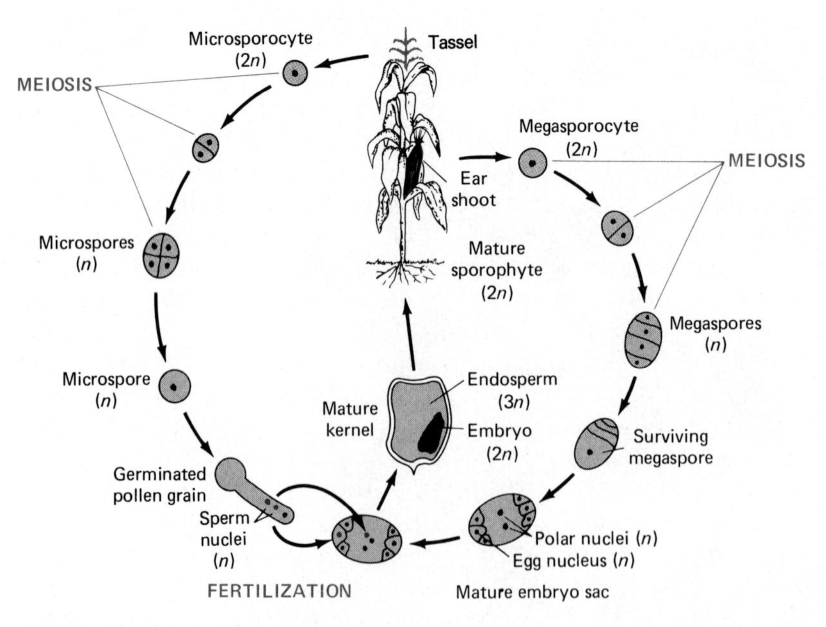


***Figure 3.17*** Life cycle of corn, or maize (*Zea mays*). Male flowers in the tassel and female flowers in the ear produce a succession of reproductive cells and structures that lead to fruit (kernel) formation. Each kernel contains a single seed, which includes the $2n$ embryo from which the new corn plant will develop and the $3n$ endosperm tissue that nourishes the growing embryo and young seedling during germination. The sporophyte is conspicuous whereas male and female gametophytes are microscopic in size.

the stem and male flowers in the *tassel* at the top of the plant. Corn is *monoecious*, that is, separate male and female flowers are produced on the same plant. In contrast, peas and most other flowering plants have perfect flowers in which both the male (pollen-bearing anthers of the stamens) and female (egg-bearing ovules of the pistils) structures occur in the same flower. Mating genetically different parents in corn is a simple matter of adding the desired pollen taken from tassels of designated male parent plants and putting a bag over the ears to prevent alien pollen from reaching the female flowers. Corn is normally cross-fertilizing, so precautions must be taken in every experimental generation, unlike peas in which self-fertilizations produce $F_2$, $F_3$, and other progenies from gametes known to come from the same flower.

The corn sporophyte produces megaspores in female structures and microspores in male flower parts, by meiosis. Megaspores undergo three successive mitotic divisions to produce eight nuclei, one of which is enclosed in the egg cell of the embryo sac, or the gametophyte produced by megaspore development during mitoses. Microspores give rise to pollen grains, the male gametophytes, each of which contains a tube nucleus and two sperm nuclei. Both the female and male gametophytes of flowering plants are, therefore, inconspicuous, and both grow within and at the expense of the green sporophyte plant.

When mature pollen is carried by wind or gravity or by the experimenter to the ear-shoot, the pollen germinates to produce a long pollen tube. The tube grows down through the silks (female receptive flower parts) to the embryo sac within the ovule in the ovary, into which the two sperm nuclei penetrate. One sperm nucleus fuses with the egg nucleus to produce the zygote, the first cell of the new sporophyte. The second sperm unites with two polar nuclei to form an **endosperm nucleus** in the embryo sac. The zygote, which is diploid, develops by many mitotic divisions into an embryo; the endosperm nucleus is triploid (three genomes) and gives rise by successive mitoses to endosperm tissue, which nourishes the embryo in the seed and the young seedling. By germination the embryo develops into the new corn plant. The endosperm is used up shortly after seedling growth begins. Each kernel on an ear of corn has developed from a separate female flower and, therefore, represents a different individual. Hundreds of kernels on a single ear can develop into comparable hundreds of individuals of the next generation. In corn one embryo sac is produced per female flower, and one seed can therefore arise. In many plants, including peas, more than one embryo sac is produced in each flower, thus more than one seed can develop within a common fruit, such as the pea pod. Each seed, however, contains a unique embryo produced from an independent fertilization between sperm from a pollen grain and the egg in an embryo sac.

In most plants and animals, we cannot verify by direct tests that pairs of alleles segregate 1:1 at meiosis; we presume this to be true on the basis of observed genotypic and phenotypic ratios of many progeny, as Mendel did. In corn, however, we can make a direct test of 1:1 segregation of alleles in pollen grains and in female gametophytes as well. For instance, the kernel phenotypes *waxy* and *nonwaxy* have different kinds of starch present and can be distinguished by simple treatment with an iodine solution. Waxy pollen stains red and nonwaxy pollen stains blue in iodine. The characters are controlled by a single pair of alleles; *Wx* is the normal allele for nonwaxy and *wx* is its mutant alternative for waxy pollen. If pollen from a heterozygous *Wxwx* plant is stained with iodine, about half stain red and half blue; 3437 blue- and 3482 red-stained pollen were recorded in one particular test. Similar direct proof of 1:1 segregation of a pair of alleles has been obtained from female gametophytes of *Wxwx* heterozygotes; this study not only demonstrates the predicted ratio but also shows that the surviving megaspore of a meiotic division may be of either allelic type, occurring equally often by chance in a large enough sampling.

While most of the inherited characters in

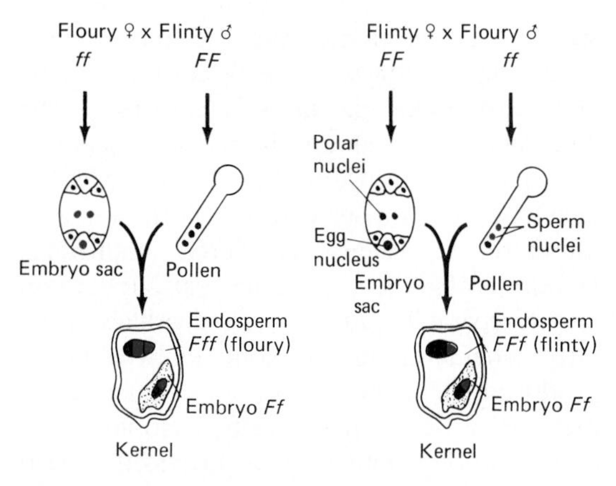


***Figure 3.18*** The triploid endosperm tissue can be examined directly on the ear of corn and be phenotypically identified. In reciprocal crosses between diploid *FF* and *ff* parents, flinty or floury endosperm develops according to the numbers of *F* and *f* alleles present. Two dominant alleles in this case are required to mask the expression of one recessive allele (*FFf* produces flinty, *Fff* produces floury endosperm).

corn are sporophytic, genetic analyses have been made of many inherited features expressed in the endosperm of the corn kernel. Since endosperm is triploid, we can analyze the interactions among three allelic copies of a gene in the same nucleus. Heterozygous endosperm receives two identical alleles from the polar nuclei of the embryo sac, and the alternative third allele is derived from the sperm nucleus with which the two polar nuclei initially fused. By crossing *flinty* (*F*) females and *floury* (*f*) males, *FFf* endosperm is produced; while the reciprocal cross gives rise to seeds with *Fff* endosperm (Fig. 3.18). The endosperm character can be observed directly in the seeds on the ear, so seeds need not be planted before the breeding results are recorded. The *FFf* genotype determines flinty endosperm, and *Fff* produces floury. Here it takes two dominant alleles to mask the expression of the *f* allele. This is not always the case since starchy (*Su*) endosperm is produced with either *Su Su su* or *Su su su* genotypes, and only the triple recessive has sugary (*su*) endosperm (Fig. 3.19).

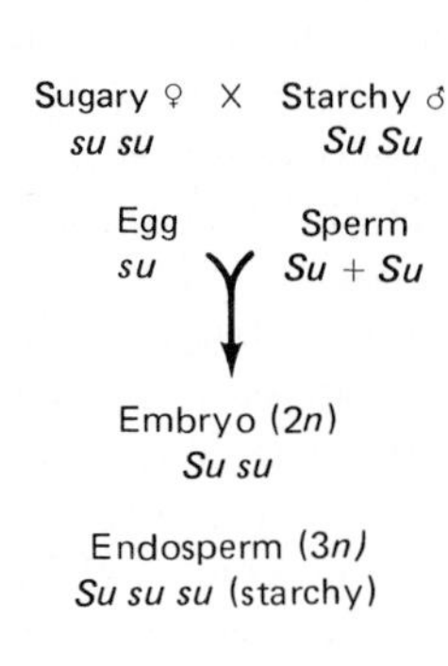


***Figure 3.19*** Starchy (plump kernels) and sugary (wrinkled kernels) endosperm segregants can be identified directly by examination of the kernels produced in ears on the female parent plant. Approximately ¾ of the kernels shown here are starchy and about ¼ are sugary. Only *sususu* endosperm develops the sugary phenotype, since one, two, or three *Su* alleles produce starchy phenotypes.

## 3.9 *Neurospora*

Important breakthroughs were made in biochemical genetics during the 1940s and afterward when the filamentous fungus *Neurospora crassa* took its place in the lineup of genetically studied species. There are four species of *Neurospora,* but virtually all studies have used *N. crassa,* and this is the species meant when no other is specified. *Neurospora* belongs to the Ascomycetes, a group of fungi characterized by

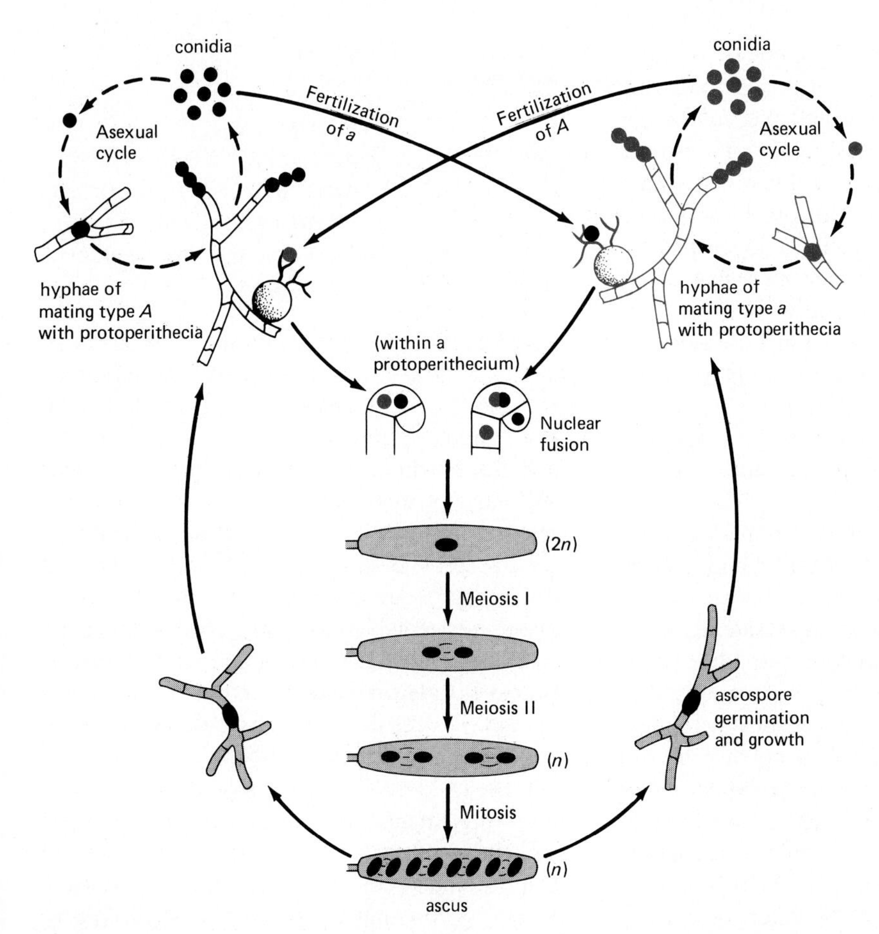


***Figure 3.20*** Life cycle of *Neurospora crassa,* showing asexual and sexual systems. In the sexual phase, fertilization occurs when haploid nuclei of opposite mating type (*A*,*a*) unite to produce the 2*n* zygote nucleus. Haploidy is reinstituted by meiosis, producing four nuclei per ascus. Each haploid nucleus undergoes mitosis, producing a final linear array of eight (four pairs) ascospores ordered precisely in the sequence in which the nuclei arose during meiosis and mitosis. Each ascospore can germinate asexually to produce a new haploid mycelium.

reproducing sexually to produce **ascospores** in a sac called an **ascus,** and reproducing asexually to produce salmon-pink spores called **conidia** (Fig. 3.20). The conidia are produced in fantastic numbers on the **mycelium,** the haploid filamentous body of the fungus, which is a mass of tubular **hyphae.** The common name of "red bread mold" comes from its color due to masses of conidia on the mycelium, and because it once was a nasty contaminant in bakeries.

*Neurospora* has two mating types, *A* and *a*, which are determined by a pair of alleles of the mating-type gene. Sexual reproduction can only take place when haploid nuclei in cells of opposite mating type unite and produce the diploid fusion nucleus, or zygote. The zygote is the only diploid stage in *Neurospora,* since meiosis begins shortly after nuclear fusion and leads to four haploid meiotic products within the ascus. These four nuclei divide again by mitosis, and thick spore walls soon form around each nucleus until every ascus has a set of eight ascospores ripe for release. The ascus is a slender sac with no space for nuclei to slip past one another. The four nuclei formed after meiosis or the eight nuclei formed when each of these has completed mitosis remain in a linear row within an ascus. While the initial four products of meiosis may be genetically distinct, depending on their alleles, the two members of a pair of ascospores are allelically identical because they arose by mitosis. When ascospores are released spontaneously or after manipulation from the ascus, each ascospore may germinate to produce a new haploid mycelium whose genotype was determined during segregation and reassortment of alleles.

The linear ordering of ascospores provides an unusual opportunity to recover and analyze all four products of meiosis (called a *tetrad*) from a single fusion nucleus. By **tetrad analysis,** we can determine genotypic and phenotypic ratios directly for *single* meiotic events rather than relying exclusively on statistical sampling. Using a microscope and a fine needle, we can dissect individual ascospores *in the order of their alignment within an ascus* and transfer them in known sequence to tubes of culture media for germination and growth (Fig. 3.21). By lining up the set of eight tubes, we can refer the phenotypes directly to segregation of alleles at meiosis. Each culture is haploid, so there is no masking or masked allele of a pair as there is in a heterozygous diploid in other species; each phenotype is developed from one genome and we can, therefore, relate it directly to the haploid genotype present in the culture.

Cultures that produce pink spores are genetically different from cultures that produce colorless spores; each breeds true in both sexual and asexual reproduction. When *pink* × *colorless* crosses are analyzed, each ascus can be shown to have 4 pink and 4 colorless products of ascospore development. The ratio of 1 pink:1 colorless is read out directly as segregation of *A* and *a* alleles at meiosis. The alignment of these alleles from one end of an ascus to the other include the arrangements shown at the top of the next column.

| arrangement | order of ascospores | |
|---|---|---|
| 1 | *AAAAaaaa* | |
| 2 | *aaaaAAAA* | (All asci have |
| 3 | *AAaaAAaa* | 4*A*:4*a*, or 1:1 |
| 4 | *aaAAaaAA* | ratio of the |
| 5 | *AAaaaaAA* | two alleles.) |
| 6 | *aaAAAAaa* | |

From Fig. 3.21 you can see how arrangements 1 and 2 can arise when homologous chromosomes segregate at the first meiotic division; then, during Meiosis II, each Meiosis I nucleus produces an identical partner nucleus and each of these, in turn, undergoes mitosis. The explanation for the ascospore ordering in arrangements 3 and 4 can be found in Fig. 3.22 (next page). An exchange of non-sister chromatid segments (crossover) between the gene and the centromere took place, giving rise to **second-division segregations.** Arrangements 1 and 2 result from **first-division segregations.** Information about the proportions of these two kinds of segregation patterns can be used to estimate the distance between the gene and the centromere of a chromosome, and to map the likely position of the centromere itself. Since it is very difficult to study the small *Neurospora* chromosomes by microscopy, genetic analysis of the ascospores is the principal means for discovering cytological information in this organism. We have not explained the chromosomal events leading to arrangements 5 and 6. Can you diagram the probable basis for these two arrangements?

In addition to tetrad analysis and its many uses, another general test method used with *Neurospora* and some other organisms is the complementation test. This test can determine whether mutations are nonallelic or are governed by mutant alleles of the same gene. The hyphae are filaments of multinucleate cells that can undergo localized fusions that allow nuclear exchanges. When mycelia of genetically marked cultures are put together in the same medium, nuclear exchanges usually lead to **heterokaryons,** whose cells have one or more

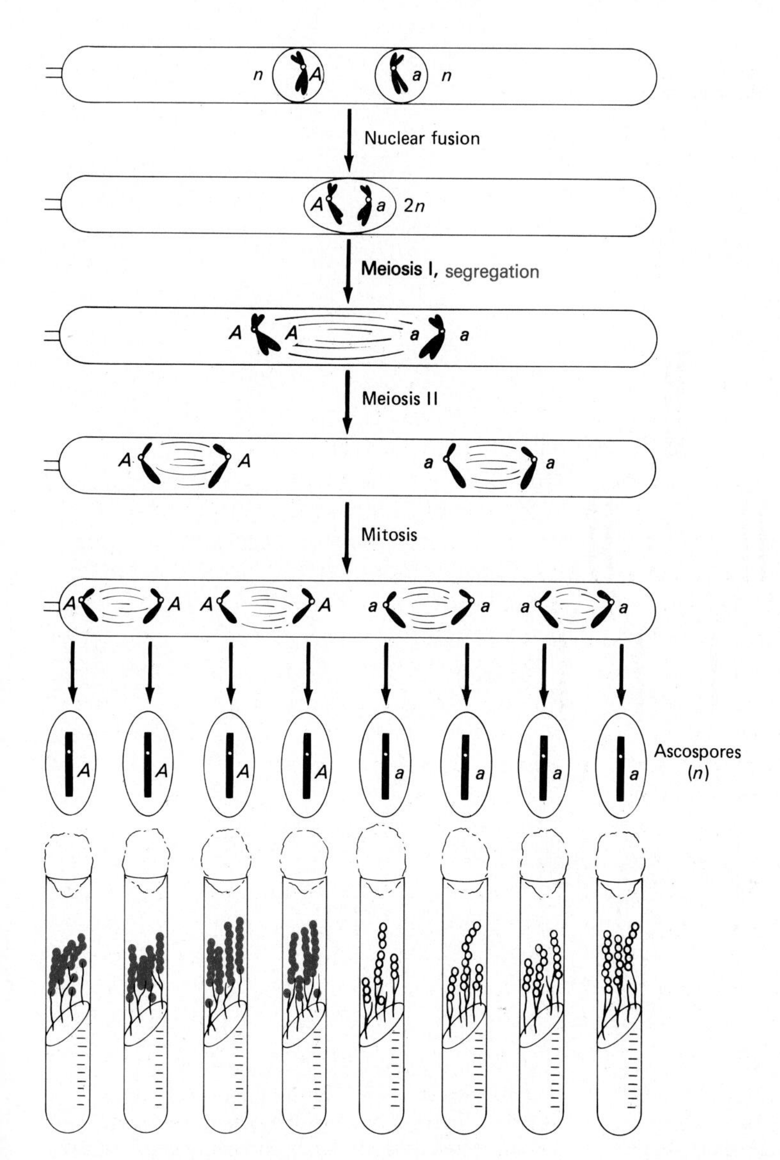


***Figure 3.21*** First-division segregation within an ascus of *Neurospora crassa.* The behavior of the pair of homologous chromosomes carrying alleles *A* and *a* (conidiospore color) provides the basis for understanding arrangements of ascospore types within an ascus. Ascospores are dissected in the order of their arrangement within the ascus, and are placed individually on growth media in separate test tubes. When the mold develops, conidial color can be related back to the particular order of alleles on chromosomes within the ascus. Knowing about chromosome behavior at meiosis, we can see that *A* and *a* segregated at Meiosis I in the heterozygous *Aa* diploid nucleus. Centromeres are shown as open circles on chromosomes.

genetically different haploid nuclei. This situation is comparable to heterozygous organisms whose homologous genomes may contain different alleles of the same genes. There is one important difference, however. The genomes are in the same nucleus and all the nuclei are identi-

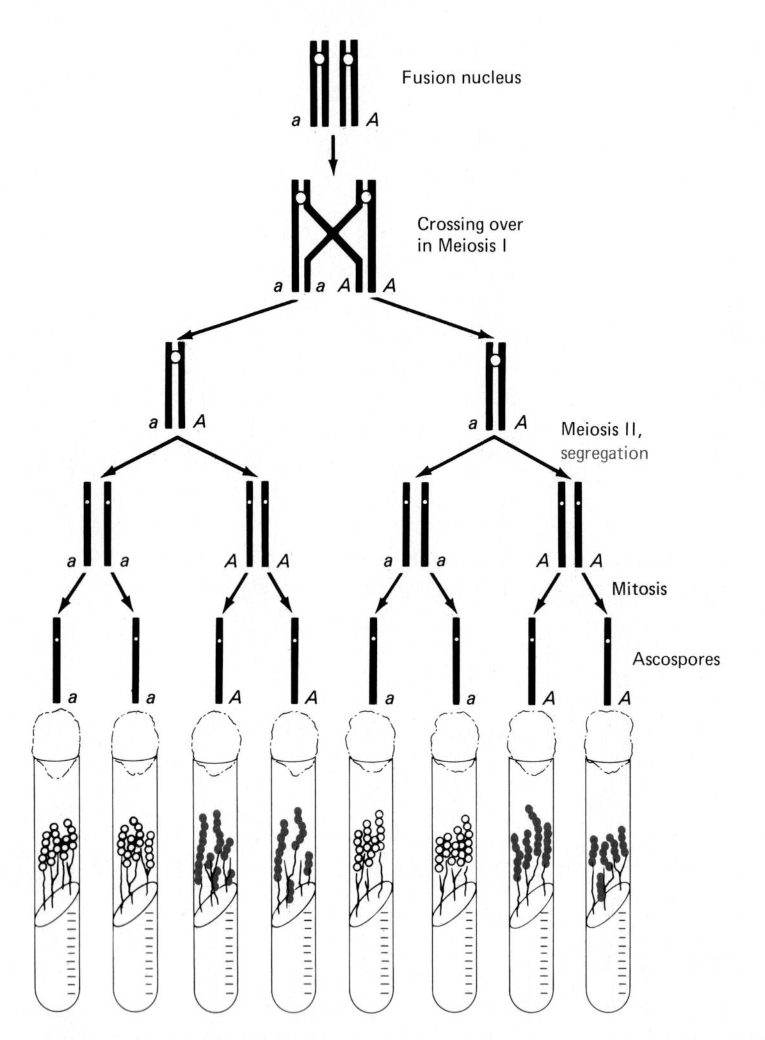


***Figure 3.22*** Second-division segregation within an ascus of *Neurospora crassa.* Each nucleus produced during Meiosis I is still heterozygous (*Aa*); *A* and *a* alleles do not segregate (assort into separate nuclei) until Meiosis II in this case. Ascospore order can be determined from culture growth in the eight ordered test tubes and be related back to the distribution of alleles during meiosis. Second-division segregations result when an exchange occurs between gene locus and centromere of two chromatids in the bivalent during meiosis.

cal in a multicellular heterozygote; heterokaryons have different alleles in different haploid nuclei in the same cells or hyphae (Fig. 3.23).

Wild-type *Neurospora* can be cultured in a minimal medium that contains all its growth requirements. Many known mutants have one or more additional growth requirements, due to modified biochemistry and metabolism governed by mutant alleles of the wild-type genes. When a mutant requiring the amino acid arginine is grown in culture, arginine must be added to the minimal medium to induce growth; a similar situation is true for a histidine-requiring mutant. When the two mutants strains are cultured together, the hyphae fuse to form a heterokaryon with different haploid nuclei in the same cells. The resulting heterokaryon is able to grow in minimal medium unsupplemented with

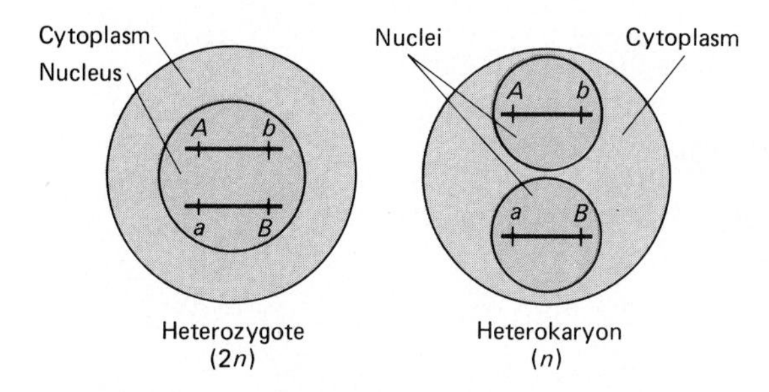


***Figure 3.23*** Different alleles of genes are present in the same nuclei in a diploid heterozygote (*AaBb*), and different alleles are present in separate nuclei in a haploid heterokaryon (*Ab* + *aB*).

arginine or histidine. The two mutants *complement each other's deficiencies.* This fact must mean that the arginine mutant has the wild-type allele for histidine and the histidine mutant has the wild-type allele for arginine (Fig. 3.24). Each kind of nucleus in the heterokaryotic hyphae governs synthesis for one of these amino acids in the cytoplasm, so that both amino acids are available in the common cytoplasm for normal development of the heterokaryon in minimal medium.

It is not particularly surprising that different phenotypes have different genotypes. But in cases of heterokaryons formed from different strains with the same phenotype, we may find one of two possible alternatives: (1) neither strain can grow either separately or together in minimal medium or (2) both can grow together in minimal medium but can grow singly only in supplemented media. Depending on the result of the complementation test—whether or not the phenotypically similar strains can complement each other and grow on minimal medium—we can make the preliminary determination of whether the phenotypes are governed by the same gene or by different genes.

When two independently isolated strains that require arginine, called arginine-1 and arginine-2, form heterokaryons that complement each other, it must mean that at least two different genes govern arginine metabolism and that each strain has the wild-type allele for one of the genes and the mutant allele for the other, or *arg*-$1^+$ *arg*-$2^-$ + *arg*-$1^-$ *arg*-$2^+$. If, on the other hand, the two arginine-requiring strains do not complement each other, we have preliminary evidence that each strain has the same mutant allele for a single gene and the deficiency cannot be remedied. The evidence in the second case is negative, that is, nothing happens. Other reasons for such negative results could exist; for example, hyphal fusions do not occur between certain strains of *Neurospora.* But further investigation is warranted in order to establish the basis for noncomplementation.

Complementation in heterozygous diploids of various species can also be tested when appropriate crosses are made. The discovery of multiple alleles, that is, more than two alleles for the same gene, was based on finding that *mutant*

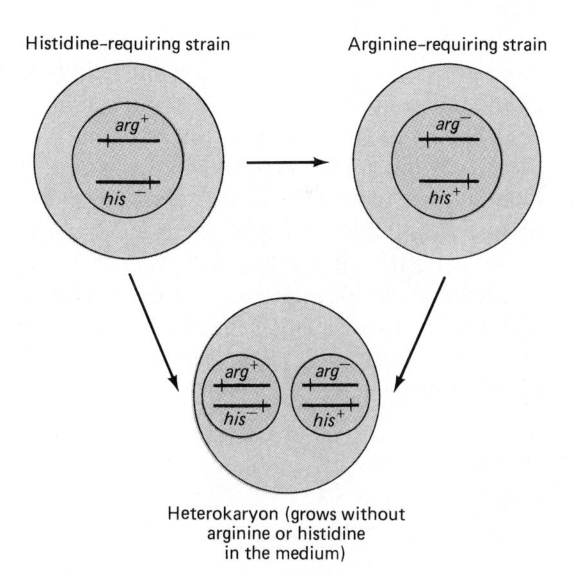


***Figure 3.24*** Mutant strains of *Neurospora* require supplements in the nutrient medium in order to grow. Heterokaryons formed from strains with different growth requirements may grow in media without supplements. Each mutant nucleus complements the other in the heterokaryon, which indicates that the *his*$^-$ nucleus must carry *arg*$^+$, and the *arg*$^-$ nucleus must carry *his*$^+$. With *arg*$^+$ and *his*$^+$ alleles in heterokaryons, growth is possible on minimal media.

× *mutant* → *mutant* phenotype, versus *mutant* × *mutant* → *wild-type* phenotype, as follows:

| **case no. 1** | **observed phenotypes** | **deduced genotypes** |
|---|---|---|
| parents | mutant × mutant | $A_1A_1a_2a_2 \times a_1a_1A_2A_2$ |
| | ↓ | ↓ |
| progeny | wild type (complementation) | $A_1a_1A_2a_2$ |
| **case no. 2** | | |
| parents | mutant × mutant | $a_1a_1 \times a_2a_2$ |
| | ↓ | ↓ |
| progeny | mutant (no complementation) | $a_1a_2$ |

In case No. 1 two pairs of alleles must govern the same phenotype, and neither one alone can produce the wild phenotype. The alleles would be $A_1$ and $a_1$ of one gene and $A_2$ and $a_2$ for the second gene. This situation resembles the inheritance of cyanide content in white clover (see section 1.7).

In case No. 2 only one gene must govern the characteristic, and expression of the wild-type phenotype requires at least one dominant allele ($A$). When a mutant allele is present on each of the two chromosomes in the nucleus, the mutant phenotype is expressed. In this case, therefore, there are multiple alleles of one gene: two recessive mutant alleles ($a_1$ and $a_2$), and one dominant allele for expression of the wild-type phenotype. This situation resembles the one described for the white eye gene in *Drosophila* (Section 1.6).

## Reproduction of Bacteria and Viruses

Bacteria are haploid unicellular microorganisms that reproduce by fission (a form of asexual reproduction in which cells divide in two). They can be grown in culture on defined nutrient media and can be studied by all the methods of biochemistry and microscopy at our disposal. When grown in a liquid medium, the cells remain in suspension and the medium becomes cloudier as more cells are formed. On a solid nutrient medium where cell movement is inhibited, individual cells continue to reproduce in place until a visible pile of cells, called a **colony,** appears on the surface of the medium. The group of identical cells in a colony is a **clone,** since each colony arises by fission from a single parent cell. Each colony can be analyzed as if it were the single cell from which the entire clone arose, but many studies can be done with colonies that would be impossible using only one cell.

Viruses can reproduce only inside a living host cell. There is never a sign of cellular organization or of a cell division process such as fission. Instead, the entering virus genes subvert the host metabolic machinery to make new virus genes and proteins and to stop making their own requirements. The newly synthesized virus molecules *assemble* into new viruses within the infected host cell, and mature infective virus progeny are released when the host cell bursts, or lyses. Each new virus can then initiate a new infection cycle in another host cell.

### 3.10 *Escherichia coli*

The common enteric bacterium *E. coli* is a normal inhabitant of the human gut. It is a typical rod-shaped species with a generation time of 20 to 30 minutes; an individual cell is about 1 or 2 μm long. Its rapid reproduction and small size make *E. coli* as useful an experimental organism as *Neurospora* but more limited in its morphological variations that are visible to the naked eye. Most of its phenotypic features are biochemical and physiological traits that can be observed in relation to colony growth on minimal versus supplemented media. Methods for counting cells in liquid culture are simple and straightforward and are based on serial dilutions or on optical readings of the cloudiness (turbidity) of the culture medium. Colonies on solid media are visible to the naked eye after one day's growth, and an apparently clear liquid culture will turn cloudy overnight as billions of cells are produced by fission.

One of the great surprises in the history of genetics was the discovery of new combinations of traits among descendants of genetically marked bacterial parent strains grown together in liquid culture and then plated out on solid media to detect new genotypes. In 1946 Joshua Lederberg and Edward Tatum, both of whom received the Nobel prize in 1960 for their work, published a brief report that opened the way to a whole new discipline of bacterial genetics. Since bacteria have no conventional means for sexual reproduction or even for mitosis as we know these processes in eukaryotic species, the way in which rearranged genotypes arose was not immediately clear and was not described until some years later. We will have ample opportunity to discuss *E. coli* genetics in various parts of this book, but for now it will be instructive to see how Lederberg and Tatum designed their experiment and how they used a general selection method that has since become a standard in microbial genetics.

Mutant cells having single nutritional deficiencies were obtained by exposing *E. coli* to X rays or ultraviolet light. Multiple deficiencies were incorporated into individual strains by successive mutagenic treatments. Two triple mutants were studied: strain Y-10 required threonine, leucine, and thiamin supplements in minimal medium in order to grow, and strain Y-24 required biotin, phenylalanine, and cysteine. Thiamin and biotin are vitamins and the other four substances are amino acids. After a period of growth in the same liquid culture medium, samples containing many millions of cells were plated on minimal media and allowed to develop into colonies. The colonies were isolated and grown in pure culture to establish their altered genetic nature and their purity of type. Only about one cell per million from the original mixed culture actually formed colonies on the solid minimal medium. Because of their small size and the huge populations that they produced within hours, the rare recombinants were detectable by this **prototroph selection** method. Prototrophs are wild-type cells and their mutant alternatives are called **auxotrophs.** The prototrophic recombinants must have contained wild-type alleles for all six of the observed nutritional characteristics, otherwise they could not have grown in minimal medium and maintained their new genotypes in subsequent generations in culture:

auxotrophic parents: $thr^-leu^-thi^-bio^+phe^+cys^+ \times$
$thr^+leu^+thi^+bio^-phe^-cys^-$

prototrophic
recombinants: $thr^+leu^+thi^+bio^+phe^+\ cys^+$

By using prototroph selection, we can analyze many millions of cells in a short time and in a small amount of space and detect rare new phenotypes and genotypes. Strains carrying numerous mutant alleles can be studied by breeding tests that are unmanageable with higher organisms, since examination of millions upon millions of flies, corn plants, or similar species in $F_1$, $F_2$, or testcross generations is neither desirable nor convenient.

## 3.11 Bacteriophages

Viruses that infect bacteria are called **bacteriophages** or, simply, **phages,** and they have been studied for more than 40 years. Like all viruses, phages can reproduce only in living host cells. The particular host species and cell type infected by a virus provide two of a number of criteria by which viruses are identified and named (Table 3.3).

Phage T2 is one of the viruses known to infect *E. coli.* When T2 infects its host, the phage attaches by its tail to the host cell and injects its DNA into the cell and leaves its empty protein coat outside (Fig. 3.25). The bacterial metabolic machinery begins to make new virus molecules instead of tending to its own needs. After a brief interval of biosynthesis, new phage particles assemble. This *vegetative phase* of the cycle ends when mature, infective phage particles emerge as the bacterial lyses, or breaks up, to release a burst of progeny viruses. Each infective virus particle can initiate a new cycle of infection by reproducing in another host cell. Phage T2 is an example of a **virulent phage,** that is, a phage that destroys its host cell after an infection has been established. The virus does not exist in the cell in a noninfective state.

Because of their small size, most viruses can be seen only by electron microscopy. They are usually studied phenotypically after growth on a lawn of bacterial cells covering the entire surface of solid nutrient medium. Under appropriate conditions, one phage infects one cell, and its progeny are released when that cell bursts. The liberated phages proceed to infect other cells around the original site, and in one day a visibly clear area, called a **plaque,** appears on the plate. Plaque type is an inherited characteristic since different phages and strains produce clear or cloudy zones, some with a sharp margin and others more jagged, some larger and others smaller, and so forth. In addition to plaque phenotypes, viruses can be identified by a number of other characteristics, such as phage morphology, biochemical variability, and the range of host types that can be infected. Each plaque is equivalent to one virus when determining phage counts, just as each colony represents one bacterial cell in quantitative studies. The entire plaque arose from lysed bacteria that can be

***Table 3.3*** Some characteristics of representative viruses.

| nucleic acid | virus | main host | comments |
|---|---|---|---|
| **DNA** | | | |
| single stranded | fd | *Escherichia coli* | |
| double stranded | T2, T4, T6 | *Escherichia coli* | |
| | P22 | *Salmonella typhimurium* | |
| | herpes simplex | human | type 1 causes "fever blisters"<br>type 2 causes genital herpes |
| | variola | human | causes smallpox |
| | Epstein-Barr | human | causes infectious mononucleosis; associated with Burkitt's lymphoma |
| | cauliflower mosaic | cauliflower | transmitted by aphids |
| **RNA** | | | |
| single stranded | $Q\beta$ | *Escherichia coli* | |
| | tobacco mosaic | tobacco | |
| | polio | human | |
| | measles | human | |
| | mumps | human | |
| | influenza A, B, C | human | |
| double stranded | reovirus | human | causes mild illness of respiratory and GI tracts |
| | wound tumor | plants | transmitted by leafhoppers |

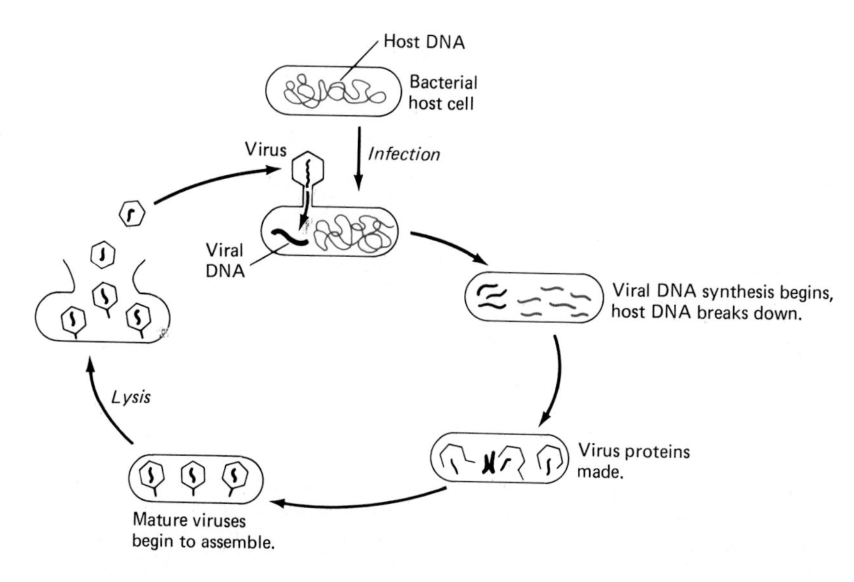


***Figure 3.25*** Life cycle of a virulent phage. Infection of host cells by the virus leads to breakdown of host DNA and subversion of host metabolic machinery to make new viral molecules according to genetic instructions in viral DNA. These new molecules assemble into new viruses. Lysis of the host allows release of mature virus progeny, which may initiate new rounds of infection in other host cells.

traced back to a single phage infecting a single bacterium on the plate.

In contrast to virulent phages such as T2, other phages, called **temperate phages,** can infect their hosts without destroying them. Temperate phages are physically integrated into the bacterial chromosome, and they are transmitted from one generation to the next as part of the host chromosome. Only the DNA of the temperate phage is integrated into the bacterial chromosome; its protein coat is not. Occasionally a temperate phage becomes infective, leaving the host chromosome and subverting host metabolism to make new viral DNA and protein. The viral molecules assemble into new phage particles that may initiate new infection cycles. Such an occasional virulent particle allows us to recognize the presence of integrated phages in various strains. The integrated phage is called a **prophage,** and its normally healthy host is said to be *lysogenic* (Fig. 3.26, next page). A lysogenic bacterial strain ordinarily carries the benign prophage, but the bacteria can be lysed under appropriate conditions of infection.

We can induce phage production in certain lysogenic strains by exposing the strain to ultraviolet light and other agents. Such studies have shown that temperate and virulent phages are essentially similar in their reproduction. A principal difference is that the *latent period* is very brief in bacteria infected with virulent phages but is suspended indefinitely before virus multiplication takes place in lysogenic bacteria.

Phages provide model systems for genetic studies of many kinds. Among the first of these studies was the work in 1952 by Alfred Hershey and Martha Chase, who showed that the molecule carrying genetic instructions is DNA and not proteins, as many had believed prior to that time. By tagging phage T2 DNA and protein differentially with radioactive isotopes, Hershey and Chase were able to show that only phage DNA entered *E. coli* cells, and phage proteins remained outside the cells. Different radioactive

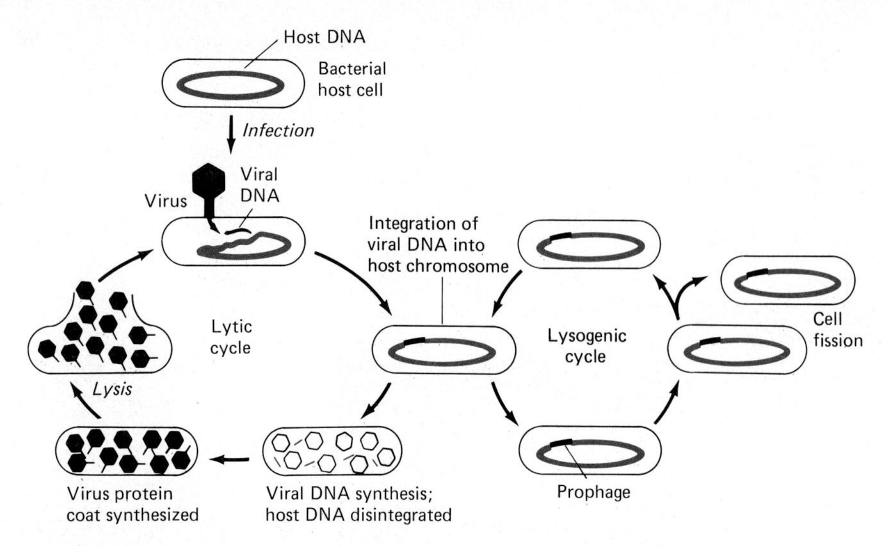


***Figure 3.26*** Life cycle of a temperate phage. Invasion of host cells by temperate viruses need not cause destruction of the host. In lysogenic bacterial strains, viral DNA is integrated into host DNA and there assumes the prophage state of viral existence. Prophage DNA replicates in synchrony with host DNA and is distributed as part of the host chromosome to descendant cells in division (fission) cycles. The destructive lytic cycle can be induced by external agents, or may occur spontaneously in occasional cells. During a lytic cycle the host is destroyed and new progeny viruses are assembled from viral molecules made before the cell bursts, according to genetic instructions in viral DNA.

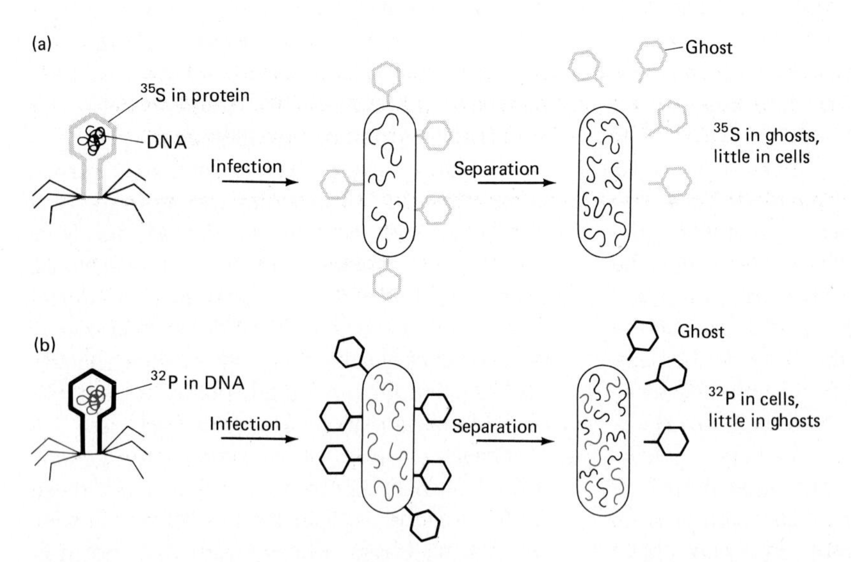


***Figure 3.27*** Experiments in 1952 by Alfred Hershey and Martha Chase showed that DNA and not protein was the genetic material, since only viral DNA entered host cells and there directed new progeny virus production. Tagging (a) viral protein with the radioactive isotope of sulfur, $^{35}$S, or (b) viral DNA with radioactive phosphorus, $^{32}$P, permitted identification of viral molecules that entered the host during infection. Labeled DNA was found inside infected cells and labeled protein was found mainly in viral ghosts that remained outside the cells. New viruses are made according to viral genetic instructions in the infected host; and since only viral DNA entered host cells, DNA must be the genetic material.

isotopes identified DNA and protein of the infecting phages, so it was clear that only the DNA had entered (Fig. 3.27). Since new phages made in infected cells are genetically identical to the infecting parent phages, the genetic instructions for making progeny T2 viruses must have been carried by the kind of molecule that entered *E. coli* cells. The molecules remaining outside the host cells are not copied and incorporated into progeny viruses; therefore, proteins must not have a genetic role in phage reproduction. We will discuss this landmark genetic study in detail in Chapter 7.

## Cultured Eukaryotic Cells

The development of cell culture methods has opened new avenues for studying complex multicellular organisms under conditions resembling those used to study microorganisms. Cell can be grown in culture as genetically pure *clones* (identical descendants of a single individual), examined for growth characteristics in defined nutrient media, and prepared for the same kinds of genetic, molecular, and biochemical analyses previously possible only with microorganisms. Studies of viruses using eukaryotic host cells grown in culture can provide information about the viruses themselves and about their interaction with hosts. Such information is difficult to obtain from studies of whole organisms. Since we will discuss all these systems in detail later in this book, we will mention only highlights of their nature and utility in the following sections of the chapter.

### 3.12 Eukaryotic Cells in Culture

Cells removed from an organism can be grown in primary culture. Such cultures are difficult to maintain because the cells have a finite life span, that is, they stop multiplying after a relatively few generations and may die. If cells can be selected from primary cultures and grown, their clonal descendants may become established in a virtually immortalized secondary culture. Secondary cultures continue to produce descendants indefinitely as long as appropriate conditions are maintained and nutrients are replenished for continued growth and multiplication.

A variety of mammalian cell types can be cultured. One kind desirable for studies of differentiated cell characteristics is the *embryonic cell.* Embryonic cells are undifferentiated, but they can give rise to descendants that develop into the differentiated state. In addition, other descendant cells retain their embryonic nature and continue to give rise to new cells indefinitely. Another kind of cell useful for these studies is the *precursor stem cell,* which is capable of cell divisions that give rise to other stem cells and to differentiated descendants whose phenotypic traits can be studied under controlled conditions. A particularly interesting type of stem cell is the erythroid precursor stem cell that can give rise to hemoglobin-producing, differentiated erythrocytes of the red-blood-cell group. Stem-cell cultures can be maintained indefinitely in secondary culture. Furthermore, they possess properties similar to those in primary culture; in contrast, secondary cultures of differentiated cells are usually altered in some properties.

Genetic studies of many kinds can be carried out using cultured mammalian cells. For more than 20 years, somatic cells in culture have been analyzed by **somatic cell hybridization.** In this procedure, induced fusion of cells carrying different allelic markers produces **somatic cell hybrids.** Hybrid cells multiply on nutrient media and give rise to various clones of descendants, each of which develops from an individual somatic cell hybrid isolated from the original mixture. The technique has been invaluable in biochemical and molecular analyses of defects that are due to mutant alleles. They are important experimental systems for making assignments of specific genes to specific chromosomes in the experimental species. Somatic cell hybridization is a useful genetic method since *cell fusions take place,* like fertilization in sexual systems, and *random losses of individual*

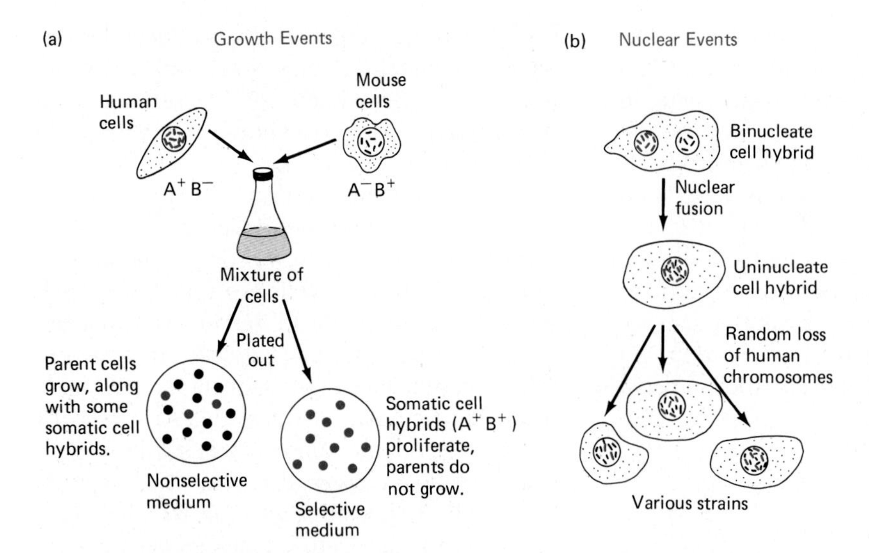


***Figure 3.28*** Somatic cell hybridization provides an alternate approach to genetic studies and uses cultured cells that reproduce asexually by mitosis and cell division. Fusions between somatic cells (all body cells exclusive of the gametes) are encouraged by appropriate treatment to alter the cell surface. (a) When mixtures of cells are plated on solid medium, each immobilized cell or cell hybrid reproduces mitotically in place and gives rise to a colony of genetically identical descendant cells. On a selective medium lacking molecules needed by $A^-$ and $B^-$ genotypes, only hybrids of $A^+B^+$ genotype can grow. Parental $A^-B^+$ or $A^+B^-$ fail to grow because of their inherited deficiencies. Cell hybrids can then be collected from selective media and can be studied further. (b) In the nuclear events characteristic of somatic cell hybridization (shown at the right), fusion of nuclei and random losses of chromosomes during growth are analogous to sexual fusion and allelic segregation typical of sexual reproduction.

*chromosomes occur* during growth of hybrid clones, thereby providing a substitute process for allelic segregation during meiosis in normal sexual reproduction (Fig. 3.28). Since segregation of members of pairs of alleles can take place and since assortment of different pairs of alleles is possible, somatic cell hybrids provide a convenient means for genetic studies of humans and other species that would be impossible or more difficult to pursue with whole organisms.

## 3.13 DNA Viruses of Eukaryotic Cells

Viruses that cause tumors to develop in their host organism are called **tumorigenic**. If *malignancy* characterizes the tumor, the viruses are described as **carcinogenic** or **oncogenic** types. Tumorigenic and oncogenic viruses can be studied in cell cultures as well as in the whole organism. Cell cultures, however, facilitate many kinds of genetic and biochemical studies that are difficult to manage with the whole organism or are considerably more expensive because of maintenance needs of animals kept in the laboratory.

The analog of malignant growths in the organism is **cellular transformation** in cultured cells. Malignant growths are *invasive,* spreading to distant sites within the organism from initial foci of growth. Transformed cells grown in culture display a number of modified properties that distinguish them from normal cells (Fig. 3.29). Normal cells in primary culture stop dividing after a finite number of generations. Virus-

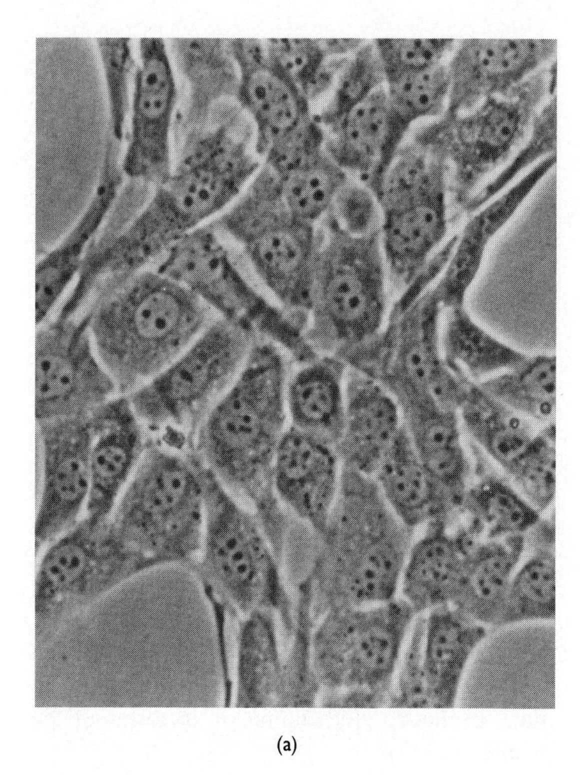

(a)

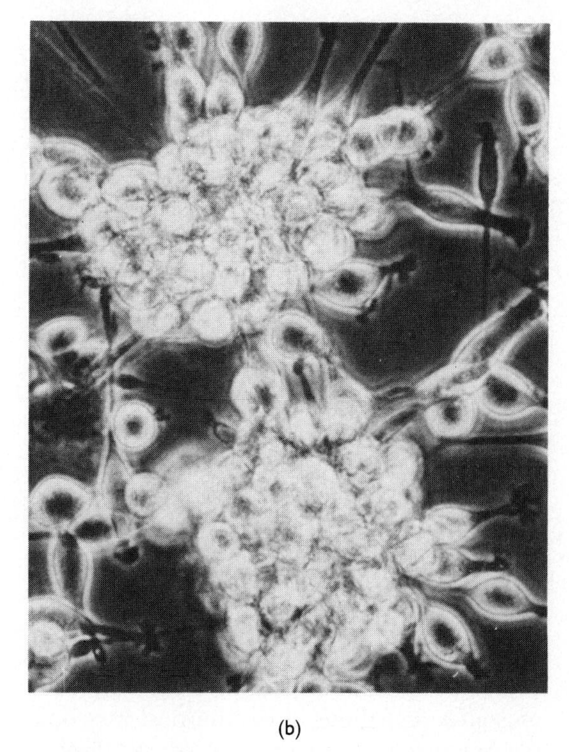

(b)

***Figure 3.29*** Mouse fibroblasts in cell culture. (a) Confluent monolayer of normal cells on solid substratum, and (b) disordered mounds of transformed cells on solid substratum.

induced transformation leads to cells that *can continue to multiply indefinitely.* In other cases, transformed cells lose *anchorage-dependence* properties and can multiply in semisolid medium. Normal cells showing anchorage dependence can multiply only when they are attached to a rigid surface. Transformed cells may be able to multiply in medium made semisolid by addition of agar or methyl cellulose; normal cells will not multiply in this substratum. In other cases transformed cells may *grow without restraint* on a rigid substratum and produce mounds of cells in disordered array. Before transformation, normal cells usually stop dividing once physical contacts are made between cells on the rigid subtratum. Normal cells produce a confluent single layer of cells, transformed cells produce multilayered mounds of descendants even on a rigid substratum.

Cellular transformation usually involves the integration of viral genes into the host chromosomes. These viral genes direct the host to synthesize one or a few viral proteins that appear to be required to maintain the transformed state. The *transformed phenotype is heritable,* since transformed cells give rise to transformed descendant cells indefinitely. Tumorigenic and oncogenic viruses can induce tumors in the organism as well as cause transformation in cell cultures. Extracts from cell cultures can be injected into special strains of nude mice, whose immunity mechanisms are defective. These strains have no thymus gland and cannot make immune substances that ordinarily act to reject foreign materials. When tumors develop they provide independent verification of the tumorigenic nature of the virus that transformed cells in culture.

## 3.14 RNA Viruses of Eukaryotic Cells

A number of viruses have RNA genomes instead of DNA genomes. RNA viruses that are oncogenic in vertebrate animals are classified as **retroviruses.** The first retrovirus that was shown to be the specific agent causing malignant growths in animals is avian sarcoma virus (ASV), also known as Rous sarcoma virus (RSV) in honor of its discoverer, Peyton Rous. The scientific community came to accept his interpretations of viral oncogenesis many years after Rous's discovery.

Retroviruses have a very unusual reproductive system compared with that of DNA viruses. On infection of the host, retroviral RNA is injected into host cells. This viral genetic material acts as a blueprint for synthesis of DNA copies of the viral RNA genome. The unique enzyme that catalyzes DNA copying from RNA instructions is called a RNA-directed DNA polymerase, or **reverse transcriptase.** These DNA copies become integrated into the host chromosomes and there direct the host metabolic machinery to make new viral molecules. Eventually, newly made viral RNA and protein molecules assemble into mature viruses, which exit from the cell and may initiate new cycles of infection. Certain specific retroviral genes can cause malignant growths in organisms or transformation of cultured cells when viral DNA copies integrate into host chromosomes. We will discuss this topic in Chapter 14.

Cultures of eukaryotic cells and of virus-infected cells provide systems for genetic, molecular, and biochemical studies that previously could only be achieved with microorganisms. Cell cultures can be grown under controlled conditions, which facilitates the design of controlled experiments. Because of the simplicity of a single cell type in a culture, experimental results are easier to interpret than are those from studies using whole organisms. In general, studies of cells in culture have made whole-animal studies more meaningful. Both kinds of studies are still necessary since cell culture data must be interpreted ultimately in terms of the whole organism.

## Questions and Problems

***3.1*** Suppose you were given an electron micrograph of a section through a eukaryotic cell, but the section happened to be made of a part of the cell excluding the nucleus. How could you verify that the cell was indeed from a eukaryote?

***3.2*** Which phases of the cell cycle take place between telophase of one mitosis and prophase of the succeeding mitosis? What is the general nature of events taking place in these particular phases of the cell cycle?

***3.3*** Distinguish between two chromatids and two homologous chromosomes. During which phase of the cell cycle are sister chromatids first produced? When do they become visible by microscopy?

***3.4*** What are the genetic consequences of mitosis? of meiosis?

***3.5*** There are 23 pairs of chromosomes in humans. What proportion of human gametes will have centromeres of the following types:

***a.*** paternal origin only.

***b.*** a mixture of maternal and paternal.

***3.6*** How many chromosomes will be found in the following kinds of cells in humans:

***a.*** secondary spermatocyte.

***b.*** spermatogonial cell.

***c.*** primary oocyte at leptonema.

***d.*** a polar body.

***3.7*** The amount of DNA in a haploid nucleus of the mouse (*Mus musculus*) is about 2.5 picograms ($2.5 \times 10^{-12}$ g). What would be the DNA content of a nucleus in each of the following:

***a.*** body (somatic) cell in $G_1$ of a cell cycle.

***b.*** spermatozoan.

***c.*** primary spermatocyte at diplonema.

***d.*** secondary spermatocyte at prophase II.
***e.*** secondary spermatocyte at telophase II.
***f.*** zygote at metaphase of the first mitotic division.

***3.8*** A synaptonemal complex forms between homologous chromosomes early in prophase I of meiosis and holds the bivalent in register throughout pachynema. Why is it unlikely that the synaptonemal complex initiates synapsis of homologues at zygonema?

***3.9*** How does chromosome behavior during meiosis help to explain Mendel's First and Second Laws of Inheritance?

***3.10*** *Aspergillus nidulans,* like *Neurospora,* is an ascomycetous fungus. During an experiment, a yellow, adenine-requiring strain and a white, proline-requiring strain were inoculated together onto minimal medium that lacked adenine and proline. Shortly afterward, growing green mycelium appeared from the inoculum. Give your interpretation of the occurrence of growth. How would you test this interpretation genetically?

***3.11*** In a cross between homozygous flinty (*FF*) females and floury (*ff*) males in corn, what are the genotypes of the following:
***a.*** endosperm of the kernels produced on the female plants.
***b.*** embryo of the kernels produced on the female plants.
***c.*** sperm nuclei of germinated pollen grains.
***d.*** any nucleus of the embryo sac (female gametophyte).
***e.*** endosperm of kernels produced on female plants of the reciprocal cross.

***3.12*** In crosses between *Neurospora* strains that produce black spores (*B*) and others that produce tan spores (*b*), the following kinds of asci were recovered:

| ascus type | order of ascospores |
|---|---|
| 1 | *b b b b B B B B* |
| 2 | *B B B B b b b b* |
| 3 | *B B b b b b B B* |
| 4 | *b b B B B B b b* |
| 5 | *b b B B b b B B* |
| 6 | *B B b b B B b b* |

***a.*** Which arrangements arise from first-division segregations and which arise from second-division segregations?
***b.*** Diagram the events during meiosis that lead to production of ascus types 5 and 6.

***3.13*** Mutation rates for genes in bacteria are usually low and occur on the average of about $1 \times 10^{-7}$. In the prototroph selection experiments conducted by Lederberg and Tatum using *E. coli*, what is the probability that spontaneous prototrophic mutants of the genotype $thr^+leu^+thi^+bio^+phe^+cys^+$ would arise in either the $thr^+leu^+thi^+bio^-phe^-cys^-$ or the $thr^-leu^-thi^-bio^+phe^+cys^+$ auxotrophic parent cultures?

***3.14*** We usually are not particularly concerned about dominance or recessiveness of allelic genes in species like *Neurospora crassa* or *E. coli,* but we do care about these designations in organisms such as corn, mice, and *Drosophila.* Why is this true?

***3.15*** Define the following terms:
***a.*** bacteriophage
***b.*** virulent phage
***c.*** temperate virus
***d.*** lysis
***e.*** plaque
***f.*** prophage
***g.*** lysogenic strain

***3.16*** What evidence did Hershey and Chase provide in experiments using phage T2 to support the hypothesis that DNA and not protein is the genetic material?

***3.17*** What are some of the differences between primary and secondary cultures of eukaryotic cells?

***3.18*** Mouse cells phenotypically $A^+B^-$ and human cells phenotypically $A^-B^+$ were mixed to produce mouse-human somatic cell hybrids, and the entire mixture of parent and hybrid cells was plated on solid media.
***a.*** How would you select clones of somatic cell hybrids from these plates?
***b.*** In what way does random loss of human chromosomes from mouse-human somatic cell hybrids resemble allelic segregation during meiosis in sexually reproducing systems?

***3.19*** One cell culture in a collection of cultures of normal human cells suddenly changes to the transformed state.
***a.*** What change has occurred in the nature of growth in this culture?
***b.*** What is the distribution of cells in this culture growing on solid substratum?
***c.*** If transformation has been induced by an oncogenic virus, where would you search in order to isolate the viral genome?

***3.20*** What is the role of the enzyme reverse transcriptase in the reproductive cycle of RNA retroviruses?

## References

Avers, C.J. 1981. *Cell Biology,* 2nd ed. Boston: Willard Grant Press.

Bishop, J.M. Mar. 1982. Oncogenes. *Sci. Amer.* **246**:80.

Campbell, A. Dec. 1976. How viruses insert their DNA into the DNA of the host cell. *Sci. Amer.* **235**:102.

DuPraw, E.J. 1970. *DNA and Chromosmes.* New York: Holt, Rinehart and Winston.

Henle, W., G. Henle, and E.T. Lennette. July 1979. The Epstein-Barr virus. *Sci. Amer.* **241**:48.

Hershey, A.D., and M. Chase. 1952. Independent functions of viral protein and nucleic acids in growth of bacteriophage. *J. Gen. Physiol.* **36**:39.

Mazia, D. Jan. 1974. The cell cycle. *Sci. Amer.* **230**:54.

Moens, P.B. 1978. The onset of meiosis. In L. Goldstein and D.M. Prescott, eds., *Cell Biology, A Comprehensive Treatise,* vol. 1. New York: Academic Press.

Pollack, R., ed. 1981. *Readings in Mammalian Cell Culture,* 2nd ed. Cold Spring Harbor, N.Y.: Cold Spring Harbor Laboratory.

Simons, K., H. Garoff, and A. Helenius. Feb. 1982. How an animal virus gets in and out of its host cell. *Sci. Amer.* **246:**58.

White, M.J.D. 1973. *The Chromosomes,* 6th ed. London: Chapman & Hall.

# CHAPTER 4

# The Chromosomal Basis of Inheritance

From genetic analysis Mendel and his successors found that members of a pair of alleles segregated from each other during reproduction and that members of different pairs of alleles underwent independent assortment during reproduction. Cytological observations using microscopy showed that chromosomes exist in pairs in meiotic cells. Members of pairs of homologous chromosomes segregated during meiosis, and members of different pairs of chromosomes underwent independent assortment during meiosis. These parallels in behavior of pairs of alleles and pairs of chromosomes provided a strong circumstantial case for locating genes on chromosomes. Experimental evidence in support of the hypothesis was obtained in studies in which known pairs of alleles could be identified with specific chromosomes and both alleles and chromosomes could be analyzed in the same system.

Once it was established that genes were located on chromosomes, it was necessary to analyze chromosomes in some detail. In particular it was essential to understand chromosome chemistry, chromosome structure and organization, and chromosome functions in order to gain information about the genes themselves. Studies of chromosomes provided and still provide the biological framework for genetic analysis of inheritance.

## Genes on Chromosomes

Chromosomes that are involved in sex determination are called **sex chromosomes,** the most familiar ones being the X and Y chromosomes. All other chromosomes are called **autosomes.** Autosomes and the alleles they carry are present in pairs in diploid cells, and their segregation and reassortment lead to Mendelian inheritance patterns based on equal contributions by each parent to the progeny. Sex chromosomes, in contrast, are not necessarily present in pairs in both sexes, and different kinds of sex chromosomes may be genetically different. These features of the sex chromosomes lead to inheritance patterns that are distinctive and different from autosomal gene inheritance patterns. The unique nature of the sex chromosomes helped to provide the evidence needed to confirm the Chromosome Theory of Heredity early in this century.

### 4.1 Discovery of the Sex Chromosomes

The first specific information on sex chromosomes came from cytological studies by C.E. McClung, E.B. Wilson, and N. Stevens between 1901 and 1905. They all studied insect chromosomes, particularly in spermatocytes of grasshoppers and other members of the order Orthoptera. These insects were favorite materials of the early microscopists. Meiotic chromosomes are particularly well displayed in these cells. McClung noted that all the chromosomes were paired except for one "accessory" chromosome in grasshopper spermatocytes (Fig. 4.1). He suggested that this chromosome was concerned with sex determination at fertilization since females had only paired chromosomes in oocyte meiosis and lacked the "accessory" chromosome. Wilson and Stevens straightened out the details of the story in 1905 and called McClung's "accessory" chromosome the **X chromosome.**

Female grasshoppers and most other female orthopteran insects have a pair of X chromosomes; males have only one X. When eggs are

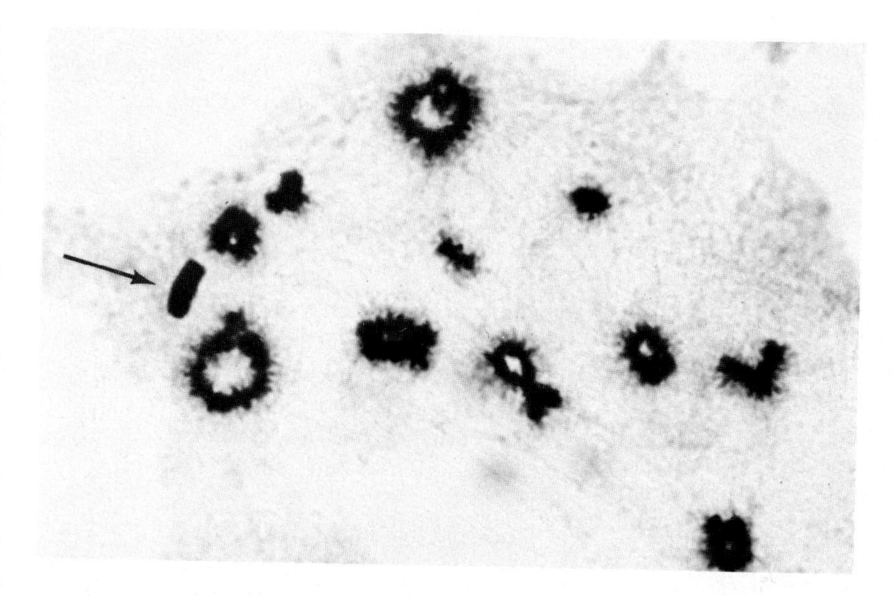

*Figure 4.1* Diakinesis in grasshopper spermatocyte. There are 11 bivalents and a highly condensed, unpaired X chromosome (arrow), making a diploid chromosome count of 23 in this XX ♀/XO ♂ orthopteran species.

formed at meiosis, each egg receives one X chromosome of the pair in the oocyte. Stevens suggested that when meiosis takes place in spermatocytes, half the sperm receive an X chromosome and half do not. When fertilization occurs, the chances are even that an egg will be fertilized by an X-carrying sperm or by a sperm without an X chromosome; fertilized eggs may be either XX or XO (pronounced "oh," signifying that no second sex chromosome is present). Because of random fertilizations, about half the zygotes will develop into females (XX) and about half into males (XO). The sex of the individual is thus determined at fertilization by the nature of the sperm that fuses with the egg (Fig. 4.2).

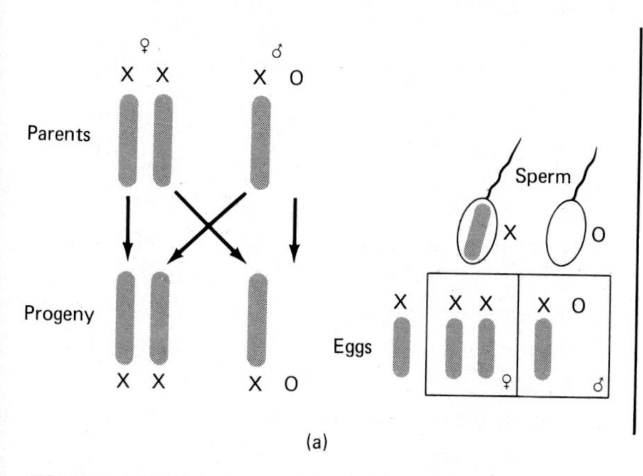


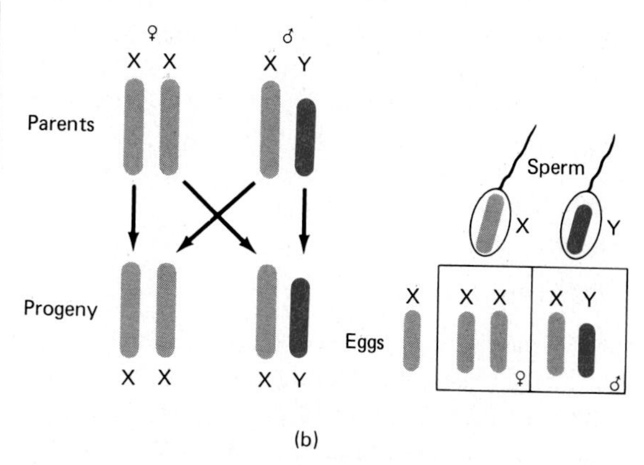


*Figure 4.2* Chromosomal mechanisms of sex determination (a) in the XX/XO system and (b) in the XX/XY system. Sex is determined at fertilization in accordance with the kind of sperm fertilizing the X-carrying egg. The sex ratio, at least at birth, should be 1 ♀:1 ♂, since in both systems each of the two kinds of sperm has a 50% chance of fertilizing the egg.

Stevens, Wilson, and others described sex chromosomes in other species, including the fruit fly, *Drosophila melanogaster*. In most of these species the X chromosome has a partner in males, and this partner is called the **Y chromosome.** In some species the Y chromosome is much smaller than the X, as it is in human beings. In other cases the Y is different in shape but not necessarily smaller, as in *Drosophila*. A sex ratio of 1 female : 1 male is obtained similarly in XX/XO and XX/XY species; sex is determined on a random basis at fertilization according to the kind of sperm that fertilizes the egg. Each gamete also contributes one set of nonsex chromosomes, or autosomes, to the zygote, and these pairs of autosomes carry many pairs of alleles in the diploid nucleus.

## 4.2 X-Linked Inheritance in *Drosophila*

In 1903 W. Sutton proposed the Chromosome Theory of Heredity in an essentially modern form. The *parallel behavior* of genes in inheritance and chromosomes in the nucleus made a powerful but circumstantial case for genes being located on chromosomes. There was no experimental evidence to support this theory until 1910, when Thomas Hunt Morgan published a brief report on the inheritance of white eyes and wild-type red eyes in *Drosophila*.

A white-eyed male appeared suddenly in true-breeding cultures of red-eyed flies. Morgan mated this male to its red-eyed sisters and found that all the $F_1$ offspring had red eyes. The $F_2$ generation consisted of the following:

2459 red-eyed females

1011 red-eyed males

782 white-eyed males

Since the $F_1$ had red eyes and since the $F_2$ showed a reasonable Mendelian ratio of 3 red : 1 white, red was dominant over white for eye color. Morgan noted the unusual feature that all the $F_2$ white-eyed flies were males.

In order to determine if females could be produced with the mutant white eye phenotype, Morgan performed a testcross of $F_1$ red-eyed females with the original white-eyed male parent. In their progeny Morgan found

129 red-eyed females

132 red-eyed males

88 white-eyed females

86 white-eyed males

Once again a deficiency in the recessive class distorted the ratio, but it was a reasonable approximation of 1 red : 1 white for both males and females. The 1 : 1 testcross ratio verified single-gene inheritance and also showed that white eyes could develop in females as well as in males.

Knowing that females are XX and males are XY, we can explain Morgan's results if we assume the white eye color gene is present on the X and absent from the Y chromosome (Fig. 4.3). All sons receive their only X chromosome through the eggs from their mother. The Y chromosome, of course, comes only from their father. Daughters, on the other hand, get one X from their mother and one X from their father. By putting the alleles on the X chromosome and following through the $F_1$ and $F_2$ generations, we can see that the pattern of **X-linked inheritance** parallels the pattern of transmission of the X chromosome. Males are **hemizygous** for their X-linked alleles, having only one X-linked allele of every kind on their one X chromosome. If the mother is heterozygous, half her eggs receive the X chromosome carrying the + allele and half the eggs receive the *w*-carrying X when these chromosomes segregate at meiosis. Half the sons of a $+/w$ mother are red eyed and half are white eyed, thereby reflecting the fact that half the eggs are of each genetic type.

In the testcross of the white-eyed male to his heterozygous $F_1$ daughters, we would predict sons with white eyes and sons with red eyes on the same basis as just described. We would further predict the production of white-eyed females in this testcross progeny since all daughters receive an X with the *w* allele from their

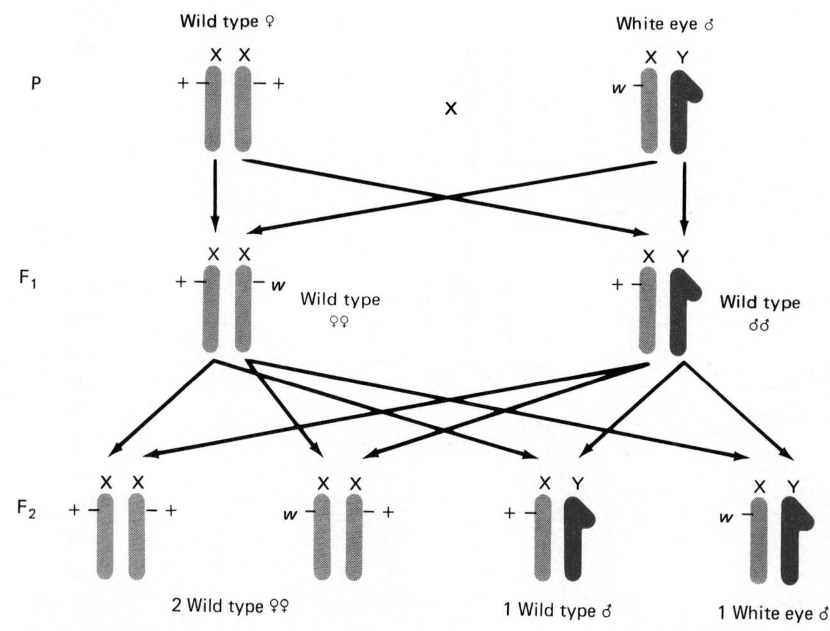


***Figure 4.3*** Illustration of the parallels between the pattern of inheritance of X-linked alleles and the pattern of transmission of the sex chromosomes in *Drosophila*.

father, and half get an X from their mother carrying the *w* allele and half receive the + allele (Fig. 4.4). We not only would predict these kinds of females, but we would be able to say that half the females would be red eyed and half would be white eyed. This is just what Morgan found.

Morgan also made the reciprocal cross between white-eyed females and red-eyed males (Fig. 4.5). These results conform to the predictions made for parallel transmission of X chromosomes and X-linked alleles on these chromosomes. They also point out two important features that distinguish X-linked inheritance:

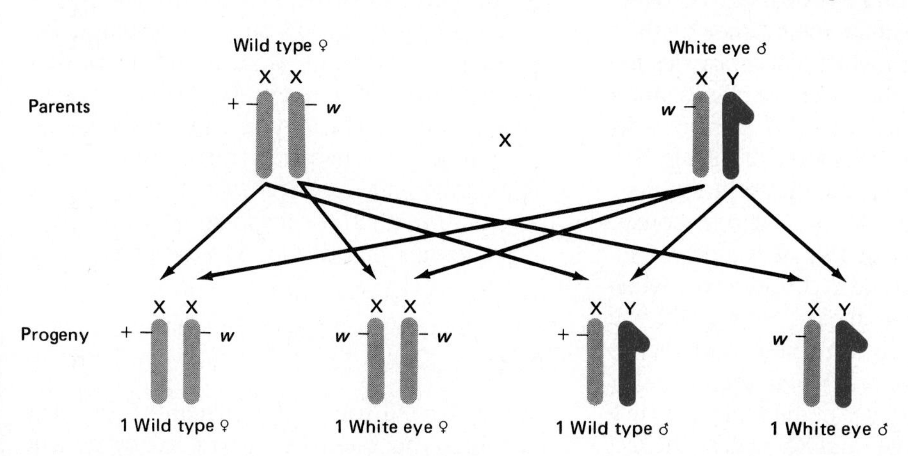


***Figure 4.4*** We can predict that equal numbers of red-eyed and white-eyed ♂♂ and ♀♀ will be produced in the testcross progeny of the white-eyed ♂ to his heterozygous $F_1$ daughters, if the eye color gene is X linked as hypothesized.

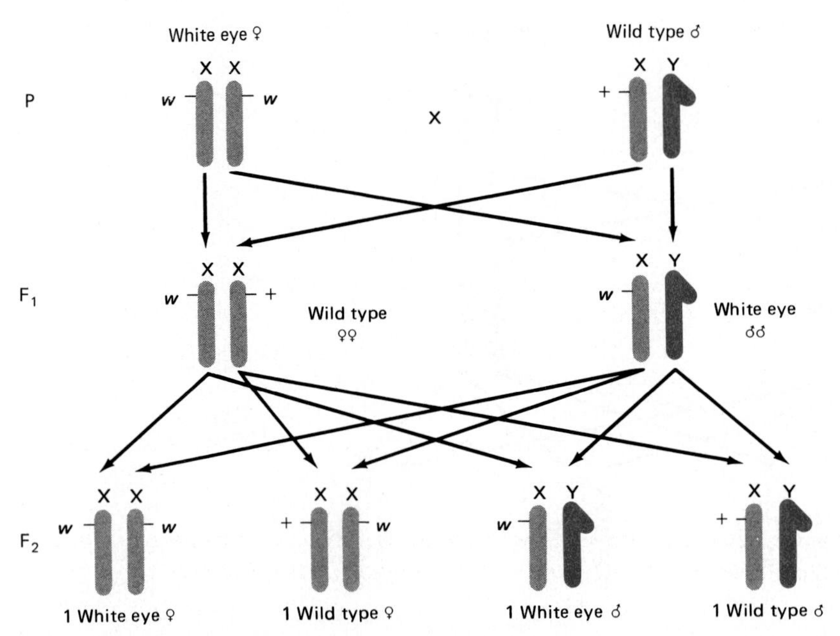


***Figure 4.5*** Reciprocal cross to one shown in Fig. 4.3 in which results conform to predictions made on the basis of parallel transmission of X chromosomes and X-linked alleles on these chromosomes.

(1) reciprocal crosses produce different progenies and (2) sons resemble their mothers more than their fathers, thus revealing a "criss-cross" inheritance pattern.

In **autosomal inheritance,** such as Mendel's studies with peas, *reciprocal crosses produce identical $F_1$ and $F_2$ progenies.* Sons and daughters resemble either their father or their mother, depending on which parent carries the dominant and which the recessive alleles. Autosomal patterns tell us that each pair of autosomes carries an equivalent set of alleles. Sex linkage, on the other hand, indicates the genetic difference between the X and Y chromosomes. What would you predict for the pattern of inheritance of genes on the Y chromosome when the X has no matching alleles? How could you distinguish X-linked and Y-linked inheritance from a pattern produced when the X and Y chromosomes carry alleles of the same gene in a species with XX females and XY males? The fact that all of these patterns have been found is a very good reason for using the more specific term X-linked inheritance for genes on the X chromosome. The more general term *sex linked* should be reserved for situations that have not yet been resolved to known sex chromosomes or for general discussions of any sex chromosome–linked inheritance pattern.

Morgan's pioneer study of the inheritance of white eyes in *Drosophila* opened the way to investigations of genes on chromosomes. Between 1911 and 1913, Morgan and his student Alfred Sturtevant located six different mutant alleles on the X chromosome in *Drosophila.* As we will see in Chapter 5, these studies of X-linked inheritance led Sturtevant to propose and develop a method for mapping genes on chromosomes, a method we use to the present day.

## 4.3 Other Inheritance Patterns Associated with Sex

Species having some other system of sex chromosomes can often be discovered by genetic analysis, even when cytological studies are difficult or uninformative. Chickens were among the first animals studied that showed a pattern different from the one found in *Drosophila* (Fig. 4.6). The inheritance of the sex-linked dominant for barred versus nonbarred feather pattern in poultry is clear from the two major features of the breeding analysis: (1) reciprocal crosses give rise to different $F_1$ and $F_2$ progenies in direct relation to the parent having two sex chromosomes of one kind versus one of that kind and (2) most of the progeny resemble one parent more than the other. In this case, however, the transmission of alleles parallels the transmission of two like sex chromosomes from the male parent rather than from the female parent.

We can tell that barred is dominant because it appears in all the $F_1$ of one of the reciprocal crosses and in the Mendelian 3:1 ratio in the $F_2$ from that cross. The 1:1 $F_1$ and $F_2$ ratios of the reciprocal cross are not informative in this respect, but they do indicate that sex linkage determines the genotypic and phenotypic distributions when both reciprocal sets of crosses are compared.

A **heterogametic** (two kinds of gametes) female and **homogametic** (one kind of gamete) male are typical of reptiles, some kinds of amphibia and fish, the moths and butterflies (Lepidoptera), and birds. The convention has been to distinguish this pattern from female XX/male XY by using different letters (WZ females and ZZ males). It is just as simple to retain the letters X and Y, as long as the sex of the XX and XY individuals is specified.

Two other patterns of inheritance are associated with the sex of the individual, but the controlling genes are located on autosomes and not on sex chromosomes. The inheritance patterns are very clear for **sex-limited** phenotypic expression since one sex expresses the two phenotypic alternatives and the other sex only develops one of the two expressions, regardless

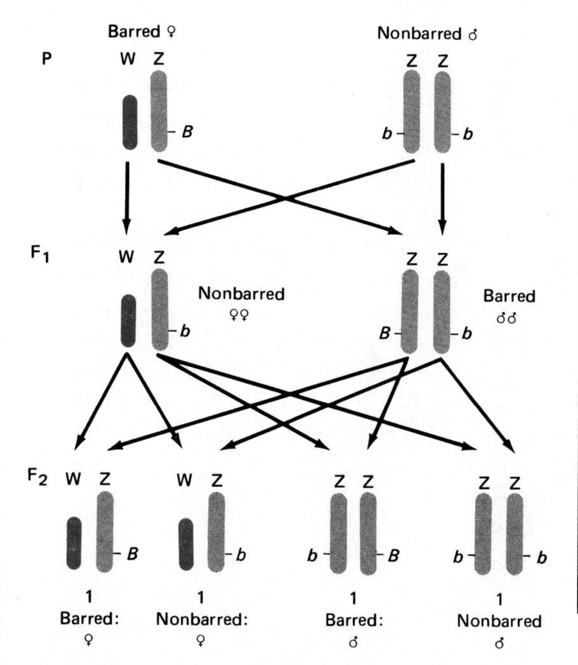


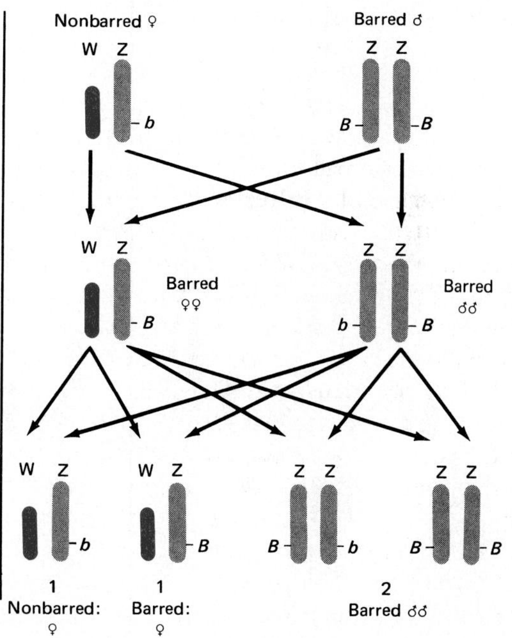


***Figure 4.6*** Sex-linked inheritance of feather color pattern in chickens. Barred feather color is inherited as a sex-linked dominant trait, according to the $F_1$ and $F_2$ progeny genotypic and phenotypic ratios. The particular phenotypic distribution found here is characteristic of organisms in which the female has a single X chromosome and is, therefore, the heterogametic sex. For species in which the female is heterogametic, the X chromosome equivalent is often referred to as the Z chromosome; and the Y chromosome equivalent, as W.

of its genotype (Fig. 4.7). Clearly, the characteristics usually are ones by which the males and females differ in their sexual development. In human beings, an inherited recessive condition called *testicular feminization syndrome* is only expressed in XY individuals. The mutant phenotype involves development of female genitalia and other anatomical features under sex-hormonal control. It can only be detected in males because these reproductive structures would develop anyway in XX females of all three genotypes. Such males are XY, and they have testes. Their insensitivity to the male sex hormone produced in the embryonic testes leads to their external physical development as females under

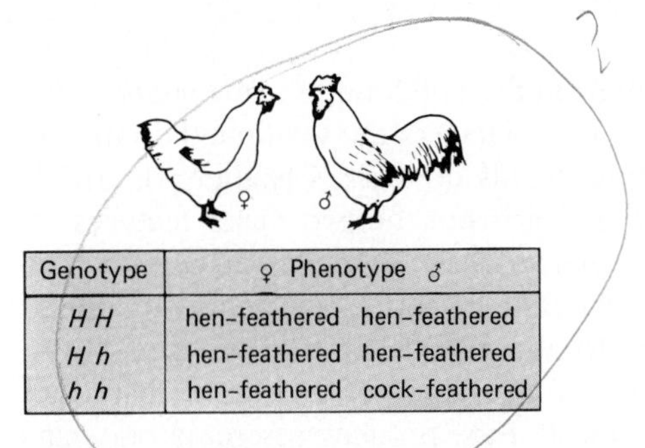


| Genotype | ♀ Phenotype | ♂ |
|---|---|---|
| *H H* | hen-feathered | hen-feathered |
| *H h* | hen-feathered | hen-feathered |
| *h h* | hen-feathered | cock-feathered |

***Figure 4.7*** Sex-limited inheritance of tail-feathering in chickens. The phenotypic ratio of 3 hen-feathered:1 cock-feathered ♂♂ in reciprocal $F_2$ progeny of heterozygous $F_1$ parents provides the basis for interpreting this characteristic as due to autosomal, single-gene inheritance.

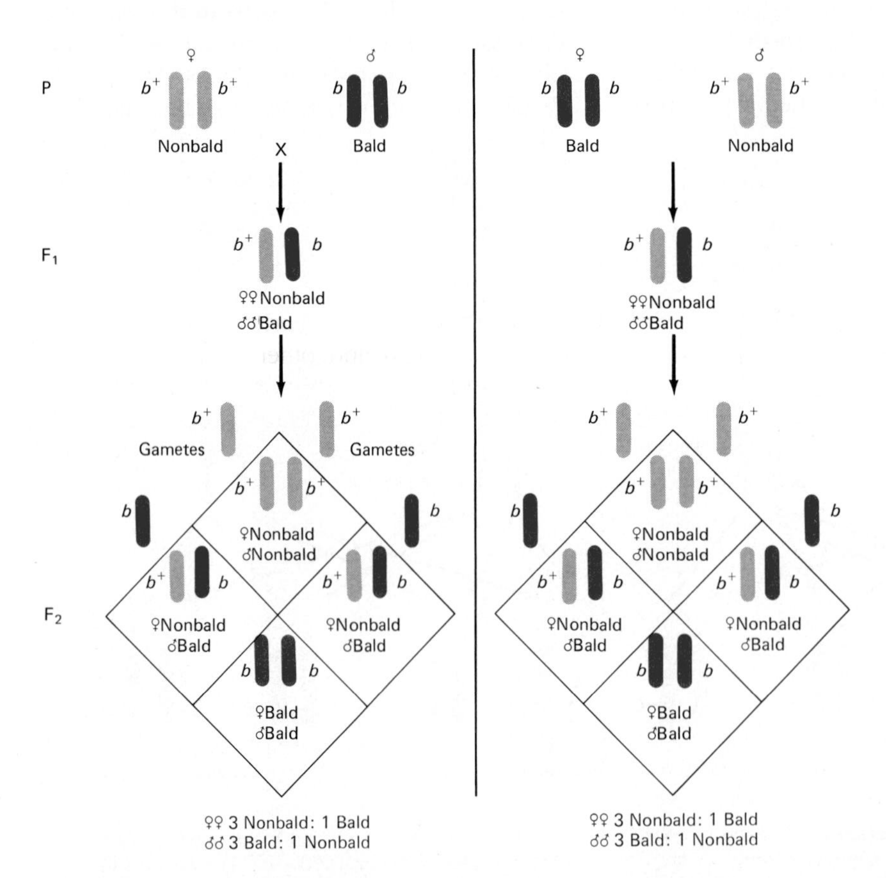


***Figure 4.8*** Sex-influenced inheritance of pattern baldness in humans. Theoretically, different phenotypes for $F_1$ ♀♀ and ♂♂ and 3:1 versus 1:3 $F_2$ ratios for the two sexes establish this inheritance pattern for a single autosomal gene. The allele that is dominant in one sex is recessive in the other. Family studies have provided the evidence for sex-influenced inheritance in the case of this human trait.

the direction of female sex hormones that are produced in the adrenal glands of all males.

In **sex-influenced inheritance** of one pair of alleles, we find the Mendelian 3:1 ratio among the $F_2$ (Fig. 4.8). Since reciprocal crosses produce identical $F_1$ and identical $F_2$ progeny, inheritance must be autosomal. The difference concerns the distribution of dominant and recessive phenotypes between the sexes. There is a reversal of dominance and recessiveness of the two alleles in relation to the sex-hormonal constitution of the individual. The allele that is dominant in one sex behaves as a recessive in the other sex. In human beings, pattern baldness is far more common among men than among women. The phenotype is expressed in men whether they are homozygous or heterozygous for the *b* allele, thereby indicating dominance of this allele in males. Women express pattern baldness only when they have the *bb* genotype, a phenomenon expected in recessive inheritance. In a large sampling of heterozygote matings, we would find that the daughters are ¾ nonbald : ¼ balding and the sons are ¼ nonbald : ¾ balding.

Another sex-influenced inheritance pattern in human beings involves comparative lengths of the index finger and the fourth finger. When the index finger is equal to or longer than the fourth finger, we find that the characteristic is female dominant and male recessive (Fig. 4.9). Can you reach this conclusion by recording the phenotype among members of your family? If your family is large enough and includes several generations, you should receive at least a hint of this inheritance pattern. A more reliable interpretation can be made if you pool your data with those from your friends and classmates; a large enough sampling should reveal the genetic pattern.

In many cases of sex-influenced inheritance, we can trace the differences in expression to hormonal differences during development. The behavior of an allele as a dominant or recessive is a function of the biological environment in which gene action takes place. Dominance and recessiveness are *relative behaviors* and are not based on some special chemical or physical characteristic of the alleles themselves. Although we can see this more clearly in some inheritance patterns than in others, the principle applies to all genes and their alleles.

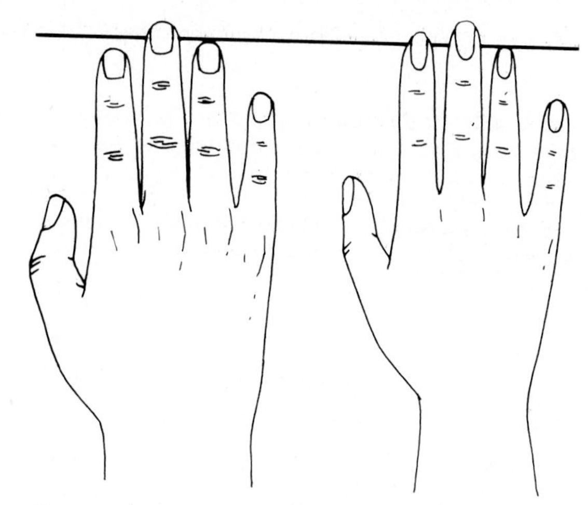

***Figure 4.9*** An autosomal sex-influenced human trait. The dominant trait in females is an index finger equal to or longer than the fourth finger; but it is the recessive condition in males. Both of these individuals are heterozygous for the trait (♀, right; ♂, left).

## 4.4 Sex Determination

Extensive genetic and chromosomal studies using *Drosophila* and other insects have permitted the definition of the sex chromosome makeup in different species and have revealed the mechanism responsible for sex determination in these insects (Table 4.1).

***Table 4.1*** Summary of the major chromosomal sex determination patterns.

| females | males | species in which these sex chromosomes occur |
|---|---|---|
| XX | XY | mammals, *Drosophila*, some flowering plants |
| XX | XO | most grasshoppers (Orthoptera) |
| XY | XX | some fishes, reptiles, birds, Lepidoptera (butterflies and moths) |
| X | Y | liverworts (Bryophyta) |

***Table 4.2*** Sex determination is based on a balance between number of X chromosomes and number of autosome sets in *Drosophila melanogaster*.

| sex chromosomes present | number of sets of autosomes | ratio of X:A | sex of individual |
|---|---|---|---|
| XX | 2 | 1.00 | female |
| XY | 2 | 0.50 | male |
| XXX | 2 | 1.50 | abnormal female (sterile) |
| XXXX | 3 | 1.33 | abnormal female (sterile) |
| XXY | 2 | 1.00 | female |
| XXX | 4 | 0.75 | intersex |
| XX | 3 | 0.67 | intersex |
| XY | 3 | 0.33 | abnormal male (sterile) |
| X | 2 | 0.50 | male (sterile) |

Since male *Drosophila* and orthopterans develop whether or not a Y chromosome is present, the Y chromosome is irrelevant to sex determination in these insect species. Through comparisons of *Drosophila* strains differing in numbers of X chromosomes and sets of autosomes (A), the **Sex Balance** theory of sex determination was conceived (Table 4.2). When the ratio of X:A was 1.0 or greater, females developed; when the ratio of X:A was 0.5 or less, males developed.

By this theory, females that were XXY and had two sets of autosomes were sexually equivalent to females that were XX:AA; males developed whether they were X:AA or XY:AA. Various intersex or other aberrant types appeared when the ratio deviated from X:A = 1.0 or 0.5. The determination of sex, therefore, depended on the ratio, or balance, between number of X chromosomes and number of autosome sets that came together at fertilization.

The same chromosomal mechanism for sex determination was presumed to characterize other XX female/XY male (or XY female/XX male) systems, including human beings and other mammals. Beginning in 1959, evidence was collected from cytological and genetic studies in mice, which showed that XO animals were female and that XXY were males (Fig. 4.10). Female mice were assumed to be XO if they showed a recessive phenotype that should have been obscured if they had also inherited the dominant allele on an X chromosome from their mothers; this was verified cytologically. Other mice were shown to be XXY males from cytological examination and from their presumed genotype which showed that one X-linked allele must have come from each parent and, therefore, one of their two X chromosomes had come from each parent.

From combined cytogenetic studies and from many chromosomal studies, the system of **chromosomal sex determination** in mammals was shown to be strongly Y-chromosome-determining (Table 4.3). The crucial observations were that XO = male in *Drosophila* but female in mammals, and XXY = female in *Drosophila* but male in mammals. These observations were amply supported by consistent findings that males always had at least one Y chromosome and one or more X chromosomes, and females had at least one X but no Y chromosome. All of these sex chromosome constitutions were found in individuals having two sets of autosomes. Mammalian species do not tolerate gross disturbances in their autosomal

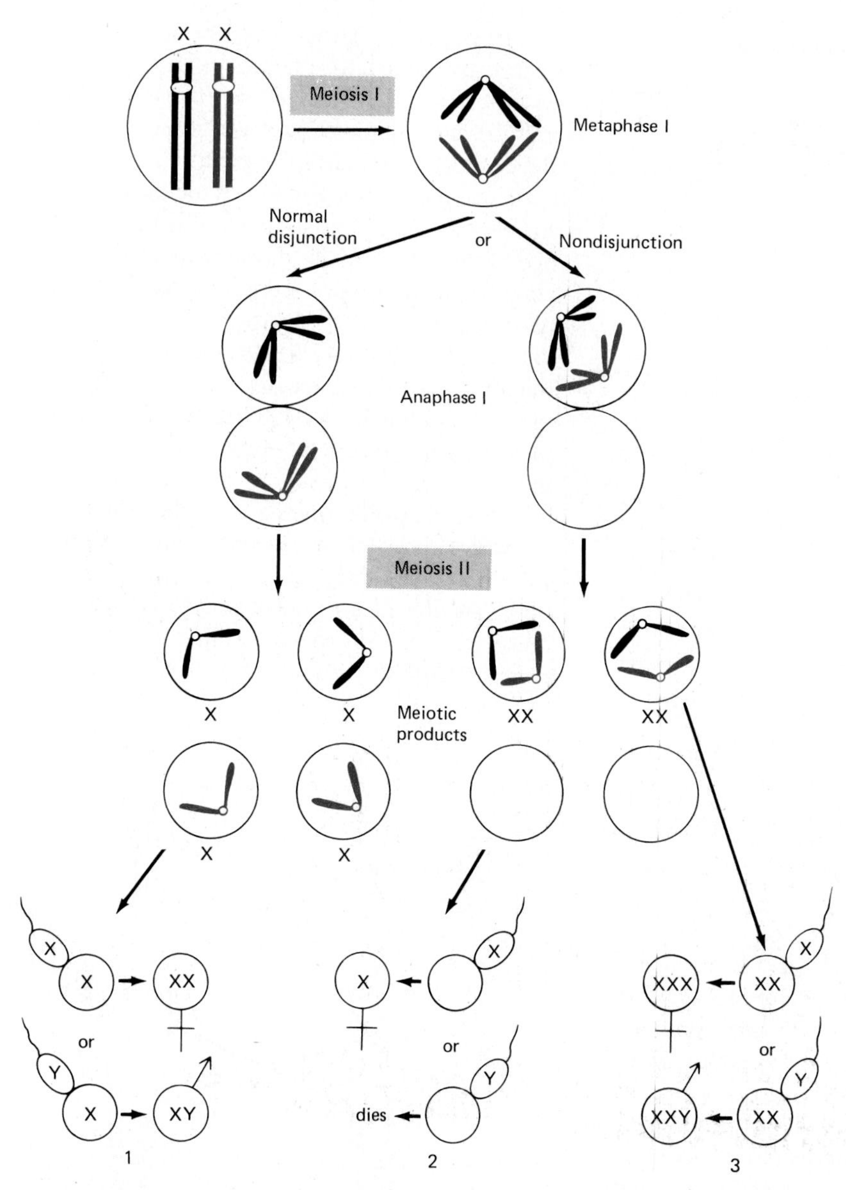


***Figure 4.10*** In mammalian species, males develop when a Y chromosome is present, along with one or more X chromosomes. There is no Y chromosome in mammalian females. The occurrence of XO ♀♀, XXY ♂♂, and other anomalous conditions is due to missing or extra sex chromosomes in one or both of the gametes that fuse to produce the new individual. The phenomenon responsible for aberrant chromosome numbers is nondisjunction, a faulty distribution event in which homologous chromosomes do not separate at anaphase and are, therefore, included or excluded together in a meiotic nuclear product.

numbers, either of whole sets or of individual chromosomes, so all possible comparisons with the *Drosophila* pattern cannot be made. There really is little need for such extensive comparison, however, since the role of the Y chromosome in mammalian sex determination is

*Table 4.3* Sex chromosome anomalies and sex determination in humans.

| individual designation | chromosome constitution* | sex |
|---|---|---|
| normal male | 46,XY | male |
| normal female | 46,XX | female |
| Turner female | 45,X | female |
| triplo-X female | 47,XXX | female |
| tetra-X female | 48,XXXX | female |
| penta-X female | 49,XXXXX | female |
| Klinefelter males | 47,XXY | male |
| | 48,XXXY | male |
| | 49,XXXXY | male |
| | 48,XXYY | male |
| | 49,XXXYY | male |
| XYY male | 47,XYY | male |

*The two-digit number indicates the total number of chromosomes in diploid cells, followed by the exact number and kinds of sex chromosomes in this complement. Two sets of autosomes are present and account for 44 of the total number of chromosomes in these cells.

quite clear from the available data.

Women who are XO show the symptoms of **Turner syndrome,** characterized in part by short stature, underdeveloped breasts and internal reproductive organs, and certain other physical traits. Men who have one or more extra X chromosomes very often display the symptoms of **Klinefelter syndrome,** characterized by relatively long limbs in proportion to torso length, larger breast development than in XY males, and smaller testes. In general, XO women are less feminized and XXY men are more feminized in their external appearance than their XX and XY counterparts. Hormonal modifications under the influence of the sex chromosomes are clearly indicated. We will discuss these and other sex chromosomal anomalies more fully in Chapter 12.

Many species, particularly among eukaryotic microorganisms, have a genic sex determination system (Fig. 4.11). As described earlier, *Neurospora* has two mating types based on two alleles of the one mating-type gene. This pattern is fairly typical among many of the algae, fungi, and protists. Many variations on this theme have been found, principally showing more than one gene for mating types. Simpler organisms may have more than two mating types, which are determined by these genes. Genic sex determination has also been found in certain fish, often operating in conjunction with sex chromosomes. Some genetic studies have shown that the genic and chromosomal systems can be separated so that only one or the other functions.

Various organisms have a pattern of noninherited sex determination, even when individuals are clearly males and females, reproductively and morphologically. The marine worm *Bonnelia* provides the usually cited ex-

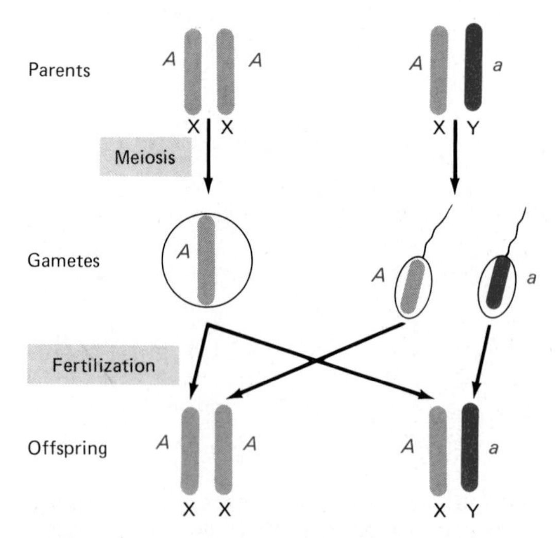


*Figure 4.11* Whether sex determination is due to two kinds of sex chromosomes or two kinds of alleles of a gene, a sex ratio of 1 ♀:1 ♂ is expected if both kinds of gametes from the heterogametic sex (shown here as the ♂) occur in equal numbers and if fertilization of the single kind of egg takes place at random with respect to male gamete type. Diploids are shown here, but the systems work similarly in haploid organisms.

ample of such a system. Larvae that hatch from fertilized eggs are sexually neutral. When a free-swimming larva settles on a rock or similar surface, it develops into a female. Larvae that settle on a mature female, however, develop into tiny male individuals. When larvae are placed in water that contains extracts of female worms, these larvae develop as males. Larvae in control media that lack female extract will develop into females after settling on the surface of the experimental container.

Many **hermaphroditic** species (having both ovaries and testes in the same individual) lack a sex determination system, as we would expect. Earthworms, snails, most of the flowering plants, and other bisexual organisms lack any sex-determining system of genes or chromosomes. In those few flowering plants that are dioecious (separate male and female flowers on separate plants), sex chromosomes may be present (Fig. 4.12). For example, the flowering plant *Melandrium* has XX female and XY male differentiation, with the Y chromosome being strongly male-determining.

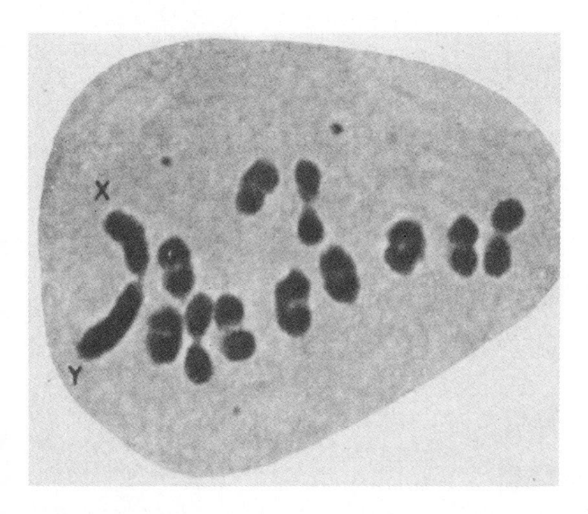


***Figure 4.12*** Metaphase I in a diploid male plant of *Melandrium* (*Lychnis*). There are 11 pairs of autosomes and one pair of sex chromosomes, seen at the left. The Y chromosome appears to be a little larger than the X chromosome. (From H.E. Warmke. 1946. *Amer. J. Bot.* **33**:648. Fig. 1.)

A relatively great variety of other patterns are known. We will gain little, however, by their detailed comparison at this stage in our discussion. When it is appropriate to some genetic question elsewhere in this book, one or more of these alternative patterns will be described at that time.

## Family Studies of Human Inheritance

Analysis of human inheritance obviously requires methods other than controlled breeding analysis. The method of choice for single-gene inheritance is to construct family histories, or pedigrees, and seek the clues that match expected patterns of allele transmission to phenotypic ratios known from genetic analysis of organisms such as *Drosophila,* mice, and garden peas.

### 4.5 Autosomal Inheritance

From what we know about other species, autosomal alleles in humans should be expressed equally in both sexes. An autosomal pair of alleles, $a^+$ and $a$, should show a 3:1 phenotypic ratio in heterozygote × heterozygote ($a^+a \times a^+a$) matings and a 1:1 ratio in heterozygote × recessive ($a^+a \times aa$) matings. For this pair of alleles, we expect the following:

$a^+a \times a^+a$

| | $a^+$ | $a$ |
|---|---|---|
| $a^+$ | $a^+a^+$ | $a^+a$ |
| $a$ | $a^+a$ | $aa$ |

$a^+a \times aa$

| | $a$ |
|---|---|
| $a^+$ | $a^+a$ |
| $a$ | $aa$ |

These patterns can be interpreted from pedigrees for the characteristic of brown versus blue eyes. We will consider a family consisting of four

(I–IV) generations (Fig. 4.13). Several family groups in this pedigree provide the necessary clues to show that (1) brown eyes and blue eyes are alternative phenotypes for one gene, (2) the allele for brown eyes is dominant to the allele for blue eyes, and (3) the eye color gene is autosomal.

*Single-gene inheritance.* We expect a testcross phenotypic ratio of 1 brown:1 blue if one parent is heterozygous and the other parent is homozygous recessive. Alternatively, we expect a ratio of 3 brown:1 blue if both parents are heterozygous. We find here that brown × blue gives a 1:1 phenotypic ratio for the children of I-3 × I-4 and II-10 × II-11. The family is too small to decide if the ratio is 3 brown:1 blue among the children of I-1 × I-2.

*Dominance of the allele for brown.* If brown is the dominant phenotype, then all the children of a homozygous dominant brown and a recessive blue should have brown eyes, as in $F_1$ progeny. This is the case for the children of II-1 × II-2, and III-5 × III-6. In addition, brown and blue segregate in the children of heterozygous parents with the dominant phenotype (I-1 × I-2), but do not segregate if both parents have the recessive phenotype (III-7 × III-8 and III-9 × III-10 are blue-eyed parents whose children are all blue eyed).

*The eye color gene is autosomal.* Both males and females have brown or blue eyes equally often, which we expect for autosomal inheritance. We also expect reciprocal matings to produce identical phenotypic ratios in the progeny, as seen in families of II-1 × II-2 and III-5 × III-6.

In cases of relatively *rare* dominant or recessive mutant alleles, we can derive additional clues from family studies (Fig. 4.14). Since these alleles are rare in the population as a whole and

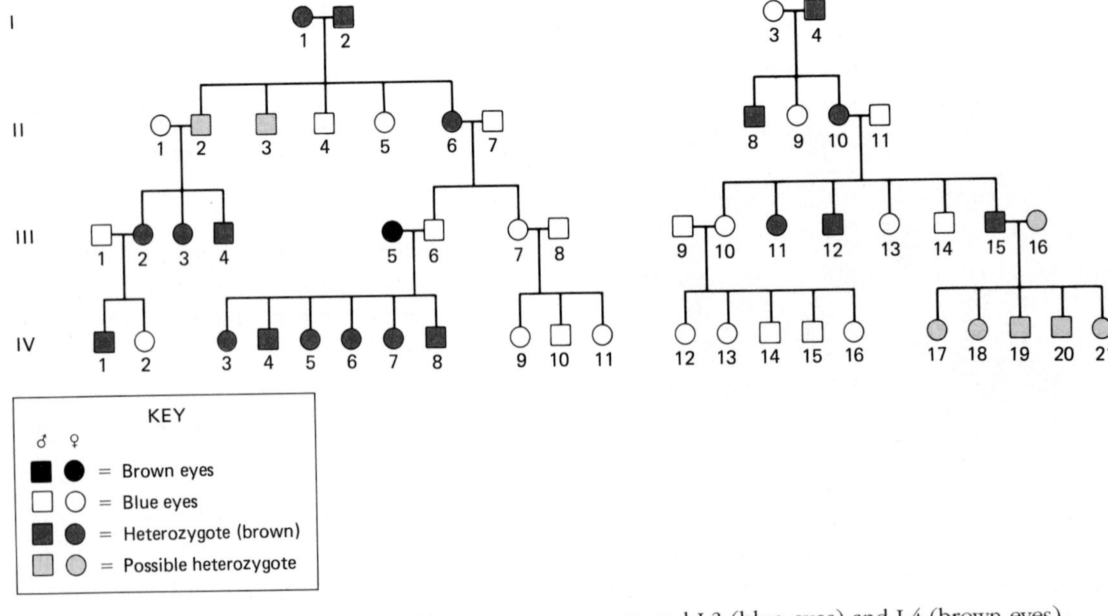


***Figure 4.13*** Inheritance of brown and blue eye color in two families is shown in pedigrees covering four generations (I–IV). Individuals are identified by generation number and member number in that generation; for example, the first generation couples are I-1 and I-2, both with brown eyes, and I-3 (blue eyes) and I-4 (brown eyes). From standard Mendelian analysis we can see that brown and blue eye color inheritance is based on one pair of alleles of an autosomal gene, with brown dominant over blue.

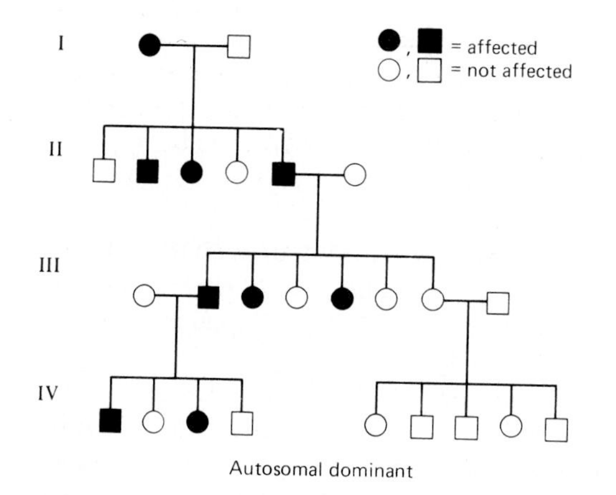


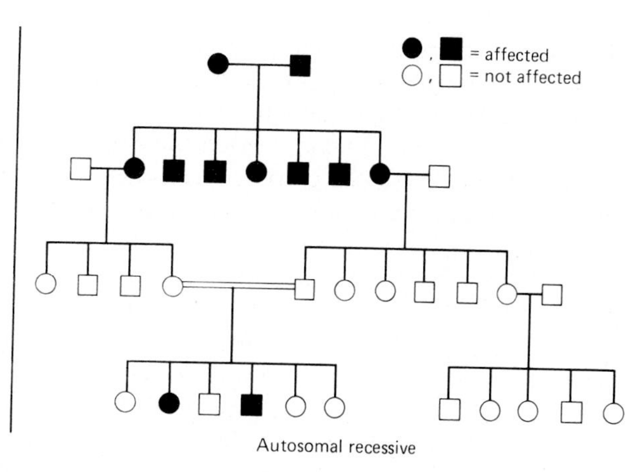


***Figure 4.14*** Pedigree analysis of inheritance of relatively rare traits. Autosomal inheritance is apparent from the occurrence of the trait approximately equally in males and females. Dominance usually is indicated by several clues, including the occurrence of the trait in some children in each generation when only one of their parents expresses the trait. Recessive inheritance, on the other hand, is usually characterized by the presence of the trait in all the children of parents who both possess the trait and by the skipping of generations. Autosomal recessive traits are expressed more often in children of parents who are related than those of unrelated parents. In the family shown at the right, first cousins in generation III produce some children with the trait. These cousins must both be heterozygous since each had one parent with the trait. Due to the rarity of the recessive allele in this example, its occurrence in both parents is more probable if these parents are related than if they are unrelated. A double horizontal bar indicates the couple are related to each other, which is clear from the family history.

since people tend to select partners who are not relatives, we expect the following patterns to appear in **autosomal dominant inheritance:**

***1.*** Most matings are between heterozygotes and recessives and lead to predominantly 1:1 phenotypic ratios for all the children, regardless of sex.

***2.*** The dominant phenotype appears in every generation if penetrance is high and if families are large enough to provide reasonable ratio approximations.

***3.*** The phenotype may be transmitted by either parent having the dominant trait, but it is never transmitted when neither parent expresses the trait (and must, therefore, be recessive).

In **autosomal recessive inheritance** for rare traits, the family patterns usually reveal the following features:

***1.*** The trait does not appear in every generation; it skips generations or appears in some family member even when it was absent in a number of previous generations within the family.

***2.*** The mutant phenotype is usually produced in children of parents with the normal phenotype (who must therefore be heterozygotes).

***3.*** All the children express the mutant trait when both parents have the recessive phenotype.

***4.*** Sons and daughters are equally likely to inherit a particular autosomal allele.

## 4.6 X-Linked Inheritance

From the analysis of red eyes and white eyes in *Drosophila,* we can more easily interpret Queen Victoria's family history of hemophilia, the "bleeder's disease," as one showing X-linked inheritance of a recessive trait (Fig. 4.15). Hemophilia occurred only in the males of her family, a fact that might seem at first glance to indicate either sex-limited or Y-chromosome inheritance. By putting the appropriate alleles on the X chromosome, we can show that only certain male relatives were hemophilic only because there were no matings between hemophilic men and either hemophilic or carrier women. In particular, we find the "crisscross" pattern of sons having this characteristic because they receive an X chromosome only from their mothers and no X from their fathers. If the characteristic were sex limited, sons could receive the mutant allele from either father or mother. It is certainly not Y-chromosome inheritance because, if it were, *every* son would express the same trait as his father, but we find hemophilic sons from normal fathers and a normal son produced by Leopold, a hemophilic child of Victoria and Albert.

In **X-linked recessive inheritance,** therefore, we expect to find that: (1) predominantly or only males are affected, depending on the parental genotypes; (2) affected sons have carrier mothers and affected male relatives (uncles, cousins, and others); (3) if the family is large enough we find affected and normal sons about equally often among the children of carrier mothers and normal fathers; and (4) affected

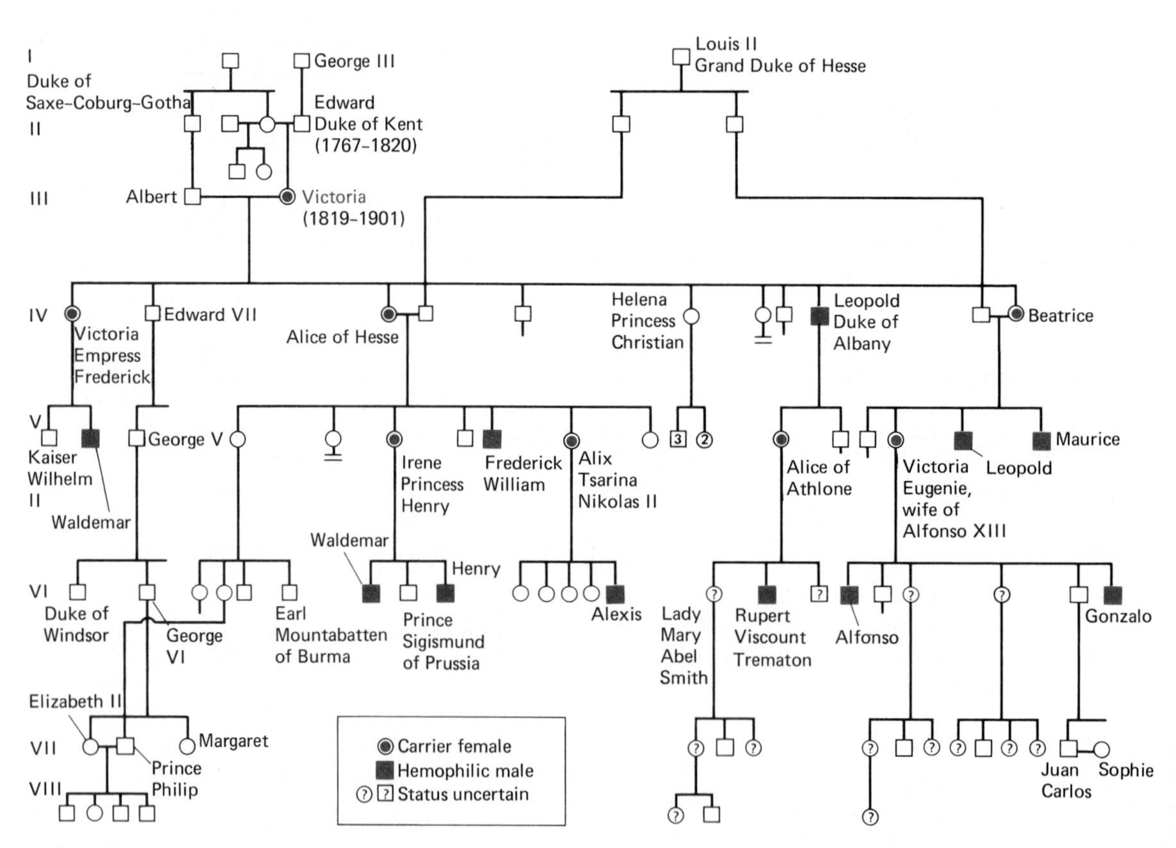


***Figure 4.15*** Hemophilia A (classical hemophilia) in royal European families related to Queen Victoria of Great Britain. The pattern of X-linked recessive inheritance can be deduced from various clues, including (1) afflicted males are born of carrier mothers, (2) afflicted males have afflicted male relatives, and (3) carrier daughters and normal sons are produced by afflicted fathers, as in the case of Leopold who was a hemophilic son of Queen Victoria. Marriage partners not shown in the pedigree were normal.

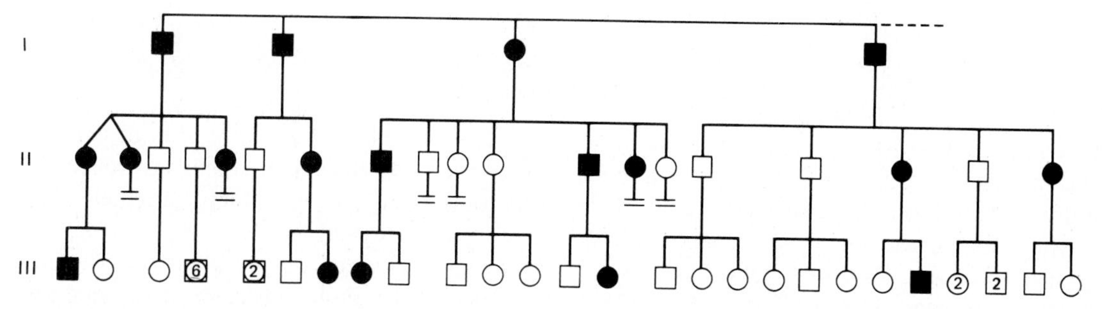


***Figure 4.16*** X-linked dominant inheritance of vitamin D–resistant rickets in one human family. Notice that both sexes may be affected. Marriage partners are not shown, but all were normal. In patterns such as this, affected males transmit the trait only to their daughters and affected mothers transmit the trait to their sons and daughters. (From *The Metabolic Basis of Inherited Disease* by J.B. Sandbury, J.B. Wyngaarden, and D.S. Frederickson, eds. Fig 65-12. Copyright © 1978 McGraw-Hill. Used with permission of McGraw-Hill Book Company.)

fathers produce all carrier daughters but all normal sons.

**X-linked dominant inheritance** is a different situation: females have twice as much chance of receiving an X chromosome carrying the dominant allele than males do (from either a carrier mother or an affected father; Fig. 4.16). Whether the female is homozygous or heterozygous for the dominant allele, the characteristic will be expressed in her. A son may receive either X chromosome through the egg from a heterozygous mother, so the son of an affected mother has a 50% chance of being affected. Sons cannot inherit the trait from their father because his contribution to the fertilized egg is a Y and not an X chromosome. Once again, for rare X-linked dominant alleles we would find few matings between unrelated people who both happen to carry the same rare mutant allele. Most matings would be between one parent with the trait and one with the normal phenotype (Box 4.1).

Through studies of this kind and others described in Chapter 5, more than 100 human genes have been shown to be X linked. Most of these genes on the human X chromosome have nothing whatever to do with sex; hemophilia is one such example of a gene related to a nonsexual phenotypic trait. Over 1100 genes have been shown by inheritance pattern analysis and by other means to be autosomal. At least one gene has been specifically assigned to every one of the 22 autosomes in the human complement. Most of the 1100 different autosomal genes identified from inheritance studies, however, have not yet been assigned to particular autosomes. In contrast with autosomes and the X chromosome, the human Y chromosome has at least one known gene, but two more may also be present. The Y-linked human genes are entirely concerned with development of sexual reproductive structures in the male embryo after approximately 6 weeks of its growth. Biochemical studies provided the information for these early-acting genes on the human Y chromosome.

## Overview of Chromosome Organization and Activity

An overall picture of chromosome morphology, types, numbers, and behavior permits us to move the gene from the realm of abstraction into the tangible world of test tubes and microscopes. Using a variety of methods, we can probe into the most intimate properties of genes as parts of the chromosome. At this stage in our discussions it will be helpful to paint in the broad outlines of the chemistry and organizational features of the chromosome. We will return to this topic in greater detail in later chapters.

## BOX 4.1 Prediction of Risk in Human Inheritance

Prediction of the risk, or chance, of a child's inheriting various single-gene defects present in a family is based on the expected ratios of phenotypes and genotypes in accordance with Mendelian rules of allelic segregation and assortment during reproduction. The specific risk predicted for any situation depends on four basic items of information: (1) the genotypes of the prospective parents, (2) the nature of the inheritance pattern, (3) the dominance and recessiveness of the alleles of the gene, and (4) whether the gene is located on the X chromosome or on an autosome.

In the case of single-gene inheritance of autosomal alleles, the risk is equal for both sons and daughters to inherit the allele causing the defect. The risk, or probability, of a child's inheriting an autosomal dominant allele is 50% if one parent is affected and 75% if both parents are affected. Because such alleles are relatively rare in populations, it is highly unlikely that either parent would be homozygous for the dominant allele causing the affliction. The inheritance of the dominant allele thus depends on whether one parent is heterozygous for the allele or both parents carry the allele. If one parent is heterozygous, the parental genotypes would be *Aa* × *aa*, and the child in each pregnancy has a 50% (½) chance of receiving the *A* allele from its affected parent. If both parents are affected, the child has a 75% (¾) chance of expressing the dominant phenotype (*Aa* × *Aa*). This outcome is based on an expected ratio of 3 dominant phenotype : 1 recessive phenotype in offspring whose parents are both heterozygous.

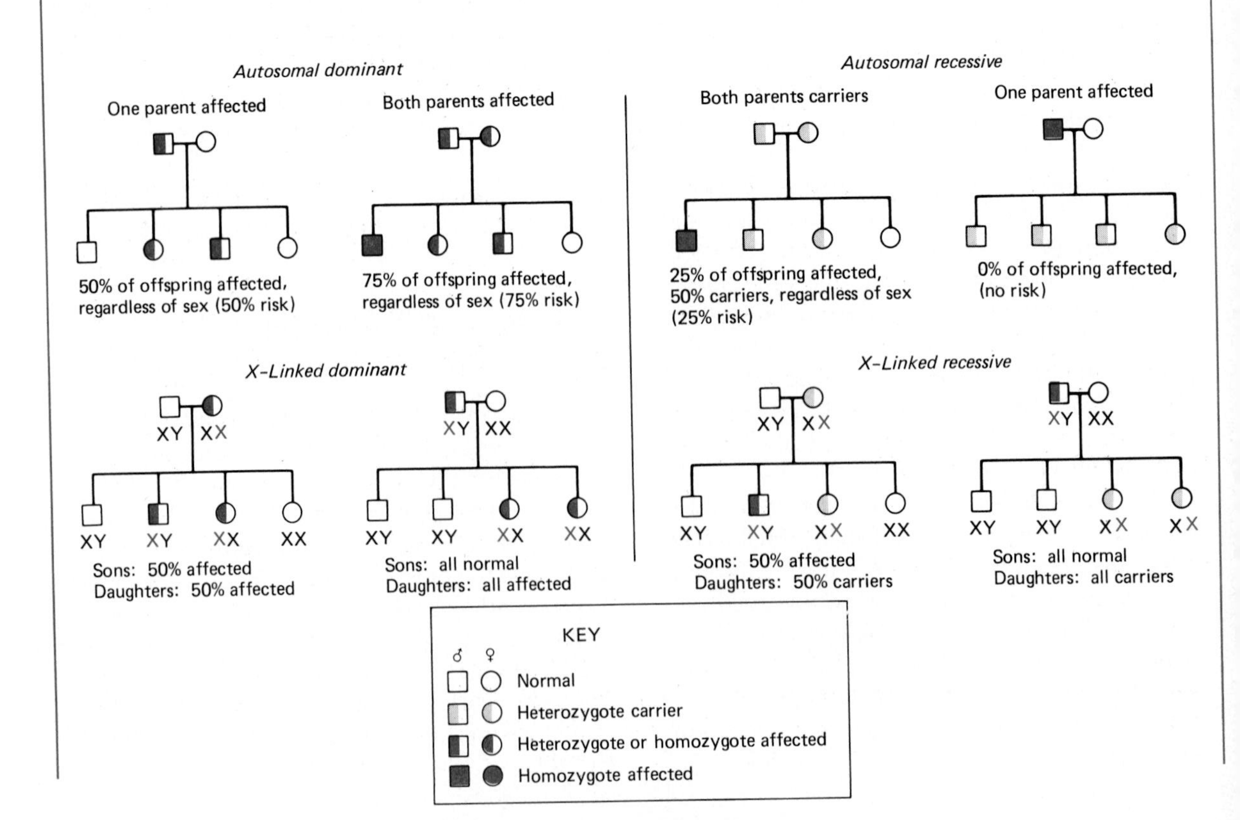

We can predict conditions due to autosomal recessive inheritance according to the genotypes of the parents involved. If both parents have the recessive phenotype (and genotype), the child in each pregnancy has a 100% risk of inheriting the defect (*aa* × *aa*). If both parents have the normal phenotype but are heterozygous for the allele, the child has a 25% risk of inheriting the recessive allele from each parent (¾ dominant phenotype : ¼ recessive phenotype). Since offspring inherit alleles from both parents, no risk would exist in families in which one or both of the parents were homozygous dominant in genotype.

X-linked inheritance involves different predictions of risk for sons and daughters since daughters receive an X chromosome from each parent, whereas sons receive their only X chromosome from their mother. By following the distribution of the X chromosomes during reproduction, we can predict risk in families in which the mother is a carrier, the father expresses the condition, or in which both parents carry the X-linked recessive allele. Since such conditions are relatively rare, the usual situation involves one parent with the recessive allele and the other parent lacking the recessive and carrying the dominant normal allele(s). Transmission is, therefore, from father to daughters and from mother to sons. The condition is expressed in hemizygous males, thereby accounting for the higher incidence of X-linked recessive conditions occurring in males than in females. Females are usually carriers of the recessive allele, which is not expressed when the normal dominant allele is also present in the female's genotype.

X-linked dominant inheritance also leads to different predictions of risk for sons and daughters, who are XY and XX, respectively. Daughters are at higher risk to inherit the dominant condition since they may receive an X chromosome carrying the dominant allele from either mother or father; in contrast, sons receive an X chromosome only from their mother. If the mother is affected, each son or daughter has a 50% chance of inheriting the mother's X chromosome that carries the dominant allele. If the father is affected, only daughters have a 50% chance of inheriting the X-linked dominant allele. Due to rarity of these alleles, it is unlikely that the mother would be homozygous or that both parents would carry the dominant allele.

## 4.7 Chemistry of the Chromosome

Chromosomes in most eukaryotes are rod-shaped structures that stain vividly and specifically with certain reagents. The stainable material was called **chromatin** by the early microscopists, and this term continues to be useful today. The unreplicated chromosome consists of a single **chromatin fiber,** which is actually one continuous, linear molecule of **DNA** (deoxyribonucleic acid) in association with basic **histone proteins,** acidic or neutral **nonhistone proteins,** minor amounts of **RNA** (ribonucleic acid), and a number of enzymes active in DNA and RNA synthesis. The chromatin fiber, therefore, is a high-molecular-weight, complex **nucleoprotein** fiber (Fig. 4.17). About 13% to 20% of the chromosome is DNA, and most of the remainder is protein. When the chromosome replicates, a new and identical chromatin fiber is synthesized so that each chromatid consists of one fiber carrying the same alleles as the original DNA before replication.

The linear integrity of the chromosme resides in its continuous linear DNA molecule. If digestive enzyme tests are conducted, we find that the only enzyme that causes the chromosome to break up into pieces is deoxyribonuclease (DNase). Enzymes that digest

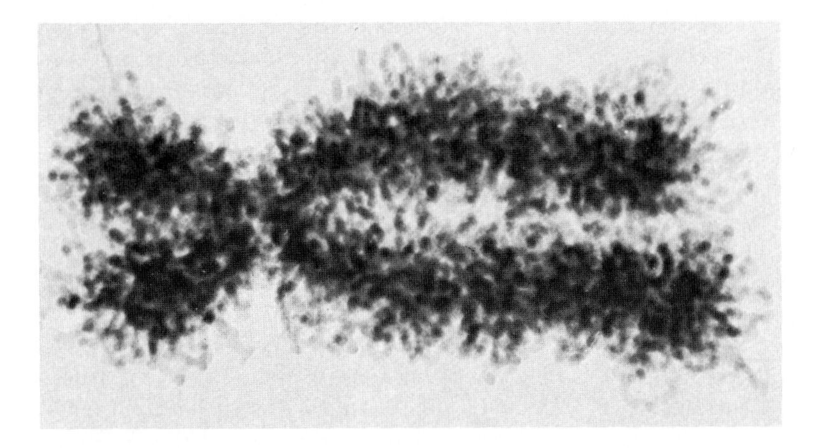

***Figure 4.17*** Electron micrograph of human chromosome 12. Each chromatid in the replicated metaphase chromosome is a single nucleoprotein, or chromatin, fiber. × 44,000. (From E.J. DuPraw. 1970. *DNA and Chromosomes.* New York: Holt, Rinehart, and Winston.)

proteins produce an eroded appearance of the chromosome, but the structure retains its original length. These general observations have been extended recently and have provided a basis for a detailed molecular analysis of chromosome structural organization. In current studies of chromosome structure, DNase is used to fragment the chromosome into small, repeating units called **nucleosomes.** Each nucleosome unit of a chromatin fiber consists of a particular length of DNA complexed with histones. These studies and their implications will be described in Chapter 8.

## 4.8 Functionally Different Parts of Chromosomes

Chromosomes are not just strings of genes lined up end to end; the structures are *differentiated* into regions with specific functions and morphology (Fig. 4.18). The centromere region, or primary constriction, is one such differentiation. Chromosomes cannot move directionally at anaphase of nuclear divisions unless a centromere is present. The centromere is a specific site for attachment of spindle fibers. Spindle fibers cannot be inserted if there is no centromere and no means by which the **acentric** (lacking a centromere) chromosome can move directionally toward the poles. If some rearrangement of chromosomes leads to a dicentric (two-centromere) chromosome and if each centromere is oriented toward a different pole, then

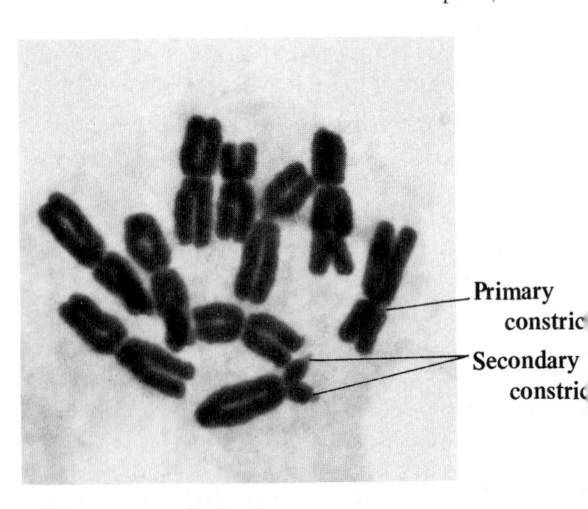


***Figure 4.18*** Ten of the sixteen chromosomes from onion root tip cell. There is an obvious primary constriction (centromere region) in each replicated metaphase chromosome. One chromosome in the haploid genome has a secondary constriction at its nucleolar-organizing region and the typical satellite knob extending beyond this region on each chromatid.

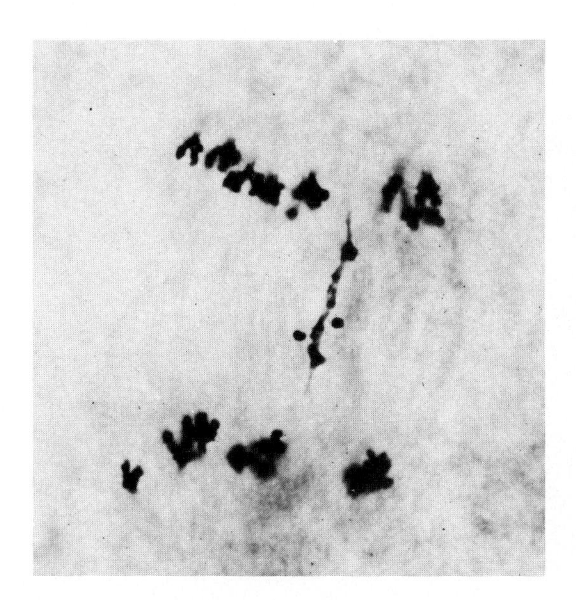

***Figure 4.19*** When the centromeres of a dicentric chromosome move to opposite poles, a bridge chromosome is produced at anaphase. Two acentric fragments lie alongside the anaphase bridge, and neither can move directionally since neither has a centromere. (Courtesy of M.M. Rhoades.)

a "bridge" chromosome is observed at anaphase (Fig. 4.19).

The centromere may be located anywhere along the length of a chromosome, even at the very tip. But each chromosome has its centromere in a fixed location, which is constant for that chromosome. Chromosomes in a particular set can be recognized according to various criteria, including their centromere location, their relative lengths, and the presence of **nucleolar-organizing regions (NOR)** in satellited chromosomes of the complement. Each haploid complement of chromosomes must have at least one nucleolar-organizing region at a fixed location on a **nucleolar-organizing,** or **NO-chromosome.**

The NO-chromosome has hundreds or thousands of repeated copies of the ribosomal RNA gene, arranged in tandem in the nucleolar-organizing region. There may be one or more NO-chromosomes in a species genome. Loss of the NOR by chromosome damage or by mutation leads to total absence of ribosomal RNA genes and is a lethal condition. Without ribosomal RNA no ribosomes can be made, and without ribosomes no proteins can be synthesized.

## 4.9 Heterochromatin and X Inactivation

For over fifty years we have been aware of two kinds of chromatin, which are distinguished by their behavior during interphase between nuclear divisions. The chromatin that is greatly extended in conformation is called **euchromatin;** in contrast, **heterochromatin** remains condensed during interphase (Fig. 4.20). Knowledge of these cytological features was supplemented by genetic studies and by biochemical analysis of DNA replication during a cell cycle. The genetic studies revealed that heterochromatin is a remarkably stable form of DNA. This discovery is based on mutation studies in which very few mutant alleles have been located in heterochromatic regions as compared with the great bulk of genes mapped in euchromatic parts of chromosomes. Biochemical analysis of DNA replication has shown that heterochromatin replicates late in the synthesis period of the cell cycle, whereas euchromatin replicates earlier in this period. Heterochromatin is therefore: (1) condensed during interphase, (2) genetically stable, and (3) late replicating.

Heterochromatin may be facultative or constitutive. **Facultative heterochromatin** contains active genes but may become condensed and genetically inactive in response to physiological and developmental conditions, and it may revert to a euchromatic state at certain times. **Constitutive heterochromatin** is permanently condensed, genetically stable, late-replicating material all of the time. In most of the species studied so far, the most common site for constitutive heterochromatin is around the centromere region of all chromosomes. This kind of stable chromatin is highly desirable in a region of the chromosome that is essential for move-

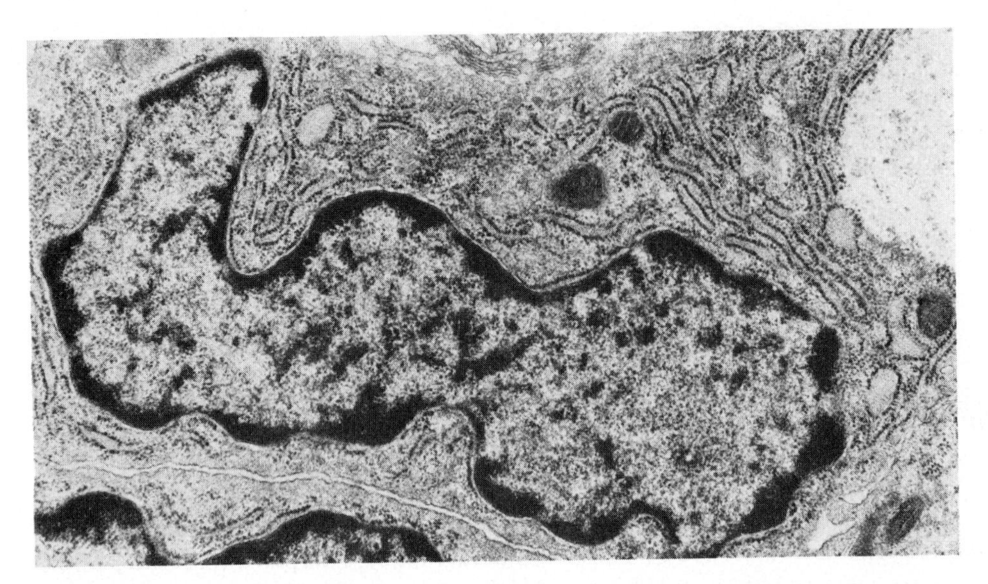

***Figure 4.20*** Electron micrograph of a thin section of rat osteoblast. The condensed heterochromatin is located mainly at the periphery, next to the nuclear envelope. Euchromatin is dispersed in the remainder of this interphase nucleus. × 24,000. (Courtesy of M. Federman.)

ment in the countless nuclear divisions during an individual's lifetime.

A well-known instance of facultative heterochromatin involves the mammalian X chromosome. The single X chromosome in males is almost entirely euchromatic. In females, one X chromosome remains largely euchromatic during the life of each cell, the second X chromosome becomes condensed heterochromatin during embryonic development. The heterochromatic X chromosome is visible as a dense blob in the interphase nucleus, whereas other chromosomes are not distinguishable (Fig. 4.21). This blob is called **sex chromatin,** or a **Barr body,** and it permits a simple test for the identification of biological sex in human beings and other mammals. Females have one Barr body per nucleus, this being the condensed second X chromosome. Normal males have no Barr body since their only X chromosome remains euchromatic throughout life.

In some patients with clinical symptoms that involve a sex-related characteristic, counts of Barr bodies have provided the starting point for a more detailed examination of the chromosome complement. Men with two X chromosomes (XXY) will have one Barr body in the nucleus, thus confirming the diagnosis of Klinefelter syndrome. As the number of X chromosomes, visible as Barr bodies, increases in a male patient, the more severe are the symptoms of Klinefelter syndrome and the greater the degree of mental retardation. Women with Turner syndrome (XO) have no Barr body since their only X remains euchromatic. The simple relationship is: one X chromosome + $n$ Barr

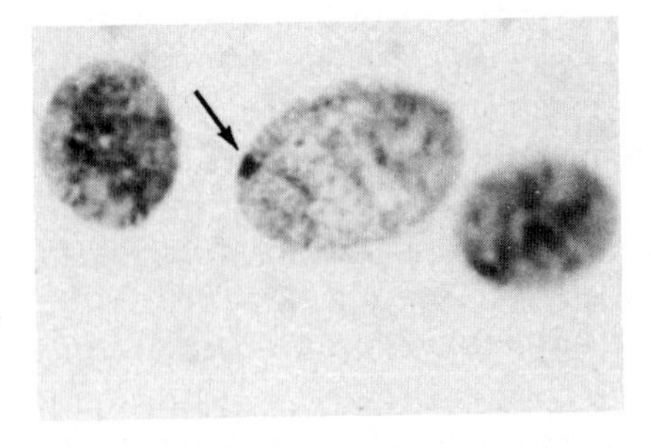

***Figure 4.21*** Barr body in the human female. One of the two X chromosomes remains condensed during interphase and is seen here at the periphery of one nucleus (arrow). The condensed X chromosome is called a Barr body, or sex chromatin. (Courtesy of T.G. Tegenkamp.)

bodies = the total number of X chromosomes in the nucleus. Only the euchromatic X is active and genetically functional.

The Y chromosome in mammals seems to be concerned primarily with sex determination. The few genes that have been postulated for the human Y chromosome seem to function only during early embryonic development. One or more of these genes contribute to differentiation of the unspecified embryonic gonads into testes. In *Drosophila*, Y-linked genes function in sperm differentiation into swimming, active gametes. Male XO flies are sterile because they produce nonmotile sperm. The X chromosome, on the other hand, carries a large number of genes needed for normal development and activities in both sexes. Total absence of X chromosomes is presumed to be lethal at very early stages in development since neither mammalian nor *Drosophila* embryos or individuals have ever been found with no X chromosomes at all. Most X-linked genes have no sex-related functions.

Since only one of the two X chromosomes remains functionally euchromatic in females, there is an equivalence of *active* copies of X-linked genes in both sexes. But which of the two X chromosomes remains genetically active in females? Is it one particular X or either X chromosome? If either X (from the mother or the father) is inactivated while the partner X remains functional, is **X inactivation** a random (either maternal or paternal) or nonrandom (only maternal or only paternal) event?

Mary Lyon first suggested and investigated the pattern of X inactivation with genetically marked mice. Thus X inactivation is often referred to as **Lyonization.** Through her analysis of mouse coat color patterns produced in heterozygous females carrying these X-linked coat color alleles, she found that *either* X could be inactivated at random. These heterozygous females developed variegated (patchy) color patterns, thus indicating that some somatic cell lineages expressed the normal allele and other lineages in the same animal expressed the mutant allele. The mechanism of X inactivation, however, remains to be discovered.

The same situation characterizes other mammalian species, as we can see from examples of two other species. In cats an X-linked gene governs black or orange fur color. Males are either black or orange, but heterozygous females develop the familiar calico (tortoiseshell) pattern of black and orange variegation. The black patches develop from cells in which the X carrying the orange allele is inactivated, and orange patches arise when the other X chromosome is inactivated in a heterozygote. On rare occasions, a male may be calico, but he invariably turns out to be XXY in sex chromosome constitution.

The consequences of random inactivation of X chromosomes can also be observed in human beings. We can assay skin biopsies from females who are heterozygous for X-linked alleles that govern synthesis of identifiably different proteins. Women who are heterozygous for the X-linked gene governing synthesis of the enzyme *glucose 6-phosphate dehydrogenase* (GPD) have two kinds of skin cells that contain either the normal or the altered enzyme but never both kinds in the same cell. These cells are cloned in culture and each clone breeds true for one or the other of the two protein types. Whether the inactivated X came from the mother or the father makes no difference.

In human beings, X inactivation is believed to occur at about the sixteenth day of embryonic development, according to Barr body observations. Differentiation of ovaries and other reproductive structures begins at about the twelfth week in a female fetus. According to various studies, fetal development is similar in XO and XX females, since ovarian structures are present in both the Turner and the normal female fetus. Apparently, in a Turner female, internal reproductive structures degenerate during development. This leads to almost total absence of internal reproductive structures in adult Turner women. The causes of these changes during development are uncertain.

X inactivation occurs very early in mouse embryos also, perhaps in the 10-to-60-cell stage of the early blastocyst. Early X inactivation is a very likely characteristic of mammalian female embryos in general.

Whole cell lineages, seen as patches of identical tissue, retain their allelic distinction throughout the life of the female. The inactivated X is therefore inherited through mitosis, and each mitotic descendant of a particular cell has the same inactivated X chromosome. *Different* cell lineages, however, may have different inactivated X chromosomes since inactivation is a random event (Fig. 4.22).

The mammalian X chromosome is facultatively heterochromatic since the condensed chromosome is restored to the euchromatic state in the egg or in cells giving rise to the egg. This must be the case because both X chromosomes, the one from the egg and the one from the sperm, are euchromatic during the first sixteen days of human embryo development. In addition, the fact that X inactivation is random in the embryo also shows that the two X chromosomes are initially euchromatic at this early stage and that either one of these becomes heterochromatic later on.

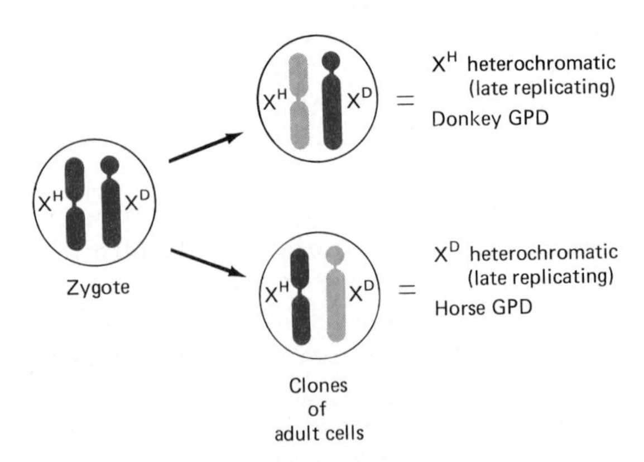


***Figure 4.22*** The X chromosome of horse and donkey are morphologically distinct, and each species produces distinctive glucose 6-phosphate dehydrogenase (GPD) protein, an X-linked trait. Female hybrids (mules) give rise to clones of body cells, about half of which show a late-replicating $X^{horse}$ and donkey GPD, and half of which show a late-replicating $X^{donkey}$ and horse GPD. These observations conform to predictions for the randomness of X inactivation and support the theory that only the euchromatic X chromosome is genetically active in the XX cell.

Although we still have much to learn about the processes of X inactivation and reactivation, we can be reasonably sure that these processes have little or nothing to do with the sex of the individual. X inactivation occurs in both males and females who have two or more X chromosomes, it does not take place in males or females with one X chromosome.

## The Chromosome Complement

Chromosomes can be identified by their morphological features from metaphase preparations of ordinary body cells, or somatic cells. Chromosome length, relative arm lengths as a function of centromere location, and unique features such as nucleolar organizers all contribute to specific identifications. The development of staining methods that produce unique band and interband patterns in somatic chromosomes provided the means to recognize each individual chromosome in a complement, or set of chromosomes, regardless of their otherwise similar morphologies.

### 4.10 Chromosome Morphology and the Karyotype

The ordered arrangement of all the somatic chromosomes in the nucleus is called a **karyotype** (Fig. 4.23). The usual preparations of metaphase nuclei are obtained from cells in culture or from actively dividing tissues such as root tips in plants. Human karyotype studies are usually made from lymphocytes (a type of blood cell) or fibroblasts (a type of connective tissue cell) maintained or established in culture. Metaphase chromosomes are the most condensed of any division stage, and it is more likely that the chromosomes will be well spread out and easier to separate and identify than in division stages, where they are longer and more tangled.

Cells in culture are stimulated to undergo mitosis in preparation for karyotyping, and the

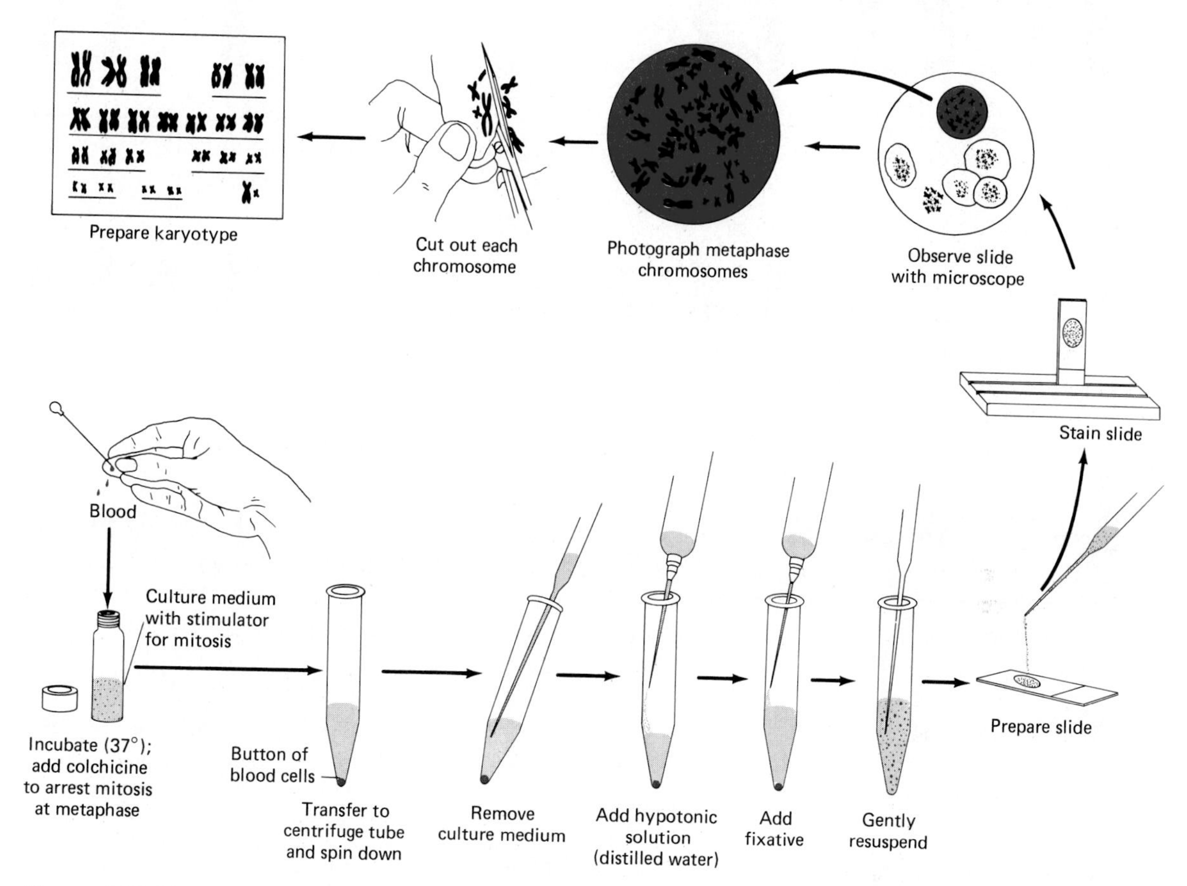


***Figure 4.23*** The procedures for karyotype preparation.

drug *colchicine* is added at the proper time to arrest mitotic nuclei in the metaphase stage. Well-spread metaphase chromosomes in arrested nuclei are stained, photographed, and then cut out of photographs. They are positioned according to an established convention, proceeding from the largest to the smallest chromosomes, with their centromeres aligned to emphasize differences in chromosome length and relative arm lengths of each chromosome.

Chromosomes of the human karyotype were first grouped into seven different classes, labeled A to G, according to chromosome length and centromere position. The three largest chromosomes, of group A, are **metacentric,** that is, they have a median centromere location and two equal-length arms (Fig. 4.24). The B and C groups of chromosomes are **submetacentric,** that is, they have one arm slightly longer than the other because of centromere location. B chromosomes are larger than C chromosomes in absolute length. Groups D and G contain the **acrocentric** chromosomes, which have one arm that is considerably longer than the other because the centromere is near one end of the chromosome. Group E consists of three small submetacentrics, and group F has two small metacentrics. While metacentric, submetacentric, and acrocentric types are represented, the human chromosome complement contains no **telocentric** type, that is, one with a terminal centromere and only a single chromosome arm. This kind of chromosome is found in some species, but it is not common. The X and Y

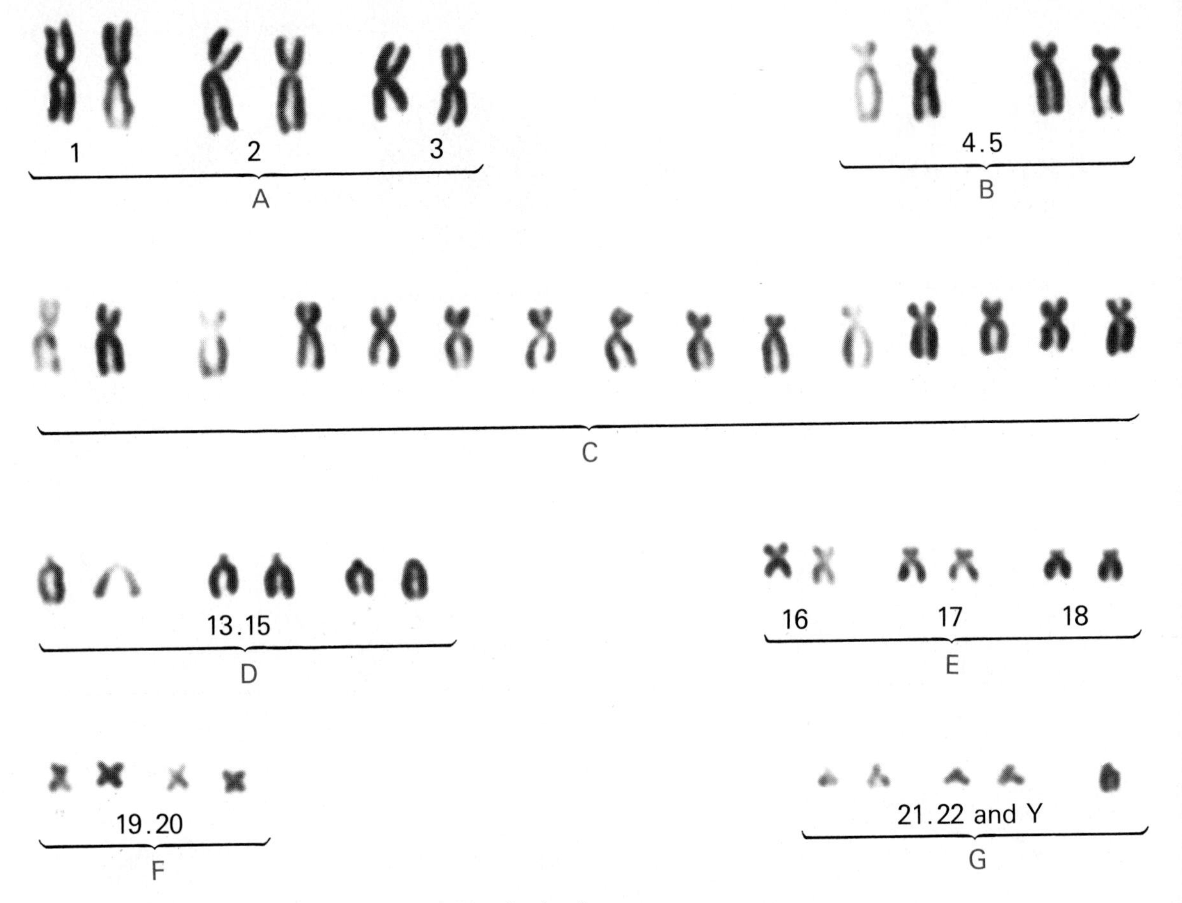


***Figure 4.24*** Karyotype of human male. Chromosomes are arranged into 7 groups (A–G) according to size and centromere location. Group A includes large metacentrics; B, large submetacentrics; C, intermediate-size submetacentrics; D, acrocentrics (and nucleolar organizing); E, smallest submetacentrics; F, smallest metacentrics; and G, smallest acrocentrics (nucleolar organizing). The X and Y chromosomes can be put into groups C and G, respectively, on the basis of morphology.

chromosomes in the human complement fall into the C and G chromosome groups, respectively, on the basis of their size and centromere location.

The five nucleolar-organizing chromosomes in the human complement include all three D-group chromosomes and both G-group chromosomes. These five NO-chromosomes are acrocentric and have satellites terminating the short arm. Satellite knobs frequently, but not always, serve as indicators of nucleolar-organizing regions of chromosomes. The satellite is simply a tiny extension of the chromosome beyond the organizer region.

Karyotype analysis of human and other mammalian species was difficult to conduct before 1956. In that year, J. Tjio and A. Levan reported their new procedure to obtain human mitotic cells with well-spread-out metaphase chromosomes. They reported the correct number of human chromosomes in diploid nuclei to be 46, as opposed to the incorrect number of 48, which had persisted in the literature for many years. Tjio and Levan's method for preparation of somatic metaphase chromosomes is used with modifications today. It opened a new era of analysis of mammalian chromosomes that simply had not been possible before.

## 4.11 Chromosome Banding

In 1969, T.C. Hsu and others introduced new methods for staining chromosomes by which distinct patterns of stained bands and lightly stained interbands became evident (Fig. 4.25). These staining methods were enormously important since they permitted each chromosome to be identified uniquely, even if the overall morphology was identical. Distinctions can now be made among the relatively similar C-group chromosomes, for example, so that we may refer to chromosome 9 or chromosome 12 instead of merely a C-group chromosome. Using the group reference is still convenient in many instances, but we now more often refer to a specific chromosome by its number as set by the karyotype conventional ordering.

The most useful chromosome banding method is **G-banding.** The earlier methods in-

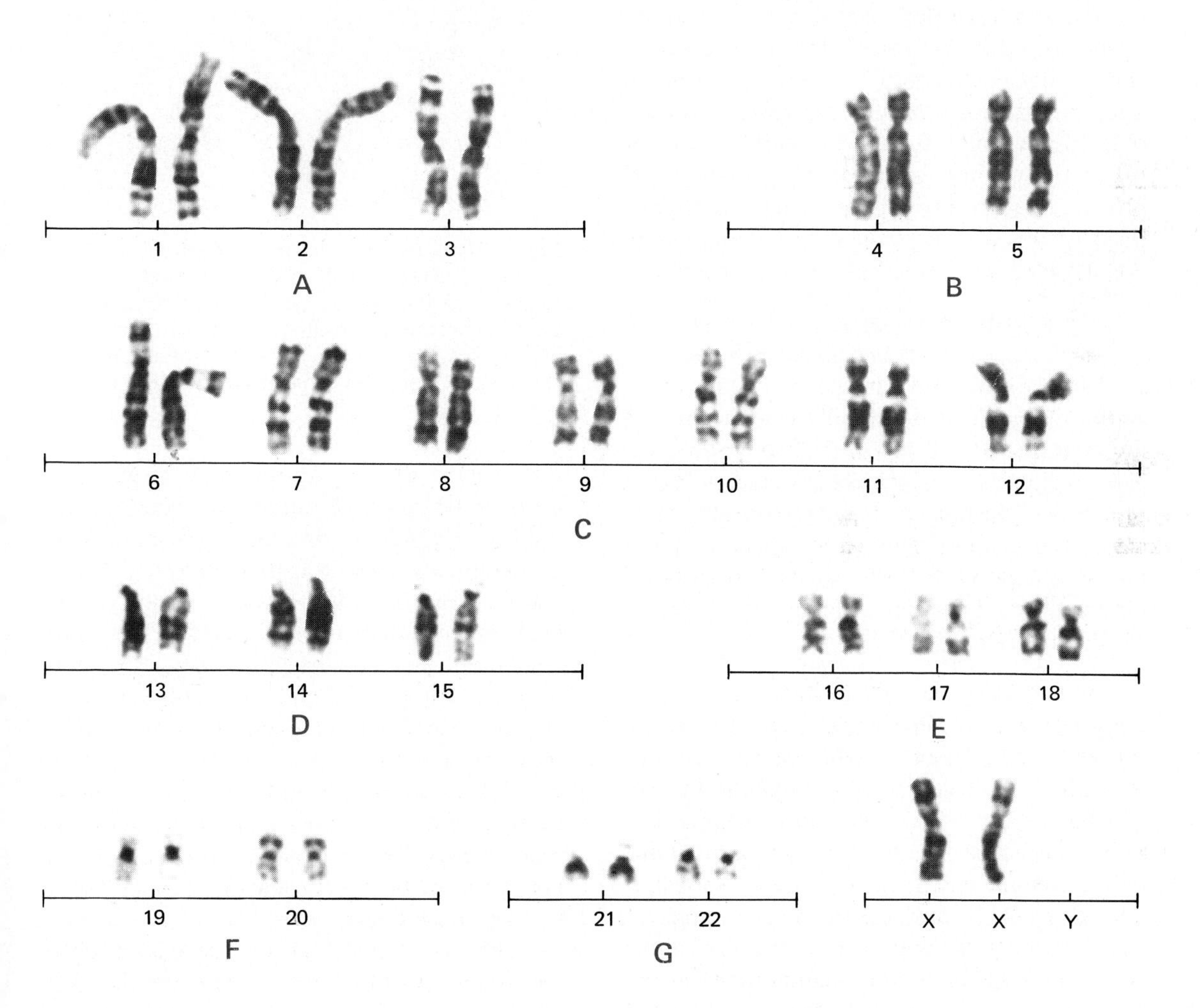


***Figure 4.25*** Karyotype of human female showing G-band staining patterns that establish the unique identity of each chromosome in the complement.

cluded some special steps in the staining procedure, but these were found to be unnecessary and no special conditions are really needed to visualize G-bands by staining with the Giemsa reagent. Giemsa staining had been used for many years to bring out contrast in nuclear material, and it was one of the first stains to delineate the bacterial nucleoid by microscopy.

Two main categories of chromosome banding patterns are recognized:

*1.* **G-bands** from Giemsa staining and **Q-bands,** which develop after staining with quinacrine and other fluorescent dyes, give relatively similar, but not identical, patterns. Fluorescent stains fade after a short time, and special microscope optics plus ultraviolet illumination are needed to see fluorescent bands. Giemsa-stained preparations are more permanent and require ordinary microscope optics and illumination. For these reasons, G-staining is used routinely.

*2.* **C-bands** are visualized by Giemsa staining after pretreatments using HCl and NaOH to partially denature the chromosomes in a preparation. C-bands are especially evident around the centromere and in other chromosome regions that contain substantial amounts of highly repetitive constitutive heterochromatin (Fig. 4.26). The Giemsa stain is not specific, but it binds to regions of DNA that have responded differently from nonbanded regions to HCl and NaOH pretreatments.

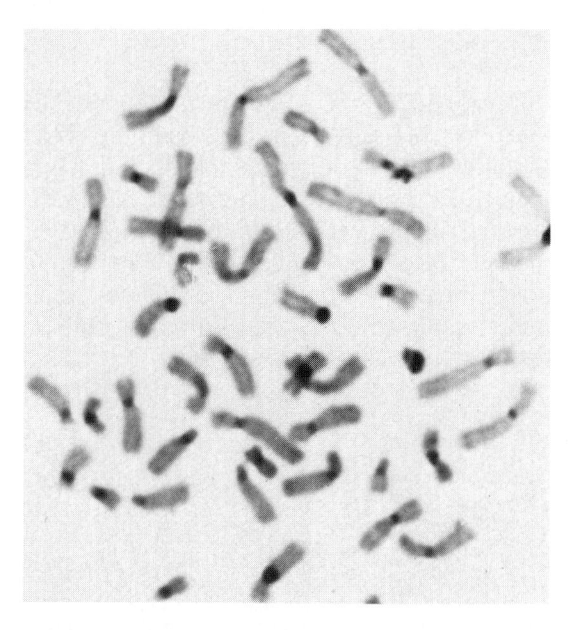

***Figure 4.26*** Human chromosome complement showing C-bands, mainly around the centromere region of each chromosome. This pattern reflects regions containing constitutive heterochromatin, which is known to occur in the centromere region and various other chromosome parts in most species.

Despite many attempts to interpret banding reactions on a molecular basis, we still know relatively little about specific interactions between DNA and any staining reagents in use. Q-bands apparently result from binding between quinacrine dye and DNA regions that are rich in adenine and thymine. Since guanine and cytosine quench fluorescence, GC-rich regions of DNA generally appear as unstained interbands. Since Q-bands and G-bands are relatively similar, it seems likely that a common mechanism of interaction exists between DNA and the dye. So far, this has not been shown to be true, and G-banding mechanisms remain unclear.

C-bands are very distinctively located and arise as the result of binding between Giemsa stain and residual chromatin remaining after pretreatments that extract nucleoproteins. More chromatin remains in areas of constitutive heterochromatin than in other parts of the chromosomes after denaturing steps, so more material is present to bind more of the stain and yield a band that is contrasted with lightly stained or unstained regions in between. (There is little or no nucleoprotein extraction in G- or Q-banding methods.) Locations of C-bands correlate very well with localizations of constitutive heterochromatin, according to several independent lines of evidence.

With new methods for G-banding chromosomes, investigators can unequivocally identify various structural and numerical changes in a normal complement. The greater certainty of identifying whole chromosomes or parts of chromosomes by G-bands often allows the

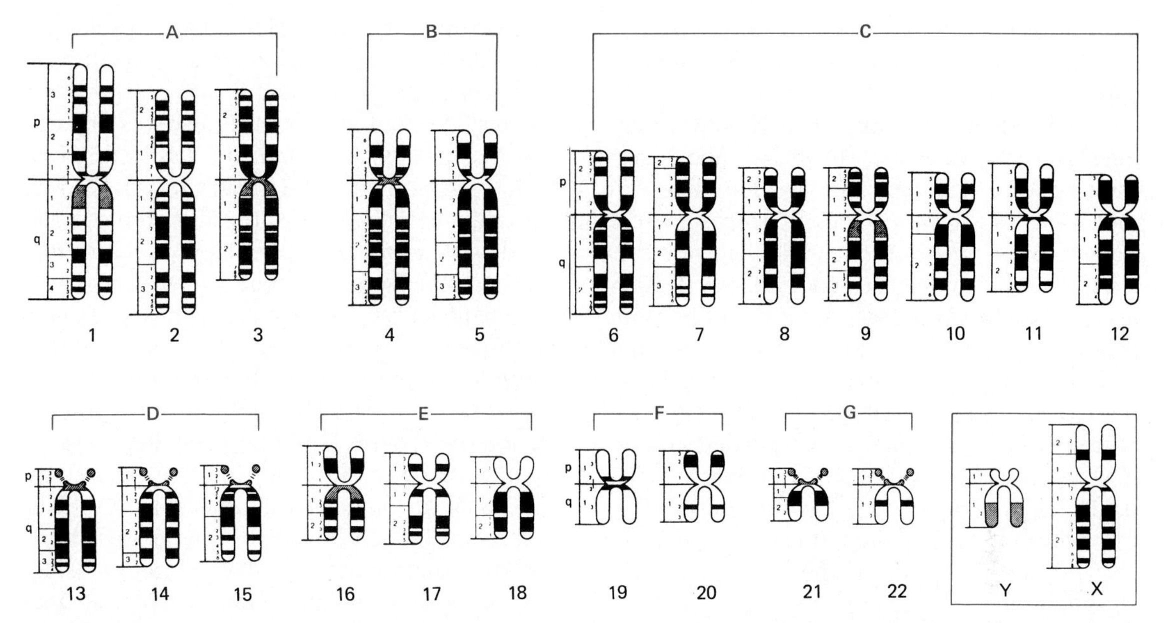


***Figure 4.27*** Standardized representation of the human complement of 22 autosomes and the X and Y sex chromosomes. Chromosomes are arranged in order of decreasing size and by position of the centromere (median, submedian, or subterminal) into groups A-G. Each chromonome, however, can now be identified unambiguously according to its banding pattern after G-staining. The nucleolar-organizing chromosomes 13, 14, 15, 21, and 22, are all shown with typical satellite knobs adjacent to the nucleolar-organizing region (secondary constriction). (From *Paris Conference (1971): Standardization in Human Cytogenetics.* Birth Defects: Original Article Series 8, No. 7, 1972. New York: The National Foundation.)

investigator to know exactly which chromosomes are present and which chromosome parts have undergone structural rearrangements (Fig. 4.27). Banding also provides a means to compare karyotypes of related species and to describe differences that apparently have an evolutionary basis.

## 4.12 Polytene Chromosomes

Typical somatic cell chromosomes have one chromatin fiber. The fiber doubles during replication and the replicated strands are distributed to daughter nuclei by mitosis. In various plant, animal, and protozoan species, giant multi-stranded or **polytene chromosomes** may develop instead of typical single-stranded chromosomes. In these nuclei, chromatin fibers replicate but are not separated into different nuclei since mitosis does not take place. Hundreds or thousands of replicated strands may remain associated in individual chromosomes. If a chromosome undergoes 10 rounds of replication, $2^{10}$ strands are produced. Such a polytene chromosome therefore exists as a unit made up of 1024 chromatin fibers. Their giant size is due in part to the numbers of chromatin fibers present and in part to the extended configuration of the chromosomes, which is typical of interphase nuclei. Interphase is the most actively metabolic stage of the cell cycle. Polytene chromosomes lend themselves very well to studies of chromosomal synthesis activities that can be observed cytologically by microscopy as well as by biochemical methods. Such activities, including synthesis of nucleic acids and proteins, can be analyzed for each chromosome individually. Each chromosome of the complement has a unique pattern of bands and interbands, and each can be identified easily by its unique pattern (Fig. 4.28).

The bands appear darker than the alternating interbands because the chromatin fibers are folded more tightly in band regions than in interbands. We can see these patterns in unstained chromosomes, but staining heightens the contrast. Combined genetic and cytological studies, or **cytogenetics,** have shown that the band patterns are constant for each chromosome and that bands can serve as markers indicating locations of particular genes on the chromosome. Studies of banded polytene chromosomes using microscopy can, therefore, provide information on the genes located on these chromosomes. We can interpret bands and band patterns for each chromosome in terms of the genes on each chromosome.

Wolfgang Beermann used larval polytene chromosome analysis in the midge *Chironomus* to determine whether differentiated larval cells with different functions and appearance all possessed the same complete set of genes of the species. He examined banding patterns of each chromosome of the complement in cells from various larval tissues and organs. If each kind of cell in the individual had the same chromosomal banding patterns, then each kind of cell in the individual possessed a complete set of genes, despite these cells being different in appearance and function. In every cell type studied, Beermann found that the complete set of bands and interbands for each chromosome was present. He interpreted this finding to mean that, despite their phenotypic differences, all the cells of an individual are genotypically the same. Cellular differentiation in these organisms was, therefore, not the consequence of different genes being lost or retained in different cell types: all the genes were present. In later studies, Beermann and others showed by similar cytogenetic meth-

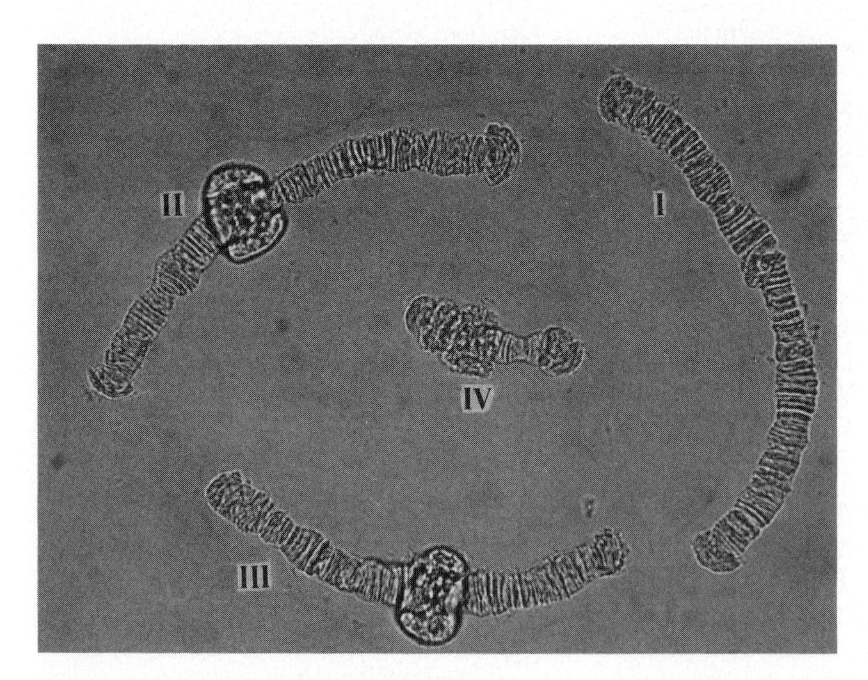


***Figure 4.28*** Photograph taken with the phase contrast light microscope of the chromosome complement of the midge *Chironomus tentans.* The unstained chromosomes are identified by roman numerals. Each chromosome is recognizable by its morphology and band pattern. The 8 polytene chromosomes are very closely paired in this diploid cell and give the impression of only 4 chromosomes. × 375. (Courtesy of B. Daneholt.)

ods that phenotypic differences among cells arose as the consequence of *differential gene action*. Some genes were turned on and others were turned off in different cells. We will discuss this important topic at length in Chapter 10.

From this survey of the nature of chromosomes and of the chromosomal basis of inheritance, it is clear that genetic studies are firmly based in the biological framework of the cell. Mendel's abstract inheritance factors were open to many kinds of studies using genetic, cytological, molecular, and biochemical methods. With the full scope of biological techniques, genes can be studied as physical entities, just like other parts of the cell.

## Questions and Problems

***4.1*** In *Drosophila* vermilion eye color ($v$) and curved wings ($c$) are recessive to their wild-type alternatives. A vermilion ♀ was crossed with a curved wing ♂, producing the following progeny: ♀♀—½ wild type, ½ curved wing; ♂♂—½ vermilion, ½ vermilion, curved wing.

***a.*** What was the genotype of the female parent?

***b.*** What was the genotype of the male parent?

***c.*** What are the genotypes of the male and female progeny?

***4.2*** Red-green color blindness in humans is recessive and X-linked. A normal woman whose mother was color-blind marries a color-blind man. They produce a son and a daughter.

***a.*** What is the probability that the son is color-blind?

***b.*** What is the probability that the daughter is color-blind?

***c.*** What is the probability that both children are color-blind?

***4.3*** Sex determination in the cockroach is based on a balance between the number of X chromosomes to autosome sets and the X:A ratio of 1.0 leads to female development. The somatic cells of a normal cockroach are examined and found to contain 23 chromosomes.

***a.*** What is the sex of this individual?

***b.*** What is the diploid number of the opposite sex?

***c.*** What is the usual chromosome constitution of male and female cockroaches?

***4.4*** In the dioecious plant *Melandrium albus* a recessive allele of an X-linked gene ($l$) is known to be lethal when homozygous in females. When present in the hemizygous condition in males, it produces patches of yellow-green color. When females are homozygous or heterozygous for the wild-type allele ($L$) or males are hemizygous for this allele, the plant develops the normal dark green color. In this species, females are XX and males are XY. Determine the expected genotypes and phenotypes in the progeny from the following crosses:

***a.*** heterozygous females × yellow-green males

***b.*** heterozygous females × dark green males

***4.5*** Females are XY and males are XX in birds. In chickens a dominant allele of an X-linked gene ($B$) produces barred plumage and the recessive allele ($b$) leads to nonbarred plumage. Removal of the ovary leads to the development of testes in the animal, and such a "male" can produce sperm. Such a "male" with nonbarred plumage is mated to a barred female.

***a.*** What sex ratio will occur in their progeny?

***b.*** What are the genotypes and phenotypes of the progeny?

***4.6*** Suppose we have two homozygous strains of *Drosophila,* one found in San Francisco (strain A) and the other in Los Angeles (strain B). Both strains have bright scarlet eyes, whereas wild-type flies have red eyes.

***a.*** When strain A ♂♂ are crossed with strain B ♀♀ you obtain 100 wild-type ♂♂ and 100 wild-type ♀♀ in the $F_1$ generation. From this result, what can you say about the inheritance of the eye color in the two strains?

***b.*** When strain B ♂♂ are crossed with strain A ♀♀ you obtain 98 scarlet-eyed ♂♂ and 100 wild-type ♀♀ in the $F_1$ generation. What can you say about the inheritance of eye color from this result?

***c.*** When you cross members of the $F_1$ progeny of part (a) you obtain the following in the $F_2$: 76 wild-type ♀♀, 24 scarlet ♀♀, 63 scarlet ♂♂, 37 wild-type ♂♂. Diagram the genotypes of the $F_1$ offspring and indicate the expected ratio of $F_2$ phenotypes for each sex.

***4.7*** The following three pedigrees represent a particular family segregating for three different traits: color blindness, glucose 6-phosphate dehydrogenase (GPD) deficiency, and XG blood group system. Use the following gene symbols: *C* for normal vision, *c* for color blindness, *Gpd* for normal GPD enzyme levels, *gpd* for GPD deficiency, *Xg* for presence of XG blood group, *xg* for absence of XG blood group.

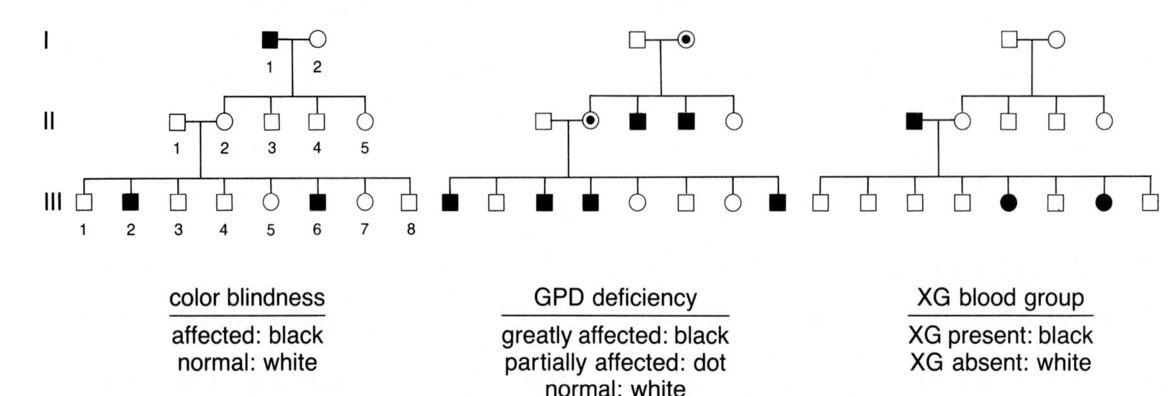


***a.*** Determine the mode of inheritance of each trait.
***b.*** Give the most probable genotypes of individuals I-1, I-2, II-1, II-2, III-1, and III-5, for all three genes.

***4.8*** The following pedigree is concerned with night blindness:

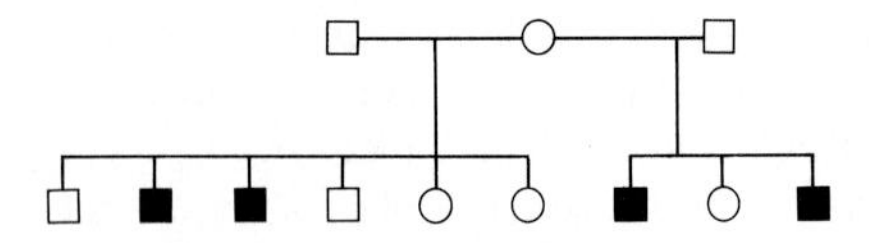

***a.*** What mode of inheritance best accounts for the transmission of this trait?
***b.*** Determine the genotypes of the members of this family according to your hypothesis.

***4.9*** The following human pedigree represents a family segregating for brachydactyly (shortened fingers) and a dental abnormality called *amelogenesis imperfecta.* Individuals with brachydactyly are indicated by a dot, and those with the dental abnormality are shown by filled squares; normal individuals are indicated by open and undotted squares and circles.

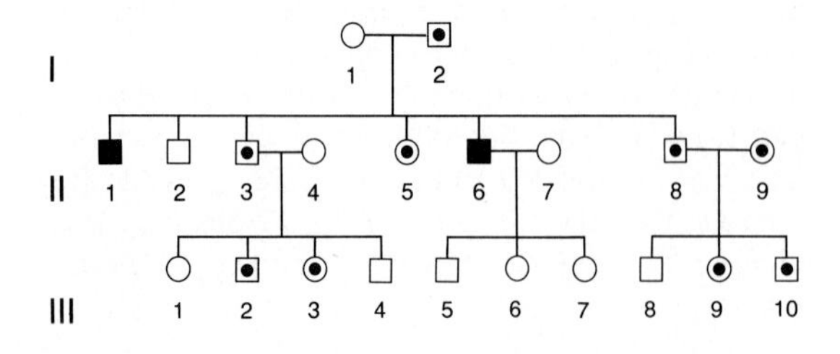


***a.*** What can you tell about the inheritance of brachydactyly?
***b.*** What can you tell about the inheritance of the dental disease?
***c.*** What is the probability that a child of III-1 will have shortened fingers or the dental disease?
***d.*** What are the genotypes of individuals I-1 and I-2?

***4.10*** Cleft lip (incomplete fusion of upper lip) in humans appears to be dominant in men and recessive in women.
***a.*** What proportion of the sons of two heterozygous parents should be expected to have cleft lip?
***b.*** What proportion of all their children should be expected to have this trait?

***4.11*** A young couple have had a hemophilic son and want to know the risk of having a second hemophilic child should they plan to enlarge their family. The genetic counselor tells them there is no risk of having a hemophilic daughter but there is a predictable risk for their producing a second hemophilic son.
***a.*** What are the genotypes of the two parents?
***b.*** What is the probability that their second child will be a hemophiliac?
***c.*** What are their chances of having two afflicted sons in a family of only two children?

***4.12*** Two cats mate and produce a litter of 8 kittens, consisting of 3 calico ♀♀, 2 orange ♀♀, 2 orange ♂♂, and 1 black ♂.

***a.*** What are the genotypes and phenotypes of the parents?
***b.*** Suppose one of the male kittens proved to be XXY. Would it be orange or black if nondisjunction occurred at Meiosis I? Diagram the possible sex chromosomes present in gametes that could produce such an XXY ♂ at fertilization if nondisjunction occurred at Meiosis I.
***4.13*** A color-blind woman and a man with normal vision have a color-blind son who is 47,XXY and shows characteristics of Klinefelter syndrome.
***a.*** What sex chromosomes were present in the egg and sperm that fused to produce their son?
***b.*** If Barr body counts are made for both parents and their son, would the son's cells resemble those from his mother or those from his father?
***c.*** If these two people have another son of the normal 46,XY chromosome constitution, what is the probability that he would be color-blind?
***4.14*** A color-blind woman of blood type O is suing a man with normal vision and blood type AB in a paternity suit over her color-blind daughter whose blood type is A. Color blindness is a recessive X-linked trait, and the major blood groups are governed by a multiple allelic series of a gene on chromosome 9.
***a.*** Can this man be the father of her child? Explain.
***b.*** If the woman had a son rather than a daughter, would that change your interpretation in this case? Explain.
***4.15*** Three women and two men had cell samples taken for karyotype analysis and Barr body counts. Each of the women had a different chromosome number, and each of the two men had a different chromosome number from the other. The slides showed that two of these people had no Barr bodies, two had one Barr body, and one person had two Barr bodies per cell. Match up the Barr body counts, chromosome counts, and sex for each of these five people.
***4.16*** How would you determine each of the following features of a chromosome complement:
***a.*** Identification of each submetacentric chromosome in a group of approximately equal length chromosomes?
***b.*** Location of constitutive heterochromatin in each chromosome?
***c.*** Identification of nucleolar-organizing chromosomes?
***d.*** Identification of all the acrocentric chromosomes?
***4.17*** Define each of the following terms: ***a.*** centromere ***b.*** nucleolar-organizing region ***c.*** heterochromatin ***d.*** euchromatin ***e.*** sex chromatin ***f.*** X inactivation ***g.*** karyotype ***h.*** polytene chromosome

## References


Allan, G.E. 1979. *Thomas Hunt Morgan: The Man and His Science.* Princeton, N.J.: Princeton University Press.

Bergsma, D., ed. 1972. Paris conference (1971): Standardization in human cytogenetics. *Birth Defects, Original Article Series* **8**(7):1.

Brown, S.W. 1966. Heterochromatin. *Science* **151**:417.

Drets, M.E., and M.W. Shaw. 1971. Specific banding patterns of human chromosomes. *Proc. Nat. Acad. Sci. U.S.* **68**:2073.

Ephrussi, B., and M.C. Weiss. Apr. 1969. Hybrid somatic cells. *Sci. Amer.* **220**:26.

Gerald, P.S. 1976. Sex chromosome disorders. *New Eng. J. Med.* **294**:706.

McKusick, V.A. Aug. 1965. The royal hemophilia. *Sci. Amer.* **213**:88.

McKusick, V.A. 1969. *Human Genetics,* 2nd ed. Englewood Cliffs, N.J.: Prentice-Hall.

Mohandas, T., R.S. Sparkes, and L.J. Shapiro. 1981. Reactivation of an inactive human X chromosome: Evidence for X inactivation by DNA methylation. *Science* **211**:393.

Silvers, W.K., and S.S. Wachtel. 1977. H-Y antigen: Behavior and function. *Science* **195**:956.

Simpson, E. 1982. Sex reversal and sex determination. *Nature* **300**:404.

Sturtevant, A.H. 1913. The linear arrangement of six sex-linked factors in *Drosophila,* as shown by their mode of association. *J. Exp. Zool.* **14**:43.

Sutton, W.S. 1903. The chromosomes in heredity. *Biol. Bull.* **4**:231.

Tjio, J.H., and A. Levan. 1956. The chromosome number of man. *Hereditas* **42**:1.

Wachtel, S.S. 1977. H-Y antigen and the genetics of sex determination. *Science* **198**:797.

# CHAPTER 5

## Linkage, Recombination, and Mapping in Eukaryotes

Genes are located on chromosomes according to patterns of gene transmission seen in genetic analysis. The phenotypic ratios of progeny generations are predictable on the basis of chance segregation and independent assortment of genes on different chromosomes. There are many more genes than chromosomes, however, and we would predict genes on the same chromosome to be transmitted according to rules other than those that involve totally independent events. Studies of genes on the same chromosome have also been interpreted according to patterns of gene transmission to progeny generations. These gene transmission studies have led to the mapping of genes on chromosomes, and to the development of a more comprehensive picture of the chromosomal basis of inheritance.

## Linkage Analysis

Through breeding studies we can distinguish between genes that assort independently into new combinations of alleles and genes that have a tendency to be transmitted together more often than to be transmitted separately into new combinations. Genes that are transmitted together more often than they are transmitted separately are called **linked genes,** and they are situated on the same chromosome. When linked genes do separate they lead to recombinant progeny whose genotypes consist of a new combination, or **recombination,** of alleles from the two parents. The parallel relationship between patterns of transmission of alleles from parents to progeny and behavior of chromosomes during reproduction was shown to hold true for linked genes as much as for independently assorting genes on different chromosomes. These consistencies provided additional support for the Chromosome Theory of Heredity.

### 5.1 Linkage versus Independent Assortment

Apparent exceptions to the Mendelian Law of Independent Assortment were discovered very soon after 1900. William Bateson and R.C. Punnett were among the first geneticists to report such exceptions in several species, including the same garden peas that Mendel had studied.

In monohybrid crosses between *purple-* and *red-*flowered peas, Bateson and Punnett found the expected ratio of 3 purple:1 red in the $F_2$. Similarly, they found a 3:1 phenotypic ratio in $F_2$ progeny of crosses between plants with *long* pollen grains and others with *round* pollen. Dihybrid crosses, however, did not produce the expected ratio of 9:3:3:1 phenotypic classes in the $F_2$ generation. Instead the results were as follows:

| | | |
|---|---|---|
| P | *PPLL* × *ppll* | |
| | purple, long | red, round |
| $F_1$ | all *P–L–* (purple, long) | |

| $F_2$ | **progeny class** | **phenotype** | **number of plants observed** | **number of plants expected (9:3:3:1 ratio)** |
|---|---|---|---|---|
| | parental | purple, long | 296 | 240 (9/16 of 427) |
| | recombinant | purple, round | 19 | 80 (3/16 of 427) |
| | recombinant | red, long | 27 | 80 (3/16 of 427) |
| | parental | red, round | 85 | 27 (1/16 of 427) |
| | | | 427 | 427 |

Statistical tests are not needed to see the distortions in the number of plants actually observed relative to the numbers expected in all the $F_2$ phenotypic classes. Particularly important is the fact that the deviation is not random. The $F_2$ progeny shows an excess of both **parental phenotypic classes** (purple, long and red, round) and a deficiency in the **recombinant phenotypic classes** (purple, round and red, long) relative to the expected number for a 9:3:3:1 $F_2$ ratio.

The dominance-recessiveness relationships between the members of the two pairs of alleles remained the same as in monohybrid crosses. Purple was still dominant over red and long was still dominant over round, as we can see from the dihybrid $F_1$ phenotype and from the two most numerous phenotypic classes in the $F_2$ progeny. Analysis of the $F_2$ progeny data shows that each pair of alleles segregated, as expected, and produced a 3:1 phenotypic ratio in each case.

purple: 296 + 19 = 315
red: 27 + 85 = 112
= 3:1

long: 296 +27 = 323
round: 19 + 85 = 104
= 3:1

The genes did not change nor did the members of each pair of alleles. The single difference is that the alleles tended to stay together in the parental combinations more often than they separated into new combinations. This tendency led to a distortion of the 9:3:3:1 $F_2$ ratio of the four phenotypic classes. These results indicate that the two pairs of alleles did not assort independently, for if they had undergone independent assortment a 9:3:3:1 $F_2$ ratio would have been observed.

The tendency of alleles of different genes to stay together more often in parental combinations than to separate into new combinations is called **linkage.** An *excess* of parental classes and a *deficiency* of recombinant phenotypic classes in the progeny, when compared with pairs of alleles that undergo independent assortment during reproduction, are evidence of the phenomenon of linkage.

Bateson and Punnett proposed a number of possible explanations of these unexpected results, but none of their proposals was supported by experimental data. They were unable to explain their results on the basis of modified ratios or on any other basis of genetic behavior known at that time. T.H. Morgan and his colleagues at Columbia University proposed the first satisfactory explanation of linkage in 1911.

## 5.2 Linkage Studies Using *Drosophila*

By 1911 Morgan had isolated a number of mutants in addition to the white eye mutant discussed in Section 4.2. Each of these mutants was assigned to the X chromosome in *Drosophila,* based on its transmission pattern. Morgan used various mutant strains in crosses involving two or more X-linked genes. He and his collaborators, particularly A. Sturtevant, provided a carefully argued analysis of inheritance patterns for these X-linked genes. They concluded from their analysis that genes on the same chromosome had a tendency to be transmitted together during the formation of gametes. Genes on different chromosomes, on the other hand, underwent independent assortment during gamete formation in sexual reproduction. We can follow their analysis by examining some of the original data reported between 1911 and 1913. In particular we will look at two of the six X-linked characters that they studied: *gray* wild-type body color, $y^+$, versus mutant *yellow, y*; and *red* wild-type eye color, $v^+$, versus mutant *vermilion, v*.

When yellow females (♀♀) were crossed with gray males (♂♂) and when vermilion ♀♀ were crossed with red ♂♂ in monohybrid tests, each pair of alleles showed typical X-linked monohybrid inheritance (Fig. 5.1). The behavior of these two pairs of alleles was then analyzed in dihybrid crosses. Three possible sets of results were predicted for the dihybrid $F_2$ progeny: (1) *independent assortment* of the two pairs of alleles would lead to a 1:1:1:1 ratio of the four $F_2$ phenotypic classes; (2) *complete linkage* of the two pairs of alleles would produce only the two parental phenotypic classes in a 1:1 ratio,

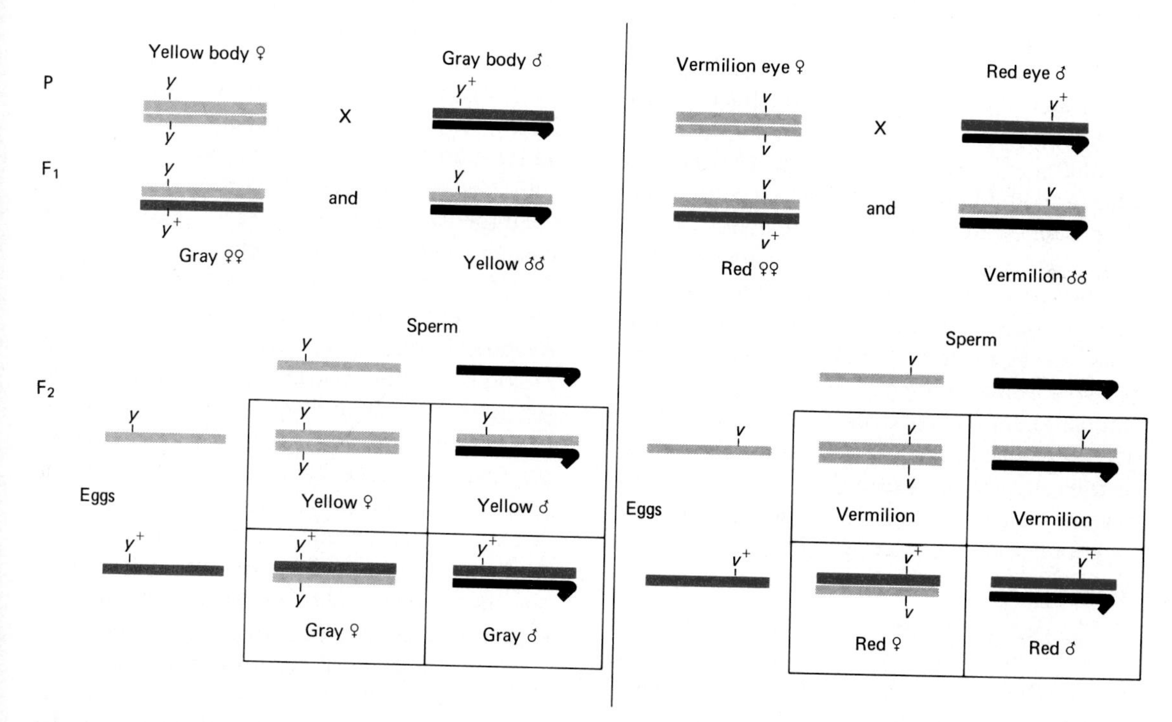


***Figure 5.1*** Monohybrid X-linked recessive inheritance characterizes yellow body color versus dominant wild-type gray and, similarly, underlies expression of mutant vermilion eye color versus wild-type red eye color in *Drosophila.* The distribution of alleles parallels the distribution of the sex chromosomes in each breeding analysis of inheritance.

***Table 5.1*** $F_2$ progeny from a dihybrid cross between parental yellow, vermilion ♀♀ $\left(\frac{y\ v}{y\ v}\right)$ and gray, red ♂♂ ($\underrightarrow{y^+ v^+}$) of *Drosophila melanogaster*.

| phenotypes | number of ♀♀ observed | number of ♀♀ expected* | number of ♂♂ observed | number of ♂♂ expected* | total progeny observed | total progeny expected* |
|---|---|---|---|---|---|---|
| gray, red | 427 | 323.5 | 385 | 271 | 812 | 594.5 |
| gray, vermilion | 240 | 323.5 | 186 | 271 | 426 | 594.5 |
| yellow, red | 213 | 323.5 | 189 | 271 | 402 | 594.5 |
| yellow, vermilion | 414 | 323.5 | 324 | 271 | 738 | 594.5 |
| totals | 1294 | 1294 | 1084 | 1084 | 2378 | 2378 |

*Expected for 1:1:1:1 ratio.

that is, the alleles were transmitted together and did not separate during reproduction; or (3) *incomplete linkage* of the two pairs of alleles such that some recombinants would be produced along with a majority of the two parental phenotypic classes, that is, the genes were capable of being transmitted separately at least some of the time.

In crosses between yellow, vermilion ♀♀ and gray, red ♂♂, Sturtevant found the expected $F_1$ phenotypes of gray, red ♀♀ and yellow, vermilion ♂♂. The $F_2$ progeny obtained by interbreeding $F_1$ males and females consisted of four phenotypic classes, but the ratio was not 1:1:1:1 (Table 5.1). These results indicated that the two genes were linked but that they did not remain together in every case of gamete formation during reproduction. Using X-linked genes in *Drosophila,* Sturtevant obtained results very similar to those obtained by Bateson and Punnett for autosomal genes in garden peas. These similarities suggest the existence of some common feature of genetic behavior that underlies the phenomenon of linkage. The similarities include the following observations:

***1.*** The inheritance pattern for linkage occurred in both a representative plant species and a representative animal species, thus implying a broad biological basis.

***2.*** The dihybrid $F_2$ ratios deviated from the expected for both autosomal genes (in peas) and X-linked genes (in *Drosophila*), thus implying the existence of some basic genic feature unrelated to specific chromosomal assignments.

Before these studies, linkage had appeared to be a phenomenon without explanation because we had discovered only that some genes do not undergo independent assortment whereas other genes do. We know that in *Drosophila,* however, all X-linked genes are on the X chromosome. Linkage behavior of X-linked genes, therefore, is predicted because the genes must be in the single kind of X chromosome that the organism possesses.

If the two genes are indeed linked on the X chromosome and if their behavior is otherwise typical for two pairs of alleles that segregate during reproduction, we can predict that the reciprocal cross (gray, red ♀♀ × yellow, vermilion ♂♂) should produce an $F_2$ progeny consisting of parental and recombinant phenotypic classes in the *same proportions* as we found in the first cross (see Table 5.1). The reciprocal cross will be somewhat different from the first cross, however, because of the presence of two X chromosomes in females and only one X chromosome in males. In the reciprocal cross each of the $F_2$ females will receive one $y^+v^+$ combination in the X chromosome contributed by its $F_1$ father so every female will be wild-type in appearance (Fig. 5.2). The $F_2$ males, on the other hand, will receive either X chromosome in eggs from $F_1$ mothers who are heterozygous for both pairs of alleles. Four classes of $F_2$ males should be produced, occurring in the same proportions found for all the $F_2$ progeny in the first cross described.

The reason we can predict the nature of the $F_2$ male phenotypic classes for the reciprocal cross is that the $F_1$ females will have the same combinations of the two pairs of alleles on their X chromosomes in both crosses. Whatever happened to those X chromosomes in the $F_1$ females in the first cross should also happen in the $F_1$ females in the reciprocal cross *if* the source of the X chromosome in the $F_1$ females is not a factor. In other words, if the process is a general one, then it should not matter if the $F_1$ female receives $y^+v^+$ or $yv$ from its mother or from its father.

As shown in Figure 5.2, the $F_2$ males of both crosses include all four phenotypic classes in the predicted proportions. This information provides support for the existence of some general mechanism that leads to recombination of parental alleles in the same frequency for the two genes, regardless of which of the parents in reciprocal crosses carried the dominant alleles and which carried the recessive alleles.

In order to know exactly what genotype is present in each individual produced in a breeding analysis, it is imperative to perform testcrosses rather than obtain $F_2$ progeny. The same $F_2$ phenotypic classes may include individuals of different genotypes and these would be undetectable unless we conducted additional crosses for an $F_3$ generation to discover the existing genotypes. The wild-type $F_2$ females produced in the gray, red ♀♀ × yellow, vermilion ♂♂ reciprocal cross are an example of the problems encountered in allelic analysis when each phenotype is not produced by a unique and different genotype. We either discard the data or have great difficulty in gaining the needed information. In testcrosses, however, each phenotype is produced by a different, single genotype,

P yellow, vermilion ♀♀ X gray, red ♂♂

$F_1$ gray, red ♀♀ and yellow, vermilion ♂♂

$F_2$

| | ♀♀ | ♂♂ | phenotypic class | percentage (♀♀ + ♂♂) |
|---|---|---|---|---|
| gray, red | 427 | 385 | parental | 34.2 |
| gray, vermilion | 240 | 186 | recombinant | 17.9 |
| yellow, red | 213 | 189 | recombinant | 16.9 |
| yellow, vermilion | 414 | 324 | parental | 31.0 |
| totals | 1294 | 1084 | | 100.0 |

P gray, red ♀♀ X yellow, vermilion ♂♂

$F_1$ gray, red ♀♀ and gray, red ♂♂

$F_2$

| | ♀♀ | ♂♂ | phenotypic class | percentage (♂♂ only) |
|---|---|---|---|---|
| gray, red | 1021 | 325 | parental | 34.6 |
| gray, vermilion | 0 | 162 | recombinant | 17.2 |
| yellow, red | 0 | 153 | recombinant | 16.3 |
| yellow, vermilion | 0 | 300 | parental | 31.9 |
| totals | 1021 | 940 | | 100.0 |

***Figure 5.2*** Reciprocal dihybrid crosses of X-linked body color and eye color traits in *Drosophila* reveal a consistent pattern of an excess of parental phenotypes and a deficiency of recombinant phenotypes in the $F_2$, instead of a phenotypic ratio of 1:1:1:1 that is typical of two independently assorting pairs of alleles. The observed distortion of the ratio is characteristic of transmission patterns for linked genes.

and we can utilize all the data in analysis without resorting to additional crosses.

If we compare the frequency of recombinant genotypes in the reciprocal crosses shown in Figure 5.2, we find very similar proportions of the two recombinant classes in the two progenies. The formula used to calculate the percentage recombination between two linked genes is

$$\frac{\text{number of recombinants}}{\text{total testcross progeny}} \times 100 = \%\ \text{recombination}$$

In this special case of X-linked genes, some or all of the $F_2$ progeny provide the same results that would have been obtained in testcrosses between heterozygotes and double recessives. For this reason we can calculate recombination in this particular example. We could not, however, calculate recombination values from the $F_2$ data reported for garden peas by Bateson and Punnett. In that case different genotypes were represented in all the $F_2$ classes except for the doubly recessive class. Some of the genotypes are recombinant and some are parental combinations, even though both kinds of genotypes may produce the same phenotype.

Sturtevant did perform testcrosses and showed that the percentage of recombinants was approximately the same as he had found in the $F_2$ progenies:

| | |
|---|---|
| P | gray, red ♀♀ × yellow, vermilion ♂♂ |
| $F_1$ | gray, red ♀♀ and ♂♂ |
| Testcross | gray, red $F_1$ ♀♀ × yellow, vermilion ♂ |

| $F_2$ progeny class | phenotype | number of ♀♀ | number of ♂♂ |
|---|---|---|---|
| parental | gray, red | 31 | 23 |
| parental | yellow, vermilion | 41 | 21 |
| recombinant | gray, vermilion | 11 | 13 |
| recombinant | yellow, red | 12 | 8 |
| | totals | 95 | 65 |
| | | 160 | |

$$\text{percentage recombination} = \frac{44}{160} = 0.28 \times 100 = 28\%$$

The discrepancy between the absolute recombination values in the $F_2$ and testcross progenies may be due to differences in progeny sizes and differences in degree of sampling errors encountered. Precise recombination frequencies must be based on many progenies from many crosses in order to reduce errors to a minimum.

Sturtevant performed another important set of crosses for these same two genes. He reasoned that if two genes are located on the same chromosome, it should not matter which of the two alleles of each pair are present in each parent. He had already shown that either parent could carry both dominants or both recessives and still produce the same proportions of phenotypic classes in the progenies of reciprocal crosses. He then was testing whether any other combinations of these two pairs of alleles in the parents made any difference in recombination frequencies for the two genes.

The crosses between yellow, red ♀♀ and gray, vermilion ♂♂ or the reciprocal crosses between gray, vermilion ♀♀ and yellow, red ♂♂ did yield about the same recombination frequencies as did the earlier crosses.

| | |
|---|---|
| P | gray, vermilion ♀♀ × yellow, red ♂♂ |
| $F_1$ | gray, red ♀♀ and gray, vermilion ♂♂ |

| $F_2$ phenotype | number of ♀♀ | number of ♂♂ | progeny class |
|---|---|---|---|
| gray, vermilion | 182 | 149 | parental |
| gray, red | 199 | 59 | recombinant |
| yellow, vermilion | 0 | 50 | recombinant |
| yellow, red | 0 | 105 | parental |
| totals | 381 | 363 | |

$$\text{percentage recombination} = \frac{109}{363} = 0.30 \times 100 = 30\%$$

These results show that the same linkage pattern and recombination frequency were produced whether the dominant alleles for both genes are carried on the same chromosome (called the **cis** or **coupling arrangement,** $\frac{y^+v^+}{y\ v}$) or on different chromosomes of a homologous pair (called the **trans** or **repulsion arrangement,** $\frac{y^+v}{y\ v^+}$). Results were consistent, regardless of the specific alleles that happened to be present on each one of the pair of homologous chromosomes. It was now necessary to relate the generalization, derived from genetic analysis and based only on behavior of alleles in crosses, to the behavior of the chromosomes on which the genes were located.

## 5.3 Chromosomal Basis for Genetic Recombination

If linkage were a consequence of genes being located on the same chromosome, as seemed to be the case from *Drosophila* studies, it would explain the tendency for linked genes to be inherited together in parental combinations. But what was responsible for recombination, which indicated that linked genes sometimes separated into new combinations of alleles in progeny? In 1911 Morgan suggested that recombinations of alleles arose as a result of exchanges between paired chromosomes during meiosis. He called this process *crossing over.*

Morgan based his hypothesis on a synthesis of data, observations, and ideas of his own and of other biologists. He was fully aware of Sutton's Chromosome Theory of Heredity and of the simple fact that each chromosome must contain more than one gene because every species has far more genes than chromosomes. Earlier F. Janssens had suggested that *chiasmata* (a term he coined) might be the sites of physical exchange between chromosomes during the time they are closely paired as bivalents in early stages of meiosis (Fig. 5.3). Morgan cited

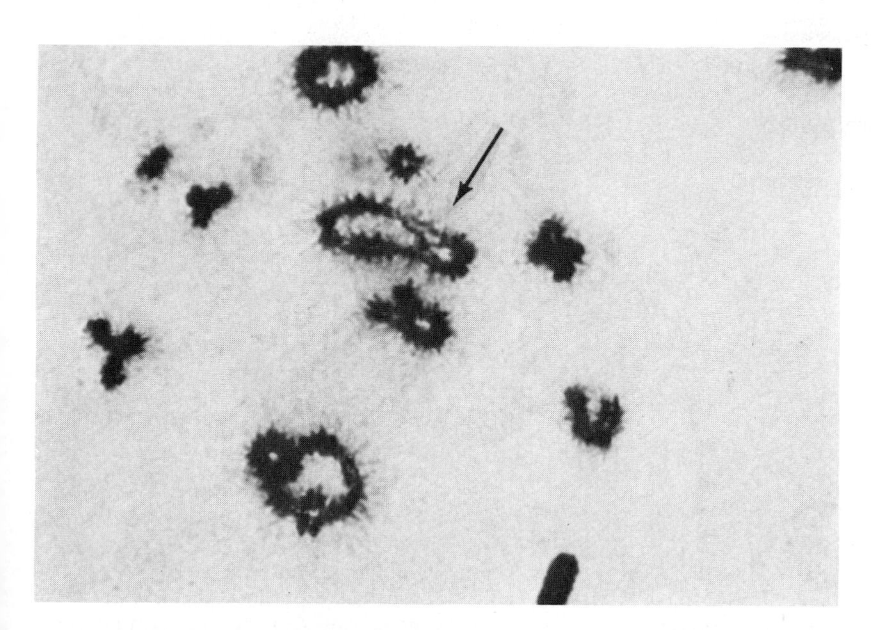

***Figure 5.3*** Diakinesis in grasshopper spermatocyte, showing chiasmata. The arrow points to a bivalent with a particularly clear display of a chiasma in which only two of the four chromatids are involved. × 3,000.

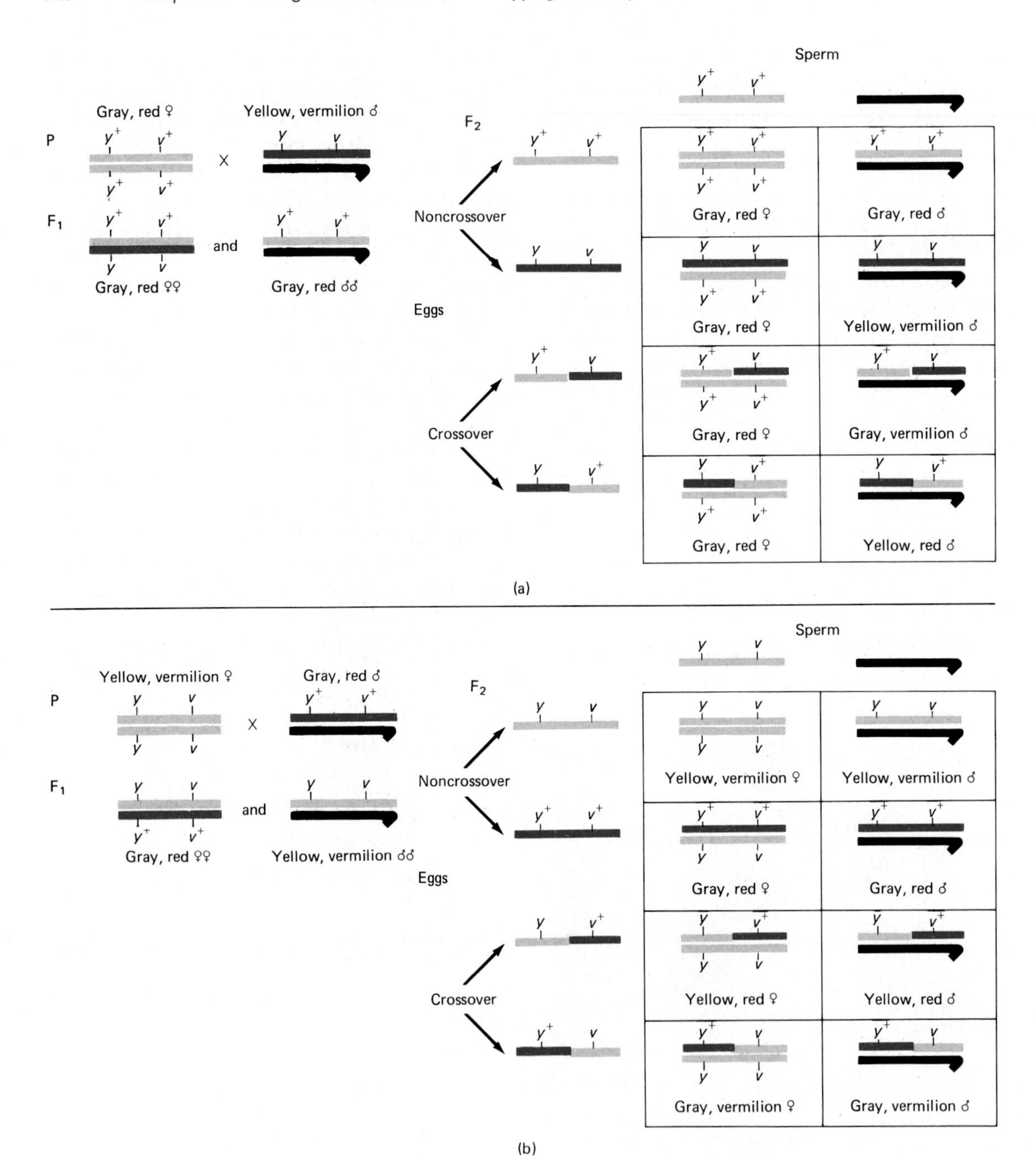


***Figure 5.4*** Reciprocal crosses demonstrate the origin of recombinant progeny classes from gametes that carry crossover X chromosomes. Linked genes are inherited together in parental allelic combinations if the chromosome has not undergone crossing over, but new combinations of alleles of linked genes will arise if crossing over has occurred between the two linked genes in a chromosome. Recombinant progeny are produced from fusions involving gametes that carry crossover chromosomes. Crossing over involves a physical exchange between parts of homologous chromosomes during meiosis.

Janssens' suggestion and supported it as the physical basis for recombinations between linked genes.

Morgan developed the concept of linkage to explain the patterns of inheritance he had observed in breeding analysis in which two or more pairs of alleles did not undergo independent assortment. The chromosomal exchanges during crossing over that Morgan postulated, however, were entirely theoretical. But this was a working hypothesis that related the known behavior of chromosomes in meiosis to the pattern of gene transmission derived from genetic analysis. However, the supporting and conclusive cytogenetic evidence that showed that physical exchanges between homologous chromosomes are responsible for gene recombinations was provided twenty years later. The power of genetic analysis was sufficiently strong, however, to make Morgan's ideas attractive and viable enough for biologists to seek evidence in their support. Just as Mendel had postulated the existence of genes on the basis of particulate patterns of transmission in inheritance, Morgan postulated chromosomal events on the basis of gene transmission patterns.

If we put the alleles on the X chromosome and follow their transmission from parents to progeny, we can see how crossing over leads to recombinations of linked genes and to recombinant phenotypic classes (Fig. 5.4). Recombinant classes of progeny arise by fertilization of eggs carrying crossover X chromosomes. Parental phenotypic classes have noncrossover chromosomes in which the original parental combinations of alleles are unchanged. We can now understand the tendency for linked genes to be inherited together, as seen in genetic analysis, on the basis of chromosome behavior during meiosis. Crossing over leads to recombinant gametes, and these in turn contribute to the recombinant phenotypic classes in the progeny. Since crossing over involves a physical exchange between homologous chromosomes, it obviously does not matter which alleles are carried in each chromosome of a pair. The same recombination frequency will be found for the same two linked genes regardless of whether dominant alleles are in coupling or repulsion, since crossing over is a chromosomal event and not a genic phenomenon. We will discuss this later in the chapter.

## 5.4 Tetrad Analysis of Linkage and Recombination

Many of the fungi and various other organisms produce a tetrad of spores as products of a single meiotic cell. Recovery and genetic analysis of such tetrads, or *tetrad analysis,* can provide detailed information about patterns of segregation and recombination taking place in individual cells. This topic was discussed in Chapter 3 (see Figs. 3.21 and 3.22) for monohybrid inheritance patterns in relation to chromosomal segregation patterns during meiosis.

Tetrads may consist of spores in *ordered* or *unordered* arrangement, and each of these types can be analyzed genetically. Ordered spores retain fixed positions reflecting their origin during meiosis, unordered spores do not. In either case, events that occurred in individual meiotic cells *can be observed directly* in individual spore tetrads. Statistical analysis of gene transmission is, therefore, not necessary.

When crosses are made between haploid parents differing in two pairs of alleles, for example $a^+b^+ \times ab$, three kinds of tetrads can be found. Each of these shows a distinctive pattern of allelic segregation. One kind of tetrad contains $a^+b^+$, $a^+b^+$, $ab$ and $ab$ spores. This tetrad is the **parental ditype** (**PD**) because only two allelic combinations are present and both are parental genotypes. The **nonparental ditype** (**NPD**) tetrads also have only two allelic combinations, $a^+b$, $a^+b$, $ab^+$, and $ab^+$, both of which are recombinant genotypes. If the original cross had been $a^+b \times ab^+$, the parental and nonparental designations would be reversed. In the third kind of segregation pattern, found in **tetratype** (**T**) tetrads, all four spores have different genotypes, two of which are parental and two are recombinant, $a^+b^+$, $a^+b$, $ab^+$, and

$ab$. We can summarize this information as follows:

P $a^+b^+ \times ab$

| $F_1$ | **spore genotypes** | **classes** | **tetrad type** |
|---|---|---|---|
| | $a^+b^+, a^+b^+, ab, ab$ | parental only | parental ditype (PD) |
| | $a^+b, a^+b, ab^+, ab^+$ | recombinant only | nonparental ditype (NPD) |
| | $a^+b^+, a^+b, ab^+, ab$ | parental and recombinant | tetratype (T) |

If the two genes are situated in different chromosomes, we expect independent assortment of the segregating pairs of alleles (Fig. 5.5). This situation is evident in the equal percentages of PD and NPD tetrads because of random segregation of homologous chromosomes of each pair during meiosis. Tetratype tetrads may be produced by recombination between linked genes or by crossing over between the centromere and the gene of one of the pairs of independently assorting chromosomes, as in the case of unlinked genes (Box. 5.1). Regardless of the T tetrads, therefore, the critical observation is of the PD:NPD proportions. An excess of PD over NPD indicates linkage, and equal percentages of PD and NPD indicate independently assorting genes on different chromosomes.

Isolated spores can provide the same data since independently assorting genes lead to a phenotypic ratio of 1:1:1:1, which occurs if PD and NPD tetrads are produced in equal proportions. If the two genes are linked, spores will show an excess of parental genotypes and a deficiency of recombinant genotypes. This occurrence is equivalent to an excess of PD tetrads over NPD tetrads. In the cross $a^+b^+ \times ab$, we may find the following results shown below.

On the basis of tetrads, we see an excess of PD over NPD types rather than equal percentages. This result indicates gene linkage rather than independent assortment. In this case the T tetrads arose by crossing over between the linked genes, and they can be included in the calculations of recombination frequency. The simple formula is the same as the one we used for *Drosophila,* namely, the proportion of the

| **spore genotypes in tetrads** | **tetrad** | | **number of parental spores** | **number of recombinant spores** |
|---|---|---|---|---|
| | **type** | **number** | | |
| $a^+b^+$ | PD | 400 | 400 | |
| $a^+b^+$ | | | 400 | |
| $a\ b$ | | | 400 | |
| $a\ b$ | | | 400 | |
| $a^+b^+$ | T | 90 | 90 | |
| $a\ b$ | | | 90 | |
| $a^+b$ | | | | 90 |
| $a\ b^+$ | | | | 90 |
| $a^+b$ | NPD | 10 | | 10 |
| $a^+b$ | | | | 10 |
| $a\ b^+$ | | | | 10 |
| $a\ b^+$ | | | | 10 |
| totals | | 500 | 1780 | 220 |
| | | | 2000 | |

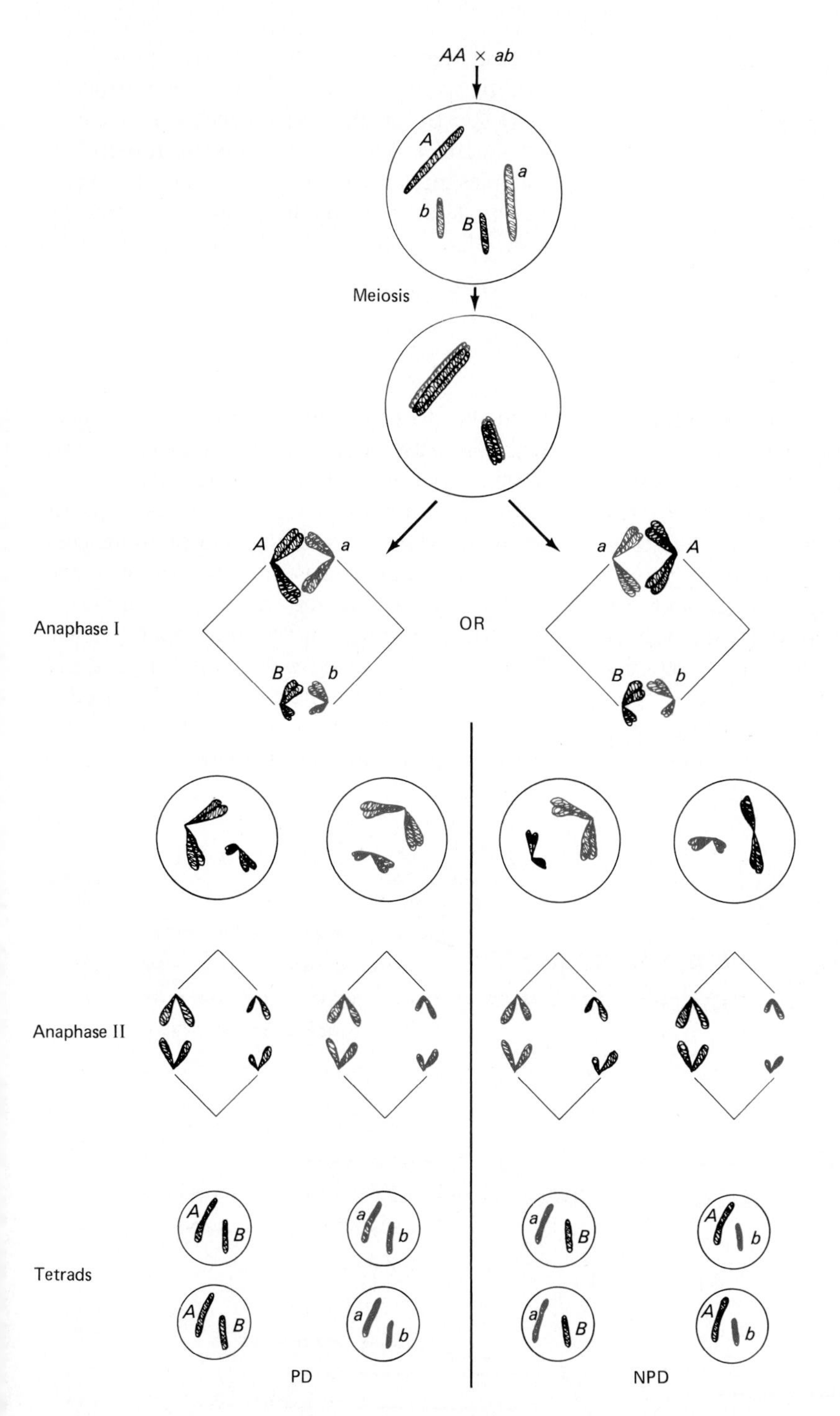


*Figure 5.5* Random segregation of homologous chromosomes at meiosis produces PD and NPD tetrads with approximately equal frequency when unlinked genes are involved. Independent assortment of two genes is, therefore, indicated when %PD = %NPD tetrads in the progeny.

total progeny that consists of recombinants:

$$\frac{\tfrac{1}{2}\text{T} + \text{NPD}}{\text{total tetrads}} \times 100 = \%\ \text{recombination}$$

or,

$$\frac{45 + 10}{500} \times 100 = 11\ \%$$

We use the term ½T in the numerator since only half the spores in T tetrads are recombinants. All NPD spores, on the other hand, are of the recombinant genotype. The numerator, therefore, includes the number of recombinants recovered in the total progeny from intact tetrads. We would have obtained the same information if we

## BOX 5.1 Origin of Tetratype Tetrads

Individuals heterozygous for two unlinked genes (+/*a* +/*b*) may produce tetratype (T) tetrads if one crossover occurs between the gene and the centromere of the chromosome. This crossover may involve either one of the chromosomes, that is, either the one carrying +/*a* or the one carrying +/*b*, as shown in (a) and (b). The same T tetrad may also be produced in cases of linked genes in heterozygotes (++/*ab*) if a single crossover occurs between the two gene loci during meiosis, as shown in (c). In linkage analysis, therefore, it is the proportion of PD:NPD tetrads that provides the necessary information to determine whether the two gene loci segregate independently (unlinked) or are linked. The proportion of T tetrads is not a useful measure of linkage or independent assortment when considering two or more gene loci. It is a useful measure of the distance separating a gene and the centromere, however, since the percentage of T tetrads depends on the frequency of crossing over between gene and centromere. The frequency of crossing over, in turn, is translated into map units of distance between gene and centromere of the chromosome.

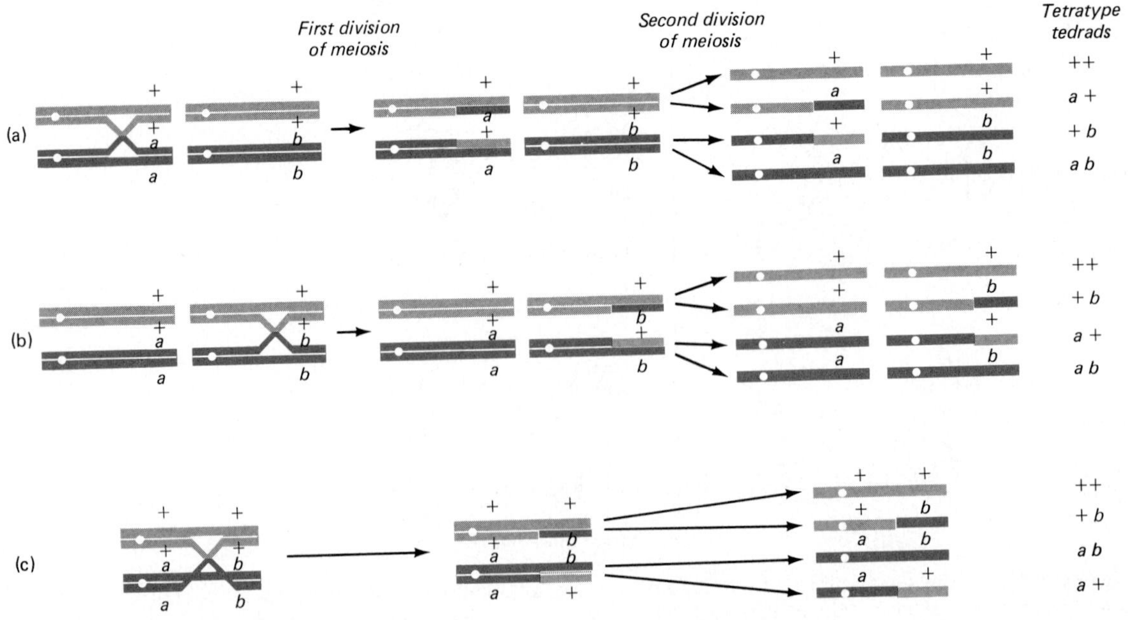

had examined only isolated spores from these 500 tetrads:

$$\frac{\text{recombinants}}{\text{total spores}} = \frac{220}{2000};$$

$$\frac{220}{2000} \times 100 = 11\% \text{ recombination}$$

But, statistical analysis would be required to evaluate the conclusions reached by *random* sampling of progeny spores.

Isolated spores or unordered tetrads can provide information for recombination frequencies, as can ordered tetrads. However, ordered tetrads allow us to analyze the particular chromosomal events giving rise to the genotype of each spore in each tetrad. On the basis of the hypothesis that crossing over between linked genes leads to recombinant chromosomes and genotypes, we predict the following:

***1.*** PD tetrads, which have only parental genotypes, arise when no crossing over has occurred between linked genes, thereby leaving parental combinations intact.

***2.*** T tetrads arise as a consequence of a single crossover event between two linked genes during meiosis, thus producing recombinant spores with crossover chromosomes. The parental genotypes in a T tetrad occur because the single crossover involves only two of the four chromatids in a meiotic bivalent. The two crossover chromatids are recombinant, but the other two chromatids were not involved in the one crossover event, and they remain unchanged genotypically (Fig. 5.6).

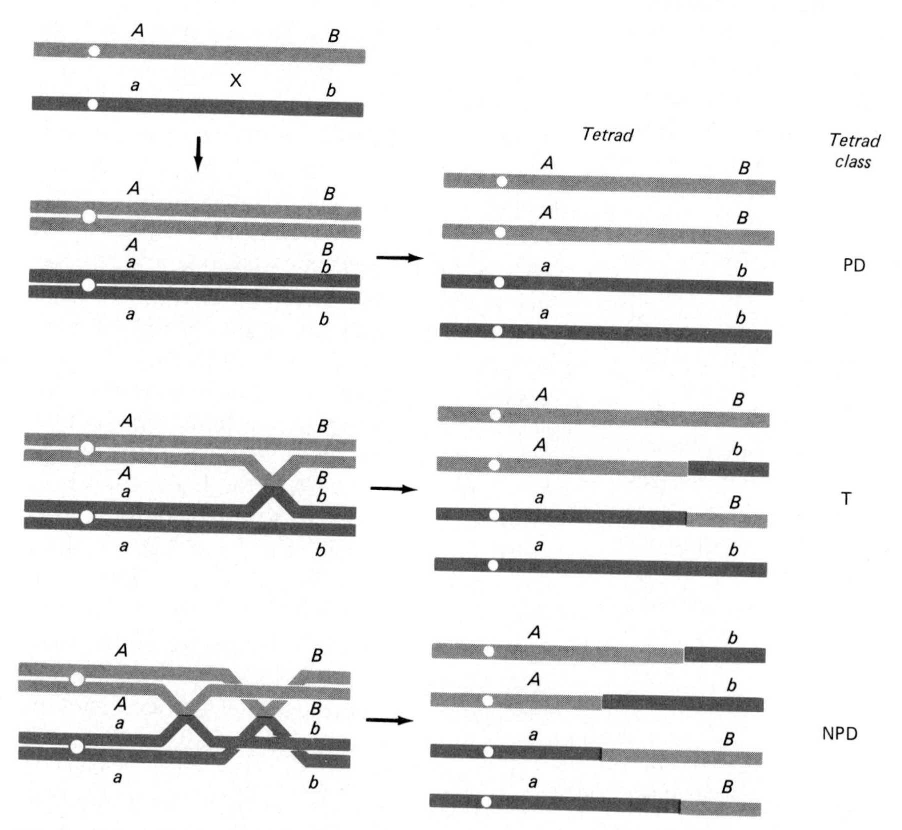


***Figure 5.6*** Tetrad analysis requires recognition of tetrad segregation patterns for two genes, which may produce three tetrad classes: parental ditype (PD), tetratype (T), and nonparental ditype (NPD). Each class arises as a consequence of crossover events, or lack of these, as shown.

***3.*** Nonparental ditype tetrads arise only when two crossovers take place between two linked genes. More specifically, each crossover involves a different pair of the four chromatids in the bivalent. The NPD tetrads will, therefore, be the rarest of the three segregation patterns for two linked genes.

The order in which the spores are linearly arranged within the spore case provides supporting evidence for these postulated chromosomal crossover events. Each pair of *adjacent* spores represents the two sister chromatids of one of the pair of homologous chromosomes. If crossing over takes place between *nonsister* chromatids—that is, one chromatid of each of the two chromosomes—this event will be reflected in the recombinant genotype of one spore in an adjacent pair of spores. Similarly, if one crossover involves two nonsister chromatids and another crossover involves the other two nonsister chromatids of the bivalent, the linear order of spore genotypes will show how these crossovers gave rise to NPD tetrads in which adjacent identical recombinant spores are found in pairs.

As we will see, tetrad analysis in *Neurospora* and other suitable organisms provided important supporting evidence for the chromosomal basis of genetic recombination. Definitive experimental evidence for the hypothesis that crossing over between homologous chromosomes gives rise to genetic recombinants was first obtained in other eukaryotic species. Nevertheless, *Neurospora* proved to be a valuable system for analyzing particular features of the crossing-over process.

## Crossing Over

We have accepted the working hypothesis that crossing over can be characterized in at least three ways:

***1.*** It is a process that involves an exchange of parts of homologous chromosomes.

***2.*** It leads to recombinations of linked genes in reproduction.

***3.*** It takes place after chromosomes have replicated, that is, in the four-strand stage when each bivalent consists of two pairs of chromatids, and each crossover event involves only two of the four chromatids.

We are now in a position to look at the experimental evidence in support of these assumptions. We will be in a better position to pursue our analysis of linkage and recombination as well as other topics once we have a basic understanding of these major features of the phenomenon of crossing over.

### 5.5 Cytogenetic Evidence for Crossing Over and Its Consequences

T.H. Morgan proposed, in 1911, that new combinations of linked genes arose by exchanges between chromosomes during meiosis and that chiasmata might be the visible evidence for such exchanges. No experimental support for this proposal existed until 1931, when two reports were published independently of each other. Each of these studies showed in an elegant way that genetic recombinations were the consequence of physical exchanges between homologous chromosomes during meiosis.

The major difficulty in obtaining evidence for physical exchanges was that homologous chromosomes ordinarily cannot be distinguished from one another even at the highest levels of magnification. In order to establish the occurrence of exchanges, the homologous chromosomes had to be *physically altered* in different ways so that each homologue was recognizable. Strains were obtained in which physically recognizable homologues were *genetically* distinguishable because they carried different alleles of suitable genes. The parallels in behavior and distribution of physically and genetically marked chromosomes provided the cytogenetic foundation for interpreting the experimental results. The cytological aspect involves microscopic observations of physically marked

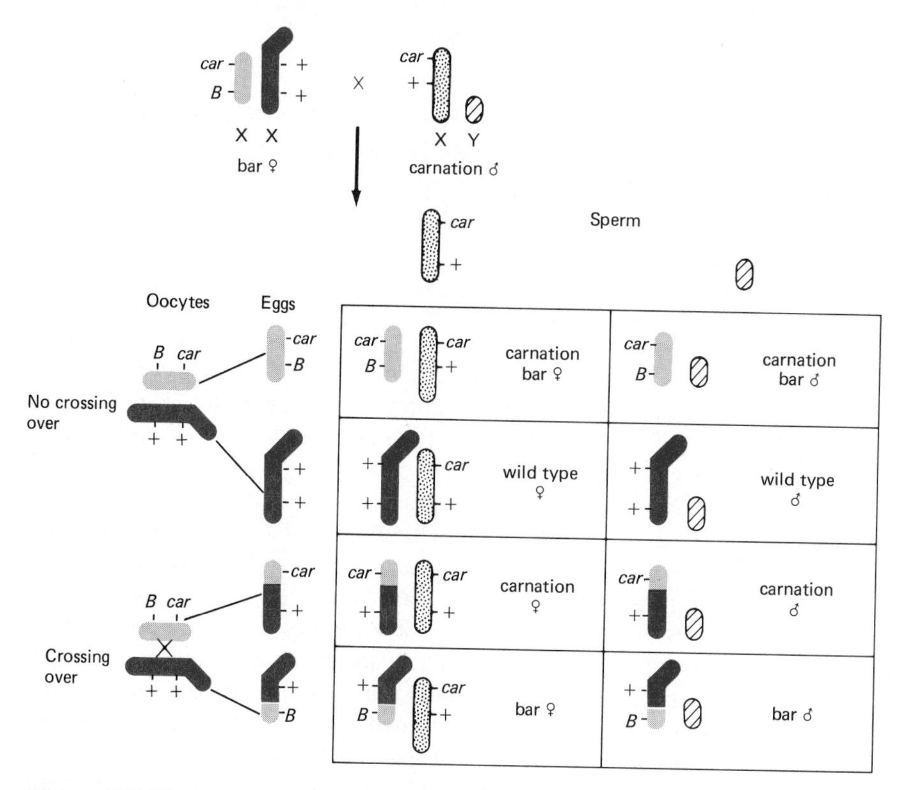


***Figure 5.7*** Diagrammatic summary of Curt Stern's 1931 cytogenetic experiments with *Drosophila melanogaster,* in which he obtained strains having physically distinguishable X chromosomes that were made allelically different for eye color and eye facet number. These classic studies relate physical exchange between homologous chromosome segments (crossing over), as seen by microscopy, to genetic recombination, as seen in progeny phenotypes and the genotypes deduced from these.

chromosomes in the progeny, and the genetic aspect involves identification of genotypic classes of parental and recombinant progeny. Taken together, the combined analysis is referred to as a study in cytogenetics.

The two independent reports made in 1931 were by Curt Stern, using *Drosophila,* and by Harriet Creighton and Barbara McClintock, using corn. Since the experimental logic and design were essentially the same in the two studies, we can illustrate the principles using either study. Stern's analysis using *Drosophila* may be better since we should be familiar with various aspects of *Drosophila* genetics by now.

Stern had come across unusual strains of *Drosophila* in his studies, including ones in which the X chromosome was physically altered and recognizably different from the normal X chromosome. In one strain the X chromosome had a piece of chromosome 4, an autosome, attached in place of its own missng piece. In another strain a portion of the Y chromosome was attached to the X chromosome so that the altered X was much longer than the normal X chromosome. Stern made the crosses required to obtain females heterozygous for these two unusual X chromosomes instead of having the normal two X chromosomes. The two physically distinguishable X chromosomes were also made allelically different by appropriate crosses with other stocks. In this way Stern could tell which alleles were present in the parents and progeny by observing the phenotypes of the flies and which chromosomes were present in parents and progeny by microscopic observations of cells in these flies (Fig. 5.7).

The special females were crossed with males whose normal X chromosome carried the recessive carnation (*car*) mutant allele for eye color and the recessive wild-type allele ($B^+$) for eye construction. The dominant mutant allele for bar eyes (*B*) leads to a reduced number of facets in the compound eye of the insect, and a "bar" of color develops in the faceted part of the eye. The dominant wild-type allele for eye color ($car^+$) governs red eyes. Females were phenotypically bar eyed with red eye color, therefore, and the parental males were phenotypically nonbarred with carnation eye color.

When Stern examined the progeny of these crosses he found that flies with unaltered X chromosomes carried the parental combinations of alleles. Flies identified by phenotype as having recombinant genotypes carried a physically altered X chromosome derived from the heterozygous female parent. This correlation between physically exchanged chromosomes and new combinations of alleles provided powerful evidence in support of the hypothesis that crossing over involves a physical exchange of homologous chromosome segments and that this exchange is responsible for recombinations involving members of linked pairs of alleles.

## 5.6 Time and Mechanism of Crossing Over

Stern's experimental results provided no particular evidence for the time when crossing over occurs, although it was generally assumed to occur during meiotic prophase. His cytogenetic analysis showed that chromosomes were physically rearranged by crossover events, but the mechanism that was responsible for chromosomal exchanges was not apparent from the experimental results. Studies made to answer these two questions in the 1930s and 1940s used *Neurospora* as the principal experimental organism because of its unique advantage of having easily analyzable ordered tetrads. Tetrad analysis could be interpreted in terms of meiotic events in individual cells, including events taking place at the first and second divisions of meiosis.

Crossing over was assumed to occur after chromosomes had replicated, in the so-called four-strand stage; but it was also possible that crossing over took place before chromosome replication, while each chromosome was a single structure rather than a pair of chromatids. The use of *Neurospora* made it possible to distinguish between these alternatives by the analyzing of ordered tetrads for the position of allelically marked spores. Spore position could then be interpreted in terms of chromosomal events during meiosis, specifically, for the time of crossing over in relation to chromosomal replication from two strands to four strands in a bivalent (Fig. 5.8).

If crossing over occurs before chromosomal replication, tetrads would carry only parental genotypes and would be of the PD type or they would have only recombinant spores and would be of the NPD type. Both parental and recombinant spores cannot occur in the same tetrad (tetratype, T) as a consequence of a *single* crossover between the two linked gene markers of the chromosome.

If crossing over occurs in the four-strand stage, after replication, then all three types of tetrads could be produced. The relative frequencies of these three kinds of tetrads would depend on the frequency with which crossing over took place between the two genes. Confirming evidence also came from studies of crossing over in trihybrid crosses involving three linked genes (Fig. 5.9). The simplest explanation for many tetrads containing spores that are recombinant for three genes is the occurrence of double crossovers involving three or four chromatids in addition to those double crossovers involving the same two chromatids of a bivalent. Exceedingly complicated interpretations would be needed to explain these ordered trihybrid recombinant spores on the basis of two-strand crossing over. Accepting the simplest interpretations possible, all these and other data were accepted as convincing evidence for the occurrence of crossing over in bivalents comprised of four chromatids.

Some years later studies showed that DNA replicates in the interphase before meiosis. Similar studies also showed that crossing over occurs in meiotic prophase, after synapsis is completed.

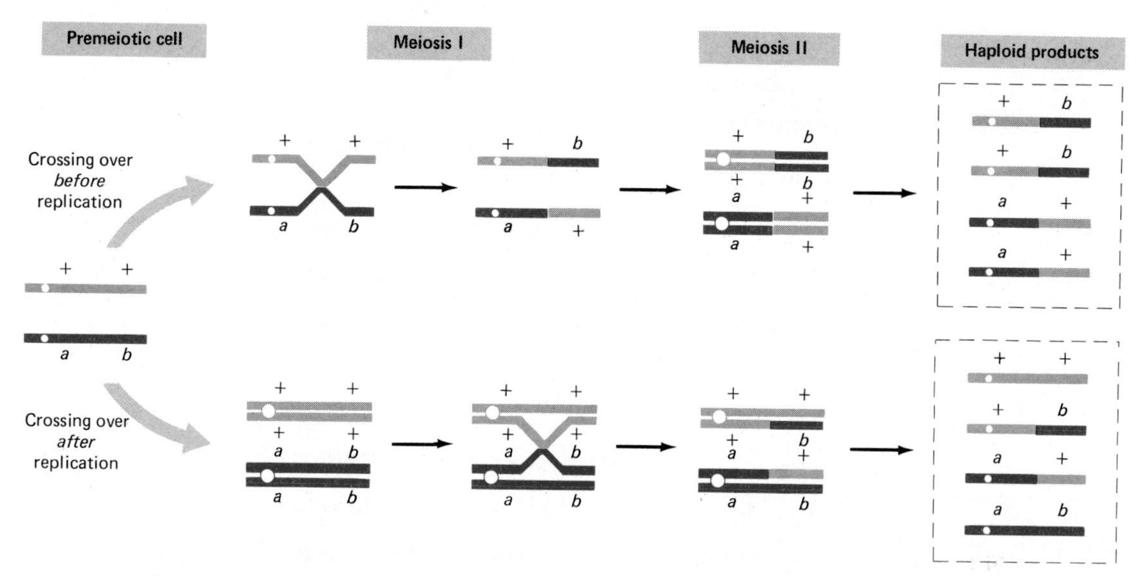


***Figure 5.8*** Different predictions are made for the two possible alternatives shown at left. If crossing over occurs before chromosome replication, NPD tetrads may be produced. If crossing over occurs after replication, in the four-strand stage, tetratype tetrads can be produced. T tetrads *cannot* be produced if crossing over between two linked genes takes place before chromosome replication. Since T tetrads are produced, crossing over probably occurs between two chromatids of a bivalent consisting of four chromatids, that is, after replication.

These observations remove all doubt that both the genetic and cytological data proved that chromosomes replicate first and that *each* crossover takes place between two of the four chromatids of a bivalent. Multiple exchange events, however, may involve two, three, or all four chromatids.

Analysis of ordered tetrads of *Neurospora* was equally useful in clarifying the question of the mechanism of exchange between homologous chromosomes. In the 1930s there were two major hypotheses for this mechanism: (1) copy-choice and (2) breakage and reunion (Fig. 5.10). The copy-choice hypothesis stated that recombinant chromatids arose during replication as the newly forming chromatids "copied" alleles partly along one chromosome and partly along the homologous chromosome of a pair. According to this idea, only two chromatids in a bivalent could be recombinant for two linked genes because the original two chromatids would remain unchanged and would not undergo gene recombination during replication ("copying"). Tetrad analysis in *Neurospora* was used in the 1940s to test the copy-choice hypothesis. According to the reasoning, the copy-choice mechanism could not produce NPD tetrads since all four spores in NPD tetrads carry the recombinant genotype on crossover chromosomes. Furthermore, since genetic analysis showed that four-strand double crossovers occur regularly, copy-choice could be ruled out as a viable hypothesis.

*Neurospora* studies apparently eliminated copy-choice, but they did not provide evidence in favor of breakage and reunion. The breakage-and-reunion hypothesis remained the only possibility, however, and was widely accepted. It was not until 1961 that direct evidence was provided to support breakage and reunion as the mechanism responsible for exchange of homologous chromosome segments in crossing over. We will return to this topic in Chapters 8 and 11, and we will see that breakage and reunion is an exceedingly precise process, and not "chop and glue" as the name might imply.

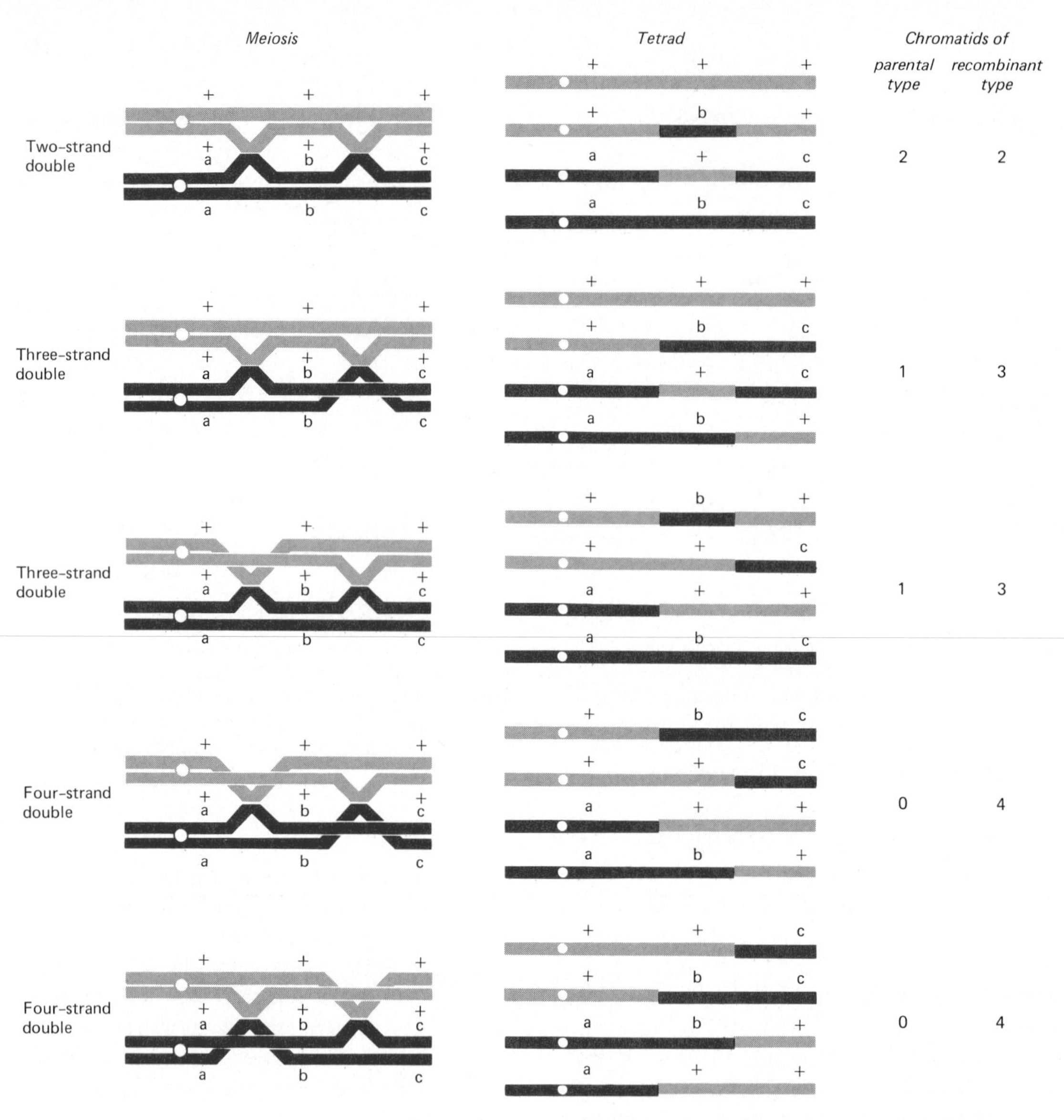


***Figure 5.9*** Some of the types of tetrads that may arise as the result of double crossovers involving two, three, or all four strands (chromatids) of a bivalent during meiosis.

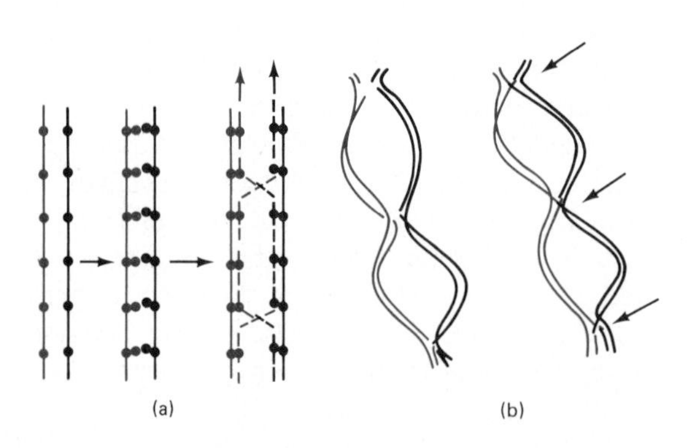


***Figure 5.10*** Diagrams showing two mechanisms postulated to explain crossing over, both proposed in the 1930s: (a) copy-choice, according to John Belling in 1931; and (b) breakage and reunion (at arrows) as described in 1937 by Cyril Darlington.

## Chromosome Mapping

Linkage studies using *Drosophila* and other organisms indicated that recombination values ranged between zero and 50% for any two genes linked on the same chromosome (Box. 5.2). Although a range of values was found for different pairs of linked genes, the *same* values were found for the same pair of genes in repeated experiments. This general observation can be interpreted to mean that each gene occupies a particular location, or **locus** (pl., loci), on a chromosome.

If genes occupy fixed positions, then the different recombination frequencies found for different pairs of genes could be a reflection of the distance separating any two genes along the length of the chromosome. The greater the distance between two genes, the greater the probability that crossing over will occur between them because there is more space for such an event to take place anywhere in that region. If two genes are situated closer together, there is less space and, therefore, less chance for an exchange to take place between them. Thus the second generalization derived from linkage analysis is that the frequency of recombination between linked genes is a consequence of the amount of space between them. These two generalizations provided the basis for mapping genes on a chromosome in fixed order and at specified distances from one another on the chromosome map.

### 5.7 Mapping by Two-Factor Testcrosses

Sturtevant summarized testcross data from a number of experiments with X-linked genes in *Drosophila* in the form of a chromosome map. The consistency of recombination values for any two linked genes showed that the gene loci were fixed and that all the genes in the same **linkage group** occurred in a linear order along the same chromosome. Their relative distances apart, or their relative locations on the chromosome, were expressed in a *chromosome map* based on recombination data from genetic analysis.

We still use Sturtevant's original proposal that the *percentage of recombinants arising from crossing over can be converted into a measurement of distance between two linked genes.* Specifically, *one map unit of distance equals the space between genes in which 1% recombinants arise by crossing over.* A **map unit** is a relative measurement in arbitrary units and not an absolute or actual measurement of chromosome length in micrometers or other physical units.

Suppose we found 6% recombinants in testcross progeny produced in both the coupling and repulsion phases of linkage.

**coupling (cis)**

| | |
|---|---|
| P | $\frac{AB}{AB} \times \frac{ab}{ab}$ |
| $F_1$ | $\frac{AB}{ab}$ |
| Testcross | $\frac{AB}{ab} \times \frac{ab}{ab}$ |

| **progeny class** | **genotype** | **number of individuals** |
|---|---|---|
| parental | $\frac{AB}{ab}$ | 390 |
| parental | $\frac{ab}{ab}$ | 410 |
| recombinant | $\frac{Ab}{ab}$ | 26 |
| recombinant | $\frac{aB}{ab}$ | 24 |
| | total progeny | 850 |

$$\text{percent recombinants} = \frac{50}{850} = 0.058$$

$$0.058 \times 100 = 5.8\%$$

**repulsion (trans)**

| | |
|---|---|
| P | $\frac{Ab}{Ab} \times \frac{aB}{aB}$ |
| $F_1$ | $\frac{Ab}{aB}$ |
| Testcross | $\frac{Ab}{aB} \times \frac{ab}{ab}$ |

## BOX 5.2 Crossing Over and Recombinant (Crossover) Gametes

By the process of crossing over, recombinant gametes are produced with combinations of alleles that are different from combinations occurring on the parental chromosomes in the meiocyte, or meiotic cell. If the diploid parent cell carries + + on one chromosome and *yv* on the homologous chromosome, recombinant gametes will have +*v* or *y*+ chromosomes, as shown in (a). Since each crossover event involves only two of the four chromatids making up the replicated homologous pair of chromosomes, the meiocyte will produce two parental (noncrossover) types of gametes as well as two reciprocal recombinant types of gametes.

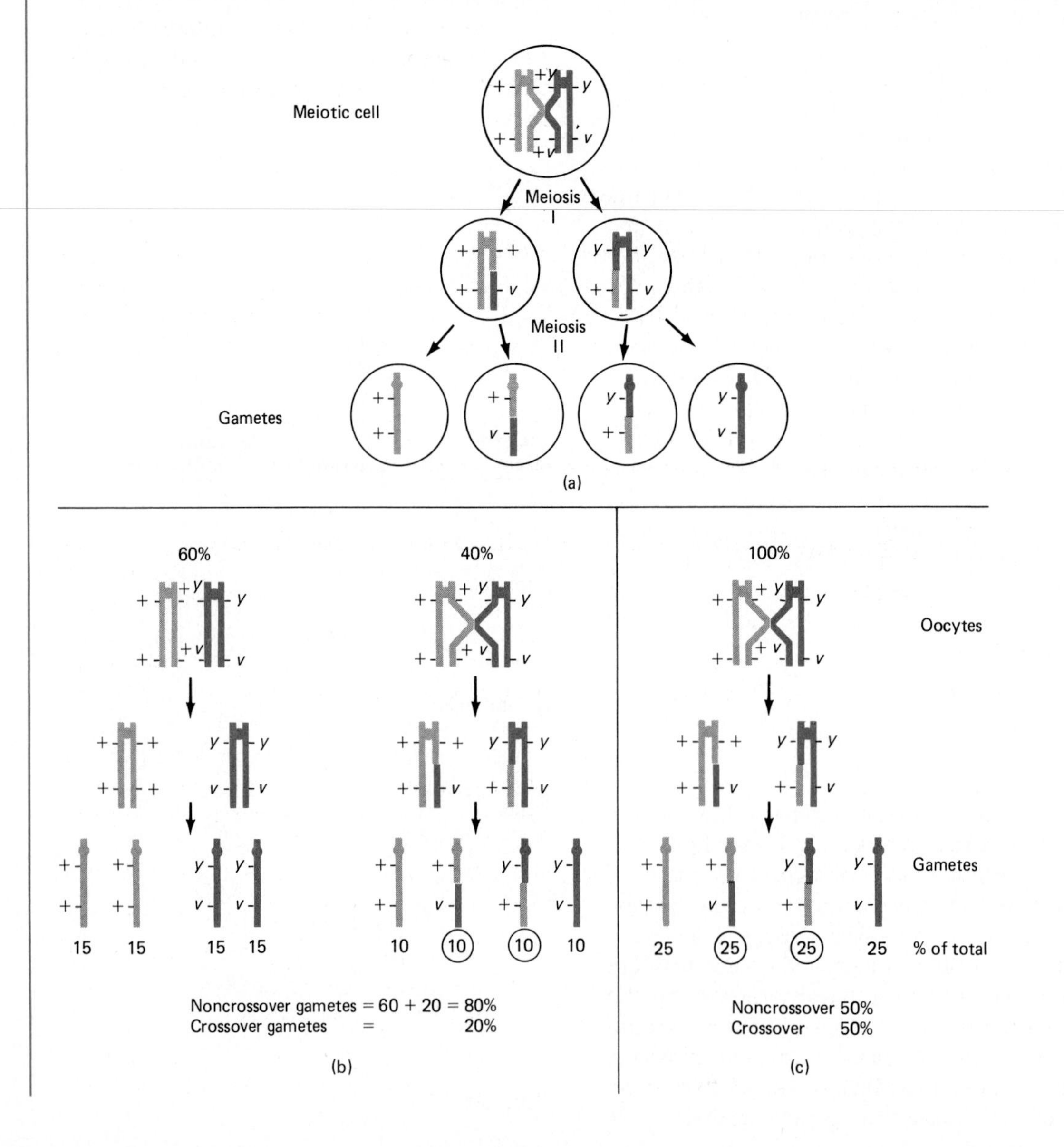

Each cell in which one exchange has taken place will, therefore, produce 50% crossover and 50% noncrossover gametes.

If a single exchange event takes place between the two gene loci in 40% of the meiocytes and no crossover event occurs in the remaining 60% of the meiotic cells, the meiocyte will produce 80% noncrossover and 20% crossover gamete types. All the gametes arising from cells in which no crossing over occurs (60%) plus half the gametes arising from cells in which crossover events take place (40%/2 = 20%) will be of the noncrossover types, as shown in (b). Therefore, in any case in which every meiocyte experiences one crossover event between two particular linked genes, a maximum of 50% crossover gamete types will be produced, as shown in (c).

In cases of 100% crossing over between two linked genes (100% of the meiotic cells have one crossover between two of the four chromatids), there can be a maximum of only 50% crossover gametes since half the gametes will receive the crossover chromatids and half will receive the noncrossover chromatids. In crosses between doubly heterozygous parents and doubly recessive types (+ +/*y v* × *y v*/*y v*) the progeny could be 50% parentals (+ +/*y v* and *y v*/*y v*) and 50% recombinants (+ *v*/*y v* and *y* +/*y v*), or 1:1:1:1, even though the two genes are linked.

| progeny class | genotype | number of individuals |
|---|---|---|
| parental | $\frac{Ab}{ab}$ | 305 |
| parental | $\frac{aB}{ab}$ | 295 |
| recombinant | $\frac{AB}{ab}$ | 18 |
| recombinant | $\frac{ab}{ab}$ | 22 |
| | total progeny | 640 |

$$\text{percent recombinants} = \frac{40}{640} = 0.062$$

$$0.062 \times 100 = 6.2\%$$

We would place these linked genes 6 map units apart on the chromosome map, as follows:

A B

6 units

In further studies we may find that gene *C* is linked to *A* and shows 10% recombination, or crossovers. Gene *C* is 10 map units distant from gene *A* in the same linkage group, but is the order of the three genes *ABC* or *CAB*? (We know it cannot be *ACB* since *A* and *B* are 6 units apart.) We can find out which of the two possibilities is correct by determining the distance between *B* and *C* in another set of two-factor testcrosses. We can predict that the gene order is *CAB* if genes *B* and *C* show 16% recombination but that the gene order would be *ABC* if we found 4% recombination between *B* and *C*, as follows:

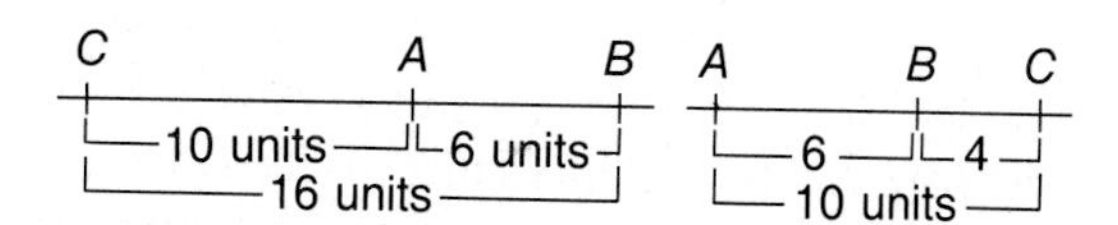


The three sets of two-factor crosses, involving *A—B, B—C,* and *A—C,* give consistent recombination values, thereby showing that genes occur in *fixed positions* relative to one another, that they occur in a *linear order* on the chromosome, and that they are found at *particular distances* from one another. We can use these results to predict the numbers of recombinant progeny that will be found any time these three genes are involved in crosses with one another.

The chromosome map (or gene map of the chromosome) has the following features:

***1.*** It summarizes the types of progeny obtained from particular crosses, and it summarizes all the linkage data.

***2.*** It contains two items of information: order of and distance between the genes, and it indicates consistency of recombination values in crosses.

*3.* It is a map of a linkage group, including all the genes found to be linked together on a single chromosome. A chromosome map can be derived for each linkage group of the genome (whole set of genes or chromosomes) in a species.

*4.* The same gene map will be derived no matter which alleles are carried by each parent in the crosses; the same results are obtained whether alleles are in the coupling (cis) or the repulsion (trans) phase of linkage.

*5.* Gene maps are reliable forecasting devices. They allow predictions about numbers and kinds of progenies in new crosses. Linkage analysis is, therefore, a powerful tool for describing the genome of a species.

By proceeding with two-factor crosses of linked genes, we can add more and more genes to the linear sequence, depending on the availability of mutant alleles for the wild-type alternatives. We cannot identify, much less map, a gene unless we know its inheritance pattern. To do this we must be able to identify segregation and recombination patterns involving members of pairs of alleles. Once a number of genes have been shown to be linked—that is, all are in the same linkage group—the order of and distances between genes are found to be *consistent* in all the combinations analyzed. The genes, therefore, are indeed in a linear arrangement. As Sturtevant pointed out, such an arrangement of genes made a very strong argument in favor of their location in the chromosome, which is the only known linear component in the cell that has a hereditary function.

We can construct an actual map of the chromosome by placing the linked genes, which have been analyzed, into arbitrary locations, based on the distances between them. When a gene is discovered to have linked genes only to its right and none to its left, that gene may be positioned at locus 0.0. The genes to its right are then located in accordance with the map units of distance found from linkage analysis. In Sturtevant's study of six X-linked characters in *Drosophila,* for example, the gene for body color was put at locus 0.0 of the X chromosome and the others were placed in relation to this gene.

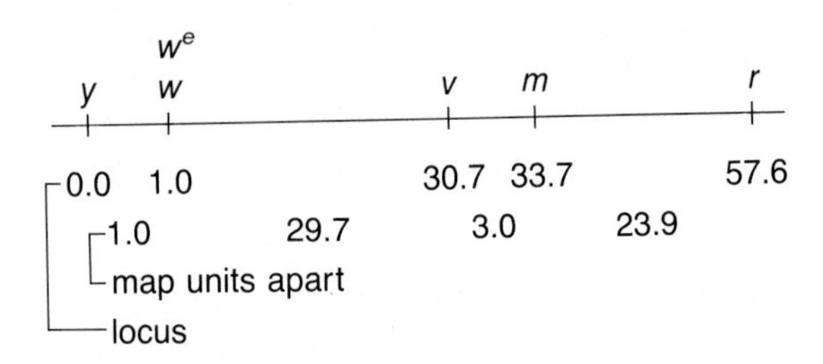


By putting the mutant allele on the map, we can more easily identify the gene, knowing at the same time that each of these alleles is recessive to its wild-type alternative. If a dominant mutant allele is mapped, then a capital-letter symbol is shown on the map (Fig. 5.11).

Sturtevant's map differs somewhat in the specific locus designations from the standard map of the X chromosome in *Drosophila.* The standard map was constructed from numerous linkage studies; it is based on vast amounts of data and can, therefore, be more refined in its details. Note that alleles *w* (white eye color) and $w^e$ (eosin eye color) are positioned at the same locus. The fact that testcrosses involving different X-linked genes and *w* or $w^e$ consistently showed about the same percentage recombinations, was preliminary evidence that *w* and $w^e$ were two mutant alleles of the same wild-type $w^+$ gene. The more critical and reliable test for multiple alleles of the same gene versus alleles of different genes verified this tentative conclusion (see Fig. 1.9). Since crosses between white-eyed and eosin-eyed flies always gave mutant progeny, these alleles were interpreted to be at the same locus on the chromosome. If the situation had been mutant × mutant → some wild-type segregants, then two different genes would have been implicated in the development of the same phenotypic character (eye color, in this case), as was found for *w* and *v*.

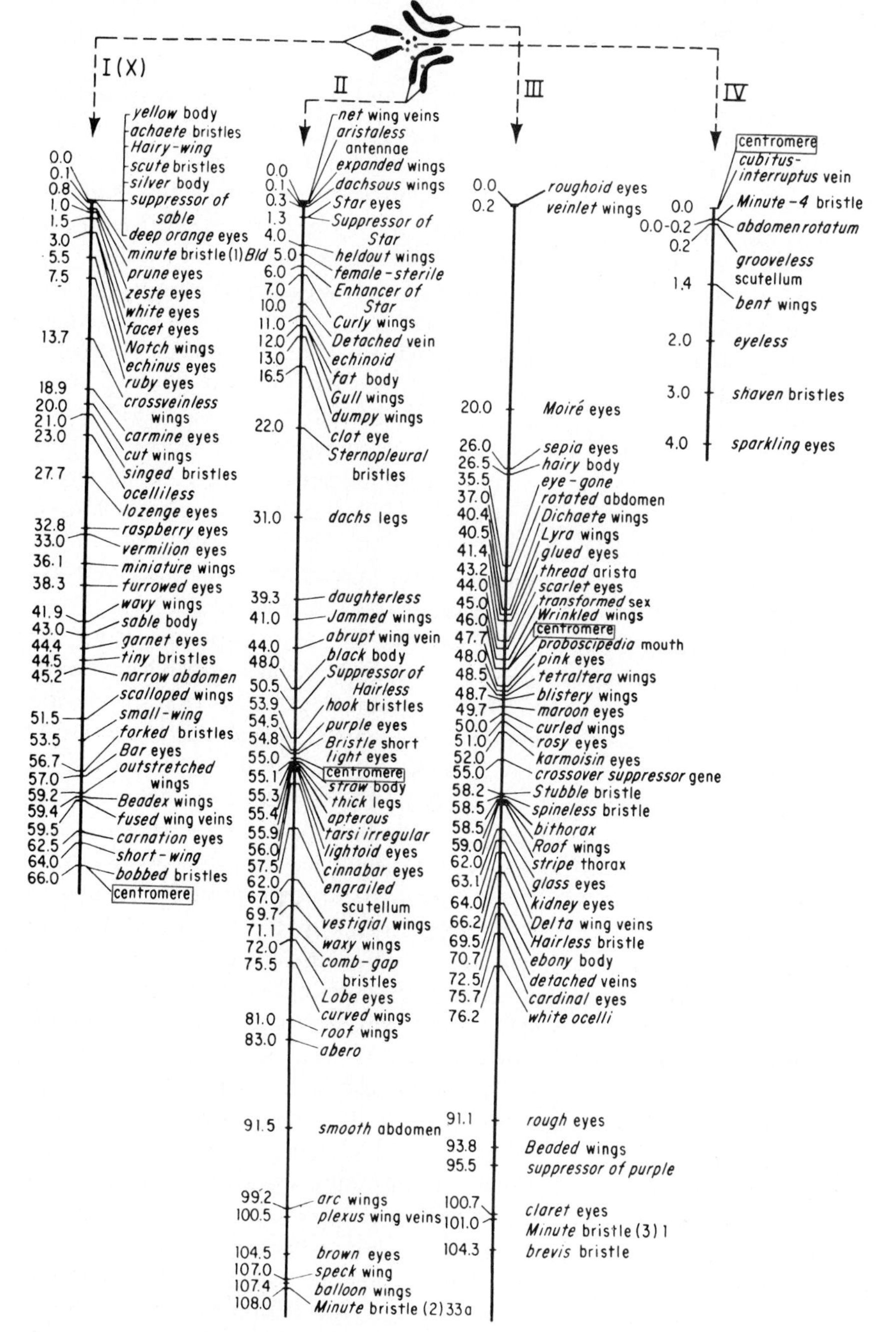


***Figure 5.11*** Linkage map of the four chromosomes of *Drosophila melanogaster,* showing some of the known genes. The *locus* is to the left and the *phenotypic mutant trait* is to the right of each chromosome. The centromere position is shown in a color box. Mutant characteristics due to dominant alleles are indicated by an initial capital letter, recessive mutant traits are shown in lowercase letters. Different genes may affect the same character, such as eye color, body color, wing shape. See D.L. Lindsley and E.H. Grell. 1968. *Genetic Variations of Drosophila melanogaster.* Carnegie Inst. Wash. Publ. No. 627, Washington, D.C., for extensive descriptions of these and other mutant traits. (Adapted with permission of Macmillan Publishing Company from *Genetics,* 2nd ed. by M. Strickberger. Copyright © 1976 by Monroe W. Strickberger.)

## 5.8 Double Crossover Problem

The percentage of genetic recombinants is converted directly into percentage of crossover gametes or chromosomes, and this latter percentage in turn is translated into map units of distance between two linked genes. This method has a pitfall, however, because recombination and crossing over are not the same thing. Crossing over is a chromosomal process involving an exchange of homologous chromosome segments; in contrast, recombinations are genetically identified from genotypic constitutions in progeny. We *infer* that a recombination arises as the result of a crossover between two genes, but the values for genetic recombination and detectable chromosomal crossovers are not always identical and, therefore, measurement of distances between genes in map units is not always entirely accurate.

The main reason that we cannot always equate genetic recombinations and chromosomal crossovers is that crossing over is a random process that takes place along the length of the chromosome. Greater distances between pairs of genes provide *more chances* for crossing over to take place, at random, between these genes. Another way of saying this is that the probability is greater that two or more crossovers will take place between genes that are farther apart than between genes that are closer together on the chromosome.

Suppose that genes *A* and *B* are really 30 map units apart on a chromosome. What are the chances that one crossover will take place between them? The answer is 30% because 30-map-unit measurement is derived from the percentage of recombinants in progenies and is translated directly into percentage of crossover gametes. What are the chances that two crossovers will take place, at random, within this distance of 30 map units, or, what is the probability of a double crossover in the region between two genes that are 30 map units apart? If the probability for one crossover is 30%, and each crossover is an independent event that takes place at random and without regard to other crossovers in the region, then the probability for two crossovers is the product of their separate probabilities. In this case, the probability for each crossover is 30% and, therefore, the probability of two crossovers between *A* and *B* is 30% × 30%, or 9% ($0.3 \times 0.3 = 0.09$). In other words, we may expect that out of every 30 gametes per 100 produced by crossing over between *A* and *B*, 9 of these will be double crossovers. What is the consequence of double crossovers? They yield alleles in parental combinations and, therefore, lead to parental, not recombinant, phenotypes. They reduce the percentage of recombinants recovered in the progeny and, therefore, lead to an *underestimate* of crossing over. This in turn leads to an underestimate of the map units of distance between two genes, as shown in Fig. 5.12.

Instead of 30% of the gametes from the heterozygous parent in a testcross having the *Ab* or *aB* recombinant genotype and producing *Ab*/*ab* or *aB*/*ab* recombinant testcross progeny, only 21% ($30 - 9 = 21$) of the gametes will lead to recombinants. The remaining 9% of these gametes will have double crossover chromosomes and will lead to the parental types *AB*/*ab* or *ab*/*ab* in testcross progeny. On finding 21% recombinants in the two-factor testcross progeny, we would assume 21% of the gametes resulted from crossing over and put only 21 map units between genes *A* and *B* on the chromosome.

The probability is lower for two crossovers to take place in the space between two genes if the distance is smaller. If genes *C* and *D* are actually 5 map units apart, then 5% of the gametes will have crossover chromosomes. Of these, only 0.25% ($0.05 \times 0.05$) will be double crossovers and 4.75% will be single crossovers, on the average. The discrepancy for small distances is, therefore, less significant and usually can be ignored until very fine detail and location are required for the standard map or for other purposes.

The double crossover problem may be remedied when a third gene is situated between *A* and *B* (Fig. 5.13). The gene in the middle serves as a "marker" for double crossover chromosomes. If we look at the double crossover

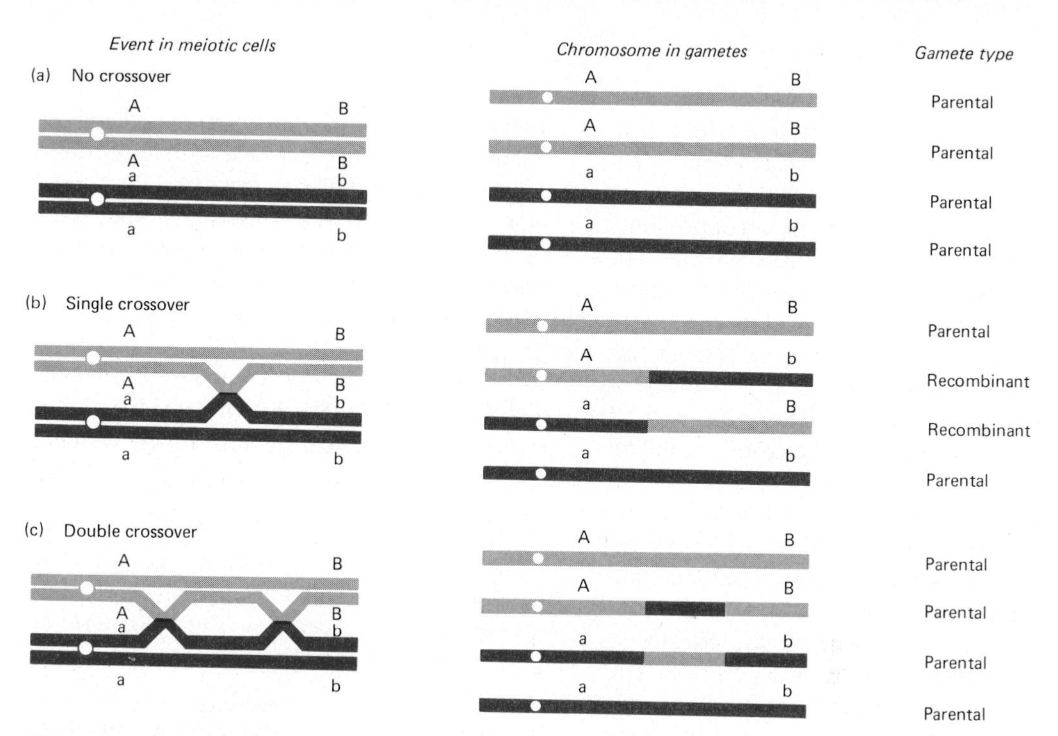


***Figure 5.12*** Gamete types arise from meiocytes according to crossing over in the region between two linked genes on a chromosome. Double crossovers lead to reduction in the observed frequency of recombinant gametes and, therefore, lead to underestimates of distances between linked genes.

chromosomes, we see that the *middle pair of alleles* is reversed when compared with the parental arrangements of alleles.

For the following reasons, the **three-point testcross** is the usual method of choice for mapping linked genes on chromosomes:

***1.*** The presence of three gene differences in testcross parents allows the identification of double crossover recombinants. This identification makes the relationship more meaningful between percentage of recombinants and percentage of gametes resulting from crossing over and provides more accurate values for determining map units of distance between linked genes.

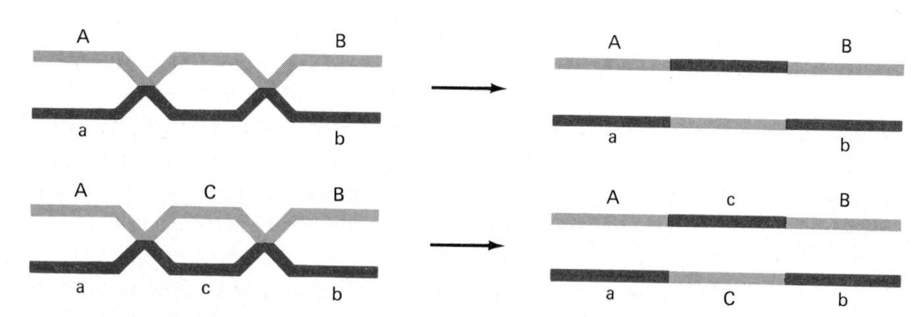


***Figure 5.13*** Double crossover chromosomes can be recognized genetically if one crossover occurs to either side of the middle "marker" gene. The middle pair of alleles is reversed relative to the parental arrangement. Such double crossover types of recombinants may not be detected in two-point testcrosses.

***2.*** A trihybrid cross yields eight phenotypic classes and each class represents a unique genotype. Eight classes is still a manageable number to analyze from reasonable numbers of progeny. We know that $2^n$ = the number of different phenotypic classes for alleles showing complete dominance, where $n$ = the number of pairs of heterozygous alleles. Thus, 16 phenotypes are produced for 4 pairs of alleles ($2^4$), 32 different phenotypes for 5 pairs of alleles ($2^5$), and so forth. Four- or five-factor testcrosses would require substantially larger numbers of progeny in order to recover all the phenotypic classes in the numbers needed for reliable samplings of genetic events observed from recombinations. From a practical standpoint, three-factor testcrosses are more convenient and equally as reliable as crosses involving more factors. The situation could be different, of course, for microorganisms that can be raised in huge numbers in a very brief time and in a relatively small space.

***3.*** It is more efficient and less time-consuming to analyze three genes at one time than to analyze three genes in several two-factor testcrosses. Instead of $A \times B$, $A \times C$, and $B \times C$ to discover the map order and distances between $A$, $B$, and $C$, we need only one three-factor testcross involving $A$, $B$, and $C$. This reduction of the number of crosses needing to be made is particularly important when we use organisms with a longer life cycle or ones requiring a great deal of space and maintenance. Anything that saves time and money and gives the same or improved results is highly desirable.

As we analyze representative three-point testcross experiments, we should try to keep these three features of experimental design and management in mind.

## 5.9 Mapping by the Three-Point Testcross Method

We can begin with a familiar example to illustrate how linkage analysis is performed when the three genes are known to be in the same linkage group. Since the goal is to learn the relative order and distance between linked genes, we can start with the three X-linked genes in *Drosophila* already mentioned: body color (gray, $y^+$, versus yellow, $y$), eye color (red, $w^+$, versus white, $w$), and wing length (long, $m^+$, versus miniature, $m$).

If $\frac{ywm}{ywm}$ females are crossed with $\underline{y^+w^+m^+}$ males, which is simply abbreviated throughout the discussion to $\underline{+\,+\,+}$ for the three wild-type alleles, $F_1$ progeny consist of heterozygous $\frac{+\,+\,+}{y\ wm}$ females and $\underline{ywm}$ males. We can either perform a testcross of $F_1$ females to triply recessive males from another stock or obtain the $F_2$ since these crosses would be identical. The $F_2$ or testcross progeny include the following:

| **phenotypic class** | **maternal chromosome present** | **number of progeny** |
|---|---|---|
| gray, red, long | $+\,+\,+$ | 1087 |
| yellow, white, miniature | $y\ w\,m$ | 1042 |
| gray, white, miniature | $+w\,m$ | 17 |
| yellow, red, long | $y\ +\,+$ | 15 |
| gray, red, miniature | $+\,+m$ | 543 |
| yellow, white, long | $y\ w\,+$ | 502 |
| gray, white, long | $+w\,+$ | 6 |
| yellow, red, miniature | $y\ +m$ | 4 |
| | | 3216 |

We must first determine the pair of parental phenotypic classes in this progeny. Since these genes are linked, we expect an excess of parentals, so we look for the largest two classes among the total of eight; these are $+\,+\,+$ and $ywm$. We already know this in this example because we know the alleles that were present on each of the X chromosomes in the $F_1$ females, and the parental, or noncrossover, genotypes are identical with these unchanged chromosomes.

The second step is to determine the pair of classes with the double crossover chromosomes since this pair will define the middle gene and, therefore, give us the order of the three genes in the chromosome. The double crossover phenotypic classes are the least numerous of all because double crossovers are less frequent than single crossover events. To produce a double crossover, one crossover must occur in the space between the middle gene and the gene to its right and another crossover between the middle gene and the gene to its left. The probability of a double crossover occurring is the product of the separate probabilities for each single crossover occurring, so it is less frequent than either of the single crossover events. By comparing the parental + + + and *ywm* with + *w* + and *y* + *m* in the two phenotypic classes present in the lowest numbers, we can see that *w* must be the middle gene and that it is flanked by *y* and *m*.

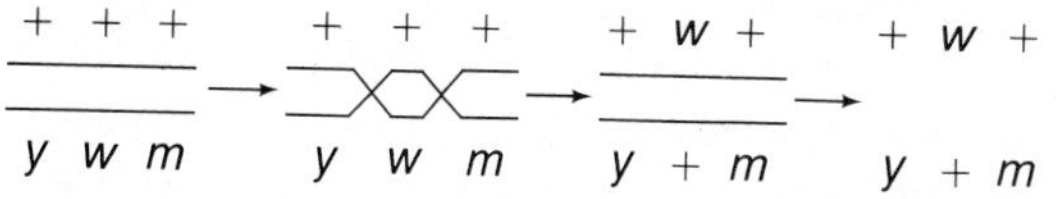

| + + + | → | + + + | → | + w + | → | + w + |
|---|---|---|---|---|---|---|
| y w m | | y w m (double crossover) | | y + m | | y + m |

It is now a simple matter to identify each of the single crossover pairs of classes. Crossing over between *y* and *w* would change the combination of $y^+/y$ in a chromosome retaining the parental combination of *wm* or + + alleles.

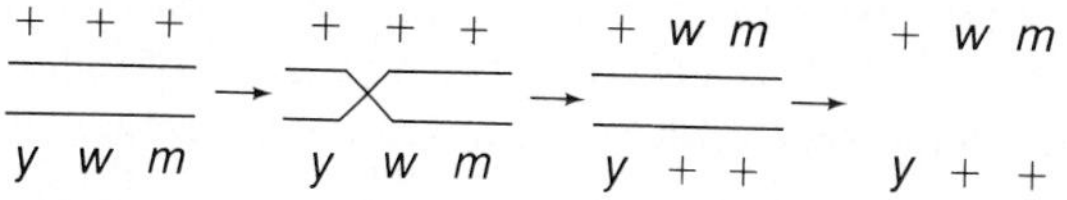

| + + + | → | + + + | → | + w m | → | + w m |
|---|---|---|---|---|---|---|
| y w m | | y w m (crossover) | | y + + | | y + + |

The remaining single crossover classes must be those involving the space between the loci for *w* and *m*.

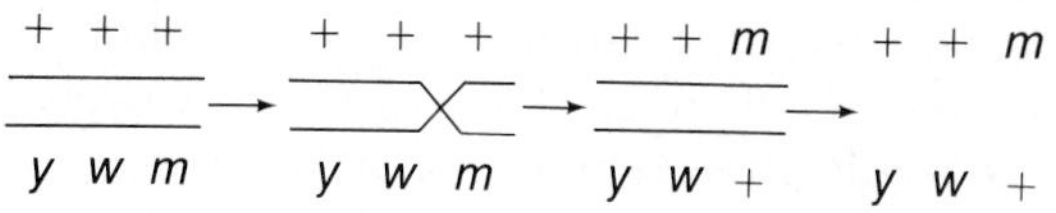

| + + + | → | + + + | → | + + m | → | + + m |
|---|---|---|---|---|---|---|
| y w m | | y w m (crossover) | | y w + | | y w + |

Notice that each of the crossover events gives rise to reciprocal classes of recombinant gametes. We expect to find approximately equal numbers of each genotype arising from two reciprocal crossover chromosomes because approximately equal numbers of gametes will have each of the crossover chromosomes due to the segregation of homologous chromosomes at meiosis.

The percentages of crossovers are calculated separately for the two regions in which crossovers occurred; we usually start at the left and label the first space region I and call the second space region II. For crosses of more than three factors, additional regions are identified in sequence by roman numerals. When we calculate crossovers in region I, we must *add* the total of the double crossovers to the total of the single crossovers for this *y*—*w* region. Similarly, to obtain the total percentage of crossovers in region II, we add the double crossovers to the single crossover events between *w* and *m*.

The reason we add double crossovers to each of the single crossover values is that crossing over took place in region I and region II in the double crossover classes. If we omit the double crossovers, we lose one of the advantages of the three-point testcross, namely, that we can detect region I crossovers or region II crossovers, even when a second independent crossover event has obscured the expected recombinations. When all the crossovers within a region are summed, the resulting percentage of crossovers (gametes or chromosomes) more accurately reflects the distance between two genes and translation into map units is closer to the actual distance (Box. 5.3).

The results of calculations from the three-point testcross are as follows:

| **region** | **gene loci** | **proportion of crossovers** | **percentage of crossovers** | **distance in map units** |
|---|---|---|---|---|
| I | *y*—*w* | $\frac{32 + 10}{3216}$ | 1.3 | 1.3 |
| II | *w*—*m* | $\frac{1045 + 10}{3216}$ | 32.8 | 32.8 |
| I + II | *y*—*m* | $\frac{32 + 1045 + 10 + 10}{3216}$ | 34.1 | 34.1 |

These three gene loci can now be mapped in their proper sequence and distances from each other on the X chromosome.

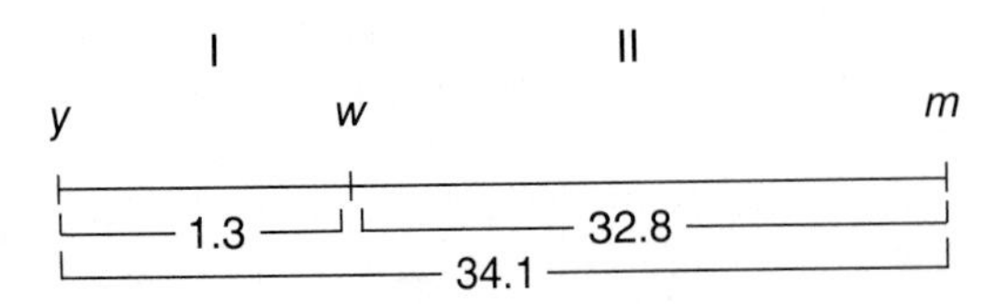


Now let us look at linkage analysis for autosomal genes in a three-point testcross. Previous studies of *Drosophila* have shown that the genes for *gray* versus *black* body color ($b^+$ and $b$), *red* versus *purple* eye color ($pr^+$ and $pr$), and *long* versus *vestigial* wing length ($vg^+$ and $vg$) are located on autosomal chromosomes since they produce identical progeny in reciprocal crosses. X-linked genes show different reciprocal progeny. The gene symbols indicate that all the wild-type alleles are dominant to their mutant alternatives, as was discovered in breeding analysis. We now want to find out if these genes are in the same linkage group, each in a different chromosome, or two in one chromosome and the third in another linkage group. To do this, homozygous flies from true-breeding stocks are crossed to obtain $F_1$ progeny that are triply heterozygous for the three pairs of alleles. The $F_1$ heterozygous ♀♀ are testcrossed to triply recessive ♂♂ and the following progeny are obtained:

| phenotype | chromosome from heterozygous parent | number of individuals |
|---|---|---|
| wild type | + + + | 59 |
| black, purple, vestigial | *b pr vg* | 51 |
| vestigial | + + *vg* | 416 |
| black, purple | *b pr* + | 402 |
| purple | + *pr* + | 23 |
| black, vestigial | *b* + *vg* | 21 |
| | | 972 |

The first question is: Are these genes linked? The $F_2$ consists of only six phenotypes instead of eight, with only two of the phenotypes present in very large numbers. These phenotypes are reciprocal (what is recessive in one phenotype is dominant in the other for each pair of alleles: + + *vg* and *b pr* +), and there is no hint of a 1:1:1:1 ratio, which means that no two genes are assorting independently. We can, therefore, assume that all three genes are linked. There is no evidence for independent assortment. The noncrossover, or parental, chromosomes from the heterozygous parent are, therefore, + + *vg* and *b pr* + since they are the most numerous classes. The missing phenotypes must be the relatively more rare double crossovers, none of which was produced, for some unknown reason. (Perhaps a crossover in region I or in region II reduces the probability for another crossover in the other region.) In any event, it is clear that the missing two genotypes are + *pr vg* and *b* + + since all the other possible combinations are present.

If we assume that + *pr vg* and *b* + + are the missing double crossover phenotypic classes, we can find the middle gene by comparing these classes with the parental types. The transposed pair of alleles, relative to the arrangement in the noncrossover chromosomes, is +/*pr*, which is the gene in the middle $\left(\frac{+\,+vg}{b\ pr+} \text{ versus } \frac{+pr\ vg}{b\ +\,+}\right)$. The sequence is, therefore, *b pr vg*, and we can now find the single crossovers in region I, between *b* and *pr*. We then calculate the percentage recombinants as $^{44}/_{972}$ = 4.5% and determine that there are 4.5 map units in region I. The single crossovers in region II, between *pr* and *vg*, include 110 recombinants out of a total progeny of 972, which is 11.3% recombination and thus the distance between *pr* and *vg* is 11.3 map units. The map is therefore:

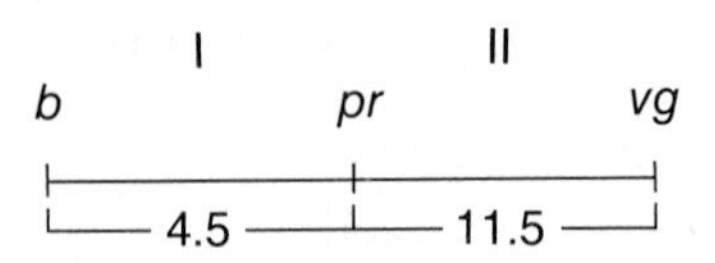


We can verify the gene order by calculating the percentage recombination between *b* and *vg*.

If these are the outside members of the chromosome segment and if *pr* is in the middle, then the percentage of crossovers between *b* and *vg* should approximate 4.5 + 11.3. A crossover between *b* and *vg* would produce the following:

| b | + | → | b | vg |
|---|---|---|---|---|
|  | vg |  | + | + |

The recombinant phenotypic classes showing *b* and *vg* in the same chromosome are *b pr vg* and *b* + *vg*; those showing $b^+$ and $vg^+$ in the same chromosome are + + + and + *pr* +. When these are totaled (51 + 21 + 59 + 23 = 154), we can calculate 154/972 = 15.8% recombinants involving *b* and *vg*. This is exactly what is expected for the two genes that are farthest apart and at each end of the chromosome segment carrying these three linked genes.

We also could have taken genes in combinations of two to determine the percentages of recombinants in each two-factor combination. By finding 4.5% for *b*—*pr*, 11.3% for *pr*—*vg*, and 15.8% for *b*—*vg* crossovers, we would have put these genes together on the map in the same logical order. It would be very similar to the example in Section 5.7 in the discussion of mapping by two-point testcrosses for genes *A*, *B*, and *C*.

As a final example of the value of the three-point testcross, it is useful to examine the progeny of a testcross with $F_1$ ♀♀ heterozygous for ebony (*e*) body color, rough (*ro*) eyes, and vestigial (*vg*) wings, as follows:

| | |
|---|---|
| *e ro vg* | 210 |
| + + + | 202 |
| *e ro* + | 198 |
| + + *vg* | 206 |
| *e* + *vg* | 47 |
| + *ro* + | 49 |
| *e* + + | 48 |
| + *ro vg* | 50 |
| total progeny | 1010 |

We first notice that some linkage is involved, because independent assortment of all three genes would have produced a 1:1:1:1:1:1:1:1 ratio. Since crossing over is infrequent, the four largest classes must be noncrossover genotypes. We can find out which two genes are linked by determining which two pairs of alleles do not assort in the noncrossover classes. For the genes *e* and *ro*:

| | |
|---|---|
| *e ro* | 210 |
| + + | 202 |
| *e ro* | 198 |
| + + | 206 |

408 *e ro*:408 + + = 1:1

Since there are only two classes, *e ro* and + +, these two pairs of alleles have not assorted to give a 1:1:1:1 ratio and must, therefore, be linked. These large classes are the noncrossover, parental classes.

We can now find out whether *vg* is linked to *e* and *ro* by checking the parental classes again. If *vg* is linked to *e* (or *ro*), there will be only two parental types, just as before.

| | |
|---|---|
| *e vg* | 210 |
| + + | 202 |
| *e* + | 198 |
| + *vg* | 206 |

Since the ratio is 1:1:1:1 for the two pairs of alleles, they must have assorted independently and are, therefore, not linked. (And *vg* is not linked to *ro* either because *ro* is linked to *e*.)

We can now calculate linkage between *e* and *ro* by looking at the smaller-sized recombinant classes.

| | |
|---|---|
| *e* + | 47 |
| + *ro* | 49 |
| *e* + | 48 |
| + *ro* | 50 |

194 recombinants/1010 total progeny = 0.192

0.192 × 100 = 19.2% recombinants = 19.2 map units

The map is:

| e | ro | vg |
|---|---|---|
| 19.2 | | |

## BOX 5.3 Interference and Coincidence

The chromosome map represents a table of probabilities and allows us to predict the frequencies of recombinants that can be expected for any group of linked genes at assigned loci on the chromosome. The map assignments themselves, after all, are based on observed frequencies of recombination due to crossing over between linked genes. Since crossing over is a random event, we can predict the probability of such a random event taking place in other meiotic cells in future crosses. The laws of probability govern the frequencies of random events.

In order to determine whether two different recombinations arose as independent events, we compare the *observed* and *expected* (predicted) frequencies of double recombinants. If each crossover is independent of other crossover events, we expect the simultaneous occurrence of the two events to be equal to the product of their separate probabilities. If one recombination frequency is 5%, and the second recombination frequency is 18%, we expect the double recombination frequency to be $0.05 \times 0.18$, or 0.009. If we find 0.9% double recombinants, we can say that each recombination was an independent event.

In many cases the observed frequency of double recombinants is *less* than the expected value. For example, according to the X chromosome map of *Drosophila,* we expect 1.5% recombination frequency between the *y* and *w* gene loci, and 34.1% recombination frequency between the *w* and *m* loci:

| gene | *y* | *w* | *m* |
|---|---|---|---|
| locus | 0.0 | 1.5 | 36.1 |

The expected double recombination frequency is $0.015 \times 0.341 = 0.005$, or 0.5%, but the observed frequency in the experiment shown in Section 5.9 is $10/3216 = 0.3\%$. The consistent occurrence of values lower than those expected is interpreted to mean that the occurrence of one crossover in a region reduces the chances of another crossover in that region (in our example, the *y—m* region). This phenomenon is called **interference** because one crossover interferes with the occurrence of another crossover. Its causes are not known at present.

The strength of interference is usually summarized as a **coefficient of coincidence,** which is simply a ratio of observed to expected frequencies of double recombinants.

$$\frac{\text{observed frequency of double recombinants}}{\text{expected frequency of double recombinants}} = \text{coefficient of coincidence}$$

In the experiment cited above for the *y—m* region, the coefficient of coincidence can be calculated to be $\frac{0.3}{0.5} = 0.6$. This ultimate value would be determined from many crosses involving this region of the *Drosophila* X chromosome. Once coefficients of coincidence have been calculated for chromosomal regions, the map values for the gene loci are adjusted appropriately. For coincidence values that are less than 1 (some interference), the map distance between two loci is *lengthened* to reflect the lower observed double recombinant frequency compared to the expected.

Coincidence varies in inverse proportion to interference, with lower coincidences for higher degrees of interference. Coincidence values generally range between 0 and 1. Complete interference of the second crossover leads to a coincidence value of 0; total lack of interference leads to a coincidence value of 1 (observed = expected). In studies of very closely linked genes, particularly in some microorganisms and bacteriophages, the number of double recombinants observed may be greater than expected. This phenomenon is called negative interference, and it is recognized by coincidence greater than 1. Its causes are not understood at present.

## Human Pedigree Analysis

One method of gene mapping in humans, called the **family method,** involves analysis of inheritance patterns and of linkage, using information obtained from family histories, or pedigrees. Such studies have a conventional genetic basis because they are concerned with patterns of gene transmission from parents to progeny during reproduction. Using the family method, however, we are restricted to analyzing existing families, which have relatively small numbers of individuals, rather than progenies obtained in breeding experiments that are conducted under controlled conditions and are based on particular experimental designs.

### 5.10 Recombination Analysis by the Family Method

A useful procedure for estimating recombination frequencies and map units of distance between X-linked genes is the **grandfather method.** By this method it is possible to determine whether a mother who is doubly heterozygous for two X-linked genes has both mutant alleles in the same X chromosome (coupling phase) or in different X chromosomes (repulsion phase) of the pair she carries. If the phenotype of the maternal grandfather is known, then we know which alleles are present on his single X chromosome. His daughter inherits this X chromosome, so we know that one of her X chromosomes carries the same combination of alleles that her father had. Her second X chromosome must then carry the alternative members of the two pairs of alleles since she is doubly heterozygous. If the maternal grandfather is Ab in phenotype, his daughter has *Ab* on the X chromosome she inherited from him. Her second X chromosome must have alleles *aB* to complete her doubly heterozygous genotype. If her father's phenotype indicates that he is genotypically *AB*, then she must carry the two pairs of alleles in the coupling phase (*AB*/*ab*).

If the location of the specific alleles is not known, it is impossible to accurately determine which of her sons carry the parental allelic combination and which sons carry a recombinant X chromosome with a recombinant genotype. When the allelic combination of the mother's X chromosomes is known, it is possible to determine which of her sons carry the recombined allelic combination and which sons do not. For example, if we know that she carries the genotype *AB*/*ab*, then her sons with the AB or ab phenotype must have a noncrossover parental X chromosome. Any sons of this woman with Ab or aB phenotype must carry a crossover X chromosome with recombined alleles.

From a number of family histories, about 1 in 20 sons of mothers who are doubly heterozygous for the X-linked genes governing *color blindness* (*Cb*) and *glucose 6-phosphate (GPD) deficiency* (*Gpd*) were found to be recombinants, according to their phenotypes (Fig. 5.14). On the basis of these combined family studies, the *Cb* and *Gpd* gene loci would be situated about 5 map units apart ($\frac{1}{20} \times 100 =$ 5% recombination) on the X chromosome map. In cases in which a minimum of detailed information is available or the genotype of the maternal grandfather is unknown, statistical methods of analysis are used to estimate recombination frequencies. Advanced methods have been developed to calculate probabilities of occurrence of particular recombinants from summed family history data.

Relatively few autosomal chromosomes have been mapped using the family method, as compared with X chromosome mapping. It is more difficult to detect segregation of pairs of autosomal alleles than to detect segregation of single X-linked alleles in hemizygous males. Autosomal gene recombination can be analyzed more easily in families in which one parent is doubly heterozygous, the other parent is doubly recessive (equivalent to a testcross), and the grandparents' genotypes are known so that coupling or repulsion phases of linkage can be determined for the doubly heterozygous parent (Fig. 5.15).

The analysis is even easier if the genes being studied have codominant alleles or if the linked marker trait is governed by a dominant allele. In

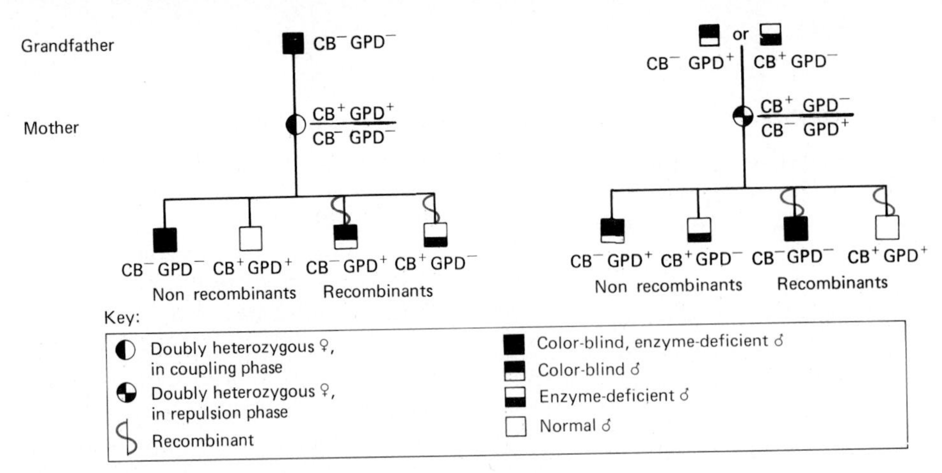


***Figure 5.14*** Determining linkage distances on the human X chromosome by the grandfather method. Recombinant genotypes can be identified and their frequency determined if it is known whether the doubly heterozygous mother of affected sons has both linked mutant alleles in *coupling* (left) or *repulsion* (right). This information can be deduced if the genotype of her father (grandfather of affected children) is known since she obtains one X from each of her parents. Since her father's X does not undergo crossing over, it is transmitted intact to his daughters. The frequency of recombinant sons, relative to the total number of sons, provides the information for determining linkage distance between the two loci.

such situations it is possible to determine which individuals carrying dominant or codominant alleles are heterozygous and which are homozygous. Allelic segregation can be determined in every generation for dominant or codominant alleles, whereas recessive alleles can be masked in heterozygotes and, therefore, will not be evident in every generation.

Once the first gene has been assigned to a particular autosome, other genes can be as-

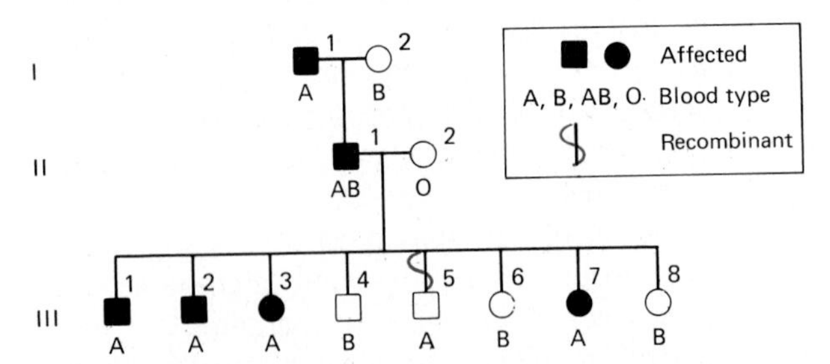


***Figure 5.15*** Determining linkage by the grandfather method for a rare affliction and blood group. The affliction is due to a dominant allele ($D$) expressed in heterozygotes and homozygotes; normal individuals are recessive ($dd$). Inheritance of the ABO major blood-group gene is governed by codominant alleles $I^A$ and $I^B$ plus recessive $i^O$ multiple alleles. The father (II-1) must be doubly heterozygous in the coupling phase of linkage ($D\ I^A/d\ I^B$), according to the phenotypes of his parents. All the children receive $d\ i^O$ from their doubly recessive mother (II-2) but they receive various allelic combinations from their father. If the two pairs of alleles assorted independently, the children would be equally likely to receive any of four possible allelic combinations ($D\ I^A$, $D\ I^B$, $d\ I^A$ $d\ I^B$) from II-1. If the two genes are linked, parental combinations $D\ I^A/d\ i^O$ and $d\ I^B/d\ i^O$ would be more frequent than recombinations $D\ I^B/d\ i^O$ and $d\ I^A/d\ i^O$. The one recombinant is III-5, who must have genotype $d\ I^A/d\ i^O$, whereas the other seven children are parental types. Linkage appears evident, but these data must be grouped with other family histories to estimate the degree of linkage between the two genes.

signed to the same chromosome if linkage is established. The first human autosome to have a gene assigned to it was chromosome 1. Roger Donahue made this discovery in 1968. He found a correspondence between the presence of a cytologically altered chromosome 1 and the expression of the Duffy blood-group gene. The assignment was based on concordance between a visibly distinct chromosome and the Duffy marker trait expression in family members. Once this first gene was assigned to chromosome 1, other genes found to be linked to the Duffy gene were also assigned to chromosome 1. Because of this head start, chromosome 1 is the best mapped of all the human autosomes at the present time.

Gene mapping by the family method, as well as by other methods to be discussed in Chapter 12, has provided a considerable amount of linkage data. The human genome is the best mapped of any mammalian species. A great deal of linkage data and other genetic information have been stored in computer banks. Geneticists can extract the information required and conduct the most complex kinds of linkage analysis, using specially derived statistical methods appropriate for these kinds of data.

## 5.11 Usefulness of the Genetic Map in Prenatal Diagnosis

In addition to many kinds of biologically important information that can be gained from detailed genetic maps, knowledge of gene location provides an immediate practical value in predicting risk of birth of an afflicted child. Two examples can illustrate the principles involved.

A number of inherited disorders cannot be detected in the fetus, but prospective parents can be informed about the risk of having an afflicted child in cases in which the gene is known to be closely linked to a useful marker gene. Suppose a pregnant woman who has had one hemophilic son wants to know the risk (chance) of having another son with this X-linked disorder. She must be heterozygous for the *hm* allele since she has no symptoms, but one of her X chromosomes was given to her hemophilic son through the egg. If she has no linked marker trait on her X chromosome, she can only be told that the chance is 50% for a son to inherit the X chromosome carrying the recessive *hm* allele (see Box. 4.1).

If the woman is known to be doubly heterozygous for hemophilia and for another X-linked trait, the grandfather method can provide information about the occurrence of the two pairs of alleles in the coupling or the repulsion phase. Under these conditions more reliable estimates of risk can be calculated. Suppose the woman is doubly heterozygous for the hemophilia gene and for the codominant alleles governing the synthesis of two distinct forms of the GPD enzyme, which we can call GPD-A and GPD-B. If we know that she carries the *hm* allele and the *Gpd-A* allele on the same chromosome and *Hm* and *Gpd-B* on the other X chromosome, a test of the GPD enzyme present in the fetus will reveal the risk that the fetus is carrying the *hm* allele.

The hemophilia gene and the GPD gene are 5 map units apart on the X chromosome. If the fetus is found to have GPD-A, the chance is 95% that it also carries the *hm* allele in a nonrecombinant X chromosome (Fig. 5.16). Since the chance is 5% for recombination between *hm* and *Gpd-A* in this chromosome, only 95%, and not 100%, of the chromosomes carrying *Gpd-A* will also carry *hm*. The risk of the male fetus developing hemophilia after birth is, therefore, 95% if it has the GPD-A trait in this case. If the fetus is found to have the GPD-B trait instead, 5% of the chromosomes with *Gpd-B* will be recombinant and will carry the *hm* allele instead of *Hm*, in this case. A male fetus with the GPD-B trait will, therefore, be at 5% risk of developing hemophilia. It is obvious that a prenatal diagnosis can provide odds of 95% or 5%, which is more specific than 50% risk for a son to carry the recessive *hm* allele.

In order to determine whether the fetus has the GPD-A or GPD-B enzyme, a physician, using

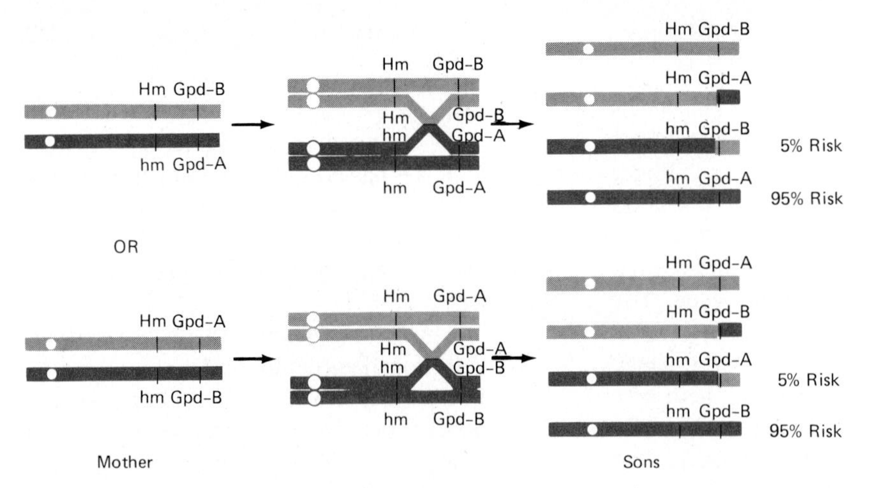


***Figure 5.16*** Estimation of risk of inheritance of the X-linked recessive disorder of hemophilia in sons of a doubly heterozygous carrier mother who has linked codominant alleles (*Gpd-A*, *Gpd-B*) for the two forms of the enzyme glucose 6-phosphate dehydrogenase. If the mother has one X chromosome with the *Gpd-A* allele, along with the recessive hemophilia allele, *hm*, the risk is 95% that the fetus will have the noncrossover X chromosome with *hm* and 5% that it will have the *hm* allele on a crossover X chromosome if GPD-A enzyme activity is found to characterize fetal cells. Alternatively, if the mother is known to have received X chromosomes from her parents with *Gpd-B/hm* on one chromosome and *Gpd-A/Hm* on the other, the risk of the fetus receiving the *hm* allele is 95% if GPD-B activity is present, or 5% if GPD-A activity is present in the amniotic fluid. Without this linked marker, the risk of a male fetus inheriting the *hm* allele from its heterozygous carrier mother can only be given as 50% (see Box. 4.1). Probabilities of 95% and 5% are based on the known value of 5 map units separating the two X-linked genes and, therefore, of 5% probability of crossover X chromosomes or 95% probability of noncrossover chromosomes being produced in meiosis.

a hypodermic syringe needle inserted through the woman's abdominal wall, removes a sample of amniotic fluid surrounding the fetus. Cells normally shed by the fetus into the surrounding amniotic fluid can be isolated from the sample and analyzed microscopically and biochemically. The first step is to look for a Y chromosome and, thereby, determine if the fetus is male. If it is a male fetus it will be at risk for hemophilia in this family. A female fetus is at no risk because the second X chromosome comes from the normal father, who carries the dominant *Hm* allele for normal blood clotting.

The GPD test is performed if the fetus is male. The presence of GPD-A in this case means that the chance is 95% that the X chromosome is a noncrossover and carries the *hm* allele. The fetus is at 95% risk of developing hemophilia. If the fetus is found instead to have the GPD-B trait, the chance is 5% that the chromosome is recombinant and, therefore, carries the *hm* allele. These risk values would be reversed, of course, if the woman carries the *hm* allele on the same chromosome as the *Gpd-B* allele, 5 map units away (see Fig. 5.16).

Knowing if the risk is 5% or 95% can be more helpful to the family in making plans than the 50% value for risk based only on the knowledge of the inheritance pattern. More specific predictions are possible with information from the genetic map.

In a similar situation, involving an autosomal disorder, similar procedures can be implemented. The dominant disorder *myotonic dystrophy,* which affects muscles, and the *secretor protein,* which is found in the saliva and in other secretions of people carrying the dominant *Se* allele, are known to be governed by genes that are linked about 10 map units apart on an unspecified autosome. Depending on whether the doubly heterozygous parent carries the dominant allele of each gene in the coupling or repulsion phase (and whether the other parent is doubly recessive), different predictions of risk for myotonic dystrophy can be made based on prenatal diagnosis. The marker trait is secretor

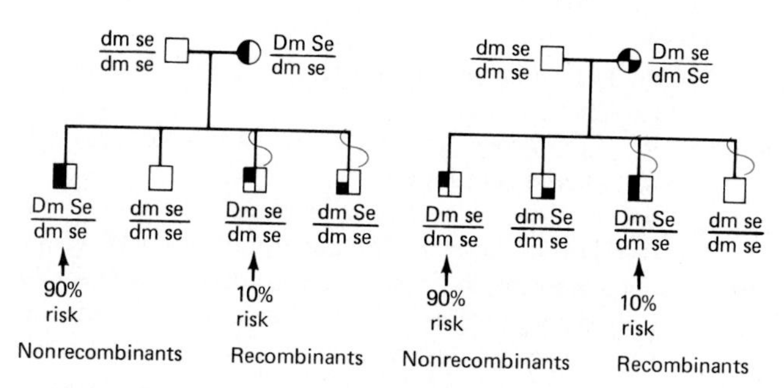


***Figure 5.17*** Estimation of risk of inheritance of myotonic dystrophy (*Dm*), an autosomal dominant muscular disorder. This gene is linked about 10 units apart from the secretor gene, whose dominant allele *Se* governs production of salivary secretor protein. Secretor protein is absent in recessives of genotype *se/se*. Risk of inheritance of allele *Dm* is 90% for nonrecombinant or 10% for recombinant genotypes if the doubly heterozygous parent is known to carry the linked genes in either coupling or repulsion. The fetal amniotic fluid would be examined for the presence or absence of secretor protein, which is the marker trait for the autosome also carrying *Dm*. In the absence of the linked marker, risk is estimated as 50%, according to standard probability determinations (see Box 4.1).

protein, and its presence or absence in fetal cells that are isolated from amniotic fluid can provide information about the probability that the chromosome also carries the dominant allele of the myotonic dystrophy gene (Fig. 5.17).

If both dominants are in the same chromosome (coupling phase) in the heterozygous parent, the chance is 90% that the two dominants will be inherited together in a noncrossover chromosome in the offspring. In 10% of the cases, the two dominant alleles will be separated by a crossover since the two genes are 10 map units apart. The presence of secretor protein means that the fetus has the chromosome carrying the *Se* allele and that the risk is 90% that this is a noncrossover chromosome and, therefore, also carries the dominant allele (*Dm*) for myotonic dystrophy.

If the doubly heterozygous parent carries both dominants in the repulsion phase of linkage, the risks are reversed. Whether the predicted risk is 10% or 90%, the value is much closer to the real situation than is the 50% risk that is based only on probabilities for inheritance of a dominant allele in this family.

In an increasing number of cases, an inherited or congenital disorder can be diagnosed directly from fetal materials obtained by amniocentesis. The need to estimate probabilities of occurrence is eliminated when direct determinations can be made during a pregnancy. In the great majority of cases, however, prenatal tests are not available or are not informative. As more genes are mapped in specific linkage groups and as more marker genes are assigned to chromosomes, more accurate predictions of risk can be calculated. In cases in which linked neighbor genes are not yet mapped, general predictions of risk can still be estimated on the basis of probabilities derived from Mendelian phenotypic ratios.

## Questions and Problems

***5.1*** In *Drosophila,* the dominant X-linked gene for bar eye (*B*) is at locus 57, and the recessive miniature wings (*m*) is at locus 36. The cross shown at right is performed:

$$\frac{+\ \ m}{B\ \ +}\ ♀♀ \times \frac{+\ \ +}{\rightharpoonup}\ ♂♂$$

***a.*** What is the expected proportion of recombinant female gametes produced by the female parents?
***b.*** What is the expected proportion of genotypes and phenotypes in the progeny?
***c.*** How many kinds of sperm would be produced relative to these two genes?

**5.2** In tomatoes, round fruit shape (*O*) is dominant over elongate (*o*), and smooth fruit skin (*P*) is dominant over peach skin (*p*). Testcrosses of double recessives to $F_1$ individuals heterozygous for these two pairs of alleles gave the following results:

| smooth, round | smooth, long | peach, round | peach, long |
|---|---|---|---|
| 24 | 246 | 266 | 24 |

***a.*** In the $F_1$, were the two pairs of alleles linked in the coupling or the repulsion phase?
***b.*** What is the percentage recombination between these two genes?

**5.3** In *Drosophila,* crossing over does not occur during meiosis in male spermatocytes but does occur in female oocytes. Taking advantage of this unusual situation, what crosses would you perform to show that the autosomal genes for aristaless antennae (*al*) and black body (*b*) are linked?

**5.4** What kinds of tetrads would be produced by doubly heterozygous *Neurospora* if the following crossovers occurred between two genes linked on the same chromosome arm: ***a.*** Single crossover. ***b.*** Three-strand double crossover. ***c.*** Four-strand double crossover. ***d.*** What kinds of tetrads would result if the single crossover took place between the centromere and the gene nearest to it?

**5.5** A strain of *Neurospora crassa* that was adenine-requiring ($ad^-$) and tryptophan-requiring ($trp^-$) was crossed to a wild-type strain (+ +) and the following tetrads were produced, as shown below.

| | tetrad classes | | | | | | |
|---|---|---|---|---|---|---|---|
| | 1 | 2 | 3 | 4 | 5 | 6 | 7 |
| | *ad trp* | *ad* + | *ad trp* | *ad trp* | *ad trp* | *ad* + | *ad trp* |
| | *ad trp* | *ad* + | *ad* + | + *trp* | + + | + *trp* | + + |
| | + + | + *trp* | + *trp* | *ad* + | *ad trp* | *ad* + | + *trp* |
| | + + | + *trp* | + + | + + | + + | + *trp* | *ad* + |
| frequency | 147 | 21 | 93 | 6 | 24 | 3 | 6 |

***a.*** Determine whether the two genes are linked. Explain your reasoning.
***b.*** If the genes are linked, draw a linkage map and include the centromere.

**5.6** In *Neurospora crassa,* synthesis of arginine and histidine are known to be under the control of different genes (*arg, his*). After a cross between two genetically different strains, the following arrangements of spore pairs are observed in the asci:

| | ascus types | | | |
|---|---|---|---|---|
| | 1 | 2 | 3 | 4 |
| spore pair 1 | *arg his* | + *his* | + + | *arg* + |
| spore pair 2 | *arg his* | + *his* | + + | *arg* + |
| spore pair 3 | + + | *arg* + | *arg his* | + *his* |
| spore pair 4 | + + | *arg* + | *arg his* | + *his* |
| frequency | 84 | 80 | 78 | 84 |

***a.*** Are the two genes linked? Explain your reasoning.
***b.*** How are the genes situated in the chromosome(s) with respect to the centromere?

**5.7** A riboflavinless strain ($rib^-$) of *Neurospora* is crossed with a tryptophanless strain ($trp^-$), with the following results:

| number of asci | tetrads | | | |
|---|---|---|---|---|
| 258 | *r* + | *r* + | + *t* | + *t* |
| 8 | *r* + | + *t* | *r* + | + *t* |
| 124 | *r* + | *r t* | + + | + *t* |
| 4 | *r* + | + + | *r t* | + *t* |
| 2 | *r t* | + + | *r* + | + *t* |

Construct a genetic map showing the arrangement of the two genes in relation to the centromere and to each other. Calcutate the map distances.

**5.8** The cross *a b c* × + + + is made using an ascomycete with unordered tetrads. From analysis of

200 asci, determine the linkage relationships between the three loci.

| | tetrad class | | | |
|---|---|---|---|---|
| | 1 | 2 | 3 | 4 |
| | *a b c* | *a b* + | *a* + *c* | *a* + + |
| | *a b c* | *a b* + | + + *c* | + + + |
| | + + + | + + *c* | *a b* + | *a b c* |
| | + + + | + + *c* | + *b* + | + *b c* |
| frequency | 80 | 84 | 20 | 16 |

***5.9*** Crossing over is known to occur in diploid somatic cells of various organisms, including *Drosophila* and the ascomycetous fungus *Aspergillus nidulans*. Patches of tissue showing recombinant phenotypes appear in a background of tissues with parental phenotype, indicating the occurrence of segregation and recombination of pairs of alleles during mitotic divisions that produce cell lineages making up the organism. A diploid somatic cell of *A. nidulans* has the following genetic constitution:

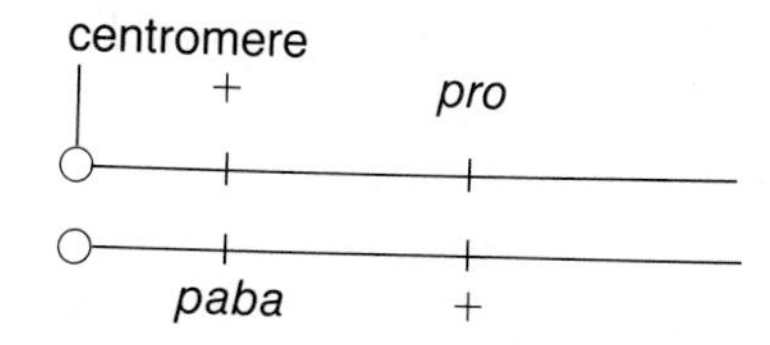


*pro* = proline requiring
*paba* = *p*-aminobenzoic acid requiring

***a.*** If crossovers occur between the *pro* and *paba* loci, what genotypes could be produced in daughter cells after mitosis?
***b.*** What patches of different phenotype might be produced in wild-type background?
(Hint: In mitosis, sister chromatids of each replicated chromosome in diploid cells move to opposite poles so that each daughter nucleus receives one chromatid of each replicated chromosome.)

***5.10*** In corn, a dominant gene *C* produces colored kernels, its recessive allele *c* produces colorless kernels. Another dominant gene, *Sh*, produces full kernels, and its recessive allele (*sh*) produces shrunken kernels. A third dominant gene, *Wx*, produces normal endosperm, and its recessive allele *wx* produces waxy starch. A testcross involving triply recessive and $F_1$ triply heterozygous plants produced the progeny listed at the top of the next column.

| | |
|---|---|
| colored, shrunken, waxy | 305 |
| colorless, full, waxy | 128 |
| colorless, shrunken, waxy | 18 |
| colored, full, waxy | 74 |
| colorless, shrunken, nonwaxy | 66 |
| colored, full, nonwaxy | 22 |
| colored, shrunken, nonwaxy | 112 |
| colorless, full, nonwaxy | 275 |
| | 1000 |

Give the following: ***a.*** Gene sequence. ***b.*** Map distances. ***c.*** Coefficient of coincidence.

***5.11*** Vermilion eyes (*v*), lozenge-shaped eyes (*lz*), and cut wings (*ct*) are recessive traits in *Drosophila*. A cross between females heterozygous at these three loci and wild-type males produced the following results:

| | | |
|---|---|---|
| ♀♀: | + + + | 1010 |
| ♂♂: | + + + | 30 |
| | + + lz | 32 |
| | + ct + | 441 |
| | + ct lz | 1 |
| | v + + | 0 |
| | v + lz | 430 |
| | v ct + | 27 |
| | v ct lz | 39 |

***a.*** What is the sequence of the three linked genes in their chromosome?
***b.*** Calculate the map distance between the genes, and the coefficient of coincidence.
***c.*** In what chromosome of *Drosophila* are these genes carried?

***5.12*** Given triply recessive and triply heterozygous *Drosophila* stocks and the genetic map shown, predict the phenotypic classes and number of offspring in each that you would obtain from 1000 progeny of a cross between triply recessive males and triply heterozygous females (assume that all mutants are recessive).

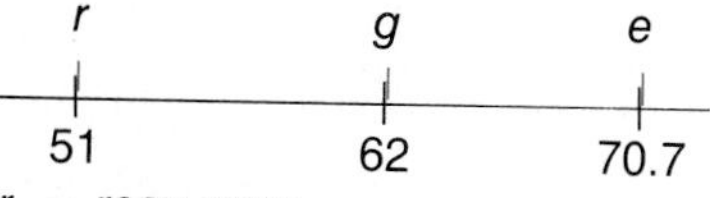


*r* = rosy eyes
*g* = glassy eyes
*e* = ebony body

***5.13*** Testcrosses of *Drosophila* females heterozygous for the recessive traits sepia eyes (*se*), blistery wings (*bl*), and spineless bristles (*ss*) with triply recessive males produced these progeny:

| | |
|---|---|
| + + + | 22 |
| + + ss | 265 |
| + bl + | 590 |
| + bl ss | 96 |
| se + + | 88 |
| se + ss | 550 |
| se bl + | 250 |
| se bl ss | 26 |
| | 1887 |

***a.*** How were the three pairs of alleles arranged in the female parent chromosomes?
***b.*** What is the order of genes and the distances between them?

***5.14*** Testcrosses of *Drosophila* females heterozygous for the recessive traits scarlet eyes (*st*), dumpy wings (*dp*), and black body (*b*) with triply recessive males produced the following progeny:

| | |
|---|---|
| + + + | 210 |
| + + b | 104 |
| + dp + | 96 |
| + dp b | 232 |
| st + + | 224 |
| st + b | 98 |
| st dp + | 102 |
| st dp b | 240 |
| | 1306 |

***a.*** What are the linkage relations for these three genes?
***b.*** What is the map distance between linked genes?

***5.15*** In *Drosophila,* hairy body (*h*) and glass eyes (*g*) are recessive traits and wrinkled wings (*W*) is dominant over wild type. Crosses of triply heterozygous females with triply recessive males produced the following progeny:

| | |
|---|---|
| + + + | 301 |
| + + W | 3 |
| + g + | 81 |
| + g W | 90 |
| h + + | 105 |
| h + W | 90 |
| h g W | 330 |
| | 1000 |

***a.*** Are these three genes linked? If so, what is their order on the chromosome?
***b.*** Calculate the map distance between linked genes.

***5.16*** Five linked autosomal recessive genes in *Drosophila* produce aristaless antennae (*al*), dachs legs (*d*), brown eyes (*bw*), waxy wings (*wx*), and hook-shaped bristles (*hk*). From the results of the following three experiments, construct a genetic map for this region of the chromosome.

| **experiment number** | **1** |
|---|---|
| **♀ parent triply heterozygous for / ♂ parent triply recessive for** | *al d hk* |
| **progeny** | 262 + + + |
| | 76 + + hk |
| | 30 + d + |
| | 131 + d hk |
| | 114 al + + |
| | 29 al + hk |
| | 82 al d + |
| | 276 al d hk |
| | 1000 |

| **experiment number** | **2** |
|---|---|
| **♀ parent triply heterozygous for / ♂ parent triply recessive for** | *d hk wx* |
| **progeny** | 84 + + + |
| | 340 + + wx |
| | 70 + hk + |
| | 20 + hk wx |
| | 20 d + + |
| | 60 d + wx |
| | 320 d hk + |
| | 86 d hk wx |
| | 1000 |

| **experiment number** | **3** |
|---|---|
| **♀ parent triply heterozygous for / ♂ parent triply recessive for** | *hk wx bw* |
| **progeny** | 15 + + + |
| | 62 + + bw |
| | 286 + wx + |
| | 140 + wx bw |
| | 148 hk + + |
| | 280 hk + bw |
| | 58 hk wx + |
| | 11 hk wx bw |
| | 1000 |

***5.17*** The ABO blood groups and the nail-patella syndrome (a rare anomaly involving abnormal fingernails, toenails, and kneecaps, together with other structural abnormalities) are controlled by different genes in humans. In the pedigree shown, the blood-group genotypes appear below the symbols and those individuals with the syndrome are represented by solid symbols.

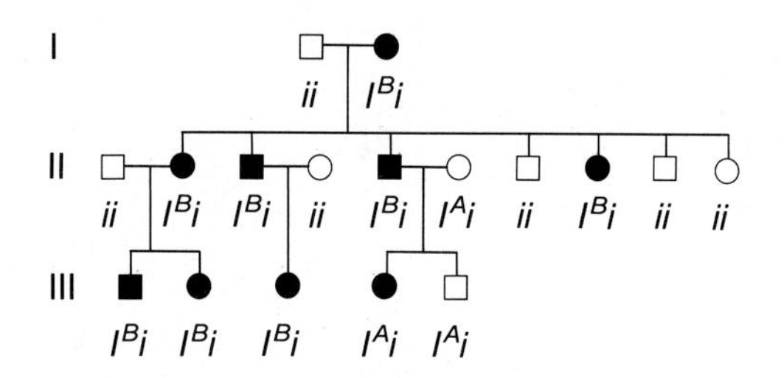


***a.*** Is nail-patella syndrome controlled by a dominant or recessive allele? Is the gene autosomal or X-linked?
***b.*** Do the above data provide any evidence for linkage between the two genes? Explain.
***c.*** Draw the linked genes on each of the two homologous chromosomes of both grandparents of the family (I-1 and I-2).

***5.18*** A woman who is doubly heterozygous for the X-linked traits hemophilia and GPD-A/GPD-B is anxious to know if the fetus she carries will inherit hemophilia, which is known to have occurred among some of her male relatives. The woman's father is normal and is known to have the codominant *Gpd-A* allele on the basis of enzyme tests. If the fetus is male, what is the chance that it is carrying the *hm* allele
***a.*** if its GPD type is unknown?
***b.*** if it proves to be GPD-A or if it proves to be GPD-B?
***c.*** if the GPD type of the woman's father was unknown?
(Note: the two gene loci are about 5 map units apart on the X chromosome.)

***5.19*** Lesch-Nyhan disease is transmitted via an X-linked mutant allele. A woman with normal vision gives birth to a son who is color-blind and has Lesch-Nyhan disease. (Note: The woman's father was color-blind.)
***a.*** Give the possible genotypes and phenotypes of the following: (1) woman's father (2) woman's mother (3) woman (4) her son
***b.*** What is the probability of her having another doubly affected son?

## References

Creighton, H.S., and B. McClintock. 1931. A correlation of cytological and genetical crossing-over in *Zea mays. Proc. Nat. Acad. Sci. U.S.* **17**:492.

Donahue, R.P., et al. 1968. Probable assignment of the Duffy blood group locus to chromosome 1 in man. *Proc. Nat. Acad. Sci. U.S.* **61**:949.

Firshein, S.I., et al. 1979. Prenatal diagnosis of classical hemophilia. *New Eng. J. Med.* **300**:937.

McKusick, V.A. Apr. 1971. The mapping of human chromosomes. *Sci. Amer.* **224**:104.

McKusick, V.A. 1978. Genetic nosology: Three approaches. *Amer. J. Hum. Genet.* **30**:105.

McKusick, V.A., and F.H. Ruddle. 1977. The status of the gene map of the human chromosomes. *Science* **196**:390.

Merritt, A.D., et al. 1973. Human amylase loci: Genetic linkage with the Duffy blood group locus and assignment to linkage group I. *Amer. J. Hum. Genet.* **25**:523.

Morgan, T.H. 1911. Random segregation versus coupling in Mendelian inheritance. *Science* **34**:384.

O'Brien, S.J., ed. 1982. *Genetic Maps,* 2nd ed. Frederick, Md.: National Institutes of Health.

O'Brien, S.J., and W.G. Nash. 1982. Genetic mapping in mammals: Chromosome map of the domestic cat. *Science* **216**:257.

Sturtevant, A.H. 1913. The linear arrangement of six sex-linked factors in *Drosophila melanogaster. J. Exp. Zool.* **14**:43.

# CHAPTER 6

# Linkage, Recombination, and Mapping in Bacteria and Viruses

In chapter 5 we saw that linkage and independent assortment of different genes in eukaryotes can be interpreted from analysis of gene transmission from parents to progeny. These data have provided the basis for genetic maps of chromosomes. Each map is a summary of the relative distances between and order of the genes located on the chromosomes. All the genetic data can be confirmed by independent studies of the chromosomes, including their behavior during meiosis and their numbers per species. In this chapter we deal with biological systems in which meiosis and fertilization are absent. In spite of this major difference between eukaryotes and other organisms in their reproductive processes, analysis of genetic recombination accomplished by other means can still be carried out and can yield useful data regarding the location and order in which genes are arranged in bacteria and viruses.

## Linkage Analysis in Bacteria

Gene transmission in bacteria involves *one-way transfer* of one or more genes from a **donor** cell to a **recipient** cell, rather than the bringing together of two whole sets of genes, which is typical of gamete fusion in sexually reproducing eukaryotes. In eukaryotes, gene transmission involves the reproductive processes of meiosis and fertilization. In bacteria and viruses, these processes are absent, and substitute reproductive processes function instead. These substitutes are called **parasexual** systems or processes since they perform functions similar to sexual processes but are not themselves sexual in nature in any conventional sense.

### 6.1 Parasexual Systems in Bacteria

Like sexual systems, parasexual systems bring together genes from different parents. They also account for segregation of alleles, which leads to haploid progeny that carry recombinant genotypes. Because gene exchange is not reciprocal in bacteria, different strategies are required to analyze gene transmission patterns. Linkage and recombination in organisms with parasexual systems can be analyzed by various experimental methods and genes can be mapped.

Gene transmission is achieved by three main avenues in bacteria:

*1.* **transformation,** in which genes enter the recipient as parts of duplex DNA fragments from the donor;

*2.* **conjugation,** a mating process by which genes from the donor enter the recipient, perhaps by way of a special conjugation structure; and

*3.* **transduction,** in which a virus acts as the vector for transferring donor genes into recipient cells.

We will discuss each of these parasexual systems. Regardless of differences in processes of gene transmission and in strategies required to analyze these processes in bacteria, all the data show that the same fundamental genetic processes are present in bacteria as in eukaryotes. Bacterial genes display linkage, undergo recombination by crossing-over processes, and can be mapped in the order in which they are present in the chromosome.

## 6.2 Features of Genetic Analysis in Bacteria

Since a bacterial cell is invisible to the naked eye, the usual phenotypes studied involve cell growth in nutrient media. Bacterial cells, either in liquid media or as colonies of cells growing on solid media, can be assayed for a variety of phenotypic traits. Among these traits are differences in the size, shape, color, and texture of colonies; sensitivity or resistance to drugs or to viral infections; and dependence on or independence of amino acids, vitamins, and other supplements in the culture medium. These and other phenotypic traits make bacteria very suitable systems for genetic analysis.

Since each bacterial individual is a haploid cell, each phenotype can be related to a particular genotype in that cell. Each colony or group of haploid descendants of a single cell can be genotypicaly identified by its expressed phenotype. Dominance relations between allelic alternatives of a gene usually can be disregarded in bacteria since each gene is usually represented by only one allele in haploids, and allelic expression will not be masked even if it is recessive. Under special conditions it is possible to construct strains that carry two alleles of a gene and then test allelic dominance and recessiveness in the strains. Most genetic analyses, however, do not involve such partially diploid strains.

The symbols that characterize alleles of a gene usually reflect the phenotype expressed when that allele is present. For example, the gene governing cellular response to the drug streptomycin is the *str* gene; it is symbolized by the resistance allele, *str-r*, and the allele conferring sensitivity will be *str-s*, phenotypes would be Str-r and Str-s, respectively. For strains that can synthesize their own metabolic requirements, the appropriate alleles are designated by a superscript +, and the alternative alleles for inability to synthesize the substances are designated by a superscript −. Strains that can synthesize the amino acid proline are thus identified as having the allele $pro^+$, and proline-requiring strains have the allele $pro^-$; phenotypes would be $\text{Pro}^+$ and $\text{Pro}^-$, respectively. No dominance or recessiveness is implied in designations for these alleles or for similar pairs. In discussions of some hypothetical gene $a$, the allele conferring independence will be shown as $a^+$. Its alternative allele will be $a^-$, indicating a need for the substance to be supplied externally.

## Bacterial Transformation

Transformation is achieved by providing fragments of donor duplex DNA *in suspension* to living recipient cells. Because many special conditions are required to accomplish successful transformation in the laboratory, this procedure has a limited general usefulness for bacterial genetic analysis. The earliest transformation studies utilized the pneumococcus species *Diplococcus pneumoniae,* but a number of other species have been added to the list in the past 40 years. It has recently become possible to achieve the process in other bacteria, such as *Hemophilus influenzae* and *Bacillus subtilis*, as well as in yeast and cultured mammalian cells.

## 6.3 Transformation Analysis

If we add pieces of duplex DNA from the donor $a^+b^+$ to a culture of recipient cells of the $a^-b^-$ genotype, we expect to find four phenotypic classes of cells in the recipient population at the end of the experiment. In addition to the unaltered recipients with the $a^-b^-$ genotype, three classes of **transformants** will be recognized by their having one or both of the donor genes in their genotype. These transformant classes are $a^+b^-$, $a^-b^+$, and $a^+b^+$. The transformants are descendants of recipient cells that have been altered genetically (transformed), and they exhibit inherited genotypic alterations.

Since donor DNA is randomly broken into fragments during cellular extraction, closely linked genes will be included in the same fragment more often than they are separated into different fragments. A single transformation event is enough to produce double transformants for closely linked genes. On the other hand, if two genes are relatively far apart in the chromosome, they will tend to separate into different fragments of donor DNA. In this situation, two transformation events are required to give rise to double transformants. Thus, (1) widely separated genes give rise to more single transformants than double transformants since any one event is more likely to happen than any two events together, and (2) the frequency of double transformants for widely separated genes is equal to the product of the separate frequencies of the two classes of single transformants, based on laws of probability. Comparison of the frequencies of double transformants with single transformants will, therefore, provide an indication of the relative nearness of the two genes in the chromosome.

Suppose the cross of donor $a^+b^+ \times$ recipient $a^-b^-$ produces the following number of transformants:

| | | |
|---|---|---|
| single transformant class 1 | $a^+b^-$ | 430 |
| single transformant class 2 | $a^-b^+$ | 180 |
| double transformant class | $a^+b^+$ | 102 |
| total transformants | | 712 |

The double transformants are obviously fewer in number than either single transformant class, thus indicating that genes $a$ and $b$ are relatively far apart on the chromosome and not closely linked. If this is true, the frequency of double transformants should equal the product of the two single frequencies, or $430/712 \times 180/712 = 0.604 \times 0.253 = 0.153$, or 15.3%. The observed frequency of double transformants is $102/712 = 0.143$, or 14.3%. This value is within the range of expectation but can be checked statistically.

Donor $a^+b^+ \times$ recipient $a^-b^-$ could have produced the following results (instead of those shown above):

| | | |
|---|---|---|
| single transformant class 1 | $a^+b^-$ | 210 |
| single transformant class 2 | $a^-b^+$ | 106 |
| double transformants | $a^+b^+$ | 400 |
| total transformants | | 716 |

We can see at a glance that the number of doubles is higher than the number of singles, thus indicating close linkage of genes $a$ and $b$. If the genes were widely separated and required two independent events to produce double transformants, the frequency of double transformants would have been about 4.3% ($210/716 \times 106/716$) rather than the observed 55.9% ($400/716 \times 100$).

In the process of transformation, homologous DNA from the donor preparation becomes integrated into the DNA of the recipient, whether the donor DNA carries one or more of the allelic markers. The recipient chromosome is a *circular* DNA molecule, so each event that leads to integration of a piece of donor DNA into the circle must involve an *even* number of crossovers (Fig. 6.1). If one crossover or some other odd number of crossovers occurs, the circle is opened, and such recombinant cells are

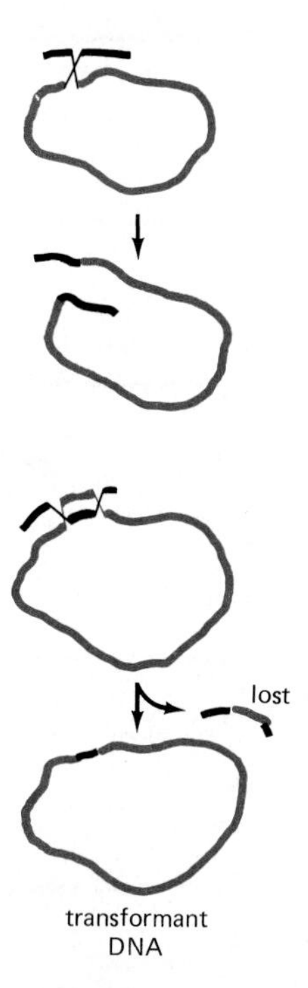


***Figure 6.1*** Integration of a piece of donor DNA (black) into an intact circle of recipient DNA (color) requires an even number of crossovers. With an odd number of crossovers, the circle is opened. Such cells would be inviable.

inviable. Inviable cells do not reproduce and, therefore, do not develop into colonies to be scored among the progeny.

In view of the requirement for even numbers of crossovers, we can see that the *single transformant classes arise through double crossovers.* One of these crossovers takes place *between* genes $a$ and $b$, and the second crossover takes place outside the $a$—$b$ region (Fig. 6.2). The double transformants also arise as the result of two crossovers; one crossover occurs on each side of the $a$—$b$ region and none takes place between the two genes. As a result of one crossover taking place on each side of the region, the whole $a$—$b$ segment is integrated into the recipient chromosome with both donor alleles unchanged and transferred together.

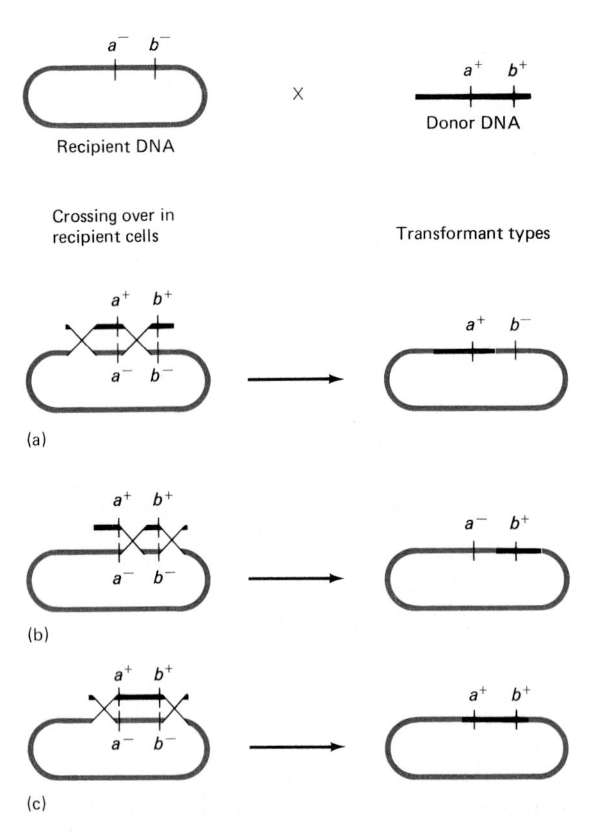


***Figure 6.2*** Origin of three classes of transformants by double crossovers. Only classes (a) and (b) provide data to estimate linkage distance between genes $a$ and $b$.

When linkage is estimated for the $a$—$b$ region, the only transformant classes included in the calculations as recombinants are those that arose as the result of crossing over in the region between genes $a$ and $b$. Although the $a^+b^+$ class is doubly transformed, no crossover took place between genes $a$ and $b$. This class, therefore, does not provide information about the percentage of recombination between the two genes, and it is classified differently for mapping purposes and for estimates of recombination frequency.

In the cross between donor $a^+b^+$ and recipient $a^-b^-$ cells, the distance between genes $a$ and $b$ is calculated from the number of recombinants in the total population, as follows:

| transformant genotype | number of transformants | number of recombinants |
|---|---|---|
| $a^+b^-$ | 130 | 130 |
| $a^-b^+$ | 70 | 70 |
| $a^+b^+$ | 800 | 0 |
| | 1000 | 200 |

$\frac{\text{recombinants}}{\text{total transformants}} = \frac{200}{1000} = 0.2$, or 20% recombinants, or 20 map units

Notice that the two recombinant classes do not occur in equal numbers. Unlike reciprocal crossovers in eukaryotes, which lead to equal proportions of reciprocal recombinant classes, prokaryotic recombinants are not reciprocal products of the same crossover event. Different double crossovers gave rise to the two recombinant classes, as shown in Figure 6.2.

## 6.4 Mapping by Transformation

Gene mapping can be accomplished by three-factor or multifactor transformations, as well as by two-factor experiments. The basic principles are the same in transformation tests as in mapping eukaryotic genes, and a single example should, therefore, be adequate. In a typical three-factor transformation experiment involving hypothetical $a^-b^-c^-$ recipient cells and DNA extracts from donors with the $a^+b^+c^+$

genotype, we might find the following transformants:

| single transformants | | double transformants | | triple transformants | |
|---|---|---|---|---|---|
| $a^+b^-c^-$ | 300 | $a^+b^+c^-$ | 150 | $a^+b^+c^+$ | 1400 |
| $a^-b^+c^-$ | 60 | $a^-b^+c^+$ | 290 | | |
| $a^-b^-c^+$ | 140 | $a^+b^-c^+$ | 10 | | |

The great excess of triple transformants indicates close linkage for all three genes. Double crossovers account for five of the recombinant classes; and quadruple crossovers will produce the class showing the lowest frequency of recombinants since four independent events are less likely to occur simultaneously than any two events (Fig. 6.3). The $a^+b^-c^+$ transformants are the least numerous and are therefore the quadruple crossover class. The gene order is determined from this class. The allele of the gene in the middle is the same as in the recipient, and the two flanking genes have alleles like those in the donor. In this case, gene $b$ is in the middle.

The preferred procedure for calculating linkage and map distances is to deal with two genes at a time, taken in all possible combinations. This permits equivalent treatment for single and double transformants. Triple transformants are not counted as recombinants since crossovers took place only in the regions outside the $a$—$b$—$c$ segment and not between genes.

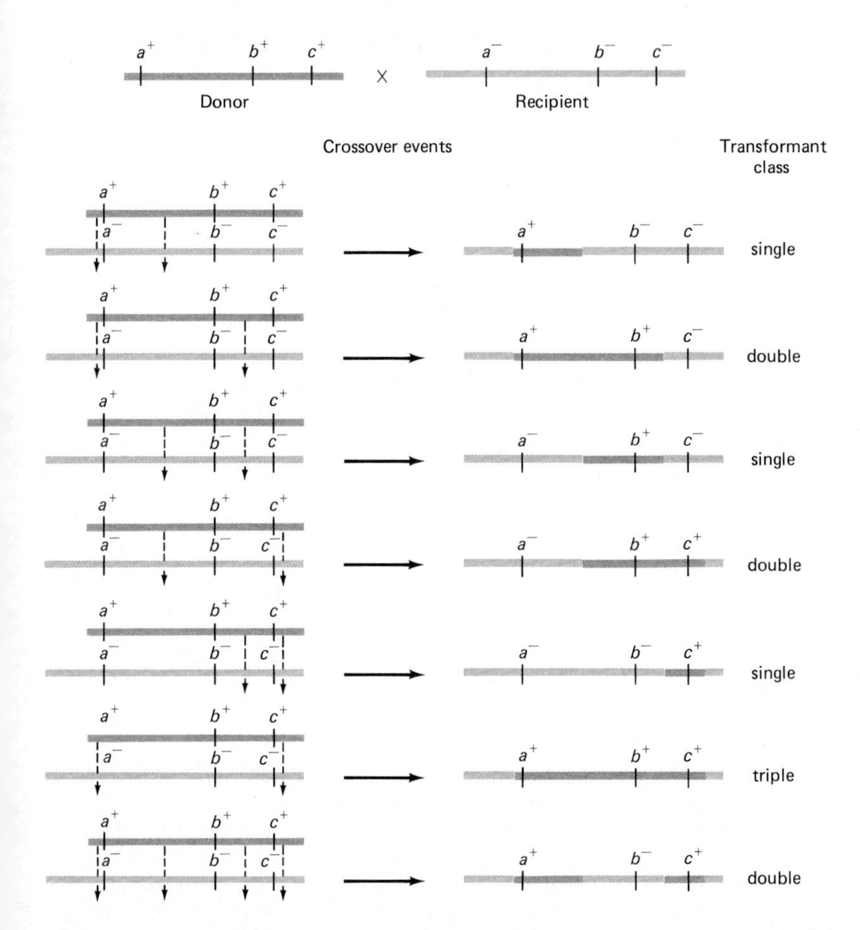


***Figure 6.3*** Crossing over between triply marked donor and recipient DNA may involve two exchange events in various regions or four exchange events in fewer cases. Depending on the number and location of crossovers, single, double, or triple transformants may be produced (having one, two, or three donor alleles in the transformant bacterial DNA, respectively).

Triple transformants are not recombinant for any of the three genes since the whole donor segment is integrated, unchanged, into the recipient chromosome.

We calculate distances between pairs of genes *a* and *b*, *b* and *c*, and *a* and *c*, disregarding the third gene in each case for the moment (Fig. 6.4). We calculate linkage distance as the proportion of recombinants among the total transformants for the two genes that we are analyzing at any one time. Upon completing these calculations, we can see that the linkage distance is greater for genes *a* and *c* than for *a* and *b* or *b* and *c*, thereby indicating that the gene order is *a*—*b*—*c*, as we had surmised by comparing the quadruple crossover class with classes showing the parental allelic combinations.

Linkage analyses by transformation and by other available methods have led to the construction of relatively detailed gene maps for bacterial species. One of the most detailed maps derived from transformation studies is that of the common soil bacterium *Bacillus subtilis* (Fig. 6.5).

## Bacterial Conjugation

In 1946 J. Lederberg and E. Tatum published the first significant genetic evidence for the existence of a mating system in *E. coli* (see Section 3.10). Their innovative method involved selecting for rare wild-type recombinants from mating mixtures of parental strains that were analyzed. Since the distance between *a* and *c* is greater than either of the other two intervals, the gene order must be *a*–*b*–*c*. Confirmation of this gene order can be obtained by noting that the quadruple transformant class (lowest frequency class) is $a^+b^-c^+$. The middle gene (in this case, *b*) is represented by a recipient allele and is flanked by two donor alleles in the lowest frequency transformant class (see Fig. 6.3).

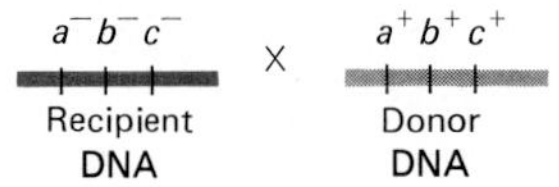

$a^-b^-c^-$ Recipient DNA × $a^+b^+c^+$ Donor DNA

| | Transformants for genes *a* and *b* (disregarding *c*) | | Transformants for genes *a* and *c* (disregarding *b*) | | Transformants for genes *b* and *c* (disregarding *a*) | |
|---|---|---|---|---|---|---|
| Single transformants | $a^+b^-(c^-)$ | 300 | $a^+(b^-)c^-$ | 300 | $(a^-)b^+c^-$ | 60 |
| | $a^-b^+(c^-)$ | 60 | $a^-(b^-)c^+$ | 140 | $(a^-)b^-c^+$ | 140 |
| | $a^-b^+(c^+)$ | 290 | $a^+(b^+)c^-$ | 150 | $(a^+)b^+c^-$ | 150 |
| | $a^+b^-(c^+)$ | 10 | $a^-(b^+)c^+$ | 290 | $(a^+)b^-c^+$ | 10 |
| | | 660 | | 880 | | 360 |
| Double transformants | $a^+b^+(c^-)$ | 150 | $a^+(b^-)c^+$ | 10 | $(a^-)b^+c^+$ | 290 |
| | $a^+b^+(c^+)$ | 1400 | $a^+(b^+)c^+$ | 1400 | $(a^+)b^+c^+$ | 1400 |
| | | 1550 | | 1410 | | 1690 |
| | TOTALS | 2210 | | 2290 | | 2050 |
| Linkage distance | Gene *a* to gene *b* $\frac{660}{2210}=0.30$ $0.30 \times 100 = 30$ | | Gene *a* to gene *c* $\frac{880}{2290}=0.38$ $0.38 \times 100 = 38$ | | Gene *b* to gene *c* $\frac{360}{2050}=0.18$ $0.18 \times 100 = 18$ | |

Gene order: *a* —30— *b* —18— *c*; *a* to *c*: 38

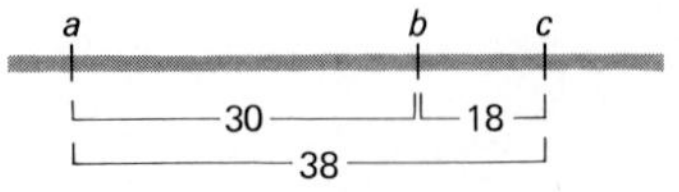

**Figure 6.4** Gene mapping by three-factor transformation. Taking two genes at a time in all possible combinations, single and double transformant classes are recorded for each combination of alleles. In this approach the nonrecombinant triple transformant $a^+b^+c^+$ class is included in the calculations. Linkage distance is calculated as the proportion of recombinants among the total transformants for each pair of genes being

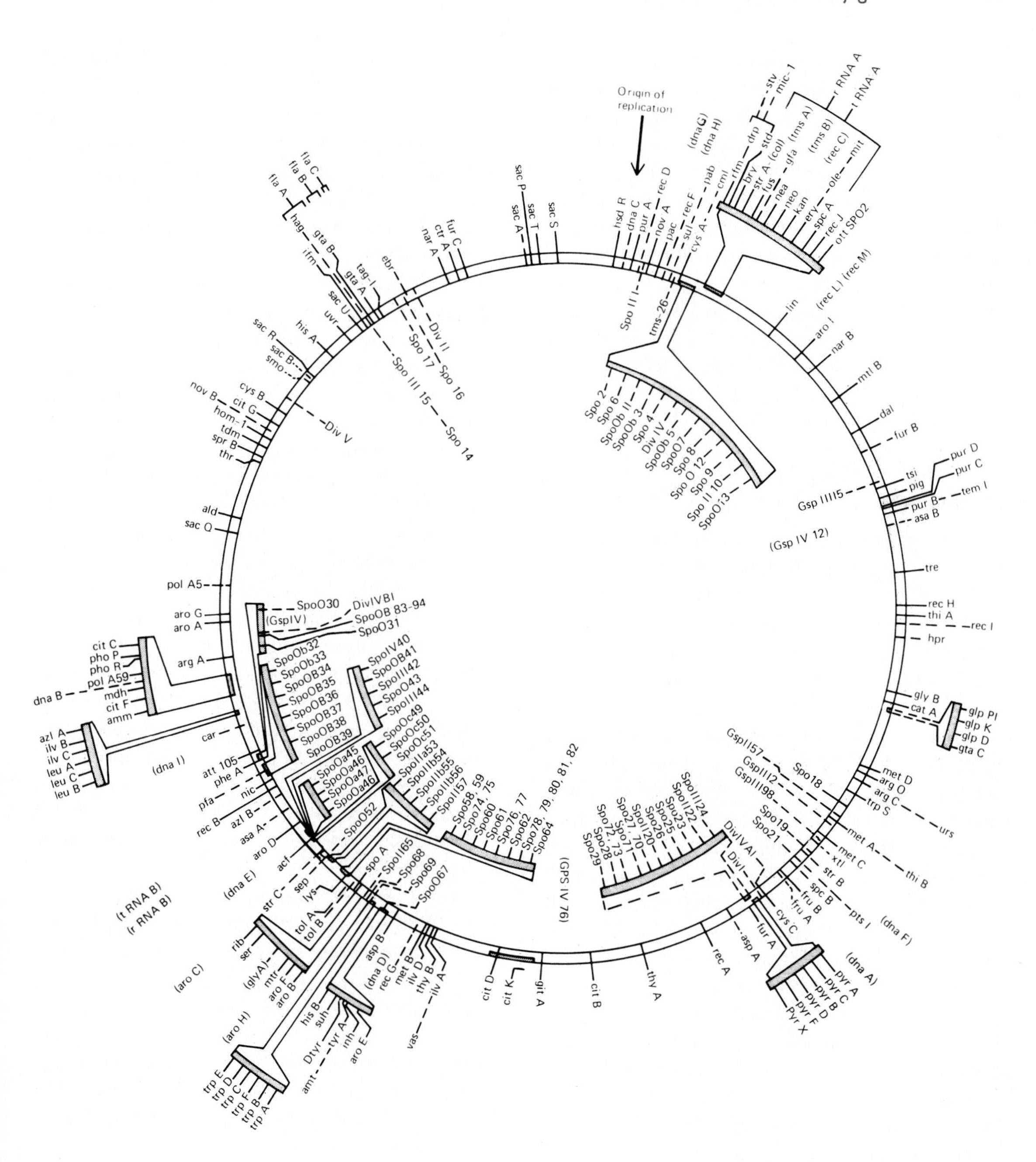


***Figure 6.5*** Gene map of the linkage group of the soil bacterium *Bacillus subtilis.* (From F.E. Young and G.A. Wilson. 1974. In R.C. King, ed., *Handbook of Genetics,* vol. 1, p. 69. New York: Plenum.)

multiply deficient in growth requirements. Even though recombinants were as rare as one per million cells, huge populations of bacteria could be screened easily and rapidly for wild-type cells that contained alleles from both parental cell types. Lederberg and Tatum opened an entirely new approach to genetic analysis of bacteria, and their approach influenced other investigators to study bacterial genetics. A Nobel prize was awarded to Lederberg and Tatum in 1960 in recognition of their milestone achievement and of their other significant genetic studies.

## 6.5 Recombination in Conjugating *E. coli*

Lederberg and Tatum's experimental design involved mixing multiply deficient strains of *E. coli* in complete nutrient medium, which contained all the substances that the mutant parental strains needed to induce growth. After incubation in the liquid medium, samples of the mixture were plated on a solid minimal medium, which lacked all the substances needed by the parental mutant strains. Only wild-type cells could grow in minimal medium; neither parental strain could grow in the absence of its nutritional substances. The genetically deficient strains are called *auxotrophs* and wild-type strains, or recombinants, are called *prototrophs*. The experimental method therefore involved selecting for rare prototrophic recombinants among millions of auxotrophic cells plated on the same medium.

After plating incubated mixtures of two different triply deficient auxotrophs in very high concentrations, Lederberg and Tatum recovered prototrophic colonies at a frequency of about one per million plated cells. All six wild-type traits were expressed in these prototrophs, apparently as the result of gene recombination between parental auxotrophs:

$$a^+b^+c^+d^-e^-f^- \times a^-b^-c^-d^+e^+f^+$$
$$\downarrow$$
$$a^+b^+c^+d^+e^+f^+$$

Although they could reasonably conclude that the prototrophic recombinants had arisen as the result of mating, or *conjugation,* between the auxotrophic parents, the investigators had to consider two other explanations for prototroph origin: (1) all three mutant alleles in some auxotrophs had undergone reverse mutation to nutritional independence or (2) prototrophs were a consequence of multiple transformations. They evaluated all three hypotheses experimentally.

We know that mutations are independent events. Suppose the chance for any one of the six parental genes to undergo random, spontaneous mutation was 1 per $10^5$ cells, a high rate for bacterial genes. The chances for all three genes to mutate independently in the *same* cell would be 1 in $10^5 \times 10^5 \times 10^5$, or 1 in $10^{15}$ (the product of their separate probabilities). Prototrophs appeared at a frequency of 1 in $10^6$, however, rather than 1 in $10^{15}$. To explain the observed prototrophic frequency by the mutation hypothesis would require the chance for reverse mutations to be 1 per $10^2$ for each gene ($10^{-2} \times 10^{-2} \times 10^{-2} = 10^{-6}$). Such mutation values were much too high for the genes studied, according to mutation analysis of the single genes. The reverse mutation hypothesis was therefore regarded as a highly improbable explanation for the observed prototroph frequencies.

Lederberg and Tatum were able to rule out the multiple transformation hypothesis by two kinds of experimental evidence. First, they failed to find prototrophs when they added DNA extracts from one strain to living recipient cells of the other strain. This result was interesting but negative, that is, something expected didn't happen. The negative results could be explained by reasons other than the absence of transformation events.

The second line of experimental evidence was more convincing in ruling out transformation. The investigators put the two parent strains in opposite sides of a U-tube, with a glass filter at the bottom of the tube. The filter had very small pores that prevented bacterial cells from passing through it from one side of the U-tube to the other side (Fig. 6.6). The fluid media mixed, but they found no prototrophs. When they removed the filter, cells were able to mingle in the common fluid medium. They removed samples of cells from the mixed culture at timed intervals and plated them out on solid minimal medium on which only prototrophs could grow. The investigators found prototrophic colonies at about the same frequency as these colonies had occurred in the original experiments. The single essential condition for prototroph formation, therefore, is *contact between living cells* that carry complementary auxotrophic allelic markers. These results indicate that the process leading to prototrophic recombinants is a mating system that permits gene exchange (conjugation) rather than mutation or transformation. Microscopic evidence for conjugation was obtained some years later (Fig. 6.7).

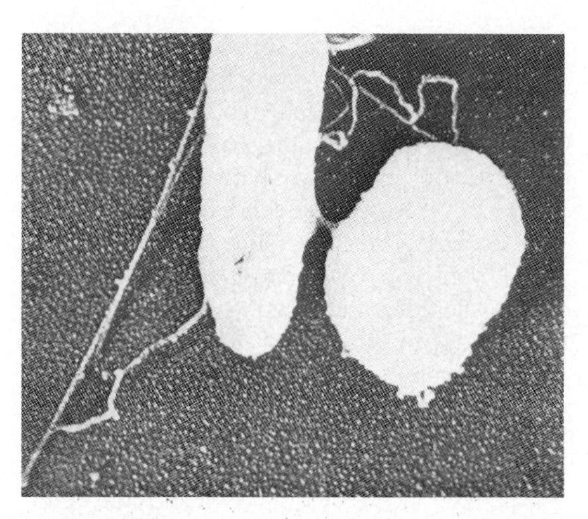

***Figure 6.7*** Electron micrograph of two conjugating *E. coli* cells from strains differing in cell shape. A slender bridge connects the elongated Hfr donor and round $F^-$ recipient cell. (From E.L. Wollman, F. Jacob, and W. Hayes. 1957. *Cold Spr. Harb. Symp. Quant. Biol.* **21**:147. Photograph by T.F. Anderson.)

Lederberg assumed that the role of the two parental strains as equal partners in gene exchange was similar to parental roles in sexual systems of organisms such as *Drosophila* or *Neurospora.* In 1953 William Hayes showed this assumption to be incorrect when he published the first of a series of studies on the conjugation system in *E. coli.* Hayes found that the success of the standard cross between different auxotrophs depends on the continued viability of only one of the two parental strains. Crosses of streptomycin-sensitive parent A (*str–s*) and streptomycin-resistant parent B (*str–r*) produced prototrophs on minimal medium that contained streptomycin. In the reciprocal cross, (B *str–s* × A *str–r*), however, no prototrophic colonies formed when streptomycin was present. Neither parent type could grow in minimal medium because of their nutritional deficiencies, but prototrophs apparently were produced only if parent B cells were viable. The conclusion was that genetic recombination took place in viable B cells, that is, B cells were the *recipients* of genetic material. The role of A cells, on the other hand, was to act as genetic *donors.* Once the A cells had performed their function, they were dispensable. Conjugation was thus shown to be a one-way transfer of genetic material from donor cells to recipient cells.

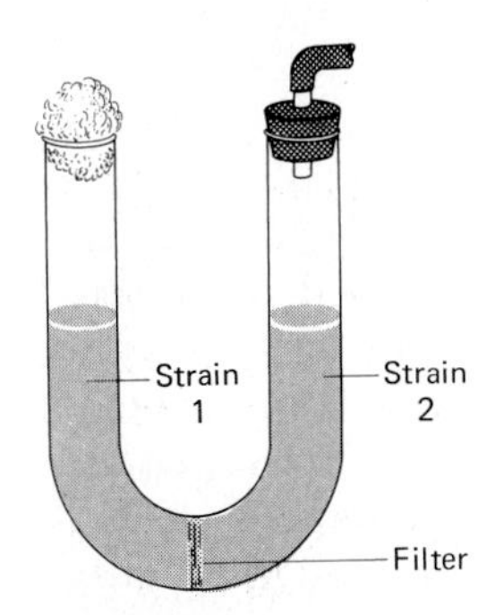


***Figure 6.6*** A U-tube. Auxotrophic parental strains are inoculated into media on each side of the U-tube, with a glass filter between, which allows the medium but not the cells to pass through.

Through other experimental studies Hayes showed that the donor state is conferred by an agent called F, the **fertility factor** (or sex factor). This fertility factor is readily transmitted to recipient cells and converts these cells to the donor state; they act as donors if used in subsequent crosses. On the basis of the transmission of allelic markers from donor to recipient, Hayes also found that F is transmitted *independently* of the bacterial chromosome. In crosses between fertile donors and sterile recipients, prototrophs were recovered at a frequency of about 1 in $10^6$ cells. Different sterile strains did not produce prototrophs when mixed together, but in some cases different fertile strains mixed together produced prototrophs, at frequencies significantly lower than in fertile × sterile crosses. The donor state is called $F^+$ and indicates the presence of the fertility factor. Recipients are $F^-$ because they lack the fertility factor. The most successful crosses, therefore, are between $F^+$ donors and $F^-$ recipients. In these crosses, $F^-$ recipients readily convert to $F^+$, thereby indicating the transfer of the fertility factor from donor to recipient during conjugation.

Recombination analysis was restricted by the low frequency of prototrophic recombinants recovered in $F^+ \times F^-$ crosses. As genetic studies continued in various laboratories, investigators found strains that were capable of high-frequency recombination. They called these strains Hfr, as an indication of their behavior. Donor Hfr strains transferred their genes to recipient $F^-$ cells, but the recipients usually remained $F^-$. In contrast, $F^+ \times F^-$ led to relatively higher rates of conversion of $F^-$ recipients to the $F^+$ state. Investigators showed that Hfr strains did have the fertility factor since Hfr cells often changed to $F^+$ (as seen by a change to low-frequency recombination in crosses with $F^-$).

These and other behavioral differences between $F^+$ and Hfr cells were studied in various experiments. Interpretation of these behaviors included the following features:

***1.*** Hfr strains carry the F factor *integrated within the bacterial chromosome,* whereas $F^+$ donors possess an independently replicating and transferrable F factor (Fig. 6.8).

***2.*** Hfr cells convert to $F^+$ when their F factor dissociates from the bacterial chromosome and maintains its existence independent of the chromosome.

***3.*** $F^+$ cells convert to Hfr at a frequency of about 1 in $10^4$ to 1 in $10^5$, as a consequence of F being integrated into the bacterial chromosome.

***4.*** $F^-$ cells change to $F^+$ fairly often because F moves readily, from the $F^+$ donor into the $F^-$ recipient, when it exists independently of the bacterial chromosome.

***5.*** $F^-$ cells only occasionally convert to $F^+$ in Hfr × $F^-$ crosses since integrated F is the last part of the bacterial chromosome to be transferred into the recipient cell.

***6.*** The F factor is a new kind of genetic component, called an **episome.** An episome is an extrachromosomal genetic element that may exist free in the cell under some conditions and may be integrated into the chromosome of the cell under other conditions. (We will say more about episomes later.)

With the advantages gained by using Hfr donors, genetic recombination analysis proceeded rapidly in *E. coli.* In the 1950s François Jacob and Elie Wollman made a very important breakthrough in interpreting many sets of results. Their experimental procedure opened the way to rapid progress in mapping genes on the *E. coli* chromosome, determining the circular conformation of the chromosome map, and showing that all the genes in *E. coli* could be mapped in a single linkage group.

## 6.6 Mapping Genes by Conjugation Analysis

Jacob and Wollman conducted *interrupted mating experiments* using Hfr and $F^-$ strains with different allelic markers of a number of genes. The Hfr strain had been shown to transfer these marker genes, which we will call $A, B, C, D$, and $E$, at high frequency. In addition, on the basis of recombination frequencies, they inferred that these genes were arranged in the order $ABCDE$. When recombinants were selected for allele $A$ in crosses of Hfr strains with recipients carrying alleles $abcde$, the remaining

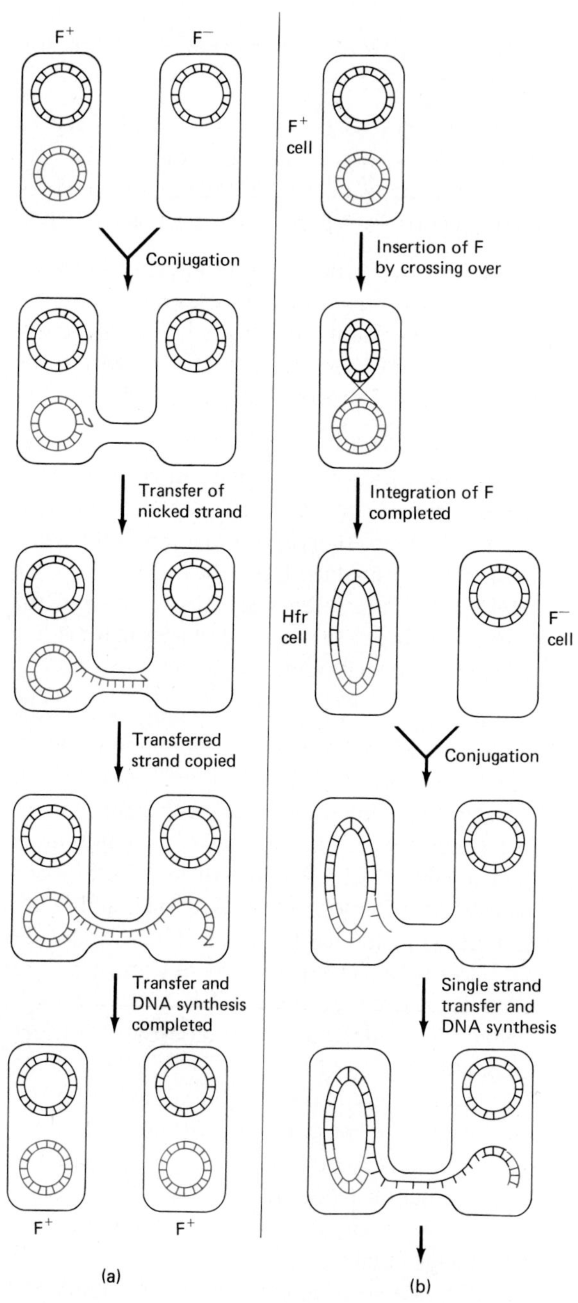


***Figure 6.8*** (a) Transfer of F factor (color) from $F^+$ donor to $F^-$ recipient occurs after a conjugation bridge has joined the cells. One strand of the F DNA duplex is nicked and transferred to the $F^-$ cell, where DNA replication (strand copying) occurs. When F has replicated, the $F^-$ cell becomes $F^+$, and the original $F^+$ donor retains at least one copy of the episome. (b) If F integrates into the bacterial chromosome, an Hfr cell arises. During conjugation, bacterial genes are transferred to the $F^-$ recipient with part of the integrated F DNA (color) leading the way. Much of the F DNA, however, remains at the far end of the large bacterial chromosome and does not enter the recipient because the cells usually break apart long before.

donor alleles, *B*, *C*, *D*, and *E*, appeared among the recombinants at frequencies of about 90%, 75%, 40%, and 25%, respectively.

In interrupted matings with *ABCDE* donors and *abcde* recipients, they mixed cells together and allowed them to conjugate. The investigators removed samples of cells from the mixture at timed intervals and subjected them to shearing forces in a high-speed blender. The violent agitation tears mating cells apart but does not otherwise damage them. They then plated cells treated in this way for subsequent scoring of recombinants.

Jacob and Wollman found that samples removed at any time less than 8 minutes after mixing produced no recombinants. At 8 minutes, recombinants appeared but they had inherited only the donor allele *A* and were genotypically *Abcde*. Samples removed about 9 and 10 minutes after mixing showed recombinants that were *ABcde* and *ABCde*, respectively. Cells removed 17 minutes after mixing gave rise to *ABCDe* recombinants, and cells removed 25 minutes after mixing gave rise to *ABCDE* recombinants. From the time a donor allele first appears in the recipient cells, its frequency among recombinants increases with time until the frequency in standard crosses is attained (Fig. 6.9).

The most reasonable interpretation of these results is that the effect of agitation in the blender is to break the chromosome during its movement from donor to recipient cell. Only those genes that had already entered the recipient at the time of agitation can be found in the recombinants. The interrupted mating experiments, therefore, show that the time at which a particular donor allele first begins to appear in the recombinants corresponds to the length of time it takes to be transferred on the donor chromosome to the recipient cell. The donor genes penetrate the recipient at precise and characteristic intervals and in the same order (*ABCDE*) in which they are arranged on the chromosome.

From these and other experimental data additional interpretations have been made concerning the *E. coli* chromosome, including the following:

***1.*** The consistency of order and entry of donor alleles into recipients indicates that the alleles are parts of a linear structure, or chromosome, on which genes are located at fixed loci.

***2.*** The difference in time intervals between successive gene transfers indicates different distances separating these genes on the chromosome, based on the assumption that chromosome transfer occurs at a relatively constant rate.

***3.*** The maximum, or plateau, values for the frequency with which each donor allele appears in recipients in interrupted matings is the same as the recombination frequency found for each allele in standard crosses. This confirms the validity of the time map in linkage analysis and chromosome mapping.

In long-term interrupted matings, Jacob and Wollman found that it took about 90 minutes for the entire donor set of known genes to be transferred to recipients. In these cases, the recombinants were Hfr. Shorter conjugation times had resulted in $F^-$ recombinants. These observations were interpreted as showing that the integrated F factor is located at the far end of the Hfr donor chromosome and that most recombinants do not receive the F factor and therefore remain $F^-$ unless the whole chromosome has penetrated.

Correspondence has been demonstrated between genetic maps based on time of transference in interrupted matings and those based on recombination frequencies in standard crosses. We can incorporate data obtained in either kind of linkage analysis into the same map of the chromosome by equating minutes for entry with map units from recombinations. At present, one minute is equivalent to 20 map units. The linkage map for *E. coli* is subdivided into 100 minutes (Fig. 6.10), which are equivalent to 2000 map units. In general practice, linkage relationships for *E. coli* are first established by interrupted matings. Greater refinement of map

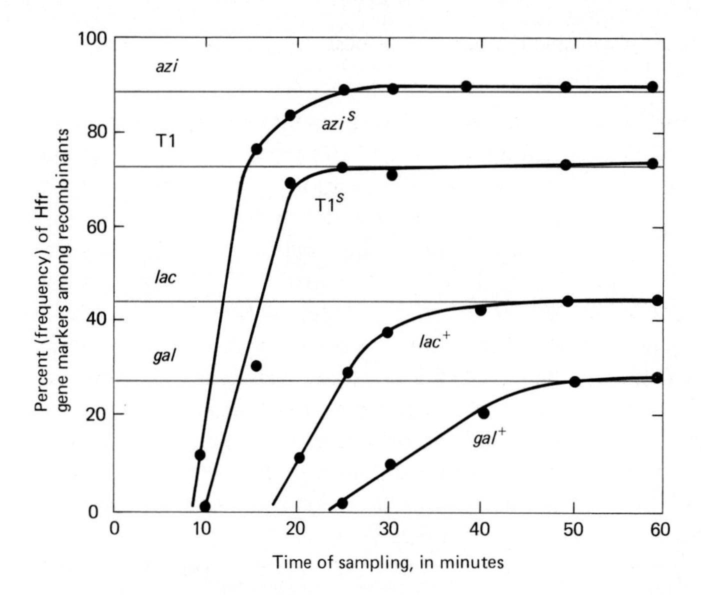


***Figure 6.9*** Kinetics of gene transfer from Hfr to $F^-$ in crosses. The transfer of Hfr marker genes recovered in recombinants sampled from the conjugation mixture, after blender treatment, is plotted as a function of time of sampling. (From F. Jacob and E.L. Wollman. 1961. *Sexuality and the Genetics of Bacteria.* New York: Academic Press.)

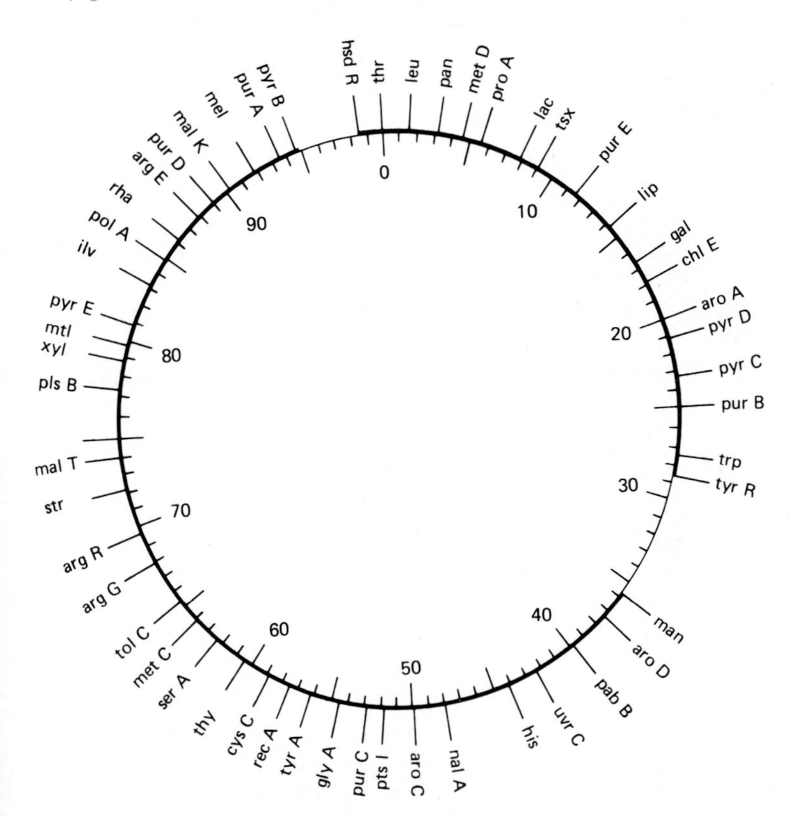


***Figure 6.10*** Gene map of the linkage group of *E. coli* strain K-12. The 100-minute map shows some of the important genes that have been located by various means, such as interrupted matings and transduction analysis. (From B.J. Bachmann, K.B. Low, and A.L. Taylor. 1976. *Bact. Rev.* **40**:116.)

distances between genes separated by less than 2 minutes is routinely obtained by transduction analysis. (We will discuss this topic shortly.)

## 6.7 Circularity of the *E. coli* Map

From linkage analysis of gene recombinations, it was found that the entire genome (all the genes of the species) was organized into one linkage group in *E. coli.* These data did not directly indicate that the chromosome was a circular structure rather than a linear structure with two free ends. The evidence for circularity of the chromosome first came from genetic studies of different Hfr strains.

In crosses involving various Hfr strains, some strains transferred the same genes in opposite directions. For example, Jacob and Wollman found that *ABCDE* donor genes entered in that order; others found that some Hfr strains donated genes in the order of *EDCBA* or that *BCDE* entered as expected but that gene *A* took a longer time to enter the recipient. In general, genes of various strains may enter recipients in different order or at times which indicate that their linkage distances vary. Comparisons of all the available data showed that all the genes enter in orders that are *circular permutations of a single sequence* (Fig. 6.11).

The explanation for these puzzling obser-

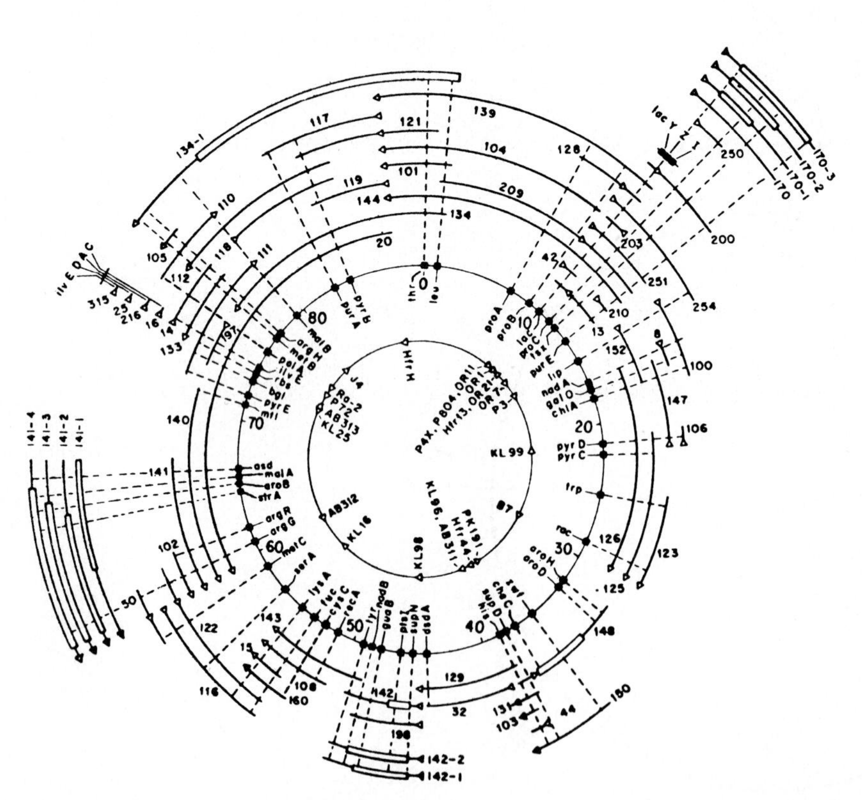


*Figure 6.11* Gene map of *E. coli* with arcs showing the approximate chromosomal regions carried into F$^-$ recipients by various F′ factors. The various orders of gene transfer appear to be circular permutations of a single sequence, leading to the summarized circular linkage map in Fig. 6.10. (From K.B. Low. 1974. In R.C. King, ed., *Handbook of Genetics,* vol. 1, p. 157. New York: Plenum.)

vations involved the F factor. The stimulus for chromosome transfer during conjugation was found to be the integration of F into the bacterial chromosome, by crossing over. The insertion of F initiates a break in the circular chromosome, and since the F factor is the last component to enter the recipient, F determines the polarity, or direction, of chromosomal transfer into the recipient. Furthermore, F may integrate at any of many sites along the bacterial chromosome. Depending upon the particular site of F integration and its orientation, therefore, different genes may lead the way into the recipient during conjugation. The sum of all these features is a single circular map of the chromosome, based on piecing together the circular permutations discovered in different genetic studies. Microscopic confirmation of the circularity of the *E. coli* chromosome was obtained by John Cairns in 1963 (Fig. 6.12).

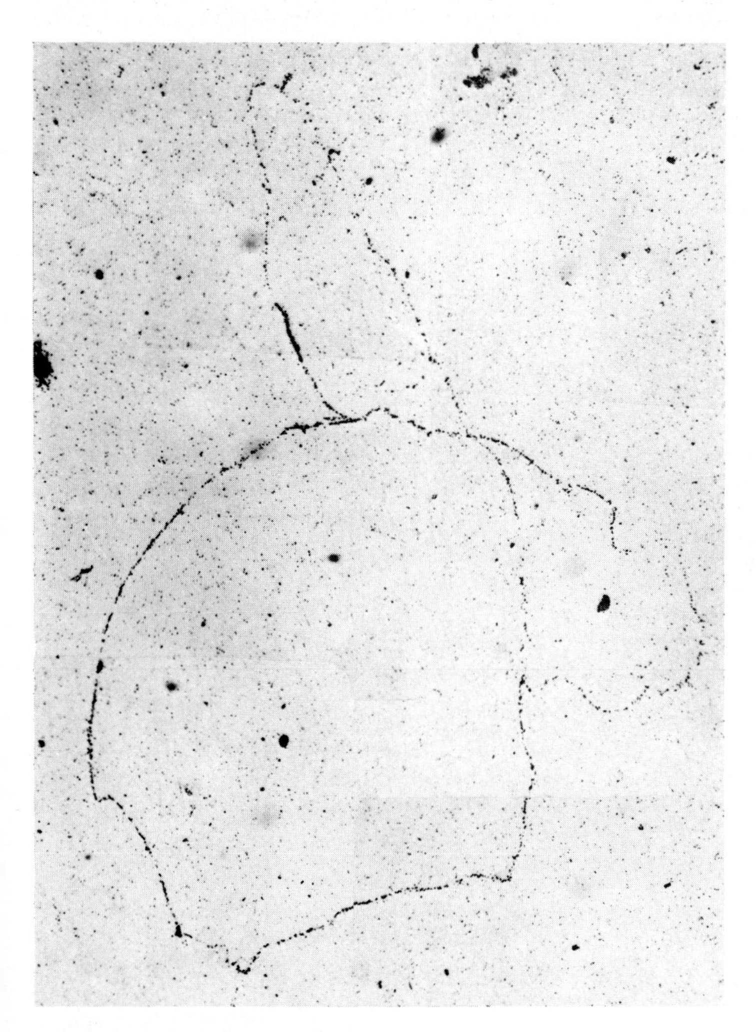

***Figure 6.12*** Replicating circular DNA of *E. coli.* (From J. Cairns. 1964. *Cold Spr. Harb. Symp. Quant. Biol.* **28**:44.)

## 6.8 Sexduction in *E. coli*

We stated earlier that the F factor could exist free in the cytoplasm in $F^+$ strains or as an integrated element in Hfr strains. Also, its genes are known to be distinct from those in the *E. coli* genome (Fig. 6.13). Under some conditions, however, the F factor has been shown to carry a variable number of *E. coli* genes along with its own genes. This type of F factor is designated F′ (F prime) in order to distinguish it from the usual fertility factor.

When F′ is transferred to a recipient during conjugation, the recipient can become a partial diploid for *E. coli* donor genes carried by F′. On occasion, recombinants arise as the result of crossover events between pairs of homologous genetic regions. Partial diploids are called **merozygotes** or **merodiploids.** The process of F′-mediated creation of merodiploids is called **sexduction.**

In studies that led to the identification of F′ factors, Edward Adelberg and François Jacob found a very high frequency of transfer of the $lac^+$ allele, which governs lactose utilization, into $F^-$ $lac^-$ recipients. This finding was unexpected because the Hfr strain that was used was known to have the *lac* locus at the far end of the chromosome, according to interrupted matings. Recipients became $F^+$ and $Lac^+$ in phenotype, but about 1 per 1000 of these recombinants produced $F^+$ $lac^-$ segregants.

The occurrence of $lac^-$ segregants from $lac^+$ recombinants was interpreted as evidence for a

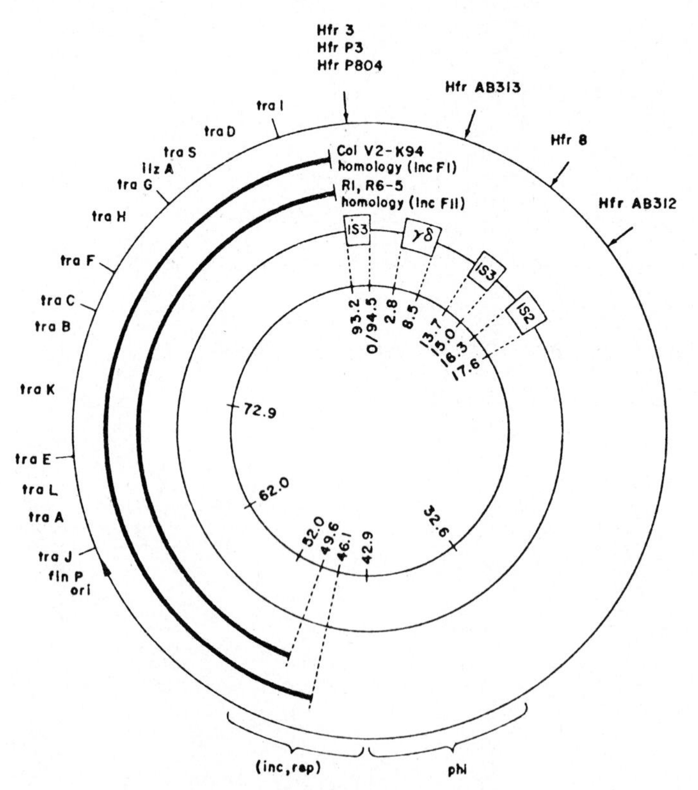


*Figure 6.13* Gene map of F factor of *E. coli.* The inner circle indicates DNA molecular length in kilobase units (1000 bases = 1 kilobase unit), and the outer circle shows locations of certain genes involved in molecular transfer (*tra*) during conjugation, replication (*rep*), and other functions. (From J.A. Shapiro. 1977. In A.I. Bukhari, J.A. Shapiro, and S.L. Adhya, eds., *DNA: Insertion Elements, Plasmids, and Episomes,* p. 671. New York: Cold Spring Harbor Laboratory.)

state of partial diploidy for the *lac* locus in the recipient. After transfer of the $lac^+$ allele into $lac^-$ recipients, the locus existed for a time in the $lac^+/lac^-$ partially diploid state. Subsequent crossover events between the two homologous *lac* regions produced $lac^-$ haploid segregants. The investigators later found $lac^+$ haploid segregants, too. The fact that F′ $lac^+/lac^-$ merodiploids exhibited the $Lac^+$ phenotype showed that $lac^+$ was dominant to the $lac^-$ allele.

In certain cases some Hfr strains may transfer F′-associated bacterial genes into recipients and thus produce merodiploids. Sexduction can be used to create various merodiploid strains for use in studies of allelic dominance and recessiveness as well as for recombination and other genetic analyses.

## Transduction Analysis

Gene transfer by **transduction** is mediated by viruses, which carry donor bacterial genes into recipient cells. The virus acts as a *vector,* or agent, for gene exchange. N. Zinder and J. Lederberg, using *Salmonella typhimurium,* a close relative of *E. coli*, reported the first evidence for this gene transfer mechanism in 1952 (Box 6.1).

### BOX 6.1 Comparative Mapping

Between 22 and 45 units on the 100-unit genetic maps of these two enteric species exhibit considerable similarity. Some of the loci (*pyr C*, *pur B*, *fla*, *his*) occupy identical positions on the two maps. The portion of the maps from 25.5 to 35.5 units (inside the triple lines) is inverted between the two species.

We can compare linkage maps to analyze the degree of genetic homology between related species. Homology of linkage maps and of DNA nucleotide sequences can hardly be expected to arise from random, unrelated events in evolution. The greater the homology, the greater are the evolutionary ties between species. Maps of bacterial and viral systems have provided some of the strongest lines of evidence for genetic relationships among different organisms.

It is now possible to compare band patterns of chromosome complements of related species and even to compare nucleotide sequences of similar genes in related species. Through all these comparative studies, the case for evolutionary descent with modification has become undeniable.

Illustration from K.E. Sanderson and P.E. Hartman. 1978. *Microbiol. Rev.* **42**:471.

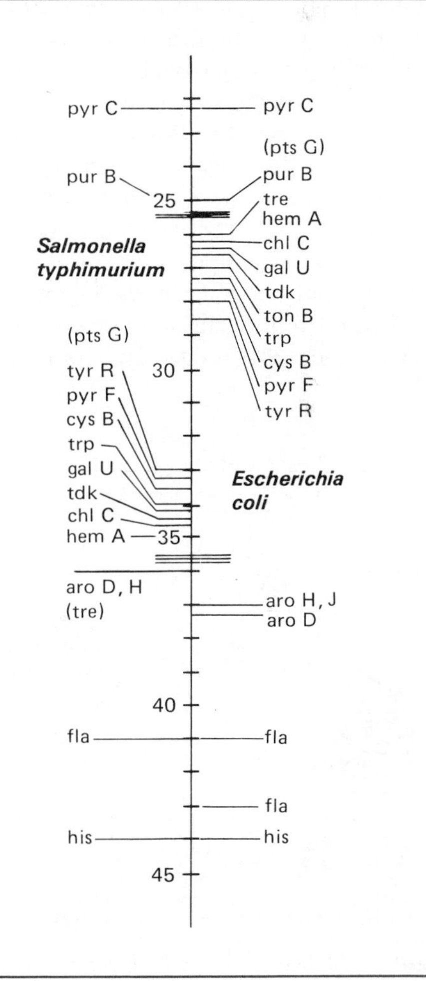

## 6.9 Transduction Analysis of Gene Transfer in Bacteria

Lederberg and Zinder were investigating whether gene transfer took place by conjugation in *Salmonella* as it did in *E. coli.* They mixed together two auxotrophic strains that differed in alleles for five genes and found prototrophic recombinants at low frequency. When the same auxotrophic strains were incubated separately on either side of a filter in U-tube experiments (see Fig. 6.6), prototrophs were recovered at about the same frequency as in the crossing experiments. This observation effectively ruled out conjugation as the process responsible for gene transfer and subsequent genetic recombinations since cell contact was not a requirement for recombination in this species.

Lederberg and Zinder also conducted transformation experiments in which they presented cells with extracts of DNA that had been degraded by the digestive enzyme *deoxyribonuclease,* as well as with extracts of intact fragments of DNA. They recovered prototrophs in *both* sets of experiments. Transformation was, therefore, also eliminated as a gene transfer mechanism in this species because degraded DNA is genetically inactive and could not have provided functional donor genes to the observed recombinant cells.

Some noncellular agent appeared to be a *transducing particle,* an agent small enough to pass through the pores of the U-tube filter and protected against degradation by deoxyribonuclease. The kind of agent they suspected was a virus because viruses can easily pass through filter pores too small for bacterial cells and because viral DNA is wrapped into a protein coat that protects the DNA from degradation by the enzyme. The transducing particle was eventually discovered to be bacteriophage P22, a virus that interacts specifically with *Salmonella.*

The interaction involves a temperate phage and a lysogenic bacterial strain. A *temperate phage* exists in its host cells primarily in the form of a *prophage,* that is, a virus that is physically integrated into the bacterial chromosome. *Lysogenic bacteria* are hosts to temperate phages and carry these viruses in a noninfective state from generation to generation. Occasionally, a bacterial cell lyses (bursts) when the integrated prophage leaves the bacterial chromosome. The free phage DNA behaves as an infecting, *virulent* component rather than as a noninfective prophage (Fig. 6.14). During the lytic, or infection, cycle, bacterial DNA is degraded. New viral mol-

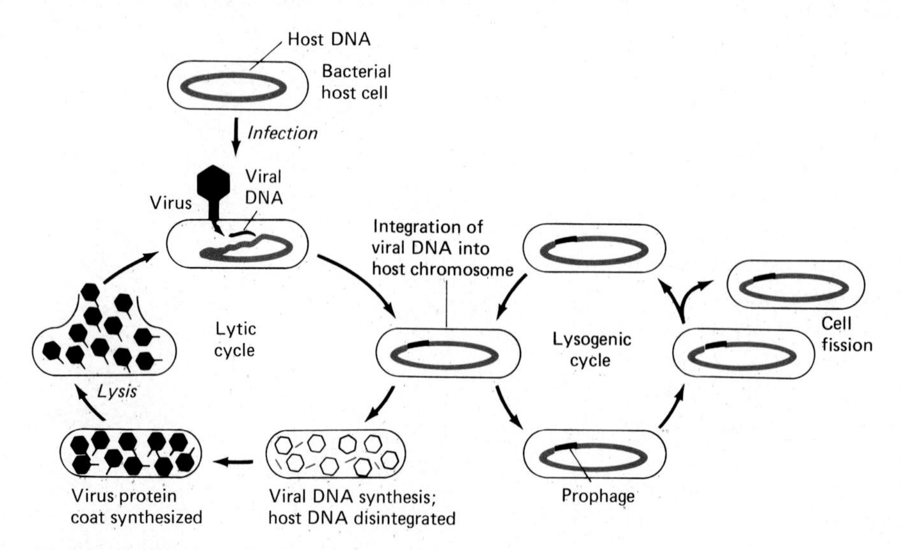


***Figure 6.14*** Life cycle of a temperate phage.

ecules are made and assembled into mature, infective viruses, which are released when the infected cell lyses.

During the lytic cycle pieces of bacterial DNA are present in the infected cell along with newly made viral DNA. Occasionally, a temperate phage, like P22, will package a headful of bacterial DNA *instead of its own DNA* within its protein coat. These accidentally packaged bacterial genes can be transferred to the new host cell in the next infection event. Once they are transferred, donor bacterial genes can be inserted into the recipient bacterial chromosome at a genetically homologous region, by crossing over (Fig. 6.15). Recombinants may arise among recipient cells when different alleles in donor DNA replace alleles of the homologous genes in the recipient chromosome. Phages that behave in this manner are called **generalized transducing phages,** and the process is called **generalized transduction,** since *any* piece of the bacterial chromosome can be transferred in this way.

## 6.10 Mapping by Generalized Transduction Analysis

Even though generalized transducing phages account for only about one per million phages that emerge during cell lysis, they provide an admirable system for mapping closely linked bacterial genes. Transduction analysis has provided the bulk of information for *detailed* genetic mapping in *E. coli, S. typhimurium, B. subtilis,* and a number of other bacteria that have transducing phage systems. Wherever possible, more refined information from transduction analysis is superimposed on cruder preliminary recombination data that is derived from conjugation or transformation analysis. Where this superimposition is not possible, transduction may be the principal or only system available for genetic mapping of the chromosome. Bacterial species of *Shigella, Pseudomonas, Proteus, Staphylococcus,* and some other genera fall into this category (Fig. 6.16).

Gene mapping by generalized transduction analysis is usually accomplished by two- or three-factor transduction tests. Since the formation of a generalized transducing phage is a relatively rare event, the **cotransduction** of two or three genes is assumed to reflect close linkage of these captured bacterial genes transferred by the phage. It would be very unlikely that two or three different pieces of bacterial DNA would happen to be packaged together in these rare transducing phages.

If two genes are cotransduced reasonably often, we can assume the existence of close linkage and can estimate the map distance between the two genes. For example, if donor cells (those giving rise to transducing phages) are $a^+b^+$ and recipient cells are $a^-b^-$, we can assume that $a^+b^-$ and $a^-b^+$ *transductants* are recombinants. These recombinants arise as the result of crossing over between genes *a* and *b*. Transductants with the $a^+b^+$ genotype are not counted among the recombinants because crossing over does not occur in the region between the two genes. The integration of the $a^+b^+$

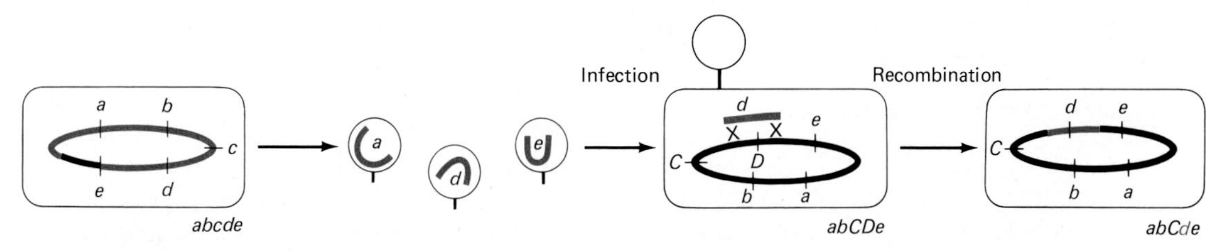


***Figure 6.15*** Generalized transducing phages may carry some bacterial genes (color) instead of their own genes from one host bacterium into another. If these genes become integrated (by crossing over) into the host genome (black), then the transduction can lead to genetically recombinant bacterial cells.

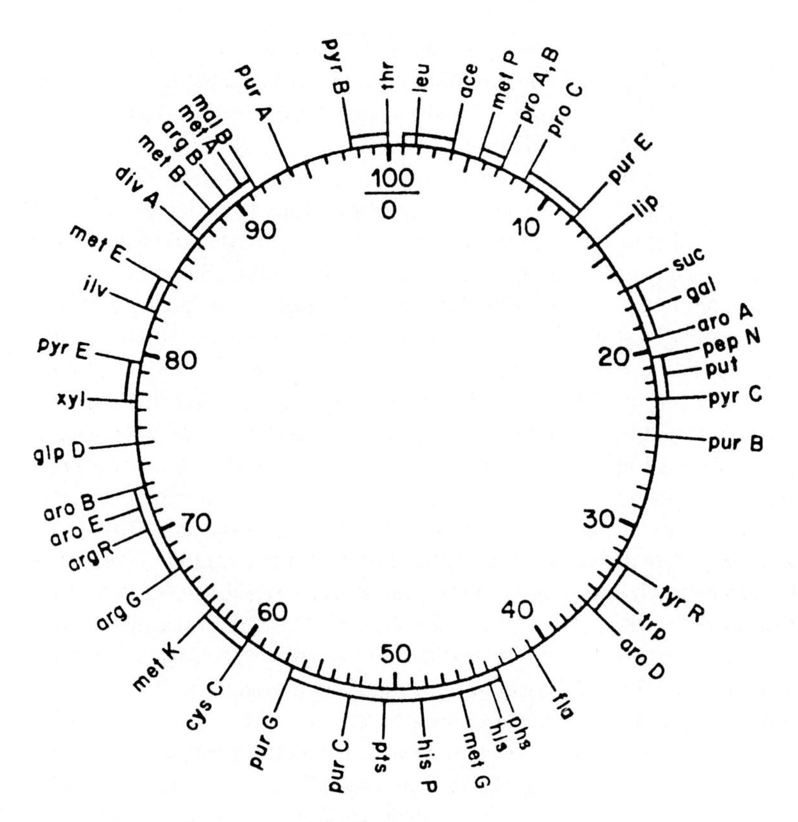


***Figure 6.16*** Gene map of the linkage group of *Salmonella typhimurium.* (From K.E. Sanderson and P.E. Hartman. 1978. *Microbiol. Rev.* **42**:471.)

intact chromosomal segment is a consequence of crossovers that take place outside the segment, one on either side of *a—b.* These methods are the same as we described for transformation analysis (see Fig. 6.3). To calculate distance between the cotransduced *a* and *b* gene loci, therefore, we determine the recombination frequency from the proportion of recombinants in the total transductant population, or

$$\frac{\text{recombinants}}{\text{transductants}} = \text{recombination frequency, or map units}$$

The total number of transductants in the denominator includes $a^+b^+$ since this class is transductant even though it is not recombinant; $a^+b^+$ transductants are excluded from the numerator.

To establish the gene order, we may compare two-factor transductions. For example, if genes *a* and *b* are cotransduced and genes *b* and *c* are cotransduced but genes *a* and *c* are cotransduced less often, the order would be *a—b—c* or *c—b—a.* Similarly, in three-factor transductions we would expect the gene order to be evident from the transductant class occurring in the lowest frequency. This class would be the result of quadruple crossovers, whereas the other classes arise through double crossovers. In this class of recombinants, the gene in the middle has the same allele as present in the recipient, and alleles of the flanking genes match those of the donor. If the gene order was *a—b—c,* we would find $a^+b^-c^+$ in the lowest frequency and would thus know that *b* was the middle gene, in crosses between $a^+b^+c^+$ donors and $a^-b^-c^-$ recipients.

Once we have identified the middle gene and determined the gene order, we can calculate map distances from recombinant frequencies for the *a—b* and *b—c* regions (see Fig. 6.4). These methods and the basic principles are the same as in transformation analysis of linkage. In fact, the basic principles of linkage analysis and mapping are identical or very similar in all the species we have discussed in this chapter and in the preceding chapters.

The great advantage of transduction analysis of bacteria lies in the relatively small size of the genome fragment that can be accidentally packaged into the transducing phage. On the average, the phage can package about 1/100 of the total bacterial genome in any single particle. This length of bacterial chromosome is roughly equivalent to the length of bacterial chromosome that can be transferred in about 1 minute during conjugation in *E. coli.* Several dozen gene loci may be present in a piece even as short as this. Therefore, conjugation analysis by interrupted matings can provide only rough outlines of the chromosome map. Transduction analysis fills in the details of gene order and linkage distances between individual gene loci throughout the length of the chromosome.

## 6.11 Specialized Transduction

Certain temperate phages transfer only certain genes from one bacterial host cell to another, *along with some of their own genes.* Since this gene package is more specific, such phages are **specialized transducing phages** and the process itself is **specialized transduction.** The bacterial genes carried by these phages are loci that flank the particular site on the bacterial chromosome where the prophage is inserted or integrated. Apparently, errors occasionally occur during prophage excision from the bacterial chromosome. The length of excised DNA is about the same as the length of prophage DNA itself, but the segment includes part of the prophage and some of the bacterial genes on one side or the other of the prophage. The excised DNA consists of a covalently linked segment of some phage genes and some bacterial genes.

Since each specialized transducing phage in its integrated prophage state occupies a particular site on the host chromosome, each kind of phage carries specific host genes. For example, in *E. coli* the integration site for lambda ($\lambda$) prophage is flanked by genes for galactose utilization (*gal*) on one side and for biotin synthesis (*bio*) on its other side. If the excision events leading to $\lambda$ prophage separation are inaccurate, the prophage may include host *gal* or *bio* genes but may leave behind some of its own genes. Specialized transducing particles formed in this way are designated $\lambda$*dgal* (deficient for some $\lambda$ genes but carrying host *gal* genes) or $\lambda$*dbio* (deficient for some $\lambda$ genes but carrying host *bio* genes (Fig. 6.17). The $\phi$80 prophage of *E. coli,* on the other hand, can carry host genes for tryptophan synthesis (*trp*) or host suppressor genes (*sup*). These specialized transducing phages would be shown as $\phi$80*dtrp* and $\phi$80*dsup*, respectively.

The deficiency of phage genes in specialized transducing particles is evident from the inability of these particles to initiate a new infection unless a fully functional phage of its type is also present in the host cell. In addition, host genes carried by these transducing phages are *added* to the host genome in the newly infected cell. This results in cells that are diploid for alleles of the genes that are transferred by the phages. These merodiploids are particularly useful in establishing dominance relationships for pairs of alleles.

Partial diploidy in specially transduced hosts is evident from their phenotype and from the production of segregating haploids in subsequent growth. If $gal^+$ is carried into recipient $gal^-$ cells, the transduced cells express the $Gal^+$ phenotype. Some haploid segregants of these transduced cells will show a $Gal^-$ phenotype, others may be $Gal^+$. Segregation of the two *gal* alleles is apparent from the $gal^-$ segregants produced by transduced $gal^+$ cells. These $gal^-$

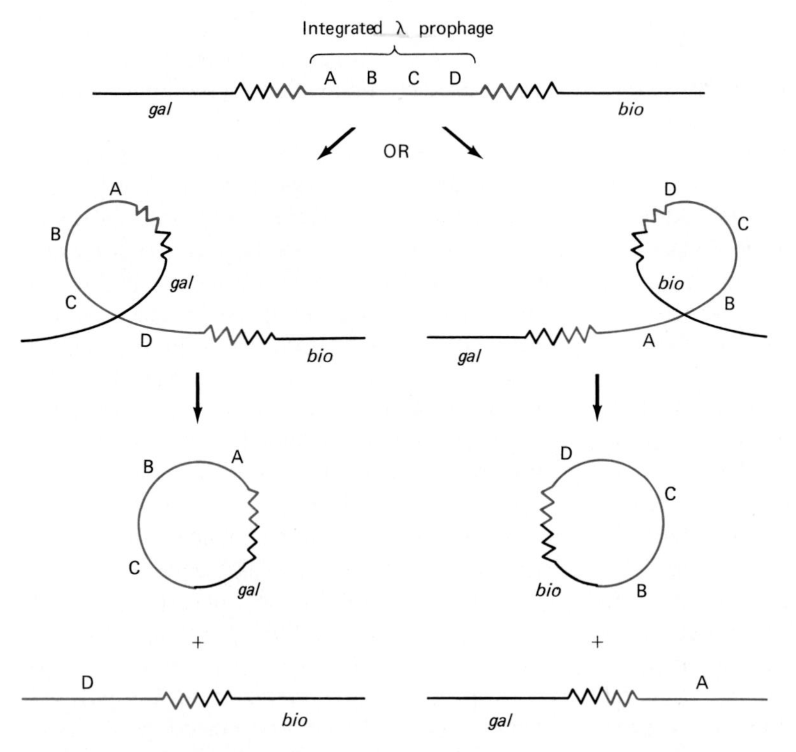


***Figure 6.17*** Specialized transduction. Formation of circular λ*dgal* (left) and λ*dbio* (right) DNA leads to transducing particles that carry particular bacterial genes (black) in place of some of their own DNA (color). In the case of the specialized transducing phage λ of *E. coli,* the prophage integrates between the *gal* and *bio* loci in the bacterial chromosome and therefore carries either *gal* or *bio* genes when inaccurate excision of the prophage takes place.

***Table 6.1*** Characteristics of generalized and specialized transduction.

| **characteristic** | **generalized transduction** | **specialized transduction** |
|---|---|---|
| viral agent | temperate phages (many known) | temperate phages (few known) |
| host bacteria | lysogenic strains | lysogenic strains |
| genes transferred by virus to host cells | host genes only | host genes and viral genes |
| host genes transferred | any part of the genome | only genes flanking integrated prophage |
| number of host genes transferred per transductional event | several dozen or more loci | a few loci |
| site of prophage integration into host chromosome | site homologous with transferred host loci | specific prophage integration site |

descendants are segregants, not newly mutated cells, because they continue to be produced for a period of time after transduction and at a higher frequency than expected for mutants.

Since transduced $gal^-$ cells display the $Gal^+$ phenotype when both alleles are present, $gal^+$ must be dominant over $gal^-$. The state of partial diploidy is due to the initial failure of the entering genes to recombine with and replace the existing alleles in the host chromosome. Integration of these genes can take place later on, as is evident from the haploid segregants that arise in subsequent generations.

Gene transfer can occur by either generalized or specialized transduction, but the two processes have significant differences as well as similarities (Table 6.1). With regard to recombination and mapping, generalized transduction is the most useful of the two processes for mapping since any part of the bacterial genome can be transduced and analyzed in genetic tests. Specialized transduction, on the other hand, provides information only on regions immediately flanking the prophage integration site. Relatively few specialized transducing phages are known in *E. coli* and other species, so that only limited regions of the bacterial genome can be mapped by specialized transduction.

## Recombination and Mapping in Viruses

Of all the organisms discussed so far, none was so different in its genetic system that we could not apply essentially similar methods to gene mapping. Gene transfer by transformation, conjugation, and transduction furnishes data for construction of maps despite the one-way transfer of donor genes and despite the requirement for double crossovers rather than single crossovers to explain recombinations. These features are really only variations on basic themes in mapping. In some ways, viruses present a different system for recombination analysis, but we can construct maps once we know and take these differences into account.

### 6.12 Recombination in Viruses

The most useful phenotypes for genetic analysis of bacterial viruses are ones that can be determined easily with the naked eye. Since viruses as units can only be seen with the electron microscope, the phenotypes of choice are ones that appear as the result of virus activity in the lysis of their bacterial hosts. When bacteria are incubated on solid medium in such large numbers that they cover the entire surface as an opaque lawn of growth, we can detect lysis by the appearance of a clear or cloudy spot called a *plaque*. Each plaque is equivalent to a colony since one original phage and all of its descendants create one plaque. *Plaque morphology* is an inherited feature of a viral strain and can serve as a useful phenotypic character. Another useful phenotypic characteristic is *host range* of the virus. Most viruses are highly restricted in host range and infect only certain species or cell types. Other viral types can infect a wider range of hosts. In either situation the particular virus can be recognized by its plaque characteristics.

A. Hershey and R. Rotman made some of the earliest studies of phage recombination in the late 1940s. Their recombination analysis of phage T2 of *E. coli* provided basic information on the genetic features of the process. They "crossed" the phages by introducing two or more phenotypically different phage strains at the same time to *E. coli* host cells. Such a *mixed infection* required careful attention to the concentrations in which these phages were added so that each cell might be infected by two or more phages. The investigators recovered the progeny phages, or **phage lysate** (viruses emerging from lysed cells), and added them to a lawn of bacterial cells in order to analyze phage phenotypes among the progeny.

The two phage parental strains differed in alleles for a plaque gene and for a host range gene. Phages carrying $r^+$ produced small plaques with irregular borders, and $r$ phages produced large plaques with sharp borders. The $h^+$ allele of the host range gene allowed a more restricted range (only *E. coli* strain B was lysed) than the *h* allele (*E. coli* strains B and B/2 were sensitive), and the plaques were different in each host. Mixed infections were made in liquid medium containing *E. coli* strains B and B/2, using $rh^+$ and $r^+h$ parental phages. The phage lysate was collected and plated on a lawn of the two kinds of bacterial cells. Four phenotypic classes were found, with parental types $rh^+$ and $r^+h$ in greater abundance than the recombinant $r^+h^+$ and $rh$ plaque types (Fig. 6.18).

When the reciprocal mixed infection was made, using $r^+h^+$ and $rh$ as parents, the same proportions of parental and recombinant classes were found as in the first experiments. Parental phage types $r^+h^+$ and $rh$ greatly outnumbered the recombinants $r^+h$ and $rh^+$. Calculation of the frequency of exchange between the *r* and *h* genes was based on the familiar proportion of total recombinants/total plaques counted.

One of the procedures used in these studies

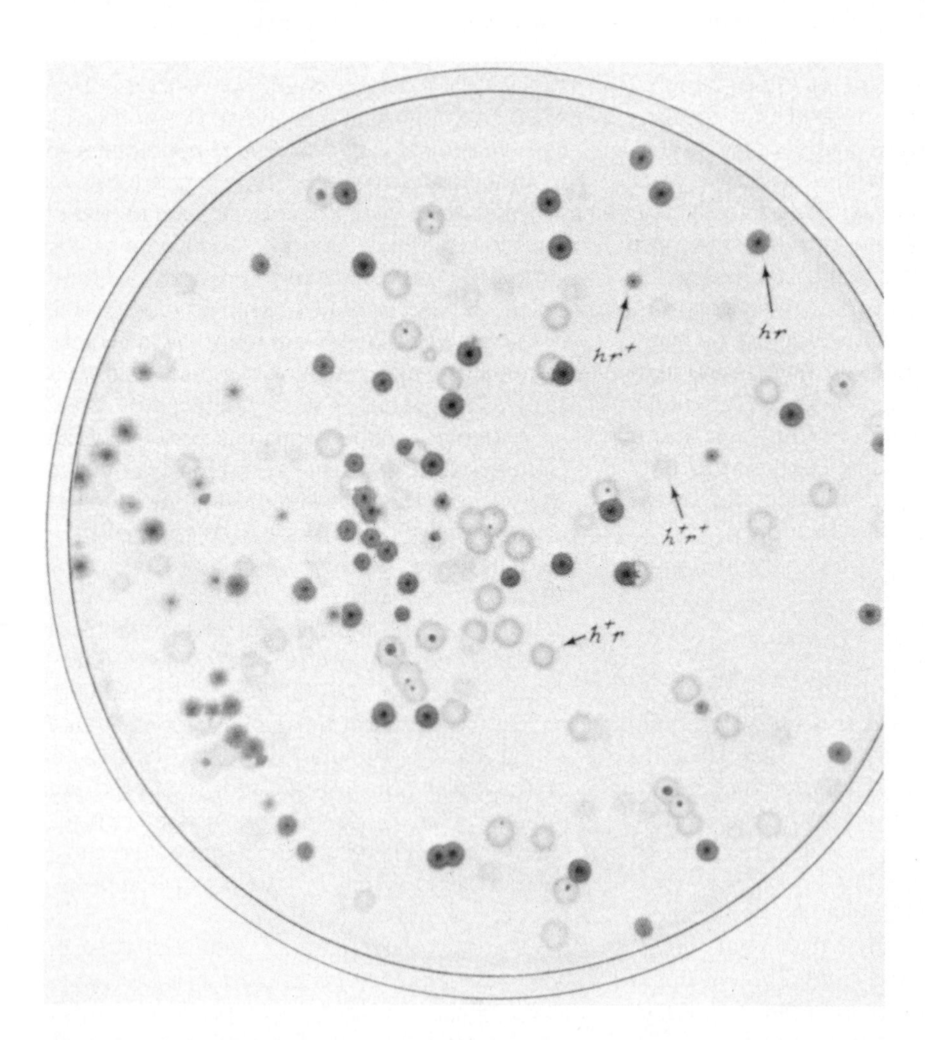


***Figure 6.18*** Plaques formed after mixed infection on *E. coli* B + B/2 cell mixtures by parental T2 phages $rh^+$ and $r^+h$ and by T2 recombinants $r^+h^+$ and $rh$. (From *Molecular Genetics: An Introductory Narrative,* Second Edition, by Gunther S. Stent and Richard Calendar. W.H. Freeman and Company. Copyright © 1978.)

was the **single-burst experiment.** In such an experiment, the bacterial culture that is exposed to mixed infection is later diluted into a series of tubes such that no more than one infected cell, on the average, is present in each tube. The phage lysate from each tube is then plated to assay phage types emerging from a single infected cell. Hershey and Rotman found that the reciprocal classes of recombinants rarely, if ever, appeared with equal frequency from any one single burst. Approximately equal amounts could be counted, however, when many such single-burst progenies were added together for a total population count. Therefore, recombination in T2 phage appeared to involve some *nonreciprocal exchange process.* The *probability* was equal for each kind of recombination event to occur in a larger sampling of phage progeny populations, but the reciprocal events did not take place in the *same* host cell with equal frequency.

The explanation of these observations was later derived from other kinds of studies. T2 DNA molecules made in host cells were found to undergo *repeated rounds of "mating"* during the latent period of the lytic cycle. In addition, these pairings between DNA molecules were indiscriminate. Any two molecules of any genotype can pair and undergo exchanges many times during the infection cycle. Since initial gene exchanges might be altered during subsequent crossover events, it is highly improbable that reciprocal recombinants would arise in equal numbers from a single infected cell. Crossing over is a random event, however, so a population of progeny from many cells may produce approximately equal numbers of reciprocal classes of recombinants, on the average.

In spite of these aspects of T2 recombination, linkage and recombination analysis of other genes as well as *r* and *h* showed a linear relationship and resulted in construction of a linear map of the chromosome. Some difficulties were encountered in constructing a *consistent* linear order for T2 genes, as linkage analysis continued. Similar problems arose in mapping genes of other T-even phages.

## 6.13 Circular and Linear Maps

From results of various three-factor crosses with T4 (and other T-even phages), it was discovered that closely linked genes could be mapped in a linear and consistent order but that widely separated loci did not behave consistently. This finding was particularly true for genes mapped at the "ends" of the chromosome. For example, in some crosses between $a^+b^+c^+$ and $abc$ parental phages, the rare double-exchange recombinants were found to be $a^+bc^+$ and $ab^+c$. This indicated that the gene order was *a—b—c*, but previous mapping of these same three genes had shown that the gene order and distances were:

To resolve this apparent inconsistency, George Streisinger and his colleagues proposed a circular rather than a linear genetic map. Several different but related linear maps could be drawn from recombination analysis, such as:

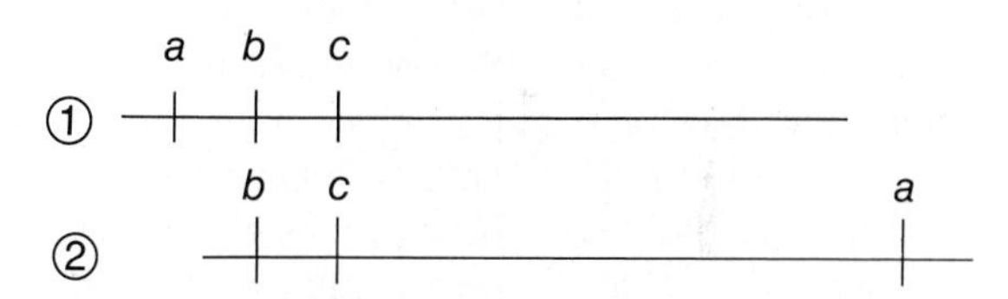


Both of these linear maps can be generated from a single circular map on the basis of where the circle is broken to yield circular permutations with two free ends, as follows:

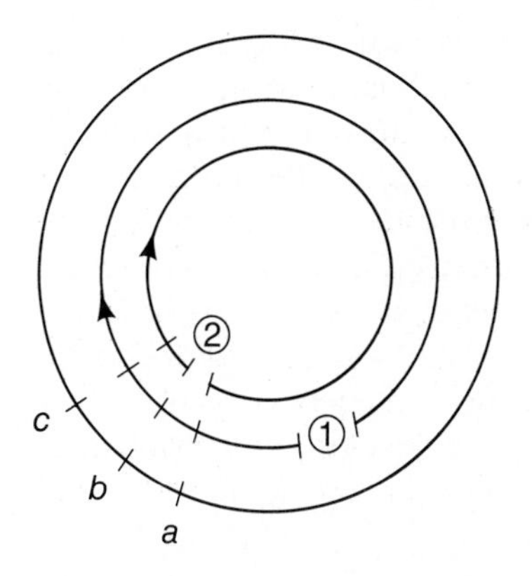

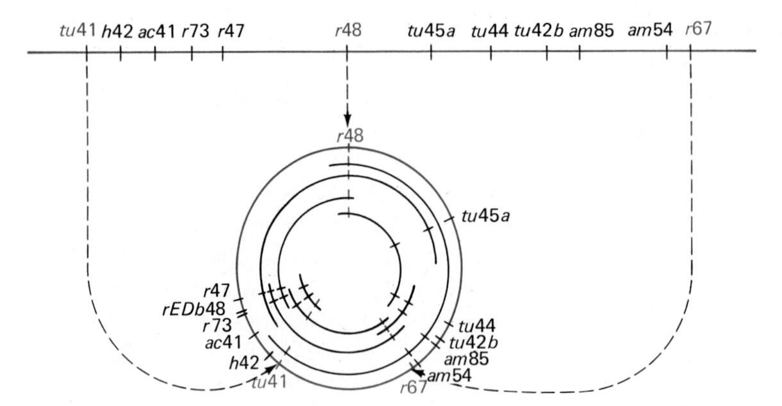


***Figure 6.19*** The linear sequence of genes established by recombinational analysis in phage T4 can be shown as a circular genetic map to accommodate linkage data from three- and four-factor crosses. The locations of markers used for any one cross are connected by an arc. The physical chromosome of T4, however, is linear, not circular, DNA. (After G. Streisinger, R. Edgar, and G. Denhardt. 1964. *Proc. Nat. Acad. Sci. U.S.* **51**:775.)

When various three- and four-factor crosses were compared, every inconsistency in linkage distances could be explained by circularity of the genetic map (Fig. 6.19). This situation was found to characterize a number of other viruses; it is, thus, a more general phenomenon and not a unique feature of T-even phages.

Later studies of T4 using molecular methods showed that the phage chromosome was actually a linear DNA molecule measuring about 53 $\mu$m from end to end. The translation of a linear genome into a circular map was explained by additional molecular studies involving DNA replication during the lytic cycle and the DNA packaging in mature phage particles that were released from lysed cells.

During DNA replication of T-even phages in infected cells, numerous large duplex DNA molecules are produced, each of which is much longer than 53 $\mu$m. These large molecules are *concatamers* of repeated sequences of the phage genome, arranged in tandem. For example, if the phage genome consists of five genes, the concatamers would be *abcdeabcdeabcdeabcdeab* or some other tandem collection. Later in the cycle, these molecules are enzymatically cut into linear pieces that are 53 $\mu$m long and include only five genes, and the pieces are packaged as a headful of DNA during viral maturation. The packaged DNA might be *abcde*, *bcdea*, *eabcd*, or some other sequence of the five genes adjacent to each other in the concatamer. Each phage has a full complement of genes, but different DNA molecules in a phage population may have different gene loci at the two ends of the linear genome. The circular map is a summary of the collection of circular permutations of the linear genome.

The long concatamers themselves probably arise by exchange events between homologous regions at the ends of two different molecules. Two molecules would be linked together as the result of a crossover event such as the following:

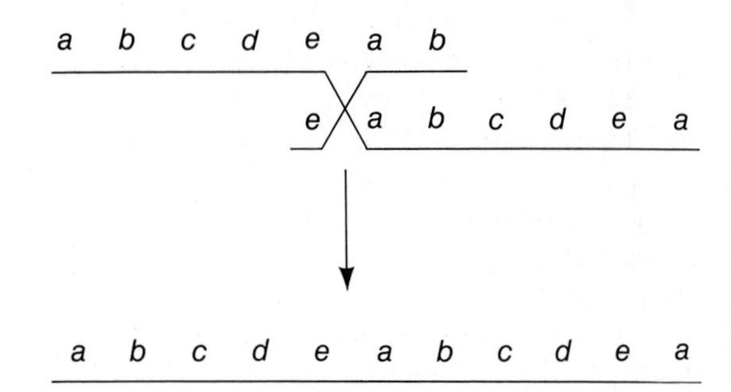

A circular genetic map is a convenient summary of recombination data that indicates the existence of permutations of a single sequence of genes. The circular genetic map is not necessarily an accurate reflection of the actual physical conformation of the chromosome. In T-even phages the genome is linear, in *E. coli* the chromosome has a circular conformation. On the basis of genetic analysis alone, we would assume that both systems have circular genomes. When we supplement genetic analysis by other, independent methods, we can determine the physical state of the genome and can compare it with the genetic map of the chromosome.

We cannot compare genetic maps of different organisms directly in terms of *physical distances* between gene loci derived from recombination analysis. In T2 phage the observed recombination frequencies reflect the occurrence of repeated rounds of matings. In lambda phage, on the other hand, DNA molecules undergo few or no exchanges during the lytic cycle. The same recombination frequencies in T2 and lambda phages may be found for some of their genes and may be translated into the same number of map units separating the loci in the two maps. The actual physical distances separating genes that are shown as the same distance apart on the genetic map may be entirely different, thus reflecting differences in recombination events. This fact is true for all map comparisons between different organisms; map units are arbitrary measurements, not absolute physical measurements. In fact, this is true even for genes on the same genetic map since different frequencies of recombination may characterize genes that are the same physical distance apart, or the same recombination frequencies may characterize genes that are different distances apart on the physical chromosome. The genetic map is a summary of recombination data and reflects the particular events that produce recombinations in each case.

## Questions and Problems

***6.1*** Donor DNA from *Bacillus subtilis* strains that are independent of the requirements for tryptophan ($Trp^+$) and phenylalanine ($Phe^+$) was used to transform $Trp^- Phe^-$ recipient cells, with the following results:

| transformant classes | number of transformants |
|---|---|
| $Trp^+ Phe^-$ | 100 |
| $Trp^- Phe^+$ | 200 |
| $Trp^+ Phe^+$ | 700 |

***a.*** How far apart on the chromosome are the genes *trp* and *phe*?
***b.*** How does the nonrecombinant $Trp^+ Phe^+$ class arise in transformation?

***6.2*** DNA extracted from wild-type *Bacillus subtilis* was used to transform mutant cells that are unable to synthesize the amino acids arginine (arg), proline (pro), and alanine (ala), with the following results:

| transformant classes | number of transformants |
|---|---|
| $Ala^+ Pro^+ Arg^+$ | 4200 |
| $Ala^+ Pro^- Arg^-$ | 420 |
| $Ala^+ Pro^- Arg^+$ | 1050 |
| $Ala^+ Pro^+ Arg^-$ | 700 |
| $Ala^- Pro^+ Arg^+$ | 210 |
| $Ala^- Pro^+ Arg^-$ | 420 |
| $Ala^- Pro^- Arg^+$ | 420 |

***a.*** What is the linkage order of these genes?
***b.*** What are the apparent map distances between these genes?

***6.3*** In a transformation experiment to determine linkage relations for genes *g*, *h*, and *z*, donor DNA of genotype $g^+h^+z^+$ was added to recipient cells of genotype $g^-h^-z^-$, with the following results:

| transformant classes | number of transformants |
|---|---|
| $g^+h^+z^+$ | 1400 |
| $g^+h^+z^-$ | 10 |
| $g^+h^-z^+$ | 150 |
| $g^+h^-z^-$ | 300 |
| $g^-h^+z^+$ | 290 |
| $g^-h^+z^-$ | 140 |
| $g^-h^-z^+$ | 60 |

What are the order and apparent map distances for the three genes?

***6.4*** Five Hfr donor strains of *E. coli* (A–E), all carrying the same wild-type alleles, are crossed to an F$^-$ recipient strain carrying the alternative set of alleles. Using the interrupted mating technique, each Hfr strain was found to transmit its genes in a unique sequence, as shown:

| Hfr strain | | | | |
|---|---|---|---|---|
| **A** | **B** | **C** | **D** | **E** |
| $mal^+$ | $ade^+$ | $pro^+$ | $pro^+$ | $his^+$ |
| *str–s* | $his^+$ | $met^+$ | $gal^+$ | $gal^+$ |
| $ser^+$ | $gal^+$ | $xyl^+$ | $his^+$ | $pro^+$ |
| $ade^+$ | $pro^+$ | $mal^+$ | $ade^+$ | $met^+$ |
| $his^+$ | $met^+$ | *str–s* | $ser^+$ | $xyl^+$ |

Draw the genetic map of the Hfr strain from which these five donors were derived.

***6.5*** In *E. coli* the following Hfr strains donate the markers shown in the order given.

| Hfr strain | marker order |
|---|---|
| 1 | Q R D H T |
| 2 | A X S T H |
| 3 | B N C A X |
| 4 | B Q R D H |

All these Hfr strains were derived from the same F$^+$ strain. What was the order of these markers on the original F$^+$ chromosome?

***6.6*** Three strains of *E. coli* have genotypes as follows:

Strain 1: $\frac{F'\ leu^+\ pro\text{-}1}{leu^+\ pro\text{-}1}$ (partially diploid for the two genes)

Strain 2: F$^-$ *leu pro*-2 *Z* (lysogenic for the generalized transducing phage Z)

Strain 3: F$^-$ $leu^+$ *pro*-1 (an F$^-$ derivative of strain 1, having lost F′)

Strains are leucine requiring (*leu*) or independent ($leu^+$), and all require proline to grow.

***a.*** How would you determine whether *pro-1* and *pro-2* are alleles of the same gene?

***b.*** Suppose that *pro-1* and *pro-2* are allelic and the *leu* locus is cotransduced with the *pro* locus. Using phage Z to transduce genes from strain 3 to 2, how would you determine the genetic order of *leu, pro-1,* and *pro-2*?

***6.7*** A bacterial strain 1 carrying two auxotrophic mutations, *a* and *b*, and the wild-type allele for gene *c*, is infected with a generalized transducing phage. The progeny phage are used to transduce strain 2 cells, which are wild type for *a* and *b* but carry an auxotrophic mutation *c*. The cells are plated on minimal media such that only those cells that are wild type for all three gene loci can survive. In a reciprocal experiment, strain 1 is transduced with phage isolated from a strain 2 infection. The following results are obtained:

| strain 1 → strain 2 | strain 2 → strain 1 |
|---|---|
| 150 wild type per $10^8$ cells | 9 wild type per $10^8$ cells |

What is the gene order for the three loci?

***6.8*** *E. coli* cells were infected with two strains of T2 virus, one being small (*s*), fuzzy bordered (*f*), and turbid (*tu*), and the other being wild type for all three traits. The lysate from this infection was plated out and classified into genotypes according to plaque morphology, as follows:

| genotype | number of plaques |
|---|---|
| *s f tu* | 6934 |
| + + + | 7458 |
| *s f* + | 1706 |
| *s* + *tu* | 324 |
| *s* + + | 1040 |
| + *f tu* | 940 |
| + *f* + | 344 |
| + + *tu* | 1930 |
| | 20,676 |

***a.*** What is the linkage order for these three genes?

***b.*** Determine the linkage distances between *s* and *f*, *f* and *tu*, and *s* and *tu*.
***c.*** What is the coefficient of coincidence?
***6.9*** *E. coli* cells were infected with two strains of T2 phage; one was wild type for three plaque traits and the other was mutant for all three traits. The lysate from this mixed infection was plated out, with the following results:

| genotype | number of plaques |
|---|---|
| + + + | 715 |
| + + *c* | 90 |
| + *b* + | 5 |
| + *b c* | 190 |
| *a* + + | 190 |
| *a* + *c* | 5 |
| *a b* + | 90 |
| *a b c* | 715 |
| | 2000 |

What are the gene order and the apparent map distances involved?
***6.10*** In another experiment using different T2 wild-type (+ + +) and triply mutant (*abc*) strains, involving the same three genes as in Question 6.9, the lysate plated from the mixed infection gave the following results:

| genotype | number of plaques |
|---|---|
| + + + | 620 |
| + + *c* | 10 |
| + *b* + | 90 |
| + *b c* | 280 |
| *a* + + | 280 |
| *a* + *c* | 90 |
| *a b* + | 10 |
| *a b c* | 620 |
| | 2000 |

***a.*** What are the gene order and the apparent map distances between genes?
***b.*** How would you explain the different results in this experiment from the results in Question 6.9 when the same three genes are involved in both cases?
***6.11*** In *E. coli* what difference in the gene transmission mechanism accounts for the following results:

$F^+ \times F^- \rightarrow F^+$ but no chromosome transfer
Hfr $\times F^- \rightarrow F^-$ but with chromosome transfer

***6.12*** What genetic tests would you use to show that a bacterial chromosome is physically circular?
***6.13*** Discuss the contribution of viral genetic studies to the Chromosome Theory of Inheritance.
***6.14*** How is it possible that a physically linear chromosome produces a circular genetic map?
***6.15*** What are the characteristics that distinguish temperate and virulent phages?
***6.16*** The generalized transducing phage P1, carrying bacterial alleles *leu*$^-$, *lys*$^+$, *gly*$^+$, was used to transduce a strain of *Salmonella typhimurium* of the genotype *leu*$^+$ *lys*$^-$ *gly*$^-$. The following transductants were produced:

| transductant classes | number of transductants |
|---|---|
| *leu*$^+$ *lys*$^+$ *gly*$^+$ | 10 |
| *leu*$^+$ *lys*$^+$ *gly*$^-$ | 90 |
| *leu*$^+$ *lys*$^-$ *gly*$^+$ | 280 |
| *leu*$^-$ *lys*$^+$ *gly*$^+$ | 1240 |
| *leu*$^-$ *lys*$^+$ *gly*$^-$ | 280 |
| *leu*$^-$ *lys*$^-$ *gly*$^+$ | 90 |
| *leu*$^-$ *lys*$^-$ *gly*$^-$ | 10 |
| | 2000 |

***a.*** What is the gene order?
***b.*** What are the apparent map distances between genes?
***6.17*** What are the differences between generalized and specialized transducing phages? In what ways are they similar?
***6.18*** Define the following terms: ***a.*** Prototroph selection. ***b.*** Auxotroph. ***c.*** Episome. ***d.*** Interrupted mating. ***e.*** Merodiploid. ***f.*** Sexduction. ***g.*** Cotransduction. ***h.*** Phage lysate.

## References


Avery, O.T., C.M. MacLeod, and M. McCarty. 1944. Studies on the chemical nature of the substance inducing transformation of pneumococcal types. *J. Exp. Med.* **79**:137.

Hershey, A.D., and M. Chase. 1952. Independent functions of viral protein and nucleic acid in growth of bacteriophage. *J. Gen. Physiol.* **36**:39.

Hinnen, A., J.B. Hicks, and G.R. Fink. 1978. Transformation of yeast. *Proc. Nat. Acad. Sci. U.S.* **75**:1928.

Hotchkiss, R.D., and J. Marmur. 1954. Double marker transformations as evidence of linked factors in deoxyribonucleate transforming agents. *Proc. Nat. Acad. Sci. U.S.* **40**:55.

Krens, F.A., L. Molendijk, G.J. Wullems, and R.A. Schilperoort. 1982. *In vitro* transformation of plant protoplasts with Ti-plasmid DNA. *Nature* **296**:72.

O'Brien, S.J., ed. 1982. *Genetic Maps,* 2nd ed. Frederick, Md.: National Institutes of Health.

Streisinger, G., R.S. Edgar, and G.H. Denhardt. 1964. Chromosome structure in phage T4. I. Circularity of the linkage map. *Proc. Nat. Acad. Sci. U.S.* **51**:775.

Zinder, N.D., and J. Lederberg. 1952. Genetic exchange in *Salmonella. J. Bact.* **64**:679.

# CHAPTER 7

# Gene Structure and Function

After the identification of DNA as the chemical basis of heredity, emphasis shifted from defining the gene according to its transmission patterns to defining the gene in molecular terms. Molecular analysis of the gene was necessarily indirect in the 1950s and 1960s since methods were not available to analyze DNA sequences directly. Studies of proteins and of RNA complementary copies of DNA were carried out, and these lines of information were then related back to the probable structure and function of the gene itself. The concept of the gene is still changing. Current information from base sequencing of DNA has revealed some unexpected features of gene structure, particularly in eukaryotes. Although we can study genes directly today using molecular methods, the earlier indirect studies provided the foundations for our current understanding of the gene.

## Genetic Properties of DNA

The chemical composition of DNA was determined over 100 years ago, shortly after the discovery of DNA in extracts of cell nuclei by Friedrich Miescher in 1871. Miescher called the material "nuclein" since it occurred only in nuclei. Some attention was paid to the compound because of its unusually high phosphorus content and its acid nature. Little interest was displayed in its possible role in inheritance. Far more was known about proteins than about nucleic acids. In fact, the prevailing ideas of the first half of this century centered around proteins being the genetic material.

Two landmark experiments provided strong evidence in favor of DNA as the genetic material. Considerable skepticism existed in the scientific community about the significance of these experiments when they were published, one in 1944 and the other in 1952. But enough interest and enough experimental data were generated to make some investigators concentrate their efforts on DNA rather than on proteins. James Watson and Francis Crick pointed out the suitability of DNA as genetic material in their two brief publications in 1953, which described a possible model for DNA molecular structure. These two publications had an immediate and powerful impact on the scientific community. Their molecular model for DNA still provides the foundation for modern genetic studies.

### 7.1 DNA Is the Genetic Material

In 1944 Oswald Avery, Colin MacLeod, and Maclyn McCarty published their seminal study on genetic transformation in bacteria. They showed that highly purified DNA extracted from one genetic strain of the pneumococcal species *Diplococcus pneumoniae* could genetically transform another strain that carried different alleles. Different types of pneumococcus initiate respiratory infections in mammalian species. These types also occur in a genetically altered noninfective, or **avirulent,** form as well as in the infective, or **virulent,** state. The virulent forms have a distinctive polysaccharide capsule surrounding the cell; the avirulent forms lack the capsule.

When purified DNA, or "transforming principle," from virulent cells of type II was added to a culture of avirulent cells of type III, colonies of virulent pneumococcus appeared among the original avirulent type III colonies on plates (Fig.

7.1). The new colonies proved to be type II virulent bacteria. When DNA was removed from the transforming principle solution, only avirulent colonies of type III cells were produced. Similar experiments were conducted, always showing that highly purified DNA extracted from one genetic strain could genetically transform a different strain to its own specificity. The transformed recipients expressed their newly inherited traits and transmitted these altered traits to all descendant generations. These features indicated that a genetic change had indeed occurred during transformation.

Since recipient cells were always transformed to the same type as the strain from which the DNA was extracted, and since transformation did not take place in the absence of DNA in the solution of transforming principle, DNA seemed to be the responsible genetic agent of transformation. Even though minute traces of protein remained in the most highly purified preparations of extracted DNA, the amount was too small to account for the genetic changes observed. We now see this evidence as highly convincing, but the scientific community did not overwhelmingly accept it at that time. One of the difficulties was the inability of other investigators to carry out successful transformation studies with other organisms. These same difficulties exist today, and transformation is not the method of choice for gene transmission studies in bacteria.

Studies did continue in many laboratories, using various genetic and biochemical methods to determine whether DNA or protein was the genetic material. In 1952 Alfred Hershey and Martha Chase reported their important experimental studies using T2 phage, which we discussed briefly in Section 3.11. They took advantage of newly available radioactive isotopes to label DNA and protein molecules differentially and to follow the fate of these labeled molecules during T2 infection of *E. coli* bacterial cells. Hershey and Chase knew that new progeny viruses were made and assembled in host cells according to viral genetic instructions. The molecules that entered the host cells after viral infection therefore must have carried viral ge-

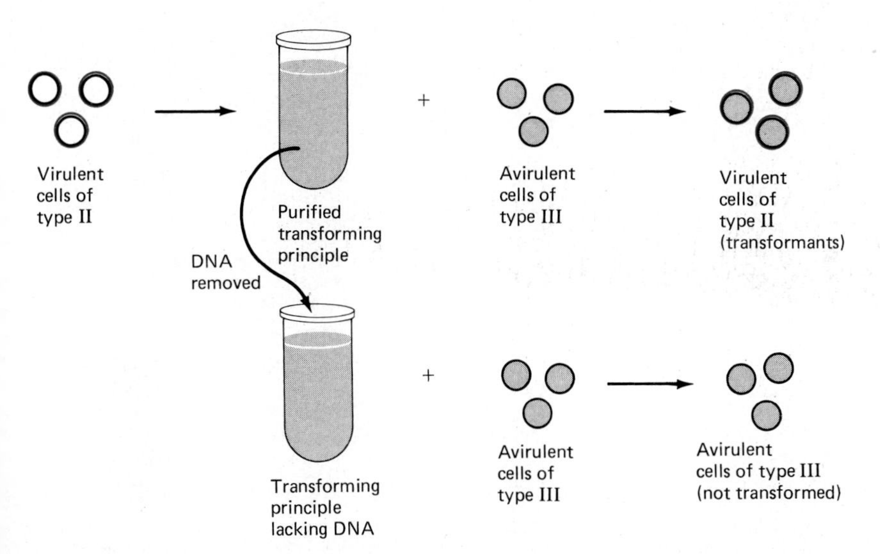


***Figure 7.1*** Diagrammatic representation of the classic 1944 experiments by O.T. Avery and coworkers, which showed that DNA is the transforming principle. This was the first significant demonstration that DNA is the genetic material. Transformation of genetically avirulent type III cells to genetically virulent type II cells was accomplished only when DNA from virulent type II cells was added to the type III culture.

netic instructions. The experiments were designed to trace the genetic continuity of DNA and protein. The results would determine which of these two types of molecules fit the requirements for the genetic material.

T2 phage consists exclusively of about equal amounts of DNA and protein. Each kind of molecule has a unique atom that is lacking in the other molecule; DNA contains phosphorus ($^{31}P$) but no sulfur ($^{32}S$), and proteins usually have S but no P atoms. DNA in one set of T2 was labeled with radioactive P atoms ($^{32}P$) by growing the viruses in *E. coli* cultured on [$^{32}P$]-labeled nutrient medium. The proteins in another set of T2 phages were labeled with radioactive S atoms ($^{35}S$) in a similar manner. The different radioactive isotopic markers would indicate which molecule—[$^{32}P$]DNA or [$^{35}S$]protein—entered the host cells and which of these molecules remained outside these cells. The molecule that entered a new host cell and directed synthesis of progeny phages must be the genetic molecule. Whichever kind of molecule remained outside the cell after infection must not be the genetic molecule.

*E. coli* host cells were grown in unlabeled nutrient medium (containing $^{31}P$ and $^{32}S$) to provide material for the infection experiments with labeled phages. In separate experiments, these unlabeled *E. coli* cells were infected with

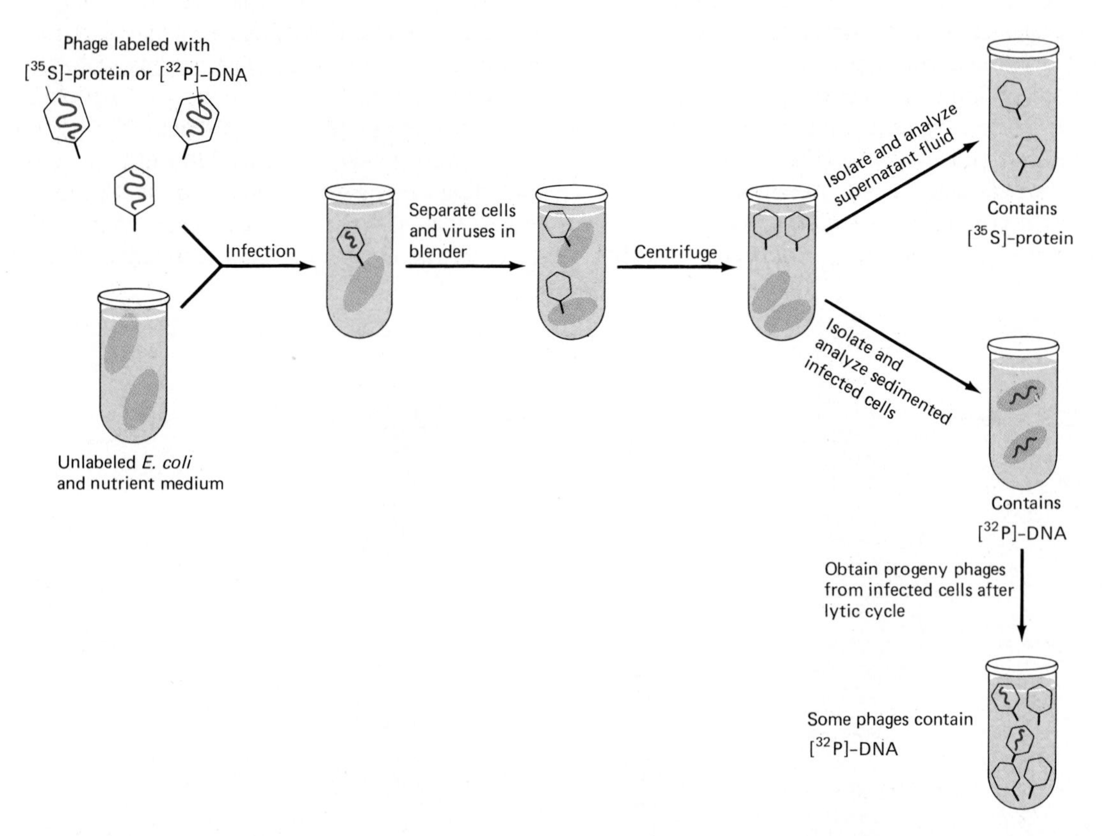


***Figure 7.2*** Illustration of the main features in the Hershey-Chase experiments, which showed that DNA was the genetic material. T2 phages carrying labeled protein *or* labeled DNA were allowed to infect *E. coli*. After infection was initiated, the viruses and their host cells were separated mechanically (blender). Labeled DNA was found inside the host cells, and labeled protein occurred primarily outside the *E. coli* cells. DNA must, therefore, be the genetic material that directs virus reproduction inside host cells during the infection cycle.

either [$^{32}$P]DNA-labeled phages or [$^{35}$S]protein-labeled phages and with both [$^{32}$P] and [$^{35}$S] viruses. After a few minutes, when infection had been initiated but before host cell lysis, the infected *E. coli* were suspended in liquid and put into a high-speed blender. The shearing forces were enough to strip away any adhering viral particles from cell surfaces but not enough to damage the *E. coli* cells themselves.

The suspended materials were removed from the blender and centrifuged to separate the bacterial sediment from the remaining supernatant fluid, which contained free viruses (viruses that had not initiated infections). Sedimented bacterial cells were examined and were found to contain [$^{32}$P]DNA and varying but much lower amounts of [$^{35}$S] protein (Fig. 7.2).

[$^{32}$P]DNA of the virus had apparently entered the bacterial cells. Most of the [$^{35}$S]protein was found in the fluid remaining after the bacterial sediment was removed, but there was little or no [$^{32}$P]DNA in this fluid. The [$^{35}$S]protein had not entered the bacterial cells. By electron microscopy in preliminary studies, Hershey and Chase could see empty ghosts of the viral protein coats that were still attached to the bacterial walls after an infection. From all these results they concluded that DNA must be the genetic material since DNA had entered the host cells and proteins had remained outside the cells for the most part. When the infection cycle was allowed to go to completion, some of the progeny phages in the lysate from burst cells were found to contain [$^{32}$P]DNA but no [$^{35}$S]protein.

On the basis of expected gene behavior—directing progeny phage synthesis and maintaining genetic continuity between generations—the only reasonable conclusion was that DNA, and not protein, was the genetic material. The entering phage DNA had directed synthesis of new phage-specific DNA, which was passed on to new generations of virus progeny.

## 7.2 Duplex DNA

DNA molecules are polymers built from **mononucleotide** monomer units (Table 7.1). The four kinds of mononucleotides differ only in

***Table 7.1*** Constituent units of nucleic acids.

| base | nucleoside* | nucleotide† | nucleic acid‡ |
|---|---|---|---|
| purines: | | | |
| adenine | adenosine | adenylic acid | RNA |
| | deoxyadenosine | deoxyadenylic acid | DNA |
| guanine | guanosine | guanylic acid | RNA |
| | deoxyguanosine | deoxyguanylic acid | DNA |
| pyrimidines: | | | |
| cytosine | cytidine | cytidylic acid | RNA |
| | deoxycytidine | deoxycytidylic acid | DNA |
| thymine | thymidine | thymidylic acid | DNA |
| uracil | uridine | uridylic acid | RNA |

*Consists of base + sugar (ribose or deoxyribose).

†Consists of base + sugar + phosphate, that is, unit is a nucleoside phosphate.

‡Polymer made up of nucleotide monomers: A, G, C, T in DNA and A, G, C, U in RNA.

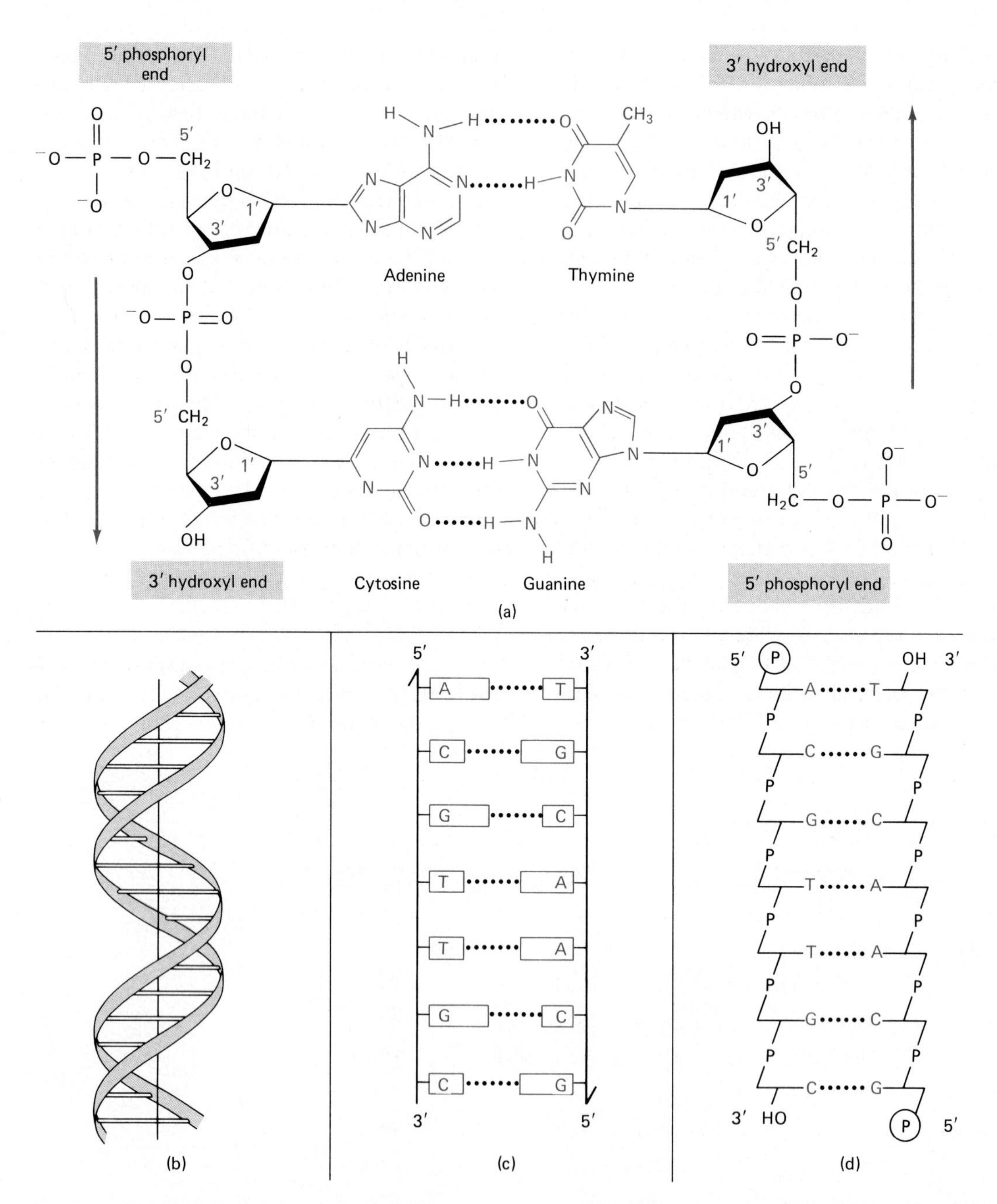


***Figure 7.3*** Molecular structure of DNA. (a) Each polynucleotide chain of the duplex consists of repeating nucleotide units held together by 3′,5′-phosphodiester bridges. The antiparallel strands are hydrogen bonded between complementary base-pairs. (b) The molecule is a double-stranded helix and must be untwisted to become two single-stranded chains. (c) The constant width of 20 Å for the duplex is due to the equivalence of width of all four possible base-pairs (A-T, T-A, G-C, C-G), consisting of a larger purine bonded to a smaller pyrimidine residue. (d) The individual chains of the duplex are often illustrated to emphasize the nature of the sugar-phosphate backbone to which the bases are covalently bonded.

their nitrogenous organic bases (**adenine, thymine, guanine,** or **cytosine**). Each organic base is covalently bonded to deoxyribose, which is bonded in turn to a phosphate residue in any deoxyribonucleotide unit (Fig. 7.3). DNA molecules are very stable in the watery environment of the cell, primarily as the result of three major kinds of chemical bonds involved in construction of the polymers.

***1.*** **Covalent bonds** link atoms within nucleotide units; and they link nucleotides together via **3′,5′-phosphodiester bridges,** extending from the 5′ carbon of one pentose sugar to the 3′ carbon of the pentose in the adjacent nucleotide. These strong bonds in the sugar-phosphate "backbone" of each strand make the polynucleotide chains relatively resistant to breakage because of the high energy required for such damage.

***2.*** The many weak **hydrogen bonds** within the DNA duplex are so arranged that most of them cannot break without many others breaking at the same time. This is energetically unlikely at cell temperatures but is easily accomplished in the laboratory by heating to 100°C or less. The separation of duplex DNA into single strands by heating provides investigators with material to analyze the unique features of each of the two complementary chains of one molecule. When native (duplex) DNA strands are separated, the molecule is **denatured.** The process of denaturation is also called **melting,** and DNA melting properties provide a useful means for identifying base composition differences between different DNAs and between various other molecular features of the gene and the genome (Fig. 7.4). Hydrogen bonding also takes place between virtually all the surface atoms in the sugar-phosphate chains and the surrounding water molecules in the cell or in solution in the test tube. These stabilizing forces help in the maintenance of molecular shape, which is essential for genetic function.

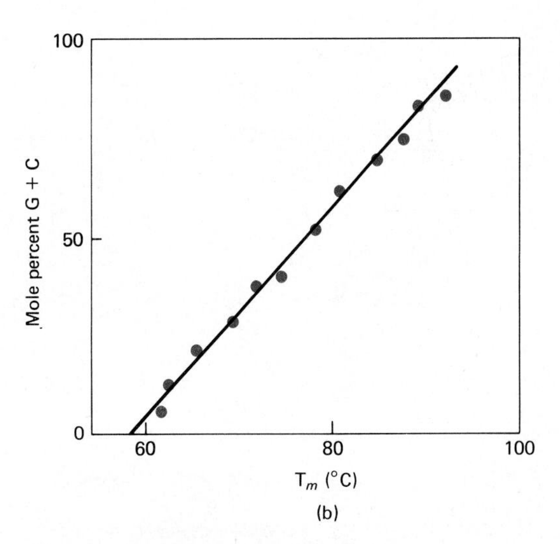


***Figure 7.4*** DNA melting. (a) Melting curve for duplex DNA in solution, with increase in absorbance (the hyperchromic shift) plotted as a function of temperature. The $T_m$ of the particular DNA is the midpoint melting temperature, at which half the denaturation has occurred in the sharp transition from double- to single-strands (at top of figure). When melting is completed there is no further change in absorbance (plateau). (b) $T_m$ is dependent on G-C content of the DNA sample, as seen by the linear relationship for various DNA sources (dots). Knowing $T_m$, we can deduce the G-C content of a DNA.

*3.* **Hydrophobic interactions** between the flat surfaces of the aromatic nitrogenous bases stacked vertically along the duplex length make an important contribution to molecular stability. These interactions lead to the exclusion of water molecules from the interior of the molecule and keep them from interfering or competing in hydrogen bonding between complementary base-pairs. Hydrophobic interactions also lend a considerable stiffness to the DNA duplex.

The cooperative result of all these kinds of bonds and interactions is stability of molecular shape and therefore maintenance of molecular properties and functions under physiological conditions.

Three important clues to the molecular structure of DNA were critical to the construction of the molecular model by Watson and Crick in 1953. Photographs of **X-ray diffraction patterns** made by bombardments of crystalline DNA preparations, from studies by Rosalind Franklin and by Maurice Wilkins, indicated that: (1) components were spaced in highly regular fashion, as opposed to a random or haphazard order in the molecule and (2) each molecule consisted of two helical chains held together as a duplex (Fig. 7.5). The third clue came from Erwin Chargaff's biochemical studies of DNA, which showed that the relative percentages of bases A, T, G, and C were consistent in any particular source of DNA but that these per-

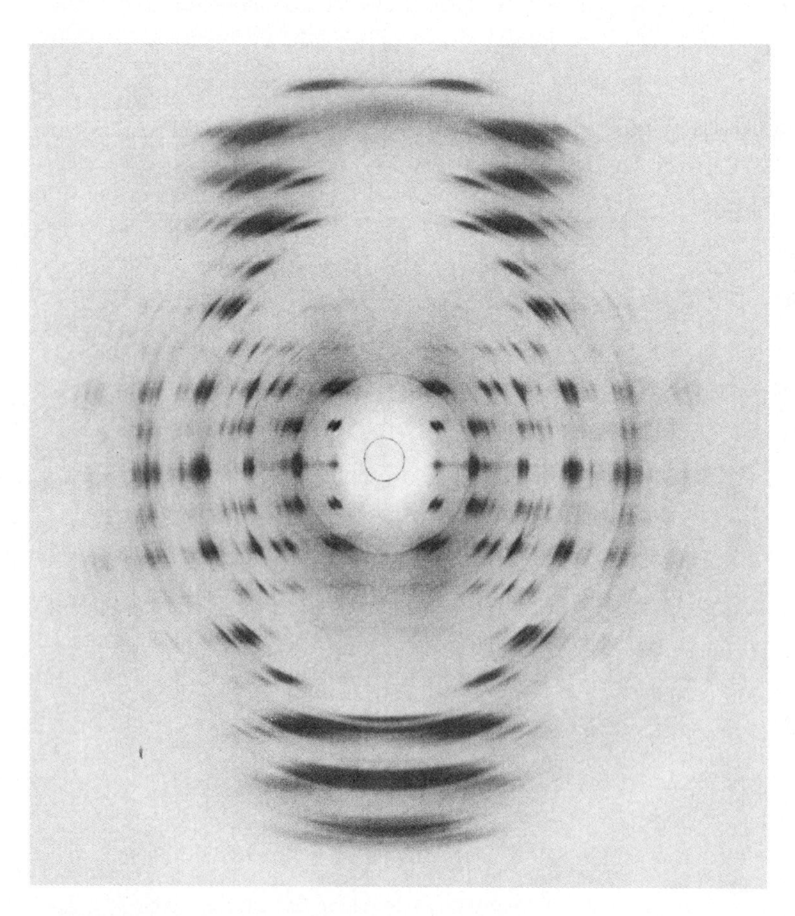

***Figure 7.5*** Photograph of X-ray diffraction pattern of crystalline DNA.

centages varied from one source to another. In particular, Chargaff had found that each source had equal amounts of A and T and equal amounts of G and C, but different sources had varying amounts of A + T and G + C (Table 7.2). The rule, therefore, appeared to be that the ratio of A:T or G:C was unity, but the ratio of AT:GC was variable. It was the brilliant conceptual achievement of Watson and Crick to build a molecular model in which all of these factors fit together logically and consistently.

By evaluating the size, shape, and theoretical bonding interactions of the four kinds of bases, Watson and Crick perceived that a double helix with a constant diameter of 2 nanometers could only be achieved if A paired with T and G paired with C across the space between the invariant sugar-phosphate chains. The ratio of 1 for A:T and for G:C was therefore due to pairing between these **complementary bases** along the length of the molecule.

The great excitement over the Watson-Crick model in 1953 was largely due to the parallels between genes and DNA, which were pointed out clearly and simply from the model. Watson and Crick noted particularly that DNA had properties that would explain four known features of genetic material: (1) stability during metabolism, (2) precise replication, (3) variety of molecular species, and (4) capacity for mutation. These properties of DNA are as follows:

***1.*** DNA does not undergo *turnover* during cell metabolism, but virtually all other molecules are made and degraded during the lifetime of a cell. DNA is stable, persisting essentially unchanged throughout the life of a cell.

***2.*** Each one of the two complementary strands of a duplex might serve as the **template** for synthesis of an exact complementary copy of a partner strand; two identical molecules can thus arise from one parent duplex molecule

***Table 7.2*** Molar proportions of bases (as moles of nitrogenous constituents per 100 g-atoms P) in DNAs from various sources.

| source of DNA | A | T | G | C | (A + T)/(G + C) | (A + G)/(T + C) | A:T | G:C |
|---|---|---|---|---|---|---|---|---|
| human liver | 30.3 | 30.3 | 19.5 | 19.9 | 1.53 | 0.99 | 1.00 | 0.98 |
| human sperm | 30.7 | 31.2 | 19.3 | 18.8 | 1.62 | 1.00 | 0.98 | 1.03 |
| human thymus | 30.9 | 29.4 | 19.9 | 19.8 | 1.52 | 1.03 | 1.05 | 1.00 |
| bovine sperm | 28.7 | 27.2 | 22.2 | 21.9 | 1.27 | 1.04 | 1.06 | 1.01 |
| rat bone marrow | 28.6 | 28.5 | 21.4 | 21.5 | 1.33 | 1.00 | 1.00 | 1.00 |
| wheat germ | 27.3 | 27.2 | 22.7 | 22.8 | 1.20 | 1.00 | 1.00 | 1.00 |
| yeast | 31.3 | 32.9 | 18.7 | 17.1 | 1.79 | 1.00 | 0.95 | 1.09 |
| *Escherichia coli* | 26.0 | 23.9 | 24.9 | 25.2 | 1.00 | 1.04 | 1.09 | 0.99 |
| *Myobacterium tuberculosis* | 15.1 | 14.6 | 34.9 | 35.4 | 0.42 | 1.00 | 1.03 | 0.98 |
| bacteriophage T2, T4, or T6 | 32.5 | 32.5 | 18.3 | 16.7* | 1.86 | 1.03 | 1.00 | 1.10 |
| bacteriophage T3 | 23.7 | 23.5 | 26.2 | 26.6 | 0.89 | 1.00 | 1.01 | 0.98 |
| bacteriophage T5 | 30.3 | 30.7 | 19.5 | 19.5 | 1.56 | 0.99 | 0.99 | 1.00 |

*The T-even phages have hydroxymethyl cytosine in place of cytosine in their DNA.

(Fig. 7.6). The rule of **complementary base pairing** underwrites the precision of DNA (and gene) replication, generation after generation.

*3.* The variety of DNA molecules is practically unlimited and matches the genetic requirement for diversity very well. Although base pairing across the molecule is restricted, the *linear* order of bases or base-pairs is not restricted. With only 4 kinds of bases or base-pairs (AT, TA, GC, CG) in a duplex of any length, the theoretical number of different molecules is $4^n$. If the average gene is 500 base-pairs from one end to the other, then $4^{500}$ different arrangements, or molecules, or genes, could be constructed from 500 units arranged in all permuted sequences of the 4 building blocks. DNA had more than enough variety to be the genetic material.

*4.* Mutations could arise as the result of *base substitution* during DNA replication. An altered base might be inserted in a replicating system by accident when adenine pairs with cytosine instead of with the usual thymine (Fig. 7.7). Upon separation in the next round of replication, the substituted cytosine will direct the incorporation of a guanine in its partner strand. The net result is substitution of an AT base-pair by a GC pair. The alteration will be perpetuated in later generations, just as mutations are transmitted through successive generations.

Together with experimental evidence from the bacterial transformation studies, the Hershey-Chase studies of T2 phage, biochemical data from Chargaff's laboratory, and other provocative information, the molecular model of DNA proposed by Watson and Crick reached a highly receptive scientific audience. Many studies were initiated after 1953 to investigate the chemical basis of heredity by seeking parallels between genes and DNA in various genetic phenomena. The age of molecular genetics had begun.

## Fine Structure Analysis of the Gene

From genetic analysis the gene appeared to be a unit of function, mutation, and recombination. After the molecular model of DNA was proposed, investigators could look at the gene more closely to determine if function, mutation, and recombination were features of the gene as an indivisible unit. During the mid-1950s and

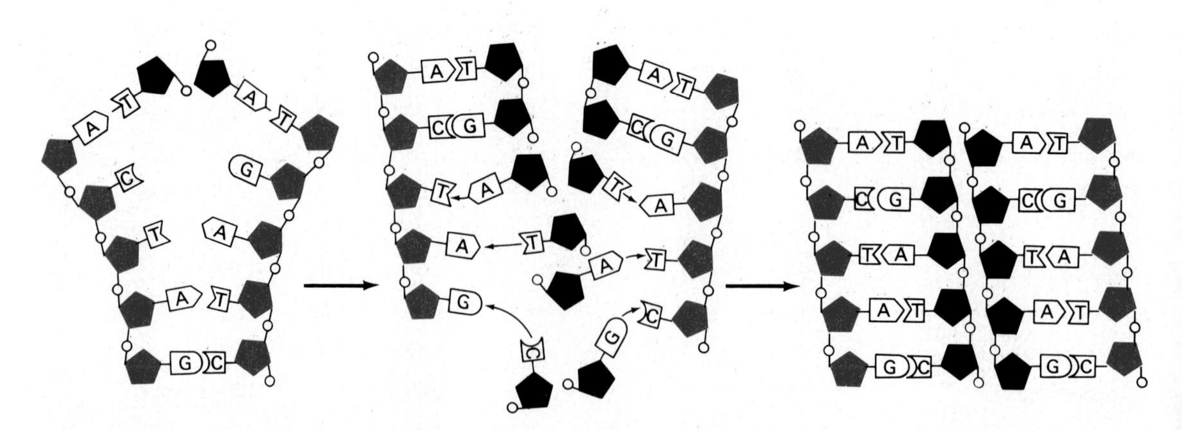

***Figure 7.6*** Each strand of the DNA duplex guides synthesis of a new complementary partner strand, thereby making two molecules that are identical to each other and to the original parental DNA. Genetic continuity is thus ensured from generation to generation.

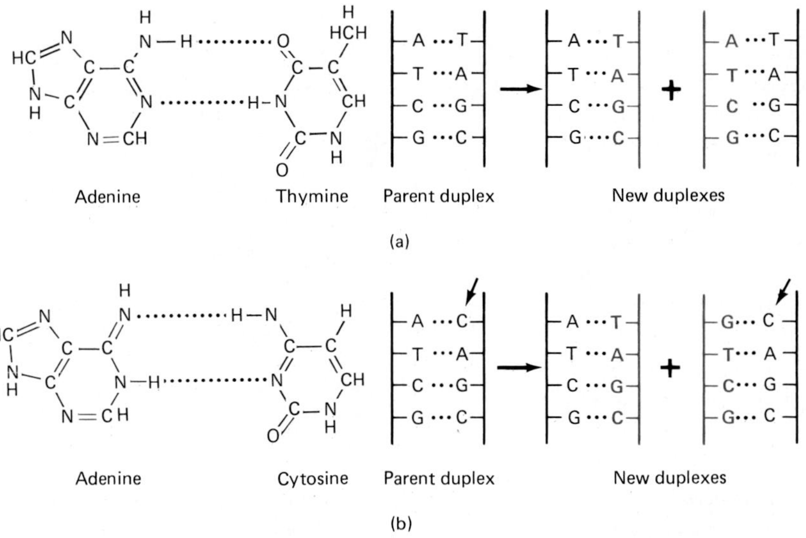

(a)

(b)

***Figure* 7.7** The shift of a hydrogen to another position produces a *tautomeric* form of adenine (as happens in other bases, too). In its more stable tautomeric form, adenine pairs with thymine; in another tautomeric form, it pairs with cytosine. Mutation by such base substitution during replication was first suggested by Watson and Crick and verified later by experiments (see Chapter 11).

early 1960s, elegant genetic analysis of the gene's fine structure provided an important bridge between the unit concept of the gene, gained from earlier studies, and the new perspective of the gene as a linear sequence of nucleotides in DNA. The experimental methods were genetic, but interpretations could be made in terms of DNA molecular structure.

## 7.3 Intragenic Recombination

The linear chromosome map of any organism was based on recombination analysis of crossing over between different genes in the linkage group. The same general principles were applied to mapping sites *within* the gene on the basis of *intragenic* recombination. Since DNA was a linear molecule composed of many hundreds of nucleotide pairs, the gene could possibly be mapped by recombination analysis just as the linear chromosome had been mapped. The detailed mapping of sites within a gene locus by genetic methods is called **fine structure mapping.** The first detailed fine structure gene maps were made of the *r*II region in *E. coli* phage T4 during the mid-1950s and 1960s by Seymour Benzer. These studies led to a new view of the gene and to later important investigations of gene function.

Wild-type phage T4 lyses host cells in 2 to 3 hours and small plaques with fuzzy edges appear 6 to 10 hours after infection of strains B and K12(λ) of *E. coli.* Mutations to the rapid lysis (*r*) trait are recognized by lysis about 20 minutes after infection and by production of large plaques with sharp edges. Several *r* genes map in different parts of the phage genome. The particular *r* mutants used by Benzer were from two adjacent genes in the *r*II region of the map, called the *r*II*A* and *r*II*B* genes. Both genes affect the same phenotypic character.

Benzer collected about 2400 different mutant strains. He made crosses between pairs of different *r*II*A* mutants and between pairs of different *r*II*B* mutant strains. To accomplish this he infected *E. coli* with an adequate number of phages of two *r*II*A* strains or two *r*II*B* mutant strains to be crossed, so that each host cell would be multiply infected by at least one phage from each of the two parental strains. Wild-type T4 can infect *E. coli* strains B and K12(λ), but *r*II mutants can only infect strain B. Since *r*II mutants can grow in *E. coli* B, recombinations can occur during infection in those host cells (Fig. 7.8). Progeny phages recovered from lysed B cells were then added to a lawn of K12 cells on nutrient media. Any wild-type recombinant phages would infect K12 and produce wild-type small plaques. The *r*II mutant parental classes could not develop in K12 nor could any doubly mutant recombinant class.

The selective assay for wild-type recombinants was very sensitive since as few as 0.0001% of the wild-type recombinants could be scored. Because some wild-type $r^+$ phages also arise by reverse mutation from the mutant state, the limit of *unambiguous* recombination that could be detected was set at a level high enough to avoid confusion with the rarer reverse mutations. A minimum frequency of 0.01% wild-type recombinants was therefore selected as the cutoff point. The actual recombination frequency observed in any cross was doubled because the reciprocal doubly mutant recombinants were expected to be present even though they could not be observed.

The astonishing result was that recombinants arose regularly in these crosses between mutants that were defective in the *same* gene. The explanation for their appearance was that crossing over had occured within the gene itself, that is, they arose by *intragenic recombination.* Since genes were composed of linear DNA, it seemed reasonable to propose that exchanges took place between different sites along the length of a single gene. The gene could be mapped in the same way that a chromosome is mapped. Crossing over is a random event, and the probability of crossing over is higher between more distant sites than between sites closer together. Higher recombination frequencies indicated greater distances between sites within the gene, and lower recombination frequencies indicated less distance separating sites. Recombination frequencies could then be equated to map units of distance in a fine structure map of the gene.

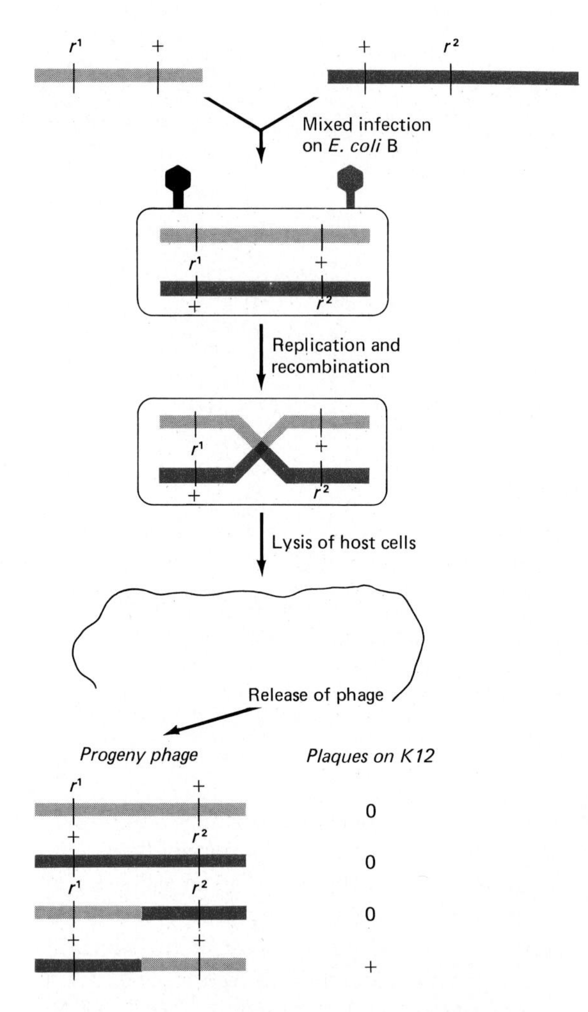


***Figure 7.8*** Recombination assay for phage T4 *r*II mutants in mixed infections on *E. coli* B. Wild-type recombinants are the only progeny class that can infect *E. coli* K12, as indicated by formation of plaques on this host strain.

The fine structure of the *r*II*A* and *r*II*B* genes revealed that they were *linear assemblies of mutable sites,* which could undergo crossing over and could thus give rise to recombinants. The gene was *not* an indivisible unit. Crossing over was *not* restricted to spaces between genes. Since crossing over took place within a gene as well as between genes, the whole chromosome seemed very likely to be a single DNA molecule or structure. A new view of the chromosome began to emerge, and new insights also opened the way to a molecular analysis of the crossing-over process, as we will see later.

## 7.4 Mapping the Gene by Overlapping Deletions

To proceed with the millions of two-factor crosses needed to map the 2400 mutant sites of the *r*II region was clearly impractical. Benzer devised an original and ingenious procedure to accomplish the task in a reasonable amount of time with a reasonable amount of effort. He assembled a reference collection of *deletion mutants.* These mutants never reverted to wild type by back mutation, which was also known to characterize deletion mutants in evolutionarily higher organisms. In each mutant a piece of the *r*II region was missing, or deleted, and these deletions were of varying lengths. The deletion regions were overlapping, so that two or more deletions in different mutants might have the same deleted segment in some part of the lesion (Fig. 7.9).

Benzer used this **overlapping deletions method** to map the *r*II region. He first established the location, length, and overlap of the deletions in the reference collection of deletion mutant strains by making crosses with a few single-site mutants that had already been mapped by recombination in the *r*II region.

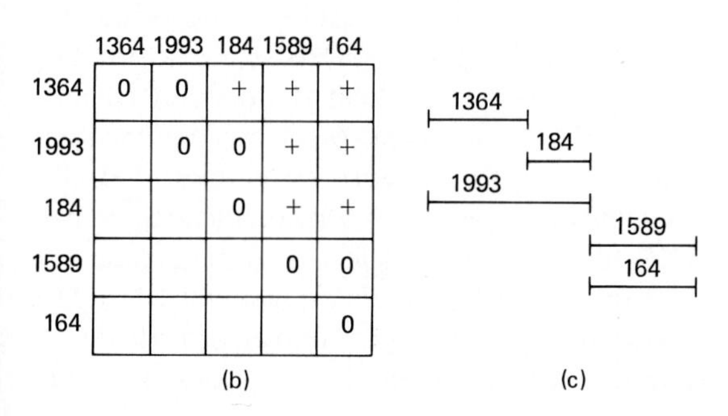


***Figure 7.9*** Overlapping deletion mapping in the *r*II region of phage T4. (a) The reference collection of overlapping deletion mutants defines the entire *r*II region. Each black bar indicates the extent of the deletion and its relative location in the *r*II*A* and *r*II*B* genes of the region. The gene lengths are subdivided into smaller segments (A1–A6 and B), and even smaller ones, which are shown in Fig. 7.11. (b) Crosses between a newly isolated *r*II mutant and reference deletion mutants permit assignment of the new mutant to an approximate map location on the basis of wild-type recombinants. The matrix of crosses between mutants in pairs can be drawn to show those crosses that produce wild-type recombinants (+) and those that do not (0). (c) The order of the mutations in the topological map can be determined from the matrix data. (From S. Benzer. 1961. *Proc. Nat. Acad. Sci. U.S.* **47**:410.)

Once he had characterized the deletion mutants, he could cross any unknown single-site mutant with known deletion mutants to locate roughly the single-site mutation at some location within the gene (Fig. 7.10). If unknown mutant *a* produced wild-type recombinants with deletion mutant strains 1, 2, 3, and 4, but not with 5, then the single-site mutation must be located somewhere in the region deleted in strain 5.

To pin down the location more closely, he then crossed mutant *a* with other reference strains that had *smaller* deletions within the same region that was missing in strain 5. After several crosses of unknown mutant *a* with a set of increasingly smaller overlapping-deletion mutants, he could make the last cross between mutant *a* and a known, mapped, single-site mutation in a nearby location. Recombination frequencies between these two single-site mutants yielded the values to map the unknown mutant site precisely and to calculate its distance from other mutant sites in the neighborhood.

In this way, 428 different sites were identified from intragenic recombination data, and the fine structure map of the *r*II region was defined (Fig. 7.11). The linear genetic map provided very strong evidence that the gene itself was linear in construction, which coincided with the known linear construction of duplex DNA.

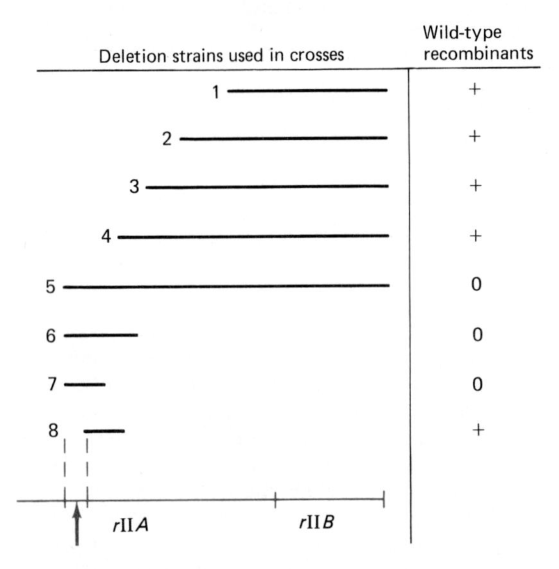


***Figure 7.10*** Mapping a single-site mutation in the *r*II region. The mutant is crossed with a series of known deletion mutants 1–5, from which an approximate location is determined (left end of *r*IIA, according to reference map shown at bottom of figure). Using small-deletion mutants 6–8, we can make a more refined location according to production of wild-type recombinants between the single-site and deletion mutants. The single-site mutation is then located (color arrow) on the reference map. Precise map location is later determined by crosses with other single-site mutants in this immediate part of the *r*II map.

## 7.5 Cis-Trans Complementation Test

The *r*II region consists of two genes, yet mutations in either *r*II*A* or *r*II*B* produce the same rapid lysis character. Why are these considered to be separate genes, one right next to the other, and not a single gene governing a single phenotype? The basis for Benzer's interpretation of two genes in the *r*II region was the **cis-trans test** (Fig. 7.12).

When *E. coli* K12 was infected with two *r*IIA or two *r*IIB mutant phages, no virus progeny developed. But when an *r*IIA mutant and an *r*IIB mutant infected strain K12 simultaneously, virus progeny were produced even though no recombination had occurred. This showed that mutant viruses could *multiply* in K12 because each parent phage provided one of the functions needed for development; two different functions were required for virus multiplication. The *r*IIA mutant provided a functional *B* gene product, and the *r*IIB mutant provided a functional *A* gene product. The two gene products, or functions, allowed growth. On the other hand, two *r*IIA mutants had a defective *A* gene function and two *r*IIB mutants had a defective *B* gene function, and neither cross allowed mutant virus growth. The rare wild-type recombinants produced in any of these three kinds of crosses were not confused with production of large numbers of mutant progeny after infection.

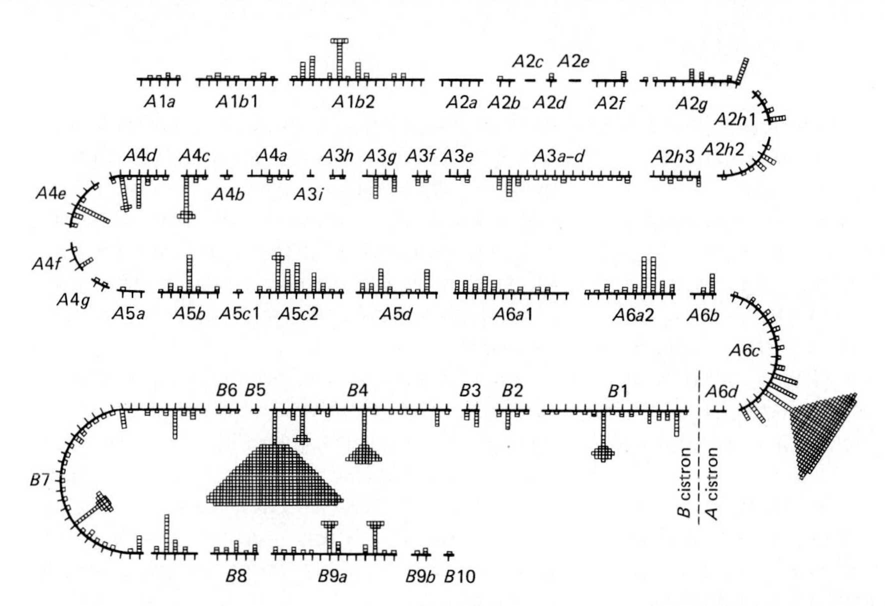


*Figure 7.11* Fine structure genetic map of the *r*II region of phage T4. The region is divided into deletion segments (A1a–B10). The single-site mutants are indicated by short vertical lines; thus segment A1a contains five sites, A1b1 has nine sites, and so on. Each square represents one occurrence of a spontaneous mutation at the site. Sites without squares have been identified by induced mutations. Some sites are "hot spots" of spontaneous mutation, such as those in segments A6c and B4. (From S. Benzer. 1961. *Proc. Nat. Acad. Sci. U.S.* **47**:410.)

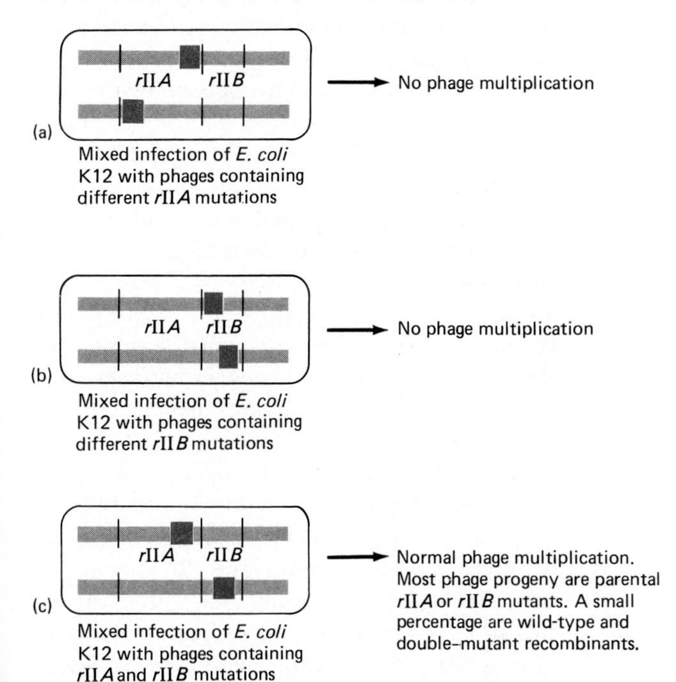


*Figure 7.12* The cis-trans test for complementation distinguishes between mutants having lesions in the same or in different genes. Simultaneous infections with two mutants in (a) *r*IIA or in (b) *r*II*B* do not complement and must, therefore, be lesions in the same gene. Complementation as shown in (c) indicates there are two different genes in the *r*II region of phage T4 since each mutant makes up for the other's defect and together they both become capable of normal multiplication in the host.

When each mutation is present in a different homologous chromosome, the alleles are in the *trans* arrangement. When both mutations are in the same chromosome and the wild-type alternatives are in the opposite homologue, the alleles are in the *cis* arrangement. The trans arrangement is the critical one to assay since the cis arrangement has one intact chromosome introduced by one parent and virus multiplication can occur. The cis test serves as a formal control to see that normal experimental conditions prevail.

Benzer coined the term **cistron** (from cis-trans test) as a substitute for gene. He redefined the gene as a cistron that directed one function in phenotypic development. Two or more cistrons might direct different but related functions in the development of a single phenotypic character, such as rapid lysis. The cis-trans test for **complementation** between mutants provides the evidence for the number of cistrons that govern a single phenotypic character. Mutations that do *not* complement each other must, therefore, be located in the same cistron. Thus, the cis-trans complementation assay is a test for functional allelism.

Using the cis-trans complementation test, we can show that regions previously assumed to contain a single gene consist of two or more genes or cistrons in viruses other than T4. Similar results have been obtained for bacteria, in which partial diploids can be produced through conjugation or transduction. It is fairly characteristic for related genes or cistrons to be clustered in viral and bacterial genomes. It is very rarely the case in eukaryotes. In fact, in eukaryotes different genes with a known related set of functions are often found in different chromosomes, such as the gene for the $\alpha$-globin chain and the gene for the $\beta$-globin chain of hemoglobin in human beings. These observed differences in genome organization are believed to be associated with different systems for regulation of gene expression but they are not fully understood at the present time.

The cistron concept was very important in the 1960s in many kinds of studies designed to identify genes with related functions in phenotype development. Since gene locations were not at all random, it soon became clear that the genome was a highly organized system and not just a randomly scattered collection of genes. This was especially true for viruses and bacteria. Of course, many genes that contribute to a single phenotype are not necessarily adjacent in these organisms.

As our information about polypeptides as products of gene expression has increased, the term cistron has lost some of its former heuristic significance. We use the term *gene* once again today, but we have a better idea of its action because of the cistron concept, which led to many modern insights about the gene and its product.

## The Genetic Code

The general hypothesis that one gene specified one enzyme had grown out of biochemical genetic studies of metabolic pathways in *Neurospora,* particularly those by George Beadle, Edward Tatum, Boris Ephrussi, and others during the 1940s. Using a variety of mutants that affect metabolism, they showed that each step in a metabolic pathway is catalyzed by an enzyme that is under genetic control (Fig. 7.13). Later studies by Benzer and others in the 1950s and 1960s led to the concept that a linear sequence of nucleotides in DNA structure specifies the linear sequence of amino acids in the structure of a protein produced under genetic instructions.

How could the relationship between gene structure and protein structure be established? In phage T4 the protein product specified by the *r*II gene was not known. The *r*II system was therefore unsuitable for the next step in analyzing gene function through its protein product. The central technical problem was to find a suitable test system in which specific nucleotides could be correlated with specific amino acids in the polypeptide or protein. The genetic code had to be deciphered. By insightful methods developed between 1961 and 1967, this spectacular accomplishment was achieved.

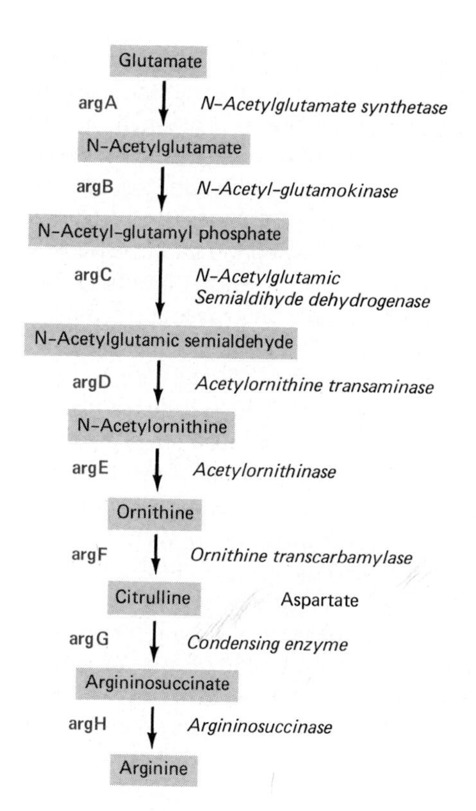


***Figure 7.13*** Pathway of arginine synthesis from glutamate in *E. coli.* Each step is catalyzed by a particular enzyme (right, italics) that is governed by a specific gene (left, color).

## 7.6 Deciphering the Genetic Code

The central problem in deciphering the **genetic code** was to identify the coding units, or **codons,** in DNA that specified amino acids in proteins. It was not possible to work directly with DNA in the 1960s, as it is today. It was possible then, however, to work with the RNA copies of the genetic instructions. These copies serve as intermediates between genes and proteins.

Information flows from DNA to the system that carries out polypeptide synthesis via an intermediate molecule. DNA information is first copied into RNA by synthesis of a *complementary strand* of RNA from the DNA template (Fig. 7.14). This RNA copy of the genetic instructions is then used as the guide for polypeptide synthesis, that is, DNA → RNA → protein. It is just as relevant, therefore, to study the relationship between RNA and protein as it is to study DNA and protein, since DNA and RNA are complementary to each other and the information in one molecule reflects the information in the other molecule.

Most of the significant information on the genetic code was obtained by the use of artificially constructed RNA in experiments and ob-

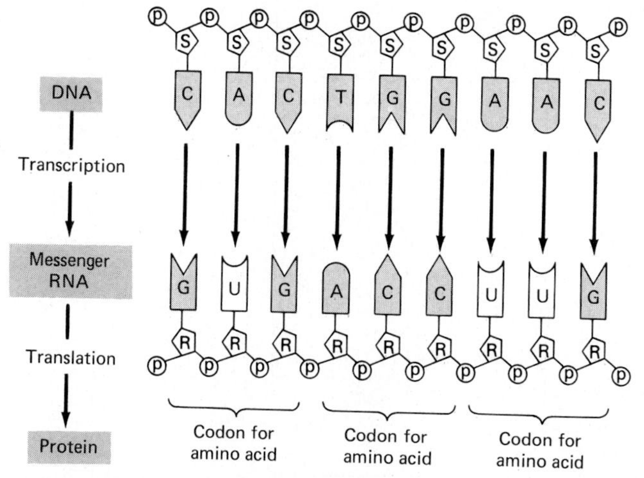


***Figure 7.14*** Genetic information in the sequence of nucleotides in DNA is copied into a complementary copy of messenger RNA in the process of transcription. The information copied into messenger RNA is then translated into a linear sequence of amino acids that make up the protein molecule. Each amino acid in the protein translation is specified in type and in position in the polymer according to 3-nucleotide code words, or codons, in DNA and its messenger RNA copy.

serving the particular amino acids incorporated into polypeptides *in vitro,* that is, in test-tube systems. Once a particular set of RNA nucleotides was associated with a particular amino acid in the polypeptide product, the DNA codon could be deduced since it would be complementary to the RNA codon defined in the experiments. For example, the first RNA codon to be identified was UUU, which specifies the amino acid phenylalanine (Fig. 7.15). The DNA codon for phenylalanine is the complementary triplet AAA.

In 1961 Marshall Nirenberg and Heinrich Matthaei showed that an artificial RNA made up only of uracil bases bound to the sugar-phosphate backbone of the polynucleotide chain, or **poly(U),** could direct synthesis of a polypeptide consisting entirely of phenylalanine residues. The *in vitro* test system contained a mixture of amino acids plus all the other ingredients known to be needed for polypeptide synthesis in a test tube. Each mixture had a different single amino acid labeled with a radioactive isotope, and all the other amino acids in that mixture were unlabeled. They had only to look for the one test tube out of the total number in which a *radioactively labeled polypeptide* had been made from poly(U) instructions. On collecting the labeled polypeptide and hydrolyzing it to its constituent amino acids, Nirenberg and Matthaei found that only phenylalanine was present in the polymer. The RNA codon for phenylalanine contained only uracil, and the complementary DNA codon contained only adenine.

A codon was assumed to be a *triplet* of nucleotides; however, the poly(U) data could not be used directly to show that UUU was the codon. There are 20 naturally occurring amino acids that are directly incorporated into polypeptides during their synthesis. It was assumed, therefore, that there were at least 3 nucleotides per codon, specifying each of the 20 amino acids. With 4 kinds of nucleotides in all permutations of threes, 64 ($4^3$) unique codons can

| First Nucleotide | Second Nucleotide: U | | C | | A | | G | | Third Nucleotide |
|---|---|---|---|---|---|---|---|---|---|
| U | UUU | Phe | UCU | Ser | UAU | Tyr | UGU | Cys | U |
| | UUC | | UCC | | UAC | | UGC | | C |
| | UUA | Leu | UCA | | UAA | Stop | UGA | Stop | A |
| | UUG | | UCG | | UAG | | UGG | Trp | G |
| C | CUU | Leu | CCU | Pro | CAU | His | CGU | Arg | U |
| | CUC | | CCC | | CAC | | CGC | | C |
| | CUA | | CCA | | CAA | Gln | CGA | | A |
| | CUG | | CCG | | CAG | | CGG | | G |
| A | AUU | Ile | ACU | Thr | AAU | Asn | AGU | Ser | U |
| | AUC | | ACC | | AAC | | AGC | | C |
| | AUA | | ACA | | AAA | Lys | AGA | Arg | A |
| | AUG | Met | ACG | | AAG | | AGG | | G |
| G | GUU | Val | GCU | Ala | GAU | Asp | GGU | Gly | U |
| | GUC | | GCC | | GAC | | GGC | | C |
| | GUA | | GCA | | GAA | Glu | GGA | | A |
| | GUG | | GCG | | GAG | | GGG | | G |

| Abbreviation | Amino acid |
|---|---|
| Ala | Alanine |
| Arg | Arginine |
| Asn | Asparagine |
| Asp | Aspartic acid |
| Cys | Cysteine |
| Gln | Glutamine |
| Glu | Glutamic acid |
| Gly | Glycine |
| His | Histidine |
| Ile | Isoleucine |
| Leu | Leucine |
| Lys | Lysine |
| Met | Methionine |
| Phe | Phenylalanine |
| Pro | Proline |
| Ser | Serine |
| Thr | Threonine |
| Trp | Tryptophan |
| Tyr | Tyrosine |
| Val | Valine |

***Figure 7.15*** The genetic coding dictionary shows messenger RNA codons and the amino acids or punctuations (starts and stops) they specify.

be assembled. With 2 nucleotides per codon, only 16 ($4^2$) combinations arise, and only 4 ($4^1$) codons are possible if one base equals one codon. Neither the doublet nor the singlet codon would be adequate to specify 20 amino acids.

Between 1961 and 1964 M. Nirenberg, Severo Ochoa, and others reported various studies in which artificially synthesized RNAs were used in *in vitro* coding tests. These synthetic RNA polymers consisted of different proportions and kinds of the four ribonucleotides (U, A, C, G). They identified about 50 codons after statistical analysis of the experimental results. They could determine the composition of codons in this way but not the sequence within the codons. For example, they found poly(UG) to be involved in coding for leucine, valine, and cysteine because polypeptides containing these amino acids were formed when the synthetic RNA contained twice as many Us as Gs. There was no way, however, to show that UUG coded for leucine, GUU for valine, or UGU for cysteine.

These studies were very important in leading the way to deciphering the genetic code, and they provided a clear demonstration that all or most of the 64 possible triplet codons were part of the dictionary that spelled out amino acids in proteins. They further showed that an amino acid could be specified by more than one codon since more than 50 triplets were used to code for only 20 amino acids.

By 1964 M. Nirenberg, H.G. Khorana, and others had invented new methods that specified the sequence of three nucleotides in most of the codons. By 1967 the last of the 64 triplet codons had been deciphered. Of this group, 61 of the codons specified amino acids and the other 3 were stop signals marking termination of a gene message. These **termination codons** are also called punctuation codons. Just as a genetic message has a terminus, so must there be a beginning or initiation signal for the first amino acid to be positioned in polypeptide synthesis. The RNA codon AUG is the major, and in some cases the only, **initiation codon,** as well as being the codon that specifies methionine.

In summary, the following features have been found to characterize the genetic code:

***1.*** The code is *triplet.* Each codon consists of a unique combination of three nucleotides.

***2.*** The code has *punctuations* (start and stop) that mark the limits of each genetic instruction.

***3.*** Most amino acids are specified by more than one codon, that is, the code is redundant. There are codon *synonyms* for 18 of the 20 amino acids included in the code. Only methionine and tryptophan are specified by a single codon.

***4.*** The code is *consistent* since each one of the 61 codons is specific for only one amino acid out of the set of 20.

***5.*** The code is essentially *universal.* Studies using many different kinds of viruses, bacteria, and eukaryotes have all shown that the same codons are translated into the same amino acids in every case. The same initiating and terminating codons have been demonstrated in all organisms that have been studied. More recently it has been shown that certain codons are translated into different amino acids from mitochondrial DNA instructions for mitochondrially specified proteins (Box 7.1). Most of the code, however, is the same as the code in nuclear genes. We will discuss mitochondrial genes in Chapter 13.

Altogether an overwhelming amount of evidence supports the conclusion that all *organisms* share a common genetic code, from viruses to human beings. That each group of organisms happened to stumble on the same genetic code by accident or coincidence is highly unlikely. The inescapable conclusion is that life forms at every level of complexity share a common ancestry and evolutionary history, as revealed in their genetic operations. We may further deduce that the present-day code must have been established early in evolution and has continued virtually unchanged for billions of years.

## BOX 7.1 The Genetic Code in Mitochondria

In 1979 it was first found that certain codons in mitochondrial DNA had different specifications from the standard ones known for nuclear genes. The great majority of codons have identical meanings, but two or more of the 64 codons do have a different meaning in mitochondria. Mammalian mitochondrial DNA is known to use two codons instead of one for the amino acid methionine. The second Met-specifying codon is AUA, which ordinarily is translated as isoleucine. The AUA codon does specify isoleucine in yeast and *Neurospora* mitochondrial genes, however, so the variation is not typical for all mitochondrial DNA. Mammalian mitochondria as well as other mitochondrial DNAs use the codon UAA to signify the end of a genetic message, just as do nuclear systems. However, every mitochondrial DNA analyzed to date appears to use the stop codon UGA to specify tryptophan instead of message termination. Mammalian mitochondria apparently do not use the arginine codons AGA and AGG, and some evidence suggests that these may be stop codons in these systems. Yeast, on the other hand, does use AGA as an arginine codon. Some variation in codon meaning and in codon usage in mitochondrial systems as compared with nuclear genes occurs. The extent of this variation, however, would be inadequate to rule out the broad generalization that the genetic code is universal. We could state that it is *virtually* universal and could emphasize in this way that some minor variation occurs but that the bulk of the genetic code serves as the common dictionary for all known genes.

| First Nucleotide | Second Nucleotide: U | | C | | A | | G | | Third Nucleotide |
|---|---|---|---|---|---|---|---|---|---|
| U | UUU | Phe | UCU | Ser | UAU | Tyr | UGU | Cys | U |
| | UUC | | UCC | | UAC | | UGC | | C |
| | UUA | Leu | UCA | | UAA | Stop | UGA | Trp | A |
| | UUG | | UCG | | UAG | | UGG | | G |
| C | CUU | Leu | CCU | Pro | CAU | His | CGU | Arg | U |
| | CUC | | CCC | | CAC | | CGC | | C |
| | CUA | | CCA | | CAA | Gln | CGA | | A |
| | CUG | | CCG | | CAG | | CGG | | G |
| A | AUU | Ile | ACU | Thr | AAU | Asn | AGU | Ser | U |
| | AUC | | ACC | | AAC | | AGC | | C |
| | AUA | | ACA | | AAA | Lys | AGA | Arg | A |
| | AUG | Met | ACG | | AAG | | AGG | | G |
| G | GUU | Val | GCU | Ala | GAU | Asp | GGU | Gly | U |
| | GUC | | GCC | | GAC | | GGC | | C |
| | GUA | | GCA | | GAA | Glu | GGA | | A |
| | GUG | | GCG | | GAG | | GGG | | G |

## 7.7 Genetic Evidence for the Code

Independent evidence in support of the proposed genetic code came from genetic analysis of the effects of mutations on amino acids in proteins of known primary structure (amino acid sequence). Allelically different mutants make slightly different versions of the same polypeptide. The goal of the studies was to determine whether the observed amino acid substitutions in the variant polypeptides agreed with DNA codon assignments for these amino acids. It was expected that one altered codon in a mutant gene would correspond to one altered amino acid in the polypeptide product of the gene and that the nature of the change in the amino acid would indicate the nature of the codon and its nucleotide alteration.

One of the best examples of correlations between altered amino acids and altered codons came from *in vivo* studies by Charles Yanofsky of the enzyme *tryptophan synthetase* in *E. coli*. Two adjacent genes govern synthesis of this enzyme, which catalyzes the final step in the biochemical pathway leading to synthesis of the amino acid tryptophan (Fig. 7.16). The enzyme consists of

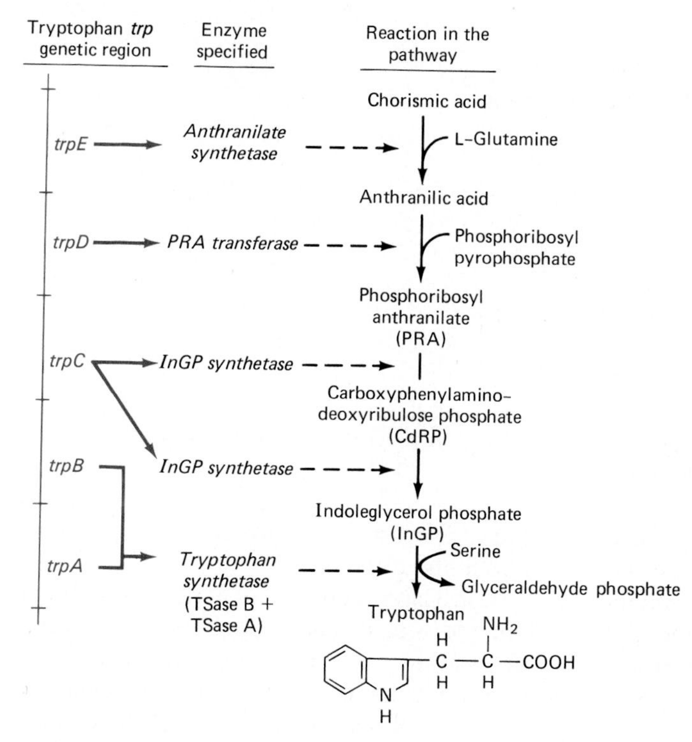


***Figure 7.16*** Some steps in the pathway of tryptophan biosynthesis, showing the *trp* genetic region in *E. coli* (color) and the enzymes specified.

two subunits, with two A polypeptide chains in one subunit and two B chains in the other. Both subunits are required for enzyme function. Gene *trpA* governs A chain synthesis and gene *trpB* governs B polypeptide synthesis.

Using strains with different mutations in the *trpA* gene, Yanofsky compared the amino acid substitutions in mutant polypeptide chains with the corresponding amino acids in the wild-type polypeptide. The sequence of 268 amino acids was known for the *trpA* wild-type polypeptide. The rationale was to see what DNA codon changes could theoretically have produced the observed amino acid substitutions in the mutants, based on the genetic code (Fig. 7.17).

An example of this approach can be shown for amino acid number 210 in the A polypeptide chain of 268 amino acids. The amino acid *glycine* is found in position 210 in the wild-type polypeptide. Mutant strain A23 has *arginine* at this location, and mutant strain A46 has *glutamic acid* instead of glycine at position 210. There are four codons that specify glycine in the genetic code, but the one codon that could give rise to either arginine or glutamic acid through a single base substitution is the DNA codon CCT. A change in the first base produces TCT, for arginine; a change in the middle base produces CTT, for glutamic acid:

**CCT** (wild type) *gly*

*arg* (mutant A23) **TCT** ↙ ↘ **CTT** (mutant A46) *glu*

If this analysis is correct, any further mutational changes in the TCT codon of mutant A23 should give rise to revertants with the original glycine at site 210 and to others with partial or full enzyme activity restored because serine (TCG or TCA),

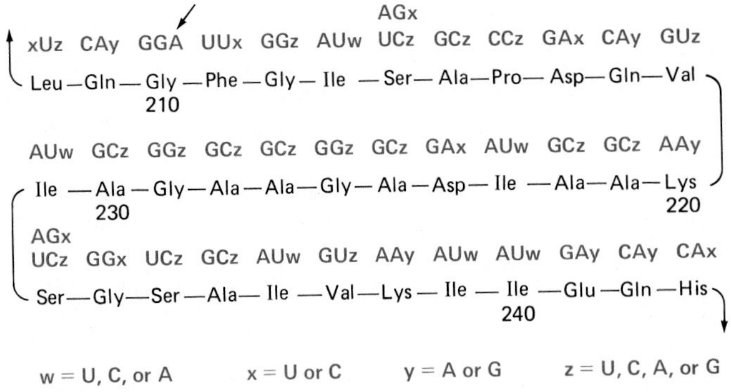

AGx
xUz CAy GGA UUx GGz AUw UCz GCz CCz GAx CAy GUz
Leu—Gln—Gly—Phe—Gly—Ile—Ser—Ala—Pro—Asp—Gln—Val
210
AUw GCz GGz GCz GCz GGz GCz GAx AUw GCz GCz AAy
Ile—Ala—Gly—Ala—Ala—Gly—Ala—Asp—Ile—Ala—Ala—Lys
230
220
AGx
UCz GGx UCz GCz AUw GUz AAy AUw AUw GAy CAy CAx
Ser—Gly—Ser—Ala—Ile—Val—Lys—Ile—Ile—Glu—Gln—His
240
w = U, C, or A
x = U or C
y = A or G
z = U, C, A, or G

***Figure 7.17*** Part of the chain of 268 amino acids in tryptophan synthetase A in *E. coli,* including amino acids 208–243. The probable codon for each amino acid is shown in color. (From C. Yanofsky et al. 1967. *Proc. Nat. Acad. Sci. U.S.* **57**:296.)

threonine (TGT), or isoleucine (TAT) is present. Using mutation-inducing agents, Yanofsky found the predicted true revertant to wild type and the other three expected types, all with the predicted amino acid at site 210. Similar results were obtained in mutation induction studies for mutant A46 (Fig. 7.18).

Although the precise codon could not always be determined, since all the amino acids involved in this particular study have codon synonyms, the substituted amino acid could always be accounted for by a *single base change in one codon.* The codons derived from biochemical studies appeared to be the same ones as those used in the living cell.

Alan Garen and co-workers studied the alkaline phosphatase gene and protein in *E. coli* to see whether one base change in a codon was responsible for observed mutations. One particular mutant made a shorter alkaline phosphatase polypeptide because the mutant codon specified "stop" instead of the amino acid tryptophan. When this premature termination mutant was subjected to chemicals that caused base substitutions, new mutants appeared as well as revertants to wild type (Fig. 7.19).

Gly
CCT
Wild type
GGA
Arg
TCT
(A23)
(A46)
Glu
CTT
Mutant strains
AGA (A23)
(A46) GAA
Gly
CCT
Ser
TCA
TCG
Thr
TGT
Ile
TAT
Ala
CGT
Val
CAT
Gly
CCT
Induced mutations
GGA
AGU
AGC
ACA
AUA
GCA
GUA
GGA
Amino acid substitutions and DNA codon changes proposed
(a)
RNA codons equivalent to altered DNA codons
(b)

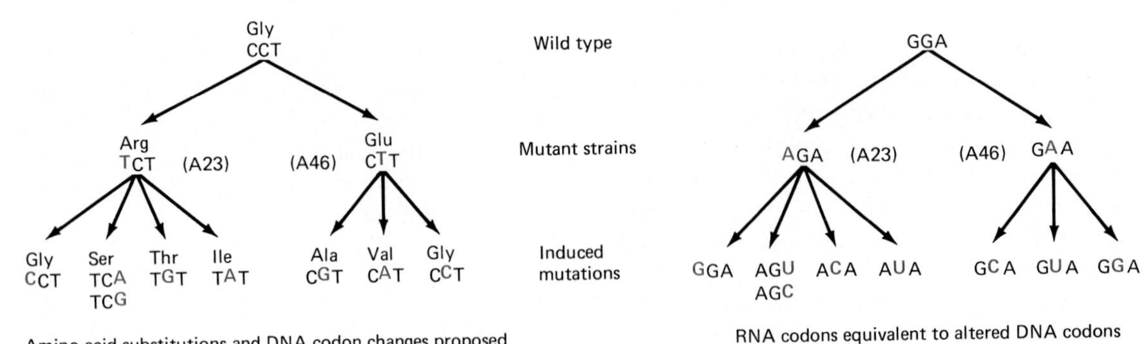

***Figure 7.18*** Revertants at amino acid site 210 in the *trpA* gene of *E. coli.* (a) The amino acid and its DNA codon in wild type, its derived mutants A23 and A46, and their induced revertants are shown. (b) The corresponding RNA codons of the genetic dictionary are shown, for each strain pictured in (a).

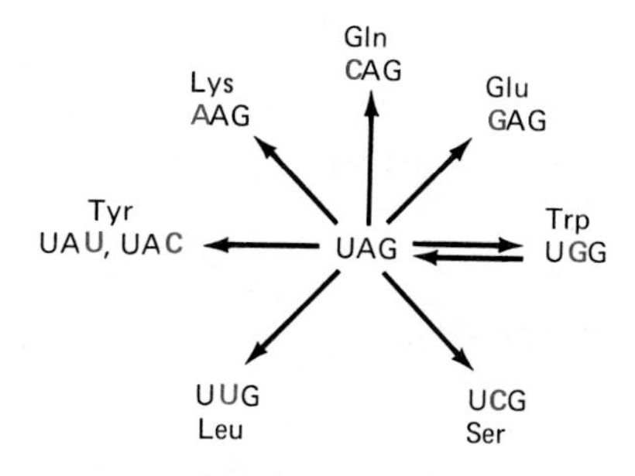


***Figure 7.19*** Codons arising by induced mutation at the termination codon UAG in *E. coli.* The probable single-base substitution is indicated in color for each of the observed reverse mutants that have an amino acid at the former termination site of the polypeptide.

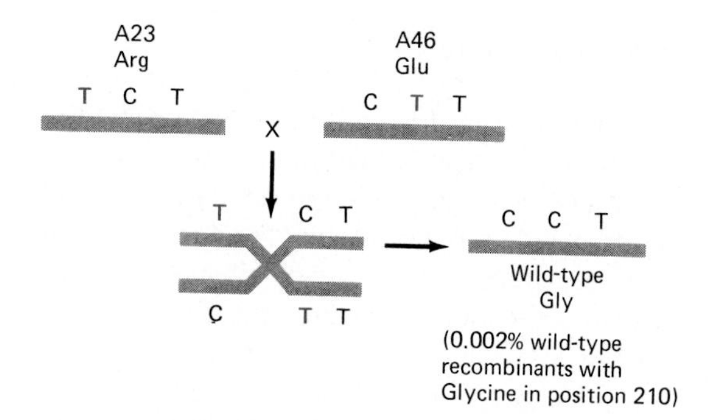


***Figure 7.20*** If a crossover occurs in *E. coli* between nucleotides in *trpA* DNA from mutant A23 × A46 so that a strand with the original CCT codon is produced, the amino acid glycine would be inserted in the protein at position 210 and the recombinant would show the wild-type phenotype. The unit of recombination (sometimes called a *recon*) could be as small as the space between two adjacent nucleotides in a chain.

Seven different amino acid substitutions were found in these induced mutants. The revertant had tryptophan restored in the proper site in the polypeptide, and the other six strains had a different amino acid present instead of being terminated. In every case a single base substitution in the DNA codon could explain the newly derived mutants if the termination codon was ATC (or UAG in the complementary RNA codon).

These *E. coli* studies and other studies therefore showed the following:

**1.** A single base substitution in one codon is responsible for mutation and may lead to a modification of one amino acid site in the polypeptide.

**2.** Any one of the three bases in a codon may be altered by mutation, thus giving rise to different alleles whose altered codons cause polypeptide alterations.

**3.** The unit of mutation can be the minimum of one base change in one codon.

Mutants A23 and A46 provided another item of information about the gene in Yanofsky's *trpA* studies. When these two mutants were crossed, some wild-type progeny were produced. The wild types were shown to have glycine restored at site 210 in the polypeptide. How did these changes arise? The simplest explanation was that a crossover had occurred between the first and second bases of the codons (Fig. 7.20). The unit of recombination, therefore, could be the minimum distance represented by the space between two adjacent nucleotides in one codon. Benzer had also come to this conclusion in the *r*II mapping studies.

## 7.8 Readout of the Genetic Message

The genetic message is a coded sequence of DNA that specifies the amino acids of a polypeptide. The message has a beginning and an end, and it is read *in sequence* from the start to the finish of the instruction, codon by codon. The instructions in DNA are copied into a complementary messenger RNA molecule, from which the codons can be translated into amino acids in the polypeptide product:

$$\text{DNA} \xrightarrow{\text{transcription}} \text{messenger RNA} \xrightarrow{\text{translation}} \text{polypeptide}$$

Crick and co-workers in England, using phage T4 mutants, obtained genetic evidence for the sequential reading of the genetic message, that is, for the existence of a *reading frame*.

They exposed T4 to a chemical agent that was believed to cause mutations by its action in adding or deleting single base-pairs in the DNA sequence. The mutant strain "FCO" obtained by these treatments of wild-type phage could be recognized by altered plaque morphology. The FCO mutant was then exposed to the same mutagenic agent, and wild-type revertants were sought and found. When the original wild type T4 and the revertant wild type were crossed, the two strains produced recombinants with mutant plaque morphology. If these had been truly wild-type revertants, they should have produced only wild-type progeny and no segregants. It was therefore obvious that the apparent revertant was a **pseudowild** (mimicking wild type) strain and not a true reverse mutant to the original wild type. The events were interpreted to mean that the pseudowild strain had both the original mutation and another base change at a different site in the same gene. FCO and other segregants could arise by crossing over between these two mutant sites in the gene (Fig. 7.21).

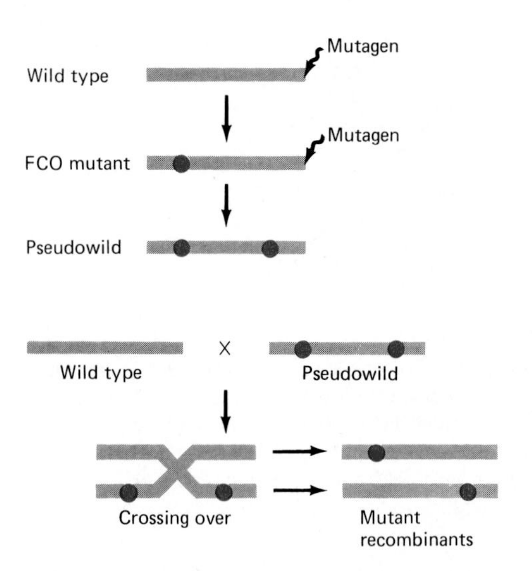


***Figure 7.21*** When the induced FCO mutant was itself treated with a mutagen, some of its progeny appeared to be phenotypically wild type. Crossing these apparent wild types (pseudowilds) with true wild types may produce mutant recombinants as a consequence of crossing over between genetically different DNAs. Crosses between true wild types will yield only wild-type progeny.

If the genetic message is read consecutively from a point of origin to the end, then we can explain the FCO mutant and its pseudowild derivative. Suppose a base had been deleted in the FCO mutant and the reading frame shifted so the rest of the message was garbled (Fig. 7.22). If a second mutation caused a base addition, the reading frame would be restored and a functional protein produced. Some garble would remain in the region between the two mutant sites in the gene, but it might not affect protein function if relatively few amino acids were affected in this block. The *polarity* in message readout demonstrated by these and other T4 mutants was most easily interpreted to mean that a genetic reading frame has a particular origin from which the readout proceeds in a particular direction.

G. Streisinger and others provided proof that such *frameshift mutations* were responsible for mutants and pseudowild strains in the case of the lysozyme protein in T4. The complete amino acid sequence is known for this enzyme, which digests bacterial cell walls. In one pseudowild strain the enzyme was found to have a cluster of five amino acids that differed from those in the wild-type enzyme (Fig. 7.23). Streisinger interpreted this finding in terms of the effects of base deletion and base addition on the reading frame. As Crick had predicted, a small part of the pseudowild protein contained a small number of altered amino acids in a block, representing the remaining garble between the site of the original codon change and the changed codon in the pseudowild mutation event. The amino acids on both sides of the remaining garbled region were identical in wild-type and in pseudowild proteins. The reading frame was altered only in between the two mutant codons in the pseudowild strain.

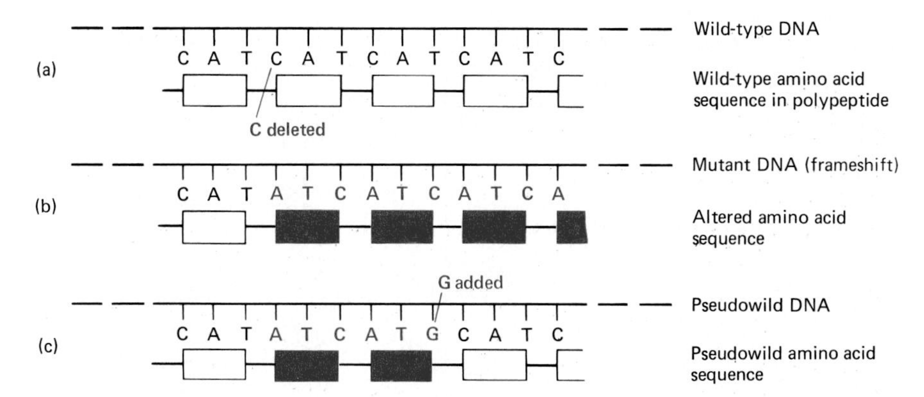


***Figure 7.22*** Restoration of the genetic reading frame in a pseudowild mutant. (a) Deletion of a base in wild type DNA alters the reading frame, as shown in (b). If a base is added near the one originally deleted, the reading frame is restored, as shown in pseudowild DNA in (c). The message readout has *polarity* (readout begins at a particular origin and proceeds in a particular direction).

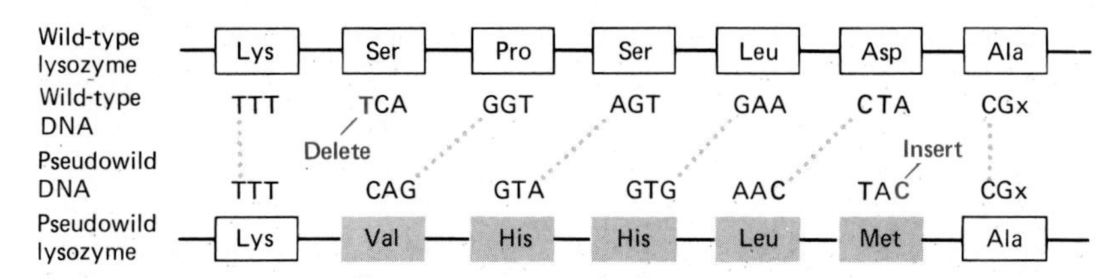


***Figure 7.23*** Frameshift mutations in phage T4. The difference in five amino acids (color) in the lysozyme protein of a pseudowild mutant compared with wild type is explained most easily by a compensating deletion and insertion of single bases into this part of the genetic reading frame. (From G. Streisinger et al. 1967. *Cold Spr. Harb. Symp. Quant. Biol.* **31**:77.)

## 7.9 Colinearity of Gene and Polypeptide

Using *E. coli,* Charles Yanofsky designed an elegant series of experiments in the 1960s to determine whether gene and protein are **colinear,** that is, whether a linear segment of DNA codes for a linear sequence of amino acids in the polypeptide. He compared the location of amino acid changes in wild-type and mutant polypeptides with the location of mutable sites in the fine structure map of the *trpA* and *trpB* genes coding for A and B subunits of the enzyme tryptophan synthetase. The *correspondence in loations* of mutable sites and amino acid substitutions was predicted if gene and polypeptide were colinear molecules.

Yanofsky isolated a large number of mutants that made defective A chains and mapped these at various sites in the *trpA* gene. He showed these mutants to be alleles of the same gene according to cis-trans complementation tests. He then mapped mutants defective in B chain synthesis in the *trpB* gene and showed them to be allelic by complementation tests. Cis-trans tests for complementation showed that *trpA* and *trpB* were functionally different genes since *trpA* mutants could complement *trpB* mutants, thus leading to wild-type tryptophan synthetase production.

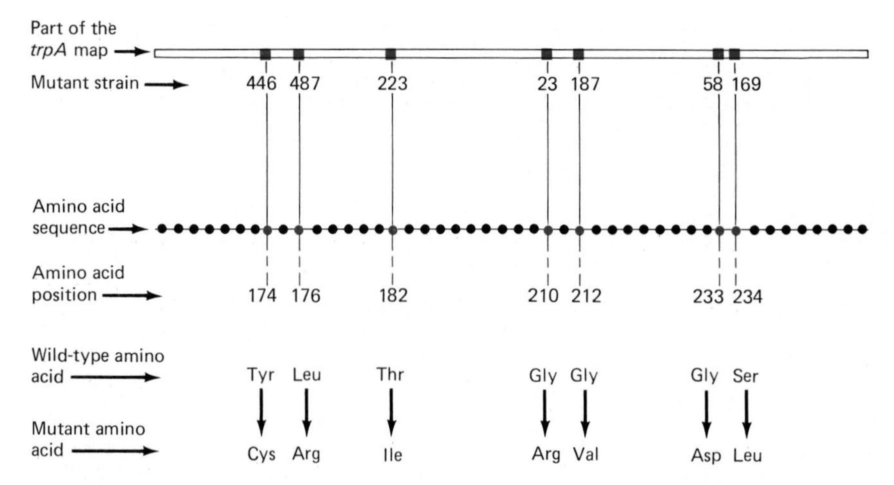


***Figure 7.24*** Colinearity of the gene and its protein product. Locations of mutations on the linear genetic map correspond to locations of amino acid substitutions in the linear polypeptides made by these *trpA* mutants of *E. coli.* Each colored square on the genetic map represents a mutant site, and each colored dot on the polypeptide represents the site of the substituted amino acid in that mutant. The position number of the amino acids, out of the total of 268, and the particular amino acids at these positions are shown. (From C. Yanofsky et al. 1967. *Proc. Nat. Acad. Sci. U.S.* **57**:296.)

The results of these genetic tests for colinearity showed that the substituted amino acid in a mutant polypeptide corresponded in location to the mutable site mapped in the gene (Fig. 7.24). Gene and protein were colinear, and the gene specified the kinds of amino acids and their sequence in the protein.

In another study using somewhat different methods, A. Sarabhai and colleagues showed colinearity between a phage T4 gene and the protein product that forms part of the phage head. In this experiment all the mutants had *termination codons* present instead of the normal codon specifying an amino acid in the protein. These mutants, therefore, made defective shortened polypeptide chains since protein synthesis terminated prematurely when the substituted termination codon was reached in the gene sequence. The investigators had mapped the mutants at different sites in the length of the gene. If gene and protein were colinear, they predicted that the length of the defective protein would correspond to the site of the mutation in the linear gene map (Fig. 7.25). The defective protein would be progressively longer for mutants whose substituted codon occurred progressively closer to the finish of the gene map, that is, progressively nearer to the end of the genetic reading frame. These results were found, thus providing support for the concept of colinearity.

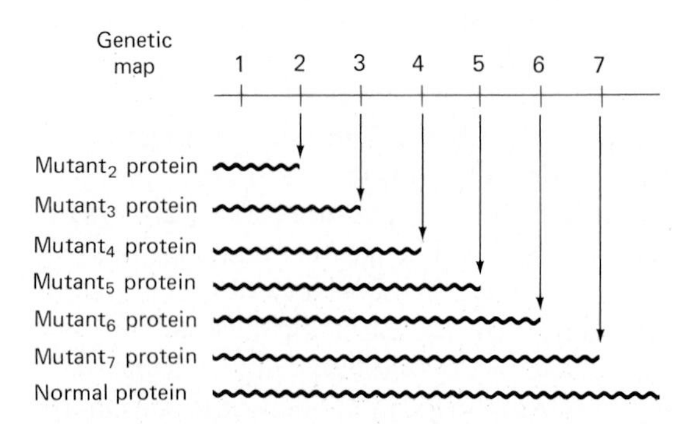


***Figure 7.25*** Demonstration of colinearity between gene and protein in phage T4. The known position of an inserted termination codon in a mutant DNA sequence (shown as 1 to 7) corresponds to the observed site of premature termination of the polypeptide chain translated from the altered sequence.

Studies of colinearity in bacteria and viruses were presumed to be equally applicable to eukaryotic organisms. The principle of colinearity was, therefore, accepted as a basic feature of gene structure and function in all organisms. A sequence of contiguous codons in DNA was translated into a sequence of contiguous amino acids in the polypeptide product of the gene, according to the established genetic code. This view was maintained until about ten years later, when DNA sequencing methods became available and revealed that all genes did not always consist of a simple linear sequence of contiguous codons spelling out the polypeptide. The concept of colinearity has been modified but not abandoned.

## DNA Sequencing and Gene Structure

The studies we have just discussed utilized indirect approaches to determine gene structure. From these studies investigators predicted that the sequence of DNA nucleotides carries the information that specifies amino acids in polypeptides and that the gene is comprised of contiguous codons. They also predicted that this uninterrupted sequence of codons is colinear with the sequence of contiguous amino acids in the specified polypeptide chain. These predictions were open to verification in the mid-1970s, when new methods were developed to determine base sequences in DNA by direct, rapid, and highly accurate techniques. Genes were organized as expected, according to base sequencing of bacterial and viral DNA. Quite unexpected results were found, however, from base-sequencing studies of genes in eukaryotes and their viruses.

### 7.10 DNA Sequencing Methods

The first published reports of complete nucleotide sequences were made for genes in *E. coli* and in the *E. coli* phage $\phi$X174. The sequencing methods involved cutting DNA into pieces of useful lengths, determining the original order in which these pieces had existed in intact DNA, and identifying each base in its proper sequence in the pieces and in the intact DNA.

Two principal sequencing methods are used today, and both of these depend on the same techniques to cut up the DNA and to identify the fragments in their original order in the DNA molecule. Duplex DNA is cut into pieces of appropriate length by restriction endonucleases, now called **restriction enzymes** (Table 7.3). Dozens of restriction enzymes are known, and each cuts the sugar-phosphate backbone of both strands of duplex DNA at the sites of particular combinations of base-pairs. With the use of various restriction enzymes singly and in combinations, collections of cut pieces were generated. These cut pieces, called **restriction fragments,** are separated according to their lengths by **gel electrophoresis.** Progressively shorter fragments move progressively faster through the gels, and entire displays of restriction fragments can be referred back to their order in the uncut DNA (Box 7.2). Different sets of restriction fragments are then isolated and the base sequences are determined by the **Maxam-Gilbert method** or by the *chain-terminating method* developed by Frederick Sanger.

In the Maxam-Gilbert chemical method of base sequencing, the 5′ end of the restriction fragment is labeled with $^{32}P$. The labeled fragments are divided into four portions. Each of the four aliquots is subjected to a different treatment that *breaks* the restriction fragment at the site of *one* of the four kinds of bases. Pieces of different lengths will therefore be produced, according to the place where the base-specific break was induced by the chemical treatment. The reaction conditions are arranged so that an average of one break is made per restriction fragment in each of the four aliquots. Otherwise the restriction fragments would be broken into tiny bits that would be useless for sequencing. The single break will occur at random along the length of a fragment but only at the site of the particular base that was sensitive to the chemical treatment. For example, in a population of restriction frag-

***Table 7.3*** Site-specific restriction endonucleases isolated from bacteria.

| enzyme | bacterial source | restriction site |
|---|---|---|
| *Eco*RI | *Escherichia coli* RY13 | 5′—G-A-A-T-T-C—3′<br>3′—C-T-T-A-A-G—5′ |
| *Eco*RII | *Escherichia coli* R245 | 5′—C-C-T-G-G—3′<br>3′—G-G-A-C-C—5′ |
| *Bam*I | *Bacillus amyloliquefaciens* | 5′—G-G-A-T-C-C—3′<br>3′—C-C-T-A-G-G—5′ |
| *Hpa*II | *Hemophilus parainfluenza* | 5′—C-C-G-G—3′<br>3′—G-G-C-C—5′ |
| *Hae*III | *Hemophilus aegyptius* | 5′—G-G-C-C—3′<br>3′—C-C-G-G—5′ |
| *Hind*II | *Hemophilus influenza* Rd | 5′—G-T-Py-Pu-A-C—3′<br>3′—C-A-Pu-Py-T-G—5′ |
| *Hind*III | *Hemophilus influenza* Rd | 5′—A-A-G-C-T-T—3′<br>3′—T-T-C-G-A-A—5′ |
| *Pst*I | *Providencia stuartii* | 5′—C-T-G-C-A-G—3′<br>3′—G-A-C-G-T-C—5′ |
| *Sma*I | *Serratia marcescens* | 5′—C-C-C-G-G-G—3′<br>3′—G-G-G-C-C-C—5′ |
| *Hha*I | *Hemophilus haemolyticus* | 5′—G-C-G-C—3′<br>3′—C-G-C-G—5′ |
| *Bgl*II | *Bacillus globiggi* | 5′—A-G-A-T-C-T—3′<br>3′—T-C-T-A-G-A—5′ |

All known restriction sites are 4 to 6 nucleotide pairs long and have a twofold rotational symmetry. Arrows indicate specific sites of cleavage on each strand of the duplex DNA.

ments treated chemically to induce breaks at adenine sites, some fragments will be broken at one adenine site and some fragments at other adenine sites. In the entire population of treated fragments, every adenine site would have experienced a break in some of the fragments. Thus a series of radioactive fragments, each containing a $^{32}$P-labeled 5′ end, is generated in each aliquot. These fragments are visible in gel autoradiographs (Fig. 7.26).

After chemical treatments, each of the four aliquots is allowed to migrate in separate, adjacent channels in the gel. Broken restriction fragments of the same size, having experienced a

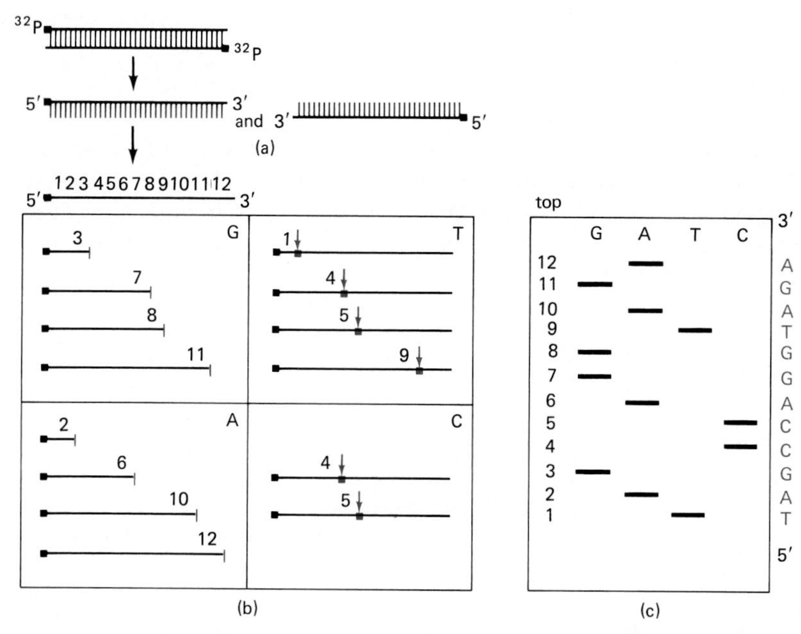


*Figure 7.26* The Maxam-Gilbert method for sequencing DNA. (a) A pure preparation of DNA, such as a population of a particular restriction fragment 12 base-pairs long, is labeled with $^{32}$P at the 5′ ends of both strands of the duplexes. Duplexes are separated into two fractions of complementary single strands, and one of these fractions is then sequenced. (b) The fraction of single strands is divided into four approximately equal portions and each of these aliquots is treated specifically to identify one of the four bases (G, A, T, C). Each treatment causes breakage of the DNA strand at the site of the base in question, but the concentration of reagents is adjusted so that each strand experiences an average of one break. The break may occur by chance at any of the sites occupied by the base involved, and a break will occur at one or another of these sites in the many fragments making up the aliquot of DNA. The different lengths of the broken 12-base-long fragments are shown in panels for G and A; the site of a break in the 12-base-long fragments is indicated by an arrow in the panels for T and C. Each usable fragment retains its original 5′ end, as determined by autoradiographic detection of the radioactive $^{32}$P marker. Fragments lacking the 5′ end are not visible in the autoradiograph, and thus confusion is avoided. (c) Each of the four aliquots migrates in a separate lane in a gel that is subjected to an electrical field in a gel electrophoresis apparatus. Fragments migrate at rates proportional to their length, with the shorter fragments migrating more rapidly toward the bottom of the gel from the origin at the top. By comparing the autoradiographic patterns of the 12 different-sized fragments in the four gel lanes, we can read the entire sequence of 12 bases directly. Reading proceeds from the shortest to the longest pieces, going from the 5′ to the 3′ end of the sequence, upward from the bottom to the top of the gel.

break at the same site in the same chemical treatment, occupy a single band. Broken restriction fragments of different sizes migrate to different places in the gel and produce $^{32}$P-labeled bands in these places in the gel. The electrophoretic method is sensitive enough to resolve chains that differ by only one nucleotide in length. Since each channel in the gel holds DNA that has been treated to reveal one of the four kinds of bases, the DNA sequence is read one band at a time across all four channels, from the bottom to the top of the gel. The reading proceeds, therefore, from the shorter 5′-labeled pieces to the longer 5′-labeled pieces.

A single run of four aliquots of some particular restriction fragment population can permit 150 or more bases to be identified in exact sequence. When all the data are pooled from

## BOX 7.2 Restriction Mapping of DNA

Physical mapping of genomic DNA can be achieved by cutting the DNA with various sequence-specific **restriction enzymes** and sizing the **restriction fragments** by **gel electrophoresis.** DNA molecules or fragments migrate through a gel, such as 1% to 2% agarose, under the influence of an applied electrical field. The smaller the molecule or fragment, the faster it will migrate in the gel. Larger molecules thus come to rest nearer to the origin (place of

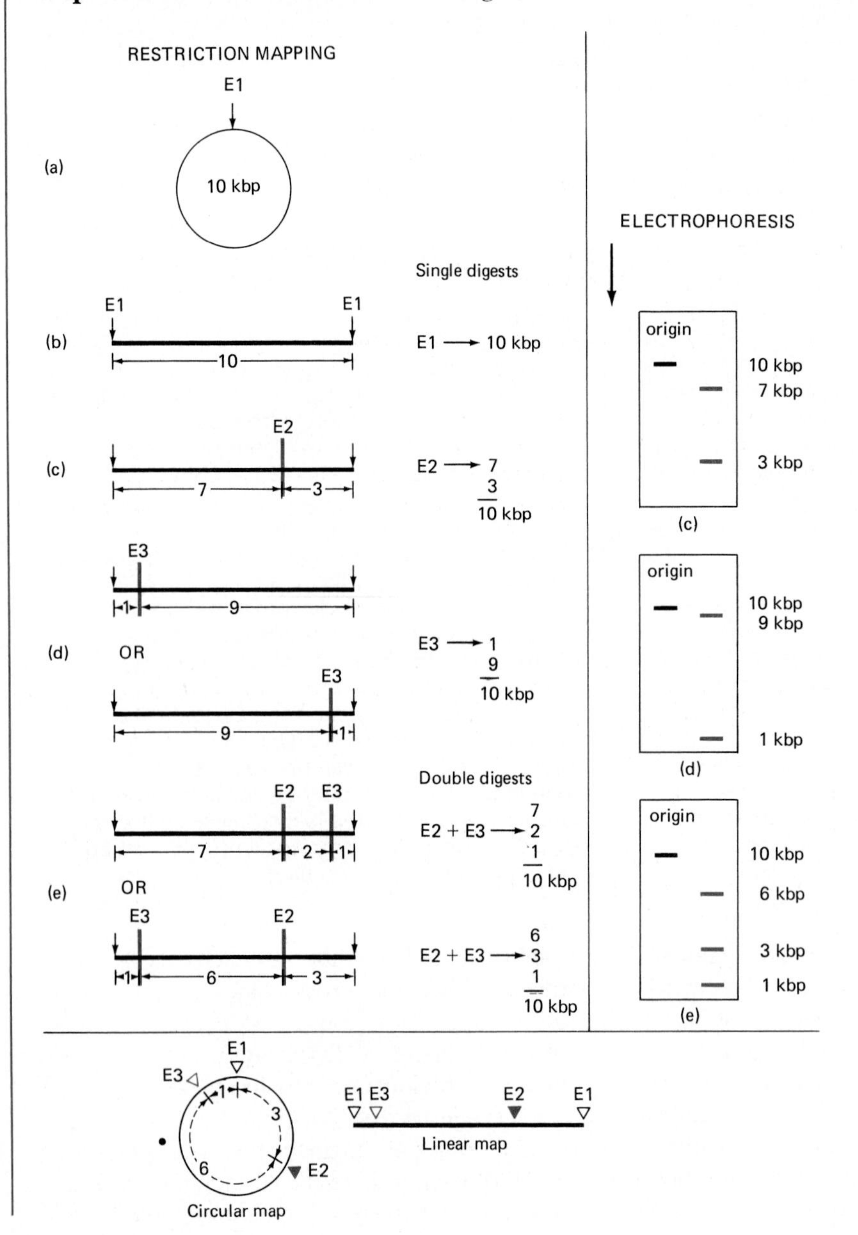

application of the DNA sample), and smaller molecules come to rest in the gel at points farther from the origin. The bands of DNA should consist of homogeneous groups of molecules, each band consisting of molecules of a particular length in the original sample. These bands can be visualized by appropriate means, such as staining, ultraviolet optics, or radioactivity. The size of molecules or fragments in the set of bands in a gel is determined by reference to fragments or molecules of known size that are run in neighboring tracks or lanes in the gel. Pieces produced by digestion with restriction enzymes may be sequenced and placed in the order in which they occurred in the original genomic DNA. An example of restriction mapping is shown on the facing page.

Starting with (a) a whole circular genome or (b) a genomic fragment 10 kilobase pairs (kbp) long, produced by restriction enzyme E1, the linear DNA fragment is cut by another sequence-specific restriction enzyme (E2). The fragment lengths are determined by gel electrophoresis. As shown in (c), the two fragments measure 7 kbp and 3 kbp, thus showing that enzyme E2 cuts at only one site on the original 10-kbp DNA sequence. All other restriction sites are located with reference to this E2 site. Another restriction enzyme (E3) also cuts the 10-kbp genomic DNA at one specific site, thereby producing fragments that are 9 kbp and 1 kbp long, as seen in (d). In order to determine the E3 site relative to the E2 restriction site, **double enzyme digests** are prepared and run, with results shown in (e). If the E3 site is to the right of E2, there will be no change in the E2 7-kbp fragment but the E2 3-kbp piece will be cut by E3 into 2-kbp plus 1-kbp fragments. If, on the other hand, the E3 site is to the left of the E2 restriction site, then the E2 3-kbp fragment will be unchanged but the E2 7-kbp fragment will be cut by E3 into 1-kbp plus 6-kbp pieces. Electrophoresis of the double digests will reveal which of the two possible alternatives is correct. Once the E3 site is located with reference to the E2 site, these restriction sites can be placed in correct order on the physical map of the genome or genomic fragment. Additional studies that use a variety of appropriate restriction enzymes can be conducted in this same way, and more details can be added to the physical map of restriction sites.

base sequencing of all the different restriction fragments obtained from the intact DNA being analyzed, the entire base sequence of that DNA can be assembled. What had taken years to accomplish by tedious methods before 1977 can now be accomplished in days or weeks by the use of rapid sequencing methods. In addition to speed, these methods provide a very high degree of accuracy.

## 7.11 Organization of Gene Sequences in Bacteria and Their Viruses

Genes in *E. coli* proved to have base sequences that were colinear with the sequences of amino acids in known protein products. These sequences were matched and showed corresponding codons in DNA and amino acids specified by these codons in the protein (Box 7.3). In *E. coli* the coding sequence of the gene started with the initiation codon TAC (equivalent to RNA codon AUG) and terminated with one of the three known termination codons, ATT, ATC, or ACT (equivalent to RNA codons UAA, UAG, UGA). In addition, other noncoding components were present at each side of the protein-specifying region. These flanking DNA segments are associated with control mechanisms in translation, a topic that we will discuss in Chapters 8 and 9.

Sanger and co-workers sequenced the entire genome of *E. coli* phage $\phi$X174. The genome included ten genes, which code for the ten

## BOX 7.3 DNA Sequence Information

In many published reports of DNA sequences, the noncoding strand of DNA is the only information given. It is actually the most informative sequence possible, if a single strand is to be shown. By deducing the complementary bases we can derive the coding strand of DNA in the appropriate 3′ to 5′ orientation from which messenger RNA is transcribed. By simply changing all T to U in the printed noncoding DNA strand, we read off the messenger RNA sequence in the correct 5′ to 3′ orientation. From such deduced mRNA codons, we know the amino acids and their sequence in the polypeptide translation, from the amino-terminus to the carboxy-terminus of the chain, according to the standard coding dictionary (See Fig. 7.15).

```
3'   TAC CTA TTT CAA AAT TTG TCT CTC CTT AGA ATT   5'   Coding strand of DNA
     ••• ••• ••• ••• ••• ••• ••• ••• ••• ••• •••
     ATG GAT AAA GTT TTA AAC AGA GAG GAA TCT TAA        Noncoding strand of DNA

5'   AUG GAU AAA GUU UUA AAC AGA GAG GAA UCU UAA   3'   mRNA transcript

H2N–Met – Asp – Lys – Val – Leu – Asn – Arg – Glu – Glu – Ser •   COOH   Polypeptide translation
```

proteins specified by the phage (Fig. 7.27). Almost the whole sequence of 5386 nucleotides is involved in coding; fewer than 217 nucleotides, contained in four short regions of the genome, have no known coding function. Gene and protein proved to be colinear in phages just as they are in bacteria. Colinearity appears to be a general phenomenon among bacteria and their viruses.

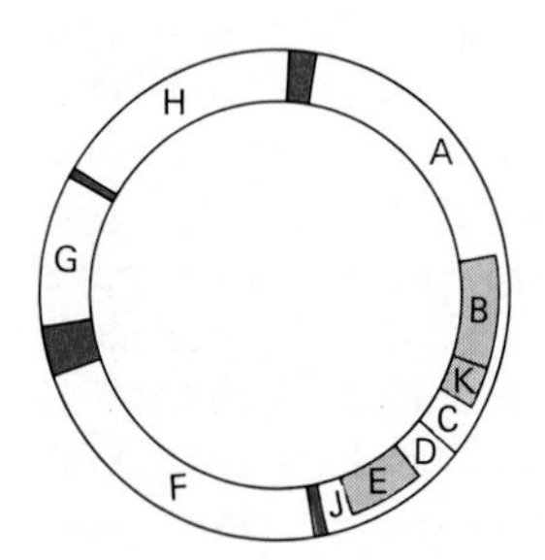


***Figure 7.27*** The genetic map of the circular genome of bacteriophage $\phi$X174 consists of ten genes (A–K) distributed among 5386 base-pairs of DNA. Genes B, K, and E overlap with other genes in the genome and share some or all of the sequence involved. Four small regions of the genome (dark color) have no known coding function. These noncoding regions occupy fewer than 217 of the 5368 base-pairs in this genome. (Data from M. Smith. 1979. *Amer. Sci.* **67**:57–67.)

Base-sequencing analysis in phages and in viruses that have eukaryotic hosts revealed an unexpected feature of genome organization. **Overlapping genes** are present in viral genomes but not in genomes of cellular organisms. The region of gene overlap may be very short or rather long. In any case, the overlapping genes on the same strand of DNA use *different reading frames* in polypeptide synthesis. Depending on the origin of the reading frame, the same nucleotides can be read in three different sequences (Fig. 7.28). This observation was unexpected since the whole concept of overlapping codons had been excluded in the 1960s when the genetic code was deciphered. It has been suggested that overlapping genes would be advantageous in viruses, which have small genomes that package a fair amount of information. Even more information can be packaged in a molecule of the same length if overlapping sequences specify different polypeptides.

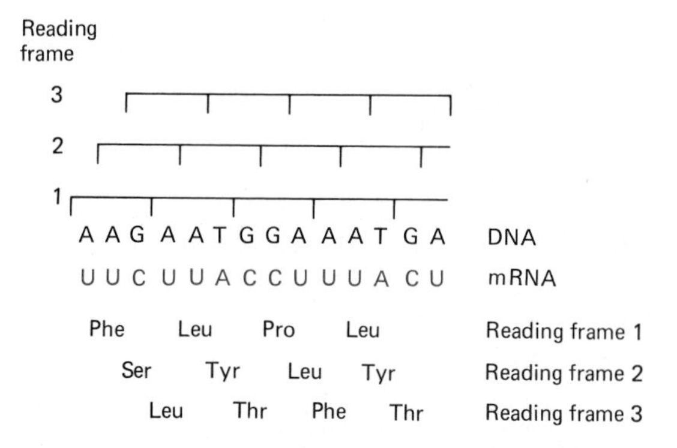


***Figure 7.28*** Overlapping genes on the same strand of DNA utilize different reading frames in polypeptide synthesis. Reading frame 1 begins with UUC and proceeds to successive mRNA triplet codons UUA, CCU, UUA, and so on, and specifies amino acids Phe, Leu, Pro, Leu, and so on, in the polypeptide translation. Reading frame 2 begins with UCU and proceeds to UAC, CUU, UAC, and so on; reading frame 3 begins with CUU, ACC, UUU, ACU, and so on to the end of the message. Each translation product is different since each reading frame constitutes a different sequence of codons.

## 7.12 Organization of Gene Sequences in Eukaryotes and Their Viruses

Compared to prokaryotic cells, gene organization in eukaryotes and their viruses was found to be quite different. For virtually every eukaryotic gene that has been sequenced, the region between the initiation and termination codons consists of *coding regions interrupted by noncoding regions.* In these **interrupted genes, coding sequences,** or **exons,** are interspersed with noncoding **intervening sequences,** or **introns,** in the genetic message. The messenger RNA transcribed from the entire gene sequence is processed before translation. Intervening sequences are cut out and coding sequences are spliced together, thus producing a continuous coding sequence in the mature messenger RNA, which is then colinear with the amino acids in the polypeptide product of the gene. As in prokaryotic genes, noncoding segments flank the genetic message on each side. These flanking segments are involved in gene expression in both eukaryotic systems and prokaryotic cells. We will discuss this topic in Chapter 8.

The number of intervening sequences in the genetic message varies considerably among different genes. Only two intervening sequences between exons have been found in genes coding for globin polypeptides of hemoglobin, but more than fifty intervening sequences are located in the gene specifying one of the collagen polypeptides (Fig. 7.29). On the other hand, some eukaryotic genes have no introns at all. Among these are genes for histone proteins, which are parts of chromosome structure, and genes for two of the three interferons, which function in immunity responses.

In view of this variation, the *number* of introns seems to have no influence on gene integrity, but it may have an influence on gene expression. In some cases, when an intervening sequence has been excised experimentally from a gene, the messenger RNA transcript of the altered gene will be nonfunctional or very unstable. Under natural conditions altered duplicates of some genes are present in the genome, but they are not transcribed. These altered duplicates are called **pseudogenes.** Comparisons between sequences of functional genes and their pseudogenes have revealed differences in their introns. Some pseudogenes have lost their introns altogether and consist only of a continuous sequence of coded information. These pseudogenes are apparently not transcribed. Other pseudogenes have deletions in parts of one or more introns and are also not transcribed. These observations indicate some importance for intervening sequences in gene expression, particularly in transcription of DNA into RNA copies.

Interrupted genes are also characteristic of viruses that infect eukaryotic hosts. The similarity in gene organization in eukaryotes and their viruses may reflect the fact that the same metabolic machinery is involved in both systems to transcribe and translate gene sequences. Cells transcribe and translate their own genes, and they also transcribe and translate viral genes brought into the cell during infection. A similar

| coding sequences (exons) | intervening sequences (introns) | location in gene (nucleotide nos.) | sequence size (no. nucleotides) |
|---|---|---|---|
| I | | 1-47 | 47 |
| | A | 48-1636 | 1589 |
| II | | 1637-1821 | 185 |
| | B | 1822-2072 | 251 |
| III | | 2073-2123 | 51 |
| | C | 2124-2704 | 581 |
| IV | | 2705-2833 | 129 |
| | D | 2834-3233 | 400 |
| V | | 3234-3351 | 118 |
| | E | 3352-4309 | 958 |
| VI | | 4310-4452 | 143 |
| | F | 4453-4783 | 331 |
| VII | | 4784-4939 | 156 |
| | G | 4940-6521 | 1582 |
| VIII | | 6522-7564 | 1043 |
| | | | 7564 |

(a)

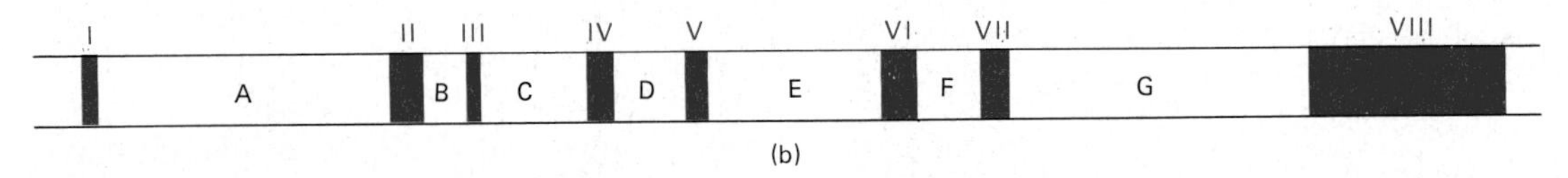


(b)

***Figure 7.29*** Organization of the ovalbumin gene in chickens. (a) The gene sequence extends across a continuous length of 7564 nucleotides in one strand of duplex DNA. From DNA sequence analysis in relation to the known amino acid sequence of 386 residues in ovalbumin, it was found that eight coding sequences were separated by seven noncoding intervening sequences, all of particular lengths. (b) The ovalbumin gene map consists of the exons (color) and the intervening sequences (A–G) in proportionate lengths corresponding to data from DNA sequencing. (Reprinted with permission from S.L.C. Woo et al. 1981. *Biochemistry* **20**:6437–6446. Copyright 1981 American Chemical Society.)

explanation may account for the existence of uninterrupted genes in prokaryotes and their viruses, namely, the same transcriptional and translational systems are utilized for host genes and for viral genes.

Viruses that infect eukaryotic hosts have overlapping genes, just as phage genomes do (Fig. 7.30). This similarity in viral genome organization may be an evolutionary consequence of similar advantages in packaging a maximum amount of genetic information into the small confines of a viral particle. It is probably not related to the prokaryotic or eukaryotic nature of the host organisms themselves.

## 7.13 Isolating Gene Sequences by Recombinant DNA Technology

In order to analyze a particular gene or DNA sequence, it must first be isolated from the genome. Once isolated, the desired DNA must be obtained in pure form and in adequate quantities for analysis. These objectives can be achieved by procedures collectively referred to as **recombinant DNA technology.**

Part or all of a genome can be cut into restriction fragments by the use of one or more restriction enzymes (see Table 7.2). The fragments are separated by gel electrophoresis and

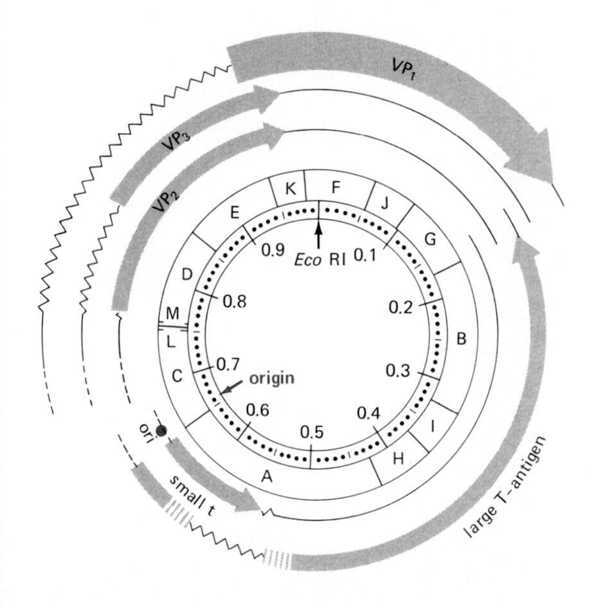


***Figure 7.30*** Standard physical map of simian virus 40 (SV40) shows the single break site for the restriction enzyme *Eco* RI in this duplex DNA molecule, which is used as a reference point for the map. The point of origin of DNA replication is indicated at position 0.663. The five virus-coded proteins are indicated relative to the parts of the map where they are encoded: small t- and large T-antigen overlap completely; virus proteins (VP) 1–3 also overlap completely with each other on the DNA map. The only portions of the genes that translate into amino acids in these proteins are shown by thick, colored arrows. The untranslated parts of the pre-mRNA are shown by thin lines and the segments excised from pre-mRNA are shown as wavy lines. The arrows point in the direction of genetic readout. (From W. Fiers et al. 1978. *Nature* **273**:113.)

collected for subsequent identification and purification. The individual restriction fragments in the genomic collection are incubated with *vector DNA,* which has also been cut by restriction enzymes. Some of the molecules in this mixture will be **recombinant DNA,** which is genomic DNA spliced to vector DNA through base pairing between sticky ends of cleaved molecules of the two kinds. The vector DNA serves several purposes: (1) it provides a means of entry for the desired gene into a host cell, (2) it provides a replicating system to make many copies of the desired gene that is spliced to it, and (3) it carries suitable gene markers to indicate the presence of a recombinant DNA molecule in the host.

Plasmids and viral DNA are the principal vectors used in recombinant DNA procedures. **Plasmids** are DNA molecules that exist and replicate independently of the host genome in the host cell. They are transferred naturally by transformation or conjugation and can be manipulated in the laboratory. Viral DNA is spliced to foreign genes and can be transferred to host cells by infection. By the use of either type of vector, foreign DNA sequences can be inserted into host cells in the form of spliced molecules of recombinant DNA. Part of the molecule consists of genomic sequences of interest for study, and part consists of the vector DNA sequences (Fig. 7.31).

Once in the host cells, the desired foreign DNA must be identified by the examination of host colonies that display phenotypes indicating that marker genes of the vector DNA are present (Fig. 7.32). When they are identified, host colonies carrying recombinant DNA must be tested to find the particular colony that is carrying the desired sequence from among many different colonies carrying many different spliced genes from the genome under study. The task can be formidable. A typical mammalian genome may consist of $5 \times 10^9$ base pairs (bp) of DNA. If restriction fragments are 20,000 to 50,000 bp long, it would be necessary to establish 100,000 to 250,000 different clones, each carrying one unique restriction fragment in a recombinant DNA molecule in a host clone. The useful size of the restriction fragment is based on the amount of DNA that can be spliced to the vector and still permit transformation or infection reactions for vector entry into the host.

The search for the desired clone in such a **genomic library** of **cloned DNA** is made simpler if *probes* can be used to identify a specific desired sequence. Such probes may consist

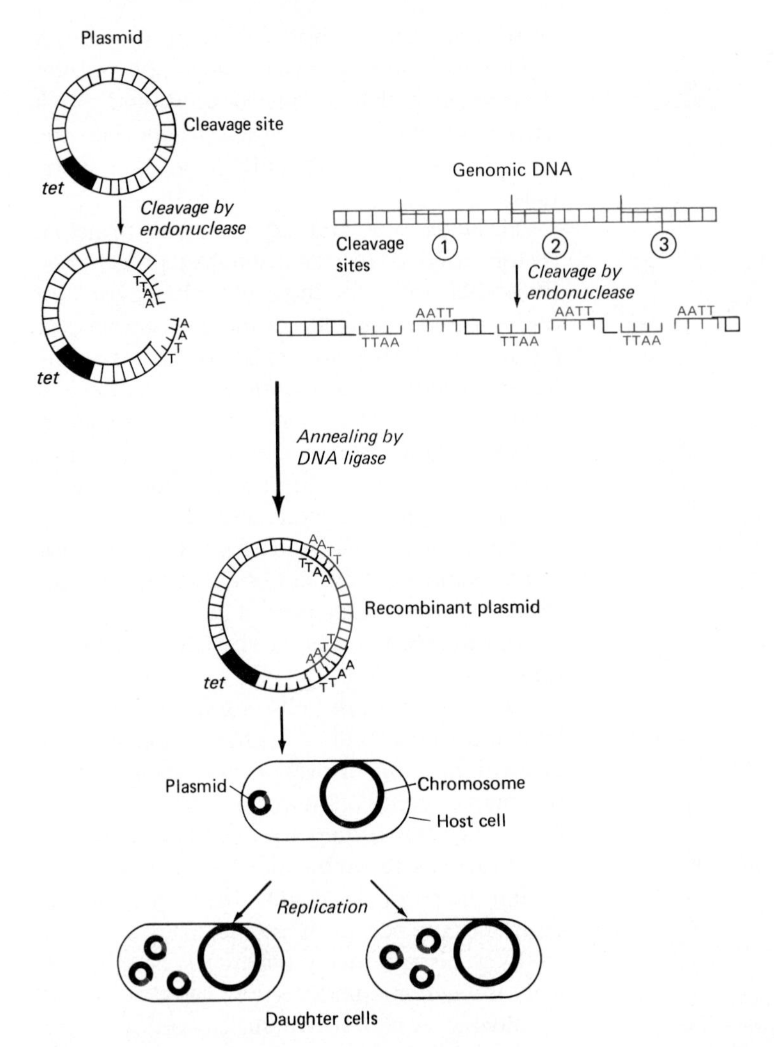


***Figure 7.31*** Recombinant DNA technology. Recombinant DNA is made by cleaving allelically marked (*tet*) plasmid DNA or other cloning-vehicle DNAs (which permit entry and multiplication of the recombinant DNA inside host cells) and genomic DNA (such as the gene for human interferon) with a suitable restriction enzyme that has endonuclease activity. Plasmid DNA and genomic DNA are annealed through complementary base pairing of their exposed single-stranded ends, and the duplex is sealed by DNA ligase action through which the sugar-phosphate backbones are covalently bonded. Once the recombinant DNA has entered the host cells, replication takes place as it does for host DNA itself. There is no synchrony of replication for recombinant and host DNAs, however, and many copies of the recombinant molecule may be made in any individual host cell.

of complementary RNA molecules that will pair with the unique gene sequence being sought. For example, large quantities of labeled messenger RNA can be obtained from cells that make an abundance of one kind of gene product, such as reticulocytes that make globins for hemoglobin or oviduct cells that make ovalbumin egg protein. Isolated messenger RNA from such cells can be applied to a genomic library to locate the colony containing the globin or ovalbumin gene sequence in recombinant DNA form. Once the clone has been identified it can be grown in bulk

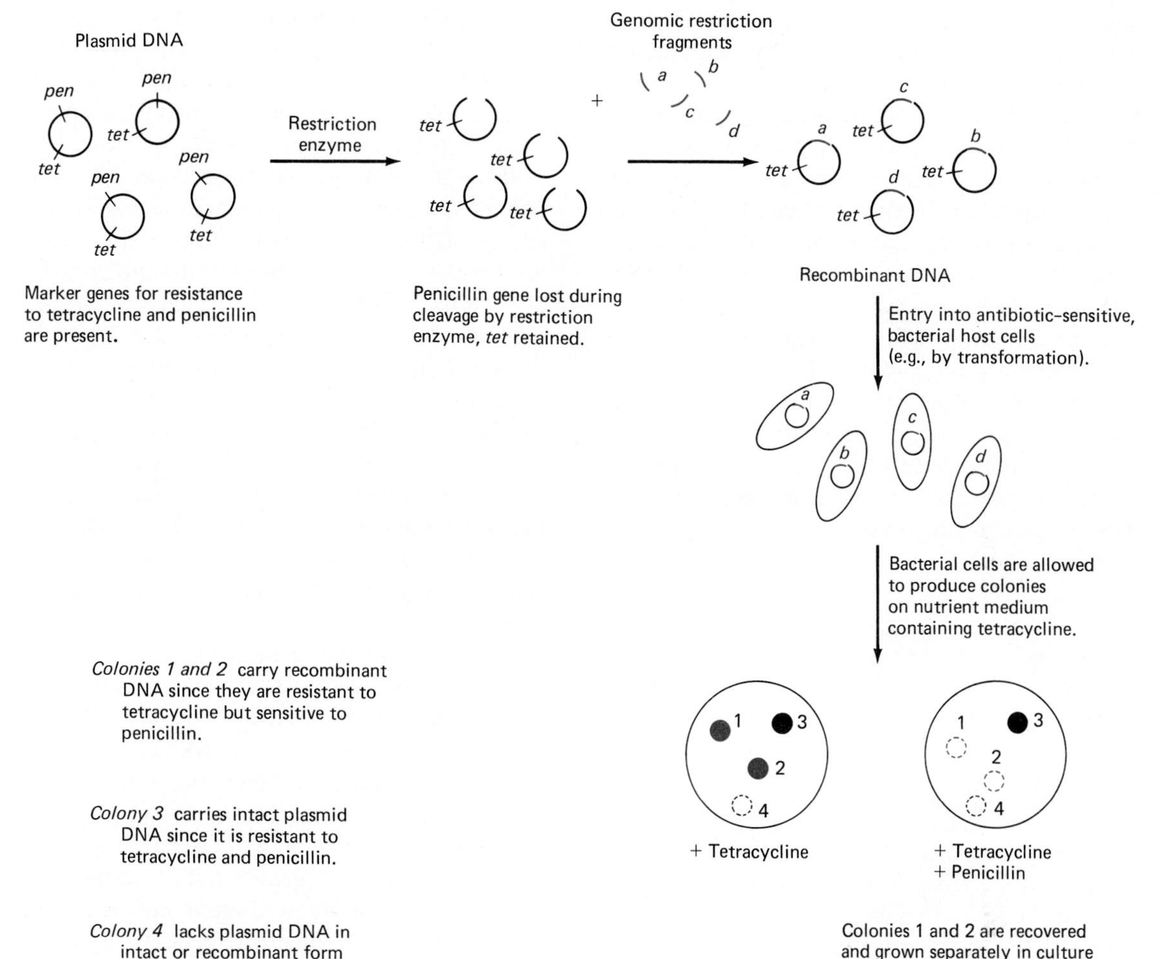


***Figure 7.32*** Isolation and identification of a genomic sequence using recombinant DNA technology. Plasmid DNA or other vector DNA carrying marker genes for antibiotic resistances are cut by a suitable restriction enzyme to selectively remove the penicillin gene. The cleaved vector DNA is incubated with genomic restriction fragments to produce spliced molecules of recombinant vector + genomic DNAs. DNA is taken into antibiotic-sensitive bacterial host cells by transformation or similar process, and the cells are incubated in selective media to identify colonies that carry marked recombinant DNA. Cells lacking recombinant DNA cannot grow in media containing either antibiotic since the host has only antibiotic-sensitive alleles. Colonies growing in media containing both antibiotics can be discarded since they must carry intact plasmid DNA with both resistance genes present. Colonies that can grow in media with tetracycline alone but not in media with both antibiotics must represent cells carrying modified plasmid DNA spliced to genomic DNA in recombinant molecules since only a circular DNA will replicate in host cytoplasm. These colonies are collected individually for further study, either immediately (see Fig. 7.33) or after growth in fresh culture medium to permit increase in numbers of desired cells carrying the desired recombinant DNA. (Adapted from "Split Genes" by P. Chambon. Copyright © 1982 by Scientific American, Inc. All rights reserved.)

amounts in culture, and the amplified desired DNA sequence can be collected for sequencing or for other studies (Fig. 7.33).

Other methods can be used to make a desired DNA sequence as a copy of the particular messenger RNA isolated. The messenger RNA can be copied into complementary DNA in reactions catalyzed by the enzyme *reverse transcriptase* (see Section 3.14). The **copied DNA (cDNA)** is then separated and allowed to synthesize its partner strand to make duplex gene sequences for recombinant DNA splicing. The recombinant DNA, which carries the desired gene sequence, is inserted into the host cells where it is allowed to replicate. The amplified sequences are then available for isolation and base-sequencing analysis.

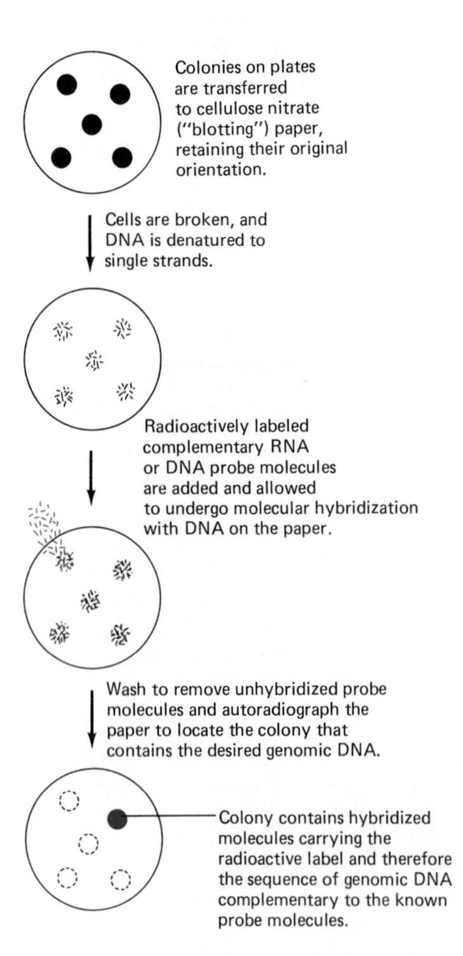


Through these and other methods of recombinant DNA technology (to be discussed in other chapters), even chromosomes that are largely inaccessible to genetic analysis can be analyzed piece by piece in chemical detail. In addition to studying gene sequences that code for polypeptides, we can alter sequences and study the effects of these changes on processing coded information in transcription and translation in host cells. Chemically synthesized gene sequences can be spliced into vector DNA for mass production of some desired protein in bulk clonal culture (Fig. 7.34). Synthetic genes for the mammalian hormones *somatostatin* and *insulin* have been introduced into *E. coli,* where the hormonal polypeptides are made from the introduced genetic instructions. The hormones are collected, purified, analyzed for biological activity and can then be prepared for therapeutic uses in medicine.

Naturally occurring desired genes can also be spliced into vector DNA, which is then inserted into a suitable host for production of large amounts of the protein product. Genes for human *interferon* have been successfully introduced into *E. coli*, and the interferon made in these host cells has been found to possess reasonably high biological activity. Interferon is a potentially important therapeutic agent for certain disorders initiated by viral infections.

***Figure 7.33 (left)*** Identification of a specific genomic DNA sequence that is carried in recombinant DNA molecules in colonies of host cells can be achieved using known specific complementary DNA or RNA probes of the desired sequence in a blotting method developed by E.M. Southern. Colonies are "blotted" onto cellulose nitrate paper and their adhering cells are treated to expose and denature their DNA contents. The radioactively labeled probe molecules are applied and allowed to undergo molecular hybridization. The colony containing the desired genomic DNA is identified by the occurrence of bound radioactive probe molecules hybridized with the DNA. The original plate with the identified colony can then provide intact cells for growth in culture and consequent amplification of the desired DNA for further study.

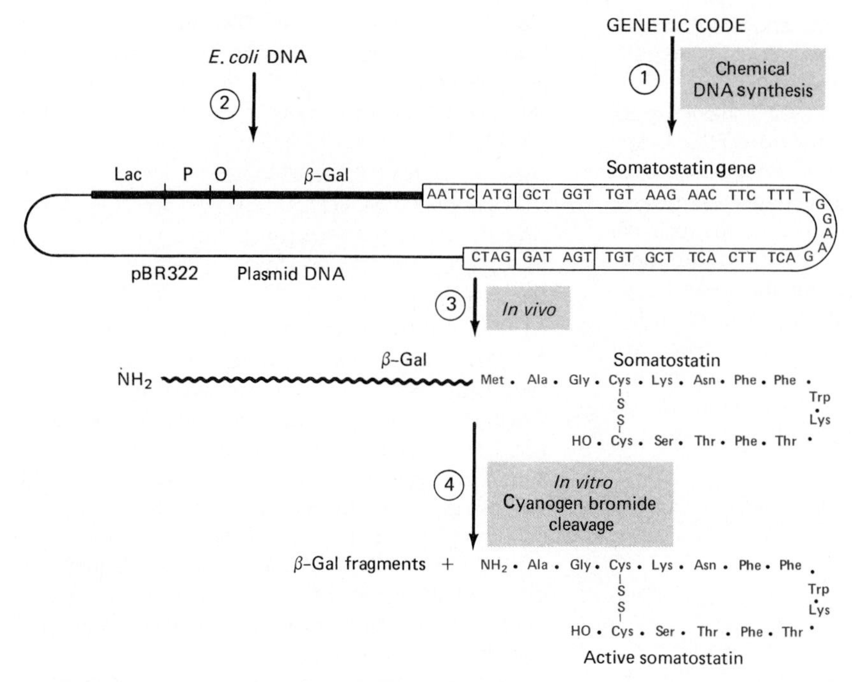


***Figure 7.34*** The somatostatin gene was synthesized in the laboratory and spliced to the *Lac* gene of *E. coli* in a carrier DNA (plasmid DNA) so transformation could be achieved. After transformation, the whole protein was cleaved *in vitro* by cyanogen bromide to yield the active mammalian hormone consisting of 14 amino acids. (From K. Itakura et al. 1977. *Science* **198**:1056. Copyright 1977 by the American Association for the Advancement of Science.)

The only source of this immunity agent has been whole blood, from which small amounts of interferon can be recovered in impure form. By **genetic engineering,** that is, introducing functioning foreign genes into host organisms or cells that otherwise do not possess these genes, it should be possible in future to exploit host organisms for human needs. The hosts act as virtual factories to make proteins that are important in medicine, research, and other human activities. Recombinant DNA technology has opened new horizons for study and exploitation of the gene and its product.

## Questions and Problems

***7.1*** In order to determine whether phage DNA or protein enters *E. coli* cells in an infection, phage molecules must be distinguished from each other and from host DNA and protein.

***a.*** Why is it desirable to use radioactive $^{32}P$ as the DNA marker and $^{35}S$ as the protein marker, instead of other radioactive isotopes such as $^{3}H$ or $^{14}C$?

***b.*** Why does entry of viral DNA but not of viral protein indicate that DNA is the genetic material of the virus?

***c.*** Why can we generalize and conclude that DNA is the genetic material in other organisms when the Hershey-Chase experiments concerned only T2 phage?

***7.2*** Avery, MacLeod, and McCarty showed that purified DNA extracted from one strain could genetically transform cells of a different strain.

***a.*** What was the importance of their control experiments, in which they used a donor extract that lacked DNA and found that transformation did not take place?

***b.*** What results would be predicted for the reverse experiment, in which DNA extracts from the recipient strain were used to transform cells from the donor strain, if DNA indeed is the genetic material?

***7.3*** The base composition and base ratios of nucleic acids from 8 different viruses are given below. For each of these viruses, indicate whether the nucleic acid is DNA or RNA and whether these molecules are single or double stranded. Explain in each case.

| | percentages | | | | | $\frac{A+T \text{ or } U}{G+C}$ | $\frac{A+G}{C+T \text{ or } U}$ |
|---|---|---|---|---|---|---|---|
| **virus** | **G** | **C** | **A** | **T** | **U** | | |
| 1 | 19 | 19 | 31 | 31 | | | |
| 2 | 19 | 19 | 31 | | 31 | | |
| 3 | 31 | 31 | 19 | 19 | | | |
| 4 | | | | | | 1.00 | 1.26 |
| 5 | 25 | 18 | 25 | 32 | | | |
| 6 | 20 | 25 | 23 | | 32 | | |
| 7 | | | | | | 1.26 | 1.00 |
| 8 | | | | | | 1.00 | 1.00 |

***7.4*** What is the base sequence of the DNA strand that is complementary to each of the following base sequences?

***a.*** T A G C A T T C C C G A G G

***b.*** G C G C T T A C G A T C A T C C A

***7.5*** During studies of DNA base composition in various laboratory strains, it was discovered that strain B had a different proportion of G+C:A+T in a later analysis than had been found several months earlier in the original assays. The data are as follows:

| | percentages | | | |
|---|---|---|---|---|
| **date of assay** | **G** | **C** | **A** | **T** |
| February | 24.8 | 24.9 | 25.2 | 25.1 |
| June | 30.1 | 29.9 | 20.1 | 19.9 |

Assume that replication error led to base substitution in strain B some time after February.

***a.*** What base substitution took place to alter strain B?

***b.*** Diagram possible errors during replication that could have been responsible for the observed change in base composition.

***7.6*** Place the three duplex DNA molecules shown below in order from highest to lowest melting temperature ($T_m$):

(1) A A A T C T G T T A G A C
T T T A G A C A A T C T G

(2) G G G C A A T C C A G G T C C
C C C G T T A G G T C C A G G

(3) C G A T T G A C T T A G A C
G C T A A C T G A A T C T G

(4) T T C A T A G C T A A A
A A G T A T C G A T T T

***7.7*** Two independently isolated small plaque mutants (*sm*-1 and *sm*-2) were analyzed to determine whether they are governed by alleles of the same gene or of different genes. Mixtures of the two mutants were used to infect *E. coli* B (mutants and wild type can grow) in one assay and *E. coli* K12 (only wild type can grow) in a second assay, with the following results: (1) *E. coli* B: about 99 mutant plaques per plate and an average of 1 wild-type plaque per plate, (2) *E. coli* K12: an average of 1 wild-type plaque per plate but no mutant plaques on any plate.

***a.*** Are *sm*-1 and *sm*-2 alleles of the same gene or of different genes? Explain.

***b.*** Why is approximately the same frequency of wild-type plaques observed in both tests?

***c.*** What is the frequency of recombination in these assays?

***7.8*** A series of six independently isolated small plaque mutants (*sm*-1 to *sm*-6) were studied in complementation tests to determine their allelic relationships and the number of genes involved in plaque size development. Mixed infections in all possible combinations of pairs of mutants were made using *E. coli* K12 as host, and observations were made of phage growth (+) or no growth (0) on lawns of bacterial cells. The results were as follows:

| 1 | 2 | 3 | 4 | 5 | 6 | |
|---|---|---|---|---|---|---|
| 0 | 0 | + | + | + | + | 1 |
| | 0 | + | + | + | + | 2 |
| | | 0 | 0 | + | + | 3 |
| | | | 0 | + | + | 4 |
| | | | | 0 | + | 5 |
| | | | | | 0 | 6 |

***a.*** How many genes govern plaque size in this set of mutant strains?

***b.*** Which strains have mutant alleles for the same gene?

**7.9** Complementation tests were performed for six independently isolated mutants governing the same phenotypic trait in phage T4, with results as follows for plaque development on *E. coli* K12 (+ = complementation, 0 = no complementation):

| 1 | 2 | 3 | 4 | 5 | 6 | |
|---|---|---|---|---|---|---|
| 0 | + | + | 0 | + | + | 1 |
| | 0 | + | + | 0 | + | 2 |
| | | 0 | + | + | 0 | 3 |
| | | | 0 | + | + | 4 |
| | | | | 0 | + | 5 |
| | | | | | 0 | 6 |

***a.*** How many different genes are involved in this phenotypic trait?
***b.*** Which mutant strains are altered in the same gene?

**7.10** Eight deletion mutants of the *rIIA* gene of phage T4 were crossed in all possible combinations and scored for wild-type recombinants, with the following results (+ = recombination, 0 = no recombination):

| 1 | 2 | 3 | 4 | 5 | 6 | 7 | 8 | |
|---|---|---|---|---|---|---|---|---|
| 0 | + | 0 | + | + | 0 | + | 0 | 1 |
| | 0 | 0 | + | + | 0 | + | 0 | 2 |
| | | 0 | + | + | 0 | + | 0 | 3 |
| | | | 0 | + | + | 0 | 0 | 4 |
| | | | | 0 | 0 | 0 | 0 | 5 |
| | | | | | 0 | 0 | 0 | 6 |
| | | | | | | 0 | 0 | 7 |
| | | | | | | | 0 | 8 |

Construct a topological map for these deletions.

**7.11** Four mutants (*a*, *b*, *c*, *d*) were crossed in all possible combinations with the eight deletion strains described in Question 7.10, with the following results (+ = recombination, 0 = no recombination):

| | 1 | 2 | 3 | 4 | 5 | 6 | 7 | 8 |
|---|---|---|---|---|---|---|---|---|
| *a* | 0 | + | 0 | + | + | 0 | + | 0 |
| *b* | + | + | + | + | 0 | 0 | 0 | 0 |
| *c* | + | + | + | 0 | + | + | 0 | 0 |
| *d* | + | 0 | 0 | + | + | 0 | + | 0 |

What is the order of genes *a*, *b*, *c*, *d* on the topological map constructed for Question 7.10?

**7.12** A number of mutations were found in the *rII* region of phage T4. From the recombination data shown in the table, determine whether each mutant is the result of a point mutational defect or a deletion. One of the four mutants has never been known to revert to wild type, the other three have been observed to undergo reverse mutation. Draw a topological map to represent your interpretations.

| 1 | 2 | 3 | 4 | |
|---|---|---|---|---|
| 0 | 0 | 0 | + | 1 |
| | 0 | + | + | 2 |
| | | 0 | + | 3 |
| | | | 0 | 4 |

**7.13** Three independently arising *rIIB* mutants were crossed in phage T4 with the following results:

| **crosses** | **percentage recombinants** |
|---|---|
| *rIIB2* × *rIIB1* | 0 |
| *rIIB2* × *rIIB3* | 0 |
| *rIIB1* × *rIIB3* | 0.8 |

What is the nature of these three mutants? How are they related to each other?

**7.14** After treatment with the mutagen nitrous acid, phage T4 produced mottled plaques (a consequence of the coexistence of two genetically different viruses in the same plaque) while ϕX174 viruses gave only nonmottled plaques on a lawn of *E. coli* cells. What do these results suggest with regard to DNA structure of the two phages?

**7.15** A synthetic RNA produced from a mixture of U and C in the proportion of 3:1, respectively, is used to direct polypeptide synthesis *in vitro*. Assume that all the monomers are incorporated randomly into poly(UC) and that all 20 amino acids are provided in the experimental system.
***a.*** Predict the relative frequencies in which all possible triplet codons would be formed in poly(UC) used in the system.
***b.*** According to the standard genetic code, predict the kinds of amino acids that would be incorporated into the polypeptide and the relative proportions of each.

**7.16** The amino acid sequence of an enzyme segment has been shown to be: —Met–Lys–Ser–Pro–Ser–Leu–Asp–Ala–Tyr— in a wild-type strain. After mutation induction, the same enzyme segment in the mutant is found to be: —Met–Lys–Ser–His–His–Leu–Met–Leu–Thr—.
***a.*** If the orignal DNA sequence encoding this polypeptide segment were 3′—TAC TTT TCA GGT AGT GAA CTA CGA ATG G—, what was the nature of the mutation leading to the mutant amino acid sequence?

***b.*** What particular change in the mutant DNA would restore all of the original codons except for the codon specifying amino acid 4 (His)?

***7.17*** An analysis of a protein shows that four different mutants contain an amino acid substitution at the site normally occupied by glycine. The origin of these mutants is

$$\text{Gly} \begin{array}{l} \nearrow \text{Val} \rightarrow \text{Met-1} \\ \searrow \text{Arg} \rightarrow \text{Met-2} \end{array}$$

Assign the respective RNA codons to each mutant and to the wild-type strain.

***7.18*** A peptide was isolated containing only four amino acids: valine, alanine, serine, and histidine. After partial degradation, three kinds of dipeptides were recovered: Val-Ala, Val-His, and Ser-His. What is the amino acid sequence in the peptide?

***7.19*** Crosses were made between a known, standard wild-type strain 1 and two other wild-type strains in order to identify possible pseudowilds. The results were

strain 1 × strain 2: all wild-type progeny

strain 1 × strain 3: 80% wild-type progeny:20% mutant progeny

strain 2 × strain 3: 80% wild-type progeny:20% mutant progeny

***a.*** Is strain 2 or strain 3 the pseudowild?

***b.*** How could a true wild type arise from a mutant strain, in contrast with the origin of a pseudowild strain?

***7.20*** Diagram the pattern of bands produced in a gel after electrophoresis of the single-stranded DNA sequence 3′—G C C A T T T A A C G T—5′, if the Maxam-Gilbert method is used.

***7.21*** With the use of the Maxam-Gilbert procedure for DNA sequencing, a particular restriction fragment was treated to identify the bases in 5′-labeled preparations of part of one strand of a duplex known to encode amino acids. The results obtained after gel electrophoresis are shown at the top of the next column.

***a.*** What is the 5′ → 3′ sequence of the five triplets of bases?

***b.*** Is the sequence in (a) for the coding or the noncoding strand? Explain.

***c.*** What is the advantage of labeling the 5′ end of the strand with $^{32}P$?

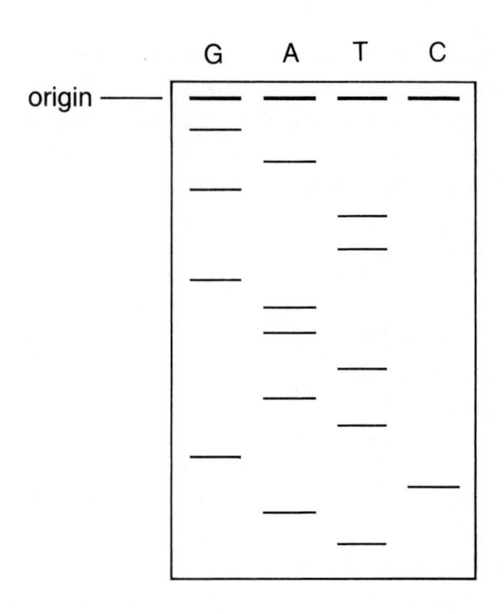


***7.22*** The circular duplex DNA plasmid p999 has one site each for the restriction enzymes *Eco*RI, *Sal*I, and *Hpa*I. After the circular DNA was linearized by digestion with *Eco*RI, these linear duplexes were digested further and fragments produced were as follows:

| **enzymes in mixture** | **fragment size (kbp)** |
| --- | --- |
| *Eco*RI | 4.5 |
| *Sal*I | 2.5, 2.0 |
| *Hpa*I | 3.4, 1.1 |
| *Sal*I, *Hpa*I | 2.5, 1.1, 0.9 |

***a.*** Draw a linear map showing the *Sal*I and *Hpa*I restriction sites relative to *Eco*RI.

***b.*** Draw the restriction map as a circle.

***7.23*** Suppose you discovered that the sequence of a coding region in a messenger RNA was 5′—U C G U C G U C G U C G U C G—3′. Specify the amino acid sequence in the polypeptide translation product if the following were the case:

***a.*** The codons were read three at a time in consecutive order.

***b.*** Three overlapping reading frames were possible.

***7.24*** You have isolated and purified the coding strand of a gene and the mature mRNA complementary to this DNA strand. After hybridizing the DNA and mRNA to obtain heteroduplexes, you note that electron micrographs of some of these heteroduplexes, made up of 2000 nucleotides of DNA and 200 nucleotides of mRNA, appeared as shown at the top of the next column.

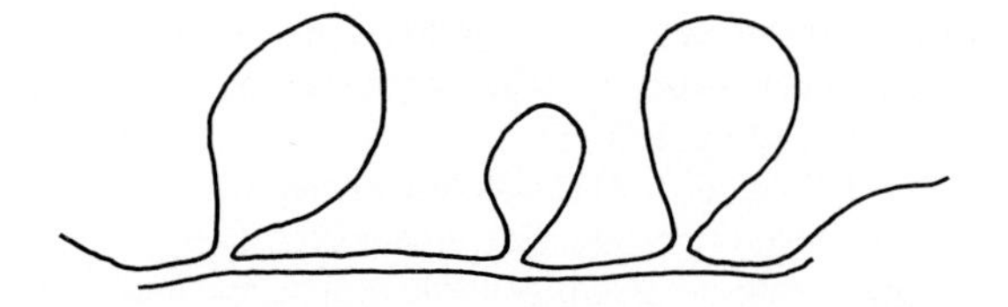

*a.* Was this gene isolated from a prokaryotic or a eukaryotic organism?
*b.* How many intervening sequences and how many coding regions (exons) occur in the DNA segment shown?
*c.* What events or processes were responsible for the shortened length of 200 nucleotides in the messenger copy of the gene sequence?

***7.25*** Plasmids carrying *pen-r* and *tet-r* genes were exposed to a restriction enzyme that excised the region carrying *pen-r* but left *tet-r* intact. The plasmid digest was mixed with genomic restriction fragments from a cloned library of mouse DNA to produce recombinant DNA, which was used to transform drug-sensitive bacterial cells. The transformed cells were plated on selective media to identify those bacterial colonies carrying recombinant DNA, with the following results:

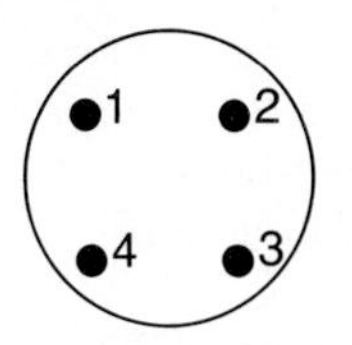


nutrient media
+ tetracycline

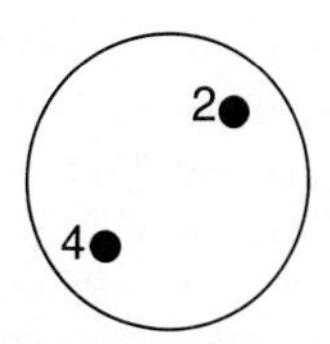


nutrient media
+ penicillin

*a.* Which colonies carry recombinant DNA? Explain.
*b.* Which colonies would grow on media containing penicillin and tetracycline? Explain.

## References

Anderson, W.F., and E.G. Diakumakos. July 1981. Genetic engineering in mammalian cells. *Sci. Amer.* **245**:106.

Beadle, G.W., and E.L. Tatum. 1941. Genetic control of biochemical reactions in *Neurospora*. *Proc. Nat. Acad. Sci. U.S.* **27**:499.

Benzer, S. 1961. On the topography of the genetic fine structure. *Proc. Nat. Acad. Sci. U.S.* **47**:403.

Chambon, P. May 1981. Split genes. *Sci. Amer.* **244**:60.

Chargaff, E. 1951. Structure and function of nucleic acids as cell constituents. *Fed. Proc.* **10**:654.

Fiers, W., et al. 1978. Complete nucleotide sequence of SV40 DNA. *Nature* **273**:113.

Garen, A. 1968. Sense and nonsense in the genetic code. *Science* **160**:149.

Gilbert, W. 1981. DNA sequencing and gene structure. *Science* **214**:1305.

Gilbert, W., and L. Villa-Komaroff. Apr. 1980. Useful proteins from recombinant bacteria. *Sci. Amer.* **242**:74.

Ingram, V.I. Jan. 1958. How do genes act? *Sci. Amer.* **198**:68.

Itakura, K. 1980. Synthesis of genes. *Trends Biochem. Sci.* **5**:114.

Khorana, H.G. 1979. Total synthesis of a gene. *Science* **203**:614.

Maxam, A.M., and W. Gilbert. 1977. A new method for sequencing DNA. *Proc. Nat. Acad. Sci. U.S.* **74**:560.

Meselson, M., and F.W. Stahl. 1958. The replication of DNA in *E. coli*. *Proc. Nat. Acad. Sci. U.S.* **44**:671.

Mirsky, A.E. June 1968. The discovery of DNA. *Sci. Amer.* **218**:78.

Motulsky, A.G. 1983. Impact of genetic manipulation on society and medicine. *Science* **219**:135.

Nathans, D. 1979. Restriction endonucleases, simian virus 40, and the new genetics. *Science* **206**:903.

Nirenberg, M.W., and J.H. Matthaei. 1961. The dependence of cell-free protein synthesis in *E. coli* upon naturally occurring or synthetic polyribonucleotides. *Proc. Nat. Acad. Sci. U.S.* **47**:1588.

Novick, R.P. Dec. 1980. Plasmids. *Sci. Amer.* **243**:102.

Reddy, V.B., et al. 1978. The genome of simian virus 40. *Science* **200**:494.

Sanger, F., et al. 1977. Nucleotide sequence of bacteriophage φX174 DNA. *Nature* **265**:687.

Sarabhai, A.S., et al. 1967. Colinearity of gene with the polypeptide chain. *Nature* **210**:14.

Smith, H.O. 1979. Nucleotide sequence specificity of restriction endonucleases. *Science* **205**:455.

Smith, M. 1979. The first complete nucleotide sequencing of an organism's DNA. *Amer. Sci.* **67**:57.

Watson, J.D. and F.H.C. Crick. 1953. A structure for deoxyribose nucleic acid. *Nature* **171**:737.

Watson, J.D. and F.H.C. Crick. 1953. Genetical implications of the structure of deoxyribose nucleic acid. *Nature* **171**:964.

Wetzel, R. 1980. Application of recombinant DNA technology. *Amer. Sci.* **68**:664.

Wilkins, M.H.F., A.R. Stokes, and H.R. Wilson. 1953. Molecular structure of deoxypentose nucleic acids. *Nature* **171**:738.

Woo, S.L.C., et al. 1981. Complete nucleotide sequence of the chicken chromosomal ovalbumin gene and its biological significance. *Biochemistry* **20**:6437.

Yanofsky, C. May 1967. Gene structure and protein structure. *Sci. Amer.* **216**:80.

Yanofsky, C., et al. 1967. The complete amino acid sequence of the tryptophan synthetase A protein ($\alpha$ subunit) and its colinear relationship with the genetic map of the A gene. *Proc. Nat. Acad. Sci. U.S.* **57**:296.

# CHAPTER 8

# The Molecular Nature of the Genome

With the development of molecular concepts of the gene and with the introduction of molecular methods for analyzing the genetic material, it became possible to study various properties of genes. One of the important properties of genes is the ability to replicate, or make exact copies, of the nucleotide sequence. DNA replication also is involved in recombination processes, in which parts of different DNA molecules are exchanged and a new sequence of nucleotides is produced. Studies of DNA replication and recombination had to be related to complex eukaryotic chromosomes as well as to the molecular behavior of naked DNA molecules that constitute prokaryotic and viral genomes.

## DNA Replication

Arthur Kornberg has identified basic rules governing **DNA replication** in all life forms. These rules include the following six:

***1.*** Replication is a semiconservative process.

***2.*** Both strands of the duplex replicate by addition of nucleotide monomers in the $5' \rightarrow 3'$ direction.

***3.*** Replication is semidiscontinuous for the most part, occurring relatively continuously on one strand (the leading strand) and discontinuously on the other (the lagging strand). The fragments thus produced later join the main body of the growing chains.

***4.*** Replication in short fragments requires initiation with a short piece of RNA primer for subsequent polymerization in DNA chain growth.

***5.*** Replication starts at a unique point, called the origin. There may be one or more origins in a DNA molecule.

***6.*** Replication proceeds from an origin in one or in both directions sequentially to the terminus, usually in both directions away from an origin.

Each of these basic rules has emerged from a variety of studies conducted during the past thirty years, beginning with the original suggestion by Watson and Crick that each strand of the DNA duplex might act as a template for making a new complementary partner strand. According to this proposal, two identical DNA duplexes would be produced from one parental duplex. The demonstration of the validity of this proposed mode of replication came within five years of Watson and Crick's suggestion in 1953.

The rest of the story is still unfolding. An ever-increasing number of genes and gene products have been found to act in DNA replication, each one required for a specific action in DNA synthesis along a template strand. The current view includes a complicated set of processes, all or most of which are set in motion by the requirement for high fidelity in making accurate, exact copies of all the genes in the genome, in every cell generation, in every organism throughout time.

### 8.1 Semiconservative Replication

One of the compelling features of the Watson-Crick molecular model for DNA was their proposal for DNA replication by a **semiconservative** mode, in which each partner strand of the duplex acts as a template guiding synthesis of a new complementary partner strand. Each of the two new duplexes consists of one parental strand and one newly synthesized strand—that is, the original duplex is partly conserved, or semiconserved, in the new duplex molecule. This idea could be experimentally tested, thus making the idea even more viable.

Before experimental supporting evidence was obtained for semiconservative replication, however, two other possible modes of replication had been proposed, namely, conservative replication and dispersive replication. In **conservative replication** the original duplex remains intact—that is, it is entirely conserved, and the whole duplex guides the synthesis of a new duplex replica of itself. In *dispersive replication* bits and pieces of newly synthesized DNA become assembled with bits and pieces of the original duplex to reconstitute two duplexes from the original template molecule (Fig. 8.1).

In 1958 Matthew Meselson and Franklin Stahl provided elegant and very convincing evidence in support of the semiconservative mode of DNA replication. The experiments were designed so that the results would match the set of predictions for only one of the three possible replication modes. Their experimental system involved a new method for high-speed centrifugation of DNA, called **equilibrium density-gradient centrifugation.** DNAs were placed in a solution of cesium chloride (CsCl) that was centrifuged to equilibrium, thus creating a density gradient of CsCl. DNAs of different densities settled at equilibrium in that part of the CsCl density gradient that corresponded to their buoyant density in CsCl (Fig. 8.2). DNAs were labeled by incorporating ordinary nitrogen

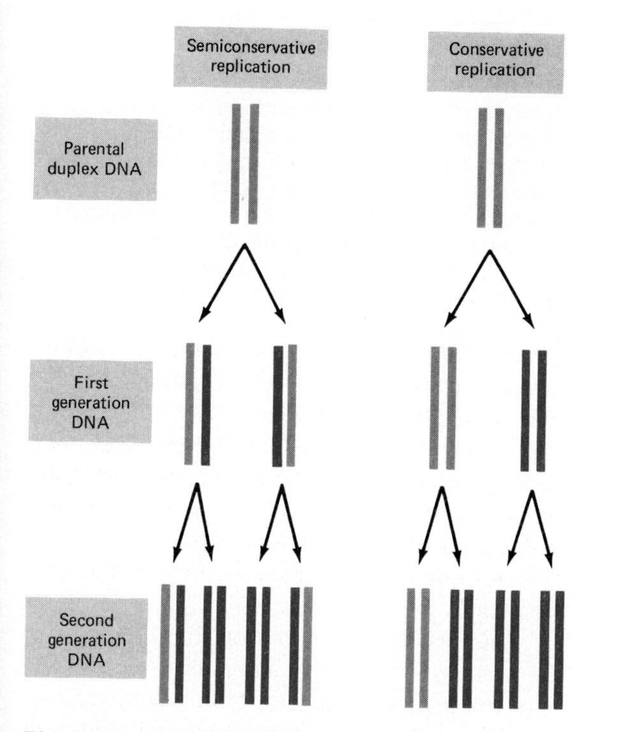


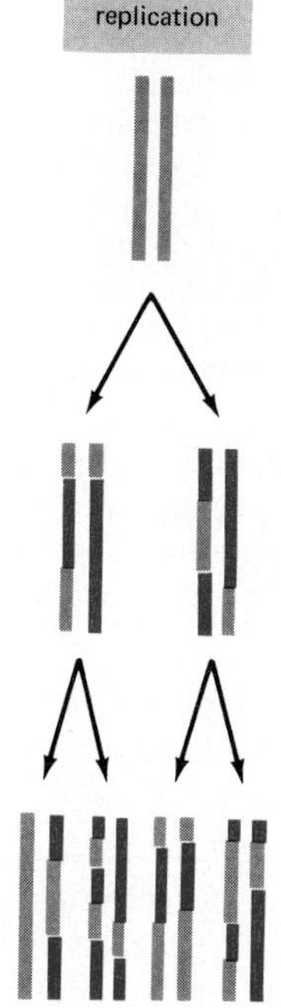


***Figure 8.1*** Three possible modes of duplex DNA replication. The predicted distribution of original parental strands (gray) and newly synthesized strands (color) are shown for two rounds of replication.

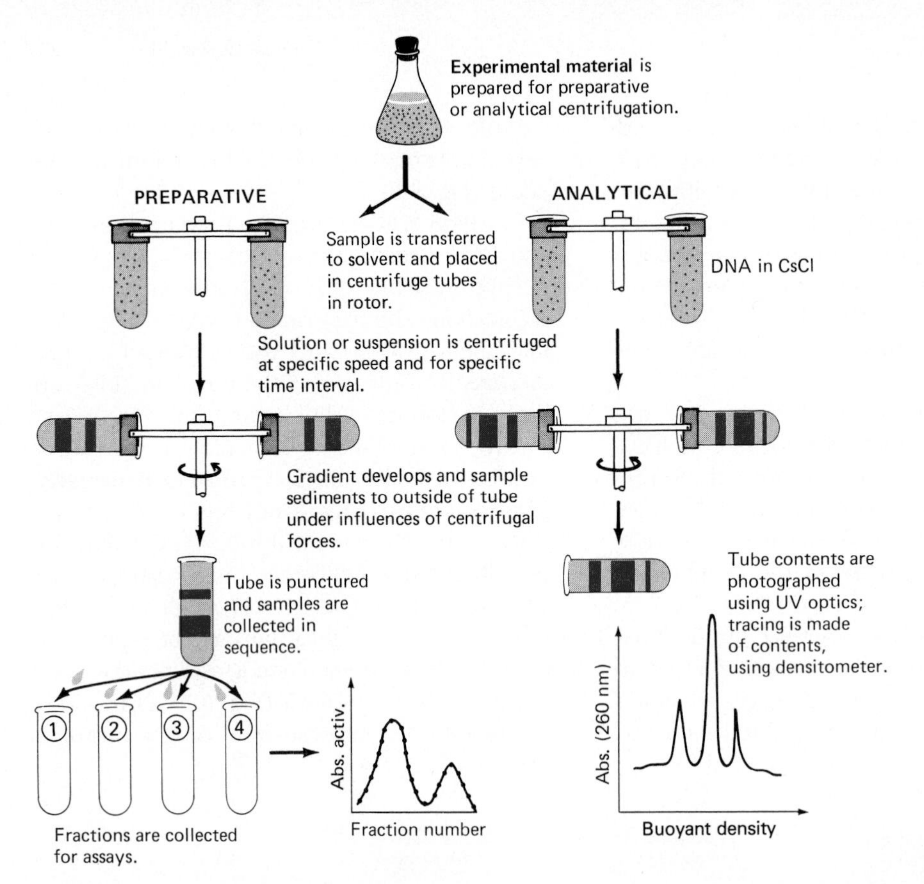


***Figure 8.2*** Equilibrium density-gradient centrifugation. Samples of experimental material can be sedimented in a preparative centrifuge, from which the fractions can be collected for analysis by vârious means. Alternatively, components may be centrifuged to the point of equilibrium in the density gradient in an analytical ultracentrifuge system and then processed directly in the centrifuge tube or cell. DNAs in different regions of the density gradient of CsCl are visualized by photographs showing UV absorption bands (shown here as colored bands) from which tracings are made using a densitometer. Each band or its trace corresponds to DNA of a particular buoyant density in CsCl, reflecting the G-C content or differences in amount of $^{15}$N incorporated into the molecules.

($^{14}$N) or the **heavy isotope** $^{15}$N at specified times during the replication cycle in *E. coli* cultures. Labeled DNAs could then be distinguished by their densities in CsCl gradients after high-speed centrifugation.

The *E. coli* cultures had been grown in nutrient media containing the heavy isotope $^{15}$N, so that both strands of all their DNA were $^{15}$N-labeled. When the experiment began, these [$^{15}$N] cells were transferred to media containing ordinary $^{14}$N. After one generation, during which cells and their DNA had doubled, the DNA was extracted and centrifuged to equilibrium in CsCl. Molecules with $^{15}$N were heavier than [$^{14}$N] DNA and settled at equilibrium in the part of the gradient that corresponded to their buoyant density in CsCl. During cell doubling, only $^{14}$N was available for synthesis of new DNA (and of other substances) since the bacteria had been transferred to unlabeled media at the start of the experiment. Any original DNA or parts of such DNA would retain the $^{15}$N label, which had been incorporated before the experimental doubling interval.

The predictions, based on the three modes of replication, were that amounts of $^{15}$N and $^{14}$N present in first-generation DNA would vary according to the mode of replication. Because all the DNAs would have different and non-overlapping buoyant densities in CsCl, their N contents could be determined by their specific positions in the gradient. If replication were semiconservative, all first-generation DNA duplexes would have one strand of original $^{15}N$-containing bases; the other strand would be newly synthesized from the $^{14}N$-containing precursors present in the unlabeled medium. Such $^{15}N$-$^{14}N$ duplexes would be half-heavy (half as much $^{15}N$ as in fully labeled DNA). According to conservative replication, 50% of the molecules would be fully labeled, or heavy ($^{15}N$-$^{15}N$), representing the conserved original duplexes, and 50% would be unlabeled, or light ($^{14}N$-$^{14}N$), representing the new duplexes made in unlabeled medium. Dispersive replication was predicted to produce DNA with varying amounts of original [$^{15}N$] bases and newly made [$^{14}N$] bases, depending on how the bits and pieces of original and new DNA might assemble in a duplex.

When DNA from first-generation cells was centrifuged, the contents of the centrifuge tubes were photographed with ultraviolet optics to make DNA bands visible in the photographic emulsion. The position of a DNA band in the gradient corresponded to its N content, and the results clearly showed a single band of DNA in the position expected for half-heavy duplexes (Fig. 8.3).

These results ruled out conservative replication but did not discriminate between semiconservative and dispersive replication. After continuing the experiments for four generations of growth, or doublings, the only reasonable interpretation of all the data was that DNA replicated semiconservatively. The predictions for this replication model were fulfilled *in every cell generation,* and the two other possibilities were eliminated. Such instances of beautifully clear experimental design, data, and conclusions are realized all too rarely, and we appreciate them all the more when they happen.

One year before the Meselson and Stahl experiments, a report by J.H. Taylor, P. Woods, and W. Hughes showed that semiconservative replication characterized chromosomal DNA in eukaryotic cells. Cells were grown in media containing thymidine labeled with the radioactive isotope **tritium,** or $^{3}H$. Since thymidine is found uniquely in DNA, original and newly made DNA molecules could be located and identified in chromosomes of these cells by the presence or absence of radioactivity in **autoradiographs** (Box 8.1).

Cells were first labeled with [$^{3}H$] thymidine by incubation in media containing this DNA precursor. Labeled cells were then transferred to a medium containing unlabeled thymidine (with ordinary $^{1}H$) and were allowed to undergo one cycle of doubling. The cells were removed to glass slides and coated with a photographic emulsion for subsequent autoradiography using the light microscope. The emulsion was developed to reveal any silver grains, which indicated a radioactive decay event at that site in the cell underneath, and the chromosomes were stained to enhance contrast. Silver grains were found along one chromatid of each replicated chromosome and were absent from the sister chromatid almost entirely (Fig. 8.4).

The pattern of silver grains conformed to the predictions for semiconservative replication since each chromosome was half-labeled. If replication had been conservative, 50% of the chromosomes would have been labeled along both chromatids and 50% of the chromosomes would not have been labeled. This was not found to be the case. Apparently, eukaryotic and prokaryotic DNA replicate by the same mode.

In 1958 Taylor further showed that semiconservative replication characterized DNA in meiotic cells. Within five years of the original suggestion by Watson and Crick, therefore, semiconservative DNA replication had been successfully demonstrated in bacteria and in both mitotic and meiotic cells of eukaryotes.

These and other experiments showed that progeny DNA duplexes were semiconserved, but they did not show *how* the new strands were made and incorporated into new molecules. The processes of DNA replication have been a focus of investigation for more than thirty years. Some of the highlights of these studies will be discussed in the next sections.

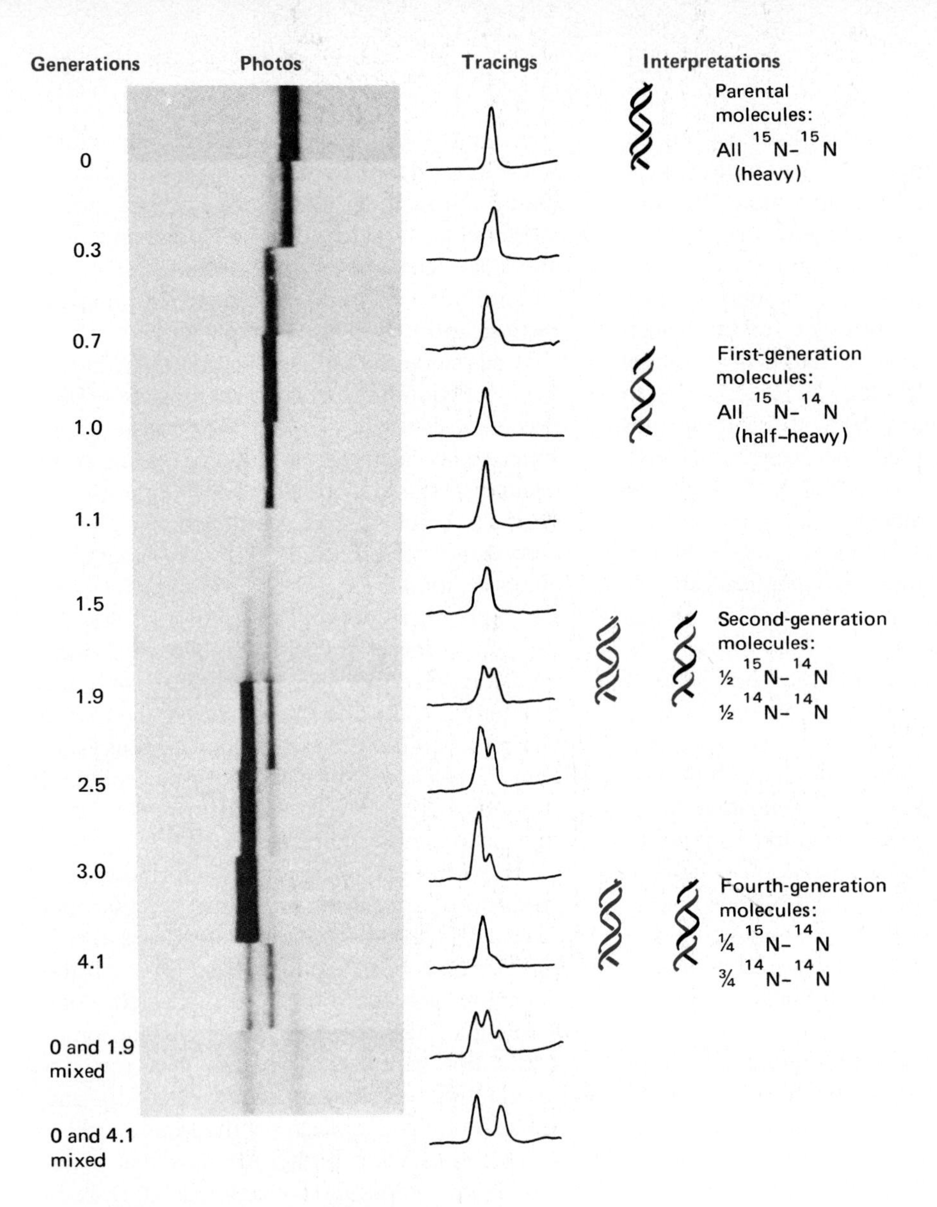


***Figure 8.3*** Experimental demonstration of semiconservative replication of *E. coli* DNA. Ultraviolet absorption photographs from analytical centrifugation and densitometer tracings of absorption bands show distribution of DNA (bands) in samples monitored at different times in the experiments. Cells were labeled initially with $^{15}N$ and grown afterward in $^{14}N$-containing medium for the number of generations indicated. The interpretations of the kinds of DNA and amounts of each (at right), according to centrifugation data, fit the predictions made for semiconservative replication of DNA. (From M. Meselson and F.W. Stahl. 1958. *Proc. Nat. Acad. Sci. U.S.* **44**:671.)

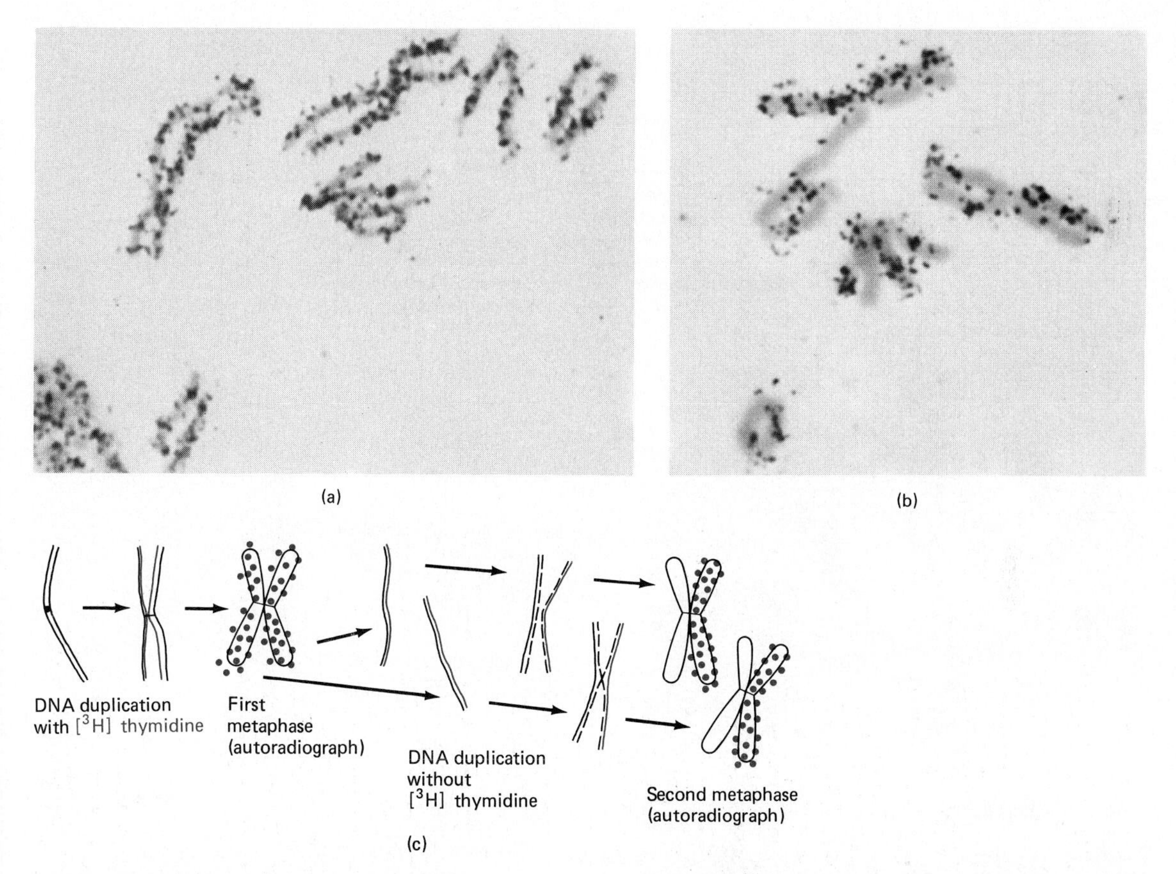


***Figure 8.4*** Semiconservative replication of eukaryotic chromosomes. (a) Autoradiograph of metaphase chromosomes of the bean *Vicia faba* after one DNA replication cycle in the presence of [$^3$H] thymidine. (b) Autoradiograph of similar root tip nucleus after one replication cycle in the presence of [$^3$H] thymidine and a second cycle in the absence of the labeled DNA precursor. (c) Distribution of DNA strands and of metaphase chromosomes during two cycles of replication. $^3$H-labeled strands are shown in color and colored dots represent the silver grains of autoradiographs such as shown in (a) and (b). Solid and dashed black lines represent unlabeled strands. (From J.H. Taylor. 1963. *Molecular Genetics,* Part I, pp. 74–75, New York: Academic Press.)

## 8.2 Synthesis of New Strands

The building blocks incorporated into DNA are deoxyribonucleoside 5′-*mono*phosphates. The active precursors are deoxyribonucleoside 5′-*tri*phosphates. In the triphosphate form, the units are added onto a growing chain and pyrophosphate is released in a hydrolysis reaction that drives the synthesis to completion (Fig. 8.5). The overall direction of chain growth is 5′ → 3′, with the addition of the nucleoside triphosphate at the 3′-hydroxyl end of the chain.

The two complementary strands of the DNA duplex are antiparallel, and chain growth proceeds along each template strand only in the 5′→ 3′ direction. Growth of both new complementary strands in a replicating duplex, therefore, must proceed in *opposite* directions from one another. It has been shown, with labeled units, that the triphosphate precursors are only added at the 3′ end of a chain. The labeled units are found only at the 3′ ends of the new chains, which means they were added most recently during synthesis. The replication enzyme, or

## BOX 8.1 Autoradiography

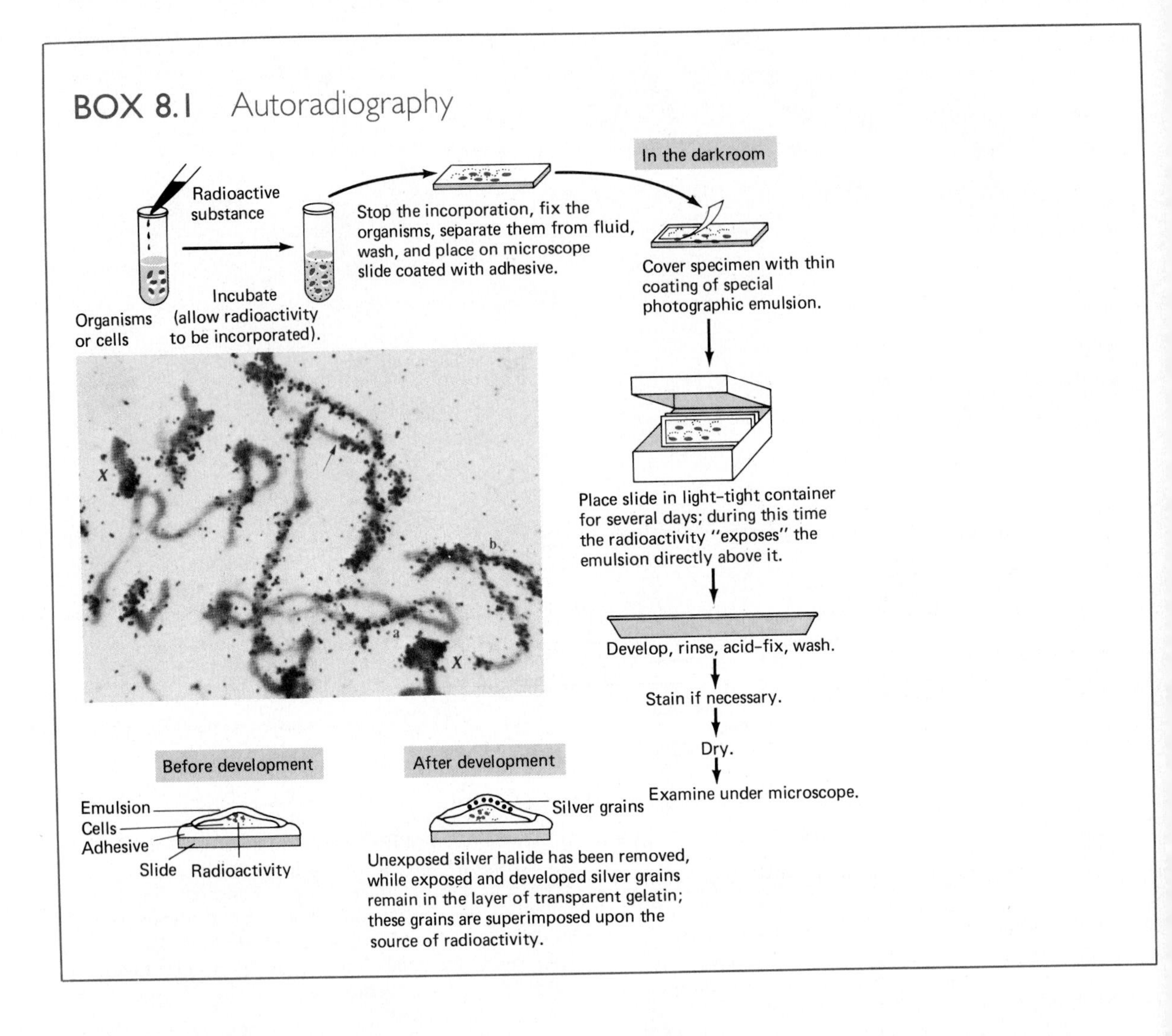


**DNA polymerase,** is only capable of adding a new monomer onto the 3′ free end of the growing chain, which further verifies the 5′ → 3′ direction of chain growth in all systems that have been studied.

Since the antiparallel chains are growing in opposite directions, their synthesis may not be simultaneous or synchronous, that is, each chain may grow by independent steps although the same enzymes catalyze all these steps. On the other hand, the DNA strand may be synthesized in short pieces that are later assembled into lengthening stretches, until both chains of a replicating duplex are completed and two daughter replica duplexes have been made from one parental double helix (Fig 8.6).

This sketchy picture of DNA replication does little justice to the actual processes involved. All these replication events require specific proteins and enzymes, and all cooperate to produce an accurate pair of replicas. How is all this possible? What are these components and how is accuracy ensured?

The first of the DNA polymerases to be identified was **DNA polymerase I,** called the "Kornberg enzyme," after its discoverer. Al-

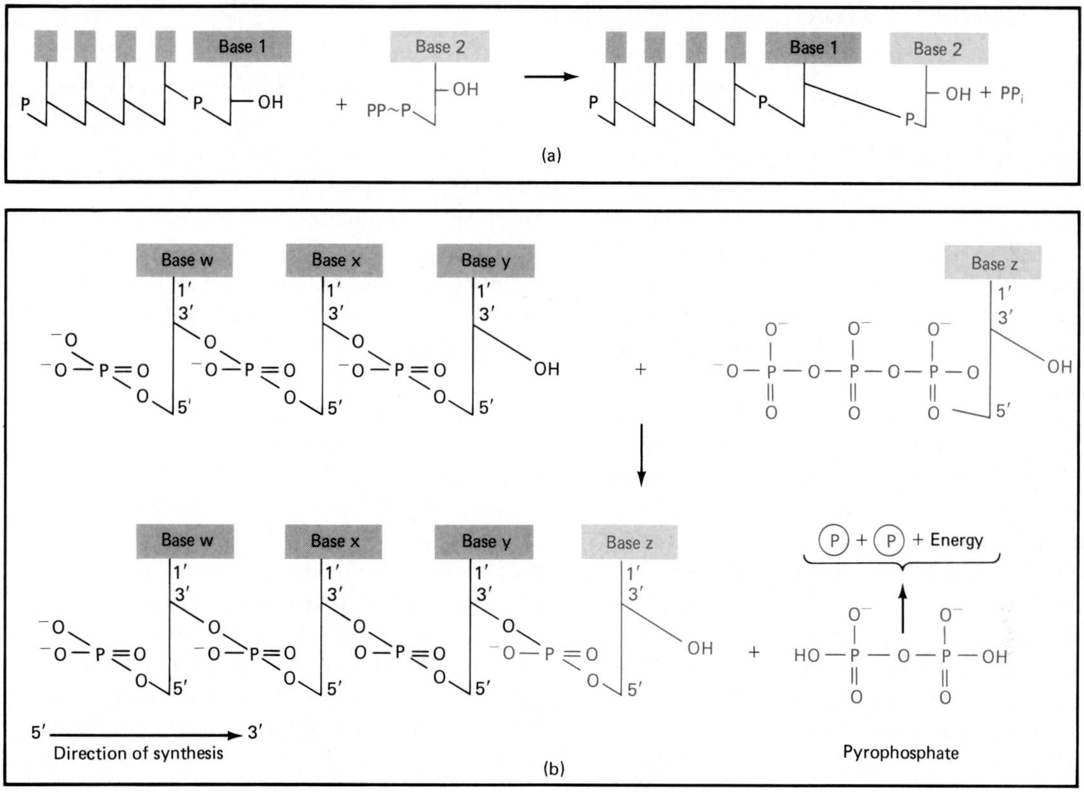

***Figure 8.5*** DNA synthesis. (a) New strand synthesis proceeds from 5′ to 3′, through the addition of mononucleotides derived from triphosphate precursors. (b) Detail of a step in synthesis showing the nucleoside 5′-phosphate precursor (color) and the addition of a mononucleotide monomer. Release of pyrophosphate drives the reaction to completion.

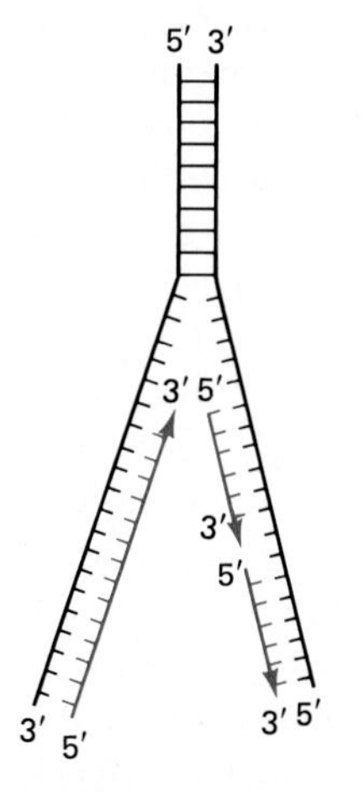


***Figure 8.6*** Replicating region of duplex DNA. We do not know whether both new strands are synthesized synchronously or not, or continuously versus discontinuously (in short pieces). We do know that two semiconserved duplexes arise from one parental duplex.

though the enzyme was shown to catalyze DNA synthesis *in vitro,* its products were often branched chains, whereas linear, unbranched DNA strands are the only form made in living systems. This enzyme was therefore considered to function in some way in replication, but it was not believed to be the main replicating enzyme because of its peculiar properties in test systems. The search for other DNA polymerases was frustrated by the dominating presence and activity of DNA polymerase I, which masked other enzymes and made them difficult to isolate. The situation was improved with the discovery of polymerase-deficient or -defective mutants in *E. coli.*

In 1969 John Cairns described the *polA*$^-$ mutant he had discovered in *E. coli.* The mutant synthesizes DNA at normal rates even though DNA polymerase I activity is barely detectable.

Using *polA*$^-$ mutants, a search was begun for other DNA polymerases, which had to be present since replication was unimpaired in *polA*$^-$ strains.

Polymerase I activity is so high in wild-type cells that other polymerase activities are masked. But *polA*$^-$ mutants do not present this problem. They were found to have **DNA polymerase II** and, later, also to have **DNA polymerase III.** DNA polymerase II was ruled out as the main replicating enzyme when *polB*$^-$ mutants were isolated and shown to synthesize DNA at normal rates despite a deficiency of this enzyme. Studies of *polC*$^-$ mutants, which are deficient in DNA polymerase III, have provided some evidence for this as the main replicating enzyme in *E. coli.* But DNA polymerase I also fulfills a vital function in DNA chain growth since mutants that totally lack this enzyme have never been found. Its function can be better understood when the nature of discontinuous chain growth is examined; it acts in reactions that lead to joining between the short pieces of DNA made by DNA polymerase III reactions. *It fills in gaps.*

R. Okazaki provided evidence that DNA synthesis might be discontinuous, proceeding by bursts of synthesis of small pieces of DNA, about 1000 to 2000 nucleotides long. Fragments of this size were isolated from mutants that were unable to hook these pieces together to make whole chains. When such *E. coli.* mutants were presented with short pulses of labeled precursor at low temperature, almost all the label was found in small pieces of DNA. Discontinuous DNA synthesis might characterize either one or both growing strands. Evidence shows that newly synthesized, single-stranded fragments bind to *both* parental template strands to an equal extent. Such highly specific binding between polynucleotides is possible if they are complementary to each other and join by complementary base pairing. These replication fragments are often called **Okazaki fragments,** after their discoverer.

The short pieces of DNA join together in wild-type *E. coli,* which indicates that an enzyme must link Okazaki fragments to each other and to the growing chain. The enzyme **DNA ligase** fulfills this function. It catalyzes the formation of phosphodiester links that make the chain continuous by splicing the sugar-phosphate backbone. The mutants used by Okazaki to show that synthesis was discontinuous were ligase-deficient and could not join the pieces together, so the pieces accumulated and could be isolated and analyzed.

The *initiation* of synthesis for each Okazaki fragment cannot be accomplished by DNA polymerases. This finding created a problem for a time, until it was discovered that chain growth was initiated or "primed" by the addition of a short stretch of RNA, catalyzed by an **RNA polymerase.** Some of these RNA polymerases have been called **primases,** but these priming enzymes vary considerably among different systems. Once primed, DNA polymerase III catalyzes the addition of DNA monomers to a short sequence of **RNA primer,** and an Okazaki fragment is made. The same events would take place for each burst of discontinuous synthesis along both growing chains in a replicating duplex (Fig. 8.7).

However, since newly made DNA duplexes do not contain RNA segments, the RNA primers must be excised from the Okazaki fragment. This leaves a gap, which must be filled in with DNA monomers. DNA polymerase I appears to be the catalyst in these reactions. The polymerase has an **exonuclease,** or digesting, function as well as a polymerase function. It can start at a free 3′ end and digest away the RNA primer in the $3' \rightarrow 5'$ direction. It can then catalyze DNA monomer additions, $5' \rightarrow 3'$, to fill in the gap. The same polymerase I is also believed to fill in gaps between Okazaki fragments and make the pieces continuous. The gap-filling activities occur throughout the period of synthesis, so chains are lengthened step by step and not all at once after the whole complementary chain has been made.

DNA polymerase I has another important function that helps to ensure accuracy of chain synthesis. It *"proofreads"* the growing chain fragments and can excise inaccurate monomer additions since it has the ability to work in the 3′

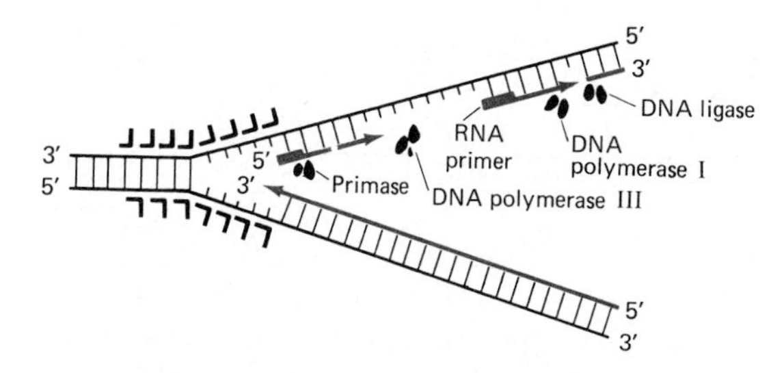


***Figure 8.7*** Overview of DNA strand synthesis at the replication fork. Synthesis is initiated by activity of an RNA polymerase, or primase, which guides production of a short segment of RNA primer. Once primed, synthesis of DNA is catalyzed by DNA polymerase III, and proceeds by addition of deoxyribonucleotides to the primer segment. The RNA primer segment is later excised and the gap is filled in with deoxyribonucleotides in reactions catalyzed by DNA polymerase I. The newly synthesized segment is then sealed into a continuous sugar-phosphate backbone by action of DNA ligase. Asynchronous, discontinuous synthesis of Okazaki fragments of the two new strands (color) is depicted in this diagram.

→ 5′ direction. If excision does occur through its exonuclease function, the DNA polymerase I can exercise its polymerase function and guide the addition of correct monomers in the 5′ → 3′ direction, according to the base sequence in the complementary template strand.

## 8.3 Proteins at the Replication Fork

A variety of catalytic and structural proteins have been associated with DNA replication, in addition to the polymerases and ligase already mentioned. Investigators have identified more than two dozen different genes that govern replication processes in phage T4, and more probably remain to be discovered even in a genome as small as the T4 system. The most dynamic events appear to take place at the **replication fork,** the site of separation of the two duplex strands and synthesis of new complementary partner strands.

Since 1970 Bruce Alberts has discovered a number of proteins that function at the replication fork in phage T4. These same classes of proteins were later found to act in prokaryotic and eukaryotic DNA replication processes. One of these proteins is not catalytic but binds tenaciously to single-stranded regions of replicating duplex DNA. It has been called the **single-strand binding (SSB) protein,** although it was known by other names in the earlier literature. Through the energy of binding, SSB proteins are able to shift the equilibrium of helically coiled DNA toward strand separation. Bound SSB proteins lead to unwinding of duplex DNA through the breaking of hydrogen bonds but without the breaking of covalent bonds.

SSB proteins bind to the sugar-phosphate backbone of DNA, so no sequence specificity is involved in its binding. The elongated shape of the protein permits it to bind tightly to single-stranded stretches of DNA about 10 nucleotides in length. Because protein-protein interactions are highly favored, SSB protein molecules bind cooperatively. They line up adjacent to one another, therefore, rather than at scattered sites on the duplex (Fig. 8.8). Once bound to DNA, SSB proteins lead to strand separation, or melting, of duplex regions. In addition, SSB proteins facilitate base-pair alignment between single strands during DNA replication and recombination.

DNA exists in *supercoiled* form as well as in the helix-coil conformation of the extended double helix. In fact, supercoiling must characterize the native state of long molecules of DNA packaged within the relatively tiny confines of a virus particle, bacterial cell, or eukaryotic chromosome. Replication must involve *relaxation* of **supercoiled DNA,** which is then made single stranded and kept taut and accessible by SSB proteins and other molecules.

Among the replication enzymes described in the 1970s are two kinds of **topoisomerases.** These enzymes were named with reference to their capacity as catalysts in interconversions of topological isomers of DNA. The first topoisomerase was discovered in *E. coli* by J.C. Wang in 1971. Enzymes of the same or similar function have since been found in other prokaryotes and in eukaryotes. The *E. coli* enzyme was shown to catalyze reactions of *nicking and closing* of du-

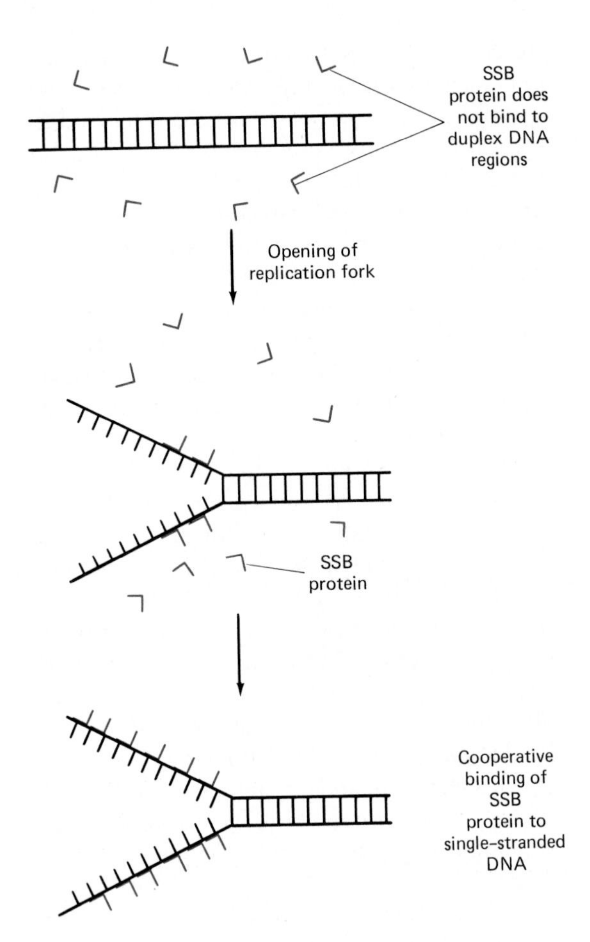


***Figure 8.8*** Action and effect of single-strand binding (SSB) protein (color) at the replication fork of duplex DNA. SSB proteins bind cooperatively to single-stranded stretches of DNA about ten nucleotides in length, this action aids unwinding of the double helix and facilitates the alignment of new monomers next to the template strand during DNA synthesis.

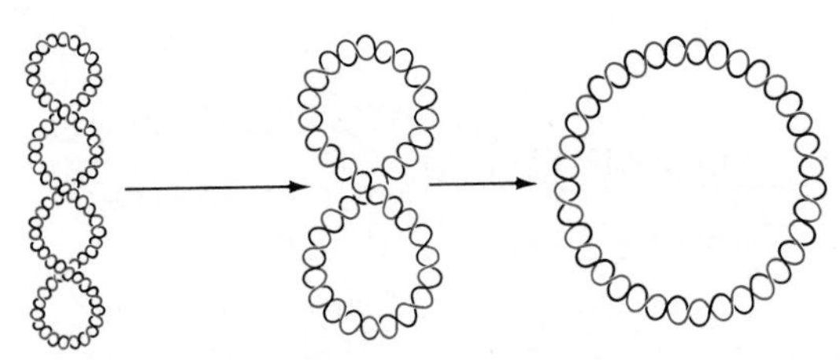

***Figure 8.9*** Untwisting of duplex supercoiled DNA involves nicking of the sugar-phosphate backbone of one strand (opening the strand) and closing of the nicked strand in reactions catalyzed by topoisomerase I. Relaxation of supercoiled duplex DNA takes place without change in coiling of individual strands around one another in the double helix.

plex DNA, which results in the untwisting, or unwinding, of supercoiled DNA (Fig. 8.9). In this way the duplex is prepared for binding with proteins that require a more relaxed topological form of DNA, such as single-strand binding proteins. This untwisting enzyme is one of a class of **topoisomerase I** enzymes, which relax supercoils. Enzymes belonging to the **topoisomerase II** class, colloquially called **gyrases,** can induce relaxed DNA to undergo supercoiling through nicking and closing reactions of the sugar-phosphate backbone of DNA.

Other catalytic and structural proteins have been identified with activities at the replication fork. One of these is a **helicase** that displays ATPase activity, that is, it catalyzes the hydrolysis of ATP and consequent energy release. The helicase snips away at hydrogen bonds in the duplex DNA region just ahead of the advancing replication fork (Fig. 8.10).

As emphasis continues to shift from studies of DNA chain growth to studies of the *initiation* of replication, more information undoubtedly will be revealed concerning the dynamic events at the growing replication fork. We know virtually nothing, however, about regulation of DNA replication. Numerous mutant strains have been isolated and used in assay systems for chain growth, but mutants defective in regulation of replication are virtually unknown. Thus, studies of replication control have lagged behind studies of the replication processes themselves.

## 8.4 Direction of Replication

J. Cairns provided the first experimental evidence in 1963 to show that DNA replication proceeded in a specific direction, from what is now called the **origin,** or site of initiation of replication. The particular origin has been mapped in some viruses (for example, at posi-

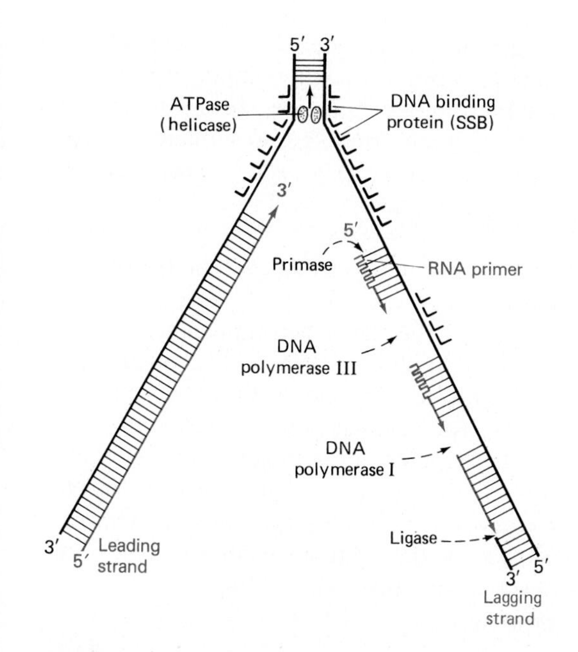


***Figure 8.10*** Dynamic events at the replication fork during synthesis of new duplex DNA molecules. Binding of single-strand binding (SSB) proteins to single-stranded regions aids unwinding of the duplex and also keeps strands taut and accessible to monomers and to enzymes involved in replication. A helicase with ATPase activity catalyzes reactions of unwinding ahead of the fork, and new strand synthesis proceeds under the guidance of various well-characterized enzymes. Asynchronous replication is shown here, in which events in the leading strand have been completed while the same set of discontinuous events is just occurring in the lagging strand. (Adapted from *DNA Replication* by A. Kornberg. Copyright © 1980 by W.H. Freeman and Company. All rights reserved.)

tion 67 in the 100-unit map of SV40) and bacteria (the 82-minute site on the 100-minute *E. coli* map).

Cairns grew *E. coli* in media containing [$^3$H] thymidine for one 30-minute generation period to label one strand of their DNA. He then allowed the cells to grow in the same labeled medium for varying lengths of time in a second 30-minute cycle before samples were removed and prepared for autoradiography. He expected label density to be twice as great where the molecule had undergone two rounds as compared to one round of replication (Fig. 8.11). In other words, some parts of the molecule had both strands labeled and other parts still had only one strand labeled in their semiconservatively replicating duplex DNA. The images of these circular replicating molecules resembled the Greek letter *theta,* and for this reason they are referred to as *θ-forms.*

These forms were interpreted as showing that replication proceeded directionally from an origin and continued around the circle until two semiconserved daughter duplexes were formed. Label densities served as markers to identify which parts of the molecule had replicated twice and which parts only once. Since two segments of the θ-form were singly labeled and the third segment was doubly labeled, all three segments were measured in molecules at different stages

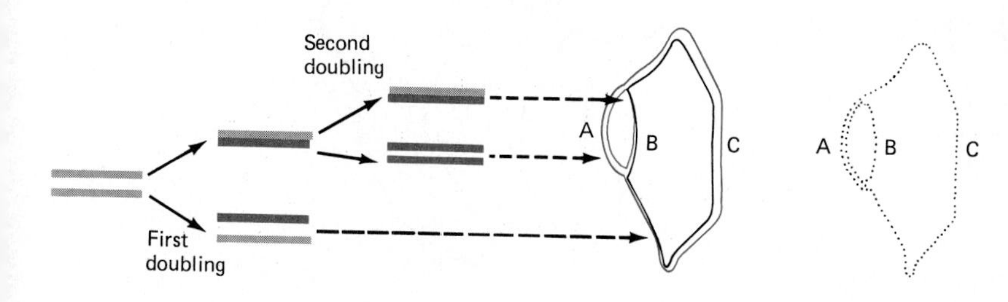


***Figure 8.11*** Distribution of silver grains in replicating circular DNA in *E. coli,* and interpretation of such autoradiographs according to semiconservative replication. Strands A and B have replicated a second time since the experiment started, strand C has only replicated once thus far.

of the second replication cycle (Fig. 8.12). In every case, two differently labeled segments were of equal length, whereas one of the singly labeled segments was longer in molecules just starting the second replication and became progressively shorter as this second cycle neared its completion. These measurements served as the basis for proposing that replication proceeded directionally around the circle, beginning at some point of origin on the molecule. In most viruses, bacteria, and eukaryotic systems that have been studied, replication proceeds in *both* directions away from the origin. There are two *replication forks* in each bidirectionally replicating "bubble" (Fig. 8.13).

There usually is one replication origin in viral and bacterial genomes and the entire genome can be considered a **replicon,** or unit of replication. Eukaryotes have many replicons per genome, as has been deduced from electron micrographs that show a number of replication "bubbles" along a single chromosomal DNA segment, each "bubble" presumably having a particular origin (Fig. 8.14). From electron micrographs and from studies of the rate of DNA synthesis the number of replicons per eukaryotic genome is estimated to be between 100 and 200.

One of the major accomplishments of biochemical studies of DNA replication has been the complete test-tube synthesis of functional, whole DNA genomes of several viruses. Not only are these molecules complete, but they start *in vitro* at the correct origin. These completely synthesized molecules can serve as templates *in vivo* for the accurate synthesis of new generations of functional, complete DNA viral genomes.

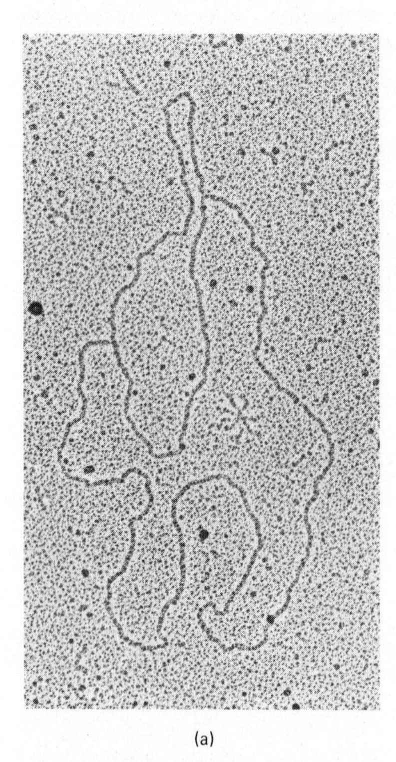

(a)

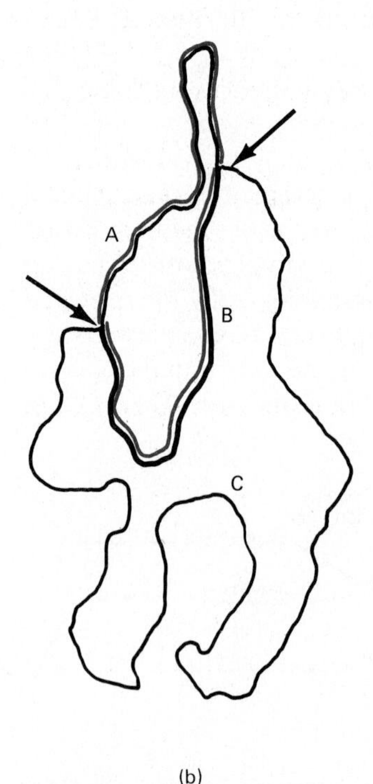


(b)

***Figure 8.12*** Replicating DNA. (a) Electron micrograph of a theta-form molecule and (b) interpretation showing unreplicated part C and replicated, equal-length parts A and B of the molecule. There are two replication forks (at arrows). × 56,000. (From D.R. Wolstenholme, et al. 1974. *Cold Spr. Harb. Symp. Quant. Biol.* **38** (1973):267.)

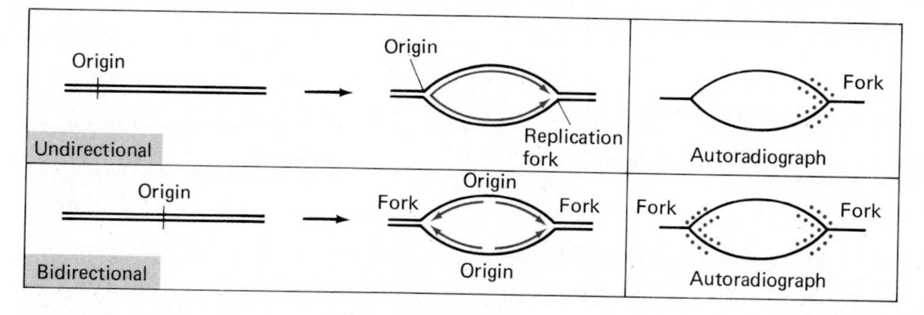


***Figure 8.13*** Expected labeling patterns, after incorporation of [$^3$H] thymidine for a few minutes following onset of DNA replication, reveal whether replication proceeds in one or in both directions away from the origin. Label would appear only in new DNA that had just incorporated the tritiated precursor.

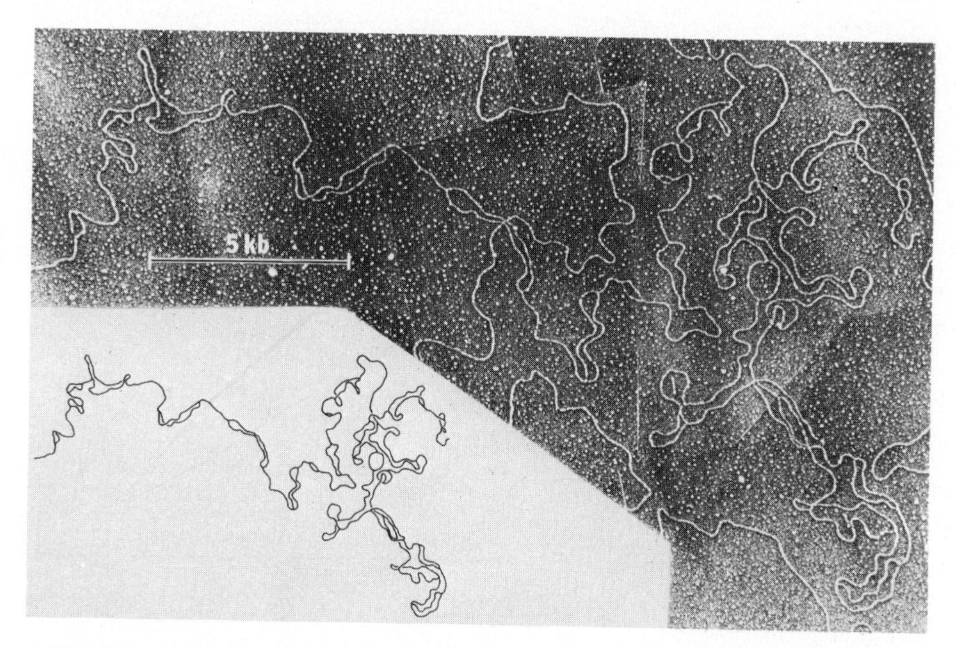


***Figure 8.14*** Electron micrograph and interpretive drawing of replicating DNA from *Drosophila melanogaster* nuclei. The portion of the molecule shown here is 119,000 base-pairs long and has 23 replication "bubbles." A kilobase (kb) is a unit of length equal to 1000 bases or base-pairs in single- and double-stranded nucleic acids, respectively. (From H.J. Kriegstein and D.S. Hogness, 1974. *Proc. Nat. Acad. Sci. U.S.* **71**:135, Fig. 1.)

## 8.5 Replication in Viruses

DNA phages may have single-stranded or double-stranded DNA, and some viruses have either single-stranded or duplex RNA genomes (Table 8.1). How do these genomes replicate? From many *in vitro* studies and from analysis of a number of mutants in either the virus or the host genetic system, it seems clear that there are a number of variations in replication. No single set of requirements serves all systems. To get some sense of the diversity among viruses, we will examine a few examples in sufficient depth to see their characteristics and their variations.

The DNA phage ϕX174 of *E. coli* is one of the best-studied single-stranded DNA systems. When the virus infects its host cell, the single-stranded molecule is converted to a double-stranded cir-

***Table 8.1*** Some characteristics of representative viruses.

| nucleic acid | virus | main host | comments |
|---|---|---|---|
| DNA | | | |
| single stranded | fd | *Escherichia coli* | |
| double stranded | T2, T4, T6 | *Escherichia coli* | |
| | P22 | *Salmonella typhimurium* | |
| | herpes simplex | human | type 1 causes "fever blisters" type 2 causes genital herpes (a venereal disease) |
| | variola | human | causes smallpox |
| | Epstein-Barr | human | causes infectious mononucleosis; associated with Burkitt's lymphoma |
| | cauliflower mosaic | cauliflower | transmitted by aphids |
| RNA | | | |
| single stranded | Qβ | *Escherichia coli* | |
| | tobacco mosaic | tobacco | |
| | polio | human | |
| | measles | human | |
| | mumps | human | |
| | influenza A, B, C | human | |
| double stranded | reovirus | human | causes mild illness of respiratory and GI tracts |
| | wound tumor | plants | transmitted by leafhoppers |

cular molecule called the *replicating form* (RF). The entering strand, which we will call the (+) strand, serves as a template for the synthesis of a complementary (−) strand, making the duplex RF. The duplex RF then acts as the template, in turn, to make new (+) single strands that are packaged in virus particles which are later released from the cell during lysis (Fig. 8.15). In this system, therefore, complementary base pairing serves the same function in ordering nucleotides into a precise sequence as it does in other replicating situations. The major difference is that only (+) strands are made, that is, only one strand of the duplex RF serves as a template for new chain synthesis.

Considerable variation exists among viruses in the particular enzymes that function in replication. Some virus-coded enzymes are essential in one species and dispensable in another species, or an enzyme coded by a host gene may function in replicating viral DNA.

In the mammalian virus SV40, duplex DNA is always present in combination with histone proteins, and it actually resembles a miniature version of a eukaryotic chromosome. Virus replication proceeds by reactions catalyzed exclusively by host cell enzymes; the virus does not code for its own replication components. The

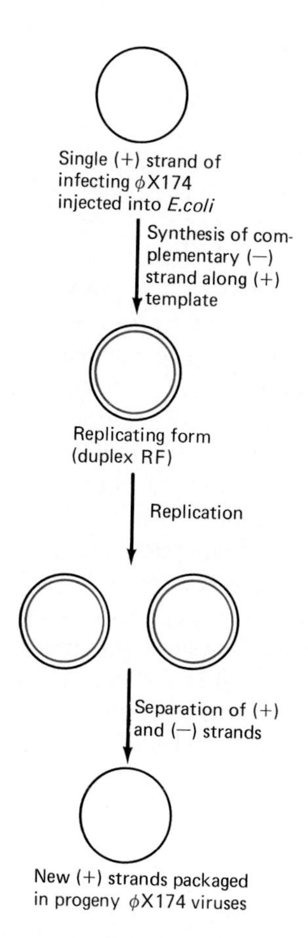


***Figure 8.15*** DNA replication of the single-stranded DNA phage $\phi$X174.

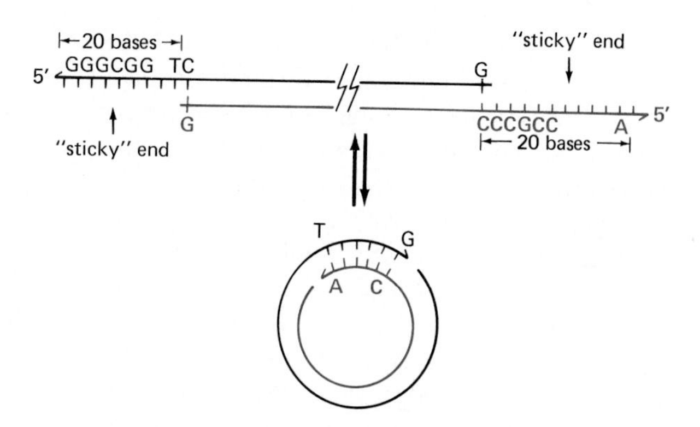


***Figure 8.16*** Interconvertibility of linear and circular forms of phage λ DNA.

circular viral nucleoprotein provides a model system to study eukaryotic DNA replication and chromosome organization. In double-stranded DNA phages, on the other hand, naked DNA molecules replicate semiconservatively to produce new naked duplex DNA.

The *E. coli* phage λ is known to occur inside the host cell as a circle during the lytic cycle but as a linear segment in the virus particle or when integrated into *E. coli* DNA while it is in its prophage state. Furthermore, λ DNA can be isolated in either circular or linear forms from *E. coli* cells. How can these two configurations be reconciled?

Apparently, the circular and linear forms of phage λ are interconvertible. The DNA circle can be linearized, or the linear molecule can be circularized (Fig. 8.16). This property of interconvertibility is made possible by the presence of "sticky" ends of the linear form, which can undergo complementary base pairing to constitute the circular form. The property is of immense importance in temperate phages such as λ since the molecule must be able to assume either configuration in different phases of the lytic cycle. If you will recall, recombination events that integrate or excise λ from the *E. coli* chromosome require circular and linear forms at different stages of the process (see Fig. 6.16) Circularity per se is not required for DNA replication, however, since phages such as T2 always exist in linear form and are never found as circles (see Section 6.13). These are only some of the variations that have been described for viruses with duplex DNA genomes.

RNA viruses also occur in double-stranded and single-stranded forms. The duplex RNA viruses apparently replicate through the activities of RNA polymerases, which are the replication enzymes in such systems. The most interesting cases, however, have been described for the single-stranded RNA tumor viruses, such as mouse mammary tumor virus and avian (Rous) sarcoma virus (ASV). The whole group of single-stranded RNA tumor viruses appears to be very closely related, and these viruses have been shown to cause tumors in many kinds of vertebrate animal species. In the case of ASV, one of the best-studied and a fairly typical virus in this group, the infective particle penetrates into the cell cytoplasm, where its RNA genome separates from its protein coat. The single-stranded RNA genome is then *transcribed into a complementary DNA strand* with the help of the virus-specific enzyme *reverse transcriptase*. This enzyme is an RNA-dependent DNA polymerase, that is, it catalyzes DNA strand synthesis from an RNA template (Fig. 8.17). The discovery of this enzyme by Howard Temin and David Baltimore came as quite a surprise. It led to the change in the dogma of information flow from the gene, from

DNA→RNA→protein

in cells and many viruses, to

DNA⇌RNA→protein

or

RNA→protein

in some viruses only.

The single-stranded DNA serves as the template for enzymatically catalyzed synthesis of a complementary DNA partner forming a circular duplex DNA, called the **provirus.** The provirus then integrates into the host chromosome, probably by a recombination process similar to the one that integrates phage λ into the *E. coli* chromosome. After integration, the proviral DNA serves as the template for synthesis of viral RNA molecules, which is catalyzed by the host cell's RNA polymerase II, the same enzyme responsible for messenger RNA transcription. Some of these RNA chains guide the synthesis of viral proteins much as a messenger RNA does. Eventually, several of these RNA strands combine with newly made viral proteins and new virus particles are formed, each with one RNA strand. These particles go through a series of finishing

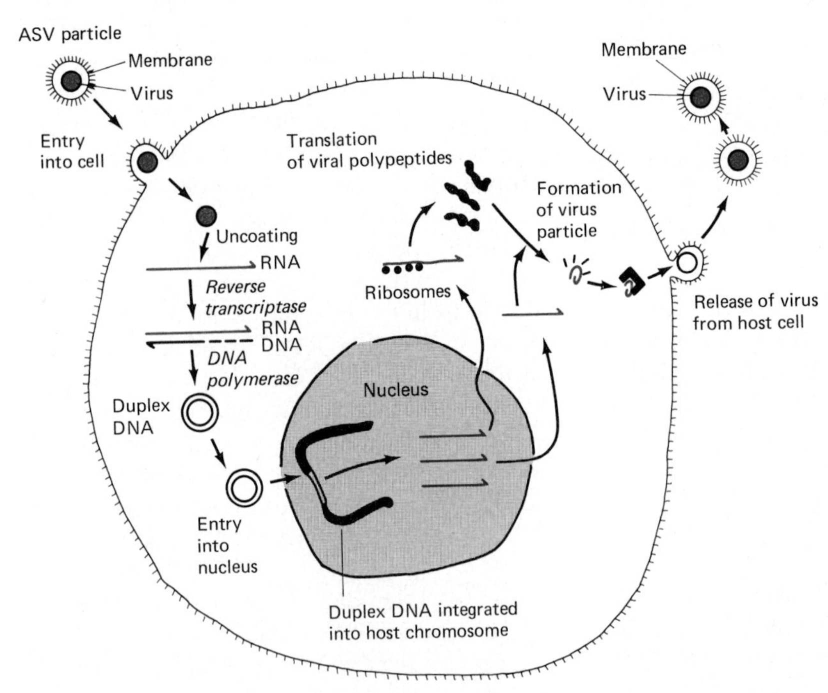


***Figure 8.17*** Major steps in replication of an RNA tumor virus, avian sarcoma virus (ASV). The enzyme *reverse transcriptase* (RNA-dependent DNA polymerase) is essential for replication in this virus.

steps before they are released from the host cell surface. None of these steps in the multiplication of RNA tumor viruses interferes with normal cellular processes and functions. The host cell is not obliged to die, as happens when DNA tumor viruses multiply in host cells. These few examples serve to show some of the variety in virus replication systems. We will discuss RNA and DNA tumor viruses again in later chapters.

# Recombination at the Molecular Level

In several previous chapters we discussed recombination that arises by crossing over, which involves an exchange of chromosome segments. Crossing over takes place after chromosomes have replicated, and each exchange involves only two of the four chromatids in a meiotic bivalent. The two general mechanisms tested in *Neurospora* by tetrad analysis were copy-choice and breakage and reunion (see Fig. 5.10). The occurrence of certain kinds of tetrads was interpreted to mean that copy-choice could not be the actual mechanism of crossing over, but the mechanism of breakage and reunion was neither proved nor disproved.

Genetic recombinants arise by crossing over in bacteria and viruses as well as in mitotic and meiotic cells in eukaryotes. The process of crossing over must involve the DNA molecule, and presumably we can study any suitable system and derive certain generalizations that may be applied to all systems where crossing over takes place. We would expect a number of enzymes to be involved in the molecular crossover events since polynucleotide chains are broken and rejoined by enzymatically catalyzed reactions. Crossing-over phenomena also play a vital role in excision and integration of viral genomes into host chromosomes as we have seen in Chapter 6. We are faced at every turn with some phenomenon involving crossing over. It is very important, therefore, to understand the molecular events behind it.

## 8.6 Breakage and Reunion

In 1961, Matthew Meselson and Jean Weigle provided convincing evidence that crossing over involved breakage and subsequent reunion of DNA molecules. They used a doubly mutant strain and a wild-type strain of phage λ; one strain had been grown in *E. coli* on media containing the heavy isotope $^{15}N$ and the other strain in ordinary $^{14}N$. These preliminary experiments were done reciprocally, so they obtained phages that were $++[^{15}N]$ and $cmi[^{14}N]$, or $++[^{14}N]$ and $cmi[^{15}N]$. Crosses were made by mixedly infecting *E. coli* cells with both viruses in $[^{14}N]$ medium. We will follow the $++[^{15}N] \times cmi[^{14}N]$ cross although similar results were obtained in the reciprocal cross.

Each phage chromosome was *genetically* marked by two gene loci governing two plaque characteristics, so recombinants could be identified by their phenotypes, and each phage chromosome was distinguished *physically* by having $^{14}N$ or $^{15}N$ in its construction. This experiment is a cytogenetic one and is comparable in every way with the *Drosophila* experiment done in 1931 by Curt Stern (see Fig. 5.7). We can interpret the results by comparing the distribution of gene markers with the distribution of isotopically distinguishable DNA molecules (chromosomes).

The experiment was designed to obtain results which showed that one proposed mechanism was correct and the other incorrect. The opposing mechanisms were *copy-choice* and **breakage and reunion.** Copy-choice had been revived as a plausible mechanism of crossing over during the 1950s, as the result of new studies using phages and bacteria in recombination experiments. It had to be reconsidered alongside breakage and reunion because of newer results obtained by that time.

If crossing over is by copy-choice, it must take place *during* DNA replication. In this case, *all* recombinants should be "light" since only $^{14}N$ is available for new strand synthesis during infection in the experiment. These $^{14}N$ viruses would be genetically recombinant as the result of the new strand switching back and forth between two parental strands as replication proceeds, but physically they would be of only one isotopic composition. If breakage and reunion is the mechanism of crossing over, at least some of the genetic recombinants would also have recombined DNA molecules, physically part $^{14}N$ and part $^{15}N$. This would result from breakage and runion involving parental and new strands.

Progeny phages were collected after cellular lysis and were centrifuged in CsCl density gradients to separate phages of different densities. Samples of these phages from different parts of the gradient were removed to determine their genotype on the basis of plaque morphology. Meselson and Weigle found that *c*+ and +*mi* genetic recombinants were also physically recombinant, according to their density positions in the CsCl gradients. They must have contained both $^{14}N$ and $^{15}N$ atoms since completely light or completely heavy DNA-containing particles were present in less dense and more dense regions of the gradient, respectively (Fig. 8.18).

Crossing over therefore took place by a mechanism involving breakage and reunion of parental and newly synthesized DNA. The molecular reactions involved in breakage and reunion could now be hypothesized, and recombination enzymes could be sought in appropriate test systems.

## 8.7 Recombination Mechanisms

The underlying assumption in recombination studies is that base pairing in single-stranded regions of duplex DNA brings molecules into the intimate association that would permit exchanges to take place. Single-stranded regions

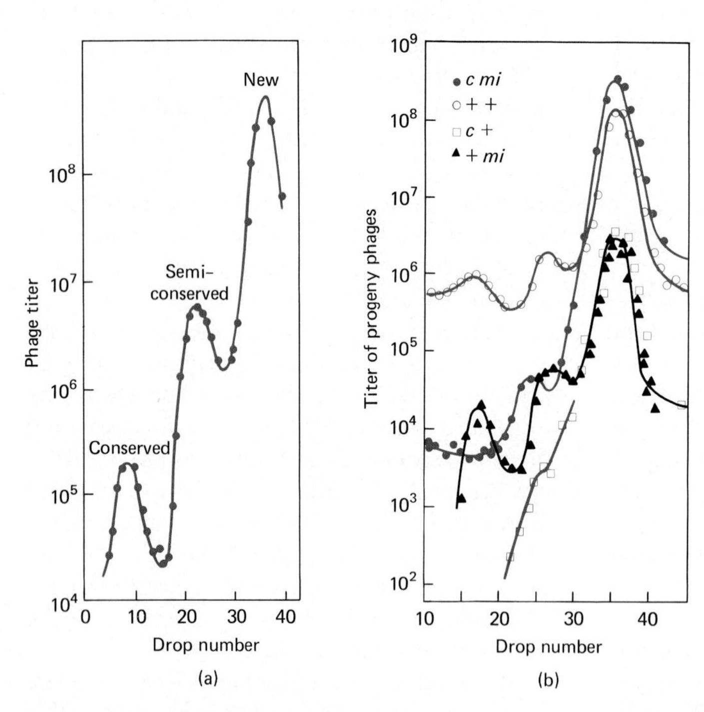


***Figure 8.18*** Crossing over occurs by breakage and reunion in phage λ and produces genetic recombinants. (a) Positions in the gradient of phages containing conserved ($^{15}N$ in both strands), semiconserved ($^{15}N$ in one strand), and new ($^{15}N$ in neither strand) duplex DNA, depending on 0, 1, or 2 cycles of replication of heavy phage infecting light *E. coli* in light medium. (b) Note the presence of conserved and semiconserved + *mi* recombinants, in particular. These must have arisen by breakage and reunion since copy-choice would lead only to new ($^{14}N$ only) recombinant phage progeny. (From M. Meselson and J.J. Weigle. 1961. *Proc. Nat. Acad. Sci. U.S.* **47**:857.)

would develop if one strand of a duplex were broken at a specific site by nuclease action. If two homologous DNAs experienced breaks at nearly the same sites, their single-stranded regions could both undergo base pairing. This occurrence would set the stage for enzymatically catalyzed reactions that would lead to rejoining of the broken molecules and to recombinant DNA. We know of no forces through which intact, homologous DNA duplexes would attract each other at specific sites. Recombination requires specific-site interaction, however, since all the genes are preserved intact within and around the recombination region. This last piece of knowledge is based on many years of genetic analysis of recombination and relies on the fact that precision must underlie recombination within the gene, as shown by Benzer, Yanofsky, and others.

There is no shortage of models for molecular recombination, but there is only a limited amount of evidence in support of any of them. However, most models share common features, and we will take a brief look at the more prominently mentioned ones.

All current models for molecular recombination in viruses and prokaryotes state that specific **endonucleases** cut or nick one strand of a duplex. *DNA polymerase* can add nucleotides onto the free end to make a single-stranded "tail" (Fig. 8.19). If homologous DNAs are near enough for base pairing along these "tails," a short duplex "bridge" could form and connect the two duplexes. Once this **recombination intermediate**, which is *heteroduplex* (containing parts of two different DNA molecules), has formed, subsequent nicks by endonucleases would lead to a recombinant DNA duplex along with fragments of the remaining strands of the parental molecules. Gaps in the nucleotide sequence of the heteroduplex would be filled in by DNA polymerase action, and the chains would be made continuous by a reaction catalyzed by *DNA ligase*. These events usually lead to *nonreciprocal* recombination—that is, if gene markers *ab* were in one parental duplex and + + in the other duplex, the single recombinant molecule would be either *a*+ or +*b*. In many such recombination events, the chance would be equal that either kind of recombinant would arise. But the products of a single virus-infected cell, for example, might show more of one recombinant than another.

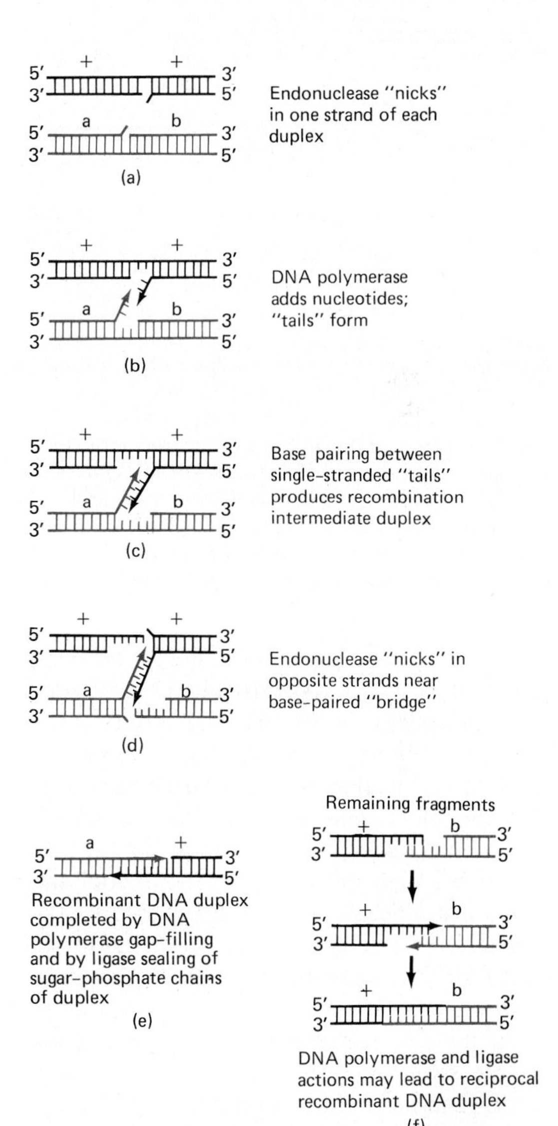


***Figure 8.19*** Steps involved in recombination by crossing over in prokaryotes and viruses may involve a sequence of enzymatically catalyzed events leading from a recombination intermediate to a completed heteroduplex DNA, shown in steps (a)–(d) leading to (e). Theoretically a reciprocal recombinant duplex can be made from remaining fragments, as shown in (f).

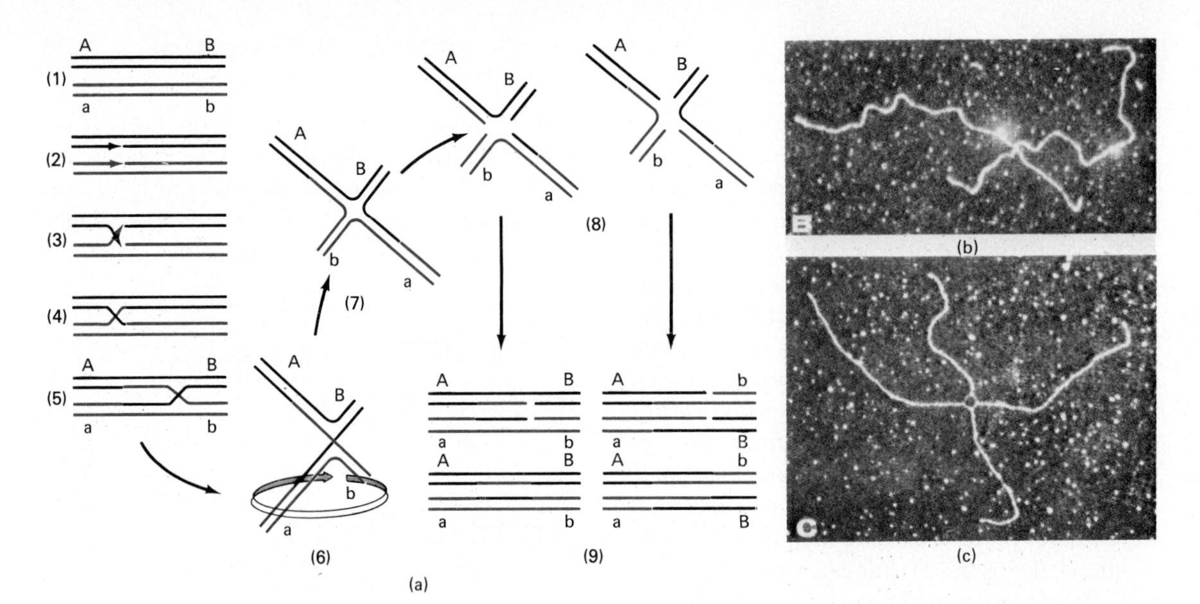


***Figure 8.20*** Structural intermediates of reciprocal recombination between two paired DNA duplex molecules, according to the prototype model developed by R. Holliday. (a) Genetic recombinants may arise after a series of steps shown in (1)–(6), and these would be reciprocal if the interlocked strands are first rotated by 180° before nucleases cleave the strands, as shown in (7). After nuclease-induced breaks, gap-filling by DNA polymerase I and ligase-catalyzed sealing would produce heteroduplex DNAs that were genetically recombinant (8)–(9). (b) and (c) Electron micrographs of *E. coli* plasmid DNA resembling the postulated structural recombination intermediates shown in (6) and (7), respectively. (From H. Potter and D. Dressler. 1977. *Proc. Nat. Acad. Sci. U.S.* **74**:4168.)

Recall that virus progeny of a single-cell burst usually does not have equal numbers of both recombinant classes, but that the sum of many such populations do. Furthermore, nonreciprocal recombination is a general characteristic of bacterial systems that involve transformation, transduction, or conjugation, as we discussed at length in Chapter 6. This model and others that differ in one or more particular details apparently coincide with observations of nonreciprocal recombination in viruses and bacteria.

All the enzymes required for recombination are the same as, or are similar to, DNA replication enzymes, which are known to be present during recombination. The kind of DNA synthesis that occurs during recombination, however, involves reactions in only small portions of the DNA molecules and is usually called **DNA repair synthesis.** Evidence for the occurrence of DNA repair synthesis during recombination has come from the identification of enzymatic defects in recombination-deficient mutants in *E. coli* and in certain phages.

In *E. coli, recA*$^-$ mutants are defective in repair of DNA that has been damaged by ultraviolet light, and they cannot undergo recombination. Mutants of the *recB*$^-$, *recC*$^-$, and *recB*$^-$*C*$^-$ genotypes are also deficient in repairing UV-induced damage to DNA, and they carry out very little recombination. The *recB* and *recC* gene products are the two subunits of a powerful endonuclease; a defect in such an enzyme could strongly influence DNA repair synthesis and recombination. Similar enzymatic defects have been found in mutants of phages such as T4 and λ, where recombination is severely affected.

Numerous electron microscopy studies have been conducted to visualize the heteroduplex recombination intermediate (Fig. 8.20). In addition, physical evidence for heteroduplex DNA has been obtained in transformation experiments and in other studies involving monitoring of isotopically marked parental DNA molecules that contain different alleles for marker genes. Genetic recombinants have been shown to arise from physically exchanged DNA, carrying the isotopic markers of both parental DNAs as well as different alleles of the two parental duplex molecules.

These lines of evidence support the breakage and reunion mechanism, but they usually can be interpreted in various ways in relation to specific models for molecular events of recombination. No one model currently satisfies everyone, and the subject seems to get more complex as new models and new data are presented.

Recombination in eukaryotic systems involves *chromatids* rather than naked DNA molecules, and a model must explain the formation of *chiasmata* in meiotic bivalents since chiasmata develop only when recombination has taken place. The problem of bringing homologous chromosomes together is not so difficult because synapsis is a regular event in the early stages of meiosis. The physical basis for exact pairing, however, is still not known. The models of molecular recombination for eukaryotic systems attempt to explain recombination leading to chiasma formation. One model that has enjoyed considerable favor since it was first presented in 1964 by Robin Holliday has also had the largest amount of experimental support. Over the years, some modifications have been introduced into the Holliday model, but the basic postulates have been retained.

The *Holliday model* for molecular recombination calls for a series of enzymatically catalyzed events in which single-strand nicks by endonuclease action arise first (Fig. 8.21). Nicks are made at corresponding sites in duplex DNAs of paired nonsister chromatids in the meiotic bivalent. The broken strands separate and then reassociate to produce a heteroduplex when ligase seals the free ends of these strands. A series of events involving digestion of DNA strand fragments by exonuclease action and gap-filling by polymerase action eventually produces two intact DNA duplex molecules that are finally made whole by DNA ligase action. The physical

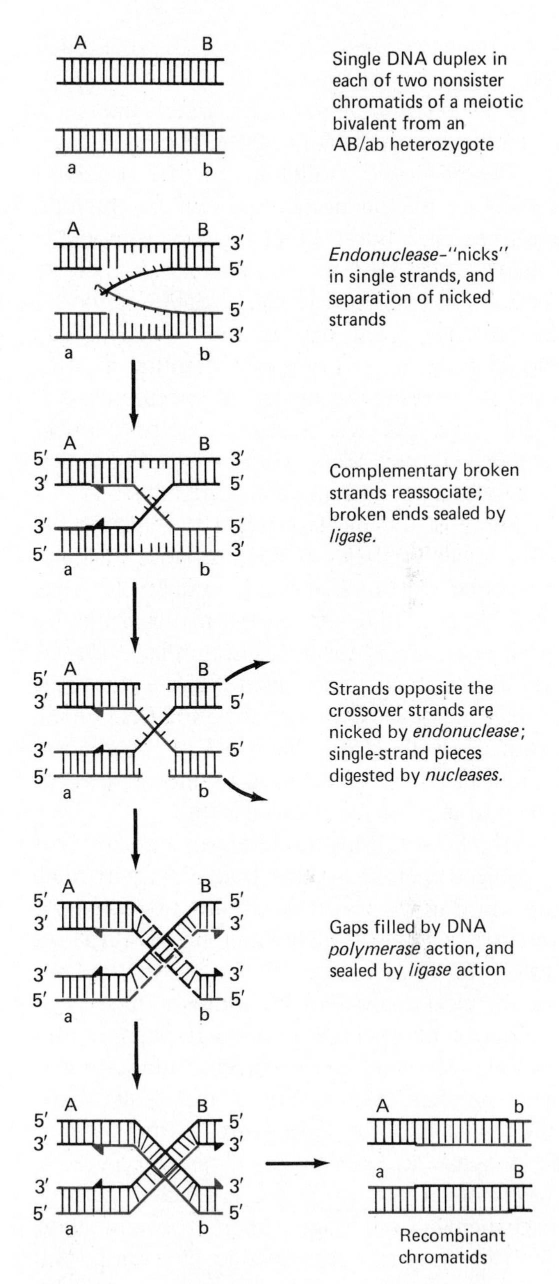


***Figure 8.21*** The Holliday model of genetic recombination applied to crossing over between nonsister chromatids in meiosis in eukaryotes. Through a series of enzymatically catalyzed steps, reciprocally recombinant chromatids arise, and a chiasma develops at the site of recombination. Only two of the four chromatids of a meiotic bivalent are shown here, and each chromatid has been represented as a duplex DNA molecule.

consequence is the development of a chiasma at the site of recombination. Two intact, *reciprocally recombinant chromatids* arise as the result of molecular events at the DNA level.

The proposed model fits all the observed recombination phenomena known for meiotic chromosomes. But whether the correct model is this particular mechanism, some variation of the mechanism, or some entirely different model has not yet been determined. The Holliday model provides a very useful working hypothesis and permits the design of specific experiments that can test various features of the proposed model. Some features have been verified, others are still under investigation.

Some of the most convincing evidence for DNA repair synthesis as the probable basis for molecular recombination in eukaryotes has come from studies of meiosis in lily plants by Herbert Stern and Yasuo Hotta. Similar evidence has also been obtained from studies of mammalian and other animal species, and DNA repair synthesis is therefore likely to be of general significance in explaining meiotic recombination in all or most eukaryotes.

Lily flower buds undergo essentially synchronized meiosis so that buds of a particular size are known to be in a particular stage of meiosis. In addition, all the meiotic cells in a bud are in the same stage. These features permit biochemical analysis of lily meiosis since large quantities of material in known parts of the meiotic sequence can be collected and prepared for enzymatic assays. Stern and Hotta have shown that endonuclease, DNA polymerase, and DNA ligase activities begin to appear in early stages of meiotic prophase I, reach a peak at pachynema (the stage when recombination probably occurs), and decline afterward (Fig. 8.22). These enzymatic events coincide with the presumed timing of synapsis, crossing over, and chiasma formation in meiotic cells.

Using radioactively labeled precursors, Stern and Hotta have shown that short pieces of DNA are made in pachynema and that some DNA digestion also occurs at this time. Because synthesis and breakdown of short pieces of DNA

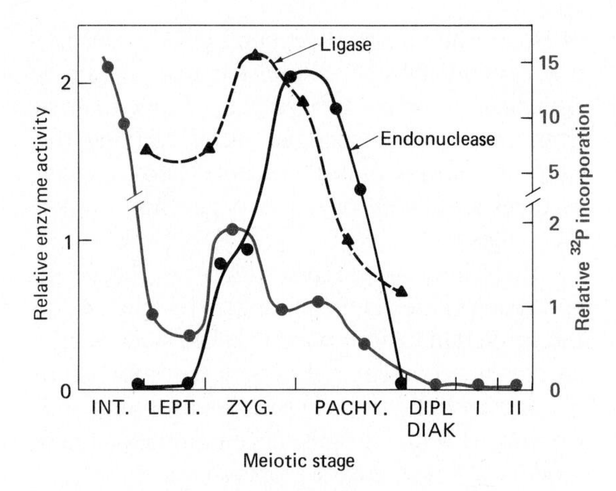


***Figure 8.22*** During early prophase I of meiosis in lily bud anthers (male floral parts). DNA synthesis occurs (colored line) as determined by incorporation of monomers labeled with radioactive $^{32}$P into high-molecular-weight DNA polymers. At the same time there is a surge of ligase and endonuclease activities. These and other data have provided experimental support for the occurrence of DNA repair synthesis during pachynema. This is the very stage of meiotic prophase I in which it is believed that crossing over takes place by mechanisms involving breakage and reunion of DNA. (From S.H. Howell. 1972. *Stadler Sympos.* **4**:57–68.)

compensate one another, there is *no net change* in the amount of DNA present before and after pachynema. This is exactly what we expect for DNA repair synthesis, namely, replacement of DNA rather than synthesis of more DNA than was present at the beginning.

Hotta and Stern have conducted many other kinds of experiments to determine the influence of inhibition of DNA, RNA, and protein synthesis on chiasma formation as an index to molecular recombination events. Thus far, all of their studies have strengthened the proposal that DNA repair synthesis is involved in crossing over and chiasma formation and, therefore, underlies molecular recombination events.

We now have the broad picture of molecular recombination. Although some of the specific details that contribute to this overall picture are still missing, we have every reason to expect these details to be discovered and the whole story of recombination to be revealed in future investigations.

## 8.8 Gene Conversion

The phenomenon of **gene conversion** refers to unusual patterns of allele segregation in which expected ratios of 1:1 for a pair of alleles may be distorted to give 3:1 and other ratios. The term gene conversion was coined to reflect the belief that somehow one allele was converted or transformed into its alternative allelic form. Gene conversion has been studied primarily in different species of fungi, such as yeast and *Neurospora,* where tetrad ratios for allele segregations can be examined directly. *Neurospora,* for example, may have a ratio of $4a^+:4a$ as expected among the eight ascospores of a single ascus, or it might have 2:6 or 6:2 or 5:3 or some similar ratio distortion. Gene conversion refers to this phenomenon of aberrant ratios for a single pair of alleles segregating at meiosis.

Gene conversion reflects a recombination event and not mutation or some other DNA modification, as is evident from crosses in which *outside* (*flanking*) *gene markers* can be followed *along* with incidents of gene conversion. When recombinants have an aberrant ratio for one pair of alleles, the alleles of genes flanking the converted region can be shown to have undergone recombination to produce 1:1 allele segregations. Because of this relationship, gene conversion itself probably arises by some recombination event that has concurrently rearranged genes on either side. Since gene conversion is confined to heterozygotes, it hardly seems possible that mutation could be responsible for the observed modifications. Mutation would affect homozygotes and heterozygotes equally.

The favored explanation for incidents of gene conversion, which occur very frequently in yeast and some other fungi, is the mismatching of nucleotide segments during DNA repair synthesis in recombinant heteroduplexes. Gene conversion is therefore believed to arise through what is called **mismatch repair** (Fig. 8.23). In fact, gene conversion has been proposed as one line of evidence for the occurrence of DNA repair synthesis in recombination.

Holliday has provided evidence showing that gene conversion rates and DNA repair are both affected in certain enzyme-deficient mutants. He found that mutants of the smut fungus *Ustilago* that were deficient in an endonuclease known to act on heteroduplex DNA showed reduced rates of gene conversion and reduced efficiency of DNA repair synthesis. Similarly, other studies with yeast mutants showed that a deficiency in repair synthesis of radiation-damaged DNA was accompanied by higher frequencies of gene conversion events in the mutants. These observations are relevant to proposed mismatch repair of a heteroduplex, which can yield 2:6 or 6:2 tetrad ratios for a pair of alleles, instead of 4:4. They are also relevant to segregation and subsequent replication of mismatched heteroduplex DNA, phenomena believed to be re-

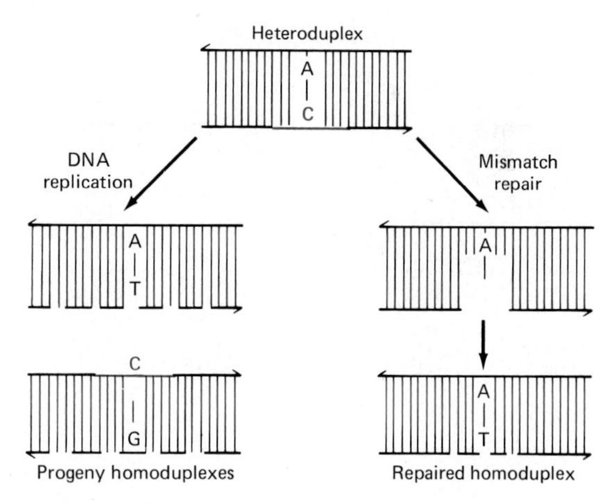


***Figure 8.23*** The mismatched base pair in a recombinant heteroduplex can be resolved in either of two ways: replication to produce semiconserved duplexes with matched base-pairs or mismatch repair of the heteroduplex itself. In the latter case, one strand is digested in the mismatched region, the intact strand is copied, and the repair is sealed. These two resolutions could yield aberrant tetrad ratios if mismatch repair occurred within a sequence with a marker gene.

sponsible for observed 3:5 and 5:3 tetrad ratios for a pair of alleles (Fig. 8.24).

At the present time we have useful models that can be tested genetically and physically. Gene conversion is a very significant component of genome change by recombination in other organisms as well as in fungi. Understanding how these events arise is important; studies with *Drosophila,* fungi, and other organisms have been increasingly informative in recent years. The complete resolution of the molecular mechanism is theoretically accessible to experimental analysis through the use of genetic, molecular, and physical methods.

## Chromosomal DNA in Eukaryotes

Eukaryotes have their DNA distributed in separate molecules among two or more chromosomes that make up the genome, and this DNA is part of a nucleoprotein complex in each chromosome. Prokaryotes and viruses have all of the genome contained in one DNA or RNA molecule. Furthermore, there is a great deal more DNA per nucleus in the eukaryotic cell than in any bacterium or virus (Table 8.2). How is all of the DNA arranged in the eukaryotic chromosome? How is the vast amount of DNA packaged into the relatively small confines of a chromosome or a nucleus? In order to understand replication, recombination, and the operations of the genetic system in eukaryotes, it is essential to know the physical ordering of the genetic material and molecules with which genes interact in transcription into RNA, regulation of gene expression, and other phenomena. Certain differences in chromosome and genome organization may provide the basis for understanding the profound differences in development and differentiation of a complex multicellular organism compared with a bacterial cell or a virus particle.

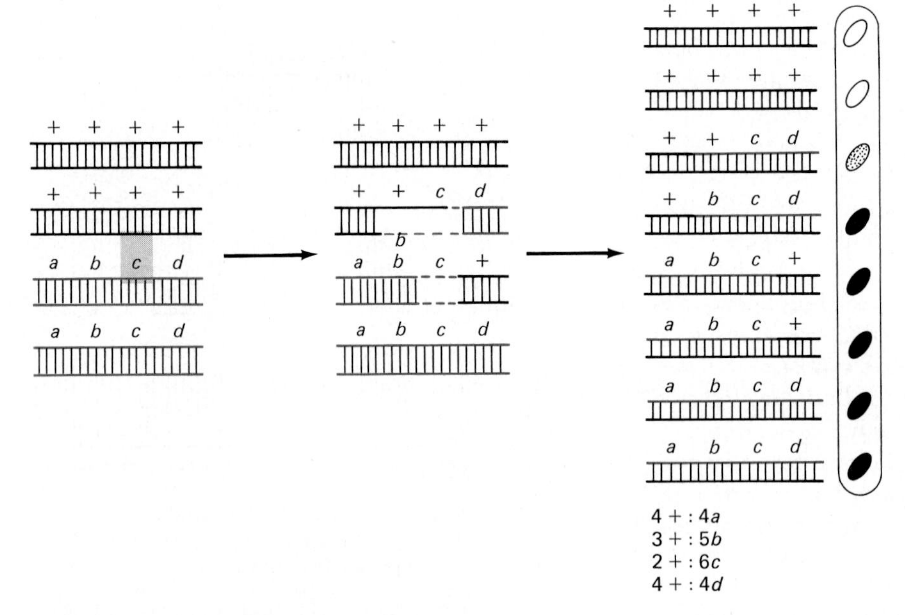


***Figure 8.24*** Following a crossover event, mismatch repair of heteroduplex DNA could produce 2 + : 6 *c*, and segregation and later replication of mismatched heteroduplex DNA could be responsible for 3 + : 5 *b* tetrad ratio. Outside markers +/*a* and +/*d* segregate 4 : 4, as expected. These 4 : 4 tetrad ratios for the outside markers emphasize the recombinational origin of gene conversion ratios for pairs of other alleles in the same tetrads.

***Table 8.2*** DNA content in various organisms (see also Box 8.3).

| organism | haploid DNA content in picograms ($10^{-12}$ g) |
|---|---|
| viruses | |
| herpes simplex | 0.00011 |
| bacteriophage λ | 0.000055 |
| bacteriophage T2 | 0.0002 |
| bacteria (prokaryotes) | |
| *Escherichia coli* | 0.0047 |
| *Staphylococcus aureus* | 0.007 |
| *Salmonella typhimurium* | 0.013 |
| eukaryotes | |
| yeast | 0.18 |
| corn | 7.5 |
| *Drosophila melanogaster* | 0.18 |
| *Rana pipiens* (frog) | 6.5 |
| *Mus musculus* (mouse) | 2.5 |
| human | 3.2 |

Various sources for data, including H.A. Sober, 1970. *Handbook of Biochemistry,* 2nd ed. Cleveland, Ohio: Chemical Rubber Co.

In the remainder of this chapter, we will examine selected features of eukaryotic chromosome organization, as analyzed at the molecular level. This information should provide some background for discussions of related topics in succeeding chapters.

## 8.9 One DNA Duplex per Chromosome

Interpretations of available genetic data as well as information from replication and recombination studies were consistent with the concept of a single duplex DNA molecule in a chromosome. Until 1973, however, there was inadequate evidence to support this working hypothesis. Two independent experimental approaches were reported in 1973, each showing that the eukaryotic chromosome contains a single duplex DNA molecule that extends from one end of the chromosome to the other.

Studies of yeast chromosomal DNA by Thomas Petes and others showed that one DNA molecule was present in each of the 17 chromosomes of a haploid nucleus. When total nuclear DNA is extracted from a known number of cells, it is a matter of simple arithmetic to figure out the amount of DNA per nucleus. There are about $10^{10}$ daltons of DNA per nucleus, and an average of $6 \times 10^8$ daltons of DNA per chromosome in yeast.

Petes used two independent methods to determine whether an average molecular weight of $6 \times 10^8$ represented one duplex DNA molecule per chromosome:

***1.*** Carefully isolated yeast nuclear DNA was centrifuged in density gradients. DNA occurred in regions of the gradient which indicated that the duplex molecules had a molecular weight of about $6 \times 10^8$. This value corresponds to one DNA duplex per chromosome.

***2.*** Nuclear DNA was examined by electron microscopy, and molecules ranging in length from 50 to 365 μm were observed. Each μm of duplex DNA has an equivalent in molecular weight of $2 \times 10^6$. Therefore, the observed molecules had a range of molecular weight between 1 and $7 \times 10^8$ and an average weight of close to $6 \times 10^8$ in this particular genome. Once again, the evidence indicates a single DNA duplex per chromosome.

In a comparable study using a different analytical method, Ruth Kavenoff, Lynn Klotz, and Bruno Zimm reported in 1973 one DNA molecule in the chromosomes of *Drosophila* species. They used a physical method in which DNA molecular weight was determined according to the rate of recoil of a stretched molecule that is undergoing relaxation. This *viscoelastic method* provides information only for the largest molecules in the DNA solution; short or broken molecules do not register. The investigators found that the longest DNA duplex in preparations of chromosomes from *Drosophila melanogaster* had a molecular weight of about $4 \times 10^{10}$, which is equivalent to a length of 20,000 μm (2 cm). Molecules of this length could be present in chromosome 2 or chromosome 3, both of which are about the same length but are much longer than the X chromosome or chromosome 4 in this species (see Fig. 5.11). Each chromosome appeared to have one DNA molecule, but another interpretation also seemed possible. Each of these chromosomes has its centromere about in the middle, so that the two arms of either chromosome 2 or 3 are approximately equal in length. Was it possible that each arm had one DNA molecule that was folded in half, and that each chromosome really contained two DNA molecules, with one chromosome-length DNA on each side of the centromere (Fig. 8.25)?

To test this possibility, Kavenoff and coworkers examined the lengths of DNA molecules from strains of *Drosophila* whose large chromosomes had been structurally rearranged. In one case the rearrangement changed the position of the centromere so that it was at one end of the chromosome, without changing the length of the whole chromosome itself. If there were only one DNA molecule in each arm, DNA should be twice as long in this particular altered chromosome since it had one arm that was twice as long as the wild-type chromosome arms. They found no change in DNA molecular weight between wild-type chromosomes and the altered chromosome. This finding suggested that DNA existed as a single continuous duplex molecule, through the centromere, from one end of a chromosome to the other. Confirming evidence for this interpretation was obtained from studies of other structurally altered chromosomes in *D. melanogaster* and in other *Drosophila* species.

The evidence from these studies in yeast and *Drosophila* as well as other evidence reported since 1973 have amply confirmed the conclusion that each eukaryotic chromosome contains one duplex DNA molecule throughout its entire length.

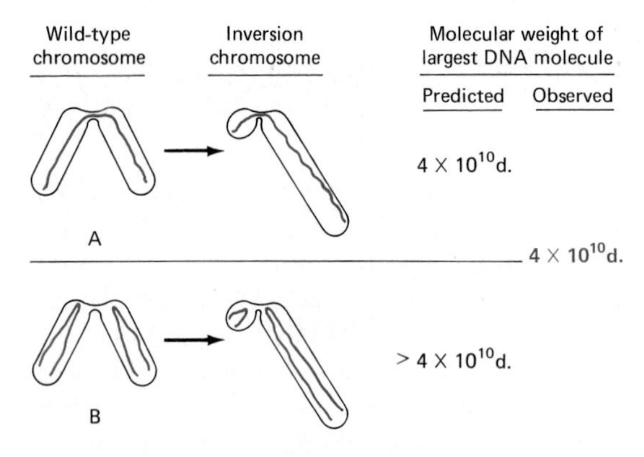


***Figure 8.25*** No increase in size of DNA (color) was found when arm lengths were modified in inversion chromosomes and compared with equal-armed wild-type chromosomes of the same overall length. This evidence supports the presence of only one DNA duplex per chromosome. The DNA molecule extends from one end of the chromosome to the other and passes through the centromere region.

## 8.10 Nucleosomes: Repeating Subunits of Chromosome Organization

Chromosomes are nucleoprotein structures in which a DNA molecule is bound to an assortment of proteins. The structural unit of the chromosome is the *chromatin fiber,* and it is best seen in interphase nuclei or metaphase chromosomes as a greatly extended and folded threadlike structure (Fig. 8.26). This fiber folds back on itself over and over again to produce the even more condensed and compacted chromosome of a typical metaphase nucleus. Each chromatin fiber contains one DNA duplex, so each chromatin fiber is equivalent to an unreplicated chromosome, but chemically it is a nucleoprotein. How are the proteins arranged in relation to DNA in the chromatin fiber? It is absolutely essential for us to understand this structural organization if we are to understand DNA activities in eukaryotic cells.

The current view of chromatin fiber organization began in 1974 with reports on biochemical analyses of chromatin by Roger Kornberg and on electron microscopic analyses by Ada and Donald Olins. Kornberg found a repeating unit consisting of about 200 base-pairs of DNA complexed to an octamer of *histone proteins*

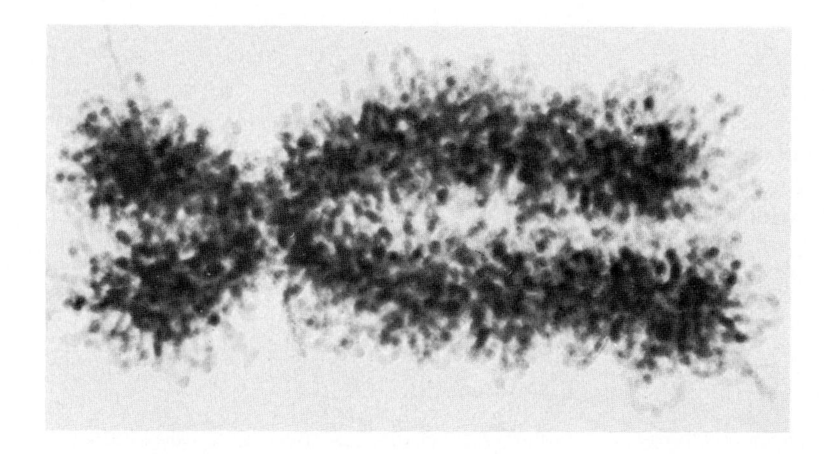

***Figure 8.26*** Electron micrograph of human chromosome 12, showing the extensively folded chromatin fiber of each of the two chromatids. (From E.J. DuPraw. 1970. *DNA and Chromosomes,* Fig. 9.10, p. 144. New York: Holt, Reinhart and Winston.)

(two each of histones H2A, H2B, H3, and H4), plus the fifth kind of histone, H1, in some undetermined association. This repeating unit of DNA plus the histone octamer was called a **nucleosome** (or nu-body). Olins and Olins produced electron micrographs showing that the chromatin fiber looked like a string of beads, each bead presumably representing a nucleosome (Fig. 8.27).

These studies created a considerable stir in the scientific community, and many laboratories became engaged in nucleosome analysis, which remains a subject of intensive study today. This breakthrough opened the way to detailed analyses of chromosome organization at the molecular level.

When isolated chromatin is partially degraded by nucleases, nucleosomes separate

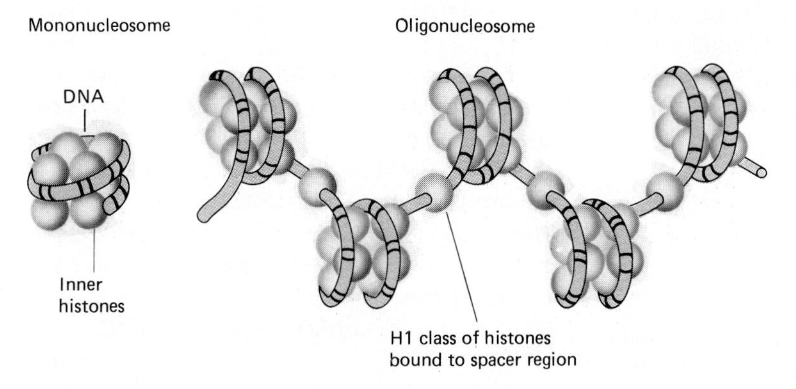


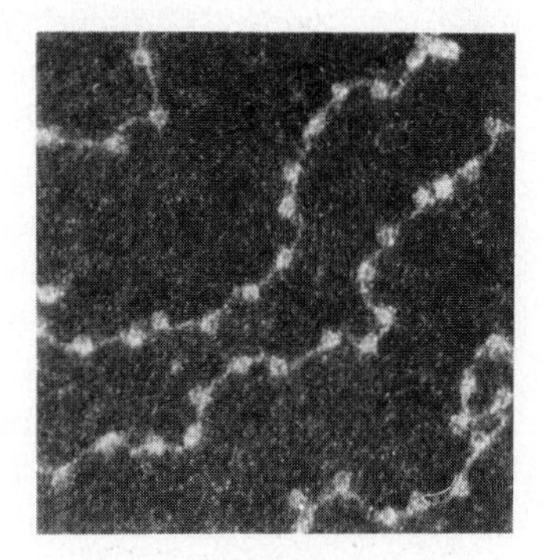

***Figure 8.27*** The nucleosome model of chromatin fiber organization. The repeating unit of the fiber is the nucleosome core, which consists of about 140 base-pairs of DNA wound around an octamer of histones (two each of H2A, H2B, H3, and H4). The nucleosomes are part of a continuous nucleoprotein (chromatin) fiber, consisting of one DNA duplex wound around histone octamers at regular intervals and in association with histone H1 in the space of about 60 base-pairs of DNA between adjacent nucleosome cores. When the chromatin fiber is digested by nuclease action, separate nucleosomes are recovered. The spacer DNA + histone H1 are removed by further digestion, thus leaving a nucleosome core particle of 140 base-pairs wrapped around the histone octamer. Electron micrograph of chicken erythrocyte chromatin shows fibers artifactually stretched out during preparation, resembling a string of beads. (From D.E. Olins and A.L. Olins. 1978. *Amer. Sci.* **66**:704.)

from the continuous chromatin fiber. Further enzymatic degradation produces a nucleosome core particle of 140 base-pairs of DNA plus the histone octamer; histone H1 and 60 or more base-pairs of DNA are lost during these last stages of nuclease action. As work has progressed in analyzing nucleosome composition and organization, the definitions have undergone various changes. We can view the nucleosome as consisting only of what was formerly called the core, that is, having 140 base-pairs of DNA and an octamer consisting of two each of the four kinds of histone proteins. The 60 or more base-pairs of DNA and histone H1 are then viewed as "linking" segments between nucleosomes, and not as parts of the nucleosome itself. Both definitions of nucleosomes remain in use today.

The whole chromatin fiber has a *constant diameter* along its length and is not really a string of wider beads linked by narrower nucleoprotein pieces. Pictures such as those produced in the mid-1970s are now believed to be of unraveled fibers, which are produced by harsh treatments during preparations for electron microscopy. Duplex DNA is coiled around the outside of the nucleosomes and gives rise to a *flexibly jointed filament of repeating units.* Each nucleosome around which the helical DNA is wound appears to be a roughly disk-shaped particle that is made up of two wedge-shaped, symmetrically arranged halves.

The nucleosome concept of chromatin fiber organization has several important implications. First, it provides a satisfactory model by which a rather stiff nucleoprotein fiber can be folded back on itself repeatedly to occupy a space that may be little more than one or a few micrometers of chromosome length. Such a fiber has enough flexibility for 20,000 μm or more to fold into the tiny condensed chromosomes typical of metaphase nuclei. Second, the proposed association with histones helps to explain how DNA is protected against attack by the abundance of nucleases known to be present in nuclei of living cells. We expect that additional information on nucleosome organization will provide the needed perspective to understand how parts of the DNA molecule are accessible to enzymatically controlled transcription into RNA while other parts of the same molecule are not engaged in transcription. We will discuss this topic again in Chapter 10.

## 8.11 Repetitive DNA

Eukaryotic DNA has two general kinds of nucleotide sequences: (1) **unique-copy sequences,** occuring in one or a few copies in the genome, and (2) **repetitive DNA** sequences of various lengths, occurring in repeated units numbering up to millions of copies per genome. Unique-copy sequences include all or most of the genes that code for polypeptides. These latter genes are the classical genes that have been analyzed by recombination and mapping and, more recently, by base sequencing in cloned DNA and by other molecular methods. At least three types of repetitive DNA are distinguishable according to their transcription into RNA and their functions:

***1.*** Informational DNA, which codes for products such as ribosomal RNAs, transfer RNAs, and histone proteins. These genes exist in *families of repeated sequences* and number in the hundreds or thousands of copies per genome. They may be clustered in particular chromosomes or scattered among two or more chromosomes in the genome. This class of DNA is referred to as **middle-repetitive DNA.**

***2.*** Noncoding DNA, which is transcribed into RNA. The transcript RNA regions are not expressed. They usually are excised from the pre-mRNA before this mRNA acts in protein synthesis during translation. The sequences may be very long, measuring hundreds or thousands of base-pairs. They are often present in regions between functional unique-copy genes and comprise much of the *spacer regions between functional genes.* This class of repetitive DNA is also middle repetitive, but the sequences do not specify a coded gene product.

*3.* Noncoding DNA, which is not transcribed into RNA. These repeated sequences, often numbering in the millions of copies per genome, are relatively short regions consisting of fewer than a dozen base-pairs. This class is referred to as **highly repetitive DNA** and is found throughout the genome. Highly repetitive DNA is characteristic of centromeric regions of all the chromosomes in the genome and of regions within genes and between genes.

Initial studies of genomic DNA involved the separation of purified cellular DNA into fractions during equilibrium density-gradient centrifugation in CsCl (Fig. 8.28). Eukaryotic DNA invariably sediments into more than one fraction, whereas prokaryotic DNA bands as a single homogeneous population of DNA molecules in CsCl density gradients. This difference reflects a fundamental distinction in the organization of eukaryotic and prokaryotic genomes, even though both are made up of DNA. The major component of eukaryotic genomes sediments in a large population of **bulk nuclear DNA** molecules. A number of minor **satellite DNA** components can be identified after sedimentation. The differences in position of the several DNA fractions in the gradient are largely due to their different G + C molar percentages in base composition.

The cellular location of satellite DNAs can be determined by comparisons between DNA fractions obtained by centrifugation of whole-cell DNA and fractions obtained from DNA extracts of isolated nuclei, mitochondria, and chloroplasts. Apart from mitochondrial and chloroplast DNAs, one or more satellite nuclear DNAs can be separated from bulk nuclear DNA in fractions obtained after centrifugation of isolated nuclear preparations. Molecular characterization of nuclear DNAs was opened to detailed analysis in the 1960s, long before recombinant DNA and base-sequencing methods were available. Roy Britten and Eric Davidson developed quan-

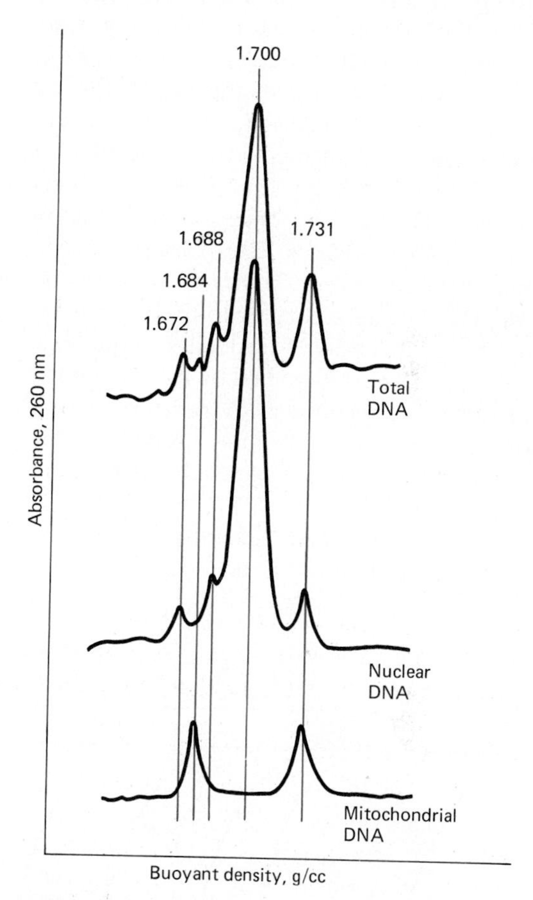


***Figure 8.28*** Densitometer tracings of purified DNAs from yeast, centrifuged to equilibrium in CsCl density gradients in an analytical ultracentrifuge. Buoyant density of DNA in CsCl is expressed in g/cc and is proportional to G + C content of the DNA. Higher G + C content produces proportionately higher buoyant density values. The reference DNA bands at 1.731 g/cc, bulk nuclear DNA at 1.700 g/cc, and three satellite DNAs at 1.688, 1.684, and 1.672 g/cc. Examination of the bottom tracing reveals that the 1.684 g/cc satellite consists of mitochondrial DNA. The other two satellites must be nuclear in origin since they appear in the middle tracing.

titative methods for DNA analysis, according to DNA **renaturation kinetics,** to obtain $C_0t$ plots (Fig. 8.29).

With the use of renaturation kinetics, different fractions of nuclear DNA were analyzed according to the rate of **reassociation,** or reannealing, of single-stranded DNA fragments derived from melted duplex DNA collected from gradient fractions. In the assays, duplex DNA is melted to single-stranded chains, and these strands are fragmented into pieces of a predetermined average size. These fragments are then allowed to undergo renaturation through complementary base pairing between single-stranded pieces. The *rate* of renaturation to duplexes varies according to the relative extent of repetitiveness of nucleotide sequences. Single-copy DNA is identified by its slow rate of renaturation. Repetitive DNA renatures much more rapidly; and the more repetitive the DNA is, the faster the fragments will renature (Box 8.2).

The observed kinetic differences have been interpreted as follows: if one copy or a few copies of a nucleotide sequence are present in a DNA preparation, complementary fragments take some time to "find" each other and pair. The more copies there are of some sequence, the better the chances for complementary sequences to "find" each other and renature. The highest renaturation rates therefore characterize highly repetitive DNA, which exists in millions of copies in the genome. Middle-repetitive DNA renatures more slowly, and unique-copy DNA sequences exhibit the slowest rates of renaturation. Renaturation kinetics defines the organization of DNA sequences in the genome and can provide quantitative estimates of the proportion of a genome that is comprised of each of the several kinds of sequences (Table 8.3).

The chromosomal locations of repetitive DNA were determined by a microscopical procedure that is based on the molecular method of DNA-RNA hybridization, which depends on base pairing between complementary sequences. This method, called ***in situ* hybridization,** was developed independently in 1969 by Mary Lou Pardue and Joe Gall in the United States and by

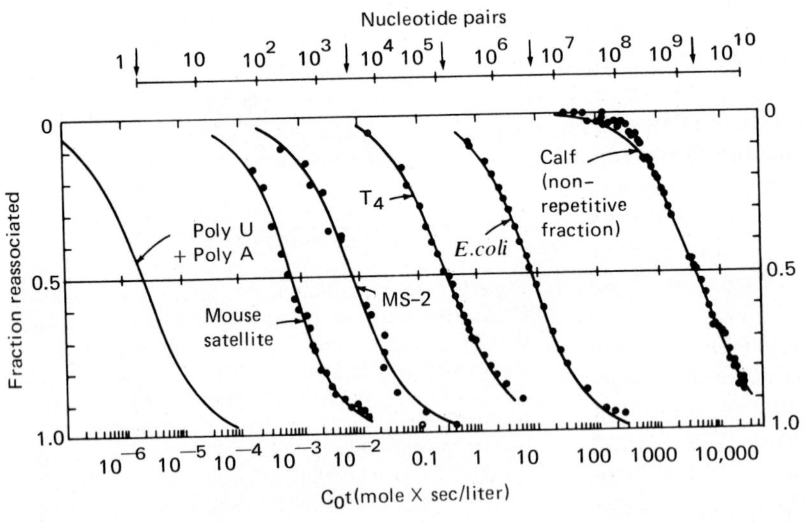


***Figure 8.29*** Kinetics of reassociation of dissociated fragments of duplex DNA from various sources. The genome size (in terms of nucleotide pairs) of each DNA source is indicated by arrows on the logarithmic scale at the top. The genome size in nucleotide pairs is proportional to the $C_0t$ time required for 50% reassociation (half-reaction), over a factor of $10^9$. The shape of each curve indicates a second-order reaction and, therefore, shows that each sample was relatively homogeneous (See Box 8.2). (R.J. Britten and D.E. Kohne, "Repeated Sequences in DNA," *Science* **161**:529–540, Fig. 9, August 1968. Copyright © 1968 by the American Association for the Advancement of Science.)

## BOX 8.2 Renaturation Kinetics of Duplex DNA

The presence of unique-copy and repetitive DNA sequences in eukaryotes contrasts with the virtual absence of repetitive DNA in prokaryotic and viral genomes. This difference becomes apparent from the renaturation kinetics of different sources of DNA, that is, from the measurements of the time course of reassociation of dissociated strands of duplex DNA. The extent of reassociation is usually monitored during an experiment by passing samples of the reaction mixture over *hydroxyapatite columns* since only duplex DNA binds selectively to hydroxyapatite crystals under appropriate conditions. Conditions for renaturation are standardized for temperature, ionic conditions, and 300–400 nucleotides per sheared fragment of single-stranded DNA from melted duplexes; this permits comparisons among different experiments and for different DNA sources.

The rate-limiting step in the reassociation reaction is one in which collisions occur between complementary single-stranded regions such that base pairing begins. Reassociation thus follows second-order kinetics, according to the equation

$$\frac{C}{C_0} = \frac{1}{1+KC_0t}$$

where $C_0$ is the total DNA concentration, $C$ the concentration of DNA remaining single stranded at time $t$ and $K$ the reassociation rate constant.

(a) The time course of an ideal, second-order reaction illustrates the features of the logarithmic $C_0t$ plot, with total DNA in the initial state being single stranded and reassociated into double strands in the final state of the reaction. The rate constant of reassociation is inversely proportional to genome size, and the $C_0t$ is therefore proportional to genome size (See Fig. 8.29). (b) The single S-shaped (sigmoidal) curve of the second-order reaction typifies total DNA from prokaryotes but not from eukaryotes. The $C_0t$ plot for calf thymus (eukaryotic) DNA indicates that some sequences are reassociating faster than others. When each component is isolated from the total DNA of eukaryotic origin, each component reassociates with second-order kinetics, but with different rates of reaction. The skewed curve for calf thymus DNA indicates that the population of DNA molecules is heterogeneous with respect to its kinetics of reassociation. Very low $C_0t$ rates characterize rapidly reassociating DNA, indicating that many copies of the same sequences are present and, therefore, they can "find" correct partners relatively easily. *E. coli* DNA is typically homogeneous, according to its second-order kinetics and shows the standard prokaryotic $C_0t$ plot. (From R.J. Britten and D.E. Kohne. 1968. *Science* **161**:529. Copyright 1968 by the American Association for the Advancement of Science.)

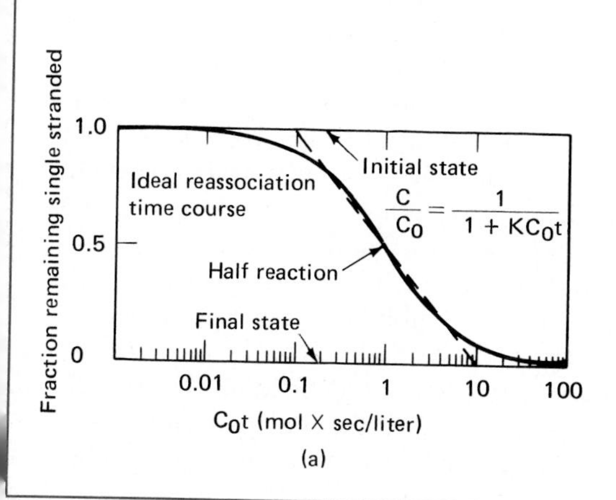


(a)

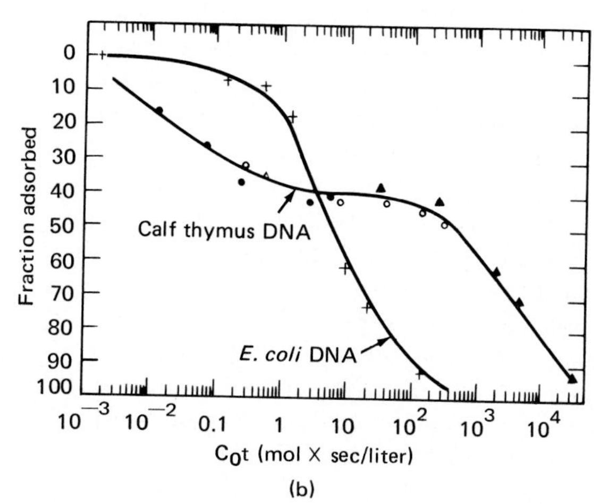


(b)

***Table 8.3*** Estimated amounts of unique-copy and repetitive DNA sequences in representative eukaryotic genomes.

| organism | percent unique-copy DNA | repetitive DNA | |
|---|---|---|---|
| | | **percent** | **number of copies** |
| *Paramecium aurelia* (protozoan) | 85 | 15 | 50–75 |
| *Dictyostelium discoideum* (slime mold) | 60 | 28 | 113 |
| *Neurospora crassa* (true fungus) | 80 | 20 | 60 |
| *Strongylocentrotus purpuratus* (invertebrate: sea urchin) | 50 | 27<br>19 | 10<br>164 |
| *Nassaria obsoleta* (invertebrate: gastropod) | 38 | 12<br>15 | 20<br>1,000 |
| *Bombyx mori* (invertebrate: silkworm moth) | 55 | 21<br>24 | 500<br>50,000 |
| *Drosophila melanogaster* (invertebrate: fruit fly) | 78 | 15<br>7 | 35<br>2,600 |
| *Xenopus laevis* (vertbrate: amphibian; African clawed toad) | 54 | 6<br>31<br>6<br>3 | 20<br>1,600<br>32,000<br>high |
| *Gallus domesticus* (vertebrate: bird; chicken) | 70 | 24<br>3<br>3 | 120<br>330,000<br>1,100,000 |
| *Bos taurus* (vertebrate: mammal; cattle) | 55 | 38<br>2<br>3 | 60,000<br>1,000,000<br>>1,000,000 |
| *Homo sapiens* (vertebrate: mammal; human) | 64 | 13<br>12<br>10 | low<br>intermediate<br>high |

Data from various sources.

Hugh John in Edinburgh. Radioactively labeled RNA or DNA is copied from satellite DNA strands *in vitro,* with RNA or DNA polymerase used to catalyze the incorporation of labeled nucleotide precursors into complementary RNA or DNA strands. The copied RNA or DNA is then applied to chromosome preparations on a glass slide, after it is pretreated to slightly denature chromosomal DNA and allow base pairing to occur. The preparation is covered with a photographic emulsion and autoradiographs are later developed to locate the silver grains, which indicate specific sites on the chromosomes to which radioactively labeled RNA or DNA has become bound through complementary base pairings.

*In situ* hybridization showed that middle-repetitive genes for ribosomal RNA are clustered at the nucleolar-organizing region of a chromosome. In virtually every species examined, highly repetitive DNA was found to occur in the centromeric region of each chromosome in the genome. (Fig. 8.30). Molecular analysis of highly repetitive centromeric DNA revealed the presence of between $10^5$ and $10^7$ tandem repeats of between two and ten base-pair sequences each. This particular highly repetitive DNA appears to be confined to regions of *constitutive heterochromatin,* which is defined by genetic stability, replication late in the cell cycle, and a highly condensed conformation in interphase chro-

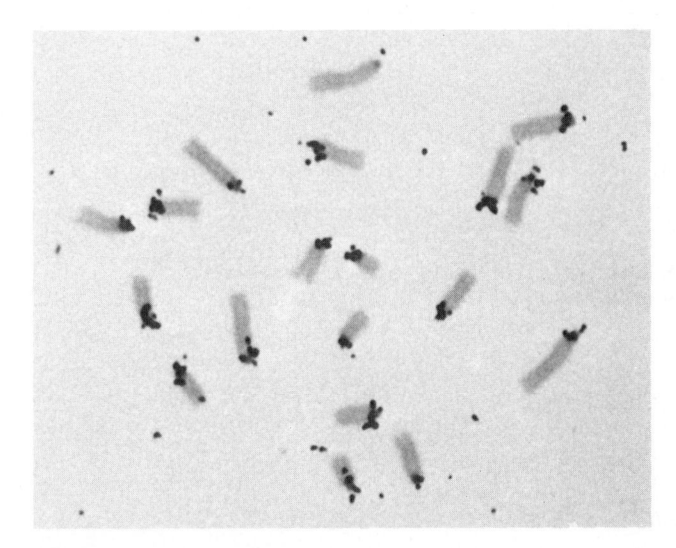

***Figure 8.30*** Autoradiograph of mouse metaphase chromosomes after *in situ* hybridization with radioactively labeled RNA copied from mouse satellite DNA. Only centromeric heterochromatin is labeled (each chromosome has its centromere near one end), as seen by the distribution of dark silver grains over the lightly stained chromosomes. (Courtesy of M.L. Pardue and J.G. Gall.)

mosomes (see Section 4.9). Furthermore, highly repetitive DNA is not transcribed into RNA, as was shown by the failure of satellite fractions of highly repetitive DNA to form DNA-RNA hybrids with *any* part of whole-cell RNA preparations. Since this DNA does not hybridize with any cellular RNA, it must not make any RNA transcripts. If it were transcribed, some hybrid duplexes would be formed in molecular hybridization assays. We would not in any case expect a sequence of two to ten base pairs to code for a polypeptide since no more than three amino acid–specifying codons could be accommodated in the short sequences.

The functions of highly repetitive DNA are uncertain for the most part. Centromeric DNA is believed to have a structural function in chromosomal organization and to provide a stable segment housing the vital region of each chromosome that is required for movement of chromosomes to the poles during mitotic and meiotic anaphase stages. Current studies using cloned DNA have shown that highly repetitive DNA exists throughout the genome in spacer regions between functional unique-copy genes and in intervening sequences within the gene itself. Specific repetitive DNAs of identical or similar sequences exist in *families of repeats dispersed throughout the genome.* Whether or not these dispersed families of repeats carry out the same or different functions is now under extensive study. Experimental evidence has implicated some families of repeats in genetic processes associated with gene expression and its regulation. In other cases families of repeats have been implicated in processes leading to genomic rearrangements by recombinational and other processes. We will have occasion to discuss these gene families in later chapters, along with families of clustered unique-copy genes.

That so much of the genome is comprised of noncoding DNA has added to the general belief that the absolute amount of DNA in species genomes has no bearing on evolutionary rankings or genetic complexity of species (Box 8.3). Molecular studies bearing on evolution have made significant contributions to our understanding of species relationships and of genomic modifications. Molecular evolutionary changes will be discussed in Chapter 14 in relation to developmental pathways of phenotype de-

## BOX 8.3 Genome Size

Genome size exhibits an enormous spread of values among animal groups, ranging from $10^8$ nucleotide pairs in an insect to $8–9 \times 10^{11}$ in some amphibians. The greatest spread within a group occurs among amphibians, whereas the other groups of land vertebrates (reptiles, birds, and mammals) are far less variable. This finding may truly reflect group differences or perhaps the fact that amphibians have been studied more intensively than the other groups.

If the number of genes were reflected in genome size and in evolutionary rankings of species, we would not expect the most highly evolved animals (mammals) to have fewer genes than fish and amphibians. There is, in fact, no evident pattern showing correlations between evolutionary ranking and increase in the genome size or gene number. Some groups are quite variable and others are not, for reasons that remain uncertain. (From R.J. Britten and E.H. Davidson. 1971. *Quart. Rev. Biol* **46**:111.)

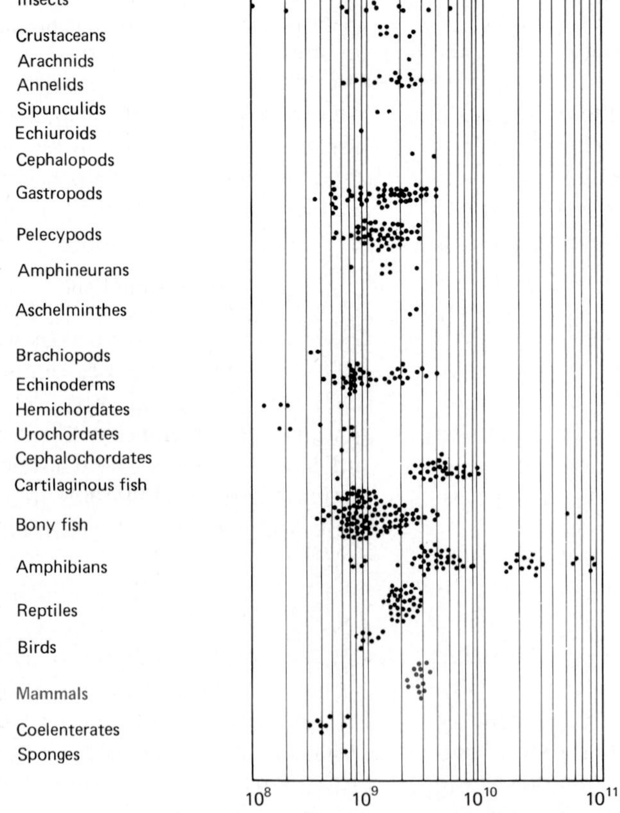

velopment and in Chapter 16 in relation to evolutionary studies at various levels of biological organization.

From this preliminary survey of chromosomal DNA, it is evident that eukaryotic genetic systems are organized into functionally distinct regions in individual chromosomes and in the genome as a whole. DNA has coding functions, which lead to proteins that underwrite phenotypic trait development, and an assortment of noncoding functions still being studied. The eukaryotic genome is thus far more than strings of genes situated in a variable number of chromosomes.

## Questions and Problems

***8.1*** In a Meselson-Stahl experiment, *E. coli* is fully labeled with $^{15}N$ by long-term growth in [$^{15}N$] medium is then transferred to [$^{14}N$] medium.

***a.*** Predict the fractions of total DNA that would be heavy, half-heavy, and light when centrifuged to equilibrium in CsCl density gradients after 0, 1, 2, and 4 generations, if DNA replicates semiconservatively.

***b.*** What predictions would you make in (a) if DNA replicates conservatively?

***8.2*** If $^{15}N$-labeled *E. coli* is transferred to [$^{14}N$] medium, diagram the DNA duplexes that would be found after 1 and 2 generations, using a straight line for a [$^{15}N$] chain and a wavy line for a [$^{14}N$] chain:

***a.*** If DNA replication is semiconservative.

***b.*** If DNA replication is conservative.

***8.3*** We believe that during DNA synthesis precursor Okazaki fragments are joined together in a series of enzymatically catalyzed reactions at the replication fork. What events take place for each of the following sequences of enzymatic reactions:

***a.*** DNA polymerase I (exonuclease), DNA polymerase I (polymerase), ligase?

***b.*** RNA polymerase (primase), DNA polymerase III? Which of these two sets of reactions precedes the other?

***8.4*** What functions are believed to be carried out at the replication fork by each of the following: ***a.*** SSB protein ***b.*** topoisomerase I ***c.*** topoisomerase II ***d.*** helicase ***e.*** ligase?

***8.5*** A ligase-deficient strain of *E. coli* (*lig*$^-$) is transferred to a medium containing radioactively labeled thymidine ($^3H$-thymidine) for a brief time, and DNA is extracted from the cells immediately afterward. What is the probable explanation for each of the following DNA fractions in the extract:

***a.*** $^3H$-labeled duplexes about 1000 to 2000 base-pairs long?

***b.*** Unlabeled duplexes tens of thousands of base-pairs long?

***8.6*** Eukaryotic cells can be grown in a medium containing bromodeoxyuridine (BrdU), an analog of thymidine. DNA in such cells would contain BrdU in place of thymidine residues in many regions of newly synthesized duplexes. A fluorescent dye that binds to DNA makes it possible to recognize duplex DNA in which one or both strands have BrdU substituted for thymidine since the dye fluoresces strongly when one strand contains BrdU (bright appearance) but weakly when both strands contain BrdU (dull appearance).

***a.*** Diagram the two strands of the duplex present in each chromatid of a metaphase chromosome after two replications in BrdU medium, using a solid line for an unsubstituted strand and a dashed line for a BrdU-substituted strand. Indicate the relative brightness of the sister chromatids.

***b.*** What would be the relative brightness of the metaphase chromosome in each daughter cell after a third replication in BrdU medium?

***c.*** Suppose the cell had 40 chromosomes. After four DNA replications in BrdU, how many brightly fluorescing *chromatids,* on the average, would be found in a given cell? (Assume random segregation of sister chromatids during mitosis.)

***8.7*** Bidirectional and unidirectional replication of DNA can be distinguished in autoradiographs of DNA molecules labeled briefly in [$^3H$] medium. Describe the predicted appearance of autoradiographed DNA as follows:

***a.*** If replication from a single origin is unidirectional.

***b.*** If replication from a single origin is bidirectional.

***c.*** If there are two unidirectional replicating units (replicons) near each other in a region of duplex DNA.

***8.8*** Duplex DNA in many phages is characterized by single-stranded cohesive termini, or "sticky" ends, similar to the molecule diagrammed below:

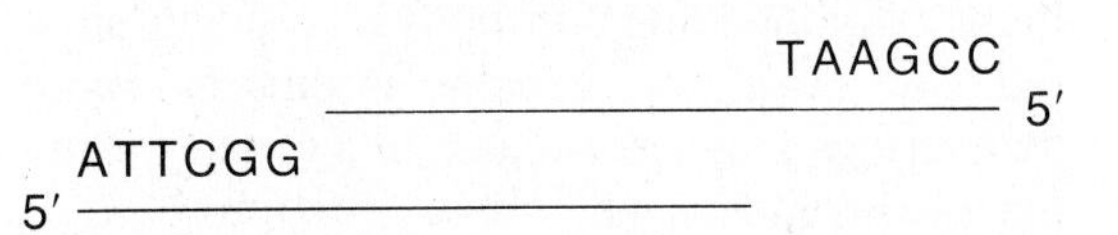


***a.*** Draw the circularized molecule formed when complementary base pairing of the two 5′ ends of the duplex takes place, before ligase seals the sugar-phosphate backbone of each strand.
***b.*** Describe the difference in centrifugation behavior of linear duplex DNA with "sticky" ends and those without "sticky" ends, after treatment to induce complementary base pairing and chain ligation.
***8.9*** Duplex DNA in some viruses is characterized by redundant termini in one strand and lack of these termini in the complementary strand, as shown below:

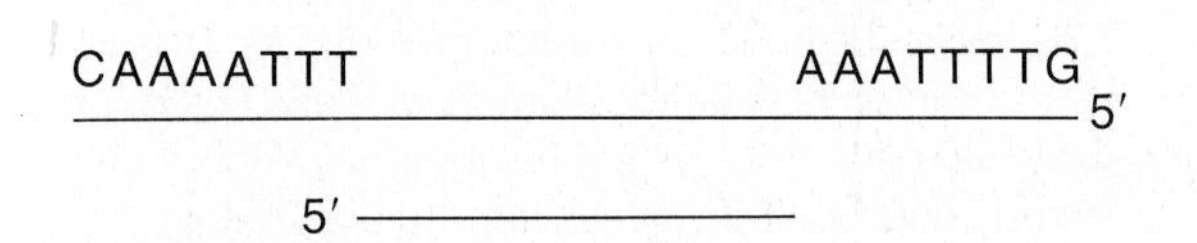


***a.*** Draw the duplex molecule that would be formed after complementary base pairing between redundant termini.
***b.*** Describe the difference in centrifugation behavior of linear duplex DNA with redundant termini and those without redundant termini, after treatment to promote complementary base pairing of single-stranded ends.
***8.10*** Electron micrographs were made of circular replicating DNA molecules isolated from a cell culture, and all were found to be theta-forms. Measurements in $\mu$m of the three segments observed in each of four molecules were: (a) 1.2, 4.3, 1.2 (b) 0.6, 0.6, 4.9 (c) 3.5, 3.5, 2.0 (d) 2.5, 3.0, 2.5
***a.*** What is the contour length of a nonreplicating circular molecule in these cells?
***b.*** List the four molecules in order of increasing progress in replication.
***8.11*** After infecting a culture of unlabeled *E. coli* with a mixture of $[^{15}N]++$ and $[^{14}N]ab$ phages from lambda stocks, the phage lysate was collected and centrifuged to equilibrium in CsCl density gradients to determine the distribution of parental and recombinant particles throughout the density gradient. Two possible results were predicted (see next column).

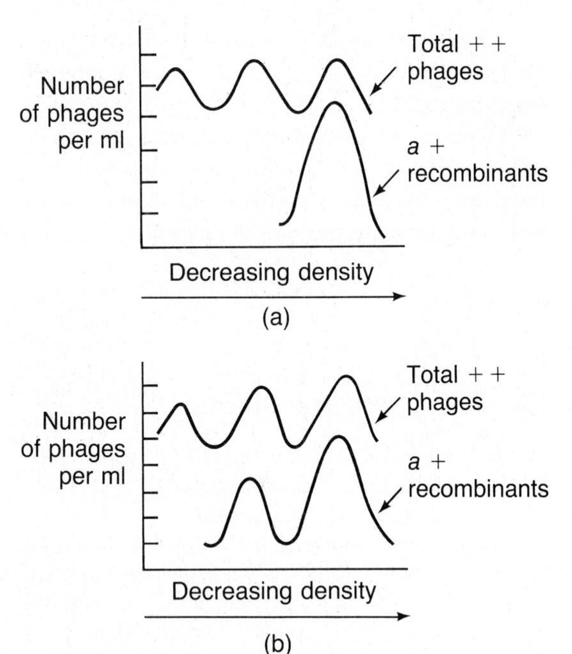


***a.*** Which of the two possibilities represents prediction for copy-choice recombination? Explain.
***b.*** What is the explanation for each of the three density classes of ++ phages shown by the upper curve in diagrams (a) and (b)?
***c.*** What is the theoretical distribution in the density gradient of $[^{15}N]++$ parental phages after 0, 1, and 2 semiconservative replications of DNA in unlabeled *E. coli?*
***8.12*** *Neurospora crassa* produces eight-spored asci since each of the four meiotic products undergoes a mitotic division. The eight spores are ordered in a linear array that indicates the origin of each spore in meiosis and mitosis. In crosses between +++ and *abc* genetic strains, some asci were found to contain spores in the following order:

+ + +
+ + +
+ *b c*
+ *b c*
*a b* +
*a b* +
*a b c*
*a b c*

***a.*** What is the unusual feature of this set of spores in an ascus?
***b.*** What is the probable explanation for the origin of such an ascus?

*c.* How does the distribution of the +/*c* gene marker support the proposition that a recombination event was involved in the development of such an ascus?

***8.13*** Define the following terms: ***a.*** heavy isotope ***b.*** radioactive isotope ***c.*** replication fork ***d.*** copy-choice ***e.*** gene conversion ***f.*** chromatin fiber ***g.*** nucleosome ***h.*** repetitive DNA ***i.*** *in situ* hybridization.

## References


Alberts, B.M. and L. Frey. 1970. T4 bacteriophage gene 32: A structural protein in the replication and recombination of DNA. *Nature* **227**:1313.

Bauer, W.R., F.H.C. Crick, and J.H. White. July 1980. Supercoiled DNA. *Sci. Amer.* **243**:118.

Britten, R.J., and D.E. Kohne. 1968. Repeated sequences in DNA. *Science* **161**:529.

Cairns, J. Jan. 1966. The bacterial chromosome. *Sci. Amer.* **214**:36

Cold Spring Harbor Symposia on Quantitative Biology. 1978. *Chromatin.* Vol 42.

Cold Spring Harbor Symposia on Quantitative Biology. 1979. *DNA: Replication and Recombination.* Vol. 43.

Davidson, E.H., et al. 1975. Comparative aspects of DNA organization in metazoa. *Chromosoma* **51**:253.

Fisher, L.M. 1981. DNA supercoiling by DNA gyrase. *Nature* **294**:607.

Freifelder, D. 1983. *Molecular Biology.* Boston: Science Books Internat., Publ.

Gall, J.G., and M.L. Pardue. 1969. Formation and detection of RNA-DNA hybrid molecules in cytological preparations. *Proc. Nat. Acad. Sci. U.S.* **63**:378.

Hall, B.D., and S. Siegelman. 1961. Sequence complementarity of T2-DNA and T2-specific RNA. *Proc. Nat. Acad. Sci. U.S.* **47**:137.

Kavenoff, R., L. Klotz, and B. Zimm. 1974. On the nature of chromosome-sized DNA molecules. *Cold Spring Harbor Symp. Quant. Biol.* **38**:1.

Kornberg, A. 1980. *DNA Replication.* San Francisco: Freeman.

Kornberg, R.D. 1974. Chromatin structure: A repeating unit of histones and DNA. *Science* **184**:868.

Kornberg, R.D., and A. Klug. Feb. 1981. The nucleosome. *Sci. Amer.* **244**:52.

Kriegstein, H.J., and D.S. Hogness. 1974. Mechanism of DNA replication in *Drosophila* chromosomes: Structure of replication forks and evidence for bidirectionality. *Proc. Nat. Acad. Sci. U.S.* **71**:135.

Lawson, G.M., et al. 1982. Definition of 5′ and 3′ structural boundaries of the chromatin domain containing the ovalbumin multigene family. *J. Biol. Chem.* **257**:1501.

Meselson, M., and C.M. Radding. 1975. A general model for genetic recombination. *Proc. Nat. Acad. Sci. U.S.* **72**:358.

Meselson, M., and F.W. Stahl. 1958. The replication of DNA in *E. coli. Proc. Nat. Acad. Sci. U.S.* **44**:671.

Meselson, M., and J.J. Weigle. 1961. Chromosome breakage accompanying genetic recombination in bacteriophage. *Proc. Nat. Acad. Sci. U.S.* **47**:857.

Olins, A.L., and D.E. Olins. 1979. Stereo electron microscopy of the 25-nm chromatin fibers in isolated nuclei. *J. Cell Biol.* **81**:260.

Pardue, M.L., and J.G. Gall. 1970. Chromosomal localization of mouse satellite DNA. *Science* **168**:1356.

Potter, H., and D. Dressler. 1976. On the mechanism of genetic recombination: Electron microscopic observation of recombination intermediates. *Proc. Nat. Acad. Sci. U.S.* **73**:2000.

Schmid, C.W., and W.R. Jelinek. 1982. The Alu family of dispersed repetitive sequences. *Science* **216**:1065.

Singer, M.F. 1982. SINEs and LINEs: Highly repeated short and long interspersed sequences in mammalian genomes. *Cell* **28**:433.

Spiegelman, S. May 1964. Hybrid nucleic acids. *Sci. Amer.* **210**:48.

Taylor, J.H. 1965. Distribution of tritium-labeled DNA among chromosomes during meiosis. I. Spermatogenesis in the grasshopper. *J. Cell Biol.* **25**:57.

Wang, J.C. July 1982. DNA topoisomerases. *Sci. Amer.* **247**:94.

# CHAPTER 9

# Gene Expression: Transcription and Translation

*Gene expression* refers to the overall phenomenon of transcribing and translating coded genetic instructions into polypeptide and RNA products. Coded information in DNA is copied into RNA, which acts as the genetic intermediate in synthesis of the gene products in the cytoplasm. Using molecular methods in concert with genetic methods where appropriate, the molecules and processes of gene expression have been analyzed in considerable detail. We must understand these processes before we can determine how they are regulated during the development of a phenotype from genotypic coded information.

## Information Flow from DNA: Transcription

The genetic information encoded in DNA is expressed in two distinct steps:

*1.* **Transcription,** in which the sequence of deoxyribonucleotides in DNA directs the synthesis of a complementary sequence of ribonucleotides in RNA, and

*2.* **Translation,** in which the sequence of ribonucleotides in messenger RNA guides the synthesis of a corresponding sequence of amino acids in polypeptides.

Proteins, which consist of one or more polypeptide chains, contribute to cellular activities in many different ways. Proteins that function as enzyme catalysts facilitate many biochemical reactions whose products contribute to the development of the phenotype.

In the last chapter we discussed replication. In this chapter we will look at the processes of transcription and translation and how we can use genetic and molecular methods to analyze these steps in **gene expression.**

### 9.1 Direction of Information Flow

In eukaryotes, DNA remains fixed in its location in the cell nucleus and protein synthesis takes place in the surrounding cytoplasm. Using specific stains or autoradiography, we can see that DNA is spatially separated from the sites of protein synthesis. Therefore, an intermediary must carry DNA instructions to these sites of protein synthesis. The logical candidate for such an intermediary, or messenger, is RNA, which is also a polynucleotide and can interact with DNA through complementary base pairing (Fig. 9.1). We would therefore expect RNA to be found in the nucleus and in the cytoplasm of actively synthesizing systems. In fact, we have known for a long time that protein synthesis will not take place in the absence of RNA.

Three kinds of RNA must interact during protein synthesis: **messenger RNA** (**mRNA**), **ribosomal RNA** (**rRNA**), and **transfer RNA** (**tRNA**). All these RNAs are complementary copies of DNA sequences, are synthesized during transcription, and all are required for translation of genetic information into polypeptides.

Each kind of RNA has a different but related function in translation. DNA codons, which specify amino acids, are transcribed into mRNA complementary codons, as we implied in our earlier discussion of the genetic code. If the DNA coding sequence is ATCGGCTA, then the

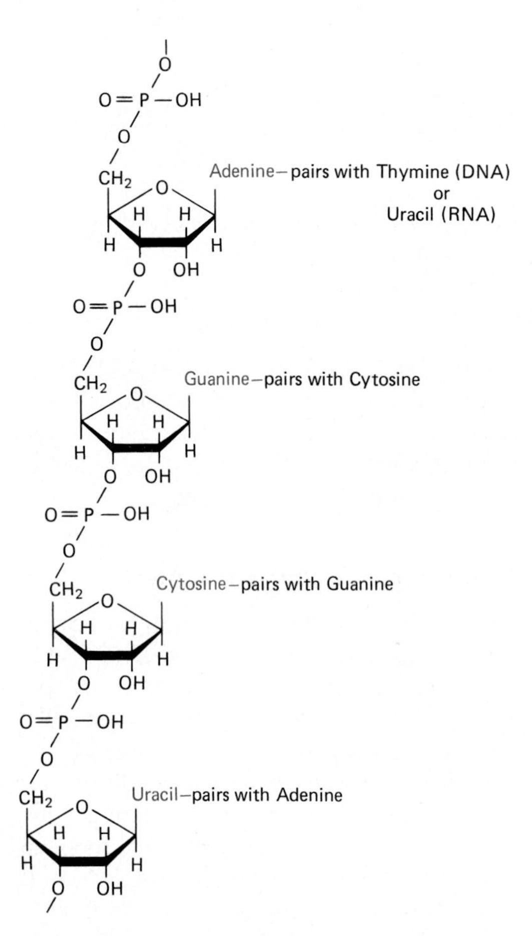


***Figure 9.1*** RNA polymers are composed of four kinds of nitrogenous bases (adenine, guanine, cytosine, and uracil) covalently bonded to a backbone of repeating phosphate and D-ribose sugar units. Pairing properties of bases in RNA depend on hydrogen bonding with particular complementary bases in DNA or RNA or with their constituent mononucleotide monomers, which contain these bases.

mRNA copy will be UAGCCGAU. Translation of the mRNA message absolutely requires the participation of ribosomes, which contain rRNA, and tRNAs, which carry amino acids to the mRNA-ribosome complexes. Complementary base pairing underwrites the accuracy of RNA interactions and, therefore, aids in the accurate translation of the genetic instructions (Fig. 9.2). The 5′ end of mRNA binds to the ribosome by base pairing with the 3′ end of ribosomal RNA. Amino acids are attached to specific transfer RNA molecules and are brought to the ribosome by tRNAs. The correct amino acid is inserted in correct sequence in the growing polypeptide chain through base pairing between the mRNA codon and the triplet **anticodon** of the tRNA structure. The recognition system, therefore, involves the fundamental molecular rule of base pairing between G and C and between A and T or U.

When the $Hb^A$ allele for the normal β-globin polypeptide of adult hemoglobin is transcribed into mRNA, the codon for *glutamic acid* is copied precisely into the messenger site corresponding to amino acid 6 in the polypeptide chain. During translation of the polypeptide, glutamic acid is correctly positioned in the growing chain because of complementary base pairing between the mRNA codon and the anticodon of the tRNA that carries this amino acid. If mRNA

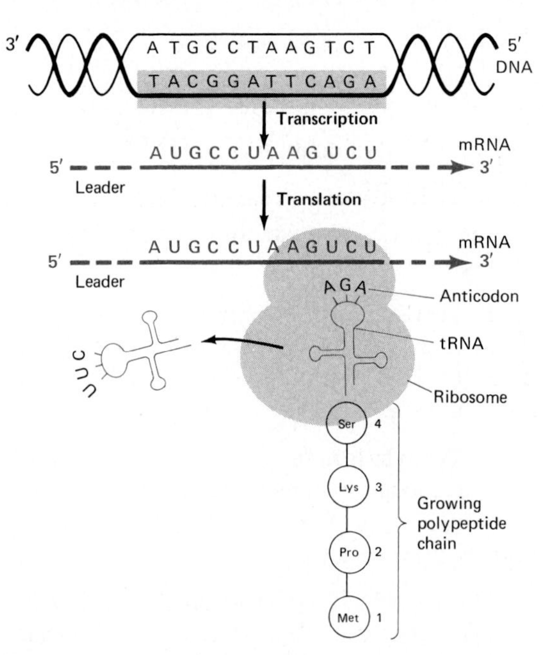


***Figure 9.2*** Complementary base pairing is the basic feature of the recognition system that provides for the high degree of accuracy in translation of the genetic message into a polypeptide at the ribosome.

is copied from the alternative $Hb^S$ allele for sickle-cell β-globin, then a different codon occupies the messenger site that corresponds to position 6 in the polypeptide. In this case, *valine* is installed at position 6 in the globin chain when a valine-carrying tRNA appears with the matching anticodon (Fig. 9.3). This seemingly simple change leads to profound alteration in protein shape and affects the oxygen-carrying capacity of sickle-cell hemoglobin molecules. The red blood cells that carry altered hemoglobin are responsible in turn for circulatory problems and other symptoms of the sickle-cell disorder in individuals who carry the homozygous recessive genotype $Hb^S Hb^S$ (see Fig. 1.13). For this disorder and for a number of other inherited disorders, we can trace the origin of a complex phenotype in the organism from a change in DNA sequence to a change in a polypeptide to a clinical set of symptoms.

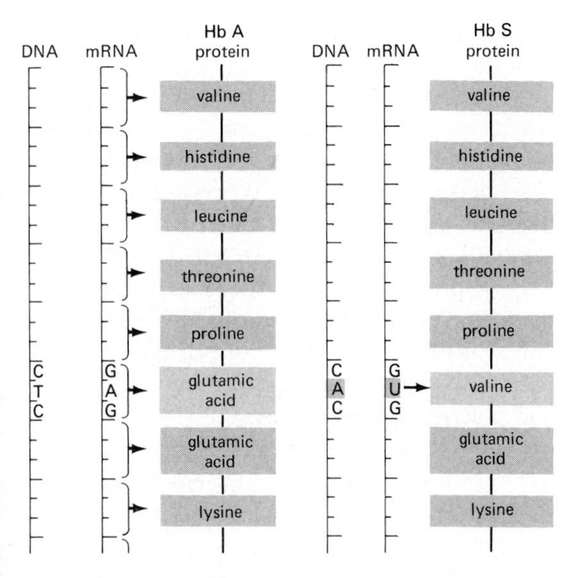


***Figure 9.3*** Glutamic acid is inserted at position 6 in the β-chain of HB A protein for normal adult hemoglobin, or valine at position 6 in Hb S (sickle-cell hemoglobin), according to the mRNA codon that is recognized by the transfer RNA (tRNA) anticodon, which carries the amino acid to the ribosome.

## 9.2 Transcription of mRNA

Enzymes that catalyze transcription are called *DNA-dependent RNA polymerases,* or usually just *RNA polymerases* for short. The template for synthesis is one strand of DNA, and the product of synthesis is an RNA polymer consisting of a single chain of covalently linked ribonucleotides that are complementary to template DNA. Of all the RNA polymerases, the *E. coli* enzyme has been characterized in the greatest detail by genetic and biochemical analyses.

RNA polymerase of *E. coli* consists of six polypeptide subunits: one each of β, β′, ω, and σ, and two α chains. If the σ subunit is dissociated from the whole enzyme *in vitro,* transcription is initiated at random along the DNA template, instead of at its special initiation site, and both DNA strands are copied instead of the usual coding strand. When σ is added back, transcription is initiated correctly and only the coding strand of DNA is copied. These data indicate that the RNA polymerase requires the σ factor for recognition of the initiation site, or promoter, on DNA.

The gene loci for *E. coli* RNA polymerase polypeptides are not clustered in the genome (Fig. 9.4). From the observed scatter between 66 and 89 minutes on the map, it appears that subunits may not be made coordinately and that they probably assemble into the whole func-

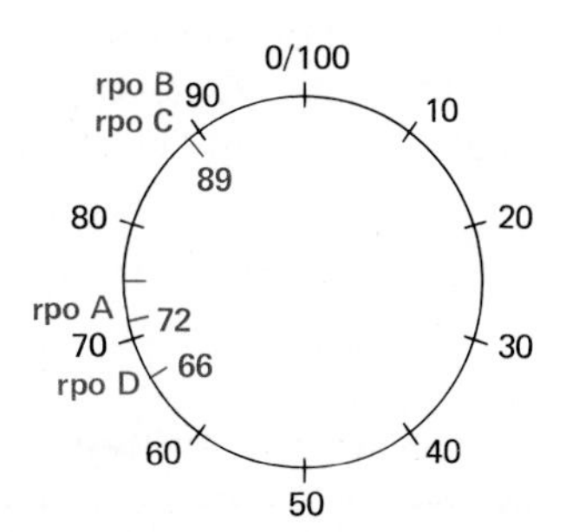


***Figure 9.4*** Genetic map of *E. coli* shows known genes that code for the polypeptide components of functional RNA polymerase protein.

tional enzyme in the cells. The *rpo* loci on the map (*rpo* stands for *R*NA *po*lymerase) are those for $\alpha$ (*rpoA*), $\beta$ (*rpoB*), $\beta'$, (*rpoC*), and $\sigma$ (*rpoD*); the gene for $\omega$ has not yet been mapped. The *rho* locus codes for the rho protein, which is essential for termination of transcription of the gene sequence; its mode of action is still uncertain.

Transcription in eukaryotes is controlled by three different nuclear RNA polymerases (Table 9.1). These enzymes can be distinguished by their subcellular location, by sensitivity to particular antibiotic drugs, and by the functional kinds of RNA transcripts that they produce.

When separated strands of duplex DNA are allowed to interact with RNA, most of the RNA binds to only one of the two DNA strands. The fact that an RNA transcript will bind mainly to one strand indicates that this is the principal template strand of DNA, or the coding **sense strand.** The other DNA strand is **antisense** and predominantly noncoding. Some genes are present in the antisense strand, but the majority of the genome is coded in the sense strand of cellular organisms. This information was obtained from **DNA-RNA molecular hybridization,** a method developed in the early 1960s by Sol Spiegelman and Benjamin Hall (Fig. 9.5).

The two DNA strands can be separated and identified because they contain different percentages of purines and the heavier A + G–rich strand settles in a different part of an alkaline CsCl gradient than its lighter partner, which has less A and G. The isolated **heavy (H)** and **light (L)** strands can then be hybridized separately with any known kind of RNA. Complementary DNA and RNA strands form **DNA-RNA hybrid duplexes.** Since base pairing is highly specific, investigators interpret these data as evidence that the RNA must have been transcribed from a DNA strand identical to the strand to which it binds. The antisense strand, therefore, is not transcribed. Molecular hybridization is a fundamental and extremely useful method in molecular genetics.

***Table 9.1*** Characteristics of RNA polymerases from animal nuclei.

| **polymerase type** | **nuclear location** | **cellular RNAs transcribed** |
|---|---|---|
| I | nucleolus | 18S and 28S rRNAs |
| II | nucleoplasm | pre-mRNA and mRNA |
| III | nucleoplasm | tRNAs, 5S rRNA |

In certain viruses and in some mitochondrial and chloroplast DNAs, parts of *both* the H and L strands may code for polypeptides or for tRNAs. In other words, different genes may be transcribed from the H and the L strands. In these cases, relatively little coded information is included in the L strand, and most of the coded genome is in the H, or sense, strand.

Whether the H or L strand is transcribed, *RNA polymerization proceeds in the* $5' \rightarrow 3'$ *direction, along the DNA template strand.* Ribonucleotides (like deoxyribonucleotides) are added *only* at the growing $3'$ end. Since the two chains are *antiparallel* and since mRNA is *translated in the* $5' \rightarrow 3'$ *direction,* the gene sequence must start at the $3'$ end of the DNA. These features are important considerations when comparing DNA and mRNA sequence complementarity and mRNA and polypeptide colinearity.

The relationship between template DNA and its mRNA transcript in prokaryotes and eukaryotes is as follows:

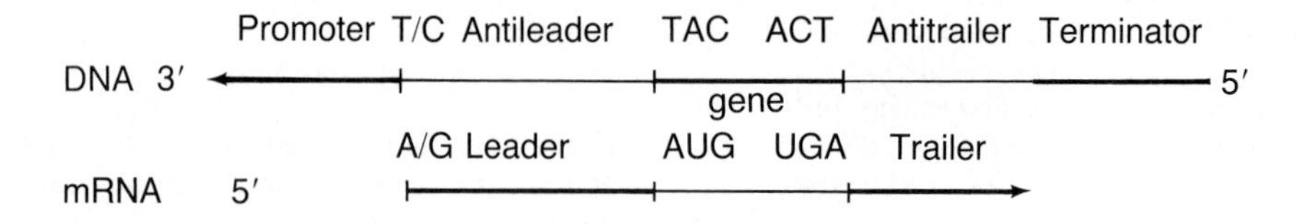

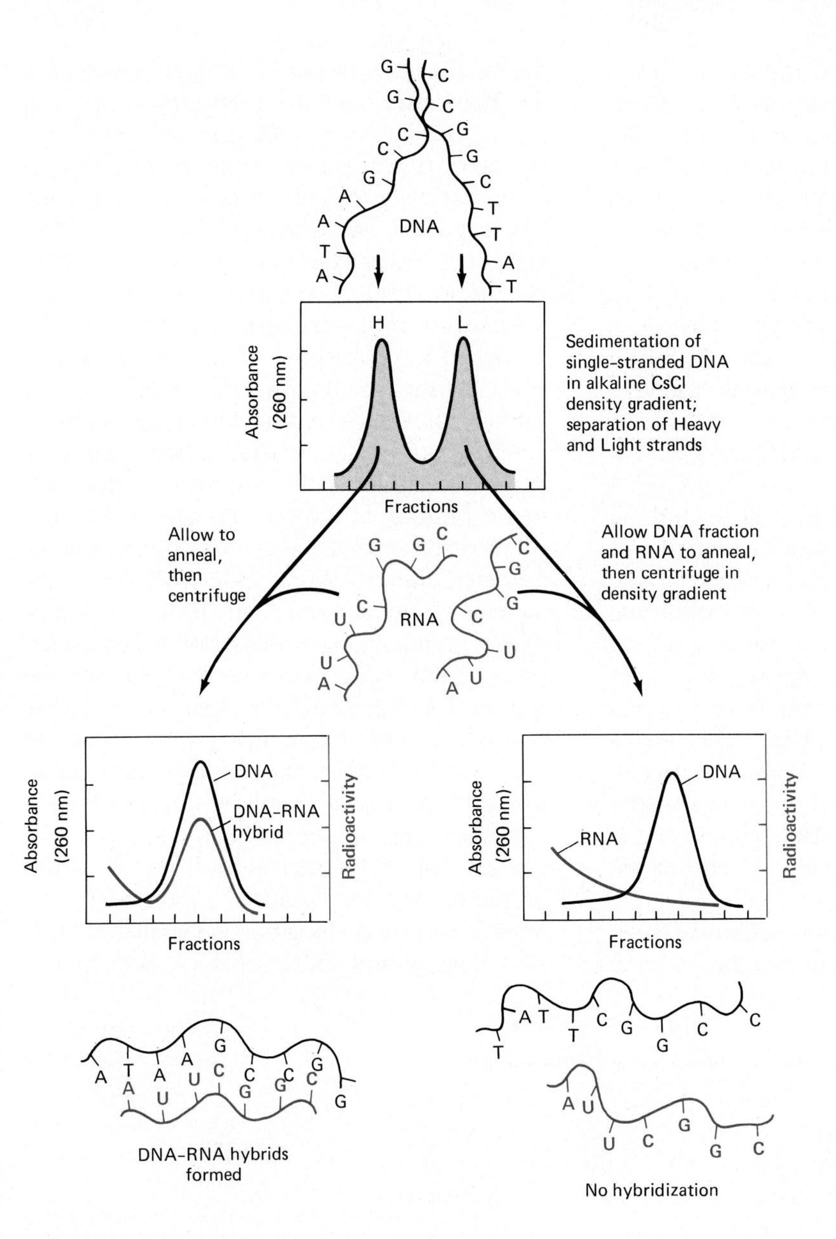


***Figure 9.5*** DNA-RNA molecular hybridization. Melted single strands of duplex DNA, separated by centrifugation, are mixed with radioactively labeled RNA in the annealing (association of complementary strands) phase of the procedure. The positions and amounts of DNA and of DNA-RNA molecular hybrids are revealed by absorbance at 260 nm and are distinguished by the radioactive labels carried. Curves such as these indicate whether or not hybridization has taken place.

Base sequencing of purified mRNAs and DNA has shown that complementary noncoding segments flank the region that codes for polypeptide. The mRNA **leader** is a sequence that is *upstream* of the coding region, extending from an invariant 5′ terminal purine (5′-pppA or 5′-pppG) to the initiation codon of the gene itself. The leader may consist of fewer than 20 to more than 600 nucleotides. *Downstream* of the mRNA coded message (toward the 3′ end of the mRNA molecule) is the noncoding **trailer** segment. The trailer extends from the termination codon of the gene (UGA, UAG, or UAA) to the 3′ end of the molecule.

Regardless of their length, all sequenced prokaryotic mRNA leaders contain a sequence of four or five bases that are complementary to the 3′ end of the rRNA exposed on the ribosome surface (Fig. 9.6). This fact suggests that at least part of the function of the leader sequence is to orient mRNA to the ribosome by promoting binding between the 5′ end of mRNA and a complementary 3′ end of rRNA. Thus, the beginning of the message would be placed in a proper position for sequential translation of the coded sequence, proceeding from the 5′ end of the message to its 3′ end.

The DNA sequence contains additional flanking regions beyond the leader and the trailer in mRNA. These *untranscribed* segments are called the **promoter** and the **terminator.** In both prokaryotic and eukaryotic gene sequences, the promoter region is involved in the *initiation of transcription by RNA polymerase.* The length and base sequence vary among promoters of different genes, except for a *heptanucleotide* (7 nucleotides). This heptanucleotide has been called a **Pribnow box,** for the person who first described it in prokaryotic DNA, and a **Hogness box,** for the person who described it in eukaryotic DNA. The heptanucleotide is also more generally referred to as a **TATA box** because of its bases (Fig. 9.7). The mechanism that promotes binding of RNA polymerase to the DNA promoter is unclear. As we mentioned earlier, however, the $\sigma$ subunit of *E. coli* RNA polymerase recognizes and binds to the DNA promoter in initiation of transcription. In eukaryotes, TATA boxes may be involved in the specific binding of RNA polymerase II to the promoter. Such binding may be one signal for the polymerase to transcribe an mRNA rather than a tRNA or rRNA sequence (Table 9.2).

Once transcription has been initiated—after binding of RNA polymerase to the DNA promoter—mRNA synthesis proceeds in the 5′ → 3′ direction along the antiparallel 3′ → 5′ DNA sense strand. Presumably, (1) DNA strands

| $\phi$X gene | 5′→3′ sequence of mRNA leader and initiator regions |
|---|---|
| A | C A A A U C U U G G A G G C U U U U U U *A U G* G U U |
| B | A A A G G U C U A G G A G C U A A A G A *A U G* G A A |
| D | C C A C U A A U A G G U A A G A A A U C *A U G* A G U |
| E | C U G C G U U G A G G C U U G C G U U U *A U G* G U A |
| F | C C C U U A C U U G A G G A U A A A U U *A U G* U C U |
| G | U U C U G C U U A G G A G U U U A A U C *A U G* U U U |
| J | C G U G C G G A A G G A G U G A U G U A *A U G* U C U |
| 16S rRNA (3′ end only) | HO A U U C C U C C A G U A G ..... 3′ → 5′ |

*Figure 9.6* Homology between 5′ leader segments and initiator codon AUG locations in different genes of the *E. coli* phage $\phi$X174. The segment in each mRNA that is complementary to the 3′ end of 16S rRNA is shown in colored letters. Such complementarity may promote binding of the 5′ end of mRNA to the exposed 3′ end of 16S rRNA of the small ribosomal subunit, just before translation begins.

**TATA Sequence Homologies**

| Eukaryotic genes | TATA sequences (5′ → 3′) |
|---|---|
| rabbit β-globin | CATAAAA |
| mouse β-globin | TATATAA |
| rat insulin | TATAAAG |
| chick insulin | TATAATT |
| chick ovalbumin | TATATAT |
| **Prokaryotic genes** | |
| *E. coli lac* | TATTGGT |
| *E. coli str* | TAAAATT |
| phage T7 *A2* | TAAGATA |
| phage λ *PRE* | AAGTATT |

***Figure 9.7*** Heptanucleotide TATA sequences, or TATA boxes, are found within the promoter sequence of the gene or operon. TATA sequences vary but all are believed to contribute to the correct initiation of transcription, perhaps by binding RNA polymerase to the DNA strand or by providing a recognition signal of some sort to the transcription enzyme. The noncoding 5′ → 3′ strand orientation is shown here, from which the complementary 3′ → 5′ coding DNA strand or the 5′ → 3′ mRNA can be deduced (substitute U for T in mRNA, see Box 7.3).

separate just ahead of the growing mRNA chain, (2) the DNA sense strand is transcribed as mRNA, (3) the mRNA is displaced, and (4) the DNA duplex closes when transcription has been completed in that segment. Transcription terminates when the RNA polymerase reaches the terminator region, beyond the antitrailer gene segment.

Most prokaryotic genes seem to have a string of GC base-pairs, followed by a string of AT pairs at the 5′ end of the antitrailer. The 3′ end of prokaryotic mRNAs usually consists of the sequence 5′–UUUUUUA–3′. It has been suggested that the physical difference between stronger bonding in GC stretches and looser bonding in AT stretches acts as a signal for dissociation of the RNA polymerase from template DNA. Whatever the signal may be, it involves an interaction between proteins and the DNA sequence. As we mentioned earlier, the presence of rho protein is required to terminate transcription in *E. coli*. Our understanding of interactions between proteins and nucleic acids lags behind our understanding of interactions between nucleic acid molecules.

## 9.3 Posttranscriptional Processing of Eukaryotic mRNA

A major difference between mRNA in eukaryotes and their viruses compared with mRNA in prokaryotes and their viruses is that the former undergoes a significant amount of *posttranscriptional modification* after dissociation from the DNA template. Eukaryotic messenger transcripts are modified posttranscriptionally in at least four ways:

***1.*** Removal of up to 90% of the length of a precursor mRNA (**pre-mRNA**) prior to production of a **mature mRNA;**

***Table 9.2*** Some characteristics of cytoplasmic ribosomes.

| source | ribosome | ribosomal subunits | rRNA in subunits | number of proteins in subunit |
|---|---|---|---|---|
| prokaryotes | 70S | 30S | 16S | 21 |
| | | 50S | 23S, 5S | 32–34 |
| eukaryotes | 80S | 40S | 18S | ~30 |
| | | 60S* | 25–28S, 5S, 5.8S | ~50 |

*The large rRNA in animal 60S subunits is 28S; whereas the molecule is 25S to 26S in 60S subunits of plant, fungal, and protist ribosomes.

*2.* **"Capping"** of the 5′ purine terminus with the unusual nucleoside **7-methylguanosine** ($m^7G$) to produce 5′-$m^7G$-ppp(leader)-gene-(trailer)-3′;

*3.* Addition of a string of adenine nucleotides, making up a **poly(A) tail** at the 3′ terminus; and

*4. Excision* of noncoding intervening sequences that were copied from the gene, followed by ligation, or **splicing,** of coded exon sequences at splice junctions to produce mature mRNA whose coding sequence corresponds to the coded amino acid sequence in the polypeptide to be made during translation (Fig. 9.8).

In most other respects, such as presence of noncoding leader and trailer segments, eukaryotic and prokaryotic mRNAs are fairly similar.

Differences between eukaryotic and prokaryotic messengers may be related partly to the timing of their transcriptional activities and to their transport within the cell. In prokaryotes, transcription and translation are coordinated processes (Fig. 9.9). While mRNA is still being transcribed from the DNA template, the first rounds of translation begin at the 5′ end of the incomplete mRNA molecule. There is relatively little time for any extensive processing of prokaryotic mRNA before translation begins. In addition, mRNA and its DNA template are in intimate association during translation. Thus, mRNA remains near its site of synthesis and does not pass through membranes since the nucleoid and cytoplasm are not separated by membranes in prokaryotic cells.

In eukaryotes, mRNA is transcribed in the nucleus along chromosomal DNA. It must then be transported across the nucleus and through the nuclear membranes in order to enter the cytoplasm. Once in the cytoplasm, mRNA may associate with ribosomes and participate in translation.

All or most of the posttranscriptional processing of eukaryotic mRNA takes place in the

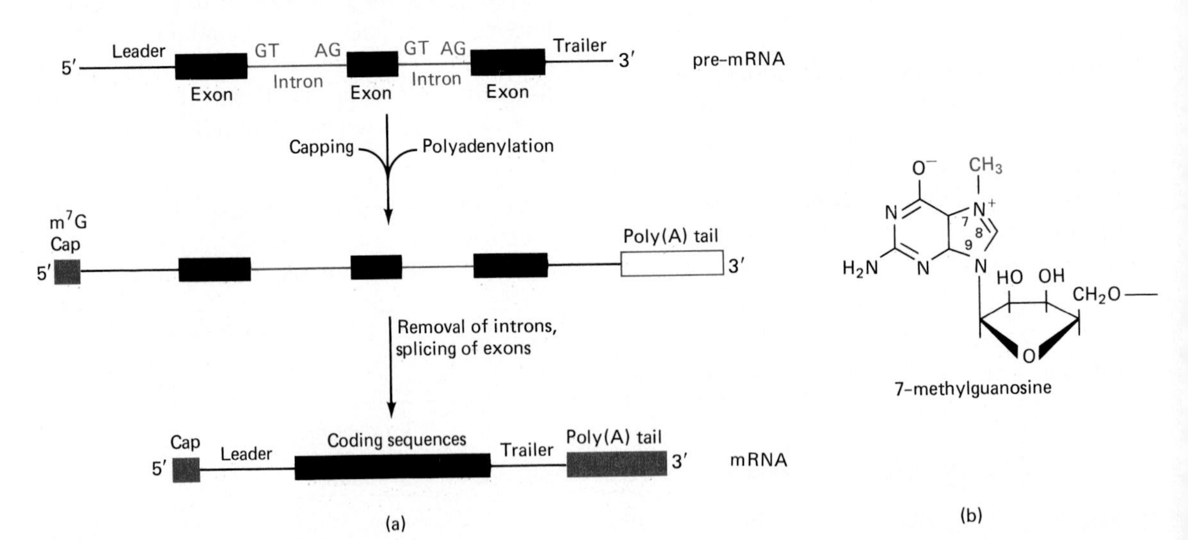


***Figure 9.8*** Posttranscriptional modification of eukaryotic mRNA. (a) The initial transcript, called pre-mRNA, includes noncoding leader and trailer segments that flank the mosaic of noncoding intervening sequences (introns) and coding sequences (exons) of the gene. Pre-mRNA is capped at its 5′ terminus with 7-methylguanosine, and its 3′ terminus is lengthened by the addition of adenine residues making up a poly(A) tail. This modified molecule is further altered in the nucleus by excision of introns at GT and AG splice sites, and splicing of exons to make a continuous reading frame for translation. (b) Chemical formula of the unusual nucleotide cap of the mRNA strand.

nucleus. Comparisons between pre-mRNA isolated from nuclei and mature mRNA isolated from cytoplasm reveal that removal of large chunks of pre-mRNA takes place in the nucleus. The excised portions of pre-mRNA are found in nuclear preparations but not in the cytoplasm. All the other processing steps take place in the nucleus; they begin with capping, proceed through addition of the poly(A) tail by polyadenylation, and culminate in excision and splicing to produce mature mRNA. The mRNA that leaves the nucleus is not a naked RNA molecule,

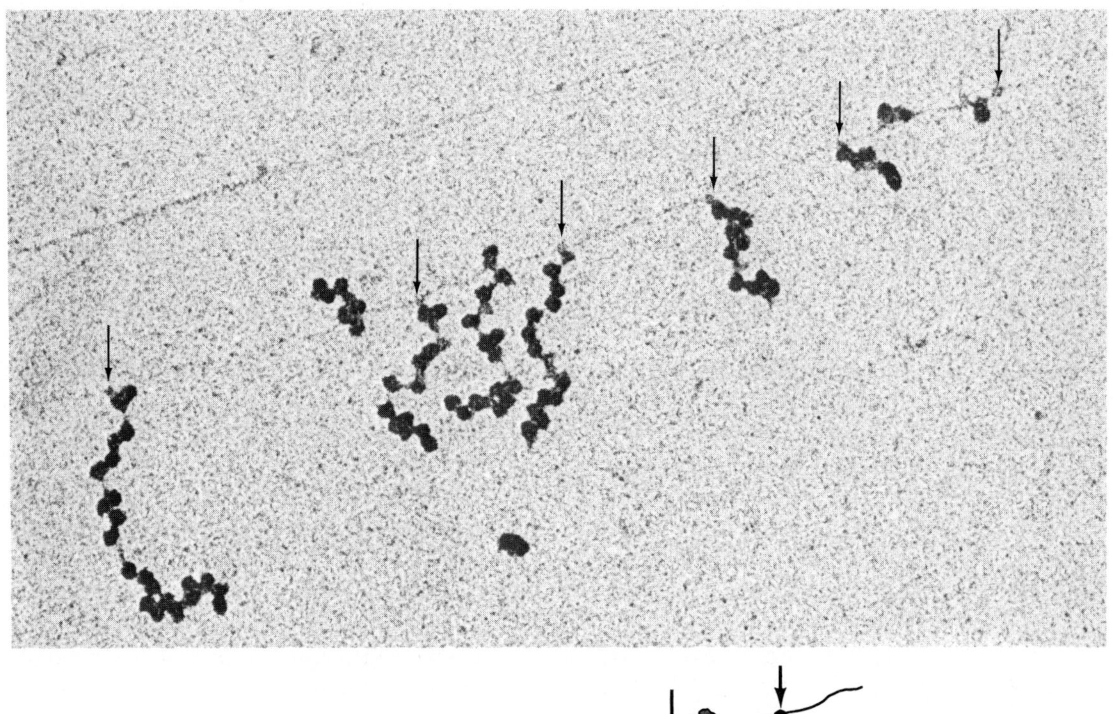

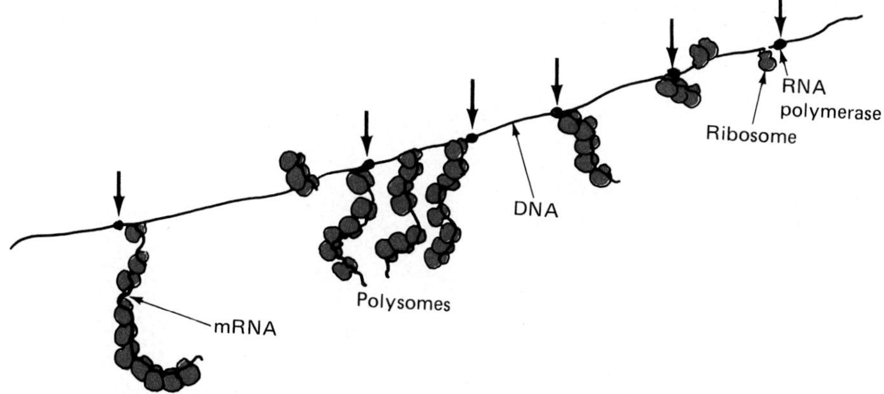


***Figure 9.9*** Electron micrograph of transcription-translation complexes from *E. coli.* Ribosomes bind to mRNA as the messenger "peels off" the template DNA during transcription, seen here by increasing length of the mRNA-ribosome aggregates when viewed from right to left along a gene segment. Molecules of RNA polymerase (arrows) catalyze transcription. An explanatory diagram is shown below the photograph. (O.L. Miller, Jr., et al. "Visualization of Bacterial Genes in Action," *Science* **169**:392–395, Fig. 3, 24 July 1970. Copyright © 1970 by the American Association for the Advancement of Science.)

however. Proteins are bound to mRNA and may aid and protect the messenger. As a *ribonucleoprotein* in the nucleus, mRNA is protected from degradation by the abundant nucleases that are present. The proteins may also facilitate penetration of the nuclear membranes. These suggestions are based more on analogies with other chemical systems than on direct experimental evidence. They do seem reasonable, however, and in line with other kinds of evidence on molecular interactions in cells.

The $m^7G$ cap on the terminal 5′ purine of the mRNA is believed to aid translation in some manner. The poly(A) tail may help bind mRNA by its 3′ end to membranes of the endoplasmic reticulum in eukaryotic cytoplasm (Fig. 9.10), thereby helping to stabilize the mRNA at one end while its 5′ end was secured by binding to ribosomes attached to the endoplasmic reticulum. Binding of the 3′ end should also protect mRNA from degradation by abundant ribonucleases in the cytoplasm. These proposals regarding the cap and tail functions remain speculative, however.

Excision of intervening sequences and ligation of coding sequences to create a contiguous sequence of codons for translation are remarkably exact processes. The presence of **consensus sequences,** such as TCAGGT, at splice junctions may function as splicing signals. In every case RNA splicing occurs immediately 5′ to a GT doublet and 3′ to an AG doublet at the termini of an intervening sequence, or intron:

```
pre-mRNA 5′ ———|GT · · · · · · · · AG|——— 3′
              Exon        Intron       Exon
```

We do not yet know whether the identity of splice junctions indicates identity or similarity of enzymatic events that catalyze the splicing processes in eukaryotic pre-mRNAs. In any case intervening sequences may occur between whole codons of a message or they may even interrupt a triplet codon. In the β-globin gene of the rabbit, 550 nucleotides are excised between the codons for amino acids 104 and 105 of the 146 amino acids in the globin polypeptide. After the intervening sequence is excised, exons are spliced together so that codons for amino acids 104 and 105 are immediately adjacent in the mature mRNA.

The *ovalbumin gene* of the chicken has been sequenced, as has its mature mRNA. The gene consists of 7564 base-pairs, including eight exons and seven introns. The mature mRNA is 1872 nucleotides long, with 64 nucleotides making up the leader at the 5′ end of the molecule and 650 nucleotides making up the stop codon

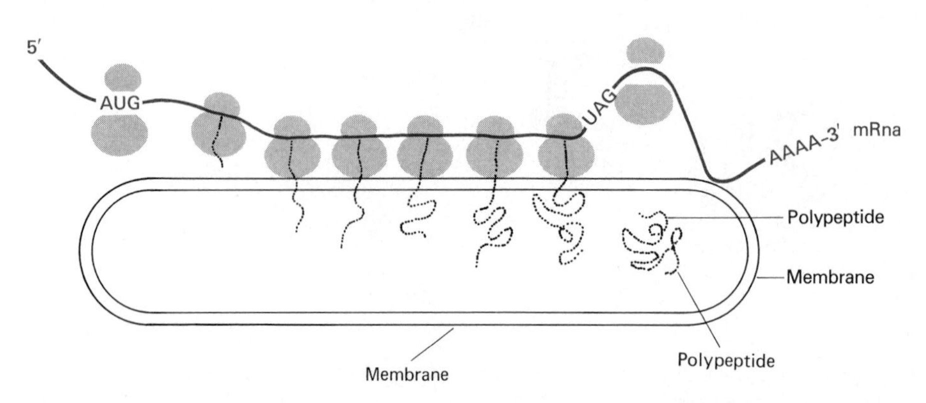


***Figure 9.10*** Assembly of bound ribosomes with mRNA (color) attached by its 3′ end to the membrane of a eukaryotic cell. Binding of the complex to the membrane may occur through interactions between growing polypeptide chains and the membrane in addition to, or instead of, mRNA binding as shown. (Reproduced from *The Journal of Cell Biology,* 1975, Vol. 65, p. 513 by copyright permission of The Rockefeller University Press.)

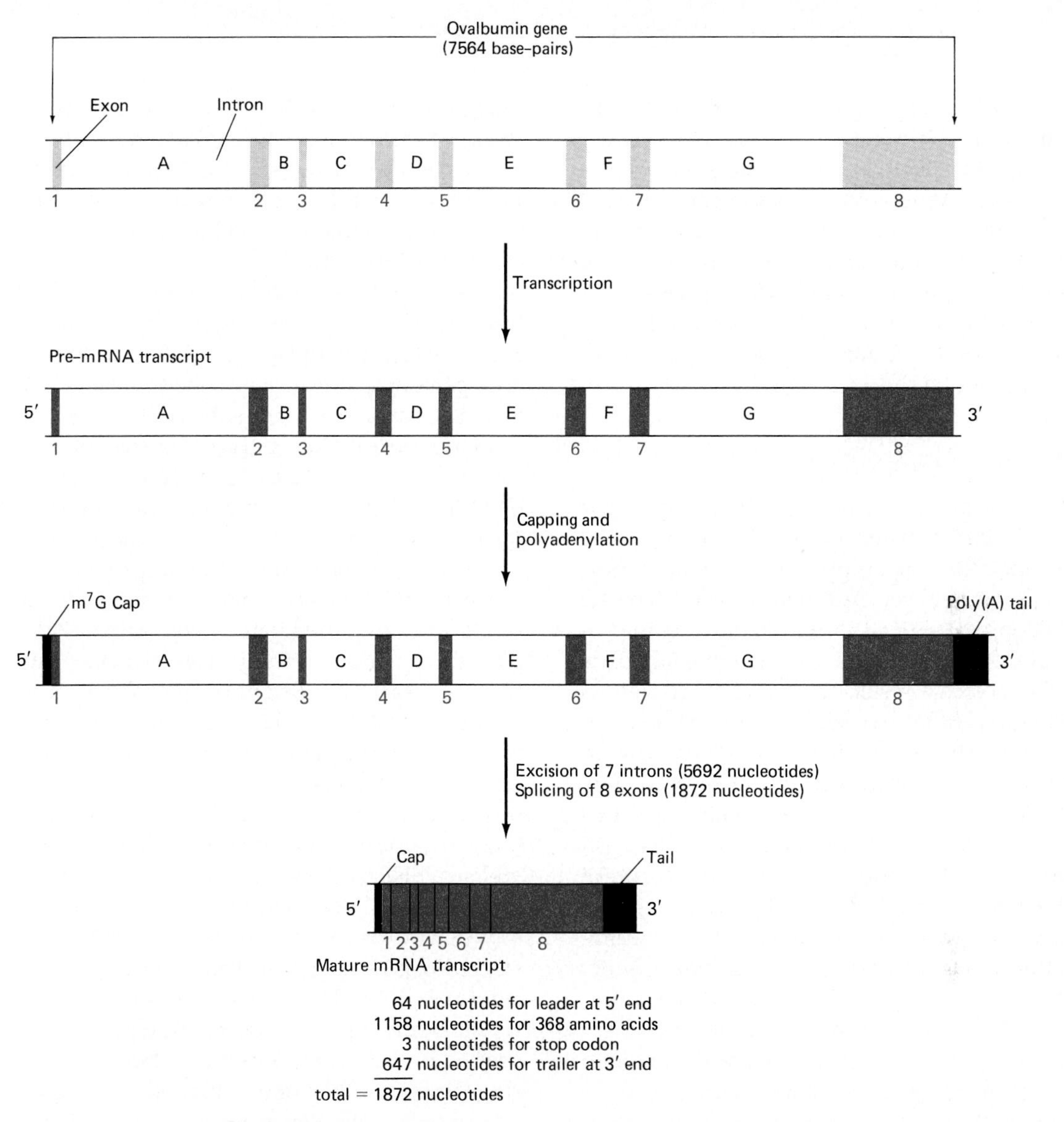


***Figure 9.11*** Posttranscriptional modification of the pre-mRNA transcript of the chicken ovalbumin gene is remarkably precise and takes place countless times in nuclei of the animal's oviduct cells. After capping and polyadenylation, the bulk of the transcript is removed when the seven introns are precisely excised. The remaining 1158 nucleotides, which code for the ovalbumin protein, are spliced into a continuous reading frame that remains flanked by untranslated cap and leader sequence at the 5′ end and by a stop codon and trailer at the 3′ terminus of the mature mRNA molecule. The processed mRNA enters the cytoplasm from the nucleus and participates in translation at the ribosomes. (Data from S.L.C. Woo et al. 1981. *Biochemistry* **20**:6437–6446. [1981]. Copyright 1981 American Chemical Society.)

and the trailer at the 3′ end. In between are 1158 nucleotides, consisting of 386 codons that specify the 386 amino acids in ovalbumin (Fig. 9.11). The seven intervening sequences in this gene range in length from 52 to 1589 nucleotides, and their sequences vary except at the splice junctions. Processing of pre-mRNA to produce the mature mRNA of the ovalbumin sequence takes place repeatedly in oviduct cells, with remarkable precision in every cell of every animal. The same accuracy characterizes the processing of eukaryotic mRNA in general. Post-transcriptional modifications characterize the processing of rRNA and tRNA transcripts from genes coding for rRNA and tRNA. Processing of these two types of RNAs differs from mRNA in some important features, however, as we will now discuss.

## 9.4 tRNA Transcription

Precursor tRNA is not much longer than the mature tRNA that functions in translation. The precursor transcript is processed so that its short 5′ leader is removed, its 3′ end is extended by the addition of 3′–ACC–5′, and its intervening sequence excised. Ligation of the remaining parts of the molecule results in mature tRNA, with its *anticodon* triplet exposed on the anticodon arm (Box 9.1). The tRNA anticodon is the recognition site, which base-pairs with the complementary mRNA codon that specifies the amino acid carried on the 3′ end of the tRNA molecule. Post-transcriptional modifications convert some of the common bases, U, A, G, and C, to unusual bases even in the anticodon (Fig. 9.12). An important modified base is *inosine,* which is derived by deamination of adenine at carbon 6.

According to the genetic code, we might expect 61 different tRNAs, each with a unique anticodon, to be needed for pairing with 61 unique mRNA codons that specify amino acids. In fact only 32 different tRNAs handle all 61 mRNA codons. Crick proposed the **wobble hypothesis** to explain the puzzling situation of fewer tRNAs than expected and the presence of modified bases in tRNA anticodons. In essence, he suggested that the 5′ base of the anticodon is a wobble base, which is able to pair with more than one complementary 3′ base in the mRNA codon in certain cases. We expect the tRNA anticodon 3′–AGU–5′ to pair with the mRNA codon 5′–UCA–3′. According to the wobble hypothesis, 5′–U in the anticodon can also pair with 3′–G in a codon. Thus, 3′–AGU–5′ would also recognize the codon 5′–UCG–3′. The 5′–U is a wobble base and it is *less restricted in its pairing properties when occupying the 5′ position in the tRNA anticodon.* Pairing is restricted to G-C and A-U for the middle base and the 3′ base of any anticodon (Fig. 9.13).

The wobble hypothesis is based on observed features of the synonymous mRNA codons for amino acids, as displayed in the conventional arrangement for the genetic code (see Fig. 7.15). Synonymous mRNA codons have the same first two bases but different 3′ bases. For example, the four valine codons are GUU, GUC, GUA, and GUG. The 5′ and middle bases are the same GU, but any of the four bases can occupy the 3′ site in these synonymous codons. The only exceptions are two amino acids, each specified only by single codons (methionine and tryptophan), plus three amino acids (serine, leucine, and arginine), each specified by six mRNA codons. In the cases of all three amino acids, four of the six codons have the same first two bases and the other two codons have their own identical doublets in the 5′ and middle base position. These observations of identical doublets at the 5′ end and different bases at the 3′ end of mRNA synonymous codons led to the suggestion that the 3′ base of the codon might have less restricted pairing specificity with the complementary (wobble) 5′ base in the anticodon.

All possible pairings do not take place, however, and only 5′-U and 5′-G in the tRNA anticodon can pair with either one of two different bases; 5′-A and 5′-C pair as expected, with only 3′-U and 3′-G, respectively. Inosine, however, can pair with 3′-A, 3′-U, or 3′-C in mRNA codons (see Fig. 9.13).

Support for the wobble hypothesis came from studies showing that the six serine codons could be handled by the predicted minimum number of three tRNAs, each tRNA with a different anticodon that carries a 5′ wobble base. Whether or not the wobble hypothesis fully explains the presence of only 32 tRNAs is still open to some question. Additional experimental evidence is needed to understand the situation completely. We do know, however, that tRNAs

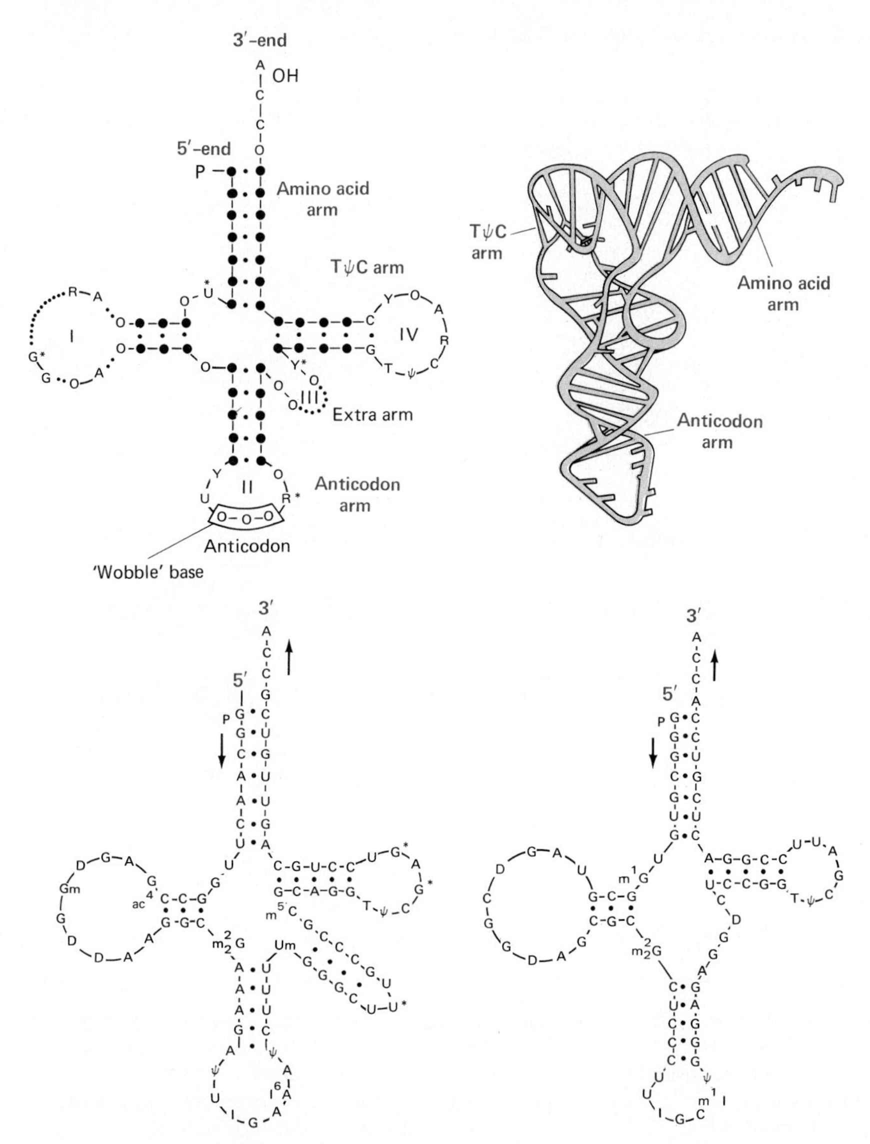


***Figure 9.12*** Molecular structure of transfer RNA (tRNA). (a) Generalized cloverleaf model of secondary structure of the single-stranded RNA molecule. Solid circles represent bases in the hydrogen-bonded helical regions, and open circles stand for unpaired bases. Unusual bases include ribothymidine (T), pseudouridine (ψ), inosine (I), and others. (b) Schematic model of tertiary structure of yeast $tRNA^{Phe}$ (From S.H. Kim, et al. 1974. *Science* **185**:435.) (c) Secondary structure of yeast $tRNA^{Ser}$ in which inosine occupies the 5′ position in the anticodon 3′–AGI–5′. (d) Secondary structure of yeast $tRNA^{Ala}$, the first nucleic acid that was completely sequenced (by Robert Holley, 1965). The 3′–CGI–5′ anticodon should pair with mRNA codons 5′–GCA–3′, 5′–GCU–3′, and 5′–GCC–3′, according to the wobble concept.

specified by mitochondrial genes may number only 22 to 25. The mitochondrial genetic code varies somewhat from the code in cellular systems, and the mitochondrial tRNA repertory is consistent with the particular codon-usage pattern that is unique to the mitochondrial genetic system. We will discuss the mitochondrial system in Chapter 13.

Eukaryotic genomes contain hundreds of copies of tRNA genes in middle-repetitive DNA

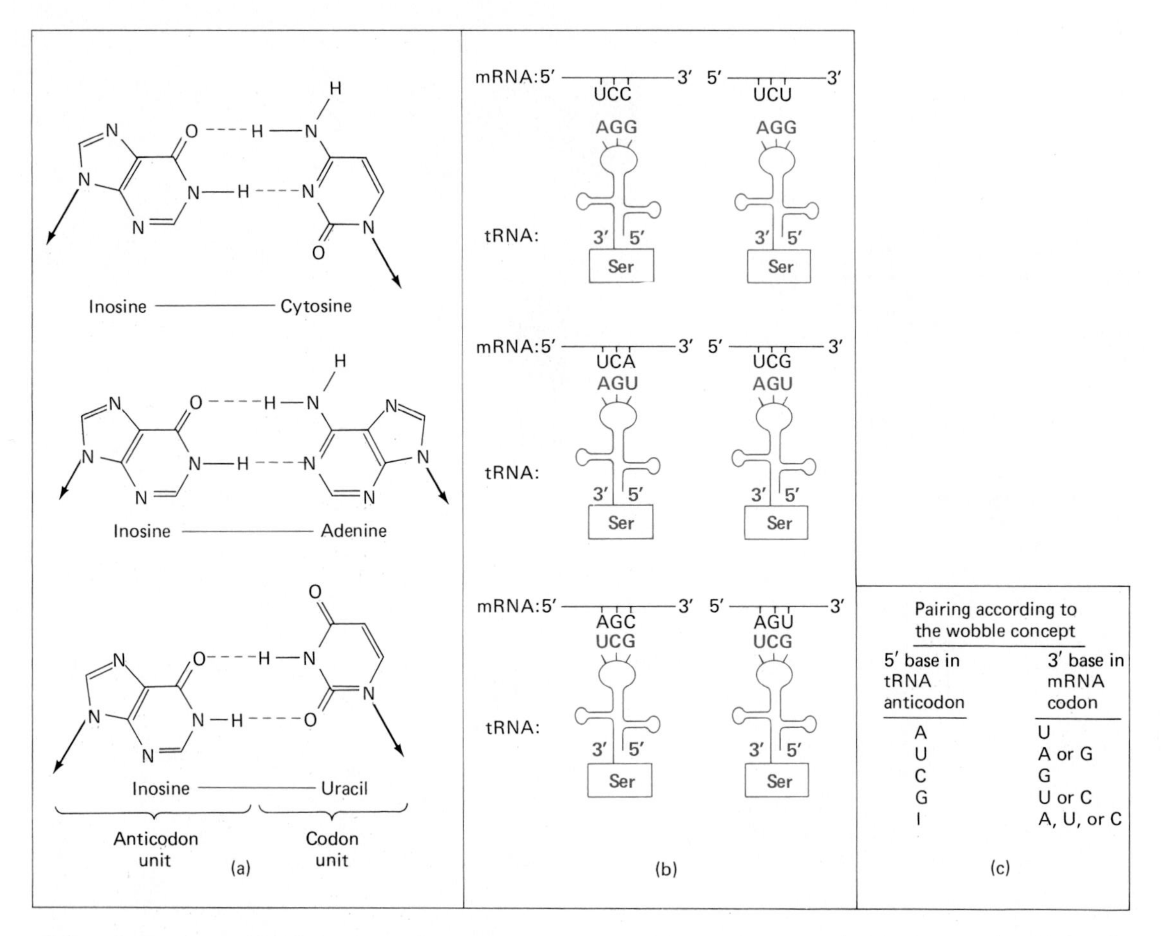


***Figure 9.13*** The wobble hypothesis. (a) Pairing of certain bases is less restricted, as seen by the example of inosine, which may pair with cytosine, adenine, or uracil, rather than with a single complementary base. (b) Fewer than 61 tRNA anticodons can interact with all 61 amino acid–specifying mRNA codons because of the less restricted pairing between certain 5′ bases in the tRNA anticodon (wobble-base position) and the 3′ base in the mRNA codon. The amino acid serine is specified by six mRNA codons, but these can interact with a minimum of three tRNA anticodons. (c) Possible pairing relationships between the 5′ anticodon wobble base and the 3′ codon base show different degrees of restriction, some of which were shown in (b).

sequences (see Section 8.11). Multiple copies of some tRNA genes are found in prokaryotic DNA, but these repeated sequences are relatively few in number. The high redundancy of eukaryotic tRNA gene copies far exceeds the level known for prokaryotic genomes.

## 9.5 rRNA Transcription

Some of the largest RNA molecules that participate in protein synthesis are *ribosomal RNAs.* There may be several thousand nucleotides in the largest rRNA chains, only 70 to 80 in tRNAs, and a variable number in mRNAs, according to the coding and noncoding sequence lengths. Even the ovalbumin mRNA, consisting of 1872 nucleotides, is only about half the length of the largest rRNA molecules.

The conventional way of describing and referring to RNA molecules and ribosomes is according to their sedimentation coefficient, expressed in *Svedberg units* (S). These coefficients

## BOX 9.1 Processing Pre-tRNA

Like mRNA and rRNA, tRNA is transcribed in a precursor form, which is processed into the mature, functional RNA that is active in trans-

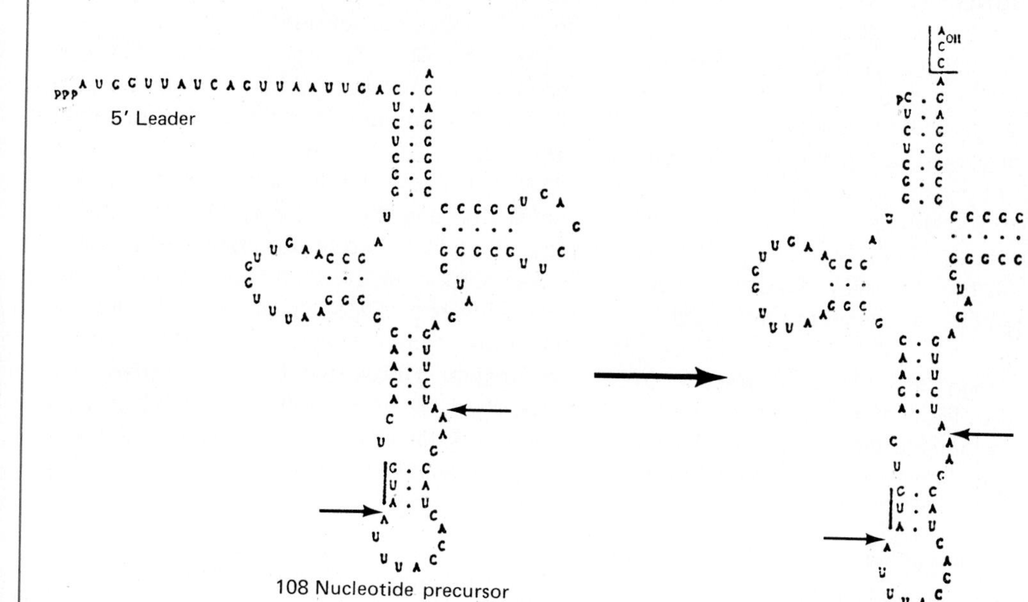


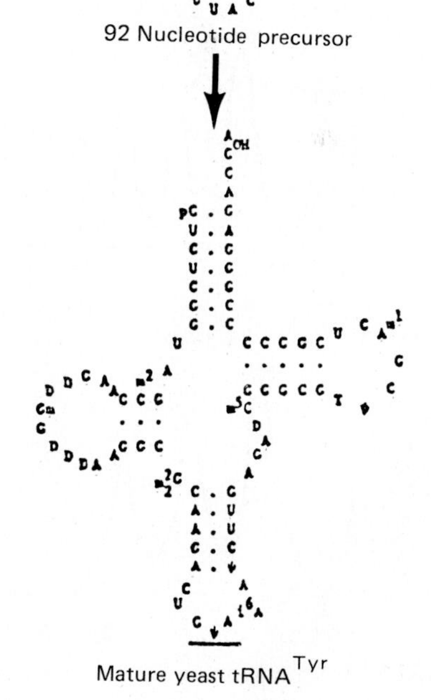


lation. The main processing stages are shown for yeast $tRNA^{Tyr}$. The primary transcript has a 5′ leader sequence, 19 nucleotides long, and an intervening sequence, 14 nucleotides long (between arrows). The 5′ leader is removed in at least two steps and the molecule is converted into a 92-nucleotide precursor form. The 3′–ACC–5′ end of the molecule is added post-transcriptionally to this precursor. The intervening sequence of 14 nucleotides is then excised and the molecule is re-ligated to produce mature $tRNA^{Tyr}$. The position of the anticodon in the various forms is underlined. All forms are depicted as cloverleaf secondary structures. (From E.M. DeRobertis and M.V. Olson. 1979. *Nature* **278**:137, Fig. 7.)

are calculated from analysis of molecule or particle sedimentation in sucrose-gradient centrifugation (Fig. 9.14).

Ribosomes and ribosomal subunits are also characterized by and discussed with reference to their S values. Each ribosome is made up of two unequal **subunits.** A 50S and a 30S subunit comprise the prokaryotic ribosome, which has a value of 70S. The eukaryotic ribosome is 80S and consists of a 60S and a 40S subunit (see Table 9.2). Since shape as well as molecular weight are determining factors in sedimentations, the sum of the S values for two subunits is greater than the determined S value for a whole ribosome in sedimentation tests. Similarly, the 23S rRNA is twice as long as the 16S rRNA, but the absolute difference in their lengths is not reflected in their S values.

In both prokaryotes and eukaryotes transcription of rRNA leads to a large precursor, or **pre-rRNA**, molecule. Certain differences characterize each type, however, so we will briefly describe the types separately. In *E. coli,* each of the seven rRNA genes (*rrn*) can be transcribed into a 30S pre-rRNA molecule, although this is rarely found because the molecule is processed *during* transcription (Fig. 9.15). The first fragment to be cleaved from pre-rRNA is 16S rRNA; and as transcription proceeds, 23S rRNA and finally 5S rRNA are released.

In eukaryotes, the size of pre-rRNA varies according to the group of species; insect pre-rRNA is 37S, amphibian is 40S, and mammalian pre-rRNA is 45S. In every case, however, pre-rRNA is processed into only two rRNA molecules: 18S and 25S to 28S rRNAs. Pre-rRNA is transcribed from **nucleolar chromatin,** which is the portion of the chromosomal nucleolar-organizing region (NOR) that extends into the nucleolus proper. Hundreds of rRNA genes, classified as **ribosomal DNA,** or **rDNA,** are transcribed simultaneously from a continuous strand of DNA (Fig. 9.16). The molecules are processed within the nucleus, where they also

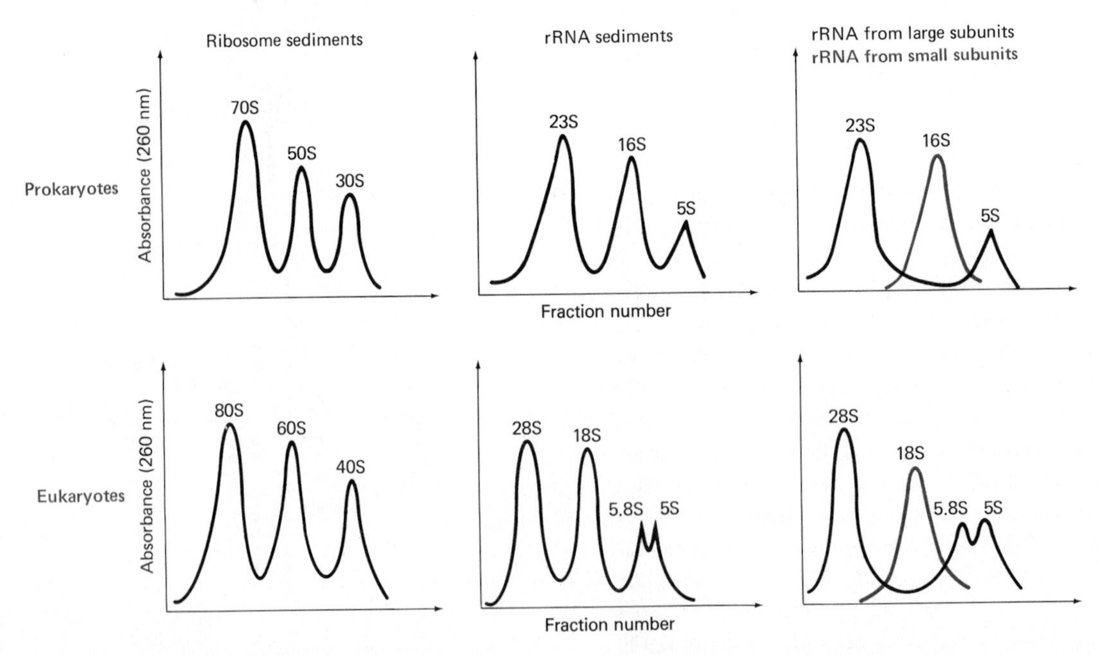


***Figure 9.14*** Separation of ribosomes and rRNAs from cytoplasm of prokaryotic (upper row) and eukaryotic (bottom row) cells by centrifugation in sucrose gradients. Whole ribosomes (70S or 80S) consist of two subunits of unequal size (50S + 30S or 60S + 40S) and several rRNA molecules of different sizes. Purified small subunits contain one rRNA (color); all the other rRNAs are present in the larger subunit of both the 70S and 80S ribosomal types.

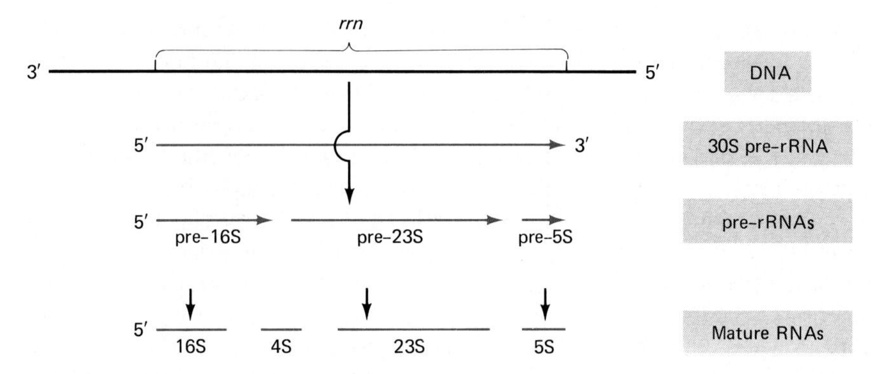


***Figure 9.15*** Processing of the pre-rRNA transcript of an *rrn* gene in *E. coli* results in 16S, 23S, and 5S rRNAs, along with one tRNA (4S) molecule). In RNAase-deficient *E. coli* it is possible to recover 30S pre-rRNA; otherwise, the molecule is processed into smaller units *during* transcription.

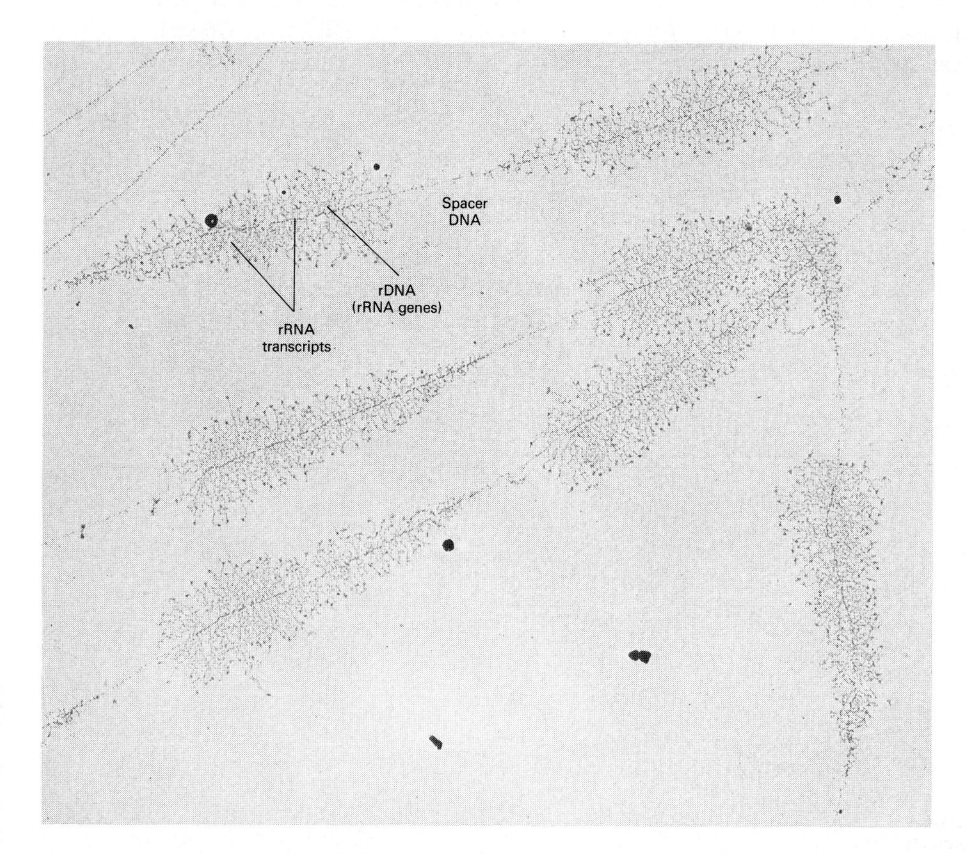


***Figure 9.16*** Electron micrograph of actively transcribing rDNA isolated from oocyte nucleoli of the spotted newt (*Triturus viridscens*), an amphibian. Each arrow-shaped region represents one rRNA gene with rRNA transcripts "peeling off" during transcription. The longer rRNAs in a group are near the 5′ end of the gene; the short transcripts are still near the 3′ end of the gene, where transcription begins. The rRNA genes (= rDNA) occur as tandem repeats, which are separated by untranscribed spacer DNA segments of uncertain function. (Miller, O.L., Jr., and Beatty, B., "Visualization of Nucleolar Genes," *Science* **164**:955–957, Fig. 2, 23 May 1969. Copyright © 1969 by the American Association for the Advancement of Science.)

may assemble with ribosomal proteins to form inactive or precursor ribosomal subunits. By processes that we do not entirely understand, inactive subunits are transformed into active ribosomal subunits, probably at the nuclear envelope or in the adjacent cytoplasm.

Time studies of pre-rRNA processing show that 18S rRNA appears first and the remainder of the original 45S pre-rRNA is cleaved to a 32S molecule (Fig. 9.17). Sometime later, the 32S molecule is processed to the finished 28S rRNA. Genes for 5S rRNA are situated elsewhere in the genome, not necessarily near rDNA of the NOR chromosome. Transcription of 5S rRNA apparently is under separate control.

In 1965 F.M. Ritossa and S. Spiegelman, using DNA-RNA molecular hybridization, first demonstrated localization of rDNA to the NOR of the eukaryotic chromosome. Earlier studies of an anucleolate mutant of *Xenopus laevis* indicated that absence of the NOR correlated with absence of rRNA and ribosome synthesis in young embryos. Ritossa and Spiegelman therefore sought molecular evidence that rRNA genes were located at the NOR.

They used four strains of *Drosophila melanogaster* that contained identical chromosomes except that each had from one to four NORs per nucleus. They supplied radioactively labeled uridine and they purified newly synthesized rRNA from isolated ribosomes. They then hybridized this labeled rRNA with chromosomal DNA from the same strains. The percentage of DNA-rRNA hybrids exactly paralleled the number of NOR per strain (Fig. 9.18). This result showed conclusively that rDNA was located at the NOR since the percentage of molecular hybrids reflected the amount of rDNA present in each strain. In later studies they showed that 5S rRNA and 4S tRNA did not increase when extra NORs were present; genes for these RNAs were situated elsewhere in the genome.

## Protein Synthesis

Most proteins are coded by single-copy genes. In addition to mRNA, tRNA, and ribosomes, translation of the genetic message re-

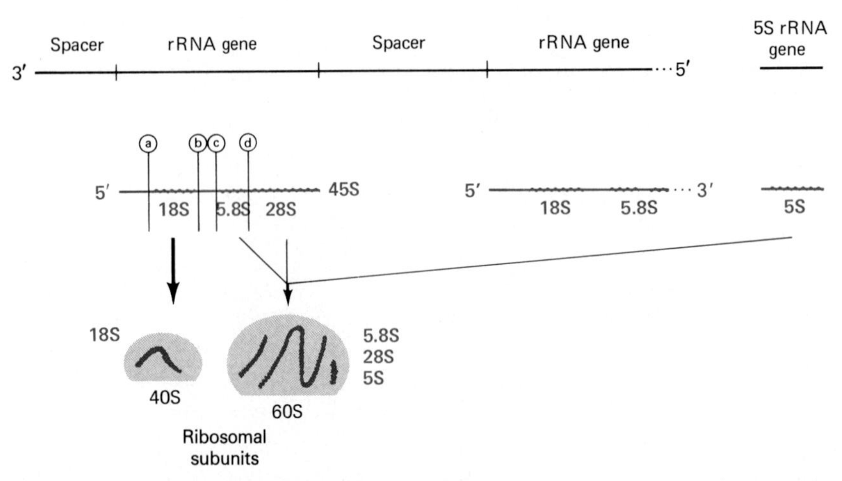


***Figure 9.17*** Processing of pre-rRNA and packaging of rRNAs into ribosomal subunits in eukaryotes. The 45S pre-rRNA is cleaved in at least four sites: (a) Removal of leader leaves a 41S fragment; (b) cleavage of 18S rRNA for the 40S ribosomal subunit leaves a 36S pre-rRNA fragment; (c) removal of a small region leaves a 32S fragment; and (d) processing of the 32S fragment yields 5.8S and 28S rRNAs of the larger ribosomal subunit. The 5.8S and 28S rRNAs are packaged together with 5S rRNA, which is encoded elsewhere in the genome.

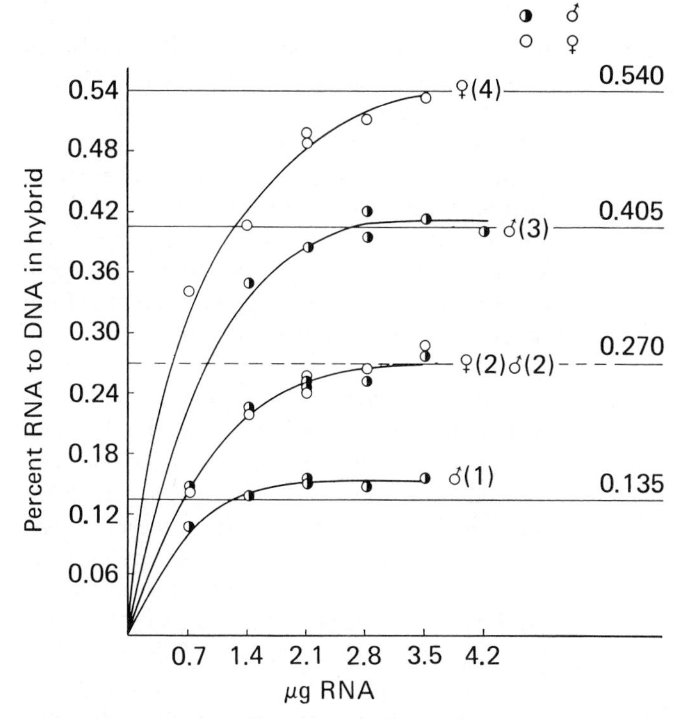


***Figure 9.18*** Demonstration that rDNA in *Drosophila melanogaster* is located at the nucleolar-organizing region (NOR) of the chromosome. Hybridization of purified rRNA to DNA from male and female flies carrying from 1 to 4 NORs per cell (2 NORs is the normal number) is shown. Horizontal lines indicate the hybridization plateau values predicted for 1, 3, or 4 NORs, on the basis of the normal plateau value of 0.270% for 2 NORs. In these saturation hybridizations, increasing amounts of RNA (μg) are added to hybridization mixtures that contain a fixed amount of DNA. Once all the complementary DNA has been saturated by pairing with RNA, no additional hybrids form since no more DNA is available for hybridization regardless of the amount of RNA present. (From F.M. Ritossa and S. Spiegelman. 1965. *Proc. Nat. Acad. Sci. U.S.* **53**:737.)

quires action of a variety of enzymes and other proteins at specific stages during translation into polypeptides. The whole translation apparatus operates through the coordination of various components and processes, an accurate recognition system between RNAs, and special conditions to initiate, elongate, and terminate the polypeptide chain.

## 9.6 Amino Acids, tRNAs, and Synthetases

Before amino acids can be linked to form a polypeptide, they must be raised to a sufficiently high energy level and a recognition mechanism must function to correctly insert these units in the sequence dictated by mRNA.

Energy for synthesis is provided by interaction of ATP with amino acids, as the first step in the preparation process. The second step is the binding of the *activated* amino acid, or *aminoacyl* residue, to its tRNA carrier. Both of these steps are catalyzed by the same enzymes, called **aminoacyl-tRNA synthetases.** There is at least one synthetase for *each* of the 20 coded amino acids.

The synthetase drives the first reaction by coupling the first reaction to the hydrolysis of ATP to form adenosine monophosphate (AMP). A high-energy intermediate is formed, called an aminoacyl-adenylate, and this complex remains bound to the enzyme until the second reaction when the aminoacyl group is bonded to the terminal nucleotide of tRNA at the 3′ end of the amino acid arm (see Fig. 9.12). The reactions can be summarized as follows:

amino acid + ATP + synthetase

① ↓

aminoacyl-adenylate-synthetase + pyrophosphate ($PP_i$)

followed by:

aminoacyl-adenylate-synthetase + tRNA

② ↓

aminoacyl-tRNA + AMP + synthetase

A particular tRNA is identified according to the amino acid it can carry; for example, $tRNA^{Met}$ is a methionine-carrying tRNA. The particular aminoacyl-tRNA is identified by its aminoacyl group and specific tRNA; for example, Met-$tRNA^{Met}$ is the methionyl-tRNA complex that participates in adding methionine to protein.

## 9.7 Subunits, Monosomes, and Polysomes

Until 1962 it was generally believed that proteins were synthesized on single, free ribosomes. The true picture was discovered by Alexander Rich and co-workers, who showed that polypeptides are made at groups of ribosomes, called polyribosomes, or **polysomes** for short. The polysome is a complex composed of a variable number of individual ribosomes, which we can refer to as **monosomes,** held together by a strand of mRNA. The length of a polysome is usually proportional to the length of the bound mRNA, or genetic message.

Their experiments involved rabbit reticulocytes which, like other mammalian red blood cell precursors, are virtual factories for hemoglobin synthesis and do little else. The advantages of using rabbit reticulocytes, therefore, were: (1) the possibility of bulk biochemical analysis from relatively homogeneous cell populations and (2) the possibility of studying almost a single polypeptide synthesis reaction without the need of screening out interference from many other polypeptides being made at the same time.

Reticulocytes were incubated in media containing radioactively labeled amino acids, and after a suitable time the cells were broken and the cell-free lysates were separated into different fractions by centrifugation. The prediction was that growing globin chains would be found in the fractions actually engaged in globin-chain synthesis and that polypeptides could be identified by their radioactive label (Fig. 9.19). The fraction in which most of the radioactive globin fragments were found consisted of 5-ribosome groups, not the monosome fraction.

In another experiment the labeled polypeptides were shown to be truly in the process of synthesis and were functionally associated with polysomes, rather than randomly associated or present by accident. When the drug *puromycin,* which acts as an analogue of tRNA, is added to an actively synthesizing system, chain growth is terminated *prematurely* and incomplete molecules are released from the polysome complex. Such *nascent* (growing) chains then disappear from the polysome fraction and are found in the fluid supernatant phase of the system and can be recognized by their radioactive label. In this way, it was verified that chain growth takes place at the polysomes since only nascent polypeptides are affected by puromycin.

Supporting evidence for globin synthesis at polysomes was obtained by Rich from electron micrographs of various gradient fractions (Fig. 9.20). In the 5-ribosome fractions, groups of 5 ribosomes were held together by a thin strand. This strand was presumed to be mRNA since it was digested by RNase, and the polysomes then

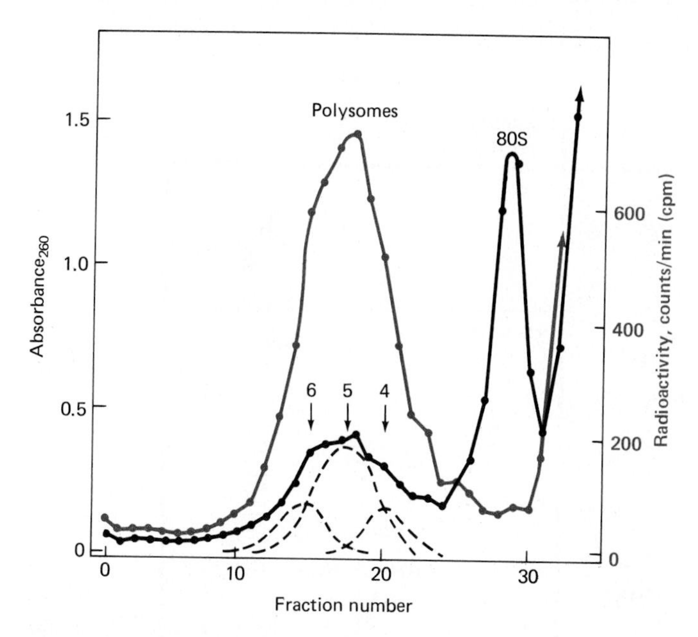


***Figure 9.19*** Sedimenting positions of ribosomal preparation from rabbit reticulocytes to which [$^{14}$C] amino acids had been added for 45 seconds to determine the location of nascent polypeptides into which these amino acids were incorporated. Black circles indicate absorbance readings (optical density) at 260 nm in the 36 fractions. The bulk of the radioactivity (color) occurs in the polysome region of the gradient where aggregates of 4, 5, and 6 ribosomes are found (170S = five-ribosome peak aggregate). Little radioactivity is present in the monosome peak (80S). (J.R. Warner et al., "Electron Microscope Studies of Ribosomal Cluster Synthesizing Hemoglobin," *Science* **138**:1399–1403, Fig. 1, 28 December 1962. Copyright © 1962 by the American Association for the Advancement of Science.)

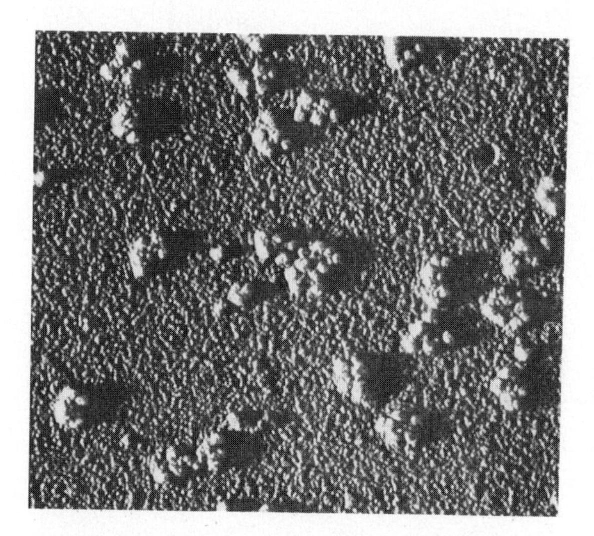

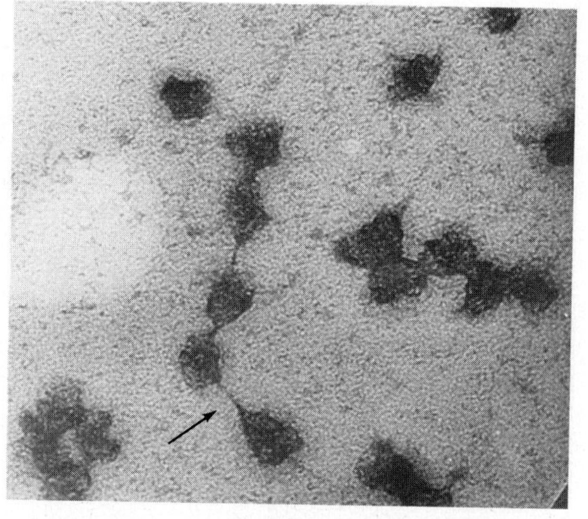

***Figure 9.20*** Electron micrographs of polysomes isolated from rabbit reticulocytes. (left) Clusters of 4, 5, and more ribosomes are evident in a preparation shadowed with gold. × 70,000. (right) When stained by uranyl acetate and photographed at higher magnification, a thin strand of RNA (presumably mRNA) is seen to connect ribosomes of a cluster. × 300,000. (Courtesy of A. Rich. With permission from H.S. Slayter et al. 1963. *J. Mol. Biol.* **7**:652. Copyright by Academic Press Inc. [London] Ltd.)

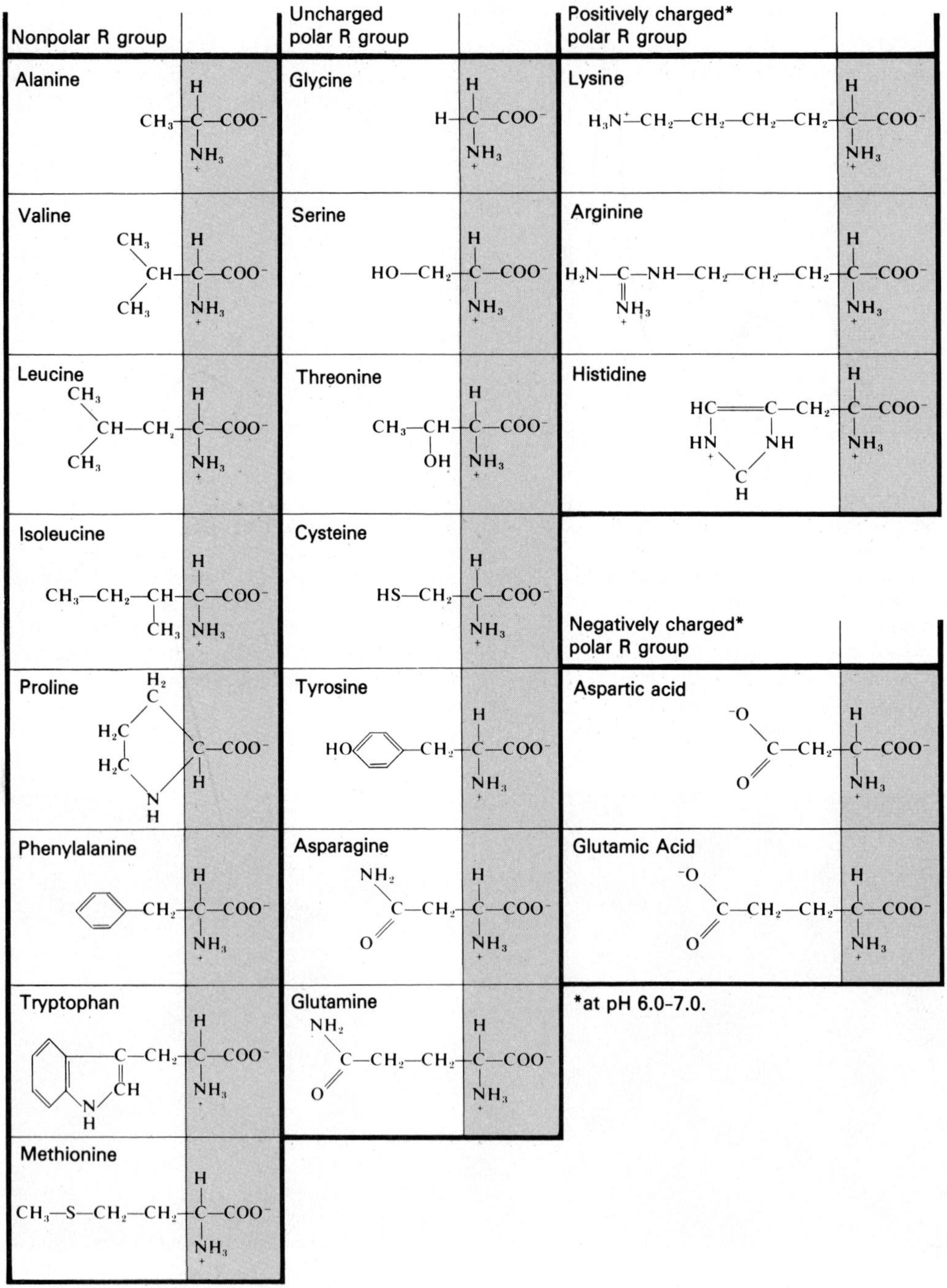

| Nonpolar R group | |
|---|---|
| Alanine | $CH_3$–$\overset{H}{\underset{NH_3^+}{C}}$–$COO^-$ |
| Valine | $(CH_3)_2CH$–$\overset{H}{\underset{NH_3^+}{C}}$–$COO^-$ |
| Leucine | $(CH_3)_2CH$–$CH_2$–$\overset{H}{\underset{NH_3^+}{C}}$–$COO^-$ |
| Isoleucine | $CH_3$–$CH_2$–$\underset{CH_3}{CH}$–$\overset{H}{\underset{NH_3^+}{C}}$–$COO^-$ |
| Proline | ring: $H_2C$, $H_2C$, $\overset{H_2}{C}$, $\underset{H}{N}$ – $\underset{H}{C}$–$COO^-$ |
| Phenylalanine | $Ph$–$CH_2$–$\overset{H}{\underset{NH_3^+}{C}}$–$COO^-$ |
| Tryptophan | indole ring ($\underset{H}{N}$, $CH$, $C$)–$CH_2$–$\overset{H}{\underset{NH_3^+}{C}}$–$COO^-$ |
| Methionine | $CH_3$–S–$CH_2$–$CH_2$–$\overset{H}{\underset{NH_3^+}{C}}$–$COO^-$ |

| Uncharged polar R group | |
|---|---|
| Glycine | H–$\overset{H}{\underset{NH_3^+}{C}}$–$COO^-$ |
| Serine | HO–$CH_2$–$\overset{H}{\underset{NH_3^+}{C}}$–$COO^-$ |
| Threonine | $CH_3$–$\underset{OH}{CH}$–$\overset{H}{\underset{NH_3^+}{C}}$–$COO^-$ |
| Cysteine | HS–$CH_2$–$\overset{H}{\underset{NH_3^+}{C}}$–$COO^-$ |
| Tyrosine | HO–$Ph$–$CH_2$–$\overset{H}{\underset{NH_3^+}{C}}$–$COO^-$ |
| Asparagine | $NH_2$–C(=O)–$CH_2$–$\overset{H}{\underset{NH_3^+}{C}}$–$COO^-$ |
| Glutamine | $NH_2$–C(=O)–$CH_2$–$CH_2$–$\overset{H}{\underset{NH_3^+}{C}}$–$COO^-$ |

| Positively charged* polar R group | |
|---|---|
| Lysine | $H_3N^+$–$CH_2$–$CH_2$–$CH_2$–$CH_2$–$\overset{H}{\underset{NH_3^+}{C}}$–$COO^-$ |
| Arginine | $H_2N$–$\underset{NH_3^+}{\overset{\parallel}{C}}$–NH–$CH_2$–$CH_2$–$CH_2$–$\overset{H}{\underset{NH_3^+}{C}}$–$COO^-$ |
| Histidine | ring: HC=C, $HN^+$, NH, $\underset{H}{C}$ –$CH_2$–$\overset{H}{\underset{NH_3^+}{C}}$–$COO^-$ |

| Negatively charged* polar R group | |
|---|---|
| Aspartic acid | $^-O$–C(=O)–$CH_2$–$\overset{H}{\underset{NH_3^+}{C}}$–$COO^-$ |
| Glutamic Acid | $^-O$–C(=O)–$CH_2$–$CH_2$–$\overset{H}{\underset{NH_3^+}{C}}$–$COO^-$ |

*at pH 6.0–7.0.

***Figure 9.21*** The twenty naturally occurring amino acids specified by the genetic code.

dissociated into monosome units. The length of the presumed mRNA was about right for a message that was long enough to code for 141 amino acids in α-globin or 146 amino acids in β-globin. Such a message would contain at least 146 × 3 nucleotides for the whole set of codons, and its length would be about 1500 Å since the Watson-Crick model indicated a dimension of about 10 Å per 3 nucleotides in a polynucleotide chain. Rich found that the mRNA was 1500 Å long, as predicted.

## 9.8 Protein Structure

With the exception of proline, the 20 amino acid building blocks of proteins have a common feature in their construction (Fig. 9.21). The first carbon, called the *α-carbon,* is asymmetric since four different components are bound to it. An *α-amino* (α-$NH_2$), *α-carboxyl* (α-COOH), and *H* are bound at three sites, and the fourth site is occupied by a residue, or *R-group.* Most of the chemical and conformational properties of proteins are due to R-group interactions between amino acids.

Interactions between amino acids during growth of the polypeptide chain uniformly involve **peptide bond** formation between the α-$NH_2$ of one unit and the α-COOH of the adjacent unit and lead to the —C—N— peptide linkage (Fig. 9.22). When two or more amino acid units are thus bonded together, the product is a **peptide.** A polypeptide contains many amino acid units, with the R-groups projecting from the zigzagging "backbone" of —C—N— linkages.

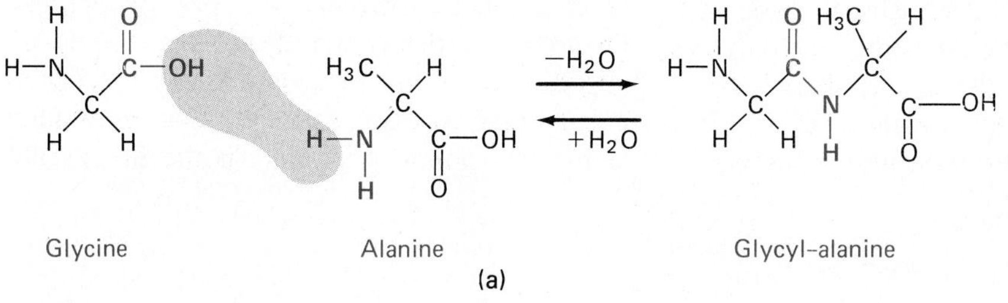

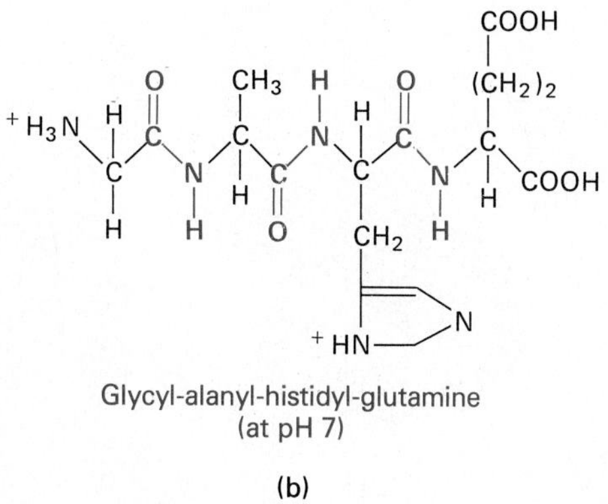

Glycine Alanine Glycyl-alanine

$-H_2O$ $+H_2O$

(a)

Glycyl-alanyl-histidyl-glutamine (at pH 7)

(b)

***Figure 9.22*** Peptide bond formation in protein synthesis. (a) Dehydration reaction produces a dipeptide from two amino acids; reactions between a peptide and a free amino acid would produce similar products. (b) Adjacent amino acid residues are joined by peptide bonds throughout a peptide, regardless of its length. Differences are largely due to amino acid side-chains, and not to the invariant zigzag polypeptide backbone.

The **primary structure** of a protein consists of the linear sequence of amino acids (Fig. 9.23). Three other orders of structure are also recognized: **secondary structure,** which develops from bonding between neighboring atoms; **tertiary structure,** which develops from interactions of R-groups at some distance from one another in the chain; and **quaternary structure,** which develops only when two or more polypeptide chains exist as subunits in a single functional protein molecule. All the orders of structure contribute to biological activity and proper functioning of the protein, whether the protein consists of one polypeptide and has only three orders of structure or consists of more than one polypeptide and has four orders of structure.

Mutations leading to amino acid substitutions in the primary structure may have severe effects if the substituted amino acid is important in the development of the three-dimensional shape of the protein. Otherwise, the new amino acid may have little or no effect on protein shape and, therefore, little or no demonstrable effect on protein function and phenotype development.

## 9.9 Polypeptide Chain Initiation

The first step in initiation of translation probably involves binding of the mRNA at its 5′ end to the 3′ end of a homologous sequence of rRNA in the small subunit of the ribosome. In prokaryotes the small ribosomal subunit contains a 16S rRNA; in eukaryotes, it is an 18S rRNA (Box 9.2). The next step involves binding of the specific initiator tRNA to the AUG codon on the mRNA sequence. This initiator tRNA carries methionine (*Met*) in eukaryotes and a modified methionine, called **N-formylmethionine,** or *fMet* for short, in prokaryotes and in mitochondria and chloroplasts. The initiator $tRNA^{Met}$ is different from the $tRNA^{Met}$ that recognizes AUG codons at *internal* sites within the mRNA sequence.

The methionine residue carried by initiator $tRNA_f^{Met}$ can be formylated by a formylase enzyme, whereas noninitiator $tRNA_m^{Met}$ cannot have its methionyl formylated *in vitro* or *in vivo.* Eukaryotic initiator Met-$tRNA_f^{Met}$ can be formylated *in vitro,* but formylase is not present in eukaryotic cytoplasm, so methionine is installed as the first amino acid. fMet is the first amino

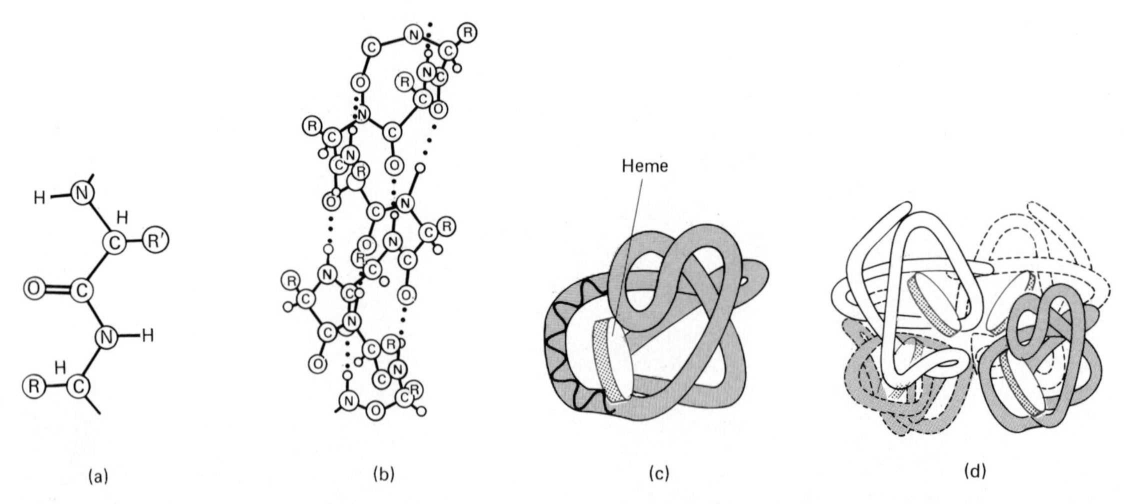


***Figure 9.23*** Four levels of structural organization in proteins, exemplified by hemoglobin. (a) Primary structure: the sequence of a chain of amino acids joined by peptide bonds. (b) Secondary structure. (c) Tertiary structure. Both (b) and (c) are responsible for the three-dimensional shape of the polypeptide. (d) Quaternary structure: the aggregation of two or more polypeptides in a single functional protein molecule. There are four globin chains in a hemoglobin molecule and each globin chain is bound to an iron-containing heme group.

## BOX 9.2 Ribosomal RNA Secondary Structure

Using information from biochemical, molecular, and electron microscopic studies, investigators have proposed a secondary structure for 16S rRNA. Development of stem-and-loop secondary structure is the result of hydrogen bonding between nearby bases in the single-stranded RNA molecule. As a consequence of secondary structure, the unpaired 5′ and 3′ ends of the rRNA strand come to be relatively near each other. The tentative locations of several ribosomal proteins that are bound to specific regions of the rRNA molecule have been proposed on the basis of protein-binding studies.

Physical studies and ultrastructural analysis have provided most of the important information concerning structural features of the small ribosomal subunit. Although investigators' interpretations of electron micrographs differ and, therefore, their proposals for the precise shape of the subunit vary, they do agree on the general form and organization of the subunit. The ultimate aim of these studies is to determine the functions of ribosomal regions and of ribosomal molecular components in translation and in regulation of translation. The 3′ terminus of 16S rRNA, for example, has been localized to the "platform" region of the 30S ribosomal subunit, near the cleft that separates the "platform" from the larger "head" region, which bulges from the body of the subunit. The 3′ end of rRNA is believed to play an important role in mRNA binding, codon-anticodon recognition, and the selection of translational initiation sites in mRNA. (From H.F. Noller and C.R. Woese. 1981. *Science* **212**:403, Fig. 3. Copyright 1981 by the American Association for the Advancement of Science.)

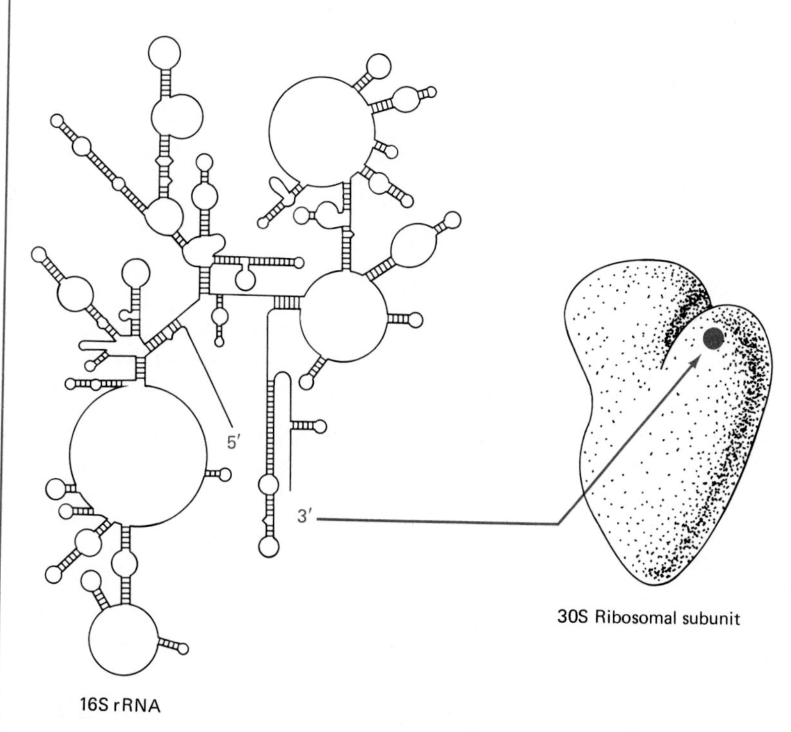

acid installed in prokaryotes and in organelles, which do have a formylase activity. The subscripts *f* and *m* indicate the initiator and non-initiator tRNAs, respectively.

Although Met or fMet is the initiating amino acid for all polypeptide chains, it may be modified or even cleaved from the polypeptide chain later on, through deformylase or aminopeptidase actions. A polypeptide, therefore, may or may not have methionine at its N-terminus. About half the polypeptides in *E. coli* have an N-terminal methionine residue.

Instead of AUG, the mRNA initiation codon may be GUG in some cases and will bind to Met-$tRNA_f^{Met}$ or fMet-$tRNA_f^{Met}$. Internal GUG codons, however, bind only valyl-$tRNA^{Val}$. If fMet is the initiating amino acid, it can only form peptide bonds through its free α-carboxyl group because its α-$NH_2$ terminus is blocked by the formyl residue. Once the initiator tRNA has been bound to the mRNA initiation codon through base pairing between anticodon and codon, the **initiation complex** is completed (Fig. 9.24). The large ribosomal subunit is not present and is not needed for chain initiation. Chain initiation can take place *in vitro* when all the components are present, even in the absence of large subunits. Among the components that are needed for formation of the initiation complex are three protein *initiation factors* (IF-1, -2, and -3), and a molecule of *GTP,* which is hydrolyzed during this formation process.

Once the initiation complex is formed, the large ribosomal subunit associates with the small subunit of the complex, through reactions that include binding with 16S rRNA. Chain elongation can then proceed along the mRNA, which is now bound to a whole ribosome. As the ribosome moves away from the 5′ end of the mRNA during polypeptide chain elongation, the mRNA can continue to participate in new initiation complex formations at its freed 5′ end. Polysome formation takes place as additional ribosomes join to a single mRNA strand, which repeatedly undergoes initiation complex formation in proportion to its length and the space taken up by bound ribosomes engaged in polypeptide synthesis.

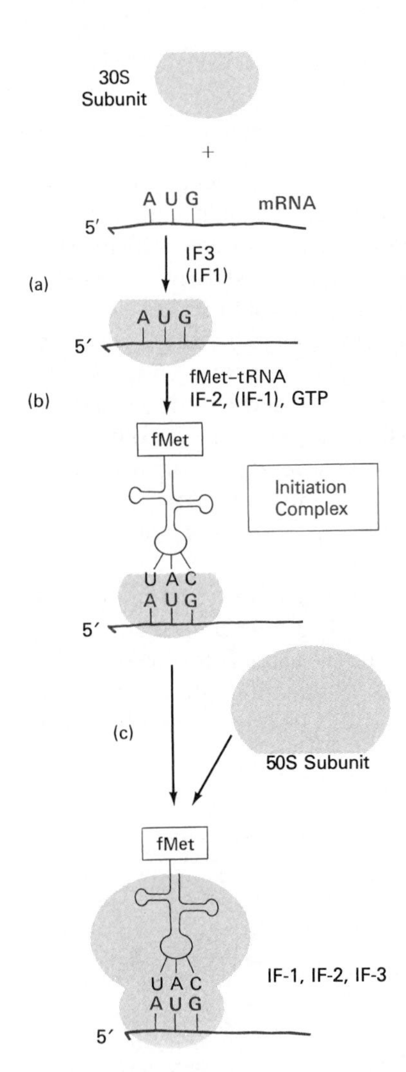


***Figure 9.24*** Formation of the initiation complex in protein synthesis in prokaryotes. (a) Binding of mRNA to 30S ribosomal subunit in the presence of one or more initiation factors (IF). (b) Binding of fMet-$tRNA^{Met}$, the initiating aminoacyl-tRNA, and formation of the initation complex. (c) Elongation of the polypeptide may proceed once the 50S ribosomal subunit binds to the initiation complex and IFs dissociate from the complex.

## 9.10 Polypeptide Chain Elongation

Once the initiating aminoacyl-$tRNA_f^{Met}$ is in place, the next aminoacyl-tRNA can be bound to the mRNA at the A site of the ribosome. The initiating aminoacyl-tRNA is the only one that enters directly at the **P (peptide) site** of the ribosome; all others enter only at the **A (amino acid) site** (Fig. 9.25).The ribosome *coordinates* and *catalytically assists* processes of polypeptide synthesis.

The first amino acid is joined to the amino acid of the second aminoacyl-tRNA in a peptide-linking reaction that is catalyzed by **peptidyl transferase,** an enzyme component of the ribosome. Once the dipeptidyl chain has been formed, the free $tRNA_f^{Met}$ is released from the P site. the entire dipeptidyl-tRNA is then translocated from the A site to the open P site, catalyzed by a **translocase** enzyme, or *G factor,* which is another component of the ribosome. The available A site can now accept the next incoming aminoacyl-tRNA, and the whole process of peptide bond formation and translocation continues repeatedly until the end of the genetic message.

During chain elongation the ribosome keeps moving from codon to codon, toward the 3′ end of mRNA. Translocation of the lengthened peptidyl-tRNA is an energy-requiring process that involves GTP hydrolysis to GDP and $P_i$. If either GTP hydrolysis or translocation from the A to the P site fails to take place, further chain elongation stops since the A site is not available for aminoacyl-tRNA entry. In addition, one or

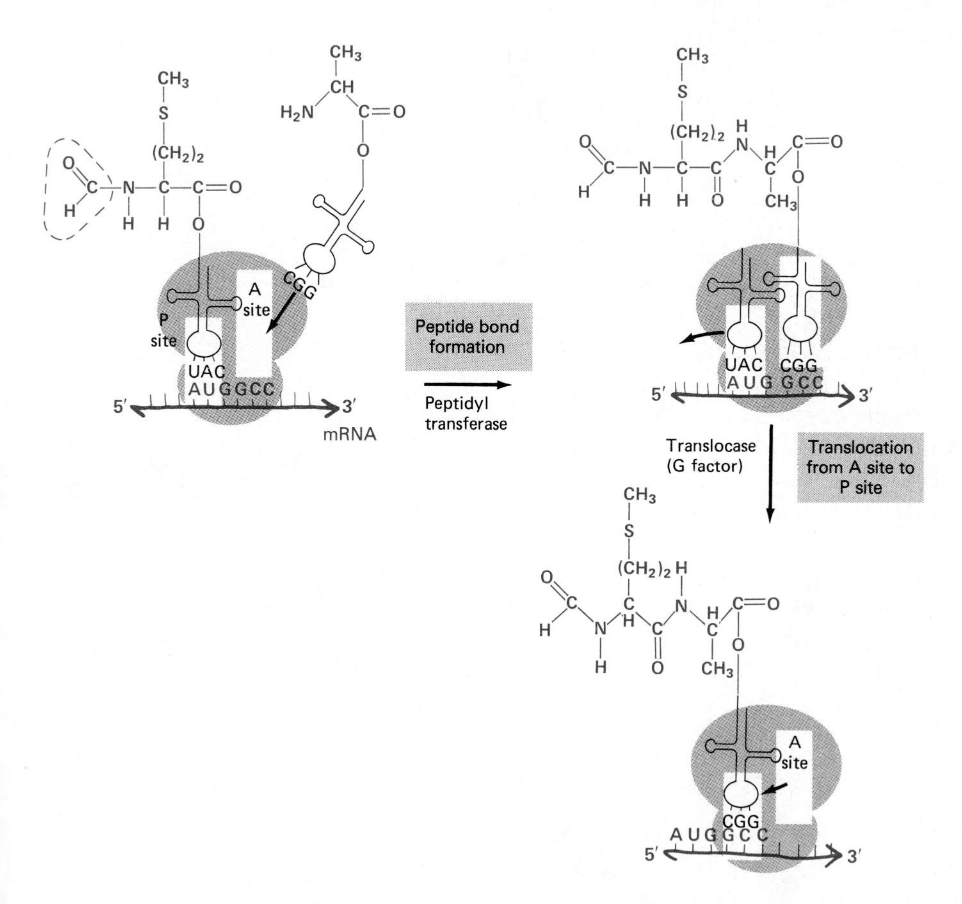


***Figure 9.25*** Polypeptide chain elongation at the ribosome. Incoming aminoacyl-tRNA enters at the A site; peptide bond formation is catalyzed by peptidyl transferase, making the peptidyl chain one unit longer. The tRNA is discharged from the P site after giving up its peptidyl chain to the incoming unit, and the new peptidyl-tRNA is translocated from the A to the P site in a reaction requiring translocase (G factor). The A site is now open for the next aminoacyl-tRNA specified by the coded sequence in mRNA. These same steps are repeated for each amino acid residue until chain termination occurs.

more *elongation factors (EF)* are required for polypeptide synthesis (Fig. 9.26). EF combines with GTP, and the EF-GTP complex combines with the incoming aminoacyl-tRNA. In this form of *EF—GTP—aminoacyl-tRNA,* the incoming unit is accurately brought into the A site, where it can then participate in chain growth. When complexed with EF and GTP, the incoming aminoacyl-tRNA will *only* bind to the A site of a ribosome that has its P site filled with a peptidyl-tRNA. Once bound to the ribosome, the GTP of this complex is hydrolyzed so that EF-GDP is released, and the *GTPase* component of the ribosome can bind G factor (translocase) and a free second molecule of GTP. This second GTP is hydrolyzed during translocation of the peptidyl-tRNA from the A to the P site. At the same time, GDP and G factor are released and the A site becomes available for the next aminoacyl-tRNA.

All these closely coordinated events of chain elongation provide greater insurance that few mistakes will take place. Enzymes and other proteins would be less effective or even nonfunctional if they were assembled inaccurately. Cell survival depends absolutely on the accuracy of protein synthesis since even one incorrect amino acid can cause protein dysfunction in some cases.

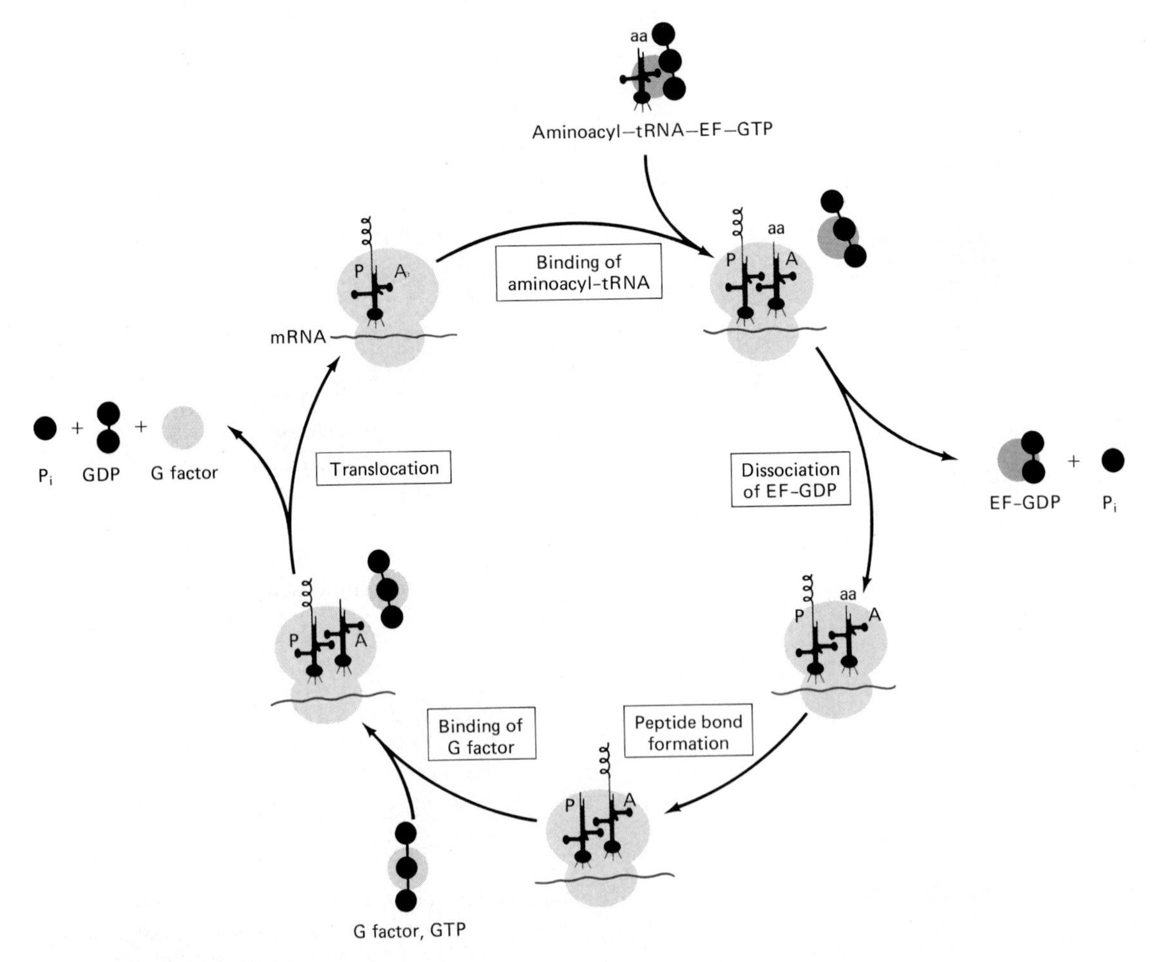


***Figure 9.26*** Components involved in lengthening the peptidyl chain by one amino acid residue during polypeptide synthesis.

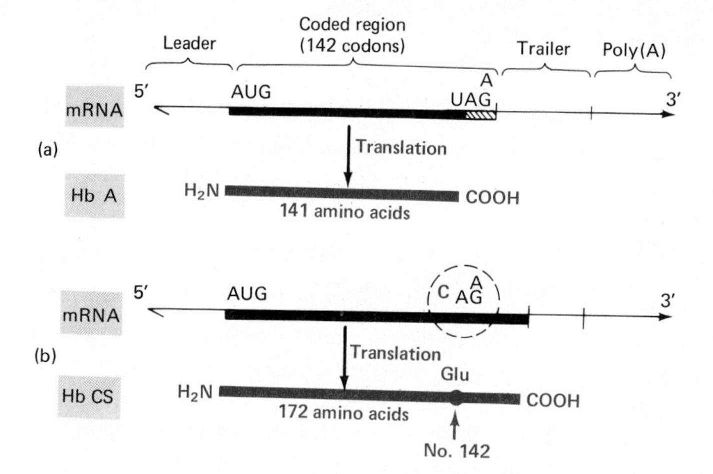


***Figure 9.27*** Effect of mutation in the termination codon of the human $Hb_\alpha$ gene. (a) The α-globin of normal Hb A is 141 amino acids long and (b) Hb Constant Spring (Hb CS) has 172 amino acids in its α chain, due to base substitution in the UAG termination codon, which now codes for glutamic acid (CAG). A portion of the trailer mRNA segment is translated and adds 31 amino acids to the normal chain length.

## 9.11 Polypeptide Chain Termination

Before the finished polypeptide can be released from the ribosome, the link between the C-terminal amino acid and its tRNA must be broken. Separation is not spontaneous, nor is it efficient even when the termination codon UAA, UAG, or UGA is exposed in the mRNA segment at the A site of the ribosome. No more amino acids can be added, however, since none of the tRNA anticodons recognize the termination codons.

Protein *releasing factors (RF)* are believed to interact at the termination-codon–A site location on the ribosome, and these RF activities apparently prevent the A site from being filled. Then a hydrolysis reaction, probably mediated by another protein, severs the link between the tRNA and the polypeptide and frees both components from the polysome. The mRNA link to the ribosome is also broken.

Upon release of the polypeptide, the two ribosome subunits *dissociate* and return to the cytoplasmic pool of subunits for further participation in protein synthesis. Whole monosomes usually are not found in an active cell, except as parts of polysome groups. In fact, if the subunits do not dissociate and the monosome persists, protein synthesis slows down and may even stop due to the deficiencies of free small subunits available for initiation complex formation and of large subunits for elongation events. Some evidence indicates that one of the initiation factors, IF-3, aids in monosome dissociation into subunits by binding to the small subunit. At the same time, therefore, that IF-3 helps to dissociate monosomes, it also becomes bound to the small subunit in preparation for another round of chain initiation.

In some of the variant hemoglobins, the particular modification can be traced to mutations in the normal termination codon, leading to chains that are longer than normal. Three particular human α-globin variants have chains longer than 141 amino acids. Hemoglobin Constant Spring (Hb CS) has 31 extra amino acids at the C-terminus, but the normal 141 amino acids are all present between the N-terminus and the extra length beginning at codon 142. Since there is a *glutamine* at position 142 in Hb CS, the most reasonable explanation is that the normal UAA or UAG terminator of α-globin mRNA has been altered to CAA or CAG, which code for glutamine. A simple base substitution, therefore, seems to have eliminated the punctuation and allowed translation of normally untranslated sequences that are in the trailer segment beyond the message (Fig. 9.27).

The other two α-globin variants are similar to

Hb CS, except that they each have a different amino acid at position 142. In Hemoglobin Icaria the first of the 31 extra amino acids is *lysine;* the amino acid *serine* is first in Hemoglobin Koya Dora. The codon change in Hb Icaria leading to lysine could have arisen by one base substitution of UAA or UAG to AAA or AAG, and would therefore involve the same base as in Hb CS, an altered first base in the triplet. A single base substitution from a termination to a serine codon in Hb Koya Dora could only involve a change in the middle base of the codon, from UAG or UGA to UCG or UCA. All other possibilities would require two substituted bases in the termination codon. Since the one termination codon common to all three variants is UAG, this was the most probable terminator for the α-globin message. This was later verified by direct sequence analysis of normal α-globin mRNA, showing that UAG is the termination codon.

## 9.12 Effects of Streptomycin on Protein Synthesis

Various chemical agents, including a number of *antibiotics,* interfere with protein synthesis at one or more stages in the process (Table 9.3). These drugs are useful as therapeutic agents in health care and as experimental probes of protein synthesis and other biological activities. We mentioned puromycin effects in Section 9.7.

In 1964 it was shown that streptomycin exerts its effect specifically on the 30S ribosome subunit in *E. coli.* In mixed reconstitution experiments using small and large subunits, in all

***Table 9.3*** Characteristic action of some inhibitors of protein synthesis.

| synthesis stage inhibited | inhibitor | mode of action of inhibitor | effective in prokaryotes | effective in eukaryotes |
|---|---|---|---|---|
| initiation | aurintricarboxylic acid | Prevents association of ribosomal subunit with messenger RNA | + | + |
| | streptomycin | Releases bound fMet-tRNA from initiation complex | + | – |
| elongation | streptomycin | Inhibits binding of aminoacyl-tRNA to ribosome; inhibits translocation on the ribosome | + | – |
| | chloramphenicol | Stops amino acid incorporation by inhibiting peptidyl transferase | + | – |
| | cycloheximide | Inhibits tRNA movement on the ribosome | – | + |
| | puromycin | Acts as amino acid analogue and causes premature polypeptide chain termination | + | + |
| termination | various drugs | Inhibit releasing factors; inhibit ribosome release from messenger RNA of polysome | + | + |

combinations, from streptomycin sensitive (*str-s*) and resistant (*str-r*) strains, protein synthesis was monitored *in vitro* using the artificial messenger poly(U) in the presence and absence of streptomycin. When streptomycin was absent, all systems synthesized polyphenylalanine, as expected. When streptomycin was present, however, the only systems that functioned were those with a 30S small subunit derived from the *str-r* strain. The source of the 50S large subunit made no difference (Fig. 9.28).

One year later further experiments showed that low concentrations of the drug caused **misreading** of codons in *str-s E. coli* strains. When poly(U) is the messenger, we expect only phenylalanines to be incorporated into polypeptide since the codon is UUU. Drug-sensitive bacteria, however, made polypeptides containing isoleucines and serines as well as phenylalanines from poly(U) messengers. Resistant strains made only polyphenylalanine. In comparisons with other studies using different artificial messengers, misreading of codons was found to involve only the pyrimidines U and C in a codon (Fig. 9.29). For example, poly(C) promoted incorporation of the usual proline and also of serine, histidine, and threonine in sensitive strains. The

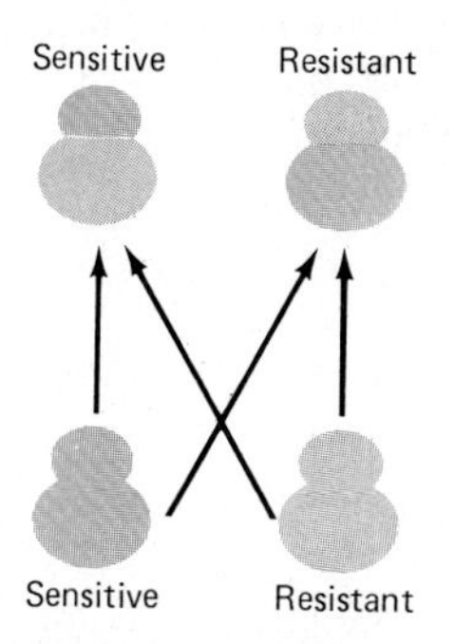


***Figure 9.28*** Mixed reconstitution experiments using ribosomal subunits of *str-s* (color) and *str-r* (gray) *E. coli* showed that resistance or sensitivity to streptomycin depended on the source of the 30S ribosomal subunit and not on the 50S subunit.

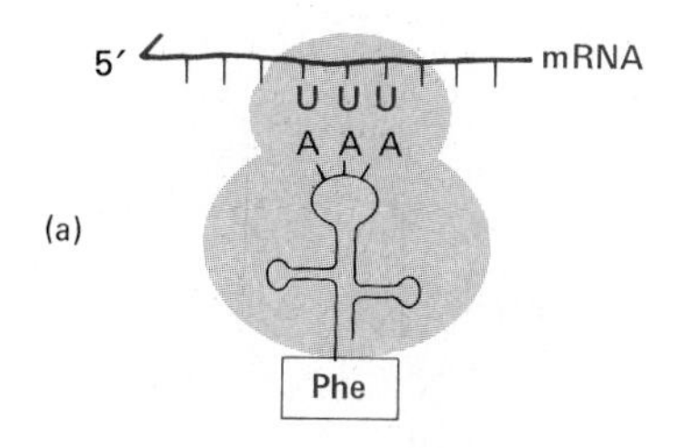


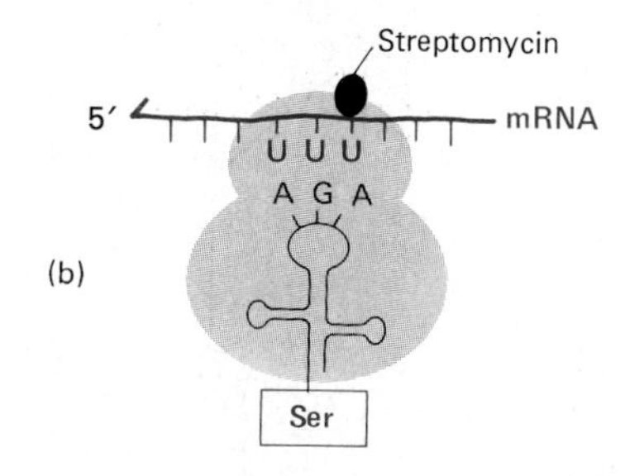


***Figure 9.29*** Misreading of codons in *E. coli*. (a) UUU is translated as phenylalanine under normal conditions, but (b) misreading of UUU as UCU is a frequent event in *str-s E. coli* in the presence of streptomycin. Serine instead of phenylalanine would be inserted into the polypeptide in the case illustrated.

U is misread as C and the C is misread as U and occasionally either is misread as A. Misreading is not known to occur in eukaryotes, for unexplained reasons.

The specific change in *str-r E. coli* was traced to one of the 21 proteins of the small ribosome subunit (proteins S1 to S21), specifically to S12. M. Nomura conducted mixed reconstitution experiments using purified rRNAs and ribosomal proteins from *str-s* and *str-r* strains. The only reconstituted small subunits that could function normally in the presence of streptomycin were ones containing protein S12 from *str-r* strains, even if all the other ribosomal proteins came from *str-s* strains (Fig. 9.30). Presumably, streptomycin binding to the small ribosomal subunit is mediated by protein S12 in sensitive strains. The mutation that confers resistance to strep-

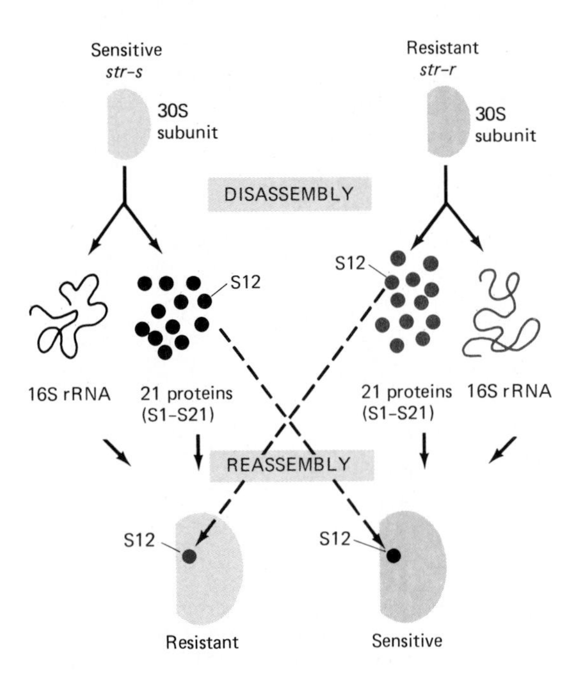


***Figure 9.30*** Summary of mixed reconstitution experiments conducted by M. Nomura in which he showed that streptomycin-sensitive *E. coli* differed from streptomycin-resistant strains in one of the 21 proteins making up the small subunit of the ribosome.

tomycin involves an amino acid alteration in protein S12, perhaps because of the substitution of one base in one codon. Protein S12 and two others in the small subunit appear to monitor fidelity of translation. They are called *fidelity proteins* and are believed to act by binding correct tRNAs rather than incorrect but similar tRNAs to the ribosome before the aminoacyl-tRNA enters the A site.

## Questions and Problems

***9.1*** Messenger RNA was transcribed *in vitro* from enzymatically synthesized DNA. The DNA was melted to its constituent single strands, and base ratios were determined for DNA and for mRNA. On the basis of the following ratios, which strand of the DNA duplex served as the template for mRNA synthesis?

| | A | T or U | G | C |
|---|---|---|---|---|
| DNA-1 | 27.0 | 32.5 | 18.5 | 22.0 |
| DNA-2 | 32.7 | 26.8 | 22.1 | 18.4 |
| mRNA | 27.0 | 33.0 | 18.0 | 22.0 |

***9.2*** The part of a gene sequence that encodes seven amino acids has been found to be the following:

T T A A T C C C T A C G G G C A T T A A T
A A T T A G G G A T G C C C G T A A T T A

***a.*** Is the upper or the lower strand the one transcribed into mRNA? Explain.
***b.*** Show the 3′ and 5′ ends of each strand of the duplex DNA.
***c.*** Show the 5′ → 3′ base sequence of mRNA complementary to this part of the coding strand.
***d.*** Starting with the first triplet in the mRNA sequence, what would be the amino acids and their sequence in the polypeptide translation?

***9.3*** A genomic restriction fragment has the following base sequence:

3′–TTTATTTTAAATATATTCGTACTTTCGGAG
1 10 20 30
GAAAAGGATCAATTTTATTT–5′
40 50

***a.*** What would be the base sequence of mRNA transcribed from this strand?
***b.*** If a short but complete peptide chain were encoded somewhere in this sequence, what would be its primary structure (kinds and sequence of amino acids)?
***c.*** According to your answer in (b), what parts of the mRNA sequence represent all or part of the leader and all or part of the trailer segments?

***9.4*** Suppose the DNA sequence in Question 9.3 included all or part of the promoter.
***a.*** Would the promoter be at the 3′ or the 5′ end of the coding strand?
***b.*** How would the presence of a promoter segment alter your answer to (a) in 9.3?
***c.*** What feature in the presumed promoter sequence helps to characterize the promoter here or in other cellular genes?

***9.5*** In a diagram, summarize the important events and their sequence leading from a pre-mRNA transcript in the nucleus to the appearance of a processed mature mRNA in the cytoplasm of a eukaryotic cell.

***9.6*** The wobble hypothesis predicts that an A in the 3′ position of a codon can pair with either a U or an I from a tRNA anticodon. Similarly, a C in the 3′ position of a codon can pair with either a G or I from a tRNA anticodon. According to the genetic code, the codon 5′–AGA–3′ codes for the amino acid arginine and 5′–AGC–3′ codes for serine. Would you expect the serine tRNA to have the anticodon 5′–ICU–3′? Why?

***9.7*** Using Crick's wobble hypothesis, suggest an explanation for the specificity of tRNA anticodons in distinguishing among the codons for cysteine (UGU, UGC), tryptophan (UGG), and the UGA termination codon.

***9.8*** Human HeLa cell chromatin can be separated into one portion with attached nucleoli and another portion without attached nucleoli. Design an experiment to test these two chromatin portions for the presence of rDNA and describe the results you would expect.

***9.9*** How would you determine that the RNA detected around a puff in a polytene chromosome was directly transcribed from the puff DNA and was not simply an accumulation of RNA that had been synthesized elsewhere in the genome?

***9.10*** You have been given three different RNA preparations, all of which have been isolated and purified from one population of eukaryotic cells. One preparation is rRNA, another is mRNA, and the third is tRNA.
***a.*** How would you unambiguously identify the tRNA preparation according to a function test?
***b.*** How would you unambiguously identify the rRNA preparation using autoradiography?
***c.*** How would you determine that the remaining preparation contained mRNA even though you know from the above experiments that it is not rRNA or tRNA?

***9.11*** In an experiment designed to examine the relationship between 45S RNA in the nucleus and 28S + 18S rRNA in cytoplasmic ribosomes, a pulse of labeled uridine is presented at time zero to experimental rat liver cells in culture. After 10 minutes the cells are washed and transferred to medium containing unlabeled uridine. Samples of cells are then removed at intervals and analyzed, with the following results:

| | | **RNA containing [$^3$H]uridine** | | | |
|---|---|---|---|---|---|
| **portion of cell examined** | **time (min)** | **45S** | **32S** | **28S** | **18S** |
| cytoplasm | 0 | – | – | – | – |
| | 10 | – | – | – | – |
| | 20 | – | – | – | – |
| | 30 | – | – | – | + |
| | 60 | – | – | + | + |
| nucleus | 0 | – | – | – | – |
| | 10 | + | – | – | – |
| | 20 | – | – | – | – |
| | 30 | – | + | – | + |
| | 60 | – | – | + | + |

State which data provide evidence for the following interpretations:
***a.*** RNA is synthesized in the nucleus and not in the cytoplasm.
***b.*** 45S RNA is the precursor to 32S RNA.
***c.*** 32S RNA is the precursor to 28S RNA.
***d.*** 28S and 18S RNAs are processed from 45S RNA in the nucleus.
***e.*** 18S RNA is processed to mature rRNA before 28S RNA is completed.
***f.*** Why is 45S RNA labeled only at 10 minutes and not afterward?

***9.12*** Ribosomal RNAs from the small subunit of *E. coli* and from chloroplast ribosomes are 16S single-stranded molecules that show a striking degree of homology according to base sequence analysis. How would you demonstrate the degree of identity or similarity between 16S RNAs isolated and purified from *E. coli* ribosomes and from chloroplast ribosomes of a species not studied previously, using only the method of molecular hybridization for the preliminary analysis since it is quicker than RNA sequencing?

***9.13*** A set of experiments was performed to determine whether globin polypeptides were synthesized at polysomes or at monosomes. [$^{14}$C]amino acids were pulsed into reticulocyte cultures for 45 seconds, after which the cells were lysed and the cytoplasmic extract was sedimented to equilibrium in a density gradient. Measurements of absorbance (solid line) and radioactivity (dashed line) were made of 35 consecutive fractions representing the entire gradient, with the following results:

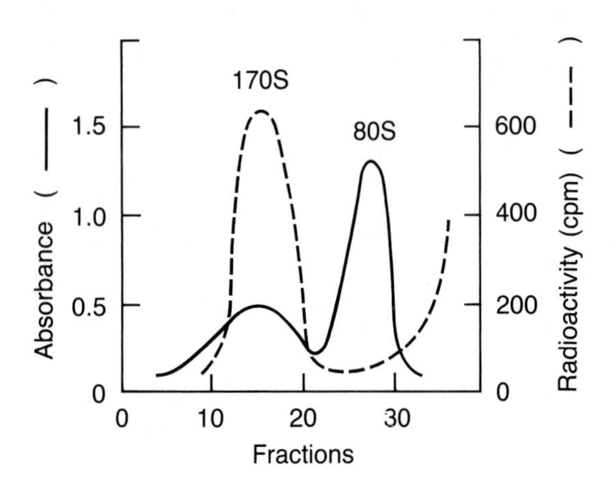


***a.*** Which fractions contain the polysomes and which contain monosomes?
***b.*** Why is the radioactivity peak higher in the 170S region of the gradient than elsewhere in the preparation?
***c.*** Why is there little or no radioactivity in fractions 1 through 10?
***d.*** Fractions 30 through 35 contain the fluid supernatant phase of the extract. Why is there so much radioactivity in this nonparticulate material?

***9.14*** An experiment similar to the one described in Question 9.13 was performed except that puromycin was added after the 45-second pulse with labeled amino acids and the cells were lysed shortly afterward. Puromycin causes premature termination of polypeptide chain synthesis and leads to early release of nascent chains from ribosomes.
***a.*** If translation takes place at polysomes, which of the results shown at the top of the next column would be predicted?

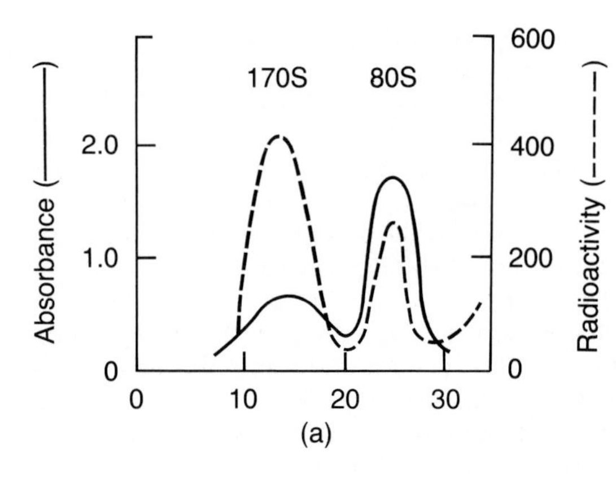


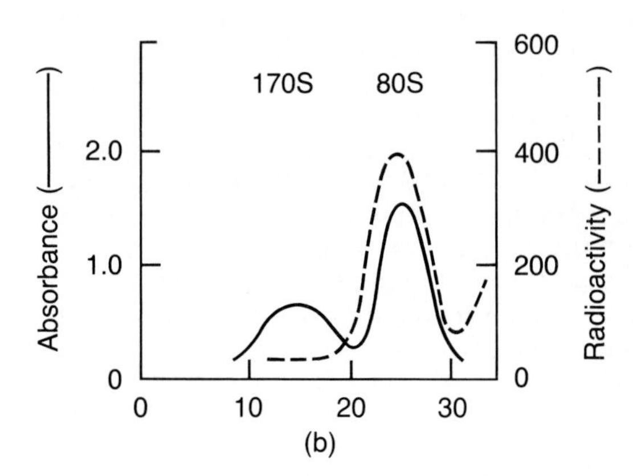


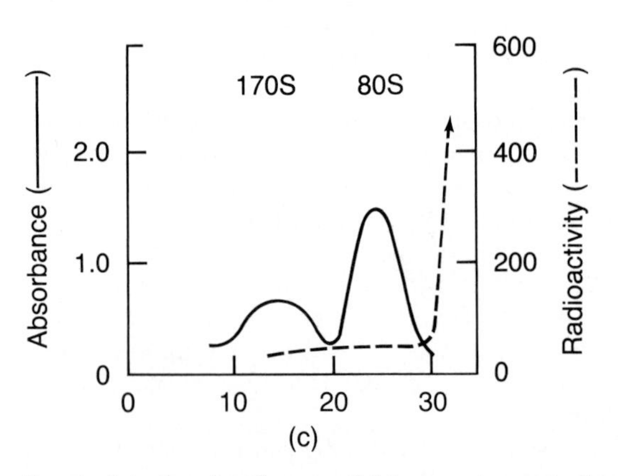


***b.*** Explain the significance of this experiment in the interpretation of the data in 9.13.

***9.15*** A messenger RNA, known to be a transcript of a gene that encodes a small peptide hormone only seven amino acids long, has been shown to have the construction shown at the top of the next page.

Leader | Trailer

5′———— AUG AAU UUG CCA UGG CAU UUU UAG ———— AAAAAAAAA–3′

***a.*** Which of the seven amino acid–specifying codons could be changed to a stop codon by a single base substitution? Give the particular stop codon in each case you have cited.
***b.*** If the stop codon at the 3′ end of the encoded sequence was altered to a codon specifying an amino acid by base substitution or by deletion or addition of a base within the triplet, what would be the effect on the polypeptide translated from such an altered sequence?

***9.16*** The α-globin chains in normal human hemoglobin (Hb A) are 141 amino acids long. The genetic variants hemoglobin Constant Spring (Hb CS) and hemoglobin Wayne 1 (Hb W1) have α-chains longer than 141 amino acids. The first 141 amino acids of Hb CS are identical to normal Hb A chains, but only the first 139 amino acids of Hb W1 are the same as normal Hb A α-chains. Assuming the simplest case of an alteration by substitution, deletion, or insertion of one nucleotide, explain the origin and consequence of each mutation:
***a.*** From Hb A to Hb CS.
***b.*** Of Hb A to Hb W1.

***9.17*** Streptomycin interferes with protein synthesis in prokaryotes in the following ways:
***a.*** Releasing fMet-tRNA from the ribosome just after the 50S ribosomal subunit has joined the initiation complex,
***b.*** Inhibiting binding of aminoacyl-tRNA to the ribosome, and
***c.*** Inhibiting translocation of peptidyl-tRNA on the ribosome.
For each of these effects, explain the reason(s) for inhibition of polypeptide chain synthesis in cells.

***9.18*** When synthetic RNA polymers were added as messenger RNAs to *in vitro* protein synthesizing systems, different amino acids were incorporated into polypeptides in the presence and absence of streptomycin, as shown below:

| synthetic RNA | streptomycin absent | streptomycin present |
|---|---|---|
| poly(U) | Phe | Phe, Tyr |
| poly(A) | Lys | Lys, Ile |
| poly(C) | Pro | Pro, Arg |
| poly(G) | Gly | Gly, Ala |

***a.*** What is the effect of streptomycin in this system?
***b.*** What is a common feature in each case of the effect stated in (a)?

***9.19*** Define the following terms: ***a.*** promoter ***b.*** TATA box ***c.*** exon ***d.*** intron ***e.*** anticodon ***f.*** monosome ***g.*** polysome ***h.*** rDNA ***i.*** initiation complex ***j.*** peptidyl transferase.

## References

Brenner, S., F. Jacob, and M. Meselson. 1961. An unstable intermediate carrying information from genes to ribosomes for protein synthesis. *Nature* **190**:576.

Brown, D.D., and J.B. Gurdon. 1964. Absence of ribosomal RNA synthesis in the anucleolate mutant of *Xenopus laevis*. *Proc. Nat. Acad. Sci. U.S.* **51**:139.

Chambon, P. May 1981. Split genes. *Sci. Amer.* **244**:60.

Crick, F.H.C. 1966. Codon-anticodon pairing: The wobble hypothesis. *J. Mol. Biol.* **19**:548.

Crick, F.H.C. 1979. Split genes and RNA splicing. *Science* **204**:264.

Dujardin, G., C. Kacq, and P.P. Slonimski. 1982. Single base substitution in an intron of oxidase gene compensates splicing defects of the cytochrome *b* gene. *Nature* **298**:628.

Federoff, N.V. 1979. On spacers. *Cell* **16**:697.

Flint, S.J. 1979. Spliced viral messenger RNA. *Amer. Sci.* **67**:300.

Gorini, L. Apr. 1966. Antibiotics and the genetic code. *Sci. Amer.* **214**:102.

Hollis, G.F., et al. 1982. Processed genes: A dispersed human immunoglobulin gene bearing evidence of RNA-type processing. *Nature* **296**:321.

Korn, L.J. 1982. Transcription of *Xenopus* 5S ribosomal RNA genes. *Nature* **295**:101.

Kozak, M. 1981. Possible role of flanking nucleotides in recognition of the AUG initiator codon by eukaryotic ribosomes. *Nucl. Acids Res.* **9**:5233.

Lake, J.A. Aug. 1981. The ribosome. *Sci. Amer.* **245**:84.

Mantei, N., and C. Weissmann. 1982. Controlled transcription of a human α-interferon gene introduced into mouse L cells. *Nature* **297**:128.

Marx, J.L. 1982. Gene scanning with block mutations. *Science* **217**:434.

Miller, O.L., Jr. Mar. 1973. The visualization of genes in action. *Sci. Amer.* **229**:34.

Noller, H.F., and C.R. Woese. 1981. Secondary structure of 16S ribosomal RNA. *Science* **212**:403.

Pribnow, D. 1975. Nucleotide sequence of an RNA polymerase binding site at an early T7 promoter. *Proc. Nat. Acad. Sci. U.S.* **72**:784.

Proudfoot, N.J. 1980. Pseudogenes. *Nature* **286**:840.

Rich, A., and S.H. Kim. Jan. 1978. The three-dimensional structure of transfer RNA. *Sci. Amer.* **238**:52.

Ritossa, F.M., and S. Spiegelman. 1965. Localization of DNA complementary to ribosomal RNA in the nucleolus organizer region of *Drosophila melanogaster*. *Proc. Nat. Acad. Sci. U.S.* **53**:737.

Tilghman, S.M., et al. 1978. The intervening sequence of a mouse β-globin gene is transcribed within the 15S β-globin mRNA precursor. *Proc. Nat. Acad. Sci. U.S.* **75**:1309.

Warner, J.R., A. Rich, and C.E. Hall. 1962. Electron microscope studies of ribosomal clusters synthesizing hemoglobin. *Science* **138**:1399.

Watson, M.D. 1981. The role of DNA topoisomerase I in transcription and transposition. *Trends Biochem. Sci.* **6**:7.

Weisbrod, S. 1982. Active chromatin. *Nature* **297**:289.

Woo, S.L.C., et al. 1981. Complete nucleotide sequence of the chicken chromosomal ovalbumin gene and its biological significance. *Biochemistry* **20**:6437.

Zalkin, H., and C. Yanofsky. 1982. Yeast gene TRP5: Structure, function, regulation. *J. Biol. Chem.* **257**:1491.

# CHAPTER 10

# Regulation of Gene Expression

Phenotype development requires gene expression, that is, transcription of coded information and its translation into proteins. Since phenotypically different cells may arise in genotypically identical populations, gene expression must be regulated so that some gene products are made in some cells and not in others. We make this observation each time we see a multicellular individual whose cells can all be traced back to a single fertilized egg. Red blood cells make hemoglobin, but skin cells do not; many kinds of cells express only certain genes at certain times in development. Various types of control mechanisms regulate gene expression. In some cases gene expression is modulated; in other cases genes themselves may be altered in specific ways that influence their expression. Some of the control mechanisms that regulate gene expression will be discussed in this chapter. Other aspects of regulation of gene expression will be the subject of Chapter 14, in which developmental pathways will be discussed.

## Control of Gene Expression in Bacteria

The expression of genetic information via protein products that contribute to phenotype development was analyzed by biochemical genetic methods in the 1930s and 1940s, principally using *Neurospora* and *Drosophila.* These studies led to the *one gene–one polypeptide concept,* which stated that each gene governs the synthesis of a polypeptide in the organism. In the 1950s and 1960s genetic studies of the genes that govern lactose utilization in *E. coli,* conducted by Francois Jacob and Jacques Monod in Paris, considerably modified this view of gene function. Their interpretations of gene function in phenotype development were believed to explain all or most of gene expression in bacteria. In the 1970s, other systems of regulating gene expression were described for bacteria. Although these control mechanisms were ultimately shown to be absent in eukaryotic organisms, the bacterial regulation studies provided a strong foundation for studies of eukaryotic mechanisms that were later described by other investigators.

### 10.1 Operon Concept of Transcriptional Control

Jacob and Monod identified *lac* genes in *E. coli* that govern synthesis of proteins needed for lactose utilization in metabolism. In addition, they discovered genes that regulated the transcription of messenger RNA for these proteins. The genes specifying proteins and the genes regulating the expression of the protein-specifying genes are parts of a coordinated gene cluster called an **operon.** They analyzed the *lac* operon by genetic methods.

From studies of isolated *lac* mutants, Jacob and Monod identified three genes that encode the amino acid sequences of three different proteins needed to metabolize lactose. These three genes were mapped adjacent to one another in the *lac* region, in the order *lacZ–lacY–lacA.* They were called **structural genes** since they coded for the primary structures of the lactose-metabolizing enzyme *β-galactosidase* (encoded by the *z* gene, or *lacZ*); of a membrane protein called *galactoside permease,* which helps lactose molecules across the membrane and into the cell (by the *y* gene, or *lacY*); and of the enzyme *thiogalactoside transacetylase,* whose function is still uncertain (by the *a* gene, or *lacA*). Mutations in any of these three genes led to amino acid changes in their specified proteins, which altered protein function and lactose metabolism. Chemical and physical changes in these proteins and their effects on the phenotype could be identified and analyzed in mutant cell cultures.

Two other classes of *E. coli* mutants were isolated and described. The mutants produced wild-type proteins, but *the amounts made of the three structural gene products in mutant cells were altered.* Both of the mutant classes exerted a regulatory effect on Lac protein synthesis. The mutants indicated the existence of **regulatory genes** that were involved in controlling the expression of the three structural genes of the *lac* operon. One class of regulatory mutants had modifications in the **operator gene** (*o* gene, or *lacO*) of the operon; the second class had a modified **repressor gene** (*i* gene, or *lacI*) of the *lac* operon. Each class of regulatory mutant was distinguishable from the other class and from structural gene mutants by the nature of its effects on structural gene expression.

Operator mutants influenced production of all three proteins at the same time and in the same way; that is, one operator mutant allele affected all three structural genes *coordinately.* Although the three proteins remained structurally unaltered, different operator mutant alleles were identified on the basis of differences in structural gene expression in response to lactose (Table 10.1). Wild-type cells with the genotype $o^+z^+y^+a^+$ made all three proteins *only* when lactose was provided in the nutrient medium. Synthesis of the three proteins was therefore *inducible* by lactose. Mutants that could not synthesize the three proteins in the presence of lactose were *operator-negative* and carried the allele $o^-$. Cells with the genotype $o^-z^+y^+a^+$ made none of the three proteins in the presence or absence of lactose, even though wild-type alleles of the three structural genes were present in the genotype. Other operator mutants made all three proteins *constitutively* (all the time), rather than inducibly, regardless of the presence or absence of lactose. These *operator-constitutive* mutants carried the allele $o^c$. Cells with genotype $o^cz^+y^+a^+$ made all three proteins all the time. Cells with genotype $o^-z^+y^+a^+$ never made the three proteins. Cells with genotype $o^+z^+y^+a^+$ made the three proteins only when lactose was present. An operator gene thus regulated the expression of structural genes in the operon.

To determine whether regulatory genes acted directly as DNA sites or through proteins made from their DNA sequence, Jacob and Monod constructed *merodiploid* strains that were partially diploid for the *lac* region (see Section 6.9). If regulatory genes interacted directly as DNA sites, they would only influence structural genes located in the same chromosome (*cis* arrangement) and not genes in another chromosome (*trans* arrangement). If, on the other hand, regulatory genes exerted their effects through circulating protein prod-

***Table 10.1*** Genotypes and phenotypes of *E. coli* strains carrying various alleles of the operator locus in the *lac* region.

| | phenotypes (proteins synthesized) | | | | | |
|---|---|---|---|---|---|---|
| | lactose present | | | lactose absent | | |
| **genotype** | **$\beta$-galactosidase** | **permease** | **acetylase** | **$\beta$-galactosidase** | **permease** | **acetylase** |
| $o^+z^+y^+a^+$ | + | + | + | 0 | 0 | 0 |
| $o^-z^+y^+a^+$ | 0 | 0 | 0 | 0 | 0 | 0 |
| $o^cz^+y^+a^+$ | + | + | + | + | + | + |

ucts, then they should have the same regulatory effects in *both* the *cis* and *trans* arrangements.

In cells merodiploid for *lac* genes, the $o^c$ allele only affected the expression of *z–y–a* structural genes located in the *same* chromosome (*cis*), and not *z–y–a* genes in the *trans* arrangement (Table 10.2). The $o^c$ allele is *cis dominant* since it only influences structural genes in the same chromosome as itself. Operator mutations therefore define a gene cluster whose closely linked units act as a coordinated unit of function; this is the operon. The operator gene site mapped adjacent to and just to the left of the *z–y–a* structural genes, which were tightly linked in the *lac* operon. The operator makes no protein product.

The second class of regulatory mutants defined the repressor gene *lacI*. These mutants did not alter structural features of the Lac proteins, but they did influence the amounts made of all three proteins coordinately. One type of repressor mutant was called superrepressor ($i^S$) since it prevented induction of the *z–y–a* proteins when lactose was present. Its action was therefore similar to the action of the operator-negative mutants. The second type of repressor mutant ($i^-$) caused all three proteins to be made constitutively in the presence or absence of lactose in the medium. Hence its action was similar to that of the operator-constitutive mutants. Repressor mutations could be functionally distinguished from operator mutations by genotype-phenotype comparisons in merodiploids. While operator alleles influenced only those genes in the same chromosome (*cis* arrangement), repressor alleles influenced *z–y–a* gene expression in the same way in *both* the *cis* and the *trans* arrangements of the merodiploid genotypes (Table 10.3).

**Table 10.2** Genotypes and phenotypes of *E. coli* strains partially diploid for genes of the *lac* region and heterozygous for the cis-dominant $o^c$ allele of the operator site.

| merodiploid genotype | phenotypes* constitutively synthesized proteins | phenotypes* inducibly synthesized proteins |
|---|---|---|
| $\frac{o^+ z^- y^- a^-}{F' o^c z^+ y^+ a^+}$ | β-galactosidase<br>permease<br>acetylase | none |
| $\frac{o^+ z^+ y^+ a^+}{F' o^c z^- y^- a^-}$ | none | β-galactosidase<br>permease<br>acetylase |
| $\frac{F' o^+ z^- y^- a^+}{o^c z^+ y^+ a^-}$ | β-galactosidase<br>permease | acetylase |

*Allele $o^c$ is cis-dominant to $o^+$ since $o^c$ causes *constitutive* production only of proteins coded by structural genes on the same chromosome as itself; $o^+$ governs *inducible* enzyme synthesis only by structural genes on the same chromosome as itself.

These and other data led Jacob and Monod to formulate the *operon concept* of gene expression and its regulation. Structural genes code for enzymes and other proteins that are utilized in metabolism and cell construction, and regulatory genes control expression of the structural genes in the same operon. All the genes in an operon constitute a functional unit of coordinated gene action.

Since operator alleles were effective only in the *cis* arrangement, Jacob and Monod proposed that the operator gene regulated *coordinated transcription* of the *z–y–a* structural-gene cluster, which is mapped right next to the operator in the same chromosome. The physical proximity required for operator influence and its coordinated effect on all three structural genes in the *cis* arrangement were important features. These features indicated that the operator determined whether or not all three structural genes would be transcribed as a unit. If the structural genes were transcribed, then the three proteins would be translated from the mRNA transcripts.

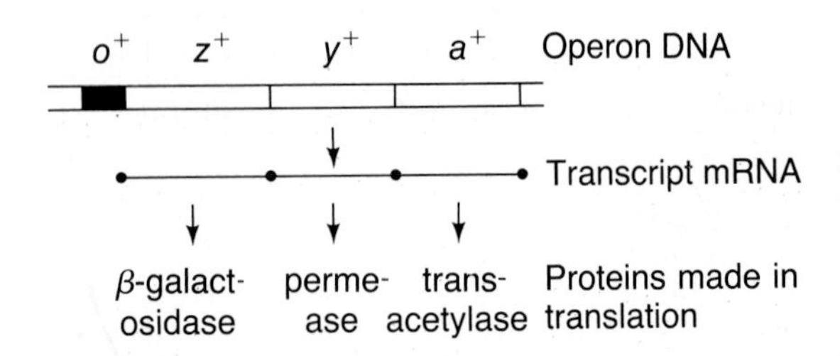

***Table 10.3*** Genotypes and phenotypes of haploids and merodiploids for the *lac* region in *E. coli* strains. Allelic behavior can be determined from phenotypes of merodiploids carrying different *i* alleles in *cis* or *trans* arrangements with *lac* operon loci.

| | **phenotypes** | | |
|---|---|---|---|
| **genotype** | **constitutively synthesized proteins** | **inducibly synthesized proteins** | **interpreted *i* allele behavior** |
| haploid | | | |
| $i^+o^+z^+y^+a^+$ | none | all three | wild type |
| $i^so^+z^+y^+a^+$ | none | none | superrepressor mutant |
| $i^-o^+z^+y^+a^+$ | all three | none | constitutive mutant |
| merodiploid | | | |
| $\frac{i^so^+z^+y^+a^+}{i^+o^+z^-y^-a^-}$ | none | none | $i^s$dominant over $i^+$ |
| $\frac{i^so^+z^-y^-a^-}{i^+o^+z^+y^+a^+}$ | none | none | $i^s$ dominant over $i^+$ |
| $\frac{i^-o^+z^+y^+a^+}{i^+o^+z^-y^-a^-}$ | none | all three | $i^+$ dominant over $i^-$ |
| $\frac{i^-o^+z^-y^-a^-}{i^+o^+z^+y^+a^+}$ | none | all three | $i^+$ dominant over $i^-$ |

If the operator did not permit transcription, none of the three proteins would be made even if lactose were present because no mRNAs would be translated.

Jacob and Monod postulated that the repressor gene acted through the cytoplasm since the *i* alleles were effective in both the *cis* and the *trans* strand arrangements in merodiploids. They postulated further that the *lacI* gene specified the production of a **repressor** protein, which interacted physically with the operator. When an *inducer* such as lactose was present, it bound to the repressor in such a way that the repressor could no longer interact physically with the operator gene. With the repressor removed, transcription could then take place and all three **inducible enzymes** would be synthesized in cells that had an active operator. In the absence of an inducer, therefore, the repressor could interact with the operator and prevent transcription of the three structural genes. None of the three inducible enzymes would be made since no mRNA transcripts would be synthesized (Fig. 10.1).

In the late 1960s Walter Gilbert and Benno Müller-Hill isolated and purified the postulated repressor gene product. They showed that repressor protein was present in $i^+$ strains but not in $i^-$ strains in which constitutive synthesis of all three structural gene products is affected. In addition, they showed that repressor protein would bind to $o^+$ but not to $o^c$ DNA. Strains with the $o^c$ allele made all three structural gene products constitutively. These studies and similar studies of a phage repressor system by Mark Ptashne plus other kinds of evidence helped to confirm the Jacob-Monod model of regulation of gene expression at the level of transcription.

In later studies of the *lac* region, a third type of regulatory genetic component was identified

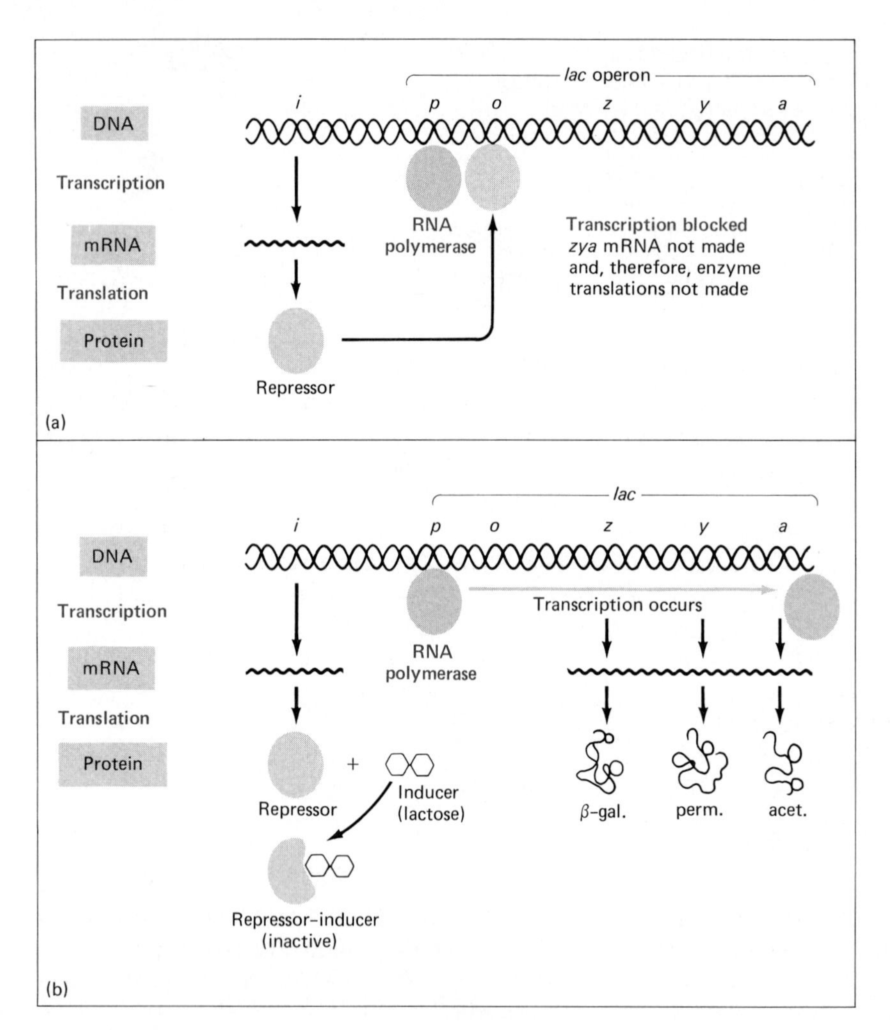


***Figure 10.1*** Operon concept of control over gene expression in the case of inducible enzymes of the *lac* region in *E. coli.* (a) Transcription is blocked and protein translations are not made in the absence of lactose inducer since the *lac* repressor blocks RNA polymerase movement past the operator site. (b) In the presence of the inducer, an inactive repressor-inducer complex forms and the operator site no longer is blocked. RNA polymerase can move from its binding site at the promoter along the DNA template and catalyze transcription of the *z–y–a* genetic region. Translation occurs once mRNA is made.

as the promoter (*p* gene, or *lacP*). The promoter mapped just adjacent to the left of the operator. Promoter mutants coordinately reduce the amounts of proteins synthesized from the *z–y–a* cluster, but these mutants continue to respond to *i* gene control. In other words, the three proteins are made in reduced amounts but only when lactose is present, which means that repressor control has not changed. The promoter is the specific DNA region to which the enzyme RNA polymerase binds. This is the enzyme that catalyzes transcription of DNA into mRNA. The polymerase can move along the operon DNA if it is not physically blocked by a repressor molecule bound to the operator, which lies between the promoter and the structural genes.

All the mutants of the *lac* region can be understood on the basis of this operon model for regulation of structural gene activities by regulatory genes. The direction of transcription appears to be distal, proceeding from the promoter to the operator and on to *z*, *y*, and *a*, in that order. This was interpreted from genetic studies of polar mutants. A *polar mutation* reduces all wild-type activity distal to that mutation, so that a polar defect in *z* influences *z*, *y*, and *a*; a polar defect in *y* affects only *y* and *a*; and a polar mutation in *a* affects only the *a* gene product and not those of the *z* and *y* genes. These different polar mutations can be explained by the fact that transcription begins at the operator and proceeds distally through *z–y–a*. The result is a *single polygenic transcript,* with all the copied information for the three structural genes in one mRNA molecule.

The repressor gene–promoter–operator complex regulates transcription of a cluster of the three structural genes in the *lac* operon, whose proteins interact in lactose metabolism (Table 10.4). The *z–y–a* genes are coordinately turned on when lactose is present, or they are coordinately turned off when lactose is absent from wild-type cells. This finding is an example of **negative control** since the repressor turns off the system.

A similar instance of negative control characterizes the synthesis of many **repressible enzymes.** Unlike inducible enzymes, repressible enzymes are made only when the reaction product is depleted and *not* when the metabolite is present in excess (Fig. 10.2). In repressible-enzyme systems the metabolite is called a **corepressor** since it helps the repressor bind to the operator and prevent transcription. Sometimes inducers and corepressors are referred to generally as *effectors,* that is, metabolites that influence repressor binding in a gene regulation system.

Jacob and Monod deduced the existence of the operon on the basis of genetic data. Some years later, biochemical and molecular genetic studies in various laboratories verified their deductions. Later studies not only described the molecular basis for the interactions of genes, proteins, and metabolites of operons in different species but provided the exact nucleotide sequences of promoter and operator sites. These data have opened the way to a detailed analysis of the physical and chemical interactions between DNA and such proteins as repressors and polymerases.

## 10.2 Attenuation of Operon Transcription

An independent mechanism controls bacterial gene expression for a number of operons that specify enzymes for amino acid synthesis. This control mechanism, called **attenuation,** regulates gene expression by *terminating tran-*

***Table 10.4*** Regulatory and structural genes of the *lac* region in *E. coli.*

| component | symbol | function | protein product |
|---|---|---|---|
| structural genes | *z* | codes for enzyme protein | β-galactosidase |
| | *y* | codes for membrane protein | galactoside permease |
| | *a* | codes for enzyme protein | thiogalactoside transacetylase |
| regulatory genes | | | |
| operator | *o* | binding site for repressor | none |
| promoter | *p* | binding site for RNA polymerase | none |
| repressor gene* | *i* | codes for repressor protein | *lac* repressor |

*Gene *i* is not part of the *lac* operon, but it has been mapped very nearby.

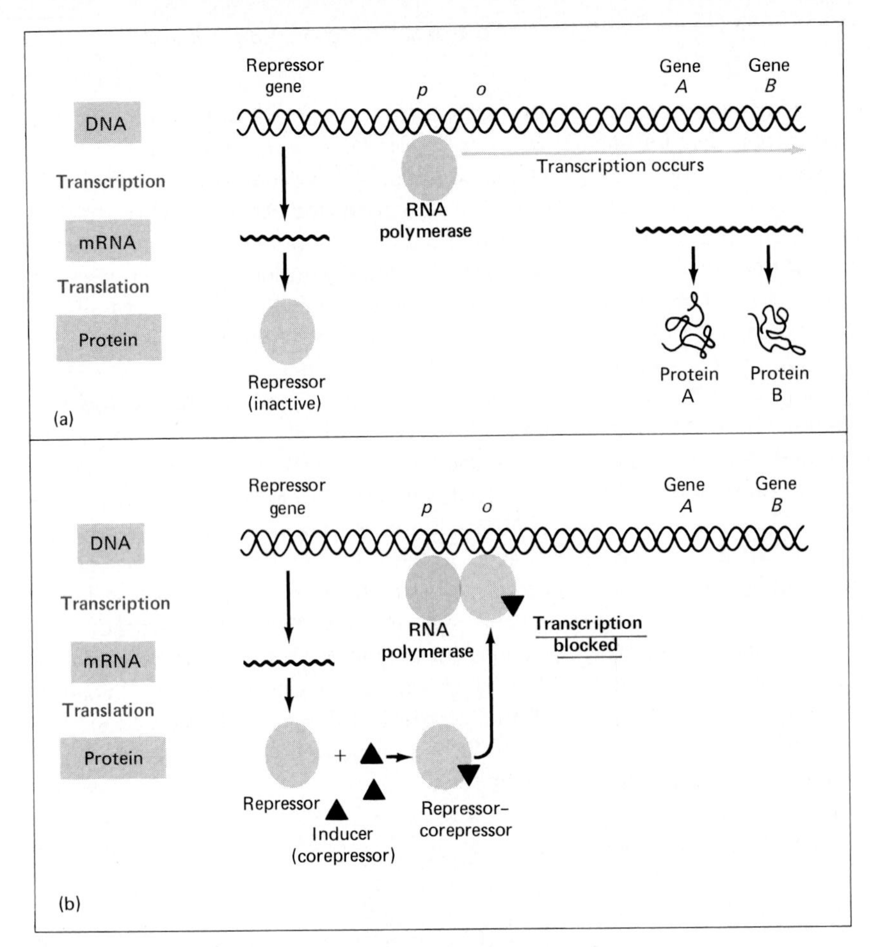


***Figure 10.2*** Differential gene expression in the case of repressible enzymes encoded by structural genes *A* and *B*. (a) Gene expression is turned on in the absence of the corepressor metabolite, but (b) it is turned off when repressor-corepressor bind to the operator site and block RNA polymerase movement to catalyze transcription.

*scription* before RNA polymerase reaches the structural genes. Bruce Ames and co-workers provided the first experimental evidence for such a mechanism for the histidine (*his*) operon. The details of the mechanism were later uncovered by Charles Yanofsky for the tryptophan (*trp*) operon and for a number of other operons concerned with amino acid biosynthesis. Similar controls operate in *E. coli* and other bacteria and in some phages, such as lambda.

The repressor-operator model for regulation of operon genes was believed to apply to all bacterial operons. Although repressors were found for *trp, arg,* and *tyr* operons, investigators could not find repressors for the *his* and isoleucine-valine (*ilv*) operons. Clues to the existence of another regulatory mechanism were obtained from analyses of *E. coli* strains that had small deletions of the *trp* operon. The deletions were near to, but distinct from, the promoter-operator region (*trpP–trpO*). Some of the deletions caused an *increased* expression of the structural genes *trpE–trpD–trpC–trpB–trpA,* which specify five enzymes in the tryptophan biosynthetic pathway (see Fig. 7.16). The effective deletions removed a region between *trpO* and the first structural gene in the operon sequence, *trpE.* Any deletion that was wholly contained within the structural genes, however, did not cause increased expression. Since all the mutant strains were capable of normal repressor-operator function, another regulatory system was postulated. The deletions presumably had removed this other, independent regulatory system. Furthermore, since gene expression was enhanced in the regulatory deletion mutants, the regulatory region appeared to be involved with some block to maximal structural gene expression. The site of this block probably occurred between *trpO* and *trpE* (Fig. 10.3).

Studies of other deletion mutants revealed the presence of the regulatory site within 30 bases of the start of *trpE*. Analysis of *trp* messenger RNA transcripts showed the presence of about 160 bases between the 5′ start of the message and the start of the *trpE* sequence. This long transcript segment preceding *trpE* is called the **leader sequence,** and its genetic designation is *trpL*. The postulated regulatory site was probably located within the leader sequence.

To determine at what level of gene expression the regulatory region acts, the frequencies were compared between transcription of the *trpL* region preceding the regulatory site and transcription of the most distal region of the operon, at genes *trpB–trpA*. It was found that *trpB–trpA* transcription was less than 15% as frequent as transcription of *trpL*. In other words, about 85% of the events that transcribe *trpL* do not continue onwards along the operon to transcribe *trpB–trpA*. Since *trpE* and *trpA* are transcribed at the same frequency, clearly most of the transcripts must terminate before *trpE*. The block to maximal expression of the *trp* structural genes, therefore, was caused by *termination of transcription before the structural genes were reached.*

The site at which transcription termination occurs is called the **attenuator.** Sequence analysis of the leader region of the mRNA transcript provided important clues to the structure and function of the attenuator in regulating termination of transcription. In particular, two observations were significant.

First, the leader sequence contained a ribosome-binding site near the 5′ start and enough codons to specify a polypeptide 14 amino acids long. Two codons in tandem sequence specified Trp residues. Beginning at the start of the leader transcript with nucleotide 1, we find an AUG initiation codon present at nucleotides 27 through 29, two Trp codons at nucleotides 54 through 59, and the stop codon UGA at nucleotides 69 through 71 (Fig. 10.4).

Second, the mRNA transcript was capable of forming extensive *stem and loop secondary structure* in the leader region immediately fol-

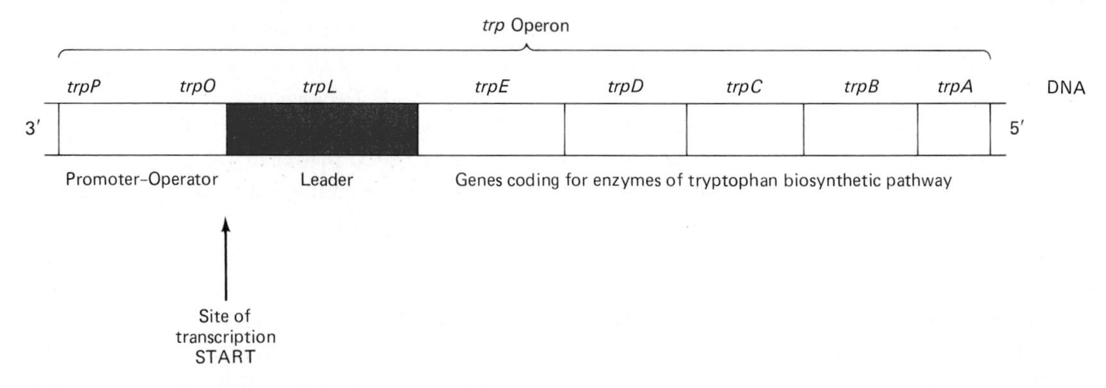


***Figure 10.3*** The regulatory and structural genes of the *trp* operon of *E. coli*. Initiation of transcription is controlled at a promoter-operator region (*trpP–trpO*), and termination of transcription is regulated by a sequence in the transcribed leader (*trpL*), which is situated just ahead of the set of structural genes (*trpE* to *trpA*). (Reprinted by permission from C. Yanofsky. 1981. *Nature* **289**:751–758. Copyright © 1981 Macmillan Journals Limited.)

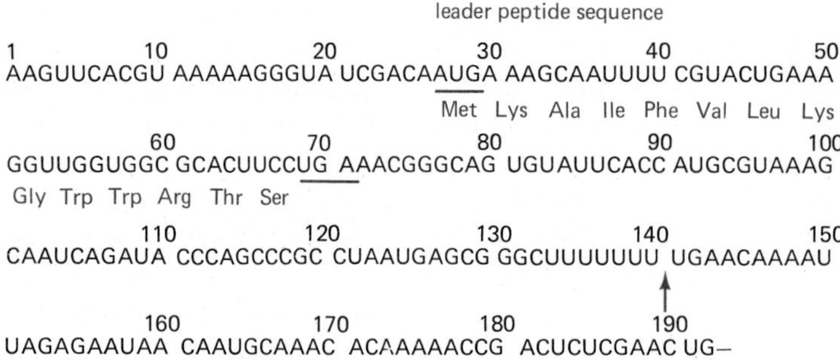


***Figure 10.4*** Nucleotide sequence of the 5′ end of *trp* messenger RNA, including the 162 nucleotides of the regulatory leader sequence and a few of the codons specifying the *trpE* polypeptide. Part of the leader sequence encodes a leader peptide 14 amino acids long, from the AUG start at nucleotides 27 to 29 through the UGA stop codon at nucleotides 69 to 71. The nonterminated transcript is shown. If transcription had been terminated at the attenuator, a transcript sequence of only 140 nucleotides would have been made (its 3′ terminus is marked by an arrow). (Reprinted by permission from C. Yanofsky. 1981. *Nature* **289**:751–758. Copyright © 1981 Macmillan Journals Limited.)

lowing (distal to) the UGA stop codon. This feature was interpreted on the basis of two regions of symmetrically inverted repeat sequences. But, since the two symmetrical regions partially overlapped, both regions could not form secondary structures at the same time; they were mutually exclusive (Fig. 10.5). The second stem and loop structure, nearest to *trpE,* is rich in guanine and cytosine and is immediately followed by eight uridine residues. Such a GC-rich stem and loop followed by poly(U) has been found near transcription termination sites in other bacterial operons that show attenuation control of gene expression.

Studies of deletion mutants, DNA sequencing, and mRNA sequencing have shown that part of the leader is translated. The translation polypeptide influences transcriptional readthrough by determining which stem and loop will be formed. One stem and loop structure decreases transcription termination; in contrast, the GC-rich stem and loop enhances transcription termination.

Many details of the model of regulation of gene expression by attenuation have been verified for the *trp* operon and other operons involved in amino acid biosynthesis. The *attenuator itself* consists of the GC-rich stem and loop structure followed by the poly(U) segment, located at the 3′ end of the leader. If the attenuator secondary structure develops, RNA polymerase will terminate transcription in the poly(U) segment and will dissociate from the transcript before reaching the structural genes. The regulation of termination of transcription is therefore concerned with whether or not the GC-rich stem and loop structure forms. If it does not develop, then the RNA polymerase will continue through the poly(U) segment and will continue transcription in the structural gene region of the operon (Fig. 10.6).

The responsiveness of the attenuation control is determined primarily by the utilization of tryptophan in protein synthesis. If little Trp is present, an insufficient amount of Trp-tRNA$^{Trp}$ will be available to complete translation of the leader peptide. The ribosome stalls at the tandem Trp codons in the leader mRNA and allows mRNA to develop a secondary structure, which prevents the mutually exclusive attenuator stem and loop from forming. The RNA polymerase continues transcription readthrough into the structural gene region. Enzymes are made from the transcript and Trp is synthesized in the cell.

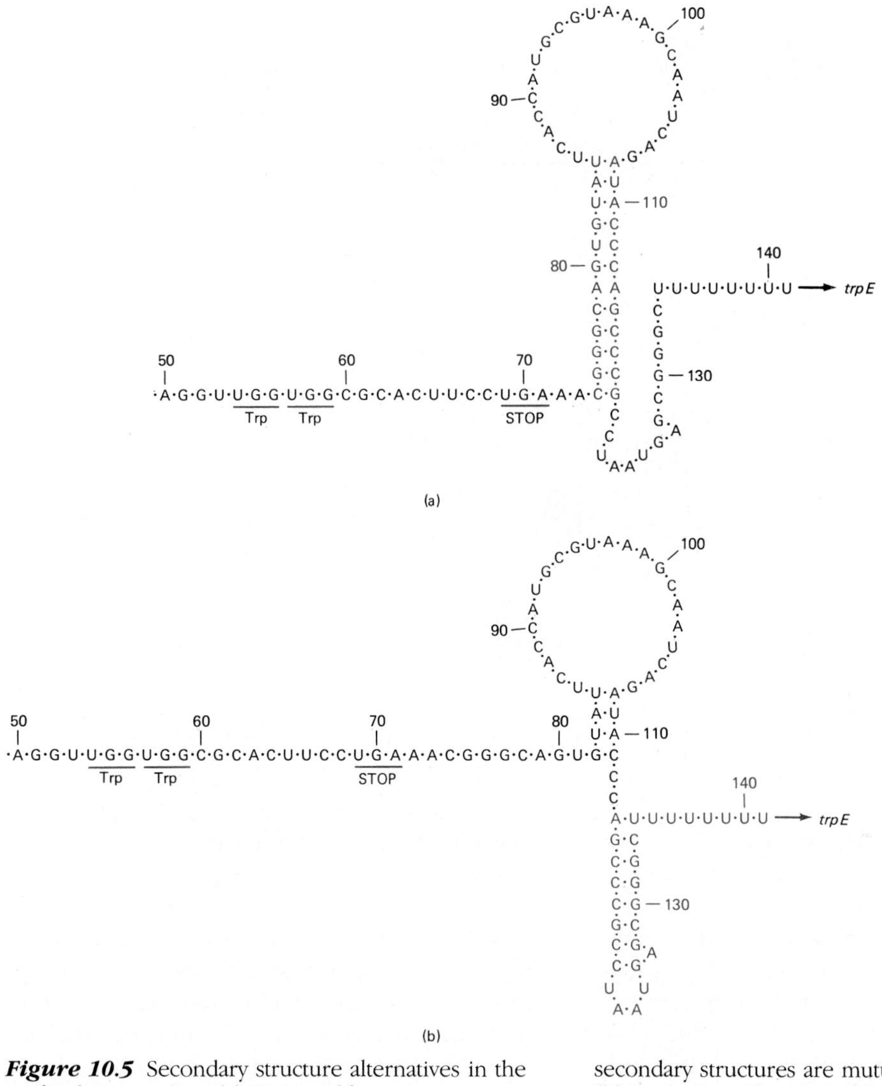

(a)

(b)

***Figure 10.5*** Secondary structure alternatives in the *trp* leader transcript. (a) Stem-and-loop structure believed to permit continued transcription beyond nucleotide 140 into the structural gene region. Formation of a locally double-bonded stem (color) near the leader peptide sequence prevents attenuator stem and loop formation since the two secondary structures are mutually exclusive. (b) Formation of stem and loop secondary structure in the attenuator region (color) leads to termination of transcription in the poly(U) segment preceding *trpE*. (Reprinted by permission from C. Yanofsky. 1981. *Nature* **289**:751–758. Copyright © 1981 Macmillan Journals Limited.)

This typical repressible enzyme synthesis occurs when the end product of the pathway is limiting.

When an excess of Trp is present, ample Trp-tRNA$^{Trp}$ is available for leader peptide synthesis. The ribosome moves along the mRNA transcript, right behind the RNA polymerase, and translates the leader sequence. The ribosome blocks the development of the first possible stem and loop secondary structure, thereby allowing the development of the attenuator stem and loop. The ribosome dissociates at the leader UGA stop codon, and the RNA polymerase dissociates from the transcript on reaching the attenuator poly(U) segment. The structural

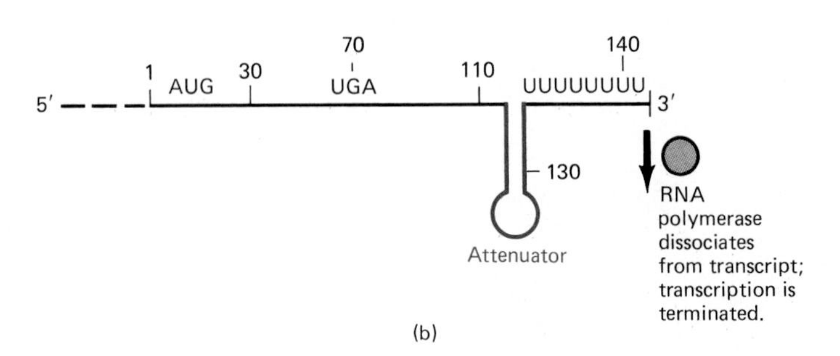


***Figure 10.6*** Regulation of *trp* operon expression by attenuation control of transcription. (a) The attenuator (color) cannot undergo stem and loop formation if a preceding region of the transcript has already developed secondary structure. RNA polymerase continues moving along the DNA sequence in this case, transcribing structural gene sequences. (b) Attenuator stem and loop structure prevents RNA polymerase from continuing its activity beyond the poly(U) segment, and transcription terminates before the enzyme reaches *trpE.* Gene expression occurs in (a) but not in (b), under attenuation control of repressible enzyme synthesis.

genes are not transcribed and their enzyme products are not made, thus halting Trp biosynthesis when Trp is present in the cell in excess (Fig. 10.7).

Some biosynthetic operons may have only attenuation control; these include the operons that code for enzymes in the pathways for histidine, threonine, and leucine synthesis. Other bacterial biosynthetic operons, such as *trp,* have both repressor-operator and attenuation controls for repressible enzyme synthesis. If RNA polymerase is blocked by the repressor control, transcripts are not made and enzymes are not synthesized, as occurs in conditions of excess tryptophan (see Fig. 10.2). If RNA polymerase escapes repressor control, it initiates transcription of the operon. Attenuation control will be activated, depending on the availability of Trp-tRNA$^{Trp}$ within the cell. In effect, the repressor-operator control is responsive to supplies of tryptophan within the cell, including molecules entering the cell from the environment. The attenuation control mechanism is responsive to the supply of Trp-tRNA$^{Trp}$, which fluctuates according to the rate of protein synthesis in the cell. Attenuation provides a control mechanism for fine-tuning gene expression for various bacterial operons and for various phage operons as well.

## 10.3 Positive Control of Gene Action

Positive controls *turn on* transcription when some molecule binds to the appropriate site of the operon DNA. Negative controls, such as repressor/operator systems, *turn off* tran-

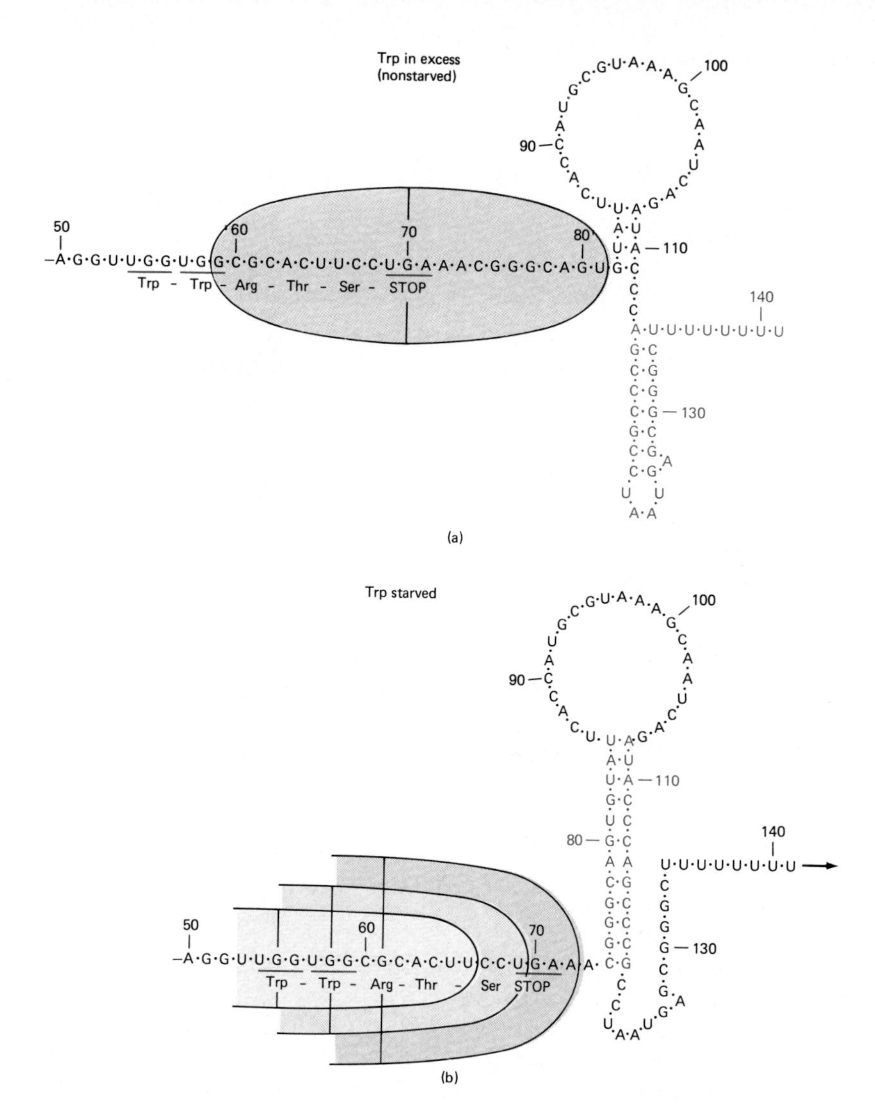


*Figure 10.7* Responsiveness of the attenuation control is determined partly by the utilization of tryptophan (Trp) in protein synthesis. (a) When Trp is in excess, leader peptides are translated as the ribosome (colored oval) moves along the mRNA transcript right behind RNA polymerase. The ribosome blocks development of the first stem and loop alternative and thereby allows formation of the attenuator stem-and-loop secondary structure (color). Trp is not made in the cell since the ribosome and the RNA polymerase dissociate from the mRNA transcript before the structural genes are reached. (b) A deficiency of Trp in cells leads to formation of the first stem and loop (color) and prevents formation of the mutually exclusive attenuator stem and loop. Transcription continues into the structural gene region and Trp is synthesized once the biosynthetic enzymes are translated from *trpE–trpA* encoded instructions. The outcome is typical of repressible enzyme synthesis, occurring when the end product of the pathway (Trp) is limiting but not when it is present in excess. (Reprinted by permission from C. Yanofsky. 1981. *Nature* **289**:751–758. Copyright © 1981 Macmillan Journals Limited.)

scription when repressor molecules bind to operator DNA. A maximum transcription response usually requires the mediation of both negative and positive controls over gene expression.

The positive control system was discovered through studies of **catabolite repression,** or the **glucose effect,** in blocking transcription of *glucose-sensitive operons,* which are involved in sugar catabolism, or breakdown. When glucose is present in the medium along with one or more other sugars, such as lactose or galactose, only the glucose is catabolized. The other sugars remain unutilized because the necessary enzymes, including $\beta$-galactosidase, are not synthesized. When glucose is depleted or is absent from the medium, the glucose-sensitive operons are transcribed and their enzyme products are made in the cells, according to the other particular sugars that are present. The products of glucose-sensitive operons are inducible enzymes, which are synthesized in the presence of their substrates (see Fig. 10.1). Catabolite repression, therefore, refers to the failure of enzyme induction when glucose is present, even though the substrates of the inducible enzymes are also present.

The key metabolite that influences catabolite repression is **cyclic adenosine monophosphate,** or **cyclic AMP** (Fig. 10.8). This molecule is required for transcription of operons that are sensitive to glucose catabolism. Cyclic AMP (cAMP) acts only when it binds with the catabolite activator protein (CAP); CAP, in turn, is only active in the cAMP-CAP complex. When the cAMP-CAP complex binds to the regulatory region of the operon, the rate of transcription of the structural genes of that operon increases. The ineffectiveness of each of these chemicals by itself was indicated by analysis of mutant strains that were deficient in CAP or that were deficient in the enzyme *adenylcyclase,* which catalyzes cAMP formation from ATP. Adenylcyclase-deficient mutants make CAP but not cAMP, and these mutants can synthesize inducible enzymes only in small amounts. CAP therefore does not act alone. CAP-deficient mutants make cAMP, and very little inducible enzyme synthesis takes place in these strains. Thus cAMP does not act alone.

Studies of repressor gene mutants of *E. coli* have shown that the cAMP-CAP positive control does not act through the agency of the repressor system. Wild-type $i^+$ strains and repressor gene mutants ($i$) of *E. coli* are stimulated to produce inducible enzymes equally well if cAMP is added to the cultures. The positive and negative controls of inducible enzyme synthesis act together, but each is a separate gene regulation system for operon transcription.

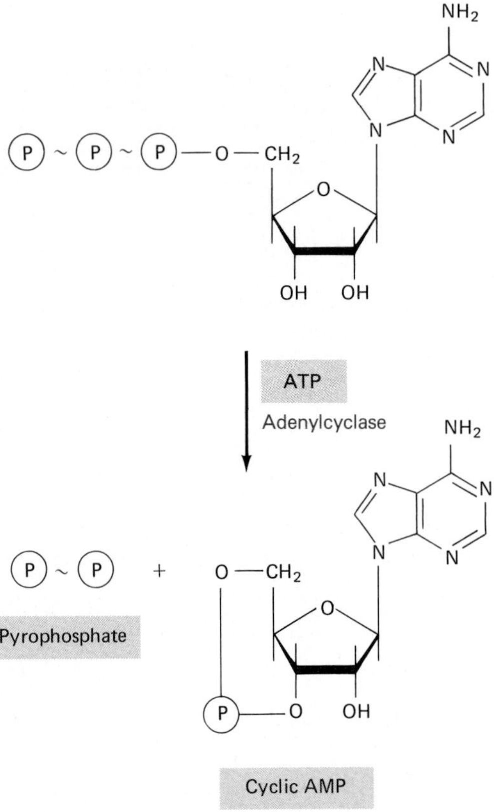

***Figure 10.8*** Cyclic AMP (adenosine 5′-monophosphate) is derived from ATP in a reaction catalyzed by adenylcyclase.

The cAMP-CAP positive control stimulates operon transcription much more efficiently than does negative control, which acts through repressor protein. Using isolated *lac* operon DNA, investigators have found that RNA polymerase binds to the promoter only infrequently, even when repressor has been removed from the adjacent operator site. Binding of RNA polymerase to the *lac* promoter is greatly enhanced, however, when cAMP-CAP is also bound to the operon regulatory region (Fig 10.9). The effect of bound cAMP-CAP is believed to provide a

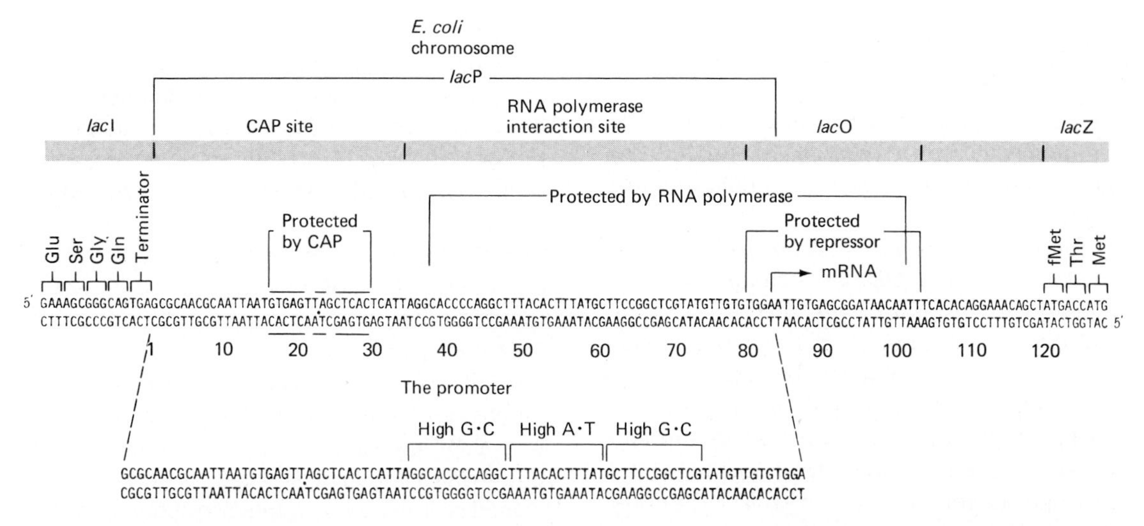


***Figure 10.9*** DNA base sequence of the regulatory region of the *lac* operon in *E. coli.* The end of the repressor gene *i* (*lacI*) and all of the promoter (*lacP*) and operator (*lacO*) are shown, along with the start of the *z* gene (*lacZ*). Binding sites for CAP, RNA polymerase, and the *lac* repressor show some overlaps. (R.C. Dickson et al., "Genetic Regulation: The Lac Control Region," *Science* **187**:27–35, 10 January 1975. Copyright © 1975 by the American Association for the Advancement of Science.)

more favorable situation for effective steric interaction between RNA polymerase and promoter DNA. The binding site for cAMP-CAP is proximal to the promoter. Bound cAMP-CAP can not only enhance binding of RNA polymerase to the promoter, but it also does not interfere with interactions between the repressor protein and the operator site just distal to the promoter.

Control over transcription *initiation* in bacterial operons appears to depend on the interactions between proteins and particular DNA binding sites. In the *lac* operon, which is one of the best-studied systems, we have seen the following:

***1.*** Repressors, made by the *i* gene, bind to the operator and turn off transcription. They are removed through binding with an inducer, such as lactose, and transcription can then proceed along the structural gene sequence of *z–y–a.*

***2.*** CAP, made by the *crp* gene, binds to a DNA site proximal to the promoter and turns on transcription of *z–y–a* genes. CAP can bind to operon DNA only if it is complexed with the metabolite cAMP. After the cAMP-CAP complex binds to regulatory DNA, coordinated transcription takes place along the structural gene sequences.

***3.*** RNA polymerase, made by the *rpo* genes, binds to the promoter and catalyzes transcription. It binds efficiently only when cAMP-CAP is also bound to operon DNA just proximal to the promoter and in the absence of repressor protein at the operator site adjacent to the promoter.

Transcriptional control appears to be the most common means of regulation of gene expression, at least in bacteria. It determines the *amounts* of gene products made in cells whose structural genes determine the *kinds* of proteins that can be made. Nongenetic regulation over phenotype development usually involves control over protein *activity,* rather than the kinds or amounts of protein. Many such controls have been described for enzyme activities, which generally fluctuate in response to a number of environmental factors, such as pH. Genetic processes put the proteins in cells, and nongenetic interactions then influence protein activities.

## Control of Gene Expression in Eukaryotes

The mechanisms for regulating gene expression found in eukaryotes are different from those found in prokaryotes. Similarly, prokaryotes lack most of the known eukaryotic regulatory mechanisms for control of gene expression. We can identify two general groups of genetic mechanisms for regulating gene expression in eukaryotes: (1) mechanisms that *modulate* gene expression through transcriptional, posttranscriptional, and translational controls and (2) mechanisms that *alter* genes through modification, amplification, rearrangement, and diminution of genomic sequences. The variety of observations and mechanisms cannot be covered adequately in the space available here, but some of the highlights and more significant mechanisms will be discussed.

### 10.4 Transcriptional Control

Transcriptional control has been amply demonstrated in prokaryotes, as we have just seen in the previous sections of this chapter. Such control, has also been demonstrated in eukaryotes, particularly for certain genes that can be studied most readily. The occurrence of transcriptional control is evident if cells expressing the gene have mRNA transcripts of the gene and cells not expressing the gene have no detectable mRNA transcripts of this same sequence. Such observations would be evidence for *differential gene activation,* that is, genes being turned on and off, but they would not necessarily reveal the mechanism by which transcription was controlled.

DNA-RNA molecular hybridization provides a specific and sensitive assay for mRNA transcripts. Hybrid DNA-RNA duplex molecules can be formed only if the two polynucleotides are complementary and can undergo base pairing. Specialized cells such as reticulocytes, which make globin polypeptides of the hemoglobin molecule, or oviduct cells, which make ovalbumin egg protein, are among the best-studied systems showing transcriptional control. Since each of these cell types makes large quantities of its particular protein product, large quantities of mRNA transcripts of the globin gene sequence or the ovalbumin gene sequences are readily available. It is a simple matter to demonstrate that these cells are actively transcribing the relevant genes, by molecular hybridization as well as by direct base sequencing comparisons of DNA copies of mRNA.

Cells that do not make hemoglobin or ovalbumin can be assayed for the presence of pre-mRNA and mRNA transcripts, which might occur but would not be translated. Clones of DNA copied (cDNA) from globin or ovalbumin mRNA can be established and used as *probes* for the relevant transcripts in other cells and tissues. When virgin oviduct tissue or reticulocytes are assayed for ovalbumin pre-mRNA or mRNA, less than one molecule of RNA hybridizes with ovalbumin cDNA. If the oviduct tissue is stimulated by hormone induction to make ovalbumin, large quantities of ovalbumin pre-mRNA and mRNA then hybridize with cDNA. Hormonal induction has no effect on reticulocytes, which do not make ovalbumin mRNA under any conditions. Similarly, if oviduct tissue is probed for the presence of globin pre-mRNA or mRNA, these molecules are not found. Globin transcripts are found only in blood-forming cells and only when they are actively synthesizing globins.

An independent assay for differential gene activation can be conducted using the eukaryotic transcription enzyme specific for mRNA synthesis, namely, *RNA polymerase II* (see Table 9.1). Binding of the enzyme to a particular DNA provides suggestive evidence that transcription can take place at that site in the system (Fig. 10.10). Pierre Chambon and co-workers have shown that substantially more RNA polymerase II binds to ovalbumin genes in induced oviduct tissue than to globin genes in these same cells. Several investigators have obtained similar evidence for a number of other species and tissues. These observations provide additional evidence for transcriptional control.

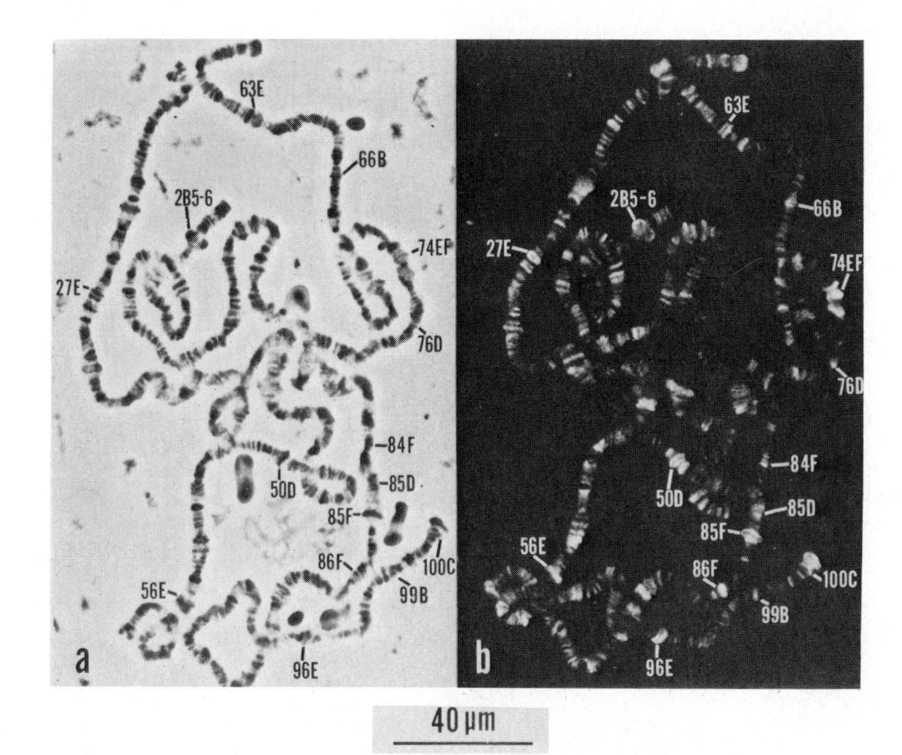


***Figure 10.10*** Immunofluorescence staining pattern obtained using antiserum prepared against the $\rho$ (rho) subunit of RNA polymerase in salivary glands of *Drosophila melanogaster*. The photographs were taken with a light microscope equipped with (left) phase contrast optics, or (right) fluorescence optics. The standard banding pattern seen on the left provides references for fluorescent bands visible on the right. (From L.M. Silver and S.C.R. Elgin. 1977. *Cell* **11**:971. Copyright © by The M.I.T. Press.)

## 10.5 Hormone-Mediated Positive Control of Transcription

In some cases positive control appears to be exerted over transcription in eukaryotic cells. Positive control is indicated by the *turning on* of gene expression. Examples of such control include the activation, or turning on, of an inactive nucleus when it is present along with an actively transcribing nucleus in a binucleate cell hybrid (Fig. 10.11). For example, if human HeLa cells that are actively transcribing RNA are fused with hen erythrocyte cells that are not transcribing, the erythrocyte nucleus soon enlarges and begins transcription. The condensed hen chromosomes become capable of synthesizing nucleic acids once nuclear enlargement occurs and the chromosomes become extended structures. Positive control action is indicated since the hen nucleus is activated, rather than the HeLa cell nucleus being inhibited in the binucleate cell hybrid. Transcription is turned on in the erythrocyte nucleus as the HeLa nucleus continues its transcribing activities. Transcription would be turned off if a negative control was operative.

Among the best examples of positive controls are those in which *steroid hormones* stimulate the onset of transcription. During larval development in *Drosophila* and other dipteran insects, specific chromosome bands expand by *puffing* and regress later on as the larva progresses toward the pupal stage of the life cycle (Fig. 10.12). **Puffs** are chromosome regions at which the chromatin is greatly extended and unfolded. Particularly large, well-developed

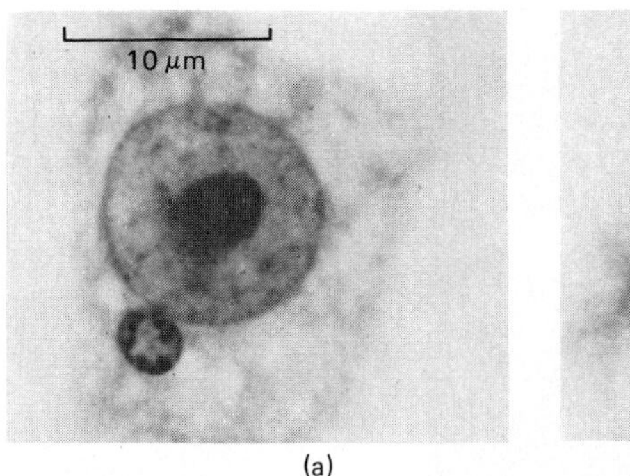


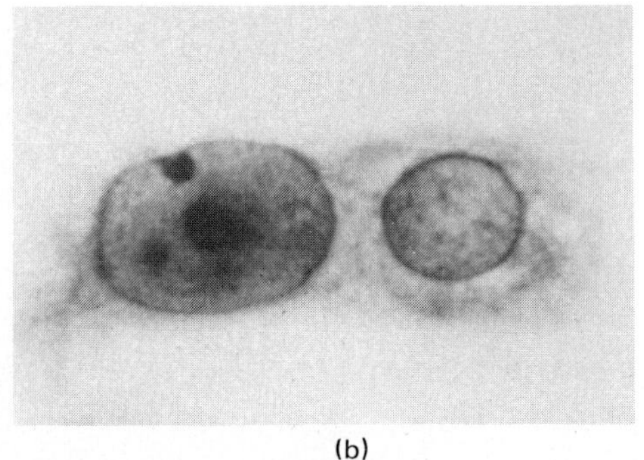


***Figure 10.11*** Binucleate somatic cell hybrid containing actively transcribing human HeLa and hen erythrocyte nuclei. (a) The large HeLa nucleus contrasts sharply with the small, condensed hen nucleus shortly after cell fusion, but (b) the erythrocyte nucleus enlarges and becomes genetically active after an interval of coexistence in the same cytoplasm with the HeLa nucleus. (From H. Harris. 1967. *J. Cell Sci.* **2**:23.)

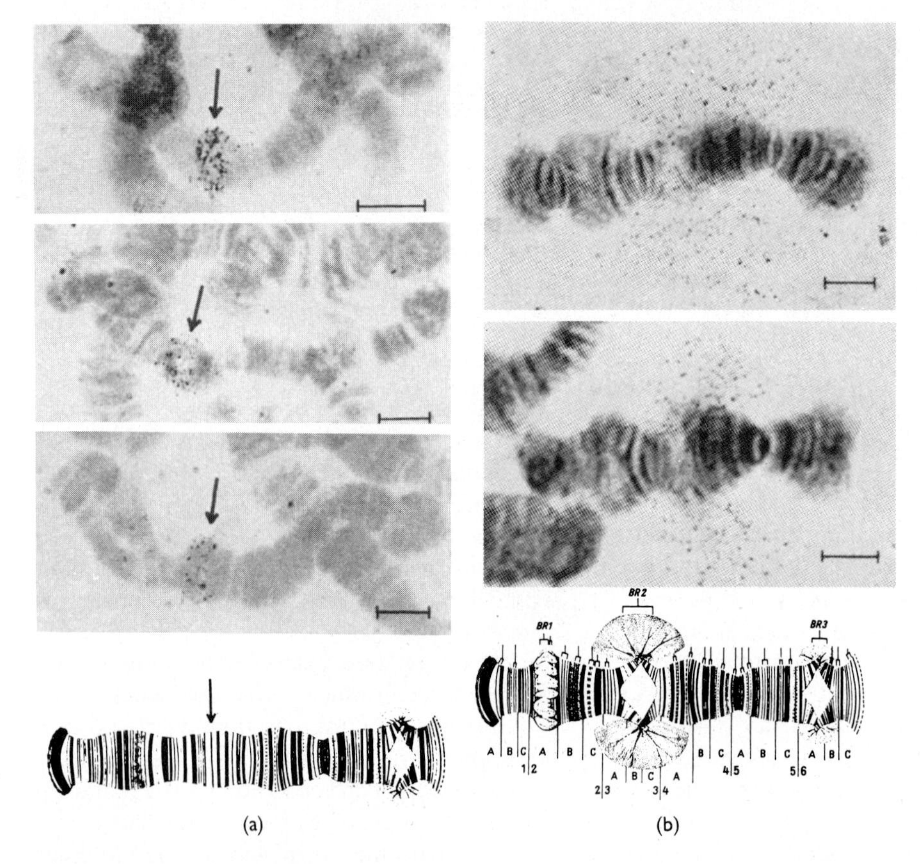


***Figure 10.12*** Differential gene action seen in puffing of chromosome IV of *Chironomus tentans*. (a) Rectum chromosomes hybridized *in situ* with BR 2 (Balbiani ring 2) RNA. Relatively little puffing or RNA is evident at the BR 2 locus (arrow). (b) Salivary gland chromosomes hybridized *in situ* with BR 2 RNA show a large, expanded Balbiani ring and a high count of silver grains. (From B. Lambert. 1975. *Chromosoma* **50**:193.)

puffs are called Balbiani rings. RNA synthesis occurs at puffs, according to evidence from autoradiography, binding of RNA polymerase II, molecular analysis, and other information. When puffs regress, RNA synthesis stops. The steroid hormone *ecdysone* plays a major role in turning on transcription, which is observed by puff development. Ecdysone is a growth hormone produced in the prothoracic gland of young larvae. The hormone is released into the bloodlike hemolymph in which it circulates throughout the organism.

When early larvae are injected with ecdysone, existing puffs regress and new puffs appear within 5 minutes. Over a period of days, about 125 different and *specific* chromosome bands undergo puffing; each puff appears and regresses at regular intevals during development. The earlier puffs in this developmental cascade may form in direct response to the hormone since puffing is not inhibited by drugs such as cycloheximide that block protein synthesis (see Table 9.3). Puffs that ordinarily develop later will not form, however, if protein synthesis is inhibited after ecdysone injection.

Highly specific *protein receptors* interact with ecdysone to turn on transcription. Each *target cell* of hormonal induction has thousands of protein receptor molecules in the cytoplasm. We know this to be the case from specific interactions between the hormone and cytoplasmic proteins extracted from *uninduced* target cells, that is, cells that have not been provided with the hormone. After ecdysone enters the target cells, the hormone combines with specific protein receptors in the cytoplasm. The hormone-receptor complexes move from the cytoplasm into the nucleus, where they bind to chromatin of the chromosomes. This is evident from the apparent absence of receptors in the cytoplasm after hormonal treatment and the presence of hormone-receptor complexes in chromatin extracts of induced cells. The protein receptors are believed to be positive control elements that turn on transcription at particular gene loci.

*Drosophila* larvae can be *ligatured* in such a way that part of the salivary glands is tied off from the remainder of the glands. In such ligatured larvae the salivary gland cells that can still receive ecdysone from the neighboring prothoracic glands will undergo puffing and tissue differentiation. Salivary gland cells on the other side of the ligature constriction do not receive ecdysone. These cells show very little puffing, and the larval tissues do not differentiate. The puffing differences between the two parts of the ligatured larvae show that ecdysone is required for transcription of specific genes that are needed for a normal program of tissue development during metamorphosis.

A more specific system in which particular molecules can be traced and analyzed is the chick or hen oviduct, which can be cultured *in vitro*. The steroid hormone *estradiol*, an estrogen compound, activates the ovalbumin gene to transcribe messenger RNA in the oviduct target cells. In immature chick oviduct or uninduced hen oviduct, ovalbumin mRNA is absent. This absence is determined by DNA-RNA hybridizations using cDNA probes of the ovalbumin gene sequence; no molecular hybrids are formed in the absence of estradiol induction. When estradiol is provided, oviduct cells transcribe the entire ovalbumin gene sequence in the form of pre-mRNA molecules, as shown by DNA-RNA molecular hybridization tests. If the drug *actinomycin* is added to induced oviducts, RNA synthesis stops. Actinomycin specifically inhibits the messenger RNA transcription enzyme RNA polymerase II.

The pre-mRNA is processed to mature ovalbumin mRNA, and translation into ovalbumin protein takes place on polysomes in oviduct cells. Assays can therefore be conducted to show that translation of the ovalbumin mRNA transcripts indeed takes place after hormonal induction of transcription. Ovalbumin synthesis is induced in oviduct cells as the consequence of mRNA transcription being turned on when estradiol-protein receptor complexes bind to chromatin.

The mechanism of positive control remains to be elucidated in the hormone-mediated oviduct system and in others of a similar nature. As molecular methods become increasingly precise and sophisticated, we expect to learn much more about protein–nucleic acid interactions and the means by which inactive chromatin is activated to begin transcription of specific gene sequences. Chromosomes are chemically and physically complex structures, however, and it has been difficult to analyze the eukaryotic chromosome in the same way that naked DNA molecules in prokaryotes and viruses have been analyzed.

## 10.6 Posttranscriptional and Translational Controls

In eukaryotes the entire gene sequence is transcribed into a large pre-mRNA transcript, which must then be processed to mature mRNA before translation can begin. *Posttranscriptional control* may be exercised through any one of the major steps in processing pre-mRNA: (1) capping the 5′ end of the mRNA with methylguanosine, (2) adding the poly(A) tail to the 3′ end of the message, and (3) excising intervening sequences and splicing together the coding sequences (Fig. 10.13).

Using cloned DNA, various modifications can be incorporated into the gene sequence to see if the modifications result in detectable effects on subsequent translation of mRNA transcribed from the altered sequences. If intervening sequences are deleted or even partially deleted from the gene, the pre-mRNA is improperly processed and does not get translated. If the methylguanosine cap is absent from the pre-mRNA, in most cases the pre-mRNA transcript will be nonfunctional. If the poly(A) tail is absent, transcription may take place but the mRNA is less stable than mRNA that has the poly(A) tail.

In the case of naturally occurring variants, independent evidence has been obtained confirming that a complete pre-mRNA transcript must be made and processed for translation. Pseudogenes are duplicates of functional genes, but they have deletions in their intervening sequences or, in some cases, no intervening sequences at all. Pseudogenes are not transcribed even though they have retained all the coding

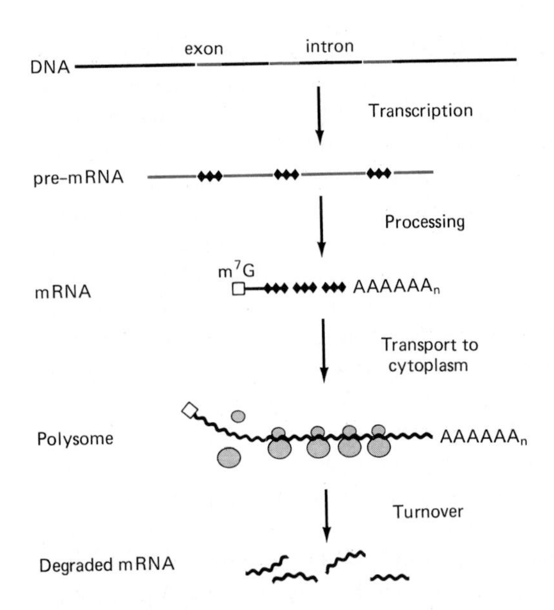


***Figure 10.13*** Posttranscriptional control may occur at any one of the steps of pre-mRNA processing (capping, polyadenylation, excision of introns and splicing of exons) and during or after transport of the processed mature mRNA from the nucleus to the cytoplasm.

sequences needed to specify the polypeptide in translation (see Section 7.12).

The situation is not quite as simple as it may appear, however. Some genes have no intervening sequences, but they make functional transcripts that act in translation. Among these are the genes for histone proteins and for two of the three interferons. Histones provide another exception since their transcripts have no poly(A) tails. A great deal obviously remains to be learned about variations in transcripts and in posttranscriptional control.

One important *translational control* in eukaryotes is the relative stability of messenger RNA. Mammalian reticulocytes lose their nucleus but continue to make globins for weeks afterward. No new transcript can form in the absence of the nucleus in which the genes were located. Globin mRNA is very stable, and continued synthesis of globins is possible because of this control at the level of translation.

The larva of the silkworm *Bombyx mori* synthesizes the protein *fibroin* in its posterior silk gland, from which strands of silk are fashioned. A single fibroin gene is responsible for synthesis of about $10^{10}$ molecules of fibroin protein in several days. This accomplishment is due to a high rate of fibroin mRNA synthesis and the efficient utilization of stable mRNA molecules. About $10^5$ mRNA molecules are transcribed from one fibroin gene, and each mRNA serves as the template for synthesis of about $10^5$ molecules of fibroin. Control over fibroin gene expression is therefore due to the stability of mRNA, which allows prolonged protein synthesis of the fibroin molecules in translation.

In at least one case the relative stability of a particular mRNA appears to be under hormonal control. Mammary glands produce *casein,* a milk protein. The synthesis of casein is decreased if the hormone *prolactin* is withdrawn from the system. Studies of the mRNA transcripts revealed a shorter half-life for transcripts made in mammary glands without prolactin than in tissues provided with prolactin. The high lability of casein mRNA, therefore, appears to be due to hormonal functions that influence translation in this system and not to hormonal induction of transcription as in oviduct tissues.

## 10.7 Gene Amplification

Stable forms of mRNA allow large amounts of a single protein to be synthesized over a considerable period of time, particularly in cells that are highly specialized for one particular function. Reticulocytes, oviduct tissue, and silk glands are systems devoted to a particular function and to synthesis of a particular protein to the exclusion of all or many other proteins. The application of this kind of amplification of protein synthesis is limited to specialized cells.

In contrast to amplification of protein synthesis via stable mRNA, **gene amplification** involves the *differential replication* of certain genes while the remainder of the genome does not replicate. The number of copies of the gene is increased, thus allowing the production of immense quantities of a specific gene product from the amplified gene(s). Among the best-studied examples of gene amplification are the oocytes of certain vertebrates and insects, which specifically amplify their genes for 18S and 28S ribosomal RNA (rRNA). The rRNA genes are usually referred to as ribosomal DNA (rDNA). One of the species that has been studied most intensively is the amphibian *Xenopus laevis,* the South African clawed toad.

Amplification of rDNA in *Xenopus* takes place only in oocytes and only during the diplonema phase of meiosis in these huge cells (Fig. 10.14). The extra copies of rDNA are synthesized at the nucleolar-organizing region (NOR) of the chromosome. Transcription of rDNA produces rRNA, which becomes incorporated into new ribosomes that subsidize high rates of protein synthesis during early embryogenesis of the fertilized egg. These rRNA genes become inactive after oogenesis. In somatic cells rRNA is transcribed from the normal complement of rRNA genes, which provides more than an adequate amount of RNA for ribosome formation and for the levels of protein synthesis carried out in development of the organism.

During amplificiation of rDNA in *Xenopus,* the original number of 900 copies of rRNA genes is increased to between 600 and 1600 *times* this number. Replication apparently proceeds by the **rolling-circle mechanism,** which was mentioned in relation to DNA replication in certain phages (Fig. 10.15). Large numbers of rRNA genes are produced rapidly, as multiple lengths of this DNA are synthesized and then cut by nuclease action. The linear molecule is circularized when its "sticky" ends undergo base pairing, and the duplex sugar-phosphate chains are ligated to restore covalent bonding. Replication continues along the rolling circle while copies of gene sequences are released on completion of their synthesis.

Only recently, however, investigators found that in eukaryotes genes other than rRNA genes undergo gene amplification episodes during normal development. In 1980, Allan Spradling and Anthony Mahowald showed that *chorion* genes in *Drosophila* were amplified tenfold in egg chambers as compared with other somatic

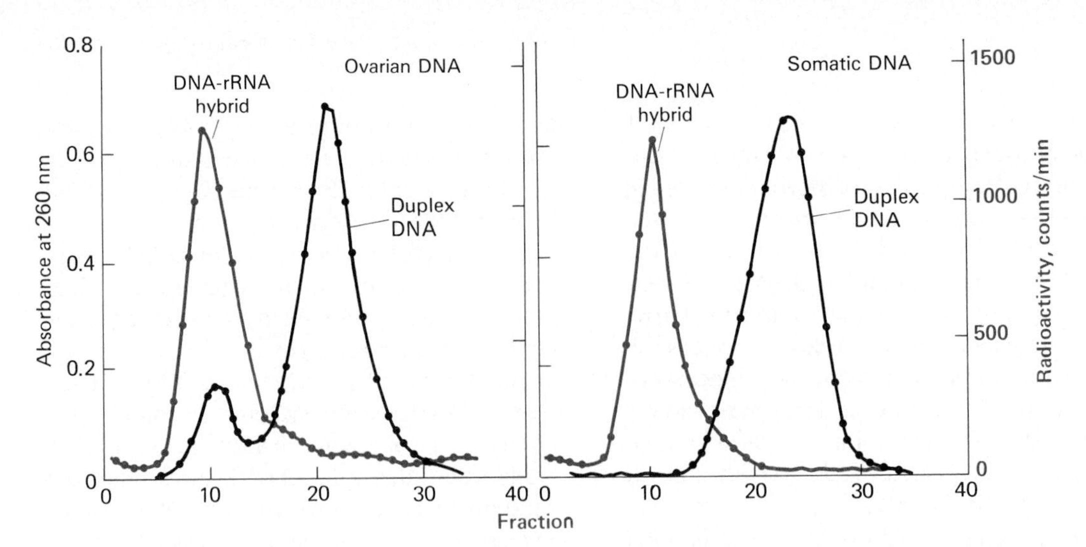


*Figure 10.14* Identification of a satellite DNA as rDNA and amplification of rDNA in oocyte nuclei. DNA-rRNA hybrids sediment in the same region of the gradient as a satellite DNA from oocytes in ovarian preparations, indicating the satellite is rDNA. Although the satellite is not evident in the absorbance curve in somatic nuclear DNA preparations, rDNA is present since molecular hybrids between rRNA and DNA from this satellite region of the DNA do form and do sediment where expected in the gradient. The increased amount of rDNA in oocytes provides evidence for gene amplification; that is, replication of certain DNA while most of the nuclear DNA does not replicate.

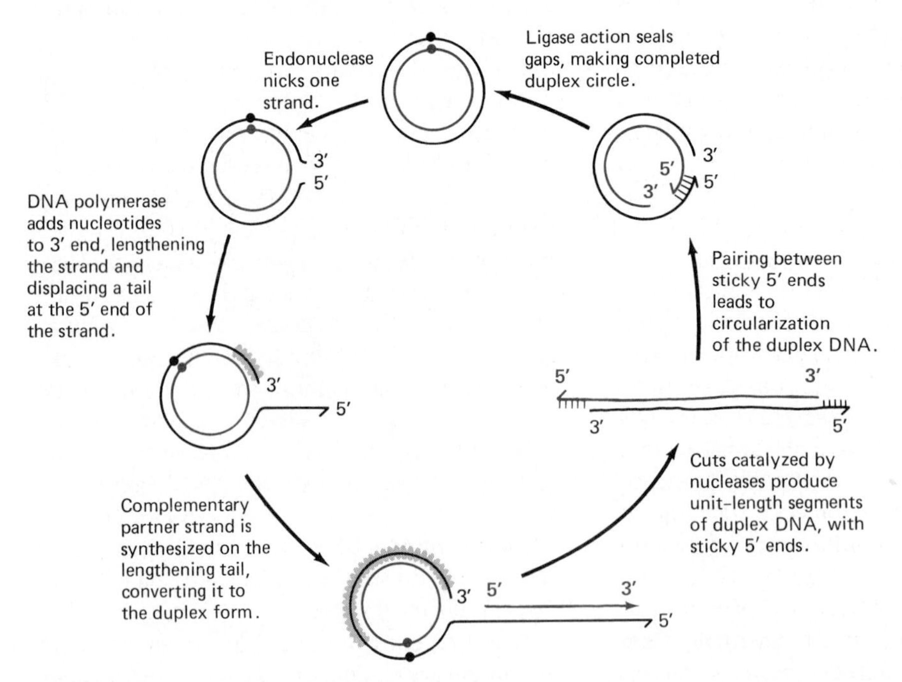


*Figure 10.15* The rolling circle mechanism of rDNA replication permits rapid synthesis of large numbers of rDNA sequences. Mulitple lengths of rDNA are cut by nuclease action, and the linear molecules are circularized by base pairing between complementary ("sticky") single-stranded ends. Ligase action seals the duplex circles as covalent bonds restore intact sugar-phosphate chains at the paired ends. Replication continues at the same time that copies of rDNA sequences are released when their synthesis is completed.

cells. Chorion proteins contribute to eggshell development.

The period of gene amplification of chorion genes in *Drosophila* is brief and takes place prior to the onset of mRNA and protein synthesis. In this situation, as with rRNA gene amplification, large amounts of a gene product are needed for limited and specified times during development. Unspecialized cells can express a different set of genes at different times in development. Thus their needs can be better met by synthesis of large amounts of RNA that function for a limited time during development. This situation is different from that of specialized cells that serve a single function and can therefore afford to make and utilize stable RNAs for their single product. In either case, whether the demand for more gene product is met by modulating gene expression through stable RNAs or by making more genes that specify a product, the needs of the developing organism can be met. The precise nature of the controls that regulate gene expression—by synthesis of stable mRNA or by synthesis of more gene copies—is uncertain. Environmental agents such as hormones undoubtedly are involved, but we know little about these factors at the present time.

## 10.8 Immunoglobulin Diversity

One of the most remarkable systems for generating diversity of gene expression through *gene rearrangements* has been described for sets of genes that code for antibodies, or **immunoglobulins,** which are synthesized in B lymphocytes. Immunoglobulins (Ig) bind to invading foreign substances, or **antigens,** in immune responses. Although the number of different genes coding for Ig molecules may only be in the hundreds, the animal can generate over a million different immunoglobins to specifically confront more than a million different antigens that may be encountered in a lifetime. Within the past few years, rapid progress has been made in analyzing Ig genetics, principally through the use of new methods in molecular genetics, such as base sequencing, cloned DNA, and molecular hybridization using specific DNA and RNA probes.

Each Ig molecule is made up of a pair of identical heavy (H) polypeptide chains and a pair of identical light (L) chains (Fig. 10.16). Each H and L chain consists of three functionally different regions: **constant** (C), **joining** (J), and **variable** (V). The J segment links the V and C segments. Each H chain also has a fourth region of great **diversity** (D) between the V and J segments. The four polypeptides are held together in the Ig molecule by disulfide bonds. Five different classes of Ig molecules can be distinguished by the components of their H chains (Table 10.5). All classes interact with antigens through the *antigen combining site* of the Ig molecule, which is located at the variable N-terminal ends of the four polypeptide chains.

The organization of Ig genes reflects the organization of the Ig molecules. In mammalian

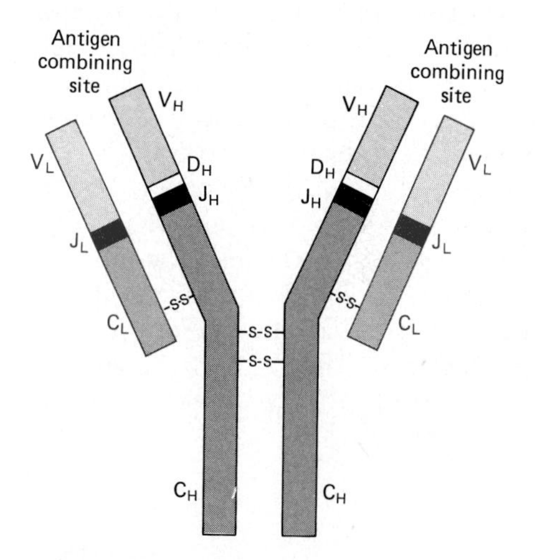


***Figure 10.16*** Immunoglobulin molecules consist of an identical pair of higher molecular weight heavy (H) polypeptide chains and an identical pair of lower molecular weight light (L) chains, all held together by disulfide bonds (—S—S—). Each L chain includes variable (V), joining (J), and constant (C) regions, and each H chain has V, J, and C segments plus a fourth diversity (D) region. Specificity of interaction between these antibody molecules and foreign antigens is localized to the two combining sites at the variable ends of the immunoglobulin.

***Table 10.5*** Types of human immunoglobulins.

| immunoglobulin class | light chain present | heavy chain present | molecular weight | functions |
|---|---|---|---|---|
| IgG | $\kappa$ or $\lambda$ | $\gamma$ (gamma) | 144,000 | main serum antibody, activates complement |
| IgM | $\kappa$ or $\lambda$ | $\mu$ (mu) | 160,000 | cell-surface receptor, serum antibody (early), activates complement |
| IgA | $\kappa$ or $\lambda$ | $\alpha$ (alpha) | 144,000 | main antibody in saliva and intestinal fluids |
| IgD | $\kappa$ or $\lambda$ | $\delta$ (delta) | 156,000 | cell-surface receptor on immature B lymphocytes |
| IgE | $\kappa$ or $\lambda$ | $\epsilon$ (epsilon) | 166,000 | antiparasitic immune response, releases histamine from mast cells |

germ-line DNA, from sperm or similar sources, Ig genes are arranged in three families of multiple genes located on three different chromosomes. Two of these *multigene families* code for the two kinds of L chains, kappa ($\kappa$) and lambda ($\lambda$); the third multigene family codes for all H-chain classes (Fig. 10.17). Each family of L-chain genes consists of *C, J,* and *V* gene clusters. The family of H-chain genes consists of *C, J, D,* and *V* gene clusters. To distinguish among these genes and their corresponding polypeptide segments in the Ig molecule, a subscript identification is used: $C_L$, $J_L$, $V_L$ for components in light chains (or, specifying these components in kappa and lambda L chains, as $C_\kappa$, $J_\kappa$, $V_\kappa$, and $C_\lambda$, $J_\lambda$, and $V_\lambda$), and $C_H$, $D_H$, $J_H$, and $V_H$ for components in heavy chains.

During somatic cell differentiation of B lymphocytes from germ-line precursor cells, two functionally and mechanistically different rearrangements of Ig genes take place:

***1.*** In *V-gene translocation* a particular *V* gene becomes associated with a *C* gene and generates a functional Ig gene that determines the antigenic specificity of that B lymphocyte and its mitotic descendants. This rearrangement occurs independently of contact with the antigen.

***2.*** In *heavy-chain class switching* the expressed $V_H$ gene is reassociated with a different $C_H$ gene. The antigenic specificity of the molecule remains unchanged by virtue of its V region, but the Ig class of antibody produced does

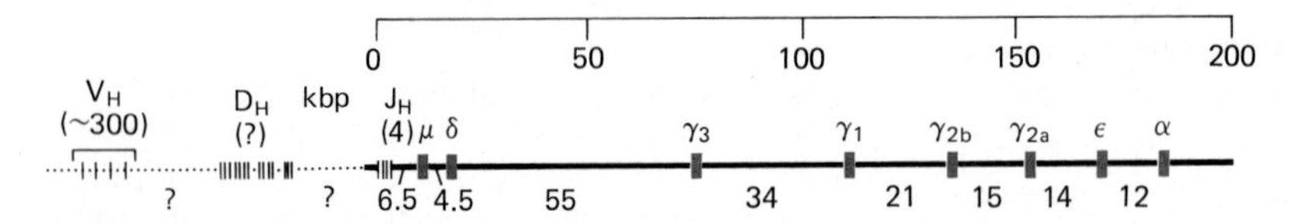


***Figure 10.17*** Organization of mouse immunoglobulin heavy chain genes in germ-line DNA. The multigene family encodes all components of all H-chain classes (IgM, IgD, IgG, IgE, IgA). There are about 300 $V_H$ genes, about a dozen $D_H$ genes, 4 $J_H$ genes, and 8 $C_H$ genes. The region encompassing $J_H$ and $C_H$ genes and spacer DNA is about 200 kbp long, but lengths of regions involving $V_H$ and $D_H$ genes are undetermined at present. (Adapted from K.B. Marcu. 1982. *Cell* **29**:719–721. Copyright © by The M.I.T. Press.)

change according to its $C_H$ chain component. The same antigenic specificity can thus be associated with a different biological activity, according to the nature of the H-chain constant region (see Table 10.5).

The ways in which these rearrangements occur have been analyzed by molecular genetic methods and by biochemical analysis of amino acid sequences in immunoglobulin molecules during different stages of development and in different B lymphocyte populations. The first steps in these processes involve modifications and rearrangements of the DNA sequences themselves. The subsequent processing involves the mRNA transcripts of the rearranged Ig genes.

During L-chain gene assembly, a *V* gene is joined to a *J* gene during lymphocyte differentiation. This joining is achieved by deletion of germ-line DNA sequences between any one of the *V* genes and any one of the *J* genes in either the $C_\kappa$ or $C_\lambda$ multigene family. In any particular cell only one of the two L-chain gene families, $\kappa$ or $\lambda$, is expressed, along with a set of H-chain genes. The DNA sequence between L-chain *VJ* and *C*, which is retained during *V-J* joining, is transcribed along with the *VJ* and *C* sequences themselves (Fig. 10.18). This intervening transcribed region is later excised from the pre-mRNA transcript and the *VJ-C* regions are spliced together. Expression of an Ig gene, therefore, requires two key processes: *V/J* joining and *VJ-C* mRNA splicing.

The formation of a complete $V_H$ sequence requires fusion of three genes: $V_H$, $D_H$, and $J_H$, which may be selected from clusters of $V_H$, $D_H$, and $J_H$ sequences in the multigene heavy-chain family (Fig. 10.19). The coding sequence for the heavy chain variable region is generated by DNA sequence deletions that join *V* to *D* segments and *D* to *J* segments. Joining of of *V-D-J* in heavy-chain variable sequences and of *V-J* in light-chain variable sequences thus produce active Ig genes for heavy and light polypeptide chains in each B lymphocyte *before interactions with antigens* take place. In addition, combinations of different gene sequences from clusters of genes that encode the variable portions of L and H chains produce a large and highly diverse population of B lymphocytes, each containing a single type of Ig molecule, from germ-line DNA.

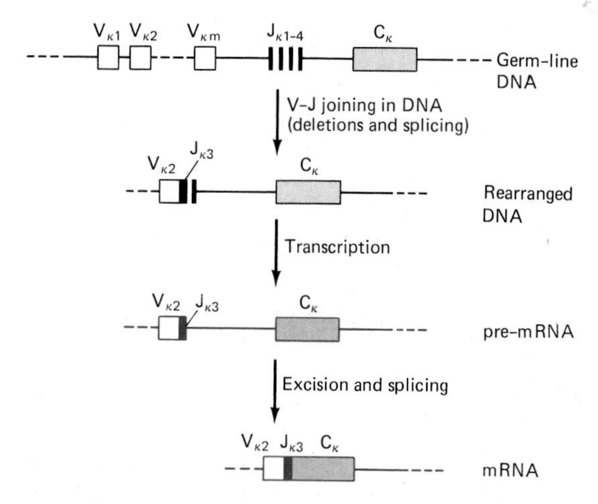


***Figure 10.18*** Light-chain gene assembly. The original arrangement of $\kappa$ chain genes in germ-line DNA is rearranged by *V-J* joining events such that any of the $V_\kappa$ genes can be joined to any of the $J_\kappa$ genes. The *V-J* region remains separate from the $C_\kappa$ region and all regions are copied into one pre-mRNA during transcription in differentiating B lymphocytes. The second process involved in gene expression is excision of intergenic sequences (spacer DNA) and splicing of the *VJ* to the *C* gene region of the transcript to produce mature mRNA that guides translation of a particular L-chain polypeptide. (J.L. Marx, "Antibodies: Getting Their Genes Together," *Science* **212**:1015–1017, Fig. 2, 29 May 1981. Copyright © 1981 by the American Association for the Advancement of Science.)

Heavy-chain class switching produces Ig molecules that retain their specific antigenic combining sites (in the variable regions) but with altered biological activity, depending on the $C_H$ gene switch that takes place (see Table 10.5). These switches from one class of Ig to another involve modifications in the DNA sequence that is contained in the genome of the differentiating B lymphocyte.

All immature B lymphocytes initially synthesize antibodies of the IgM class, which func-

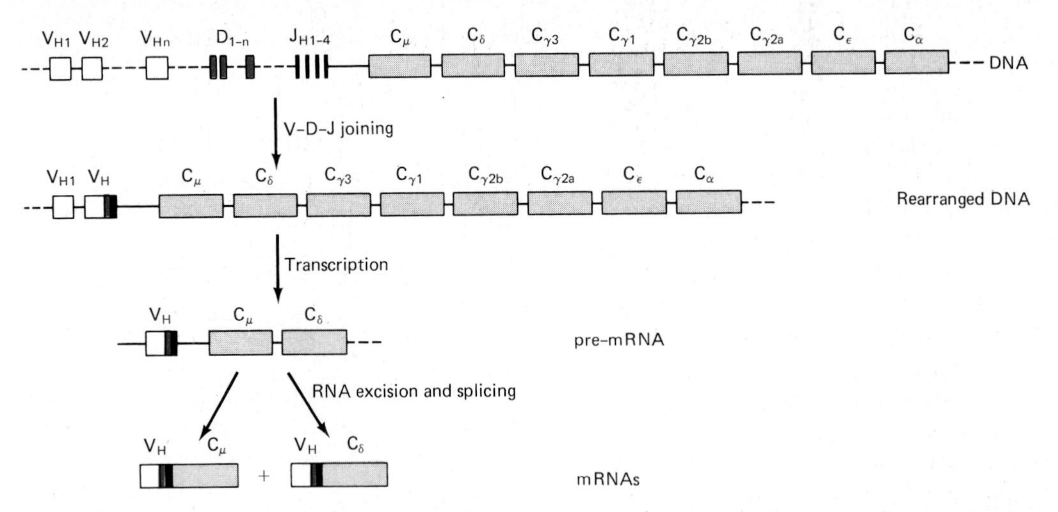


***Figure 10.19*** Heavy-chain gene assembly. Germ-line heavy-chain genes are rearranged by joining of one each of the V, D, and J genes in the multigenic sets to the set of C genes. Transcription of rearranged DNA into pre-mRNA is followed by differential excision and splicing reactions to produce two classes of heavy-chain mRNA. The messengers guide synthesis of either IgM or IgD heavy chains in differentiating B lymphocytes. Any one lymphocyte makes only one type of heavy chain. (J.L. Marx, "Antibodies: Getting Their Genes Together," *Science* **212**:1015–1017, Fig. 3, 29 May 1981. Copyright © 1981 by the American Association for the Advancement of Science.)

tion as cell-surface antigen receptors. These cells differentiate to lymphocytes that carry both IgM and IgD receptors on the cell surface. These developing but still immature B lymphocytes differentiate into active antibody-secreting *plasma cells* (activated B lymphocytes, or B cells) *after antigenic stimulation* (Fig. 10.20). Plasma cells either actively secrete IgM or change to produce Ig molecules of one of the other three classes (IgG, IgA, IgE). *Antigenic specificity is retained throughout these switches.*

Heavy-chain class switching involves gene deletions and relocations in the heavy-chain multigene family. Since $J_H$ genes occur only at one locus 5′ to the $C_\mu$ gene, which specifies the IgM constant region, $C_\mu$ gene expression must accompany the initial $V_H$ gene translocation event. IgM is therefore the first Ig class to be synthesized in immature B lymphocytes. To switch expression from IgM to another class, DNA must be deleted in order to effect re-

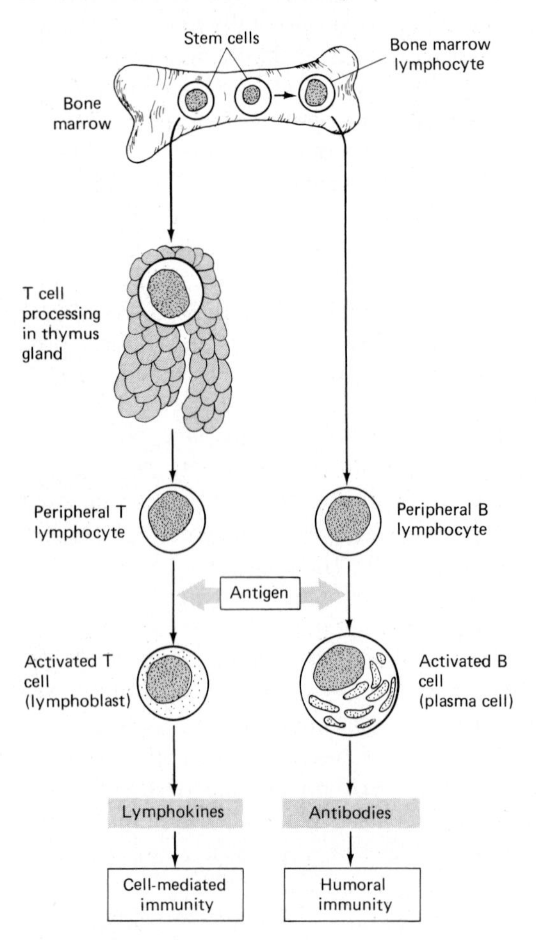


***Figure 10.20*** (right) Stem cells of the bone marrow are hematopoietic, that is, they give rise to blood cells. Different kinds of blood cells, some of which are shown, all originate and differentiate from the single kind of precursor stem cell.

location of $V$ and $C$ genes in the gene cluster (Fig. 10.21). For example, to switch from IgM to IgA expression, most of the DNA of the $C_H$ locus is deleted. The $V_H$ gene is thus relocated next to the $C_\alpha$ gene. DNA sequencing has shown tandemly repeated homologous sequences in regions 5′ to each $C_H$ gene in the cluster. These are called *multiple switch sites.* In the case of certain gene switches, sequence homology between switch sites is sufficient to permit conventional recombination between the homologous sequences. In other cases, homology is inadequate and some mechanism other than recombination between homologous paired sequences must be involved. These mechanisms have not yet been elucidated.

Heavy-chain class switching may become impossible in particular B lymphocyte lineages if large deletions of multiple switch sites occur. For example, some B lymphocyte populations secrete only IgM and undergo no class switching. In these populations a large segment of DNA 5′ to the $C_\mu$ genes has been deleted. The deleted region encompasses the $C_\mu$ switch sites.

In summary, therefore, an enormous variety of Ig molecules can be generated by deletions and rearrangements of genes in L-chain gene clusters of *V, J,* and *C* sequences and in H-chain gene clusters of *V, D, J,* and *C* sequences. These DNA sequences are transcribed and further processing of the pre-mRNA molecule produces a mature mRNA transcript. This mRNA codes for *one each* of the multiple genes that code for V, J, and C regions of the light-chain polypeptide and for V, D, J, and C regions of the heavy-chain polypeptide. Once the genome has been rearranged to generate only certain combinations of sequences, thus establishing the antigenic specificity of the B lymphocyte and its descendants, additional diversity can be achieved in the biological specificity of these specified molecules through heavy-chain gene switching in the genome. The gene sequences will be processed at the pre-mRNA level, just as they were

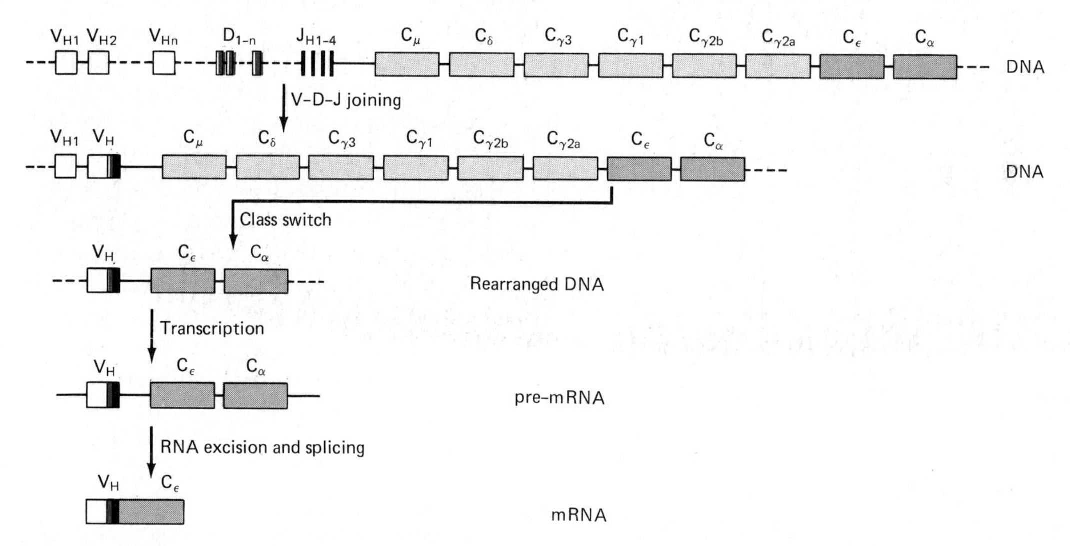


***Figure 10.21*** Heavy-chain class switching in developing B lymphocytes, after antigenic stimulation, leads to secretion of IgG, IgA, or IgE in cells originally active in secreting IgM. To switch expression from IgM to another class, deletions of C genes and splicing of remaining C segments to VDJ segments must take place. In the example shown, six C genes are excised and the two remaining C genes are spliced to the VDJ segment in a rearranged DNA. After pre-mRNA transcription, the transcript is processed further to produce a functional mRNA that will guide translation of IgE heavy-chain polypeptide. (J.L. Marx, "Antibodies: Getting Their Genes Together," *Science* **212**:1015–1017, Fig. 3, 29 May 1981. Copyright © 1981 by the American Association for the Advancement of Science.)

before gene switching. Heavy-chain gene switching provides a mechanism for altering the biological specificity, or function, of the Ig molecule, without altering its antigenic specificity.

Through gene rearrangements by DNA sequence deletions and by excision and splicing of RNA transcripts, the animal can generate over a million different and specific kinds of B lymphocytes. Each B lymphocyte produces a single kind of Ig molecule, carrying a particular antigenic combining site that permits interaction with a single kind of invading antigen in the body. Antigenic specificity is determined before antigen confrontation occurs; but once stimulated by antigen, a particular and specific B lymphocyte proliferates and synthesizes quantities of the relevant Ig molecule to produce an immune response (Fig. 10.22). In this way the animal can generate an enormous and diverse repertory of antibodies from coded information in fewer than a thousand gene sequences. The animal is thus genetically prepared to interact with virtually any antigen it may encounter in a lifetime, long before these encounters even take place.

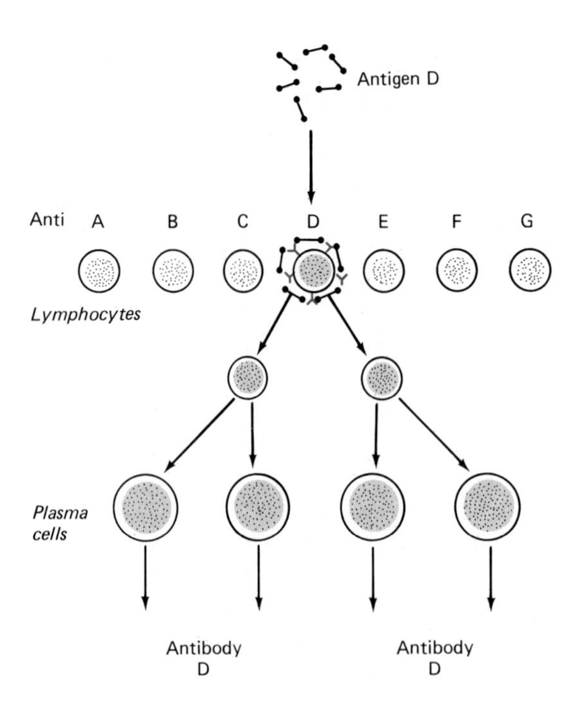


*Figure 10.22* According to the clonal selection theory, for which there is ample evidence, various lymphocytes in the body have become sensitized to particular antigens through previous exposures. When an antigen, such as D, again invades the body, sensitized B lymphocytes of the anti-D variety are stimulated to develop into activated B or plasma cells. These plasma cells proliferate and secrete anti-D antibodies that interact with the invading D antigens.

## Transposable Genetic Elements

**Transposable genetic elements** are unique DNA segments that can insert into several sites in a genome. Transposable elements in eukaryotic cells are similar to those of transposons in bacteria, plasmids, and phages, and to the DNA proviruses of vertebrate retroviruses. These transposable genetic sequences influence gene expression, lead to rearrangements of genomic sequences, and may be responsible for mutations near the insertion site of the transposable element. There are important differences among all these kinds of movable genetic elements, but they all contribute to the current picture of a dynamic genome rather than of the relatively stable genome indicated by classical genetic analysis.

### 10.9 Insertion Sequences and Transposons in Bacteria

Simple **insertion sequences** (**IS**) and the more complex **transposons** (**Tn**) are transposable elements that can be distinguished from each other by at least two criteria: (1) IS elements are less than 2000 base pairs (or 2 kilobase pairs) long, whereas Tn elements are usually much longer than 2 kbp, and (2) IS elements only contain genes involved in the function of DNA insertion, whereas Tn elements contain genes unrelated to insertion as well as insertion genes. A number of transposons also contain IS components. Both IS and Tn elements are transposable, that is, they can move onto and off the bacterial chromosome and from place to place on the chromosome with considerable frequency. They were called "jumping genes" by some biologists.

IS elements were discovered through studies of mutants at the galactose (*gal*) operon in *E. coli,* and they have been studied in *lac* operon mutants as well. All the mutant strains showed a *polar* effect on structural gene transcription; all genes were turned off distal to the integration site between the *gal* operator-promoter and the three *gal* structural genes, which control three enzymes of galactose metabolism (Fig. 10.23). Strains of these polar mutants also gave rise to rare wild-type revertants (1 in $10^7$ cells) and to more frequent types that had undergone some alteration in *gal* operon genes and sometimes in nearby genes as well.

These genetic alterations were explained on the basis of insertion and excision of IS elements in the control region between the *gal* operator-promoter and the three structural genes of the operon. When an IS was inserted, it turned off transcription of all structural genes distal to its integrated site. When the IS was excised precisely, the wild-type gene expression was restored to its original state. When the IS was excised inaccurately, some parts of the *gal* operon were also removed and sometimes nearby genes were deleted as well, thereby producing altered mutant strains.

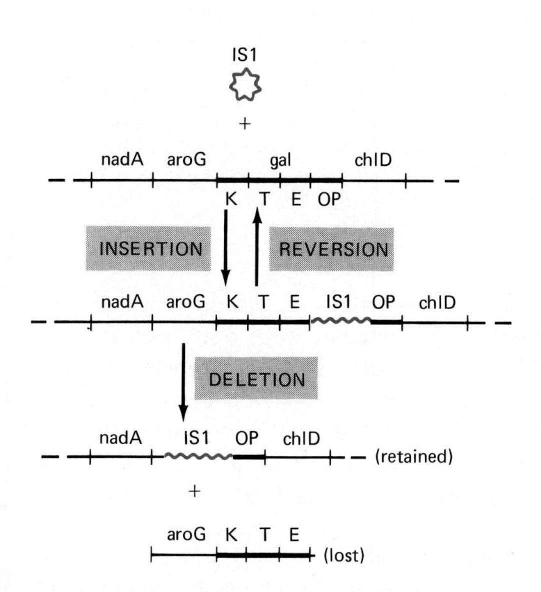


***Figure 10.23*** Insertion of the IS*1* element in the *gal* region of *E. coli* shuts off the *K*, *T*, and *E* structural genes distal to the integration site next to operator-promoter (*OP*); transcription is restored in all three genes when IS*1* leaves the site. Excisions and insertions of IS*1* may lead to deletion of the *gal*K, *gal*T, and *gal*E genes and to other effects on the genome.

The particular element IS*1* exerted a negative control over *gal* operon transcription, turning it off by insertion and turning it on by excision from the integrated site. This kind of negative control, that is, turning off transcription when IS inserts in the genome, may occur normally in *E. coli* since the normal *E. coli* genome contains about eight copies of IS*1* at various sites.

The presence of IS*1* within the *gal* operon was deduced from genetic studies and was verified by electron microscopy. DNA was extracted and purified from specialized transducing phages carrying the normal *gal* operon and from specialized transducing phages carrying the *gal* polar mutant operon. These DNAs were melted to single strands and then mixed to permit reannealing of duplex DNA. Reannealed duplex segments carrying one copy of normal *gal* and one copy of mutant *gal* are ***heteroduplex*** molecules. Electron microscopy of heteroduplexes reveals a loop of single-stranded, unpaired DNA extending from the mutant DNA (Fig. 10.24). This loop consists of the IS element, which has no partner on the normal *gal* DNA. Similar heteroduplex mapping has revealed IS*1* elements in other parts of the *E. coli* chromosome. The method is generally used to analyze the extent of pairing between DNAs and deletions and insertions of DNA sequences. The length of a deleted or inserted segment can be estimated by the length of the unpaired region of the heteroduplex molecule since the segment can be measured directly in micrometers. The segment can then be described in terms of its molecular weight ($1\ \mu m = 2 \times 10^6$ daltons) or length in kilobases ($1\ \mu m =$ about 3kb).

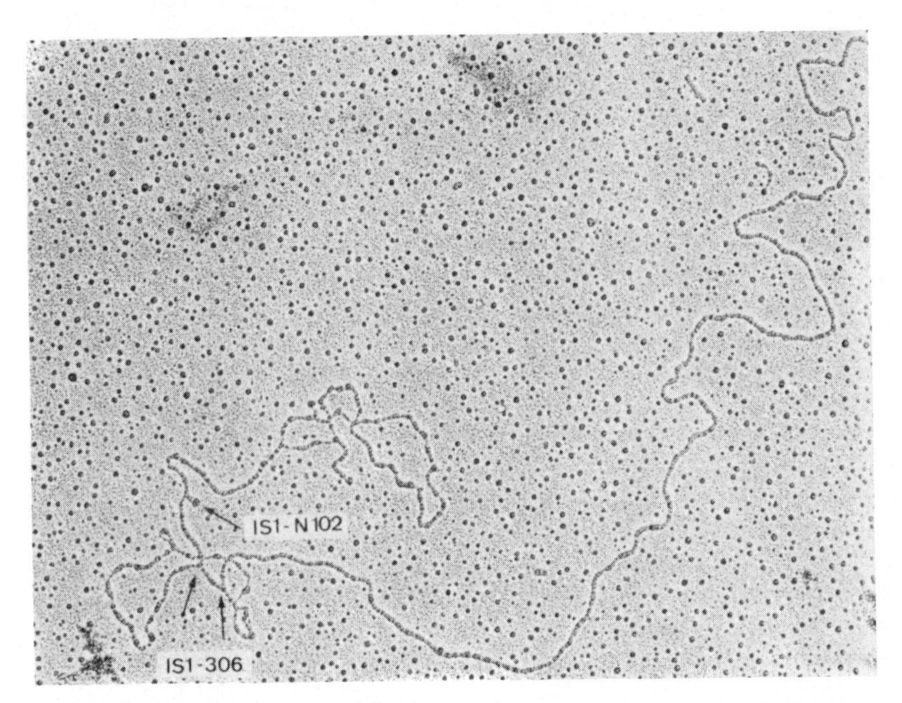


***Figure 10.24*** Electron micrograph of a heteroduplex molecule obtained by hybridizing a specialized transducing *λdgal* phage with a *λdgal* phage carrying an IS*1*-induced deletion. If IS*1* were still present after the deletion had occurred, there would be two single-stranded loops in the heteroduplex: one due to the nondeleted region, which has no partner in the deleted strand, and another nearby, which would be the IS*1* element still present. If the IS*1* sequence were deleted too, only one single-stranded loop would be present since neither would have IS*1*. The two loops shown here mean that IS*1* has been retained near the deleted region, as was shown in Fig. 10.23. (From H.J. Reif and H. Saedler. 1977. In *DNA : Insertion Elements, Plasmids, and Episomes.* A.I. Bukhari, J.A. Shapiro, and S.L. Adhya, eds., p. 81. New York: Cold Spring Harbor Laboratory.)

IS*2* is a very interesting element; it has the ability to implement positive control over transcription in *E. coli.* It was found that IS*2* turns on constitutive synthesis of *gal* structural genes if the element inserts in the *same* direction as operon transcription (Fig. 10.25). If IS*2* integrates in the *gal* operon in the direction *opposite* to transcription, it causes a reduction in structural gene transcription distal to the insertion site. The two orientations of IS*2* can therefore act as a simple on-off switch for transcription. Similar "flip-flop" control mechanisms have been demonstrated to regulate flagellar type in the bacterium *Salmonella* and mating type expression in the eukaryotic species *Schizosaccharomyces,* a fission yeast. **Inversions** of IS elements, from one orientation to the other and back again, may represent a fundamental mode of regulation of gene expression. IS*2* apparently may contain nucleotide sequences that act as strong promoters (start signals) and other sequences that recognize the rho protein, which terminates transcription in *E. coli* (stop signals). These sequences help to explain the different behavior of IS*2* and other IS elements since IS*1* and IS*4* do not have promoter or rho-recognition sequences. IS*1* turns off transcription in genes distal to its integrated site whether it is oriented in the same or the opposite direction as operon transcription.

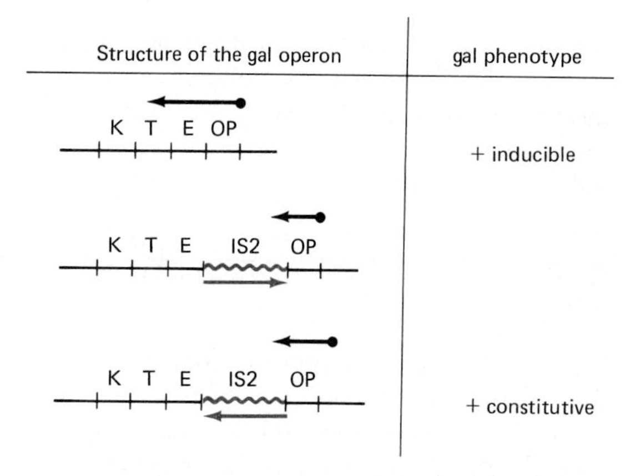


***Figure 10.25*** When insertion element IS*2* sits between *gal*OP and *gal* K-T-E either of two events takes place. The *gal* enzymes remain inducible but are made in reduced amounts if IS*2* is oriented opposite the direction of *gal* transcription; but constitutive enzyme synthesis characterizes the operon when IS*2* is oriented in the same direction as operon transcription. The IS*2* orientations act like a simple on-off switch for transcription in this case.

Transposons have at least two genes of their own: (1) a gene that codes for an enzyme, called a *transposase,* which is responsible for transposition onto and off DNA molecules, and (2) a gene that codes for a repressor substance, which regulates transcription into RNA of the transposase gene and of the repressor's own synthesis. The discovery of transposons provided an explanation for the puzzling phenomenon of rapid spread of multiple resistances to different antibiotics among pathogenic bacteria. This phenomenon has serious implications for public health since antibiotic therapy is a major theme in health care. The genes conferring antibiotic resistance in pathogenic enteric bacteria are carried in a class of plasmids called **R plasmids.** Transposons play a major role in generating these plasmids. Like other plasmids, R plasmids exist as circular DNA molecules within cells. They may exist and replicate independently, or they may be integrated into or dissociated from other plasmids, from viral genomes, and from the host bacterial chromosome itself. This accounts for the rapid spread of drug-resistant phenotypes. Plasmids that can move onto and off the bacterial chromosome are called *episomes.* The fertility factor F in *E. coli* is an example of an episomal type of plasmid (see Section 6.5).

R plasmids consist of two kinds of DNA sequences: **r determinants,** which carry genes that confer antibiotic resistance and other genes, and **resistance transfer factors** (RTF), which mediate transfer of the R plasmid from cell to cell (Fig. 10.26). The *r* determinants are transposons that have IS elements at their ends. Different antibiotic-resistance genes are incorporated into the same R plasmid molecules by insertion of transposons, each carrying one or more of these genes. There is no homology between transposons of different types, according to base sequencing analysis. Little or no homology has been found between transposons or R plasmids and the bacterial chromosome or other genetic molecules. In view of observed nonhomologies, transpositional integration and excision events must be due to some unconventional or "illegitimate" recombination process. Molecular models have been proposed to explain illegitimate recombination, but many details remain unclear at present.

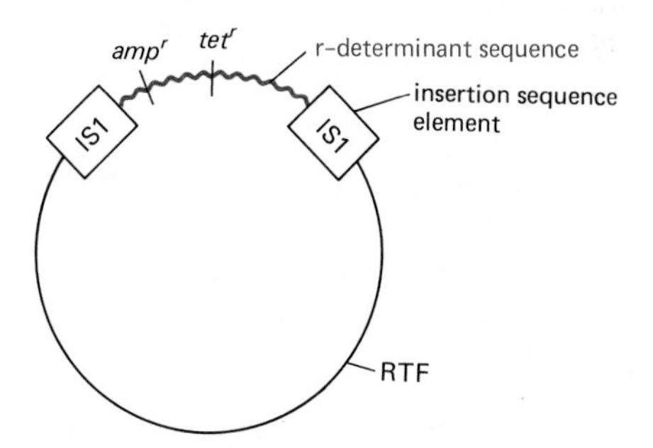


***Figure 10.26*** Organization of R plasmids. Resistance transfer factor (RTF) segments, which mediate transfer of the plasmid from cell to cell, are covalently bonded to *r* determinants, which carry antibiotic resistance genes. The *r* determinants are transposons that are inserted into RTF sequences through integration of IS elements at each end of the *r* determinant sequence. One or more *r* determinants may be bonded to an RTF in any individual R plasmid DNA molecule.

Through transposition, *r* determinant portions of R plasmids can acquire a variable number of different antibiotic resistance genes (Fig. 10.27). The *r* determinant portions of R plasmids have therefore evolved as collections of transposons. R plasmids may be transferred from cell to cell by conjugation, transformation, or transduction. Thus they may be exchanged between distantly related or even unrelated bacterial species, as well as within a single species or population of bacteria. Multiple resistances may appear suddenly and disappear suddenly from particular populations of pathogenic bacteria. The sudden appearance of bacterial strains carrying multiple resistance genes poses a very serious threat to health care. Pathogenic bacteria that can be controlled or eradicated by antibiotics one week may become resistant to one or more antibiotics the next week and remain intractable to treatment.

## 10.10 Mu Phage Transpositional Effects in *E. coli*

Mu is a temperate phage of *E. coli.* Mu phages may exist as integrated DNA sequences in the bacterial chromosome and replicate in synchrony with it, or they may be infectious and replicate independently in the cytoplasm of lysogenic host cells (see Fig. 6.14). Heteroduplex

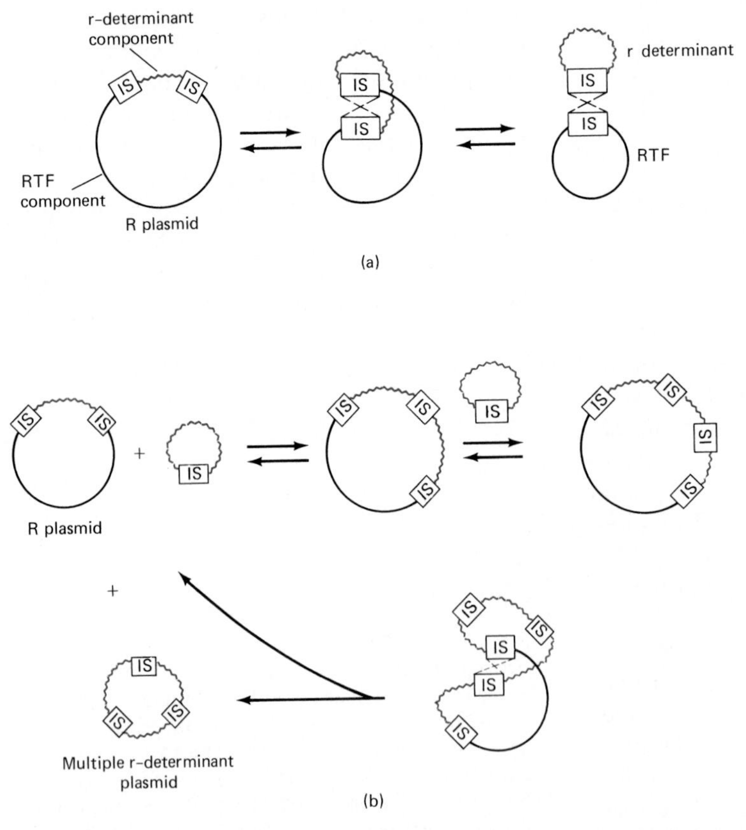


***Figure 10.27*** Proposed mechanism for reversible dissociation of R plasmids at the sites of IS*1* insertion. (a) Independent *r* determinants and RTF units may arise or may cointegrate to become an R plasmid. (b) Through multiple integrations, two or more transposons may be incorporated into the same R plasmid, and multiple *r* determinant plasmids may arise in this way and be transferred eventually to bacterial cells, which gain multiple antibiotic resistances in a single step. (From K. Ptashne and S.N. Cohen. 1975. *J. Bact.* **122**:776.)

analysis, using electron microscopy, and base sequencing analysis have revealed the presence in the Mu phage genome of *invertible* DNA sequences flanked by repeated IS elements (Fig. 10.28). These IS elements may be oriented in the molecule in such a way that the two neighboring invertible segments face in opposite directions or in the same direction. When the two segments are oriented in the same direction, Mu phages are viable. When the two segments face in opposite directions, the phages are inviable and cannot replicate in the host cell. Mu genes are therefore turned on and off according to the strand orientation in the invertible DNA segment. This situation resembles the one we described earlier for IS*2* effects in the *gal* operon

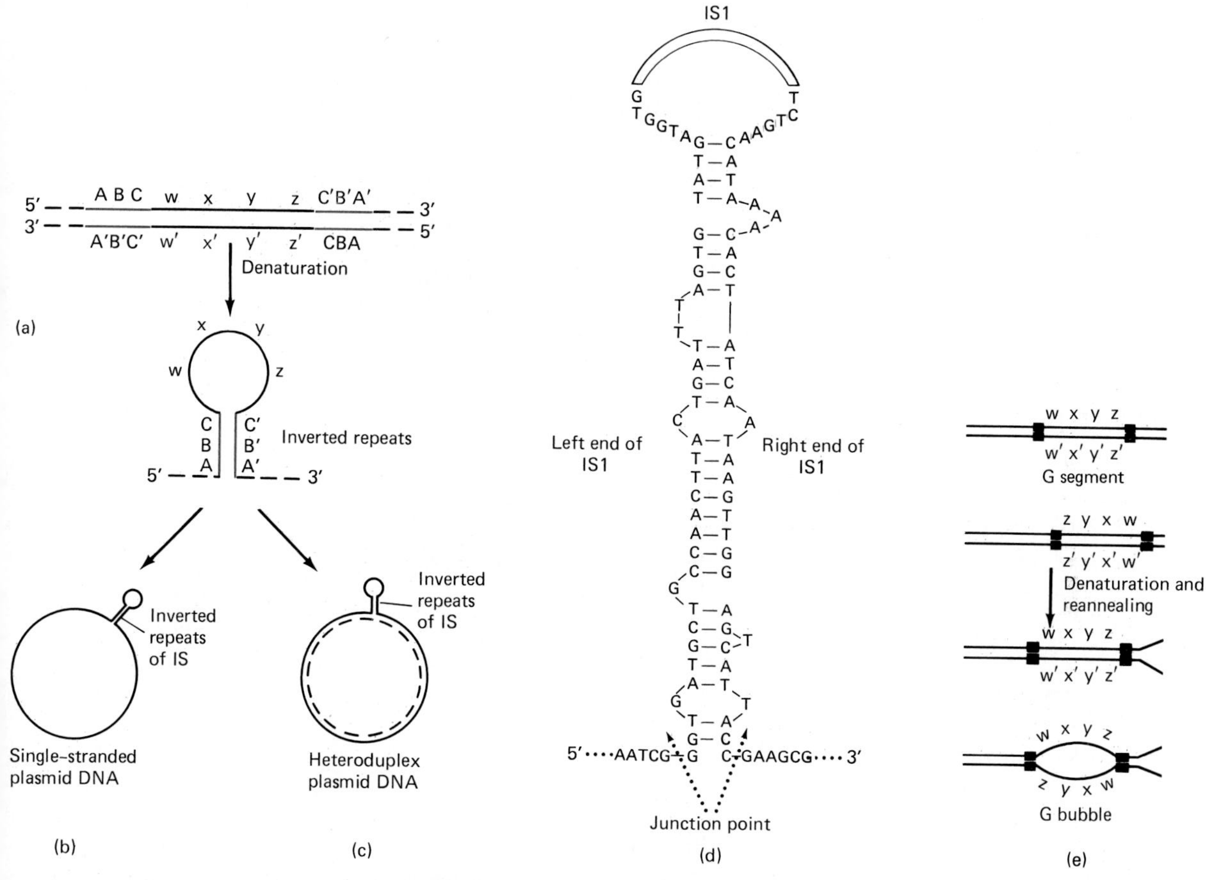


***Figure 10.28*** Transposons flanked by inverted repeats of IS elements provide the basis for heteroduplex analysis and mapping of these regions. (a) Transposon *W X Y Z* is flanked by inverted IS repeats, *A B C* and *C B A* in duplex DNA, which produce predicted stem-loop conformations in single strands from denatured duplexes. (b) The stem loop is evident in electron micrographs of single-stranded plasmid DNA or in (c) heteroduplex DNA, where one strand loops out in the lollipop form. (d) The sequence of IS*1*, showing the inversely repeated stem of the element. (e) Structure of part of Mu DNA as seen in electron micrographs of denatured and reannealed duplexes. Genes *w x y z* and complementary *w' x' y' z'* (hypothetical) are flanked by inverted repeat sequences (black boxes) that serve as sites of recombination for inversion. When Mu DNA mixtures reanneal after denaturation, the two orientations of the gene segment are obvious from electron micrographs which show closely paired duplexes and duplexes with the "G bubble" formation due to the heteroduplex inversion in this renatured region (the G segment).

and "flip-flop" control in *Salmonella* and *Schizosaccharomyces*. The genes in the invertible DNA segment include ones that code for proteins needed by the phage for adsorption to the host cell, an essential prerequisite for phage entry into the host.

Mu is highly unusual because it behaves like a transposable genetic element but is capable of initiating infection like a conventional virus. In addition, Mu integrates indiscriminately into the host genome or into DNA molecules of plasmids and other phages. Other viruses integrate at specific sites in the host chromosome. Phage lambda, for example, integrates at the λ attachment site on entering its prophage state in the *E. coli* chromosome. There is considerable interest in Mu phage behavior because of its eccentricities and because it may provide a model system for studying transposable genetic elements.

## 10.11 Transposable Genetic Elements in Eukaryotes

The first descriptions of transposable genetic elements were made by Barbara McClintock in the 1940s. She described sporadic occurrences of altered gene expression as seen in variegated phenotypic patterns in kernels on ears of corn. From the results of complex but incisive genetic analysis, McClintock attributed the phenomenon to the influences of transposable genetic elements, which she called *controlling elements* (Fig. 10.29). These elements cause mutations and chromosomal rearrangements as well as influencing gene expression.

A particular controlling element could turn off the activity of a marker gene when it was transposed within or near that gene. The change in marker gene expression was observed as patches of pigment-producing kernel tissue interspersed with colorless patches. A controlling element could later be transposed to another site in the genome, thus restoring marker gene activity but possibly influencing the activity of another marker gene in the same cells. Transposable elements could turn on gene action as well as turn it off after being integrated in the region. These significant experimental studies by McClintock were poorly understood by others at the time and were given little or no consideration in explanation of gene expression controls. On discovery of bacterial transposons and the genetic and molecular analyses of these more convenient experimental systems, investigators finally realized their importance in the 1970s. Whether eukaryotic transposable genetic elements and bacterial transposons are similar in their biochemical features is still unclear but they are certainly analogous in genetic behavior and in their influence over gene expression and genomic rearrangements.

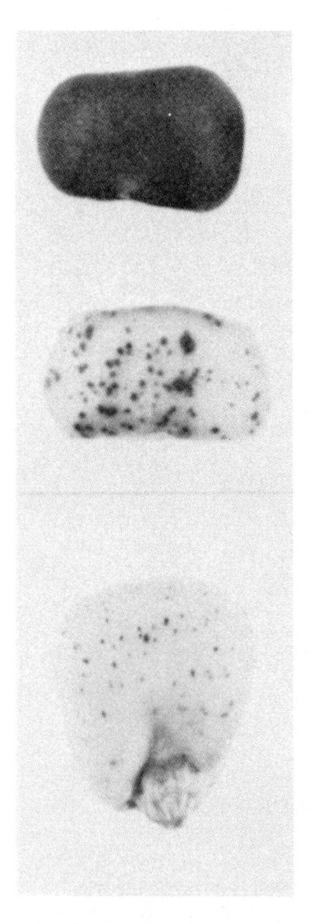

***Figure 10.29*** The pattern and size of spots produced in these kernels of corn are the result of transpositions of genetic elements to various sites in the genome at different times in development and in relation to marker genes for color. (Photo courtesy of B. McClintock.)

Transposable genetic elements have been described in other eukaryotes as well as in corn. In the 1960s it was evident from genetic analysis of various *Drosophila* strains that transposable genetic elements influenced gene expression and led to rearrangements of genomic segments. Transpositions of gene sequences coding for mating types in yeast underlie mating type determination and other genetic expressions in the organism.

With the introduction of recombinant DNA methods and rapid base sequencing in the 1970s, molecular analysis of transposable genetic elements in eukaryotes became possible. Among the best-studied systems are those in *Drosophila,* particularly of the *copia* gene. *Copia* and other transposable genetic elements exist in the genome as *families of dispersed repeated sequences.* Members of the *copia* family are located at 20 to 40 different sites in the *Drosophila* genome. There are believed to be 20 or 30 different families of dispersed repeats in *Drosophila,* accounting for about 5% of the genomic DNA in the organism.

*Copia* and other transposable genetic elements resemble bacterial transposons in genetic behavior and in the organization of their DNA sequences. The elements range in size from 5000 to 7000 base pairs. The sequence of bases at one end of each molecule is a long direct repeat of the sequence at the other end. These long direct repeats may be 300 to 500 base pairs in length. At the ends of each long direct repeat is a short inverted repeat of a few base pairs. An intriguing current observation is that integrated proviral forms of vertebrate retroviruses (RNA viruses that infect animal cells, see Section 3.14) resemble eukaryotic transposable genetic elements like *copia* (Fig. 10.30). These similarities have been proposed to be reflections of the evolutionary origin of proviruses from transposable elements. Whether or not this is true is uncertain at present. Base sequencing and other molecular and genetic methods may shed more light on the similarities among transposable genetic elements.

Some types of transposable genetic elements integrate only at specific sites in the genome and influence specific genomic rearrangements. Most of these genetic elements, however, seem to wander from place to place in the genome and influence genetic events randomly, depending upon their site of integration and the genomic regions flanking them. In most of the known eukaryotic transposable systems, but not in all, the elements do not seem to carry genes that encode any specific characteristic. *Copia* elements transcribe copious quantities of RNA (thus, the name *copia*), which are translated *in vitro* to produce polypeptides of no known functions. Since transposable elements influence gene expression and genomic organization, their significance may lie more in control of gene expression than in contributing proteins that determine particular traits of the organism. At the present time we are uncertain about the functional and evolutionary significance of

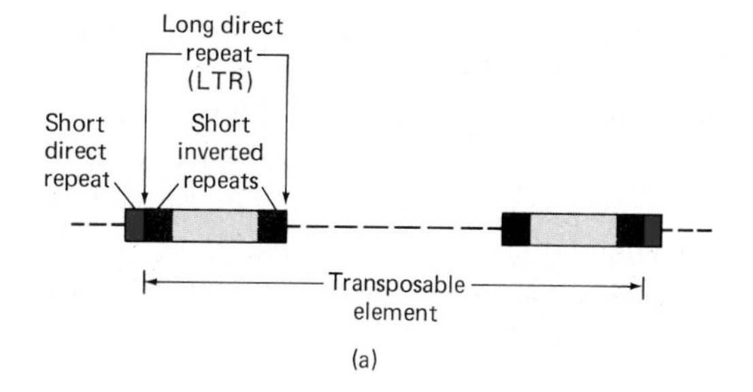


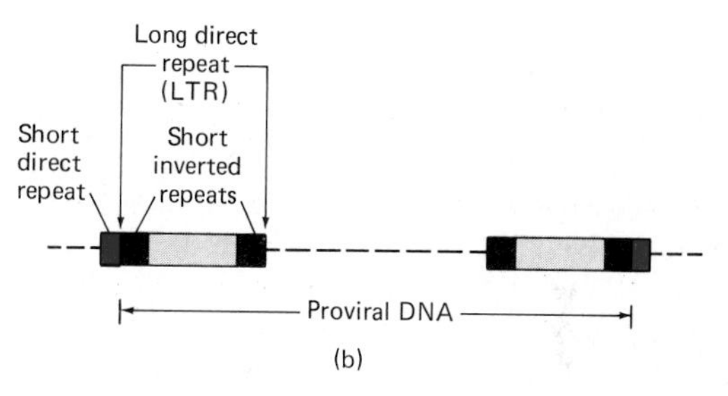


***Figure 10.30*** Similarities in organization of (a) genomic transposable elements and (b) integrated proviral DNA of vertebrate retroviruses. Each of these DNA sequences is flanked at each end by short direct repeat segments (color) and long direct repeat segments. The long direct repeats consist of a central sequence and flanking short inverted repeats. (Reprinted by permission from D.J. Finnegan. 1981. *Nature* **292**:800–801. Copyright © 1981 Macmillan Journals Limited.)

transposable genetic elements. We are certain, however, that the existence of these elements contributes to a dynamic state of flux of the genome. Despite this genomic lability, wholesale rearrangements do not take place. Genes still retain their order in linkage groups, and this same order exists in both germ-line and somatic cells throughout development, according to molecular studies and observations of the constancy of banding patterns in polytene chromosomes.

## Questions and Problems

***10.1*** State whether β-galactosidase synthesis is inducible or constitutive in the following haploid *E. coli* strains.

***a.*** $i^- o^c z^+$
***b.*** $i^+ o^+ z^+$
***c.*** $i^- o^+ z^+$
***d.*** $i^+ o^c z^+$

***10.2*** The allele $i^s$ codes for a superrepressor, which is unable to combine with the inducer and, therefore, is able to bind to the operator. An allele $o^0$ has been found, which turns off the operon in the presence or absence of inducer. With this in mind, complete the following table (use + to indicate enzyme synthesized, − to indicate enzyme not synthesized):

| | | lactose present | | lactose absent | |
|---|---|---|---|---|---|
| **strain** | **genotype** | **β-galactosidase** | **permease** | **β-galactosidase** | **permease** |
| 1 | $i^+ o^+ z^+ y^+$ | | | | |
| 2 | $i^s o^+ z^+ y^+$ | | | | |
| 3 | $i^- o^+ z^+ y^+$ | | | | |
| 4 | $i^+ o^0 z^+ y^+$ | | | | |
| 5 | $i^+ o^c z^+ y^+$ | | | | |
| 6 | $i^- o^0 z^+ y^+$ | | | | |
| 7 | $i^- o^c z^+ y^+$ | | | | |
| 8 | $i^s o^0 z^+ y^+$ | | | | |
| 9 | $i^s o^c z^+ y^+$ | | | | |

***10.3*** State whether β-galactosidase synthesis is inducible or constitutive in each of the following merodiploid strains of *E. coli:*

***a.*** $i^+ o^+ z^+ / i^+ o^c z^+$
***b.*** $i^+ o^c z^+ / i^+ o^c z^+$
***c.*** $i^+ o^+ z^+ / i^- o^+ z^+$
***d.*** $i^- o^+ z^+ / i^- o^+ z^+$
***e.*** $i^+ o^+ z^+ / i^+ o^+ z^+$

***10.4*** A strain of *E. coli* exhibits constitutive synthesis of β-galactosidase. This strain remains constitutive after it is made partially diploid by the addition of an F′ phenotype of $i^+ z^+$. Explain the basis of the genetic problem in this strain.

***10.5*** In an investigation of the genetic control of enzyme synthesis, several strains of *E coli* were constructed with different genotypes and β-galactosidase activity was measured as a function of time. If lactose (inducer) was added at 15 minutes, there could be four possible results:

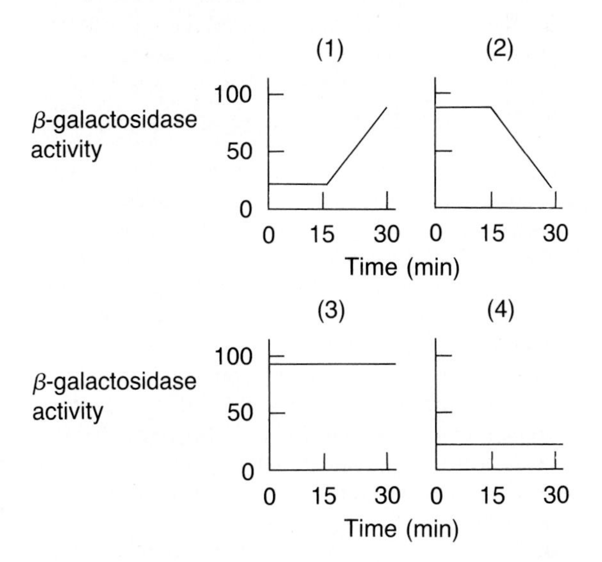


Which result would you predict for each of the following:

***a.*** $i^+o^+z^-$
***b.*** $i^-o^+z^+$
***c.*** $i^so^+z^+$
***d.*** $i^+o^cz^+$
***e.*** $i^-o^+z^+/i^+o^+z^+$
***f.*** $i^+o^+z^+/i^+o^cz^+$

***10.6*** In *Salmonella typhimurium* three linked loci were found; their linkage order is given in the following table. The mutant alleles are symbolized by *a*, *b*, and *c*, and their wild-type alternatives by +. In the presence or absence of the inducer, the cells exhibit high enzyme or no enzyme production (high and none, respectively, in the table). Which of the three mutant alleles is the structural gene, which the operator, and which the repressor gene?

| | phenotype | |
|---|---|---|
| **genotype** | **inducer** | **no inducer** |
| *a* + + | high | high |
| + *b* + | none | none |
| + + *c* | high | high |
| + + +/*abc* | high | none |
| + + *c*/*ab* + | high | high |
| + *b* +/*a* + *c* | high | high |
| *a* + +/+ *bc* | high | none |

***10.7*** Tyrosine aminotransferase (TAT) activity in liver cells grown in culture increases when the cells are exposed to steroid hormones. Assume this activity is genetically regulated in a manner analogous to operon control of β-galactosidase in *E. coli.*

***a.*** Describe the phenotype of liver-cell mutants altered in the structural gene for TAT enzyme.

***b.*** Describe the phenotype of a liver-cell mutant that was obviously altered in a regulatory gene for TAT enzyme.

***c.*** How could you tell whether a mutation in this system was dominant or recessive?

***10.8*** The enzymes of the tryptophan biosynthetic pathway are repressibly synthesized under genetic control of the *trp* operon in *E. coli.* Proceeding from the 3′ to the 5′ region of the operon, the order of genes is *trpP–trpO–trpL–trpE–trpD–trpC–trpB–trpA.*

***a.*** Which are the regulatory genes or segments, and what is the function of each?

***b.*** Which regulatory components in the operon function in repressible enzyme synthesis under operator-repressor control and which are under attenuation control?

***c.*** In wild-type strains, will Trp enzymes be made in the presence of low or high levels of tryptophan? Explain on the basis of operator-repressor control.

***10.9*** A wild-type strain of *E. coli* and one with *trpL* deleted are incubated in media containing low concentrations of tryptophan and media with excess tryptophan. Fill in the following table for each strain under each condition of growth, indicating no by 0, yes by +, and enhancement by ++.

| | leader peptide made | |
|---|---|---|
| **medium** | **wild type** | ***trpL^0*** |
| low Trp | | |
| excess Trp | | |

| | attenuator formed | |
|---|---|---|
| **medium** | **wild type** | ***trpL^0*** |
| low Trp | | |
| excess Trp | | |

| | ***trpE*** **transcribed** | |
|---|---|---|
| **medium** | **wild type** | ***trpL^0*** |
| low Trp | | |
| excess Trp | | |

***a.*** Explain your predictions for the wild-type strain in both media.

***b.*** Explain your predictions for the *trpL*$^0$ strain in both media.

***10.10*** What are the differences between the operator and the leader sequences of an operon in each of the following cases:

***a.*** Transcription into mRNA.

***b.*** Translation into peptide chains.

***c.*** Regulation of gene expression by negative control.

***10.11*** Eight strains of *E.coli* were incubated in media with and without lactose, and $\beta$-galactosidase synthesis was assayed. The strains differed in alleles of the *lac* operon repressor gene *i*, the adenylcyclase gene *cyc,* and the catabolite activator protein gene *crp.* Results are tabulated as follows (0 = no enzyme, + = low enzyme levels, ++ = medium enzyme levels, +++ = high enzyme levels):

| **strain** | **genotype** | **lactose present** | **lactose absent** |
|---|---|---|---|
| 1 | $i^+ cyc^+ crp^+$ | +++ | 0 |
| 2 | $i^+ cyc^+ crp^-$ | ++ | 0 |
| 3 | $i^+ cyc^- crp^+$ | ++ | 0 |
| 4 | $i^+ cyc^- crp^-$ | + | 0 |
| 5 | $i^- cyc^+ crp^+$ | +++ | + |
| 6 | $i^- cyc^- crp^-$ | + | + |
| 7 | $i^s cyc^+ crp^+$ | 0 | 0 |
| 8 | $i^s cyc^- crp^-$ | 0 | 0 |

***a.*** How is $\beta$-galactosidase synthesis regulated in each of the eight strains?

***b.*** If cyclic AMP is added to any of these culture media, which strains will show enhanced enzyme synthesis to the +++ level? Explain.

***10.12*** Eukaryotic gene expression may be modulated through transcriptional, posttranscriptional, and translational controls. For the two following examples of gene expression, state (1) which control is involved, (2) how the control regulates gene expression, and (3) one line of evidence in support of the statements in (1) and (2):

***a.*** Ovalbumin synthesis in chick oviduct.

***b.*** Hemoglobin synthesis in mammalian reticulocytes.

***10.13*** *Chironomus tentans* is a useful insect for studies in the regulation of gene expression because it has large polytene chromosomes that are suitable for microscopic studies.

***a.*** When [$^3$H]thymidine was injected into developing *Chironomus* larvae, radioactive thymidine was found to be distributed evenly all along the length of all the chromosomes in every tissue, according to autoradiography. Explain the significance of these results.

***b.*** When [$^3$H]uracil was injected into developing larvae, only certain regions of chromosomes in only some of the tissues contained labeled uracil a short time afterward. Explain.

***c.*** After a longer time, [$^3$H]uracil was found in the cytoplasm of those larval cells whose chromosomes had previously been labeled in (b). Explain.

***10.14*** What would be the distribution of nuclear DNA in CsCl density gradients for cells engaged in rDNA amplification and for nonamplifying cells? How would you identify the "satellite" DNA as rDNA? Diagram your expected results.

***10.15*** Diagram the immunoglobulin protein molecule and indicate the kind and location of the functioning regions in each type of polypeptide chain.

***10.16*** During immunoglobulin light-chain gene assembly, one gene from the $V_L$ gene cluster joins with one gene from the $J_L$ gene cluster.

***a.*** What alterations take place in the genome during *V/J* joining?

***b.*** If the *V* gene is a kappa ($\kappa$) type, what type would the *J* gene be?

***c.*** What steps occur between joining of genes to make *V-J* and the production of mRNA sequences used in translation of the polypeptide chain?

***d.*** In what cell type does this rearrangement of DNA take place?

***10.17*** In what respect does heavy-chain gene assembly differ from light-chain gene assembly in germline DNA?

***10.18*** During antigenic stimulation of immature B lymphocytes, IgM is produced. After a second stimulation and the development of antibody-secreting plasma cells, or B cells, the same lymphocytes may be found to secrete IgE instead of IgM. Explain how this occurs at the molecular level.

***10.19*** Patients being treated with tetracycline to alleviate a bacterial infection suddenly became insensitive to the antibiotic, even though they appeared to be improving earlier. The patients were given ampicillin, but they were insensitive to this drug, too. When they were given streptomycin as an alternative, the patients improved and eventually recovered from the infections. Explain the basis for these events in terms of R plasmids.

***10.20*** State at least one similarity between bacterial transposons and transposable elements in eukaryotes in relation to the following aspects:

***a.*** DNA sequence organization.

***b.*** Genetic behavior.

***10.21*** Define the following terms: ***a.*** structural gene ***b.*** repressor ***c.*** operon ***d.*** attenuation control ***e.*** catabolite repression ***f.*** target cell ***g.*** gene amplification ***h.*** immunoglobulin ***i.*** transposable genetic element ***j.*** insertion sequence ***k.*** transposon.

## References


Alt, F.W., et al. 1982. Immunoglobulin heavy-chain expression and class switching in a murine leukaemia line. *Nature* **296**:325.

Brown, D.D. 1981. Gene expression in eukaryotes. *Science* **211**:667.

Brown, D.D., and I.B. Dawid. 1968. Specific gene amplification in oocytes. *Science* **160**:272.

Calabretta, B., et al. 1982. Genome instability in a region of human DNA enriched in *Alu* repeat sequences. *Nature* **196**:219.

Clowes, R.C. Apr. 1973. The molecule of infectious drug resistance. *Sci. Amer.* **229**:18.

Cohen, S.N., and J.A. Shapiro. Feb. 1980. Transposable elements. *Sci. Amer.* **242**:40.

Cold Spring Harbor Symposia on Quantitative Biology. 1981. *Movable Genetic Elements.* Vol 45.

Gall, J.G. 1968. Differential synthesis of the genes for ribosomal RNA during amphibian oogenesis. *Proc. Nat. Acad. Sci. U.S.* **60**:553.

Hourcade, D., D. Dressler, and J. Wolfson. 1973. The amplification of ribosomal RNA genes involves a rolling circle intermediate. *Proc. Nat. Acad. Sci. U.S.* **70**:2926.

Kolata, G. 1981. Z-DNA. *Science* **214**:1108.

Kurosawa, Y., et al. 1981. Identification of D segments of immunoglobulin heavy-chain genes and their rearrangement in T lymphocytes. *Nature* **290**:565.

Landy, A., and W. Ross. 1977. Viral integration and excision: Structure of the lambda *att* sites. *Science* **197**:1147.

Leder, P. May 1982. The genetics of antibody diversity. *Sci. Amer.* **246**:102.

McClintock, B. 1965. The control of gene action in maize. *Brookhaven Symp. Biol.* **18**:162.

Maurer, R.A. 1982. Estradiol regulates the transcription of the prolactin gene. *J. Biol. Chem.* **257**:2133.

Milstein, C. Oct. 1980. Monoclonal antibodies. *Sci. Amer.* **243**:66.

Nevers, P., and H. Saedler. 1977. Transposable genetic elements as agents of gene instability and chromosomal rearrangements. *Nature* **268**:109.

Pastan, I. Aug. 1972. Cyclic AMP. *Sci. Amer.* **227**:97.

Ptashne, K., and S.N. Cohen. 1975. Occurrence of insertion sequence (IS) regions on plasmid deoxyribonucleic acid as direct and inverted nucleotide sequence duplications. *J. Bact.* **122**:776.

Ptashne, M., A.D. Johnson, and C.O. Pabo. Nov. 1982. A genetic switch in a bacterial virus. *Sci. Amer.* **247**:128.

Razin, A., and A.D. Riggs. 1980. DNA methylation and gene function. *Science* **210**:604.

Royal, A., et al. 1979. The ovalbumin gene region: Common features in the organization of three genes expressed in chicken oviduct under hormonal control. *Nature* **279**:125.

Schimke, R.T. Nov. 1980. Gene amplification and drug resistance. *Sci. Amer.* **243**:60.

Shapiro, J.A. 1979. Molecular model for the transposition and replication of bacteriophage Mu and other transposable elements. *Proc. Nat. Acad. Sci. U.S.* **76**:1933.

Shimotohno, K., S. Mizutani, and H.M. Temin. 1980. Sequence of retrovirus provirus resembles that of bacterial transposable elements. *Nature* **285**:550.

Stronowski, I., and C. Yanofsky. 1982. Transcript secondary structures regulate transcription termination at the attenuator of *S. marcescens* tryptophan operon. *Nature* **298**:34.

Yanofsky, C. 1981. Attenuation in the control of expression of bacterial operons. *Nature* **289**:751.

Yanofsky, C., and R. Kolter. 1982. Attenuation in amino acid biosynthetic operons. *Ann. Rev. Genet.* **16**:113.

# CHAPTER 11

# Mutation

Alternative forms of a gene arise by mutation, and mutant alleles are inherited in subsequent generations. The study of mutations and mutational processes is important to our understanding of the gene and its behavior in organisms. In addition, we are profoundly interested in relationships between mutation and disease or other aspects of health. Studies of mutations by gene transmission analysis were very important in earlier years, as they are today. The addition of molecular methods for analyzing mutations and mutational processes, however, has brought us much closer to a fundamental understanding of the gene and its contribution to phenotype development.

## General Nature of Mutations

**Mutations** are sudden, heritable changes in the genetic material. These changes become evident when a new phenotype appears and is passed on to new generations. Inherited changes typically arise as the result of alterations in single genes; these changes are defined as **point mutations.** In addition, **chromosomal mutations** may arise by gross alterations of chromosome structure and genome restructuring. We will confine ourselves to point mutations in this chapter and will discuss chromosomal mutations in Chapter 12.

Point mutations are a major source of hereditary variation in evolution, and they provide the only *new* information to subsidize species changes. They are the "raw materials" of the evolutionary pathways in past, present, and future species of organisms.

### 11.1 Spontaneous Mutations

Mutations can arise spontaneously in any gene at any time in any cell and are, therefore, *random* changes in the genes. The causes of spontaneous mutations are largely unknown, but they are generally believed to be the result of occasional mistakes in base pairing during DNA replication. Radiation or chemicals in the environment contribute to DNA damage, but the amounts of these agents in the environment have been calculated to be insufficient to account for observed mutation rates.

Mutations are *recurrent* events and we can thus determine their rates of occurrence (Table 11.1). We can see from the data that mutations are *rare* events and that different genes and different alleles mutate at different rates. In sexual organisms, the **mutation rate** is expressed as the number of mutations per gamete per generation; in asexual organisms such as bacteria, mutation rate is expressed as the number of mutations per cell-generation. The **mutation frequency,** or the number of mutations in a population, will be much higher in *E. coli* than in human beings if we count up the mutants over a period of time, such as one year. When we make comparisons on the basis of generations, regardless of their absolute duration, we can judge the general nature and properties of gene mutations in any organisms as observed in gene mutation rates.

Mutations from one allelic form to another, such as wild type to mutant, usually yield a different mutation rate than mutations back to the original form. The rate of **forward mutation** and the rate of **reverse mutation** back to some original form can be calculated by the

***Table 11.1*** Mutation rates of specific genes in various organisms.

| organism | gene | trait | mutations per $10^6$ cells or gametes |
|---|---|---|---|
| Bacteriophage T2 | *h* | host range | 0.003 |
| *Escherichia coli* | *str* | streptomycin resistance | 0.0004 |
| (colon bacillus) | $lac^-$ | lactose fermentation (to $lac^+$) | 0.2 |
| *Neurospora crassa* | $ad^-$ | adenine requirement (to $ad^+$) | 0.04 |
| (red bread mold) | $inos^-$ | inositol requirement (to $inos^+$) | 0.08 |
| *Zea mays* (corn) | *Sh* | shrunken seeds | 1 |
| | *Y* | yellow seeds | 2 |
| | *I* | color inhibitor | 106 |
| *Drosophila melanogaster* | $y^+$ | body color (to *yellow*) | 100 |
| (fruit fly) | $w^+$ | eye color (to *white*) | 40 |
| | $e^+$ | body color (to *ebony*) | 20 |
| *Mus musculus* (mouse) | $a^+$ | coat pattern (to *nonagouti*) | 30 |
| | $c^+$ | coat color (to *albino*) | 10 |
| *Homo sapiens* (human) | | Huntington's disease | 1 |
| | | aniridia (absence of iris) | 5 |
| | | retinoblastoma (tumor of retina) | 20 |
| | | hemophilia A | 30 |
| | | achondroplasia (dwarfness) | 40–140 |
| | | neurofibromatosis (tumor of nervous tissue) | 130–250 |

same methods. Mutations are therefore *reversible,* and the rates of spontaneous forward and reverse mutation of a gene are often different. But spontaneous mutation in either direction occurs at a rate that is *characteristic* for any particular gene or allele and varies from one gene or allele to another.

Mutations are essential tools for genetic analysis since we cannot study the gene unless we can compare allelic differences in relation to gene structure, function, and regulation. We recognize the existence of a gene through detection of its variants, from which we can then backtrack to determine the nature of the unaltered or wild-type genetic material. Nonmutated genes are undetectable by observations of phenotypes, but they can be identified by molecular methods.

Since mutations are random and rare but recurrent events, we can search for evidence of mutations in families, progenies, and both laboratory and natural populations. In human beings and other diploid organisms, the easiest mutations to detect are dominants because the mutant allele is expressed even in heterozygotes (Fig 11.1). If the altered characteristic is heritable, then it must have arisen in the generation in which it first appeared. In human beings and other species in which one sex is hemizygous, sex-linked recessive mutations can be detected in the hemizygous sex because the unpaired allele is expressed directly. We believe that Queen Victoria must have been the source of the X-linked recessive mutation for hemophilia since none of her male ancestors displayed the condition and many of her male descendants did (see Fig. 4.15).

Autosomal recessive mutations usually can be detected only under special conditions because the phenotype is expressed only in homozygotes for the recessive allele. In human beings the sudden appearance of a new recessive inherited trait in a family can sometimes be observed, but it usually cannot be traced back to

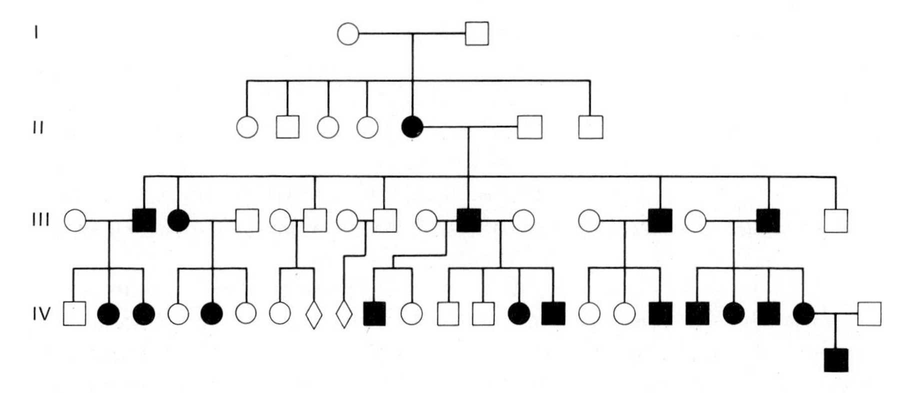


*Figure 11.1* Pedigree of a genetic mutation. Solid black symbols indicate family members with severe foot blistering, open symbols indicates individuals free of this affliction. The circles designate females; the squares, males; diamonds, more than one individual or unknown (s) of either or both sexes. II-5 must have inherited a mutant allele through the egg or sperm of her parents, and the allele was then transmitted to her descendants in the next generations. Autosomal dominant inheritance is apparent. (Data from J.B.S. Haldane and J. Poole. 1942. *J. Hered.* **33**:17.)

the person or generation in which the mutation first occurred. We cannot determine when the rare recessive allele first occurred because it may remain undetected in the heterozygotes for any number of generations (Fig. 11.2).

New autosomal recessive mutations can be detected in diploid species that can be studied by breeding analysis. Investigators can construct highly homozygous inbred stocks in which homozygous recessives segregate regularly thus producing new phenotypes that can be identified on a regular basis. Other kinds of special test stocks are also available in mice, *Drosophila,* and other species. These stocks can be used as a parent in crosses to detect newly arisen recessive alleles observed in the progenies. Otherwise, new mutations would be difficult to distinguish from recessives already present and

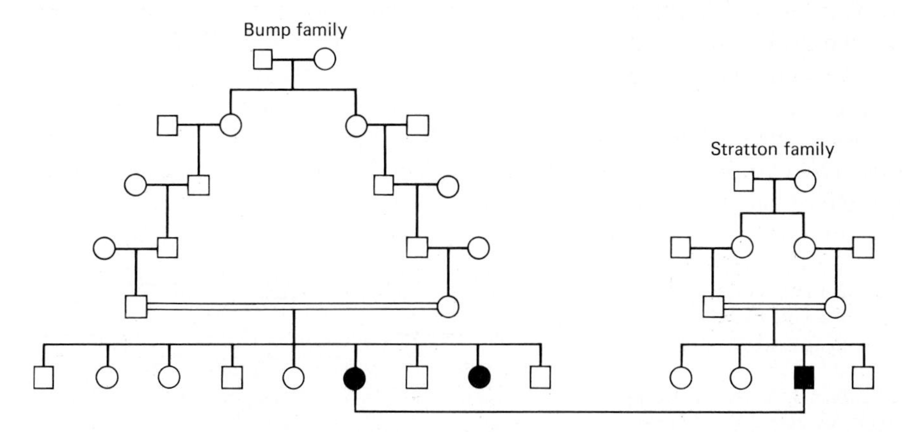


*Figure 11.2* Lavinia Bump and Charles Stratton ("General Tom Thumb") were midgets in the famous P.T. Barnum circus in the late 1800s. They were from unrelated families, but each of them was the child of parents who were cousins (note double horizontal line connecting the parents in each case). Such autosomal recessive inheritance may not be phenotypically evident for generations, but a consanguineous marriage often produces recessive offspring. (From D.L. Rimoin. 1973. In *Medical Genetics,* p. 149. V.A. McKusick and R. Claiborne, eds. New York: HP Publishing Co.)

carried in the heterozygous state. Such existing alleles would be expressed only in the occasional recessives that appear among the progeny of two parents who happen to be heterozygous for the same rare allele, as shown in Figure 11.2.

## 11.2 Detection of Spontaneous Mutations

By experimental methods for analyzing spontaneous X-linked and autosomal mutations, investigators can (1) detect new mutations, (2) calculate their frequencies in progenies and derive mutation rates for particular genes or for classes of gene changes, and (3) determine the nature of the effect of mutations on the organism. The **ClB method** was devised by H.J. Muller to detect spontaneous X-linked mutations in *Drosophila.*

The ClB method involves a specially constructed stock of *Drosophila:* heterozygous *ClB* females carry one normal X chromosome and one X chromosome with the following markers: *C,* which stands for inverted regions on the X chromosome that prevent crossing over and maintain this *ClB* chromosome intact and unchanged during the experiment; *l,* which stands for a lethal allele expressed only in hemizygous males or in homozygous females; and *B,* which stands for Bar eye, a dominant phenotypic marker showing that the *ClB* chromosome is present in an individual. (Fig. 11.3).

*ClB* females are crossed with wild-type males. Each $F_1$ *ClB* female carries one X chromosome contributed by the male parent, and all surviving $F_1$ males are wild type. Each $F_1$ *ClB* female is mated singly in culture with a wild-type $F_1$ male. The expected ratio of $F_2$ progeny is 2♀ : 1♂ since *ClB* males die. Males can be detected by simple inspection of culture bottles. Any variation from the expected proportion of males indicates that the original sperm cell carried a spontaneous mutation.

If no males are present in an $F_2$ culture, the X chromosome of the original sperm cell must have carried a spontaneous **lethal mutation.** The frequency of all-female cultures among the total $F_2$ cultures is used to calculate the mutation rate. Data summed up from many experiments indicates that the mutation rate for all X-linked, recessive spontaneous lethal mutations in *Drosophila* is about two per thousand per generation (0.2% or $2 \times 10^{-3}$). The mutation rate for any one specific gene that produces a lethal allele may, of course, vary from this average rate.

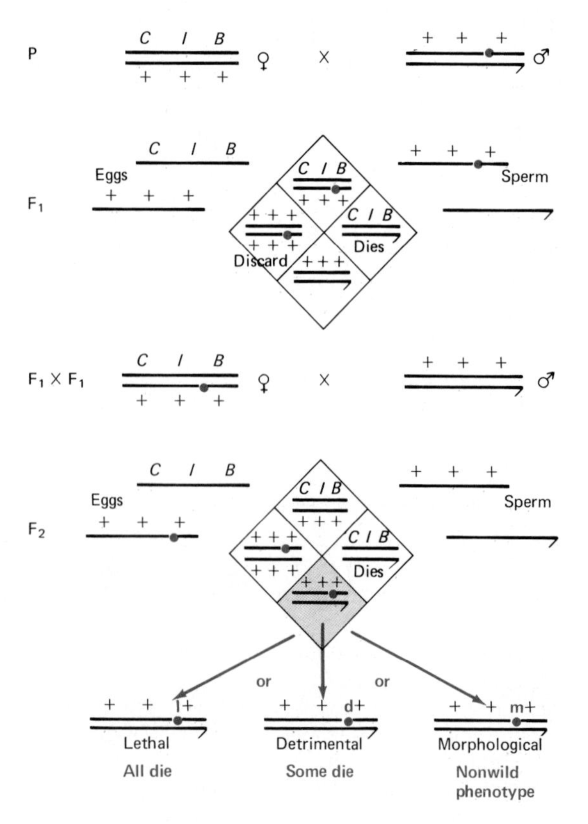


***Figure 11.3*** Through the ClB method the frequency of spontaneous or induced mutations on the X chromosome can be analyzed quantitatively and qualitatively. In the $F_2$ progeny, the ratio expected is 2 ♀ : 1 ♂, since *ClB* ♂♂ die. Depending on the ratio actually observed, we may identify the X-linked mutation (locus shown by colored dot) as a lethal, detrimental, or morphological type. From data obtained in such experiments, we can also calculate the spontaneous mutation rate for a particular X-linked locus or for a category of mutations, such as recessive lethals.

If $F_2$ males are present but have mutant phenotypes, then the original sperm cell must have carried a **morphological mutation.** The percentage of $F_2$ progeny with a particular mutant phenotype among the total $F_2$ progeny in many experiments serves as a basis for determining the mutation rate of a particular gene that yields a morphological mutation. Summed data can also be used to calculate the overall mutation rate for this class of gene mutations.

A third class of mutations, called **detrimental mutations,** can also be detected by the ClB method. In these cases the $F_2$ progeny include males with the wild phenotype but in numbers significantly lower than the expected 33% of a progeny. We have no simple way to assess the kind of genetic change that has taken place, except that *on the average* there is a reduction in the life expectancy or in the relative success among these mutants in reaching the adult stage. They may be more susceptible to parasites or disease or may succumb earlier because of any one of a number of reasons. This class of mutations is of considerable importance in the long-term evolutionary success of a species and is of immense concern in considering the damaging effects of excessive radiation or chemical agents on human populations and their descendants. Detrimental mutations are much more frequent than lethal or morphological mutations, and they contribute substantially to the **genetic load** of harmful mutant alleles carried in every diploid sexually reproducing species.

In addition to these classes of mutations, others have been identified in a variety of organisms according to the phenotypic effect they produce. For example, physiological, or **biochemical, mutations** alter the phenotype as the consequence of some metabolic defect but do not necessarily lead to a visible morphological change in the structure, color, or other outward characteristics of the individual. In the large and diverse class of **conditional mutations,** the alteration may or may not be expressed, depending on the environment. Mutants that require some nutrient will be perfectly normal as long as the nutrient is supplied but may grow slowly or be inviable when the nutrient is not available. Many mutants are *temperature sensitive,* developing normally at one temperature (permissive) but abnormally or not at all at a nonpermissive (restrictive) temperature. Many *conditional lethal* mutations fall into this category of changes that can be detected in certain environments and not in others (Table 11.2).

Different experimental systems have been used to detect mutations in various organisms. Drug-resistant bacterial mutants can be detected as colonies that grow in media containing a drug. Cells that carry the sensitivity allele will grow only in the absence of the chemical. Auxotrophic mutants can likewise be recognized by their growth on supplemented media and their inability to grow in minimal media. Prototrophic revertants can be detected by plating auxotrophic strains on minimal media and observing growth of wild-type colonies. Several of these procedures are equally as applicable to mammalian cells or fungi as to bacteria. They can also be used to identify mutations in viruses, which lead to altered host range, plaque differences, and other phenotypic characteristics.

Because spontaneous mutations are relatively rare events, **enrichment methods** have been devised to detect and isolate mutants on a more predictable and controlled basis. For example, penicillin kills only actively growing bacterial cells. If a population of bacteria is

***Table 11.2*** Examples of different categories of mutations in human beings.

| category of mutation | mutant characteristic |
|---|---|
| lethal | Tay-Sachs disease |
| morphological | achondroplasia (dwarfness) |
| detrimental | sickle-cell anemia |
| physiological (biochemical) | glucose 6-phosphate dehydrogenase deficiency (GPD deficiency) |
| conditional | phenylketonuria (PKU) |

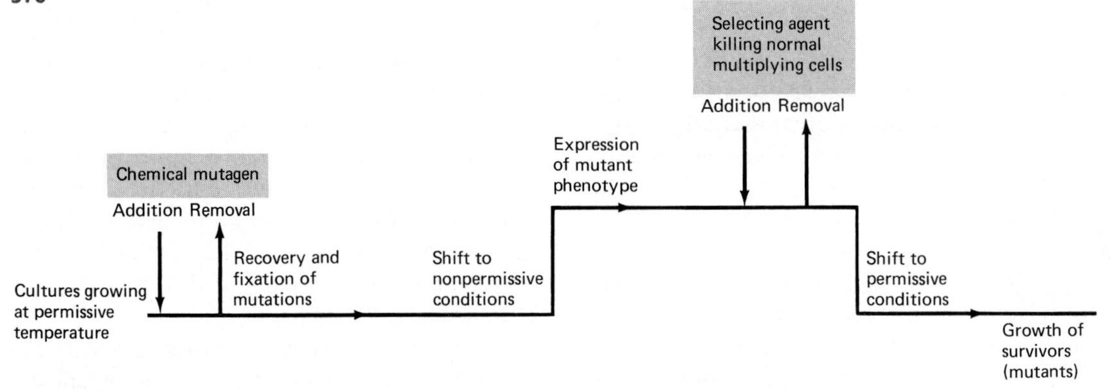


***Figure 11.4*** Enrichment method for isolation of temperature-sensitive mutants in mammalian cell cultures. Mammalian cells can multiply in a temperature range between 32°C and about 40°C. It is possible to isolate temperature-sensitive mutants that differ from wild-type cells because mutants cannot grow at 38°–39°C (the nonpermissive temperature), but they can grow at 33°–34°C (permissive temperature). The desired mutants are more likely to be detected if the illustrated enrichment method is used. (From N.R. Ringertz and R.E. Savage. 1976. *Cell Hybrids.* New York: Academic Press.)

incubated in minimal-nutrient media containing this antibiotic, growing prototrophs will die. Auxotrophic mutants that may be present in small numbers in these wild-type populations can be recovered because they do not grow in minimal media and are therefore not killed by the antibiotic. After the penicillin is washed away, the population of surviving viable cells is transferred to media supplemented with a variety of nutrients. The auxotrophic cells that grow into colonies can be tested later for specific nutritional requirements. Similar methods can be used with fungi, such as *Neurospora.* By these and other enrichment methods, a population originally containing relatively few mutants among many nonmutants can be amplified into a population consisting largely of mutants that can be easily detected, isolated, and characterized. Similar procedures are widely used today to isolate mammalian mutant cells in culture and provide important materials for many kinds of experiments (Fig. 11.4).

## 11.3 Spontaneous versus Directed Mutational Changes

In the 1930s and 1940s most biologists believed that mutations arose spontaneously, without regard to benefit or harm, in any cell anywhere at any time. Some microbiologists, however, thought that bacterial mutations arose in direct response to a need in a particular environment. For example, when a population of sensitive bacteria was incubated with an antibiotic, the resulting population consisted mostly of antibiotic-resistant bacteria. This result *appeared* to be an example of the inheritance of acquired characteristics, as proposed by Lamarck in 1809 but long discredited by many studies with higher organisms. These same results, however, could also be explained by selection.

In 1943, Salvador Luria and Max Delbrück designed the **fluctuation test** to determine whether bacterial mutants were *selected* over nonmutants in certain environments, or were *induced* to mutate by the environmental factor. They incubated a number of bacterial cultures and plated these cultures onto solid nutrient media in plates containing phage T1. If preexisting mutants resistant to the phage happened to be present in these cultures, they could grow on the plates; phage-sensitive nonmutants, however, were killed. The number of mutant colonies present in each culture would depend on how early in the growth period the mutational event had occurred. There would, therefore, be a large and random fluctuation in colony numbers among the separate cultures (Fig. 11.5). If,

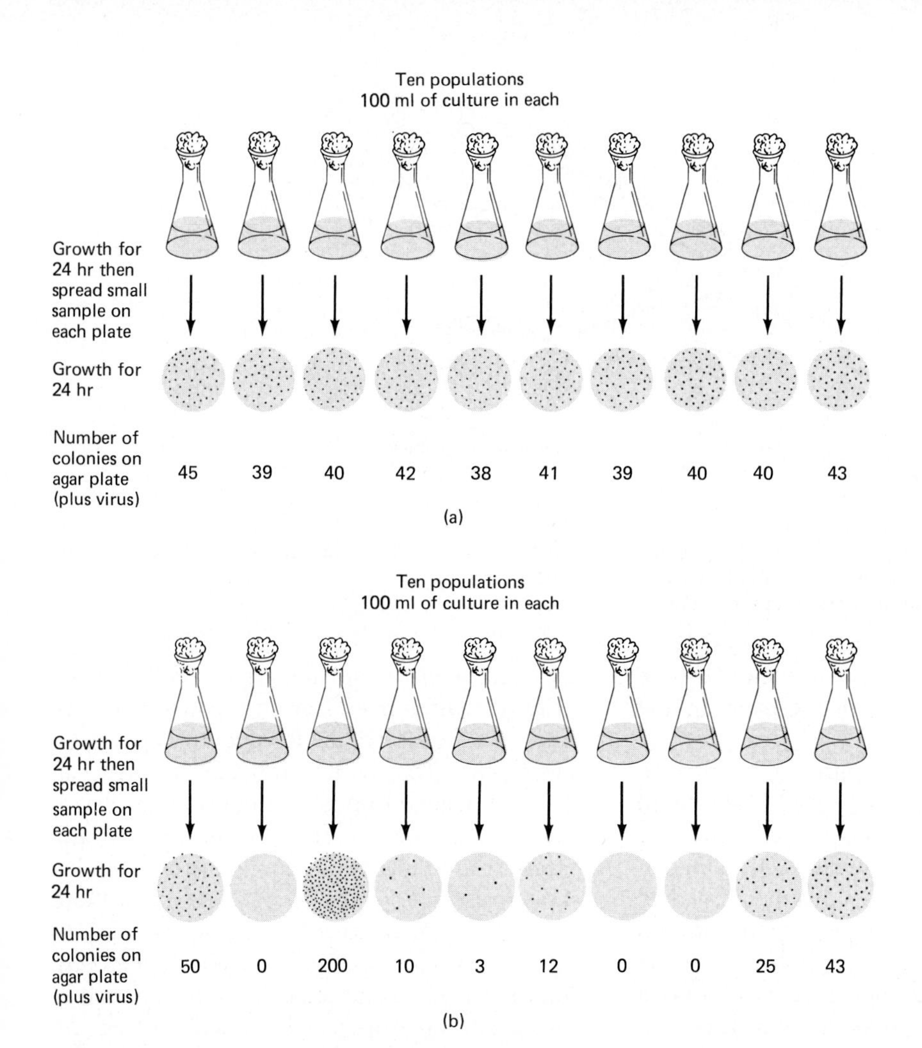


***Figure 11.5*** The Luria-Delbrück fluctuation test to distinguish between induction of new mutants or selection of preexisting mutants in populations. The population is subdivided and each subpopulation is analyzed after growth in the presence of viruses or other agents to which the cells are sensitive. The number of colonies of resistant cells sampled from each subpopulation (a) will be more or less consistent if the agent induced mutations to resistance since the probability of induced mutation would be the same in each subpopulation or (b) will fluctuate between none and hundreds of resistant cells per sampled subpopulation, depending on the numbers of preexisting mutants that happened to be present. Higher numbers of resistant colonies are expected if one or more mutants multiplied during the 24 hours of growth in liquid culture before plating a sample of cells; lower numbers are expected if a mutation arose shortly before plating and only a few descendants were produced by the spontaneous mutant; no mutants would be present if the genetic alteration did not take place in any cell in a subpopulation during the growth period.

on the other hand, cells were induced by the phage to mutate to resistance, each cell in each culture would have an equal chance to undergo such a directed change, and approximately equal numbers of mutants should be found in each of the separate cultures. When the tests were performed, wide fluctuations in the numbers of resistant bacterial mutants were found, thus showing that selection of preexisting, spontaneous mutations was the correct explanation for mutations in bacteria as it was for other organisms. The fluctuation test is used as a general method to identify agents that induce mutations in contrast with agents that permit

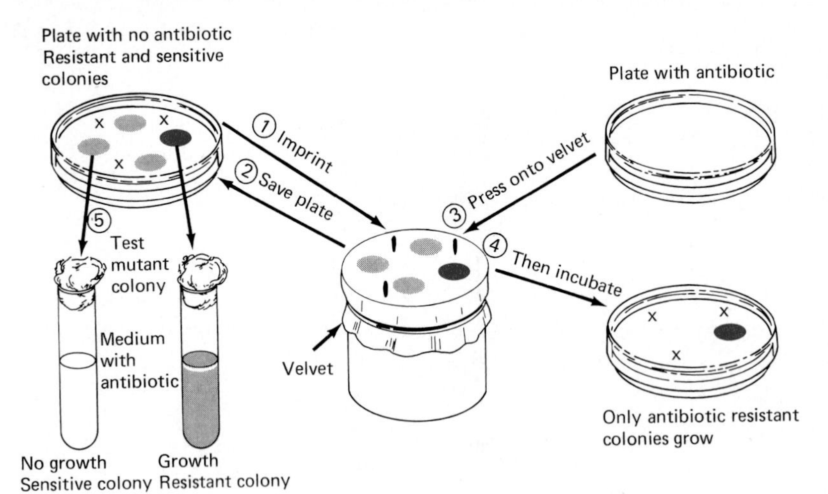


***Figure 11.6*** The method of replica plating permits indirect selection of mutants that have not been exposed to the environment containing the agent (drug, virus) to which they respond differently from wild type.

spontaneous mutants to grow in their presence and merely mediate selective growth in the environment.

There were some difficulties at the time in accepting Luria and Delbrück's statistically based fluctuation test as adequate evidence for the selection of spontaneous mutants. In addition, there was the lingering doubt that the phage had somehow induced the change since the bacteria were always incubated in the presence of the virus during the experiments. In 1952, Joshua and Esther Lederberg provided convincing evidence for the selection of spontaneous mutants: resistant mutants appeared in cultures that had not been exposed to the virus. Their method was the **replica plating test,** which has since become a standard tool in microbial genetics (Fig. 11.6).

In replica plating experiments, large numbers of bacteria are spread on solid nutrient media so that they cover the surface as a lawn of growth. This serves as the *master plate.* Next, a velvet-covered cylinder is gently pressed to the surface of the bacterial lawn so that some cells are caught in the nap of the velvet. The velvet is then pressed onto a fresh plate such that bacterial cells are inoculated onto its surface. This second plate contains a drug, virus, or some other environmental agent to which the master-plate bacteria are sensitive, and it is a *replica* of the master plate. Only resistant cells will grow on the replica plate, and their sister cells on the original master plate can be identified according to the location of the colonies that appear on the replica. The master and replica plates can then be matched up and samples of sister colonies on the master plate can be isolated and transferred to media containing the drug so that their resistance can be tested. Growth of these cells in the presence of the drug or virus, to which they had not been previously exposed, indicates that resistance arose in the absence of the specific environmental factor. Therefore, the drug, virus, or other agent leads to differential survival and multiplication of mutants in a mixed population; it does not induce the mutational change.

Replica plating is a very useful and timesaving method in mutation and other microbial studies. For example, after the use of the penicillin enrichment procedure to isolate spontaneous mutants, (as described in the preceding section), surviving cells can be grown in culture and then replica-plated to many different kinds of media to determine the exact nature of the mutational change. A nutritional mutant needing some amino acid can be identified when it grows on the supplemented medium but not on replica plates lacking amino acids. Further replica plating to media containing all amino acids but one eventually shows the specific amino acid requirement for a specific mutant. By very simple means, it is possible to identify almost any kind of auxotrophic mutant on a routine basis by replica plating.

## 11.4 Mutation at the Molecular Level

Most gene mutations apparently arise as the result of mispairing or similar mistakes in DNA replication, sometimes due to the occurrence of an unstable or tautomeric form of a base and subsequent incorrect base pairing (see Fig. 7.7). On other occasions one or another of the replication enzymes may guide the wrong base into position on the new strand and this strand will give rise in later replications to an altered set of descendant strands. In general, alterations in the base sequence of DNA lead to two types of mutations, which are classified according to the nature of the molecular change: *base substitution* and *frameshift* classes of heritable changes.

Base substitutions may alter a codon so that it specifies a different amino acid; these are called **missense mutations.** The meaning or sense of the message has been altered, but the whole polypeptide is translated (Fig. 11.7). If, on the other hand, the substituted base changes an amino acid–specifying codon to a termination codon, it is called a **nonsense mutation.** Premature termination of polypeptide synthesis at the mutant codon results in fragments of polypeptide that do not contribute to cell functions.

Frameshift mutations alter the reading frame through addition or deletion of an existing base in the gene sequence. Since messenger RNA is read sequentially from the initiating AUG to the terminating UAG, UAA, or UGA codon in translation, the readout is garbled (see Fig. 7.23).

Mutants can be restored to the wild phenotype by *reverse mutations* or by **suppressor mutations.** True reverse mutations arise when the base alteration in a codon reverts back to the original sequence. When the codon is restored, the original amino acid is specified once again and the wild phenotype will be expressed because the wild-type protein is made. The nucleotide sequence of the wild-type gene and the revertant gene are identical.

Suppressor mutations are categorized in two general classes, depending on whether the second mutation occurs in the same gene (*intragenic*) or in another gene (*intergenic*) in the mutant strain. In either case, suppression of the mutant phenotype and the appearence of the wild phenotype are the result of a second mutation. Suppressor mutants are double mutants that have a *pseudowild* phenotype, that is, the phenotype looks like the wild type but the suppressor strain has a different nucleotide sequence from that of the wild-type strain.

Intragenic suppression may arise in any one of several ways. In base-substitution mutants, a nonsense mutation can be suppressed if the mutant termination codon is altered to an amino

| Hemoglobin Type | Residue Number 143 | 144 | 145 | 146 | TERMINATOR |
|---|---|---|---|---|---|
| Normal β-chain | | | | | |
| nucleotide sequence | – CAC – | AAG – | UAU – | CAC – | UAA |
| amino acid sequence | – His – | Lys – | Tyr – | His – | COOH |
| Hb Rainier (Missense) | – CAC – | AAG – | UGU – | CAC – | UAA |
| | – His – | Lys – | Cys – | His – | COOH |
| | | | A | | |
| Hb McKees Rocks (Nonsense) | – CAC – | AAG – | UAG | | |
| | – His – | Lys – COOH | | | |

***Figure 11.7*** Human hemoglobin variants illustrating missense and nonsense mutations that affect the carboxy terminus of the β-globin chain. In Hb Rainier a different amino acid is specified than in wild type, and in Hb McKees Rocks the β-chain is shortened by two amino acids due to a change from a codon specifying the amino acid tyrosine to a termination codon (UAA or UAG).

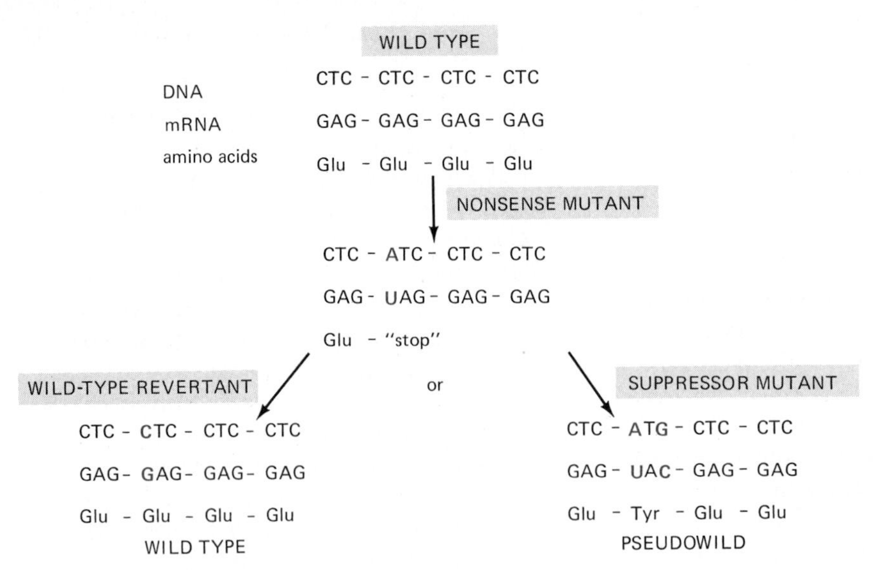


***Figure 11.8*** Intragenic suppression may arise through base substitution in a nonsense mutant, altering the termination codon to a code word that specifies an amino acid. The suppressor mutant then has a pseudowild phenotype.

acid–specifying codon (Fig. 11.8). The suppressor mutant has the pseudowild phenotype, but the composition of the protein product of the gene will differ by one amino acid at the location corresponding to the nonsense codon in the mutant gene. Missense mutations can also be suppressed by base substitutions in the mutant codon. In each case, a functional protein is made and a pseudowild phenotype is expressed. Frameshift mutations can be suppressed when a base is added or deleted to compensate for the first mutational deletion or addition, respectively, as described in Section 7.8. Once again, the suppressor strain makes a slightly different protein than the wild-type strain, but their phenotypes may be identical or very similar. Molecular analysis of the protein or of the gene sequence can reveal the difference between the original mutant, the suppressor-mutant strain, and the wild type. Genetic analysis would show that segregation and recombination took place when suppressors were crossed with wild types. A true revertant would not produce segregant progeny types when crossed with a wild-type stock.

Intergenic suppression involves a second mutation in a different gene from the first mutation. Suppression may be *indirect* if the second mutation alters a metabolic pathway that cancels out the effect of the first mutation on a different metabolic pathway. For example, a *Neurospora* mutation causing sensitivity to high concentrations of arginine within the cell can be suppressed indirectly by a second mutation that causes very little arginine to be synthesized in the cells. The original mutation remains unchanged, but the suppressor strain will grow well because the suppressor gene has modified intracellular metabolism relating to arginine. Single-mutant, double-mutant, and wild-type progeny will segregate if the suppressor strain is crossed to a wild-type tester strain and if recombination takes place.

*Direct intergenic suppression* usually involves a second mutation that functions in the process of translation from mRNA. In many cases the second mutation affects a tRNA gene so that its anticodon is altered. The suppressor tRNA can then recognize a different mRNA codon. It can either insert a different amino acid from the one specified by the first mutation, or it can insert an amino acid at a mutant codon that would have caused premature termination of protein synthesis. In other words, the direct suppressor mutation can correct a missense or a nonsense mutation through direct interaction at the site of the first mutation when translation is in progress. The altered codon caused by the first mutation remains unchanged by the mutation in the tRNA gene, but its mutational effect is corrected or suppressed by the second mutant gene product.

In suppressor mutation phenomena, therefore, the effects of the first mutation are modified by the effects of the second mutation in the same gene or in a different gene. The first mutation could have arisen by base substitution or by frameshifts, causing missense, nonsense, or garbled messages. The second mutation directly allows a sense message to be made or it dampens the influence of a mutant message by indirect means. Suppressor mutants can be distinguished from wild-type revertants by genetic analysis and by molecular analysis, as described above.

All the varieties of mutations that have been described in this section have been found in spontaneous mutants in all kinds of organisms (Table 11.3). These same kinds of mutations have been induced in experiments using particular mutagenic substances or agents. There is no essential difference between spontaneous and induced mutations either at the molecular level or at the phenotypic level of observation.

***Table 11.3*** Some of the kinds of mutations recognized according to operational or molecular criteria.

| kind of mutation | observations and interpretations |
|---|---|
| *Operational classes* | Altered phenotype; altered inheritance pattern |
| point (single site) | Change in one nucleotide |
| deletion | Loss of gene(s) or parts of genes |
| forward | Change of wild-type allele to mutant alternative |
| reverse (back) | Change of mutant allele to wild-type alternative |
| dominant/recessive | Mutant allele masks or is masked by wild-type allele |
| sex-linked/autosomal | Location of mutated site determined by inheritance pattern |
| lethal | Mutant dies before age of reproduction |
| detrimental | Mutant has reduced average life expectancy |
| morphological | Visible change detected in some body feature |
| physiological (biochemical) | Altered metabolism; may or may not show altered morphology |
| conditional | Phenotypic expression varies according to environment |
| suppressor | Second-site alteration in presence of first-site change, leading to pseudowild phenotype and inheritance pattern |
| *Molecular classes* | Alteration in nucleotide sequence of gene(s) |
| base substitution | One nucleotide replaced in one codon |
| missense | Alteration in a codon causing different amino acid |
| nonsense | Change from codon specifying an amino acid to codon specifying termination of the polypeptide |
| transition | Substitution of a pyrimidine by another pyrimidine, or of one purine by another purine in the codon |
| transversion | Substitution of a pyrimidine by a purine, or vice versa, in a codon |
| frameshift | One or more nucleotides added or deleted in a codon, causing altered reading frame |

Studies of the genetic material by induced mutagenesis under controlled experimental conditions can therefore provide detailed information about the mechanisms and consequences of spontaneous mutation.

## Induced Mutagenesis

In 1927 H.J. Muller reported that X rays induced mutations in *Drosophila,* and in 1928 L.J. Stadler reported similar results using barley plants. Up to that time, only occasional spontaneously occurring mutants were available for genetic studies. The discovery of a useful mutagenic agent was therefore of great practical importance. Even more significantly, it opened the way to studies of the molecular nature of the gene. Induced mutagenesis by chemicals was first reported in the 1940s by Charlotte Auerbach. She found that nitrogen mustard, the "poison gas" that had produced such horrible effects in World War I, induced the same kinds of mutations in *Drosophila* that had been found after radiation exposure or had arisen spontaneously. Investigators had made some progress in elucidating the properties of genetic material before 1953, but Watson and Crick's molecular model of DNA permitted a more systematic and primary analysis of the effects of mutagens on genes and a better understanding of the molecular nature of the gene itself.

### 11.5 Mutations Induced by Chemical

There are four general groups of **chemical mutagens** whose modes of action on DNA have been studied intensively: (1) base analogues, (2) chemicals that act directly on DNA bases, (3) alkylating agents that remove purines from DNA, and (4) acridine dyes that cause deletion and addition of bases from DNA.

A **base analogue** is a moleucle whose structure mimics the naturally occurring base so that the analogue may be incorporated instead of the usual base when DNA replicates (Fig. 11.9).

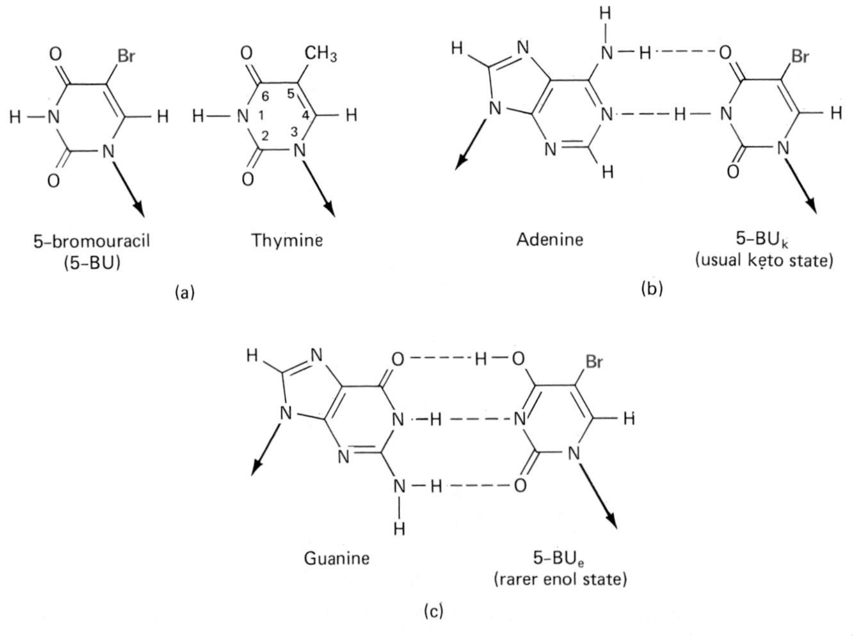

5-bromouracil (5-BU) Thymine (a)

Adenine 5-BU$_k$ (usual keto state) (b)

Guanine 5-BU$_e$ (rarer enol state) (c)

***Figure 11.9*** Pairing properties of the mutagenic base analogue 5-bromouracil (5-BU): (a) 5-BU closely resembles thymine, and (b) pairs with adenine as thymine does; but (c) 5-BU may pair with guanine under certain conditions, such as being in the tautomeric enol state.

Once incorporated, however, the base analogue leads to different pairings so that a base substitution may occur at the next replication of a partner strand opposite the strand that carries the analogue. The mutation results from a base-pair substitution at the site where the analogue was originally incorporated.

For example, 5-bromouracil (5-BU) is similar to thymine (5-methyluracil) except for substitution of a bromine atom for the methyl group in thymine. When 5-BU is present in the DNA strand, it sometimes pairs with guanine, whereas thymine would have paired with adenine. The difference in pairing properties is due to the influence of the bromine atom in effecting a frequent tautomeric shift such that 5-BU may exist in an enol form, leading to altered pairing properties. If we think of thymine and 5-BU as being essentially the same molecule except for the difference at carbon atom 5, we can see how a spontaneous mutation could arise when a normal thymine undergoes a tautomeric shift and a cytosine-guanine pair substitutes for the nonmutant thymine-adenine pair in DNA (Fig. 11.10).

When a purine replaces a purine or a pyrimidine replaces a pyrimidine in the mutational event, such a base-pair substitution is called a **transition;** if a purine replaces a pyrimidine, or vice versa, then a **transversion** is responsible for the mutation.

| **transitions** | **transversions** |
|---|---|
| AT → GC | AT → TA |
| GC → AT | AT → CG |
| TA → CG | CG → AT |
| CG → TA | CG → GC |

In the case of 5-BU, transitions arise in either direction. When 5-BU is incorporated opposite adenine, it may lead to a transition from an AT to a GC base-pair in later replications if 5-BU pairs with guanine. The analogue may cause a GC → AT transition if it is initially incorporated opposite guanine and pairs with adenine at a later replication.

Several other base analogues, such as 5-bromodeoxyuridine, which mimics thymidine, or 2-aminopurine, which mimics adenine, also cause mutations by transitions that arise because of tautomeric shifts and altered pairing proper-

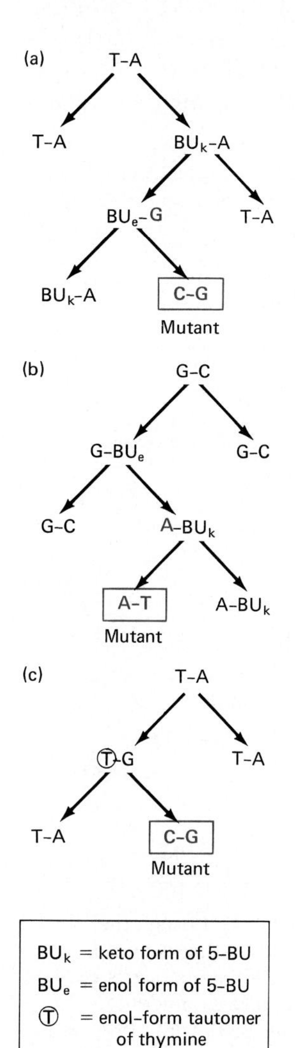


***Figure 11.10*** (right) Origin of mutation during DNA replication. (a) 5-$BU_k$ pairs with adenine and DNA replicates as usual; but 5-$BU_e$ pairs with guanine and leads to base-pair substitution during replication. (b) A transition may arise because of 5-BU pairing in a G-C base-pair, producing an A-T base pair, which is the reverse transition of the one shown in (a). (c) A base-pair substitution of the transition type may also arise if altered pairing takes place when the enol tautomer of thymine appears in the DNA sequence.

ties during DNA replication. All of these mutagens help us understand spontaneous mutations in normal DNA, which arise presumably through altered pairing of tautomers.

*Direct-acting chemicals* act on the existing base structure of DNA rather than being incorporated into the molecule. Nitrous acid ($HNO_2$) is a mutagen that causes oxidative deamination of adenine, cytosine, and guanine (Fig. 11.11). When oxygen replaces the amino group at carbon atom 6, differences in hydrogen bonding properties lead to transitions in either direction. For example, deaminated cytosine pairs like thymine and causes a GC→AT transition, whereas deaminated adenine assumes the bonding properties of guanine and an AT→GC transition results. Since neither thymine nor uracil has an amino group in the molecule, these bases are unaffected by $HNO_2$.

Other chemicals that act directly on DNA bases, such as hydroxylamine or nitrosoguanidine, also cause base-pair transitions. These chemicals are potent mutagens since they are effective in very low concentrations and since they deaminate bases very efficiently. These mutagens can act on replicating or nonreplicating DNA.

*Alkylating agents* are very reactive compounds that can add an alkyl group (such as ethyl or methyl) at various positions on DNA bases and thereby alter base-pairing properties. These mutagens cause both transitions and transversions in either replicating or nonreplicating DNA. The nitrogen and sulfur mus-

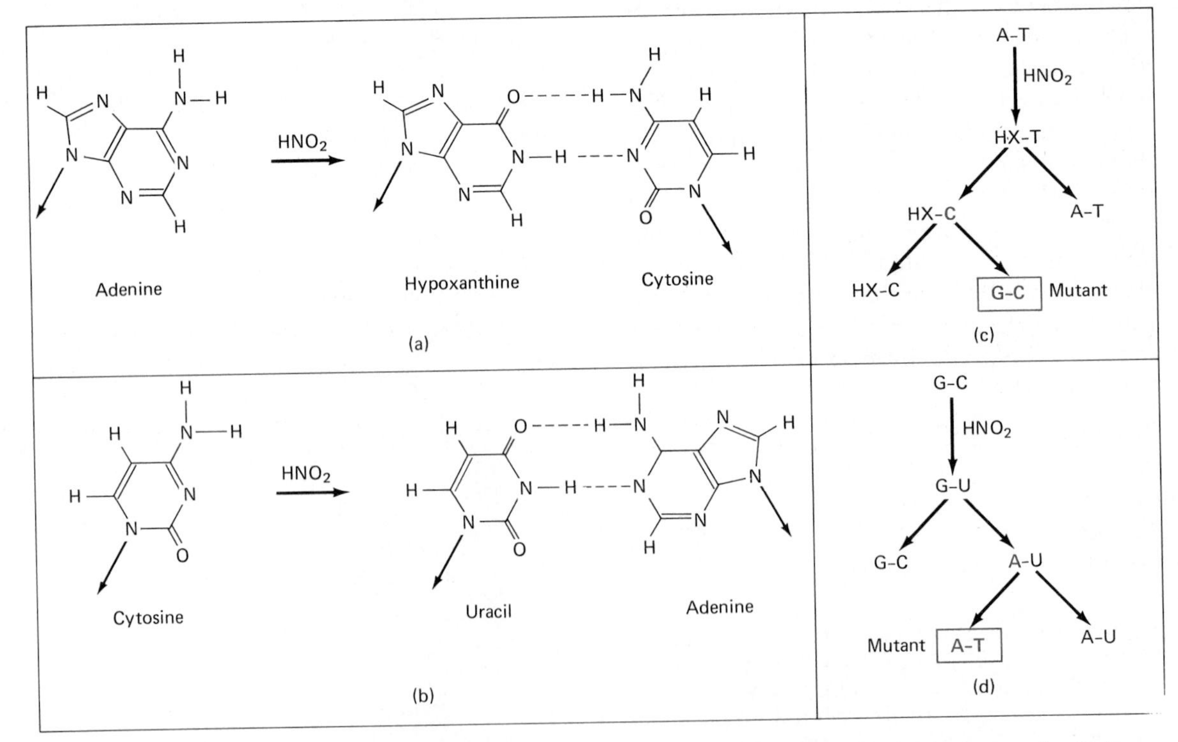


***Figure 11.11*** Induced mutation through the agency of nitrous acid, a direct-acting chemical. Changes are instituted in a DNA sequence as the result of (a) oxidative deamination of adenine, producing hypoxanthine; this pairs with cytosine and not with thymine; and (b) oxidative deamination of cytosine, producing uracil, which pairs with adenine and not guanine. The results of these changes in DNA structure are (c) an A-T → G-C transition and (d) a G-C → A-T transition, respectively.

tards were the first chemical mutagens to be studied, and they were shown to have a delayed mutagenic effect. For example, phage mutants arose several replications after treatment with alkylating agents such as the mustards or ethyl ethanesulfonate. *Drosophila* progeny included flies with variegated or mosaic phenotypes, indicating that the mutational effect arose several generations after the sperm had been treated and only became evident in certain mitotic cell lineages in the progeny flies.

Ethyl ethanesulfonate (EES) and ethyl methanesulfonate (EMS) are less toxic than the nitrogen and sulfur mustards and are frequently used as mutagens. Both EES and EMS act directly on guanine, adding an ethyl or methyl group at carbon 7. This weakens the linkage of guanine to deoxyribose and the guanine is lost from DNA, leaving a gap. Depending on which of the four bases fills in this gap, transitions or transversions may eventually arise (Fig. 11.12). If alkylated guanine remains in the DNA strand, it pairs like adenine and a GC→AT transition can occur. Alkylating agents therefore modify DNA even in nonreplicating molecules, but the effects appear only after subsequent DNA replications, which yield altered codons.

*Acridine dyes* bind to DNA and intercalate (insert themselves) between adjacent bases. They distort the DNA strand at the insertion site and cause either addition of an extra base or deletion of a base when the DNA replicates. Acridines such as proflavine and ethidium bromide are sometimes responsible for frameshift mutations and are very often responsible for gross deletions or complete loss of DNA because bound dye molecules interfere with new strand synthesis during DNA replication. Phages are very sensitive to acridines, as is mitochondrial DNA in yeast and other species. There is some doubt that acridines are mutagenic for eukaryotic nuclear genes since the experimental evidence is either negative or ambiguous.

As we will see in Chapter 13, ethidium bromide is a powerful and very useful mutagen for mitochondrial genes in yeast. If exposure to the agent is prolonged, yeast mitochondrial DNA will be lost completely from the organelles. The mutant yeast can then be used in many kinds of genetic analysis of the mitochondrial genome. By the adjusting of the concentration of the mutagen and the duration of exposure, deletions can be induced in mitochondrial DNA, which will affect the mitochondrial phenotype.

Through chemical, physical, and genetic studies of the action of base analogues on DNA and of the several classes of mutagens that directly modify DNA, we have come to an understanding of the molecular basis for spontaneous mutations that would have been difficult or impossible to obtain from direct investigations of spontaneous mutants. Induced mutagenesis also provides the abundance of mutants that are necessary for genetic analysis.

Seymour Benzer provided one of the most striking demonstrations of gene structure analy-

Guanine — EMS → 7-ethylguanine

(a)

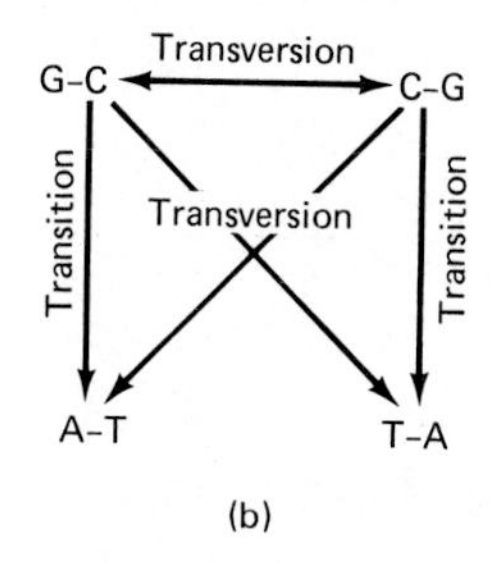


(b)

***Figure 11.12*** Alkylating agents lead to altered base-pairing properties. (a) EMS alters guanine at carbon 7, weakening binding to deoxyribose and leading to loss of the base from the DNA sequence. (b) When the gap left by the lost guanine is filled, transitions or transversions may arise, depending on which of the four bases replaces the guanine in the sequence.

sis using induced and spontaneous mutants. We discussed his analysis of the *r*IIA and *r*IIB genes in phage T4 in Section 7.3. Over two thousand independent mutations in these two adjacent genes were used in crosses to construct fine structure maps for these loci (Fig. 11.13). These studies showed that:

***1.*** numerous sites of mutation occur within a gene, perhaps between 1000 and 1500 for the *r*IIA and *r*IIB genes; some sites are "hot spots" of mutation;

***2.*** the gene map is linear, implying that the gene has a linear construction;

***3.*** most mutations are changes at a single site and are reversible by back mutation.

Mapping the genetic fine structure was a significant accomplishment twenty years ago and provided important insights into the nature of the gene. Investigators have continued to report similar fine structure analysis, although modern experiments are more concerned with the actual sequence of nucleotides within the gene. Both approaches to gene mapping revealed the important feature that numerous mutations, and therefore numerous alleles, can arise in a single gene and produce an altered phenotype.

## 11.6 Radiation-induced Mutagenesis

Radiations at wavelengths too short to be visible to us include cosmic rays, alpha and beta particles, X rays, $\gamma$ rays, and ultraviolet radiation (Fig. 11.14). The most widely used *physical* mutagens are X rays and ultraviolet (UV) light, each of which has a different mode of action on DNA. X rays were the first mutagens to be identified;

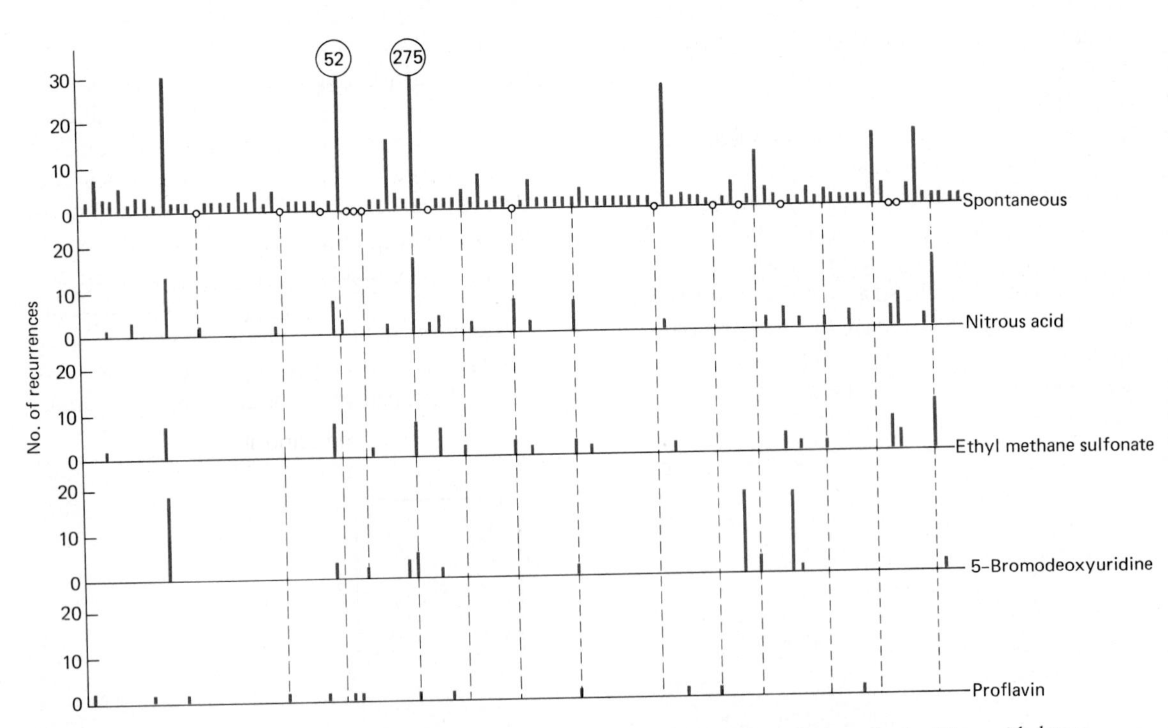


***Figure 11.13*** Sites of spontaneous and induced mutations in the *r*IIB gene of phage T4. Small circles on the base line in the distribution of spontaneous mutations indicate sites known only from induced mutations, primarily after treatment with the acridine dye proflavin. Sites with large numbers of recurrences ("hot spots") are evident, as well as the difference in distributions of mutations induced by different agents. (From S. Benzer. 1961. *Proc. Nat. Acad. Sci. U.S.* **47**:403.)

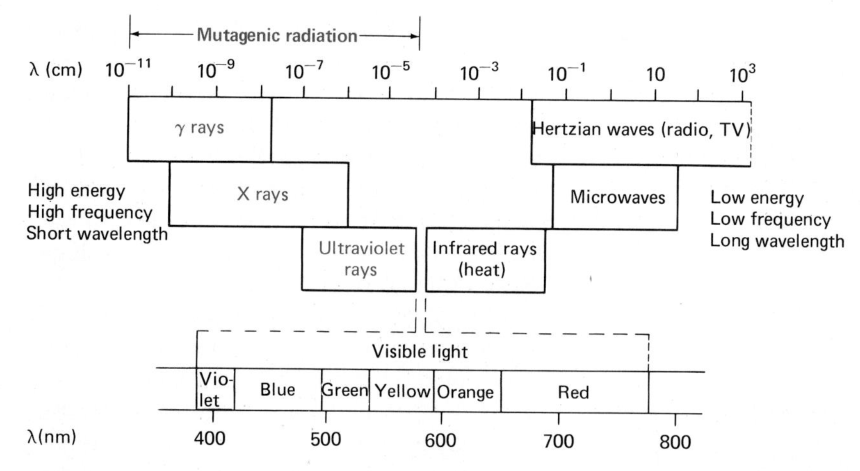


***Figure 11.14*** Electromagnetic radiation. The approximate ranges of radiations are plotted on a logarithmic scale of wavelength (λ) in cm, in the upper part of the diagram. The spectrum of visible light, however, is shown in nm on an arithmetic scale of wavelengths.

although we have studied X rays since the late 1920s, we know far less about their mode of action than we do about UV. The main reason for this is that UV irradiation effects can be studied more easily, using biochemical methods to complement genetic analysis. In fact, we have acquired an extensive understanding of enzymatic mechanisms of UV damage and its repair by an ever-increasing repertory of enzymes and proteins.

**X rays** are *ionizing radiations,* that is, they cause an electron to be ejected from the atom, thus producing ionized atoms, radicals, and molecules. These ionized substances are often highly reactive in cell chemistry and undoubtedly cause a variety of effects that lead to very complex changes in irradiated systems. X rays generally produce gross chromosomal damage, such as breaks and rearrangements of the DNA. They also apparently cause point mutations, according to studies of inheritance patterns of X-irradiated mutants. Some interpretations of X-ray mutagenesis results are conflicting, however, because some evidence indicates that many apparent point mutations may actually be tiny deletions of genetic material, which produce phenotypic effects similar to those caused by base changes in DNA.

The *dosage,* or amount of X rays received by the organism, is measured in **roentgens** (*r*). One roentgen is defined as the amount of radiation that yields $2.08 \times 10^9$ ion-pairs per cubic centimeter of air under standard conditions of temperature and pressure. In a biological perspective one roentgen produces two ionizations per cubic micron of tissue or water. It has been estimated that in thirty years (one human generation) we each receive a dose of about 3*r* from background radiation and perhaps an additional 0.1*r* as the result of radioactive fallout from testing of atomic weapons in the atmosphere.

There is a directly proportional relationship between X-ray dosage and its effect in inducing lethal X-linked mutations in *Drosophila* (Fig. 11.15). This same relationship is found whether the same total dosage is delivered in bursts for short periods or in small amounts over extended periods of time. We may conclude from this observation that any amount of ionizing radiation is potentially mutagenic since a given dose will yield the same number of mutations whether it is delivered at high intensities for short times or in low intensities over months or years. The effects of ionizing radiations are therefore *cumulative,* at least in some species. In mice, however, the effect from chronic low-dose

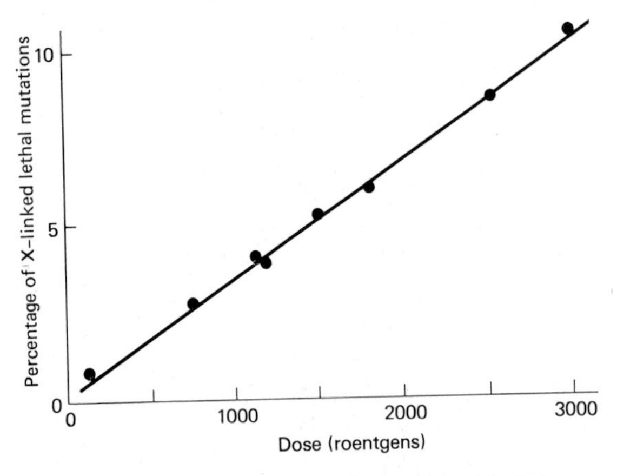


***Figure 11.15*** The percentage of X-linked lethal mutations in *Drosophila* induced by X rays is directly proportional to the dosage of ionizing radiation received by the organism.

radiation seems to be less than from a short, high-intensity exposure to the same total amount of radiation. Apparently, in mice and other mammals, DNA can be healed by repair mechanisms if the damage that has been inflicted at any given time has not been very extensive.

The "target theory" states that the gene is a target that is "hit" by an ionizing particle and that one hit yields one mutation. This theory was proposed on the basis of the observed linear relationship between X-ray dosage and its induction of X-linked lethal mutations in *Drosophila*. As the dosage increases, however, a two-hit or even a multiple-hit relationship appears since the plot yields a curve instead of a straight line. This result is probably due to the fact that more than one mutation is induced but all are recorded only as lethals. The same lethal phenotype will arise whether one or more than one mutation was induced. In the case of chromosomal damage, however, two-hit plots are usually found for radiation doses that cause breaks in chromosomes. These relationships are often complicated because a variety of events may take place between the time of exposure to X rays and the time the phentoypic or chromosomal effect is observed.

**Ultraviolet light** (UV) at the 254 nm wavelength has been used for many years as a sterilizing agent to kill bacteria. UV also induces gene mutations, primarily through photochemical changes in DNA. The wavelength that is germicidal is also the most potently mutagenic, and the two phenomena are related by the fact that DNA absorbs UV more intensively at 254 to 260 nm than at other wavelengths (Fig. 11.16).

The lower-energy wavelengths of the UV spectrum penetrate solid materials rather poorly, unlike the higher-energy (shorter wavelength) X rays and other ionizing radiations, which penetrate solids far more effectively. UV, therefore, is effective only for thin layers of cells or for dispersed cells.

UV induces various kinds of cellular damage, but the best-known effect of UV is its induction of **pyrimidine dimer** formation in DNA, whereby pairs of adjacent pyrimidine bases in a strand become linked together by carbon-carbon bonding (Fig. 11.17). **Thymine-thymine** dimers, abbreviated as $\hat{TT}$, are far more common than cytosine-cytosine or thymine-cytosine dimers. Dimerization causes a distortion, or bulge, in the DNA duplex in the altered region so that hydrogen bonds are broken and cross-linkages form. A large part of the lethal effect of UV irradiation is apparently due to interference with DNA synthesis because of the newly formed cross-links after dimerization. The mutagenic action of UV has been clarified in recent years through studies of mutation repair systems, which we will discuss next.

## 11.7 Mutation Repair

Many years ago, when UV-irradiated bacteria were incubated in order to locate and isolate mutants, it was found that very few mutants appeared in plates that had accidentally been left in the light when compared with the plates incubated in the dark. The phenomenon was called **photoreactivation,** and such mutation repair was later shown to be the result of **DNA repair** by light-dependent photoreactivating enzyme action. The most effective photo-

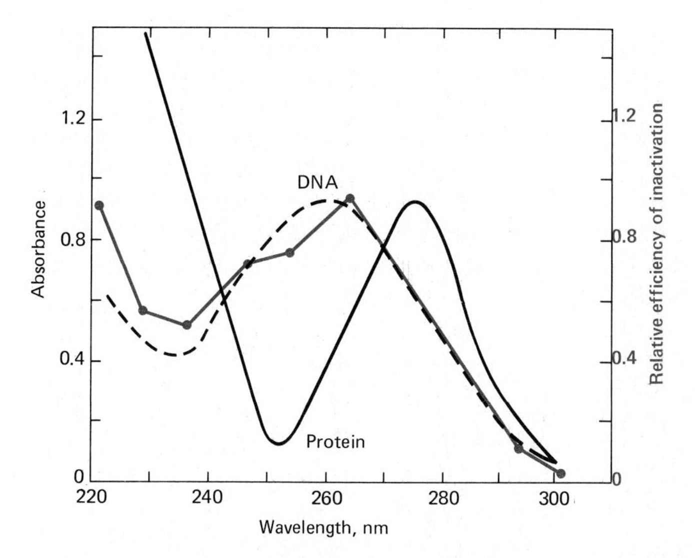


*Figure 11.16* Action spectrum of UV inactivation of phage T2. The action spectrum (color) parallels the absorption spectrum for DNA very closely, but not the absorption spectrum for a typical protein. These results indicate that phage inactivation is a consequence of UV radiation damage to phage DNA.

reactivation wavelengths are in the range between 310 and 400 nm. In this "light repair" process, the photoreactivating enzyme(s) splits the dimer so that single pyrimidines are restored at the altered site.

Photoreactivation is an efficient mutation repair process and is highly accurate; few deaths and few mutants occur after UV-irradiated cells are incubated in the light. Some DNA repair, however, can also go on in the dark. The exis-

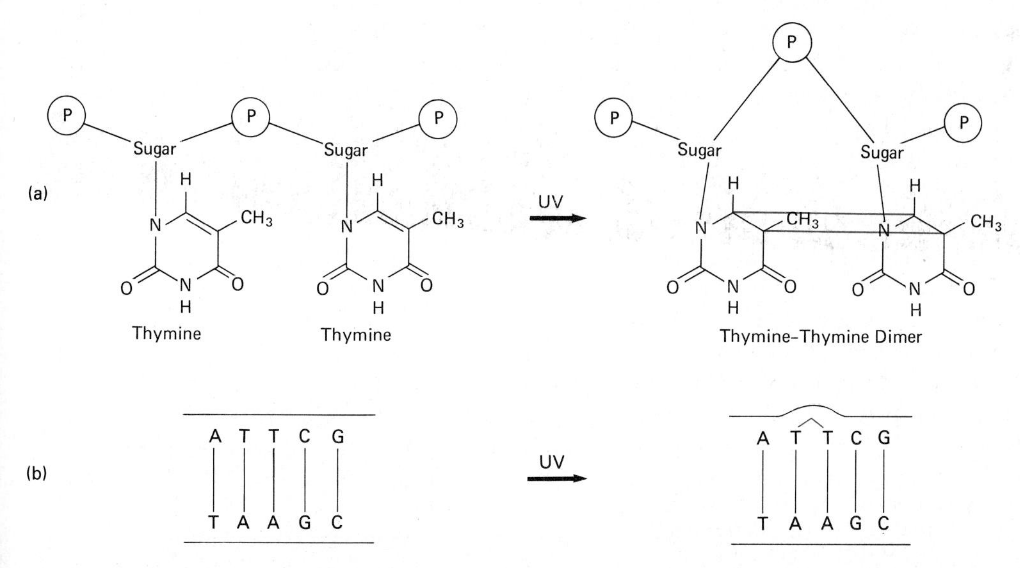


*Figure 11.17* Pyrimidine dimer formation by UV. (a) Thymine-thymine dimer formation through carbon-carbon bonding (color lines) in DNA. (b) Dimerization causes a bulge in the altered region of the DNA duplex, where the dimer $\widehat{TT}$ occurs.

tence of "dark repair" ability was deduced from studies of *uvr* (*u*ltra*v*iolet *r*epair) mutants in *E. coli.* These *uvr* strains give rise to many more mutations after UV irradiation than do $uvr^+$ cells that undergo identical exposure. The gene products of a number of *uvr* mutants (*uvrA, uvrB,* and others) are all needed for efficient and accurate **excision repair** of irradiated DNA in several enzymatically catalyzed steps (Fig. 11.18).

The two UV-repair systems we have mentioned so far are both enzymatically catalyzed, but by different enzymes encoded by different genes. In addition, the process of dimer correction is not the same in the two pathways. In photoreactivation, the dimer itself is split to restore the single-base condition, and the DNA strand regains its original intact state. In excision repair, the entire dimer and some adjacent sequences are removed from the DNA strand, and the gap is accurately patched in by the action of DNA polymerase and then sealed by ligase action. The patch sequence is dictated by the complementary strand opposite the gap. Both repair processes produce undamaged DNA as an end result, and the correction processes are essentially error free: far fewer mutants arise in $uvr^+$ than in $uvr^-$ strains and almost no mutants appear in photoreactivated cell cultures. Multiple errors in the repair pathways would result in significantly fewer surviving cells and significantly more mutants than are typically recovered in irradiation experiments.

The rare recessive hereditary disease called *xeroderma pigmentosum* (XP) is largely due to inefficient repair of UV damage, including defects in the first step in the excision-repair pathway. People with this disease are characterized by a high sensitivity to sunlight, which includes the UV wavelengths, and a very high incidence of skin cancer on those body surfaces exposed to sunlight. An XP patient may have more than 100 skin cancers at one time, including malignant melanomas, which are rare in the general population. Cultures of XP cells from different patients show a heterogeneous assortment of defects in UV-induced mutation repair, but some kinds of chemical agents as well as X rays can be handled in the normal way by these mutant cells.

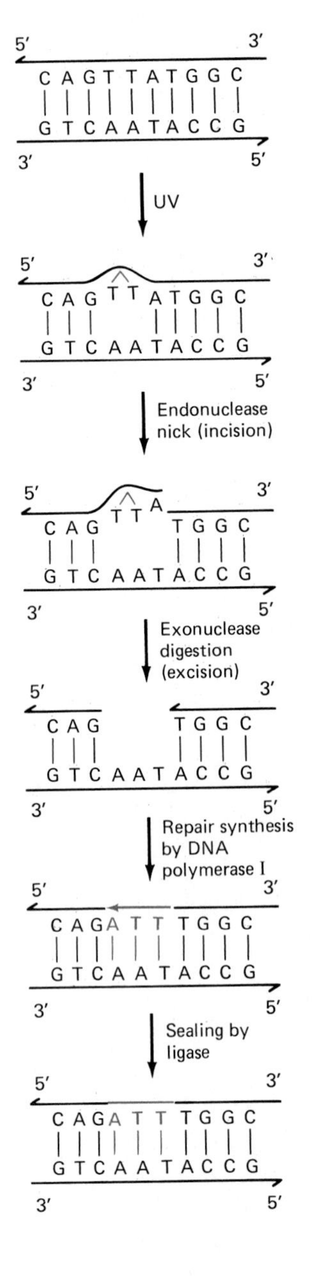


***Figure 11.18*** Excision repair of UV radiation–damaged DNA. Dimers are excised and the accurate repair is catalyzed by DNA polymerase I using the intact strand as the template. After the gap is filled, DNA ligase seals the sugar-phosphate backbone of the strand.

***Table 11.4*** Some damaging agents or products for which genetically repair-defective cells are repair proficient or deficient.

| cell type | proficient | deficient |
|---|---|---|
| Xeroderma pigmentosum | ionizing radiation<br>strand breaks<br>anoxic<br>ethyl methanesulfonate (EMS)<br>proflavin + light<br>mitomycin C | ultraviolet light<br>dimers<br>strand breaks<br>ionizing radiation—anoxic<br>chlorpromazine + light<br>EMS<br>$HNO_2$ |
| Ataxia telangectasia | ionizing radiation<br>strand breaks<br>endonuclease sites<br><br>mitomycin C<br>methyl methanesulfonate (MMS)<br>ultraviolet | ionizing radiation<br>chromosomes<br>survival<br>endonuclease sites<br>mitomycin C<br>MMS<br>actinomycin D |
| Fanconi's anemia | MMS<br>ultraviolet | mitomycin C<br>ultraviolet (high dose)<br>$\gamma$ rays |

From R.B. Setlow. 1978. *Nature* **271**:713–717. Table 1.

These and similar studies indicate that several repair pathways probably exist in mammalian systems by which DNA damage caused by chemical and physical agents can be restored to relatively undamaged and functional DNA. Different kinds of repair deficiencies have also been found in cells cultured from patients with other recessive hereditary diseases (Table 11.4).

A third major enzymatic repair system that has been studied in *E. coli* is **postreplication repair.** In this stystem the pyrimidine dimers are retained rather than being split or excised. An important clue to the existence of this third repair system was the discovery that *uvr* mutants could still carry out UV repair in the dark even when the other two systems were not functioning.

After pyrimidine dimers have formed as a primary lesion of UV irradiation in parental DNA, replication of new daughter strands leads to polynucleotide chains containing large gaps that may be 1000 nucleotides long (Fig. 11.19). These gaps in daughter strands are found at locations that correspond to the dimers in parental DNA strands, and gaps and dimers are found in approximately the same quantities. The gaps are secondary lesions that apparently result from problems in base pairing between parental and daughter strands during replication, due to distortions caused by dimers in the parental strands. Postreplication repair is the process by which these "daughter strand gaps" are filled in by polymerase-directed synthesis. The filled gaps are joined to the main DNA strand segments on either side, probably by ligase action, and a continuous informational DNA is thereby produced.

The pathways by which postreplication repair and other events occur are still under investigation. In UV-irradiated *E. coli,* postreplication repair can be achieved by several distinct pathways. One pathway requires recombination between parental and daughter strands, so that daughter strands ultimately become whole and intact while the dimer remains in a parent strand that may be left behind in the cell. Other postreplicational repair systems do not require recombinational events for the UV-induced damage to be corrected. These pathways may be either error free or error prone, as seen by the

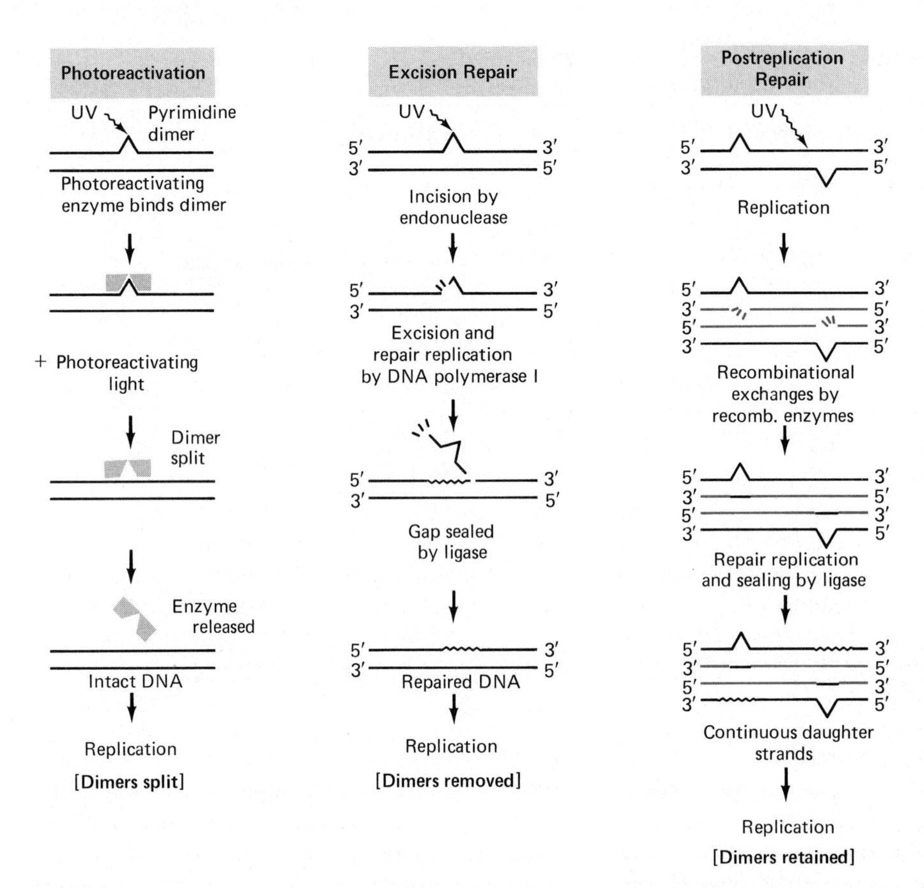


***Figure 11.19*** Three modes of UV repair, each leading to a different specific end result, although repair has been achieved in each case. (From E.M. Witkin. 1976. *Bact. Rev.* **40**:869.)

relative frequencies of mutants obtained after irradiation. Because a variety of specific repair-deficient mutants are available in *E. coli,* it is the best-known system for the complex genetic analyses of radiation repair.

Through biochemical and genetic analysis of UV-induced mutations and mutation repair systems in viruses, bacteria, and eukaryotes, a fascinating story of biological adaptations has emerged. Cells clearly have a variety of mechanisms to clean up after damage from radiation or chemical agents. In the case of UV hazards, it is obvious that life could not exist for very long in the open air and sunlight without efficient, error-free repair systems. Since strong photoreactivating wavelengths arrive along with the dangerous UV wavelengths from the sun, photoreactivation appears to be the first line of defense against UV-irradiation dangers. The example of XP patients who cannot tolerate sunlight provides a graphic demonstration of the problems of species survival that would have been encountered without evolutionary adaptations to overcome UV effects.

Cells must obviously be able to tolerate UV, and it makes evolutionary sense that backup systems have developed that can neutralize harmful UV effects. If one system fails, other systems can take over. These repairs must be relatively error free, otherwise the damage caused by numerous mutations arising during repair would be equally dangerous since most mutations are harmful to some degree. Since a variety of repair mechanisms have been found in even the simplest organisms, genetic controls over the potentially harmful effects of mutagenic agents probably evolved early in the history of life on this planet. These ancient genetic solutions were then transmitted to descendants, along with new and embellished repair systems as evolution proceeded during billions of years.

## Mutagens and Carcinogens

All of the agents that cause cancer are collectively called **carcinogens,** just as mutation-causing agents are called mutagens. Recent studies have shown that some (but not all) mutagens may act as carcinogens and that many (but not all) carcinogens are mutagenic agents. Relationships between mutagenesis and **carcinogenesis,** as phenomena, are similar to the relationships between agents that act as mutagens and those that act as carcinogens. In both kinds of phenomena, a primary effect can be traced to perturbations in DNA. Once a mutation has occurred in a cell, the mutation is transmitted to descendants. Once a cell has been transformed from a normal type into a cancerous type, the transformation is also inherited in descendants of the altered cell. We are still tragically far away from understanding, much less curing or preventing, the many different diseases that come under the single heading of cancer. We have achieved some basic progress, however, from studies of the relationships between mutagens and carcinogens and between the phenomena of mutation induction and cancer induction.

### 11.8 Tests for Mutagenicity

Many carcinogens have also been shown to possess mutagenic activity, and it is of considerable medical importance to identify possible carcinogenic agents in the environment. The *in vivo* tests to demonstrate whether an agent causes cancer are expensive and time-consuming. They require large numbers of laboratory animals, animal care facilities, maintenance, personnel, and other costly items. These tests are essential, but they need not be the *first* series of tests to determine whether or not a substance is carcinogenic. The ideal situation would be to make a quick and inexpensive test showing mutagenicity, and then select such a compound for subsequent tests to determine its carcinogenicity *in vivo.* The Ames test and the Sister Chromatid Exchange (SCE) test are two tests for mutagenicity, among others, that are widely used in preliminary screening for potentially carcinogenic compounds.

The bacterium *Salmonella typhimurium* is the test organism used in the **Ames test,** devised by Bruce Ames. The bacterial strain carries a cell wall mutation that permits most chemicals to enter the cells readily, a *uvr* mutation that abolishes most excision repair, a plasmid carrying some unknown factor that exerts mutator activity in *Salmonella* so that DNA damage is converted into mutations with high frequency, and a genetic requirement for the amino acid histidine. The $his^-$ mutation can be reverted back to $his^+$ by either base substitution or frameshift mutations. A mixture of cytoplasmic ingredients obtained as a cell-free extract from rat liver is also added to the culture dishes. The enzymes in this rat liver fraction can convert test chemicals to other products, some of which may be mutagenic even if the original substance is not. Many carcinogenic chemicals are not harmful until they have been metabolized to other products in the mammalian system, and the rat liver extract can effect these changes in the culture medium. The assay consists of scoring for $his^+$ revertants (or suppressor mutants that cause a pseudowild phenotype), which appear as colonies on minimal media.

The Ames test has been used to test hundreds of chemicals (Table 11.5). In a 1975 summation of data, Ames and co-workers reported that 87% of 179 known carcinogenic chemicals in animals were also mutagenic, whereas 86% of the apparently noncarcinogenic compounds were not mutagenic (101 out of 117 tested). Because of the high correlation between carcinogenicity and mutagenicity, the Ames test has

***Table 11.5*** Correlation between carcinogenicity in animals and mutagenicity in *Salmonella* (Ames test strains).

| category of compounds | carcinogens detected as bacterial mutagens | noncarcinogens not mutagenic to bacteria |
|---|---|---|
| aromatic amines | 23/25 | 10/12 |
| alkyl halides | 17/20 | 1/3 |
| polycyclic aromatics | 26/27 | 7/9 |
| esters, epoxides, carbamates | 13/18 | 5/9 |
| nitro aromatics and heterocycles | 28/28 | 1/4 |
| nitrosamines | 20/21 | 2/2 |
| fungal toxins and antibiotics | 8/9 | 5/5 |
| cigarette smoke condensate mixture | 1/1 | — |
| azo dyes and diazo compounds | 11/11 | 2/3 |
| common laboratory biochemicals | — | 46/46 |
| miscellaneous organics | 1/6 | 13/13 |
| miscellaneous heterocycles | 1/4 | 7/7 |
| miscellaneous nitrogen compounds | 7/9 | 2/4 |
| | 156/179 | 101/117 |

From J. McCann, E. Choi, E. Yamasaki, and B.N. Ames. 1975. *Proc. Nat. Acad. Sci. U.S.* **72**:5135–5139.

excellent *predictive* value. In other words, if a substance is found to be mutagenic there is a very good chance that it is a carcinogen. The substance is tested in animals such as mice to determine whether it will produce tumors. At that stage, the substance is definitely identified as a carcinogen, and foods, cosmetics, or other consumer products containing the carcinogen will be banned by the Food and Drug Administration (FDA) according to law.

Another preliminary screening test involves a comparison between the frequency of **sister chromatid exchange** (**SCE**) in control cells and in cells treated with a test substance (Fig. 11.20). The basis for chromatid exchanges remains uncertain, except that breaks and rejoinings must occur to produce the altered pattern seen in stained preparations. Many mutagens induce chromosomal breakage that leads to structural rearrangements, but we do not yet know whether or not SCEs arise by the same processes. The value of the SCE test is its speed, low cost, and high correlation between increased frequency of SCEs and known mutagenicity of a chemical compound or radiation source.

Cultured mammalian cells are allowed to undergo two rounds of replication in the presence of 5-bromodeoxyuridine (BrdU). The test substance or treatment is provided during the second cell cycle, and colchicine is added to arrest the cells in metaphase of the subsequent mitosis. Colchicine disrupts spindle formation so the metaphase chromosomes are spread out through the cell in a condensed state and can be seen and photographed easily. The chromosomes are stained with ordinary Giemsa stain and examined with the light microscope. The

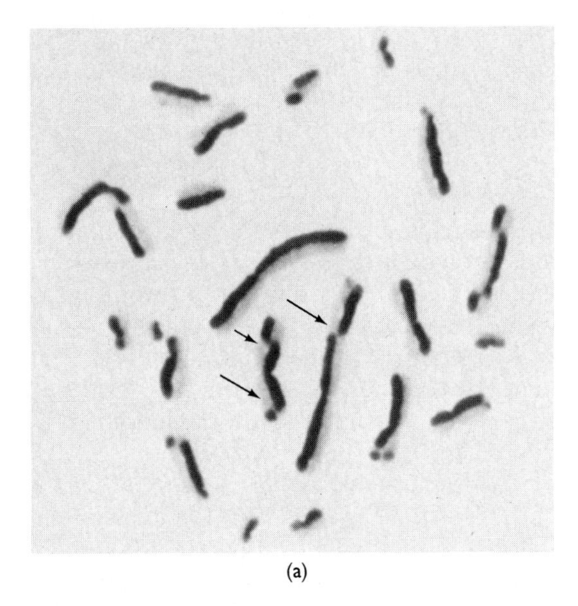

(a)

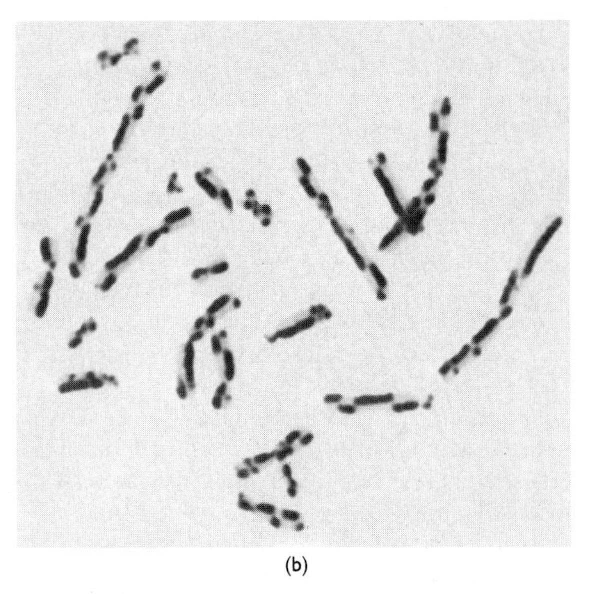

(b)

***Figure 11.20*** Metaphase Chinese hamster chromosomes prepared and stained to show sister chromatid exchanges (SCEs) in chromosomes whose two chromatids are differentially stained: (a) untreated control, with 12 SCEs (3 are shown at arrows) among 20 chromosomes and (b) chromosomes from a cell exposed to the mutagenic agent nitrogen mustard, with approximately a tenfold increase in SCEs compared with the controls. (Reprinted by permission from P. Perry and H.J. Evans. 1975. *Nature* **258**:121–125. Copyright © 1975 Macmillan Journals Limited.)

sister chromatids of a metaphase chromosome stain differently because one chromatid has both its strands of duplex DNA loaded with BrdU, and the sister chromatid, which is produced in the second cell cycle, has only one brominated strand in the duplex. The stain interacts more strongly with the doubly brominated molecules than with the molecules that have only one brominated strand, making the sister chromatids visibly distinct.

Any exchange between sister chromatids will be evident by a change in staining pattern of the pair of chromatids making up a metaphase chromosome. SCEs appear in normal cells. About 5 to 15 SCEs per chromosome complement has been established as the baseline frequency for untreated human lymphocytes in culture. The addition of certain mutagenic alkylating agents, such as nitrogen mustard, to the lymphocyte cultures causes a considerable increase in the number of SCEs per cell at concentrations of the test substances that produce no other visible chromosomal changes. At very high concentrations, these alkylating agents also break chromosomes. Since they seem to be effective at much lower concentrations in the SCE test, the SCE test apparently is a more *sensitive* assay for chromosomal breakage than are tests used previously to detect gross structural rearrangements of whole chromosomes.

One of the difficulties with the SCE test was the uncertainty regarding the amount, if any, of genetic damage that arose in cells with high numbers of SCEs. In other words, did the presence of SCEs indicate that mutations and other genetic effects had been induced, or did it simply mean that chromatid exchanges had occurred but that no genetic or inherited alteration necessarily accompanied SCE increases? In the past few years, stronger lines of evidence have indicated that an increase in the frequency of SCEs is accompanied by an increase in genetic damage and in mutation rate.

Anthony Carrano and co-workers applied the SCE test to Chinese Hamster Ovary cells (CHO cells) that were resistant to the drug 8-azaguanine. The resistancy was due to mutations at the X-linked locus for the enzyme hypoxanthine-guanine phosphoribosyl transferase (HPRT). Cells were exposed to one of four mutagens, and the results were similar for all four experiments. Each mutagen that caused increased mutation rate from resistancy to sensitivity also increased SCEs. The frequency of induced SCEs was linearly related to the frequency of induced mutations (Fig. 11.21). A number of conditions must be established, however, before the SCE test system for CHO cells can be used as a simple and direct quantitative assay of mutagenicity. We must determine whether CHO cells can behave *in vitro* and *in vivo* in a similar manner and whether CHO results can be extrapolated to human cells. As with the Ames test, agents that are identified as mutagens by the SCE test can be considered prime suspects for carcinogens. Such substances can then be tested with live animals for their ability to sponsor tumors.

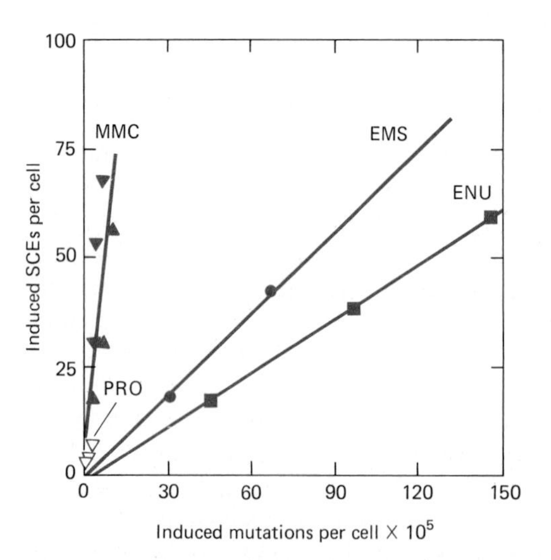


***Figure 11.21*** There is a linear relationship between the frequency of induced sister chromatid exchanges and the frequency of induced mutations in Chinese hamster ovary cells in culture after exposure to a mutagen. Four mutagens were tested: MMC (mitomycin C, a cross-linking agent), PRO (proflavin, an intercalating agent), EMS (ethyl methanesulfonate, an alkylating agent), and ENU (*N*-ethyl-*N*-nitrosourea, an alkylating agent.) (Reprinted by permission from A.V. Carrano et al. 1978. *Nature* **271**:551. Copyright © 1978 Macmillan Journals Limited.)

## 11.9 Cancer and DNA Perturbations

There is relatively little direct evidence showing that modifications in DNA can initiate **cancer,** that is, cause (1) uncontrolled growth, (2) invasion of other tissues, and (3) spread to distant sites (metastasis) where new tumors can be initiated. However, indirect lines of evidence suggest that much if not most carcinogenesis is the result of changes in DNA.

Cancer cells divide to give rise to other cancer cells like themselves. This fact immediately suggests that chromosomal DNA is responsible for the heritability of the cancer phenotype or at least that gene action has been altered in the environment of the cancer cell and that as long as the altered environment is maintained the altered gene action will be maintained. That gene action rather than gene structure is altered has been well demonstrated in one particular cancer by several groups working with rare teratocarcinoma tumors in mice. These tumors, called embryoid bodies, arise in the gonads and contain a large variety of differentiated and undifferentiated tissues in an unorganized mass. If certain embryonal carcinoma cells are removed from the teratocarcinoma and injected into mouse blastocysts, some blastocysts proceed through normal embryo development and give rise to genetic mosaic (chimeric) mice with normal cells derived from embryonal carcinoma cells (Fig. 11.22). If gene structure had been altered, all descendants from parental tumor cells would be tumor cells; whereas gene action is subject to environmental influences and is reversible when the environment changes.

As we described in the preceding section, most carcinogens have mutagenic activity. The induction of cancer by physical and chemical agents is thus linked to alterations in DNA. The nature of the genetic changes that lead to the cancerous state, however, remains unknown. It seems unlikely that changes in a single gene

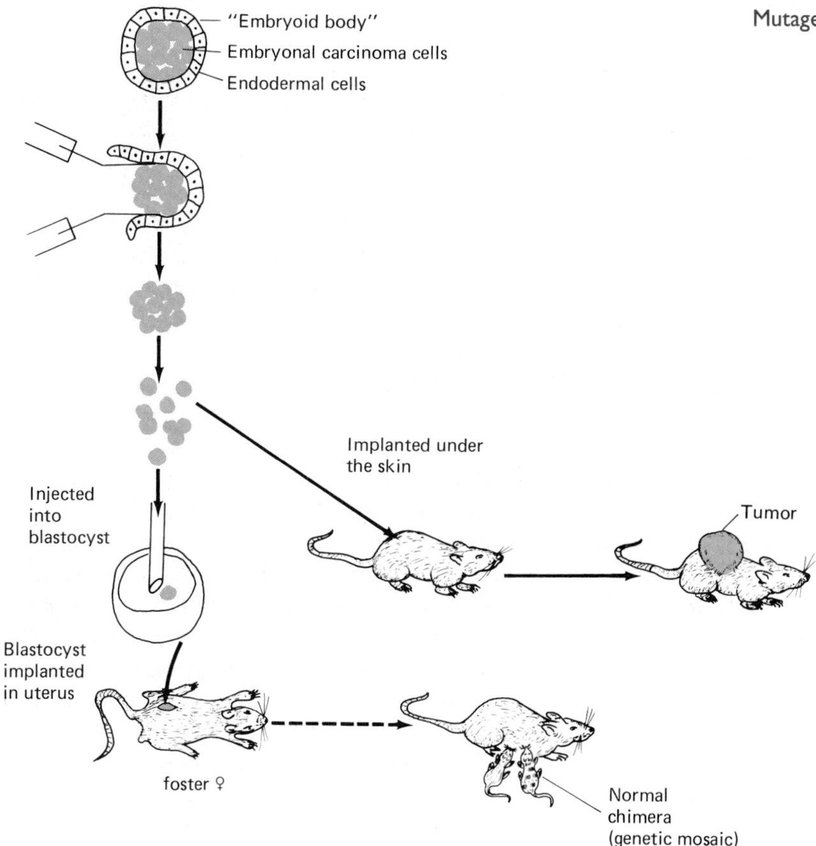


***Figure 11.22*** Demonstration that gene action rather than gene structure has been altered in embryoid-body tumors in mice. Embryonal carcinoma cells can be injected singly into mouse blastocyst and these can give rise to genetic mosaic, or chimeric, mice containing normal cells derived from the cancer cell. The cancerous nature of the embryonal carcinoma cell is shown in control mice, which develop subcutaneous tumors from such cell implants. Reversible gene action may explain these results; the same cell type behaving differently in different environments.

could be responsible. In addition, DNA alterations are difficult to analyze because only some exposed cells become cancerous while many others in the same population remain unaffected, or the organism succumbs if it suffers much cell death due to various other causes. Some carcinogens provoke changes in cell behavior only after they have been metabolized to one or more products. For example, nitrates and nitrites (food preservatives) are harmless until they are metabolically converted to powerfully mutagenic and carcinogenic *nitrosoamines* in the body (Fig. 11.23). Many known carcinogenic-mutagenic agents, such as X rays, UV, nitrogen mustard, and others, cause base substitution and frameshift mutations, chromosome breaks, and other genetic damage. But whether any of this genetic damage leads to the induction of cancer is still uncertain.

Three recessively inherited human diseases—xeroderma pigmentosum (XP), ataxia telangiectasia (AT), and Fanconi's anemia (FA)—show defects in mutation repair. The symptoms of these diseases are very different and the repair defects also are different (see Table 11.4). Each of these diseases, however, makes the patient cancer prone. It would seem, therefore, that damage to DNA can be carcinogenic and that there is a causal connection between mutagenic and carcinogenic agents.

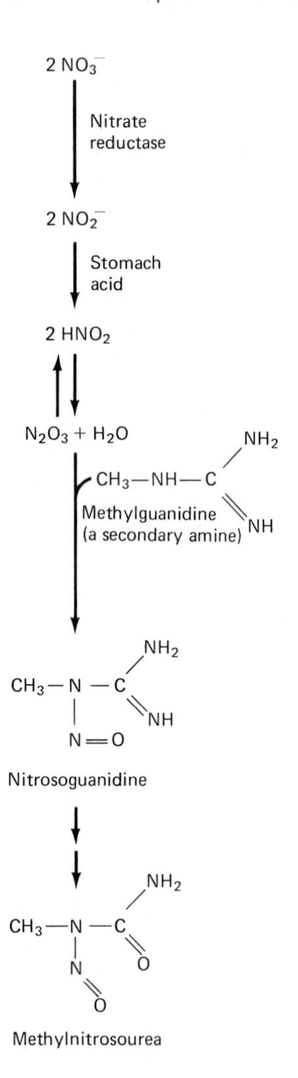


***Figure 11.23*** Conversion of harmless food preservatives to powerful nitrosamines, which may induce mutagenic or carcinogenic changes in the body.

The clinical symptoms of these disorders are characterized by extreme sensitivity of the skin to sunlight in XP patients, problems with motor coordination and immune systems functions in AT patients, and blood disorders in FA patients, which lead to death as the result of hemmorrhage or other blood system failure. Approximately one in several hundred thousand births produces a child with any one of these rare diseases. One in every several hundred people, however, may be heterozygous for the diseases. XP and AT heterozygotes are not cancer prone, but risk of leukemia is higher in FA heterozygotes than in people who are homozygous for the dominant normal allele. This inference is based on data that about 5% of all people dying from acute leukemia are FA heterozygotes, which is far higher than the heterozygous representation in the general population.

In addition to seeking relationships between repair deficiencies and proneness to cancer in other hereditary disorders, experiments have been done in attempts to correlate *ageing* with changes in DNA repair. The results are difficult to interpret directly, and comparisons among the different experiments are also difficult to make. The connection remains plausible, but the hypothesis is supported by little evidence. The same can be said for the hypothesis that **somatic mutations** lead to ageing, cancer induction, and other medical problems of the aged. The idea that body cells accumulate mutations over the years or that the cellular and extracellular environments undergo changes that induce mutations leading to cancer and other problems in old age is unproven, but this idea recurs regularly as a hypothesis to be tested. Basic problems in designing appropriate experiments and test systems are numerous, not the least of which is our poor knowledge of the normal sequences of developmental events during a lifetime. If we do not know the normal kinds of cellular changes taking place, it is hard to detect the abnormal, much less relate some observed difference to diseases of such a varied nature as cancer.

All of these lines of study are bringing us closer to a better understanding of normal and abnormal development of cells, tissues, and organisms. Unrepaired damage to DNA clearly seems to have a high carcinogenic potential, but it seems unlikely that all cancers arise from defects in the repair of DNA. Nevertheless, we can now examine the rate and processes of DNA repair as one parameter in the steps leading to carcinogenesis. A number of cancer-causing genes, called *oncogenes,* have recently been described. These genes will be discussed in Chapter 14. Progress may seem slow, but it is steady.

## Questions and Problems

***11.1*** L.J. Stadler compared spontaneous mutation rates of seven genes in corn, with the following results:

| gene | number of gametes tested | number of mutations |
|---|---|---|
| *R* | 554,786 | 273 |
| *I* | 265,391 | 28 |
| *Pr* | 647,102 | 7 |
| *Su* | 1,678,736 | 4 |
| *Y* | 1,745,280 | 4 |
| *Sh* | 2,469,285 | 3 |
| *Wx* | 1,503,744 | 0 |

What was the spontaneous mutation rate, expressed in mutations per $10^6$ gametes, for each of the seven genes?

***11.2*** A study revealed that among 735,00 children born to normal parents in a certain country, 14 of these children were brachydactylic (a dominant trait involving shortened fingers and toes). Estimate the mutation rate to brachydactyly.

***11.3*** In corn the gene for aleurone color has a spontaneous mutation rate of $11 \times 10^{-6}$ and the gene for color of the corn plant has a spontaneous mutation rate of $492 \times 10^{-6}$. What is the probability of a wild-type plant producing a gamete with mutations at both of these gene loci?

***11.4*** The percentage mutations induced by X rays has been shown to be directly proportional to dosage. If 50 mutants were detected among 500 progeny of males that received 1000 roentgens (*r*) and 80 mutants among 400 progeny of males that received 2000 *r*, what percentage of mutants would be predicted to appear among progeny of males that receive 3000 *r*? Plot the three values for percentage mutants induced as a function of X-ray dosage.

***11.5*** The following results were obtained in a ClB test to detect X-linked mutations in *Drosophila* males irradiated with X rays (refer to Fig. 11.3 for method):

| | phenotypes of $F_2$ progeny | | | |
|---|---|---|---|---|
| cross | number of bar eye females | number of wild-type females | number of wild-type males | number of mutant males |
| 1 | 100 | 100 | 100 | 0 |
| 2 | 100 | 100 | 0 | 0 |
| 3 | 100 | 100 | 25 | 0 |
| 4 | 100 | 100 | 0 | 100 |
| 5 | 100 | 100 | 0 | 30 |

What was the effect of irradiation on the original male parent in each cross?

***11.6*** Self-fertilized pale green barley plants heterozygous for a gene determining chlorophyll production yield twice as many pale green progeny as full green progeny. Explain.

***11.7*** In the fluctuation test designed by Luria and Delbrück, bacteria were plated onto media containing the virulent phage T1 to isolate T1-resistant bacterial mutants. Had they used phage lambda, which is temperate, how would the experimental results have been different from those observed for T1?

***11.8*** You wish to isolate a bacterial strain that cannot produce either leucine or tryptophan. How would you select for double mutants?

***11.9*** In each of three different mutants, glycine at position 66 in the protein is replaced by cysteine, aspartic acid, and alanine, respectively. The codons for these three amino acids include respectively, UGU, GAU, and GCU. If each mutation only involves the change of a single base, what must be the codon for glycine at position 66 in the wild-type protein? Explain.

***11.10*** A wild-type strain of *E. coli* has histidine as amino acid 30 in a polypeptide chain, and a mutant induced in this strain is found to have glutamine substituted as amino acid 30. When induced to mutate again, the mutant strain produced only wild types with His at position 30, plus the original mutant type.

***a.*** Assuming the simplest case of a single-base substitution in the original mutant and its wild-type revertants, what bases in the mRNA codon remained unchanged in the mutants induced to mutate again?

***b.*** How would you explain the appearance only of wild-type revertants in the mutant induced to mutate again?

***11.11*** A mutation occurs in structural gene *Z* of the *lac* operon (*lacZ*). What would you expect if this was a nonsense mutation? a missense mutation? a frameshift mutation?

***11.12*** Suppose you have determined the partial amino acid sequence of a wild-type protein in *E. coli* to be—Pro-Trp-Ser-Glu-Lys-Cys-His—. You recover a series of mutants that have lost the function performed by this protein, and you find the mutant proteins to include the following partial sequences:

mutant 1 —Pro-Trp-Arg-Glu-Lys-Cys-His—

2 —Pro—

3 —Pro-Gly-Val-Lys-Asn-Cys-His—

***a.*** What is the molecular basis for each of these three mutations?

***b.*** What is the DNA sequence for the wild-type protein in this particular region of the gene?

***11.13*** The wild-type coat protein of tobacco mosaic virus (TMV) contains proline at position 20. Treatment with nitrous acid, which is known to deaminate cytosine to uracil, produced variants with amino acid substitution at this position in the polypeptide, according to the following scheme:

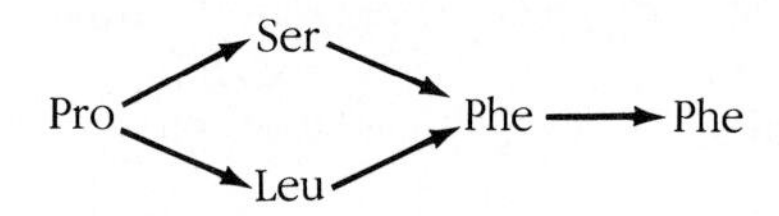


Treatment of the ultimate Phe-mutant using nitrous acid cannot induce further amino acid substitutions at this position.

***a.*** List all possible codons for these four amino acids (using the coding dictionary in Fig. 7.15), and identify the most probable codons for these amino acids based on the assumption that each mutational change shown above was due to a single-base substitution.

***b.*** What kinds of mutational changes were induced by nitrous acid treatment?

***11.14*** Each of four strains of *Neurospora crassa* have a single-gene mutation that results in the deficiency of a different enzyme in a biosynthetic pathway. Each strain can grow in minimal media supplemented with substances A, B, C, D, or E, but cannot grow in minimal media alone. The results of these assays are tabulated below (+ means growth, 0 means no growth):

| | medium supplement | | | | |
|---|---|---|---|---|---|
| **strain** | **A** | **B** | **C** | **D** | **E** |
| 1 | + | 0 | + | + | 0 |
| 2 | 0 | 0 | + | 0 | 0 |
| 3 | + | + | + | + | 0 |
| 4 | + | 0 | + | 0 | 0 |

***a.*** What is the sequence of reactions in the biochemical pathway?

***b.*** Identify the step in the pathway at which each mutant strain is blocked.

***11.15*** Three strains of *E. coli* have been isolated. Each strain is mutant for a different enzyme active in a particular metabolic pathway. When supplements of growth factors A, B, C, and D are added to the media, the bacteria respond by growth (+) or no growth (0). Using the results tabulated below, diagram a pathway that could explain these data and show the steps in the pathway at which each mutant strain is genetically blocked.

| | growth factor added | | | |
|---|---|---|---|---|
| **strain** | **A** | **B** | **C** | **D** |
| 1 | 0 | 0 | + | + |
| 2 | 0 | 0 | 0 | + |
| 3 | + | 0 | + | + |
| 4 | 0 | + | + | + |

***11.16*** Four strains of *E. coli* have been found to have a photoreactivation system for repair of ultraviolet-damaged DNA in the light. To determine whether excision repair and postreplication repair systems are also present in these strains, they were incubated in the dark for a long period after UV irradiation. The results of the assays performed afterward are as follows (0 = none, + = low, +++ = high):

| **strain** | **growth** | **mutation increase** | **thymine dimers remaining in DNA** |
|---|---|---|---|
| (1) $uvr^+$ $rec^+$ | +++ | 0 | + |
| (2) $uvr^+$ $rec^-$ | +++ | 0 | 0 |
| (3) $uvr^-$ $rec^+$ | +++ | + | +++ |
| (4) $uvr^-$ $rec^-$ | +++ | +++ | 0 |

***a.*** Which strains carry out normal excision repair? Explain.

***b.*** Which strains carry out postreplication repair? Explain.

***c.*** How would you account for the differences between strains 3 and 4?

***d.*** Which gene governs excision repair and which governs postreplication repair capacity?

***11.17*** You have been asked by a community to determine whether their drinking water supply is contaminated by pollutants with mutagenic activity. The Ames test is your method of choice for the task.

***a.*** How will you conduct the test?
***b.*** What results would indicate the presence of mutagens in the water?
***c.*** What subsequent test(s) would you suggest to determine whether these same agents are also carcinogenic?
***d.*** Why would you suggest such a follow-up procedure if the only data you have obtained concern mutagenicity?

***11.18*** Define the following terms: ***a.*** mutation rate ***b.*** mutation frequency ***c.*** spontaneous mutation ***d.*** replica plating ***e.*** reverse mutation ***f.*** base analogue ***g.*** transition ***h.*** transversion ***i.*** thymine-thymine dimer ***j.*** photoreactivation ***k.*** mutagen ***l.*** carcinogen.

## References

Allen, J.W., and S.A. Latt, 1976. Analysis of sister chromatid exchange formation *in vivo* in mouse spermatogonia as a new test system for environmental mutagens. *Nature* **260**:449.

Allen, J.W., C.F. Shuler, and S.A. Latt. 1978. Bromodeoxyuridine tablet methodology for *in vivo* studies of DNA synthesis. *Somatic Cell Genet.* **4**:393.

Ames, B.N. 1979. Identifying environmental chemicals causing mutations and cancer. *Science* **204**:587.

Axelrod, D.E., R. Terry, and F.G. Kern. 1979. Cell differentiation rates of Friend murine erythroleukemia variants isolated by sib selection. *Somatic Cell Genet.* **5**:539.

Barrett, J.C., T. Tsutsui, and P.O. Ts'o. 1978. Neoplastic transformation induced by a direct perturbation of DNA. *Nature* **274**:229.

Bunn, H.F., B.G. Forget, and H.M. Ranney, eds. 1977. *Human Hemoglobins.* Philadelphia: Saunders.

Cairns, J. Nov. 1975. The cancer problem. *Sci. Amer.* **233**:64.

Cairns, J. 1975. Mutation selection and the natural history of cancer. *Nature* **255**:197.

Capecchi, M.R., S.H. Hughes, and G.M. Wahl. 1975. Yeast supersuppressors are altered tRNAs capable of translating a nonsense codon *in vitro.* *Cell* **6**:269.

Carrano, A.V., et al. 1978. Sister chromatid exchange as an indicator of mutagenesis. *Nature* **271**:551.

Cleaver, J.E. 1968. Defective repair replication of DNA in xeroderma pigmentosum. *Nature* **218**:652.

Coulondre, C., et al. 1978. Molecular basis of base substitution hotspots in *Escherichia coli.* *Nature* **274**:775.

Demple, B., and S. Linn. 1980. DNA *N*-glycosylases and UV repair. *Nature* **287**:203.

Howard-Flanders, P. Nov. 1981. Inducible repair of DNA *Sci. Amer.* **245**:72.

Illmensee, K., and L.C. Stevens. Apr. 1979. Teratomas and chimeras. *Sci. Amer.* **240**:120.

Klein, G. 1981. The role of gene dosage and genetic transpositions in carcinogenesis. *Nature* **294**:313.

Lederberg, J. and E.M. Lederberg. 1952. Replica plating and indirect selection of bacterial mutants. *J. Bact.* **63**:399.

Luria, S.E., and M. Delbrück. 1943. Mutations of bacteria from virus sensitivity to virus resistance. *Genetics* **28**:491.

McKnight, S.L., and R. Kingsbury. 1982. Transcriptional control signals of a eukaryotic protein-coding gene. *Science* **217**:316.

Muller, H.J. 1927. Artificial transmutation of the gene. *Science* **66**:84.

Perutz, M.F. 1976. Fundamental research in molecular biology: Relevance to medicine. *Nature* **262**:449.

Reif, A.E. 1981. The causes of cancer *Amer. Sci.*. **69**:437.

Setlow, R.B. 1978. Repair deficient human disorders and cancer. *Nature* **271**:713.

Stadler, L.J. 1928. Mutations in barley induced by X rays and radium. *Science* **68**:186.

Stanbury, J.B., et al. 1983. *The Metabolic Basis of InheritedDisease,* 5th ed. New York: McGraw-Hill.

Streisinger, G., et al. 1967. Frameshift mutations and the genetic code. *Cold Spring Harb. Symp. Quant. Biol.* **31**:77.

Waldren, C., C. Jones, and T.T. Puck. 1979. Measurement of mutagenesis in mammalian cells. *Proc. Nat. Acad. Sci. U.S.* **76**:1358.

Witkin, E.M. 1976. Ultraviolet mutagenesis and inducible DNA repair in *Escherichia coli. Bact. Rev.* **40**:869.

Wolff, S., and B. Rodin, 1978. Saccharin-induced sister chromatid exchanges in Chinese hamster and human cells. *Science* **200**:543.

Wolff, S., B. Rodin, and J.E. Cleaver. 1977. Sister chromatid exchanges induced by mutagenic carcinogens in normal and xeroderma pigmentosum cells. *Nature* **265**:347.

Zakour, R.A., and L.A. Loeb. 1982. Site-specific mutagenesis by error-directed DNA synthesis. *Nature* **295**:708.

# CHAPTER 12

# Classical and Molecular Cytogenetics

Cytogenetics is the study of the physical structure of genetic material and its correlations with genetic functions. Studies conducted between the 1920s and 1940s utilized microscopy to study chromosome structure and gene transmission analysis to study genetic functions. Modern cytogenetic studies still rely heavily on correlations observed between microscopic analysis and genetic analysis, but another dimension has been added by the application of molecular methods of analysis. DNA can be studied by electrophoresis of restriction fragments, and genetic functions can be studied by biochemical assays of proteins and other gene products. Regardless of the specific methods used, the principles of seeking independent evidence of some phenomenon by finding correlations between the physical structure and the genetic functions of genetic material still underlie cytogenetic analysis.

## Changes in Chromosome Number

Eukaryotic organisms have the haploid or the diploid chromosome number in nuclei of different cells in different phases of the sexual cycle. Many species have more than one or two sets of chromosomes, and the general term **euploidy** refers to the presence of any multiple of whole sets of chromosomes. Sometimes a species with more than two sets of chromosomes is a typical member of a group of related diploid species, and in other species only an occasional aberrant individual has extra sets of chromosomes.

The presence of at least one chromosome more or one less than the characteristic diploid chromosome number is called **aneuploidy.** In most species an aneuploid individual is different in appearance from the typical diploid individual, and aneuploids characteristically show some degree of infertility or inviability because of chromosomal imbalance.

Studies have shown that changes in chromosome numbers have been important in species evolution. Aberrant euploid and aneuploid types have also been important source materials for studies of genome organization and of certain human disorders. We will review some of these studies in this part of the chapter.

### 12.1 Polyploidy

When a euploid individual or species has more than two sets of chromosomes the condition is called **polyploidy** and the individual or species is a **polyploid.** The actual number of genomes present is the basis for naming particular kinds of polyploids: triploids ($3n$) have three chromosome sets, tetraploids ($4n$) have four sets, hexaploids have six sets ($6n$), and so forth. The fact that about half the known species of flowering plants are polyploids suggests that polyploidy has been a significant factor in the evolution of higher plants. Relatively few bisexual animal species are polyploids, but a number of asexual animals and hermaphroditic species can sustain the polyploid condition.

Polyploids that have an even number of chromosome sets (two, four, six, eight, and so on) are far more likely than polyploids with an

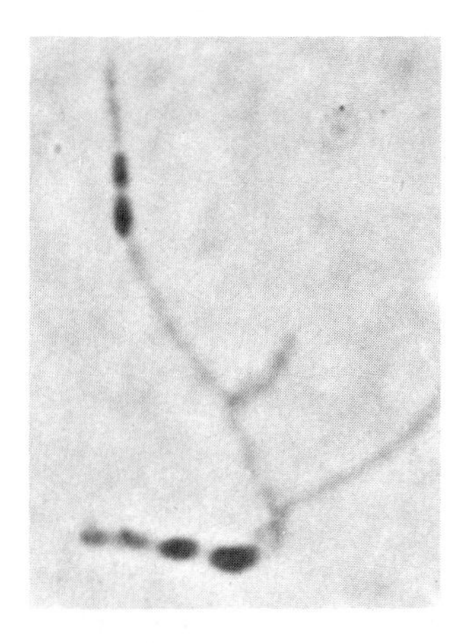

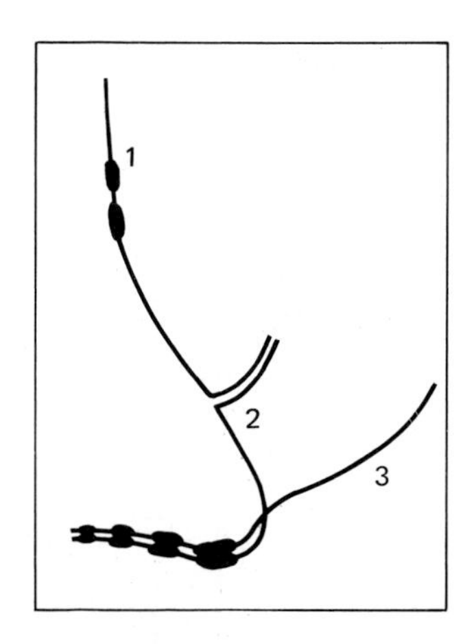


***Figure 12.1*** Pairing is two by two among the three copies of the chromosome in this trivalent from a pachytene nucleus of castor bean, *Ricinus communis.* (Courtesy of G. Jelenkovic.)

odd number of sets to be fully fertile or at least partially fertile. The problems leading to partial sterility are usually confined to meiosis and to synapsis between homologous chromosomes in prophase I. If there are more than two homologues in the meiocyte nucleus, synapsis may involve all the homologues, although each pairing event is confined to two homologue segments in any one region (Fig. 12.1). If the synapsis events lead to associations among all the homologues, then complex meiotic chromosome configurations will be formed and abnormalities will usually arise during the meiotic divisions. A variable number of the homologous chromosomes will be distributed to the gametes, and chromosomal imbalance may cause inviability of the gametes or inviability of a zygote produced by fusions between gametes with unbalanced chromosome numbers.

Sterility is almost ensured when the polyploid has an uneven number of chromosome sets, such as in triploids or pentaploids, because segregation at meiosis will lead to variable numbers of each of the kinds of chromosomes in the gametes. There is some probability in tetraploids and other polyploids with an even number of genomes that an equal distribution of chromosomes may take place at meiosis to produce some gametes with the same number of copies of each chromosome. Some fertility, therefore, can usually be expected in these kinds of polyploids. For example, tetraploids can produce diploid gametes and these diploid gametes can fuse to restore the tetraploid adult stage. Triploids, on the other hand, may give rise to gametes with two copies of some chromosomes and one copy of others, thus producing a highly unbalanced chromosomal constitution in the gametes themselves. Since these gametes usually are not functional, triploids are sterile.

Polyploidy can arise because of a failure of meiosis in a diploid individual, which then produces diploid gametes. When diploid gametes unite, as in a self-fertilizing species, a tetraploid is produced. Such an individual or species is called an *autotetraploid* since four copies of the same genome are present. Any **autopolyploid** has multiple sets of a single genome. If, however, two different genomes come together when different species interbreed and if such a hybrid becomes a polyploid when meiosis fails, it is called an **allopolyploid.** Two copies of each of two different genomes are carried by an allotetraploid, and multiple sets of two or more genomes by any allopolyploid (Fig. 12.2).

Autotetraploids generally experience problems at meiosis because of multivalents that form when more than two homologues synapse

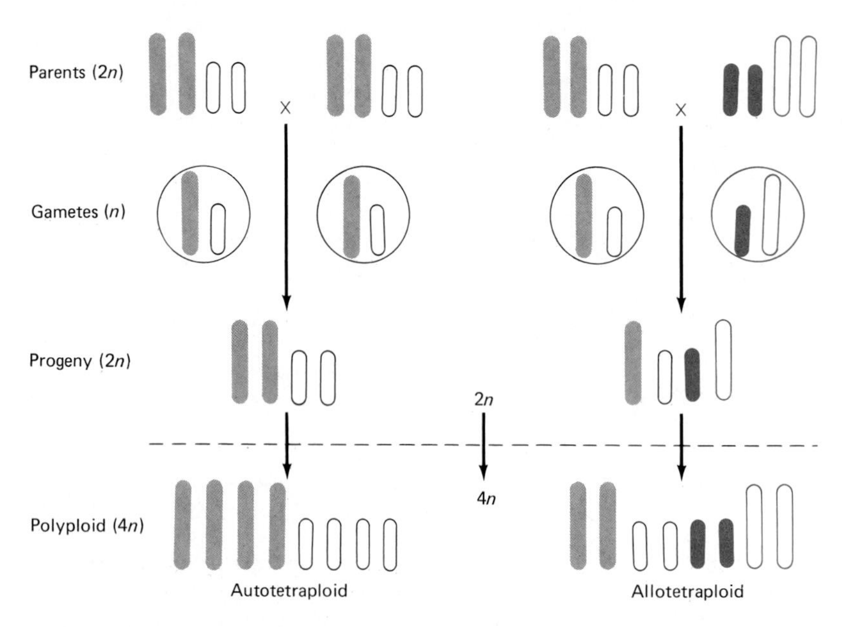


***Figure 12.2*** Origin of the genomes in an autotetraploid versus an allotetraploid.

at various places along their lengths. Distorted segregations lead to inviable gametes or zygotes. Allopolyploids, on the other hand, generally are fertile because each genome has a partner genome and chromosomes generally pair in twos to produce bivalents. Little or no meiotic irregularity should result from such regular synapsis and from segregation of pairs of homologues at metaphase.

Polyploid plants have relatively little physiological disturbance because sex chromosomes are usually absent. Most plants are hermaphroditic anyway. Even if some physiological difficulty was present, many plants can be propagated asexually, so sterile or fertile polyploids may continue to exist for many years. In animals, on the other hand, disturbances in the numbers of sex chromosomes and in sex chromosome-autosome balance are the usual result of breeding between polyploids or between a polyploid and a diploid. These chromosomal imbalances and the disruption of the normal sex-determining mechanisms make it very unlikely for sexual animals to give rise to persisting polyploids. If polyploids are produced, they usually are not perpetuated because of sexual abnormalities. These problems, of course, have little effect on asexually reproducing animals or on animals that are hermaphroditic or that have no genetic sex-determining mechanism.

A number of important agricultural species are polyploid, such as wheat, cotton, tobacco, strawberries, and many of the fruit trees. In some of these species it has been possible to trace back to the probable diploid ancestors from which the polyploids arose. Wheat species are diploid, tetraploid, and hexaploid, but their wild ancestors in Southwest Asia are diploids. Tobacco is a tetraploid species whose diploid ancestors grew in South America, and the tetraploid potato also originated from diploids in South America.

One of the great advantages of polyploidy in plants is the larger size of the plant and of its flowers, fruits, and seeds. There is better quality from the consumer standpoint and much higher yields, which helps everybody from the farmer to the consumer. In the case of ornamental plants, whose desirable features usually are in foliage and flowers, polyploids are created deliberately for larger, more luxuriant, and more decorative plants. In many cases a sterile polyploid, such as a triploid or pentaploid, is most desirable because the flowers are larger and they last longer since no seeds are produced in these sterile plants. Prize orchids are often sterile polyploids. Polyploids can be created at will by applications of *colchicine,* which prevents spindle formation and leads to retention of all the chromosomes within a single nuclear membrane. By the adjusting of drug concentration and duration of exposure, polyploids can be made at any desired level of chromosome sets.

## 12.2 Aneuploidy

*Aneuploids* have at least one more or one less chromosome than the diploid number, but they do not have multiples of chromosome sets. If a diploid has one extra copy of a chromosome, the individual is a **trisomic** and the condition itself is called **trisomy** (three bodies or chromosomes of one kind); the chromosome constitution is shown as $2n + 1$. A diploid with two extra copies of a particular chromosome is **tetrasomic**, shown as $2n + 2$. The condition of having one less than a complete set of chromosomes in a diploid is called **monosomy** and the individual is **monosomic;** it is $2n - 1$. Other specific conditions of chromosome gain or loss have been found and each has been given a specific term for identification. Since trisomics and monosomics are the types encountered most often, we will discuss only these.

Aneuploids usually arise by nondisjunction of homologous chromosomes at meiosis or by nondisjunction of sister chromatids at mitosis (Fig. 12.3). The failure to disjoin or separate accurately can occur at any nuclear division, and its consequences vary according to the division in which the event occurs and the time of occurrence. Nondisjunction at meiosis gives rise to gametes with one more or one less chromosome than usual. If such gametes are viable and fuse to produce a zygote, the zygote will be trisomic or monosomic for the nondisjoined chromosome. Nondisjunction may involve any chromosome of the complement.

We have already discussed the consequences of human aneuploidy involving the sex chromosomes (see Section 4.4). You may recall that nondisjunction can give rise to viable aneuploid individuals who may be monosomic (XO, $2n - 1$), trisomic (XXX, $2n + 1$; XXY, $2n + 1$), or tetrasomic (XXXX, $2n + 2$; XXXY, $2n + 2$; XXYY, $2n + 1 + 1$). Each of these conditions probably arose as the result of meiotic nondisjunction to produce a gamete containing one more or one less chromosome or an even greater imbalance. In some cases an individual may be a *sex mosaic*, having patches of tissues carrying different sex chromosome complements. This situation is almost certainly due to mitotic nondisjunction, with different cell lineages produced by chromosomally different daughter cells after the nondisjunctional event. The later the nondisjunction, the smaller the patches of aberrant tissue since fewer divisions and fewer descendant cells would be produced late in development.

Autosomal aneuploidy in human beings has been described for several specific chromosomes, but the most common is **trisomy-21** in which chromosome 21 is present in three copies (Fig. 12.4). Such an individual has 47 chromosomes and shows the clinical symptoms of Down syndrome. Other trisomies are lethal either early in development or in childhood. No cases of monosomy have been reported, except for monosomy of the X chromosome in Turner females, who are XO in sex chromosome constitution. Almost no species can tolerate monosomy and produce a viable individual with a missing chromosome. Monosomy is a serious genetic imbalance, and almost certainly a number of harmful recessive alleles would be expressed on the one remaining chromosome of a pair in a diploid species. An extra chromosome is better tolerated by a diploid species if it is relatively small and, presumably, carries fewer genes. But tolerance varies considerably from species to species.

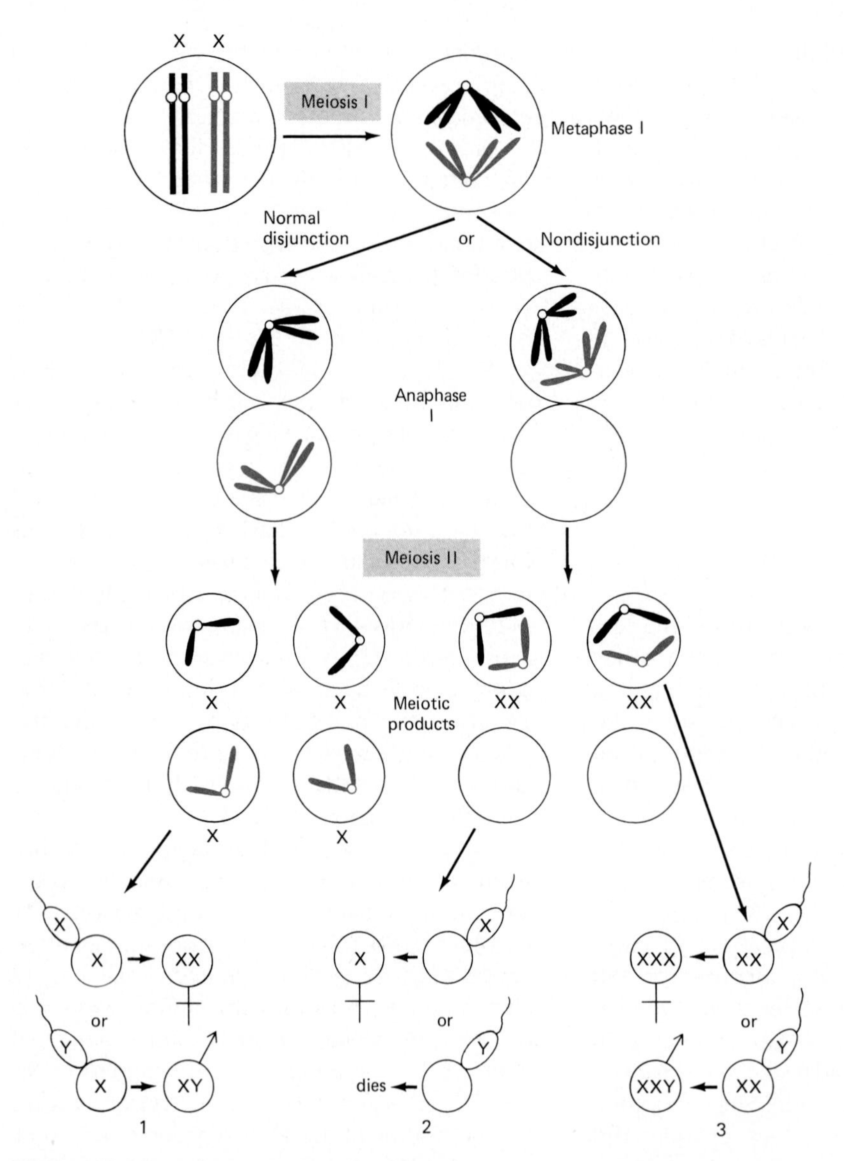


*Figure 12.3* Nondisjunction at meiosis can lead to gametes with too few or too many chromosomes and to aneuploid individuals produced by fusions involving such gametes. Nondisjunction may involve sex chromosomes, as shown here, or any of the autosomes.

Trisomy-21 leads to the nonfamilial pattern of Down syndrome in human beings. The risk of a woman giving birth to a child with this disorder increases with increasing age of the mother (Fig. 12.5). The increase in risk is presumably due at least in part to the increasing age of the oocyte, which was produced long before the woman herself was born. As she ages, the oocytes encounter a variety of changing internal environments. These physiological factors may be responsible for higher probability of meiotic nondisjunction when the oocyte matures.

Since trisomy-21 is an autosomal disorder, it occurs equally often in both sexes. It is characterized by a number of physical abnormalities as well as mental retardation, which may be very

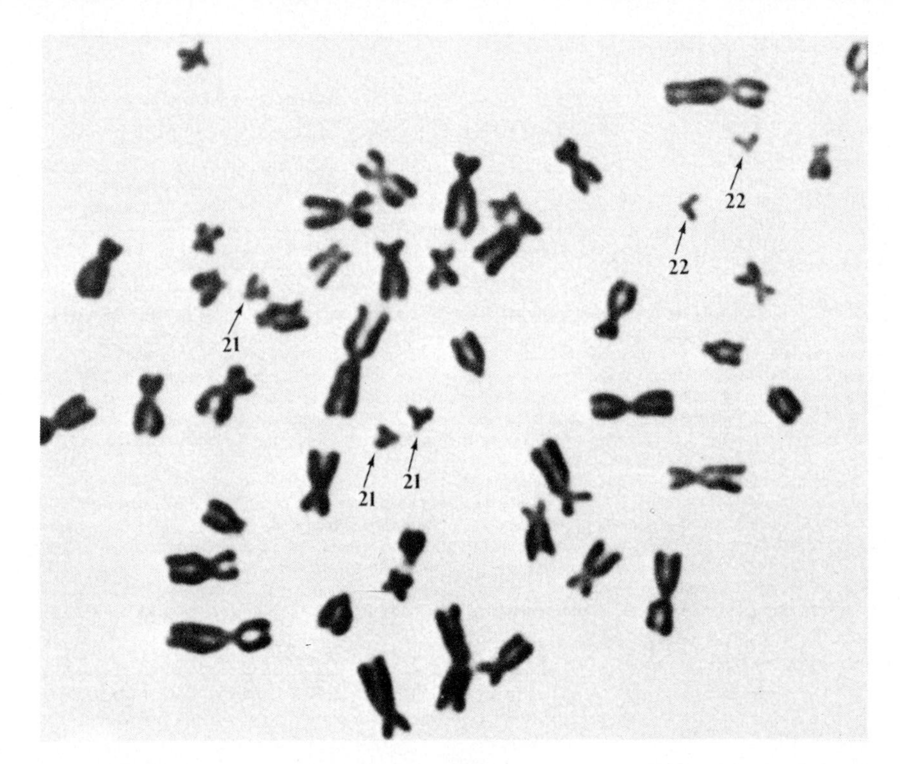


*Figure 12.4* Aneuploid chromosome complement of a female patient with trisomy 21, or Down syndrome. There are five G-group chromosomes instead of four, and a karyotype is not really necessary to see the chromosomal imbalance in this case of $2n + 1$ aneuploidy. (Courtesy of T.R. Tegenkamp.)

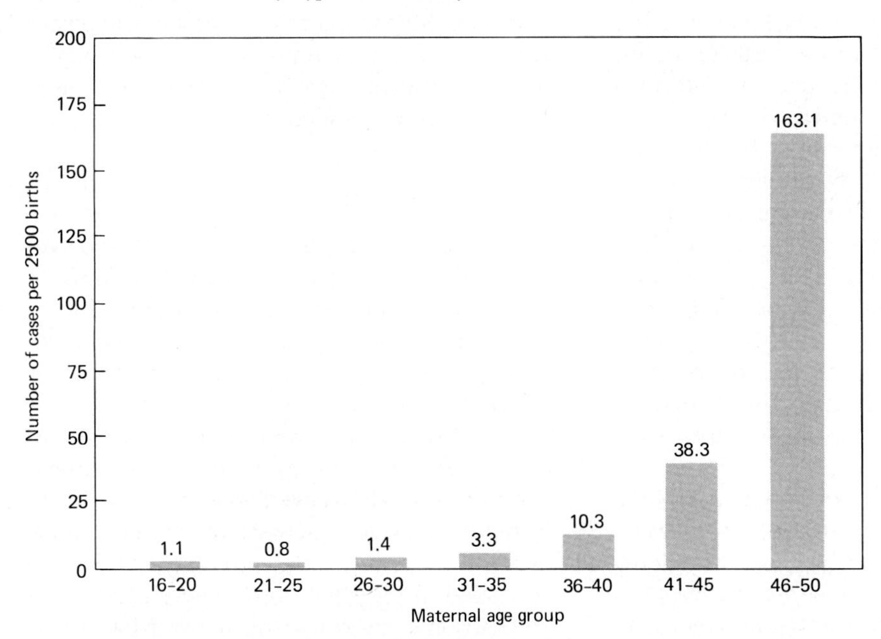


*Figure 12.5* The frequency of children with Down syndrome born to older women is considerably higher than for women in the younger age groups. The risk is particulary high for women in the 46 to 50 age group.

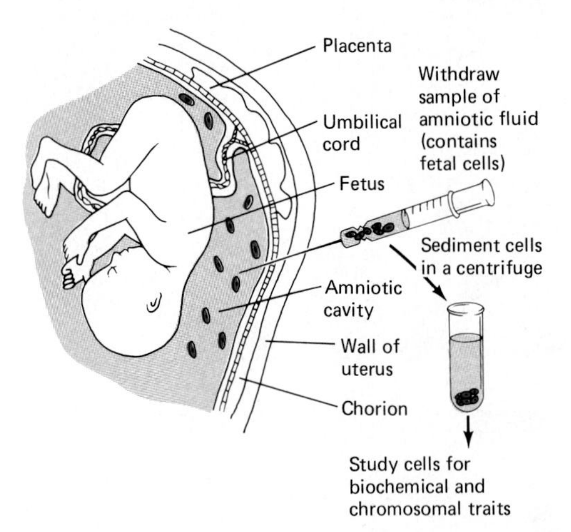


***Figure 12.6*** Amniocentesis. A sample of amniotic fluid is withdrawn from the uterus and is processed for cytological and biochemical analysis of fetal traits.

***Table 12.1*** Frequency of selected aneuploidies in humans.

| aneuploidy | frequency of live-born |
|---|---|
| sex chromosome anomalies | |
| Turner syndrome (45,X) | 1/2500 females |
| triplo-X (47,XXX) | 1/1250 females |
| tetra-X (48,XXXX) | very low (about 20 cases known) |
| Klinefelter syndrome (all types) | 1/800 males |
| XYY male | 1/900 males |
| autosomal anomalies | |
| trisomy-13 | 1/4,000 to 10,000 |
| trisomy-18 | 1/8,000 |
| trisomy-21 | 1/700 |

Data from J. de Grouchy and C. Turleau. 1977. *Clinical Atlas of Human Chromosomes.* New York: John Wiley & Sons.

severe in some cases. The average life expectancy has been lengthened with improved medical care, but relatively few individuals survive into their twenties.

If an older woman wishes to know whether her pregnancy will lead to the birth of a baby with Down syndrome, she may be informed specifically after **amniocentesis** (Fig. 12.6). After a sample of amniotic fluid has been removed from the uterus by the physician, the fetal cells that are shed normally into this fluid can be examined. The presence of an extra chromosome 21 in a nucleus that has 47 chromosomes is direct evidence that the child will be born with the disorder. The prospective parents may then make preparations for this birth or elect to have the fetus aborted. The decision is theirs alone, once they have been apprised of the situation.

The incidence of aneuploidy in human births is relatively high (Table 12.1). Some aneuploidies are lethal early in life, such as trisomy for chromosomes 13 or 18. Some aneuploidies have not been observed so far but may be found as more studies are performed. Down syndrome, the first autosomal aneuploidy discovered in human beings, was not reported until 1959. One reason for this relatively late discovery on human chromosomes and karyotypes is that suitable preparations were not achieved until new techniques of chromosome preparation were developed in 1956.

## 12.3 Centric Fusion

Aneuploidy may also arise without gain or loss of chromatin by the process of **centric fusion,** in which two smaller chromosomes fuse to form one larger one. Usually, two smaller acrocentrics fuse to produce a large metacentric or submetacentric chromosome (Fig. 12.7).

It was originally believed that one of the two centromeres was lost during centric fusion and that the larger derivative had only a single centromere. Electron microscopy revealed, however, that two centromere structures are contained in the single centromere region of a chromosome known to be derived by centric fusion. Some investigators suspected these results on the basis of their observations of various invertebrate species. In some of these populations, individuals had different chromosome

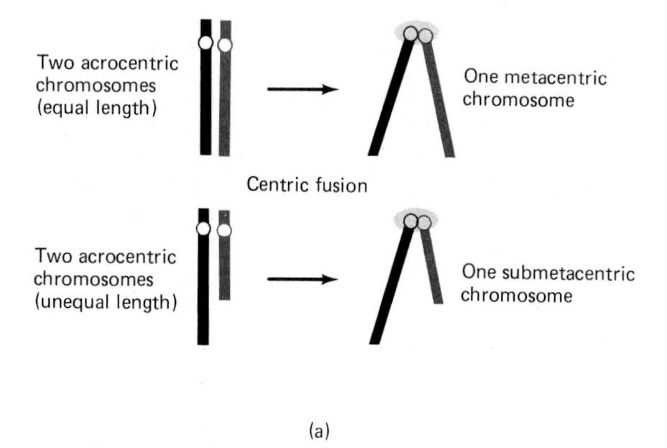


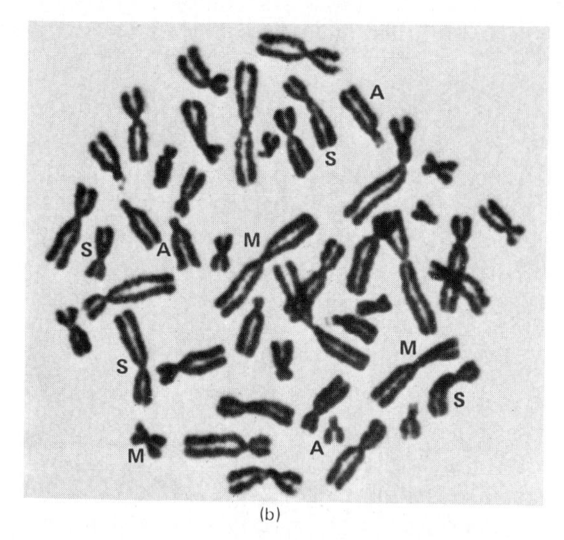


***Figure 12.7*** Centromere location in chromosomes. (a) Centric fusion may lead to one larger metacentric or submetacentric chromosome from two smaller acrocentrics. Both centromeres are usually retained in a common centromere region in the fusion chromosome product. (b) Human chromosomes (metacentrics, M; submetacentrics, S; and acrocentrics, A) in a metaphase cell.

*numbers* but they had the same number of chromosome *arms* in a genome. It has since been found that chromosomes which do undergo centric fusion, reducing the chromosome number but not the number of arms, may also dissociate into acrocentrics. The chromosome number is increased in such individuals but the number of chromosome arms remains unchanged. Centromere retention explains these reversible changes since new centromeres cannot be created from other parts of a chromosome. Neither the mechanism for centric fusion nor the mechanism for dissociation is known.

Centric fusion has played an important role in genome evolution in many animal groups and in some of the flowering plants (Fig. 12.8). The usual trend is toward reduction in chromosome number as species groups evolve and retention of most of the genome in structurally rearranged chromosomes of the complement. In *Drosophila*, the reduction from a haploid complement

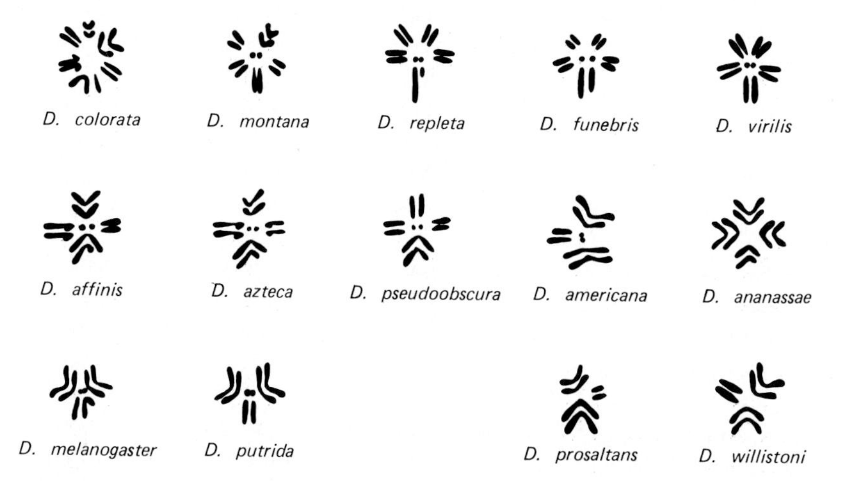


***Figure 12.8*** Chromosome complements of fourteen *Drosophila* species showing the aneuploid series of $n = 6$ to $n = 3$, which has arisen through centric fusions and other events in evolution of the group. The X and Y chromosomes of these male genomes are shown at the bottom of each drawing.

of 6 to species with only 3 chromosomes in a set has been examined in some detail from polytene chromosome band comparisons. From these detailed comparisons, it is clear that particular metacentric chromosomes had been derived from centric fusion of particular acrocentrics during evolution since matching band patterns provided unmistakable evidence of these events.

The evolutionary significance of aneuploidy by centric fusion lies in the alterations in recombination frequencies and in reduced production of new genotype combinations. In species with more chromosomes (linkage groups), independent assortment accounts for a high proportion of new genotype combinations in populations. If the same genes are present in fewer chromosomes, crossing over will produce recombinations for these genes. Fewer new genotype combinations will arise by crossing over than by independent assortment. Species with fewer chromosomes tend to be less variable than related species with the same genes in more chromosomes. Some environmental conditions favor low genetic variability. Centric fusions, if they occur, would be retained in such genetically constant populations because of the selective advantages of the slower release of inherent genetic variability. We will discuss these phenomena in more detail in Chapter 16.

## Changes in Gene and Chromosome Structure

Changes in structure may involve whole chromosomes or individual genes; both types of changes involve alterations in the linear ordering of DNA nucleotide sequences. Cytogenetic analyses of **structural aberrations** have been extended from observations at the gross level of light microscopy to the more detailed levels of DNA molecules and gene sequences. Molecular cytogenetic analyses require new tools of electron microscopy, biochemical tests, and physical methods, but they are cytogenetic *in principle* regardless of the level of observation that can be achieved.

Four general classes of structural chromosomal or genetic changes are generally recognized: (1) deficiencies or deletions, (2) duplications, (3) inversions, and (4) translocations of genetic material. Cytogenetic studies have contributed substantially to our view of the genome as a coordinated system rather than as just a random collection of genes or strings of genes in chromosomes.

### 12.4 Deficiencies or Deletions

The loss of one or more genes or parts of genes is called a **deficiency** or a **deletion;** the terms are synonymous. Chromosomal deficiencies arise through breaks caused by one or more agents, including radiation, viruses, and chemicals. The broken region may be healed or restored with little or no subsequent effect, or all or some of the genetic material may fail to be incorporated back into the chromosome. The loss of genetic material usually has a phenotypic consquence, although its magnitude will vary according to the amount of material involved, the relative need of the material for viability and function, and the particular species in question.

When the lost piece includes the centromere of the chromosome, the *acentric* chromosome or fragment will usually not be incorporated into daughter nuclei and will eventually be degraded or eliminated from the cell. By and large, deletions are lethal when present in the homozygous state, which must mean that very little genetic material does not contain some essential nucleotide sequence for the organism. Heterozygous deletions very often are phenotypically detectable, although phenotype alteration depends on the genetic lesion involved.

In human beings a number of disorders are caused by chromosome deletions. These are detectable only in heterozygotes since the homozygous condition is lethal. The *cri-du-chat* disorder is characterized by severe mental retardation and other abnormalities. The infant makes a peculiar mewing cry, which accounts for the name of the disorder (cat cry). This disorder is associated with a deletion of part of the short arm of chromosome 5 (Fig. 12.9). A few

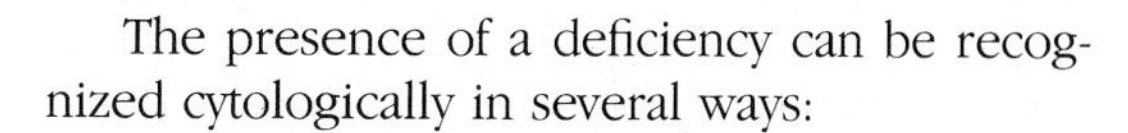


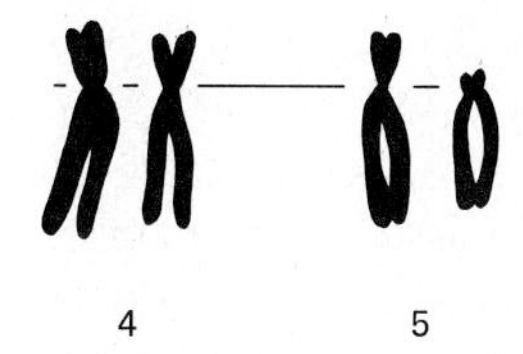


***Figure 12.9*** Group B chromosomes from a patient with *cri-du-chat* syndrome showing partial deletion of the short arm of one chromosome 5.

other disorders that are associated with chromosomal deletions generally result in mental retardation and various physical abnormalities. Apparently human beings do not readily tolerate losses of genetic material, even as heterozygotes. Whether this is due to the unmasking of harmful recessive alleles on the unpaired chromosome segment whose partner has been deleted or to genetic imbalances in the genome or to both is uncertain.

The presence of a deficiency can be recognized cytologically in several ways:

***1.*** From karyotype analysis showing the mismatched pair of chromosomes, one shorter than the other;

***2.*** From changes or differences in banding pattern in giant polytene chromosomes of *Drosophila* and other dipteran insects when homologous chromosomes are compared in larval cells, or when banded chromosomes in a human karyotype are compared; and

***3.*** From paired chromosomes in the pachynema stage of meiotic prophase, as one chromosome segment loops out because its partner has a deleted segment; or from similar conformations in polytene chromosomes, or in heteroduplex DNA photographed using the electron microscope (Fig. 12.10).

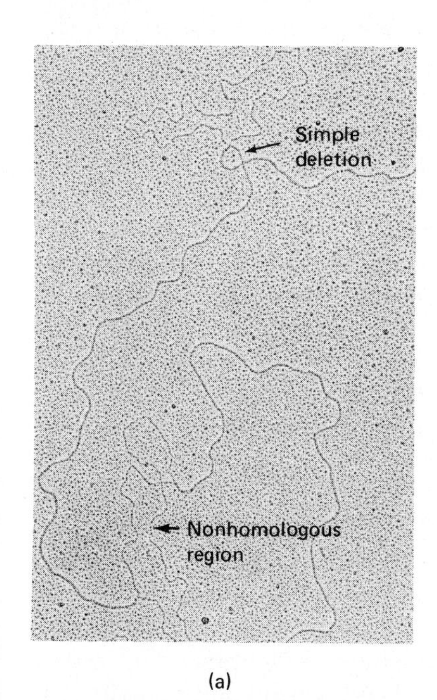


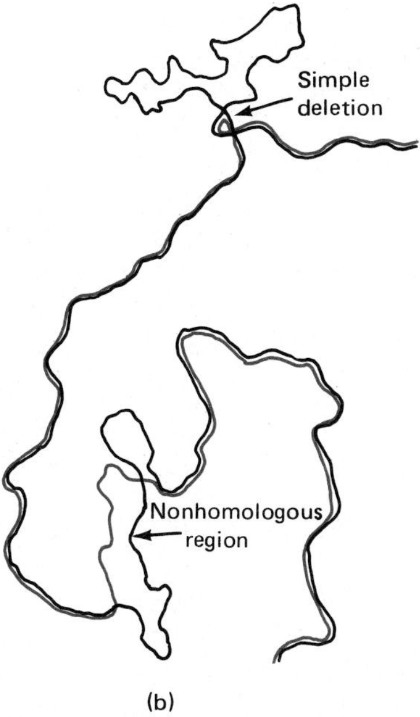


(a) (b)

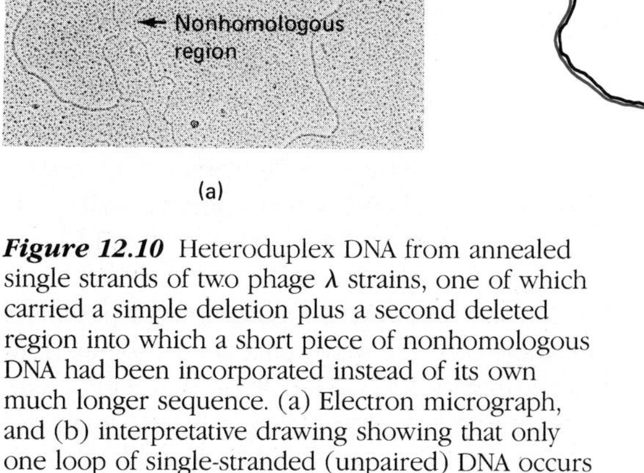

***Figure 12.10*** Heteroduplex DNA from annealed single strands of two phage λ strains, one of which carried a simple deletion plus a second deleted region into which a short piece of nonhomologous DNA had been incorporated instead of its own much longer sequence. (a) Electron micrograph, and (b) interpretative drawing showing that only one loop of single-stranded (unpaired) DNA occurs in the simple deletion site, but two unmatched single strands loop out in the second altered site. (B. Westmoreland et al., "Mapping of Deletions and Substitutions in Heteroduplex DNA Molecules of Bacteriophage Lambda by Electron Microscopy," *Science* **163**:1343, 21 March 1969. Copyright © 1969 by the American Association for the Advancement of Science.)

Deletion of all or part of a single gene can be detected genetically by the appearance of the recessive phenotype in a heterozygote, whose recessive allele has been unmasked when the dominant allele has been deleted entirely or in sufficient amount to be nonfunctional. This method was standard in searching for mutant alleles in earlier years. Organisms were irradiated to induce chromosome breaks and deletions, and recessives found in the irradiated wild-type strains were examined in detail to locate the expressed recessive allele. This approach was particularly valuable in *Drosophila,* since the giant polytene chromosomes could be examined to find the deleted band(s) and to locate the gene in that chromosome at that band (Fig. 12.11). The search was made easier because of the looped-out, unpaired segment in which the deletion occurred.

Through comparisons among deletion mutants, many of which contained large, overlapping deletions, *Drosophila* chromosome mapping was conducted with greater certainty. Mapping by overlapping deletions was originally developed for studies using *Drosophila* polytene chromosomes. This method was extended by Seymour Benzer to map the fine structure of individual genes, as we discussed in Chapter 7. When crosses between deletion mutants produce only mutant progeny, both parents must have overlapping deleted regions. Since a particular segment is deleted in both parents, wild-type progeny will not be produced. If deletion mutants produce some wild-type progeny, it indicates that the parental deletions are nonoverlapping. Recombination would give rise in some cases to progeny carrying intact chromosomal segments, derived partly from one parent and partly from the other parental genome. Once a deletion map is constructed, finer genetic distinctions can be made in subsequent crosses and in mapping of small deletions and point mutations (see Fig. 7.10).

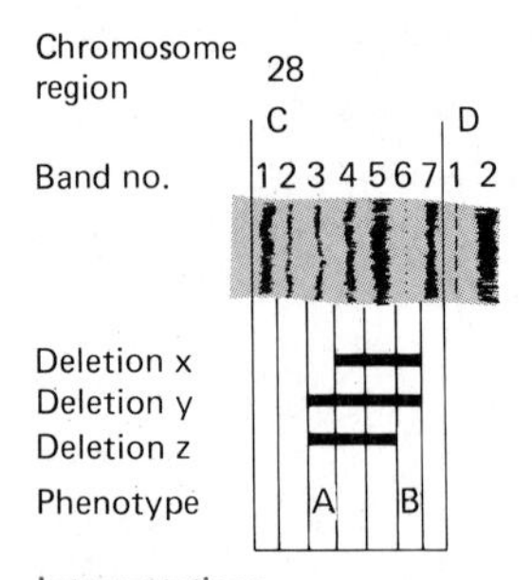


***Figure 12.11*** Localization of a gene to a particular band of *Drosophila melanogaster* polytene chromosome by deletion mapping. Recessive phenotype A is expressed when heterozygous with deletion z but not with deletion x, and recessive phenotype B is expressed when heterozygous with deletion x but not z; both A and B recessive phenotypes are expressed when heterozygous with deletion y. Such expression of the recessive phenotype in a deletion heterozygote is termed *pseudodominance.*

## 12.5 Duplications

**Duplications** are repeats of chromosome segments or genes. They affect the organism in a variety of ways, ranging from negligible to considerable effects. Duplications may even be lethal if a very large amount of chromosomal material is repeated. The primary evidence for duplicated genes came from polytene chromosome studies in *Drosophila,* in which repeated band patterns indicated repeated genes or gene segments. These may be tandem repeats, such as ABCABC, or reverse repeats, such as ABCCBA. Although duplications have usually been found on the same chromosome and rather close together because they are easier to spot, other repeats may be present in different chromosomes. Breakage and rejoining in the same chromosome or insertion into another chromosome after breaking and rejoining has occurred are often the causes for duplicated chromosome segments.

Through **unequal crossing over,** in which breakage and rejoining is not precise when nonsister chromatids exchange segments, one chromatid may acquire two copies of a gene and

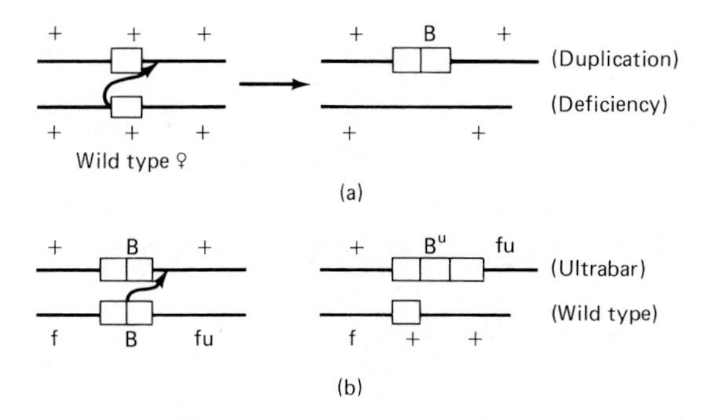


***Figure 12.12*** Origin of (a) Bar and (b) ultrabar through unequal crossing over in *Drosophila.* The locus is shown as a box on the X chromosome. Recombination of the outside markers +/*f* and +/*fu* in the Bar × Bar cross giving rise to ultrabar and wild-type segregants provides evidence in support of crossing over as the source of *B* and $B^u$ mutations.

its nonsister may have none (Fig. 12.12). A duplication and a deletion may therefore arise by unequal crossing over. The first case of presumed unequal crossing over was analyzed in Bar-eyed *Drosophila* females in the 1930s. Alfred Sturtevant and Calvin Bridges, from T.H. Morgan's group at Columbia University, provided cytogenetic observations of females having one, two, and three copies of this X-linked dominant allele in one chromosome. The Bar eye phenotype varies according to the number of duplicate genes in the cell and the number of copies on each chromosome. For example, females with two gene copies on each X chromosome have more eye facets (68) than females with three duplicates on one chromosome and a single copy on the homologous X chromosome (45 facets). Wild-type females have almost 800 facets per eye. This is an example of a **position effect,** that is, a phenomenon through which the degree of expression of a given gene (or genes) is modified in relation to its (or their) physical location within the genome.

With the advent of recombinant DNA technology to clone selected DNA sequences, extensive studies have been made of molecular duplications. Families of repeated genes may include hundreds, thousands, or millions of repeats (duplicated sequences), as we discussed in Section 8.11. These must have originated as duplications since it is virtually impossible for so many repeats to have arisen by chance in the same genome. Molecular analysis of cloned genes has also revealed that protein-specifying genes of unique-copy sequences occur in small clusters on one or two chromosomes. The extensive homologies among functionally similar genes can be explained most easily by postulating gene duplication at various times in the past history of the species. Among the best-known of these systems are the mammalian genes coding for various globin polypeptides of hemoglobin molecules.

Hemoglobin molecules in the adult consist of two $\alpha$ globin and two $\beta$ globin chains. The homology between $\alpha$ and $\beta$ globins is evident from their amino acid sequences, when these are aligned in the most favorable way (Fig. 12.13). These sequences have diverged substantially over the 500 million years estimated to have elapsed since the original duplication occurred, at the dawn of vertebrate history. While divergences have appeared in codons specifying $\alpha$ globin and $\beta$ globin amino acids, an astonishing identity is found in the *organization* of the gene sequences. Both globin genes are organized into three coding segments (exons) separated by two intervening sequences (introns). The exons in the human genes correspond to amino acid residues 1–31, 32–99, and 100–141 in $\alpha$ globin and to amino acid residues 1–30, 31–104, and 104–146 in $\beta$ globin. The exon-intron splice junctions are the same in the two genes (Fig. 12.14). Despite divergence in coding sequences so that only 63 amino acids are in the same locations in the 141 residues in $\alpha$ globin and the 146 residues in $\beta$ globin, the overall organization of these interrupted genes has remained virtually the same for 500 million years.

The five functional genes in the $\beta$-like globin cluster are spread over a DNA segment about 60 kilobases long. About 95% of the segment is made up of noncoding DNA, interspersed between genes and in the genic introns themselves. This entire 60-kb segment is also found in

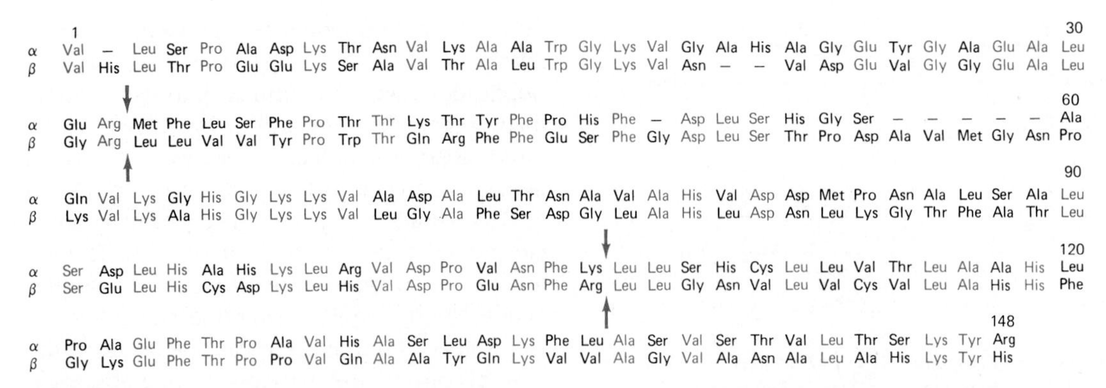


***Figure 12.13*** Amino acid sequences of alpha ($\alpha$) and beta ($\beta$) globin chains in human hemoglobin. The 141 amino acids of $\alpha$ globin and 146 amino acids of $\beta$ globin have been aligned over 148 positions to enhance the similarities present. The exon-intron splice junctions (at arrows) are essentially the same in the two globins even though the original gene duplicated and began to diverge about 500 million years ago. Only 63 amino acids (color) occupy the same locations in the two polypeptide chains. (Adapted from *Evolution* by T. Dobzhansky, F.J. Ayala, G.L. Stebbins, and J.W. Valentine, Fig. 9-17. W.H. Freeman and Company. Copyright © 1977.)

the genome of apes and Old World monkeys, indicating conservation during the past 40 million years of primate evolution (Fig. 12.15). Comparisons with more ancient primate types, such as lemurs, show that duplication of the $\gamma$ locus coding for a fetal globin and duplication of the $\beta$ locus with subsequent divergence of one of these $\beta$ duplicates to a $\beta$-like $\delta$ globin gene, must have occurred sometime between 70 million and 40 million years ago.

Duplication of DNA sequences and subsequent divergence of these sequences contributes to the richness of genetic variation and diversity in biological evolution. We will discuss these topics in Chapter 16.

## 12.6 Inversions

An **inversion** is a structural change in which a segment of a chromosome is cut out and then reinserted in an inverted order, having been rotated 180° from the original sequence. If the

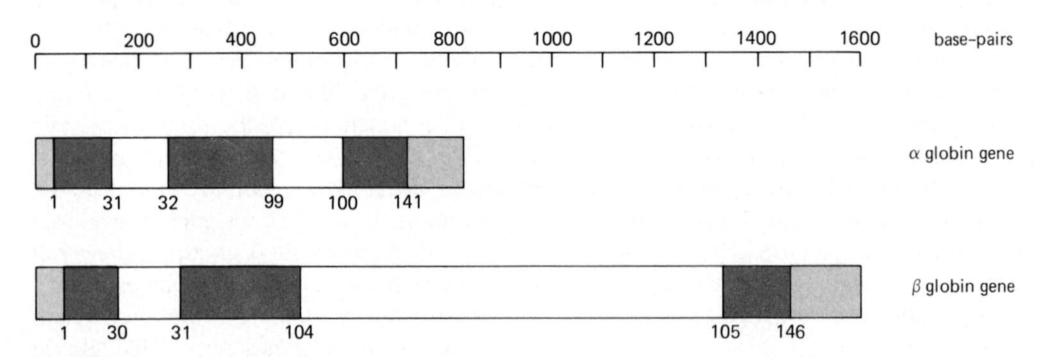


***Figure 12.14*** Organization of human $\alpha$ and $\beta$ globin genes is virutally the same despite 500 million years of evolution since the ancestral gene duplicated. The exon-intron splice junctions are the same in the two genes. The second exon is slightly longer and the second intron is considerably longer in the $\beta$-globin gene, presumably due to additions/deletions in one or both gene sequences. (Reproduced, with permission, from the *Annual Review of Genetics,* Volume 14. © 1980 by Annual Reviews, Inc.)

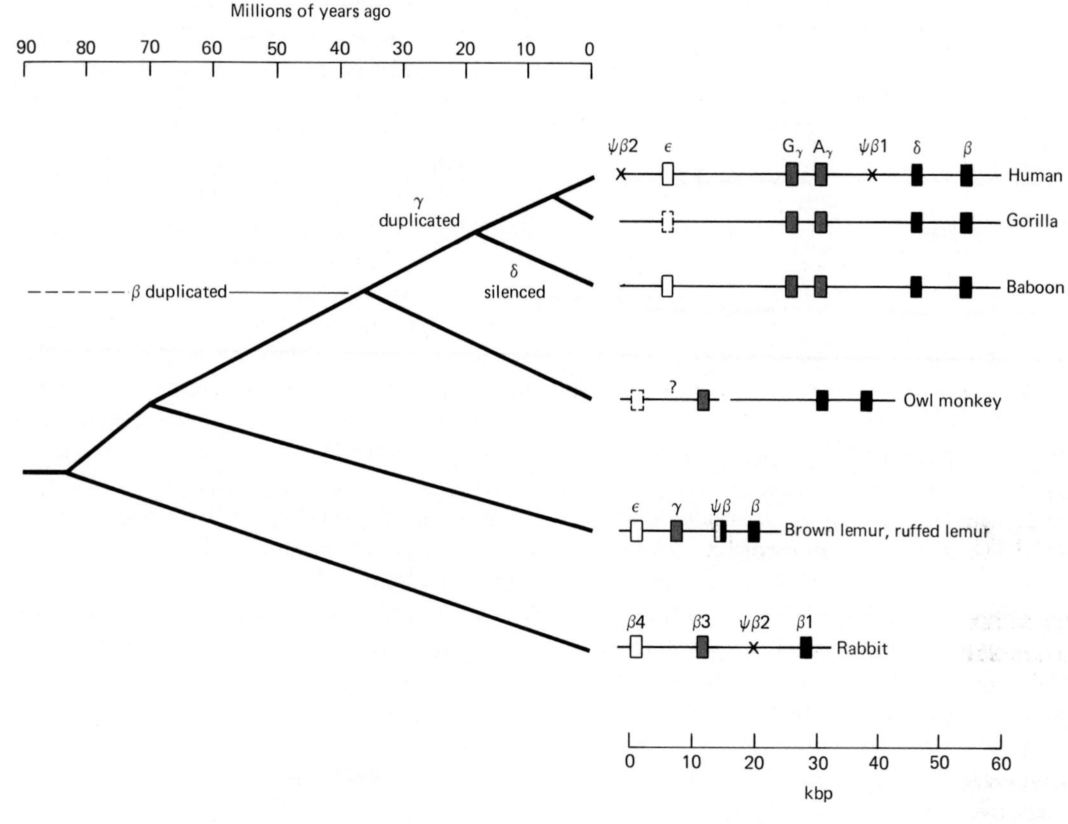


***Figure 12.15*** Changes in the cluster of $\beta$-like globin genes are shown for the past 90 million of the 180 million years of mammalian evolution. The organization of this gene cluster in human, ape, and baboon (Old World monkey) is strikingly similar even after 40 million years of divergent evolution. Genes for gamma ($\gamma$) globins are shown in color, and pseudogenes are indicated by crosses. All these genes arose by duplication and subsequent evolutionary divergence of nucleotide sequences. (R. Lewin, "Evolutionary History Written in Globin Genes," *Science* **214**:426–429, Maps, 23 October 1981. Copyright © 1981 by the American Association for the Advancement of Science.)

original sequence is *abcdefgh* and breaks occur between *b* and *c* and *f* and *g,* the inverted chromosome would have the new sequence *abfedcgh*. Usually no genetic material is lost.

Inversions can be detected genetically by altered linkage relations between genes within the inverted segments and flanking genes on both sides of the inversion. Since recombination frequency reflects the distances between genes, any alteration in their distances will show up in altered recombination percentages and different map units of distance when compared with the standard gene maps.

Cytological detection of an inversion can be accomplished by examination of chromosomes in meiosis or by comparisons between banded chromosomes in polytene nuclei or in human and other species' karyotypes after staining to show bands. In meiotic cells the inversion can be seen in individuals or cells heterozygous for the inverted segment. Karyotype analysis of somatic chromosomes can be carried out for either homozygotes or heterozygotes in comparison with standard banding patterns.

Inversion heterozygotes are of two general types, depending on whether the centromere is

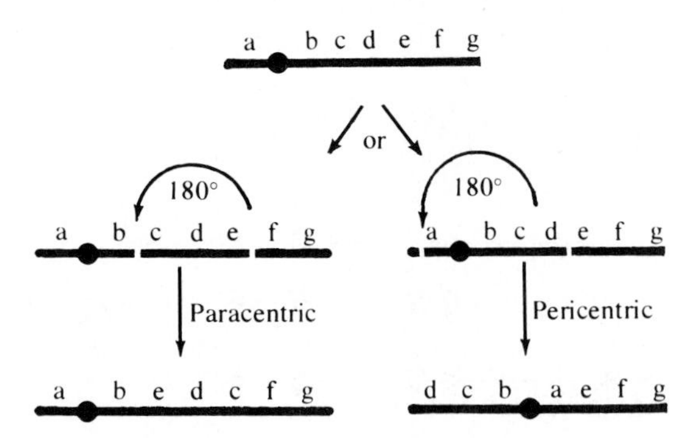


***Figure 12.16*** Origin of paracentric and pericentric inversions.

included in the inverted segment. In **paracentric inversions** the centromere occurs outside the inverted segment and is not included in the altered sequence; in **pericentric inversions** the centromere is contained within the inverted segment (Fig. 12.16). The characteristic meiotic configuration is evident in paracentric or pericentric inversion heterozygotes during the pachynema stage of prophase I. During synapsis, pairing between homologous regions takes place gene for gene as usual. Since the gene order is different in the inverted and noninverted chromosomes of a pair, a characteristic *loop* develops at the inversion region (Fig. 12.17). The relative size of the loop is one indication of the extent of the inversion.

Crossing over within the inverted region leads to aberrant chromatids. Depending on the site of the crossover and sites of other crossovers occurring within and around the inverted region, chromatids may be produced with two centromeres (dicentric) or with no centromere (acentric). Even if pachytene nuclei cannot be studied, meiotic cells in metaphase or anaphase of the first or the second divisions may have chromosome bridge and fragment formation, which can be seen and related to crossover events in prophase I (Fig. 12.18). Whether or not these bridge and fragment aberrations persist after metaphase I, the usual results of crossing over in inversion heterozygotes are gametes or spores with deficiencies and duplications of chromosome segments. For this reason, inversion heterozygotes are usually partially sterile since genetically aberrant gametes or spores often are dysfunctional.

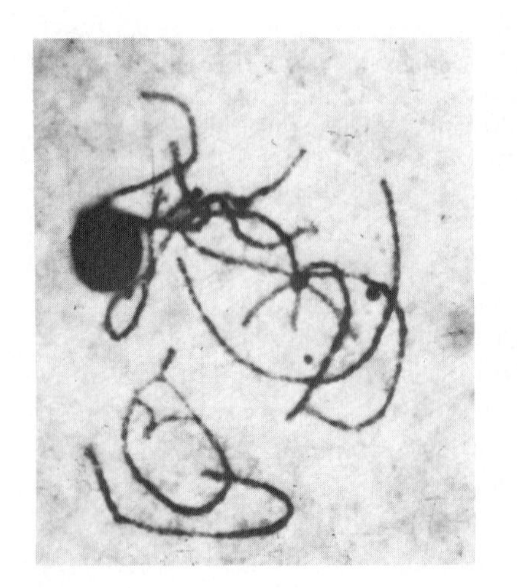

(a)

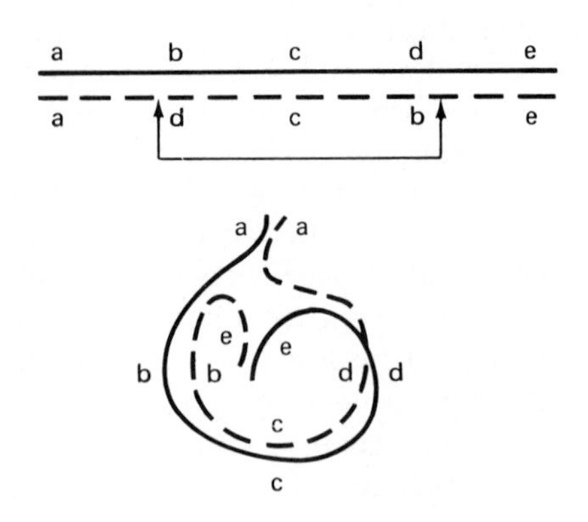


(b)

***Figure 12.17*** Pachytene nucleus from corn (*Zea mays*) meiocyte heterozygous for an inversion in chromosome 2. (a) Inversion loop and (b) interpretive drawing of the inverted and noninverted chromosomes and their alignment after pairing in pachynema. (Photo courtesy of M.M. Rhoades.)

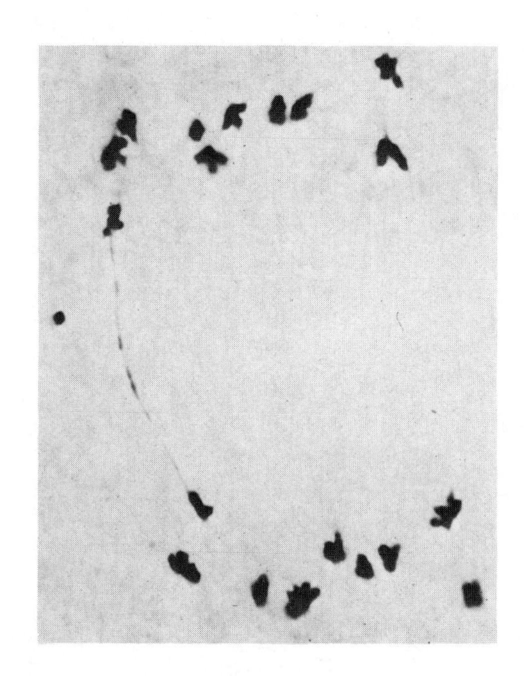

***Figure 12.18*** (right) Chromosome bridge between disjoining parts of a dicentric chromosome during anaphase I of meiosis in corn. An acentric fragment lies next to the bridge. (Photo courtesy of M.M. Rhoades.)

The actual fate of acentric and dicentric chromosomes and of cells containing these aberrations may vary from one species to another. For example, in corn the dicentric chromosome usually breaks before metaphase I so that spores are formed with deficiencies and duplications. These give rise to inviable pollen and to partial sterility in the inversion heterozygote. In *Drosophila,* however, the aberrant chromosomes often are not included in the functional egg nucleus because of the way in which the egg and the three polar bodies form. The usual result is formation of only noncrossover egg cells (Fig. 12.19). It is this particular set of events that

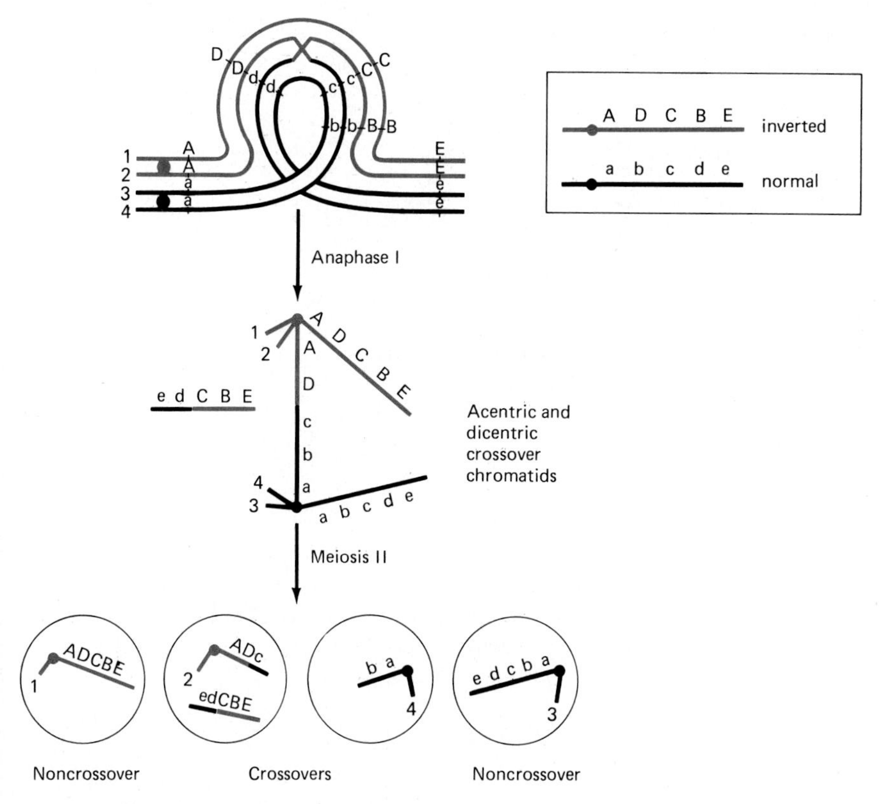


***Figure 12.19*** Consequences of crossing over within a paracentric inverted region in an inversion heterozygote. Gametes with crossover chromosomes have deletions or duplications, but noncrossover gametes have a complete set of the genes involved.

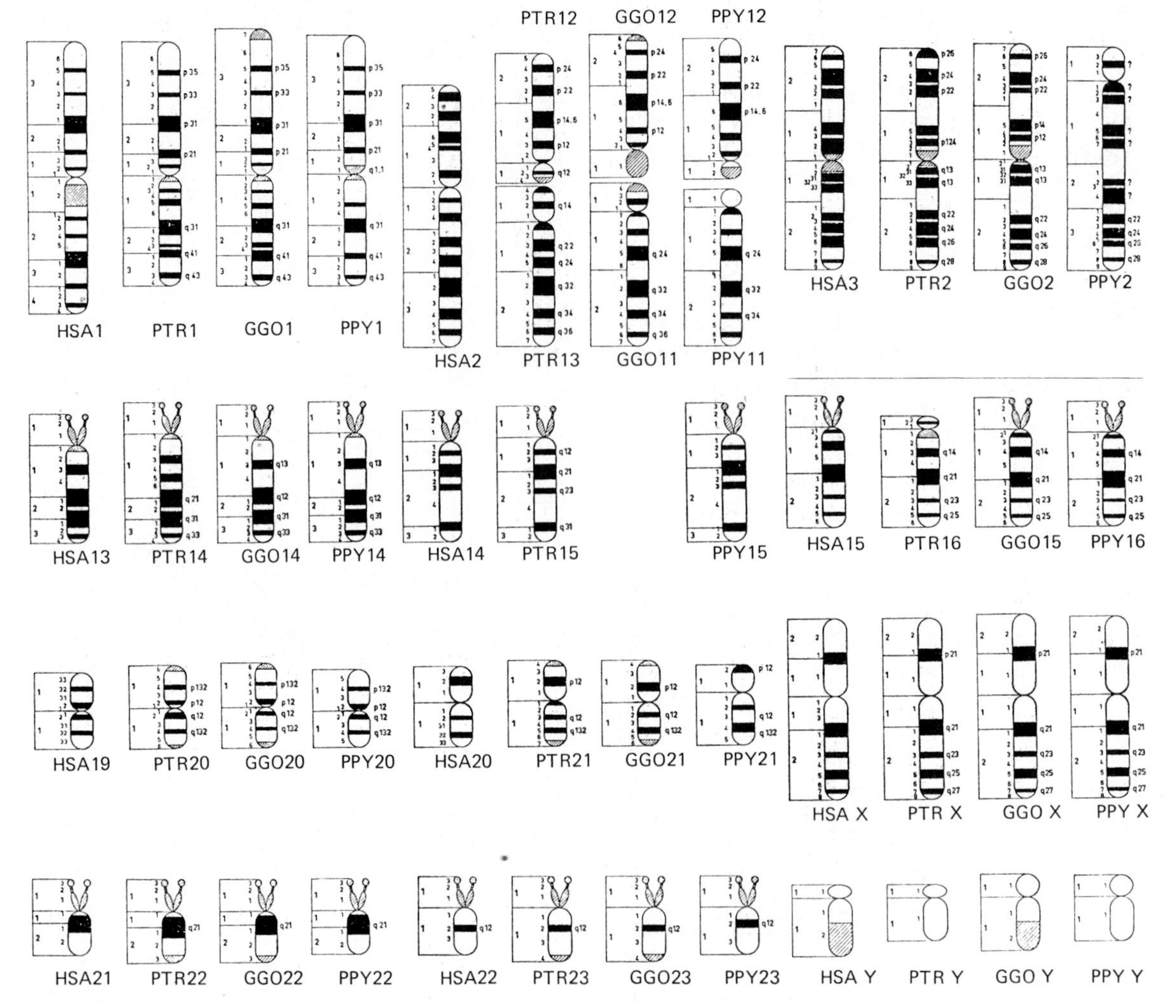


***Figure 12.20*** Diagrammatic representation of chromosome bands of selected chromosomes of human beings (HSA, *Homo sapiens*) and the chimpanzee (PTR, *Pan troglodytes*), gorilla (GGO, *Gorilla gorilla*), and orangutan (PPY, *Pongo pygmaeus*). The chromosomes have been arranged to show the similarities in banding pattern; solid bands are positive Q or G bands, and crosshatched areas depict variable bands. Notice that human chromosome 2 is closely matched by two acrocentrics in all three great ape species, and that some chromosomes are astonishingly similar in all four species. (From *Paris Conference* (1971): *Standardization in Human Cytogenetics*. Birth Defects: Original Article Series **14**:9, 1975. The National Foundation, New York.

allows the use in genetic studies of females with heterozygous inversions, when only non-crossover gametes are required for an analysis. Crossing over does take place, but recombinant egg cells are not produced. The practical effect, therefore, is suppression of crossover eggs and almost exclusive transmission of the non-crossover chromosome. We described such a system in the *ClB* test for mutation detection in *Drosophila* (see Section 11.2). The crossover suppressor (*C*) was an inversion that led to transmission of noncrossover *ClB* chromosomes to the progeny, since crossover chromosomes generally are not included in the *Drosophila* egg cell formed at meiosis.

Karyotype analysis from banded chromosome complements in primate species has revealed the existence of a number of paracentric and pericentric inversions that distinguish human chromosomes from those of the chimpanzee and other great apes, our closest living relatives (Fig. 12.20). The chromosomes are relatively similar except for the inversions and the chromosome number. Apparently, two of the short acrocentric chromosomes in the ape genome have fused to become the large, metacentric chromosome 2 in the human complement. Comparisons of the relative lengths of these chromosomes and of their banding pattern in metaphase and late prophase preparations have led to this interpretation.

Similarities in karyotypes of related species are matched by similarities in the protein products of a number of genes that have been analyzed. The amino acid compositions of human and ape proteins, including globins, enzymes, and other gene products, are either identical or virtually identical. Since the protein is a reflection of nucleotide sequence in DNA, we can see that the higher primates are genetically very closely related, and the karyotype similarities provide another line of evidence in support of this conclusion. We will discuss this general subject more thoroughly in Chapter 16.

## 12.7 Translocations

A **translocation** is the result of a transfer of part of one chromosome to another, nonhomologous chromosome. If the transfer involves only the attachment of part of one chromosome to an intact nonhomologue, it is called a **simple translocation.** If nonhomologous chromosomes exchange segments, it is called a **reciprocal translocation.** In either case, genes in one linkage group are transferred to another linkage group. The genetic identification of simple and reciprocal translocations involved the discovery that linked genes become independently assorting, whereas other independently assorting genes become linked as a direct consequence of the redistribution of chromosome segments.

Translocations require breakage and rejoining of chromosome segments, just like the other three classes of chromosome aberrations we have discussed. Like these, too, translocations can be detected cytologically in meiotic prophase (Fig. 12.21). Since synapsis is a relatively

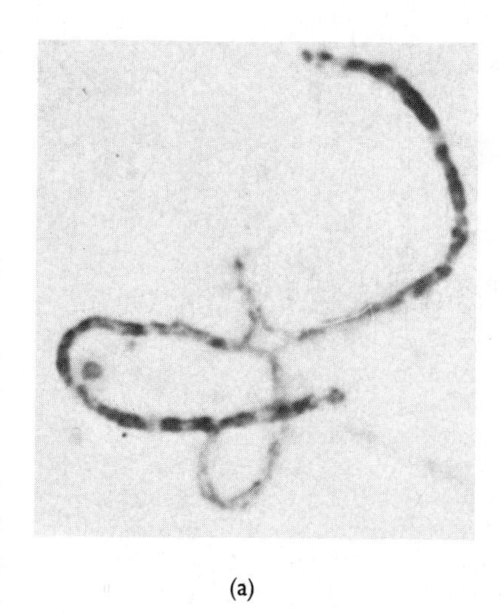

(a)

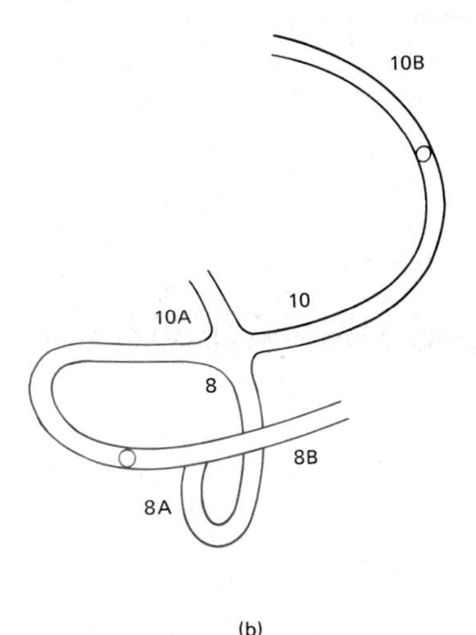


(b)

***Figure 12.21*** Pairing in pachytene nucleus of corn (*Zea mays*) heterozygous for a reciprocal translocation involving chromosomes 8 and 10. (a) Photograph showing two-by-two pairing within the complex of four chromosomes; and (b) interpretive diagram of the nontranslocated chromosomes 8 and 10 and the translocated 8A/10B and 8B/10A chromosomes. (Photo courtesy of M.M. Rhoades.)

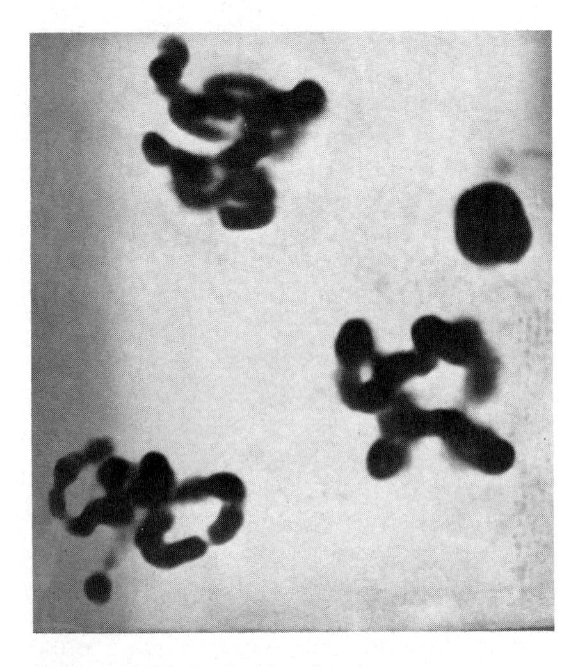

***Figure 12.22*** Circle of twelve chromosomes in *Rhoeo discolor,* babe-in-the-cradle. All six pairs of chromosomes have undergone reciprocal translocations.

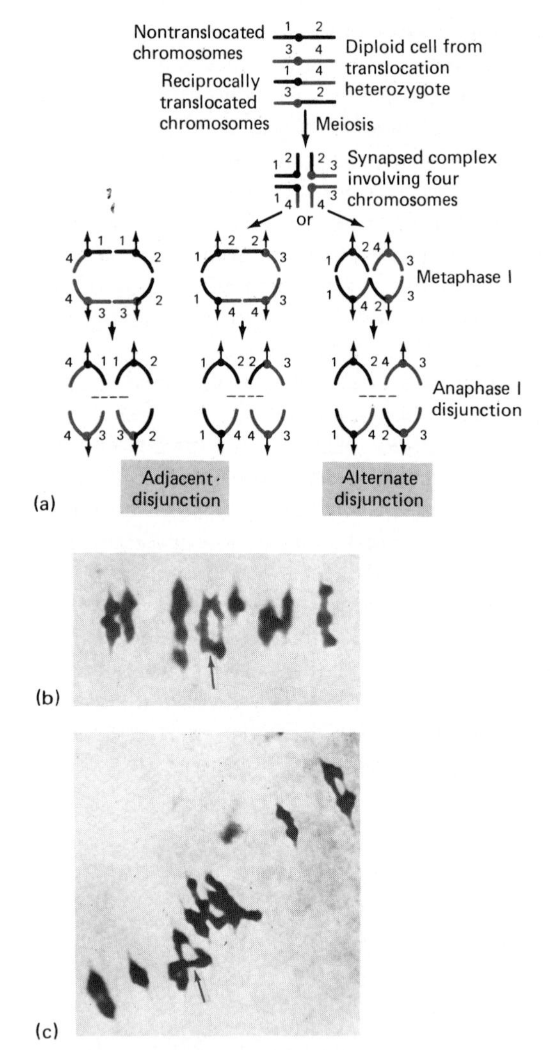


***Figure 12.23*** Meiosis in a translocation heterozygote. (a) Diagrammatic illustration of the consequences of three alignments at metaphase I, producing adjacent or alternate disjunction at anaphase I. Chromosome arms are numbered, showing that deficiencies and duplications arise by adjacent but not by alternate disjunction. (b) Alignment of a ring of four chromosomes at metaphase I in corn will lead to adjacent disjunction but (c) the "figure-8" alignment leads to alternate disjunction at anaphase I. (Photos courtesy of M.M. Rhoades.)

precise pairing process, homologous parts of translocated and nontranslocated chromosomes will pair in a heterozygote and produce complexes of chromosomes that can be identified easily by microscopy. Translocation configurations can be identified in pachynema as "cross-shaped" figures, and in diplonema, diakinesis, or metaphase I, as groups of chromosomes in chains or rings (Fig. 12.22).

Different patterns of chromatid and gene segregation occur in translocation heterozygotes depending on the occurrence of crossing over, the location of the crossovers, and the orientation of the translocation chromosome complex on the equatorial plate at metaphase I of meiosis (Fig. 12.23). If a ring of chromosomes is twisted so that both translocated chromosomes move to one pole and both normal chromosomes go to the opposite pole (**alternate segregation**), functional gametes or spores will be produced since each will have a full set of genes. If, however, one translocated and one normal chromosome move to the same pole at each end of the anaphase I cell (**adjacent segregation**), deficiencies and duplications will arise and the meiotic products will be nonfunctional. Adjacent segregation is responsible for partial sterility in translocation heterozygotes. If the aberrant gametes or spores do participate in reproduction, they will give rise to inviable zygotes or to aberrant individuals. The reduced frequency of viable offspring in crosses involving a translocation heterozygote can be understood in relation to these processes of crossing over and chromatid segregation at meiosis. Translocation homozygotes, on the other hand, would experience no difficulties of this kind. Their chromosomes only pair two by two, thus ensuring that meiosis will be normal and will give rise to functional gametes.

Translocations can be identified in human karyotypes if a simple translocation is present or if a reciprocal translocation has involved unequal lengths of chromosome segments. In either case, one or more unpaired chromosomes can be seen. About 2% or 3% of patients with Down syndrome have a translocated chromosome that contains parts of chromosome 21 (a G-group chromosome) and one of the three D-group chromosomes (chromosome 13, 14, or 15). Such a D/G translocated chromosome can also be found in one or more relatives of the patient. There is no effect on the phenotype of the relative as long as only two copies of the genes of chromosome 21 are present. Such individuals have 45 chromosomes, one of which is the D/G translocation, instead of 46 chromosomes. The second D and the second G chromosome will also be unpaired since their homologous regions are incorporated into the larger D/G translocation chromosome. Patients with Down syndrome who have an apparently normal count of 46 chromosomes actually have three copies of chromosome 21 genes. They have a normal pair of chromosome 21, plus a third copy of these genes in their D/G translocated chromosome. The homologous D chromosome is unpaired, as is the D/G translocated chromosome (Fig. 12.24).

The vast majority of Down syndrome individuals do not have a translocated chromosome. Instead, they have a total count of 47 chromosomes, with chromosome 21 present in triplicate. These cases arise by nondisjunction of homologous chromosomes during meiosis or mitosis, as we discussed earlier. Such cases do not show a familial pattern of transmission of the

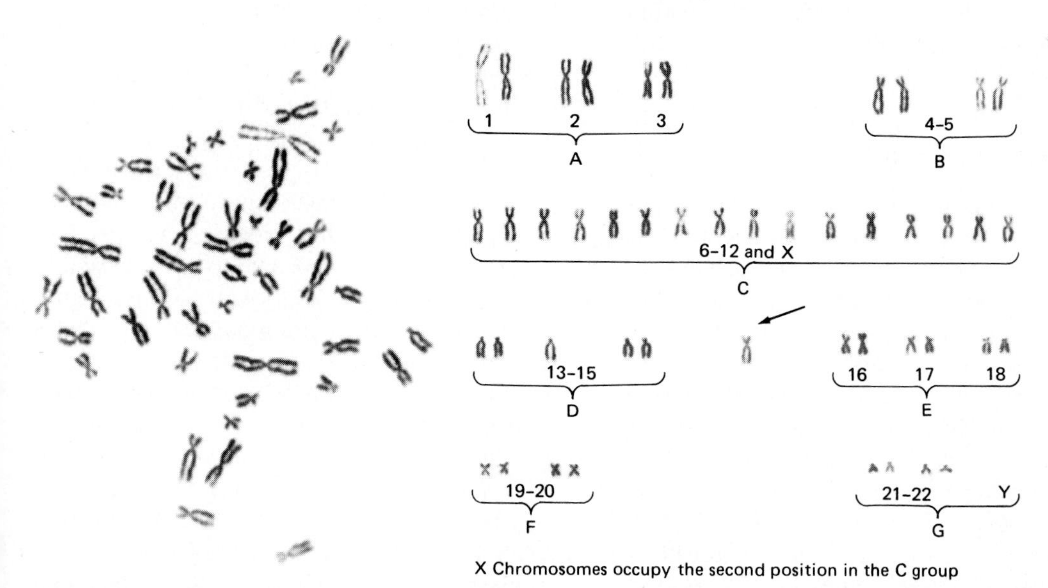


***Figure 12.24*** D/G translocation chromosome in a female patient with Down syndrome. There are three copies of chromosome 21, even though the total chromosome count is only 46. One copy of chromosome 21 genes is present in the 14/21 translocation chromosome and the other two copies exist as whole chromosomes 21. The karyotype was prepared from the metaphase spread, where only 5 chromosomes of the D group can be found in this trisomy-21 individual. (Courtesy of T.R. Tegenkamp.)

syndrome. On the other hand, families with individuals having a D/G translocated chromosome do show a familial transmission pattern since the aberrant chromosome can be transmitted through the gametes to the offspring. Years ago there was considerable confusion in interpreting familial and nonfamilial patterns of occurrence of Down syndrome. The situation became clear after human chromosome studies became possible in 1956, with the development of new methods for preparing human chromosomes for microscopic study.

All the classes of chromosomal aberrations that are due to structural changes in the genome can be identified by independent genetic and cytological criteria. The correlations between these sets of evidence strengthened the theory that genes were situated in chromosomes. These same parallels serve as well today whether studying genes at the chromosomal or the molecular level of analysis. Certain types of aberrations can be used directly to map genes or genetic regions to specific chromosomes and even to specific regions of a chromosome as we will see in the next sections. In some species, map construction virtually depends on cytogenetic rather than on classical genetic transmission methods.

## Somatic Cell Genetics

We can map human genes by genetic transmission analysis using the family method, but there are limitations in the method such that few genes have been assigned to specific chromosomes (see Section 5.10). Great advances in mapping have been made in the past twenty years by the addition of **somatic cell genetic** methods, by which limitations of the family method have been overcome. Somatic cells may fuse (mimicking fertilization) and chromosomes may later be lost from the cell hybrid (mimicking homologous chromosome segregation). This *parasexual cycle* permits us to exploit somatic cell hybrids in assigning genes to particular chromosomes of the genome. The artful utilization of deletions and translocations of human chromosomes in somatic strains has led to more specific assignments of genes to regions of the chromosome, not merely to the chromosome itself. These studies have made the human genome the best mapped of any mammalian species.

### 12.8 Somatic Cell Hybridization

When somatic cells are incubated together in a culture medium, cell fusions occasionally take place. If cells are pretreated with a chemical agent such as *polyethylene glycol*, or with *Sendai viruses* inactivated by ultraviolet light, the frequency of cell fusions increases significantly. These treatments modify the cell surface. As a result, cell-cell contacts are more frequent and sufficiently intimate so that cell membranes fuse to produce single hybrid cells. These **somatic cell hybrids** are initially binucleate, but nuclei may fuse shortly afterward to yield a *synkaryon* (*same nucleus*). If the cells differ in alleles for particular genes, the cell hybrid will be heterozygous for such pairs of alleles. Cell fusion leading to cell hybrids is a substitute process for fertilization typical of sexual systems.

During subsequent mitotic divisions, somatic cell hybrids increase in numbers and give rise to colonies, each consisting of the descendants of a different single cell. In many cases, random loss of chromosomes of one of the parents takes place. Some of the remaining chromosomes may be unpaired, if their homologues have been lost. These unpaired chromosomes provide a haploid genotype for the genes located on them; other chromosome pairs and their pairs of alleles remain in the diploid state in the same cell. Random loss of chromosomes leading to *haploidization* of some parts of one of the parental genomes provides a substitute process for meiosis in sexual systems. During haploidization, members of pairs of alleles undergo segregation. One member of a pair of alleles may be retained on a single chromosome in some cell hybrids, and the other allele of the pair may be retained on the homologous chromosome in other cell hybrids in the culture.

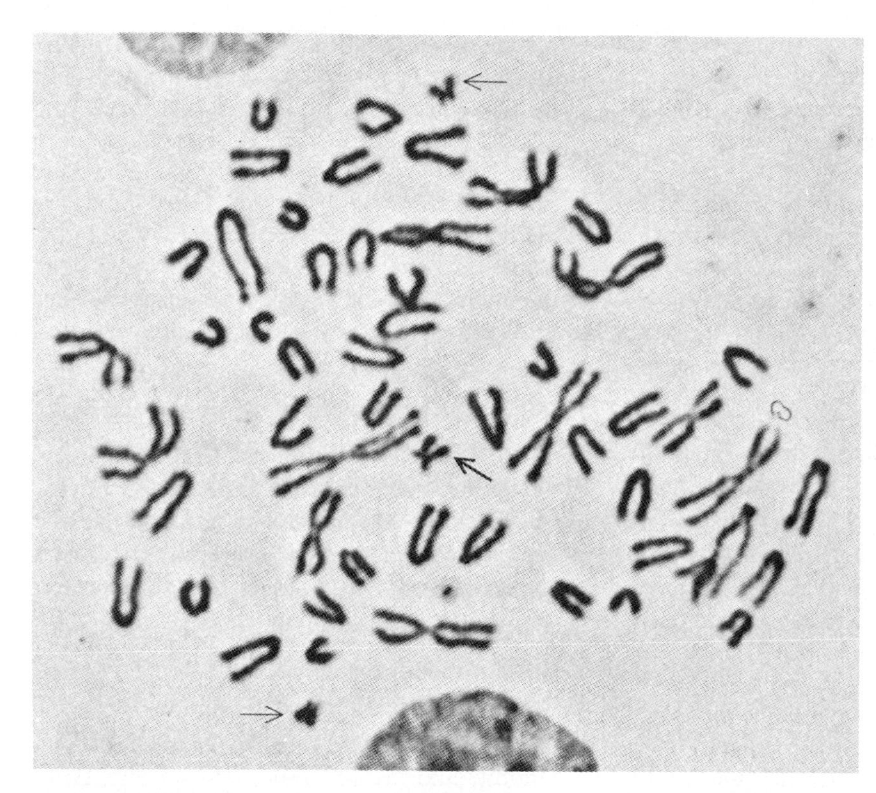

***Figure 12.25*** Human-mouse somatic cell hybrid retaining only 3 human chromosomes. The light arrows (top and bottom) indicate G-group human chromosomes (Nos. 21 and 22), and the heavier arrow (center) points to a human chromosome of group E (No. 17 or 18). (From "Hybrid Somatic Cells" by B. Ephrussi and M.C. Weiss. Copyright © 1969 by Scientific American, Inc. All rights reserved.)

The most useful cell hybrid systems are those in which rodent cells have been fused with human cells. During haploidization, human chromosomes are lost preferentially while the diploid set of rodent chromosomes is retained in the somatic cell hybrid. Instead of finding 86 chromosomes (40 mouse + 46 human), about 41 to 55 chromosomes are usually found in the haploidized cells. These remaining chromosomes of the original 86 consist of the 40 mouse chromosomes plus 1 to 15 or more of the human diploid complement of 46 chromosomes (Fig. 12.25).

The reasons for using rodent (usually mouse) cells as one of the parent types in hybridizations with somatic human cells include the following:

***1.*** Many different mutant cell-lines are available, thus making genetic analysis more comprehensive and more flexible.

***2.*** Rodent and human chromosomes have different sizes and shapes, and each can be recognized and easily identified in hybrid nuclei.

***3.*** Rodent chromosomes tend to be retained, whereas the human chromosomes in the same cells tend to be lost gradually and randomly by haploidization. This result is particularly true for mouse-human and hamster-human somatic cell hybrids, but less so in rat-human combinations.

***4.*** Rodent genes and human genes are expressed at the same time in cell hybrids, each producing *functionally similar* proteins that can be distinguished and identified by molecular assays. For example, the enzyme thymidine kinase (TK) catalyzes the same reaction in mouse and human or mouse-human systems, but the mouse enzyme and the human enzyme can be distinguished by specific molecular features assayed in a simple test.

Since somatic cell hybridization is an infrequent event, the parental cells in the incubation mixture would swamp the less numerous cell hybrids when plated on solid medium to allow colony development. A *selective growth medium* is therefore employed to encourage growth of the cell hybrids and simultaneously to discourage or inhibit growth of the parental mouse and human cells in a mixture. The usual selective medium contains the drug *aminopterin*, which blocks the major pathways for synthesis of DNA building blocks (Fig. 12.26). The essential purines can be made in the so-called "salvage" pathway if *hypoxanthine* is provided, and pyrimidines can be made if *thymidine* is present in the medium, provided the specific enzymes to catalyze these reactions are present in the cells. The enzymes are **hypoxanthineguanine phosphoribosyl transferase** (**HPRT**) and **thymidine kinase** (**TK**), respectively. The selective medium is called the **HAT medium** because it contains Hypoxanthine, Aminopterin, and Thymidine.

When TK-deficient mouse cells (phenotypically $TK^-$, carrying the mutant *tk* allele) are fused with human cells that are phenotypically $HPRT^-$ (because they carry the mutant *hprt* allele) or vice versa, cell hybrids that are phenotypically $TK^+/HPRT^+$ may be produced. When the cell mixture is plated on HAT medium after incubation, neither the $TK^-/HPRT^+$ nor the $TK^+/HPRT^-$ parents can grow because each lacks one of the two essential enzymes for the salvage pathway. Somatic cell hybrids that are phenotypically $TK^+/HPRT^+$ can grow. Such hybrids have received a functional *Tk* gene from the human parent cell and a functional *Hprt* gene from the mouse parent cell, or vice versa (Fig. 12.27).

Once somatic cell hybrids have multiplied and become visible as colonies on the HAT medium, they can be removed and each colony can be incubated separately in culture to provide materials for biochemical assays and genetic analysis. When suitable hybrid types have been identified, they can be used in gene mapping studies.

## 12.9 Gene Assignments to Chromosomes

A major goal of somatic cell genetics has been to identify and assign genes to specific chromosomes in the human haploid complement of 22 autosomes and 2 kinds of sex chromosomes (Fig. 12.28). Procedures for assigning a gene to a particular chromosome depend on finding the protein made by the gene in the *same* cell that has retained the particular chromosome. If human TK is only made when human chromosome 17 is present and never made when chromosome 17 is absent, then the *Tk* gene can be assigned to chromosome 17.

This procedure would be simple if it were possible to isolate 24 unique and different

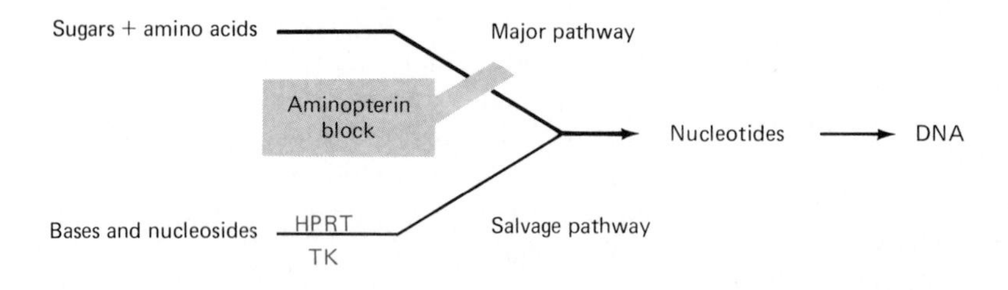


***Figure 12.26*** Pathways of synthesis of deoxyribonucleotides. If the major pathway is blocked by the drug aminopterin, nucleotides can still be made in the salvage pathway if the cells contain functional HPRT and TK enzymes.

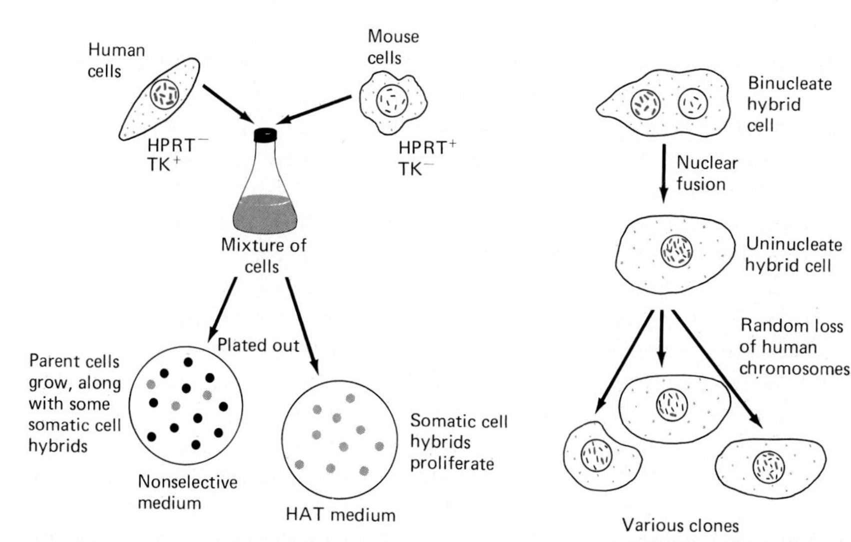


***Figure 12.27*** Somatic cell hybrids with HPRT$^+$/TK$^+$ phenotype can grow and produce colonies on selective HAT medium, whereas enzyme-deficient parent cells do not grow. After nuclear fusion, the hybrid cells experience random losses of human chromosomes. The different hybrid clones (different human chromosomes lost and retained) can then be used in studies to map human chromosomes.

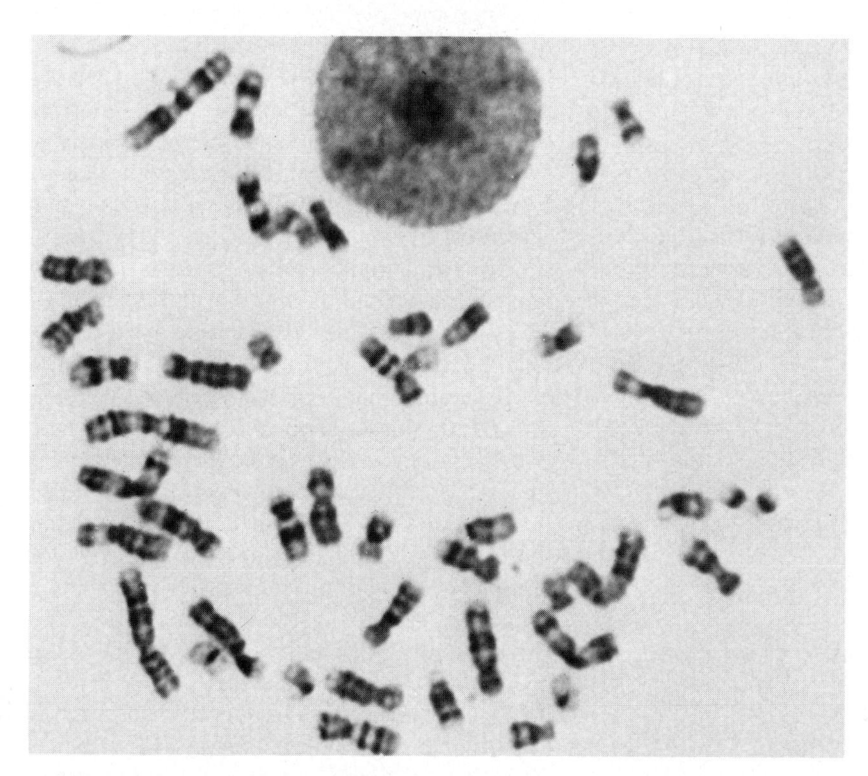

***Figure 12.28*** Human metaphase chromosomes stained to show G-banding. Each chromosome can be uniquely identified by its G bands, once the chromosome has been identified generally according to its size and centromere location (as a member of one of the seven groups, from A to G). See Fig. 12.29 for reference.

clones of somatic cell hybrids, each having retained only a single different human chromosome. Technical difficulties make this ideal situation almost impossible to attain. However, a clone panel consisting of only five different clones of somatic cell hybrids has 32 ($2^5$) possible ways to distribute a chromosome among the clones. Thus, by choosing five hybrid clones we can construct a panel in which each of the 24 chromosomes has a unique distribution pattern among the five clones. Such a clone panel can be used to identify genes for all 24 kinds of human chromosomes. In addition, it can be used to find genes that are associated in the same chromosome.

The principles can be illustrated using a simpler clone panel with a group of three different clones constituting a unique distribution for each of 8 ($2^3$) different chromosomes, as follows:

| hybrid clone | human chromosome (+ = present) | | | | | | | |
|---|---|---|---|---|---|---|---|---|
| | **X** | **2** | **3** | **4** | **5** | **16** | **17** | **18** |
| A | + | + | + | + | − | − | − | − |
| B | + | + | − | − | + | + | − | − |
| C | + | − | + | − | + | − | + | − |

Using this clone panel for reference to each of the eight unique +/− distribution patterns, we can look for chromosome assignments for four (or more) different genes. Each gene governs synthesis of a particular enzyme that can be identified by biochemical tests in each of the three hybrid clones. We would test each hybrid clone to find the pattern of presence or absence of the enzyme activities sponsored by each gene. Suppose we obtained the following results:

| hybrid clone | human enzyme (+ = present) | | | |
|---|---|---|---|---|
| | **HPRT** | **GPD** | **PGK** | **TK** |
| A | + | + | + | − |
| B | + | + | + | − |
| C | + | + | + | + |

We now compare the +/− pattern for each enzyme, in its vertical column, with the +/− pattern for each of the eight chromosomes, in their vertical columns in the clone panel. The vertical columns for the HPRT, GPD, and PGK enzymes match the vertical column for the X chromosome. This fact is evident from the presence of the three enzymes in clones A, B, and C, and the concordant presence of the X chromosome in clones A, B, and C. None of the other seven chromosomes is present in all three hybrid clones. These *concordant results* in retention or loss of a chromosome with the presence or absence of the gene product serve as the primary indicators of gene association with a particular chromosome. In addition, our results showed that three different genes were all present on the same chromosome (X).

Genes that are associated with the same chromosome are said to be **syntenic**, and the term **synteny** refers to genes associated with the same chromosome according to chromosome-phenotype concordance in somatic cell hybrids. This term is preferred to the term *linkage*, which is reserved for conclusions reached from recombination data in gene transmission studies such as pedigree analysis. Syntenic genes may or may not exhibit linkage by recombination. Unlinked genes may become syntenic if the somatic cell hybrid clone experiences rearrangements of parts of its chromosomes, as happens occasionally.

Returning to the results shown above, we would conclude that the genes for the enzymes hypoxanthine-guanine phosphoribosyl transferase (*Hprt*), glucose 6-phosphate dehydrogenase (*Gpd*), and phosphoglycerate kinase (*Pgk*) were syntenic on the X chromosome. The thymidine kinase gene (*Tk*) is not syntenic with the other three genes. Comparing the clone panel for chromosomes with the +/− pattern for the enzymes, we can see that the *Tk* gene is located on chromosome 17 since the two +/− patterns are concordant.

One or more genes have now been assigned to each of the 24 different human chromosomes (Fig. 12.29). By early 1983 over 200 genes had been assigned to particular human chromosomes by gene mapping in somatic cell hy-

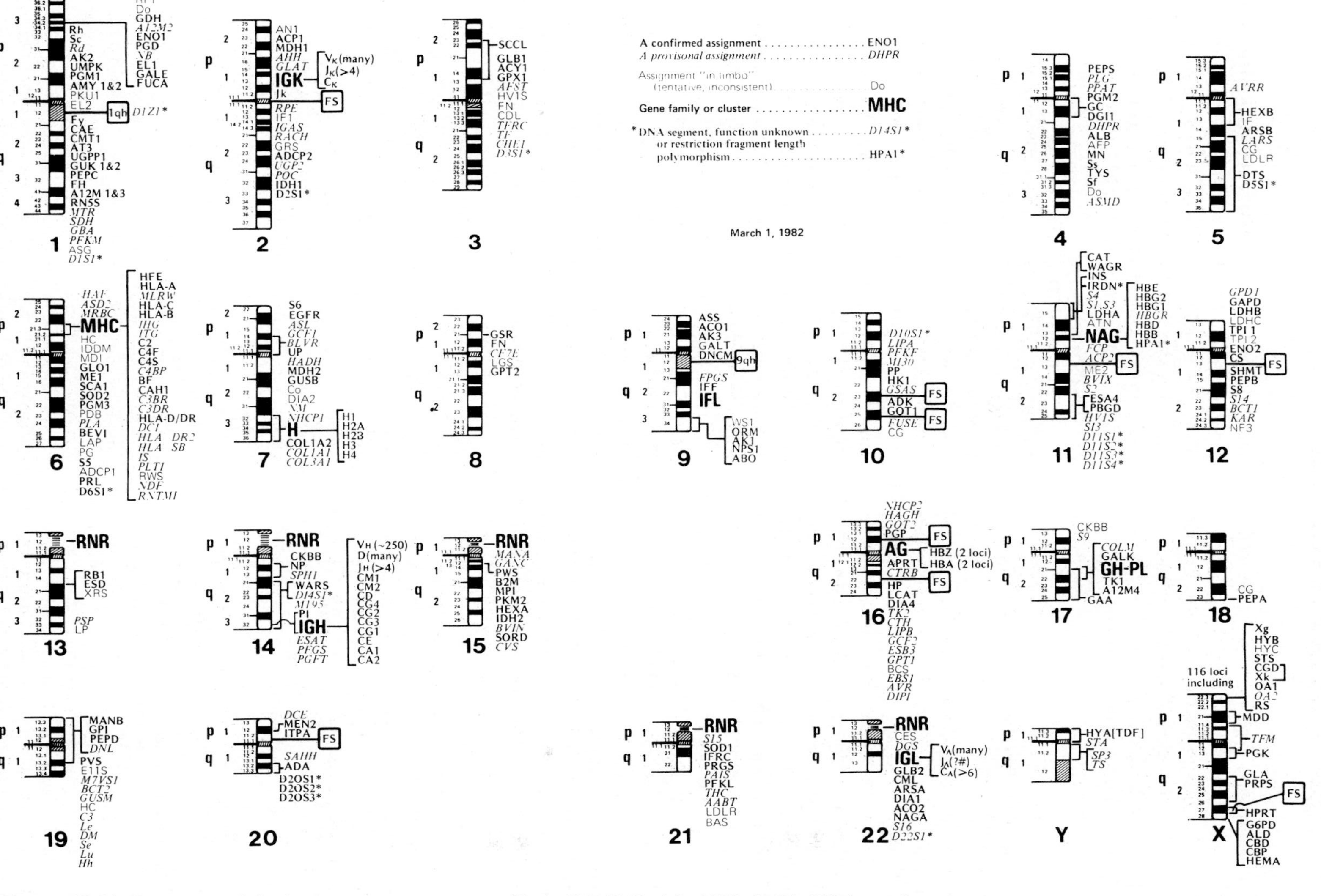


*Figure 12.29* Gene map of the human chromosome complement, showing many of the important gene assignments.

(From V.A. McKusick. 1982. *In* S.J. O'Brien, ed. *Genetic Maps.* **2**:327–350, Fig. 1.)

brids. On the basis of past performance, it has been estimated that about four human genes can be mapped per month, on the average. This rate is certain to be increased with the development of additional methods and the continued use of somatic cell genetics as an approach to human gene mapping.

## 12.10 Regional Mapping of the Chromosome

In some situations structurally rearranged chromosomes may be used to assign a gene to a particular region of a particular chromosome. A part of a chromosome may be *translocated* from its normal location and become attached to a different, nonhomologous chromosome. In one case involving a mouse-human somatic cell hybrid clone, a translocation had occurred in which the long arm of human chromosome 17 had broken off and become translocated to one of the mouse chromosomes (Fig. 12.30). The kinds of chromosomes present in this translocated clone and in other clones, which had only the mouse chromosome and not human chromosome 17 or which had both human chromosome 17 and the mouse chromosome, could be determined by ordinary light-microscopic examination of the mitotically dividing nuclei in each clone. Each of these clones was then assayed for TK activity since the *Tk* gene was known to be assigned to human chromosome 17. The experiment was designed to reveal the region of chromosome 17 in which the *Tk* gene was located.

Human TK activity was found in a normal clone carrying a normal human chromosome 17 and in the clone with the translocated chromosome. Since TK activity was found, it meant that the *Tk* gene was located in the long arm of human chromosome 17. If the *Tk* gene had been in the short arm of the chromosome, it would have been lost along with that region of chromosome 17. As a control, it was shown that human TK activity was not associated with any part of the mouse chromosome since clones lacking human chromosome 17 but retaining the mouse chromosome displayed no TK activity.

In another case, it was possible to assign the three X-linked genes governing HPRT, GPD, and PGK enzymes to the long arm of the X chromosome and to determine their order in this chromosome arm. Clones were obtained in which progressively longer sections of the long arm of the X chromosome were *deleted* as the result of breaks induced in the chromosome by

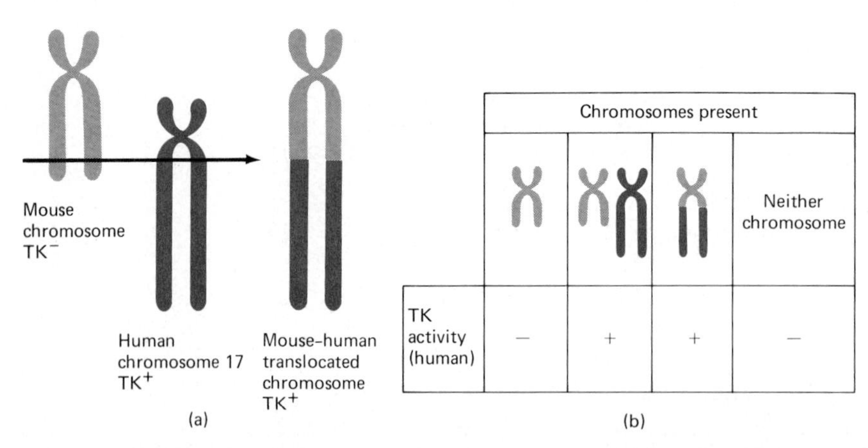


**Figure 12.30** Regional mapping within a chromosome. The gene for TK must be located on the long arm of human chromosome 17, since TK enzyme activity occurs only when the cell contains that particular region of the chromosome. TK activity is expressed whether the whole chromosome is present or only the long arm of 17 in the mouse-human translocated chromosome.

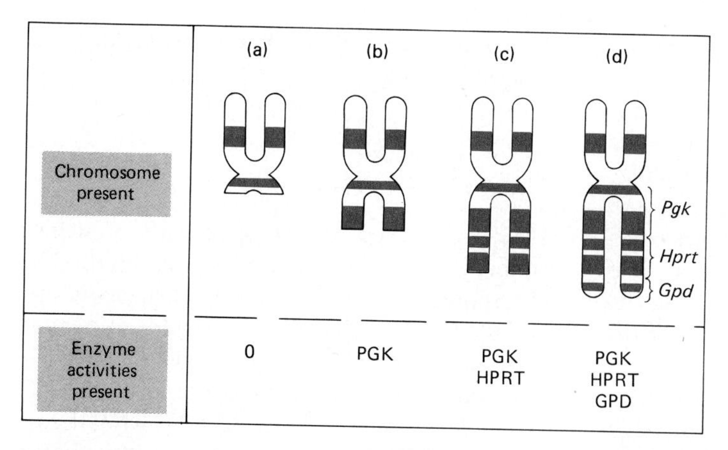


***Figure 12.31*** Regional mapping of human X-linked genes. By the comparison of the expressed cellular activities of the enzymes phosphoglycerate kinase (PGK), hypoxanthine-guanine phosphoribosyl transferase (HRPT), and glucose 6-phosphate dehydrogenase (GPD) with the missing and retained parts of the X chromosomes in the cells, the *Pgk, Hprt,* and *Gpd* genes can be assigned regionally to the chromosome. (a) All three genes must be present in the long arm of the chromosome since all three enzyme activities are absent in chromosomes with an intact short arm and a greatly deleted long arm. (b)–(d) The relative locations of the three genes are evident from enzyme activities correlated with regions retained and regions deleted in the long arm of the X chromosome. Chromosome breaks leading to deletions can be induced by particular viruses presented to the cells in culture.

particular viruses added to the cells (Fig. 12.31). The whole chromosome possessed all three genes, as seen by the presence of all three human enzyme activities. When the tip of the long arm was deleted, GPD activity was lost. When a larger segment was deleted, HPRT activity was lost along with GPD. When most of the long arm was deleted, all three enzyme activities were lost. By examining the concordance of progressively greater deletions with progressive loss of enzyme activities, the gene order could be determined. Gene *Gpd* must be at the tip since the GPD enzyme was missing in all clones with deletions. Gene *Hprt* must be the next gene in the sequence since HPRT was the next enzyme to be lost, and gene *Pgk* must be the farthest from the tip of the long arm because PGK enzyme activity was not lost until virtually all of the long arm had been deleted. These concordant results provided the basis for regional assignment of all three genes and for their order in that region of the chromosome.

More than 100 genes have been assigned to the human X chromosome by the combined methods of gene mapping by somatic cell genetics and by X-linked inheritance patterns found in family studies. More than 1100 autosomal genes have been identified but relatively few have as yet been assigned to particular autosomes.

Analysis of hybrid cells must be extended to a molecular level to a greater degree than is currently employed. Certain genes are not expressed in fibroblasts or other cell types used in somatic cell genetics. The genes for human hemoglobins, for example, are present but inactive in most kinds of somatic cells. In order to identify the gene or its chromosomal location in the absence of the gene product, it will be necessary to shift biochemical analysis from the product to the gene itself. Where such molecular methods can be used, unique segments of DNA can be identified and located even in the absence of the protein product of the gene. Globin genes can be recognized in any cell, even when globins are not made. We will discuss such studies later.

Somatic cell genetics does have limitations. Many genes have protein products that have not yet been identified. We wouldn't know what phenotype to seek without knowing the protein sponsored by the gene. We would have no molecular probes to find the gene itself. In other cases, the phenotype associated with a particular gene may be visible only in the organism and not in individual cells. Developmental processes may lead to the formation of more than five digits in persons having the gene for polydactyly. There is no way at pesent, however, for us to know how to identify this gene or its product in somatic cell hybrids since the phenotype associated with the polydactyly gene is recognized at the level of the organism and not of the individual cell.

Apart from these limitations of the moment, somatic cell genetics has been a very fruitful approach to human gene mapping. Considerable information has been gained about the human genome and many of its genes. Important data have also been obtained for prenatal diagnosis in human health care (see Section 5.11). There is every reason to expect somatic cell genetics to be a productive approach in the future.

## Molecular Cytogenetics

Cytogenetic analysis at the molecular level is an invaluable method for determining sequence organization in gene clusters and in individual genes. Some of these studies have been discussed or alluded to in this and in earlier chapters. Molecular cytogenetics provides detailed information because of the high resolution that can be achieved using electron microscopy and molecular methods such as hybridization and restriction fragment analysis. The principles are identical in any cytogenetic analysis, whether light microscopy, electron microscopy, or recombinant DNA and restriction enzyme methods are used to describe chromosomes and DNA molecules. Parallel lines of information are obtained from microscopic or molecular studies and from genetic studies. The different sets of information should confirm and support the interpretations made from any one set of information by itself.

### 12.11 Molecular Heteroduplex Analysis

Hybrid duplexes consisting of pairs of single-stranded nucleic acid molecules can be used to determine whether or not nucleic acids are complementary. The failure of duplex formation in all or part of a base sequence is evidence for noncomplementarity. Hybrid duplexes or heteroduplexes, may consist of DNA-DNA, DNA-RNA, or RNA-RNA components. Direct observation of duplexes and duplex segments is made by electron microscopy of the molecular preparation.

One of the clearest and earliest examples of genome analysis using molecular cytogenetic methods was provided by studies of rRNA repeated gene clusters in the toad *Xenopus laevis*. Classical cytogenetics had shown that loss of the nucleolar-organizing region (NOR) in an anucleolate mutant was paralleled by loss of the capacity to make rRNA and new ribosomes during embryogenesis of fertilized eggs that were homozygous for the NOR deletion. The localization of rRNA genes (rDNA) at the NOR was confirmed by molecular hybridization between rRNA and DNA from NOR-amplified sequences (see Fig. 10.14). The existence of about 450 copies of the rRNA gene per genome was also calculated from molecular studies. These data did not provide information, however, concerning the sequence organization of the tandem gene cluster or of any nontranscribed DNA regions.

The first information showing that tandemly arranged rRNA genes were separated by nontranscribed DNA spacer sequences came from electron microscopic studies by Oscar Miller, Barbara Hamkalo, and co-workers (Fig. 12.32). They obtained stunning photographs showing isolated rDNA in the process of actively transcribing rRNA molecules in preparations from amphibian oocyte nuclei. Many rRNA transcripts were being made at the same time from *each* gene in the cluster. All the rRNA genes appeared

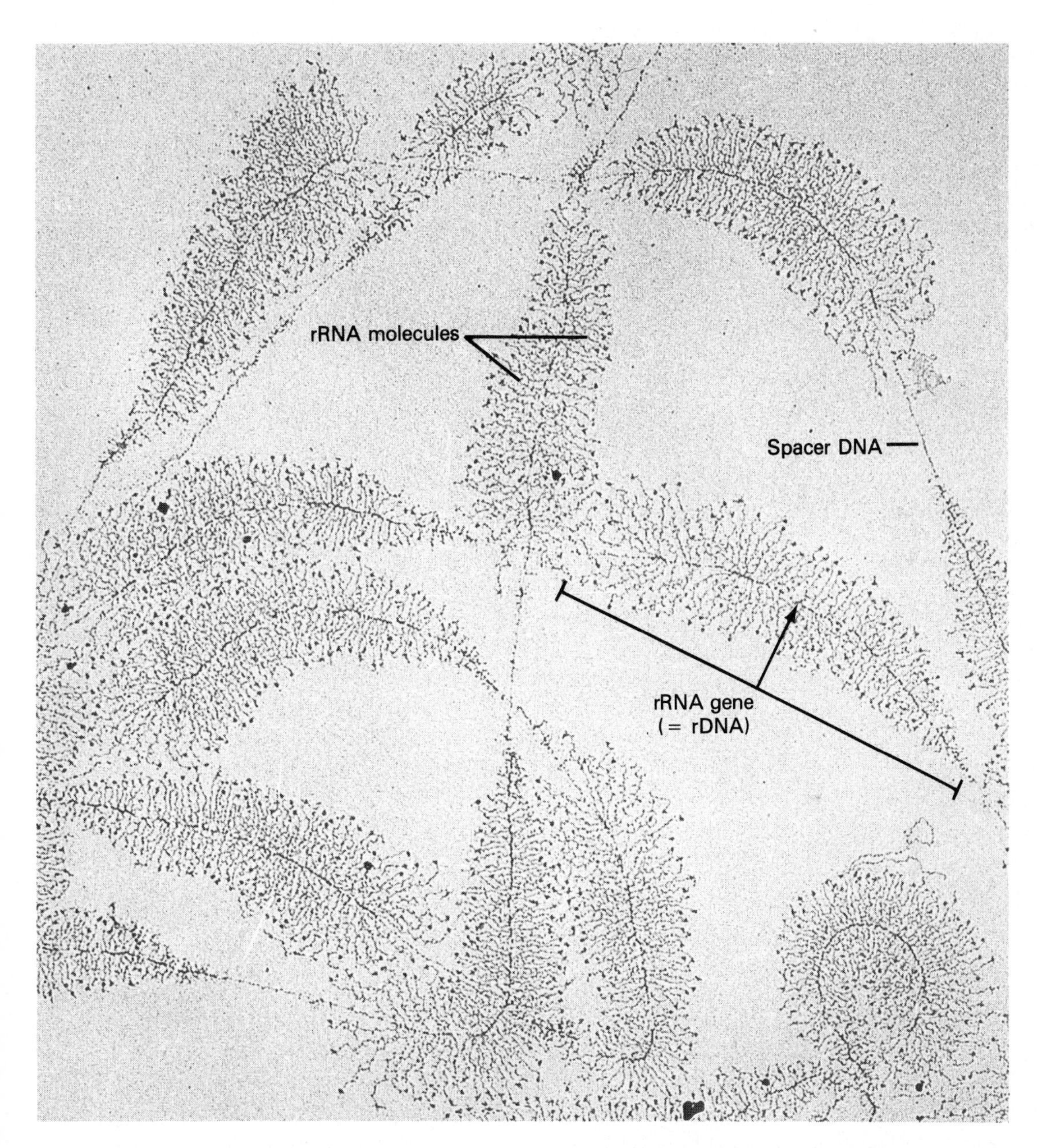


***Figure 12.32*** Electron micrograph of active ribosomal DNA (rDNA, or rRNA genes) in the process of transcribing rRNA, which appear here as fine fibrils in each arrow-shaped grouping. The tandemly repeated rRNA genes in this newt oocyte nucleus are separated by nontranscribing spacer DNA regions. (From O.L. Miller, Jr., and B. Beatty. 1969. *J. Cell Physiol.* **74**, Suppl. 1:225.)

to be transcribing in the same direction along the double-stranded DNA molecule, thus indicating that all the genes being copied were present in the same DNA strand. The measured length of each transcribing rDNA segment (= one rRNA gene) corresponded to the length expected for a 40S amphibian rRNA transcript (45S rRNA in mammals; see Section 9.5). DNA spacers between repeated rRNA genes remained bare, indicating that they were not being transcribed or at least that they were not being transcribed at the same time as rDNA.

Donald Brown, Igor Dawid, and others analyzed the organization of rRNA gene clusters by **denaturation mapping.** Duplex rDNA was exposed to mild denaturation conditions that led to strand separation in regions rich in A+T, while regions rich in G+C remained paired (Fig. 12.33). A+T regions are less stable than G+C regions, due to double hydrogen bonding between A and T residues but triple hydrogen bonding between G and C residues. Mild de-

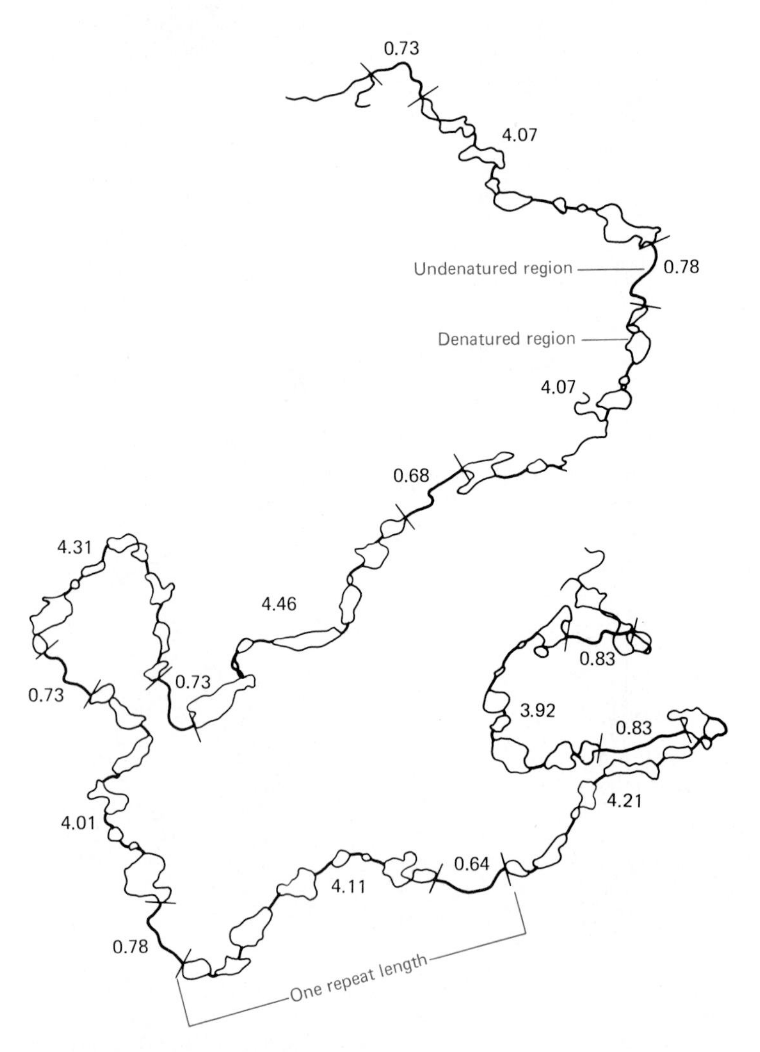


***Figure 12.33*** Tracing of an electron micrograph of one partly denatured segment of amplified rDNA from the toad *Xenopus mulleri.* This piece is 42 μm long and contains 8 complete repeats of the sequence coding for 40S pre-rRNA. Undenatured regions appear as solid, thicker lines, and denatured regions appear as open loops. Each repeat gene sequence contains one mainly denatured region (3.92–4.46 μm long) and one mainly undenatured region (0.64–0.83 μm long). (With permission from D.D. Brown, P.C. Wensink, and E. Jordan. 1972. *Journal of Molecular Biology* **63**:57, Plate 11b. Copyright © 1972 Academic Press Inc. [London] Ltd.)

naturation using heat or alkali will lead to differential strand separation, and partially denatured DNA can be photographed, measured, and assigned a particular predominating base composition in different segments of the molecule.

Measurements of partially denatured rDNA revealed alternating duplex gene regions, rich in G+C, and denatured spacers, rich in A+T. The gene was about 0.8 μm long and the denatured spacer regions were about 4 μm long. These values corresponded to Miller's measurements of transcribing rDNA molecules and to DNA-rDNA molecular hybridization data. Sequence organization of rDNA was thus verified as being tandem arrays of repeated rRNA-coding sequences alternating with noncoding spacers.

Base composition analysis of rRNA from different *Xenopus* species showed a very high degree of homology. This observation indicated that coding regions of rDNA in different species would be similar or even identical, according to base complementarity rules. Was the entire rDNA sequence homologous, including nontranscribed spacer DNA? Heteroduplex analysis provided the answer to this question and also revealed the existence of transcribed spacers between the 18S and 28S portions of the rRNA gene (Fig. 12.34).

When rDNAs from *Xenopus laevis* and *X. mulleri* were melted to single strands and then allowed to reanneal to the duplex state, heteroduplexes could be analyzed by electron microscopy to determine regions of homology and nonhomology. As predicted, 18S and 28S sequences of the two species paired; they were homologous. But nontranscribed long spacers between rRNA gene sequences remained unpaired; they were not homologous. In addition, a spacer between 18S and 28S sequences was also shown to be nonhomologous. Nonhomologous sequences remain unpaired and appear as bubbles or loops of single-stranded DNA. Homologous paired sequences are recognized by thicker duplex segments. This was one of the earliest demonstrations of the evolutionary conservation of related coding sequences. Considerable divergences in base sequence characterized spacer DNA evolution in the *same molecules*. Similar observations have since been made for many other tandemly repeated gene clusters, including genes coding for histone proteins and genes coding for 5S rRNA.

Heteroduplex mapping of individual genes is achieved by the same procedures as we just discussed for gene clusters. The initial discovery that eukaryotic mRNA is transcribed as a long

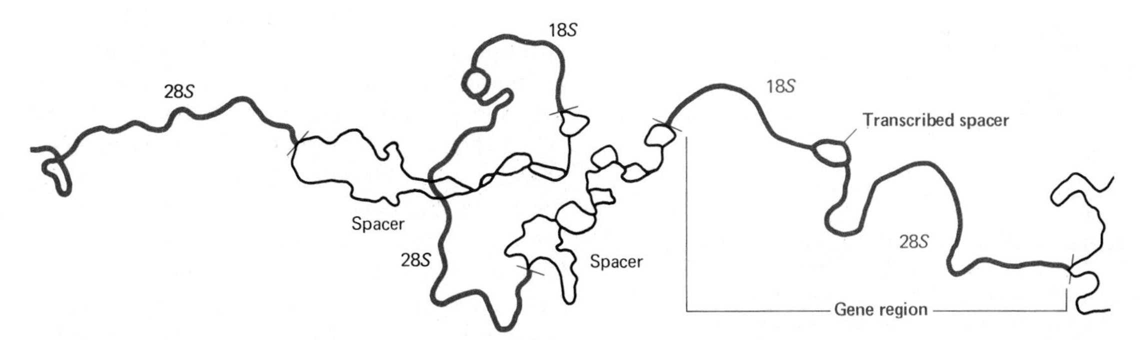


***Figure 12.34*** Tracing of an electron micrograph of heteroduplex DNA consisting of one strand of rDNA from *Xenopus laevis* and one from *X. mulleri*. The rRNA gene regions are shown in color and the untranscribed spacer DNA between genes is shown in black. Pairing (matching of DNA sequences of the two species) is very good in the gene region specifying the 18S and 28S rRNA sequences. Little base-pair matching is found in the spacers between genes, as evident from the many loops present. The rRNA sequence has been preserved in evolution of these species, but the spacer DNA has diverged almost completely *in the same time*. (From "The Isolation of Genes" by D.D. Brown. Copyright © 1973 by Scientific American, Inc. All rights reserved.)

precursor containing interspersed coding and noncoding segments was obtained by hybridizing cloned genic DNA with pre-mRNA transcripts and with mature mRNA transcripts (see Section 9.3). It was found that pre-mRNA was a copy of the entire gene sequence, but mature mRNA was considerably shorter than pre-mRNA. Heteroduplex cloned-DNA–mature-mRNA molecules revealed regions of single-stranded loops emanating from the DNA strand. These looped-out regions were later shown to correspond to noncoding intervening sequences that alternate with coding seqences, or exons, in the gene (Fig. 12.35). The number of loops corresponds exactly to the number of intervening sequences, and the paired regions in the heteroduplex correspond exactly to the number of coding sequences in a gene. These observations hold true for all the eukaryotic genes that have been analyzed in this way, including globin genes and ovalbumin genes. Through molecular analysis of base sequences, investigators could verify the lengths of coding and noncoding segments and determine the specific sequences.

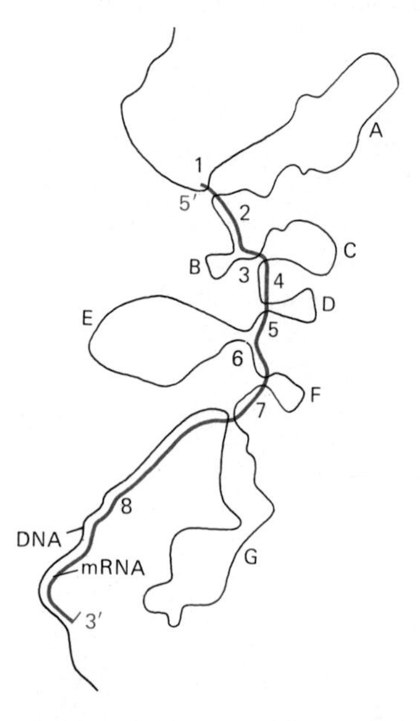


***Figure 12.35*** Tracing from an electron micrograph of a heteroduplex molecule formed by hybridization of the DNA (black) and mature mRNA (color) strands that encode the chicken ovalbumin protein. The mRNA sequence only includes the regions corresponding to the eight exons (1–8) of the gene. The regions corresponding to the seven introns (A–G) of the gene are evident from the size and location of the loops emanating from the encoded DNA strand (see Fig. 7.29). (Reprinted by permission from F. Gannon et al. 1979. *Nature* **278**:428–434. Fig. 3b. Copyrignt © 1979 Macmillan Journals Limited.)

Another method for analyzing base complementarity using the electron microscope is **R-loop mapping.** The method utilizes duplex DNA and single-stranded RNA molecules (Box 12.1). In addition to determining sequence organization in DNA and copied complementary RNA, R-loop mapping is very important in analyzing homologies between cellular DNA and viral RNA. As we will see in Chapter 14, homologies between normal cellular genes and the base sequence of RNA in vertebrate retroviruses became apparent from R-loop mapping as well as from base-sequencing studies.

## 12.12 Molecular Cytogenetics of Human Hemoglobin Disorders

Family studies and somatic cell genetics have provided the information on nonallelism of different globin genes and on gene assignments to human chromosomes. Two families of clustered genes, $\alpha$ and $\beta$, code for the globin polypeptides of hemoglobin molecules: $\alpha$ globin and $\alpha$-like globin genes are situated on chromosome 16, and $\beta$ globin and $\beta$-like globin genes are clustered in the distal portion of the short arm of chromosome 11 (see Fig. 12.29). With newly available methods for recombinant DNA technology, cloning DNA sequences in genomic libraries of restriction fragments, and base sequencing, a considerable amount of information has surfaced in a few years concerning aberrant globin genes and their contribution to human disorders involving hemoglobins. Molecular analysis of hemoglobin disorders began in the mid-1950s with Linus Pauling's discovery of electrophoretic differences between normal and sickle-cell hemoglobin molecules and with Vernon Ingram's discovery that these two hemo-

## BOX 12.1 R-Loop Mapping

R-loop analysis was first described by Thomas, White, and Davis (1976. *Proc. Nat. Acad. Sci. U.S.* **73**:2294). This method is used to map regions of the genome and deduce sequence organization and includes direct visualization of RNA-DNA hybrids in the electron microscope. RNA-DNA hybrids are more stable than the corresponding DNA-DNA duplexes, so that under certain incubation conditions RNA-DNA hybrids can form at regions of partial denaturation of the DNA duplex, as shown in (a). When such incubation is performed, a single-stranded R-loop is formed in the DNA region complementary to the added RNA, the loop having been displaced on formation of the RNA-DNA double-stranded segment just opposite the loop.

In the electron micrograph shown in (b) and the interpretive drawing shown in (c), 28S and 18S rRNA were added to duplex rDNA from *Drosophila melanogaster*. Each rRNA has hybridized to the complementary rDNA template strand, thus leading to looping out of displaced single-stranded DNA (the anticoding strand of rDNA). These molecules can be measured precisely to show the sequence organization of 28S and 18S coding regions in rDNA and the length and number of spacers (Sp 1, Sp 2) for which there is no mature rRNA transcript. (Photograph and drawing from P.K. Wellauer and I.B. Dawid. 1977. *Cell* **10**:193. Copyright © by the MIT Press.)

R-loop mapping has proven useful in many kinds of studies involving cloned DNA and in studies of RNA processing. THe R-loop pattern seen in β-globin DNA preparations to which pre-mRNA and mRNA of globin had been added clearly showed that pre-mRNA was transcribed from the entire DNA region and was later processed to a spliced mRNA complementary only to specific, identified parts of the β-globin gene sequence.

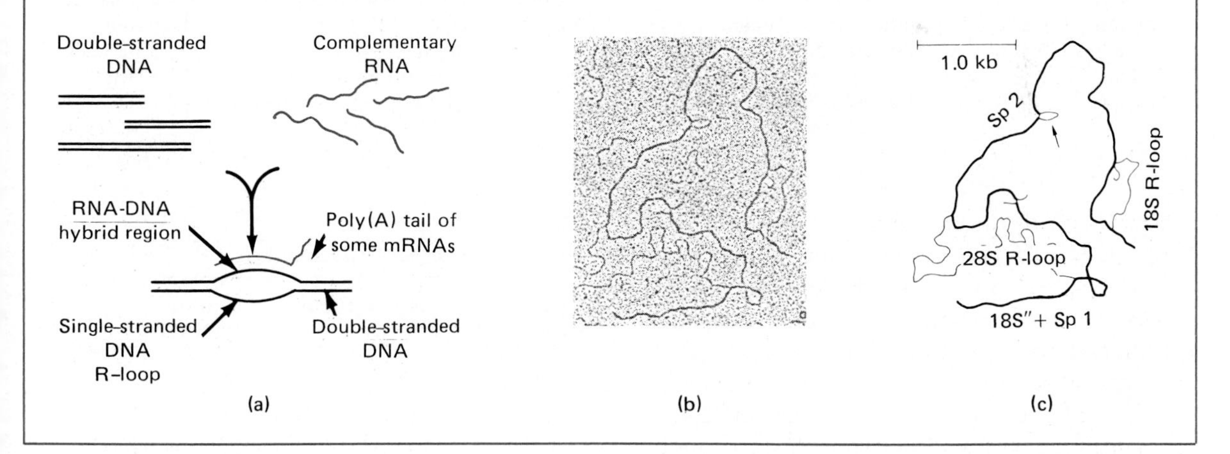


globins differed in only one amino acid in the β globin chain of 146 amino acids (Fig. 12.36).

Normal adult hemoglobin is a tetramer molecule comprised of two identical α globin chains and two identical β globin chains, signified by $\alpha_2\beta_2$. Stable hemoglobin (Hb) molecules formed during every stage of development, from the embryo to birth of the individual, must have a pair of α-like and a pair of β-like globins. Hb molecules made of four chains of the same type are unstable, and red blood cells carrying these molecules will precipitate out of the circulation

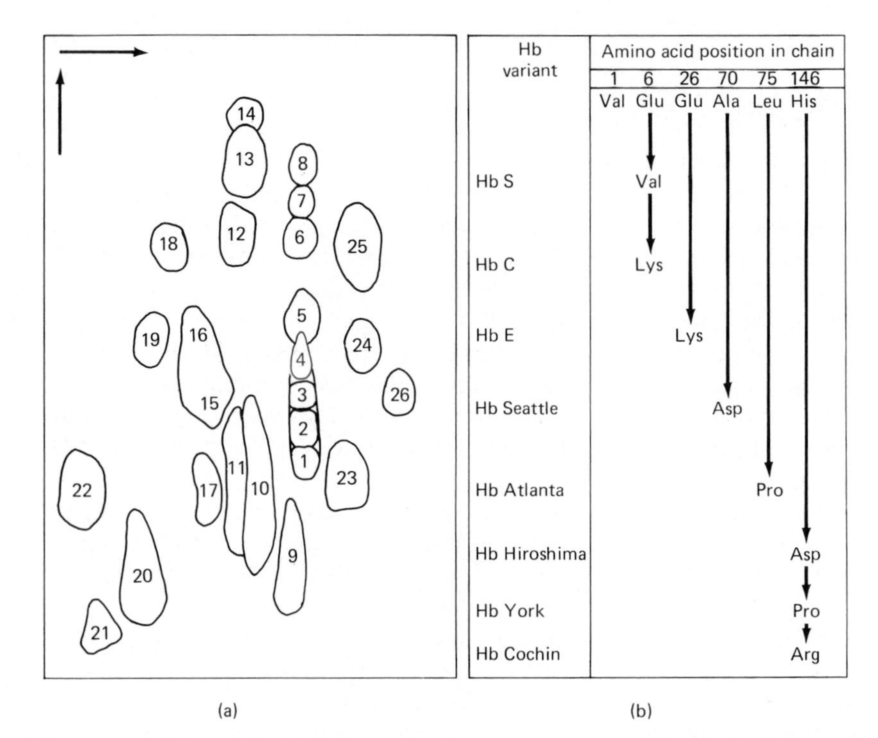


***Figure 12.36*** Molecular analysis of human hemoglobin variants. (a) Chromatographic separation of 26 fragments of the $\beta$-chain of hemoglobin produced by tryptic digestion. Fragment no. 4 in this protein digest contains the amino-terminus of the chain, including amino acid 6. Analysis of amino acids only in the one fragment (out of 26) which chromatographs differently in a mutant compared with this fingerprint of normal $Hb_{\beta}$ reduces the amount of work required to identify the molecular alteration. (b) Specific amino acid substitutions in human hemoglobin variants. Almost 100 of the 146 sites in the $Hb_{\beta}$ molecule have been found to vary in one or more individuals from one or more human populations. The $Hb_{\beta}$ gene has been mapped on chromosome 11.

stream, usually causing hemolytic (destruction of red blood cells) symptoms. Embryonic $\zeta_2\epsilon_2$ and fetal $\alpha_2\gamma_2$ hemoglobins are stable, as are $\alpha_2\delta_2$ or $\alpha_2\beta_2$ hemoglobins which are produced after birth and throughout life (Table 12.2).

A number of inherited disorders, such as sickle-cell anemia, are due to gene mutations that lead to amino acid substitutions. In addition, another class of variants includes hemoglobin disorders that are characterized by quantitative changes in the *amounts* of normal $\alpha$ and $\beta$ globins produced; certain of these disorders appear to involve alteration in globin-gene expression due to deletions within the cluster of $\alpha$ globin genes or the cluster of $\beta$ globin genes.

***Table 12.2*** Human hemoglobins.

| type of hemoglobin | globin chains present | amount synthesized |
|---|---|---|
| embryonic | $\zeta_2\epsilon_2$ (Gower I) | uncertain |
| | $\alpha_2\epsilon_2$ (Gower II) | uncertain |
| | $\zeta_2\gamma_2$ (Portland) | uncertain |
| fetal | $\alpha_2\gamma_2$ (Hb F) | major hemoglobin of fetus |
| adult | $\alpha_2\beta_2$ (Hb A) | major hemoglobin (98%) |
| | $\alpha_2\delta_2$ (Hb $A_2$) | minor hemoglobin (2%) |

*Table 12.3* Characteristics of $\beta$ thalassemias in humans.

| type of $\beta$ thalassemia | clinical symptoms | genetic defect | hemoglobins in adults |
|---|---|---|---|
| $\beta^+$ thalassemia | mild to severe anemia* | little processing of globin pre-mRNA to mature mRNA | low Hb A ($\alpha_2\beta_2$) low Hb F ($\alpha_2\gamma_2$) |
| $\beta^0$ thalassemia | mild to severe anemia* | improper regulation of gene expression | no Hb A low Hb F |
| $\delta\beta$ thalassemia | mild anemia | deletion of $\beta$ globin gene and part of $\delta$ globin gene | no Hb A or Hb $A_2$ Hb F only, in low amounts |
| HPFH disorder | none | deletion of $\beta$ and $\delta$ globin genes | no Hb A or Hb $A_2$ Hb F only, in adequate amounts |

*The relative severity of clinical symptoms varies considerably and is presumed to be due to variability in the nature and/or extent of the genetic defect in the individual.

These disorders are referred to as ***α* thalassemias** and ***β* thalassemias,** respectively. Their particular DNA defects have been discovered through molecular cytogenetic analysis, particularly within the past few years.

The two main types of $\beta$ thalassemias are $\beta^+$ and $\beta^0$ thalassemias (Table 12.3). The production of small amounts of $\beta$ globin in **$\beta^+$ thalassemics** appears to be due to mutations in the $\beta$ globin gene that alter processing of $\beta$ globin pre-mRNA transcripts. There is no $\beta$-globin synthesis at all in **$\beta^0$ thalassemics** homozygous for the structural gene defect. The symptoms of $\beta^+$ and $\beta^0$ thalassemia are somewhat variable, as are the specific genetic lesions in the $\beta$ globin gene cluster. Two other $\beta$ globin–related disorders are much rarer but have been more thoroughly characterized than other $\beta$ thalassemias. In **$\delta\beta$ thalassemic** individuals, no adult hemoglobin (Hb A, $\alpha_2\beta_2$, or $\alpha_2\delta_2$) is made at all; only fetal hemoglobin (Hb F, $\alpha_2\gamma_2$) is made. Hb F is made in insufficient amounts to compensate for missing Hb A, however, leading to mild anemia in $\delta\beta$ thalassemics. In individuals with **hereditary persistance of fetal hemoglobin** (**HPFH disorder**), only Hb F is made. The amounts of Hb F are adequate to compensate for the lack of Hb A molecules, and individuals with HPFH do not develop anemia.

From molecular hybridizations using cloned globin gene probes and from base sequencing, it has been found that only a segment of the 5′ end of the $\delta$ globin gene is present and that the remainder of the $\delta$ gene, spacer regions, and all of the $\beta$ globin genes are deleted in DNA from $\delta\beta$ thalassemic individuals (Fig. 12.37). DNA ob-

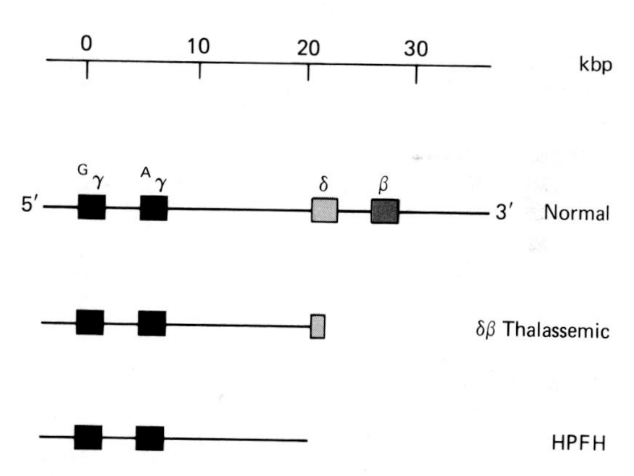


***Figure 12.37*** Deletions of $\delta$ and $\beta$ globin genes are responsible for phenotypic expressions of human blood disorders of the $\beta$-thalassemia type. Individuals having $\delta\beta$ thalassemia retain both $\gamma$ globin genes, part of $\delta$, and none of the $\beta$ globin gene. In individuals with HPFH disorder, the entire $\delta$-$\beta$ gene region is deleted. In both disorders the $\gamma$ globin genes guide synthesis of $\gamma$ chains of fetal hemoglobin, Hb F, ($\alpha_2\gamma_2$), but normal Hb A ($\alpha_2\beta_2$ or $\alpha_2\delta_2$) is not made because $\beta$ and $\delta$ globin genes are missing. (A. Bank et al., "Disorders of Human Hemoglobin," *Science* **207**:486–493, 1 February 1980. Copyright © 1980 American Association for the Advancement of Science.)

tained from individuals with HPFH disorder shows deletion of the *entire region* that carries δ and β globin genes. Production of γ globins continues in the absence of δ-β genes, but it is more restricted if some of the δ gene sequence remains (in δβ thalassemics) than if it is all deleted (HPFH). The relationship between γ globin gene expression and δβ globin genes is unclear at present, but the δβ globin genes must influence γ gene expression according to these cited observations. The role of δβ gene expression in the developmental switch from fetal globins to adult-type globins is under extensive study, utilizing variant DNAs obtained from patients.

There are two α globin genes on each one of the pair of chromosome 16, making a total of four α globin genes per individual. The range of symptoms that had been observed in patients with α thalassemias has now been shown to correspond to deletion of from one to four of these four α globin genes (Figure 12.38). The α thalassemias are characterized by a reduced rate of α globin synthesis, resulting in the presence of excess β-like chains that associate to form unstable and abnormal tetramer hemoglobin molecules, such as Hb Bart's ($\gamma_4$) and Hb H ($\beta_4$) tetramers. The clinical severity of α thalassemias increases with increasing numbers of deleted α globin genes. Individuals carrying only one of the four α globin genes develop Hb H disease, which is characterized by a mild hemolytic anemia. When all four α globin genes are deleted, the fetus dies *in utero* of a disorder called hydrops fetalis. All these cases of α globin gene deletions have been determined by restriction fragment analysis after restriction enzyme cleavage of DNA from patients, by molecular

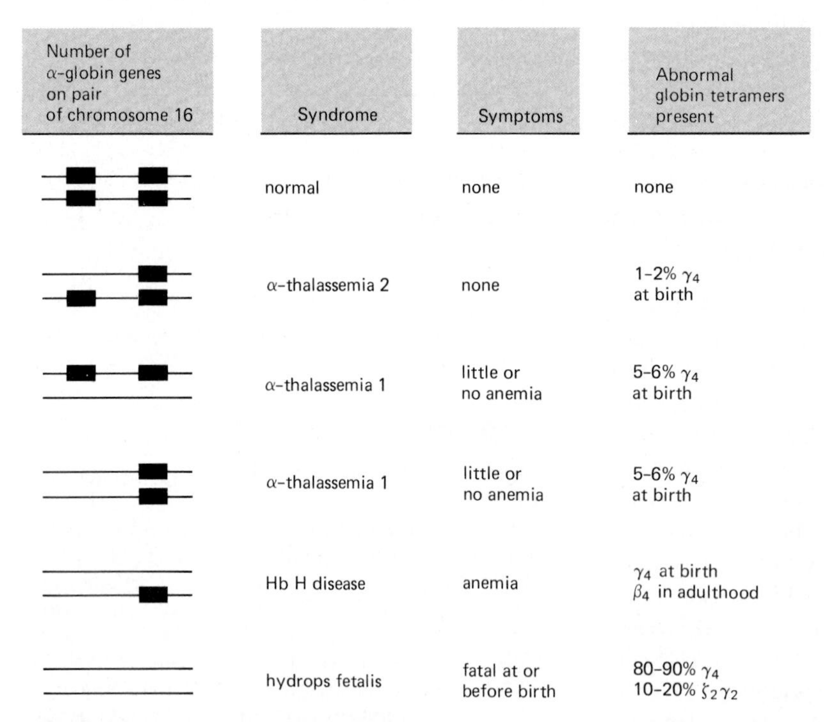


***Figure 12.38*** Schematic representation of genotypes associated with α-thalassemia syndromes and related human blood disorders. The consequences of the α-globin gene deletion depends on how many of the four genes are missing in an individual. (Reproduced, with permission, from the *Annual Review of Genetics,* Volume 14. © 1980 by Annual Reviews, Inc.)

***Table 12.4*** Associations between the size of *Eco*RI restriction fragments from the $\alpha$ globin gene region of human chromosome 16 and the number of $\alpha$ globin genes present on each chromosome 16 in normal and thalassemic individuals.

| | chromosome 16A | | chromosome 16B | |
|---|---|---|---|---|
| **condition** | **number of $Hb_{\alpha}$ genes** | **fragment size (kbp)** | **number of $Hb_{\alpha}$ genes** | **fragment size (kbp)** |
| normal | 2 | 23 | 2 | 23 |
| $\alpha$ thalassemia 2 | 2 | 23 | 1 | 19 |
| $\alpha$ thalassemia 2 | 1 | 19 | 2 | 23 |
| $\alpha$ thalassemia 1 | 1 | 19 | 1 | 19 |
| $\alpha$ thalassemia 1 | 0 | — | 2 | 23 |
| $\alpha$ thalassemia 1 | 2 | 23 | 0 | — |
| Hb H | 0 | — | 1 | 19 |
| Hb H | 1 | 19 | 0 | — |
| hydrops fetalis | 0 | — | 0 | — |



hybridizations between patients' DNA and normal cloned globin DNA, and by base sequencing.

Prenatal diagnosis of a fetus at risk for thalassemias can be carried out using DNA isolated from fetal cells in amniotic fluid obtained by amniocentesis. If the family has a history of thalassemia, DNA from both parents can be analyzed to see if globin gene deletions are present. Comparison of the size of restriction fragments carrying $\alpha$ globin genes and fragments in which the $\alpha$ globin gene cluster is present can provide information on the existence of deletions in the parental genomes. Gene sequencing is not necessary (Table 12.4). The prenatal test would be particularly important for carrier parents who both have one chromosome 16 with the two $\alpha$ globin genes deleted. The fetus would be at 25% risk to inherit one of each of these deleted parental chromosomes and develop hydrops fetalis. The parents would wish to consider whether or not to continue a pregnancy that must inevitably terminate with death of the fetus *in utero*. All other $\alpha$-thalassemics are viable, and prenatal diagnosis would not be necessary.

Individuals with easily detected $\delta\beta$ deletions would not produce a fetus at risk since $\delta\beta$ thalassemia and HPFH thalassemia produce either no clinical symptoms or only mild anemia. Restriction fragment comparisons for a fetus at risk for the more severe $\beta^+$ or $\beta^0$ thalassemias would not necessarily be informative since neither of these conditions is associated with detectable deletions.

It has been possible to utilize restriction fragment analysis in the case of a fetus at risk for sickle-cell anemia, even though deletions are not involved in this $\beta$ globin gene expression. Yuet Kan and Andrée Dozy discovered that $\beta$ globin restriction fragments produced by restriction enzyme *Hpa* I were usually 7 kbp or 7.6 kbp long. In many of the patients with sickle-cell anemia ($\beta^S\beta^S$ homozygotes), this $\beta$ globin restriction fragment was 13 kbp long. Parents of these patients were heterozygous for the $\beta^S$ *allele* ($\beta^A\beta^S$), and their globin DNA was recovered in fragments of two sizes: 7 or 7.6 kbp and

13 kbp. They were heterozygous for the alteration in *Hpa* I restriction site in this chromosomal region.

If the family has a history of sickle-cell anemia, prospective parents can be tested to see if each one makes both normal and sickle-cell hemoglobin molecules. This would indicate their heterozygous status. If each parent is then found to have two kinds of globin restriction fragments that correspond to their $\beta^A\beta^S$ genotype, a prenatal test can be conducted to determine whether the fetus is heterozygous or homozygous for the sickle-cell allele (Fig. 12.39). This test is accomplished by obtaining fetal DNA in amniotic fluid, performing restriction enzyme reactions using *Hpa* I, and determining whether the fetus carries two 13-kbp globin gene fragments. If only 13-kbp fragments are found, the fetus must be $\beta^S\beta^S$. If one 13-kbp fragment is found along with 7- or 7.6-kbp fragments, the fetus is heterozygous for the sickle-cell allele and is not at risk. Homozygous $\beta^A\beta^A$ genotypes would be evident in a fetus if the fetal DNA produced only 7- or 7.6-kbp restriction fragments of the $\beta$ globin gene cluster.

Prenatal diagnosis by restriction fragment analysis provides the safest means of determining the hemoglobin characteristics of the fetus. Removal of blood from the fetus is a tricky procedure and has a 5% to 10% failure rate, leading to death or injury to the fetus. Furthermore, fetal blood contains little or no adult hemoglobin needed to diagnose $\beta$ globin defects. A fetus carries predominantly Hb F ($\alpha_2\gamma_2$). Defects in $\alpha$ globins can be determined in fetal hemoglobin production directly, but the test introduces a risk of injury to the fetus. Analysis of fetal DNA in cases of suspected hydrops fetalis would be preferable to analysis of fetal blood.

Through classical cytogenetic and molecular cytogenetic approaches, considerable progress has been made in clinical practice and in our understanding of basic features of genetic organization and expression. We can look forward to learning a great deal more about the genome and genetic processes through the application of molecular methods to genetic and cytogenetic research.

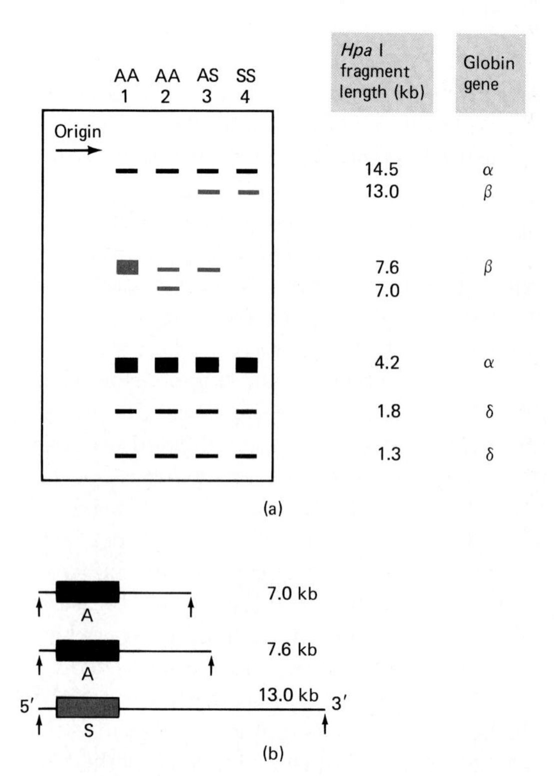


***Figure 12.39*** Restriction fragment analysis of $\beta$-globin DNA carrying the sickle-cell allele. (a) Homozygous normal ($\beta^A\beta^A$, lanes 1 and 2), heterozygotes ($\beta^A\beta^S$, lane 3), and homozygous recessives for the sickle-cell allele ($\beta^S\beta^S$, lane 4) can be distinguished by the particular $\beta$-globin restriction fragments produced by treating DNA with the restriction enzyme *Hpa* I. All other globin DNA fragments for all genotypes show no differences in the gel electrophoretogram. (b) Schematic representation of the three types of *Hpa* I fragments carrying the $\beta$-globin gene (A, normal allele; S, sickle-cell allele). *Hpa* I restriction sites are indicated by arrows. (From Y.W. Kan, and A.M. Dozy. 1978. *Proc. Nat. Acad. Sci. U.S.* **75**:5631–5635, Figs. 2 and 5.)

## Questions and Problems

***12.1*** When plants of the two species *Primula verticillata* and *P. floribunda* are crossed, they produce hybrid offspring that are vigorous but sterile.

***a.*** After many vegetative propagations of the hybrid primrose, a branch with fertile seeds developed. This was the origin of *Primula kewensis*. Explain the origin of this new, fertile primrose species.

***b.*** If the two parent species have 18 chromosomes each in their somatic cells (that is, $2n = 18$), how many bivalents can you expect in *P. kewensis* during meiosis?

***12.2*** *Raphanobrassica* is a hybrid derived from a cross between radish (*Raphanus*) and cabbage (*Brassica*), but it has no economic value because the plant has the roots of a cabbage and the leaves of a radish. If the hybrid has a somatic chromosome number of 36 and its cabbage parent produces gametes with 9 chromosomes, what is the haploid chromosome number in the radish parent? Can you explain the origin of the *Raphanobrassica* hybrid?

***12.3*** The recessive gene for bent wings (*bt*) occurs on the tiny fourth chromosome of *Drosophila melanogaster*. If a fly disomic for chromosome 4 and carrying the recessive *bt* allele on both chromosomes is crossed to a fly monosomic for this chromosome but carrying the $bt^+$ allele, what would be the phenotypes of the $F_1$ progeny?

***12.4*** In tomatoes a cross is made between a normal female plant trisomic for chromosome 6 and a disomic male having compound inflorescence (*s/s*).

***a.*** Assuming the gene *s* is located on chromosome 6, give the kinds and ratio of phenotypes in the progeny of a trisomic $F_1$ female testcrossed to a disomic *s/s* male.

***b.*** What would be the result of the same cross if gene *s* is not located on chromosome 6?

***12.5*** A person is diagnosed for Turner syndrome. Her brother and two uncles had glucose 6-phosphate dehydrogenase (GPD) deficiency, an X-linked condition. The family pedigree is shown below. Which X chromosome is missing in the individual with Turner syndrome (GPD deficient = shaded, GPD normal = unshaded)?

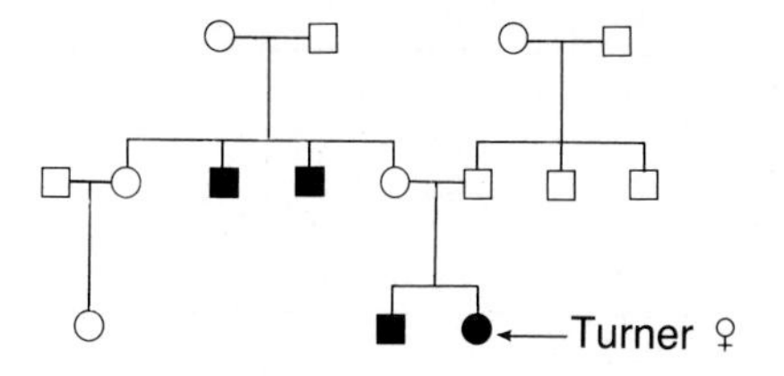


***12.6*** On chromosome 2 in *Drosophila melanogaster*, the normal linkage map includes purple (*pr*) 54.5; vestigial (*vg*) 67; Lobe (*L*) 72; arc (*a*) 99.2; brown (*bw*) 104. A cross

$$\frac{+ + L + +}{pr\ vg + a\ bw} ♀ \times \frac{pr\ vg + a\ bw}{pr\ vg + a\ bw} ♂$$

produced the following recombination percentages; *pr–vg*, 10%; *vg–L*, 0.2%; *L–a*, 1.4%; and *a–bw*, 5%. If the heterozygous female was an inversion heterozygote, what regions of chromosome 2 were involved in the inversion?

***12.7*** Backcrosses between doubly heterozygous individuals from corn strains with purple, shrunken kernels and doubly recessive strains with white, smooth kernels usually produce progeny segregating into four phenotypic classes in a 1 : 1 : 1 : 1 ratio. One cross, however, produced 300 purple, shrunken : 40 purple, smooth : 43 white, shrunken : 295 white, smooth progeny.

***a.*** Explain the probable basis for the different results in the one cross.

***b.*** What cytological test would you perform to verify the interpretation in (a)?

***c.*** How many map units apart are the two genes in the unusual heterozygous purple, shrunken parental plant?

***12.8*** A corn plant homozygous for a reciprocal translocation between chromosomes 2 and 5 and for the *Pr* allele, which governs seed color, was crossed with a normal *pr/pr* plant. The $F_1$ progeny were semisterile and phenotypically Pr. A testcross with the normal parent gave the following results:

1528 semisterile Pr

290 semisterile pr

372 normal Pr

1454 normal pr

How far is the *Pr/pr* locus from the translocation point?

***12.9*** Inversion heterozygosity reduces the number of recombinants recovered due to selective elimination of products of crossing over within the inversion loop.

***a.*** If pairing of normal chromosomes at meiosis is represented by

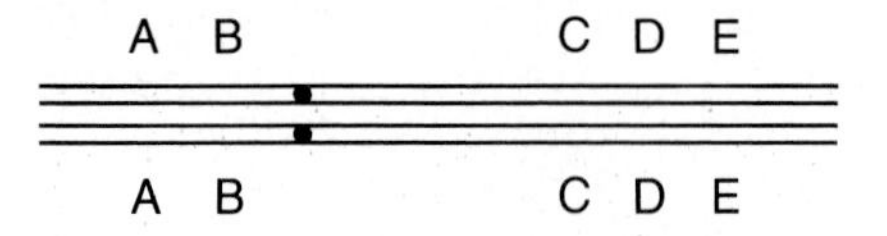


diagram chromosome pairing at pachynema, without crossing over, where one chromosome has a pericentric inversion due to break points between A and B and between C and D.

***b.*** Diagram pairing in the inversion heterozygote in (a), with one crossover between the centromere and C.

***c.*** Diagram the chromosomes shown in (b) at the end of Meiosis II, and indicate which chromosomes probably would not result in viable gametes.

***12.10*** Salivary chromosome preparations for *Drosophila melanogaster* are particularly useful for cytogenetic studies since these chromosomes show specific band patterns and homologous chromosomes are paired in the interphase nucleus. Bands designated by letters on the long arm of chromosome 2 can be represented simply in a diagram as

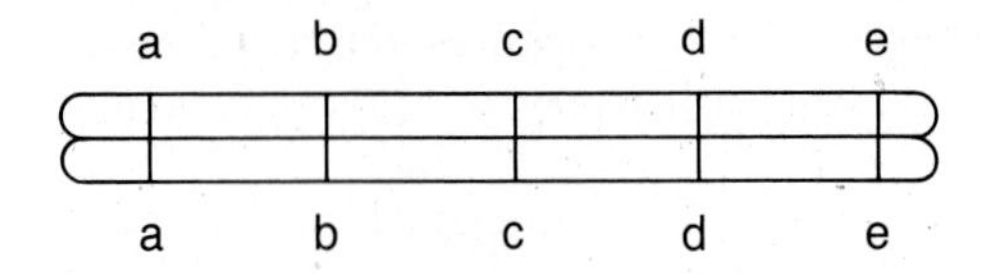


Draw a simplified diagram for paired chromosomes in which one member of the pair has the following:

***a.*** A large deletion.

***b.*** A large duplication.

***c.*** An inversion.

***d.*** A nonreciprocal translocation with chromosome 3.

***e.*** A reciprocal translocation with the X chromosome.

***12.11*** Three mouse-human somatic cell hybrid clones retained particular human chromosomes, as shown in the following clone panel (+ = present, − = absent):

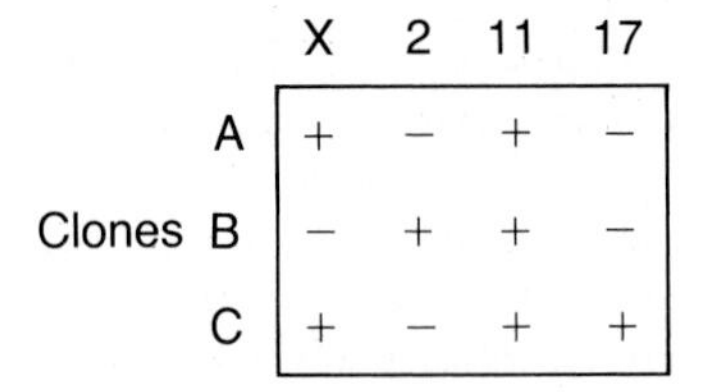

| | | Human chromosomes | | | |
|---|---|---|---|---|---|
| | | X | 2 | 11 | 17 |
| | A | + | − | + | − |
| Clones | B | − | + | + | − |
| | C | + | − | + | + |

Assays for four human enzymes were made with all three clones, with the following results (+ = present, − = absent):

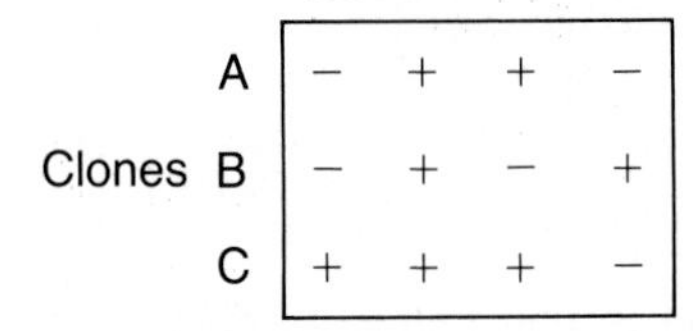

| | | Enzyme activity | | | |
|---|---|---|---|---|---|
| | | TK | LDH | PGK | AHH |
| | A | − | + | + | − |
| Clones | B | − | + | − | + |
| | C | + | + | + | − |

***a.*** Identify the human chromosome carrying the gene for each enzyme.

***b.*** Are any of these genes syntenic?

***12.12*** Three mouse-human somatic cell hybrid clones retained particular human chromosomes, as shown in the following clone panel (+ = present, − = absent):

| | | Human chromosomes | | | | | | | |
|---|---|---|---|---|---|---|---|---|---|
| | | X | 3 | 5 | 7 | 9 | 17 | 20 | 21 |
| | A | + | + | + | + | − | − | − | − |
| Clones | B | − | + | − | + | − | + | − | + |
| | C | − | − | + | + | − | − | + | + |

Assays for eight human enzymes were made with all three clones, as follows (+ = present, − = absent):

| | | Enzyme activity | | | | | | | |
|---|---|---|---|---|---|---|---|---|---|
| | | $\alpha$ | $\beta$ | $\gamma$ | $\delta$ | $\epsilon$ | $\zeta$ | $\eta$ | $\theta$ |
| | A | + | − | + | + | − | + | − | − |
| Clones | B | + | − | − | + | + | − | + | + |
| | C | + | + | + | − | − | + | − | + |

*a.* Identify the human chromosome carrying the gene for each enzyme.
*b.* Which genes are syntenic?
***12.13*** Ribosomal DNA and rRNA were isolated and purified from oocytes of the toad species *Bufo americanus* and *B. fowleri,* and studies of DNA homology were carried out using heteroduplex analysis.
*a.* How would you obtain single-stranded rDNA for hybridizations?
*b.* Which rDNA strands from each species would you incubate to obtain DNA-DNA molecular hybridizations? Explain.
*c.* Which rDNA strands would you incubate with which rRNAs to obtain molecular hybridizations? Explain.
*d.* What would be the appearance of heteroduplex DNA from the two species if divergence had occurred in noncoding spacers but not in coding regions? Draw a possible heteroduplex figure to illustrate your answer.
***12.14*** Isolation of genes *in vitro* has been achieved in various laboratories.
*a.* What properties of the major ribosomal RNA gene made it simpler to isolate than other genes?
*b.* Describe the assay system that made it possible for D. Brown and co-workers to detect a given gene in a mixture of related DNA molecules?
*c.* It is possible to introduce the gene coding for rRNA in *Xenopus* into a bacterial cell-line. How can you detect the particular bacterial clones containing the *Xenopus* rRNA gene?
***12.15*** Two people known to have $\alpha$-thalassemia 1 syndrome have had two children, one with no symptoms and no abnormal globin tetramers in her hemoglobin and another with Hb H disease.
*a.* How many $\alpha$ globin genes does each parent have?
*b.* If their normal child had four $\alpha$ globin genes, what would be the number of $\alpha$ globin genes on each chromosome 16 in each parent?
*c.* If their normal child actually has undetected $\alpha$ thalassemia 2, what would be the number of $\alpha$ globin genes on each chromosome 16 in each parent?
*d.* Since their second child has Hb H disease, is (b) or (c) the more likely situtation for the parental $\alpha$ globin gene distribution?
***12.16*** In 1983 several patients with sickle-cell anemia were treated with a drug that induced synthesis of fetal hemoglobin (Hb F, $\alpha_2 \gamma_2$), which ordinarily is not made in adults. One or more of the patients experienced relief from the anemia as a result of the treatment, and these patients did have reasonable amounts of Hb F in their blood.
*a.* What was the probable effect of the drug at the molecular level?
*b.* Why would Hb F contribute to relief of symptom in individuals with sickle-cell anemia?
***12.17*** Restriction fragment analysis of $\beta$ globin–encoded DNA revealed that DNAs from individuals with normal Hb A and of genotype $\beta^A\beta^A$ were cut by restriction enzyme *Hpa* I to produce restriction fragments 7.0 or 7.6 kbp long, as determined by gel electrophoresis. *Hpa* I restriction fragments from individuals with sickle-cell anemia (Hb S, genotype $\beta^S\beta^S$), however, were found to be 13 kbp long in many cases.
*a.* What would be the gel pattern of restriction fragments for $\beta^A\beta^S$ heterozygotes?
*b.* If two prospective parents, each heterozygous for $\beta^S$, wished to know whether the fetus carried by the woman was at risk for sickle-cell anemia, what test could be performed with fetal material obtained by amniocentesis?
*c.* What gel pattern would indicate the fetus was at risk for sickle-cell anemia?
***12.18*** Define the following terms: *a.* polyploid *b.* aneuploid *c.* trisomic *d.* structural chromosome aberration *e.* pericentric inversion *f.* somatic cell hybrid *g.* HAT medium *h.* synteny *i.* thalassemia

## References


Bank, A., J.G. Mears, and F. Ramirez. 1980. Disorders of human hemoglobin. *Science* **207**:486.

Brown, D.D. Aug. 1973. The isolation of genes. *Sci. Amer.* **230**:20.

Chang, J.C., and Y.W. Kan. 1982. A sensitive new prenatal test for sickle-cell anemia. *New Eng. J. Med.* **307**:30.

Cohen, S.N. July 1975. The manipulation of genes. *Sci. Amer.* **233**:24.

Drets, M.E., and M.W. Shaw. 1971. Specific banding patterns of human chromosomes. *Proc. Nat. Acad. Sci. U.S.* **68**:2073.

Ephrussi, B., and M.C. Weiss. Apr. 1969. Hybrid somatic cells. *Sci. Amer.* **220**:26.

Epstein, C.J., and M.S. Golbus. 1977. Prenatal diagnosis of genetic diseases. *Amer. Scientist* **65**:703.

Fuchs, F. June 1980. Genetic amniocentesis. *Sci. Amer.* **242**:47.

Jeffreys, A.J., and S. Harris. 1982. Processes of gene duplication. *Nature* **296**:9.

Kan, Y.W., and A.M. Dozy. 1978. Polymorphism of DNA sequence adjacent to human $\beta$-globin structural gene: Relationship to sickle mutation. *Proc. Nat. Acad. Sci. U.S.* **75**:5631.

Klein, G. 1981. The role of gene dosage and genetic transpositions in carcinogenesis. *Nature* **294**:313.

Knudson, A.G., Jr., et al. 1976. Chromosomal deletion and retinoblastoma. *New Eng. J. Med.* **295**:1120.

Lejeune, J., R. Turpin, and M. Gauthier. 1959. Le mongolisme, premier exemple d'aberration autosomique humaine. *Ann. Génét.* **1**:41.

Lewin, R.A. 1981. Evolutionary history written in globin genes. *Science* **214**:426.

McKusick, V.A. 1982. The human gene map. *Genetic Maps.* **2**:327.

Maniatis, T., et al. 1980. The molecular genetics of human hemoglobins. *Ann. Rev. Genet.* **14**:145.

Miller, O.L., Jr. Mar. 1973. The visualization of genes in action. *Sci. Amer.* **229**:34.

Omenn, G.S. 1978. Prenatal diagnosis of genetic disorders. *Science* **200**:952.

Orkin, S.H. 1978. The duplicated human $\alpha$ globin genes lie close together in cellular DNA. *Proc. Nat. Acad. Sci. U.S.* **75**:5950.

Orkin, S.H., and A. Michelson. 1980. Partial deletion of the $\alpha$-globin structural gene in human $\alpha$-thalassemia. *Nature* **286**:538.

Orkin, S.H., et al. 1982. Improved detection of the sickle mutation by DNA analysis. *New Eng. J. Med.* **307**:32.

Phillips, R.L., and C.R. Burnham, eds. 1977. Benchmark Papers in Genetics, Vol. 6: *Cytogenetics.* Stroudsberg, Pa.: Dowden, Hutchinson, & Ross.

Rowley, J.D. 1983. Human oncogene locations and chromosome aberrations. *Nature* **301**:290.

Ruddle, F.H. 1981. A new era in mammalian gene mapping: Somatic cell genetics and recombinant DNA methodologies. *Nature* **294**:115.

Ruddle, F.H., and R.S. Kucherlapati. July 1974. Hybrid cells and human genes. *Sci. Amer.* **228**:82.

Swanson, C.P., T. Merz, and W.J. Young. 1981. *Cytogenetics,* 2nd ed. Englewood Cliffs, N.J.: Prentice-Hall.

Van der Ploeg, L.H.T., et al. 1980. $\gamma$-$\delta$-Thalassemia studies showing that deletion of the $\gamma$- and $\delta$-genes influences $\beta$-globin gene expression in man. *Nature* **283**:637.

Wellauer, P.K., and I.B. Dawid. 1978. Ribosomal DNA in *Drosophila melanogaster.* II. Heteroduplex mapping of cloned and uncloned rDNA. *J. Mol. Biol.* **126**:769.

Yunis, J.J., and O. Prakash. 1982. The origin of man: A chromosomal pictorial legacy. *Science* **215**:1525.

# CHAPTER 13

# Extranuclear Genetics

Genes situated outside the nuclear genome may influence phenotypic expression of the cell or organism in which they occur. Extranuclear factors can be identified by unique patterns of inheritance, which differ in one or more ways from nuclear inheritance patterns. In particular, extranuclear genes are often transmitted by only one parent during sexual reproduction. Extranuclear genes segregate rapidly during mitotic divisions, whereas nuclear genes segregate during meiosis.

The principal extranuclear genes in eukaryotic cells are parts of the DNA genome in mitochondria and in chloroplasts. These organelle genomes code for rRNAs and tRNAs specific for their own translation apparatus, on which organelle-synthesized mRNAs are translated into organelle-specific polypeptides. Virus and plasmid genomes also provide extranuclear genetic information that directly influences both host and viral phenotypic expression. Unlike mitochondrial or chloroplast DNAs, viral and plasmid nucleic acids may exist as free extranuclear systems in the host cell or as integrated components of the host linkage group(s).

## Extranuclear Inheritance

The non-Mendelian pattern of inheritance of extranuclear characteristics reflects the difference in transmission and segregation between these genes and genes on chromosomes in the nucleus of the same cells. The general criteria that describe **extranuclear inheritance,** therefore, are ones which emphasize the lack of correspondence between transmission and segregation patterns expected for genes on nuclear chromosomes and genes that are located elsewhere. Genetic analysis of extranuclear transmission patterns includes experiments that are designed to *eliminate the possibility of nuclear-gene inheritance.* If the nucleus can be ruled out as the location and sponsor of the inheritance pattern, the conclusion can be made that some extranuclear system is responsible for the observed genetic phenomena.

Experimental results must, of course, provide evidence showing that such extranuclear traits are indeed inherited and not simply the result of transient changes in phenotype. The principal evidence showing that such traits are inherited is that the traits persist indefinitely in populations and are transmitted in sexual and asexual reproduction.

### 13.1 Extranuclear Inheritance Patterns

The nature of extranuclear inheritance can be demonstrated by the pattern of transmission from parents to progeny and by one or more additional criteria. Although the criteria vary from one system to another, the basic extranuclear pattern can be demonstrated. A number of useful experimental tests or observations can be applied to identify extranuclear patterns; we will briefly describe five of these.

***1.*** *Reciprocal crosses yield different progenies.* When wild-type and presumptive extranuclear mutants are crossed reciprocally, progeny of each cross may resemble only the female parent (Fig. 13.1). For example, in *Neurospora crassa,* the progeny of *poky* ♀ × wild-type ♂ are *poky,* but wild-type ♀ × *poky*

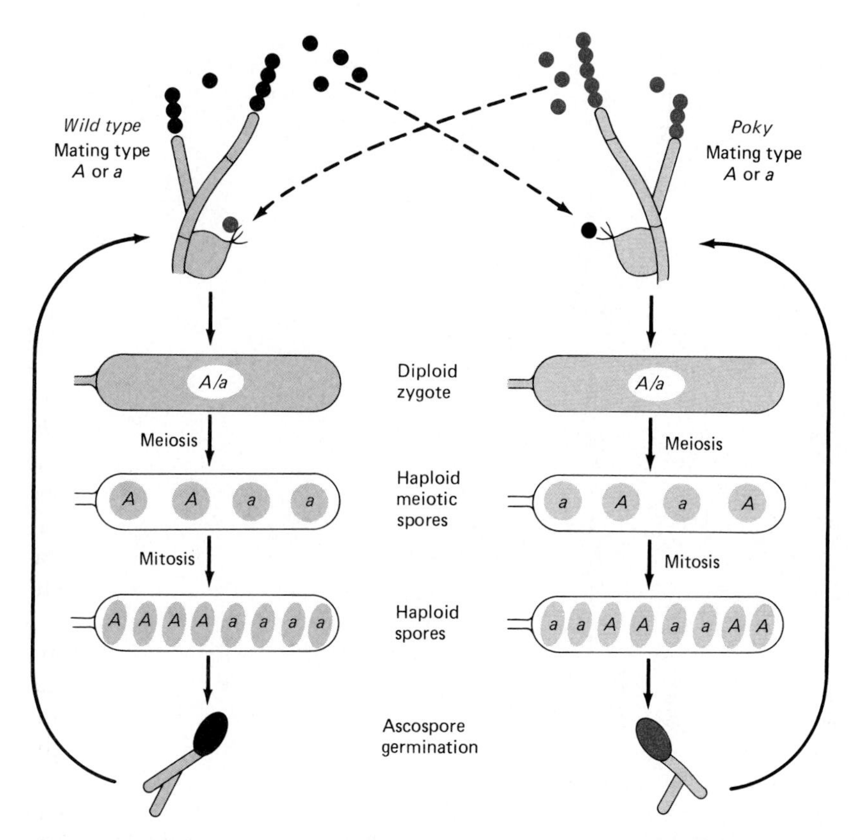


***Figure 13.1*** Reciprocal crosses usually yield different progenies in relation to extranuclear traits, but nuclear-encoded traits show the expected Mendelian pattern of segregation in the same or in different crosses. The mating type alleles *A* and *a* segregate 4:4, but all the ascospores are either wild type or *poky,* depending on the female parent or strain of *Neurospora crassa.*

♂ yields wild-type progeny. In this example of **uniparental inheritance,** only maternal zygotes are produced. In other species genes may be transmitted from both parents to produce biparental zygotes or, rarely, from only the male parent to produce paternal zygotes. Whatever the nature of the progeny, reciprocal crosses give different results. By and large these differences are seen as different proportions of maternal, biparental, and paternal zygotes in different progenies due to different transmission frequencies of alleles from the two parents.

One reason for transmission mainly through the female parent is that the female gamete supplies all or most of the cytoplasm to the zygote, whereas the male often contributes only a nucleus. This situation appears to be the case in *Neurospora* and in many plants and animals that produce a large egg and a relatively small sperm. The relative contributions by parents is probably not the explanation for uniparental inheritance in other species, such as the isogamous alga *Chlamydomonas reinhardi* (Fig. 13.2). We have not yet identified all of the mechanisms responsible for uniparental inheritance or for variable proportions of different kinds of zygotes in a single progeny.

In the budding yeast *Saccharomyces cerevisiae,* extranuclear traits may be transmitted through either the *a*- or *α*-mating type parent or from both. Biparental zygotes are produced in relatively high frequencies from parents that differ in alleles of structural genes. Uniparental zygotes are usually produced in crosses

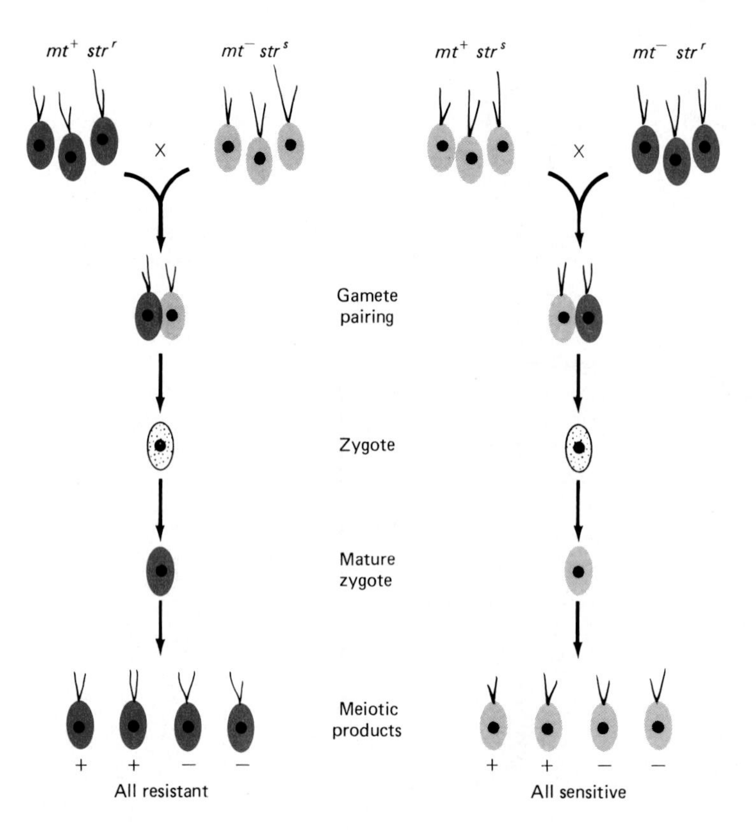


***Figure 13.2*** Although each parental cell seems to contribute an equal amount of nuclear and cytoplasmic materials in *Chlamydomonas* zygote formation, extranuclear traits generally show uniparental inheritance through the mating-type + ($mt^+$) parent. Nuclear alleles, such as $mt^+/mt^-$ segregate as expected in the 2:2 tetrad ratio, whereas the extranuclear alleles for resistance or sensitivity to streptomycin segregate 4:0 or 0:4.

between wild-type, respiration-sufficient, grande (*rho*-plus, or $\rho^+$) strains and *rho*-minus ($\rho^-$) petite strains which carry deletions that are responsible for the extranuclear trait of respiration deficiency (Fig. 13.3). Some petites give no petite zygotes in crosses with grande; these are called *neutral petites.* Most petites will produce some percentage of petite zygotes in crosses with grande, and up to 99% of the progeny may be petite in cases involving *suppressive petite* strains. We will discuss these mutants more fully later in the chapter.

The pattern of transmission through only one parent is fairly typical, but it has no consistency with regard to the particular parent that transmits; transmission may be always or usually through the same mating type or sex or through either mating type or sex. Because of this variability, no rule of inheritance or universal prediction will serve for all patterns of extranuclear transmission. Each case must be analyzed by its own features.

**2.** *Progeny show a non-Mendelian segregation ratio in a tetrad of meiotic products.* In each of the three species we have mentioned, the zygote undergoes meiosis to produce a tetrad of spores. When the four products of a single meiotic cell are examined, all four cells often have the same phenotype, that is, the phenotypic ratio of 4 : 0 (all like one parent) or 0 : 4 (all like the other parent). Pairs of alleles for

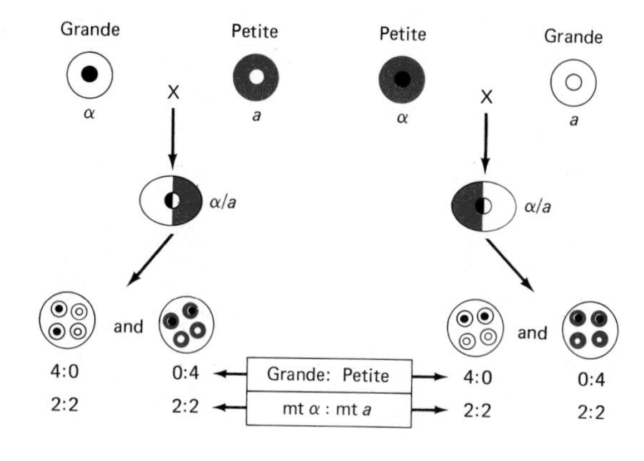


***Figure 13.3*** In the yeast *Saccharomyces cerevisiae*, extranuclear petite deletion mutants show 0:4 and 4:0 tetrad ratios in crosses with grande strains, whereas nuclear alleles for mating type segregate 2:2 in the same tetrads. Suppressive petite × grande yields varying percentages of 4 grande:0 petite tetrads, whereas neutral petite × grande yields 100% 4:0 tetrads.

nuclear-gene markers, however, segregate 2 : 2 in these same tetrads (see Fig. 13.3). These results indicate at least two features of the system : (1) the nucleus behaves normally and meiosis is not aberrant since known nuclear pairs of alleles segregate in the expected 2 : 2 ratio and (2) extranuclear alternatives may not segregate at meiosis. From these results we can infer that extranuclear factors are not located on chromosomes in the nucleus; otherwise they would behave in the same way as pairs of marker alleles, which we know are located on chromosomes. We cannot, however, infer where the extranuclear factors are located until more experimental evidence is obtained.

**3.** *Extranuclear factors cannot be mapped on any chromosome in the nuclear genome.* Confirming evidence can be obtained to verify that extranuclear factors are not situated in the nuclear genome, by tests showing that these factors are not linked to any known genes in any chromosome in the nucleus. Extranuclear factors assort independently of nuclear genes in every chromosome, and they therefore cannot be a part of any nuclear linkage group. Extranuclear genes will, however, show linkage to each other within an extranuclear genetic system.

**4.** *Extranuclear factors are transmitted through the cytoplasm and not through the nucleus.* The *heterokaryon test* can provide evidence for a cytoplasmic rather than a nuclear location of extranuclear factors in species that can maintain genetically different marker nuclei in a common cytoplasm. When haploid, multinucleate mycelia of *Neurospora* wild-type and *poky* strains fuse together, the heterokaryon that forms will have two genetically different kinds of haploid nuclei which remain separate in the mixed cytoplasm. When spores develop asexually from different hyphae of the mycelium, a single haploid nucleus is enclosed along with some cytoplasm (Fig. 13.4). When these spores are incubated and allowed to develop, the mycelium of these asexual progeny will be either wild type or *poky*. Either one of the two kinds of haploid nucleus may be present, as determined by gene markers. This shows that the source of the nucleus has no influence on poky phenotypic development; either nucleus can be found in poky progeny. Since the nucleus does not control the poky phenotype, the cytoplasm is the only other possible source of the inheritance factor transmitted from the heterokaryon to its asexual spore progeny.

**5.** *Extranuclear genes in general, and organelle genes in particular, usually segregate rapidly during mitotic divisions.* Such *vegetative segregation* can be seen quite clearly in variegated plants, in which green, white, and mixed cell lineages occur in variable amounts in different plants of a single progeny.

By taking these separate lines of evidence into consideration, we see that extranuclear factors apparently are not located on chromosomes in the nucleus. They probably are situated in the cytoplasm, which has been demonstrated to influence inheritance. The exact cytoplasmic location can be determined by other tests, which we will discuss.

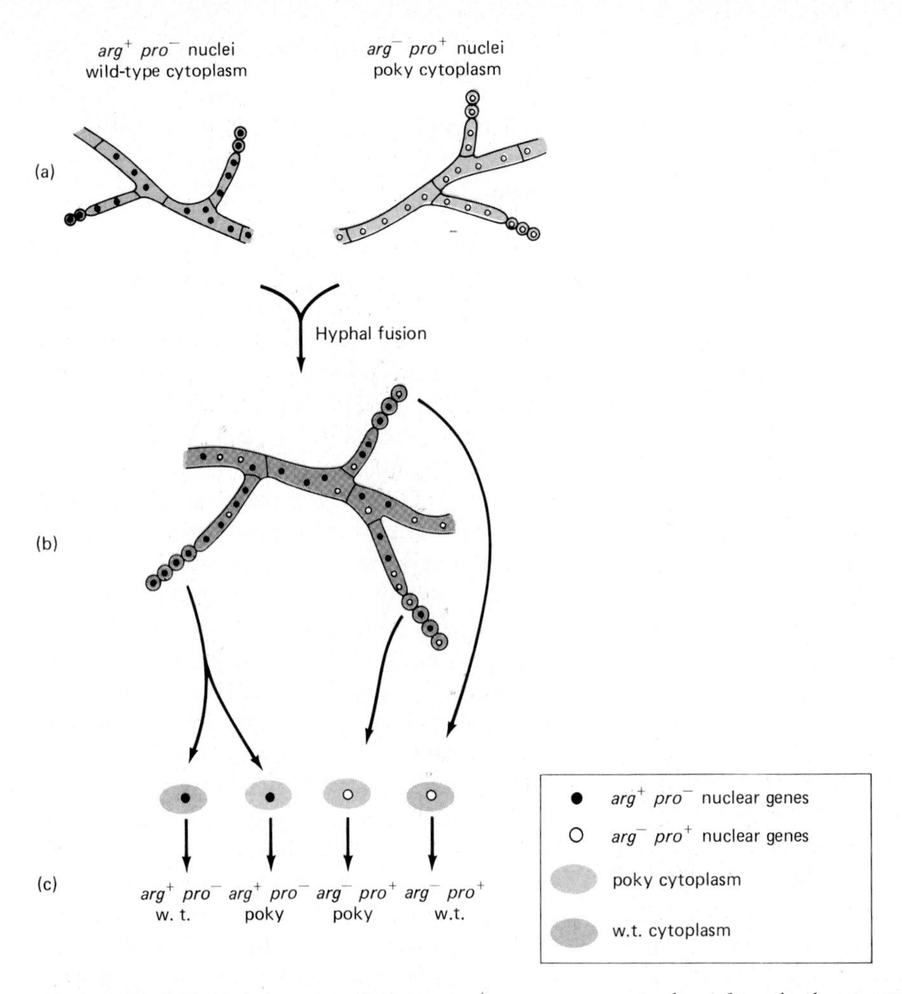


***Figure 13.4*** Heterokaryon test for extranuclear factor *poky,* in *Neurospora crassa.* (a) Allelically marked nuclei and cytoplasm in parent strains provide guidelines to assess genetic locations in heterokaryons. (b) Haploid nuclei carry different nuclear markers in heterokaryotic mycelium, with cytoplasm from both parental sources. (c) Germination of uninucleate, haploid spores produced by the heterokaryon reveals that the nucleus has no influence on *poky* phenotypic development, therefore, the extranuclear cytoplasm must be the source of the *poky* genetic factor.

## 13.2 Extranuclear Genes

Are extranuclear factors equivalent to genes that are located on chromosomes? If these factors are genes that happen to be located somewhere in the cytoplasm, we would predict that they would have genetic characteristics: (1) genes should have the capacity to mutate to alternative allelic forms; (2) different alleles of the same gene should segregate, and allelic types should breed true in the segregant populations; and (3) such genes should be physically identified with nucleotide sequences of DNA or RNA, the only known genetic molecules.

Although extranuclear mutants had been described according to their non-Mendelian patterns since 1909, two major problems had prevented their further analysis for more than 40 years. First, only an occasional, single kind of extranuclear mutant had been described in various species. Second, and most importantly, only one alternative was transmitted by one of the parents to the progeny in each cross; alternatives did not segregate at meiosis. There was no way to determine if these extranuclear alternatives were unit factors or not. The situation can be compared to the problem Mendel would have faced if he had crossed only tall plants and short plants and the progenies were either all tall or all short in every case. The principles of genic inheritance probably could not have been deduced from such inheritance patterns.

Ruth Sager began the first systematic analysis of extranuclear inheritance in the early 1950s, using *Chlamydomonas reinhardi.* She discovered that streptomycin acted as a mutagen, producing extranuclear mutations that affected resistance, sensitivity, and dependence on streptomycin itself in the growth medium. She also isolated extranuclear mutations influencing cell response to other antibiotics, as well as mutations affecting photosynthesis. By fluctuation test analysis streptomycin was shown to be a mutagen in this system (see Fig. 11.5).

Sager was able to conduct genetic analyses of these mutants when she found that some biparental zygotes appeared among the vast majority of uniparental zygotes, that is, some of the zygotes had received extranuclear factors through both parents. When these biparental zygotes underwent meiosis, extranuclear alternatives did not segregate. But segregations did take place during mitotic divisions of the haploid meiotic products. Segregant progeny types were shown to be true breeding (Fig. 13.5). These results showed that extranuclear factors existed

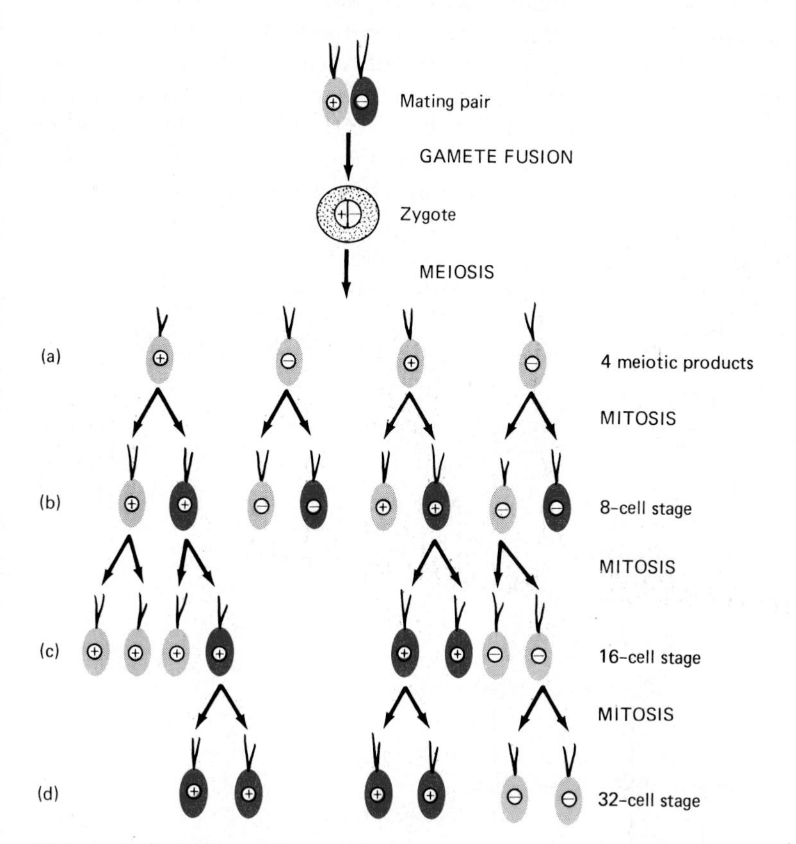


***Figure 13.5*** Analysis of biparental zygotes in *Chlamydomonas* revealed that segregation of extranuclear factors took place in postmeiotic mitotic divisions. (a) The four products of meiosis multiply asexually to produce (b) cell pairs segregating for extranuclear (gray/color) but not for nuclear (+/−) traits. (c) Some cells continue to segregate color and gray, while others have become homozygous and produce two gray or two colored mitotic products. (d) Segregation of extranuclear traits may be completed as early as the 16-cell stage in this species. (Only some of the 16 or 32 cells have been drawn.)

as pairs of allelic alternatives, which *segregated postmeiotically* during mitotic divisions of the haploid progeny. Extranuclear factors, therefore, behaved like genes, and they could be studied by available genetic methods.

But where were these extranuclear genes located? Sager proposed that the extranuclear genes she studied were located in the chloroplast of *Chlamydomonas* because some of the mutations affected chloroplast characteristics. The evidence in support of this inference was not particularly convincing, and many years elapsed before certain of these extranuclear genes were generally acknowledged to be located in the chloroplast.

The development of extranuclear genetics was also aided substantially by studies of the petite extranuclear mutant in yeast, primarily by Boris Ephrussi and Piotr Slonimski in Paris. They found that the acridine dye *proflavin* acted as a specific mutagen, and up to 99% of a grande respiration-sufficient strain could mutate to become petite (Fig. 13.6). The petite mutant was respiration deficient since it lacked some of the cytochrome enzymes required for aerobic respiration. This condition is not lethal in yeast because the organism can gain enough energy through glycolysis to sustain growth and reproduction. It grows more slowly, however, and produces petite (small) colonies on plates that also contain grande (large) wild-type colonies that can metabolize sugars through aerobic respiration, a far more efficient process than glycolysis.

Petites were recognizable by colony growth, enzyme analysis, respiration tests, and by electron microscopy. Mitochondria in the mutant cells were generally abnormal in appearance because the inner mitochondrial membrane was not organized into the typical invaginated cristae conformations (Fig. 13.7). Although many differences could be recorded to identify petites and their inheritance was clearly extranuclear, no other mutants were available and genetic tests for recombination could not be conducted as they had been in *Chlamydomonas*. When it was discovered in the early 1960s that mitochondria and chloroplasts contained DNA, Slonimski analyzed **mitochondrial DNA** (mtDNA) in grande and petite yeast. He showed that petite mtDNAs and grande mtDNAs were very different in base composition and therefore in sedimentation characteristics in CsCl (Fig. 13.8). This study in 1966 provided the first evidence that

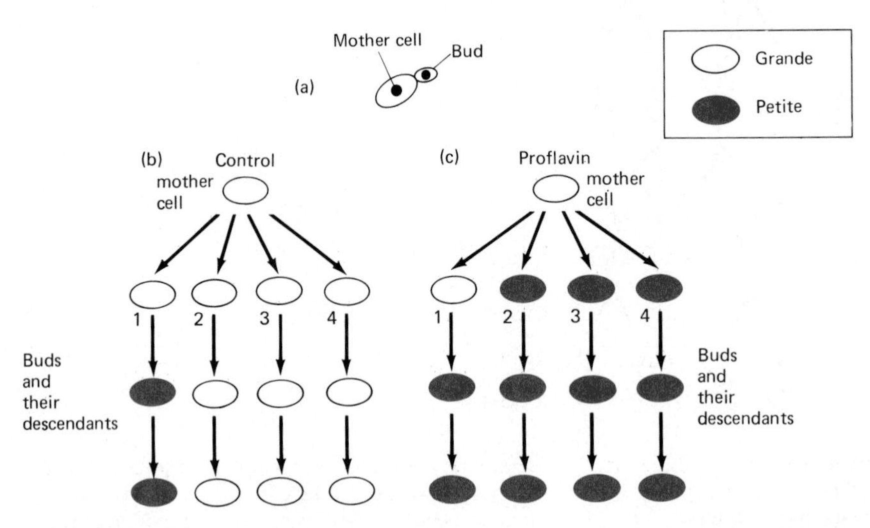


***Figure 13.6*** Lineage study of grande yeast mutation to petite. (a) Budding cells produce grande descendants, except for (b) occasional spontaneous petite mutants, which breed true thereafter. (c) In the presence of the acridine dye proflavin, up to 99.9% of the cells may become petite. The original mother cells do not mutate in the case of this particular mutagen.

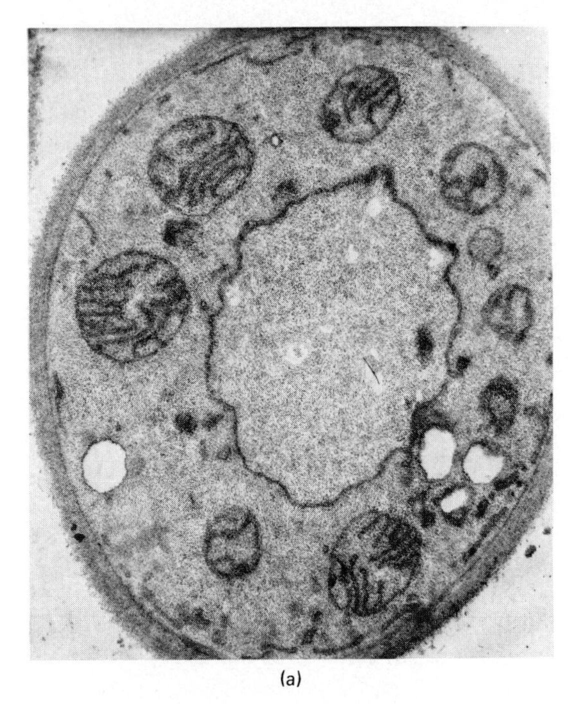

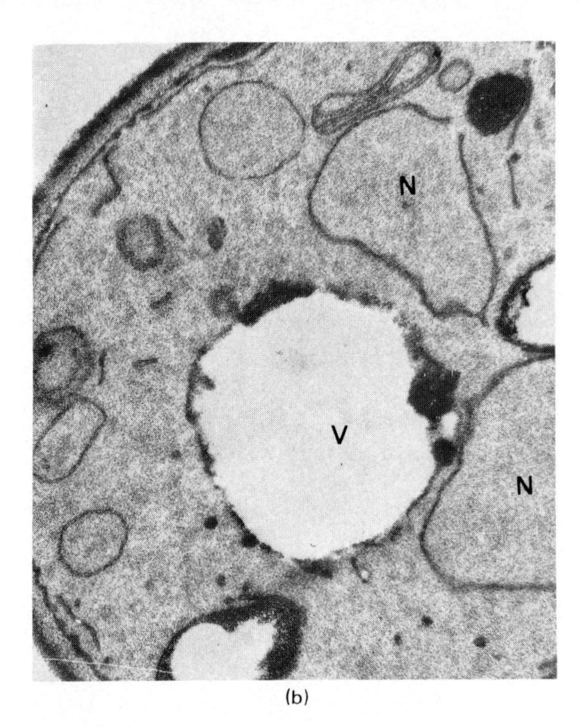


(a) (b)

***Figure 13.7*** Electron micrographs of permanganate-fixed yeast cells. (a) Mitochondrial profiles in grande yeast show typical cristae, invaginations of the inner membrane; whereas (b) petite mitochondria either lack cristae or show aberrant inner membrane conformations. The nucleus (N) in (b) appears in two separate parts due to the plane of sectioning. Vacuoles (V) are characteristic of wild-type as well as petite yeast cells. (Photos courtesy M. Federman.)

related an alteration in extranuclear DNA with mutation of a non-Mendelian inherited characteristic. The alteration in DNA paralleled the alterations in inheritance and in phenotype for petite mutants. This marked the beginning of studies of the physical basis for extranuclear inheritance.

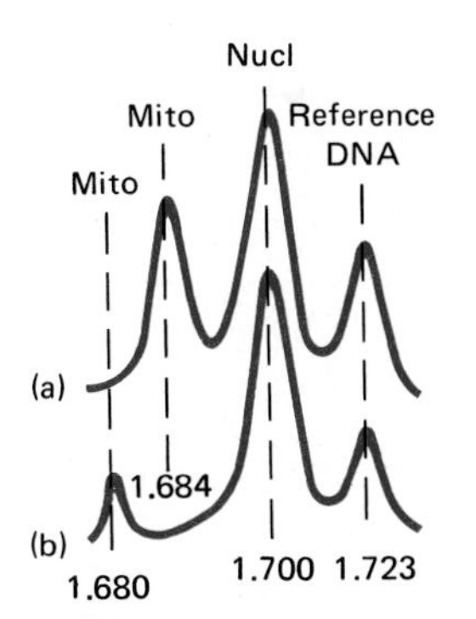


***Figure 13.8*** Densitometer tracings of yeast DNAs centrifuged to equilibrium in CsCl density gradients. Higher buoyant density (1.684) for (a) grande mitochondrial DNA than for (b) petite mtDNA reflects differences in base composition of the two DNAs. This difference parallels differences in the pattern of extranuclear inheritance and in phenotypic development between such strains.

## 13.3 Organelle DNA

The first convincing evidence for organelle DNA came from electron microscopy studies in 1963. Within only one year the occurrence of mitochondrial DNA and chloroplast DNA (ctDNA) had been reported for every species studied. At first these studies were confined to chemical and physical descriptions, which demonstrated that organelle DNA was duplex. Base composition analysis showed that A = T and G = C, as expected for base-paired duplex DNA. Physical studies confirmed the interpretation since organelle DNA melted at higher temperatures and the single strands renatured to the duplex form when the preparation was cooled.

In 1966, individual mtDNA molecules were visualized by electron microscopy of purified

DNA or of DNA released from organelles that had been subjected to osmotic shock during preparations for microscopy (Fig. 13.9). The direct observation of organelle DNA revealed the astonishing fact that both mtDNA and ctDNA were usually circular molecules. Furthermore, every species of metazoan animal appeared to have mtDNA that measured between 4.5 and 5.9 $\mu$m in contour length (Table 13.1). Protozoan mtDNA varied from one species group to another; mtDNA in fungi was four to five times larger than the average metazoan 5-$\mu$m-long circle, and mtDNA in flowering plants measured 30 $\mu$m or more. Later, it was found that ctDNA molecules were about 40 to 45 $\mu$m in contour length in many species examined, from unicellular protists such as *Euglena* to flowering plants.

Only two studies were reported in the 1960s in which mtDNA in *Neurospora* and ctDNA in *Chlamydomonas* were shown to replicate semiconservatively. Technical problems were encountered in studies of this kind using other organisms, due to large pools of DNA precursor molecules that persisted indefinitely in experimental populations. It was difficult to show that the distribution of $^{14}N$ and $^{15}N$ isotopes followed the predicted semiconservative pattern of distribution, as Meselson and Stahl had shown in *E. coli* because $^{15}N$ was not used up completely when experiments began. Because these $^{15}N$-labeled precursors remained available for synthesis of new strands over several generations, recognizing original and newly synthesized strands of DNA in the proportions expected was difficult (see Fig. 8.1).

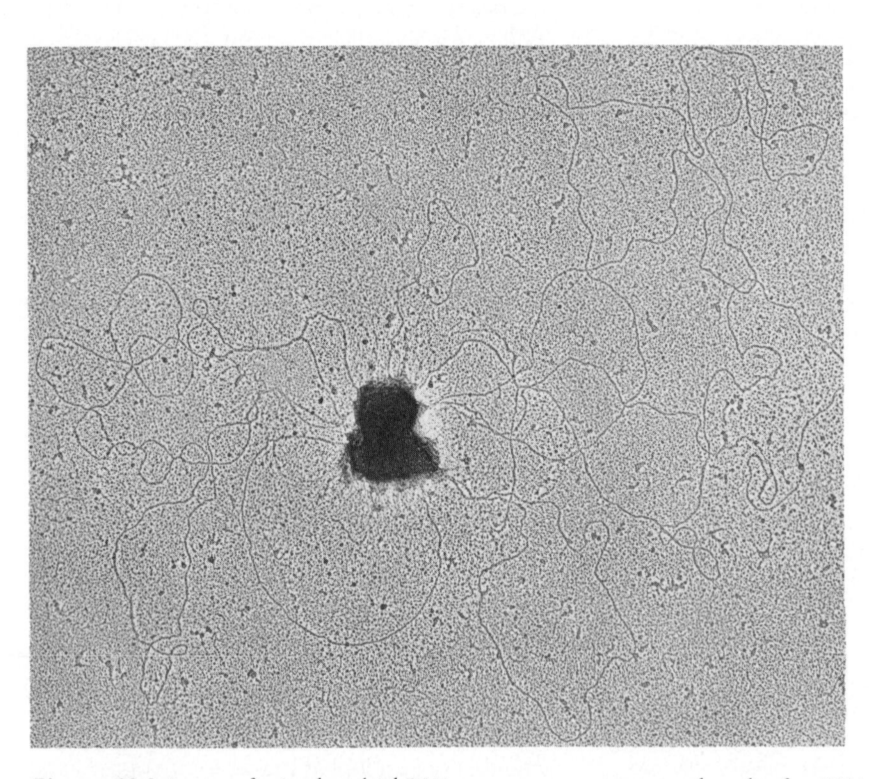

***Figure 13.9*** Loops of mitochondrial DNA are seen emerging from a piece of membrane from osmotically shocked yeast mitochondria photographed with the electron microscope. The contour length of mtDNA shown here is 48.5 $\mu$m, which represents two 25-$\mu$m circular molecules. × 29,000. (From C.J. Avers et al. 1968. *Proc. Nat. Acad. Sci. U.S.* **61**:90.)

***Table 13.1*** Size of circular mitochondrial DNAs in various organisms.

| organism | contour length (μm) |
|---|---|
| animals | |
| vertebrates | 4.7–5.9 |
| invertebrates | 4.5–5.9 |
| plants | |
| flowering plants | 30–34 |
| fungi | |
| true fungi | |
| *Podospora* | 31 |
| *Saccharomyces* | 25–26 |
| *Neurospora* | 19 |
| *Aspergillus* | 10 |
| *Saprolegnia* | 14 |
| slime molds (*Physarum*) | 19 |
| protists | |
| protozoa | |
| ciliates | 14–15* |
| amebae (*Acanthameba*) | 13 |
| trypanosomes | 0.2–0.8† |
| | 6–11† |
| *Chlamydomonas* | 4.6 |

*Molecules found in *Paramecium* and *Tetrahymena* are linear, not circular.

†Minicircles and maxicircles occur together in a meshwork of DNA, which is called kinetoplast DNA (kDNA).

Replication of mtDNA has been analyzed by electron microscopy since 1968, and mtDNA has been shown to replicate semiconservatively. In 1968, electron microscopy of rat liver mtDNA revealed typical θ-forms similar to those described in *E. coli* by Cairns (Fig. 13.10). In 1972, however, Jerome Vinograd and co-workers reported a modified semiconservative replication pattern, which is called **D-loop synthesis.** According to electron microscopy and biochemical tests, both template strands of the duplex did not replicate simultaneously. Synthesis began along the light (L) strand and some time later the heavy (H) strand initiated synthesis of a new complementary partner (Fig. 13.11).

When replication begins along the L strand, a single-stranded displacement loop, or **D-loop,** is produced. Molecular hybridizations between isolated single-stranded D-loops and L-strand parental DNA revealed the source of these single-stranded segments. The D-loop becomes larger as replication proceeds around the circular L strand, in the 5′→ 3′ direction. H-strand repliction is initiated later, and proceeds in the 5′→ 3′ direction along the antiparallel strand. Ultimately, both parental strands complete their replication and two new, identical mtDNA duplexes are produced.

In addition to replication by the theta-form and D-loop processes, mtDNA and ctDNA may

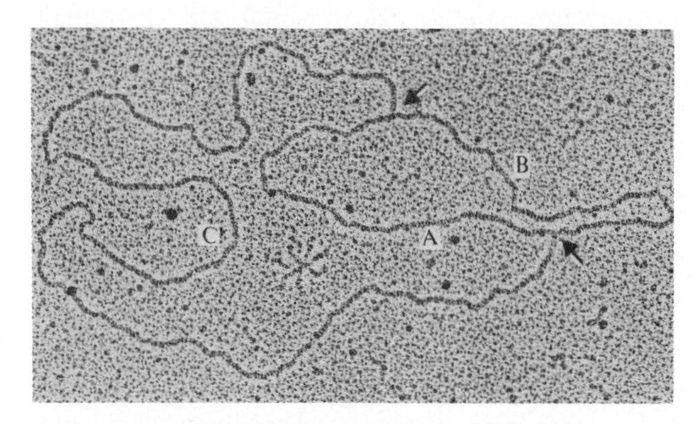


***Figure 13.10*** Electron micrograph of replicating mtDNA from rat liver, showing a typical theta-form molecule. Arrows indicate the replication forks. Segments A and B have replicated, and segment C remains unreplicated in this 5-μm-long circle. (From D.R. Wolstenholme et al. 1974. *Cold Spr. Harb. Symp. Quan. Biol.* **38**(1973):267.)

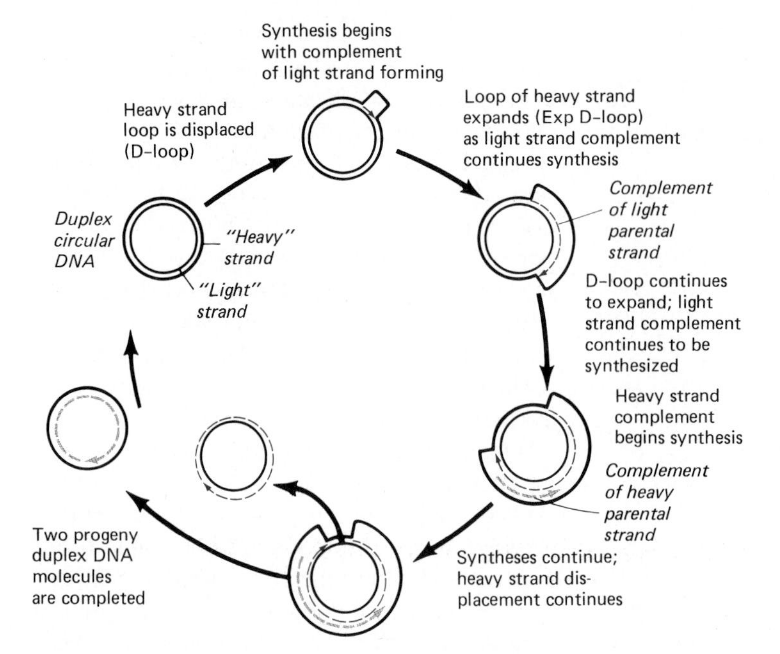


***Figure 13.11*** D-loop synthesis of mitochondrial DNA in diagrammatic summary. The heavy and light parental strands are shown in thick and thin black lines, respectively. The light strand complement is synthesized first (color) in the D-loop region, and the heavy strand complement (gray) is synthesized somewhat later. Each new strand is synthesized in the $5' \rightarrow 3'$ direction along the antiparallel template strands. Two identical semiconserved duplexes are produced from the parental DNA.

sometimes replicate by the rolling-circle mechanism (see Fig. 10.15). It may be that any or all of these modes of duplex DNA replication can occur in the same cells, regulated by different cellular and environmental conditions, such that *slower rates* of replication by theta-form and D-loop synthesis and *faster rates* of replication of rolling circles occur at different times in different cells. The mechanism of such regulation is not known.

## Organelle Transcription and Translation

Genetic studies had provided evidence of extranuclear genes in mitochondria and chloroplasts, and it seemed very likely that these genes were physically located in organelle DNA molecules. How were these genes expressed? Was organelle DNA transcribed and translated within the organelle iself, or was the cellular nucleocytoplasmic system required to process mtDNA and ctDNA information? It was well known that hundreds of different nuclear genes coded for various proteins and processes that were parts of mitochondrial and chloroplast phenotypes. It was not known whether organelle-encoded information was also processed in eukaryotic cytoplasm.

In the 1960s and 1970s, various lines of evidence revealed that mitochondria and chloroplasts were indeed *semiautonomous* components of eukaryotic cells. Not only did they contain their own DNA, which replicated and was transmitted in subsequent generations, but they also were found to transcribe coded information in their DNA molecules and to translate messenger RNA into organelle-specified polypeptides. Organelles are not totally autonomous or independent structures. They cannot increase outside the living cell and they depend on the nucleus for most of the genetic information specifying their structure, function, and regulation.

## 13.4 Protein Synthesis

In 1966, Anthony Linnane and co-workers showed that protein synthesis in the cytoplasm and in mitochondria could be selectively shut off by specific drugs. In the presence of *cycloheximide,* cytoplasmic protein synthesis is inhibited and mitochondrial activities proceed as usual; *chloramphenicol, erythromycin,* and certain other antibiotics inhibit mitochondrial protein synthesis and have little or no effect on cytoplasmic ribosome activities.

With these newly available methods for studying organelle protein synthesis *in vivo,* investigators showed that polypeptides made in mitochondria and chloroplasts were different from those made in the surrounding cytoplasm. In a typical experiment of this kind, [$^3$H] amino acids were provided during cycloheximide inhibition to label organelle-synthesized polypeptides, after [$^{14}$C] amino acids were provided to label polypeptides made in the presence of chloramphenicol (Fig. 13.12). By selectively turning off one compartment of the cell and then the other, and distinctively labeling polypeptides made in each part of the cell, it was very clear that organelles were capable of synthesizing proteins. But were these proteins coded by the organelle genes or by nuclear genes? To answer this question it was necessary to analyze polypeptide synthesis in extranuclear mutants whose defective proteins were known. The petite mutant of yeast proved to be a useful system for such a specific analysis since its cytochrome oxidase was defective when compared with the functional grande enzyme.

Cytochrome oxidase, the terminal enzyme of aerobic respiration, is a protein composed of seven identifiable polypeptide subunits in yeast. Petites can make four of these polypeptides but not the other three. When grande yeast was incubated in the presence of erythromycin or

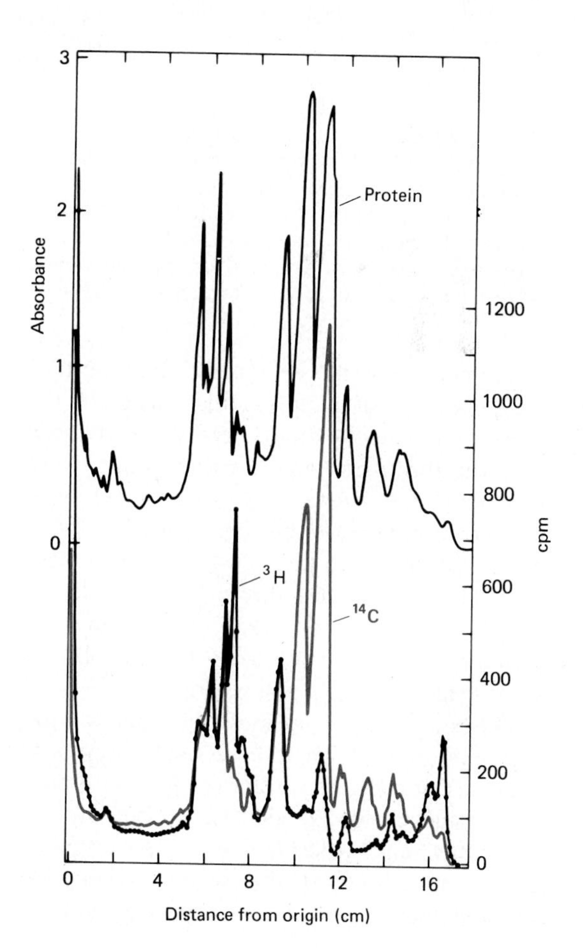


***Figure 13.12*** Double labeling experiment using differential inhibitors to identify location of synthesis of chloroplast membrane proteins in *Chlamydomonas reinhardi.* Absorbance of electrophoretically separated proteins from chloroplast membranes serves as the reference for the lower portion of the figure. Proteins that were made in the cytoplasm are labeled with [$^{14}$C] arginine, which was present during the time the cells were growing in medium containing chloramphenicol. Proteins that were made in the chloroplast are labeled with [$^3$H] arginine, which was provided later in the experiment, when cycloheximide was the inhibiting drug for protein synthesis. Cycloheximide inhibits protein synthesis at cytoplasmic ribosomes, and chloramphenicol inhibits protein synthesis at organelle ribosomes. (From J.K. Hoober. 1970. *J. Biol. Chem.* **245**:4327, Fig. 6.)

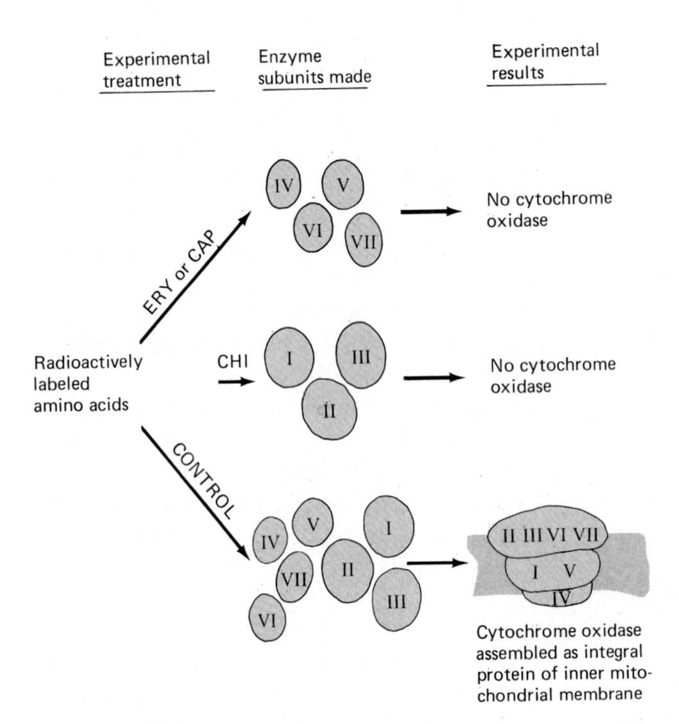


***Figure 13.13*** By labeling cytochrome oxidase polypeptides in wild-type yeast exposed to different inhibitors of protein synthesis, we can determine which polypeptides fail to be synthesized in enzyme-deficient experimental cells. In the presence of erythromycin (ERY) or chloramphenicol (CAP), mitochondrial protein synthesis is inhibited and components I–III are not made. Comparison with cycloheximide-inhibited cells (CHI), which make no proteins at cytoplasmic ribosomes, and with controls that are not inhibited, reveals the different origins for the seven polypeptides of this yeast enzyme.

chloramphenicol, the cells made only the four polypeptides that petites could make. In the presence of cycloheximide, grande yeast made only the three polypeptides that petites were unable to synthesize (Fig. 13.13). In other words, the correspondence between drug-inhibited grande activities and the activities in petite cells showed that the three larger polypeptides of cytochrome oxidase were made in mitochondria, whereas the four smaller subunits were made in the cytoplasm. These seven polypeptides assemble inside the grande mitochondrion to form a functional respiratory enzyme that catalyzes aerobic respiration. Petites are unable to oxidize glucose aerobically because their mtDNA lesion involves coded information for three of the seven polypeptide subunits; they make nonfunctional polypeptides.

If chloroplasts and mitochondria can synthesize polypeptides from coded DNA information, these polypeptides must be translated on some ribosomal machinery. The first convincing experimental evidence of ribosomes in mitochondria and chloroplasts was reported in 1970. Not only were these particles isolated from organelles and purified, but their chemical, physical, and functional characteristics were described. Mitochondrial ribosomes vary in size from 55S to 80S particles, but all chloroplast ribosomes appear to be 70S (Table 13.2). These monosomes consist of two subunits of unequal size, just like cytoplasmic ribosomes (cytoribosomes), and each subunit has one rRNA molecule. Chloroplast ribosomes apparently also have 5S rRNA, but there is no convincing evidence for the occurrence of 5S rRNA in mito-

***Table 13.2*** Some characteristics of organelle ribosomes*.

| organelle and organism | ribosome monomer | small subunit | large subunit | rRNA from small subunit | rRNA from large subunit |
|---|---|---|---|---|---|
| mitochondrion | | | | | |
| animals | 55–60S | 30–35S | 40–45S | 12–13S | 16–17S |
| flowering plants | 78S | 44S | 60S | 18S | 26S, 5S |
| fungi | 80S | 40S | 52S | 14–17S | 21–24S |
| protists | | | | | |
| *Tetrahymena* | 80S | 55S | 55S | 14S + | 21S[†] |
| *Euglena* | 71S | 32S | 50S | 16S | 23S |
| chloroplast | | | | | |
| various species | 70S | 30S | 50S | 16S | 23S, 5S |

*The reference standards are components and monomers of *E. coli* ribosomes and rat or yeast cytoplasmic ribosomes (see Table 9.2).

†There is no way to tell which rRNA comes from the two 55S ribosome subunits. It is possible that both rRNAs are present in each 55S particle, which might therefore be the actual ribosome monomer.

chondrial ribosomes in any species other than flowering plants.

Ribosome function was first established by *in vitro* tests, which showed that these particles were active in making polyphenylalanine from poly(U) messenger RNA. More stringent tests were also conducted; one such test showed that oganelle ribosomes participated in making correct virus proteins when natural viral RNA was provided *in vitro* as a messenger molecule. In suitable preparations, polysomes were observed by electron microscopy (Fig. 13.14). The thin strand connecting monosomes in the polysome group was digestible with ribonuclease, and it was therefore inferred that it was organelle mRNA. Direct evidence for specific RNA transcripts was obtained in later studies.

## 13.5 Transcription in Organelles

Using molecular hybridization assays, it has been shown that mitochondria and chloroplasts transcribe their own rRNA and all or most of their unique tRNAs. Ribosomal RNA can be purified from isolated organelle ribosomes and hybridized with H and L strands of mtDNA or ctDNA. In most cases, rRNA for both ribosome subunits is coded by the H-strand of DNA, and the two coding sites are very close together. In other cases, such as yeast, 15S and 21S rRNAs are transcribed from widely separated regions of

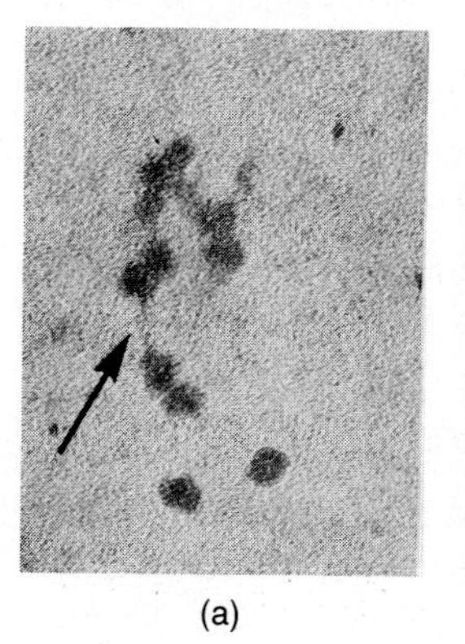

(a)

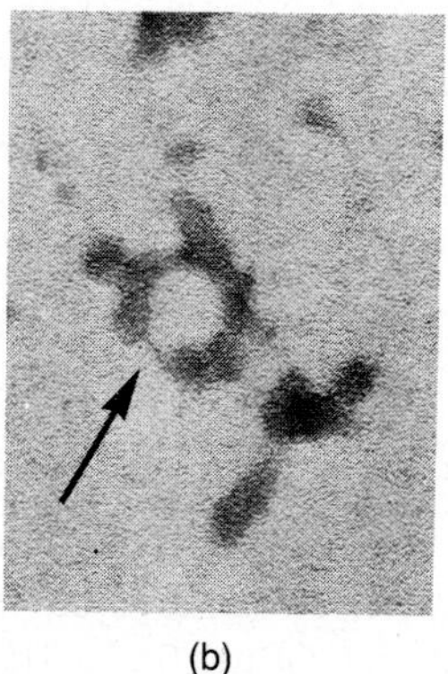

(b)

***Figure 13.14*** Electron micrographs of polysomes isolated from purified yeast mitochondria after preparative centrifugation in sucrose gradients. The presumptive mRNA strand is shown at the arrows. (a) × 160,000; (b) × 255,000. (From C.S. Cooper and C.J. Avers. 1974. In *The Biogenesis of Mitochondria,* p. 289; A.M. Kroon and C. Saccone, eds. New York: Academic Press.)

mtDNA. In chloroplasts, coding sites for rRNAs are arranged in the sequence of 16S–23S–5S (in exactly the way they occur in *E. coli*). From these molecular hybridizations, it appears that each kind of rRNA has only one coding site in both mtDNA and ctDNA in most species studied.

Twenty or more kinds of tRNA are transcribed from mtDNA and ctDNA. These tRNAs can be isolated from organelles and purified for molecular hybridizations and other assays. By molecular hybridizations, tRNA transcripts can be identified and located in organelle DNA (Fig. 13.15). Genes for tRNAs occur in several different regions of the mapped DNA, and usually at least one tRNA gene exists within the rRNA coded region. We will discuss the significance of this consistent observation later in the chapter.

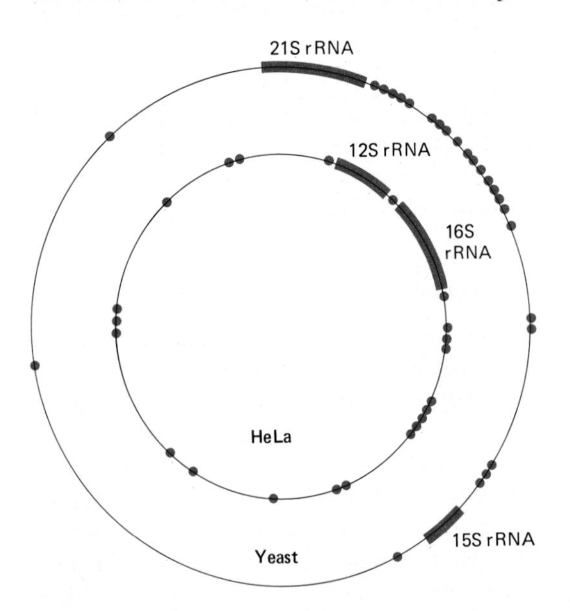


***Figure 13.15*** Sites of ribosomal RNA genes (colored bars) and transfer RNA genes (colored dots) in the circular genomes of HeLa (human) and yeast mitochondria. In yeast (outer circle), the two rRNA genes are far apart; the two rRNA genes in human mtDNA are close together. The 25 tRNA genes in the yeast mitochondrial genome are found mainly in the region between the two rRNA genes, whereas the 22 tRNA genes in human mtDNA are scattered around the genome map. (Reprinted by permission from P. Borst and L.A. Grivell. 1981. *Nature* **290**:443–444, Fig. 1. Copyright © 1981 Macmillan Journals Limited.)

Functional tests for organelle-encoded tRNAs have also been conducted. Purified organelle tRNAs can be bound to radioactively labeled amino acids for which each tRNA is specific, to identify polypeptides that are made with such labeled amino acids incorporated into the polymers. In this way, it has been shown that every tRNA made by mitochondria or chloroplasts can participate in organelle translation processes, along with organelle ribosomes and other components needed for syntheses.

By electron microscopy and biochemical methods, transcribing mtDNA complexes from mitochondria of human cells (HeLa cell cultures) and from *Drosophila* have been isolated and studied. In both cases, 80% to 100% of the 5–μm–long circular mtDNA molecule was covered with polysome groups (Fig. 13.16). It was inferred that each of these polysomes represented translation events in progress, just as they had in *E. coli* (see Fig. 9.9). Translation appears to begin while mRNA transcription is still under way, and transcription and translation are therefore coordinated processes in mitochondria, as they are in bacterial cells. Whether the number of polysomes along mtDNA can be equated to the number of different structural genes coding for polypeptides, however, remains to be determined.

Genetic and molecular studies of organelle gene expression have been expanded in the 1980s by the addition of high-resolution methods for base sequencing and for detailed analysis of cloned DNA. Comparisons can more readily be made now between organelle DNAs from related species, such as various mammalian species, and between distantly related species, such as yeast, *Neurospora, Drosophila,* toad, and human mtDNAs. These and other studies will be discussed in the following sections.

## Organelle Genomes

Gene maps for several mitochondrial and chloroplast genomes have been constructed by a variety of methods. The large number of extra-

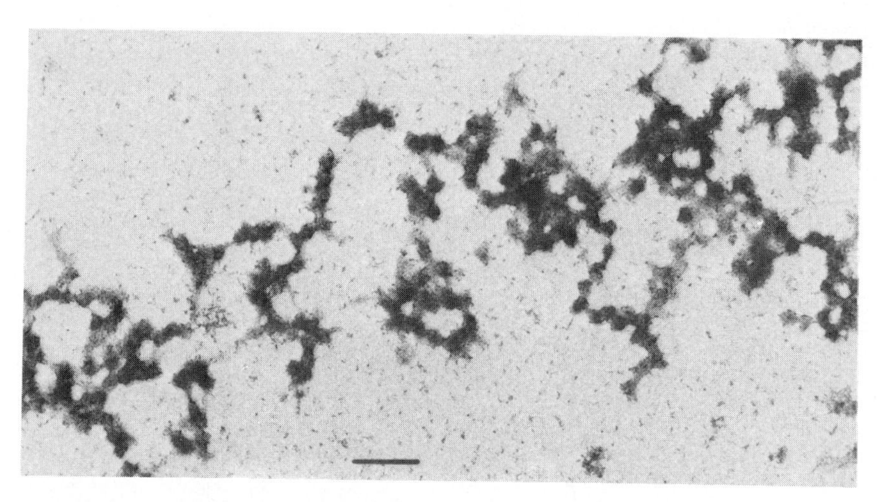

***Figure 13.16*** Electron micrograph of a string of polysomes bound to mtDNA, from osmotically ruptured mitochondrion of *Drosophila melanogaster*. × 90,000. (With permission from W.Y. Chooi and C.D. Laird. 1976. *J. Mol. Biol.* **100**:493. Copyright © 1976 Academic Press Inc. [London] Ltd.)

nuclear gene mutations known in yeast mtDNA and in *Chlamydomonas* ctDNA have facilitated the identification of organelle genes and their alleles. These genes have been mapped by genetic methods to a certain extent but more extensively by physical mapping of restriction fragments or deletion mutants and by molecular methods. These two genomes in particular have served as models for organelle genomic analysis in other organisms.

The complete sequence of 16,569 base-pairs in human mtDNA was reported in 1981 by a group of investigators in England. Their interpretations of the human mitochondrial genome sequence relied greatly on infomation concerning coding sequences and gene products that had been obtained in molecular hybridization analysis and from studies of other mtDNAs from various species. The base sequence by itself could not have been accurately interpreted without these comparative studies, as we will see in a later section.

## 13.6 Mitochondrial Mutations in Yeast

Until 1970, the petite mutation was the only extranuclear mutation known in yeast. Petites are deletion mutants of two general types: (1) *suppressive petites,* in which varying amounts of the mitochondrial genome are missing, and (2) *neutral petites,* most of which lack mtDNA altogether (Fig. 13.17). In addition to a relatively high rate of spontaneous mutation, the petite mutation can be induced by exposure to *ethidium bromide* and other acridine mutagens in 99% to 100% of treated grande cells. Both kinds of petites have the same phenotypic characteristics of slow growth, deficient aerobic respiration, and enzymatic defects. They can be distinguished by their mtDNA content and by their progeny when crossed with grande tester strains. In grande × suppressive petite, a variable proportion of the meiotic tetrads have 4 petite : 0 grande spores or 0 petite : 4 grande tetrad types. In grande × neutral petite, all tetrads are 4 grande : 0 petite. The mechanism leading to variable proportions of the two kinds of tetrads in crosses involving petites is not known.

Little could be done to map the organelle genome until extranuclear point mutations were found and were analyzed genetically in 1970. These point mutations can mutate back to the wild-type allele, whereas petite deletion mutants are not revertible. Different gene loci and alleles

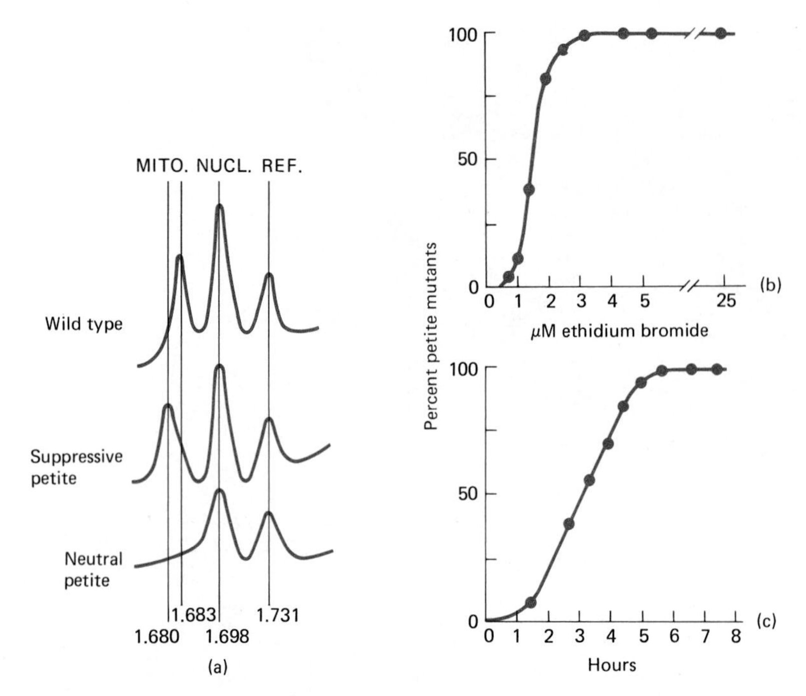


***Figure 13.17*** The petite mutation in yeast. (a) The three strains of yeast have the same nuclear DNA, according to its buoyant density in CsCl, but they differ in their mtDNA. The mtDNA base composition is altered in suppressive petites, and mtDNA is absent from neutral petite strains. (b) Mutation induction from grande to petite by ethidium bromide is dependent on drug concentration, and 100% of the cells can become petite. (c) The time course of appearance of petite mutants exposed to ethidium bromide. In 2.5 hr (time for one cell cycle), the petites have increased by mutation from less than 5% to more than 90% of the population.

were identified by recombination and complementation tests (Fig. 13.18). These genes were shown to be located in mtDNA by using ethidium bromide mutagenesis of mutant strains and by following the frequencies with which different mutant genes were *retained together or lost together* in petites induced by the mutagenic treatment. The closer together two genes were, the higher the frequency of their coordinated loss or retention in the induced-deletion mutants. In addition, any genes that were lost when neutral petites were induced had to be genes physically located in mtDNA, which is known to be lost entirely in these particular petites. Although petites were of limited value before 1970, they proved to be the crucial components in later studies such as the ones just mentioned. In addition, mapping by the deletion method could be accomplished using different petite strains with point mutation markers incorporated into their genotype.

Yeast mtDNA is about 26 μm in contour length. Although very few intact molecules have been isolated and photographed, the size of the genome has been established by restriction enzyme analysis (Fig. 13.19). According to the usual equivalents, a 26-μm-long molecule consists of 78,000 base pairs of DNA. The entire DNA molecule derived from restriction enzyme analysis has about 78,000 base-pairs of DNA and must therefore be about 26 μm long. We can determine the order of these restriction enzyme fragments in the intact molecule by comparing overlaps. The same restriction maps of the grande genome can then be used to determine the locations of gene loci by the use of petites

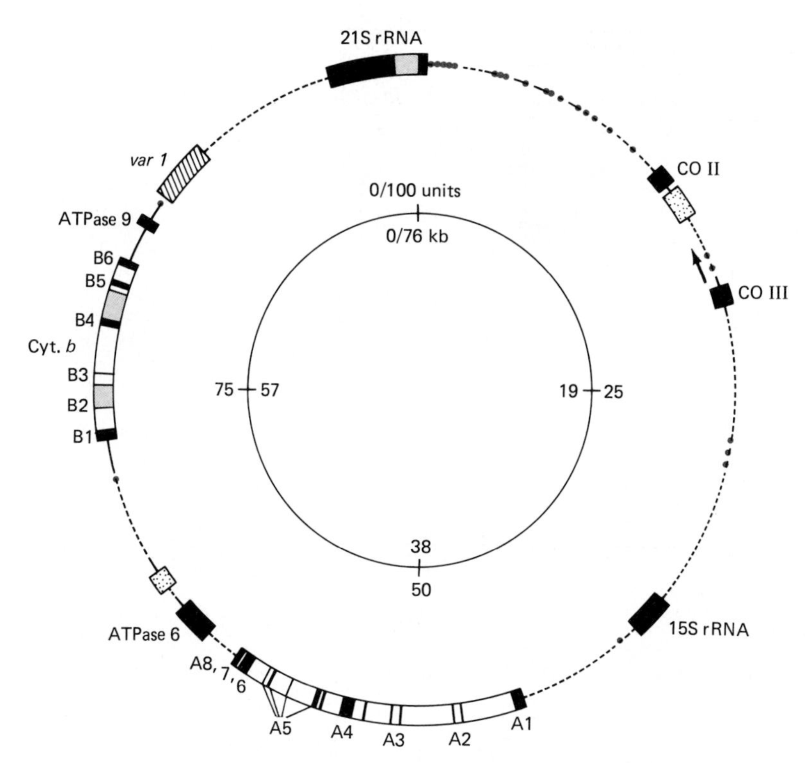


***Figure 13.18*** Organization of the mitochondrial genome in yeast. In addition to two rRNA genes and 25 tRNA genes (colored dots), genes encoding seven known polypeptides are distributed within the 76,000 to 78,000 base-pairs (76 to 78 kb) of the circular mtDNA molecule. The three largest polypeptides (COI, COII, COIII) of the respiratory enzyme *cytochrome oxidase* are each specified by a gene; polypeptide subunit chains 6 and 9 of the respiratory ATPase are each encoded by a gene; the respiratory enzyme component cytochrome *b* is encoded by a gene; and *var 1* encodes a protein of the mitochondrial ribosomes. Three of these 34 known genes are organized into introns and exons namely, the genes for COI (A1–8) Cyt *b* (B 1–6), and 21S rRNA. Considerable noncoding DNA occurs in spaces between genes. (Reprinted by permission from P. Borst and L.A. Grivell. 1981. *Nature* **290**:443–444, Fig. 1. Copyright © 1981 Macmillan Journals Limited.)

with overlapping deletions and with marker genes in the pieces of mtDNA that are retained in the petite strains.

In studies of this kind, the pieces of retained mtDNA in the petite deletion mutant strain can be identified according to the restriction fragments that are generated. The genes carried in the petites can then be ordered on the whole 100-unit map (Fig. 13.20).

Examination of the yeast mtDNA map reveals why virtually any deletion will produce a respiration-deficient petite. The loci coding for polypeptides of different respiratory enzymes and loci for both rRNA genes are widely spaced around the entire genome. The loss of almost any part of this genome will lead to some enzymatic defect or to the inability to carry out protein synthesis. The action of ethidium bromide and other acridines is through intercalation of the drug with DNA. When the drug binds to mtDNA, replication of the genome is inhibited and distorted, leading to gross alterations in the mtDNA molecule in suppressive petites. If the drug acts long enough in the system, treated cells or their descendants will lose all of the genome and become neutral petites.

All of the known loci apparently code for structural gene products. At present, we know

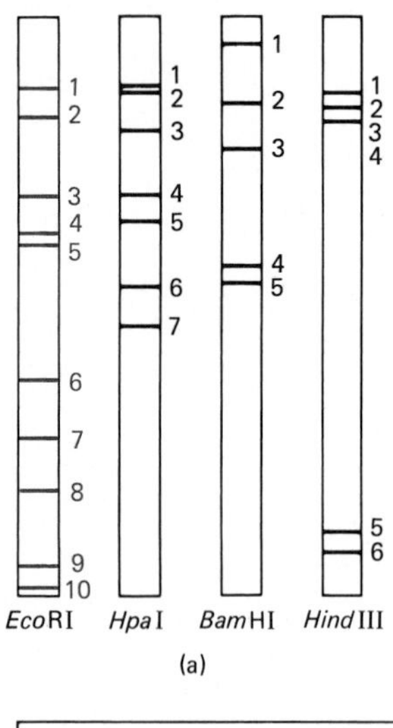


| Fragment number | Enzyme cleavage products (kb) | | | | |
|---|---|---|---|---|---|
| | *Eco*RI | *Hpa*I | *Bam*HI | *Hind*III | *Hha*I |
| 1 | 23.7 | 23.7 | 33.0 | 24.0 | 19.1 |
| 2 | 17.3 | 20.6 | 19.5 | 19.5 | 15.0 |
| 3 | 10.0 | 14.7 | 13.4 | 15.8 | 12.0 |
| 4 | 8.3 | 7.1 | 5.4 | 15.6 | 8.0 |
| 5 | 7.8 | 6.5 | 4.8 | 0.5 | 5.3 |
| 6 | 3.5 | 3.2 | 76.1 | 0.3 | 4.4 |
| 7 | 2.4 | 2.3 | | 75.7 | 4.1 |
| 8 | 1.7 | 77.2 | | | 3.5 |
| 9 | 0.9 | | | | 2.3 |
| 10 | 0.2 | | | | 1.1 |
| | 75.8 | | | | 74.8 |

(b)

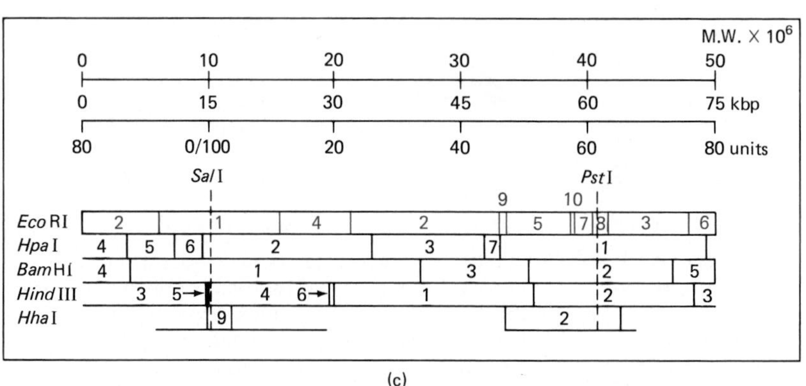


(c)

***Figure 13.19*** Molecular size of the yeast mitochondrial genome according to restriction fragment analysis. (a) Using restriction enzymes *Eco*RI, *Hpa*I, *Bam*H1, and *Hind* III, among others, sets of restriction fragments are generated from mtDNA. These are separated according to length by gel electrophoresis (numbered bands), with the shorter fragments migrating most rapidly from the origin at the top of the gel. (b) Molecular lengths (in kilobases) of the cleavage products of each restriction enzyme add up to about 76 to 78 kb in each case. This must therefore be the size of the intact mtDNA molecule in cells. (c) Restriction mapping involves ordering of the restriction fragments generated by different enzymes, with reference to each other and to the single cleavage site induced by enzyme *Sal*I. The *Sal*I site is used as the zero reference point of the 100-unit map. See Box 7.2 for further reference. (From R. Morimoto et al. 1978. *Mol. gen. Genet.* **163**:241–255, Fig. 1.)

little about regulatory genes in the yeast mitochondrial genome.

The ctDNA molecule should code for a larger number of gene products than mtDNA since most ctDNA is 40 to 45 $\mu$m in contour length, but mtDNA is only 5 $\mu$m in metazoan animals and 26 $\mu$m in yeast. Relatively few loci have been identified in ctDNA, however, and only some of these have been related directly to any feature of the chloroplast phenotype. In contrast, all of the known mitochondrial genes in yeast code for known molecular components of the mitochondrion.

Information on genes in ctDNA has come primarily from studies of *Chlamydomonas reinhardi* using genetic analysis and from iso-

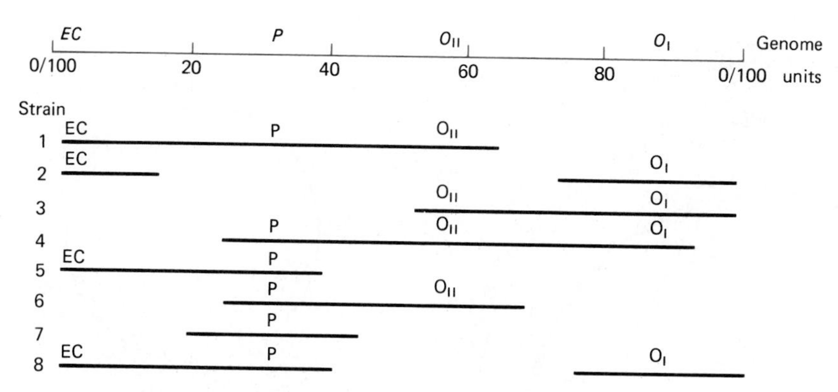


***Figure 13.20*** Physical mapping of the yeast mitochondrial genome by molecular hybridizations using mtDNA from 8 petite strains, which have selective deletions so that only certain combinations of antibiotic resistance markers are retained. (Compare with the grande genome, top.) The petite genomes are cleaved by restriction enzymes and the fragments allowed to hybridize with grande restriction fragments. By comparing molecular hybrids formed, genetic markers present, and relationships of overlapping deletions, we can derive the physical map. Each petite strain retained only the genome segments shown, relative to the 100-unit grande genome. The marker alleles *E*, *C*, *P*, $O_{II}$, and $O_{I}$ are shown in this example of mapping.

lated molecular analyses of some other chloroplast systems. There is some conflict between Sager and other investigators in designating the precise order of extranuclear loci in the single ctDNA linkage group of *C. reinhardi.* Sager has correlated information from different kinds of recombination mapping methods to assign the order of gene loci and their relative distances apart. When these data are compared, an internal consistency is evident when various linear maps are incorporated into a single circular linkage group (Fig. 13.21). Since ctDNA is a 62-$\mu$m-long, circular molecule in *Chlamydomonas,* the genetic data indicating circularity of the linkage map are consistent with the physical conformation of ctDNA molecules.

## 13.7 Gene Functions in Organelles

Transcripts of genes coding for rRNAs and tRNAs remain within the organelle, where they are synthesized and where they participate in organelle protein synthesis. Most or perhaps all of the ribosomal proteins, however, are encoded by nuclear genes. These organelle ribosomal proteins are translated in the cytoplasm and the molecules are then transported into mitochondria or chloroplasts, where they assemble with rRNAs and emerge as functional ribosomal subunits. One or more chloroplast ribosomal proteins may be specified by ctDNA, according to the limited evidence that is available. For example, Sager showed that one of the two ribosomal subunits in certain drug-resistant strains of *Chlamydomonas* was responsible for drug-resistance when mixed reconstitutions were conducted using 30S and 50S subunits from resistant and sensitive strains to make 70S monosomes (Fig. 13.22). These experiments were similar to those showing that the 30S ribosome subunit in *E. coli* was altered in streptomycin-resistant strains (see Fig. 9.28).

The polypeptides of a number of respiratory enzymes in yeast mitochondria are encoded by mitochondrial genes, and these enzymes are incorporated into the structure of the mitochondrial inner membrane. In all cases, however, some of the polypeptides of the enzyme are also encoded by nuclear genes. If the gene products in either cell compartment are defective, the entire enzyme has a defective or abnormal function and aerobic respiration is deficient or lack-

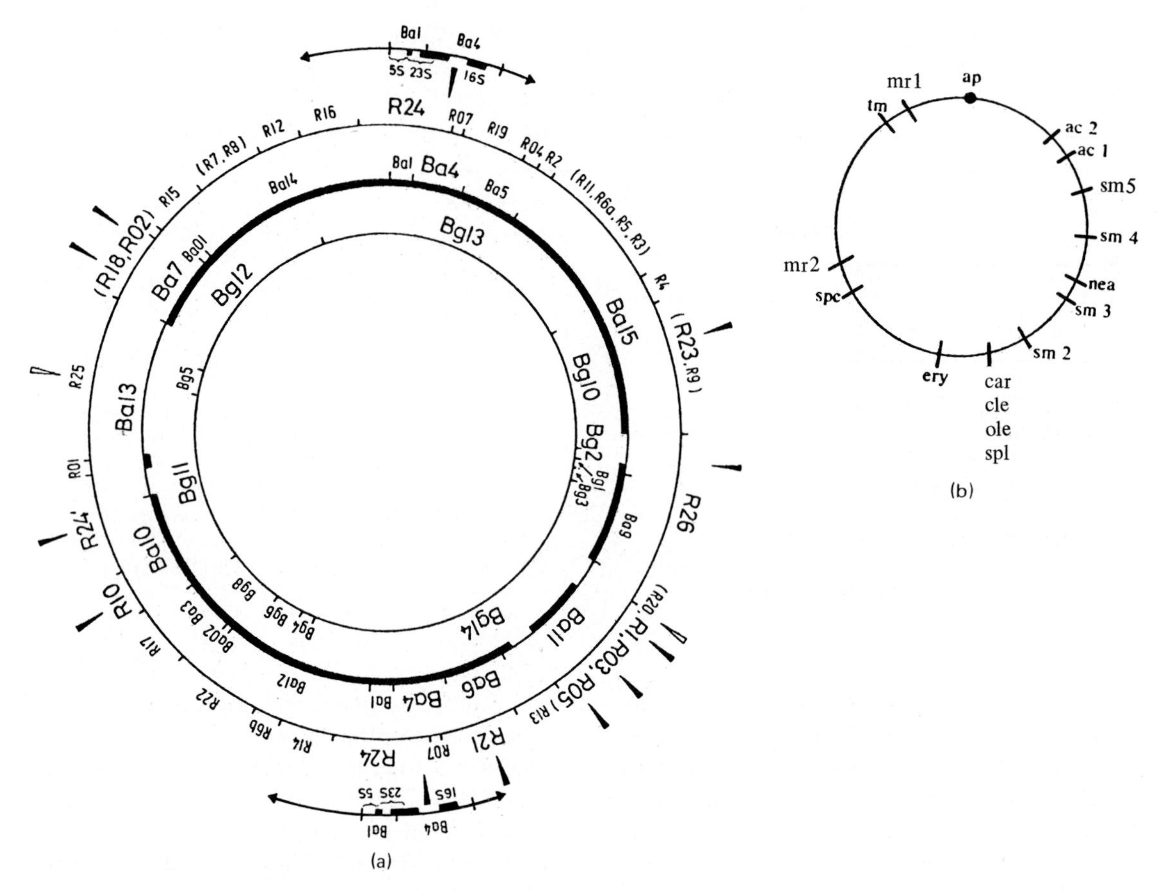


***Figure 13.21*** Maps of the chloroplast genome in *Chlamydomonas reinhardi.* (a) Physical map showing three concentric circles generated by three different restriction enzymes (*Eco*RI, *Bam*HI, and *Bgl*2). Sequences of tRNAs (4S RNAs) are shown by large letters and black arrows. The two rDNA units are indicated on the outside. Sequences coding for tRNAs and rRNAs were located by molecular hybridizations with restriction fragments. (From P. Malnoë and J.-D. Rochaix. 1978. *Mol. gen. Genet.* **166**:269.) (b) Genetic map derived from recombination analysis and other means, showing antibiotic resistance loci and other markers. (From B. Singer, R. Sager, and Z. Ramanis. 1976. *Genetics* **83**:341.)

ing. This situation may be lethal for most aerobic cells, but cells that can exist by fermentative or glycolytic carbohydrate metabolism are still able to function at a reduced level. Enzymes of electron transport and of oxidative phosphorylation are made through the cooperative actions of nuclear and extranuclear genetic systems in aerobic cells.

None of the membrane polypeptides in chloroplasts has been identified specifically, but some of these are made in the chloroplast and some are made in the cytoplasm (see Fig. 13.12). The best-known protein in chloroplasts is the enzyme ribulose 1,5-biphosphate carboxylase (RuBP carboxylase), which catalyzes the first step in reduction of $CO_2$ to carbohydrates during the "dark reactions" of photosynthesis. The enzyme consists of a number of polypeptide chains in two subunits of the functional catalyst, and one of these subunits is encoded by nuclear

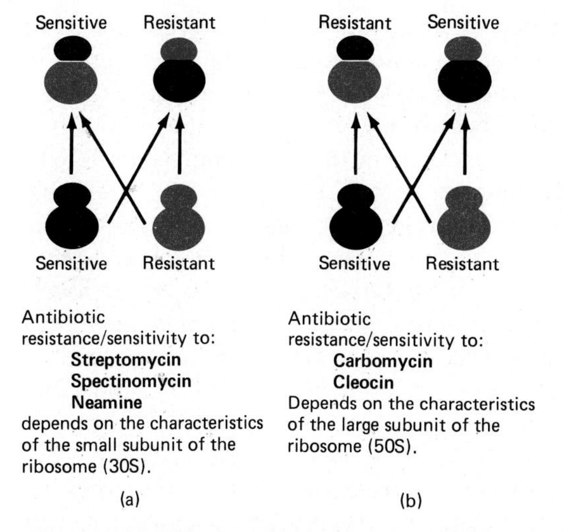


***Figure 13.22*** Diagram summarizing results from Sager's studies, showing that one of the chloroplast ribosomal subunits is altered in extranuclear antibiotic-resistant mutants in *Chlamydomonas.* The subunit involved was identified from mixed reconstitutions.

genes and the other by chloroplast genes in flowering plants. RuBP carboxylase is loosely associated with photosynthetic membranes within the chloroplast, but it is not a structural component of these membranes. The enzyme serves as a vital link between photosynthetic reactions, in which light energy is transformed into chemical energy, and the "dark reactions," in which this chemical energy is utilized in carbohydrate synthesis from $CO_2$ and $H_2O$.

In both mitochondria and chloroplasts, therefore, some of the polypeptides of the vital enzymes of aerobic respiration and of photosynthesis are made within the organelle. These polypeptides assemble with others made in the cytoplasm, and the entire functional enzyme is completed in the organelle. The protein-synthesizing machinery of mitochondria and chloroplasts serve as the centers for synthesis of organelle-specified polypeptides needed for the existence of the cell and the organism.

What is the advantage of having two sets of polypeptides made in different parts of the cell? One reason that has been suggested is that organelle-synthesized polypeptides are highly hydrophobic in nature as a consequence of very high levels of hydrophobic amino acids used in their construction. Such molecules would move across organelle membranes with great difficulty if they were synthesized in the cytoplasm. Since they are made within the organelle, problems of their transport from the cytoplasm are avoided. Cytoplasmically synthesized polypeptides of these organelle enzymes are far less hydrophobic and pass across the mitochondrial or chloroplast membranes with less difficulty. This answer raises other questions, such as, Why aren't all the enzyme polypeptides made within the organelle? At present we have very little information, and most of the discussions have been highly speculative.

## Mitochondrial Genomes Compared

With the introduction of methods for base sequencing and the continued use of molecular hybridizations, restriction fragment comparisons, and other methods, enormous strides have been made in analyzing mtDNA in the past few years. Many important questions remain to be answered, but we have learned much about mitochondrial genome organization and mitochondrial gene expression processes. Comparisons of mitochondrial genetic systems have been particularly useful in determining the underlying similarities among different genomes and in discovering interesting differences among them. These data should prove to be important in the ultimate determination of the evolutionary origin of mitochondria in the eukaryotic cell.

### 13.8 Genome Organization

The complete sequence of 16,569 base pairs in the human mitochondrial genome was published in 1981 by a group of 14 investigators in England. The entire sequence could be interpreted because of the vast amount of information that had been obtained in studies of human and other mtDNA by mapping, molecular hybridization, RNA sequencing and partial DNA sequencing, and by other molecular and genetic analyses of mitochondrial genes, RNAs, and polypeptide products of these genes.

Human mtDNA includes genes for 12S and 16S rRNA, for 22 tRNAs, and for 13 polypeptides (Fig. 13.23). Only five of the polypeptides have been identified; these genes encode three cytochrome oxidase subunits, one ATPase subunit, and cytochrome *b*. All five genes are also found in other mitochondrial genomes. In addition, yeast mtDNA carries a gene for ATPase subunit 9 that is not found in the human genome.

The eight *unassigned reading frames* in the human mitochondrial genome are presumed to be polypeptide-specifying genes, possibly for polypeptides in the ribosome or membrane. These sequences have initiation and termination codons at the termini of a contiguous array of amino acid–specifying codons. The same unassigned reading frames are present in other mammalian genomes, and their conservation in evolution is considered to be indicative of their functional nature.

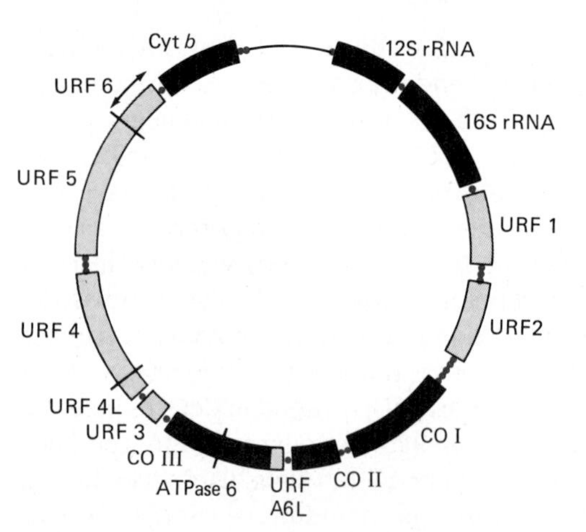


***Figure 13.23*** Organization of the human mitochondrial genome, based on studies using HeLa cell mtDNA. The genes for two rRNAs, 22 tRNAs (colored dots), and 13 polypeptides are distributed economically among virtually all of the sequence of 16,569 base-pairs of the circular mtDNA molecule. Five of the polypeptide-specifying genes have been identified (COI, COII, COIII, ATPase 6, and Cyt. *b*), but eight unassigned reading frames (URF) remain to be analyzed. (Reprinted by permission from P. Borst and L.A. Grivell. 1981. *Nature* **290**:443–444, Fig. 1. Copyright © 1981 Macmillan Journals Limited.)

Great economy of organization characterizes the human mitochondrial genome. The only noncoding region appears to be located in the D-loop segment, where the origin of replication is presumed to be located. No spacers are found between genes, and no intervening sequences occur within genes. Coding sequences are located principally on the heavy (H) strand, but eight of the 22 tRNA genes are situated on the light (L) strand of the duplex DNA (Fig. 13.24). Yeast mtDNAs have long A+T−rich spacers between many of the genes, and at least some yeast mitochondrial genes are comprised of intron and exon alternating regions. Interrupted gene organization in yeast and its absence in mammalian mitochondria provides one of the significant distinctions between these genomes.

The distribution of genes is different in human and yeast mitochondria. In the human mitochondrial genome the two rRNA genes are very near each other, separated only by a tRNA gene. In yeast the two rRNA genes are located tens of thousands of base-pairs apart on the circular mtDNA map. Yeast tRNA genes are situated in at least four regions of the genome, whereas tRNA genes in human mitochondria are scattered around the genome. This distribution of tRNA genes has led to a proposal of transcript processing in human mitochondria, which we will discuss shortly.

The size of mtDNAs differs considerably among organisms (see Table 13.1). The yeast mtDNA molecule is five times the contour length of mammalian mtDNA. We are not certain whether this difference in size of mtDNAs is sufficient explanation for differences in genome organization, such as the occurrence of intergenic spacers and interrupted genes in yeast but not in mammalian mitochondria. These size differences have contributed, however, to variations in processing RNA transcripts and other features related to gene expression in mitochondria of different organisms.

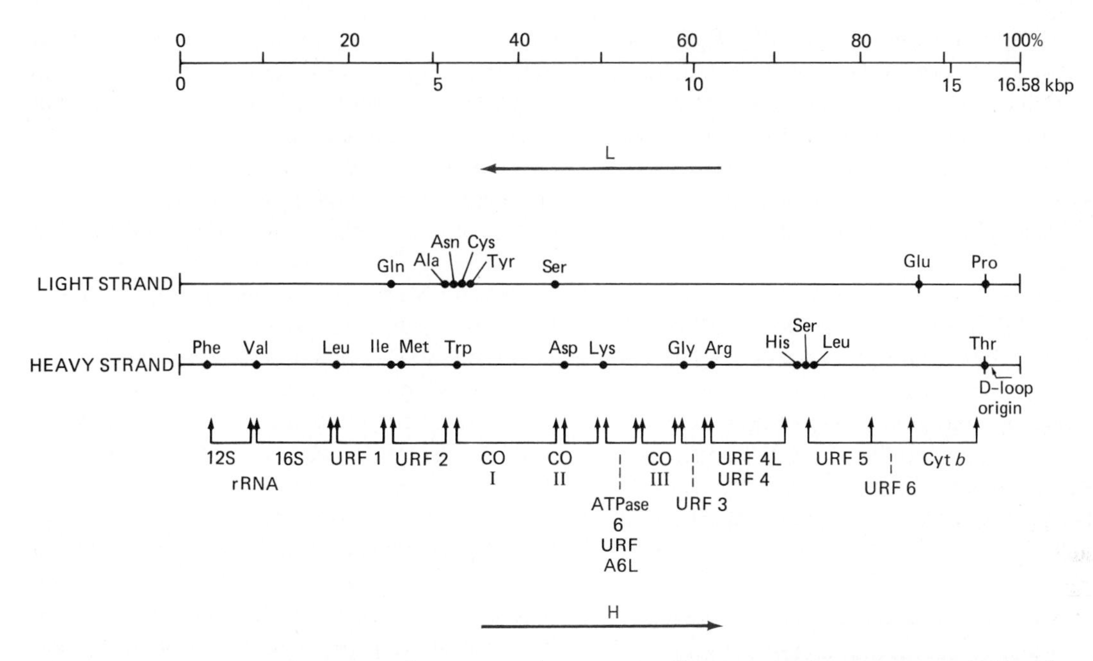


***Figure 13.24*** Location of genes on the two strands of duplex mtDNA in human (HeLa) cells. The circular strands are linearized for easier reference, according to the 16.58-kbp map shown at the top. Eight tRNA genes occur on the light strand and the remaining 14 tRNA genes, along with rRNA- and polypeptide-specifying genes (shown below the strand), are situated in heavy-strand DNA. The only noncoding sequence is found in the region of D-loop origin of replication (far right). Direction of transcription for each strand is indicated by a red arrow. (Reprinted by permission from D. Ojala, et al. 1981. *Nature* **290**:470–474, Fig. 1. Copyright © 1981 Macmillan Journals Limited.)

## 13.9 Transcription Processing in Mitochondria

The observed scatter of tRNA genes in human mtDNA led Guiseppe Attardi and co-workers to propose a model for RNA processing. Based on extensive molecular analyses of RNA transcripts and of their alignments with DNA coding sequences, Attardi suggested that wherever tRNA transcript regions are present they act as recognition signals for cleavages by nucleases of a polygenic RNA molecule copied from the H strand. The entire human mtDNA molecule appears to have a single promoter so that the entire H strand may be transcribed as a single RNA transcript. The promoter is believed to occur in the D-loop noncoding region, where the replication origin is situated.

The *tRNA punctuation model* of RNA processing in human mitochondria stipulates that tRNA-looped secondary structure may signal cleavages at each terminus of the tRNA sequence. Once the tRNA is cut out, the mRNA and rRNA transcripts flanking the tRNA will be separated out at the same time. Further processing of the transcripts produces mature RNAs that function in translation processes. One of these processing steps is *polyadenylation of transcripts:* poly(A) is added posttranscriptionally to the 3′ end of human RNA transcripts. This observation contributed to the suggestion that polyadenylation was an important aspect of RNA processing in human mitochondria.

Base sequencing of human mtDNA and its RNA transcripts has revealed that a number of genes have incomplete termination codons at their terminus. A gene may have 5′-A or 5′-AU as a terminus, immediately abutting its neighbor gene initiation codon. By posttranscriptional addition of poly(A), the RNA transcript terminating in the complementary 3′-U or 3′-UA can be extended to include 3′-UAA as a termination codon (Fig. 13.25). Yeast mitochondrial genes not only carry complete termination codons, but these are followed by noncoding trailer sequences at the 5′ end. Yeast RNA transcripts are not polyadenylated, according to indirect lines of evidence.

Interrupted genes in yeast are transcribed into pre-mRNAs that must be processed to mature mRNAs. Intervening sequences are excised by nucleases, and coding sequences are ligated to produce a contiguous array of codons colinear with the gene product. Piotr Slonimski and co-workers have suggested that the excised intervening sequences play a role in subsequent splicing of exons to produce a functional RNA transcript. By studies of yeast that carry altered mitochondrial genes, it has been shown that intervening sequences in the gene must be present and must be cut out precisely in order to generate a functional gene product, such as a cytochrome oxidase subunit polypeptide (Fig. 13.26). This is reminiscent of eukaryotic nuclear mRNA processing and its importance in functional gene expression.

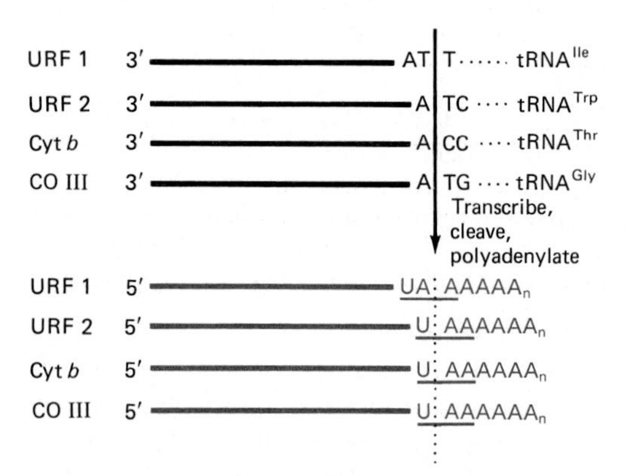


***Figure 13.25*** The transcriptional processing and polyadenylation model for termination of translation in mammalian mitochondria. Incomplete termination codons in the four genes shown were verified by comparison of DNA sequences with their complementary mature mRNA transcripts. Termination codons are posttranscriptionally completed by addition of adenine residues in polyadenylation reactions. The UAA stop codon of the processed transcripts is underlined. Secondary structure development of tRNA regions in a polygenic transcript may provide recognition signals for cleavage of a gene sequence at a particular encoded terminus. This in turn is followed by polyadenylation of the cleaved gene transcript at its 3′ terminus. Such a sequence of events has been called *punctuation processing* of a polygenic transcript. (Reprinted by permission from S. Anderson, et al. 1981. *Nature* **290**:457–465, Fig. 3. Copyright © 1981 Macmillan Journals Limited.)

In mammals, mitochondrial tRNA genes may be involved in transcriptional processing, but in yeast they apparently are not involved. Yeast mtDNA has about five different promoters, so the 26-$\mu$m-long genome is not transcribed into a single polygenic RNA. Transcription and posttranscriptional processing in yeast mitochondria resembles eukaryotic RNA systems and is quite different from the processes that contribute to economical human mitochondrial gene expression.

## 13.10 Codon Usage in Translation

One of the astonishing discoveries in 1980 was the difference in codon usage between mitochondria and the cytoplasmic translation systems of prokaryotes and eukaryotes. The whole dictionary of triplet codons has been analyzed in yeast, in *Neurospora,* and in human mitochondria. In all three species the mRNA codon UGA is read as tryptophan instead of a stop signal (Fig. 13.27). In human mitochondria, but not in yeast or *Neurospora,* the RNA codon AUA codes for methionine instead of isoleucine. In yeast only, all four codons beginning with CU specify threonine instead of leucine. All the remaining codons appear to specify the same amino acids or stop signals for mitochondrial translation as they do for cellular cytoplasmic translation.

Many tRNA anticodons in mitochondria can recognize and interact with as many as four different mRNA codons. Mitochondrial tRNA anticodons almost always have U in the 5′ wobble position for unmixed codon families, those in which the same initial doublet followed by any one of the four bases in the 3′ position of

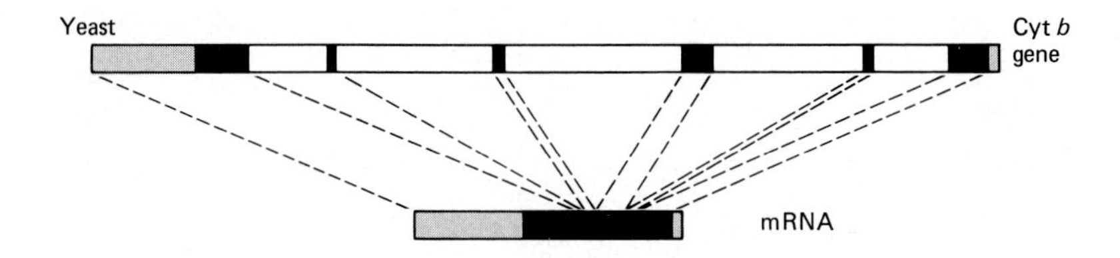

***Figure 13.26*** Comparison of the organization of the cytochrome *b*–specifying gene and its transcript in yeast and human mitochondrial systems. The yeast gene has leader and trailer sequences (hatched), six coding segments (black), and five noncoding intervening sequences (white) that are excised from pre-mRNA to make a continuous reading frame in the mature mRNA transcript. The human Cyt *b* gene is colinear with its pre-mRNA transcript, and a poly (A) tail (hatched) is added afterward to complete the mature mRNA that guides translation. The human mitochondrial gene lacks leader, trailer, and intervening sequences, but the coding regions are very similar in human and yeast mRNAs. (Reprinted by permission from P. Borst and L.A. Grivell. 1981. *Nature* **290**:443–444, Fig. 2. Copyright © 1981 Macmillan Journals Limited.)

the mRNA codons code for the same amino acid. The exception appears to be the family of four arginine-specifying codons, which are all used by anticodon 3′-GCA-5′ (Fig. 13.28). Mixed codon families are read by tRNA anticodons with U or G generally present in the anticodon 5′ wobble position. A few tRNA anticodons have 5′-C, and only the arginine-specifying codons are read by tRNA anticodons carrying 5′-A.

From observed codon-anticodon sequences in sequenced tRNAs and sequenced mRNAs and from deductions based on coding sequences in DNA we know that as few as 24 different tRNAs can recognize all available mRNA codons. This number is substantially less than the 32 different tRNAs needed in cytoplasmic translation systems (see Fig. 9.13). The unusual nature of mitochondrial tRNAs involves other parts of their se-

| | Encoded amino acid | | | |
|---|---|---|---|---|
| | Cytoplasm | Mitochondria | | |
| mRNA codon | | Mammalian | Yeast | *Neurospora* |
| CUU<br>CUC<br>CUA<br>CUG | Leu | Leu | Thr | Leu |
| AUA | Ile | Met | Ile | Ile |
| UGA | Stop | Trp | Trp | Trp |
| AGA<br>AGG | Arg | ? | Arg | – |

***Figure 13.27*** Differences in codon usage in mitochondrial and standard cytoplasmic translation systems. Variations in usage are shown in color. Specification of AGA and AGG codons in mammalian mitochondria is uncertain, but some evidence indicates that these may function as stop codons.

| Codon | Amino acid | Anticodon | Codon | Amino acid | Anticodon | Codon | Amino acid | Anticodon | Codon | Amino acid | Anticodon |
|---|---|---|---|---|---|---|---|---|---|---|---|
| UUU | Phe | AAG | UCU | Ser | AGU | UAU | Tyr | AUG | UGU | Cys | ACG |
| UUC | | | UCC | | | UAC | | | UGC | | |
| UUA | Leu | AAU | UCA | | | UAA | Ter | | UGA | Trp | ACU |
| UUG | | | UCG | | | UAG | | | UGG | | |
| CUU | Thr | GAU | CCU | Pro | GGU | CAU | His | GUG | CGU | Arg | GCA |
| CUC | | | CCC | | | CAC | | | CGC | | |
| CUA | | | CCA | | | CAA | Gln | GUU | CGA | | |
| CUG | | | CCG | | | CAG | | | CGG | | |
| AUU | Ile | UAG | ACU | Thr | UGU | AAU | Asn | UUG | AGU | Ser | UCG |
| AUC | | | ACC | | | AAC | | | AGC | | |
| AUA | | | ACA | | | AAA | Lys | UUU | AGA | Arg | UCU |
| AUG | Met | UAC | ACG | | | AAG | | | AGG | | |
| GUU | Val | CAU | GCU | Ala | CGU | GAU | Asp | CUG | GGU | Gly | CCU |
| GUC | | | GCC | | | GAC | | | GGC | | |
| GUA | | | GCA | | | GAA | Glu | CUU | GGA | | |
| GUG | | | GCG | | | GAG | | | GGG | | |

***Figure 13.28*** mRNA codons ($5' \rightarrow 3'$, left) and tRNA anticodons ($3' \rightarrow 5'$, color) of the yeast mitochondrial genetic code. The 5′ wobble base of each anticodon is underlined. Unmixed codon families (all four codons specify the same amino acid and all have the same initial doublet of bases) generally interact with a single anticodon whose 5′ wobble base is U. The unmixed codon family for arginine is the only exception in yeast. (From S.G. Bonitz et al. 1980. *Proc. Nat. Acad. Sci. U.S.* **77**:3167–3170, Fig. 2.)

quences in addition to anticodons. In yeast and *Neurospora,* and probably in other mitochondria as well, at least some tRNAs have eight nucleotides instead of seven in the anticodon arm. This may facilitate the broader range of codon-anticodon base pairings in translation. Certain other features known to be invariant in cytoplasmic tRNAs are variable in mitochondrial tRNAs. For example, mitochondrial tRNA secondary structures derived from sequence analysis may have a guanine residue instead of pseudouridine in the T$\psi$C arm (Fig. 13.29).

Certain codons are either missing altogether or rarely present in a number of mitochondrial genes that have been sequenced. In a survey of codon usage in five mitochondrial genes that specify polypeptides in yeast, members of Alexander Tzagoloff's laboratory discovered that 15 of the 62 amino acid–specifying DNA codons were not present at all. Certain DNA codons were heavily used whereas alternates specifying the same amino acid were rarely used (Fig. 13.30).

In general, mitochondrial translation utilizes a curtailed set of tRNAs and fewer codons than cytoplasmic systems. A set of 24 tRNAs appears to be adequate for mitochondrial translation since fewer codons are used and the same tRNA anticodon can interact with as many as four mRNA codon types. Codon usage studies have provided

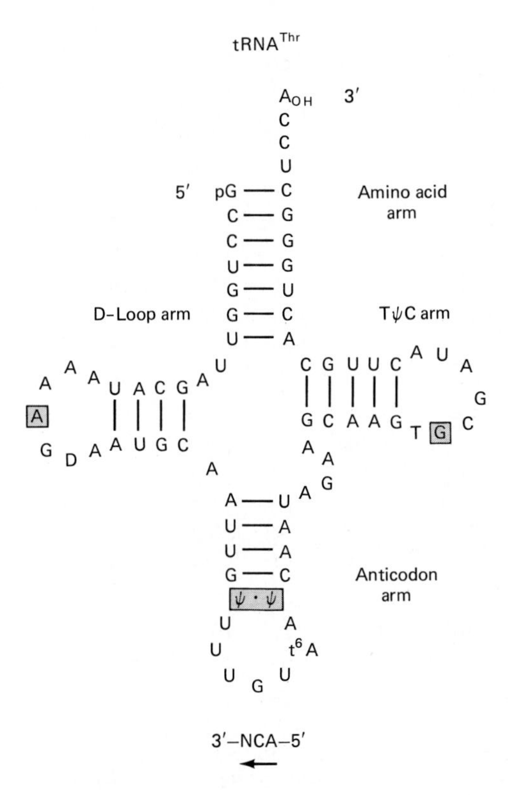


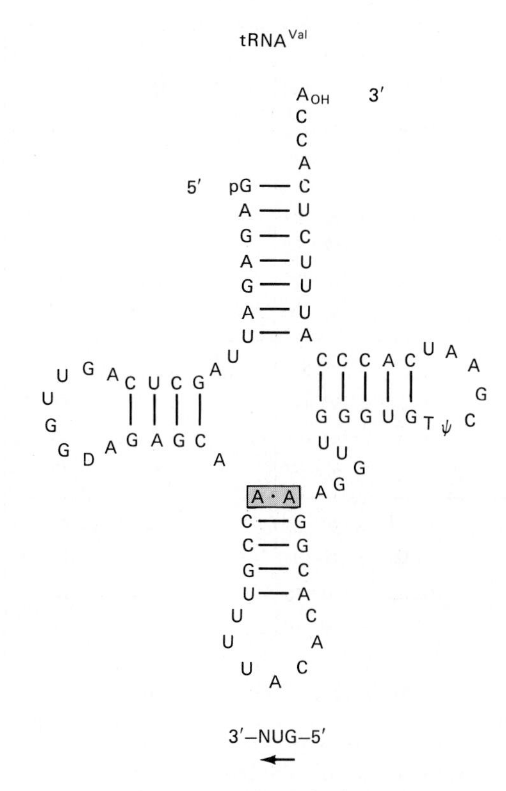


***Figure 13.29*** Secondary structure of tRNA$^{\text{Thr}}$ and tRNA$^{\text{Val}}$ sequences specified by mitochondrial genes in *Neurospora crassa*. Unusual features that distinguish these mitochondrial tRNAs from most nuclear-encoded tRNAs are shown in boxed regions (color). Underneath each tRNA anticodon portion of the sequence is the set of mRNA codons (N = A, C, U, or G) that could interact through base pairing. The mRNA codons appear in $3' \rightarrow 5'$ orientation instead of the more usual $5' \rightarrow 3'$ orientation seen in codon dictionaries (refer to Fig. 13.28). (From J.E. Heckman et al. 1980. *Proc. Nat. Acad. Sci. U.S.* **77**:3159–3163, Fig. 1.)

important information about mitochondrial systems themselves and about the minimum requirements for any translation system. The evolutionary implications as well as the nature of the molecular processes are under extensive study in a number of laboratories.

## 13.11 Evolutionary Origins Of Organelles

The question of the origin of mitochondria and chloroplasts that carry some of their own genetic machinery has stimulated the proposal of a number of speculative hypotheses. The principal basis for these hypotheses is comparison of organelle components with components in prokaryotes and eukaryotes. There is little or no evidence other than these genomic comparisons. The lack of significant experimental support for these hypotheses and the absence of living intermediate evolutionary forms have obstructed the search for widely accepted and verifiable models.

The **endosymbiont theory** revived by Lynn Margulis in 1967 is an idea that has attracted many enthusiastic proponents, despite the lack of substantial evidence in its support. Margulis proposed that the ancestral cell type was anaerobic, prokaryotic, and capable of ingesting solids through its mobile cell-surface

| | | | | | | | | | | | | | | | | | | | |
|---|---|---|---|---|---|---|---|---|---|---|---|---|---|---|---|---|---|---|---|
| Phe | UUU | 34 | 77 | 109 | Ser | UCU | 46 | 32 | 43 | Tyr | UAU | 56 | 46 | 65 | Cys | UGU | 11 | 5 | 11 |
| | UUC | 38 | 141 | 132 | | UCC | 0 | 99 | 47 | | UAC | 6 | 89 | 59 | | UGC | 1 | 17 | 20 |
| Leu | UUA | 164 | 73 | 131 | | UCA | 23 | 83 | 148 | Ter | UAA | – | – | – | Trp | UGA | 26 | 93 | 97 |
| | UUG | 2 | 16 | 16 | | UCG | 0 | 7 | 4 | | UAG | – | – | – | | UGG | 0 | 11 | 7 |
| Leu (Thr) | CUU | 12 | 65 | 87 | Pro | CCU | 24 | 41 | 30 | His | CAU | 36 | 18 | 34 | Arg | CGU | 0 | 7 | 8 |
| | CUC | 0 | 167 | 65 | | CCC | 0 | 119 | 34 | | CAC | 0 | 79 | 63 | | CGC | 0 | 25 | 18 |
| | CUA | 2 | 276 | 266 | | CCA | 26 | 52 | 139 | Gln | CAA | 22 | 81 | 79 | | CGA | 0 | 29 | 36 |
| | CUG | 0 | 45 | 26 | | CCG | 1 | 7 | 2 | | CAG | 3 | 9 | 3 | | CGG | 0 | 2 | 3 |
| Ile | AUU | 6 | 125 | 234 | Thr | ACU | 34 | 51 | 58 | Asn | AAU | 49 | 33 | 60 | Ser | AGU | 12 | 14 | 14 |
| | AUC | 108 | 196 | 140 | | ACC | 0 | 155 | 80 | | AAC | 4 | 131 | 108 | | AGC | 0 | 39 | 35 |
| Met (Ile) | AUA | 21 | 167 | 218 | | ACA | 27 | 133 | 157 | Lys | AAA | 23 | 85 | 100 | Arg | AGA | 28 | 0 | 0 |
| Met | AUG | 45 | 40 | 31 | | ACG | 1 | 10 | 6 | | AAG | 1 | 10 | 2 | | AGG | 0 | 0 | 0 |
| Val | GUU | 54 | 30 | 53 | Ala | GCU | 32 | 43 | 47 | Asp | GAU | 31 | 15 | 31 | Gly | GGU | 15 | 24 | 36 |
| | GUC | 3 | 49 | 34 | | GCC | 1 | 124 | 82 | | GAC | 1 | 51 | 43 | | GGC | 3 | 88 | 39 |
| | GUA | 38 | 70 | 76 | | GCA | 47 | 80 | 97 | Glu | GAA | 31 | 64 | 83 | | GGA | 64 | 67 | 109 |
| | GUG | 1 | 18 | 10 | | GCG | 2 | 8 | 7 | | GAG | 2 | 24 | 11 | | GGG | 0 | 34 | 29 |

***Figure 13.30*** Genetic code and codon usage in yeast, human, and mouse mitochondrial translation systems. In yeast, 15 of the 62 codons specifying amino acids are not used in at least the five mitochondrial genes analyzed (left column, color); only codons AGA and AGG are not used to specify amino acids in human (middle column) or mouse (right column) mitochondrial translation systems. Stop codons UAA and UAG were not included in the tabulations. Certain mRNA codons are used to very different degrees in these three systems, especially mRNA codons with C in the 3′ (third) position. Codons with 3′-G are used infrequently in all three species, except for the start codon AUG, which also specifies methionine. (Data from S.G. Bonitz et al. 1980 *Proc. Nat. Acad. Sci. U.S.* **77**:3167–3170, Table 1 [yeast]; F. Sanger, 1981. *Science* **214**:1205–1210, Fig. 9. [human]; and M.J. Bibb, et al. 1981. *Cell* **26**:167–180, Table 1 [mouse].)

activities. Such a cell is postulated to have engulfed respiring prokaryotic (bacterial) cells, leading to a mutually beneficial, or symbiotic, relationship between the host and its *endo*symbionts within the cell (Fig. 13.31). The host provided nutrients and protection, and the endosymbionts provided the energy-efficient aerobic respiratory pathway to the anaerobic host. In a world undergoing an atmospheric change from anaerobic to aerobic conditions, the host benefitted greatly. Other episodes of endosymbiosis occurred later and gave rise to modern flagella from a spirochetelike prokaryote, and to chloroplasts from blue-green algal cells in certain but not in all lineages. According to this theory, the mtDNA and ctDNA we observe today are the remnants of endosymbiont genomes, as are their translational systems of ribosomes and tRNAs.

The principal support for endosymbiosis are certain observed similarities between prokaryotes and eukaryotic organelles. For example:

***1.*** Genome organization is similar in bacteria and in organelles: a single circular duplex DNA molecule that is not associated with histone proteins. Eukaryotes, on the other hand, have nucleoprotein chromosomes, and their genes are distributed among two or more linkage groups separated from the cytoplasm by a nuclear envelope. No membrane separates bacterial or organelle DNA from its surroundings.

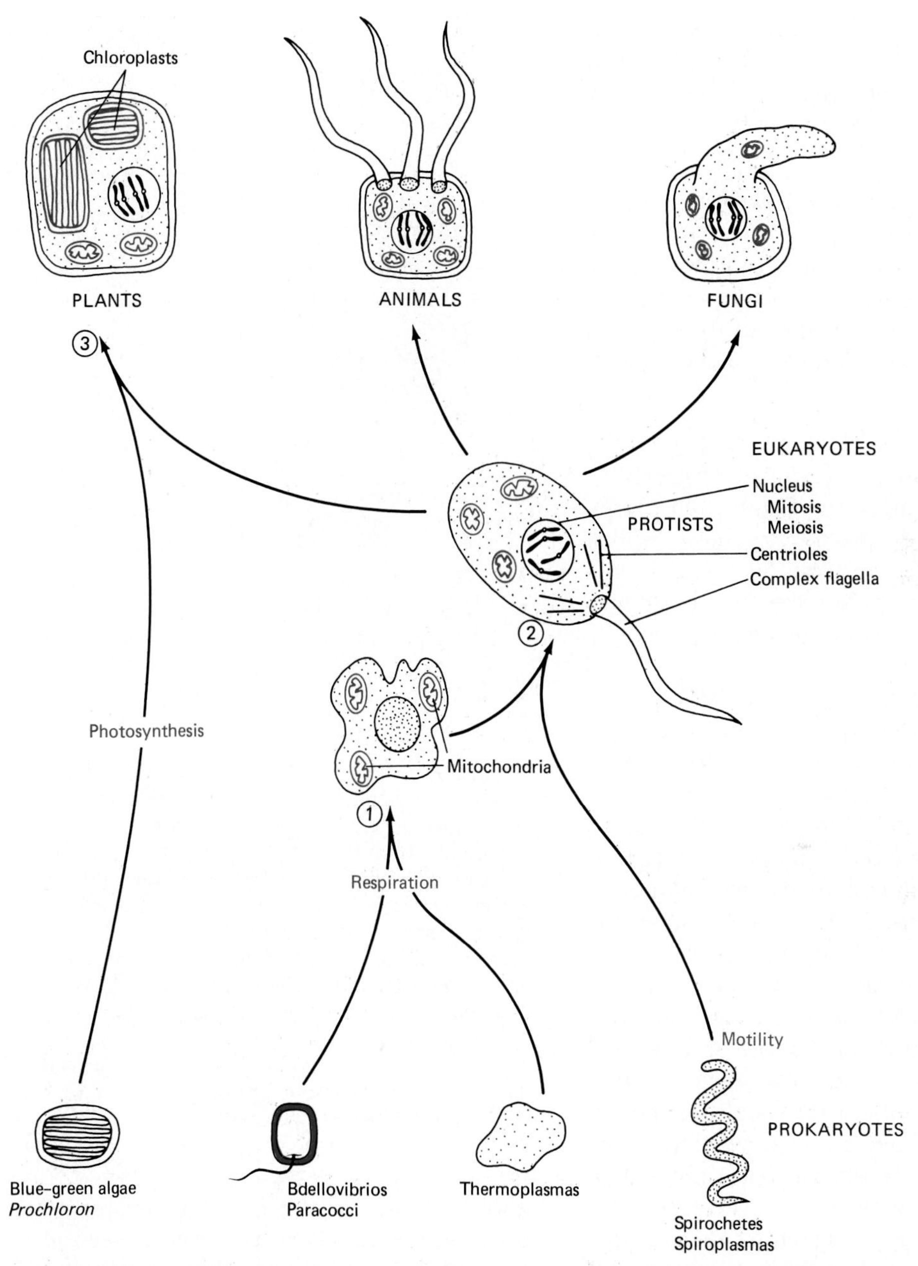


***Figure 13.31*** Model for the origin of eukaryotic cells according to the endosymbiont theory. (1) Ingestion or invasion of aerobic respirer bacteria similar to modern *Paracoccus* or *Bdellovibrio* species by a *Thermoplasma*-like bacterial host with a mobile cell surface may have led to mitochondria during evolution. (2) Prokaryotic spirochete or *Spiroplasma* species may have been the progenitor of modern flagella and cilia after a second endosymbiotic episode involving eukaryotes. (3) A third endosymbiosis leading to chloroplasts from photosynthetic prokaryotic blue-green algae and *Prochloron* is assumed to have occurred only in eukaryotic lineages leading to plants. (From *Symbiosis in Cell Evolution* by L. Margulis. Copyright © 1981 by W.H. Freeman and Company. All rights reserved.)

*2.* Bacterial and organelle ribosomes respond similarly to drugs that affect the ribosomal machinery for protein synthesis; eukaryotic cytoribosomes are inhibited by drugs that do not affect bacterial or organelle systems.

*3.* Specific enzymes of aerobic respiration and of photosynthesis are physically a part of the bacterial plasma membrane and of internal membranes in the two organelle types. The eukaryotic cell has no equivalent of the enzymes for these processes other than the ones found in its organelles.

Endosymbiosis is a well-known phenomenon and often involves a symbiotic bacterial or blue-green algal organism within a eukaryotic host. We know of no prokaryotes, however, that act as hosts for other prokaryotic organisms in a symbiotic association. Nor are any prokaryotes known to have the capacity of ingesting solids or other cellular organisms.

The alternatives to endosymbiosis involve evolutionary changes by which some existing parts of the cell have been altered so that they now exist as separate membranous compartments bathed in eukaryotic cytoplasm, much like other membranous compartments such as lysosomes, endoplasmic reticulum, and the nucleus itself (Fig. 13.32). Views differ about the exact nature of the prokaryotic ancestor of eukaryotes and about the events that led to the present-day organization of eukaryotic cells. Most of the theories, however, propose that some piece of the genome was separated from the bulk of the cellular DNA and became enclosed within membranes. Some of the suggested hypotheses postulate that a plasmid or plasmidlike DNA molecule carrying genes for organelle structure and function was enclosed within membranes making up the mitochondrial or chloroplast boundaries.

One of the central problems in the endosymbiosis theory is that the hundreds of genes now known to be encoded in the nucleus for organelle traits must somehow have moved from the original endosymbiont genome into the host nucleus. Some recent studies have shown that gene transfer between mitochondria and nucleus can occur. However, at present all the theories remain as possible explanations for the evolutionary origin of mitochondria and chloroplasts.

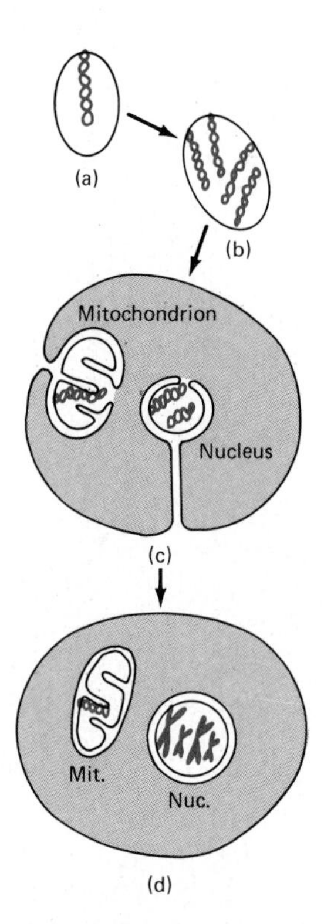


***Figure 13.32*** The evolutionary hypothesis for the origin of organelles postulates genome duplication and invagination of membrane to form double-membrane systems around each genome (color) (a) Prokaryotic cell; (b) duplication of prokaryotic genomes; (c) membrane invagination and formation of double membranes around genomes, all of which is destined to evolve into mitochondrion and nucleus (and chloroplast); and (d) eukaryotic cell, whose nuclear genome evolves toward greater complexity while the organellar genomes lose many duplicated genes. (Reprinted by permission of *American Scientist,* journal of Sigma Xi, The Scientific Research Society.)

On the basis of new information obtained from base sequencing of RNAs and DNAs from mitochondria and chloroplasts, it appears that the two kinds of organelles may have evolved from different origins. The remarkable similarities between chloroplast and blue-green algal rRNAs and ribosomes as well as a growing list of DNA sequence similarities for other genetic components in these two systems have strengthened the case for chloroplast origins from symbiotic blue-green algae. Mitochondria, on the other hand, seem to differ in important ways in different species and from both prokaryotes and eukaryotes in genomic and gene organization (Table 13.3). Whether different species of endosymbionts were ancestral to different mitochondrial types in animals, plants, fungi, and protists, remains to be determined. If not, then mitochondria from some endosymbiont(s) or some cellular compartment(s) have undergone substantial evolutionary divergence in the billion years or more since eukaryotes first appeared on Earth. Because we know little or nothing about

***Table 13.3*** Comparison of selected features of the genomes in human mitochondria, yeast mitochondria, *E. coli* (prokaryote), and *Drosophila melanogaster* (eukaryote).

| **feature** | **human mitochondria** | **yeast mitochondria** | ***E. coli* (prokaryote)** | ***D. melanogaster* (eukaryote)** |
|---|---|---|---|---|
| genomic DNA | 1 duplex; circular | 1 duplex; circular | 1 duplex; circular | 4 duplexes; linear |
| contour length | 5.5 μm | 25–26 μm | 1300 μm | >50,000 μm |
| number of kilobase pairs | 16.57 | 75–78 | ~4,000 | ~150,000 |
| intergenic spacer regions | absent | present | absent | present |
| genes | | | | |
| polypeptide-specifying genes | 5 known + 8 URFs | 7 known + URFs | 3000–4000 | 5,000–10,000 |
| tRNA genes | 22 | 25 | 32 + multiple copies | 32 + multiple copies |
| rRNA genes* | 2; adjacent | 2; far apart | 3 (5–10 copies of each) | 2 (hundreds of copies) |
| noncoding leader and trailer segments | absent | present | present | present |
| stop codons | absent in some genes | present | present | present |
| exon-intron organization | absent | present in some genes | absent | present |
| transcription | within mitochondria | within mitochondria | within nucleoid | within nucleus |
| number of promoters | 1 | 5 or more | numerous | numerous |
| pre-mRNA | absent | present | absent | present |
| mRNA | transcribed directly | processed from pre-mRNA | transcribed directly | processed from pre-mRNA |
| poly(A) tail | posttranscriptional | absent | absent | posttranscriptional |
| translation | within mitochondria | within mitochondria | within cytoplasm | within cytoplasm |
| ribosomes | 55–60S | 80S | 70S | 80S |
| codon usage | UGA = Trp | UGA = Trp | UGA = Stop | UGA = Stop |
| | AUA = Met | AUA = Ile | AUA = Ile | AUA = Ile |
| | $AG^{A}_{G}$ = not used | $AG^{A = Arg}_{G = not\ used?}$ | $AG^{A}_{G}$ = Arg | $AG^{A}_{G}$ = Arg |
| | CUN = Leu | CUN = Thr | CUN = Leu | CUN = Leu |

*Human mitochondrial rRNA genes =12S and 16S; yeast mitochondrial rRNA genes = 15S and 21S; *E. coli* rRNA genes = 16S, 23S, and 5S; *D. melanogaster* rRNA genes = 38S (processed to 18S and 28S rRNAs) and 5S.

selection factors that influence organelle evolution today, much less in the remote past, we cannot fully reconstruct evolutionary sequences. We are still in the stage of data collection, and we may be able to speculate more specifically or judiciously in the future than we can at the present time.

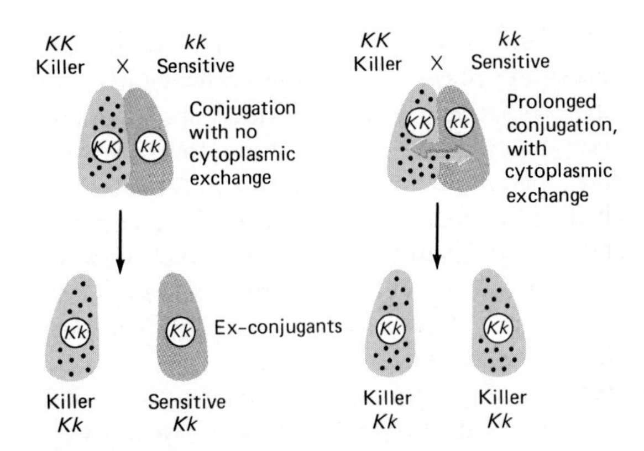


***Figure 13.33*** Infectious inheritance in *Paramecium.* Genetic "killers" are *KK* or *Kk* and maintain kappa (colored dots) in the cytoplasm. Kappa bacteria can be transmitted only through the cytoplasm, so *KK* or *Kk paramecia* remain sensitive unless they receive kappa during conjugation.

## Infectious Inheritance

Extranuclear genes or genomes can be carried in host cells in the form of plasmids, viruses, and bacterial DNA. These extranuclear genetic components influence host phenotypes, just as organelle genomes do, but they are not native to the cell. The host can survive and flourish in the absence of extranuclear plasmids, viral genes, and bacterial genes. Organelle genes, on the other hand, are vital for the expression of aerobic respiratory and photosynthetic activities in eukaryotic organisms, regardless of their evolutionary origin.

### 13.12 Kappa, Sigma, and Other Infectious Agents

Many viruses or bacteria can be transmitted in eukaryotes during sexual reproduction of the host. The gene transmission patterns in the host therefore give the impression that extranuclear traits are passed from parent to offspring in non-Mendelian fashion. In certain *Paramecium* strains, the small bacterium called **kappa** is transmitted from parents to progeny through the cytoplasm and not through parental nuclear contributions (Fig. 13.33). The maintenance of kappa in these cells, however, is dependent on the dominant nuclear allele *K.* Thus *kk* paramecia cannot harbor kappa even if these symbionts are introduced into *kk* cells by conjugation with *KK* or *Kk* partners that do have kappa. Cells with kappa bacteria are resistant to the toxin produced by these kappa elements. Kappa-containing *KK* or *Kk* paramecia are "killers." Kappa therefore influences phenotypic expression in sensitive cells and in resistant cells. The presence of kappa bacteria confers immunity to the toxin in *KK* or *Kk* paramecia.

There are two examples of infectious elements transmitted by the female parent to progeny in *Drosophila,* each demonstrating an apparently extranuclear pattern of inheritance. Flies may be phenotypically resistant to anesthesia by $CO_2$ and recover quickly from such exposure, or they may be sensitive and remain permanently paralyzed by exposure to this gas. The trait is transmitted through the female and only rarely through the male parent. Sensitive flies carry a virus called **sigma,** which alters the organism so that it becomes sensitized to $CO_2$; resistant organisms become sensitive to $CO_2$ when sigma virus is introduced during reproduction or by injection of extracts from sensitive flies. Sigma is an infectious agent that has the ability to alter the phenotype of the host organism to $CO_2$ sensitivity. It can be transmitted from generation to generation through the eggs of the female parent.

A trait called sex ratio (SR) has been traced to the effects of a symbiotic spirochete bacterium in several strains of *Drosophila.* Progeny of females harboring this spirochete are virtually all daughters because male embryos are killed very early in development if the spirochete is present in the cytoplasm of these cells. The SR trait can be introduced into normal flies through reproduction or through injections of extracts from SR individuals. As with kappa, sigma virus and the SR spirochete can be maintained in *Drosophila* only with appropriate nuclear genes, which influence the host response to the agent in the expression of a resistant or sensitive phenotype.

Certain strains of mice are much more likely to develop mammary tumors than other strains. In reciprocal crosses between mice from both kinds of strains, the proneness to cancer appeared to be transmitted through the female parent and not through the male parent. The pattern therefore resembled one due to extranuclear inheritance. Later, it was found that normal mice from low-cancer strains would develop mammary tumors with much higher frequency if they were *nursed* by females from high-cancer strains instead of by their own mothers. Such progeny mice would then transmit increased proneness to tumor development in subsequent generations. The infectious agent, *mouse mammary tumor virus,* was eventually isolated and identified. The virus was transmitted through mother's milk, but the expression of the virus depended in part on nuclear genes in the mouse. Individuals from some mouse strains are less likely to develop these tumors even though the virus may be present.

There are a number of other RNA tumor viruses in addition to mouse mammary tumor virus. Various DNA tumor viruses are known to be responsible for cancers in vertebrate animals, from fish to mammals. With all of them, the tumor virus can multiply only in certain hosts and host cell types, which are called **permissive** hosts or cells. When the same virus is introduced into host or cell type that is **nonpermissive,** the virus is unable to multiply but may *transform* the host cell into a cancerous state.

Tumor viruses induce a *lytic infection* in permissive host cells, during which virus multiplication takes place. Cell death usually follows shortly after infection. In cells kept in culture, usually derived from embryonic tissues, a lytic response is observed by the formation of a zone of dead cells somewhat resembling plaques produced by bacterial viruses. The **transforming response** is identified in cultured cells by the development of a disordered mass of cells that are heaped on top of one another (Fig. 13.34). In the living animal, the lytic response also leads to cell death and the animal may die if the infection is widespread. The development of tumors serves as the indicator of the transforming response in nonpermissive host animals.

Transformed cells transmit their new phenotypes to subsequent generations indefinitely, that is, the transformed phenotype is inherited. Where studies have been possible, the virus genome has been shown to become integrated into one or more chromosomes in the transformed host cell nucleus. As the host cell reproduces, copies of the viral genome are synthesized along with copies of the host genome, and both sets of genes are transmitted to progeny generations. In these cases, therefore, the inheritance of the transformed phenotype can be traced to transmission of viral genes.

Only a part of the viral genome is expressed in nonpermissive cells, and this part of the genome codes for certain viral products that

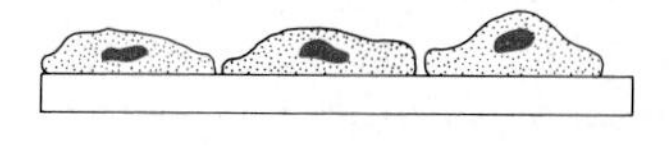

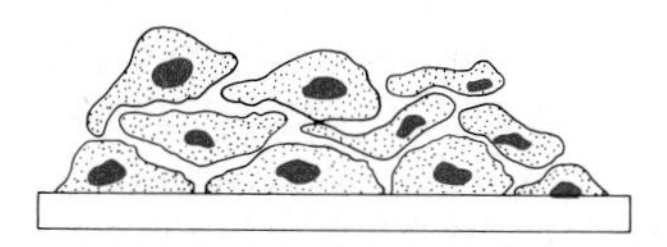

***Figure 13.34*** Diagrammatic illustration of the growth pattern of normal cells (top) and transformed cells on solid surfaces (bottom). Normal cells stop multiplying and moving when a confluent monolayer has formed; transformed cells continue to multiply (and move) and thereby give rise to a disorderd pile of cells.

induce the cancerous state in the host. Such viral genes are often called **oncogenes,** since they are responsible for the *oncogenic* (cancer-causing) potential of the virus in a suitable host. The integrated viruses do not multiply in nonpermissive cells since other genes needed for virus multiplication and infectious particle formation are turned off. In permissive host cells all the viral genes are turned on, thus allowing multiplication to occur and a lytic response to take place instead of transformation.

We will discuss oncogenic viruses in more detail in the next chapter. These systems influence host development, and they also provide admirable models to investigate gene expression during eukaryotic development.

## 13.13 Plasmids as Extranuclear Genomes

Plasmids are excellent examples of extranuclear genomes that are transmissible from cell to cell and that influence the cellular phenotype. Most of our information has come from plasmids in bacterial cells, but they are known to occur in eukaryotic cells. For example, a 2-$\mu$m-long, circular DNA plasmid is present in many yeast cells. Plasmid functions, however, are barely known in the few eukaryotic systems where they have been found.

The fertility factor **F** in *E. coli* confers a number of distinct phenotypic traits on cells in which it occurs. In addition to making such cells capable of transferring F plasmids themselves, integrated F in Hfr strains also causes transfer of the host genome into conjugant partner cells. *E. coli* $F^+$ cells are sensitive to infection by single-stranded RNA phages and certain single-stranded DNA phages, but they are resistant to other phages such as T3 and T7. These phenotypic characteristics are outcomes of phage gene action within the host, and they are inherited when F is inherited by *E. coli.*

**R plasmids** carry antibiotic-resistance genes that confer resistance on host cells harboring these plasmids. The plasmid genes can be incorporated into the host genome, so the resistance may remain even when the plasmid DNA has apparently been eliminated from the host cell. This kind of genome modification shows extranuclear inheritance when the resistance genes are in the plasmid DNA but shows alteration to the host inheritance pattern if the genes become integrated as a part of the host linkage group.

Bacterial cells that contain *Col plasmids* can synthesize proteins called *colicins,* which kill sensitive cells of their own species or of other species. Cells that harbor Col plasmids are immune to the effects of the toxic proteins since they also synthesize *immunity proteins.* The Col plasmid genome contains separable regions, much like the R plasmids. The genome segment coding for colicin and immunity proteins can be separated from the transfer component. In this case, the colicin–immunity protein coding segment may be inherited along with *E. coli* genes in the single host linkage group. A pattern of extranuclear inheritance, however, would characterize *E. coli* with Col plasmids as separate genomes coexisting in the same cells with host DNA.

As with viruses, inheritance patterns of plasmid genes reflect the physical location of the extranuclear DNA. In both instances, the host phenotype is influenced by genes that are not native to the cellular linkage group but that may be integrated into this linkage group. These situations appear to be quite different from mitochondrial and chloroplast inheritance since organelle DNA remains within the organelle and is not physically integrated into the host chromosomes. Whether the difference is only of one degree across a spectrum of variations in extranuclear inheritance or whether these all represent different kinds of evolutionary events is uncertain at present. The general observation that ties all these systems together is that each extranuclear genome codes for some of its own phenotypic traits and also influences phenotype expression of the cell or organism in which it occurs. By analyzing the patterns of gene transmission, we can determine whether or not these genes exist in a physically separate nucleic acid molecule or as an integrated part of host DNA.

## Maternal Effect

The influence of maternal substances on the phenotype of the developing organism is referred to as **maternal effect.** These maternal substances, such as messenger RNAs, are made in the premeiotic oocyte under the direction of maternal genes, and they therefore reflect the maternal genotype and not the zygotic genotype. When the zygote and developing organism are observed, however, zygotic genes seem to be responsible for the expressed phenotype. If the inheritance of such characteristics can be followed over several generations, maternal effect is usually clearly demonstrated and distinguished from true extranuclear inheritance.

Several classic examples of a maternal effect, which show different results in reciprocal crosses, can be shown to be instances of *delayed Mendelian inheritance.* In the snail *Limnea peregra,* right-handed (dextral) coiling of the shell is determined by a dominant allele *D,* and left-handed (sinistral) coiling by the recessive allele *d* of a single gene. The animal is hermaphroditic and is capable of self-fertilization as well as cross-fertilization. When homozygous dextral (*DD*) females are crossed with sinistral (*dd*) males, all the progeny are dextral; but in the reciprocal cross of sinistral (*dd*) females and dextral (*DD*) males, all the progeny show sinistral coiling (Fig. 13.35). If *Dd* $F_1$ individuals in each of these reciprocal progenies undergo self-fertilization, each one produces an $F_2$ progeny entirely composed of dextrally coiled snails. This seems to be non-Mendelian inheritance, but its true nature is clearly revealed when each $F_2$ individual undergoes self-fertilization to produce $F_3$ progeny. Three-fourths of the $F_2$ snails produce dextral progeny and one-fourth of the $F_2$ snails produce an $F_3$ generation composed entirely of sinistral types. The appearance of the same phenotypic ratio of 3 dextral : 1 sinistral in reciprocal progenies points to single-gene Mendelian autosomal inheritance, but this phenotypic ratio appears in the $F_3$ generation instead of the $F_2$ generation.

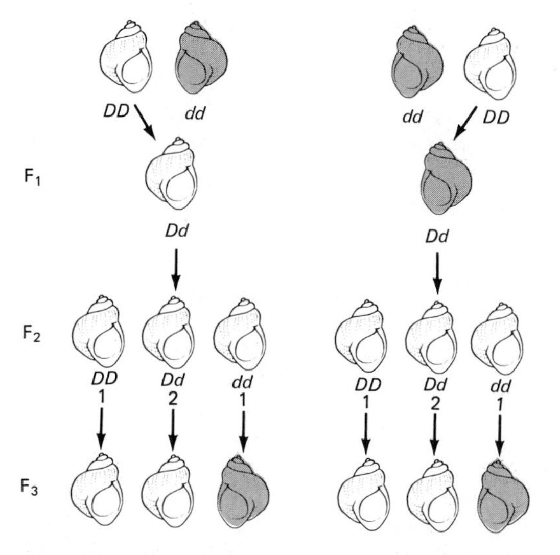


***Figure 13.35*** Maternal effect in the snail *Limnea peregra.* Direction of shell coiling depends on the mother's genotype rather than on the genotype of the individual itself. Dextral, or right-handed, coiling is dominant over sinistral, or left-handed, coiling; but the expected $F_1$ and $F_2$ phenotypic ratios are delayed one generation and appear in the $F_2$ and $F_3$ generations, respectively.

The reason for the delayed expression of Mendelian phenotypic ratios is that substances produced in the egg cytoplasm by the maternal genotype govern the symmetry of the first cleavage division of the fertilized egg, or zygote. Once this division has taken place, the direction of coiling (and of the whole body) is established and lasts the lifetime of the animal. If the maternal genotype is *DD,* all zygotes will be dextral; if the maternal genotype is *dd,* all zygotes will be sinistral, regardless of the male parental genotype. Each $F_1$ *Dd* female in reciprocal progenies, whether dextral or sinistral in phenotype, has the dominant *D* allele. Every egg produced by these *Dd* females has the *D* gene product, which directs dextral coiling in $F_2$ zygotes developed when these eggs are fertilized. When the $F_2$ snails self-fertilize, those with the *dd* genotype give rise to sinistral progeny whereas females with *DD* or *Dd* genotypes will produce dextral $F_3$ progeny. The overall result shows classical Mendelian inheritance, delayed one generation. If the pattern had been truly extranuclear, differences would have been evident in reciprocal progenies in every generation and not just in one or two generations of a lineage. In extranuclear inheritance the progeny continue to resemble only one of the two parents and the extranuclear trait is transmitted indefinitely in subsequent generations.

The inheritance of phenotypic traits can therefore be judged by various criteria using genetic methods and molecular analyses. If the pattern of inheritance is extranuclear, the responsible extranuclear genome can be identified in many cases as belonging to organelles, such as mitochondria and chloroplasts, or to viruses or plasmids. Investigation of some new extranuclear inherited trait in an organism is therefore guided by available methods through which the extranuclear factor can be located. It can then be explored in greater detail in that host system.

## Questions and Problems

***13.1*** Results of reciprocal crosses between strains of *Chlamydomonas reinhardi* that were streptomycin sensitive and streptomycin resistant are given below for tetrads of meiotic products:

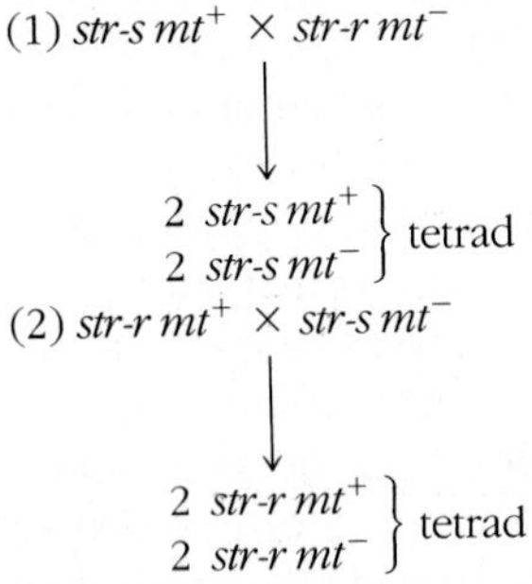


***a.*** What are the tetrad allelic ratios for the *str* and *mt* genes in (1) and (2)?
***b.*** What is the probable mode of inheritance for each gene? Explain.
***c.*** What results would be predicted if $F_1$ *str-s* $mt^+$ were backcrossed to the parental (1) *str-r* $mt^-$ strain, and (2) *str-s* $mt^-$ strain?
***d.*** What results would be predicted if $F_1$ *str-r* $mt^+$ were backcrossed to the parental (1) *str-r* $mt^-$ strain, and (2) *str-s* $mt^-$ strain?

***13.2*** Reciprocal crosses were made between large colony strains of *Neurospora crassa* that were $Arg^+$ $Pro^-$ and small colony strains that were $Arg^-$ $Pro^+$, with the following eight-spore asci produced:

(1) $Arg^+$ $Pro^-$ large ♀ × $Arg^-$ $Pro^+$ small ♂
↓

| Tetrads: | | |
|---|---|---|
| | + − large | + + large |
| | + − large | + + large |
| | + − large | − + large |
| | + − large | − + large |
| | − + large | + − large |
| | − + large | + − large |
| | − + large | − − large |
| | − + large | − − large |

(2) $Arg^-$ $Pro^+$ small ♀ × $Arg^+$ $Pro^-$ large ♂
↓

| Tetrads: | | |
|---|---|---|
| | + + small | − + small |
| | + + small | − + small |
| | − − small | + − small |
| | − − small | + − small |
| | − + small | − + small |
| | − + small | − + small |
| | + − small | + − small |
| | + − small | + − small |

***a.*** What are the phenotypic ratios for the three traits?
***b.*** What is the probable inheritance pattern for each phenotypic trait?

***13.3*** Certain crosses of grande × petite yeast will yield normal respiration-sufficient diploids. When these diploid cells undergo meiosis and produce ascospore tetrads, each ascus is found to contain 2 grande and 2 petite spores. On the basis of these results, explain the inheritance pattern.

***13.4*** Petites such as those described in Question 13.3 are called segregational petites. In crosses between segregational petites and a neutral extranuclear petite strain, what percentage of the ascospores will have the petite phenotype after meiosis has occurred in the zygotes?

***13.5*** A biologist using haploid *Neurospora* noticed a patch of orange mycelium on a plate containing wild-type mold. The biologist constructed a heterokaryon using a non-orange, arginine-requiring haploid and the arginine-independent, orange variant, and found some arginine-requiring, orange segregants, among several other phenotypes. Explain the probable origin of the orange phenotype.

***13.6*** In corn a form of male sterility is inherited extranuclearly, so the normal ♂ × male-sterile ♀ gives male-sterile offspring and the reciprocal cross gives normal offspring. Some strains of corn carry a dominant restorer gene (*Rf*), which restores pollen fertility in a male-sterile line.

***a.*** If pollen from a genotypically *Rf*/*Rf* plant is used in a cross with a male-sterile plant, what are the genotype and phenotype of the $F_1$?

***b.*** If a testcross is performed using $F_1$ plants described in (a) as females and pollen from a normal *rf*/*rf* plant, what would be the genotypes and phenotypes of these testcross progeny? What would be the nature of their cytoplasm?

***13.7*** Two species of *Drosophila,* A and B, produce interesting results when crossed. If female A is crossed with male B, only female progeny occur; if female B is crossed with male A, mostly male progeny are produced and there are few or no female offspring. Suggest an explanation for these observations.

***13.8*** Make a list of the major differences that distinguish extranuclear inheritance from conventional inheritance of nuclear genes in eukaryotes, using the following format:

| | inheritance | |
|---|---|---|
| **characteristic** | **nuclear** | **extranuclear** |
| 1. | | |

***13.9*** Use diagrams to show how replication of mitochondrial DNA can proceed by the mechanisms of D-loop synthesis, Cairns theta-forms, and rolling circle. What electron-microscopic figures would clearly permit distinction of each one of these mechanisms in populations of replicating mitochondrial DNA circular molecules?

***13.10*** Presumptive ribosomal particles have been isolated from purified mitochondria obtained from rat liver cells.

***a.*** How would you demonstrate that these were ribosome monomers using a test for function?

***b.*** How would you demonstrate that these functional particles were composed of two subunits, both of which were needed for monosome function?

***c.*** What antibiotic inhibitors would you use to determine whether these ribosomes were truly mitochondrial and not merely contaminating cytoplasmic ribosomes from the same cells?

***d.*** List the major differences between ribosomes from mitochondria and those from the cytoplasm of animal cells.

***13.11*** How would you test whether rRNA in organelle ribosomes was transcribed from organelle DNA and not from nuclear DNA? How could you tell whether organelle rRNA was transcribed from one or from both strands of its template DNA?

***13.12*** Yeast cells were incubated in media containing the mitochondrial protein-synthesis inhibitor erythromycin (ERY) and in media with the cytoplasmic protein-synthesis inhibitor cycloheximide (CHI). Assays were made to determine which of the six subunits of a respiratory enzyme were translated in mitochondria and which in the cytoplasm, with the following results:

| | enzyme subunits made | | | | | |
|---|---|---|---|---|---|---|
| **experiment number** | $\alpha$ | $\beta$ | $\gamma$ | $\delta$ | $\epsilon$ | $\zeta$ |
| (1) control | + | + | + | + | + | + |
| (2) + ERY | 0 | 0 | + | 0 | + | + |
| (3) + CHI | + | + | 0 | + | 0 | 0 |
| (4) + CHI + ERY | 0 | 0 | 0 | 0 | 0 | 0 |

***a.*** Which subunits are encoded in mtDNA and which in nuclear DNA?

***b.*** If another experiment were performed in which synthesis proceeded with [$^3$H]amino acids for the first two hours in the presence of ERY and with [$^{14}$C]amino acids for the next two hours in the presence of CHI, which enzyme subunits would be labeled with $^3$H in the first two hours and which with $^{14}$C in the next two hours?

***13.13*** Describe an experiment in which you could demonstrate that one subunit of a chloroplast enzyme is encoded in chloroplast DNA and the other subunit of the functional enzyme is encoded in nuclear DNA.

***13.14*** Heavy- and light-strand fractions from HeLa cell mitochondrial DNA were incubated with [$^3$H]labeled $tRNA^{Lys}$, $tRNA^{Pro}$, $tRNA^{Val}$, and $tRNA^{Leu}$ to obtain molecular hybridizations. H-strand mtDNA hybridized exclusively with $tRNA^{Lys}$, $tRNA^{Val}$, and $tRNA^{Leu}$, and L-strand mtDNA hybridized only with $tRNA^{Pro}$. On which strands of the mtDNA duplex are the four genes located?

***13.15*** Clones of yeast mtDNA restriction fragments carrying the gene for cytochrome *b* were prepared for denaturation, and the melted H and L strands were separated by centrifugation. H strands and L strands from each of three different clones, each from a different yeast strain, were hybridized and heteroduplexes were photographed with the electron microscope, with the results shown at right.

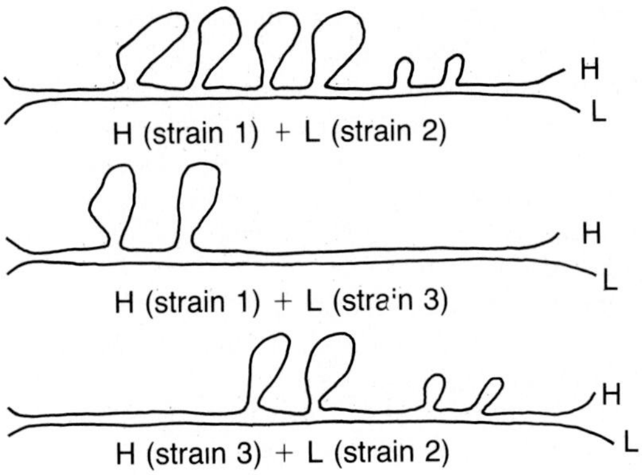

H
L
H (strain 1) + L (strain 2)
H
L
H (strain 1) + L (strain 3)
H
L
H (strain 3) + L (strain 2)

***a.*** If there is no difference in base sequences of the same introns and exons, and all the exon sequences of the cytochrome *b* gene are known to be present in the three strains, how many introns are present in the strain 1 gene?

***b.*** Which strain 1 introns are missing in the other two strains? (Label each detected intron of strain 1 with a letter of the alphabet, beginning with A at the left.)

***c.*** How many exon coding segments are present in this gene in strain 1?

***d.*** Would you predict that the polypeptide subunit of cytochrome *b* would have the same amino acid sequence in all three strains or not? Explain.

***13.16*** Aliquots of a preparation of mRNA strands 30 nucleotides long and with the base sequence

5′—CCAGAGUUU UACUGAAUCA UACUUAAAUA A—3′
(positions 1, 10, 20, 30 marked)

were incubated in protein-synthesizing systems extracted from four different sources. The amino acid sequences of polypeptides or peptides made in each system were:

| source of extract system | polypeptides and peptides synthesized *in vitro* |
|---|---|
| HeLa cytoplasm | $H_2N$-Pro-Glu-Phe-Tyr-COOH +$H_2N$-Ile-Ile-Leu-Lys-COOH |
| yeast mitochondria | $H_2N$-Pro-Glu-Phe-Tyr-Trp-Ile-Ile-Thr-Lys-COOH |
| HeLa mitochondria | $H_2N$-Pro-Glu-Phe-Tyr-Trp-Ile-Met-Leu-Lys-COOH |
| *Neurospora* mitochondria | $H_2N$-Pro-Glu-Phe-Tyr-Trp-Ile-Ile-Leu-Lys-COOH |

***a.*** What are the differences in codon usage in translation between cytoplasm and all the mitochondrial types?
***b.*** What differences exist in codon usage in translation among the three mitochondrial systems?
***13.17*** For each of the following six strains of paramecia, indicate whether the phenotype is killer or sensitive: (1) *KK,* kappa present; (2) *KK,* no kappa; (3) *Kk,* kappa present; (4) *Kk,* no kappa; (5) *kk,* kappa present; (6) *kk,* no kappa.
***13.18*** The following crosses were made between paramecia from the six strains shown in Question 13.17: 1 × 2, 1 × 6, 2 × 6.
***a.*** What would be the genotypes and phenotypes of the ex-conjugants if no cytoplasmic exchange took place during conjugation?
***b.*** What would be the genotypes and phenotypes of the ex-conjugants if prolonged cytoplasmic exchange took place during conjugation?
***c.*** What would be the genotypes and phenotypes of the progeny produced by fission from each ex-conjugant in crosses 1 × 2 and 1 × 6 from (a)?
***d.*** What would be the genotypes and phenotypes of the progeny produced by fission from each ex-conjugant in crosses 1 × 2 and 1 × 6 from (b)?
***13.19*** The snail *Limnea peregra* shows right-handed (dextral) or left-handed (sinistral) coiling of its shell; dextral is dominant over sinistral (*D, d* alleles). A snail produces only dextral progeny after self-fertilization. When these progeny snails undergo self-fertilization, however, they produce 25% sinistral and 75% dextral offspring. What is the genotype of the original snail?
***13.20*** Define the following terms: ***a.*** uniparental inheritance ***b.*** heterokaryon ***c.*** postmeiotic segregation of alleles ***d.*** D-loop synthesis of DNA. ***e.*** unassigned reading frame ***f.*** punctuation processing of RNA ***g.*** endosymbiont ***h.*** infectious inheritance ***i.*** maternal effect.

## References


Anderson, S., et al. 1981. Sequence and organization of the human mitochondrial genome. *Nature* **290**:457.

Attardi, G. 1981. Organization and expression of the mammalian mitochondrial genome: A lesson in economy. *Trends Biochem. Sci.* **6**:86, 100.

Bernardi, G. 1979. The petite mutation in yeast. *Trends Biochem. Sci.* **4**:197.

Bibb, M.J., et al. 1981. Sequence and gene organization of mouse mitochondrial DNA. *Cell* **26**:167.

Birky, C.W., Jr. 1978. Transmission genetics of mitochondria and chloroplasts. *Ann. Rev. Genet.* **12**:471.

Bonitz, S.G., et al. 1980. Codon recognition rules in yeast mitochondria. *Proc. Nat. Acad. Sci. U.S.* **77**:3167.

Borst, P., and L.A. Grivell. 1981. Small is beautiful—portrait of a mitochondrial genome. *Nature* **290**:443.

Dujardin, G., C. Jacq, and P.P. Slonimski. 1982. Single base substitution in an intron of oxidase gene compensates splicing defects of the cytochrome *b* gene. *Nature* **298**:628.

Farrelly, F., and R.A. Butow. 1983. Rearranged mitochondrial genes in the yeast nuclear genome. *Nature* **301**:296.

Gellissen, G., et al. 1983. Mitochondrial DNA sequences in the nuclear genome of a locust. *Nature* **301**:631.

Gillham, N.W. 1978. *Organelle Heredity*. New York: Raven.

Grivell, L.A. 1982. Restriction and genetic maps of yeast mitochondrial DNA. *Genetic Maps* **2**:221.

Grivell, L.A. Mar. 1983. Mitochondrial DNA. *Sci. Amer.* **248**:78.

Gruissem, W., et al. 1982. Transcription of *E. coli* and *Euglena* chloroplast tRNA gene clusters and processing of polycistronic transcripts in a HeLa cell-free system. *Cell* **30**:81.

Hoffmann, H.-P., and C.J. Avers. 1973. Mitochondrion of yeast: Ultrastructural evidence for one giant, branched organelle per cell. *Science* **181**:749.

Kolodner, R., and K.K. Tewari. 1975. Chloroplast DNA from higher plants replicates by both the Cairns and the rolling circle mechanism. *Nature* **256**:708.

Leaver, C.J., and M.W. Gray. 1982. Mitochondrial genome organization and expression in higher plants. *Ann. Rev. Plant Physiol.* **33**:373.

Margulis, L. 1981. *Symbiosis in Cell Evolution.* San Francisco: Freeman.

Morimoto, R., and M. Rabinowitz. 1979. Physical mapping of the yeast mitochondrial genome. *Mol. gen. Genet.* **170**:25.

Novick, R.P. Dec. 1980. Plasmids. *Sci. Amer.* **243**:102.

Ojala, D., J. Montoya, and G. Attardi. 1981. tRNA punctuation model of RNA processing in human mitochondria. *Nature* **290**:470.

Preer, L.B., and J.R. Preer, Jr. 1978. Inheritance of infectious elements. In *Cell Biology, A Comprehensive Treatise,* Vol. 1 (L. Goldstein and D.M. Prescott, eds.), p. 319. New York: Academic Press.

Rochaix, J.D. 1978. Restriction endonuclease map of the chloroplast DNA of *Chlamydomonas reinhardi. J. Mol. Biol.* **126**:597.

Sager, R., C. Grabowy, and H. Sano. 1981. The *mat-1* gene in *Chlamydomonas* regulates DNA methylation during gametogenesis. *Cell* **24**:41.

Sanger, F. 1981. Determination of nucleotide sequences in DNA. *Science* **214**:1205.

Slonimski, P.P., P. Borst, and G. Attardi. 1982. *Mitochondrial Genes.* Cold Spring Harbor, N.Y.: Cold Spring Harbor Laboratory.

Wurtz, E.A., J.E. Boynton, and N.W. Gillham. 1977. Perturbation of chloroplast DNA amounts and chloroplast gene transmission in *Chlamydomonas reinhardtii* by 5-fluorodeoxyuridine. *Proc. Nat. Acad. Sci. U.S.* **74**:4552.

# CHAPTER 14

# Developmental Genetics

Genetic and molecular methods have been applied successfully to the analysis of differentiation and development in viruses and cellular organisms. Although differential gene expression can account for cellular differentiation, the development of biological form and function are dependent on temporal ordering of events and on the spatial framework of developing systems. Differential gene expression takes place in different cells at different times during the orderly progression of a program of developmental events. Developmental genetics brings together information on genes and gene products with information on gene expression in tissues, organs, and organisms from the time of their inception until the time that form and function are completed.

## Temporal Ordering of Gene Action in Development

The development of form and functions in the organism, or **morphogenesis,** is one of the major areas of biological study. Form and function are the ultimate outcome of gene action, which leads to the production of proteins and nucleic acid products that contribute to biosynthesis and metabolism in general. Various external factors, such as ions, pH, metabolites, and other substances, contribute to orderly development of structures and activities unique to each organism. Developmental genetic studies have been conducted using a range of life forms, from the simplest viruses to the complex mammalian multicellular organism. In many cases information gained from studies of simpler life forms have provided models by which we can analyze and interpret more complex developmental systems. One of the more significant principles gained from developmental genetic studies of viruses has been the importance of timing of gene action to achieve an orderly progression of events that culminates in a fully developed individual.

### 14.1 Early Genes and Late Genes in Viral Morphogenesis

During phage morphogenesis in infected host cells, certain proteins encoded by viral genes are synthesized early in the infection cycle, and others are synthesized later in the cycle. These "early" and "late" proteins are products of "early" and "late" genes, that is, of genes transcribed early or late in the infection cycle.

During the 25-minute infection cycle initiated by phage T4 in *E. coli,* host metabolic machinery is subverted from making host molecules to making proteins and nucleic acids specified by viral genes. Viral enzymes appear within 1 minute after infection, and viral DNA is made in reactions catalyzed by these enzymes within the first 5 minutes after infection. The first structural proteins of the viral head and tail begin to appear about 8 minutes into the cycle, and new complete phages first begin to appear 4 to 5 minutes later. In the remaining half of the infection cycle, about 200 new phages accumulate within the host cell. The phage-specified enzyme *lysozyme,* made late in the cycle, attacks the host cell wall; free infectious phages are liberated as the host cell wall breaks down and the cell bursts (see Fig. 6.14).

These events comprise an orderly sequence of development since "early" proteins consist mainly of enzymes that catalyze the synthesis of "late" proteins, which are used in construction of new phage particles. Gene expression appears to be under positive transcriptional control since T4 does not make repressor proteins, which are part of negative controls (see Section 10.1). Experimental evidence indicates that positive control is exerted through interactions between protein-specificity factors encoded in host and viral genes and host RNA polymerase polypeptides. Different host and viral specificity factors may bind to core components of the host RNA polymerase, which is made up of five polypeptide chains (see Section 9.2). Once a specificity factor is bound to the core of the polymerase, it will recognize and bind to one or more of the various promoters in the viral genome. Different specificity factors recognize different promoters, thereby determining which genes will be transcribed in the viral genome by the same host polymerase transcribing enzyme. This selectivity leads to a *sequence of transcriptional readout of the phage genome* during the infection cycle (Fig. 14.1).

Early in infection, host-specified **sigma ($\sigma$) factor** binds to the host polymerase core and determines the specificity of transcription of the *early strand* of phage duplex DNA. Later in infection, $\sigma$ factor dissociates from the polymerase core, and other specificity factors subsequently bind to and dissociate from the host core enzyme. The activated polymerase thus recognizes and binds to different promoters in the phage genome. Transcription later in the

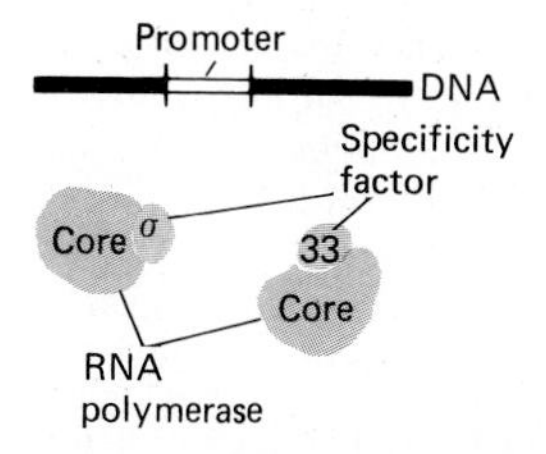


***Figure 14.1*** Various classes of promoters are recognized by different specificity factors (color) bound to the RNA polymerase core enzyme. For example, factor $\sigma$ recognizes early promoters whereas factor 33 recognizes late promoters, thus leading to orderly readout of the phage T4 genome.

cycle takes place largely from the partner *late strand* of duplex phage DNA, guided by particular viral-encoded specificity factors that were made early during infection. The order of transcriptional readout is thus specified by promoter-polymerase interactions that involve both strands of phage DNA at different times during infection (Fig. 14.2). A population of early proteins guides synthesis of a population of late proteins, leading to phage morphogenesis in accord with the temporal ordering of gene expression.

Genes in phage T7 are clustered according to their functions. The linear viral genome consists of three distinct regions: (1) genes that encode early functions which involve primarily regulation of transcription; (2) genes that specify T7 DNA replication enzymes; and (3) genes that encode structural proteins involved in T7 morphogenesis (Fig. 14.3). Unlike T4, in which different specificity factors associate with host RNA polymerase, T7 gene 1 codes for a T7 RNA polymerase. The switch in T7 from transcription of early genes to transcription of late genes is due to T7's own polymerase gene, which is located in the early region of the genome map. As infection proceeds, polymerase molecules accumulate until the T7 enzyme is the only one that acts late in the infection cycle. The exclusivity of the T7 polymerase is also due to turning off of the host RNA polymerase through proteins specified by early T7 genes. The T7 RNA polymerase recognizes and binds only to late promoters of the phage genome. The order of transcriptional readout is governed in T7 as in T4 by promoter-polymerase interactions, but T7 makes a new polymerase in addition to specificity factors that bind to the host polymerase and regulate its binding properties. Transcription is under positive control in both phages since transcription is turned on when an appropriate specificity factor binds to the host enzyme or when the viral enzyme is functioning in transcription.

Phage $\lambda$ is more complex than the T-phages since $\lambda$ is lysogenic. Phage $\lambda$ can initiate a virulent infection, or it can become integrated into

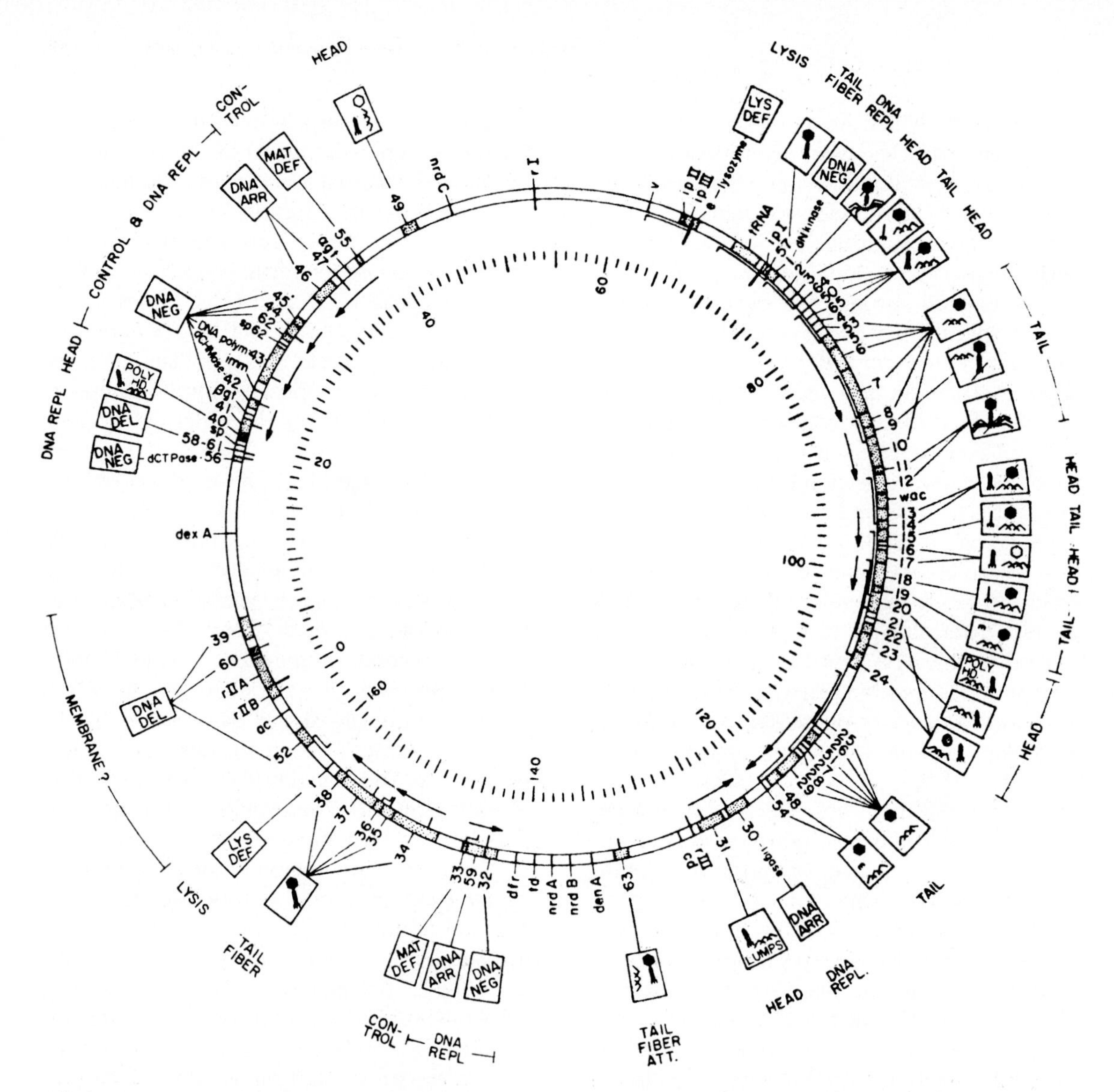


***Figure 14.2*** Map of the known genes of phage T4. The numbers on the inner circle "ticks" are map distances in recombination units. The gene symbols and numbers are just outside the outer circle, and many are known to direct functions that are indicated in the rectangles by drawings of phage parts or by abbreviations (NEG, negative; DEL, delayed; ARR, arrested; HD, head; LYS, lysis; MAT, maturation; DEF, defective). The arrows within the genetic map show the direction of transcription of the known genes in that segment of the map. (From W.B. Wood. 1974. In *Handbook of Genetics,* vol. 1, p. 327; R.C. King, ed. New York: Plenum.)

the bacterial genome in the noninfectious *prophage* state. Phage λ must be able to accomplish an orderly sequence of transcriptional readout in its infectious state, as well as limited transcriptional activity in its prophage state when new phage particles are not being made.

The major component that blocks transcription in the prophage state is the λ *repressor* encoded by phage gene $C_1$. Virtually no λ-specific mRNA is transcribed when λ repressor protein is present except gene $C_1$ transcripts for the repressor protein itself. The repressor binds

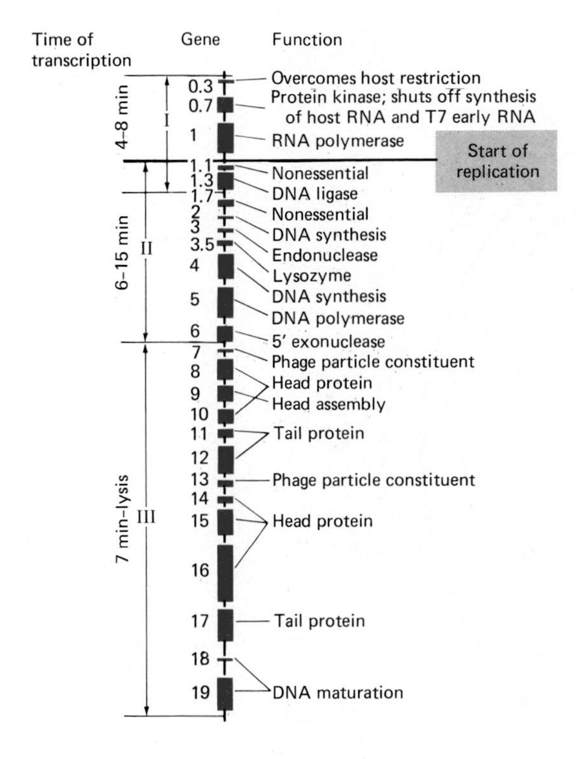


***Figure 14.3*** Genetic map of phage T7. The gene number is just left of each colored block, which indicates the size of the gene, and the gene function is on the right. The roman numerals I–III mark off the early gene region, DNA replication genes, and the late genes, respectively. (From *DNA Synthesis* by Arthur Kornberg. W.H. Freeman and Company. Copyright © 1974.)

specifically to operators of two of the early operons in the phage genome. Since these early genes are blocked, later gene functions are not expressed because the necessary early gene products are not present. If the repressor gene is inactivated by UV irradiation or some other inhibitor, the two early operons become accessible to transcription since repressor protein no longer blocks RNA polymerase from moving past the operator toward the structural genes.

One of these early operons contains genes that primarily encode proteins required for dissociation of the λ prophage DNA from the host chromosome. The other early operon includes genes needed for several early functions, including phage DNA replication. In addition to the negative control of the repressor-operator system of λ operons, positive controls must also be exerted to achieve transcription of all the early phage operons.

About half of the genes in the λ genome encode late proteins (Fig. 14.4). Synthesis of λ head and tail proteins is accomplished by translation of mRNA of a single operon, which includes about 20 genes for λ structural proteins. Transcription of this late operon is turned on by the protein product of an early gene, *Q*, whose precise nature is uncertain. The coordinated syntheses of all the λ structural proteins is ensured by the simple device of having all these late genes in a single operon, governed by a single promoter. After RNA polymerase binds to this promoter, all the gene sequences in the operon are transcribed into a single polygenic mRNA transcript, which is translated into a variety of proteins that are needed to build the viral particles. New phages are built from DNA and structural proteins, and these infective particles are released after phage lysozyme weakens the host cell wall and lysis takes place. The lytic cycle activities are all turned off, however, as long as λ repressor is being made.

## 14.2 Building a Virus

Viral morphogenesis takes place when appropriate gene products become available in the host cell. Although functionally different viral proteins are made in temporal order during infection, we have not evaluated whether *assembly* of the viral particle is itself temporally ordered. In order to analyze morphogenesis itself, it is necessary to determine the timing and nature of the processes leading to development of structures and functions.

Relatively simple viruses and cellular structures apparently become organized through **self-assembly** of component molecules. In self-assembly systems, the information required to build the structure is part of the component molecules themselves, and no additional information is needed to specify size, shape, and

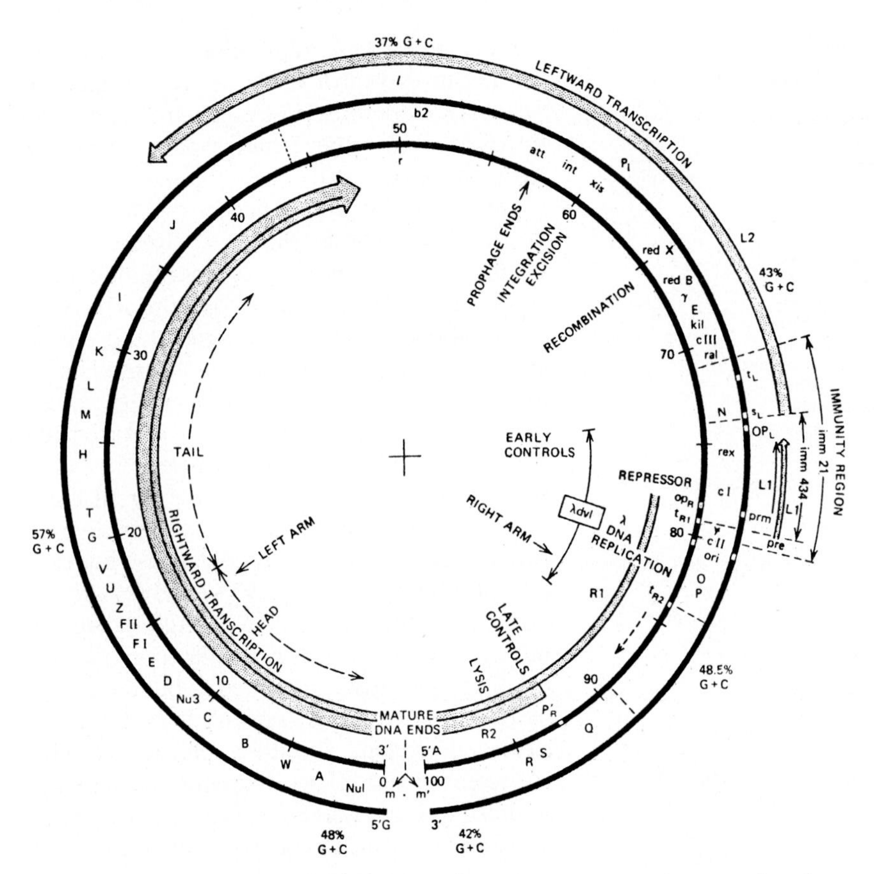


***Figure 14.4*** Genetic map of phage λ, showing the two complementary DNA strands (black) and the direction of transcription of regions along each strand (gray arrows). Functions of certain regions are shown within the circles. (From W. Szybalski. 1974. In *Handbook of Genetics,* vol. 1, p. 309; R.C. King, ed. New York: Plenum.)

organization. This fact has been demonstrated *in vitro* for certain simple viruses and for particular cellular structures. In eukaryotic cells, microfilaments assemble from actin monomers, microtubules assemble by orderly aggregations of tubulin dimers, and ribosomes assemble from rRNAs and ribosomal proteins. Assembly of these same subunits to produce functional structures can be achieved *in vitro* as well as in the cells of the organism.

In the case of tobacco mosaic virus (TMV), the constituent RNA and protein molecules can be dissociated from active viruses and then recovered for *in vitro* tests. When the separated molecules are incubated in appropriate conditions, the molecules reassemble into infective TMV rods with the same size, shape, and properties as viruses made in the living host cell. A single-stranded RNA molecule binds together with thousands of molecules of a single kind of coat protein to produce typical TMV rods (Fig. 14.5). In this self-assembly system as in the ones mentioned above, the shapes of the individual molecules guide specific interactions that lead to assembly of the final structure during morphogenesis. These molecular associations develop by formation or relatively weak chemical bonds, and not by covalent linkages. Systems capable of self-assembly therefore need no additional genetic guidance for morphogenesis,

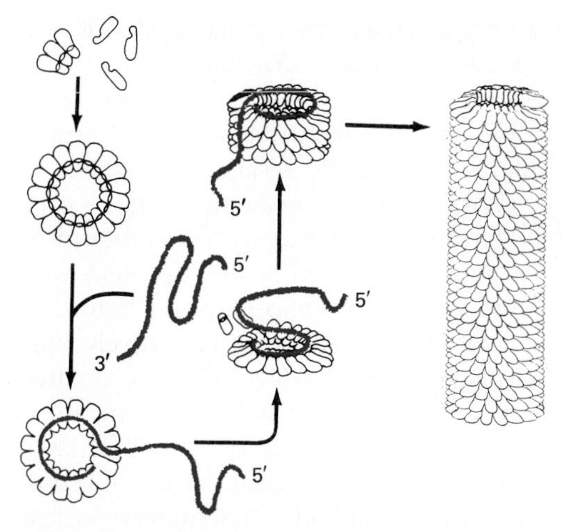


***Figure 14.5*** Reconstitution of tobacco mosaic virus (TMV) particle from dissociated coat protein molecules and single-stranded RNA (color). The order of assembly and the orientation of the RNA strand are shown. The final assembled particle consists of 130 turns of the helically ordered protein subunits, with 16⅓ subunits per turn, completely enclosing the RNA strand. (From *Molecular Genetics: An Introductory Narrative,* Second Edition, by Gunther S. Stent and Richard Calendar. W.H. Freeman and Company. Copyright © 1978.)

once the structural molecules are available. Self-assembly is essentially a spontaneous process.

When component molecules of T4 viruses are placed together *in vitro,* virus particles do not assemble spontaneously. Phage T4 is a more complex virus than TMV; of approximately 100 T4 genes, about 50 encode components involved in morphogenesis. TMV has only five or six genes in its genome, and its construction involves little more than binding of the coat protein subunits to RNA and to each other in a particular geometry. Phage T4, on the other hand, has several distinctive structural regions that make up the mature virus particle. T4 has a polyhedrally shaped *head* consisting of a protein coat wrapped around a linear duplex DN molecule. A short *neck* connects the head to a springlike *tail* consisting of a contractile sheath surrounding a central core and attached to a *base plate,* from which there protrude six short *spikes* and six long, slender *tail fibers* (Fig. 14.6). When T4 attaches by its spikes and tail fibers to the bacterial cell wall, the sheath contracts and drives the tubular core of the tail through the cell wall. This action provides a passageway for DNA to pass from the phage head into the bacterial cytoplasm.

During the 1960s, Robert Edgar, William Wood, and others provided unambiguous evidence showing that T4 morphogenesis is the outcome of *interactions between gene products* and not of a sequence of ordered transcriptional events. They isolated a large number of developmental mutants that made altered and often nonfunctional gene products. Development stopped prematurely in these mutants at the point at which the altered protein was needed in phage assembly; these developmental mutants induced abortive infections. By complementation tests between different mutant strains and by other methods, investigators identified about 40 genes that contribute to phage T4 morphogenesis. These tests distinguished different genes, but they did not provide information on the nature of phage assembly. In other words, the tests did not detemine whether phage morphogenesis was temporally ordered or whether assembly proceeded in some random fashion once the necessary gene products were pro-

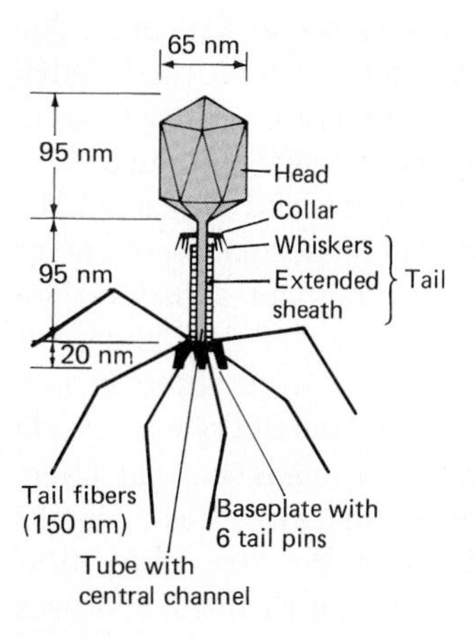


***Figure 14.6*** Structure of T4 and the other T-even phages.

vided by complementing mutants in a mixed infection.

To determine whether an infection was aborted because of a particular block in development at some particular time in an ordered morphogenetic sequence, Edgar and Wood studied T4 mutants under *permissive* and *nonpermissive* (*restrictive*) conditions. When a mutant infected permissive *E. coli* cells, the phage developed to the infective particle stage. In nonpermissive *E. coli* cells, development of the mutant ceased at the time that the altered mutant gene product was needed. The particular morphogenetic functions of the genes involved in phage morphogenesis were determined by electron microscopy of mutant extracts that had been isolated from nonpermissive host cells, in which development was blocked at a certain stage that was characteristic for each mutant strain. For example, T4 phages that carried mutant gene 23 produced tails and tail fibers but no heads. This fact indicated that gene 23 is involved in head assembly. Gene-34 mutants produced heads and tails but no tail fibers during abortive infections in nonpermissive hosts, thereby indicating that gene 34 is involved with tail fiber assembly. Similarly, gene 27 must control a step in tail formation since mutant extracts contained heads and tail fibers but no tails. The gene map for T4 reflects the morphogenetic functions of various genes, as derived from analyses using these *conditionally lethal* mutants (see Fig. 14.2). The development of such mutants is blocked (lethal to the phage) in nonpermissive cells but not in permissive cells (virulent phages produced).

Edgar and Woods' results and similar studies indicated that a block in the formation of one of the phage structural components had no influence over the formation of the other two structures. Heads and tails were made in tail-fiberless mutants, tails and tail fibers were made in headless mutants, and heads and tail fibers were made in tailless mutants. These results were interpreted to mean that the morphogenetic pathway leading to mature viruses had three separate and distinct branches. Each of these branches led to the formation of a different one of the three major structures.

Having identified the branching morphogenetic pathway, the next step was to determine whether one assembly sequence involved the three major structures or whether all three structures aggregated all at once to produce the virus. The experiments aimed at this problem were designed for *in vitro* analysis using mixtures of extracts originally isolated from abortively infected nonpermissive cells (Fig. 14.7). These extract mixtures were incubated and later examined by electron microscopy for aggregations of separate structures. The extract mixtures were also used to infect *E. coli* in an independent *in vivo* test of phage function. If virulent phages had been formed in the mixtures, infections could be produced and phage lysates would be observed.

From such *in vitro* and *in vivo* tests of morphogenesis, it was found that heads and tails assemble even in the absence of tail fibers, but that tail fibers only bind to head-tail aggregates and not to tails alone. The virus is infective only when all three structures are combined into a phage particle, as seen by comparisons of *in vitro* and *in vivo* results. These observations indicate that viral assembly apparently takes place in a two-step sequence; heads and tails assemble first, and tail fibers are added afterward in a separate step.

With this information, Wood, Edgar, and others next looked for information on the nature of the assembly of the individual structures. Essentially, they wanted to know how many steps were involved in generating the head or tail or tail-fiber structures themselves. To determine this, they conducted two-factor **extract complementation tests.** These tests are *in vitro* procedures that depend on *interactions between gene products* rather than on direct gene action, as with *in vivo* complementations.

Using paired mixtures of extracts from abortively infected cells, they found that many of the tests gave strongly positive or strongly negative results. Infective phages in a mixture either increased at least tenfold or showed no detectable

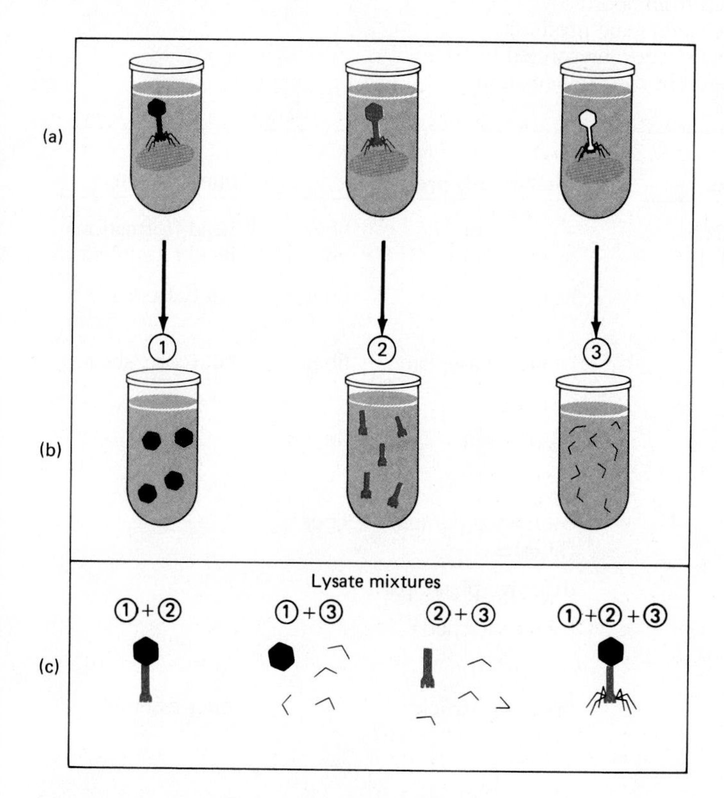


***Figure 14.7*** Results of mixing extracts of T4 mutant lysates from abortively infected *E. coli* cells. (a) Different T4 mutants infect *E. coli* and (b) lysates of the abortively infected cells are collected for (c) mixing of these lysates in various combinations. These experiments showed a sequence of morphogenetic assembly in phage T4, with tails assembling onto heads and then tail fibers assembling onto the head-tail particle.

increase when compared with either extract alone. On the basis of these tests, 14 different gene complementation groups were found to accommodate all of the paired results. Extracts from different groups do complement each other, but extracts of the same group do not complement each other *in vitro* (Table 14.1). Each complementation group defines a functional component for assembly, and all 14 of these components or reactions are needed to produce functional viruses.

With the tests already mentioned and additional studies of T4 morphogenesis, it was possible to construct the steps involved in each of the three branches of the morphogenetic pathway leading to mature viruses (Fig. 14.8). Some of the details remain to be clarified, but the overall picture of morphogenesis in T4 is clear. A number of significant features emerged from these studies:

***1.*** The morphogenetic process has a stringent sequential order. If a step in one pathway is blocked, characteristic structural intermediates accumulate. The block cannot be bypassed. Such a block in one pathway, however, does not interfere with the morphogenetic sequence leading to the other two structures.

***2.*** The sequence of morphogenetic events depends on interactions between gene products and not on sequential transcription or induction of structural gene action. All of the late proteins appear in cells at about the same time, and not in

**Table 14.1** Complementation groups determined from results of paired mixtures of extracts from lysates taken from abortively infected *E. coli* cells. Positive interactions between gene products, seen by tenfold increase in infective T4 phages, were interpreted to mean that the responsible genes were in different complementation groups.

| extract complementation group | mutant genes | components present* | | | inferred defect |
|---|---|---|---|---|---|
| Ia | 20,21,22,23,24,31 | — | tail | fiber | head (formation) |
| Ib | 49,2,64,50,65,4,16,17 | head | tail | fiber | head (completion) |
| II | 53,5,6,7,8,10,25,26, 51,27,28,29 | head | — | fiber | tail (baseplate) |
| III | 48,54 | head | baseplate | fiber | tail (core, sheath) |
| IV | 13,14 | head | tail | fiber | ? |
| V | 15 | | | | |
| VI | 18 | | | | |
| VII | 9 | contracted particle[†] (fiberless) | | fiber | ? |
| VIII | 11 | defective phage particle (fibers attached) | | | ? |
| IX | 12 | | | | |
| X | 37,38 | fiberless particle | | — | fiber assembly |
| XI | 36 | | | | |
| XII | 35 | | | | |
| XIII | 34 | | | | |
| XIV | 63 | fiberless particle | | fiber | fiber attachment |

(Groups IV–VI, VIII–IX, and X–XIII are each bracketed together and share one entry.)

*All structural components listed are unattached to each other unless otherwise indicated. Description of these as heads, tails, and so on implies only that they are identifiable in electron micrographs, not that they are complete structures.

†Heads with attached tails are designated particles.

From W.B. Wood et al. 1968. Bacteriophage assembly. *Fed. Proc.* **27**(5):1160–1166.

sequential order as would be expected for sequential transcription of structural genes. This inference is supported by the observation that several extract complementation groups correspond to single-gene products (see Table 14.1). These gene products interact *in vitro,* and presumably also interact *in vivo* in normal cells and in permissive cells infected with mixtures of mutants from different gene complementation groups, as shown by the production of infective phages in all of these systems.

Similar interactions between gene products that are synthesized coordinately or independently in various phages, but appear at about the same time during infection, probably underlie morphogenesis in all or most viruses. Similar processes very likely are responsible for orderly morphogenesis in cellular organisms, that is, development is not simply a matter of turning genes on and off through transcriptional controls. Morphogenesis also requires interactions between protein products of differential gene action, after transcripts have been translated. These interactions between gene products are also under genetic control since mutations can block particular steps in developmental sequences.

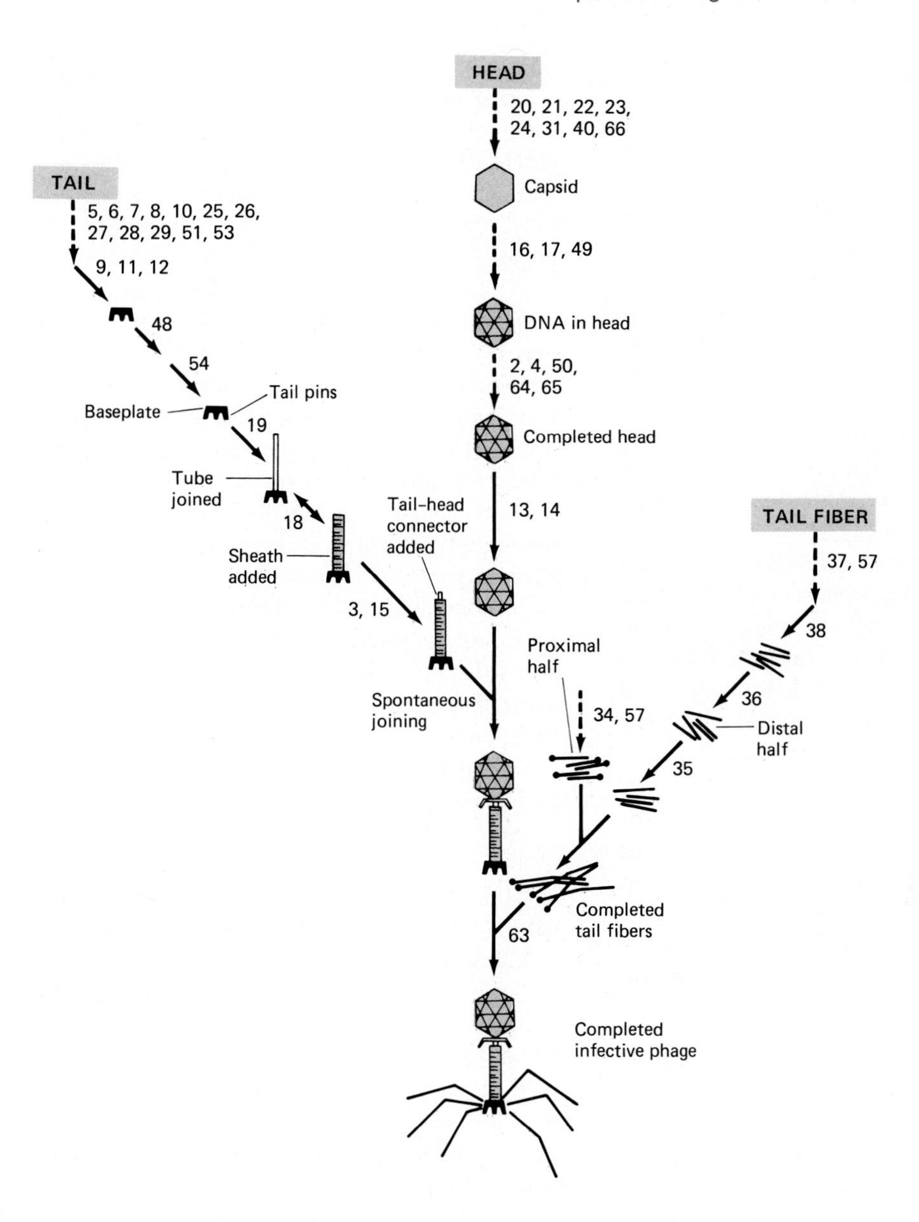


***Figure 14.8*** There are three main branches in the morphogenetic pathway of phage maturation, such that heads, tails, and tail fibers are formed independently and afterward combine to constitute the infective phage particle. The numbers refer to the genes whose products are involved at each morphogenetic step (see Fig. 14.2 for T4 map). (From W.B. Wood. 1973. In *Genetic Mechanisms of Development,* p. 29; F.H. Ruddle, ed. New York: Academic Press.)

## 14.3 Sequential Gene Action in Yeast

A useful genetic approach to the analysis of morphogenesis involves reciprocal shifts of conditionally lethal mutants between permissive and nonpermissive conditions. By this method, investigators can analyze morphogenetic events in the *same cell* under different, controlled conditions and can thus determine the time of action of gene products in development. The genetic method was first described by Jonathan Jarvik and David Botstein in 1973 in their studies using temperature-sensitive mutants of phage P22. The same approach has proven useful in the analysis of other systems, including eukaryotic organisms such as yeast.

Yeast *(Saccharomyces cerevisiae)* can exist indefinitely in either the haploid or diploid state, producing new bud cells from uninucleate mother cells by conventional mitosis. Mitosis is followed by cell division, which separates bud and mother cells by a new cell wall. The bud and mother cells may remain associated for a time after cell division and may even give rise to new bud cell generations. During each **cell division cycle,** which produces the next generation of cells, specific events lead through the succession of cell-cycle stages referred to as $G_1$, S, $G_2$ and mitosis (Fig. 14.9). In addition to molecular events, which characterize DNA replication and protein synthesis of these cell-cycle stages, a number of morphogenetic events take place. In each cell cycle a new bud cell is produced, the newly divided nucleus migrates into the bud, and a newly made cell wall closes off the bud from the mother cell before separation can occur.

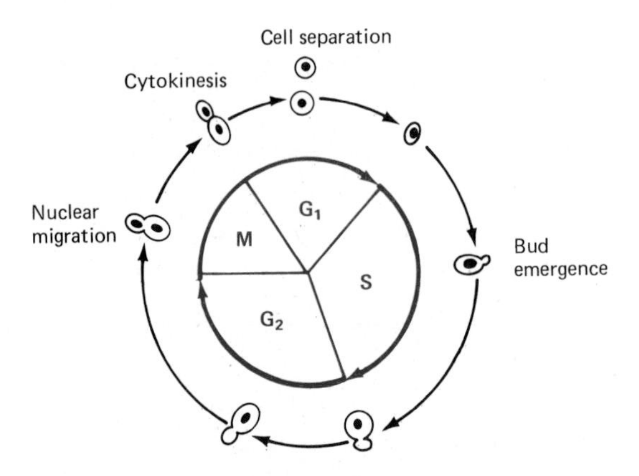


***Figure 14.9*** The cell cycle in yeast includes the usual four stages of $G_1$, S, $G_2$, and M. Reproductive events are shown in the outer circle, in relation to their timing in the cell cycle.

The $G_1$ phase of the cell cycle is a significant interval in yeast development; at this time a cell may be directed into any one of three different developmental programs. During $G_1$ the cell may initiate DNA synthesis and thus enter the S phase of the cell cycle; the cell, if haploid, may fuse with another cell of the opposite mating type in sexual reproduction; or the cell may become noncycling and enter the quiescent stationary phase of the population growth cycle. Upon initiating DNA synthesis, a cell apparently becomes committed to the cell division cycle and will not undertake either of the other two alternative pathways. The initiation of DNA synthesis may therefore be viewed as an act of determination, or commitment, and the subsequent steps of the cell division cycle can be viewed as elements of a developmental program. Since these are basic features of eukaryotic growth and development, information gained from studies of yeast can also be applied to other systems.

Leland Hartwell and co-workers used genetic methods to conduct an extensive series of studies of the yeast cell division cycle. They were able to identify about 35 different temperature-sensitive cell-division-cycle *(cdc)* mutants. Each of these mutants could be associated with some event during a cycle, on the basis of an observed block in an expected morphological event (Fig. 14.10). For example, cells carrying the *cdc* 4 mutation cease activity in $G_1$ and do not initiate DNA synthesis. In addition to assays which show that DNA synthesis fails to occur, there is a morphological *landmark* that characterizes the *cdc* 4 mutation; the spindle pole body (SPB) duplicates but fails to separate as it would normally prior to mitosis. Cells carrying the *cdc* 7 mutation are also arrested in $G_1$, but the morphological landmark, or terminal phenotype expressed by the mutant, is a later event; the

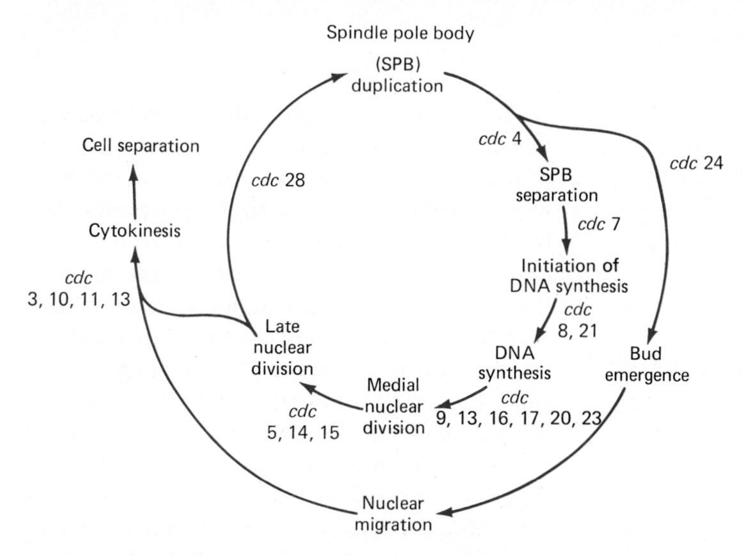


***Figure 14.10*** Dependent pathway of landmarks in the cell cycle of the yeast *Saccharomyces cerevisiae,* derived from phenotypes of cell-division-cycle (*cdc*) mutants. The designations for *cdc* genes are immediately preceding their diagnostic landmark. The diagram relates to mutant phenotypes as follows: Upon a shift to the restrictive temperature, mutant cells arrest synchronously at the position designated by the *cdc* number; all events that flow from this point do not occur while all other cell cycle events do occur. (From L.H. Hartwell. 1978. *J. Cell Biol.* **77**:627.)

spindle pole body not only duplicates but also separates. The *cdc* 28 mutation precedes both of these steps since arrested $G_1$ cells do not even duplicate the spindle pole body. Since all three of these *cdc* mutations block DNA synthesis, all three mutant types appear as unbudded cells when incubated at the nonpermissive temperature of 36° or 38°C. When these temperature-sensitive *cdc* mutants are incubated at the permissive temperature of 23°, successive cell cycles proceed normally.

Hartwell and co-workers conducted experiments to determine the *execution point* of the mutation, that is, the time at which the mutation acts to arrest the cell at a morphogenetic landmark. Temperature-sensitive *cdc* mutants were shifted from the permissive temperature, at which they were growing, to incubation at the nonpermissive temperature. The mutants were then photographed through an ordinary light microscope and were photographed again about 6 hours after the temperature shift. The size of the bud on the mother cell serves as a guide to the time or interval of the cell cycle in progress and thus indicates the execution point of the mutant. Mutant cells that were affected in an earlier part of the cycle (before the execution point) were arrested in the first cycle. Mutant cells that were affected in a later part of the cycle (past the execution point in the first cycle) were arrested during the second cycle. Since mother and bud cells remain associated during the entire six-hour interval, first-cycle arrests and second-cycle arrests can be distinguished (Fig. 14.11). Observation of mother-bud cell aggre-

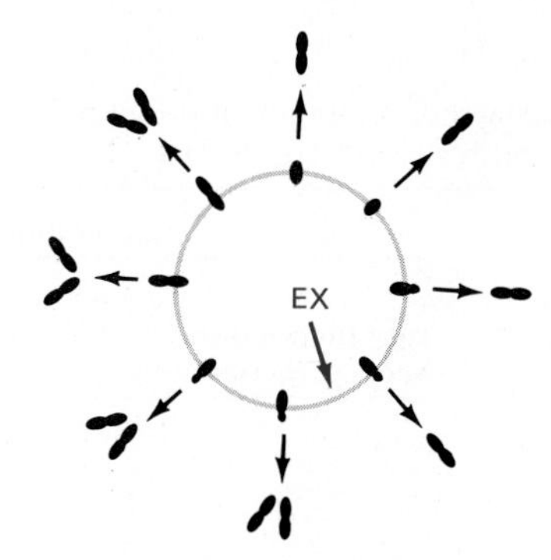


***Figure 14.11*** Determination of the execution point (EX) in temperature-sensitive mutants. First- and second-cycle arrests can be distinguished by one versus two mother-and-bud cells observed at the time of the temperature shift and 6 hours later. (From L.H. Hartwell. 1978. *J. Cell Biol.* **77**:627.)

gates can be used to analyze all the cells in the culture, even though they are not undergoing synchronous growth. Each cell will be arrested, either in the first or the second cycle, after being shifted from permissive to nonpermissive temperatures.

The visible effect of the mutation may be evident at or before the time that the execution point is reached. For example, *cdc* 8 and *cdc* 21 mutants stop synthesizing DNA immediately when shifted to the nonpermissive temperature. Both mutants have execution points near the end of the phase of DNA synthesis, and both gene products are probably involved in this stage-specific morphogenetic event. A search was undertaken to determine the biochemical nature of the gene products, which were strongly suspected of being enzymes involved in DNA replication. In the case of *cdc* 21, the mutant was found to be defective in the enzyme *thymidylate synthetase,* which catalyzes reactions in which the DNA precursor thymidylate is synthesized. Similarly, *cdc* 9 mutants were found to be defective in the ligating enzyme *DNA ligase.* After the execution point of the mutation is found, investigators can design more specific tests to seek the biochemical or molecular nature of the gene product involved in the developmental process or event.

In addition to observing developmental landmarks and determining the execution points of mutants, the sequential ordering of developmental events can be determined by performing *reciprocal shift experiments using doubly mutant strains* that are sensitive to different conditionally lethal growth conditions. One mutant may be sensitive to high temperatures and the other to low temperatures, or they may be sensitive to different inhibitors. In each case, the second mutant is not conditionally lethal for the condition to which the first mutant is sensitive. Suppose we wish to know whether mutations governing events A and B are *dependent* on one event (A or B) occurring before the other (B or A). If doubly mutant A/B is first incubated under conditions permissive for A and restrictive for B and then shifted to conditions restrictive for A and permissive for B, development will be completed for both events only if event B is dependent on event A occurring first (Table 14.2). This sequence can be verified by reversing the reciprocal shifts such that the first incubation is permissive for B and restrictive for A and the second incubation is restrictive for B and permissive for A. The developmental program would not be completed in the reversed reciprocal shifts since event A was restricted and event B could not take place without A having

***Table 14.2*** Sequencing a developmental program by reciprocal shifts to permissive and restrictive conditions.

| | | completion of developmental program | |
|---|---|---|---|
| **relationship between the two events in a pathway** | **first incubation:**<br>**second incubation:** | **restrict B, permit A**<br>**permit B, restrict A** | **restrict A, permit B**<br>**permit A, restrict B** |
| dependent $\underrightarrow{A}\ \underrightarrow{B}$ | | + | − |
| dependent $\underrightarrow{B}\ \underrightarrow{A}$ | | − | + |
| independent $\genfrac{}{}{0pt}{}{\underrightarrow{A}}{\underrightarrow{B}}$ | | + | + |
| interdependent $\underrightarrow{A,B}$ | | − | − |

With permission from L.M. Hereford and L.H. Hartwell. 1974. *J. Mol. Biol.* **84**:445-461, Table 1. Copyright: Academic Press Inc. (London) Ltd.

occurred. Whether the dependent sequence is A,B or B,A can be determined by this experimental test. Similarly, the results of these reciprocal shift experiments can reveal whether events A and B are *independent* of each other since development would be completed in both sets of conditions. If events A and B were *interdependent,* that is, if both events had to take place together, development would not be completed under either set of incubation conditions.

By comparing the results in reciprocal shift experiments, investigators could determine whether pairs of mutations governed dependent, independent, or interdependent developmental sequences since each of these situations leads to different experimental results that conform to one of the model relationship schemes. When all the data were assembled, it was apparent that developmental events occurring during the yeast cell division cycle were temporally ordered, gene-controlled steps (Fig. 14.12). These same observations have been made in studies using various organisms, amply confirming the basic principles of temporally ordered genetic control over morphogenesis.

A great deal remains to be done, however, since very few gene products have been identified for these sequential morphogenetic events. Once events have been defined and arranged into the orderly progression in which they occur, it becomes possible to search for the responsible genes and gene products using molecular and biochemical methods. Once these data have been collected, it will be essential to determine how these events and pathways are regulated in development. Clearly, remarkable progress has been made, and much remains to be done in this demanding area of biological research. Basic experimental designs are available, however, and information gained from model systems has been and can be applied to analysis of more complex developmental programs.

## Genetic Aspects of Development

Eukaryotes, and multicellular organisms in particular, are genetically and structurally complex. Eukaryotic genes are not clustered into early and late functions and, in fact, different genes on different chromosomes may specify polypeptides that assemble into a single functional protein molecule. We have described examples of this type in the cases of immunoglobulins (see Section 10.8) and hemoglobins (see Section 12.12). Furthermore, in contrast with viruses, bacteria, and unicellular eukaryotes such as yeast, many different cell types of varied

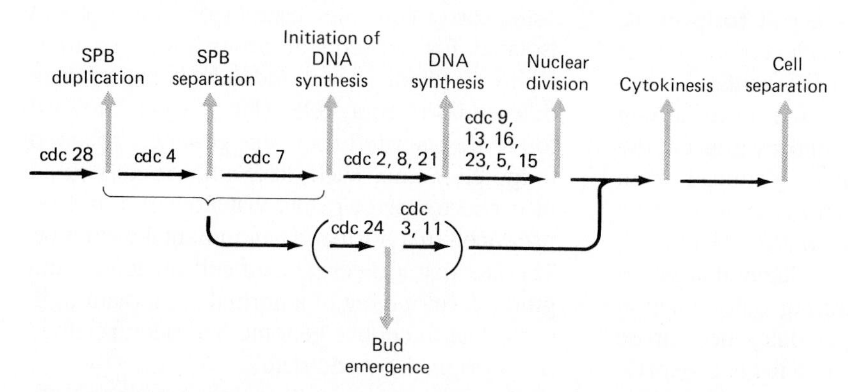


***Figure 14.12*** Summary of dependent relationships between gene-controlled steps in yeast cell cycle, according to data from reciprocal temperature-shift experiments. Compare with Fig. 14.10. (From L.H. Hartwell. 1978. *J. Cell Biol.* **77**:627.)

structure and function arise during development in multicellular organisms.

In addition to temporal order of gene-controlled events, we expect morphogenesis in multicellular organisms to be guided by the location in which different events take place. Spatial, or positional, ordering of development, therefore, is another dimension of morphogenetic events that must be studied.

## 14.4 Totipotency of Differentiated Cell Nuclei

Development of a limb or a brain differs in magnitude from building a virus or a new yeast cell. Studies of specialized cells, such as reticulocytes or oviduct cells as well as studies of viruses and simple organisms, indicate that differential gene expression underwrites morphogenetic changes. But, can we be sure that turning genes off and on also accounts for complex morphogenetic events in multicellular organisms? Perhaps other processes account for developmental programs in complex systems; for example, genes may be lost or may undergo mutation in different cells at different times during development. These possibilities had to be explored.

In the 1950s Robert Briggs and Thomas King reported on one approach taken to analyze the genetic capacity of differentiated cells. They designed experiments to see whether or not nuclei from differentiated cells were still **totipotent,** that is, whether or not such nuclei could support the development of an entire organism in the same way that the nucleus of the fertilized egg can provide all the genetic information for development of the individual. If nuclei from differentiated cells could provide genetic information for *every cell type* in the developing organism, then it would be unlikely that genes had been lost or altered during differentiation and more likely that cellular differences arose through processes of differential gene expression.

Briggs and King devised methods for **nuclear transplantation** of somatic cell nuclei into enucleated or irradiated egg cells from frogs (Fig. 14.13). They found that nuclei that were removed from somatic embryonic cells in the preblastula stage of development and transplanted into egg cells could often guide normal frog development. Nuclei from postblastula embryonic cells could not sponsor an entire developmental program leading to adult frogs, that is, postblastula nuclei were no longer totipotent.

Their detailed and extensive studies with frogs showed that some irreversible change had occurred during embryonic development and that somatic cell nuclei remained totipotent only up to the gastrula stage. John Gurdon and others showed that in toads, on the other hand, nuclei from embryonic intestinal cells remained totipotent long after gastrulation. Normal adult toads developed in many instances, and these animals were fully fertile and capable of producing normal progeny. The differences between these amphibian species is not well understood. In either case, however, a complete set of functional genes must have been retained for a considerable time during differentiation since a differentiated cell nucleus could guide development of an animal once the nucleus was placed into the egg cell environment. Clearly, interactions between the transplanted nucleus and the egg cytoplasm are a necessary feature of development. Individual somatic cells can be grown in culture, but they do not give rise to whole organisms.

Frederick Steward conducted similar studies using carrot cells dissociated from the root and isolated for subsequent observations of their ability to produce new individuals from single differentiated root cells (Fig. 14.14). Steward found that isolated root cells generally retained totipotency since such differentiated cells could give rise to carrot plants that flowered and reproduced in a normal developmental sequence. The fact that a differentiated cell nucleus could guide development of a normal adult plant indicated that the entire genome had been retained in its original genetic state.

Even more compelling data have been obtained to show that phenotypically different cells in the organism carry identical or nearly identical sets of genes. All the bands of polytene

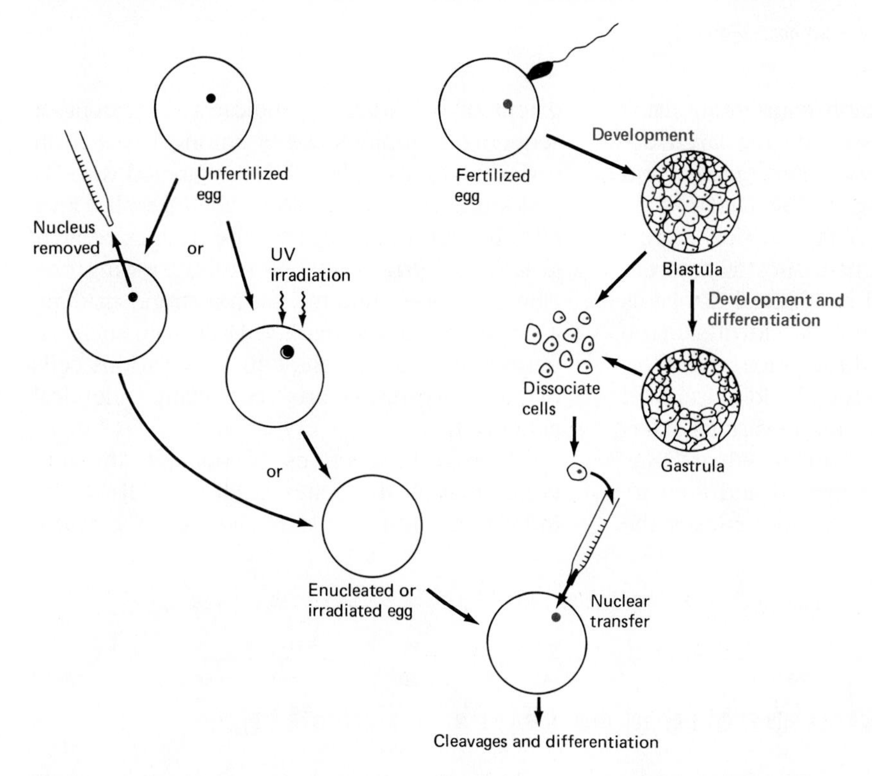


***Figure 14.13*** Nuclear transplantation experiments have shown that nuclei taken from differentiated cells and placed into enucleated or irradiated eggs can support normal amphibian development. Such evidence indicates that development and differentiation are not the consequences of gene losses or gene mutations in cells of the multicellular organism.

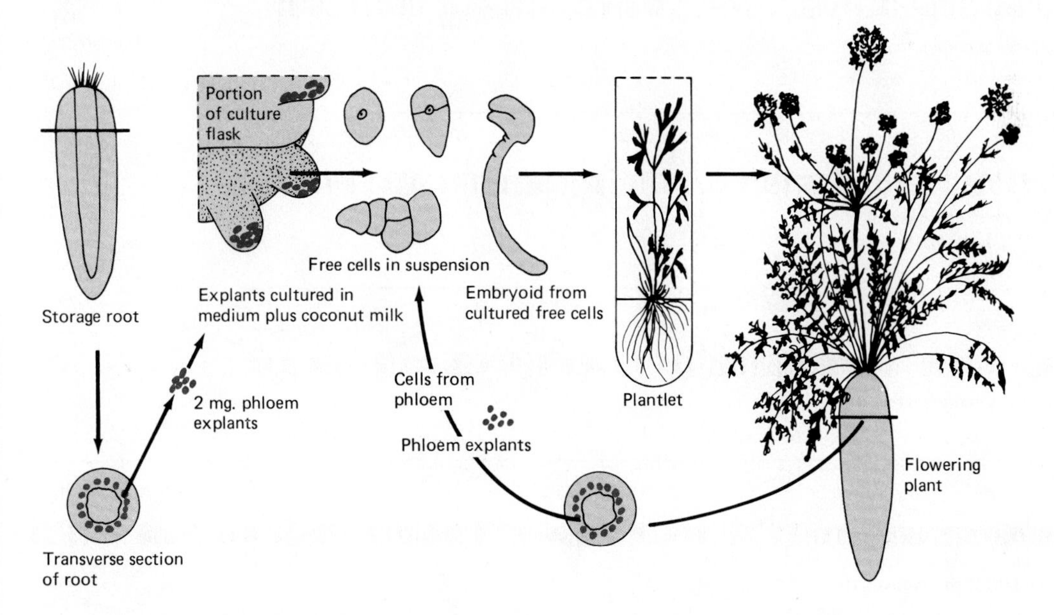


***Figure 14.14*** Cells taken from carrot root and grown in culture can give rise to normal flowering carrot plants, thus indicating that the cells are totipotent and genetically complete. (From F.C. Steward, "Growth and Development of Cultured Plant Cells," *Science* **143**:20–27, 3 January 1964. Copyright © 1964 by the American Association for the Advancement of Science.)

chromosomes are remarkably constant in different somatic cells and tissues during larval development in *Chironomus, Drosophila,* and other dipteran insects (Fig. 14.15). Different sequences of puff development and puff regression take place in different tissues at different times, and each temporal and spatial display is specific and predictable in each individual (see Fig. 10.12). Molecular hybridizations have repeatedly shown that essentially identical and complete genomes are present in various differentiated tissues of the individual. When DNAs from different tissues are melted and then allowed to reanneal to form hybrid duplexes, the extent of hybridization indicates the extent of sequence complementarity among DNAs from various sources. If liver DNA is labeled with $^{3}H$ and kidney or brain DNA is labeled with $^{32}P$ or $^{14}C$, hybrid duplexes can be recognized by ***double labeling,*** with one strand carrying one isotopic marker and the partner strand carrying a different isotopic marker. Data from such experiments routinely show that the various cells of an organism are essentially identical genetically.

There are exceptions, of course, to the generality that all the genes in all the cells of the individual continue to exist and retain their orig-

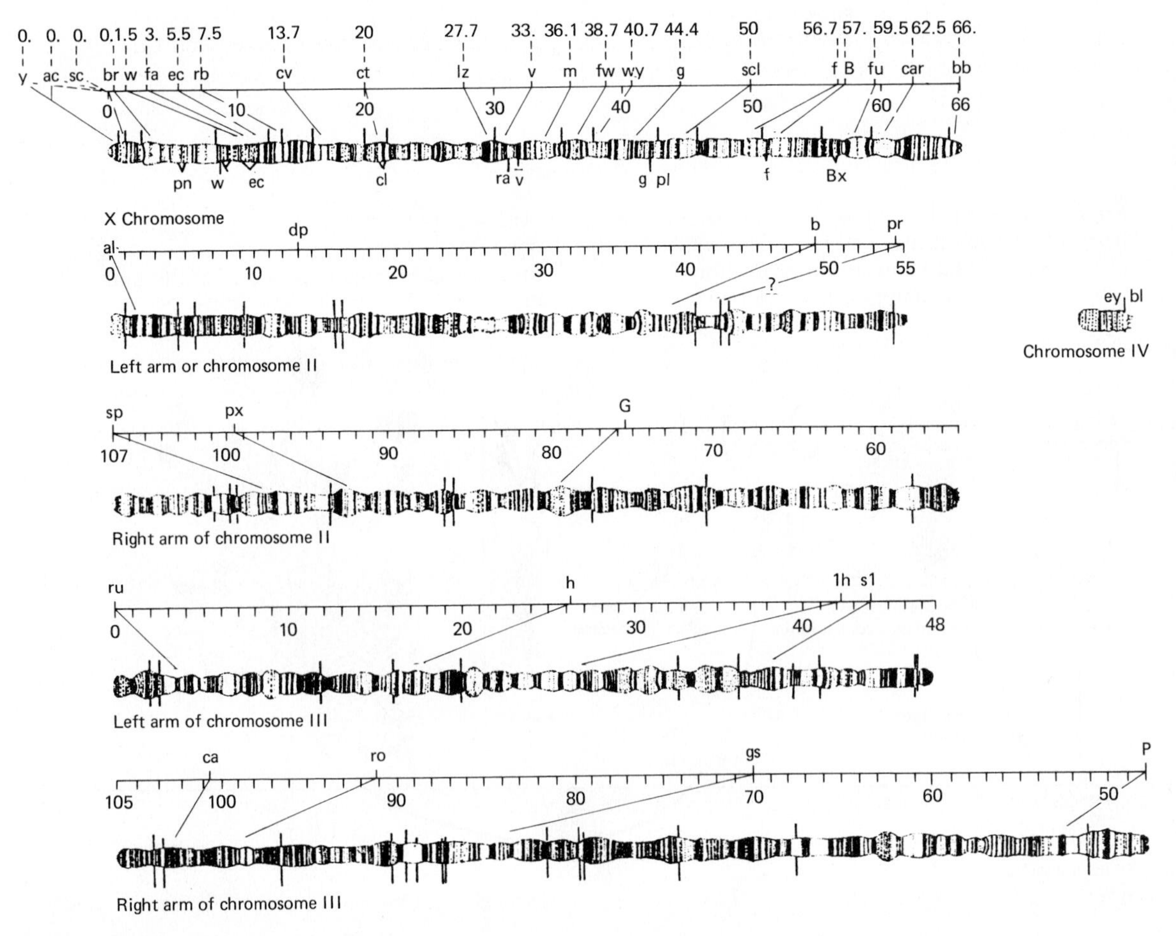


***Figure 14.15*** Salivary gland chromosome maps of *Drosophila melanogaster,* and locations of certain genes according to loci on the genetic map and bands on the cytological map. (From T.S. Painter. 1934. *J. Hered.* **25**:465.)

inal allelic forms. For example, an increasing body of evidence indicates that some immunoglobulin genes undergo mutations in somatic cells, giving rise to new kinds of immunoglobulin molecules in somatic cell lineages. In the toad *Xenopus,* ribosomal RNA genes become methylated in somatic cells but not in oocytes, thus showing that some genes may undergo different modifications during differentiation and development. Various invertebrate species undergo gene losses via chromosome diminution in somatic cell development, whereas all the chromosomal material is retained in germ-line cells set apart during development in these organisms. These novel evolutionary solutions to expression of developmental programs undoubtedly serve their purposes in the cases described. For the great majority of organisms and cell types, however, regulation of gene expression underwrites the many and varied developmental pathways in biological species.

## 14.5 Gene Activation during Development in *Chironomus*

As we have just seen in the previous section, genes apparently are neither lost nor mutated during differentiation. We can thus assume that cellular differentiation in complex organisms is the outcome of differential gene expression, for which there is abundant evidence in many different cases involving particular kinds of cells. A classic example of differentiation due to differential gene expression was described by Wolfgang Beermann in 1961 for the midge *Chironomus.* Giant polytene chromosomes of dipteran insects, such as *Drosophila* and *Chironomus,* can provide direct evidence of differential gene activation at specified times during a developmental sequence. Apparently identical cells in a tissue or an organ can be examined to see whether or not chromosomal and genic activities are the same during the progress of a developmental program.

During the prepupal stage of larval development, four cells in the salivary gland begin to make a secretion that serves to "glue" the pupa to a solid surface, where it remains during its metamorphosis into the adult insect. In *Chironomus pallidivittatus* a granular protein is made in these same cells, and the protein becomes part of the secretion substance. In these four cells and in none of the other cells of the salivary gland, a **puff** appears near one end of chromosome 4 at the time the protein is synthesized. The related species *Ch. tentans* produces a clear secretion without the protein granules. No puff is formed near the end of chromosome 4 in any of the salivary gland cells in this species. Crosses between *Ch. pallidivittatus* and *Ch. tentans* indicated that protein synthesis was under control of the *sz* gene. Strains or species that made the protein were $sz^+$ and strains that made a protein-free clear secretion were homozygous for the $sz^-$ allele.

Since puff formation on chromosome 4 appeared to be correlated with the presence of the $sz^+$ allele and with synthesis of the protein granules, Beermann conducted a genetic analysis to obtain evidence for this observed correspondence between puffing and gene expression. He crossed *Ch. pallidivittatus* and *Ch. tentans* to obtain hybrid $sz^+/sz^-$ progeny. The banded polytene chromosome 4 from each parent was recognizable in the hybrids by the presence of a large inversion in the *tentans* chromosome. Since polytene chromosomes pair in somatic larval cells, each chromosome 4 was loosely paired to the other because of inversion heterozygosity (see Section 12.6).

Microscopic examination of banded polytene chromosomes in salivary gland cells of the hybrids revealed that only the *pallidivittatus* chromosome developed a puff in the usual site, whereas no puff formed on the loosely paired *tentans* chromosome 4 at this site (Fig. 14.16). The $sz^+/sz^-$ hybrids produced half as much granular protein in their secretion as did the $sz^+/sz^+$ *pallidivittatus* parent cells. These results indicated that the correspondence between specific puff development and specific gene product synthesis was the consequence of differential

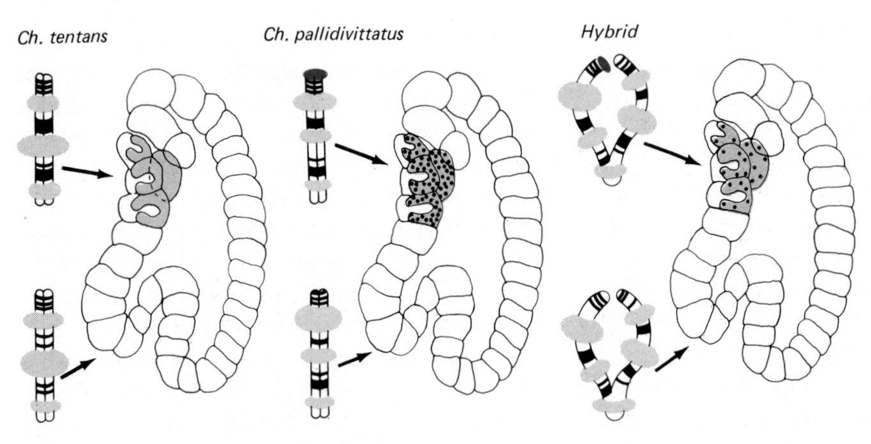


***Figure 14.16*** Diagrammatic summary of Beermann's experiments with *Chironomus.* Four cells in the salivary gland of *Ch. pallidivittatus* produce a granular secretion. A puff (color) is formed at one end of chromosome 4 only in these four cells and not in the other cells of the gland (chromosomes shown to left of gland). The corresponding four cells in *Ch. tentans* produce a clear secretion lacking the granules, and no puff forms at the end of chromosome 4 in any of its gland cells. In hybrids between these two species, half as many protein granules occur in the secretion as in the *Ch. pallidivittatus* parent. Since there are structural differences in chromosome 4 in the two parent species, the hybrid chromosome pair is not closely held together and the source of each of the two chromosomes in the pair can be identified. Only the chromosome in the hybrid that was derived from the *Ch. pallidivittatus* parent forms a puff in the four crucial gland cells. These experiments show that puffing is associated with gene action in directing synthesis of a protein that characterizes a phenotypic trait in the organism. (From W. Beermann. 1963. *Amer. Zool.* **3**:23.)

gene activation during development. The puff developed, the gene began to transcribe RNA according to autoradiographic data, and the secretion protein was translated from the copied genetic information contained in the $sz^+$ allele.

The failure of puff development in the *tentans* chromosome was the consequence of the presence of the inactive $sz^-$ allele. The absence of puffing in all the other salivary gland cells (other than four that secreted granular protein), whether of the hybrid or of the *pallidivittatus* parent, represents differential activation of $sz^+$ alleles in these cells. The control over transcription remains uncertain, but speculations are possible on the basis of other, similar instances of differential gene action and puffing. As we mentioned in Section 10.5, protein receptors may be present in some cells and not in others that are subject to hormonal induction of transcription. Puffing is almost certainly the result of unfolding of packed chromatin fibers, which makes the chromatin DNA accessible to transcription by RNA polymerase binding (Fig. 14.17). Puffing or unfolding of tightly folded chromatin may be influenced by the physical conformation of the DNA, by chemical modifications of chromosomal DNA or histone proteins, or by other factors. The mechanisms of regulation of conformational changes in chromatin remain unclear at present, but the phenomenon is currently under extensive investigation by means of molecular methods. In addition to understanding mechanisms of regulation of genes and chromosomes, we must also analyze the basis for differential gene activation in genotypically identical cells. We have no explanation at present for the activation of the $sz^+$ allele in four of the salivary gland cells during the final stages of larval development. There are many avenues of investigation, however, by which these fundamental questions can be analyzed and ultimately explained.

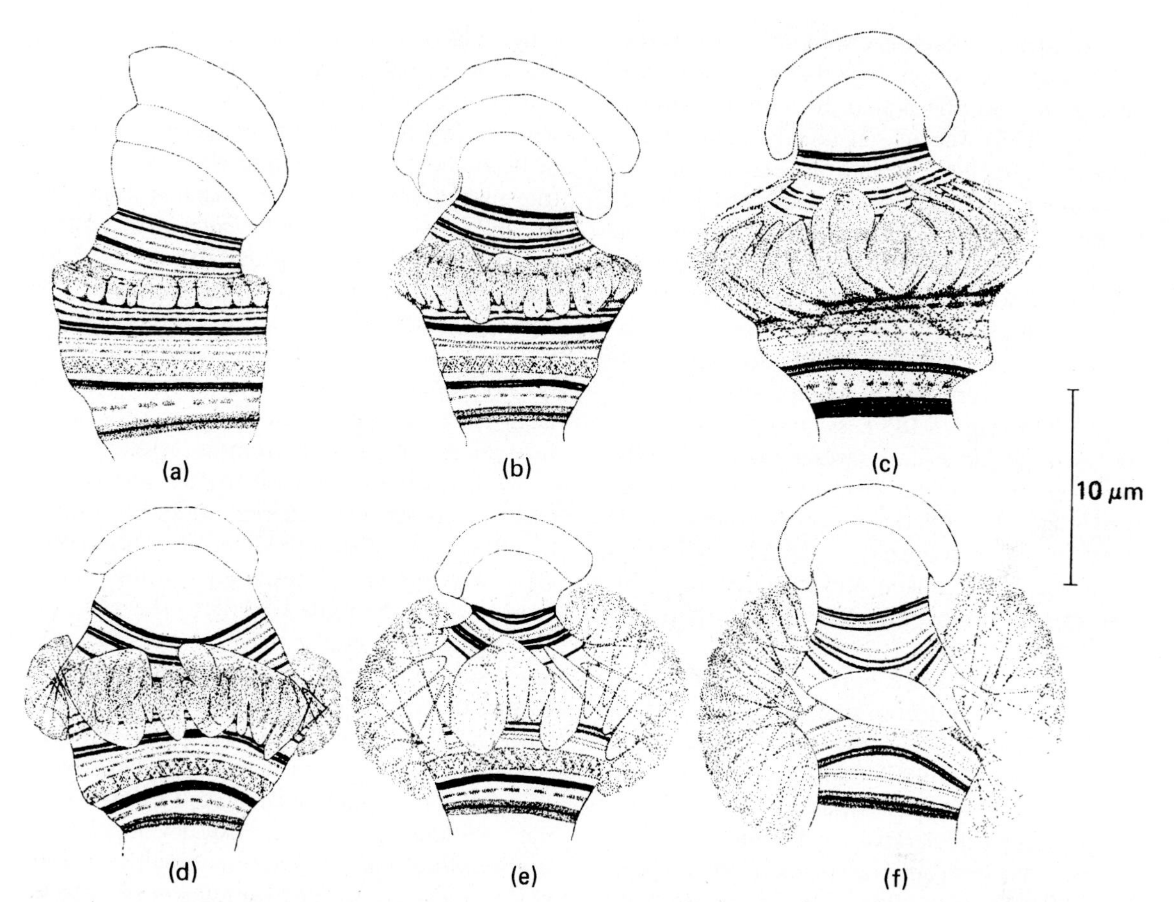


***Figure 14.17*** Diagrammatic representation of puff development in a polytene chromosome of *Chironomus tentans* larval salivary gland. The puff appears as a local chromosome region (probably one band) undergoes a gradual decondensation (a–f). (From W. Beermann. 1952. *Chromosoma* **5**:139.)

## 14.6 Disk Determination in *Drosophila*

We have seen that cellular differentiation during development is the outcome of differential gene expression. We may also ask whether the determination of a particular developmental program is due to differential gene expression. In other words, are genetic processes responsible for establishing a particular developmental program **(determination)** as well as for carrying out the instructions of that program **(differentiation)** at a later time? We might suspect that this is the case, but we must know whether or not the determined state as well as the differentiated state is the result of differential gene expression. With such information we can be more confident that the processes of morphogenesis are guided by gene action in temporally and spatially organized pathways.

During development of *Drosophila* larvae, clusters of cells are set aside in groups called **imaginal disks.** These disks do not contribute at all to larval development or function. Instead, they develop into external structures, such as head, legs, wings, and antennae of the *imago,* or adult fly. While larval cells grow and differentiate all around them, disk cells remain in an embryonic state throughout larval development. Disks only begin to differentiate into adult structures during pupation, when larval tissues break down during metamorphosis into the adult individual (see Fig. 3.15). Larvae can function quite well without these disks, as can be seen in various *Drosophila* mutants in which some or all of the imaginal disks are missing. Diskless mutants flourish during the larval stage, but they die in pupation when adult structures fail to appear.

Each disk starts out as a small group of 10 to 50 cells or less, and each disk is *determined* at this stage to later develop into a specific part of the adult fly (Fig. 14.18). Disk cells proliferate in the growing larva, and several thousand cells may comprise each disk by the *third instar,* or final stage of larval development. The various disks can be recognized in third-instar larvae according to size, shape, and location, and each disk can be removed in whole or in part for experimental studies.

From earlier work by George Beadle and Boris Ephrussi and from extensive later studies by Ernst Hadorn, we know that imaginal disk cells are programmed for specific development long before their phenotypes are expressed in the adult. A particular disk can be removed from a third-instar larva and transplanted into another larva, which is then allowed to develop into an adult fly. The transplanted disk will differentiate into the same structure that would otherwise have appeared had it been left in place in the original larva. If an eye disk from a larva carrying the white eye allele *(w)* is transplanted into the abdomen of a larva which is $w^+$, the adult will develop two red eyes in their usual location on its head and a white eye in its abdomen. Similar transplantations using other imaginal disks lead to similar results. The adult structure that differentiates is the same structure that would have differentiated in the original case, whether it is legs, antennae, or other external parts of the fly. When a disk is removed from a larva, other cells do not replace the lost disk cells. The donor larva, if it develops into an adult, lacks the structure programmed by the removed disk.

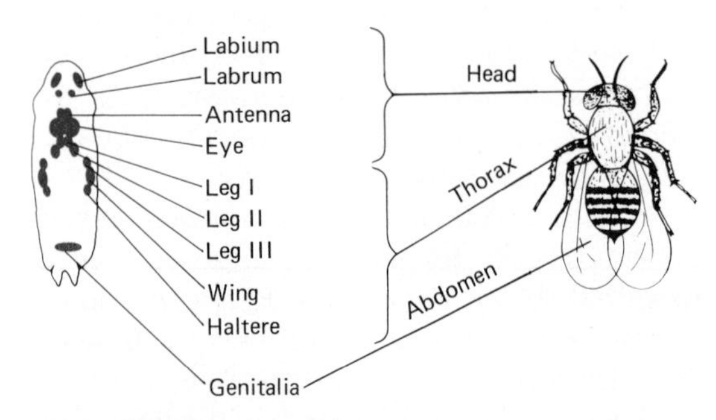


***Figure 14.18*** Location and identification of imaginal disks (color) in the larva of *Drosophila melanogaster,* and their structural derivatives in the adult, or imago.

Through transplantation studies such as the ones just described, different disks have been identified. Each disk is predetermined for a specific developmental program during metamorphosis in the pupa leading to the adult fly. Transplantation studies have shown that the larval surroundings have no apparent influence on disk determination or on the eventual phenotype. Disks will differentiate into the determined structures regardless of where they have been inserted in the recipient larvae and regardless of the recipient's phenotype. Disk development is *autonomous,* but development will not take place unless the proper stimulus is present or provided.

The stimulus for differentiation of the determined disks is the molting hormone *ecdysone,* which begins to be made in quantity in third-instar larvae. Isolated disks that are treated with ecdysone will differentiate into adult structures. In addition, if determined disks are transplanted from third-instar larvae into adult flies, cell proliferation continues but the cells do not differentiate. Ecdysone is not present in adult flies. These experiments show that disk cells are determined to develop into specific gene-encoded structures at a very early stage in development. Cellular differentiation will not take place in the absence of ecdysone, which is the stimulus for gene expression, or in the absence of other important factors that influence gene expression. Differentiation proceeds normally during pupation since ecdysone is made in the late stages of larval development. There is a considerable lag, therefore, between the time that cells are programmed to develop into certain structures and the time that development actually takes place. Even though the cells are already determined, development may never take place if the environment is unsuitable (for example, surrounded by adult instead of larval tissues).

Once determination has occurred, can disk cells retain their specific potential for differentiation indefinitely? Using *serial transplantations,* Hadorn has shown that disk cells do not de-differentiate, that is, they retain their determined state indefinitely (Fig. 14.19). By transferring part of a disk into an adult, disk cells increase in numbers and can be used over many generations as a supply of cells for transfer into larvae, where disk development can then be observed in the adult that emerges. For more than 150 such serial transfers over several years, disk cells continued to increase without differentiating in the adult hosts, but they did differentiate into the expected determined structures in the larval hosts. The determined states were regenerated and maintained since disk specificities were transmitted from generation to generation. *The determined state was clonally heritable.*

In some of the serial transfers, Hadorn found that a disk might be altered to a different determined phenotype. For example, a leg disk might differentiate into wings in a percentage of the larval hosts. This change in determined state, or **transdetermination,** would then be regenerated and maintained in subsequent generations of serial transfer adults and larvae. The new determined state was heritable. Hadorn found that transdeterminations were not random; certain transdeterminations never occurred directly. For example, haltere disks transdetermine directly to wing disks but not to genitalia or leg disks; leg disks transdetermine to wing but not to haltere disks (Fig. 14.20). The pathway of transdeterminations includes specific sequences, therefore, and the alterations seem to

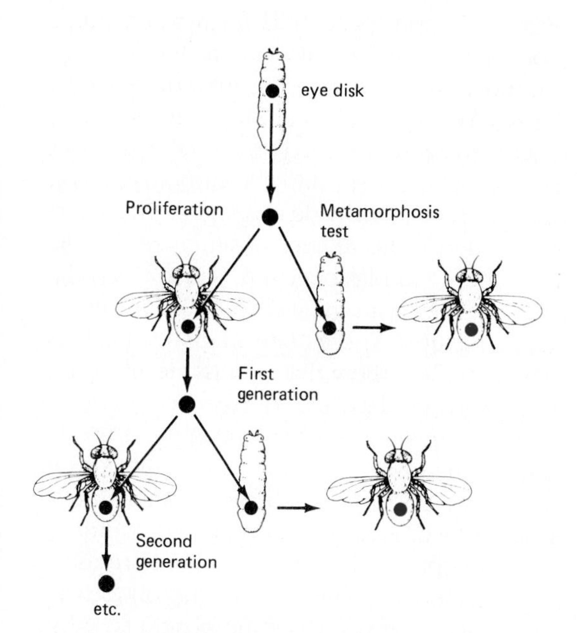


***Figure 14.19*** Serial transplantation procedure in *Drosophila* imaginal disk studies. Disks proliferate but do not differentiate in adult flies, but when they are transplanted into larvae the disk potential can be identified in the adult that emerges, in a metamorphosis test for disk specificity.

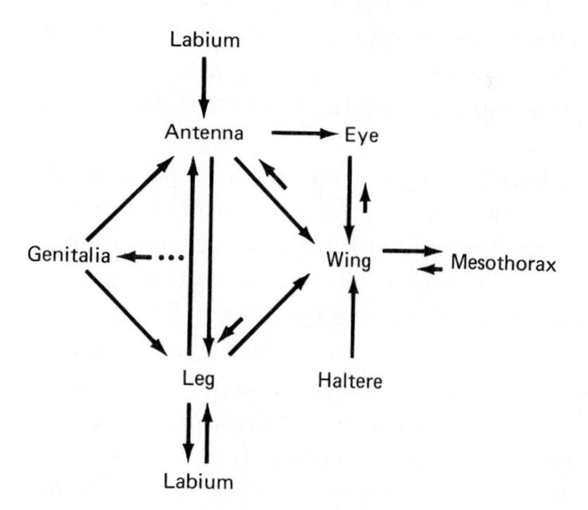


***Figure 14.20*** Pathways of transdeterminations in *Drosophila.* Certain disks transdetermine only into certain other potentials, resulting in apparent sequences of transdetermination rather than a range of possible alterations from one determined state to any of all others. The length of each arrow indicates the relative frequency of each transdetermination event. (Based on studies by E. Hadorn.)

represent switches between particular determination pathways rather than a graduated range of possible alternatives.

Transdetermination is not due to mutations in somatic cells, according to two particular observations: (1) they arise too frequently and (2) groups of clonally unrelated, adjacent cells may transdetermine at the same time. Each transdetermination event has a particular probability, however; some events occur more frequently than others. In addition, certain transdeterminations are reversible, but the rates of change are different for the pair of changes. For example, leg disks transdetermine to wing much more often than wing disks transdetermine to leg.

In one class of inherited changes, called **homeotic mutations,** alterations resembling transdeterminations have been observed. In homeotic mutants, a normal structure or part of a structure develops in an abnormal location. For example, the *tetraptera* mutant has wings in place of halteres so that four-winged flies develop; in the *tumerous head* mutant, genital or leg structures develop in place of antennae; other mutants of a similar nature have been described. The relationship between homeotic mutations and transdeterminations, however, is not known. Homeotic mutations may alter some control system, rather than causing a change in a structural protein.

When does disk determination take place in development? In experiments with embryonic systems, determinative events occurred in early blastoderms shortly after the first cleavage divisions of the fertilized egg. Blastoderms from mutant strains were cut in half at the midline, and cells from the anterior and posterior halves were put into separate suspensions. These cell suspensions were centrifuged, and clusters of sedimented cells were transplanted into wild-type adult host flies, where they multiplied. The nature of these cells was then examined by transferring some of this material into wild-type larvae to see what would develop when these larvae metamorphosed into adults.

Cells that were originally taken from the anterior half of the blastoderm developed into anterior adult structures, and posterior blastoderm cells gave rise to posterior adult structures. These experiments showed that cell determination had already occurred in the blastoderm stage. Furthermore, the position of a disk and the ultimate location of the structure in the adult appears to be established at the time that blastoderm cells are determined. Form and function develop in orderly fashion long before tissues and organs actually differentiate in the organism. From many years of experimental embryological studies of many species, we know that various factors in the egg act as stimuli for determinative events in the early cleavage divisions after fertilization. The nature of most of these factors, however, remains virtually unknown at present.

Through various methods it is possible to ascertain the number and position of disk precursor cells as they appear on the blastoderm surface (Fig. 14.21). This surface is only one cell layer thick, and these cells form when nuclei produced by mitotic divisions in the fertilized egg move outward from the interior. Seymour Benzer, Yoshiki Hotta, and others, using a technique originally invented by Alfred Sturtevant, have analyzed genetic and chromosomal mosaic flies in order to map development. They were able to derive the relative positions of and distances between blastoderm precursor cells that later give rise to imaginal disks and ultimately to adult structures. These **"fate maps"** of the blastoderm surface show that orderly development in *Drosophila* depends in part on *positional information in the embryo.* Imaginal disks develop in the blastoderm and later in the larva in places that correspond to the positions these structures will eventually occupy in the adult fly. The developmental blueprint is therefore established quite early, long before morphogenesis takes place. The nature of the positional information is poorly understood at present, but this information must ultimately be under genetic control since positional information is modified in homeotic mutants.

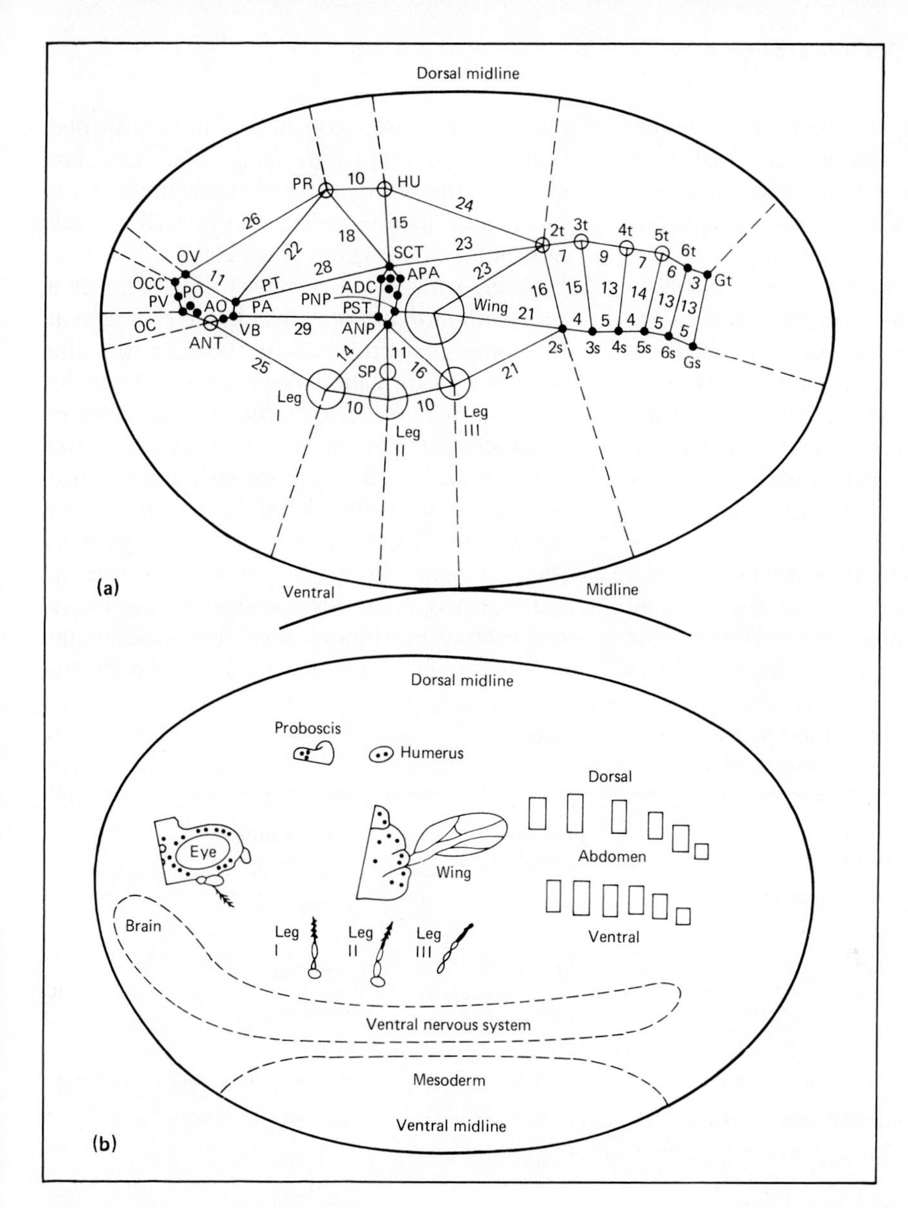


***Figure 14.21*** Fate map of disk precursors of *Drosophila melanogaster* blastoderm. (a) Fate map showing cells that will eventually develop into the indicated external body parts of the adult fly. The map is a projection of the right half of the blastoderm as seen from the inside of the egg. The abbreviations refer to parts shown in a pictorial sketch in (b), which is based on the fate mapping data. (From Y. Hotta and S. Benzer, 1972. *Nature* **240**:527.)

## 14.7 Oncogenes and Tumor Development in Vertebrates

Tumors are abnormal growths in which differentiated cells revert to an undifferentiated state, undergoing rapid rates of multiplication and sometimes migrating to new locations in the organism. Tumors may be *benign,* increasing in size in their place of origin in the organism. Tumors become *malignant* when they invade neighboring tissues and subsequently undergo *metastasis,* that is, when they spread to distant sites in the organism and there initiate new tumors. Metastasizing tumors are malignant growths called *cancers*. The initiation of abnormal growths is referred to as *tumorigenesis* or *oncogenesis* (Gr. *onkos,* tumor; *genesis,* to be born). A number of particular genes are able to initiate and maintain the tumorous state in the organism. Such genes are generically called **oncogenes,** and they are designated by the general symbol *onc.* Specific oncogenes are symbolized according to a recently adopted system that indicates their origin and the type of tumor under control of the oncogene (Table 14.3).

Both DNA and RNA viruses have been shown to be oncogenic in plants and animals. Like all viruses, oncogenic viruses are relatively host specific and tissue specific in their activities. A number of viruses may induce tumors in one host and be innocuous in another host. For example, the DNA virus SV40 (simian virus 40) causes little apparent effect in its normal monkey host but is tumorigenic in rodents.

Many kinds of oncogenic viruses can elicit a *transformation response* in cultured cells as well as a tumorigenic response in the organism. The cellular transformation response is evident by *uncontrolled growth in culture.* Instead of forming a single confluent layer of cells on a solid substratum, such as a culture dish, transformed cells grow into disordered heaps on a solid substratum. Transformed cells can also grow in semisoft agar, whereas normal cells cannot. Transformed cells often *lose their anchorage dependence,* that is, they can no longer adhere to a solid surface. Uncontrolled cellular multi-

***Table 14.3*** Retroviral strains, probable origin, and tumorigenic effects of retroviral *onc* genes (oncogenes) in host animals.

| retroviral *onc* gene* | retroviral strain | probable animal origin | type of tumor produced |
|---|---|---|---|
| v-*rel* | avian reticuloendotheliosis virus-T | turkey | lymphoma |
| v-*src* | Rous sarcoma virus | chicken | sarcoma |
| v-*myb* | avian myeloblastosis virus<br>avian leukemia virus | chicken<br>chicken | leukemia<br>leukemia |
| v-*myc* | avian myelocytomatosis virus | chicken | sarcoma, leukemia, carcinoma |
| v-*erb* | avian erythroblastosis virus | chicken | sarcoma, leukemia |
| v-*fps* | Fujinami sarcoma virus | chicken | sarcoma |
| v-*yes* | avian sarcoma virus (Y73) | chicken | sarcoma |
| v-*ros* | avian sarcoma virus (UR2) | chicken | sarcoma |
| v-*mos* | Moloney murine sarcoma virus | mouse | sarcoma |
| v-*ras* | Harvey murine sarcoma virus | rat | sarcoma |
| v-*abl* | Abelson murine leukemia virus | mouse | leukemia |
| v-*fes* | feline sarcoma virus | cat | sarcoma |
| v-*fms* | McDonough feline sarcoma virus | cat | sarcoma |
| v-*sis* | simian sarcoma virus | wooly monkey | sarcoma |

*The homologous cellular gene or gene equivalent in the animal host is designated by the lowercase letter prefix c, e.g., c-*src* is the cellular gene equivalent of the retroviral v-*src* oncogene.

Adapted with permission from J.M. Coffin et al. 1981. *J. Virology* **40**:953–957.

plication and migrations of transformed cells in culture are analogous to tumor development in the organism. Cultured cells are very convenient systems for studies of oncogenic viruses and their activities in the cell since a broader range of experimental manipulations can be accomplished that are either difficult to accomplish with animals or are much more expensive and time consuming.

One particular group of oncogenic viruses is of surpassing interest at present because of important new information gained about these viruses and about their relationship to normal cellular genes and activities. This is the group of **retroviruses,** which are animal viruses whose genomes consist of RNA instead of DNA. Retroviruses are oncogenic in a large number of vertebrate species, from fish to mammals, including humans. Considerable interest in retroviruses was first aroused in 1970 when David Baltimore and Howard Temin independently described the existence of an unusual replication enzyme in these RNA viruses. Baltimore and Temin described reverse transcriptase, and RNA-dependent DNA polymerase, that is, a DNA polymerase that uses RNA as a template to copy. The usual polymerases use DNA templates, either in replication of DNA (DNA polymerases) or in transcription of DNA into RNA (RNA polymerases). The "backward" process of copying RNA into DNA—that is, by reverse transcriptase action—was the reason for naming these RNA animal viruses *retro*viruses.

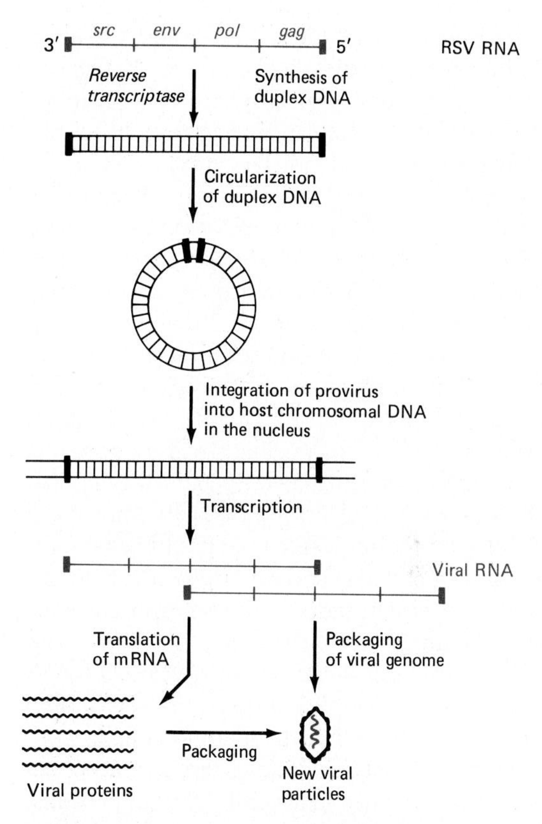


***Figure 14.22*** Infection cycle of the retrovirus RSV (Rous sarcoma virus) in chicken cells. The small RNA viral genome, consisting of four genes, enters the host cell and is copied there into DNA in reactions catalyzed by reverse transcriptase (an RNA-dependent DNA polymerase). The circularized duplex DNA is integrated into host chromosomal DNA in the nucleus as a provirus sequence. Proviral DNA replicates along with host DNA. Some of the viral mRNA transcripts are translated into virally specified proteins and some transcripts are packaged in progeny viruses along with viral proteins. The protein product of the *src* gene is an enzyme called a protein kinase, and it is not incorporated into new virus particles.

The retrovirus infection cycle is unusual because the viral RNA genome must be copied into DNA before viral replication and gene expression can take place in the host. After penetrating the host cell, viral RNA is copied into DNA in reactions catalyzed by reverse transcriptase (Fig. 14.22). The copied DNA synthesizes a partner strand to become a duplex molecule. The viral DNA duplex circularizes and becomes integrated into the host chromosome as a provirus. Once part of the host chromosome, proviral DNA is replicated along with host DNA and its sequences are transcribed into mRNA by host RNA polymerase. Proviral mRNA transcripts are translated into several kinds of proteins on host polysomes. Some of these viral proteins encapsulate some of the viral RNA transcripts to produce new viruses that can exit the cell and initiate new infections.

A number of retroviruses have been found to contain oncogenes. The initiation and maintenance of the abnormal growth response by retroviruses is dependent only on the activity of an oncogene in the viral genome. If the oncogene is deleted or incomplete, the remainder of the viral genome can infect the cell but abnormal growth will not take place. When the oncogene is present, most of the viral activity is directed to maintain the transformed state since relatively few new viruses are made and released from infected cells. Retroviruses do not ordinarily kill their host upon infection or upon the initiation of an oncogenic response in the host.

In some retroviruses the oncogene specifies a catalytic phosphoprotein called a **protein kinase.** The kinase specified by Rous sarcoma virus (RSV) was designated pp60 $^{src}$. It is a phosphoprotein (having phosphorylated amino acid residues in the molecule) of about 60,000 molecular weight and is encoded by the viral RSV-*src* oncogene, which induces sarcomas in chickens and certain other birds, such as quails. The action of this and similar virally specified protein kinases is to add phosphoryl ions ($PO_3^{2-}$) to tyrosine residues in at least some cellular proteins. These protein kinases appear to phosphorylate tyrosines in several different proteins, that is, they have a range of substrate specificities rather than acting on a single protein species. From electron microscopy and other data, it has been found that the protein kinase molecules bind to the inner surface of the cellular plasma membrane.

From molecular hybridizations, base sequencing, heteroduplex and R-loop analysis using electron microscopy, and from various molecular genetic methods, it has been shown that particular cellular genes and particular viral oncogenes have highly homologous coding sequences and that they specify the same protein kinase gene product. The viral oncogene v-RSV-*src,* for example, has a normal cellular gene equivalent designated c-RSV-*src;* the v and c symbols indicate the source of the gene as viral or cellular.

The *src* gene equivalents were discovered by isolating and purifying v-RSV-*src* from viruses and the use of cDNA of this material as a probe to search for homologous sequences in chicken DNA (Fig. 14.23). The presence of heteroduplexes indicated sequence homologies between v-*src* and normal DNA in the organism. Furthermore, looped regions of the chicken DNA strand in the heteroduplexes indicated that the viral oncogene lacked introns that were present in the normal eukaryotic gene equivalent. The cellular gene, called a *proto-oncogene,* has the usual interrupted eukaryotic gene organization of interspersed coding and noncoding sequences. The viral oncogene has only coding sequences in a contiguous array and no intervening sequences. Base sequencing has confirmed the interpretations made from heteroduplexes.

The equivalence of the viral and the cellular versions of an oncogene has been demonstrated in several ways, but two of these studies were particularly informative. In one study Hidesaburo Hanafusa and co-workers injected strains of Rous sarcoma virus carrying large deletions of v-RSV-*src* into chickens. These viruses could not initiate an oncogenic response, but they could infect cells. When progeny RSV were collected from the infected animals, Hanafusa found that they contained an intact v-*src* gene. The reconstituted v-*src* gene must have arisen through recombination between the deleted v-*src* and the host c-*src* gene sequences. These reconstituted v-*src* genes were fully oncogenic in action since they initiated and maintained an oncogenic response in animals into which the progeny viruses were injected.

In another approach, George Vande Woude and Edward Scolnick engineered recombinant DNA by splicing cellular proto-oncogenes from cloned DNA onto viral DNA that carried a promoter sequence, which contributed to regulating the expression of a nearby gene. When the *src*-promoter recombinant DNA was introduced into cultured cells, a transforming response was initiated and maintained in some of the cells. The cellular response to recombinant DNA was the same as when a viral oncogene is integrated into a cell. These observations of equivalent sequences and equivalent genetic response induced by cellular and viral oncogenes raise the

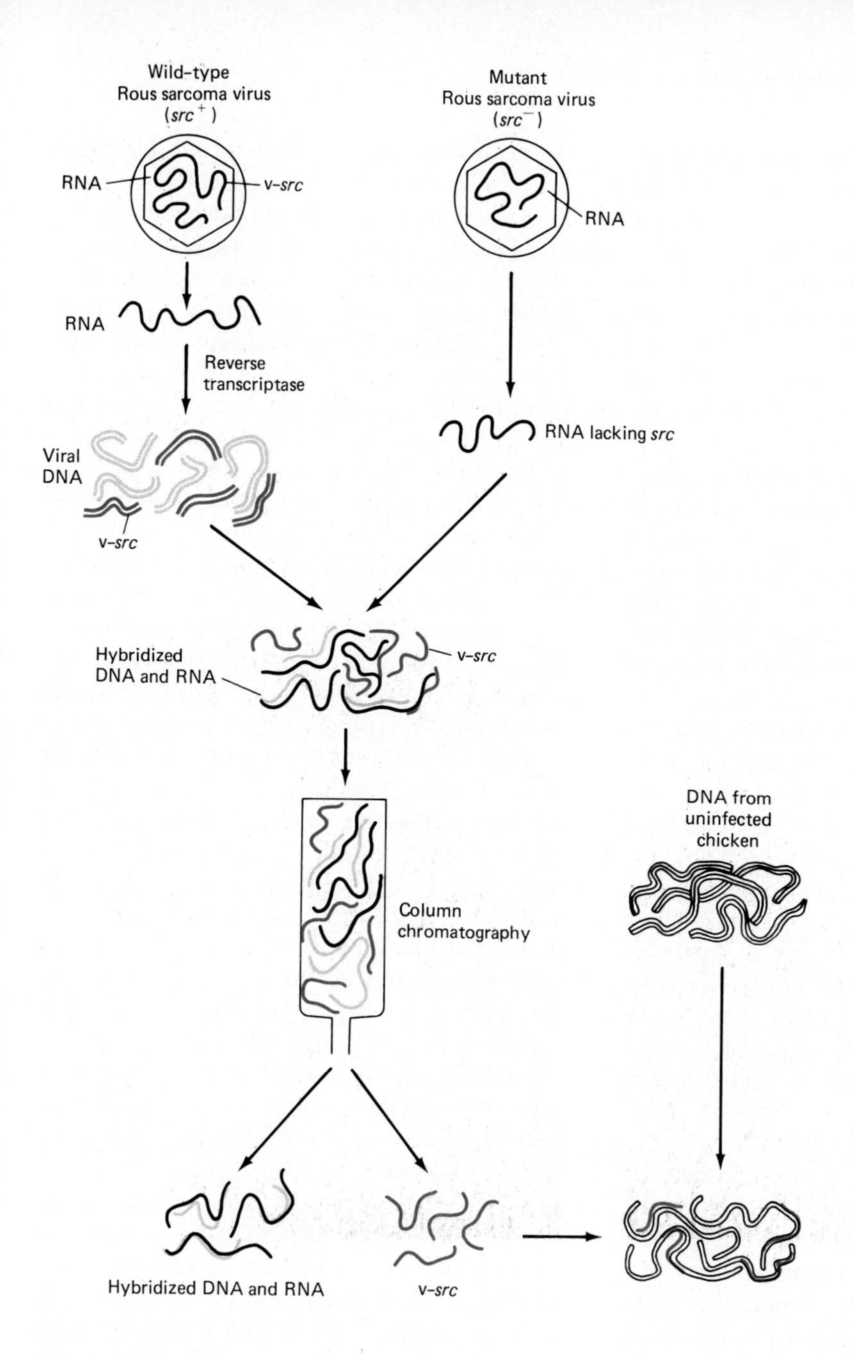


***Figure 14.23*** The occurrence of a normal cellular gene equivalent of the viral *src* oncogene was demonstrated using a radioactive DNA probe that carried v-*src*. Isolated viral RNA from wild-type RSV was copied into radioactively labeled DNA carrying v-*src*. Melted single-stranded fragments of this DNA were mixed with RNA from a mutant virus lacking *src*. Molecular hybridizations occurred between complementary DNA and RNA fragments, except for v-*src* fragments, which had no complementary sequences in the mutant RNA. The radioactive unhybridized *src* DNA probes were separated from DNA-RNA hybrid duplexes by column chromatography. When the v-*src* probes were mixed with normal chicken DNA, hybridization revealed the presence of complementary c-*src* sequences of the cellular gene equivalent. (From "Oncogenes" by J.M. Bishop. Copyright © 1982 by Scientific American, Inc. All rights reserved.)

important question of how the viral oncogene causes an oncogenic response while the cellular gene equivalent sponsors only normal cellular activities. Some evidence concerning this differential response has come from studies of the gene product in transformed and in normal cells.

The particular protein kinases that have been found to be encoded by viral and cellular oncogene equivalents are enzymes that specifically phosphorylate tyrosine residues in cellular proteins. It has been proposed that retrovirus oncogenes work harder than their cellular equivalents, thus leading to a significant and excessive amount of proteins with phosphorylated tyrosines. This **dosage effect** may swamp the cell with these kinds of proteins, leading to abnormal cellular responses. Tony Hunter and co-workers have shown that approximately ten times the amount of phosphorylated tyrosines are present in proteins of transformed cells as compared with normal cells.

Two particular effects of phosphorylated tyrosines have been demonstrated in cells, which bear on modifications from normal to abnormal cellular behavior. Stanley Cohen discovered that once bound to the cell surface, "epidermal growth factor" molecules stimulated normal DNA synthesis and cell division. The binding of epidermal growth factor specifically elicits phosphorylation of tyrosine in cellular proteins. Since some of the same proteins can also be phosphorylated by the v-*src* protein kinase, $pp60^{src}$, tyrosine phosphorylation apparently plays some role in regulating normal cell growth. Perhaps an excessive amount of phosphorylated tyrosines in particular proteins leads to swamping of normal controls and to uncontrolled cell multiplication in transformed cells in culture and in animal tumors.

A second effect of phosphorylated tyrosines concerns the loss of anchorage dependence, which leads to the inability of transformed cells to adhere to solid substratum. *Adhesion plaques* are focal contacts in the cell surface that mediate cellular attachment to solid substratum. One of the prominent proteins in adhesion plaques is *vinculin,* and this same protein has been found to contain elevated levels of phosphotyrosine in transformed cells. Adhesion plaques do not develop in transformed cells, and vinculin is no longer found to be concentrated in regions where adhesion plaques otherwise might develop. Vinculin is believed to act in some fashion in a *transmembrane cytoskeletal control system* that links the bundles of microfilaments just under the cell surface with the extracellular matrix coating the outer surface of the cell membrane (Fig. 14.24). Disruption of such a transmembrane control system could have profound effects that lead to altered cell morphology, reduced cell adhesion, and other disruptive changes owing to changes in cell surface mobility and consequent abnormal interactions with other cells or with molecules.

Cellular DNA sequences for c-*src* and for other normal gene equivalents of oncogenes have been found in many vertebrate genomes. These normal gene sequences appear to have been conserved to a large degree during 500 million years of vertebrate evolution, which implicitly indicates their importance in normal cellular functions. Retroviruses are believed to have picked up such normal cellular genes, perhaps in relatively recent times, and to have incorporated these normal genes in the viral genome. By chance, we discovered the retroviral oncogenes in earlier studies and their normal cellular equivalents in later studies. We can see now that the evolutionary sequence was quite the reverse. Retroviruses can perform quite well without their captured *onc* genes since oncogenes are not involved in the viral infection. Cellular *onc* genes, on the other hand, appear to be intimately involved in normal cellular functions. It appears at present that cellular *onc* genes may run amok when their structure or control is disturbed, but they guide normal activities in most cells most of the time.

Aberrant expression of a normal cellular *onc* gene may be the result of various perturbing factors. Carcinogens, mutagens, transpositions of genetic material, inserted viral genes, and other

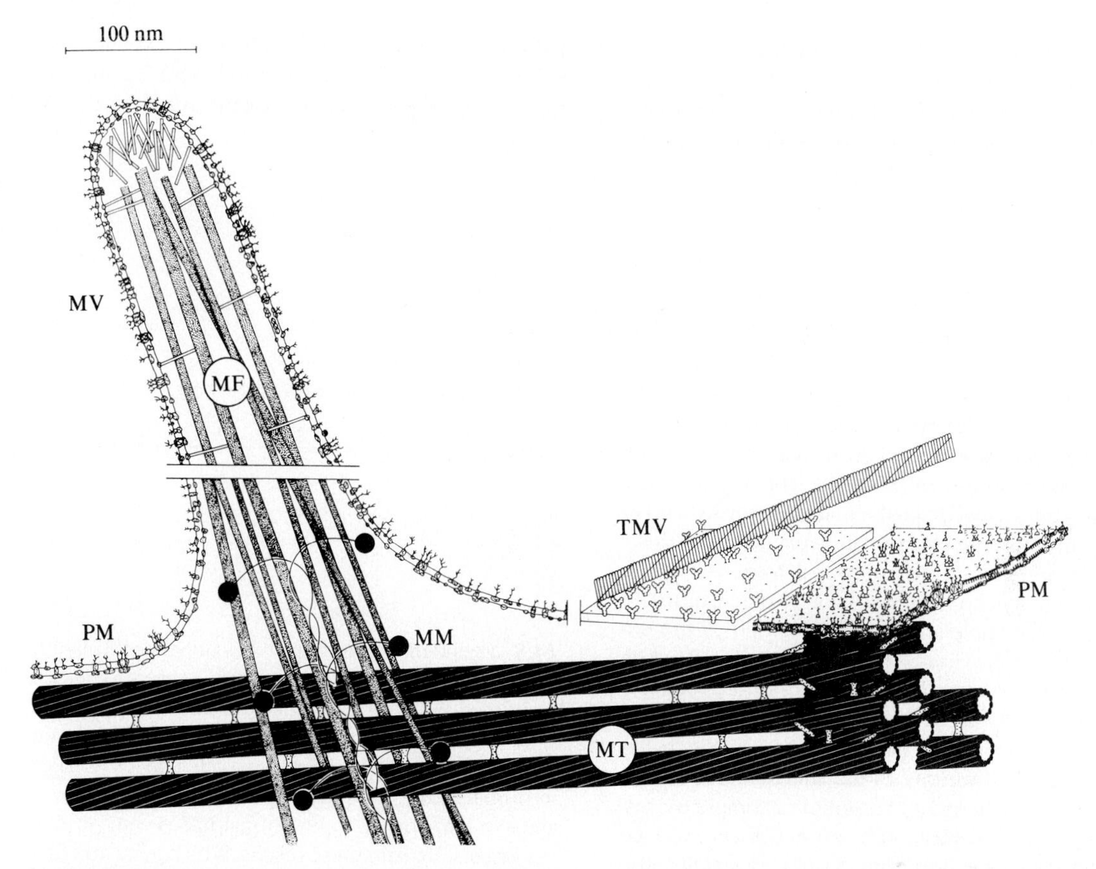


***Figure 14.24*** Model of the cytoskeletal framework associated with the plasma membrane (PM). Large microtubles (MT) occur along with microfilaments (MF) and myosin molecules (MM), which usually extend into a microvillus (MV) projection of the surface membrane. The cytoskeleton underlies the membrane in areas lacking microvillus projections, as well as where these occur. A tobacco mosaic virus (TMV) particle provides a basis for size comparisons. Note the relatively enormous dimensions of the cytoskeleton in relation to the protein and glycoprotein molecules on and in the plasma membrane. The cytoskeletal components are believed to comprise a transmembrane control system governing membrane protein mobility. (Reprinted by permission from F. Loor. 1976. *Nature* **264**:272. Copyright © 1976. Macmillan Journals Limited.)

agents have been implicated as factors contributing to altered genetic behavior. Recently, William Hayward and co-workers showed that insertion of a viral promoter sequence caused activation of a cellular *onc* gene in chickens, thus leading to formation of a characteristic lymphoma type of cancer. It appears likely that "cancer genes" may be little more than normal genes gone haywire as a consequence of some disturbance or modification in gene structure or in gene expression, induced by any one of an increasing number of external agents and factors.

If this should prove to be true, then the required cure will necessarily involve measures to restore the uncontrolled nondifferentiating cancer cells to their original controlled differentiated state in the organism. Therapy to alleviate the condition may be difficult to achieve, apart from selectively destroying abnormal cells and preserving normal cells in the individual. Reversions from transformed state to normal can be achieved at will in cell cultures by temperature shifts. Whether these reversals are due to effects on the *onc* genes themselves or on their protein products is still uncertain. These and other avenues of basic research will be important areas of investigation in the future, as they are today.

## Questions and Problems

***14.1*** When *E. coli* is infected by $\lambda$ phage during its lytic cycle, two observations can be made: (1) the early mRNA is transcribed from the middle of the phage chromosome to the right, and (2) the later mRNA is transcribed from the middle to the left end of the $\lambda$ chromosome. Suggest an explanation for this.

***14.2*** A viral extract was prepared from T4 phages defective in a gene for production of either heads or tails. When complete heads were added to the extracts, infectious viruses were produced; but when complete tails were added, no viruses were produced. Where is the defect in this T4 mutant assembly?

***14.3*** Extract complementation tests were conducted using extracts of various T4 mutants defective in one or more genes. In order to assemble a complete virus, genes for normal head, tail, and tail fibers must be present. From the following matrix, determine the nature of the defect in complementation groups B, D, E, and F. It is known that group A mutants only produce heads and complementation group C mutants only produce tails (+ indicates phage production, 0 indicates no phages are produced).

| | A | B | C | D | E | F |
|---|---|---|---|---|---|---|
| **A** | 0 | + | 0 | 0 | 0 | 0 |
| **B** | | 0 | 0 | + | + | 0 |
| **C** | | | 0 | + | 0 | 0 |
| **D** | | | | 0 | + | 0 |
| **E** | | | | | 0 | + |
| **F** | | | | | | 0 |

***14.4*** Complementation tests were performed by constructing $F_1$ hybrids between five strains of *Drosophila,* each of which had an abnormal wing structure. The matrix obtained is shown at the top of the next column (+ indicates normal wing structure).Diagram the complementation map. How many different genes are defined by this set of experiments? (Assume the mutants all are recessive to wild type.)

| | 1 | 2 | 3 | 4 | 5 |
|---|---|---|---|---|---|
| **1** | 0 | 0 | + | + | 0 |
| **2** | | 0 | + | + | 0 |
| **3** | | | 0 | + | + |
| **4** | | | | 0 | 0 |
| **5** | | | | | 0 |

***14.5*** A gene mutation in fruit flies prevents formation of functional sperm storage organs (spermothecae) and functional ovaries in females. This mutant allele, *z,* does not affect males. At another locus a mutant allele, transformer (*t*), is known to change XX females into males. Flies that are XX and homozygous for *z* and *t* (*z*/*z t*/*t*) prove to be normal males with normal sex organs. Explain these results. Which gene acts first in this developmental sequence?

***14.6*** George Beadle and Boris Ephrussi conducted very interesting disk transplantation experiments in the 1930s, using *Drosophila melanogaster,* thus marking a beginning of biochemical genetics. They transplanted eye disks reciprocally between vermilion and cinnabar larvae and found that cinnabar disks maintained their cinnabar phenotype when developing in vermilion hosts, but vermilion disks developed into wild-type eyes in cinnabar hosts. Wild-type eye color developed from wild-type disks in vermilion or cinnabar hosts, and wild-type eyes developed from vermilion or cinnabar disks in wild-type hosts. These genes act in a pathway involving synthesis of hormonelike substances needed for eye pigment production:

$$\text{tryptophan} \xrightarrow{v^+} \text{formylkynurenin} \xrightarrow{cn^+} \text{hydroxykynurenin} \rightarrow \rightarrow \text{pigment}$$

Use this information to explain the results observed in the eye disk transplantation experiments described above.

**14.7** A group of yeast cell cycle mutants were grown in restrictive and permissive conditions according to the reciprocal shift method. Their ability to complete the developmental program was scored (+ and −, for completed and not completed, respectively) as follows:

| **cdc mutants shown as** | | **completion of developmental program** | | |
|---|---|---|---|---|
| **A** | **B** | **1st incubation:** <br> **2nd incubation:** | **restrict A, permit B** <br> **permit A, restrict B** | **restrict B, permit A** <br> **permit B, restrict A** |
| *cdc* 28 | *cdc* 7 | | − | + |
| *cdc* 28 | *cdc* 4 | | − | + |
| *cdc* 7 | *cdc* 4 | | + | − |
| *cdc* 2 | *cdc* 24 | | + | + |

***a.*** What is the developmental sequence involving *cdc* 4, 7, and 28?
***b.*** How are the developmental steps governed by *cdc* 2 and *cdc* 24 related to each other?

**14.8** Four different temperature-sensitive phage mutants are known, each of which has a defect in head protein assembly. Two of the mutants have an altered allele at locus *A*, one of which is heat sensitive ($A^{hs}$) and the other is cold sensitive ($A^{cs}$). There is a heat-sensitive allele at locus *B* ($B^{hs}$) in a third mutant and a cold-sensitive allele ($C^{cs}$) at locus *C* in the fourth mutant. After *E. coli* is infected with these mutant phage strains the culture is divided into two populations—one is grown for 15 minutes at low temperature and then for 15 minutes at high temperature, and the other population is grown for 15 minutes at high temperature and then for 15 minutes at low temperature. The cultures are then scored for phage production (0 = none, + = phages produced). Explain the results shown in the table:

| | **low → high temp.** | **high → low temp.** |
|---|---|---|
| $A^{hs}\ C^{cs}$ | 0 | 0 |
| $A^{cs}\ B^{hs}$ | 0 | 0 |
| $C^{cs}\ B^{hs}$ | + | 0 |

**14.9** Two independent temperature-sensitive mutations were recovered in phage T4. Both mutants formed plaques at 25° but not at 42 °C. If one mutant was defective in DNA replication and the other was defective in tail formation, how would you determine that these phages were mutant for early and late functions?

**14.10** During serial transplantations of larval imaginal disks, occasional transdeterminations can be observed when the structure develops in the adult host. From the following information, diagram the probable pathways of transdetermination for the four kinds of disks listed.

| ***imaginal disk transplanted*** | ***structure formed in adult host*** |
|---|---|
| antenna | antenna, leg, eye |
| eye | eye, wing |
| leg | leg, wing, antenna |
| wing | wing |

**14.11** A number of tumorigenic DNA viruses occur in the mouse. Molecular hybridizations were conducted to study one of these viruses; pure preparations of virus DNA, tumor-cell DNA, and normal-cell DNA were used. Viral DNA hybridized with tumor-cell DNA but not with normal-cell DNA. Explain these results.

**14.12** Assume that a normal cell is transformed into a tumor cell as a result of a retrovirus integrating next to a cellular oncogene in the chromosome. Restriction maps of the DNA from normal and tumor cells were constructed and are shown below (arrows indicate the DNA sites cut by the restriction enzyme):

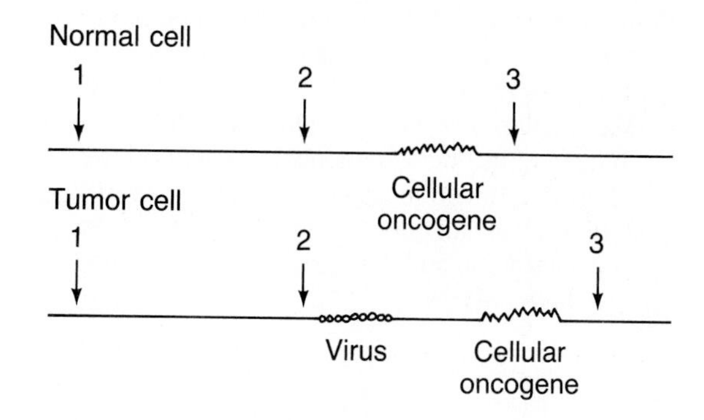

When the restriction fragments of normal and tumor cell DNA are separated by gel electrophoresis, the fragments containing the cellular oncogene and/or viral DNA sequences can be located by hybridizing radioactive copies of oncogene or viral DNA (probes) to the gel and obtaining an autoradiograph of the gel. Using the following diagram, show where bands of restriction fragments would appear on the autoradiograph:

DNA source: Normal Normal Tumor Tumor

Direction of electrophoresis

DNA probe: Oncogene Virus Oncogene Virus

***14.13*** Define the following terms: ***a.*** morphogenesis ***b.*** self-assembly ***c.*** extract complementation test ***d.*** execution point ***e.*** totipotent nucleus ***f.*** puff ***g.*** determination ***h.*** differentiation ***i.*** homeotic mutant ***j.*** oncogene ***k.*** transformation response ***l.*** retrovirus.

## References


Beermann, W., and U. Clever. Apr. 1964. Chromosome puffs. *Sci. Amer.* **210**:50.

Bishop, J.M. 1980. The molecular biology of RNA tumor viruses: A physician's guide. *New Eng. J. Med.* **303**:675.

Bishop, J.M. Mar. 1982. Oncogenes. *Sci. Amer.* **246**:80.

Briggs, R., and T.J. King. 1957. Changes in the nuclei of differentiating endoderm cells as revealed by nuclear transplantation. *J. Morphol.* **100**:269.

Butler, P.J.G., and A. Klug. Nov. 1978. The assembly of a virus. *Sci. Amer.* **239**:62.

Coffin, J.M., et al. 1981. Proposal for naming host cell–derived inserts in retrovirus genomes. *J. Virol.* **40**:953.

de Martinsville, B., et al. 1983. Oncogene from human EJ bladder carcinoma is located on the short arm of chromosome 11. *Science* **219**:498.

DeRobertis, E.M., and J.B. Gurdon. Dec. 1979. Gene transplantation and the analysis of development. *Sci. Amer.* **241**:74.

Eva, A., et al. 1982. Cellular genes analogous to retroviral *onc* genes are transcribed in human tumour cells. *Nature* **295**:116.

Gluecksohn-Waelsch, S. 1979. Genetic control of morphogenetic and biochemical differentiation: Lethal albino deletions in the mouse. *Cell* **16**:1.

Hadorn, E. Nov. 1968. Transdetermination in cells. *Sci. Amer.* **219**:110.

Hartwell, L.H. 1978. Cell division from a genetic perspective. *J. Cell Biol.* **77**:627.

Hayward, W.S., B.G. Neel, and S.M. Astrin. 1981. Activation of a cellular *onc* gene by promoter insertion in ALV-induced lympoid leukosis. *Nature* **290**:475.

Hynes, R. 1982. Phosphorylation of vinculin by $pp60^{src}$: What might it mean? *Cell* **28**:437.

Jarvik, J., and D. Botstein. 1973. A genetic method for determining the order of events in a biological pathway. *Proc. Nat. Acad. Sci. U.S.* **70**:2046.

Josephs, S.F., et al. 1983. 5′ Viral and human cellular sequences corresponding to the transforming gene of simian sarcoma virus. *Science* **219**:503.

Lewis, E.B. 1978. A gene complex controlling segmentation in *Drosophila*. *Nature* **276**:565.

Markert, C.L., and R.M. Petters. 1978. Manufactured hexaparental mice show that adults are derived from three embryonic cells. *Science* **202**:56.

Payne, G.S., J.M. Bishop, and H.E. Varmus. 1982. Multiple arrangements of viral DNA and an activated host oncogene in bursal lymphoma *Nature* **295**:209.

Santos, E., et al. 1982. T24 human bladder carcinoma oncogene is an activated form of the normal human homologue of BALB- and Harvey-MSV transforming genes. *Nature* **298**:343.

Shepard, J.F. May 1982. The regeneration of potato plants from leaf-cell protoplasts. *Sci. Amer.* **246**:154.

Stent, G.S. Sept. 1972. Cellular communication. *Sci. Amer.* **227**:42.

Ursprung, H., and R. Nothiger, eds. 1972. *The Biology of Imaginal Disks*. New York: Springer-Verlag.

Wong-Staal, F., et al. 1981. The v-*sis* transforming gene of simian sarcoma virus is a new *onc* gene of primate origin. *Nature* **294**:273.

Wood, W.B., et al. 1968. Bacteriophage assembly. *Fed. Proc.* **27**:1160.

# CHAPTER 15

# Population Genetics

The application of Mendelian principles to the analysis of the genetic structure of natural populations comes under the general heading of population genetics. The major focus in genetic analysis at the population level is to gain perspective on those factors and forces that shape the evolutionary patterns in biological systems. Population genetics provides models and analytical tools by which the genetic composition of populations can be measured and described in quantitative terms based on mathematical analysis. Although population genetics is based in sophisticated mathematics, we can gain an appreciation or flavor of this approach to understanding evolutionary processes in biological systems.

In this chapter we will first examine the conditions for equilibrium gene frequencies, and then we will see how specific factors influence changes in these frequencies and thereby influence evolution. In the next chapter we will pursue some of these same topics in more depth and will introduce more qualitative features of the evolutionary process.

## Gene Frequencies

At the beginning of this century there were various misconceptions about the relationship of Mendelian ratios found by genetic analysis and the actual proportions of different phenotypes and genotypes observed in natural populations. If brachydactyly is inherited as a dominant trait, why isn't 75% of the human population characterized by this condition and only 25% normal? If brown eyes is the dominant phenotype and blue eyes the recessive phenotype, why are some populations entirely brown-eyed and others mostly blue-eyed? Should we expect populations eventually to consist of 75% dominants and 25% recessives for inherited traits under single-gene control? If that were the case, how could Mendelian genes be responsible for biological evolution when changes in inherited traits obviously occurred over long periods of time? The Mendelian 3:1 phenotypic ratio did not seem to explain events or observations in natural populations.

These perplexing questions had to be resolved to determine whether Mendelian inheritance could be applied to events in the natural world as well as in the laboratory and to see if Mendelian genes were the factors responsible for biological variability during the course of evolution. The methods and concepts of population genetic analysis are concerned with measurements of gene frequencies in populations and with interpreting the meaning of these frequencies in evolutionary terms. The core of population genetics has been developed mathematically, and models that embody population theory have been tested in experimental and natural populations.

### 15.1 Populations and Gene Pools

We shift our perspective from individual organisms and their individual combinations of alleles to studies of *populations,* or collections of organisms as units, and of their combined genetic information content, or **gene pool.** Broadly defined, populations are aggregates of individuals sharing one or more features, such as living space. In genetic terms, we restrict our concept of a population to a collection of freely interbreeding individuals. In this sense we confine ourselves to groupings of members of the same species or even to the entirety of a species since members of different species ordinarily do not or cannot interbreed. Similarly, the sum total of genes shared by a population makes up the dynamic reservoir of information that is dealt out during reproduction to produce genotypes that lead to phenotypes, by which we can measure the population. Gametes of one generation produce zygotes of the next generation, and these zygotes develop into individuals who in turn produce new gametes and new zygotes. The gene pool is reconstituted in each generation as alleles are sorted and recombined into gametes at meiosis and into zygotes at fertilization. The population and the species persist generation after generation as long as individuals draw from and contribute to the gene pool through their reproductive activities.

Through studies of the gene pool or of particular genes within this reservoir, we take the first steps in analyzing the evolution of the population. Evolution involves genetic change with time and depends on the transmission of these inherited features to successive generations. Evolution therefore depends on genetic variability; and by analyzing genetic variability within and between populations we gain leads with which to assess the occurrence and nature of biological evolution.

## 15.2 Equilibrium Frequencies

When we analyze a laboratory population, we study phenotypes and describe the genetic situation according to ratios or proportions of phenotypes and genotypes. Suppose we study homozygous parents *AA* and *aa*, which mate and produce a uniform $F_1$ progeny all of which have the genotype *Aa*. If these $F_1$ individuals interbreed at random to produce $F_2$ progeny whose genotypic and phenotypic ratios are 0.25 *AA*:0.50 *Aa*:0.25 *aa*, we can deduce that the inheritance pattern is based on a pair of codominant alleles of an autosomal gene. The 1:2:1 ratio of genotypes and phenotypes is an outcome of chance combinations of gametes carrying the *A* or *a* allele, and we assume that all possible matings occur at random when we find this particular ratio. We don't always pay attention to the fact, however, that the ratio of genotypes also reflects the particular frequencies, or proportions, of the two alleles in the parental genotypes and in the gametes they produce (Fig. 15.1).

The frequencies of the two alleles in the gene pool are equivalent to the probabilities of gametes carrying these alleles. If the alleles are present in equal frequencies in the gene pool, then, on the average, half the gametes carry *A* (0.5) and half carry *a* (0.5). When all possible gamete fusions take place between *A* and *a* gametes produced by both males and females, the next generation will consist of the three possible genotypes in the proportions predicted by the binomial $(0.5 + 0.5)^2$; the $F_2$ ratio will be 0.25 *AA*:0.50 *Aa*:0.25 *aa*. Using the same binomial method, we can calculate that the $F_2$ gene pool with 0.5 *A* and 0.5 *a* frequencies will give rise in turn to an $F_3$ population in which the same three genotypes will be produced in the same ratio in which they appeared in the $F_2$. We can verify these calculations using the more tedious method of listing all the possible $F_3$ genotypes that can be produced (Table 15.1).

These calculations reveal that an equilibrium has been established after one generation of random mating among *Aa* individuals of the $F_1$ generation, that is, genotypic ratios show no

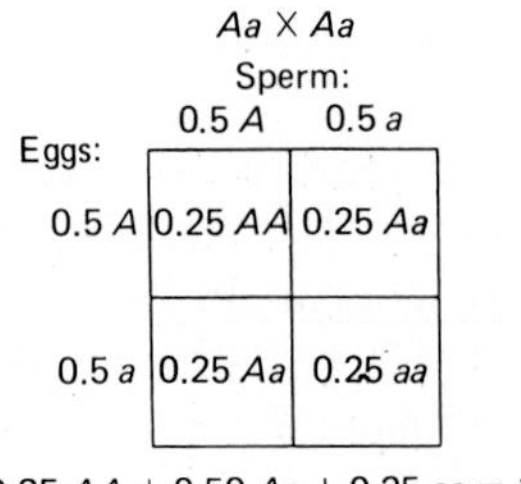


***Figure 15.1*** The frequency of genotypic classes of progeny is an outcome of the relative frequencies of gametes carrying alleles *A* and *a* in the cross *Aa* × *Aa*. In this case the frequency of $A = a = 0.5$.

***Table 15.1*** Frequencies of matings and offspring in a randomly mating $F_2$ population where $p(A) = q(a) = 0.5$ and the genotypic frequencies are 0.25 *AA*:0.50 *Aa*:0.25 *aa*.

| type of $F_2$ mating | frequency of $F_2$ mating | frequency of $F_3$ genotypic classes *AA* | *Aa* | *aa* |
|---|---|---|---|---|
| *AA* × *AA* | ¼ × ¼ = 1⁄16 | 1⁄16 | 0 | 0 |
| *AA* × *Aa* | ¼ × ½ = ⅛ | 1⁄16 | 1⁄16 | 0 |
| *Aa* × *AA* | ½ × ¼ = ⅛ | 1⁄16 | 1⁄16 | 0 |
| *AA* × *aa* | ¼ × ¼ = 1⁄16 | 0 | 1⁄16 | 0 |
| *aa* × *AA* | ¼ × ¼ = 1⁄16 | 0 | 1⁄16 | 0 |
| *Aa* × *Aa* | ½ × ½ = ¼ | 1⁄16 | 2⁄16 | 1⁄16 |
| *Aa* × *aa* | ½ × ¼ = ⅛ | 0 | 1⁄16 | 1⁄16 |
| *aa* × *Aa* | ¼ × ½ = ⅛ | 0 | 1⁄16 | 1⁄16 |
| *aa* × *aa* | ¼ × ¼ = 1⁄16 | 0 | 0 | 1⁄16 |
| | | 4⁄16 | 8⁄16 | 4⁄16 |

$$F_3\ p^2 + 2pq + q^2 = 1$$

$$0.25\ AA + 0.50\ Aa + 0.25\ aa = 1$$

$$p(A) = 0.5 \qquad p + q = 1$$

$$q(a) = 0.5$$

further change in later generations. The frequencies of the two alleles and the three genotypes they produce will remain unchanged through an infinite number of generations since $(0.5 + 0.5)^2$ will continue to define the gene pool for this gene. The population will remain in equilibrium as long as the same conditions prevail: (1) large population size, minimizing sampling errors; (2) random mating; (3) equal success in survival and reproduction of all the genotypes; (4) no mutation at the gene locus; and (5) no genotypes entering or leaving the population.

Any large, randomly mating population should theoretically attain equilibrium frequencies for any pair of alleles of a gene as long as the specified conditions exist since meiosis and fertilization events will lead to segregations and assortments for any pairs of alleles in the same manner. Since this is the case, we can generalize these observations and use the principle established to characterize a population in genetic terms, according to the frequencies of alleles and genotypes produced in random matings (Fig. 15.2).

If we have only two alleles for a gene locus, their combined frequencies must add up to 100%, or 1.0. The frequency of one allele can

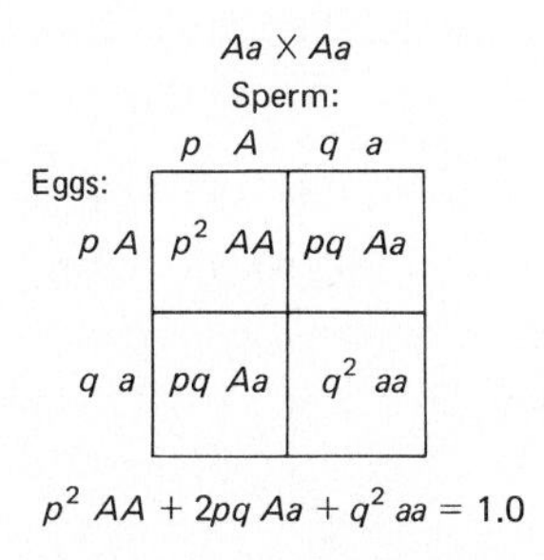


***Figure 15.2*** In general terms, the relative frequencies of genotypic classes arising from crosses in a population depends on the frequency of the alleles at that gene locus in the population.

then be designated as $p$ and the frequency of the other allele must be $1.0 - p$ or $q$ for easier reference. The genotypic class frequencies arise from $(p + q)^2 = 1.0$ and are the terms of the expanded binomial, or

$$p^2 + 2pq + q^2 = 1.0$$

If we analyze a population and find that $(p + q)^2 = p^2 + 2pq + q^2 = 1.0$, we can be reasonably sure that the observed genotypic frequencies provide a true indication of the frequencies of the two alleles in the gene pool of the population. This principle is the **Hardy-Weinberg law,** named for the two men who first pointed out the generalization in independent publications in 1908. Their theoretical proposition marked the first step toward population analysis in genetic terms, or **population genetics.**

Using $(p + q)^2 = 1.0$, we can show theoretically that an equilibrium will be established after one generation of random mating between individuals that produce both kinds of gametes in any pair of frequencies that add up to unity (Table 15.2). We can substitute any values other than 0.5 and 0.5 for $p$ and $q$. An equilibrium will be established regardless of the particular allelic

***Table 15.2*** Frequencies of random matings and progeny types produced in a population at equilibrium for alleles $A$, $a$, where $[p\,(A) + q\,(a)]^2 = p^2\,(AA) + 2pq\,(Aa) + q^2\,(aa) = 1$

| **type of mating** | **frequency of mating** | **frequencies of progeny types** | | |
|---|---|---|---|---|
| | | ***AA*** | ***Aa*** | ***aa*** |
| $p^2\,(AA) \times p^2\,(AA)$ | $p^4$ | $p^4$ | | |
| $p^2\,(AA) \times 2pq\,(Aa)$ | $2p^3q$ | $p^3q$ | $p^3q$ | |
| $2pq\,(Aa) \times p^2\,(AA)$ | $2p^3q$ | $p^3q$ | $p^3q$ | |
| $p^2\,(AA) \times q^2\,(aa)$ | $p^2q^2$ | | $p^2q^2$ | |
| $q^2\,(aa) \times p^2\,(AA)$ | $p^2q^2$ | | $p^2q^2$ | |
| $2pq\,(Aa) \times 2pq\,(Aa)$ | $4p^2q^2$ | $p^2q^2$ | $2p^2q^2$ | $p^2q^2$ |
| $2pq\,(Aa) \times q^2\,(aa)$ | $2pq^3$ | | $pq^3$ | $pq^3$ |
| $q^2\,(aa) \times 2pq\,(Aa)$ | $2pq^3$ | | $pq^3$ | $pq^3$ |
| $q^2\,(aa) \times q^2\,(aa)$ | $q^4$ | | | $q^4$ |
| | | $p^4 + 2p^3q + p^2q^2$ | $2p^3q + 4p^2q^2 + 2pq^3$ | $p^2q^2 + 2pq^3 + q^4$ |

| $AA$ | $Aa$ | $aa$ |
|---|---|---|
| $p^4 + 2p^3q + p^2q^2$ | $2p^3q + 4p^2q^2 + 2pq^3$ | $p^2q^2 + 2pq^3 + q^4$ |
| $= p^2(p^2 + 2pq + q^2)$ | $= 2pq(p^2 + 2pq + q^2)$ | $= q^2(p^2 + 2pq + q^2)$ |
| $= p^2(p + q)^2$ | $= 2pq(p + q)^2$ | $= q^2(p + q)^2$ |
| $= p^2 \times 1^2$ (since $p + q = 1$) | $= 2pq \times 1^2$ | $= q^2 \times 1^2$ |
| $= p^2$ | $= 2pq$ | $= q^2$ |

$(p + q)^2 = p^2 + 2pq + q^2 = 1$ in each generation afterward

and, $p\,(A) + q\,(a) = 1$ in each generation afterward

***Table 15.3*** Frequencies of random matings and progeny types for a population at equilibrium for alleles $A$, $a$, where $(p(A) + q(a))^2 = (0.9 + 0.1)^2 = 0.81\ AA + 0.18\ Aa + 0.01\ aa = 1$.

| | | **frequencies of progeny types** | | |
|---|---|---|---|---|
| **type of mating** | **mating frequency** | ***AA*** | ***Aa*** | ***aa*** |
| 0.81 *AA* × 0.81 *AA* | 0.6561 | 0.6561 | 0 | 0 |
| 0.81 *AA* × 0.18 *Aa* | 0.1458 | 0.0729 | 0.0729 | 0 |
| 0.18 *Aa* × 0.81 *AA* | 0.1458 | 0.0729 | 0.0729 | 0 |
| 0.81 *AA* × 0.01 *aa* | 0.0081 | 0 | 0.0081 | 0 |
| 0.01 *aa* × 0.81 *AA* | 0.0081 | 0 | 0.0081 | 0 |
| 0.18 *Aa* × 0.18 *Aa* | 0.0324 | 0.0081 | 0.0162 | 0.0081 |
| 0.18 *Aa* × 0.01 *aa* | 0.0018 | 0 | 0.0009 | 0.0009 |
| 0.01 *aa* × 0.18 *Aa* | 0.0018 | 0 | 0.0009 | 0.0009 |
| 0.01 *aa* × 0.01 *aa* | 0.0001 | 0 | 0 | 0.0001 |
| | | 0.81 | 0.18 | 0.01 |

and $p(A) = 0.9$

$q(a) = 0.1$

$p(A) + q(a) = 0.9 + 0.1 = 1.0$

frequencies as long as the specified conditions prevail in the population as listed earlier (Table 15.3).

According to this principle, we can theoretically describe the genetic structure of a population according to allelic frequencies in the gene pool and the genotypic frequencies produced in each generation. If these frequencies are in equilibrium, we can predict that the population will not change in subsequent generations. But simply finding that the three genotypes exist in the proportions that fit the Hardy-Weinberg formula is not adequate evidence that the population exists in an unchanging equilibrium. It shows that the allelic frequencies can be calculated, but the allelic frequencies themselves may change in later generations if every specified condition for an equilibrium population is not satisfied. In other words, we have a method by which we can characterize the gene pool at some moment in time, but whether this same gene pool will continue to exist in subsequent generations can only be determined by examining the population in detail. The population must be large, randomly mating, experience little or no mutation, have no selective advantage among the several genotypes, and be closed to migration of genotypes into or out of the population. Before we proceed to distinguish changing and unchanging populations, however, we should see whether the Hardy-Weinberg principle can be used to describe natural populations.

## 15.3 Allelic Frequencies in Natural Populations

We can look at the distribution of the MN blood groups in human populations to see how the Hardy-Weinberg principle can be applied. The MN blood group gene exists in two codominant allelic forms, $L^M$ and $L^N$, so all three genotypes

***Table 15.4*** Observed and expected numbers and frequencies of MN blood types in a sample population of 613 white Americans.

| | | **numbers and percentages of blood types** | | | **allelic frequencies calculated** | |
|---|---|---|---|---|---|---|
| | | **M** | **MN** | **N** | $p(L^M)$ | $q(L^N)$ |
| observed | No. | 179 | 304 | 130 | 0.54 | 0.46 |
| | % | 29.2 | 49.6 | 21.2 | | |
| expected | No. | 178 | 306 | 129 | | |
| | % | 29 | 50 | 21 | | |

can be identified unambiguously according to serological cross-reactions with anti-M and anti-N serum preparations applied to a small sample of blood. In addition to the advantage of recognizing each genotype from phenotypes of members of the population, we expect people to select mates at random with regard to their MN blood group. In fact, most people don't know their MN blood type, so matings will take place at random in relation to these alleles and genotypes. As far as we know, there is no particular selective advantage of one genotype over another in the case of this gene locus.

Since the alleles are codominant, we can calculate the frequencies of the $L^M$ and $L^N$ alleles directly from genotypic (phenotypic) frequencies in a sample of one or more human population groups. In a sample of 613 white Americans, 179 people were of blood group M, 304 of blood group MN, and 130 of blood type N (Table 15.4). We can estimate the frequencies of the $L^M$ and $L^N$ alleles in this population sample in several ways:

***1.*** People of blood type M have two $L^M$ alleles in their genotype and people with blood type MN have only one $L^M$ allele, so the frequency of the $L^M$ allele in this group of 613 individuals can be calculated as

$$\frac{\text{number M individuals} + \tfrac{1}{2}\text{ number MN individuals}}{\text{total individuals in sample}}$$

$$= \frac{179 + \tfrac{1}{2}(304)}{613} = \frac{331}{613} = 0.54, \text{ or } 54\%\ L^M$$

and the percentage of the $L^N$ allele = 100% − 54% = 46%. The frequency of the $L^N$ allele, of course, can be calculated in the same way as that for $L^M$.

***2.*** The calculated percentages of the three genotypes can be used instead of the actual numbers of individuals:

$$\text{percentage M} + \tfrac{1}{2}\text{ percentage MN} = \text{percentage } L^M \text{ allele}$$

$$29.2\% + \tfrac{1}{2}(49.6\%) = 29.2 + 24.8 = 54\%\ L^M$$

$$100\% - 54\% = 46\%\ L^N$$

***3.*** The frequency of each genotype is the product of the separate frequencies of alleles $L^M$ and $L^N$, or the product of the probability that a gamete carrying one allele will fuse with a gamete carrying the same or a different allele, expressed by $(p + q)^2 = 1.0$. Since $p^2$ is the product of $p \times p$, we can take the square root of $p^2$ and derive the value for $p$. Similarly, $q^2 = q \times q$, so we can find $q$ from the square root of $q^2$. Once we know $p$ or $q$, we simply subtract that value from 1.0 to find the value of the other allele.

$p^2\ (L^M L^M) = 0.292$, and $p = \sqrt{0.292} = 0.54$;
and $q = 1.0 - 0.54 = 0.46$; or
$q^2(L^N L^N) = 0.212$ and $q = \sqrt{0.212} = 0.46$

Using $p = 0.54$ and $q = 0.46$, we can calculate the percentages of the three genotypes expected in a sample of 613 people, according to $p^2 + 2pq + q^2 = 1.0$.

$$p^2 = (0.54)^2 = 0.29;$$
$$613 \times 0.29 = 178\ L^M L^M \text{ individuals}$$
$$2pq = 2(0.54 \times 0.46) = 0.50;$$
$$613 \times 0.5 = 306\ L^M L^N \text{ individuals}$$
$$q^2 = (0.46)^2 = 0.21;$$
$$613 \times 0.21 = 129\ L^N L^N \text{ individuals}$$

The observed and expected genotypic frequencies are essentially identical. These values are usually compared using the chi-square test, although the test is unnecessary in this case since the numbers are so similar.

From this analysis we can conclude that the frequencies of the two alleles in the gene pool of this population are 54% $L^M$ and 46% $L^N$, since the observed genotypic frequencies were those expected from the Hardy-Weinberg formula. Since we are not entirely certain that every specified condition for equilibrium actually characterizes the population, we cannot be certain that these allelic frequencies are in equilibrium. We can predict, however, that the same genotypic frequencies will be found in subsequent generations sampled if the population is in equilibrium for these allelic frequencies. If we found a different pair of frequencies, however, we would know that not all the conditions specified for an equilibrium population were satisfied. We could then look for those factors that influenced a change in allelic frequencies. Since changes in allelic frequencies characterize biological evolution, we would direct our analysis toward the factor or factors influencing such evolutionary change in the population. Population genetics therefore serves as a first step in the genetic characterization of a population and leads to further studies of its evolutionary pattern.

We can also see from this analysis that a genotypic ratio of 0.25:0.50:0.25 would arise only if the two alleles were present in equal frequency. The genotypic ratio of 0.29:0.50:0.21 is the direct outcome of the allelic frequencies $(0.54\ L^M + 0.46\ L^N)^2 = 1.0$. Will these same allelic frequencies characterize any population of human beings with regard to the MN blood group gene? We can analyze different groups of people to answer this question (Table 15.5).

We all know that different groups within a species, country, or region tend to choose partners from within their group more often than from the rest of the population. The human species actually consists of many subpopulations, or *Mendelian populations,* according to data for allelic frequencies of a gene in various subpopulations. In fact, variations in the absolute frequencies of alleles for the same gene in different populations of a species have provided the quantitative measurements for variations within the gene pool shared by a whole species. All members of a species are *potentially* interbreeding, but truly random mating does not usually characterize the whole species.

The frequencies of the $L^M$ and $L^N$ alleles in the Navaho Indian and the Australian Aborigine population samples are strikingly different from one another, and both are different from the 54% and 46% frequencies found in the white

***Table 15.5*** Numbers and frequencies of MN blood types observed in samples of two human populations.

| | | | numbers and percentages of blood types | | | allelic frequencies | |
|---|---|---|---|---|---|---|---|
| **population** | **number in sample** | | **M** | **MN** | **N** | **$p(L^M)$** | **$q(L^N)$** |
| aborigines (Australia) | 730 | No. | 22 | 216 | 492 | 0.178 | 0.822 |
| | | % | 3.0 | 29.6 | 67.4 | | |
| Navaho Indians (U.S.) | 361 | No. | 305 | 52 | 4 | 0.917 | 0.083 |
| | | % | 84.5 | 14.4 | 1.1 | | |

American population sample. Yet, each satisfies the Hardy-Weinberg formula. What do these differences in genotypic and allelic frequencies mean? We can find out by applying the Hardy-Weinberg formula to each of these two groups as we did for the first population.

If the observed genotypic frequencies coincide with frequencies estimated by expansion of $(p + q)^2 = 1.0$, where $p$ and $q$ have been calculated from the observed phenotypes, we will know that the calculated allelic frequencies represent the true proportion of the two alleles in the gene pools of these populations. The different proportions of the two alleles produce the different genetic compositions of the three human population groups. We expect, of course, that mating is random with respect to the alleles or genotypes of the individuals, and no genotype has a selective advantage in any of these cases.

For the Aborigine population we expect to find that

$$(0.178\ L^M + 0.822\ L^N)^2 = 0.0317\ L^M L^M + 0.2926\ L^M L^N + 0.6757\ L^N L^N$$

By comparing the numbers or percentages of the three genotypes (phenotypes) in the sample of 730 individuals with the expected percentages, we find a very close agreement between the two sets of values (Table 15.6).

The same procedure will show that the numbers expected in the Navaho population sample are in very close agreement to those actually observed, based on the equation:

$$(0.917\ L^M + 0.083\ L^N)^2 = 0.8409\ L^M L^M + 0.1522\ L^M L^N + 0.0069\ L^N L^N$$

Once again, the observed and expected numbers of genotypes can be evaluated by the chi-square test to judge the extent of numerical agreement according to the Hardy-Weinberg principle. The numbers in this case are sufficiently close, however, to see by inspection that such agreement does in fact exist.

In these populations we have seen that the absolute genotypic or phenotypic ratio may vary across a rather broad range and that each ratio is a reflection of the frequencies of the two alleles of the gene locus. These relationships can be plotted in the form of a graph. From such a graph we can see at a glance what genotypic ratios to expect for a given pair of allelic frequencies or what allelic frequencies we can expect to find for a particular genotypic ratio (Fig. 15.3). Indeed, if we know only one of the allelic frequencies or only one of the genotypic frequencies, we can find the other values directly from the graph. The graph is another way of expressing the Hardy-Weinberg formula of $(p + q)^2 = p^2 + 2pq + q^2 = 1.0$. The validity of these relationships has been established through studies of many gene loci in many populations for different species of sexually reproducing diploid organisms.

Upon close examination of this graph, we see that different genotypic frequencies can arise for the same pair of allelic frequencies. We already

***Table 15.6*** Observed and expected frequencies of $L^M$ and $L^N$ alleles and the three possible phenotypes (= genotypes) in two human populations (see also Table 15.5).

| | | **number of phenotypes observed** | | | **allelic frequencies** | | **number of phenotypes expected** | | |
|---|---|---|---|---|---|---|---|---|---|
| **population** | **number in sample** | **M** | **MN** | **N** | $p(L^M)$ | $q(L^N)$ | **M** | **MN** | **N** |
| aborigines (Australia) | 730 | 22 | 216 | 492 | 0.178 | 0.822 | 23 | 214 | 493 |
| Navaho Indians (U.S.) | 361 | 305 | 52 | 4 | 0.917 | 0.083 | 304 | 55 | 2 |

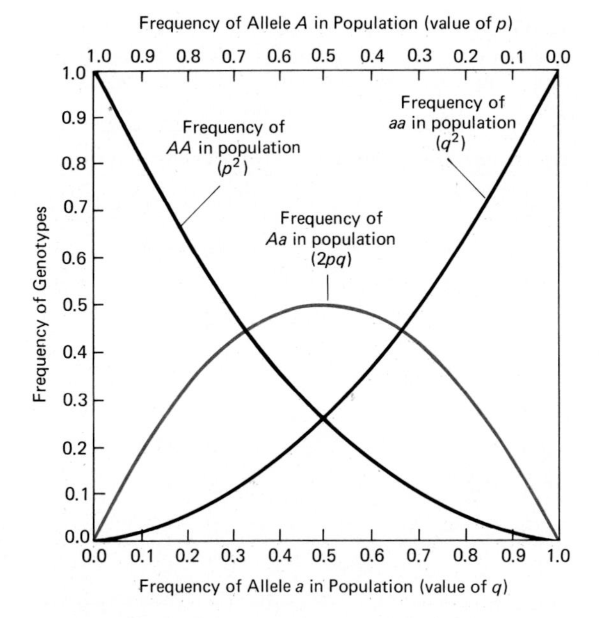


***Figure 15.3*** Relationships between the frequencies of alleles $A$ and $a$ (in values of $p$ and $q$, respectively) and the frequencies of the genotypes $AA$, $Aa$, and $aa$ as predicted by the Hardy-Weinberg formula.

know this from our first example of an initial parental population made up only of $AA$ and $aa$ individuals, which produced $F_1$, $F_2$, and $F_3$ generations.

| | genotypic frequencies | | | allelic frequencies | |
|---|---|---|---|---|---|
| | $p^2(AA)$ | $2pq(Aa)$ | $q^2(aa)$ | $p(A)$ | $q(a)$ |
| P | 0.50 | 0 | 0.50 | 0.5 | 0.5 |
| $F_1$ | 0 | 1.00 | 0 | 0.5 | 0.5 |
| $F_2$ | 0.25 | 0.50 | 0.25 | 0.5 | 0.5 |
| $F_3$ | 0.25 | 0.50 | 0.25 | 0.5 | 0.5 |

If observed and expected genotypic frequencies coincide, we know that these proportions can be used to describe the allelic frequencies in the gene pool. From an evolutionary viewpoint the frequencies of alleles and genotypes, however, can only be interpreted as being in equilibrium or undergoing change if we have information about more than one generation. If we can characterize the population in terms of its size, mating pattern, selective advantages of the several genotypes, mutation at the gene locus, and migration of genotypes, then we can use gene frequencies to describe the evolutionary status of the population. Since we do not expect most populations to conform to the specified equilibrium conditions, we do not expect most populations to be in a state of evolutionary equilibrium or state of no change in allelic frequencies. We can therefore characterize populations from an evolutionary standpoint according to the nature of the gene pool in different generations or according to the particular factors that lead to changes in the gene pool over time. Most of these studies are theoretical and are based on mathematical models that permit us to make predictions about genetic changes taking place in populations during their evolution. These points will be illustrated shortly. The main point to be made now is that we can use mathematical models to describe a particular population feature genetically. We must know a great deal more about the population, however, before we can describe it in evolutionary terms or even in complete genetic terms.

Although we have looked at only one particular gene locus in one species and discussed others in general, it should be clear that we cannot determine dominance, codominance, or recessiveness of alleles from studies of their frequencies in natural populations. In the case of the MN blood group gene, we saw that very different frequencies existed for the same two alleles in different populations. It would be nonsense to conclude that the two alleles were codominant in the white American population because their frequencies were approximately equal but that $L^M$ was dominant among Navahos and recessive among Australian aborigines on the basis of the observed frequencies. We can only discover the inheritance pattern from breeding or family studies of gene transmission and allelic behavior in phenotypic expression, not from population genetics. Population genetics is not an alternative to conventional genetic analysis of inheritance; it is, instead, an area of study in which we can *apply* known Mendelian principles to an analysis of the gene pool in natural populations. We can derive conclusions about the evolutionary situation based partly on assumptions from the mathematical model and partly on assumptions and observations using additional models in subsequent studies.

As both Hardy and Weinberg pointed out in 1908, the frequencies of dominant and recessive phenotypes depend on the frequencies of the alleles of genes in a population. We only expect 75% dominant and 25% recessive phenotypes in large, randomly mating populations where the two alleles occur with equal frequency and one allele is dominant to the other. The population will come to equilibrium for any pair of allelic frequencies, producing a particular proportion of genotypes and phenotypes accordingly, if factors do not intervene to alter these frequencies. Equilibrium can be reached after only one generation of random mating, as long as specified factors do not change. Hardy and Weinberg therefore showed theoretically that Mendelian inheritance was as characteristic of natural populations as it was of controlled populations in the laboratory and that the same rules could be applied in both situations. Many years later, others used the basic Hardy-Weinberg principle to extend the application of Mendelian genetics to analyze biological evolution.

## 15.4 Extensions of Hardy-Weinberg Analysis

We can use the Hardy-Weinberg law just as well in the case of multiple alleles at a gene locus as for two alleles, as long as all other factors remain the same. If we have three alleles at a locus, $A_1$, $A_2$, and $A_3$, with frequencies of $p$, $q$, and $r$, respectively, then $p + q + r = 1.0$. The expected frequencies of all possible genotypes can be derived by expanding the trinomial $(p + q + r)^2$ to predict the kinds and frequencies of genotypes that will be produced by random combinations of all kinds of gametes contributing to the next generation (Table 15.7).

***Table 15.7*** Equilibrium frequencies of genotypes produced by random matings involving three alleles of a gene, occurring in the frequencies of $(p(A_1) + q(A_2) + r(A_3)) = 1$.

| type of mating | frequency of progeny genotypes |
|---|---|
| $p\,A_1 \times p\,A_1$ | $p^2$ $A_1A_1$ |
| $p\,A_1 \times q\,A_2$ | $pq$ $A_1A_2$ |
| $q\,A_2 \times p\,A_1$ | $pq$ $A_1A_2$ |
| $p\,A_1 \times r\,A_3$ | $pr$ $A_1A_3$ |
| $r\,A_3 \times p\,A_1$ | $pr$ $A_1A_3$ |
| $q\,A_2 \times q\,A_2$ | $q^2$ $A_2A_2$ |
| $q\,A_2 \times r\,A_3$ | $qr$ $A_2A_3$ |
| $r\,A_3 \times q\,A_2$ | $qr$ $A_2A_3$ |
| $r\,A_3 \times r\,A_3$ | $r^2$ $A_3A_3$ |
| | $p^2 + 2pq + 2pr + q^2 + 2qr + r^2 = 1$ |

The same principle of establishing equilibrium populations from initial nonequilibrium populations applies to multiple alleles as well as to a pair of alleles at a gene locus. If we find three alleles with the frequencies of $p = 0.3$, $q = 0.5$, and $r = 0.2$, an equilibrium state will be established after one generation of random mating, providing all other conditions support such an equilibrium in a large, randomly mating population.

If all the genotypes are phenotypically distinguishable, we can estimate the frequencies of multiple alleles directly from the observed phenotypic (= genotypic) frequencies. The frequency of the homozygote of any one of the alleles is added to half of the proportion of each class of heterozygotes involving that allele. For three alleles of a gene, we would derive their separate frequencies by

$$p = p^2 + pq + pr$$
$$q = q^2 + pq + qr$$
$$r = r^2 + pr + qr.$$

The major blood group ABO gene in human beings exists in a number of allelic forms, but the three major alleles are $I^A$ and $I^B$, which are

***Table 15.8*** Phenotypes and genotypes of the ABO blood groups in human populations.

| phenotype | genotype | genotypic frequencies |
|---|---|---|
| A | $I^AI^A$ | $p^2$ |
| | $I^Ai^O$ | $2pr$ |
| B | $I^BI^B$ | $q^2$ |
| | $I^Bi^O$ | $2qr$ |
| AB | $I^AI^B$ | $2pq$ |
| O | $i^Oi^O$ | $r^2$ |

codominant, and $i^O$, which is recessive to the other two. For gene frequency analysis we can let

$$p = \text{frequency of } I^A$$
$$q = \text{frequency of } I^B$$
$$r = \text{frequency of } i^O$$
$$p + q + r = 1.0$$

For random mating, the equilibrium frequencies of genotypes are $(p + q + r)^2 = 1.0$ (Table 15.8). We can find the value for $r$ immediately by taking the square root of the frequency of the blood-type O. The values for $p$ and $q$ are derived less directly since homozygotes and heterozygotes are not distinguishable among people of either blood type A or B. Using the symbols, $\overline{A}$, $\overline{B}$, $\overline{AB}$, and $\overline{O}$ to denote the *frequencies* of A, B, AB, and O phenotypes, respectively,

$$\overline{A} + \overline{O} = (p^2 + 2pr) + r^2$$
$$= p^2 + 2pr + r^2$$
$$= (p + r)^2$$
$$\sqrt{\overline{A} + \overline{O}} = p + r$$

and $p + r = 1 - q$, since $p + q + r = 1.0$.
Therefore, $1 - q = \sqrt{\overline{A} + \overline{O}}$

$$q = 1 - \sqrt{\overline{A} + \overline{O}}$$

and similarly, $p = 1 - \sqrt{\overline{B} + \overline{O}}$

In a white American population sample of 1849 individuals, allelic frequencies as follows were calculated from the proportions of the four blood types that were observed:

| | individuals in blood groups | | | |
|---|---|---|---|---|
| | O | A | B | AB |
| No. | 808 | 699 | 259 | 83 |
| % | 43.7 | 37.8 | 14.0 | 4.5 |

| allelic frequencies | | |
|---|---|---|
| $p(I^A)$ | $q(I^B)$ | $r(i^O)$ |
| 0.24 | 0.10 | 0.66 |

The frequency of $i^O$, or $r$, is quickly determined by taking the square root of the blood group O class, $i^Oi^O$, and $\sqrt{0.437} = 0.66$. The calculation of $p$, the frequency of $I^A$, and of $q(I^B)$, are as follows:

$$p = 1 - \sqrt{\overline{B} + \overline{O}} = 1 - \sqrt{0.14 + 0.437} = 0.24$$

$$q = 1 - \sqrt{\overline{A} + \overline{O}} = 1 - \sqrt{0.378 + 0.437} = 0.10$$

and $p + q + r = 0.24 + 0.10 + 0.66 = 1$

These frequencies are not typical of all human populations that have been tested (Table 15.9). Once more we see the subdivision of the

***Table 15.9*** Average frequencies (%) of the ABO blood groups in various human populations.

| | blood groups (% of sample) | | | |
|---|---|---|---|---|
| population | A | B | AB | O |
| English | 42.4 | 8.3 | 1.4 | 47.9 |
| French | 42.3 | 11.8 | 6.1 | 39.8 |
| German | 42.5 | 14.5 | 6.5 | 36.5 |
| Italian | 33.4 | 17.3 | 3.4 | 45.9 |
| Russian | 34.4 | 24.9 | 8.8 | 31.9 |
| American | 37.8 | 14.0 | 4.5 | 43.7 |
| Japanese | 38.4 | 21.9 | 9.7 | 30.0 |
| Chinese | 30.8 | 27.7 | 7.2 | 34.3 |
| Hawaiian | 60.8 | 2.2 | 0.5 | 36.5 |
| aborigines (Australia) | 57.4 | 0.0 | 0.0 | 42.6 |
| Ute Indians (U.S.) | 2.6 | 0.0 | 0.0 | 97.4 |

human species into Mendelian populations, according to the characterizations based on blood group frequencies. These also presumably represent equilibrium frequencies since mating is random with regard to the ABO alleles and, as far as we know, there is little or no selective difference among the four types. To determine whether or not these populations really are in equilibrium with regard to the blood group alleles, we would have to obtain additional information on several generations and on factors such as mutation, migration, and the effects of chance in modulating allelic frequencies.

If we plot these blood group frequencies geographically, we find an interesting gradient of frequencies for the $I^B$ allele in the east-west direction across Europe and Asia (Fig 15.4). The distribution has been correlated with historical information and with other data for human migrations. Allele $I^B$ very likely arose in central Asia by mutation and was spread westward during migrations and invasions by Mongols and other central Asian groups between the sixth and sixteenth centuries.

American Indian populations migrated from Asia over ten thousand years ago, but $I^B$ is rarely found in present-day Indian groups. We could interpret this to mean that the $I^B$ mutation arose after the ancestors of American Indians had left the Asian mainland. Through these kinds of analyses, patterns of human migration and other anthropological interpretations have been made. Population genetics, therefore, provides a tool with which to analyze species history, including human history, as well as more ancient evolutionary events.

In addition to multiple alleles, we can apply the Hardy-Weinberg formula to analyze the genetic structure of populations for two other special cases: (1) alleles of different gene loci and (2) sex-linked genes. In effect, we can show that allelic frequencies follow the Hardy-Weinberg law in virtually any genetic situation we may wish to analyze, regardless of the number of

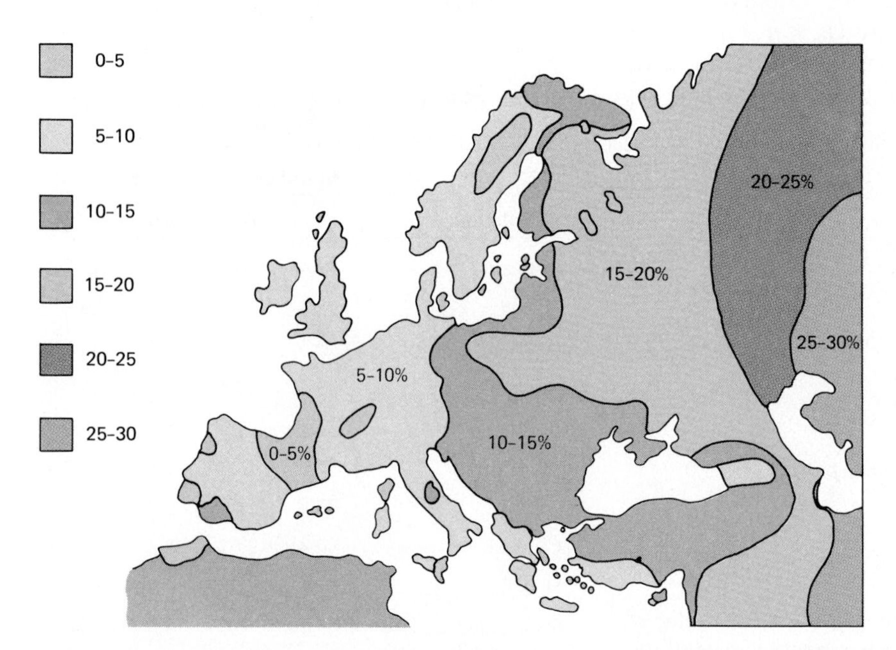


***Figure 15.4*** Frequency of the $I^B$ allele of the ABO blood group locus in populations living in Europe and parts of Africa and Asia. The allelic frequency shows a clearcut east-west gradient, which has been correlated with human migrations and invasions from Asia westward. (From A.M. Winchester. 1979. *Human Genetics,* 3rd ed., p. 197. Columbus, Ohio: Charles E. Merrill Publishing Co.)

genes we study or the particular chromosomes in which these genes may be located.

For alleles of two different genes we simply determine the frequencies of the two pairs of alleles separately for the various genotypes and then multiply these values together to obtain the frequency of any genotypic class. For example, suppose alleles $A$ and $a$ occur in the frequencies of $p_A = 0.6$ and $q_a = 0.4$, respectively, whereas the frequencies for alleles $B$ and $b$ at another gene locus are $p_B = 0.2$ and $q_b = 0.8$, respectively. In each case, since there are only two alleles at a locus, $p + q = 1$. With the separate allelic frequencies calculated, any genotypic class frequency can be determined, as in the following three examples of the nine possible genotypes that can be produced in the population:

$$AABB: (p_A^2)(p_B^2) = (0.6)^2(0.2)^2 = 0.0144 = 1.44\%$$

$$AaBb: (2p_Aq_a)(2p_Bq_b) = 2(0.6 \times 0.4) \times 2(0.2 \times 0.8) = 0.1536 = 15.36\%$$

$$aaBb: (q_a^2)(2p_Bq_b) = (0.4)^2 \times 2(0.2 \times 0.8) = 0.0512 = 5.12\%$$

From appropriate kinds of calculations, it has been shown that equilibrium will be reached at different speeds (numbers of generations) for linked genes and for unlinked genes. But once two pairs of alleles are in equilibrium, we cannot distinguish whether or not the pairs are linked. In other words, traits produced by alleles at linked loci do not show any particular association in equilibrium populations. Linkage cannot be determined directly from allelic frequencies in natural populations, any more than dominance and recessiveness can. Breeding analysis and family studies are required to interpret gene transmission and allelic behavior patterns. Using these known systems, we can analyze allelic frequencies at the population level. If we find that two known linked genes are associated more often than expected for the equilibrium frequencies of the two pairs of alleles, we may assume that a nonequilibrium situation exists and we can search for the factor(s) responsible for the situation.

The frequencies of X-linked alleles can also be analyzed using the Hardy-Weinberg formula, with the difference that females having two X chromosomes may produce three different genotypes for one pair of alleles at a locus, whereas hemizygous males display only two possible genotypes. The expected proportion of X-linked genotypes is $p^2$ $(AA)$ + $2pq(Aa)$ + $q^2(aa) = 1.0$ in populations for XX females and $p(A) + q(a) = 1.0$ for males in these populations. If the population under study is in equilibrium for a pair of X-linked alleles, the allelic frequencies will be the same in both sexes, and $p + q = 1.0$ in both sexes (Fig 15.5). Each sex is treated as a separate subpopulation in these calculations.

In the example shown, where $p = 0.8$ and $q = 0.2$, we can see that the frequency of males with the $a$ genotype and, therefore, expressing the "a" trait is considerably higher than the frequency of homozygous $aa$ females. This explains the fact that many more males than females express X-linked recessive traits, such as color blindness or hemophilia. In a population in equilibrium, the frequency of males with the $a$ genotype should be equal to the square root of the frequency of females showing the $aa$ genotypic condition. Similarly, we should be able to estimate the frequency of females expected to be $aa$ according to the squared frequency of males expressing the $a$ genotype. If 1% of males have

Sperm:

| Eggs: | X-carrying $p(A)$ 0.8 | Y carrying $q(a)$ 0.2 | |
|---|---|---|---|
| $p(A)$ 0.8 | $p^2(AA)$ 0.64 | $pq(Aa)$ 0.16 | $p$ $A$ 0.8 |
| $q(a)$ 0.2 | $pq(Aa)$ 0.16 | $q^2(aa)$ 0.04 | $q$ $a$ 0.2 |
| | Daughters | | Sons |

Genotype frequencies
Daughters: $0.64\,AA + 0.32\,Aa + 0.04\,aa = 1.0$
Sons: $0.8\,A + 0.2\,a = 1.0$

***Figure 15.5*** Hardy-Weinberg equilibrium frequency for two X-linked alleles is evident from the fact that $p(A)$ and $q(a)$ are the same in both sexes.

the recessive genotype, then we would expect $(0.01)^2 = 0.0001$, or 0.01% of females to be of the recessive genotype. If the values for $q$ in males and females are not equal, the population is not in equilibrium. Notice that for $q = 0.01$ in an equilibrium population, 1 in 100 males express the trait, but only 1 in 10,000 females have the recessive phenotype.

In the case of X-linked alleles, as for autosomal alleles which we discussed earlier, dominance and recessiveness cannot be determined directly from allelic frequencies in natural populations. Whether a particular allele is common or rare is controlled by factors other than the dominance or recessiveness of the allele. A dominant allele will not necessarily be more common nor is it destined to become more common by virtue of its dominance. A recessive allele may be common or rare for various reasons, but not because of recessiveness itself. This was part of the concept stated by Hardy and Weinberg and formulated in their simple algebraic equation.

## Changes in Gene Frequencies

Populations whose gene frequencies do not change are nonevolving since we define evolution in biological systems as genetic change with time. The Hardy-Weinberg equilibrium exists only when certain specified conditions exist. Changes or modulations in one or more of these conditions should lead to changes in allelic frequencies and to evolution of the population. The major factors or forces that influence allelic frequencies in the gene pool are:

**1.** *Mutation,* which is recurrent change from one allelic form to another at a gene locus;

**2.** *Selection,* which is responsible for differential reproduction of diverse genetic types in the population;

**3.** *Mating system or pattern,* which may be random or nonrandom and which therefore influences the proportions of genotypes produced in each generation;

**4.** *Migration* of genotypes into and out of the population, which influences the proportions of alleles contributed to the gene pool; and

**5.** *Random genetic drift,* which consists of fluctuations in allelic frequencies as the result of sheer chance, particularly in small populations.

We will see how each of these influences can modulate the gene pool.

### 15.5 Mutation

Mutations are *recurrent* heritable changes from one allelic state to another; forward and reverse mutations usually occur at different rates. Mutation is a force affecting allelic frequencies since mutations of allele $A$ to allele $a$ will slightly reduce the frequency of $A$ and slightly increase the frequency of $a$ in the gene pool; similarly $a \rightarrow A$ will reduce the frequency of $a$ and increase the frequence of $A$. The rate of forward mutation, $u$, and the rate of reverse mutation, $v$,

$$A \xrightarrow{u} a$$

$$A \xleftarrow{v} a$$

are expressed in proportions of alleles mutating per generation. Therefore, the actual proportion of total alleles changing in one generation from $A$ to $a$ will be $pu$, and from $a$ to $A$ it will be $qv$.

The relationship between mutation rates and the frequencies of a pair of alleles in the gene pool can be illustrated as follows.

Suppose the rate of $A \rightarrow a = u = 2 \times 10^{-5}$, and
the rate of $a \rightarrow A = v = 1 \times 10^{-5}$

The increase in $A = pv$ and the decrease in $A = pu$ since the extent of the mutation effect is dependent on the frequency of $A$ and $a$ alleles undergoing mutation. If the initial frequency of the two alleles is

$$p_0 = 0.6 \text{ and } q_0 = 0.4,$$

and the net change in the frequency of $A$, or $\Delta p$, is

$$
\begin{aligned}
\Delta p &= qv - pu \\
\text{then } \Delta p &= (0.4)(1 \times 10^{-5}) - (0.6)(2 \times 10^{-5}) \\
&= (4 \times 10^{-6}) - (12 \times 10^{-6}) \\
&= -8 \times 10^{-6}
\end{aligned}
$$

Therefore, the new frequency of allele $A$, or $p_1$, is the sum of the initial frequency, $p_0$, plus the change in the frequency of $A$, or $\Delta p$,

$$
\begin{aligned}
p_1 &= p_0 + \Delta p \\
\text{then } p_1 &= 0.6 + (-8 \times 10^{-6}) \\
&= 0.599992 \\
\text{and } q_1 &= 1.0 - 0.599992 \\
&= 0.400008
\end{aligned}
$$

Even though the mutation rate from $A$ to $a$ is twice the rate of the reverse mutation, the effect on the gene pool has been very slight in one generation, causing an increase in $a$ of only 8 per million genes in the population.

An equilibrium point will exist when there is no further change in allelic frequencies by mutation alone. This is expressed as

$$\Delta p = pu - qv = 0, \text{ or } \hat{p}u = \hat{q}v$$

where $\hat{p}$ and $\hat{q}$ are equilibrium values. Solving this equation to obtain the equilibrium value of $p$ ($\hat{p}$),

$$
\begin{aligned}
\hat{p}u &= \hat{q}v \\
\hat{p}u &= (1 - \hat{p})v \\
\hat{p}u &= v - v\hat{p} \\
\hat{p}(u + v) &= v \\
\hat{p} &= \frac{v}{u + v}
\end{aligned}
$$

Similarly,

$$\hat{q} = \frac{u}{u + v}$$

The equilibrium value, $\hat{p}$, therefore, depends only on the mutation rates and is independent of the initial value, $p_0$. The same equilibrium point will be reached from any initial value of $p_0$ and $q_0$, including 0 and 1.

For the example given above, the equilibrium point is

$$\hat{p} = \frac{1 \times 10^{-5}}{(2 \times 10^{-5}) + (1 \times 10^{-5})} = 0.333$$

$$\text{and } \hat{q} = 1.0 - 0.333 = 0.667$$

At equilibrium, therefore, there are twice as many $a$ alleles as $A$ alleles in this population, but since $a$ mutates half as often as $A$, the total number of mutations is the same in each direction in this particular case. The *equilibrium* due to opposing mutation rates is dependent on these rates and is independent of initial gene frequencies. The *change* in allelic frequencies, however, is dependent both on the gene frequencies and on the mutation rates.

In general, the effect of mutation *alone* is negligible in directing changes in allelic frequencies and, therefore, on evolution. The significant feature of mutation processes is that *new genetic information* arises and increases variability in the gene pool. Mutation provides the raw material for evolution, but other factors usually provide the significant influences on the speed of evolutionary change.

## 15.6 Selection

The basic concept of **natural selection** is that of *differential reproduction* in genetically diverse populations. The concept was proposed by Charles Darwin and Alfred Russell Wallace in 1858 and was overwhelmingly documented and expounded by Darwin in 1859 in his book *On the Origin of Species*. Any trait that has a genetic component and allows the genotypes expressing this trait to leave proportionately more progeny than other genotypes will tend to increase in frequency in the population as the genes responsible for the trait increase in frequency. The effects of natural selection on gene frequencies can be measured, at least in simple genetic situations.

Suppose we initially have equal numbers of two homozygous types of mice—$AA$ types with black fur and $aa$ types with white fur—and for every 100 black mice that survive and reproduce, only 80 white mice survive and reproduce. We may describe this situation by saying that the **fitness,** $W$, of the white mice is only 80% that of the black mice. If the fitness of black mice is equal to one, the fitness of white mice is

$$W = 1 - s$$

where $s$ is the **selection coefficient.** In this example

$$0.8 = 1 - s \quad \text{or,} \quad s = 0.2$$

From this we see that $s$ is a measure of **selective disadvantage** of the less fit type.

To follow this example further we can include *Aa* heterozygotes as well as the two homozygous genotypes. We will stipulate that *AA* and *Aa* genotypes are equally fit, so that the following situation exists:

| | Genotype | | |
|---|---|---|---|
| | ***AA*** | ***Aa*** | ***aa*** |
| $W$ | 1.0 | 1.0 | 0.8 |
| $s$ | 0 | 0 | 0.2 |

The genotypic frequencies will change after selection since only a fraction of the *aa* genotypes contribute to the gene pool of the next generation whereas all the *AA* and *Aa* genotypes contribute progeny (Table 15.10). The total of genotypic frequencies has changed from $p^2 + 2pq + q^2 = 1$ to $p^2 + 2pq + q^2 - sq^2 = 1 - sq^2$. The fraction $sq^2$ of all the $a$ gametes (alleles) is eliminated by selection against *aa* genotypes. In our example of black and white mice, the initial frequencies, $p_0(A) = 0.6$ and $q_0(a) = 0.4$, have been changed to new frequencies in the new generation, or $p_1(A)$ and $q_1(a)$. To find these new frequencies, we can calculate the change in $p$, or $\Delta p$:

$$\Delta p = \frac{sp_0q_0^2}{1 - sq_0^2} = \frac{(0.2)(0.6)(0.4)^2}{1 - (0.2)(0.4)^2} = \frac{0.0192}{1 - 0.032} = \frac{0.0192}{0.968}$$

$$\Delta p = 0.02$$

$$\text{Thus } p_1 = p_0 + \Delta p = 0.60 + 0.02 = 0.62$$

$$\text{and } q_1 = 0.38$$

Therefore, the genotypic frequencies in the new generation will be derived from the new allelic frequencies, so that $(p_1 + q_1)^2 = p_1^2 + 2p_1q_1 + q_1^2 = 1.0$, or

$$(0.62 + 0.38)^2 = (0.62)^2 + 2(0.62 \times 0.38) + (0.38)^2 = 0.384\,AA + 0.471\,Aa + 0.144\,aa$$

The increase in *AA* occurs at the expense of *aa* genotypes for the most part since the genotypic frequencies were $(p_0 + q_0)^2 = 0.36\,AA + 0.48\,Aa + 0.16\,aa$ in the previous generation.

The effect of selection is not necessarily very great in one generation, but the *effects are cumulative.* We would therefore expect a consistent decline in the frequency of allele $a$ and a proportionate increase in $A$ through a number of generations if conditions do not change to alter the fitness of the three genotypes and the selective disadvantage of genotype *aa.*

To gain some appreciation of the speed of change in allelic frequencies under selection we can look at an extreme case in which the selective disadvantage of genotype *aa* is one. In this case, therefore, $W = 0$ and $s = 1$ for the homo-

***Table 15.10*** Change in genotypic frequencies in one generation, where $s > 0$ for recessive (*aa*) genotypes.

| **genotype** | **initial frequency** | **fitness (W)** | **contribution to the gene pool** |
|---|---|---|---|
| *AA* | $p^2$ | 1 | $1 \times p^2$ |
| *Aa* | $2pq$ | 1 | $1 \times 2pq$ |
| *aa* | $q^2$ | $1 - s$ | $(1 - s)\,q^2 = q^2 - sq^2$ |
| sum: | $p^2 + 2pq + q^2 = 1$ | | $p^2 + 2pq + q^2 - sq^2 = 1 - sq^2$ |

zygous recessives, but both $AA$ and $Aa$ genotypes have a fitness of 1 and no selected disadvantage. Allele $a$, therefore, is a recessive lethal. This time we will start with a population where $p_0 = q_0 = 0.5$, and the three genotypes occur in the proportion of 0.25 $AA$:0.50 $Aa$:0.25 $aa$. The recessives make no contribution to the gene pool, since they are completely selected against. They produce no gametes for the next generation, and therefore, the new frequency of heterozygotes in the next generation arises only from matings between $AA \times Aa$ and $Aa \times Aa$, and the new heterozygote frequency is dependent only on $p_0^2 + 2p_0q_0$ ($q_0^2$ is eliminated). The new heterozygote frequency in the next generation is

$$\frac{2p_0q_0}{p_0^2 + 2p_0q_0}$$

or,

$$\frac{2p_0q_0}{p_0^2 + 2p_0q_0} = \frac{2q_0}{p_0 + 2q_0} = \frac{2q_0}{p_0 + q_0 + q_0} = \frac{2q_0}{1 + q_0}$$

The probability of two heterozygotes mating and producing $aa$ offspring is one-fourth the product of the separate probabilities, or frequencies of the heterozygote, which is

$$\frac{2q_0}{1 + q_0} \times \frac{2q_0}{1 + q_0} \times \tfrac{1}{4} \text{ (proportion of } aa \text{ expected)}$$

$$= \frac{q_0^2}{(1 + q_0)^2}$$

This is the new frequency of $aa$ genotypes, or $q_1^2$

$$q_1^2 = \frac{q_0^2}{(1 + q_0)^2}$$

so that

$$q_1 = \frac{q_0}{1 + q_0}$$

The value for $q_2$ in the second generation produced in this population is

$$q_2 = \frac{q_1}{1 + q_1}$$

or

$$q_2 = \frac{q_0/(1 + q_0)}{1 + [q_0/(1 + q_0)]} = \frac{q_0}{1 + q_0} \times \frac{1 + q_0}{1 + q_0 + q_0}$$

$$= \frac{q_0}{1 + 2q_0}$$

and after $n$ generations,

$$q_n = \frac{q_0}{1 + nq_0}$$

and

$$n = \frac{q_0 - q_n}{q_0q_n}$$

Using these last two equations, we can make predictions about our population as it undergoes change guided by selection against the $aa$ genotypes. If we start with $p_0 = q_0 = 0.5$, we can predict the value of $q$ in 10 generations:

$$q_{10} = \frac{0.5}{1 + 10(0.5)} = 0.083$$

In 10 generations of complete selection against the $aa$ genotypes, the frequency of allele $a$ has declined from its initial value of 0.5 to a value of 0.083. If we use this formula to calculate events over a period of 100 generations, we observe a rapid initial decrease in $q(a)$, but this decrease occurs more and more slowly with time (Table 15.11). If we plot the changes in frequencies of

***Table 15.11*** Changes in the frequency of $q(a)$ over a period of 100 generations, when $s = 1$ for $aa$ genotypes and $s = 0$ for $AA$ and $Aa$ genotypes. The population begins with $p_0 = q_0 = 0.5$ and with 0.25 $AA$ + 0.50 $Aa$ + 0.25 $aa$ as genotypic frequencies.

| **generation($n$)** | **frequency of $q^2(aa)$** | **frequency $q(a)$** |
|---|---|---|
| 0 | 0.25 | 0.50 |
| 1 | 0.11 | 0.33 |
| 2 | 0.06 | 0.25 |
| 3 | 0.04 | 0.20 |
| 4 | 0.03 | 0.17 |
| 5 | 0.02 | 0.14 |
| 10 | 0.007 | 0.08 |
| 20 | 0.002 | 0.05 |
| 50 | 0.0004 | 0.02 |
| 100 | 0.0001 | 0.01 |

all three genotypes over selected intervals for 100 generations, we see that fewer and fewer *aa* genotypes are exposed to selection and most of the *a* alleles are protected against selection in the heterozygous genotypes, which express the dominant phenotype (Fig. 15.6).

Similarly, we can predict how many generations it will take for the initial frequency of the *a* allele to reach some new frequency. For example, if the initial frequency is 0.5, how many generations are required to reduce this frequency to 0.01?

$$n = \frac{0.5 - 0.01}{(0.5)(0.01)} = 98$$

If we round off the number, it will take about 100 generations of complete selection against *aa* genotypes to reduce the frequency of *a* gametes from 50% of the gene pool in the initial population to 1% of the gene pool.

On the basis of these calculations we can make some general observations concerning the role of selection in changing the frequencies of alleles in the gene pool. Selection is not effective if $p$, $q$, or $s = 0$ or very low values. Selection is most effective at intermediate values of $p$, $q$, and $s$. But even when $s = 1$, and every recessive genotype is eliminated from contributing to the gene pool, the effectiveness of selection becomes reduced over time. This is because $q$ becomes smaller and smaller and the lethal recessive allele is protected from selection in the heterozygotes. It should also be obvious that any program of elimination of some harmful allele from the population, as by sterilization or some other eugenic method, will be relatively ineffective if it is based only on elimination of homozygous recessives. This is especially true because most harmful alleles already exist in the population in relatively low frequencies and most of the harmful alleles are carried by heterozygous members of the population.

From Table 15.12, we can also see that relatively rare recessive genotypes are present in populations along with an astonishingly high proportion of heterozygous carriers of the reces-

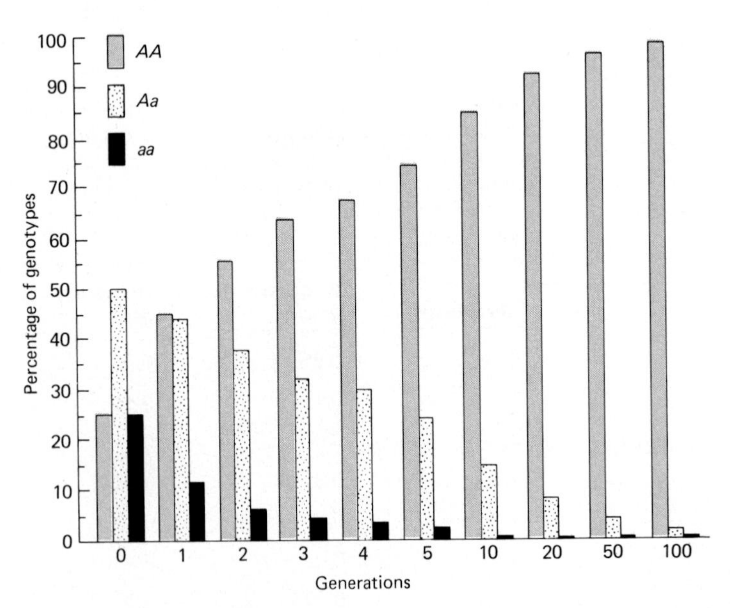


***Figure 15.6*** Course of complete selection against *aa* genotypes in a population that initially was composed of 0.25 *AA*, 0.50 *Aa*, and 0.25 *aa* genotypes. The decrease in frequency of allele *a* becomes slower with time, as most of the *a* alleles are protected in heterozygous genotypes and fewer *aa* genotypes are exposed to selection.

***Table 15.12*** Frequencies of homozygotes and of heterozygous carriers for various recessive disorders in human populations.

| **inherited disorder** | **frequency of homozygotes (*aa*)** | **frequency of carriers (*Aa*)** | **ratio of carriers to homozygous recessives** |
|---|---|---|---|
| sickle-cell anemia | 1 in 400 | 1 in 10 | 40:1 |
| cystic fibrosis | 1 in 1600 | 1 in 20 | 80:1 |
| phenylketonuria (PKU) | 1 in 40,000 | 1 in 100 | 400:1 |
| Tay-Sachs disease | 1 in 100,000 | 1 in 160 | 625:1 |
| alkaptonuria | 1 in 1,000,000 | 1 in 500 | 2000:1 |

sive allele. It is because of this high proportion of heterozygous carriers that selection becomes less and less effective against a harmful allele.

Up to now we have completely ignored the effect of mutation in relation to selection against *aa* genotypes. Although selection removes *a*, mutation from *A* to *a* replaces the recessive allele in the gene pool. The opposing forces of mutation and selection should balance out at some time in the history of the population, when an equilibrium is established between the effects of the two processes. We would expect this intuitively, but we can also provide a mathematical basis for such an equilibrium.

At equilibrium, the change in $p(A)$ can be shown as

$$\Delta p = spq^2 + vq - up = 0$$

If we ignore $vq$, then

$$spq^2 = up$$
$$sq^2 = u$$
$$\hat{q}^2 = \frac{u}{s} \text{ and } \hat{q} = \sqrt{\frac{u}{s}}$$

We can ignore the term $vq$ since the rate of mutation from $a$ to $A$ is usually much lower than the rate of forward mutation and $vq$ is negligible in these cases. The balance at equilibrium depends on $spq^2$, the selective disadvantage of *aa*, and on $up$, the rate of mutation that adds more $a$ alleles to the gene pool. So the frequency at equilibrium of recessive genotypes, $\hat{q}^2$, is a function of $u$ and $s$. Where $s = 1$, the frequency of the recessive genotypes at equilibrium will be equal only to the mutation rate of $A \rightarrow a$, that is, $\hat{q}^2 = u/1 = u$. Values of $s < 1$ will result in higher frequencies of *aa* genotypes and of the $a$ allele in the gene pool at equilibrium. For example, if a recessive lethal genotype ($s = 1$) occurs in 1 per 100,000 births, then

$$\hat{q}^2 = u/s = 10^{-5}/1 = 10^{-5}$$
$$\text{and } \hat{q} = \sqrt{u} = 0.003$$

But, if $s = 0.1$, then

$$\hat{q}^2 = 10^{-5}/10^{-1} = 10^{-4}$$
$$\text{and } \hat{q} = \sqrt{10^{-4}} = 10^{-2}, \text{ or } 0.01$$

By reducing the numbers of *aa* genotypes that are eliminated by selection, there has been a significant increase at equilibrium in the frequency of *aa* genotypes in the population (from $10^{-5}$ to $10^{-4}$) and in the frequency of the $a$ allele (from 0.003 to 0.01). By increasing the frequency of *aa* genotypes tenfold, there has been a 33-fold increase in the frequency of $a$ in the gene pool. Calculations of this kind provide a basis for estimating the results of medical improvements in the treatment of hereditary disorders and for estimating the added load of harmful alleles in the population, or the *genetic load.* Also, since the increase in the frequency of $a$ leads to an increase in the frequency of heterozygotes, there are more carriers of the harmful allele in a population and a higher chance of heterozygotes mating and producing recessive offspring, of whom 10% die before reproduction and the remaining 90% require medical care to live and be able to reproduce.

This can be shown by calculating the frequencies of heterozygotes at equilibrium for $\hat{q} = 0.003$ versus $\hat{q} = 0.01$,

$$2pq = 2(0.997 \times 0.003) = 0.006$$
$$\text{versus } 2pq = 2(0.99 \times 0.01) = 0.02$$

and the frequency of recessives from heterozygote × heterozygote matings is

$$(0.006 \times 0.006) \times 1/4 \text{ (proportion } aa \text{ produced)} = 0.000009$$
$$\text{versus } (0.01 \times 0.01) \times 1/4 = 0.000025$$

In other words, where $s$ has been reduced from 1 to 0.1 the chance for heterozygotes mating and producing recessive offspring has increased from 9 per million to 25 per million in the new equilibrium population.

In these calculations it is obvious that changes in the frequency of recessive genotypes or of the recessive allele are a function of $s$ alone since the mutation rate, $u$, does not change. Furthermore, changes in the frequencies of alleles in the gene pool are influenced predominantly by selection rather than by mutation when $p$, $q$, or $s$ is not zero or some negligible value (Fig. 15.7).

Selection against harmful dominant alleles is obviously much quicker than it is for harmful recessive alleles since the dominant allele is expressed in both the homozygote and the heterozygote. Complete selection against a dominant allele could theoretically be achieved in one generation, provided each allele was expressed phenotypically. All dominant alleles are not necessarily expressed in every genotype since some genes interact in the phenomenon of epistasis and others show reduced penetrance, by which we mean that every genotype containing the allele(s) does not produce the expected phenotype (see Chapter 1). In addition, new dominant alleles arise by recurrent mutation.

Changes in allelic frequencies due to selection against the dominant allele are shown in Table 15.13. Of the alleles eliminated, the proportion of $A$ is $sp^2 + pqs$. At equilibrium, the sum

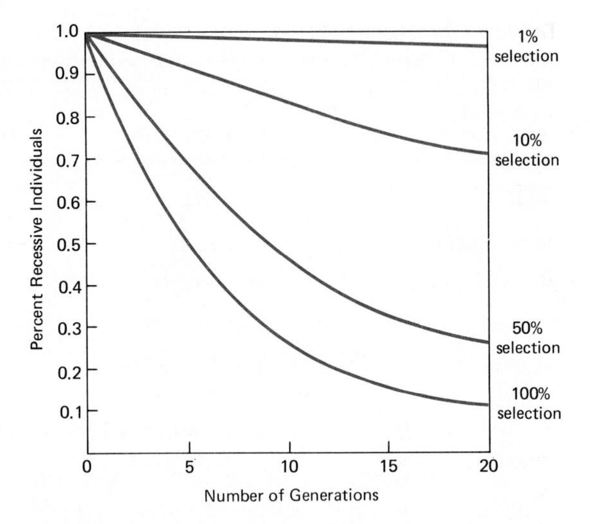


***Figure 15.7*** Individuals with the recessive genotype are eliminated more and more effectively as the intensity of selection increases. Beginning with a population that includes 1% recessives, their frequency is reduced to about 0.1% in 20 generations under a regime of complete selection, but the change is considerably less when the selection pressure is lower.

of these two terms must equal the $A$ alleles gained by mutation, or $vq$. If $A$ is rare and the frequency of $a$ approaches 1, then $sp^2 + pqs = v$, or rearranged, the equation is $p^2 + pq = v/s$. Because $q$ is nearly 1, $p$ must be very small and $p^2$ must be negligible; therefore, we can conclude that the equilibrium state is reached when

$$pq = \frac{v}{s}$$

***Table 15.13*** Changes in genotypic frequencies in populations when there is selection against the dominant phenotypes.

| genotype | initial frequency | fitness (W) | contribution to the gene pool |
|---|---|---|---|
| $AA$ | $p^2$ | $1 - s$ | $p^2 - sp^2$ |
| $Aa$ | $2pq$ | $1 - s$ | $2pq - 2pqs$ |
| $aa$ | $q^2$ | 1 | $q^2$ |
| sum: | 1 | | $1 - (sp^2 + 2pqs)$ |

***Table 15.14*** Changes in genotypic frequencies in populations after one generation of selection against one codominant allele in diploids or against a haploid genotype.

| allele or genotype | frequency | fitness (W) | contribution to the gene pool |
|---|---|---|---|
| $a_1$ | $p$ | 1 | $p$ |
| $a_2$ | $q$ | $1 - s$ | $q - qs$ |
| sum: | 1 | | $1 - qs$ |

In the cases of codominant alleles or haploid genotypes, the force of selection acting against the genotype or the allele is shown in Table 15.14. The loss of allele $a_2$, $qs$, is balanced at equilibrium by the rate of mutation from $a_1 \rightarrow a_2$, which is $pu$, the product of the frequency of allele $a \times$ mutation rate of $a_1$ to $a_2$. For a rare $a_2$ allele, $qs = u$, or

$$q = \frac{u}{s}$$

In these cases, as in the case of recessive and dominant alleles in diploids, equilibrium allelic frequencies can be reached when the forces of selection and mutation are opposable, that is, working in opposite directions. Changes in allelic frequencies, therefore, take place only in nonequilibrium, evolving populations.

## 15.7 Nonrandom Mating

Since one of the basic assumptions of the Hardy-Weinberg law is that of random mating, we would expect alternate mating patterns to contribute to nonequilibrium frequencies of alleles in the gene pool. In many populations individuals show a decided preference for particular partners. For example, tall people may prefer partners who are also tall, or short people may prefer one another, or tall people may choose partners shorter than themselves and vice versa. One obvious example of nonrandom mating is self-fertilization, which is a common mating pattern among plants.

In strictly self-fertilizing plant species, there is a very high level of homozygosity at many gene loci. We can see how this can come about by looking at an example of an initial population composed of *Aa* heterozygotes, which undergo self-fertilizations for five generations (Table 15.15). After an infinite number of generations,

***Table 15.15*** Distribution of genotypes from five generations of self-fertilization, beginning with one *Aa* individual.

| | genotypic frequencies | | | | |
|---|---|---|---|---|---|
| generation | AA | Aa | aa | $q(a)$ | $F^*$ |
| 0 | | 1 | | ½ | 0 |
| 1 | ¼ | ½ | ¼ | ½ | ½ |
| 2 | ⅜ | ¼ | ⅜ | ½ | ¾ |
| 3 | 7/16 | ⅛ | 7/16 | ½ | ⅞ |
| 4 | 15/32 | 1/16 | 15/32 | ½ | 15/16 |
| 5 | 31/64 | 1/32 | 31/64 | ½ | 31/32 |
| $n$ | $\frac{1-(1/2)^n}{2}$ | $(1/2)^n$ | $\frac{1-(1/2)^n}{2}$ | ½ | $1-(1/2)^n$ |
| $\infty$ | ½ | 0 | ½ | ½ | 1 |

*$F$ is the inbreeding coefficient, representing the proportion of heterozygosity lost.

the frequency of the heterozygotes will decrease to zero and the frequencies of the two classes of homozygotes will be 0.5 each. The overall effect of **inbreeding** for any number of generations is a decrease in heterozygosity and an increase in homozygosity. Note that the allelic frequencies do not change; only the genotypic frequencies are altered directionally.

A useful measure of the degree of inbreeding in a population is provided by the **inbreeding coefficient *F*,** proposed by Sewall Wright. This value expresses the proportion of heterozygosity that has been lost and is used to calculate the different genotypic proportions that will arise in different populations depending on the amounts of inbreeding that have occurred (Table 15.16). In human populations as well as those of other species having two separate sexes, inbreeding occurs if the mating partners are related to one another, such as first cousins or second cousins. There is a loss of heterozygosity in systematic matings between relatives, but the loss is much less per generation than under self-fertilization (Fig. 15.8).

Human societies have a general taboo against matings between close relatives. Various factors in human history have led to this observed societal pattern: one of the reasons may have been the higher frequencies of inherited defects appearing among the children of close relatives than among the general population. In western societies, about one marriage in two hundred is between cousins; pairings between other kinds of relatives are found in insignificant frequencies.

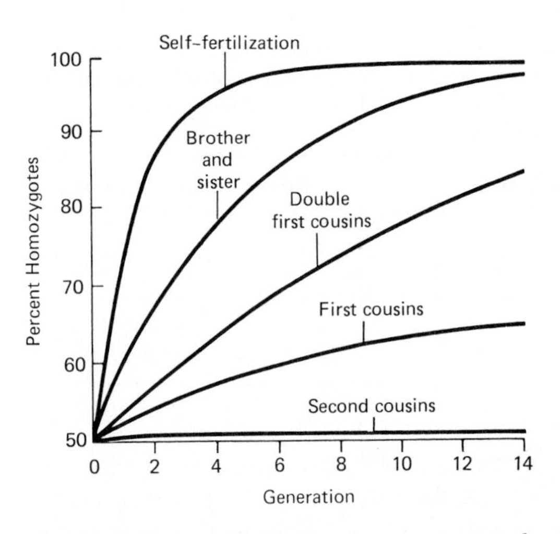


***Figure 15.8*** Graph showing the percentage of homozygous individuals in successive generations under different degrees of inbreeding, according to closeness of relationship of the parents. (From S. Wright. 1921. *Genetics* **6**:172.)

We all carry a number of harmful alleles in our genotypes—our genetic load—and these alleles are usually relatively rare in populations. If two unrelated people have children, these children will have four different grandparents and eight different great-grandparents. If these two people are first cousins, however, their children will have only six different great-grandparents (Fig. 15.9). If one of the great-grandparents in this family happened to be heterozygous for a rare recessive defect, then both first cousins might have inherited the defective allele. The chance of being heterozygous is ¼ for each of these two people. This is due to

***Table 15.16*** Genotypic frequencies in a randomly mating population for different degrees of inbreeding.

| generations of inbreeding | $F$ | genotypic frequencies | | |
|---|---|---|---|---|
| | | $AA$ | $Aa$ | $aa$ |
| none (Hardy-Weinberg) | 0 | $p^2$ | $2pq$ | $q^2$ |
| one or more | $0 < 1$ | $p^2 + Fpq$ | $2pq - 2Fpq$ | $q^2 + Fpq$ |
| infinite number | 1 | $p^2 + pq$ | 0 | $q^2 + pq$ |

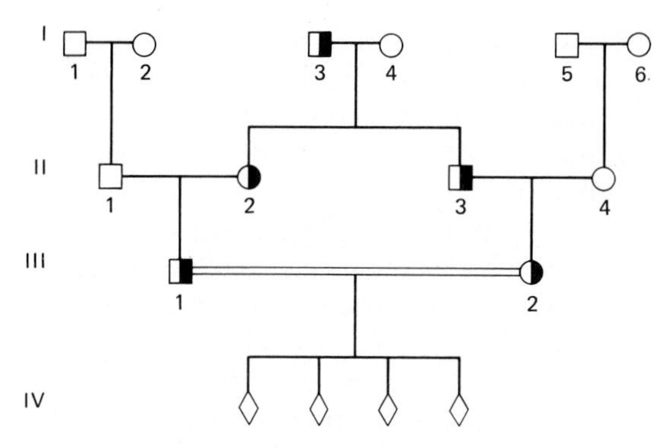


***Figure 15.9*** Pedigree of a family with consanguineous matings. Great-grandparent I-3 was heterozygous for a rare recessive defect, so III-1 and III-2, who are first cousins, each may have inherited the recessive allele from I-3 since he was a grandfather to both of them. If there is ¼ (= ½ × ½) chance for heterozygosity at this locus for III-1 and also for III-2, then there is 1/16 (= ¼ × ¼) chance for the birth of a child with the recessive genotype in generation IV from such first-cousin parents.

the fact that the chance is ½ that the allele was transmitted to a grandparent and, if it was, the chance is ½ that the allele was transmitted in the next generation to one member of our couple. There is the same ¼ chance that the other partner also is heterozygous, if we use the same line of reasoning in analyzing transmission of the allele to the second partner from the common great-grandparent. If the ¼ chance is that each partner is heterozygous for the harmful allele, then the chance is ¼ × ¼ = 1/16 that they will produce a defective child. If these two people were not related, their chances of producing such a defective child would be $2pq \times \frac{1}{4}$. Since we are discussing a rare allele, occurring perhaps with a frequency of $10^{-6}$, the chances would be about 1 in 2,000,000. In fact, a good method for estimating the frequency of a rare recessive allele is to determine the incidences of cousin marriages among parents of children homozygous for the inherited disorder.

Marriages between cousins are not necessarily harmful, as can be attested by numerous examples of prominent men and women whose parents were cousins. Many different, rare recessive defects in the gene pool of the human species, however, are concealed for the most part in heterozygous genotypes. At the same time, desirable recessive traits are more likely to be expressed in children of related individuals. In the long run, people choose their partners for important reasons that may far outweigh the likelihood of producing defective children as a hindrance to their choice. Of course, if some defective trait is known in the family history, cousins contemplating parenthood would have to take into consideration the greater probability of their producing a defective child than if they had selected unrelated partners.

## 15.8 Migration and Gene Flow

Virtually every species is subdivided into a number of breeding populations, each tending to interbreed within itself more than exchanging genes with other subpopulations of the same species. If these subpopulations remain totally isolated from one another for a sufficiently long time, they may undergo enough change in their gene pools through mutation and selection to become genetically incapable of interbreeding even if the opportunity were to arise. This is the direction presumably taken toward **speciation,** that is, toward new species arising from some ancestral population or populations. Most of the time, however, a certain amount of gene exchange, or **gene flow** between populations of a species, takes place as individuals move into or out of a population. Migration counteracts the tendency toward divergence into new species. Movements of individuals between different breeding groups also leads to changes in allelic frequencies when genotypes leave or when genotypes arrive and interbreed with resident genotypes.

If there is an influx of individuals from one population into another population, we can determine the extent of the effect of the immigrants on the gene frequencies in the recipient population from the equation for change in gene frequency shown on the next page.

$$p_1 = p_0(1 - m) + p_m m$$
$$\Delta p = p_1 - p_0 = p_0(1 - m) + p_m m - p_0$$
$$= -m(p_0 - p_m)$$

where $p$ is the frequency of the recipient or resident population, $p_m$ is the frequency of the same allele among the migrants entering the recipient population, and $m$ is the *migration coefficient.* The migration coefficient, or proportion of migrants, is a measure of the proportion of gametes that are derived from the immigrants and are contributed to the next generation.

Suppose a population with allelic frequencies of $p(A) = 0.1$ and $q(a) = 0.9$ receives migrants from a different population whose allelic frequencies are $p_m(A) = 0.9$ and $q_m(a) = 0.1$; and $m$ is 10%; the change due to the migration will be

$$\Delta p = -0.1(0.1 - 0.9)$$
$$= 0.08$$
$$p_1 = p + \Delta p = 0.18$$

The frequency of the $A$ allele is considerably greater in the recipient population, increasing from $p = 0.1$ to $p_1 = 0.18$, or almost twice the original frequency in just one generation.

When $p = p_m$, that is, when allele $A$ (or $a$) occurs in equal frequency in the resident population and in the migrants, an equilbrium can be established. It is possible for very rapid changes in gene frequency to take place as the result of migration pressure, particularly when $m$ is fairly large and $p$ is greatly different from $p_m$. By migration, favorable genes or genotypes can be spread throughout the species when gene flow takes place during interbreeding between residents and migrants.

## 15.9 Random Genetic Drift

Mutation, selection, and migration can be referred to as *pressures* because we can predict the magnitude and direction of the changes in gene frequency that these events produce. Each of these processes, however, has a random aspect. Mutation as a process is random, of course, in the sense that any gene may mutate at any time in any cell without regard to advantage or disadvantage. Another random aspect of evolution is **genetic drift,** that is, the loss of an allele or its fixation in a population by sheer chance, unrelated to selection. The most important factor involved in this phenomenon is *population size*; the smaller the population, the greater the potential for genetic drift. Since chance is the main feature underlying drift, the chance for a sampling error in constructing genotypes from the gene pool is higher if fewer individuals contribute to and draw from the gene pool.

We can illustrate this principle with an extreme example. Suppose a single heterozygous plant (*Aa*) that is self-fertilizing produces only one offspring that survives. The population size, $N$, is one and is consistent in each generation. Since three genotypes could be produced by self-fertilization of the *Aa* plant, the chance is 50% that the single surviving offspring will be *Aa*, in which case the frequencies of alleles $A$ and $a$ do not change. But it is equally possible that the single offspring might be one of the homozygous genotypes; if the offspring is *AA* then the gene frequency changes to $p = 1$ and $q = 0$, while an *aa* survivor would produce a change to $p = 0$ and $q = 1$. In either case, the allelic frequencies differ from the initial frequencies of $p = q = 0.5$.

In this way, the $A$ allele may become fixed and the $a$ allele lost, or the $a$ allele may become fixed and the $A$ allele lost from the gene pool, only because of chance and not because of the relative fitness of the genotypes. Although this is an extreme example, many populations consist of relatively small numbers of breeding individuals, at least during some periods in their history. It is important to note that the effective size of the population, $N$, is the *number of breeders* and not the total census of the population. $N$ is a measure of the number of individuals actually contributing to the gene pool, from which the next generation of genotypes arises through sexual reproduction. Each diploid individual has two alleles for each autosomal gene, and in populations with approximately equal numbers of reproductively active males and females, the rate of decrease in heterozygosity due to random genetic drift ($k$) has been shown to be

$$k = \frac{1}{2N}$$

In a population with 100 breeding individuals having a number of heterozygous gene loci, the expected rate of decrease in heterozygosity is 1/200. The rate of decrease in heterozygosity should also be viewed as the rate of fixation, that is, the rate at which homozygosity is being achieved. In this population of 100 individuals, one locus in 200 would be expected to become homozygous for one allele or the other in the next generation.

As $N$ becomes large, genetic drift becomes an insignificant component in influencing change in the gene pool. There is no clear agreement on the limits of a "small" versus a "large" population in absolute numbers. The effective size of the breeding population is a statistical concept that allows us to estimate the amount of genetic drift.

Small, self-contained breeding units, or *isolates*, in the midst of larger populations provide the most likely situations in which genetic drift could occur. The Dunkers of eastern Pennsylvania are a very small religious sect that have descended from a small group of West Germans who came to the United States in the early eighteenth century. Bentley Glass studied the Dunker community, which has about 300 members. The number of parents in each generation has remained relatively stable at about 90. Although the Dunkers are members of a much larger farming population in the region, they intermarry only within their own group and they are essentially genetically isolated by rigid marriage customs.

Glass compared the frequencies of certain traits among the Dunkers with those of the surrounding general population and with those of the West German population from which the Dunkers had emigrated over 200 years ago. By comparing the small isolate with its larger neighboring population and with its parent population, the effectiveness of genetic drift should be evident. The relative frequencies of the ABO, Rh, and MN blood groups were determined for the three populations, along with the incidences of four inherited external characteristics: left- or right-handednesss, attached or free-hanging ear lobes, the presence or absence of hair on the middle segments of the fingers, and normal versus extensive flexibility of the thumb ("hitchhiker's thumb").

These studies showed that the frequencies of some traits were strikingly different in the Dunker community when compared with the United States and West German populations. Blood group A is much more frequent, type O is somewhat rarer, and types B and AB are exceptionally infrequent in the Dunker community (Table 15.17). The $I^B$ allele was almost lost, and most of the carriers of the $I^B$ allele were converts who entered the isolate by marriage but were not born in the community.

Blood type M has increased and type N has decreased in frequency as compared with the other two populations. Only the Rh blood groups occur in frequencies that are similar to those in the surrounding population. Similarly, in physical traits only right- and left-handedness occur in frequencies resembling those of the surrounding or parental populations. The other three external physical traits are significantly lower in the Dunker community than in the other two populations.

From these data, we may assume that particular gene frequencies, some very high, some very low, and others essentially unchanged, can best be attributed to genetic drift, that is, to

***Table 15.17*** Frequencies of ABO and MN blood group alleles in three human populations.

| | allelic frequencies | | | | |
|---|---|---|---|---|---|
| **population** | $I^A$ | $I^B$ | $i^O$ | $L^M$ | $L^N$ |
| United States | 0.26 | 0.04 | 0.70 | 0.54 | 0.46 |
| West German | 0.29 | 0.07 | 0.64 | 0.55 | 0.45 |
| Dunker | 0.38 | 0.03 | 0.59 | 0.65 | 0.35 |

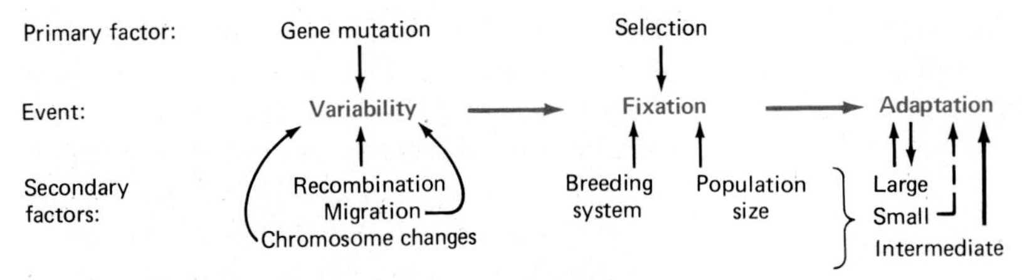


***Figure 15.10*** Summary of events and factors that influence evolution in the direction of adaptation.

chance fluctuations in gene frequencies. In these isolates, at least, drift appears to have been operative. For natural populations in general, we are less certain whether random genetic drift has helped to shape the gene pool.

What we have discussed so far has shown that various factors influence gene frequencies and thereby sustain populations in nonequilibrium conditions, that is, in the state of evolutionary change. Of these factors, certain ones may be particularly important in certain populations at certain times and under certain conditions, but the main force that shapes biological evolution is considered to be natural selection. Natural selection acts on genetic diversity, or variability of the gene pool, and guides the evolution of the population toward greater fitness of its genotypes, that is, in the direction of *adaptation* (Fig 15.10). The various processes that increase genotypic diversity in populations are mutation, recombination, chromosomal changes in number and structure, and migration. Selection is usually effective, but in very small populations, evolutionary changes may be either adaptive or nonadaptive, due to chance. Random fixation or loss of advantageous or disadvantageous alleles can occur when random genetic drift is in operation as an effective force. We will consider some of these concepts in more detail in the next chapter.

## Questions and Problems

*(Note: Unless otherwise indicated, assume that the populations are in equilibrium.)*

***15.1*** Coat color in cats is governed by a pair of codominant X-linked alleles, which give the following phenotypes: $C^BC^B$ ♀♀ or $C^B$♂♂ are black, $C^YC^Y$ ♀♀ or $C^Y$♂♂ are yellow, and $C^BC^Y$ ♀♀ are calico (mixture of yellow and black fur). A population of cats in Rome was analyzed for these alleles, with the following results:

| | black | yellow | calico | total |
|---|---|---|---|---|
| females | 554 | 14 | 108 | 676 |
| males | 622 | 84 | 0 | 706 |

***a.*** What are the allelic frequencies?

***b.*** Do the genotypic frequencies for females fit the Hardy-Weinberg formula?

***c.*** Do the genotypic frequencies for males fit the Hardy-Weinberg formula?

***15.2*** Vermilion eye color in *Drosophila* is due to an X-linked recessive gene $v$, and wild type (red eyes) is due to its dominant allele $v^+$. A population of fruit flies reared in the laboratory has 85 red-eyed males and 15 males with vemilion eye color.

***a.*** Estimate the frequencies of the two alleles.

***b.*** What percentage of females in this population would be expected to have vermilion eyes?

***15.3*** MN blood groups are determined by a pair of codominant alleles ($L^M$ and $L^N$). A sample of 426 Bedouins in the Sinai peninsula was typed for MN blood groups, with these results: 238 M, 152 MN, and 36 N.

***a.*** Calculate the allelic frequencies of $L^M$ and $L^N$.

***b.*** If the frequency of $L^N = 0.3$, how many individuals in a population of 1000 would be expected to belong to blood group MN?

***15.4*** Phenylketonuria is a metabolic disorder due to an autosomal recessive gene. If frequency of affected individuals in the population is 1 per 10,000, what is the probability that two unrelated, heterozygous parents will produce a phenylketonuric child?

***15.5*** In *Drosophila*, black and ebony body color phenotypes are due to recessive alleles (*b* and *e*, respectively) of gene loci on different chromosomes. A large population screened for these phenotypes gave these results: 9.69% wild type, 9.31% ebony, 41.31% black, and 39.69% ebony, black. Calculate the frequencies of the ebony and black alleles of the two body color genes.

***15.6*** Ocular albinism is governed by an X-linked recessive allele, and 1% of the gametes in the gene pool of a particular human population has this allele.

***a.*** What is the expected frequency of ocular albinism among males of this population?

***b.*** What is the expected frequency among females of this population?

***15.7*** Lesch-Nyhan disease is an X-linked, recessive disorder that causes neurological damage in human beings. A survey of 500 males from a Caucasian population revealed that 20 were affected with this disorder.

***a.*** What is the frequency of the normal allele in this population?

***b.*** What percentage of the females in this population would be expected to be normal?

***15.8*** In England 70% of the population are *tasters*, who can detect the bitter taste of the chemical PTC (phenylthiocarbamide), and 30% are *nontasters*. A single autosomal gene is involved, and taster is dominant over nontaster.

***a.*** What proportion of all marriages between tasters and nontasters have no chance (except by mutation) of producing a nontaster child?

***b.*** If mating is random, what proportion of marriages will be between nontasters?

***15.9*** The ABO blood group locus is on chromosome 9 in humans. In a particular population the frequency of the $i^O$ allele is 0.6. What would be the expected frequency of blood type O among individuals with trisomy-9?

***15.10*** A sample of 2000 Italians was typed for ABO blood groups, with these results: A = 640 B = 300, AB = 80, and O = 980. Calculate the frequencies of alleles $I^A$, $I^B$, and $i^O$.

***15.11*** curly fur versus straight fur is governed by a single pair of autosomal alleles. A large, randomly mating animal population consists of 23% with curly fur and 77% with straight fur. Which allele is dominant?

***15.12*** In a sample of 2400 births at an area hospital, 6 babies died shortly after birth from the effects of colonic obstruction, an autosomal recessive lethal disorder.

***a.*** What is the frequency of the recessive *co* allele in the population?

***b.*** What proportion of the population is heterozygous for the *co* allele?

***c.*** What proportion of the population is homozygous for the normal $co^+$ allele?

***d.*** What is the rate of mutation of $co^+$ to *co*?

***15.13*** A sheep rancher in Iceland finds that the recessive allele *y*, which causes yellow fat, has become established in his flock of 1000 sheep, and that 1 in 25 sheep express this trait. Assume that the population is randomly mating and that all genotypes have the same reproductive fitness.

***a.*** How many sheep express the normal trait of white fat?

***b.*** How many normal sheep carry the recessive allele?

***c.*** Since only animals with white fat are selected for breeding explain why the recessive allele has not been eliminated from the population.

***d.*** How long would it take to eliminate the recessive allele if selective breeding is continued, using only sheep with white fat?

## References

Cavalli-Sforza, L.L. Aug. 1969. "Genetic drift" in an Italian population. *Sci. Amer.* **221**:30.

Cavalli-Sforza, L.L. Sept. 1974. The genetics of human populations. *Sci. Amer.* **231**:80.

Cavalli-Sforza, L.L., and W.F. Bodmer. 1971. *The Genetics of Human Populations.* San Francisco: Freeman.

Crow, J.F. Feb. 1979. Genes that violate Mendel's rules. *Sci. Amer.* **240**:134.

Dobzhansky, T. 1968. *Genetics of the Evolutionary Process.* New York: Columbia University Press.

Dobzhansky, T., et al. 1977. *Evolution.* San Francisco: Freeman.

Eckhardt, R.B. Jan. 1972. Population genetics and human origins. *Sci. Amer.* **226**:94.

Fisher, J. 1978. *R.A. Fisher: The Life of a Scientist.* New York: Wiley.

Fraikor, A.L. 1977. Tay-Sachs disease: Genetic drift among the Ashkenazim Jews. *Social Biol.* **24**:117.

Friedman, M.J., and W. Trager. Mar 1981. The biochemistry of resistance to malaria. *Sci. Amer.* **244**:154.

Hardy, G.H. 1908. Mendelian proportions in a mixed population. *Science* **28**:49.

Hartl, D.L. 1981. *A Primer of Population Genetics.* Sunderland, Mass.: Sinauer Associates.

Lerner, I.M. 1968. *Heredity, Evolution, and Society.* San Francisco: Freeman.

Lewontin, R.C., ed. 1968 *Population Biology and Evolution.* Syracuse, N.Y.: Syracuse University Press.

Lewontin, R.C. 1974. *The Genetic Basis of Evolutionary Change.* New York: Columbia University Press.

Mayr, E. 1970. *Populations, Species, and Evolution.* Cambridge, Mass.: Harvard University Press.

Mettler, L.E., and T.G. Gregg. 1969. *Population Genetics and Evolution.* Englewood Cliffs, N.J.: Prentice-Hall.

Nei, M. 1975. *Molecular Population Genetics and Evolution.* New York: Elsevier.

Weinberg, W. 1908. Über den Nachweis der Vererbung beim Menschen. (English translation in S.H. Boyer, IV, ed. 1963. *Papers on Human Genetics,* p.4. Englewood Cliffs, N.J.: Prentice-Hall.)

Wills, C. Mar. 1970 Genetic load. *Sci. Amer.* **222**:98.

Wright, S. 1978. *Evolution and the Genetics of Populations,* vol. 4: *Variability within and among Natural Populations.* Chicago: University of Chicago Press.

# CHAPTER 16

# Evolutionary Genetics

Evolution is the fundamental theme that ties together the many aspects of biological systems. Through genetic changes acted upon by selection, all the varied forms of life have arisen on Earth and all the varied processes that maintain and diversify life have arisen. Biologists can study individual topics in different ways, using methods of genetics, physiology, systematics, molecular biology, and others. Evolutionary biology provides a synthesis of all this information and places the various aspects of evolution into a unified framework of events and processes.

## Natural Selection

The basic theme of biological evolution developed in the last chapter was that natural selection acting on genetic diversity leads to changes in gene frequencies in the gene pool. These events rest on three fundamental principles: (1) the gene pool must contain variation to select from, (2) this variation must be heritable since there is progressive change from generation to generation, and (3) different variants leave different numbers of offspring either immediately or in later generations.

Natural selection involves differential reproduction of genetically diverse types so that some types leave more offspring than others. Eventually the gene pool will be composed of a higher proportion of some types than of some others. If we stop at this point, what have we explained? To say that inherently fitter individuals leave more offspring and that individuals who leave more offspring can be defined as fitter is to state a tautology. We have provided no meaningful information or explanation. In order to gain insight into the biological meaning of natural selection and to understand the nature of evolutionary changes in biological systems, we must also consider the nature of *adaptation,* or the *relative fitness* of the genotype in the environment. In addition, we must examine the nature and wealth of genetic diversity since this is the raw material for evolution.

### 16.1 Adaptation

Darwin introduced a fourth principle underlying biological evolution, the principle of the "struggle for existence." He made it clear that the term was metaphorical in the widest sense and the struggle was not simply a matter of bloody combat to the death between individuals. Fitter individuals and populations are those that compete successfully for the finite resources of life, such as space, light, food, shelter, and mates. Less fit individuals and populations are less successful in competition, secure fewer of their requirements for existence, and therefore decrease in frequency relative to the fitter members of the population.

Fitter individuals are better adapted to their environments, that is, they are inherently better at solving their ecological problems than are other individuals with genotypes less fit for the same environments. The concept of adaptation, that is, of the fitness of the individual or population in its environment, puts us into the real world of space and time. Once we introduce the relative fitness, or adaptedness, of the individual to its environment, we can begin to predict which individuals and which populations are more likely to leave more offspring and which will leave fewer or no offspring. It is possible to predict which individuals will be fitter according to the biological characteristics by which they solve ecological problems.

We can also see more clearly why the amount of diversity in the biological world is finite, despite the random nature of mutation. In other words, considerable diversity arises by the random process of mutation, but only a proportion of this diversity becomes incorporated into the genetic structure of a population. The proportion of diversity that becomes incorporated is either *neutral,* conferring neither advantage nor disadvantage, or *positive,* enhancing the fitness or adaptedness of the population in relation to environmental challenges in different places at different times in its history.

With these concepts in mind, we can expect that evolutionary changes in the gene pool will arise in different ways according to the particular nature of interactions between populations and their environments. Furthermore, environments are rarely uniform at any one time, and they also change in various ways over time. Within this space-time mosaic of the environment, we can look at ways in which adaptation is achieved. Adaptation is relative rather than absolute, so the nature of adaptations in populations will depend partly on the genetic diversity that occurs and partly on the selection strategy that acts on this diversity to modulate its quantity and quality. In some cases diversity is maintained in the absence of selection, although the significance of this observation is controversial.

## 16.2 Selection Stategies

Natural selection influences the gene pool in a number of ways in addition to **normalizing natural selection,** which acts to reduce or eliminate disadvantageous alleles, as we discussed in the previous chapter (Fig. 16.1). In the

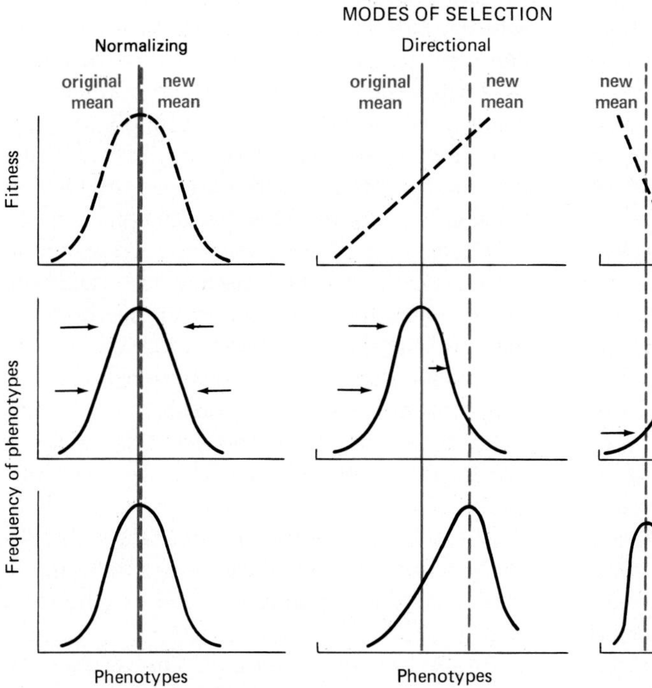


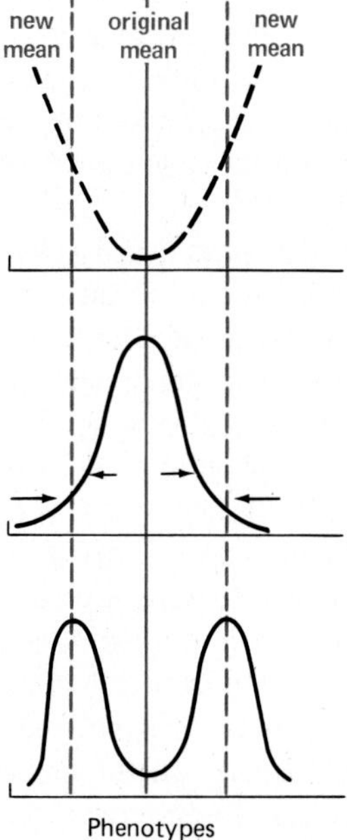


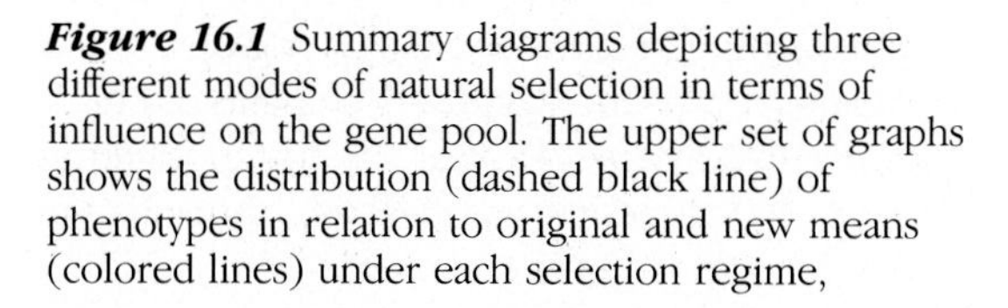

***Figure 16.1*** Summary diagrams depicting three different modes of natural selection in terms of influence on the gene pool. The upper set of graphs shows the distribution (dashed black line) of phenotypes in relation to original and new means (colored lines) under each selection regime, according to fitness of the phenotype. The influence of selection on the phenotypic frequency is indicated by arrows in the middle set of graphs, and the consequences of such selection are shown in the bottom set of graphs.

case of normalizing natural selection, genotypes with lowered fitness ($W < 1$) are reduced gradually, and genotypes with a fitness of 1 will increase proportionately in each generation until an equilibrium is reached. Normalizing natural selection results in less genotypic diversity since selection acts against one of the homozygous genotypes. For dominant, codominant, or recessive alleles that are lethal ($W = 0, s = 1$), the rate of elimination of the disadvantageous genotype is faster than if $s < 1$ (see Fig. 15.7). The price paid for increasing population fitness is a diminution of its gene-pool diversity.

In **directional natural selection,** the population is shifted in the direction of its greatest adaptedness, or fitness, in relation to changing environmental conditions. One genotype may be favored under one set of conditions and another under a different set of conditions. In either case, the proportion of favored genotypes increases and the frequency of the less fit genotype decreases. Since one homozygous genotype is fitter than the other, the frequency of heterozygotes decreases as one allele is reduced or eliminated and as the alternative allele may even increase to become the only variant fixed in the gene pool. The genotypic diversity is therefore reduced.

Examples of the action of directional selection are among the most spectacular because we can see evolution in action in our own lifetimes. The evolution of resistance of insect species to pesticides and of bacterial species to antibiotics in recent years has provided ample demonstrations of the ability of living species to undergo genetic changes in response to environmental challenges. In each case the scenario is practically the same: when a new insecticide or a new antibiotic is introduced, a relatively satisfactory level of control can be achieved with low concentrations or infrequent treatment. As time goes by an increasing dosage or more frequent applications are required until the results are no longer effective, economical, or safe for human beings. The problems involved in human health and survival as well as in management of crops and domesticated animals are of great concern and of immense practical importance to us.

Directional selection has also been observed in the spread of industrial melanism, or black pigmentation, in many species of moths, and particularly detailed studies have been made of the peppered moth (*Biston betularia*) in Great Britain. Before the middle of the nineteenth century the predominant form was a "light" variant and occurrences of the "dark" or melanistic form were rare in these populations (Fig. 16.2). The appearance and spread of the darkly pigmented form has been recorded in regions where pollution and soot have darkened the vegetation. In these industrial regions the lightly pigmented form has become greatly reduced in frequency and may even have disappeared almost entirely. In other regions of little or no pollution, the light form still occurs with high frequency and the dark form is rare. According to many observations and experiments, the main selective factor promoting directional changes in genotypic frequencies is predation of the moths by birds. The light varieties are protectively colored when they rest on nonpolluted vegetation; they are conspicuous on darkened vegetation. In contrast, dark varieties are protectively colored on darkened vegetation.

Genetic tests have shown that resistance to insecticides in insects, resistance to certain antibiotics in bacteria, and the difference between dark and light forms of many moth species are due to either a single gene with a major effect or to a set of genes with individually minor effects. In the moths, the allele producing dark color is usually dominant to the allele for light pigmentation. Also, further genetic changes have involved a system of *gene modifiers* with relatively minor phenotypic effects, but these modifiers intensify the phenotypic effect of the main gene.

In the peppered moth, directional selection has led to an increase in the dominant homozygous genotype in industrialized regions, but to relatively few homozygous dominants in nonindustrialized parts of Great Britain. In meeting these environmental challenges, the diverse

***Figure 16.2*** Example of industrial melanism in the moth *Biston betularia.* Directional selection has modified the moth populations in industrialized regions of Great Britain toward a higher frequency of the dark form than of the originally prevailing light types. Camouflage against predators is effective for the light moths in unpolluted environments (left), but not in sooty environments (right). (Reprinted from Colin Patterson: *Evolution.* Copyright © Trustees of the British Museum (Natural History) 1978. Used by permission of the publisher, Cornell University Press.)

gene pool has been altered so that fewer heterozygotes and fewer homozygotes of one of the two types are produced in polluted and nonpolluted regions. Each population type is better adapted to its immediate environment, but the chances remain high for losing one of the two alleles since adaptation has proceeded directionally toward establishing one of the two alternative homozygotes.

In contrast with these types of selection some selection strategies maintain a relatively high level of genotypic diversity despite forces acting to eliminate or reduce the frequency of a disadvantageous allele. The main types of strategy under this heading are *balancing, diversifying,* and *frequency-dependent* natural selection. While normalizing and directional selections are conservative programs in adaptation, that tend to make and keep the species relatively constant, these other types of selection act to retain diversity and, therefore, to maintain a higher degree of genetic flexibility in the species. In all of these kinds of selection, the relative fitness of the individual is a determining feature of genetic changes in the gene pool.

The classical example of the action of **balancing natural selecton** is that of sickle-cell hemoglobin (Hb S), which occurs in fairly high frequency in some parts of the world despite the low fitness of homozygous recessive indi-

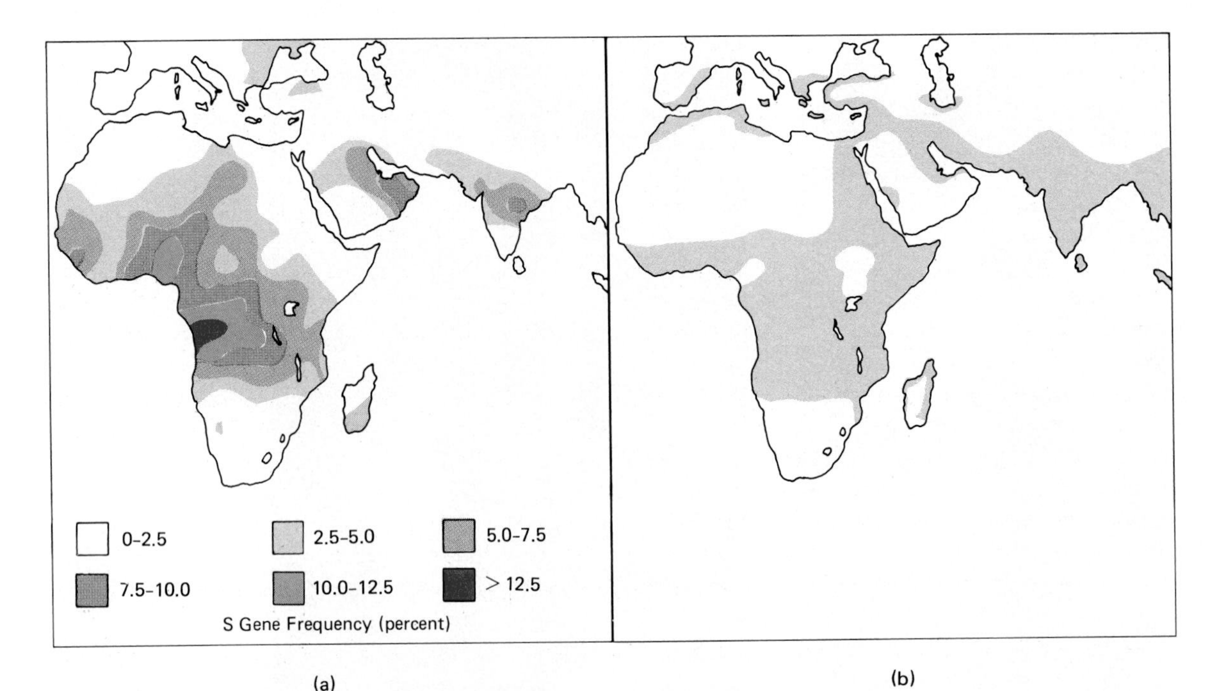


***Figure 16.3*** Distribution of the sickle-cell allele and of falciparum malaria in Africa and parts of Europe and Asia. (a) The sickle-cell allele is most common in central Africa, but it also occurs in considerable frequency in other areas; and (b) those areas where the falciparum malaria is prevalent coincide with the areas of occurrence of the sickle-cell allele. (From *The Genetics of Human Populations* by L.L. Cavalli-Sforza and W.F. Bodmer. )

viduals (Fig. 16.3). The $Hb^S$ allele is maintained at a high frequency because of the selective advantage enjoyed by heterozygotes and not because of a high rate of mutation from the $Hb^A$ to the $Hb^S$ allele. In parts of the world where malarial infection caused by *Plasmodium falciparum* is endemic, $Hb^A/Hb^A$ homozygotes often succumb to the infection, whereas $Hb^A/Hb^S$ heterozygotes are not adversely affected by the pathogen. Individuals with sickle-cell anemia, $Hb^S/Hb^S$ recessives, are probably resistant to malarial infection, but a very high percentage of these people die of their anemia disorder before reaching adulthood and are, therefore, relatively unfit. Because heterozygotes are fitter and, therefore, more likely to reproduce, a high frequency of recessive genotypes is maintained in these populations despite their selective disadvantage.

If we assume that the selective disadvantage of genotype $Hb^A/Hb^A$ is $s_1 = 0.2$ and that of $Hb^S/Hb^S$ is $s_2 = 0.9$, then the equilibrium frequency of the $Hb^S$ allele would be $\hat{q} = s_1/(s_1 + s_2) = 0.2/(0.2 + 0.9) = 0.18$. The equilibrium frequency of the $Hb^A$ allele would then be 0.82. In every generation, although the percentage of infants born who will develop anemia and probably die in childhood is $(0.18)^2 \times 100 = 3.24\%$, the *average fitness* of the population will be greater than that of a population consisting only of $Hb^A/Hb^A$ homozygotes since the $Hb^A/Hb^S$ heterozygotes have a fitness of 1.

In nonmalarial regions, normalizing natural selection will come into operation and there will be a reduction in frequency of the $Hb^S$ allele since $Hb^A/Hb^S$ heterozygotes enjoy no advantage over $Hb^A/Hb^A$ homozygotes. In the United States, sickle-cell anemia occurs mostly in the black population, with a frequency of about 0.25% (1 in 400). The initial frequency of the $Hb^S$ allele among blacks brought to the United States from

Africa 300 years ago has been estimated at about 22%. It has been calculated that over approximately 15 generations the frequency of the sickle-cell allele would have declined to about 5% and that about 10% of the current black population would be heterozygotes. These predictions agree with the values actually found. Since the three genotypes occur in the expected Hardy-Weinberg proportions, we can infer that heterozygotes are no more fit than $Hb^A/Hb^A$ homozygotes. We can predict a continued decline in the frequency of the $Hb^S$ allele in subsequent generations until the two alleles come to exist in equilibrium. The calculations are complicated by the fact that an increasing percentage of people with sickle-cell anemia do live longer and contribute to the gene pool so that the precise equilibrium frequencies are uncertain.

Populations such as those just described for malarial regions are said to exist in a *balanced polymorphism.* In such a population several different phenotypes occur and the rarest of these phenotypes exists with a higher frequency than can be accounted for by mutation alone. Polymorphisms of various types have been described, involving morphological, physiological, biochemical, and chromosomal characteristics. We will discuss these in the next section since other issues arise in relation to polymorphisms not maintained by balancing natural selection.

**Diversifying natural selection** (also called **disruptive selection**) operates to diversify the gene pool so that two or more classes of genotypes have high adaptiveness in different subdivisions of a heterogeneous environment. This selection strategy operates in favor of the homozygous genotypes and against the heterozygotes, or intermediate types. Diversifying selection is believed to be a major force in the evolution of *mimicry* (also called "Batesian mimicry" in honor of the English naturalist H.W. Bates). In mimicry an organism that is sought by predators comes to develop the phenotypic pattern of a distasteful species and is thereby somewhat more protected from predators that select their prey visually (Fig. 16.4). The predator learns to avoid the pattern of the distasteful species and will also avoid the mimic species if the two patterns are sufficiently alike. For mimicry to be successful, the distasteful species must far outnumber the edible species. Otherwise the predators will not be adequately conditioned to learn to avoid the mimics.

It has been postulated that one or a few major alleles have become predominant through diversifying selection and that subsequent selection of modifier genes has contributed to the development of several different mimetic phenotypes in individual species. The hybrids formed experimentally between different mimetic groups are intermediate, and we therefore believe that any hybrids in nature would be eliminated by predators. Elimination of less fit intermediates and selection of better-adapted extreme phenotypes therefore produces a polymorphic situation in which different phenotypes are actively maintained by selection pressures in varied environments. The maintenance of diversity is an outcome of selection against heterozygotes and leads to the preservation of several fitter genotypes that belong to homozygous classes.

Cases of Batesian mimicry also provide examples of the action of **frequency-dependent selection,** whereby a phenotype is selected against when it is more common and is favored when it is rare. Selection is exerted on mimics or nonmimics depending on their relative frequencies in the population. In populations where unpalatable models are common and nonmimics are more common than mimics, the nonmimics will suffer heavier predation than mimics. However, where distasteful models are rare and mimics are more common than nonmimics, the mimics will suffer heavier predation than nonmimics.

The above examples of selection in natural populations have been confirmed by evidence obtained using a variety of experimental approaches, such as studying samples captured in natural populations or samples of laboratory stocks simulating natural populations. In addition, selection is practiced by plant and animal

***Figure 16.4*** Batesian mimicry in African butterflies. Three distasteful species shown at the left are mimicked by strains of the species *Papilio dardanus* (center) and *Hypolimnas* (right). (From *Mimicry in Plants and Animals* by W. Wickler. Copyright © 1968. Used with permission of McGraw-Hill Book Company.)

breeders, which has produced improved strains of food crops, better breeds of domesticated animals, and numerous varieties of ornamental plants.

## 16.3 Molecular Polymorphisms

We have already discussed some polymorphisms that involve molecular differences due to the actions of different alleles of a single gene, such as those for normal and sickle-cell hemoglobins. Although recessives can be recognized by clinical symptoms, homozygous $Hb^A/Hb^A$ and heterozygous $Hb^A/Hb^S$ individuals are generally indistinguishable phenotypically. By gel electrophoresis we can see the different patterns for the two kinds of hemoglobins, and we therefore recognize a molecular difference among all three genotypes; the situation is one involving a **molecular polymorphism** (Fig. 16.5). Using gel electrophoresis to obtain preliminary information on charge differences, which lead to

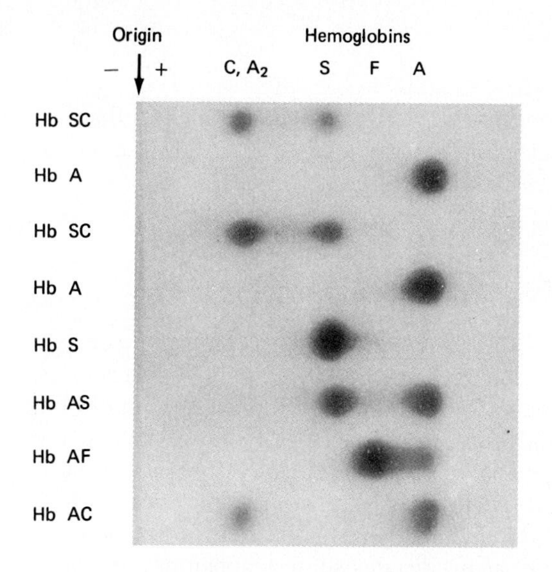


***Figure 16.5*** Electrophoretic mobilities of hemoglobins (from left to right) from individuals (vertical list) homozygous for Hb A, homozygous for Hb S, and heterozygous for S and C, A and S, and A and C. The sample designated Hb AF is from umbilical cord blood, showing a high percentage of fetal hemoglobin. Hb $A_2$ is a normal adult hemoglobin present in small amounts (about 2%) in most persons and is indistinguishable from Hb C under electrophoretic conditions used. All the hemoglobins consist of four globin chains in tetramer molecules: Hb A = $\alpha_2\beta_2$, Hb $A_2$ = $\alpha_2\delta_2$, Hb F = $\alpha_2\gamma_2$. The modification in Hb S and Hb C is at amino acid 6 in the $\beta$ chains of Hb A.

differences in migration of proteins in gels, an astonishing number of allelic variants have been described for hemoglobins and for other gene products.

In the case of human hemoglobins, for example, more than 100 different allelic variants have been identified by gel electrophoresis and by subsequent amino acid composition and sequence analysis of the $\beta$-globin chain alone. In many of these cases, the allelic variation is very rare, but some variations are relatively common in particular populations. Most of these molecular variations produce no clinical symptoms at all, but some produce a mild to moderate anemia. Interestingly, the geographical regions where some $\beta$-chain variants exist in high frequency are also regions where the falciparum malaria is endemic (Fig. 16.6). In addition to

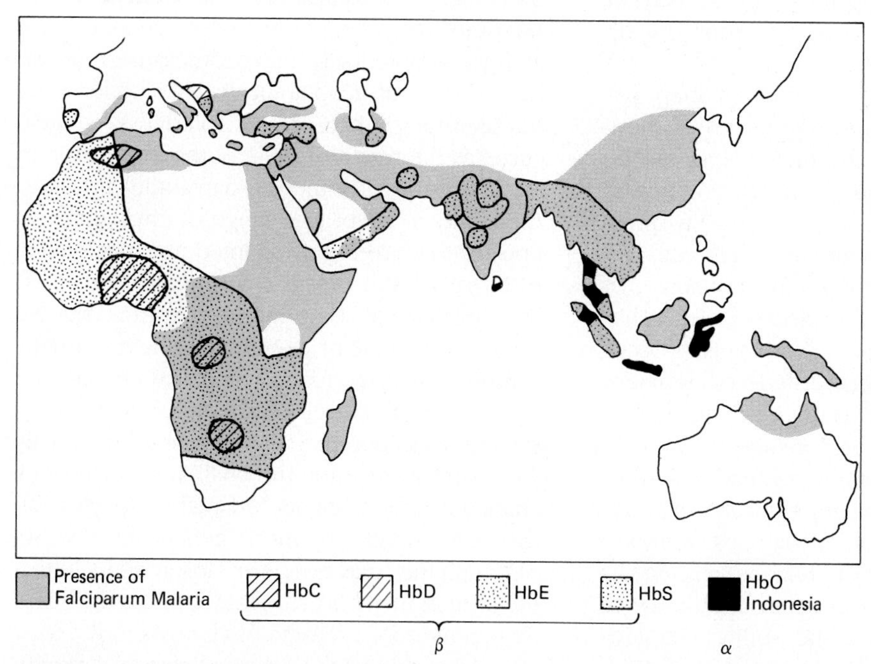


***Figure 16.6*** Distribution of falciparum malaria and of human hemoglobin variants with modified $\beta$ chains (Hb C, Hb D, Hb E, Hb S) or $\alpha$ chains (Hb O Indonesia). (From *Evolutionary Biology* by Stanley N. Salthe. Copyright © 1972 by Holt, Rinehart and Winston, Inc. Reprinted by permission of Holt, Rinehart and Winston, CBS College Publishing.)

$\beta$-globin polymorphisms, some polymorphisms involve the $\alpha$-chain gene as well as blood-group antigenic determinant genes, such as those coding for the MN, Rh, and other factors.

We know of a large number of enzyme polymorphisms for many species, including 80 allelic variants for the human gene that encodes glucose 6-phosphate dehydrogenase. In many cases the polymorphism is identified by differences in certain properties of the *isozymes,* or molecular variants of an enzyme, but all variants have the same function. From such enzyme studies it has become apparent that molecular polymorphisms are common. They had gone undetected for the most part because individuals carrying these alleles exhibit little or no phenotypic differences. Futhermore, since each organism carries only two alleles for a particular gene locus, analysis of large samplings of populations is required in order to obtain molecular data for each series of multiple alleles. Taken altogether, molecular polymorphisms provide undeniable evidence showing surprisingly high levels of genotypic diversity in sexually reproducing species.

How is this diversity maintained in populations? By definition, the frequencies of these polymorphic genotypes are higher than can be accounted for by mutation alone. In some cases selection clearly operates to maintain polymorphisms, as we saw in the case of $Hb^A$ and $Hb^S$ alleles. In other cases it seems unlikely that diversity is maintained by selection, and it has been suggested that some polymorphisms occur because of random genetic drift. In cases where drift has been proposed as the mechanism, particular allelic variants are apparently *neutral mutations,* that is, mutations conferring neither advantage nor disadvantage. Mutation rate and random genetic drift, rather than selection, are most likely to influence the relative frequencies of neutral mutant alleles. Despite the abundance of experimental studies, the subject remains controversial and there is no consensus of opinion about the overall significance of random genetic drift in maintaining diversity in the gene pool. The occurrence of neutral mutations and their preservation by chance rather than by selection has been referred to as *non-Darwinian evolution.*

## Evolution of Genetic Systems

The *genetic system* includes all of the genetic mechanisms and controls operative in a species, such as mutation, recombination, regulation, and other components directly related to the genetic material. The genetic system itself is subject to hereditary modification, as much as the individual genes contained in the genome. We can predict that some adaptations are related directly to the genetic system itself as an outcome of mutation and selection, just as we find adaptations for the multitude of phenotypic traits in a species.

The major function of the genetic system is related to factors that provide a compromise between **fitness,** or **adaptedness** (which requires genetic constancy), and **flexibility,** or **adaptability** (which requires genetic variability). Adaptedness can be measured, as we have seen, but adaptability is difficult or even impossible to measure. Adaptedness is measured as the relative fitness of the population in relation to environment. Adaptability refers to the ability to adapt to a range of environments. Species that are highly adapted are more likely to be successful in the environment for which they are fitter at some moment in time, but we have seen that the price of adaptedness is usually a reduction in genetic diversity. With reduced diversity and, therefore, higher genetic constancy, such populations may not be equally successful in meeting the challenges of new or changing environments. Adaptable species, on the other hand, are more genetically diverse. Although they may not be as closely adapted to a life-style in the short run, they possess the genetic potential for a variety of changes in the long run and in this way are more likely to succeed in new selection situations.

Different kinds of genetic systems manage to balance the short-term advantage of genetic constancy and the long-term advantage of genetic flexibility. We will look at some of these systems and the ways in which a balance has been achieved.

## 16.4 Flow of Variability

We have seen that a considerable amount of genetic diversity can characterize the gene pool of a sexual species. We must also consider, however, how this allelic diversity is doled out when new genotypes arise in each generation. In particular, we want to know *how much* genotypic diversity is generated and *how evenly* it is generated in successive generations; that is, we want to know how variability flows from the gene pool to actual genotypes produced in individuals making up the population. Thus in sexual species we must examine features related to meiosis, during which alleles segregate into gametes and recombination and reassortment take place.

One of the factors which influences the amount of genotypic diversity that can be generated in a species is the *number of chromosomes in the genome*. Genes that are on different chromosomes assort independently, but linked genes remain associated unless crossing over takes place between them. The higher the chromosome number, therefore, the more kinds of gametes will be produced since alleles of different unlinked genes segregate at random during meiosis. If each pair of chromosomes in the diploid nucleus contained just one pair of heterozygous alleles, then $2^n$ different kinds of gametes would be produced, where $n$ is the number of different chromosomes. Five chromosome pairs would produce $2^5 = 32$ kinds of gametes; 6 pairs would produce $2^6 = 64$ gamete genotypes; 7 pairs, $2^7 = 128$ gamete types, and so on. For each chromosome added to the genome, the number of different kinds of gametes is doubled, based on only one heterozygous locus per chromosome pair. Similarly, reduction by just one chromosome pair will reduce gamete variety by one-half. When a species with 7 chromosome pairs becomes tetraploid and then has 14 chromosome pairs, the number of kinds of gametes increases theoretically from $2^7 = 128$ to $2^{14} = 16{,}384$ kinds of gametes with only one heterozygous locus per chromosome pair. Increases in chromosome number by polyploidy are very common in the flowering plants, with about 50% of these species estimated to be polyploid today.

Reduction in chromosome number during evolution has been characteristic of a great many animal species-groups, such as *Drosophila* (see Fig. 12.8), and a substantial number of plants as well (Fig. 16.7). In these cases little or no genetic

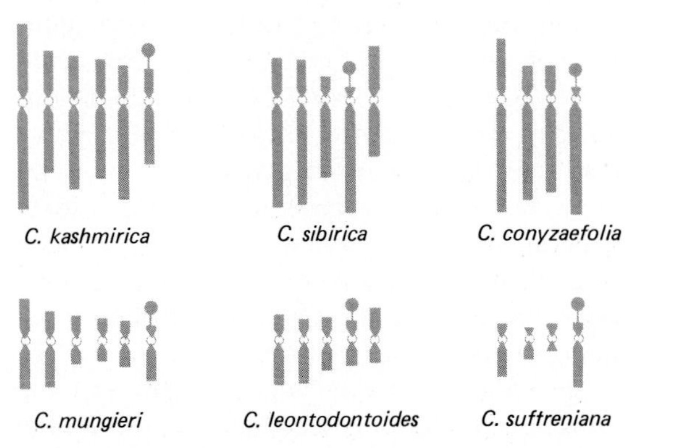


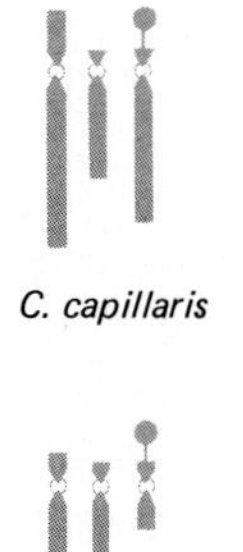


***Figure 16.7*** Diagrams of the karyotype of eight species of *Crepis,* a genus in the sunflower family (Compositae), showing reductions in chromosome number that have occurred during the evolution of this group. The centromere region is shown by a circle.

material has been lost since centric fusions have been responsible for numerical reduction. Genes formerly assorting independently are segregated into different gametes only after crossovers, and fewer gametes will, therefore, contain recombinant genotypes in species with fewer chromosomes in the genome. With fewer new combinations of alleles in the gametes, fewer kinds of genotypes will be produced in each generation, despite the equivalence of diversity in the gene pool of a species with 3, 4, 5, or 6 linkage groups, as in *Drosophila* or *Crepis.*

In addition to genotypic diversity arising from gametes containing alleles that assort independently, the amount of crossing over between linked genes must also influence the actual diversity of genotypes that appear in successive generations. The relative amount of crossing over can be estimated by breeding analysis where marker genes are available and where breeding is possible under controlled conditions. In addition, direct cytological observations of meiotic chromosomes can provide information about crossover frequencies. Each chiasma observed in a meiotic bivalent is believed to represent the site of a previous crossover event. By counting the number of chiasmata per bivalent, it has been found that crossover frequencies vary considerably among different species. Such frequencies are taken together with chromosome number to estimate the **recombination index** for a population or a species. The recombination index is derived from

$$\frac{\text{chiasma (crossover) frequency}}{\text{chromosome pairs}}$$

In a species with 5 pairs of chromosomes and an average of 10 chiasmata per meiotic nucleus, the recombination index is 2; with only 5 chiasmata for 5 pairs of chromosomes, the recombination index is 1. For the same genome size, therefore, more gamete variety and more genotypic variety will be produced when there is a higher crossover frequency, or recombination index.

Some genes reduce crossing over in various species; in any species the occurrence of chromosome inversions can dampen recombination. Inversion heterozygotes produce fewer kinds of gametes since many of the crossover gametes are inviable (see Section 12.6). Because of their huge salivary chromosomes, *Drosophila* species are admirable systems for analysis of chromosome inversion frequencies in natural populations. Different chromosome band patterns provide unequivocal evidence for inversions, and populations may be sampled repeatedly to see if inversion chromosomes are retained intact or if they become altered by crossing over. In the genome of *D. pseudoobscura,* for example, chromosome 3 is characterized by four different inversion types. Inversion heterozygotes and inversion homozygotes are present in various populations, and these same inversion chromosomes are retained year after year in the populations. The proportions of each inversion chromosome type vary from one population to another in accordance with different adaptive values (Fig. 16.8). For us the main point is that blocks of genes in chromosome 3 remain intact; such blocks of genes are called **supergenes.** In spite of a high level of allelic heterozygosity for chromosome 3, genetic constancy is maintained by inversions that preserve supergenes through their effect in reducing crossover gametes. Even without banding patterns to identify inversions, it is possible to look for inversion loops in meiotic divisions (see Fig. 12.17). Following this line of analysis, we can be sure that chromosomal inversions exist in natural populations and that they influence the amount of diversity that is released in each generation. We don't know, however, how significant inversions are in general since only a few species have been studied for inversion heterozygosity. Homozygous inversions, of course, do not affect crossing over and recombination.

In addition to the influences of chromosome number and crossover frequency on diversity at the level of meiosis, various factors act at the stage of zygote formation during fertilization.

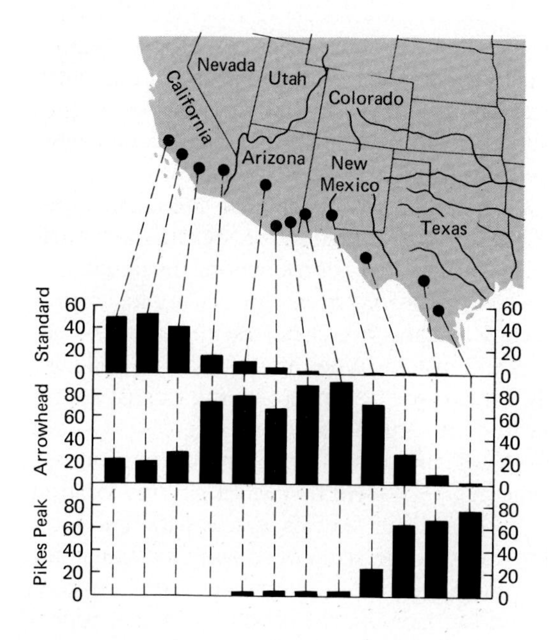


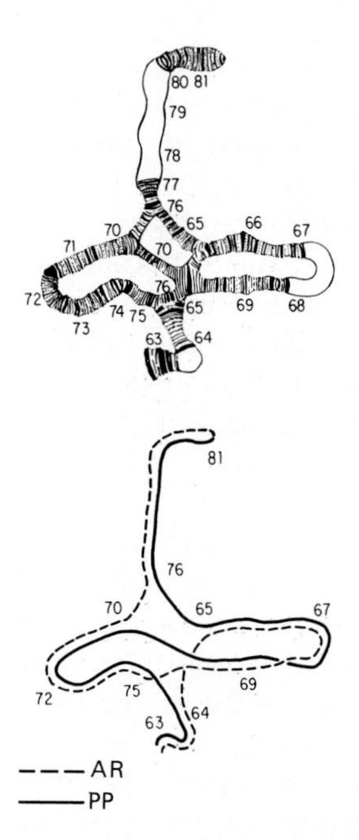


***Figure 16.8*** Relative frequencies of three different gene arrangements for chromosome 3 in *Drosophila pseudoobscura* in the southwestern United States. There are inverted blocks of genes in the Arrowhead (AR) and Pikes Peak (PP) chromosome types relative to each other and to the standard pattern. The three chromosome types can be identified according to band patterns, and inversion heterozygotes in these populations can be identified by inversion loops and bands of the paired chromosomes. Pairing in a nucleus heterozygous for AR and PP types of chromosome 3 is shown at the upper right, and an interpretive drawing of the two chromosomes is just below it. The numbers refer to bands on chromosome 3. (After studies by Dobzhansky.)

Among these are the mode of dispersal of gametes or spores, the preferences or barriers of mating between individuals, habitat differences, and other features of the life-style of the species.

## 16.5 Variations in Reproductive Systems

Self-fertilizing species usually are highly homozygous, regardless of the degree of crossing over or of independent assortment that may take place. Inbreeding leading to homozygosity is therefore a mating system that minimizes genetic variability and enhances genetic constancy in populations. Similarly, asexual reproduction also preserves genetic constancy from generation to generation, regardless of diversity levels in the gene pool. In contrast with these extreme mechanisms by which genetic constancy can be maintained, heterozygosity promoted by outbreeding serves to increase genotypic diversity in general. Heterozygosity and outbreeding or cross-fertilization are theoretically adaptive characteristics since diversity increases in amount and is incorporated into many new genotypes in each generation. Yet some species or groups of species have dispensed with such potential for variability and have still remained successful. In most cases, however, some compromise is expressed between genetic constancy and genetic variability. In other words, the life-styles of many species reveal patterns of reproduction and other features through which each species attains some degree of adaptedness (constancy) and some degree of adaptability (flexibility) at the same time. Certain situations illustrate the overall consequences of compromises in genetic systems, and only a few examples will be described to emphasize the major features involved.

Many flowering plant species have evolved from cross-fertilizing ancestors to become self-fertilizing or even asexually reproducing by some kind of apomictic mechanism. *Apomixis* refers to an asexual substitite for sexual reproduction. One apomictic mechanism, for example, involves the production of embryos from vegetative diploid cells instead of from fusion between eggs and sperm. Apomictic plants, such as the common dandelion, produce seeds, but the embryo has the same genotype as the parent plant since the embryo develops from a maternal diploid cell. What is the advantage of asexual reproduction in a species like dandelion? If we examine this species, we find that it usually inhabits temporarily suitable spaces that undergo rapid changes, such as lawns, open fields and roadsides, or areas disturbed by human activities of other sorts. Dandelion populations fluctuate rapidly in size and in location as well as in time. Dandelion seeds arriving at a suitable locale germinate quickly, produce an abundance of seeds apomictically, and can very quickly produce a large adult plant population. Genotypic constancy is highly adaptive for such a species since one or a few suitable genotypes can multiply quickly and produce many individuals with the same suitable genotype. One or a few adapted immigrants can therefore produce a large descendant population in one or two generations in some new locality, all of which are equally well adapted to the particular environment at a particular time.

Numerous protist and invertebrate species are successful colonizers, just like the common dandelion. They produce huge numbers of progeny, often by some asexual mechanism, live for a relatively brief time, inhabit transient environments, and experience considerable fluctuations in population size at frequent intervals. In all of these cases adaptedness and rapid spread of similar or identical adapted genotypes are highly desirable. High levels of diversity would be relatively less adaptive for such a life-style since only a few genotypes out of many would usually prove to be best adapted and any delay in establishing the population could be disastrous in the brief times available for colonizing a new, transient location.

New genotypes do arise in colonizing species either by mutation or by an occasional episode of sexual reproduction. In the ciliated protozan *Paramecium,* for example, reproduction usually occurs by asexual fissions. On occasion, usually when the habitat or food supply declines or becomes limiting, sexual reproduction is stimulated. Many genotypes are produced, and as the sexual progeny disperses, some genotypes will be suitable in one location or another. The species spreads to new locations, and its spectrum of suitable locations may become expanded as genotypic diversity is produced by occasional episodes of sexual reproduction. Genetic constancy is maintained by asexual reproduction most of the time, but genetic flexibility is provided by periodic bursts of sexual activity. These same properties also characterize many of the fungi and certain insects that are capable of achieving large populations very quickly when they acquire access to new and suitable locations and are equally capable of disappearing rapidly as these habitats change suddenly.

Colonizing species exhibit relatively little diversity of genotypes within any one population, but substantial levels of diversity may be found among different populations. Occasional interbreeding between members of different populations also leads to greater genotypic diversity and to overall diversity in the entire gene pool of the species.

Species with a large component of genetic flexibility and a relatively smaller component of genetic constancy have different features from the species we have described. Adaptability in changing environments would be most likely in highly heterozygous species that release genetic diversity steadily and at a slow rate. In this way new genotypes would be produced continually, and the chances would be good that one or more of these genotypes would prove to be suitable in a new or altered environment or life situation. A higher level of adaptedness or relative fitness would not be as advantageous in a diversified environment, in an environment with limited living space, or in one in which some particular genotype out of many would prove to be better under some specified set of conditions.

Adaptability, or genetic diversity, would be predicted to occur in species that are highly heterozygous and outbreeding, are relatively long-lived, exist in relatively stable habitats, and produce a relatively large number of gametes. We would predict that such species would have a high recombination index and probably would have higher chromosome numbers than related species that had opted for adaptedness in evolution. Such adaptable populations might produce many descendants, but only a few would survive to perpetuate the species. Good examples of such an adaptable genetic system can be found in various forest tree species, such as the oaks.

Oak populations undergo relatively little fluctuation in size; and once established, the trees continue to live for a long time in the same place. Progeny as fit or fitter than the successful parents would be rare, but one or a few acorns might develop into seedlings that could become established in the limited amount of space available in the forest. Oak species have relatively high chromosome numbers as well as high recombination index values, and they are almost exclusively cross-fertilizing.

Annual desert-dwelling plants provide many examples of compromise between fitness and flexibility components in the genetic system. Most of these species are highly heterozygous and cross-fertilizing, qualities that underwrite the potential for genetic diversity. These plants live in hostile terrain, however, and we would expect them to have some element of genetic constancy since they must be highly adapted to their environments and since their populations fluctuate in size a great deal. Many desert species of annual plants (they live for only one year) have fewer chromosomes and lower chiasma frequencies than related species living in friendlier environments. The lower chromosome numbers of the desert annuals lead to less diversity by independent assortment. The lower chiasma frequencies mean that favorable combinations of linked genes tend to be preserved as supergenes more often than not. By means of a low recombination index, therefore, such plants may maintain a highly heterozygous gene pool, but only a small portion of this diversity is actually generated in new genotypic combinations during reproduction. Furthermore, times of little or no rainfall are marked by little or no reproduction, and genetic constancy prevails in the populations at such times. When rain does come, the annuals have a burst of reproductive activity and new genotypes are produced along with a high proportion of genotypes that closely resemble the parents.

Species of *Haplopappus,* a member of the sunflower family (Compositae), may have 2, 4 or 8 pairs of chromosomes (Fig. 16.9). *Haplopappus gracilis* has only 2 pairs of chromosomes, and at least one of these has arisen by centric fusion between smaller chromosomes of an ancestral species. *Haplopappus* species with higher chromosome numbers have considerably smaller chromosomes than *H. gracilis.* Although all of these species are heterozygous and cross-fertilizing, *H. gracilis* exhibits less genotypic diversity, presumably because of its lower recombination index and related features. By reducing genetic flexibility and enhancing genetic constancy, *H. gracilis* has survived as a more

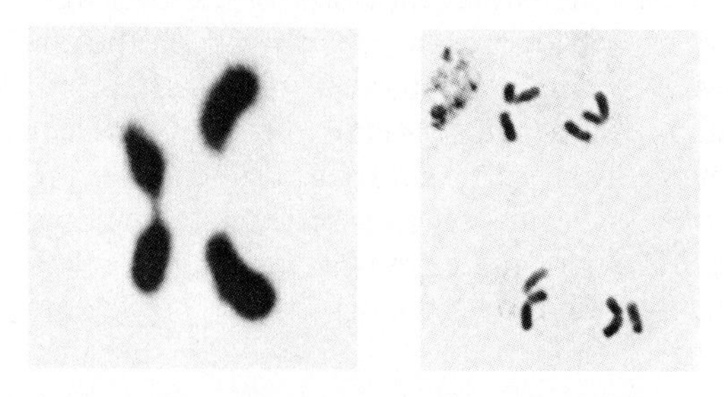

***Figure 16.9*** Meiosis in *Haplopappus gracilis*; $n = 2$. The two pairs of chromosomes are beginning to disjoin in very early anaphase I (left), and each of the four meiotic products (right) clearly has two chromosomes in its nucleus. This is the lowest known chromosome number for eukaryotic species. (From R.C. Jackson. 1959. *Amer. J. Bot.* **46**:550.)

highly adapted species in its desert environment. Should the environment change, this species may not be able to survive if it has evolved closer to adaptedness than to adaptability. In the short run, however, its adaptedness has been one basis for its success in the desert.

From many studies, at least for flowering plants, we can predict a relationship among the characteristics of the phenotype and the genetic system that is in operation. Specifically, we predict that trees and other long-lived perennials would probably exist in stable habitats, would be sexually reproducing and cross-fertilizing, and would have higher chromosome numbers and higher crossover frequencies. These features permit maximum recombination potential and, therefore, the highest amounts of diversity and the steadiest rates of release of this diversity in each generation. Species that occupy temporary habitats and have shorter life spans would predictably have at least one of the following mechanisms by which diversity is limited in amount and in rate of flow: (1) apomixis or asexual reproduction; (2) self-fertilization; or (3) a low recombination index due to lower chromosome number and fewer chiasmata per genome at meiosis. These features lead to reduced amounts of diversity in the gene pool and to an erratic flow of variability into each new generation. Through a number of different compromises between fitness and flexibility, a broad range of genetic systems becomes possible, and species may evolve toward any point between the extremes of high adaptedness and high adaptability. Since adaptability involves greater genetic diversity, it is more likely to characterize populations with a higher probability for long-term evolutionary success in meeting the challenges of the ever-changing world.

## 16.6 Sex as an Adaptive Complex

The origins of sexual reproduction are unknown, but we believe that the earliest organisms were asexual and that sexual species first appeared a little over one billion years ago. The enormous advantages of sexual reproduction center around its property of producing new genotypes by recombination in every generation. Through sexual reproduction, high levels of genotypic diversity can be attained if genetically different parents mate and produce both recombinant and parental types of progeny. The most reliable means for ensuring that parents will be genetically different is for species to produce different male and female individuals. This situation characterizes all the vertebrate animals. Cross-fertilization is mandatory if the sexes are represented by different individuals.

Once separation of the sexes has been achieved during evolution, the continued production of unequivocally separate male and female individuals is secured through the development of a genetic sex determination mechanism. Determination of sex by sex chromosomes rather than by one or a few genes makes the system less prone to alterations through point mutations. And once sex chromosomes are present, genetic differentiation between the X and Y chromosomes must be substantial or complete so that the sex chromosomes become and remain nonhomologous. Nonhomology will insure little or no pairing between the X and Y chromosomes, and crossing over will not take place between them. The genetic constitutions of both the X and Y will remain intact, and each chromosome will be transmitted as a complete X or a complete Y unit to every gamete. Blocks of sex-determining genes will be maintained as supergenes. If crossing over were possible between the X and Y chromosomes, some recombinant sex chromosome types would be produced and such X-Y recombinant sex chromosomes would direct development of sexual intermediates, leading to an eventual breakdown of the sex chromosome–determining mechanism.

In mammals, the Y chromosome is sex determining. Mammalian X and Y chromosomes have little or no homology, and no crossing over has been observed between them. In mammals, therefore, each individual with a Y chromosome is male and each individual without a Y chromosome is female. The separate sexes persist as two distinct types, and every generation is assured of being produced by parents from genetically different lineages. Heterozygosity is essentially assured, and diversity will continue to be produced and maintained in the gene pool of every vertebrate species. In this way, some degree of adaptability is guaranteed, and some degree of long-term evolutionary potential is built into the genetic system.

## 16.7 Species Formation

A species is generally considered to be a group of organisms that can freely interbreed and produce healthy, fertile offspring in each generation. Members of different species cannot or do not interbeed, and gene exchange ordinarily takes place within a species but not between species. The origin and maintenance of separate species, therefore, depends on the origin and maintenance of separate gene pools. Intuitively, we would predict that speciation involves two major steps, one that prevents interbreeding between different populations of the same species so that no gene exchange occurs, and each gene pool diversifies independently of the others, and another step that prevents gene exchange even if populations of different species inhabit the same living space. As a further corollary, we would also expect recently diverged, closely related species to be more similar genetically and distantly related species to be more distinct genetically because of their longer histories of separate evolution.

We generally accept that the initial event in speciation is the physical separation of subpopulations so that gene exchange cannot take place even though the potential exists. Once they are *spatially isolated,* subpopulations continue to diversify through mutation and selection, but their pathways of evolution will almost certainly not be identical. Mutations are random and we expect different mutations to arise by chance in different populations. Environments are varied, and we expect different selection pressures to operate in different physical spaces and on different sets of mutations (Fig. 16.10).

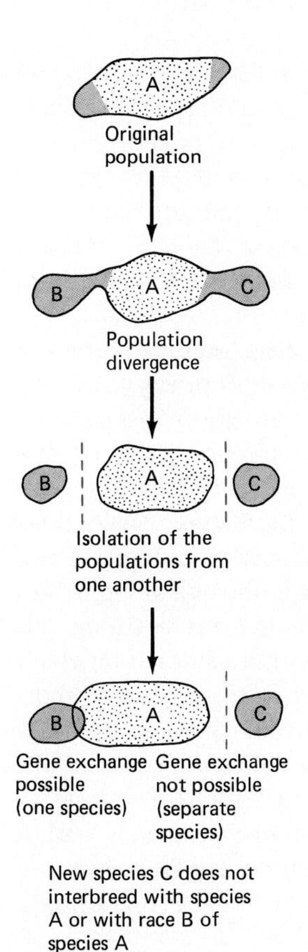


***Figure 16.10*** Hypothetical population A undergoing racial divergence and speciation. If time in isolation is sufficiently long, gene exchange may no longer be possible between subpopulations A and C, which implies they are separate species. When gene exchange is still possible, as between A and B, the two races constitute a single species.

As the separated populations become genetically differentiated during their evolution, they become *reproductively isolated,* that is, they become incapable of fruitful gene exchange. Even if the divergent populations were to inhabit the same location in the future, interbreeding would be minimized and gene exchange would be abortive. The populations, now separate species, would maintain their separate gene pools and would continue to diversify into still more distantly related entities with time. In general, therefore, genetic divergence occurs in **allopatric** populations (ones separated spatially) and may continue when these populations become **sympatric** (occupying the same space) if divergence has led to some measure of reproductive isolation during the allopatric stage. Speciation continues to the stage of different and noninterbreeding populations if natural selection leads to complete reproductive isolation of the two groups.

**Reproductive isolating mechanisms** are often divided into two groups: **prezygotic** and **postzygotic.** Prezygotic isolating mechanisms include any ecological, physiological, or behavioral barriers that prevent or interfere with mating and gamete fusions. Mating usually does not occur at all in these cases. Postzygotic isolating mechanisms include those leading to sterility or inviability of $F_1$ hybrids between different groups or to hybrid breakdown in which $F_2$ and backcross progeny are sterile or otherwise unfit. Once achieved, reproductive isolation is essentially irreversible and species will continue to diversify along separate evolutionary pathways. These pathways may ultimately lead to new categories of organisms, such as new genera, families, orders, and even higher taxonomic groups.

Numerous examples of reproductive isolating mechanisms come under the heading of prezygotic isolation. Different species of flowering plants, such as goldenrods and asters, may bloom at different seasons or times during a season. Since eggs and pollen are not available at the same time for the different species, gamete fusions will not take place. Similarly, animals that court and mate at different times, such as frog and toad species, do not engage in reproduction since potential partners from another species are not sexually receptive at the same times. Species that inhabit different ecological zones in an area are unlikely to meet and, therefore, are unlikely to mate. Behavioral differences between species are also very effective in keeping apart potential partners from different species, even if the species breed at the same times in the same place.

Postzygotic isolation is not observed as often as prezygotic isolation in natural populations, but numerous examples have been documented. The classic example of a sterile $F_1$ hybrid is the mule, which is produced by mating a horse and a jackass. Such matings are unlikely to take place in nature, but if they did, no further gene exchange would occur between the two gene pools since the $F_1$ hybrid is incapable of breeding either with others of its own kind or with either parent type.

In general, therefore, a variety of patterns of genetic differentiation characterize reproductive isolation. Genetic differences arise in allopatric populations and are usually maintained and strengthened even when the populations become sympatric. Some cases of sympatric speciation have been noted. One of the more obvious is the situation in which polyploids arise within a population and are unable to breed successfully with the parent types because any zygotes will have an unbalanced number of chromosomes. In tetraploid × diploid matings, the progeny will be triploid, and such individuals are sterile because of meiotic abnormalities arising from pairings and separations of three genomes during gamete formation. If the newly arisen tetraploid is self-fertilizing, seeds will be produced and the polyploids may establish their own gene pool and go on to develop into separate species isolated from the parental diploids by postzygotic isolating mechanisms. If two different diploid species produce a diploid hybrid that is sterile, and if the sterile hybrid becomes tetraploid by doubling of its chromosome number, the allotetraploid will be reproductively isolated from both diploid species because of triploid sterility resulting from any mating (see Section 12.1).

## Molecular Evolution of the Genome

It is not possible to interbreed organisms that belong to different kingdoms, phyla, classes, orders, or families. It is possible, however, to use various molecular methods, such as molecular hybridization, gel electrophoresis, or nucleotide sequencing, as substitutes for genetic breeding analysis and thus to study phylogenetic relationships from a genetic perspective. The extent of similarity and divergence in nucleotide sequences and amino acid compositions of proteins is remarkably close to predictions based on inferred genetic relationships and times of evolutionary divergence. Although individual genes or proteins sometimes have undergone significant divergence from an ancestral form, some features of the genome have changed remarkably little over tens of millions or hundreds of millions of years of evolution.

## 16.8 Molecular Phylogenies

By molecular hybridization it has been shown that DNAs isolated from representative vertebrate species are more similar among members of the same taxonomic group than among members of different groups. DNAs from animals of the same order or class, such as primates of the class Mammalia, resemble each other in many respects and are less similar to members of other classes, such as birds, reptiles, and amphibians. The same general similarities and differences characterize RNAs from ribosomes of a range of eukaryotic organisms. If the G + C/A + U ratios of rRNAs are plotted, ratios for related species are clustered together around similar values. These clustered values are distinct from clusters of values obtained for other groups of related organisms.

In Figure 16.11, ratios for the six protist representatives (numbers 1 to 6) are scattered along a gradient of values, indicating considerable divergence with respect to the G + C/A + U rRNA ratios. The highest value among the

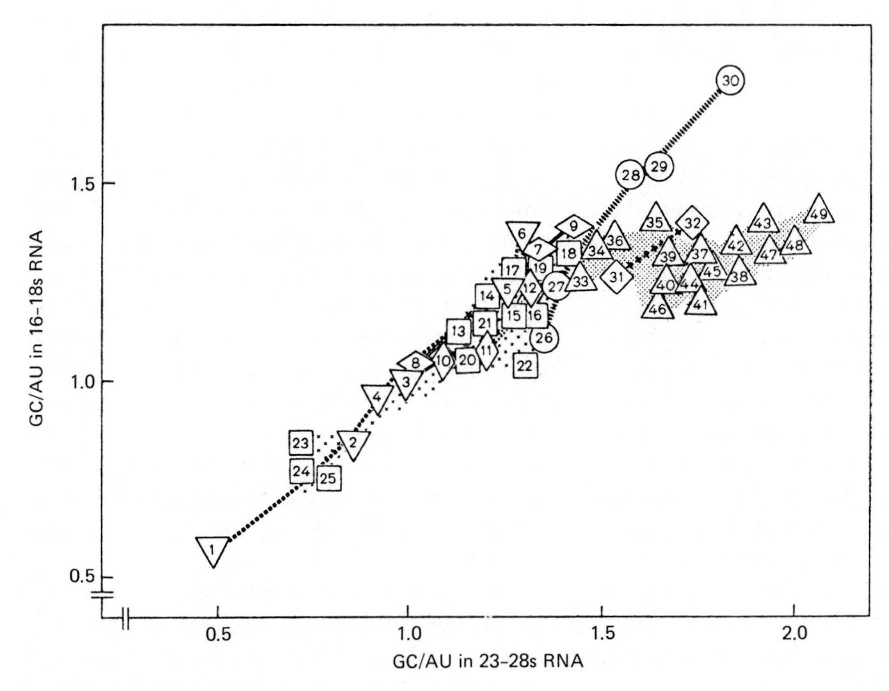


***Figure 16.11*** Evolution of the G + C/A + U ratio of rRNA in animals and certain protists. Numbers 1–6 are the protists (*Euglena* = 6); 7–9, sponges and coelenterates; 10–12, nematodes and annelids; 13–25, arthropods; 26–30, molluscs; 31–32, echinoderms; 33–49, vertebrates (*Homo sapiens* = 49). The evolutionary pattern for rRNA base composition shown here reflects the known phylogenies of these groups. (From P.A. Lava-Sanchez, F. Amaldi, and A. La Posta. 1972. *J. Mol. Evol.* **2**:44, Fig. 2.)

six is for *Euglena,* whose ratio value is very similar to that of the sponge, the single representative of the simple group called Porifera (number 7). This similarity in GC/AU values is in accord with the view that *Euglena* is evolutionarily very close to the lower multicellular animals. Looking at the central area of the plot of GC/AU values, we can see that divergences appear to have proceeded separately from the simpler molluscs (numbers 26 and 27) toward more highly evolved molluscs (numbers 28 to 30) on one pathway, and toward the echinoderms (numbers 31 and 32) and vertebrates (numbers 33 to 49) along another pathway. The known phylogenetic relationships among the taxonomic groups, based on comparative anatomy and other studies, are well reflected by the evolutionary pattern deduced from rRNA base composition. Similar relationships have been shown to characterize bacterial, protist, fungal, and land plant groups, in confirmation of phylogenetic relationships established from morphological and other parameters.

Comparative studies of amino acid composition and sequence in particular proteins common to diverse groups of organisms have also provided confirmations of phylogenetic relationships deduced from studies using a variety of independent analytical methods. Cytochrome *c* is a respiratory protein that transfers electrons toward molecular oxygen in all aerobic species, both prokaryotic and eukaryotic. We presume that the gene that encodes cytochrome *c* arose almost 1.5 billion years ago, when the Earth's atmosphere was becoming increasingly aerobic. The gene must have arisen in a prokaryotic ancestor of the eukaryotes, which evolved sometime later. We infer a common origin for the gene and its protein product in all present-day species since about half the amino acids occupy the same positions in cytochrome *c* molecules obtained from any aerobic species.

During evolution, about half the amino acid residues in cytochrome *c* have undergone substitutions, reflecting mutations in codons of the original and subsequent cytochrome *c* gene sequences during 1.5 billion years of evolution. An **evolutionary** or **phylogenetic tree** can be constructed to indicate the probable number of base substitutions in the gene that would lead to amino acid substitutions observed in the protein (Fig. 16.12). Similar phylogenetic trees have been constructed for other proteins known or believed to be homologous in origin and function in various groups of organisms.

The rate of protein change during evolution can be estimated by determining the number of amino acid changes in a protein in relation to the time of divergence of various organisms, based on the fossil record. The rates of evolutionary change appear to be relatively constant for any given protein type, but rates of change vary widely for functionally different proteins (Fig. 16.13). The slowest rates of change characterize molecules that can least tolerate amino acid substitutions and still carry out their vital functions. Histones bind to DNA at many sites along the protein. Modifications almost anywhere along the histone primary structure would interfere with binding and, therefore, would interfere with DNA packaging in the chromosomes. On the other hand, fibrinopeptides are short polypeptides within the fibrinogen molecule that are enzymatically excised from fibrinogen when it is converted to fibrin during blood clotting. Fibrinopeptide molecules have far fewer restraints in carrying out their function since only a small portion of the molecule is specifically required for enzyme recognition and action. Amino acid substitutions in most of the fibrinopeptide chains would have little or no effect on clotting reactions.

Hemoglobin and cytochrome *c* show rates of change that are intermediate between those for fibrinopeptides and histones. Hemoglobin and cytochrome *c* have some tolerance for amino acid substitutions while still retaining function, but portions of each of these molecules cannot be modified by gene mutations without altering

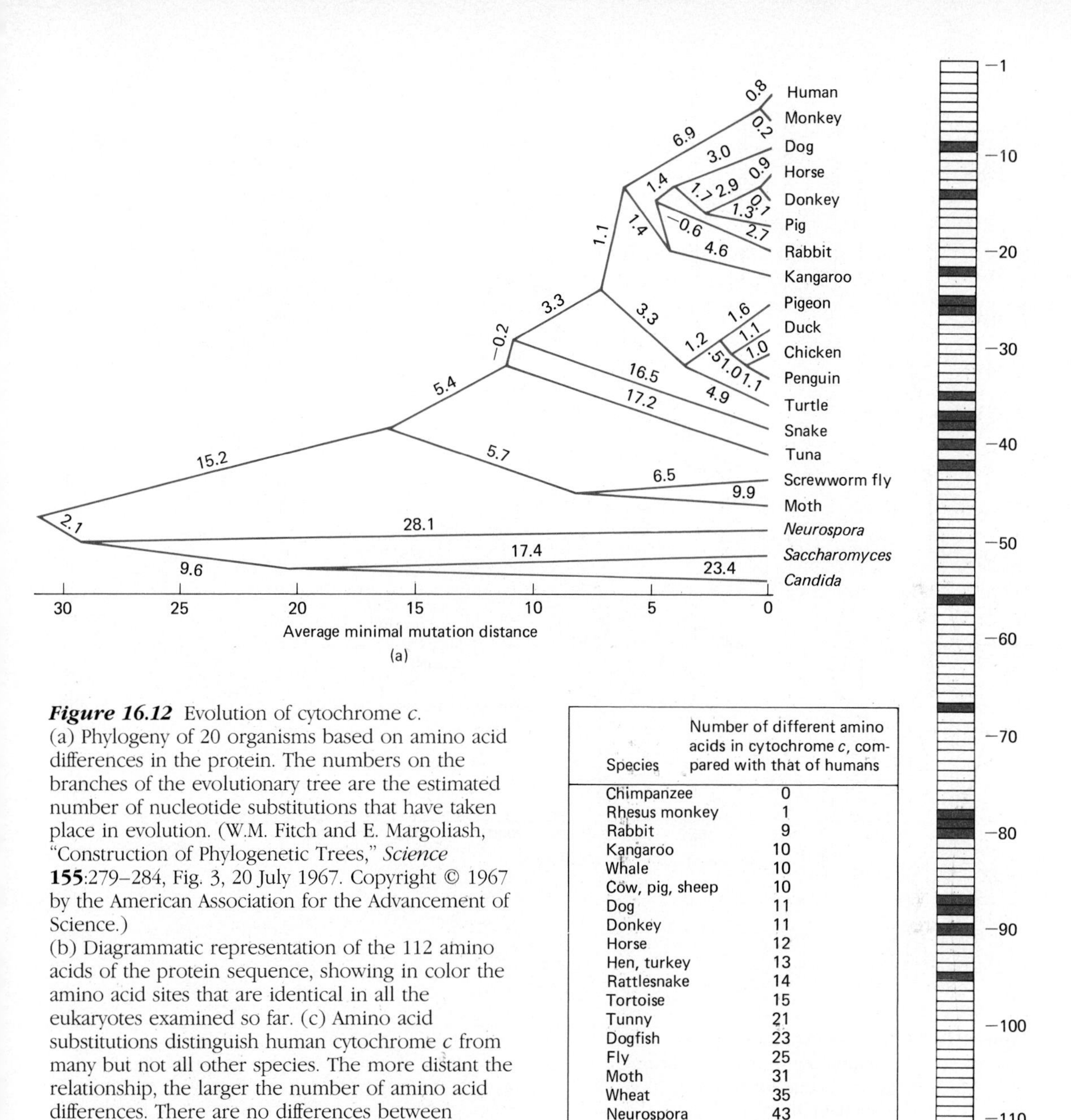


| Species | Number of different amino acids in cytochrome c, compared with that of humans |
|---|---|
| Chimpanzee | 0 |
| Rhesus monkey | 1 |
| Rabbit | 9 |
| Kangaroo | 10 |
| Whale | 10 |
| Cow, pig, sheep | 10 |
| Dog | 11 |
| Donkey | 11 |
| Horse | 12 |
| Hen, turkey | 13 |
| Rattlesnake | 14 |
| Tortoise | 15 |
| Tunny | 21 |
| Dogfish | 23 |
| Fly | 25 |
| Moth | 31 |
| Wheat | 35 |
| Neurospora | 43 |
| Yeast | 44 |

***Figure 16.12*** Evolution of cytochrome *c*.
(a) Phylogeny of 20 organisms based on amino acid differences in the protein. The numbers on the branches of the evolutionary tree are the estimated number of nucleotide substitutions that have taken place in evolution. (W.M. Fitch and E. Margoliash, "Construction of Phylogenetic Trees," *Science* **155**:279–284, Fig. 3, 20 July 1967. Copyright © 1967 by the American Association for the Advancement of Science.)
(b) Diagrammatic representation of the 112 amino acids of the protein sequence, showing in color the amino acid sites that are identical in all the eukaryotes examined so far. (c) Amino acid substitutions distinguish human cytochrome *c* from many but not all other species. The more distant the relationship, the larger the number of amino acid differences. There are no differences between cytochrome *c* from great apes and human cells, and only one difference between our cytochrome *c* (isoleucine at position 66) and that from the Rhesus monkey (threonine at position 66).

protein function. Substitutions in these critical regions of the protein would be disadvantageous and, therefore, would be selected against. Since many amino acid substitutions have occurred during evolution without altering protein function, some investigators consider the observed substitutions to be the result of neutral mutations. Neutral mutations may account in part for their relatively slow but constant rates of evolutionary change.

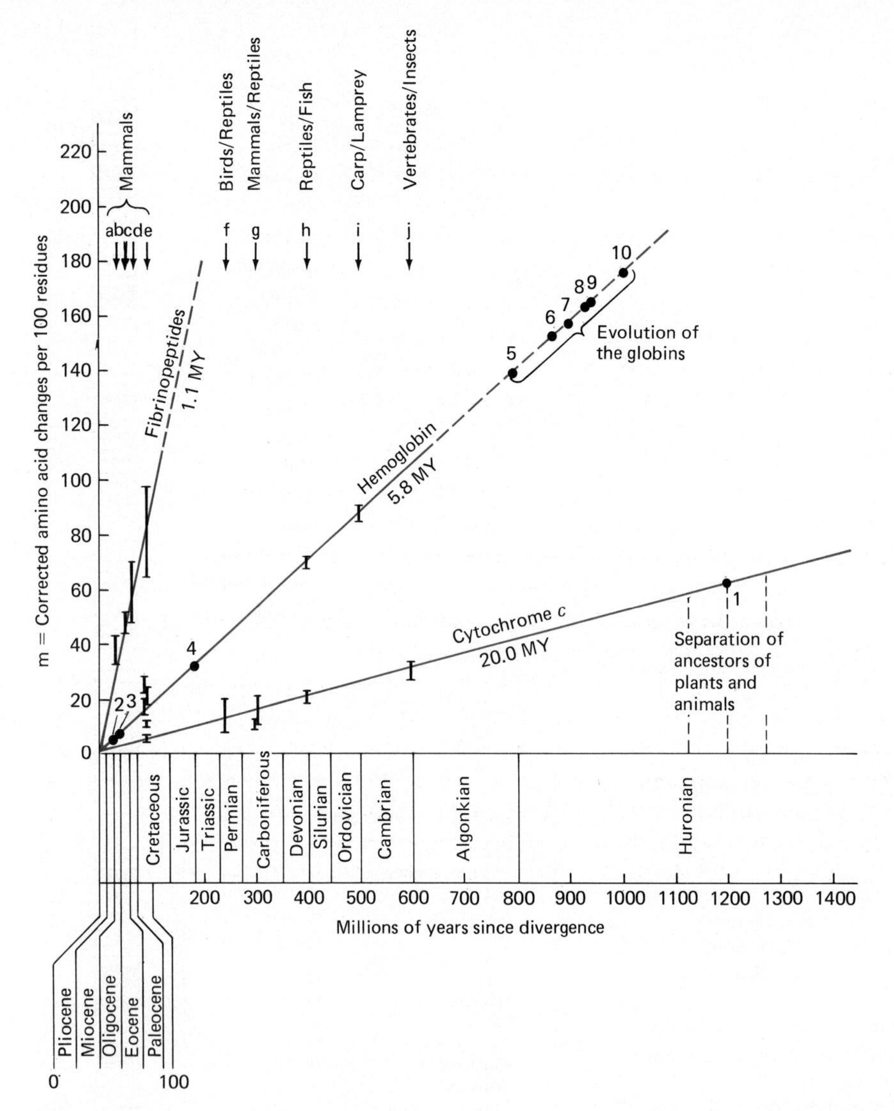


***Figure 16.13*** Rates of evolution in the fibrinopeptides, hemoglobins, and cytochrome *c* proteins. The rate of change for histones (not shown) would essentially be plotted as an almost horizontal line across the bottom of the graph. Faster rates of change characterize the molecules most able to tolerate amino acid substitutions and still maintain vital functions. Point 1 represents a date of 1200 ± 75 MY (million years) for the separation of plants and animals, based on cytochrome *c* data; points 2–10 refer to events in hemoglobin-gene evolution. The time in millions of years for a 1% change in amino acid sequence to show up between evolutionary lineages is shown just beneath each curve. (From R.E. Dickerson. 1971. *J. Mol. Evol.* **1**:26.)

## 16.9 Conservation of Gene Structural Organization

When coding sequences in eukaryotic genes are analyzed, there appears to be a striking similarity in the number of exons retained by related genes that specify homologous proteins. We mentioned this important feature earlier in discussions of the cluster of $\beta$ globin and $\beta$-like globin genes in higher primates (see Section 12.4). Despite increasing divergence of exon nucleotide sequences and amino acid substitutions in the polypeptide products of the genes, the *organization* of exons and introns remained similar or even identical in interrupted genes that had evolved over tens or hundreds of millions of years. This observation immediately suggests that the "meaning" of the exons is related in some way to the structural organization and function of the protein product. In an increasing number of cases we find that the number of exons in the gene corresponds to compact structural regions (**domains**) of the protein product. In many cases each exon or cluster of exons apparently does indeed encode each domain in the protein (Fig. 16.14).

Two important implications arise from the observed concordance between exons and protein domains. First, the concordance provides a framework for analysis of genes and proteins that we suspect are related but for which amino acid sequences alone have been uninformative or for which nucleotide sequences are largely nonhomologous. The second implication is that gene evolution in eukaryotes may involve generating new proteins by generating new combinations of exons from the same or from different genes. This suggestion may help explain the speed of eukaryotic evolution. Each of these implications has opened new avenues of investigation and has led to new insights into evolutionary processes.

One study of functionally related proteins that revealed their hitherto unknown genetic relationship was reported in 1981 by Shirley Tilghman and co-workers. They sought relationships between two proteins that constitute a major component of the blood serum. *Serum albumin* becomes the major component at birth and replaces *alpha-fetoprotein* (AFP), which is synthesized in the developing mammalian fetus. Both proteins help regulate the osmotic pressure of the intravascular fluid, and they are structurally similar in size and in amino acid composition. They act at different developmental stages in the organism, just as $\gamma$ globins and $\delta$ globins or $\beta$ globins do in hemoglobins of the fetus or in the individual after birth. To test whether the AFP and albumin genes arose from a common ancestral gene, Tilghman used two criteria. She first determined the amino acid sequences of the two proteins to compare compositional homologies and organizational homologies in protein structure. She next determined the structure of the genes themselves in order to compare nucleotide sequence homologies and organizational homology.

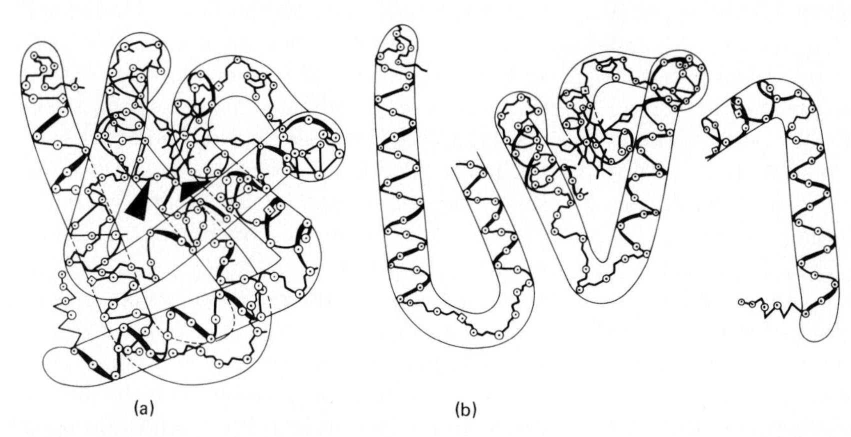


***Figure 16.14*** Domains in globin polypeptide appear to correspond to exons in globin gene organization. (a) Two arrows show the points at which the structure of a globin chain is interrupted by the two introns in the corresponding globin gene sequence. (b) The exon-intron splice junctions occur in the alpha-helical region of the polypeptide, shown here in three domains corresponding to the three exons of the globin gene. The product of the central exon surrounds the heme component of the globin chain. (W. Gilbert, "DNA Sequencing and Gene Structure," *Science* **214**:1311, Fig. 11, 18 December 1981. Copyright © 1981 by the Nobel Foundation.)

Serum albumin amino acid sequence had been analyzed in previous studies of bovine and human albumins. The human albumin consists of 584 amino acid residues, and the bovine albumin is 2 amino acid residues shorter. AFP amino acid sequence was deduced from base sequencing of cDNA clones, which were derived from mouse AFP mRNA, according to standard codon usage in specifying amino acids in polypeptides. Base sequencing is currently easier, faster, and more accurate than direct amino acid sequencing. Tilghman found that mouse AFP consisted of 584 amino acids and that amino acid sequences were strikingly similar in organization in AFP and the two albumins (Fig. 16.15). The three closely analogous domains in each protein are very similar in extent and organization; and 180 out of 584 conserved amino acid residues are common to all three proteins. These data provide strong evidence for the common ancestry shared by AFP and mammalian serum albumins.

To gain genetic evidence for this postulated common ancestry, Tilghman and co-workers compared the organization of mouse AFP and serum albumin genes, using cloned DNA from genomic libraries. Restriction enzyme mapping and heteroduplex and R-loop analysis using electron microscopy revealed that the AFP and albumin genes were organized similarly into 15 coding segments interrupted by 14 intervening sequences. The sizes of the corresponding coding segments were identical, but no nucleotide sequence homology was detected between corresponding coding segments in hybridization tests (Table 16.1). Hybridizations and heteroduplex mapping also revealed no detectable sequence homologies for intervening sequences between exons or for flanking sequences of the AFP and albumin genes.

Despite nonhomologies between the two gene sequences, the conserved organizational features of the two interrupted genes are strong evidence for their common origin. It is very unlikely that fifteen exons of similar lengths would have arisen by chance in two genes coding for proteins of similar function. Furthermore, about one-third of their amino acids are the same in kind and in location within the protein structures. These concordant features in AFP and serum albumin genes and protein products, together with similar results found for globins and other molecules subjected to genetic and biochemical comparisons, provide strong evidence for conservation of genetic organization in the evolution of homologous genes coding for a variety of homologous proteins. The relationships between gene structure and protein structure thus extend far beyond individual codons that specify individual amino acids. Conservation of functional domains in homologous proteins is achieved by conservation of a corresponding organization of exons and introns in the interrupted gene sequence. In view of observed *organizational homologies* for gene and proteins showing minimal nucleotide and amino acid *compositional* homologies, other homologous proteins showing weak compositional homology should be reevaluated after organizational analysis has been accomplished. Only in this way will it be possible to trace the ancestry and evolution of related genes, even for genes that specify proteins with some functional differences. For example, comparisons between hemoglobins and cytochromes can reveal whether the heme-binding domains of these functionally distinct molecules share a common ancestry.

***Figure 16.15*** (opposite page) The amino acid sequence of mouse alpha-fetoprotein (AFP). Where any of the 584 amino acids in the mouse AFP sequence is the same as in either human or bovine serum albumin, the amino acid circle is colored above (human) or blackened below (bovine) the label. Amino acids are numbered to the left and right of the large loops. A total of 180 out of 584 amino acids are the same in all three protein sequences, as seen in circles colored above and blackened below the amino acid name. (From M.B. Gorin et al. 1981. *J. Biol. Chem.* **256**:1954–1959, Fig. 3.)

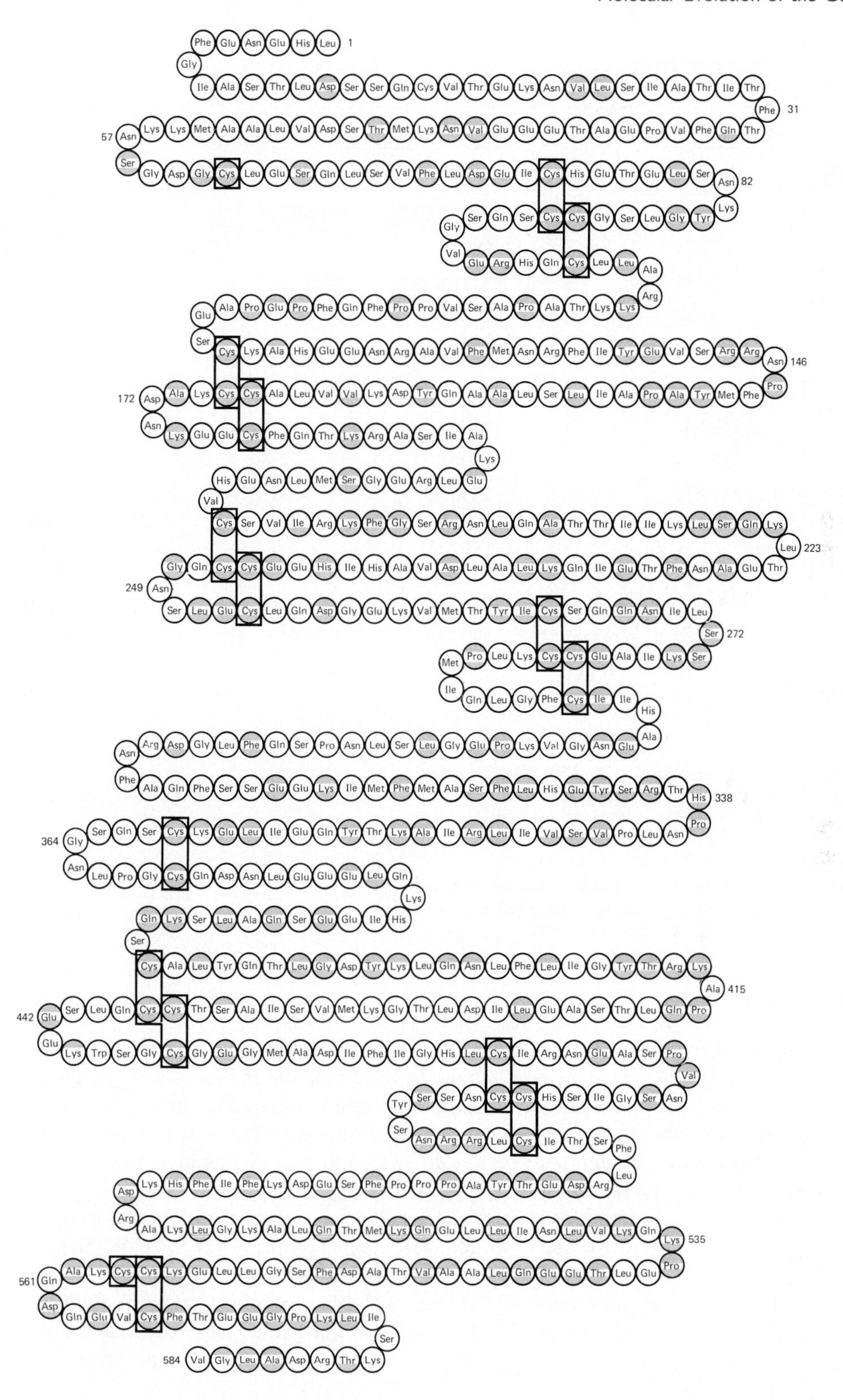
Phe Glu Asn Glu His Leu 1
Gly
Ile Ala Ser Thr Leu Asp Ser Ser Gln Cys Val Thr Glu Lys Asn Val Leu Ser Ile Ala Thr Ile Thr
Phe 31
57 Asn Lys Lys Met Ala Ala Leu Val Asp Ser Thr Met Lys Asn Val Glu Glu Glu Thr Ala Glu Pro Val Phe Gln Thr
Ser
Gly Asp Gly Cys Leu Glu Ser Gln Leu Ser Val Phe Leu Asp Glu Ile Cys His Glu Thr Glu Leu Ser
Asn 82
Lys
Gly Ser Gln Ser Cys Cys Gly Ser Leu Gly Tyr
Val
Glu Arg His Gln Cys Leu Leu
Ala
Arg
Glu Ala Pro Glu Pro Phe Gln Phe Pro Pro Val Ser Ala Pro Ala Thr Lys Lys
Ser
Cys Lys Ala His Glu Glu Asn Arg Ala Val Phe Met Asn Arg Phe Ile Tyr Glu Val Ser Arg Arg
Asn 146
Pro
172 Asp Ala Lys Cys Cys Ala Leu Val Val Lys Asp Tyr Gln Ala Ala Leu Ser Leu Ile Ala Pro Ala Tyr Met Phe
Asn
Lys Glu Glu Cys Phe Gln Thr Lys Arg Ala Ser Ile Ala
Lys
His Glu Asn Leu Met Ser Gly Glu Arg Leu Glu
Val
Cys Ser Val Ile Arg Lys Phe Gly Ser Arg Asn Leu Gln Ala Thr Thr Ile Ile Lys Leu Ser Gln Lys
Leu 223
Gly Gln Cys Cys Glu Glu His Ile His Ala Val Asp Leu Ala Leu Lys Gln Ile Glu Thr Phe Asn Ala Glu Thr
249 Asn
Ser Leu Glu Cys Leu Gln Asp Gly Glu Lys Val Met Thr Tyr Ile Cys Ser Gln Gln Asn Ile Leu
Ser 272
Met Pro Leu Lys Cys Cys Glu Ala Ile Lys Ser
Ile
Gln Leu Gly Phe Cys Ile Ile
His
Ala
Asn Arg Asp Gly Leu Phe Gln Ser Pro Asn Leu Ser Leu Gly Glu Pro Lys Val Gly Asn Glu
Phe
Ala Gln Phe Ser Ser Glu Glu Lys Ile Met Phe Met Ala Ser Phe Leu His Glu Tyr Ser Arg Thr
His 338
Pro
364 Gly Ser Gln Ser Cys Lys Glu Leu Ile Glu Gln Tyr Thr Lys Ala Ile Arg Leu Ile Val Ser Val Pro Leu Asn
Asn
Leu Pro Gly Cys Gln Asp Asn Leu Glu Glu Glu Leu Gln
Lys
Gln Lys Ser Leu Ala Gln Ser Glu Glu Ile His
Ser
Cys Ala Leu Tyr Gln Thr Leu Gly Asp Tyr Lys Leu Gln Asn Leu Phe Leu Ile Gly Tyr Thr Arg Lys
Ala 415
442 Glu Ser Leu Gln Cys Cys Thr Ser Ala Ile Ser Val Met Lys Gly Thr Leu Asp Ile Leu Glu Ala Ser Thr Leu Gln Pro
Glu
Lys Trp Ser Gly Cys Gly Glu Gly Met Ala Asp Ile Phe Ile Gly His Leu Cys Ile Arg Asn Glu Ala Ser Pro
Val
Tyr Ser Ser Asn Cys Cys His Ser Ile Gly Ser Asn
Ser
Asn Arg Arg Leu Cys Ile Thr Ser
Phe
Leu
Asp Lys His Phe Ile Phe Lys Asp Glu Ser Phe Pro Pro Pro Ala Tyr Thr Glu Asp Arg
Arg
Ala Lys Leu Gly Lys Ala Leu Gln Thr Met Lys Gln Glu Leu Leu Ile Asn Leu Val Lys Gln
Lys 535
561 Gln Ala Lys Cys Cys Lys Glu Leu Leu Gly Ser Phe Asp Ala Thr Val Ala Ala Leu Gln Glu Glu Thr Leu Glu
Pro
Asp
Gln Glu Val Cys Phe Thr Glu Glu Gly Pro Lys Leu Ile
Ser
584 Val Gly Leu Ala Asp Arg Thr Lys

***Table 16.1*** Sizes of coding and intervening sequences in the alpha-fetoprotein (AFP) and serum albumin genes in the mouse.*

| | coding | | | intervening | |
|---|---|---|---|---|---|
| **region†** | **AFP** | **albumin** | **region** | **AFP** | **albumin** |
| 1 | 114±32 | 112±31 | 1-2 | 890±166 | 804±96 |
| 2 | 53±14 | 62±26 | 2-3 | 882±165 | 1016±77 |
| 3 | 148±37 | 122±36 | 3-4 | 3324±470 | 1277±129 |
| 4 | 218±52 | 222±42 | 4-5 | 1045±126 | 558±51 |
| 5 | 144±26 | 153±31 | 5-6 | 954±198 | 1200±118 |
| 6 | 104±22 | 125±28 | 6-7 | 1260±107 | 1380±93 |
| 7 | 133±29 | 136±30 | 7-8 | 1610±120 | 1045±147 |
| 8 | 230±23 | 218±65 | 8-9 | 1830±160 | 826±72 |
| 9 | 154±57 | 137±35 | 9-10 | 755±96 | 1077±120 |
| 10 | 125±35 | 118±33 | 10-11 | 635±62 | 1343±83 |
| 11 | 135±45 | 170±70 | 11-12 | 1603±143 | 308±51 |
| 12 | 280±70 | 240±49 | 12-13 | 1072±122 | 2021±102 |
| 13 | 175±50 | 121±48 | 13-14 | 548±102 | 582±69 |
| 14 | 69±27 | 75±27 | 14-15 | 230±61 | 1138±61 |
| 15 | 149±33 | 110±29 | | | |

*Data are presented as numbers of base-pairs ± standard deviation. There is a marked similarity between the genes in the organization of coding sequences and size of the sequences, and a general lack of similarity between the genes in the size of the fourteen intervening sequences between coding segments.

†Coding segments are numbered in the 5′→3′ direction, and the intervening sequences are indicated by the coding segments between which they occur. Brackets indicate the three domains in each protein molecule.

With permission from D. Kioussis et al. 1981. *J. Biol. Chem.* **256**:1960–1967, Table 1.

## 16.10 Gene Duplication in Evolution

In a number of cases it is clear from gene sequencing and amino acid sequencing that duplicated structural genes exist in the genome of eukaryotes. Globin genes are presumed to have duplicated and subsequently diverged from an ancestral gene, beginning with the appearance of invertebrate animals about 650 million years ago. The ancestral gene may have encoded a polypeptide similar to the muscle protein *myoglobin,* which exists as a single polypeptide chain bound to a heme group. We believe that about 400 million years ago, globin gene duplication took place in the vertebrate lineage and subsequent genetic divergence produced $\alpha$ globin and $\beta$ globin genes or their equivalents (Fig. 16.16). On the basis of estimated nucleotide substitutions leading to divergent amino acid compositions, the $\beta$ globin gene continued to duplicate and diverge to produce $\epsilon$ globin, $\gamma$ globin, and, most recently, $\delta$ globin. The $\gamma$ globin gene must have duplicated again very recently since only a single nucleotide

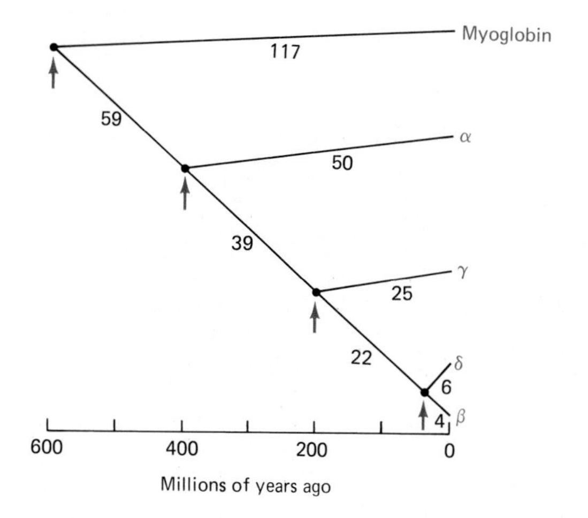


***Figure 16.16*** Evolutionary lineage of globin genes. Duplication of the ancestral myoglobin gene and subsequent divergence of the duplicates gave rise to $\alpha$ globin and $\beta$ globin genes, beginning about 400 million years ago in a vertebrate lineage. Subsequent duplications and divergences gave rise to the other known genes in the $\beta$ globin gene cluster, including $\gamma$ globin and $\delta$ globin genes. The estimated numbers of nucleotide substitutions leading to the modern human globin genes from the ancestral duplicates are shown for each branch of the evolutionary sequence.

difference distinguishes $^{A}\gamma$ globin and $^{G}\gamma$ globin genes. Only 10 amino acid differences characterize $\delta$ globin and $\beta$ globin, which is far fewer than the differences between the $\beta$ globin gene and either $\epsilon$ globin or $\gamma$ globin genes. From data such as these, phylogenetic trees can be constructed for other genes that have duplicated and subsequently diverged during evolution or for divergent sequences of the same gene in different organisms. The phylogenetic tree for cytochrome *c* was derived in this way (see Fig. 16.12).

Divergence of the same gene in different organisms or of initially identical duplicated genes in the same and in different organisms is expected to occur because of the random nature of mutation and the differences in selection pressures acting on the organism. The same mutations or other genetic modifications are highly unlikely to occur by chance in different organisms at different times and places during their evolution. Constraints are imposed on the degree of evolutionary modification, however, according to the tolerance of the protein in incorporating new amino acids or in altering domain structure and still being able to carry out a vital function (see Section 16.8).

In addition to duplication of whole-gene sequences, a number of recent studies indicate that individual coding sequences within genes may also undergo duplication and rearrangement. Walter Gilbert's suggestion that novel proteins might be generated through additions and rearrangements of exon segments of the same or different genes has now been verified by several studies. The organizational analysis of mouse alpha-fetoprotein and serum albumin genes by Tilghman provided evidence for the origin of one of the two genes by triplication of an ancestral coding sequence (Fig. 16.17). Since the three domains of these proteins are analogous in function and composition, a coding sequence probably duplicated and then underwent one more duplication of a segment to produce a gene with three similar regions coding for a protein with three similar domains. *After* the gene had evolved to a tripartite organization, the tripartite gene must have duplicated and subsequently diverged to the present-day AFP and serum albumin genes. The alternative possibility is that the single coding sequence duplicated and each single sequence then triplicated independently. This possibility is more remote because of the observed similarities in all three domains of both proteins and the similarities in organization of the AFP and albumin genes. More coincidences or identical events would have had to take place to produce two homologous tripartite genes from an original pair of duplicated coding segments.

The structure of one of the cluster of collagen genes (chick $\alpha$-2 (type I) collagen gene) consists of at least 52 exons interspersed with 51 introns over a length of 38,000 base-pairs. Structural analysis of the gene has shown that many of these 52 exons are 54 base-pairs long, which suggests that this collagen gene evolved by many

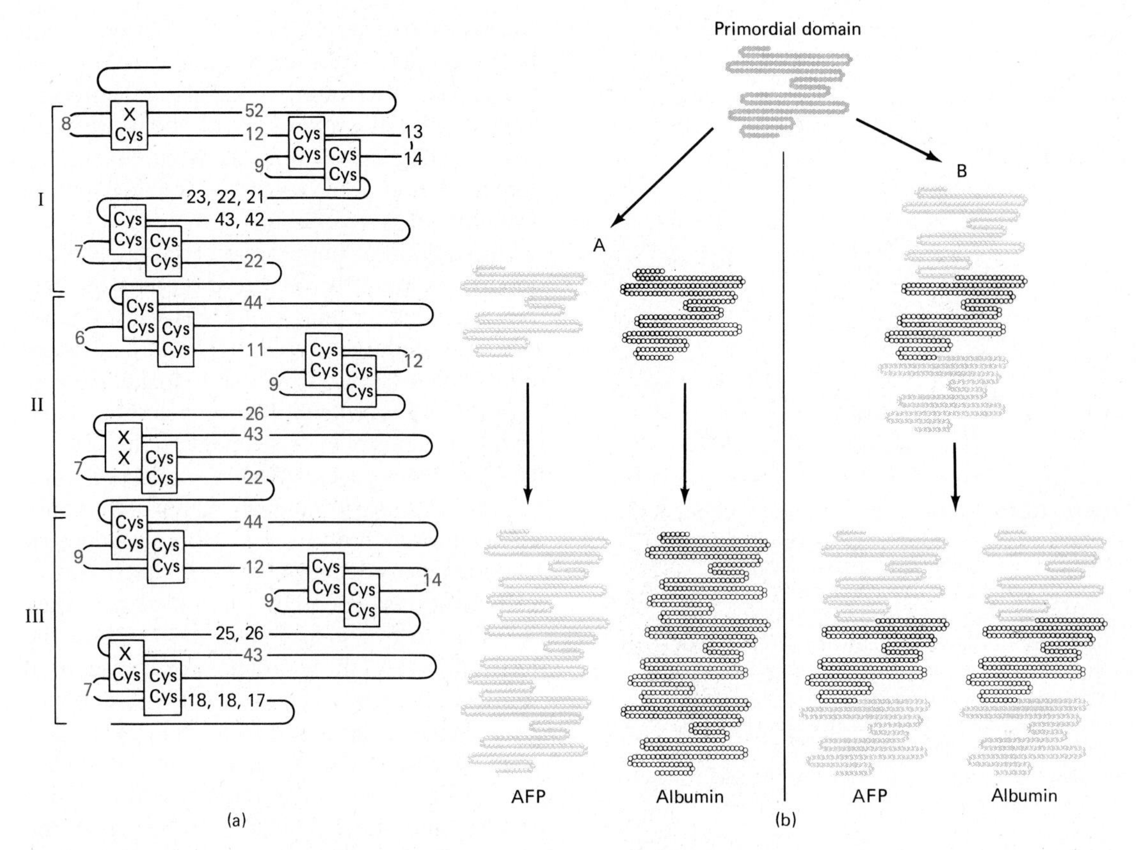


*Figure 16.17* Comparison of alpha-fetoprotein (AFP) and serum albumin sequences. (a) The lengths of amino acid sequences between cysteine-cysteine bonded residues are given for mouse AFP (numbers in color) and for human and bovine serum albumins. A single number indicates that all three proteins are identical, two numbers indicate the difference between AFP and the two albumins, and three numbers indicate the residues in AFP, human albumin, and bovine albumin, respectively, from left to right. The approximate borders of the three domains in these proteins are outlined on the left. (b) Two models to explain the evolution of AFP and albumin genes from a single primordial gene, according to analysis of the proteins, are shown in A and B. Each of the models postulates the evolution of a tripartite organization from a single coding sequence. The data best fit model B, in which AFP and albumin genes diverged *after* the ancestral gene had evolved to a tripartite organization. (From M.B. Gorin et al. 1981. *J. Biol. Chem.* **256**:1954–1959, Figs. 4 and 5.)

duplications of an ancestral gene or coding sequence that was 54 bp long (Fig. 16.18). The coincidence of many exons of identical length is explained more reasonably by duplications of an ancestral coding segment than by chance events alone producing such similar exon segments in the present-day gene.

The processes leaking to duplications of exons or of complete interrupted gene sequences are not entirely known. *Unequal crossing over* remains the most probable explanation for duplication of genetic material (see Fig. 12.12). The existence of small and large flanking repeats within the gene and between genes provides a means by which complementary base pairing can take place, thus leading to unequal recombination products. Although an extra sequence may become incorporated into one re-

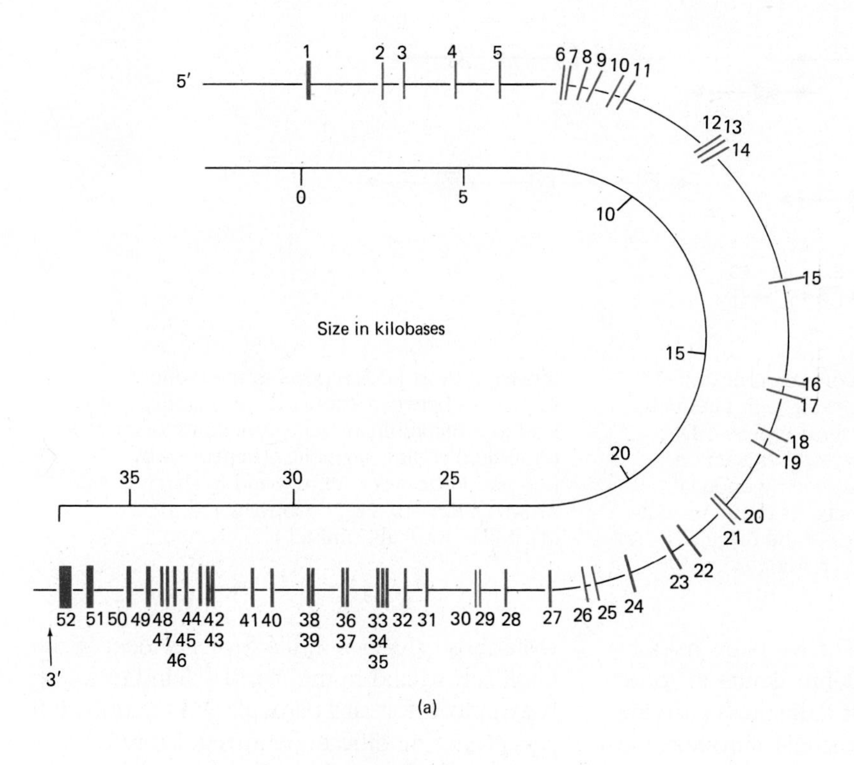


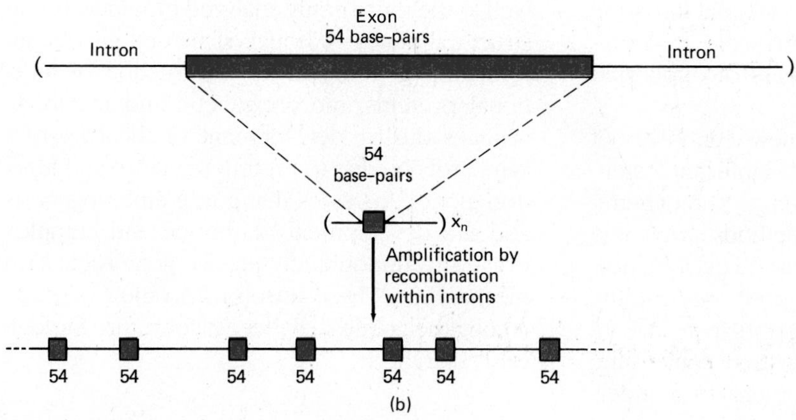


***Figure 16.18*** Organization of the chick $\alpha$-2 (type I) collagen gene. (a) The gene consists of 52 exons (color bars) interspersed with 51 introns over a length of about 38,000 base-pairs (38 kb). (b) Many of these exons are 54 base-pairs long, which may represent the length of the ancestral coding sequence that duplicated many times during the evolutionary history of the gene. (From B. de Crombrugghe and I. Pastan. 1982. *Trends Biochem. Sci.* 7:11–13, Figs. 1 and 2.)

combinant molecule, its reciprocal DNA product would be missing a sequence (Fig. 16.19). Some evidence does exist for such differences arising in reciprocal recombinant molecules involving duplicated genes. For example, the $\alpha$ globin gene exists in two copies on chromosome 16 in human genome. We know that patients with $\alpha$ thalassemias may have none or one gene rather than two $\alpha$ globin genes per chromosome (see Section 12.12). Of course, deletions, rather than

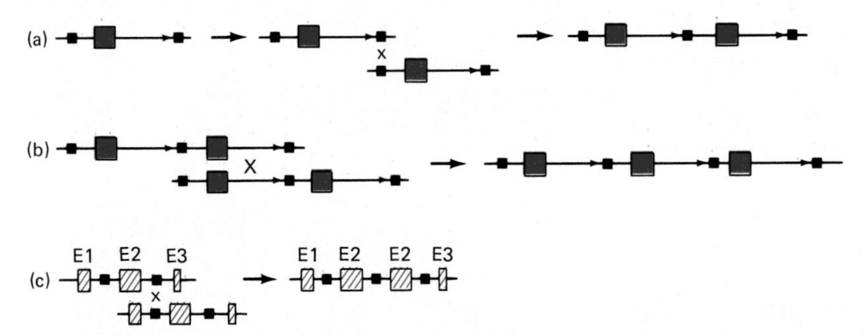


*Figure 16.19* Consequences of recombination of coding sequences (color boxes) through unequal crossing over at homologous short repeated DNA sequences (black boxes) dispersed between or within genes. Duplication of a gene by unequal crossing over may occur (a) once or (b) repeatedly to produce two or more copies of the original coding sequence. (c) Unequal crossing over involving short DNA repeats in intervening sequences between exons (E1–E3) similarly may lead to a recombinant gene containing one or more duplicated coding segments. (Reprinted by permission from A.J. Jeffreys and S. Harris. 1982. *Nature* **296**:9–10, Fig. 1. Copyright © 1982 Macmillan Journals Limited.)

unequal crossing over, may have been responsible for the missing $\alpha$ globin genes in these situations. The discovery of individuals carrying *three* $\alpha$ globin genes on a single chromosome 16, however, indicates that unequal crossing over has occurred at least in some cases involving changes in the numbers of duplicated genes in the genome.

Base sequencing and the new technology of DNA analysis have provided significant extensions to the repertory of genetic and biochemical analysis. From these new methods have come new and often unexpected data and exciting new insights into the fundamental processes guiding biological evolution. We can anticipate a continuing stream of new information concerning genetic processes and phenomena that underwrite those changes in life forms evident in the fossil record and in the world around us today. We can look forward particularly to information concerning modifications in regulatory DNA as well as the more easily analyzed modifications in structural genes. Although structural genetic information is essential to construction of functional proteins, morphogenetic and metabolic changes during development of the organism can hardly be due to protein structure and function alone. Processes that guide embryogenesis and the development of unique and complex organisms undoubtedly involve gene regulation mechanisms. These unsolved problems remain among the greatest challenges for future biological research.

## Questions and Problems

***16.1*** In parts of the world where falciparum malaria is endemic, individuals who are $Hb^A/Hb^S$ have a selective advantage over individuals with either homozygous genotype for these hemoglobin alleles.

***a.*** If the selective disadvantage of genotype $Hb^A/Hb^A$ is 0.1 and that of $Hb^S/Hb^S$ is 0.7, what will be the equilibrium frequency of the $Hb^S$ allele and of the $Hb^A$ allele?

***b.*** What is the expected percentage of $Hb^S/Hb^S$ infants born each year in the equilibrium population where $s_1 = 0.1$ and $s_2 = 0.7$?

***16.2*** Describe the changes occurring in the gene pool of a population that is undergoing normalizing, directional, and disruptive natural selection at different times in its history. What changes will occur in genotypic frequencies under each of these three programs of selection?

***16.3*** When comparing the genetic systems in *Drosophila colorata* ($n = 6$) and *D. willistoni* ($n = 3$), predict what the situation might be for each of the following characteristics (use *higher* and *lower*, comparatively):

| | ***D. colorata*** | ***D. willistoni*** |
|---|---|---|
| recombination index | | |
| frequency of inversions | | |
| frequency of supergenes | | |
| habitat stability | | |

Which species is probably more highly adapted (genetically constant)?

***16.4*** What is the evolutionary advantage of genetically differentiated X and Y sex chromosomes in species with chromosomal sex determination?

***16.5*** In a population of butterflies, individuals with brown and yellow spots are inconspicuous in their surroundings and have a relative fitness of 1. On the other hand, insects with solid yellow wings have a relative fitness of only 0.6 since they are more likely to experience predation. The solid yellow phenotype is due to a dominant allele *Y*, and the spotted phenotype is due to the recessive allele *y* in homozygotes (*yy*). If the initial frequency of allele *y* is 0.8, what are the frequencies of alleles *Y* and *y* after one generation of selection (predation)? After 10 generations?

***16.6*** Cross fertilizing plants isolated from a natural population produced an average of 32 leaves per plant. Breeding only plants with high leaf numbers for 15 generations had raised the mean to 58 leaves per plant. Selection was stopped for several generations and the mean leaf number decreased to about 42. This remained the mean for the next 16 generations, during which time selection was still not practiced.

***a.*** Suggest an explanation for these observations.

***b.*** Why didn't the mean number drop to the original value of 32?

***16.7*** In comparing cytochrome *c* from various species, we can calculate a minimum number of single nucleotide-pair changes necessary to get from one cytochrome *c* molecule to another by mutations. The number of nucleotide changes for a group of organisms might be as follows:

| | Species A | B | C | D | E |
|---|---|---|---|---|---|
| Species A | 0 | 20 | 20 | 20 | 20 |
| B | | 0 | 10 | 10 | 10 |
| C | | | 0 | 8 | 8 |
| D | | | | 0 | 2 |
| E | | | | | 0 |

***a.*** Draw an "evolutionary tree" for these species so that each line represents mutational distance.

***b.*** Which species is probably most ancient in this group? Why?

***16.8*** During the 1930s an English scientist released a large number of *Drosophila* that were heterozygous for the recessive allele for ebony body color (+/*e*) near a locality where the flies did not normally occur. After 8 generations the frequency of allele *e* had decreased to 0.11. How would you explain this change from 0.5 to 0.11?

***16.9*** Upon landing on an island that had not been explored previously, you find three populations of frogs that can be distinguished by spotting pattern.

***a.*** How would you determine whether they are members of one, two or three species?

***b.*** Suppose you discover that all these populations represent a single species of frog. How would you explain the absence of any frog hybrids on the island?

***16.10*** The number of amino acid differences between the four kinds of human globins are as follows:

| | $\alpha$ | $\beta$ | $\delta$ | $\gamma$ |
|---|---|---|---|---|
| *(141) $\alpha$ | 0 | 84 | 85 | 89 |
| (146) $\beta$ | | 0 | 10 | 39 |
| (146) $\delta$ | | | 0 | 41 |
| (146) $\gamma$ | | | | 0 |

*Total amino acids in the chain shown in parentheses.

***a.*** Construct an "evolutionary tree" for these globins.

***b.*** The gene for $\alpha$ globin is located on chromosome 16 and the loci for $\beta$ and $\delta$ globins are situtated on chromosome 11 in the human genome. How would you explain these locations in view of the postulated origin of all these globins?

***16.11*** Suppose that each of three linked genes specified a different keratin protein (A,B,C) in horse hair and each of two linked genes specified a different keratin (X, Y) in pigs.
***a.*** Suppose that amino acid compositions were largely nonhomologous between the horse keratins and the pig keratins. What might you conclude from this information?
***b.*** What would you predict for base sequence homologies of the genes, knowing about amino acid compositional nonhomologies? Explain.
***c.*** Suppose that base sequencing revealed considerable similarity between the numbers of exons and introns in the horse A keratin gene and the pig X keratin gene. From this information what might you conclude concerning the ancestry of the genes in the two mammalian species?

***16.12*** Suppose that genes for A and B keratins in the horse, as discussed in Question 16.11, arose by duplication in relatively recent times. Using the diagram below, indicate the band pattern of a gel autoradiograph if a single restriction fragment of 20 kb was obtained from DNA carrying the A gene, and a single restriction fragment of 30 kb was obtained from DNA carrying the B gene, and each of these was hybridized with radioactive A gene probe and with radioactive B gene probe?

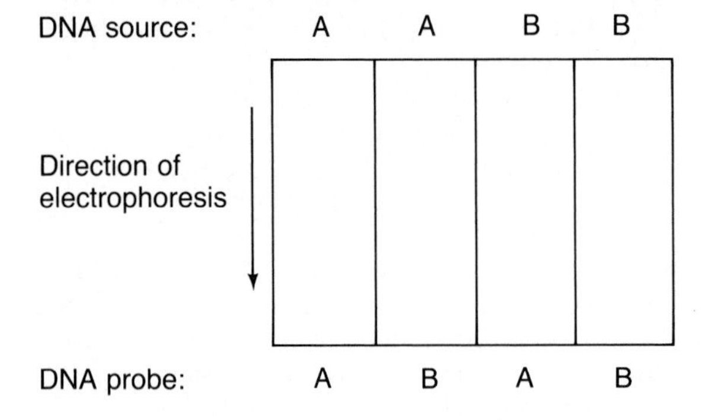


***16.13*** Suppose that base sequencing of the C keratin gene in the horse showed four exons and three introns and that two of these exons were highly homologous but were quite different from the other exons, which were very different from each other.
***a.*** Suggest a possible evolutionary origin for two homologous exons within a gene.
***b.*** Diagram a possible sequence of molecular events that might account for the origin of the two homologous exons in a gene now comprised of four exons.

***16.14*** Question 16.11 discussed three hypothetical keratin genes in the horse genome and only two keratin genes in the pig genome.
***a.*** How would you explain the absence of a third keratin gene in the pig genome and its presence in the horse genome?
***b.*** Suppose that base sequencing revealed the existence of a third keratin gene in the pig genome, but that it is a nonfunctional pseudogene. What would you predict to be the intron-exon organization of such a pseudogene? Explain.

***16.15*** Define the following terms: ***a.*** adaptation ***b.*** directional natural selection ***c.*** molecular polymorphism ***d.*** fitness ***e.*** adaptability ***f.*** recombination index ***g.*** reproductive isolation between species ***h.*** protein domain ***i.*** unequal crossing over.

## References

Ayala, F.J. Sept. 1978. The mechanism of evolution. *Sci. Amer.* **239**:56.

Beadle, G.W. Jan. 1980. The ancestry of corn. *Sci. Amer.* **242**:112.

Bishop, J.A., and L.M. Cook. Jan. 1975. Moths, melanism, and clean air. *Sci. Amer.* **232**:90.

Blake, C. 1983. Exons and the evolution of proteins. *Trends Biochem. Sci.* **8**:11.

Clarke, B. Aug. 1975. The causes of biological diversity. *Sci. Amer.* **233**:50.

de Crombrugghe, B., and I. Pastan. 1982. Structure and regulation of a collagen gene. *Trends Biochem. Sci.* **7**:11.

Dickerson, R.E. Mar. 1980. Cytochrome *c* and the evolution of energy metabolism. *Sci. Amer.* **242**:136.

Eiferman, F.A., et al. 1981. Intragenic amplification and divergence in the mouse α-fetoprotein gene. *Nature* **294**:713.

Feldman, M., and E.R. Sears. Jan. 1981. The wild gene resources of wheat. *Sci. Amer.* **244**:102.

Gilbert, W. 1978. Why genes in pieces? *Nature* **271**:501.

Gilbert, W. 1981. DNA sequencing and gene structure. *Science* **214**:1305.

Gō, M. 1981. Correlation of DNA exonic regions with protein structural units in haemoglobin. *Nature* **291**:90.

Gould, S.J., and R.C. Lewontin. 1979. The spandrels of San Marco and the Panglossian paradigm: A critique of the adaptationist programme. *Proc. Roy. Soc. London,* B **205**:581.

Graham, J.B., and C.A. Istock. 1979. Gene exchange and natural selection cause *Bacillus subtilis* to evolve in soil culture. *Science* **204**:637.

Holland, J., et al. 1982. Rapid evolution of RNA genomes. *Science* **215**:1577.

Jeffreys, A.J., and S. Harris. 1982. Processes of gene duplication. *Nature* **296**:9.

Kimura, M. Nov. 1979. The neutral theory of evolution. *Sci. Amer.* **241**:98.

Kioussis, D., et al. 1981. The evolution of $\alpha$-fetoprotein and albumin. II. The structures of the $\alpha$-fetoprotein and albumin genes in the mouse. *J. Biol. Chem.* **256**:1960.

Leder, P., et al. 1980. Mouse globin system: A functional and evolutionary analysis. *Science* **209**:1336.

Levin, D.A. 1979. The nature of plant species. *Science* **204**:381.

Lewontin, R.C. Sept. 1978. Adaptation. *Sci. Amer.* **239**:212.

Lucchesi, J.C. 1978. Gene dosage compensation and the evolution of sex chromosomes. *Science* **202**:711.

Maynard Smith, J. 1977. Why the genome does not congeal. *Nature* **268**:693.

Maynard Smith, J. 1978. *Evolution of Sex*. New York: Cambridge University Press.

Mayr, E. Sept. 1978. Evolution. *Sci. Amer.* **239**:46.

Milkman, R., ed. 1983. *Perspectives in Evolution.* Sunderland, Mass.: Sinauer Associates.

Roninson, I.B., and V.M. Ingram. 1982. Gene evolution in the chicken $\beta$-globin cluster. *Cell* **28**:515.

Schwarz, Z., and H. Kössel. 1980. The primary structure of 16S rDNA from *Zea mays* chloroplast is homologous to *E. coli* 16S rRNA. *Nature* **283**:739.

Thompson, J.N., Jr., and R.C. Woodruff. 1978. Mutator genes—pacemakers of evolution. *Nature* **274**:317.

White, M.J.D. 1978. *Modes of Speciation.* San Francisco: Freeman.

Yunis, J.J., and O. Prakash. 1982. The origin of man: A chromosomal pictorial legacy. *Science* **215**:1525.

# Answers to Questions and Problems

## Chapter 1

***1.1*** ***a.***

| | expected $F_1$ ratios | |
|---|---|---|
| **matings** | **genotypes** | **phenotypes** |
| (1) *BB* × *BB* | all *BB* | all black |
| (2) *BB* × *Bb* | ½ *BB*:½ *Bb* | all black |
| (3) *BB* × *bb* | all *Bb* | all black |
| (4) *Bb* × *Bb* | ¼ *BB*:½ *Bb*:¼ *bb* | ¾ black:¼ white |
| (5) *Bb* × *bb* | ½ *Bb*:½ *bb* | ½ black:½ white |
| (6) *bb* × *bb* | all *bb* | all white |

***b.*** *Bb,* since his phenotype is black, but the progeny segregate white as well as black offspring. If he were *BB*, all the progeny would be black.

***1.2*** ***a.*** red.

***b.*** *RR* × *RR* or *RR* × *Rr*, *Rr* × *Rr*, *RR* × *rr*, *rr* × *rr*,

| *RR* × *RR* or *RR* × *Rr* | *Rr* × *Rr* | *RR* × *rr* | *rr* × *rr* |
|---|---|---|---|
| ↓ | ↓ | ↓ | ↓ |
| all *RR* or 1 *RR*:1 *Rr* | 1 *RR*: 2 *Rr*: 1 *rr* | all *Rr* | all *rr* |

*Rr* × *rr*
↓
1 *Rr*:1 *rr*

***1.3*** ***a.*** red. ***b.*** 3 red:1 scarlet. ***c.*** 1 red:1 scarlet.

***1.4*** ***a.*** black, dominant; white, recessive. ***b.*** *BB* × *bb*. ***c.*** *Bb*. ***d.*** *Bb*. ***e.*** 1 black:1 white.

***1.5*** Cross to recessive (white): if it is heterozygous the progeny will segregate 1:1, and if it is homozygous the progeny will all be black.

***1.6*** ***a.*** I-1 must be *Bb* since segregation occurs in generation II; I-2 is *bb*. ***b.*** Heterozygous black guinea pigs include I-1; II-1, II-3, II-6, III-2, III-3, III-4 since each had one recessive *bb* parent; IV-2 and IV-3 might be homozygous *BB* since both their parents were black, but they might be *Bb* since both their parents must be heterozygous *Bb*. ***c.*** White, since III-1 and III-5 are both white and are genotypically *bb*.

***1.7*** ***a.*** 9 green, starchy:3 green, waxy:3 yellow, starchy:1 yellow, waxy. ***b.*** Phenotype ratio is 1 green, starchy (*GG Wxwx* and *Gg Wxwx*):1 green, waxy (*GG wxwx* and *Gg wxwx*); genotype ratio is 1:1:1:1. ***c.*** *GG Wxwx*; *GG WxWx*; *Gg Wxwx*; *Gg WxWx*. ***d.*** Backcross to doubly recessive yellow, waxy since phenotypic ratios will be different in backcross progeny using *GG WxWx*, *GG Wxwx*, *Gg WxWx*, or *Gg Wxwx* as a parent.

***1.8*** ***a.***

Parents: +/+ +/+ × *b*/*b se*/*se*
$F_1$: all wild type; all +/*b* +/*se*
$F_2$

| | **allelic combinations** | **ratio** | **genotypes** | **phenotypes** | **ratio** |
|---|---|---|---|---|---|
| ¼ +/+ | ¼ +/+ | = 1/16 | +/+ +/+ | gray, red | 1/16 |
| | 2/4 +/*se* | = 2/16 | +/+ +/*se* | gray, red | 2/16 |
| | ¼ *se*/*se* | = 1/16 | +/+ *se*/*se* | gray, sepia | 1/16 |
| 2/4 +/*b* | ¼ +/+ | = 2/16 | +/*b* +/+ | gray, red | 2/16 |
| | 2/4 +/*se* | = 4/16 | +/*b* +/*se* | gray, red | 4/16 |
| | ¼ *se*/*se* | = 2/16 | +/*b* *se*/*se* | gray, sepia | 2/16 |
| ¼ *b*/*b* | ¼ +/+ | = 1/16 | *b*/*b* +/+ | black, red | 1/16 |
| | 2/4 +/*se* | = 2/16 | *b*/*b* +/*se* | black, red | 2/16 |
| | ¼ *se*/*se* | = 1/16 | *b*/*b* *se*/*se* | black, sepia | 1/16 |
| | | | | | 9/16:3/16:3/16:1/16 |

**b.** Parents: $b/b$ $+/+$ × $+/+$ $se/se$
$F_1$: all wild type; all $+/b$ $+/se$
$F_2$ same as answer in (***a***) above

**c.** (1) $+/+$ $+/+$ × $b/b$ $se/se$

| | $b$ $se$ | |
|---|---|---|
| $+$ $+$ | $+/b$ $+/se$ | all wild type |

(2) $+/b$ $+/se$ × $b/b$ $se/se$

| | $b$ $se$ | |
|---|---|---|
| $+$ $+$ | $+/b$ $+/se$ | ¼ wild type |
| $+$ $se$ | $+/b$ $se/se$ | ¼ gray, sepia |
| $b$ $+$ | $b/b$ $+/se$ | ¼ black, red |
| $b$ $se$ | $b/b$ $se/se$ | ¼ black, sepia |

(3) $+/+$ $+/se$ × $b/b$ $se/se$ ⟶ ½ red : ½ sepia
$+/b$ $+/+$ × $b/b$ $se/se$ ⟶ ½ gray : ½ black

***1.9*** ***a.*** both $Rr$, both pink (incomplete dominance of one pair of alleles). ***b.*** (1) 1 red:2 pink:1 white, (2) all red, (3) all pink, (4) 1 pink:1 white.
***1.10*** ***a.*** (1) all roan, all $C^RC^W$; (2) 1 red ($C^RC^R$):1 roan ($C^RC^W$); (3) 1 red ($C^RC^R$):2 roan (2 $C^RC^W$):1 white ($C^WC^W$); (4) 1 roan ($C^RC^W$):1 white ($C^WC^W$). ***b.*** (1) none; (2) 50%, reds only; (3) 50%, all the reds and all the whites; (4) 50%, whites only.
***1.11*** (1) 1 wild type ($T^+/T^+$):1 tetraptera ($T/T^+$), (2) 1 wild type ($T^+/T^+$): 2 tetraptera ($T/T^+$) since $T/T$ is lethal.
***1.12*** ***a.*** It is possible. Since his genotype is $I^AI^B$ and the mother's is $i^Oi^O$ they could produce a child with the genotype $I^Ai^O$. ***b.*** A man of blood type B or blood type O could not be the father of this child since neither would have an $I^A$ allele. ***c.*** Only type O would rule out a man as the possible father since men of type AB ($I^AI^B$), type B ($I^BI^B$ or $I^Bi^O$), or type A ($I^AI^A$ or $I^Ai^O$) could father a type AB child whose mother is also type AB.
***1.13*** ***a.*** According to the possibilities shown below, only parents A × B could produce a child with blood type AB. The parents of the twins must therefore be AB × O, and the parents of the remaining baby must be B × O.

**b.**

| | | | |
|---|---|---|---|
| parents: | $I^AI^B$ and $i^Oi^O$ | $I^A$– and $I^B$– | $I^Bi^O$ and $i^Oi^O$ |
| | ↓ | ↓ | ↓ |
| babies: | $I^Ai^O$ and $I^Bi^O$ | $I^AI^B$ | $I^Bi^O$ |

***1.14*** ***a.*** four; $RRPP$, $RRPp$, $RrPP$, $RrPp$. ***b.*** $RrPp$ × $Rrpp$ since the 3:3:1:1 ratio indicates one pair of alleles segregates 3:1 and the other pair 1:1:

¾ $R$– < ½ $Pp$ = ⅜ $R$–$Pp$ walnut; ½ $pp$ = ⅜ $R$–$pp$ rose
¼ $rr$ < ½ $Pp$ = ⅛ $rrPp$ pea; ½ $pp$ = ⅛ $rrpp$ single

***c.*** Pea ($rrPp$) × rose ($Rrpp$) will produce ¼ singles; other combinations would produce ⅛, 1/16, or no singles in a progeny.
***1.15*** ***a.*** White × yellow $F_2$ ratio of 12:3:1 indicates dihybrid inheritance with dominant epistasis of one gene over the other; white × red $F_2$ ratio of 9:3:4 indicates dihybrid inheritance with recessive epistasis of one gene over the other:

¾ $A$– < ¾ $B$– = 9/16 white; ¼ $bb$ = 3/16 white
¼ $aa$ < ¾ $B$– = 3/16 red; ¼ $bb$ = 1/16 yellow

¾ $B$– < ¾ $C$– = 9/16 red; ¼ $cc$ = 3/16 white
¼ $bb$ < ¾ $C$– = 3/16 yellow; ¼ $cc$ = 1/16 white

Dominant allele $A$ inhibits color development governed by alleles $B$ (red) and $b$ (yellow) of another gene. Homozygous recessive $cc$ prevents expression of dominant red and recessive yellow of alleles $B$,$b$, and these colors are expressed only if the dominant $C$ allele of the color gene is present. ***b.*** Crosses between white strain $AABBCC$ and white strain $aabbcc$ produce white $F_1$ progeny $AaBbCc$. Interbreeding $F_1$ plants will produce $F_2$ progeny consisting of 52/64 white:9/64 red:3/64 yellow since both dominant and recessive epistasis will govern red and yellow color development.

| parents | | babies | |
|---|---|---|---|
| **phenotypes** | **genotypes** | **phenotypes** | **genotypes** |
| B × O | $I^BI^B$ × $i^Oi^O$, or $I^Bi^O$ × $i^Oi^O$ | B,O | $I^Bi^O$, $i^Oi^O$ |
| A × B | $I^AI^A$ or $I^Ai^O$ × $I^BI^B$ or $I^Bi^O$ | A, B, AB, O | $I^Ai^O$, $I^Bi^O$, $I^AI^B$, $i^Oi^O$ |
| AB × O | $I^AI^B$ × $i^Oi^O$ | A, B | $I^Ai^O$, $I^Bi^O$ |

***1.16*** Dihybrid inheritance involving two interacting genes: genotypes with one or two dominant alleles of both genes produce red, genotypes with one or two dominant alleles of one gene and recessive for the other gene produce sandy-colored swine, and the doubly recessive genotype produces white animals.
***1.17*** ***a.*** Dihybrid inheritance in which all genotypes carrying the dominant allele ($A$) produce white fruit, and the $aa$ genotypes produce either yellow ($B$–) or green ($bb$). The white gene is epistatic to the color gene.

***b.*** $AABB \times aabb \rightarrow$ all white ($AaBb$)
$AABb \times aabb \rightarrow$ all white (1 $AaBb$:1 $Aabb$)
$AaBB \times aabb \rightarrow$ 1 white ($AaBb$):1 yellow ($aaBb$)
$AaBb \times aabb \rightarrow$ 2 white (1 $AaBb$:1 $Aabb$):
1 yellow ($aaBb$):1 green ($aabb$)
$AAbb \times aabb \rightarrow$ all white ($Aabb$)
$Aabb \times aabb \rightarrow$ 1 white ($Aabb$):1 green ($aabb$)

***c.*** backcross to green ($aabb$) since
$aaBB \times aabb \rightarrow$ all yellow ($aaBb$)
$aaBb \times aabb \rightarrow$ 1 yellow ($aaBb$):1 green ($aabb$)

***d.*** *two*; $AABB$ and $AAbb$

***1.18*** Dihybrid inheritance since a 9:7 phenotypic ratio is modified 9:3:3:1, in which the dominant allele of one gene governs color development and the other gene governs red (dominant) or white (recessive) kernels. Either recessive homozygote is epistatic to the effects of the other gene.
***1.19*** ***a.*** At least one parent but possibly both must be $Pp$. If only one is $Pp$, the other is $pp$. ***b.*** IV-5 must be $Pp$ since he is polydactylous. ***c.*** IV-5's mother (III-6) is more likely to be $Pp$ since the trait was expressed in her father (II-9) and in other relatives, and it is a rare trait. ***d.*** The trait is not expressed in every individual of genotype $Pp$, including I-1 or I-2 and III-6. ***e.*** The trait is expressed in every generation, and afflicted parents usually have some afflicted and some nonafflicted children. ***f.*** If different polydactylous individuals have varying numbers of fingers and/or toes, the trait shows variable expressivity.
***1.20*** ***a.*** $(1/4)^n$ progeny are expected to express a homozygous genotype, and 1 per 200 recessives indicates $(1/4)^4$, or four pairs of alleles governing color. ***b.*** About 1 million progeny to find $(1/4)^{10}$ recessives.
***1.21*** One pair of alleles, since the ratio is 1 small:2 intermediate:1 large in the $F_2$. Each $A$ allele adds an increment of size, so $AA$ = large and $Aa$ = intermediate. We can also interpret this as incomplete dominance since only one gene is involved.

***1.22*** ***a.*** Two pairs of alleles, $A_1/a_1$ and $A_2/a_2$, with the dominants having equal (5 cm) and additive effects on the phenotypes. ***b.*** Cross $A_1a_1A_2a_2 \times A_1a_1A_2a_2$ produced 1:4:6:4:1 phenotypic ratio in progeny:

| tail length class | number of dominant alleles | genotypes present |
|---|---|---|
| 1/16 2.5 cm | 0 | $a_1a_1a_2a_2$ |
| 4/16 7.5 cm | 1 | $A_1a_1a_2a_2$, $a_1a_1A_2a_2$ |
| 6/16 12.5 cm | 2 | $A_1A_1a_2a_2$, $A_1a_1A_2a_2$, $a_1a_1A_2A_2$ |
| 4/16 17.5 cm | 3 | $A_1A_1A_2a_2$, $A_1a_1A_2A_2$ |
| 1/16 22.5 cm | 4 | $A_1A_1A_2A_2$ |

***c.*** $a_1a_1a_2a_2 \times A_1A_1A_2a_2$ or $A_1a_1A_2A_2$ would produce 50% 7.5 cm and 50% 12.5 cm offspring types.
***1.23*** ***a.*** Strain A is $aabbcc$, and strain B is $AABBCC$. ***b.*** 180 g. ***c.*** 7 classes with weights of 120 g, 140 g, 160 g, 180 g, 200 g, 220 g, and 240 g. ***d.*** 256 (only 1 in 64, on average, is $AABBCC$).
***1.24*** The proportion of individuals in any homozygous genotypic class in $F_2$ populations is $(1/4)^n$, where $n$ is the number of genes (allele pairs) involved. If 4/4000 had seeds weighing 30 g (one of the homozygous classes), there must be 5 genes (allele pairs) involved since 1/1000 is approximately $(1/4)^5$.
***1.25*** $2 \times (1/4)^5 = 2 \times 1024 = \$2048$.
***1.26*** ***a.*** Simple monohybrid inheritance with incomplete dominance. ***b.*** One pair of alleles would produce only three phenotypic classes, so these data must indicate two pairs of alleles producing $(1/4)^2 = 1/16$ in a homozygous class (5/80). ***c.*** $aabb$ (50 mm); $Aabb$ and $aaBb$ (62 mm); $AAbb$, $AaBb$, $aaBB$ (75 mm); $AABb$, $AaBB$ (88 mm); and $AABB$ (100 mm). Each $A$ or $B$ allele adds an increment of 12–13 mm to tail length.
***1.27*** Obtain a new supply of hybrid corn seed since seed from harvested plants will produce a highly variable $F_2$ generation of plants, many of which will be weaker than the $F_1$ type.

## Chapter 2

***2.1*** ***a.*** (1) $1/4 \times 1/4 \times 3/4 = 3/64$, (2) $3/4 \times 1/4 \times 3/4 = 9/64$, (3) $(3/4)^3 = 27/64$. ***b.*** (1) $3/4 \times 3/4 \times 1/4 \times 1/4 = 9/256$; (2) $6a^2b^2 = 6(3/4)^2\,(1/4)^2 = 54/256$.
***2.2*** ***a.*** 1/4 since each birth is an independent event. ***b.*** $(1/4)^2 = 1/16$ ***c.*** $3/4 \times 1/4 \times 3/4 = 9/64$.
***2.3*** ***a.*** $(1/2)^5 = 1/32$. ***b.*** $(1/2)^5 + (1/2)^5 = 2/32$. ***c.*** $(1/2)(1/2)^4 = 1/32$.

**2.4** ***a.*** $1 \times \frac{1}{2} \times \frac{1}{2} \times \frac{1}{2} \times \frac{1}{2} = \frac{1}{16}$. ***b.*** $\frac{1}{2} \times \frac{1}{4} \times \frac{1}{2} \times \frac{1}{2} = \frac{1}{32}$.
**2.5** ***a.*** $7a^6b = 7(\frac{1}{2})^6(\frac{1}{2}) = \frac{7}{128}$. ***b.*** $35a^3b^4 = 35(\frac{1}{2})^3(\frac{1}{2})^4 = \frac{35}{128}$. ***c.*** $\frac{1}{2} \times \frac{1}{2} \times \frac{1}{2} \times \frac{1}{2} \times \frac{1}{2} \times \frac{1}{2} \times \frac{1}{2} = \frac{1}{128}$.

**2.6** ***a.*** $\frac{8!}{(6!)(2!)} \times (\frac{1}{2})^6(\frac{1}{2})^2 = 10.9\%$. ***b.*** $10.9\% + 10.9\% = 21.8\%$.
**2.7** ***a.*** (1) $\chi^2 = 20.55$, (2) $\chi^2 = 7.85$, (3) $\chi^2 = 2.52$, (4) $\chi^2 = 3.60$, (5) $\chi^2 = 18.78$; results in experiments 3 and 4 fit the 9:3:3:1 expected ratio. ***b.*** Experiments 1, 2, and 5.
**2.8** ***a.*** $\chi^2 = 0.4$, $P = 0.7$–$0.5$, good fit. ***b.*** $\chi^2 = 4.0$, $P = 0.05$–$0.01$, poor fit; coin may be deformed and give unreliable results. Repeat 100 tosses using another coin and compare results. Larger samples are usually more reliable.
**2.9** ***a.*** Expect $(a + b)^4 = a^4 + 4a^3b + 6a^2b^2 + 4ab^3 + b^4$, or

$\frac{1}{16}(208) + \frac{4}{16}(208) + \frac{6}{16}(208) + \frac{4}{16}(208) + \frac{1}{16}(208) =$

| | | | | | |
|---|---|---|---|---|---|
| 13 | 52 | 78 | 52 | 13 | number expected |
| 10 | 51 | 80 | 54 | 10 | number observed |

$\chi^2$ test shows good fit between expected and observed distributions. ***b.*** Five times the numbers shown in (a), namely, 65 + 260 + 390 + 260 + 65, distributed among 1040 families.
**2.10** ***a.***

| experiment number | $\chi^2$ for 3:1 ratio | $\chi^2$ for 13:3 ratio |
|---|---|---|
| 1 | 1.00 | 1.23 |
| 2 | 1.52 | 1.06 |
| 3 | 27.00 | 0.41 |

***b.*** Greater reliance should be placed on the larger sample in expt. 3, which showed good fit to expected 13:3 ratio but $P<0.01$ for a 3:1 ratio. Two interacting genes with dominant epistasis govern flower color.
**2.11** ***a.*** $\chi^2 = 0.46$, $P = 0.5$ for 1:1 segregation of stem color alleles. $\chi^2 = 2.31$, $P = 0.2$–$0.1$ for 1:1 segregation of stem hair alleles. ***b.*** $\chi^2 = 2.80$, $P = 0.5$–$0.3$ for 1:1:1:1 ratio due to independent segregation of two pairs of alleles.
**2.12** $\bar{x} = 2.06$ mm; $\sigma = 0.46$; $S.E._{\bar{x}} = 0.10$.
**2.13** ***a.*** A normal curve can be plotted, and the data fit a normal distribution. ***b.*** $\bar{x} = 172.0$ cm, $\sigma = 8.5$ cm, $S.E._{\bar{x}} = 0.19$, or 0.2. ***c.*** $1\sigma$ range = 163.5–180.5 and includes 82.4% of the students, $2\sigma$ ranges from 155.0 to 189.0 cm and includes 99.5% of the students, and $3\sigma$ ranges from 146.5 to 197.5 and includes 100% of the sample. ***d.*** Yes.
**2.14** ***a.***

| | $N$ | $\bar{x}$ | $\sigma$ | $S.E._{\bar{x}}$ |
|---|---|---|---|---|
| parent 60 | 57 | 6.63 | 0.82 | 0.11 |
| parent 54 | 101 | 16.80 | 1.89 | 0.19 |
| $F_1$ (60 × 54) | 69 | 12.12 | 1.52 | 0.18 |
| $F_2$ | 401 | 12.89 | 2.25 | 0.11 |

***b.***

| | $S.E._D$ |
|---|---|
| Parent 60 and Parent 54: | 0.22 (significant) |
| $F_1$ and $F_2$: | 0.77 (not significant) |
| Parent 54 and $F_2$: | 3.91 (significant) |

***c.***

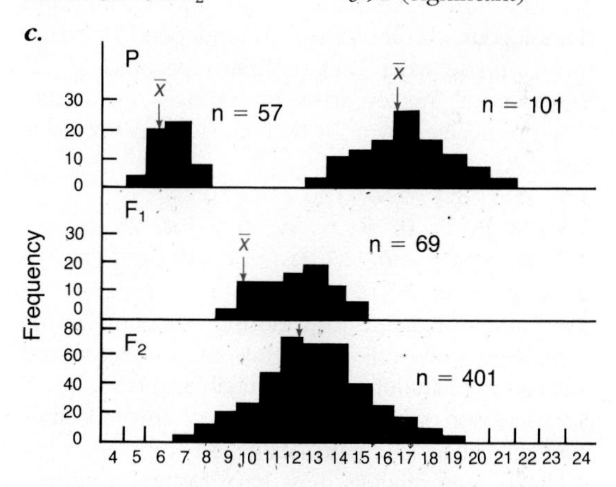


**2.15** ***a.*** *AABBCC*, 6 cm + (6 × 2 cm) = 18 cm; *aabbcc*, 6 cm + (0 × 2 cm) = 6 cm.
***b.*** $F_1$ (*AaBbCc*), 6 cm + (3 × 2 cm) = 12 cm.
***c.*** $F_2$ involves segregation of six pairs of alleles:

| number of dominant alleles | ear length | proportion of total genotypes |
|---|---|---|
| 6 | 18 cm | 1/64 |
| 5 | 16 cm | 6/64 |
| 4 | 14 cm | 15/64 |
| 3 | 12 cm | 20/64 |
| 2 | 10 cm | 15/64 |
| 1 | 8 cm | 6/64 |
| 0 | 6 cm | 1/64 |

**2.16** ***a.*** Traits 2, 3, and 5 since concordance is substantially higher for monozygotic than for dizygotic twins. ***b.*** Trait 2 since concordance is 100% for monozygotic twins.

***2.17*** Probably signifies genetic factors contributing to susceptibility to tuberculosis infection in an environment conducive to disease induction and development. Genetic influence in immunity exists against many kinds of diseases, whether caused by infection or other factors.

## Chapter 3

***3.1*** The section would contain membranous organelles such as mitochondria, chloroplasts, and endoplasmic reticulum in the cytoplasm.
***3.2*** $G_1$, $S$, and $G_2$; macromolecular syntheses and intermediary metabolism.
***3.3*** Chromatids are parts of one chomosome and homologous chromosomes are independent structures; *S* phase, when DNA replicates; prophase.
***3.4*** Identical nuclei arise by mitosis, genetically different nuclei may arise by meiosis (if genotype is heterozygous).
***3.5*** ***a.*** $(\frac{1}{2})^{23}$. ***b.*** 1 to 2 × $(\frac{1}{2})^{23}$.
***3.6*** ***a.*** 23. ***b.*** 46. ***c.*** 46. ***d.*** 23.
***3.7*** ***a.*** 5 pg. ***b.*** 2.5 pg. ***c.*** 10 pg. ***d.*** 5 pg. ***e.*** 2.5 pg. ***f.*** 10 pg.
***3.8*** The SC begins to form during zygonema and is completed when this stage ends. SC formation and synapsis are parallel, not sequential, events.
***3.9*** Behavior of homologous pairs of chromosomes parallels behavior of pairs of alleles.
***3.10*** The two mutants must have formed a heterokaryon, whose haploid nuclei from both strains were able to complement each other's deficiencies and produce wild-type mycelium. Genetic tests would include isolating individual haploid uninucleate spores and seeing whether they produce the two parental kinds of phenotypes. Each strain had the wild-type alleles that are missing in the other strain.
***3.11*** ***a.*** *FFf.* ***b.*** *Ff.* ***c.*** *f.* ***d.*** *F.* ***e.*** *Fff.*
***3.12*** ***a.*** Types 1 and 2 are first-division segregations, types 3 through 6 are second-division segregations. ***b.*** See Figs. 3.21 and 3.22.
***3.13*** $(1 \times 10^{-7})^3 = 1 \times 10^{-21}$ since each mutation of each of the 3 genes is an independent event.
***3.14*** Haploid organisms have only one axlele of each gene per nucleus. Dominance or recessiveness is irrelevant since the one allele is expressed in each case.
***3.15*** ***a.*** A virus that infects bacteria. ***b.*** A bacterial virus that destroys its host cell during infection. ***c.*** A virus that need not destroy its host cell during or after infection. ***d.*** Breakdown or dissolution of cells. ***e.*** A clear area in a confluent lawn of cells caused by lysis following viral infection. ***f.*** Phage DNA integrated into the bacterial chromosome. ***g.*** Bacterial strain that carries temperate phages in their noninfective state and transmits prophage DNA to successive cell generations.
***3.16*** Labeled DNA was found inside host cells and labeled protein remained outside these cells for the most part. Genetic material must enter the host in order to direct synthesis of new viruses identical to the infecting viral particles. Since only DNA entered host cells, DNA must be the genetic material carrying instructions for synthesis of new viral molecules.
***3.17*** Primary cultures consist of cells removed from the host organism and the descendants of these cells in the culture. Secondary cultures consist of the clonal descendants of cells removed from a primary culture and transferred to new media for independent growth.
***3.18*** ***a.*** Only $A^+B^+$ mouse-human somatic cell hybrids will grow on selective media lacking substances A and B required by the parent strains for growth. Each colony of somatic cell hybrids represents a clone descended from one original cell, and cells picked from such colonies can initiate new cultures representing different clones of hybrid cells. ***b.*** During random chromosome losses, different combinations of human chromosomes will be retained in different clones. Some clones will retain one chromosome of a pair, and other clones will retain the homologous chromosome of a pair. This situation is analogous to allelic segregation since one allele or the other may be lost, thereby leaving only one of the two alleles in the clone.
***3.19*** ***a.*** The transformed cells may continue to multiply indefinitely, may become capable of growth in semisolid medium, and may produce disordered heaps of cells on solid substratum. Normal cells have a finite life span in a culture, grow only on solid substratum if they are anchorage dependent, and stop dividing once they have produced a confluent single layer of cells. ***b.*** Transformed cells are found in multilayered mounds on solid substratum, normal cells are found in a confluent single layer. ***c.*** In host chromosomes, since oncogenic viral DNA becomes integrated into host DNA as prophage.
***3.20*** Reverse transcriptase catalyzes synthesis of DNA copies of viral RNA genes, and the copied DNA

becomes integrated into host chromosomes as prophage DNA. Prophage DNA then directs synthesis of new viral RNA and proteins, using host metabolic machinery.

## Chapter 4

**4.1** ***a.*** *v/v* $c^+/c$. ***b.*** $v^+/$ *c/c*. ***c.*** ♀♀ are $v^+/v$ $c^+/c$ and $v^+/v$ *c/c*, ♂♂ are *v/* $c^+/c$ and *v/ c/c*. Gene *v* is X-linked, gene *c* is autosomal.
**4.2** Parents are *Cc* ♀ × *c* ♂. ***a.*** ½. ***b.*** ½. ***c.*** ½ × ½ = ¼.
**4.3** ***a.*** male. ***b.*** 24. ***c.*** XX ♀, XO ♂.
**4.4** ***a.*** *Ll* ♀♀ × *l* ♂♂ → ⅓ *Ll* ♀♀ (dark green):⅓ *L* ♂♂ (dark green):⅓ *l* ♂♂ (yellow-green). The *ll* ♀♀ are lethals and are not produced. ***b.*** *Ll* ♀♀ × *L* ♂♂ → ½ *LL* ♀♀ and *Ll* ♀♀ (dark green):¼ *L* ♂♂(dark green):¼ *l* ♂♂ (yellow-green).
**4.5** ***a.*** 1♂:2♀. ***b.*** ⅓ $X^BX^b$♂♂:⅓ $X^BY$♀♀ :⅓ $X^bY$ ♀♀ (males all are barred, ½ the females are barred and ½ are nonbarred; cross was $X^bY$ "male" with $X^BY$ female).
**4.6** ***a.*** A different eye color gene is mutant in each strain since the progeny are wild type ($a^+a^+bb \times aab^+b^+ \rightarrow a^+ab^+b$ progeny). ***b.*** The mutant gene is X linked in strain A and autosomal in strain B, and the scarlet allele is recessive to its wild-type alternative at each locus. ***c.*** $F_1$ $a^+/a$ $b^+/b$ ♀♀ × $F_1$ $a^+/$ $b^+/b$ ♂♂ produces $F_2$ of 3 wild-type:1 scarlet ♀♀ and 3 wild-type:5 scarlet ♂♂.
**4.7** ***a.*** Each trait is X-linked; color blindness is fully recessive to normal, enzyme levels are determined by incompletely dominant alleles, and XG blood group presence is dominant over its absence. ***b.*** I-1 = *c Gpd xg*; I-2 = *C/C Gpd/gpd xg/xg*; II-1 = *C Gpd Xg*; II-2 = *C/c Gpd/gpd xg/xg*; III-1 = *C gpd xg*; III-5 = *C/C Gpd/Gpd Xg/xg* or *C/c Gpd/Gpd Xg/xg*.
**4.8** ***a.*** The trait is an X-linked recessive, transmitted from carrier mother to about half her sons.
***b.***

$X^NY$ — $X^NX^n$ — $X^NY$

$X^NY$ $X^nY$ $X^nY$ $X^NY$ $X^NX^{N/n}$ $X^NX^{N/n}$ $X^nY$ $X^NX^{N/n}$ $X^nY$

(The daughters could have received either $X^N$ or $X^n$ from their mother, but must have received $X^N$ from their father in either marriage.)

**4.9** ***a.*** Autosomal dominant. ***b.*** X-linked recessive trait. ***c.*** none, none. ***d.*** I-1 is *D/d b/b*, I-2 is *D/ B/b* (He is free of the dental disease and some of his children are not short fingered, thus indicating that he is heterozygous for this gene.)
**4.10** ***a.*** ¾. ***b.*** (¾ × ½) + (¼ × ½) = ½.
**4.11** ***a.*** *Hm/hm* ♀ and *Hm/* ♂ (Every daughter receives his X chromosome that carries a normal, dominant allele.) ***b.*** ½ × ½ = ¼. ***c.*** ¼ × ¼ = 1/16.
**4.12** ***a.*** $C^BC^O$ (calico) ♀ × $C^O/$ (orange) ♂. ***b.*** Calico or orange, if it received one of its X chromosomes from each parent; calico if it received both X chromosomes from its mother and only the Y chromosome from its father. It couldn't be black since at least one of the two Xs must carry the $C^o$ allele.
**4.13** ***a.*** The egg must have been XX, carrying recessive alleles on each chromosome, and the sperm carried a Y chromosome. ***b.*** His mother, having one Barr body in the nucleus. ***c.*** 100% since he must get his X from his mother and she is color-blind.
**4.14** ***a.*** No. His daughters would receive his only X chromosome along with one X chromosome through the egg. Since his X chromosome must carry the dominant allele for normal vision and the child is color-blind, her father must also have been color-blind. Blood group compatability is irrelevant in this case. ***b.*** Yes. The man could be the father, but there is no certainty that he was the father in this case. Sons inherit their only X chromosome from their mothers and their Y chromosome from their fathers. A color-blind woman would produce only color-blind sons since no compensating allele is carried on the Y chromosome.
**4.15**

| Barr bodies | chromosomes | sex |
|---|---|---|
| 0 | 45, X | ♀ |
| 0 | 46, XY | ♂ |
| 1 | 46, XX | ♀ |
| 1 | 47, XXY | ♂ |
| 2 | 47, XXX | ♀ |

**4.16** ***a.*** G-banding would distinguish the unique banding pattern of each submetacentric in the complement. ***b.*** C-banding. ***c.*** Presence of satellites at the terminus of the chromosome arm in which the NOR is located. ***d.*** Each acrocentric has one arm much shorter than the other arm.

***4.17*** ***a.*** Specific chromosomal site of attachment of spindle fibers, the primary constriction of a chromosome. ***b.*** Site of nucleolar attachment to a chromosome and site of repeated copies of ribosomal RNA genes. ***c.*** The portion of chromatin that remains condensed during interphase, is genetically stable, and replicates late in the cell cycle. ***d.*** The portion of chromatin that is greatly extended and not condensed during interphase, is subject to mutation, and replicates throughout the synthesis period of the cell cycle. ***e.*** The Barr body, or condensed X chromosome that is inactivated when more than one X chromosome is present in a nucleus. ***f.*** Condensation and consequent genetic inactivity of one or more X chromosomes in excess of one X chromosome per nucleus. ***g.*** Ordered arrangement of all the chromosomes in a photographed nucleus, with the largest first and the smallest last in the order, and grouped according to centromere location and chromosome length. ***h.*** Multistranded chromosome produced by repeated replications of chromatin strands without their separation from each other afterward.

## Chapter 5

***5.1*** ***a.*** The two genes are 21 map units apart, so 21% of the eggs will be recombinant + + and *B m* (10.5% each). ***b.*** The ♀ progeny will be 50% bar and 50% wild type, and ♂ progeny will consist of 39.5% miniature (+ *m*), 39.5% bar (*B* +), 10.5% wild type (+ +), and 10.5% bar, miniature (*B m*). ***c.*** Two (X-carrying + +, and Y-carrying).
***5.2*** ***a.*** repulsion. ***b.*** 8.6%.
***5.3*** In the cross *al b*/*al b* ♀♀ × *al* +/+ *b* ♂♂, the progeny will consist of aristaless and black individuals if the genes are linked but will consist of four phenotypic classes (aristaless; black; wild type; aristaless, black) if the genes are on different chromosomes and, therefore, assort independently to produce four kinds of sperm (*al* +, + *b*, + +, and *al b*).
***5.4*** ***a.*** T. ***b.*** T. ***c.*** NPD. ***d.*** PD. (see Figs. 5.6 and 5.9 for reference).
***5.5*** ***a.*** The two genes are linked since PD tetrads (147 + 24) far outnumber NPD tetrads (21 + 3); independently assorting genes would be found in crosses producing equal proportions of PD and NPD tetrads. ***b.*** The order of the linkage map is *ad*—centromere—*trp* since only 39 tetrads (6 + 24 + 3 + 6) show second-division segregations for *ad*/+ while 126 tetrads (93 + 24 + 3 + 6) show second-division segregations for *trp*/+. The distance between *ad* and the centromere is ½ second-division tetrads/total tetrads × 100 = 6.5 map units, between centromere and *trp* = 21 map units, and between *ad* and *trp* the distance is (½ T + NPD)/total tetrads × 100 = 27.5 map units.

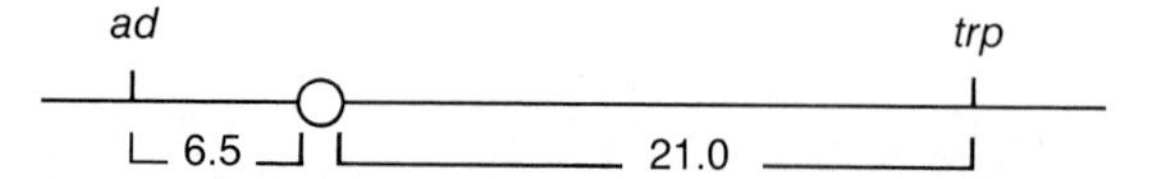


***5.6*** The genes assort independently and must, therefore, be on different chromosomes; and, since there are no second-division segregations, each gene must be very close to the centromere on its respective chromosome. (This information leads to the conclusion that the genes are in different linkage groups since they could not be 50 or more map units apart on the same chromosome and at the same time each be near the single centromere.)
***5.7***

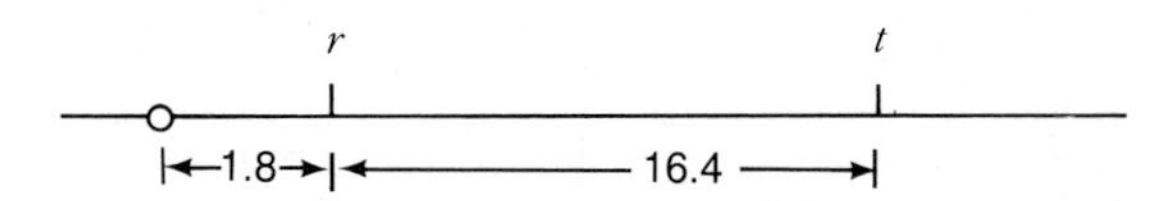


***5.8*** Class 1 is PD for all genes taken two by two (*a b*, *a c*, *b c*); class 2 is PD for *a b*, but NPD for *b c* and *a c*; class 3 is T for *a b* and *a c*, and NPD for *b c*; class 4 is T for *a b* and *a c*, and PD for *b c*. Using these observations, we can arrange the data as follows:

| gene pair | PD | NPD | T |
|---|---|---|---|
| | **tetrad types** | | |
| *a b* | 164/200 = 0.82 | 0 | 36/200 = 0.18 |
| *b c* | 96/200 = 0.48 | 104/200 = 0.52 | 0 |
| *a c* | 80/200 = 0.40 | 84/200 = 0.42 | 36/200 = 0.18 |

For the *a b* pair, PD and NPD are not equal and, therefore, genes *a* and *b* are linked. Gene *c* assorts independently of *a* and of *b* (PD = NPD), so it is on a different chromosome from the other two genes. The distance between genes *a* and *b* is (½ T + NPD)/Total tetrads = ½(36)/200 × 100 = 9 map units.
***5.9*** ***a.*** *paba pro*/+ *pro* and + +/*paba* +, or *paba pro*/+ + and *paba* +/+ *pro*, depending on the orientation of the centromeres of sister chromatids toward the poles at mitosis. ***b.*** Patches of $pro^-$ cells among wild-type cells would be produced if the cellular genotype after crossing over and chromatid segregation is *paba pro*/+ *pro*.

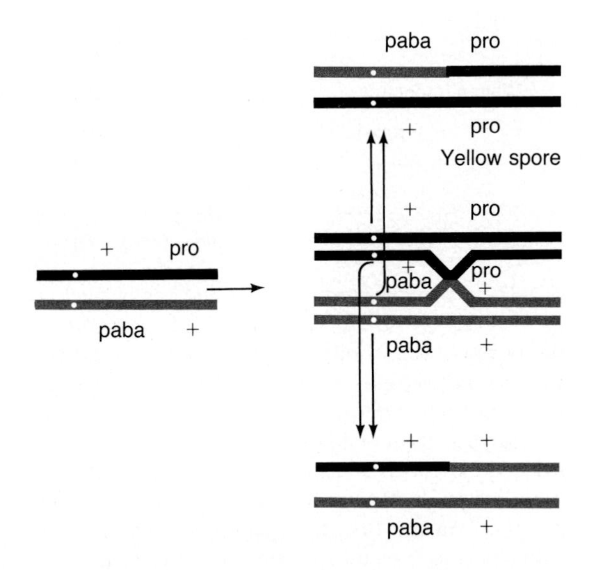


**5.10** ***a.*** *wx—c—sh.* ***b.*** *wx—c, 28* map units; *c—sh, 18* map units. ***c.*** 0.8.

**5.11** ***a.*** *v—lz—ct.* ***b.*** *v—lz*, 6 map units; *lz—ct*, 7 map units; coefficient of coincidence is 0.238. ***c.*** The genes are in the X chromosome.

**5.12** Progeny expected from

$$\frac{r\,g\,e}{+\,+\,+}\ ♀♀ \times \frac{r\,g\,e}{r\,g\,e}\ ♂♂:$$

| phenotypic classes | number of offspring |
|---|---|
| r g e | 406.5 |
| + + + | 406.5 |
| r + + | 50 |
| + g e | 50 |
| r g + | 38.5 |
| + + e | 38.5 |
| r + e | 5 |
| + g + | 5 |
| | 1000 |

**5.13** ***a.*** + *bl* +/*se* + *ss*. ***b.*** $se \underset{29.8}{—} bl \underset{12.3}{—} ss$.

**5.14** ***a.*** *dp* and *b* are linked, but *st* assorts independently of *dp* and of *b* and must be on another chromosome. This is apparent from the 1:1 ratio of *dp b*:+ + in the four noncrossover classes (largest classes), whereas these classes include 1 *dp st*:1 + *st*: 1 *dp* +:1 + +, and 1 *b st*:1 + *st*:1 *b* +:1 + +. ***b.*** *dp* and *b* are 30.6 map units apart [(96 + 98 + 102 + 104) / 1306].

**5.15** ***a.*** All three genes are linked since two phenotypic classes are much larger than any of the others. One double recombinant class is missing. The gene order is *h—W—g*. ***b.*** $h \underset{20.4}{—} W \underset{17.4}{—} g$

**5.16** Gene orders and map distances in each experiment are: (1) $al \underset{30.4}{—} d \underset{21.7}{—} hk$, (2) $d \underset{17.0}{—} hk \underset{21.0}{—} wx$, and (3) $hk \underset{14.6}{—} wx \underset{31.4}{—} bw$. Putting together all this information, the order of all five genes and the map distances calculated or averaged is:

$al \underset{30.4}{—} d \underset{19.4}{—} hk \underset{17.8}{—} wx \underset{31.4}{—} bw$.

**5.17** ***a.*** Autosomal dominant. ***b.*** The trait seems to be linked to the ABO gene since the $I^B$ characteristic appears together with the trait in all afflicted members of the three generations shown. ***c.*** I-1 (grandfather is: $\frac{i\ n}{i\ n}$; grandmother is: $\frac{I^B\ N}{i\ \ n}$. (Note: the recessive allele of the ABO gene is symbolized *i* or $i^O$.)

**5.18** ***a.*** 50%, the chance of receiving the X chromosome carrying *hm* from the mother. ***b.*** The mother's genotype is *Hm Gpd-A*/*hm Gpd-B* since she inherited her father's *Hm Gpd-A* X chromosome. If the fetus is GPD-A the chance is 5% that it has a recombinant *hm Gpd-A* chromosome and 95% that it has a noncrossover X chromosome carrying *Hm Gpd-A*. If the male fetus is GPD-B, the risk is 95% of inheriting a noncrossover *hm Gpd-B* X chromosome and 5% of inheriting a recombinant X chromosome carrying the *Hm* allele with the *Gpd-B* allele. ***c.*** 50%, the chance of receiving the X chromosome carrying *hm* from the mother.

**5.19** ***a.*** The woman's father is color-blind (*cb* +/Y); the woman's mother is normal (+ +/+ *ln*); the woman is normal (*cb* +/+ *ln*); her son has inherited a crossover X chromosome from his mother, and he is (*cb ln*/Y). ***b.*** Probability would be equal to one-half the frequency of crossing over between the two X-linked loci.

## Chapter 6

**6.1** ***a.*** (100 + 200)/1000 = 0.30 × 100 = 30 map units. ***b.*** One crossover on each side of the *trp—phe* gene segment (double crossover).

**6.2** ***a.*** The linkage order is *pro—ala—arg*.
***b.*** *pro—ala*, (420 + 1050 + 210 + 420)/7000 = 0.3(100) = 30 map units
*pro—arg*, (1050 + 700 + 420 + 420)/7000 = 0.37(100) = 37 map units
*ala—arg*, (420 + 700 + 210 + 420)/7000 = 0.25(100) = 25 map units

***6.3*** $g$—$z$, (660/2210)/(100) = 29.9 map units
$z$—$h$, (360/2050)/(100) = 17.6 map units
$g$—$h$ (880/2290)/(100) = 38.4 map units
Gene order is $g$—$z$—$h$.

***6.4***

mal, str, ser, ade, his, gal, pro, met, xyl

B, Q, R, D, H, T, S, X, A, C, N

***6.6*** ***a.*** Perform a complementation test. Allow strains 1 and 2 to conjugate and plate on minimal medium and observe colonies. Strain 1 is proline-requiring and will die. Strain 2 is leucine-requiring and will die. Strain 2 cells that have received F′ from strain 1 during conjugation will be merodiploid (*F′ leu*$^+$ *pro*-1/*leu*$^-$ *pro*-2), and will live to produce colonies if *pro*-1 and *pro*-2 are not alleles. If *pro-1* and *pro-2* are allelic, the merodiploids will not be viable and colonies will not be produced. ***b.*** Infect strain 3 with phage Z and use the progeny phase to infect strain 2 (which is immune). Plate these cells on two different growth media: minimal and minimal + leucine. The relative frequency of prototrophs on minimal media and of leucine-requiring recombinants on minimal + leucine media will indicate whether the order is *leu—pro-1—pro-2* or *leu—pro-2—pro-1,* according to:

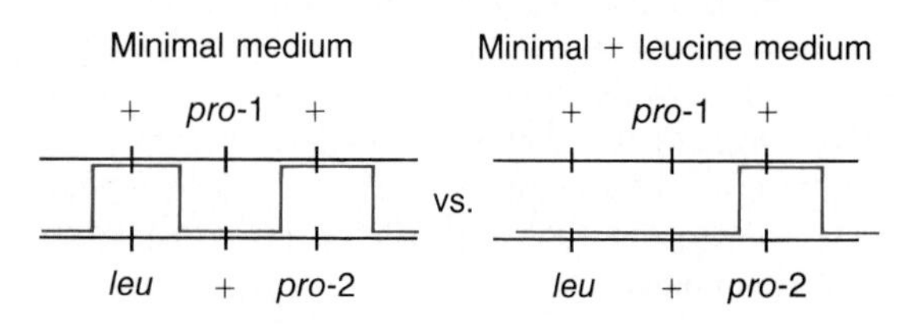


or,

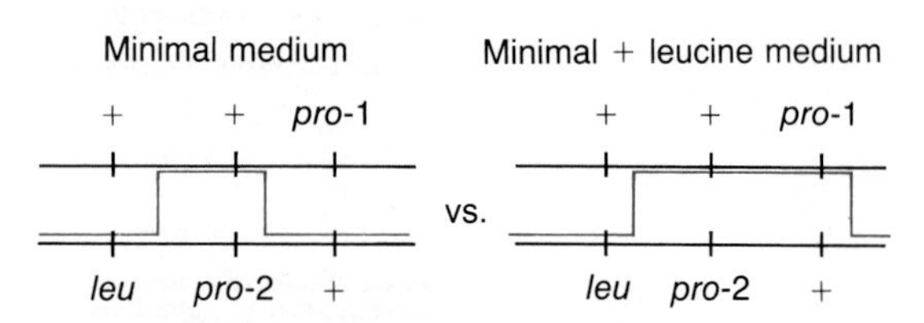


***6.7*** The gene order is $a$—$c$—$b$. Strain 1 transduction of strain 2 ($a^-$ $b^-$ $c^+$ into $a^+$ $b^+$ $c^-$) produces prototrophs when only $c^+$ is transferred; in the reciprocal, prototrophs arise if transfer $a^+$ and $b^+$ and not $c^-$ of strain 2 into strain 1. If $c$ is the middle gene, one predicts more wild-type recombinants ($a^+b^+c^+$) from strain 1 → strain 2, than from strain 2 → strain 1 transduction. It takes only a double crossover to get $a^+$ $b^+$ $c^+$ in the first, but quadruple crossover for wild-type recombinants in the second case. Double crossovers occur with higher frequency than quadruple crossovers.

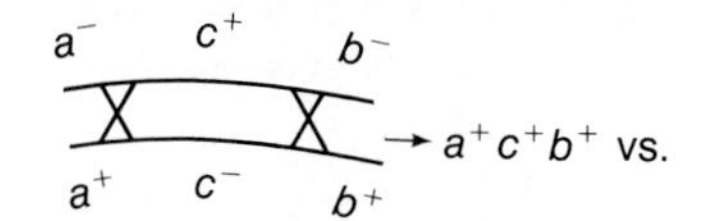


strain 1 → strain 2

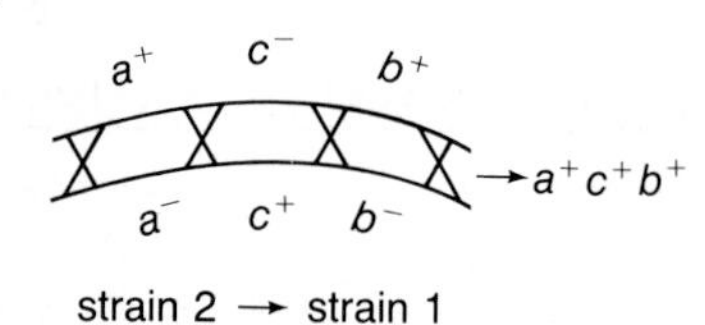


strain 2 → strain 1

***6.8*** ***a.*** The gene order is $s$—$f$—$tu$ ***b.*** $s$—$f$, 12.8 map units; $f$—$tu$, 20.8 map units; $s$—$tu$, 33.6 map units.
***c.*** Coefficient of coincidence = observed double crossovers/expected doubles = (668/550) = 1.2.

***6.9*** The gene order is $a$—$b$—$c$.
$a$—$b$, (400/2000) (100) = 20.0 map units
$b$—$c$, (190/2000) (100) = 9.5 map units
$a$—$c$, (560/2000) (100) = 28.0 map units

***6.10*** ***a.*** The gene order is $a$—$c$—$b$.
$a$—$b$, (740/2000) (100) = 37.0 map units
$b$—$c$, (200/2000) (100) = 10.0 map units
$a$—$c$, (580/2000) (100) = 29.0 map units
***b.*** The two linkage orders can be resolved by arranging the genes in a circular map, since $a$—$b$—$c$ and $b$—$c$—$a$ (or, $a$—$c$—$b$) are linear permutations of a circular gene map, as shown at the top of the next page.

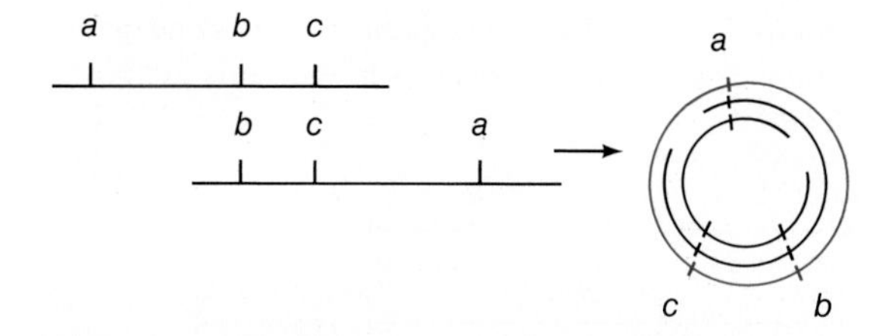


*6.11* $F^+$ donors have F factors in the free, episomal state; Hfr strains have F integrated into the bacterial chromosome. In $F^+ \times F^-$, the episome is transferred and the recipient becomes $F^+$. In Hfr $\times$ $F^-$, the bacterial chromosome leads the way and F is the last component to enter the recipient. Conjugation usually terminates before F enters, and the recipient is $F^-$.
*6.12* Conjugation experiments using Hfr and $F^-$ strains.
*6.13* Viral genetics showed that DNA was the genetic material, that genetic and physical mapping could be correlated, that various means could be used to produce comparable or almost identical genetic maps, and that various mechanisms exist as substitutes for sexual reproduction in the generation of genetic diversity among bacteria and viruses.
*6.14* Physically linear chromosomes produce circular genetic maps if the chromosomes are terminally redundant and linearly permuted molecules.
*6.15* Temperate phages can enter into a lysogenic relationship with the host cells; virulent phages have a very brief latent period and always proceed through the lytic cycle.
*6.16*

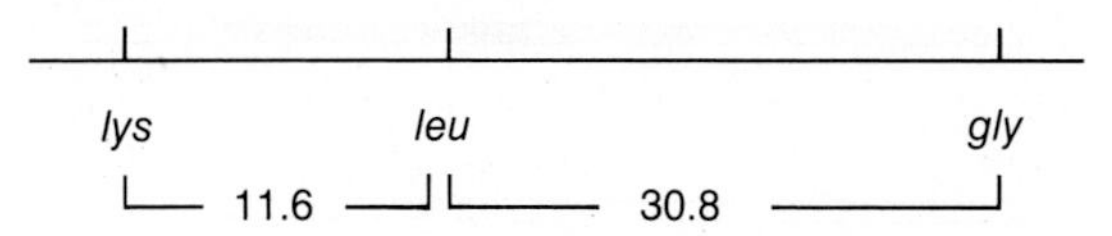


*6.17* Generalized transducing phages carry a piece of bacterial DNA instead of their own genes in the viral particle, and specialized transducing phages package certain (flanking) bacterial genes along with some of the phage genes in the mature virus particles. Both kinds of transducing phages can transfer bacterial genes from one host cell to another host cell, often leading to recombinant bacterial DNA and recombinant phenotypes in the transductant host cells. In addition, all transducing phages are temperate rather than virulent types.
*6.18* ***a.*** An experimental design in which wild-type recombinants produced from different auxotrophs can be detected by growth of prototrophic colonies on minimal media that cannot support auxotrophic growth. ***b.*** A genetically deficient type that requires supplements in minimal media to be able to grow. ***c.*** An extrachromosomal genetic element that may exist free in the cell or may be integrated into the host chromosome, depending on prevailing conditions. ***d.*** An experimental design in which entry of the donor genome into recipient cells is halted at various times, usually by mechanical agitation to break apart the conjugating cells, and the data are used to construct a map of the genes in sequential order. ***e.*** A cell or strain of bacteria carrying two copies of part of the genome (diploid for the segment) and one copy of the remainder of the genome. ***f.*** The process of producing merodiploids through the mediation of *F′* factors carrying one or more bacterial genes into cells that have a complete genome. ***g.*** The simultaneous introduction of two or more genes by a transducing phage, producing multiply recombinant recipient cells. ***h.*** A suspension of phages present or released from broken (lysed) host cells.

## Chapter 7

*7.1* ***a.*** P is unique to DNA and S is unique to protein, so each isotope unambiguously identifies a single type of molecule; C and H are common elements in all organic molecules and would not be incorporated exclusively into one type of molecule. ***b.*** The molecule entering host cells must carry the genetic instructions for making progeny viruses identical to the infecting parental viruses; the molecules remaining outside the host would be incapable of directing virus reproduction inside the cell.
***c.*** DNA is found in all cellular organisms and in many viruses, and it would have the same function in any life form just as other organic molecules of a particular kind have the same function in all or most of the life forms in which they occur.
*7.2* ***a.*** If DNA is the genetic material, we would predict that control extracts lacking DNA would be ineffective in transformation. The control results verified experimental interpretations of DNA function; the control did not falsify the hypothesis that DNA was genetically functional material. ***b.*** If DNA is the genetic material, we would predict that the reciprocal experiment would show that DNA from either source could transform the other strain, which would act as recipient.

***7.3*** (1) Double-stranded DNA; (2) double-stranded RNA; (3) double-stranded DNA; (4) single-stranded DNA if T is present or single-stranded RNA if U is present; (5) single-stranded DNA; (6) single-stranded RNA; (7) double-stranded DNA if T is present or double-stranded RNA if U is present; (8) either single- or double-stranded, which cannot be determined from these data alone.

***7.4*** ***a.*** A T C G T A A G G G C T C C
***b.*** C G C G A A T G C T A G T A G G T.

***7.5*** ***a.*** G-C or C-G base-pair substituted for A-T or T-A base-pair.
***b.***

| A→G | A→C | T→G | T→C |
|---|---|---|---|
| T A | T A | T A | T A |
| G C | G C | G C | G C |
| G C | G C | G C | G C |
| C G | C G | C G | C G |
| ↓ | ↓ | ↓ | ↓ |
| T G | T C | G A | C A |
| G C | G C | G C | G C |
| G C | G C | G C | G C |
| C G | C G | C G | C G |
| ↓ | ↓ | ↓ | ↓ |
| C G | G C | G C | C G |
| G C | G C | G C | G C |
| G C | G C | G C | G C |
| C G | C G | C G | C G |

***7.6***

| duplex | (2) | (3) | (1) | (4) |
|---|---|---|---|---|
| % G + C: | 67 | 43 | 31 | 25 |
| $T_m$: | highest → lowest | | | |

***7.7*** ***a.*** *sm-1* and *sm-2* must be alleles of the same gene since they did not complement each other on *E. coli* K12. If they were alleles of different genes, then *sm-1* + and + *sm-2* could grow on K12 and produce mutant plaques. ***b.*** The wild-type plaques are due to recombinant viruses, and the same frequency of crossing over and recombination should be evident in all crosses involving the same two genes. ***c.*** 1/100 wild-type recombinants + 1/100 reciprocal doubly mutant recombinants (not observed) = 2/100, or 2% recombination.

***7.8*** ***a.*** Four (one in strains 1 and 2, one in strains 3 and 4, one in strain 5, and one in strain 6), based on complementations. ***b.*** Strains 1 and 2 have mutations in one gene, and strains 3 and 4 have mutations in a different gene.

***7.9*** ***a.*** Three. ***b.*** Strains 1 and 4 have mutations in the same gene since they do not complement; strains 2 and 5 have mutations in the same gene; strains 3 and 6 have mutations in the same gene.

***7.10***

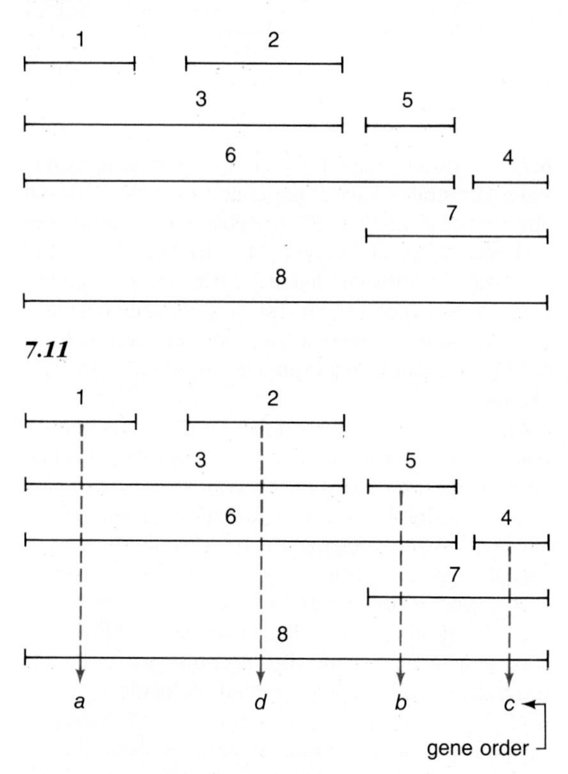


***7.12*** Strain 1 is a deletion mutant and strains 2, 3, and 4 have point mutations.

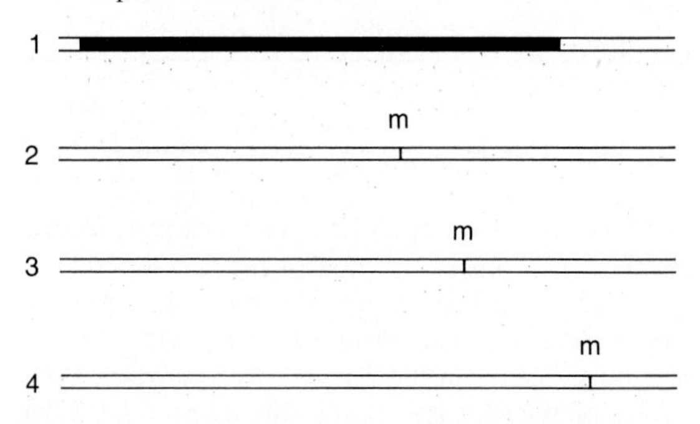


***7.13*** *r*II*B1* and *r*II*B3* are point mutations 0.8 units apart; *r*II*B2* appears to be an overlapping deletion mutant since it produces no recombinants with the other strains.

***7.14*** T4 is a duplex DNA phage, and nitrous acid induces mutations in one of the strands, which, upon replicating, segregates one normal and one mutant daughter duplex after semiconservative replication.

Phage $\phi$X174 is a single-stranded DNA virus and mutations in its single strand yield only mutant phage progeny.

**7.15** ***a.***

UUU = $(3/4)^3$ = $27/64$ = 42.2%

2U + 1C = $(3/4)^2(1/4)$ = $9/64$ each of UUC, UCU, CUU

1U + 2C = $(3/4)(1/4)^2$ = $3/64$ each of UCC, CUC, CCU

CCC = $(1/4)^3$ = $1/64$

***b.***

$27/64$ UUU + $9/64$ UUC = $36/64$ phenylalanine

$9/64$ UCU + $3/64$ UCC = $12/64$ serine

$9/64$ CUU + $3/64$ CUC = $12/64$ leucine

$3/64$ CCU + $1/64$ CCC = $4/64$ proline

**7.16** ***a.*** Base deletion causing frameshift;
3′—TAC TTT TCA ~~G~~GTA GTG AAC TAC GAA TGG—. ***b.*** Add A to alter the fifth codon from GTG to AGT, which will restore Ser as amino acid 5 and all the others in the single frameshift mutation. The new reading frame would be

3′—TAC TTT TCA GTA AGT GAA CTA CGA ATG G—.
↑

**7.17**

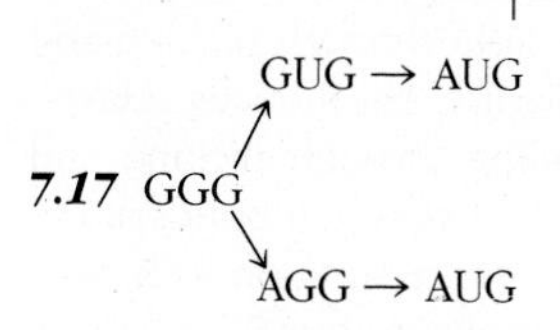


**7.18** Ala–Val–His–Ser

**7.19** ***a.*** Strain 3 is pseudowild since mutant and wild-type progeny segregate in crosses with the standard strain 1. ***b.*** Reverse mutation from a mutant to a wild-type allele at the same site in the gene, rather than a second mutation at a different site as in pseudowilds.

**7.20**

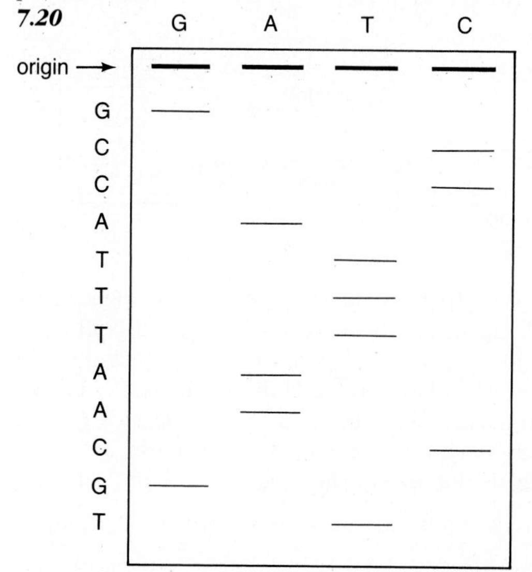


**7.21** ***a.*** 5′—TAC GTA TAA GTT GAG—3′ (3′→5′ = GAG TTG AAT ATG CAT). ***b.*** Coding strand; a complementary mRNA from this strand would be 5′—CUC AAC UUA UAC GUA—3′; if this were the noncoding strand mRNA would be 5′—UAC GUA UAA GUU GAG—3′ and the middle codon would be a stop signal rather than a triplet specifying an amino acid in this polypeptide-encoded sequence. ***c.*** Each fragment has the same 5′ reference point, and the sequence can be read from the shortest 5′-labeled piece (at the bottom of the gel) to the longest 5′-labeled piece (at the top of the gel).

**7.22** ***a.***

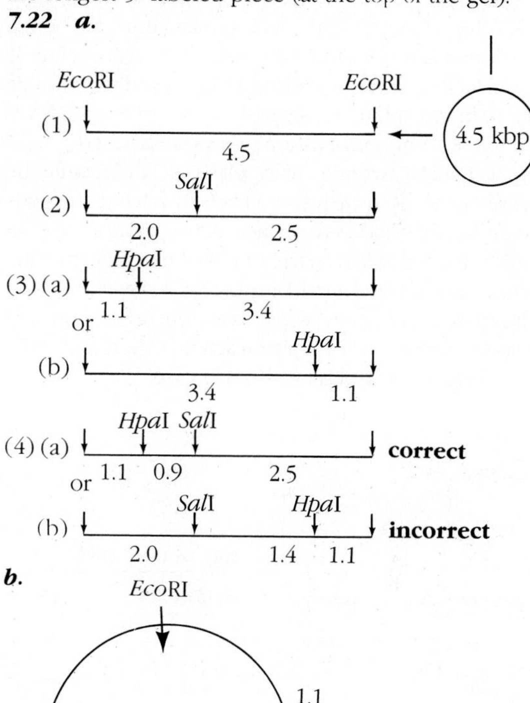


***b.***

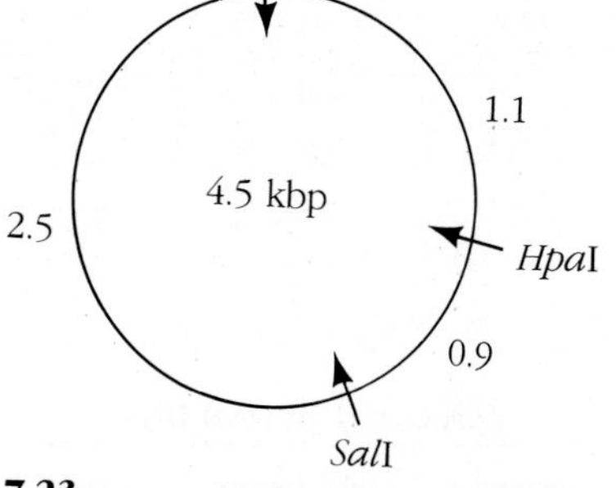


**7.23**

***a.*** —Ser–Ser–Ser–Ser–Ser—

***b.*** —Ser–Ser–Ser–Ser–Ser—
translated from 5′—UCGUCGUCGUCGUCG—3′
—Arg–Arg–Arg–Arg—
translated from 5′ —CGUCGUCGUCGU— 3′
—Val–Val–Val—
translated from 5′ —GUCGUCGUC— 3′

**7.24** ***a.*** Eukaryotic organism. ***b.*** Three intervening sequences (indicated by loops) and two complete exons between the loops plus all or parts of the two flanking exons. ***c.*** Pre-mRNA intervening sequences are excised and the remaining coding sequences are spliced together, thus making a continuous reading frame of encoded information in mature mRNA; all of this takes place in the nucleus.

**7.25** ***a.*** Colonies 1 and 3 must carry recombinant DNA since the *pen-r* gene is missing as seen by the lack of growth of colonies 1 and 3 on penicillin-containing media. Any linearized plasmid DNA would not replicate, so the only explanation for drug-sensitive cells growing on media with tetracycline is that such cells carry plasmid DNA ligated to genomic DNA in recombinant molecules. Recombinant DNA is amplified by replication in transformed cells. ***b.*** Colonies carrying intact plasmid DNA would be resistant to both drugs by virtue of having plasmids carrying *tet-r* and *pen-r* genes. A preparation of plasmids treated with restriction enzyme may include some undamaged circular plasmid DNA along with linearized molecules whose *pen-r* encoded sequence has been removed by enzyme action. Colonies 2 and 4 must carry intact plasmids in this case.

## Chapter 8

**8.1** ***a.***

| generation | heavy | half-heavy | light |
|---|---|---|---|
| | **percentage of total DNA** | | |
| 0 | 100 | 0 | 0 |
| 1 | 0 | 0 | 0 |
| 2 | 0 | 50 | 50 |
| 4 | 0 | 87.5 | 87.5 |

***b.***

| generation | heavy | half-heavy | light |
|---|---|---|---|
| | **percentage of total DNA** | | |
| 0 | 100 | 0 | 0 |
| 1 | 50 | 0 | 50 |
| 2 | 25 | 0 | 75 |
| 4 | 6.25 | 0 | 93.75 |

**8.2** ***a.*** ***b.***

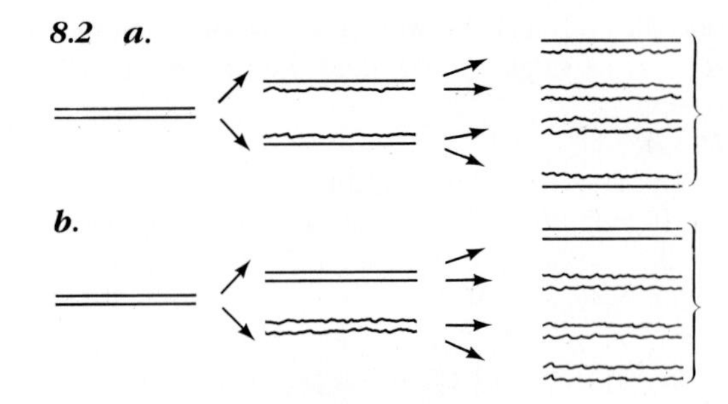

**8.3** ***a.*** Removal of RNA primer, gap-filling with deoxyribonucleotides, sealing of the sugar-phosphate backbone of the chain. ***b.*** Synthesis of RNA primer, synthesis of DNA strand segment. Reactions in (b) precede reactions in (a).

**8.4** ***a.*** Noncatalytic protein that binds cooperatively to single-stranded DNA regions, leading to unwinding of duplex DNA and holding taut the opened single-stranded regions. ***b.*** Untwisting enzyme that relaxes supercoils in duplex DNA through nicking and closing reactions. ***c.*** Gyrase that induces relaxed DNA to undergo supercoiling through nicking and closing reactions. ***d.*** Cuts hydrogen bonds at the duplex region just ahead of the replication fork, generating energy through hydrolysis of ATP by ATPase action. ***e.*** Joins exposed ends of sugar-phosphate chains by catalyzing covalent linkages, which produce ligated strands.

**8.5** ***a.*** Newly synthesized Okazadi fragments that remain unjoined in the absence of ligase action in cells. ***b.*** DNA originally present in the cells that has not yet replicated.

**8.6** ***a.***

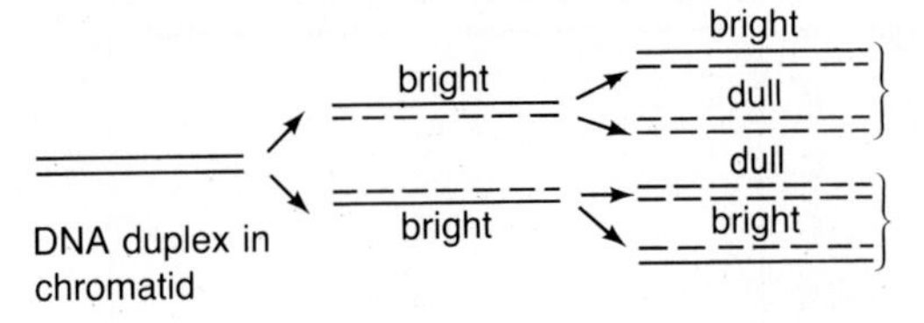


***b.*** Two daughter nuclei would have 1 bright and 1 dull chromatid in the chromosome, and both chromatids would be dull in the other two daughter nuclei after 3 replications in BrdU medium. ***c.*** Ten chromatids, since there are 80 chromatids (40 replicated chromosomes) and ⅛ of these would be brightly fluorescent after 4 generations in BrdU.

**8.7** ***a.*** ***b.*** ***c.*** **8.8** ***a.***

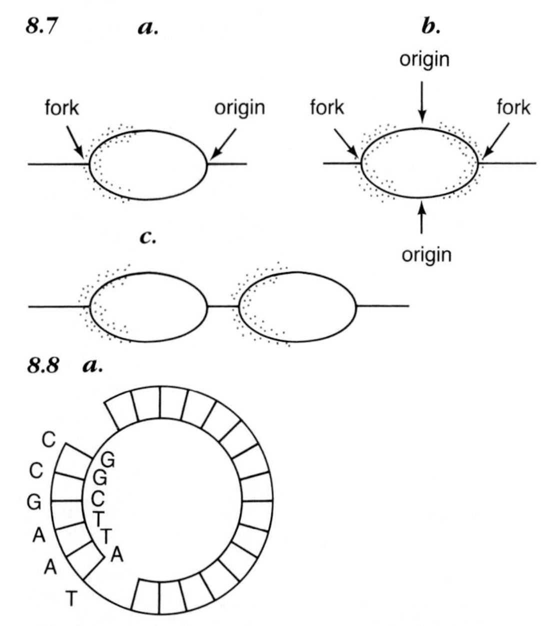


***b.*** Molecules with "sticky" ends would circularize and move more rapidly through the gradient than molecules that remain linear because they have no cohesive termini.

**8.9** ***a.***

AAATTTG 5′
TTTAAAC 3′

***b.*** Circularized molecules would move more rapidly through the gradient than those molecules remaining linear because they have no redundant ends.

**8.10** ***a.*** 5.5 μm. ***b.*** (b) (a) (d) (c), with (c) having progressed the most in replication.

**8.11** ***a.*** (a); copy-choice requires recombination during DNA replication, which would mean that all recombinant molecules would be new and would contain only $^{14}$N. ***b.*** DNA replication would take place whether or not recombination had occurred. The upper curve represents unreplicated, semiconserved, and wholly new DNA molecules; unreplicated molecules retain their original density, semiconserved molecules have $^{15}$N + $^{14}$N in their construction, and wholly new molecules contain only $^{14}$N in their construction. ***c.*** See Fig. 8.18a.

**8.12** ***a.*** Alleles $+/b$ segregated $1+:3b$ instead of $1+:1b$ as expected in haploid products of meiosis. ***b.*** Gene conversion, presumably due to mismatch repair during DNA repair synthesis of recombinant DNA. ***c.*** The distribution of $+/c$ alleles in the linear sequence of spores could only have arisen by crossing over during meiosis; the alleles segregating in an aberrant ratio are usually immediately adjacent to other allele pairs that have obviously experienced recombination.

**8.13** ***a.*** A stable alternative form of an element, characterized by a larger number of neutrons in the atomic nucleus than occurs in the common form of the element; e.g., $^{15}$N and $^{14}$N, or $^{2}$H and $^{1}$H. ***b.*** An unstable alternative form of an element, characterized by a different number of neutrons in the atomic nucleus than occurs in the common form of the element; e.g., $^{3}$H versus $^{1}$H, or $^{32}$P versus $^{31}$P; unstable isotopes undergo beta decay, releasing electrons. ***c.*** The region of duplex DNA that undergoes synthesis of new strands. ***d.*** A mechanism of crossing over suggested to be a process of switching back and forth of newly made strands along the template strands during DNA replication. ***e.*** The process responsible for aberrant allelic ratios, deviating from the expected 1:1, presumably due to mismatch repair of recombined DNA. ***f.*** The structural nucleoprotein component of the chromosome in eukaryotes, composed predominantly of DNA, histones, and nonhistone proteins. ***g.*** The repeating structural unit of the chromatin fiber, consisting of about 140 base-pairs of duplex DNA wound around an octamer of histone proteins, these being two each of histones H2A, H2B, H3, and H4. ***h.*** DNA sequences occurring in many copies per genome. ***i.*** Method of hybridizing radioactively labeled complementary DNA or RNA directly to a chromosome preparation on a microscope slide (*in situ*) and detecting specific chromosomal DNA regions by the radioactivity of bound molecules as seen in autoradiographs of the chromosomes.

## Chapter 9

**9.1** DNA-2, since it is complementary to mRNA.

**9.2** ***a.*** The lower strand; the upper strand includes stop codons ATC and ATT at the second and sixth triplets, respectively, and could not be the coding strand.

***b.*** 5′–TTAATCCCTACGGGCATTAAT–3′
3′–AATTAGGGATGCCCGTAATTA–5′

***c.*** 5′–UUAAUCCCUACGGGCAUUAAU–3′

***d.*** —Leu–Ile–Pro–Thr–Gly–Ile–Asn—

***9.3*** ***a.*** 5′–AAAUAAAAUUUAUAUAAGCAUGAAAGCCUCCUUUUCCUAGUUAAAAUAAA–3′.
1 10 20 30 40 50

***b.*** Beginning with the AUG start codon at base number 20 and proceeding toward the 3′ end of the message in translation until the UAG stop codon at bases 38–40, the amino acid sequence would be: $H_2N$—Met–Lys–Ala–Ser–Phe–Ser—COOH.

***c.*** Leader sequence, bases 1–19; trailer sequence, bases 41–50.

***9.4*** ***a.*** 3′ end. ***b.*** The promoter is not transcribed, and the mRNA would be shorter since only the leader segment of the 3′ end of the gene would be in the mRNA transcript. ***c.*** Long sequence of thymine and adenine residues, including possible TATA box.

***9.5***

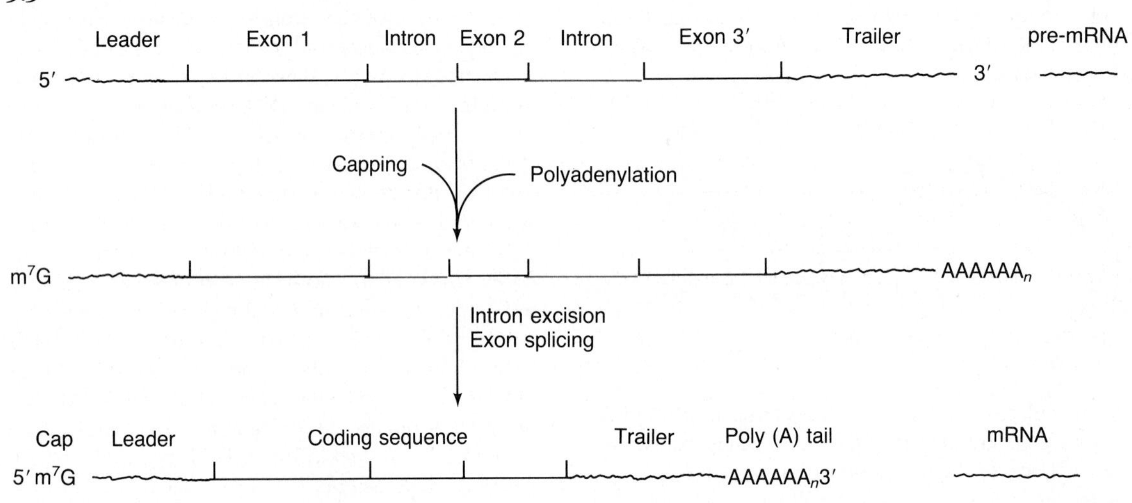


***9.6*** No. According to the wobble hypotheses, ICU anticodons recognize either AGA or AGC codons. Therefore, if ICU were the anticodon of $tRNA^{Ser}$, both AGA and AGC codons would code for serine. AGC is a serine codon, but AGA codes for arginine.

***9.7*** The cysteine codons UGU and UGC will pair with ACG (wobbly), and UGU will pair with anticodon ACA as well. The tryptophan codon UGG pairs only with tRNA anticodon ACC, since 5′-C does not wobble. There is no tRNA for the UGA termination codon, so this codon is distinct from the codons for cysteine and tryptophan, which interact specifically with different tRNA anticodons. Hence, no tRNA has an anticodon of either ACU or ACI that could pair with the stop codon and allow read-through.

***9.8*** Obtain single-stranded DNA by denaturing each chromatin fraction and then hybridize with rRNA. The fraction producing DNA-rRNA molecular hybrids is the rDNA-containing chromatin. It should be the fraction with attached nucleoli.

***9.9*** Conduct *in situ* hybridization, using labeled RNA isolated from chromosome puffs identical to the ones under study. Silver grains will appear over the puffed chromosome regions if the template DNA is located there and has hybridized with the added RNA.

***9.10*** ***a.*** tRNA will accept aminoacyl-group addition in reactions catalyzed by aminoacyl-tRNA synthetase *in vitro* or *in vivo,* whereas other RNAs will not. ***b.*** *In situ* hybridization should be carried out to see if the presumptive labeled rRNA binds to the NORs of the chromosomes, where rDNA is localized. ***c.*** mRNA binds to small subunits of ribosomes, and the mRNA-subunit aggregate should bind initiator aminoacyl-tRNA in turn. Polypeptide chain elongation will proceed in a preparation containing this RNA and other components of protein synthesis.

***9.11*** ***a.*** RNA appears first in the nucleus and later in the cytoplasm. ***b.*** 45S RNA is labeled first and this label later appears in 32S RNA; this is the expected precursor-product relationship. ***c.*** Labeled 32S RNA appears first, and it disappears as labeled 28S appears (precursor-product). ***d.*** 45S RNA is found only in the nucleus, not in the cytoplasm, and the labeling sequence shows precursor-product relationships. ***e.*** 18S rRNA appears in the cytoplasm at 30 minutes, whereas 28S rRNA takes 60 minutes to appear. ***f.*** 45S RNA loses its label after 10 minutes because initially labeled 45S RNA is processed to other RNAs (which appear as labeled components later in the experiment), and newly made 45S RNA is synthesized from unlabeled precursors during the remainder of the experiment.

***9.12*** Prepare complementary DNA copies of *E. coli* 16S rRNA in a DNA-synthesis system containing reverse transcriptase and radioactively labeled thymidine, and incubate *E. coli* labeled cDNA with chloroplast 16S rRNA to obtain molecular hybridization. The percentage molecular hybrids formed, compared with the percentage molecular hybrids formed in mixtures of labeled cDNA and *E. coli* 16S rRNA, would indicate the degree of homology between the two 16S rRNAs.

***9.13*** ***a.*** Polysomes in fractions 10 to 20, monosomes in fractions 20 to 30. ***b.*** The 170S region contains nascent polypeptides labeled with $^{14}C$, thus indicating that polypeptide chains (bound to polysomes) are made only at polysomes in the cells. ***c.*** There is little or no particulate material of greater density than polysomes averaging 170S, and little or no polypeptide synthesis would take place in the absence of ribosome groups larger than 170S. ***d.*** Fractions 30 to 35 contain unbound labeled chains, nascent chains broken during preparation, unincorporated [$^{14}C$]amino acids, and similar materials that are not associated with particulate ribosome and polysome components.

***9.14*** ***a.*** (c). ***b.*** If polypeptide chains are being made on polysomes, we predict that puromycin would strip these chains from polysomes. Removal of labeled nascent chains would be observed by lack of radioactivity in the polysome region and accumulation of large amounts of radioactivity (in stripped nascent chains) in the supernatant fluid. The experimental results confirm the original interpretations; they do not falsify the hypothesis based on data shown in Question 9.13.

***9.15*** ***a.*** UUG → UAG, UGG → UGA or UAG. ***b.*** The message would be lengthened and triplets in the trailer would be translated into amino acids until a new stop codon was reached, leading to a lengthened polypeptide chain that might be nonfunctional or might result in an altered protein that is functionally different.

***9.16*** ***a.*** Hb A → Hb CS by base substitution, changing the stop codon to one that specifies an amino acid; triplets of bases in the trailer would be translated up to a new stop codon, thus lengthening the $\alpha$-chain but not altering the first 141 amino acids. ***b.*** Hb A → Hb W1 by frameshift mutation in codon 140, thus causing a shift in the reading frame and translation of base triplets 3′ to codon 139 until a new stop codon is reached; the first 139 codons would be unchanged, thereby making these identical in Hb A and Hb W1.

***9.17*** ***a.*** Streptomycin prevents initiation and elongation of polypeptide chain synthesis: 70S monosomes would accumulate in drugged cells since IF-3 is no longer bound and dissociation into subunits cannot take place, thus leading to a deficiency of 30S subunits for initiation complexes needed to initiate chain synthesis; removal of fMet-tRNA from initiation complexes prevents chain elongation since other aminoacyl-tRNAs ordinarily do not have an initiating capacity. ***b.*** Chain elongation cannot continue in the absence of aminoacyl monomers. ***c.*** If peptidyl-tRNA is not translocated from the A site to the P site of the ribosome the A site will not be open for incoming aminoacyl-tRNA, and chain elongation will stop. (Note: (b) and (c) are probably the consequences of a single effect, namely, inhibition of translocase (G factor) activity. If peptidyl-tRNA is not translocated from the A site to the P site, the A site is not available for incoming aminoacyl-tRNA.)

***9.18*** ***a.*** Misreading of codons. ***b.*** The middle base in the codon is misread in each case, but the first and third bases are read correctly: UUU read as UAU (Tyr), AAA read as AUA (Ile), CCC read as CGC (Arg), GGG read as GCG (Ala); if other bases in the codon were also misread, a greater variety of amino acids in the polypeptides would be made in the presence of streptomycin.

***9.19*** ***a.*** An untranscribed sequence located 3′ to the coding region of the gene and involved in binding RNA polymerase and initiating transcription. ***b.*** A heptanucleotide within the promoter sequence, believed to aid in binding RNA polymerase and correctly initiating mRNA transcription. ***c.*** A coding region of a gene that is comprised of interspersed coding and noncoding segments. ***d.*** A noncoding region of a gene that is comprised of interspersed coding and noncoding segments. ***e.*** The $3' \rightarrow 5'$ base triplet in tRNA that binds to a complementary mRNA $5' \rightarrow 3'$ codon during translation. ***f.*** A single ribosome composed of a small subunit and a large subunit, part of the cytoplasmic translation machinery. ***g.*** A group of ribosomes bound to a strand of mRNA and actively engaged in polypeptide chain synthesis in the cytoplasm. ***h.*** Repeated copies of genes that specify rRNA in eukaryotes, usually localized in the nucleolar-organizing region (NOR) of one or more chromosomes of the genome. ***i.*** Combination of a small ribosomal subunit, mRNA, and the initiating aminoacyl-tRNA in polypeptide chain synthesis. ***j.*** The ribosomal enzyme that catalyzes peptide bond linking between the existing peptidyl-tRNA at the P site and the newest aminoacyl-tRNA bound to the A site of the ribosome, hence making the growing peptidyl chain longer by one aminoacyl residue.

## Chapter 10

***10.1*** ***a.*** constitutive. ***b.*** inducible. ***c.*** constitutive. ***d.*** constitutive.

***10.2***

| | lactose present | |
|---|---|---|
| **strain** | **β-galactosidase** | **permease** |
| 1 | + | + |
| 2 | 0 | 0 |
| 3 | + | + |
| 4 | 0 | 0 |
| 5 | + | + |
| 6 | 0 | 0 |
| 7 | + | + |
| 8 | 0 | 0 |
| 9 | + | + |

| | lactose absent | |
|---|---|---|
| **strain** | **β-galactosidase** | **permease** |
| 1 | 0 | 0 |
| 2 | 0 | 0 |
| 3 | + | + |
| 4 | 0 | 0 |
| 5 | + | + |
| 6 | 0 | 0 |
| 7 | + | + |
| 8 | 0 | 0 |
| 9 | + | + |

***10.3*** ***a.*** constitutive. ***b.*** constitutive. ***c.*** inducible. ***d.*** constitutive. ***e.*** inducible.

***10.4*** It has an $o^c$ mutation.

***10.5*** ***a.*** (4). ***b.*** (3). ***c.*** (4). ***d.*** (3). ***e.*** (1). ***f.*** (3).

***10.6*** Structural gene = *b*, operator locus = *c*, repressor gene = *a*.

***10.7*** ***a.*** Altered protein synthesized, perhaps nonfunctional TAT. ***b.*** The mutant would make TAT constitutively, in presence and absence of steroid hormones. ***c.*** If both homozygotes and heterozygotes for the mutant allele produce the same phenotype, the mutant allele is dominant over its wild-type alternative. If only homozygous mutant genotypes produce the mutant phenotype, the mutant allele is recessive to the wild-type allele. If the genotype is unknown, then fusion of the mutant with wild-type cells would indicate if the mutant was dominant or recessive.

***10.8*** ***a.*** *trpP* = promoter, site of RNA polymerase binding and initiation of transcription; *trpO* = operator, site of binding for repressor, which can turn transcription off and on; *trpL* = leader segment, which functions in attenuation control of transcription. ***b.*** *trpP* and *trpO* function in operator-repressor control; *trpP, trpO, trpL* function in attenuation control. ***c.*** Low Trp; Trp is a corepressor, which binds to repressor and together they block *trpO* and the transcription of structural genes; repressor alone, under conditions of low Trp cannot bind to *trpO* and cannot block transcription.

***10.9***

| | leader peptide made | |
|---|---|---|
| **medium** | **wild type** | ***trpL*$^{o}$** |
| low Trp | 0 | 0 |
| excess Trp | + | 0 |

| medium | attenuator formed: wild type | attenuator formed: *trpL*$^O$ |
|---|---|---|
| low Trp | 0 | 0 |
| excess Trp | + | 0 |

| medium | *trpE* transcribed: wild type | *trpE* transcribed: *trpL*$^O$ |
|---|---|---|
| low Trp | + | ++ |
| excess Trp | 0 | (+) |

***a.*** Wild-type strain: (1) Low Trp: insufficient Trp to translate leader codons, no attenuator formed, RNA polymerase proceeds to *trpE* and transcribes; repressor cannot block transcription in absence of corepressor; (2) Excess Trp: sufficient Trp to make leader peptide, attenuator forms, *trpE* not transcribed; Trp corepressor helps repressor block *trpO* and transcription. ***b.*** *trpL*$^0$ strain: (1) Low Trp: attenuation control absent in deletion strain, allowing enhanced transcription since repressor is also ineffective in absence of corepressor; (2) Excess Trp: transcription proceeds in the absence of attenuation control, but repressor-corepressor control severly reduces transcription; RNA polymerase that escapes repressor control can proceed beyond *trpO* and deleted *trpL* to *trpE*.

***10.10*** ***a.*** Operator is not transcribed, leader is transcribed. ***b.*** Operator is not transcribed, so its translation is not possible; leader has a reading frame in the transcribed mRNA sequence, which can be translated. ***c.*** Operator-repressor and attenuation controls are both negative since sequences must be unblocked to be transcribed.

***10.11*** ***a.*** 1 = normal inducible synthesis through operator-repressor ($i^+$) and cAMP-CAP positive control. 2 = normal negative control ($i^+$) but reduced positive control since CAP is deficient. 3 = normal negative control ($i^+$) but reduced positive control since cAMP is deficient. 4 = normal negative control only ($i^+$), no positive control ($cyc^-$ $crp^-$). 5 = no negative control since $i^-$ leads to constitutive rather than inducible synthesis, positive control only ($cyc^+$ $crp^+$). 6 = neither control is functional, enzyme synthesized constitutively. 7 = no negative control due to superrepressor $i^s$; positive control functional but ineffective since $i^s$ repressor is bound to the operator. 8 = both controls nonfunctional, enzyme not synthesized. ***b.*** Strain 3; added cAMP would overcome $cyc^-$-related insufficiency of cAMP.

***10.12*** ***a.*** (1) Transcriptional control; (2) hormone turns on transcription of ovalbumin gene; (3) using cDNA probe copied from ovalbumin gene sequence, find cDNA-mRNA hybrid molecules in hormone-induced oviduct but not in uninduced oviduct. ***b.*** (1) Translational control; (2) stable mRNA continues to guide globin translation long after nucleus has been lost (new mRNA cannot be made); (3) cDNA probe reveals presence of mRNA in cells making globins.

***10.13*** ***a.*** DNA replication occurred, without subsequent separation by mitosis. ***b.*** Transcription occurred only at certain gene loci and not at others. ***c.*** RNA transcribed in the nucleus (mRNA, rRNA, tRNA) enters the cytoplasm and functions there. Product RNA obtains label later, when pre-RNA is processed into product.

***10.14*** There would be a large satellite rDNA peak in amplifying nuclear DNA, and not in nonamplifying material. If molecular hybrids form with rRNA, then the "satellite" DNA must be complementary and must therefore be rDNA template (See Fig. 10.14).

***10.15***

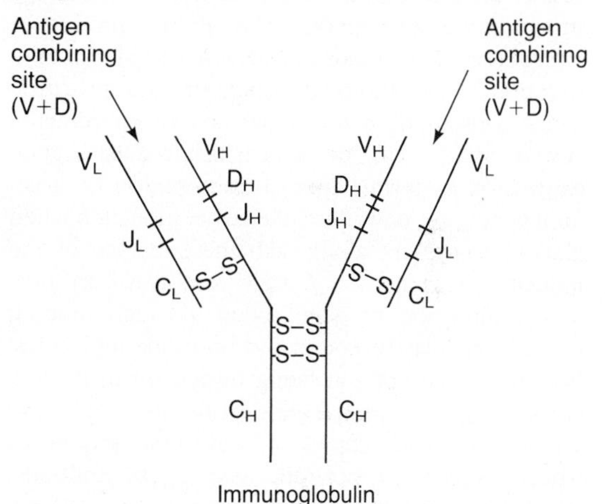


***10.16*** ***a.*** Deletions of all but one *V* and all but one *J* gene between those selected from each gene cluster, and splicing of *V/J—C* to make rearranged germ-line DNA. ***b.*** Kappa; a sequence *V-J—C* is either κ or λ, not a mixture. ***c.*** Rearranged *V-J—C* is transcribed into pre-mRNA, which is altered to mRNA by excision of the sequence between *J* and *C* and splicing to make *V-J-C* continuous. ***d.*** Immature B lymphocytes.

***10.17*** *V/D/J* joining to the $C_H$ gene cluster; there is no D segment in the light chain.

***10.18*** Heavy-chain class switching; deletion of the intervening $C_H$ genes in the cluster removes all of them between *V-D-J* and $C_\epsilon$, resulting in transcription of a different kind of pre-mRNA from the previous DNA arrangement and, ultimately, to processed mRNA with the V-D-J-$C_\epsilon$ sequence for translation to IgE.
***10.19*** R plasmids carrying resistance alleles of the *tet* and *amp* genes must have been transferred between bacterial cells, leading to resistance to both antibiotics in bacteria responsible for the infection. The R plasmids must not have included *str-r* alleles, and the infecting bacteria must have been *str-s*, allowing effective antibiotic therapy with streptomycin.
***10.20*** ***a.*** In transposable elements and transposons the coding sequence is flanked on both ends by inverted repeats, which are IS elements in the transposons. ***b.*** Transposable elements and transposons can be relocated to different sites in the genome through removal and integration events involving repeated ends; both may influence behavior of adjacent genes in different ways in their different locations.
***10.21*** ***a.*** Encoded sequence specifying the primary structure of a polypeptide. ***b.*** Protein product of a regulatory gene, binds to operator and blocks transcription. ***c.*** Promoter, operator, and structural genes constituting a functional unit of coordinated gene action. ***d.*** Mechanism that regulates gene expression by termination of transcription of structural genes. ***e.*** Failure of enzyme induction when glucose is present along with the substrate of the inducible enzyme. ***f.*** Cell responsive to hormonal induction of gene action, through binding between cellular receptors and hormone molecules, leading to transcription being turned on in the nucleus. ***g.*** Differential gene replication, leading to synthesis of many copies of some DNA sequences whereas others are not replicated. ***h.*** Antibody, which binds specifically to antigens. ***i.*** DNA sequence that may move from place to place in the genome and influence neighbor gene action or cause DNA rearrangements. ***j.*** Transposable genetic element, less than 2 kbp long, that contains only genes involved in insertion into DNA. ***k.*** Transposable genetic element, longer than 2 kbp, that contains genes unrelated to insertion as well as insertion genes.

## Chapter 11

***11.1*** $R = 273/554,786 = 0.000492 = 492/10^6 = 492 \times 10^{-6}$; $I = 105.5 \times 10^{-6}$; $Pr = 10.8 \times 10^{-6}$; $Su = 2.4 \times 10^{-6}$; $Y = 2.3 \times 10^{-6}$; $Sb = 1.2 \times 10^{-6}$; $Wx = 0.0$.
***11.2*** 14 children/1,470,000 gametes $= 9.5 \times 10^{-6}$ mutation rate.
***11.3*** $(11 \times 10^{-6})(492 \times 10^{-6}) = 5412 \times 10^{12}$. Since each mutation is an independent event, the chance that both occur together, by chance, is the product of the separate probabilities (mutation rates).
***11.4*** Since percentage mutations is directly proportional to dosage, there should be a constant value for % mutants/dosage ($r$). 10%/1000 = 0.01 and 20%/2000 = 0.01, which is a constant value that represents Difference in %/Difference in $r$ for each experiment. To find the expected percentage mutants at 3000 $r$, multiply 3000 $r$ by the 0.01 constant; $3000 \times 0.01 = 30\%$.

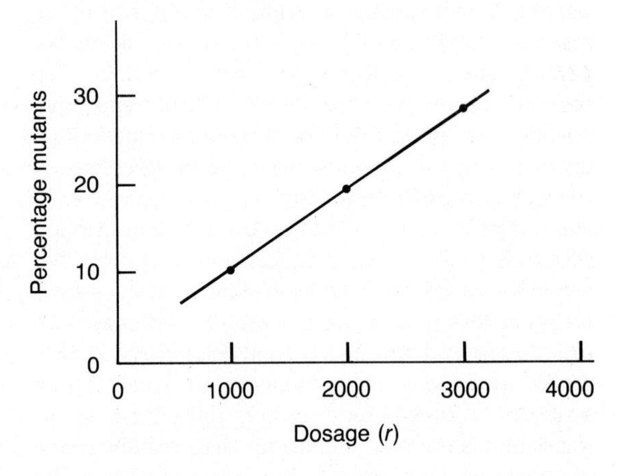


***11.5*** (1) No mutations induced; (2) X-linked lethal mutation induced; (3) detrimental mutation induced, no visible phenotypic change; (4) morphological mutation induced, no detrimental effect; (5) morphological mutation induced, detrimental effect.
***11.6*** The recessive genotype must be lethal, so $Gg \times Gg$ produces only *GG* and *Gg* offspring in the usual proportion of ⅓:⅔ for monohybrid inheritance of one pair of alleles showing incomplete dominance or codominance.
***11.7*** Phage λ is temperate, not virulent. Once phage λ exists in its host in prophage form, further infection by other λ phages is inhibited. Resistance to temperate phages arises after the bacteria have established contact with these viruses, so the results would have shown no greater variation between samples of the individual cultures than of samples taken from the mass culture. This indicates environmental induction of mutations instead of selection of preexisting, randomly arising mutations.

***11.8*** Isolation of double mutant *leu*$^-$ *trp*$^-$ can be performed using a two-step penicillin selection method: (1) Inoculate bacteria into complete media lacking tryptophan and containing penicillin. Only the *trp*$^-$ mutants survive since they cannot divide and, therefore, are not killed by the drug. Plate these mutants on tryptophan-supplemented media lacking penicillin in order to obtain stock culture. (2) Plate the above *trp*$^-$ strain in media with penicillin but lacking tryptophan and leucine. The double mutants will be the only survivors, and these *leu*$^-$ *trp*$^-$ mutants will grow on media only if supplemented with both amino acids they cannot produce.

***11.9*** The codon for glycine is GGU. In Gly → Cys, a transversion occurred (GGU → UGU); in Gly → Asp, a transition occurred (GGU → GAU); in Gly → Ala, a transversion occurred (GGU → GCU).

***11.10*** ***a.*** His = CA$^{U}_{C}$, Gln = CA$^{A}_{G}$; first and second codon bases unchanged, since the only amino acids at position 30 were His or Gln for all strains. ***b.*** No UAx, AAx, or GAx encoded amino acids (first base change), no CUx, CCx, or CGx encoded amino acids (second base change); changes in third base produce His or Gln exclusively; codon alteration for Gln resembles mutant and codon alteration for His resembles wild type.

***11.11*** A nonsense mutation would cause a shortened protein and could lead to nonfunctional enzyme; missense mutations cause nonfunctional or weakly effective enzymes if the amino acid substitution occurs in critical regions of the protein, but many substitutions have little detectable effect on protein function. Frameshift mutations alter the reading frame and usually cause the enzyme to be defective since many different amino acids occur in frameshift polypeptides.

***11.12*** ***a.*** Mutant 1 has a missense mutation: Ser (AGU) → Arg (AGA or AGG); mutant 2 has a nonsense mutation: Trp (UGG) → UGA or UAG (stop codons); mutant 3 has two frameshift mutations (the fourth base is deleted, and U or C is inserted) making Asn codon:

(−) deletion before UGG; (+) insertion after AAA

5′-CCx UGG AGU GAA AAA UG$^{U}_{C}$ CA$^{U}_{C}$-3′
(wild-type mRNA)
↓
5′-CCx GGA GUG AAA AA$^{U}_{C}$ UG$^{U}_{C}$ CA$^{U}_{C}$-3′
(mutant-3 mRNA)

***b.*** Wild-type DNA in this region is:
3′-GGx ACC TCA CTT TTT AC$^{A}_{G}$ GT$^{A}_{G}$-5′

***11.13*** ***a.*** Pro: CCU, CCC, CCA, CCG; Ser: UCU, UCC, UCA, UCG, AGU, AGC; Leu: UUA, UUG, CUU, CUC, CUA, CUG; Phe: UUU, UUC. The probable sequence of mutational changes is

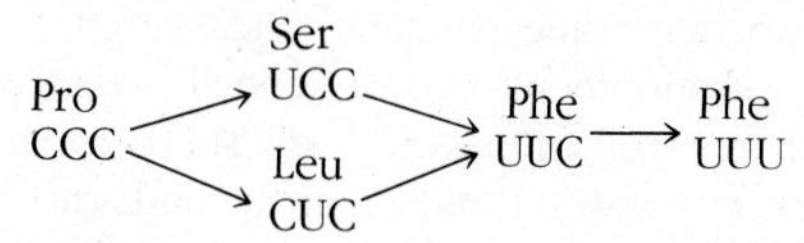


***b.*** Nitrous acid caused transitional changes in TMV (replacement of one pyrimidine by another pyrimidine).

***11.14***

E —③→ B —①→ D —④→ A —②→ C

All strains can grow on C, so this must be the final product in the pathway which, if provided, allows all the mutants to grow regardless of blocks in previous steps. Strain 3 can grow on all supplements except E, which indicates that E precedes the other substances in the pathway; in addition, none of the strains can grow with E alone, so all blocks must come afterward. Strain 2 is the only one requiring the final product exclusively, so that its block is in the last reaction leading to C. Strain 4 uses A or C, so its block precedes A and C. Strain 1 blocks B to D; and strain 3 blocks the first step.

***11.15***

precursor —③→ A
precursor —④→ B
A + B —①→ C —②→ D

***11.16*** ***a.*** Strains 1, 2, and 4 must have carried out normal excision repair since these grew well after irradiation and few or no dimers remained; strain 3 showed mutation increase, a known feature of error-prone excision repair mutants. ***b.*** Dimers are not removed in postreplication repair. Strains 1 and 3 have dimers remaining, so these must have postreplication repair systems. ***c.*** Strain 3 has error-prone excision repair but much of the damaged DNA is restored by postreplication repair; strain 4 has only excision repair of the error-prone mutant type. ***d.*** *uvr* governs excision repair, *rec* governs postreplication repair capacity.

***11.17*** ***a.*** Incubate special *Salmonella typhimurium his*$^-$ strain in media containing water sample and in control media without the sample. Record frequency of wild-type mutants in both kinds of media. ***b.*** A significant increase in wild-type mutants ($his^+$ revertants and pseudowilds) in media with water compared with control mutant frequency indicates the presence of mutagens in the water sample. ***c.*** *In vivo* test using animals to determine whether water samples induce tumor development in test animals when compared with control animals. ***d.*** There is a strong correlation between mutagenicity and carcinogenicity of chemical and physical agents. The predictive value of the Ames test for carcinogenicity is one of its principal features, along with speed and low cost compared with *in vivo* studies.

***11.18*** ***a.*** Number of mutations per gamete or cell per generation. ***b.*** Number of mutations in populations of individuals. ***c.*** Inherited change arising in the absence of known mutation-inducing agents. ***d.*** Imprinting of cells on (replica) plates containing different factors, by transfer from a master plate of cells via a velvet-covered cylindrical holder. ***e.*** Genetic change restoring the original phenotype by restoration of the original codon(s) or by suppressor mutation(s) at other sites. ***f.*** Molecule whose structure mimics the naturally occurring base for which it may substitute. ***g.*** Purine-for-purine or pyrimidine-for-pyrimidine replacement in nucleic acid. ***h.*** Purine-for-pyrimidine or pyrimidine-for-purine replacement in nucleic acid. ***i.*** Pair of adjacent thymine residues linked by carbon-carbon bonding in a strand of DNA. ***j.*** Process of repair of UV-damaged DNA by action of light-dependent enzymes. ***k.*** An agent that induces mutation. ***l.*** An agent that induces tumor or cancer development.

## Chapter 12

***12.1*** ***a.*** Allopolyploidy: a sterile diploid species-hybrid became tetraploid and could undergo normal meiosis to produce viable gametes. ***b.*** 18.

***12.2.*** $n = 9$; *Raphanus* ($n = 9$) × *Brassica* ($n = 9$) produce hybrid *Raphanobrassica* ($2n = 18$), which is sterile. Spontaneous doubling of the chromosome number gave rise to $2n = 36$ and to the fertile allotetraploid hybrid.

***12.3*** 1 wild type (+/*bt*):1 bent (*bt*/–).

***12.4*** ***a.*** If the wild-type allele *S* is dominant over *s* in any dosage, the expected results are 5 wild type:1 compound inflorescence. ***b.*** standard testcross ratio of 1 wild type:1 compound inflorescence.

***12.5*** The Turner female is hemizygous (XO) and GPD deficient, so she must have the X chromosome that carries the mutant *gpd* allele. She and her brother are identical in X-chromosome constitution. She is lacking her father's X chromosome since only her mother carries the GPD deficiency allele on one X chromosome.

***12.6*** The intervals showing lowered recombination frequencies are *vg–L* and *L–a,* so the inversion probably includes these two segments of the chromosome. Pairing at pachynema in the inversion heterozygote would lead to an inversion loop at the segment carrying gene *L.* Since *pr–vg* and *a–bw* show expected recombination frequencies, their distances are unchanged and, therefore, the inversion cannot include the two genes flanking *L.*

***12.7*** ***a.*** Reciprocal translocation between the chromosomes, each carrying one of the two genes; genes were independently assorting and are now linked in the unusual genotypic heterozygote, which is probably a translocation homozygote. ***b.*** Look for a complex of four chromosomes that shows homologies between two normal chromosomes and two translocated chromosomes in progeny during prophase I of meiosis in pollen mother cells. ***c.*** $\frac{83}{678} \times 100 = 12.2$ map units apart.

***12.8***

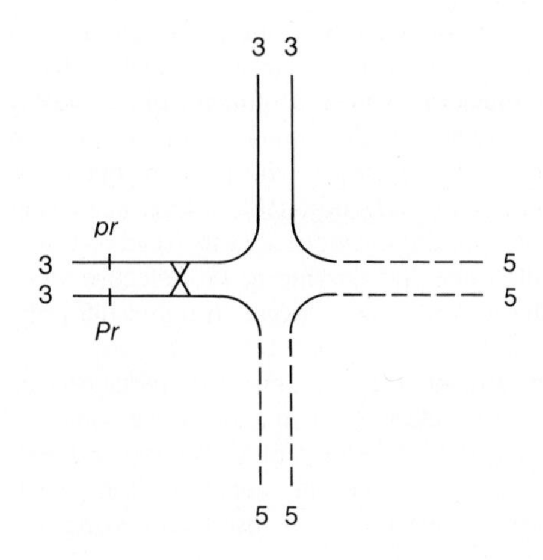


No crossover at X: $\frac{3}{3} + \frac{5}{5} \rightarrow$ 1454 normal pr

Crossover at X: $\frac{3}{5} + \frac{3}{5} \rightarrow$ 1528 semisterile Pr

$\frac{3}{3} + \frac{5}{5} \rightarrow$ 372 normal Pr

$\frac{3}{5} + \frac{3}{5} \rightarrow$ 290 semisterile pr

The locus is $\frac{662}{3644} \times 100 = 18.2$ map units from the translocation point.

***12.9*** ***a.***

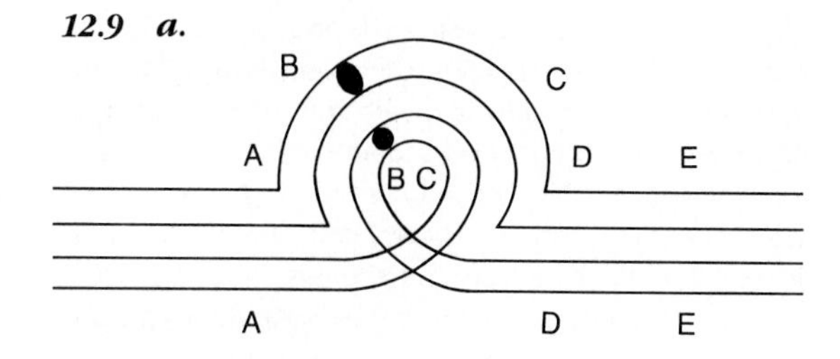


***b.***

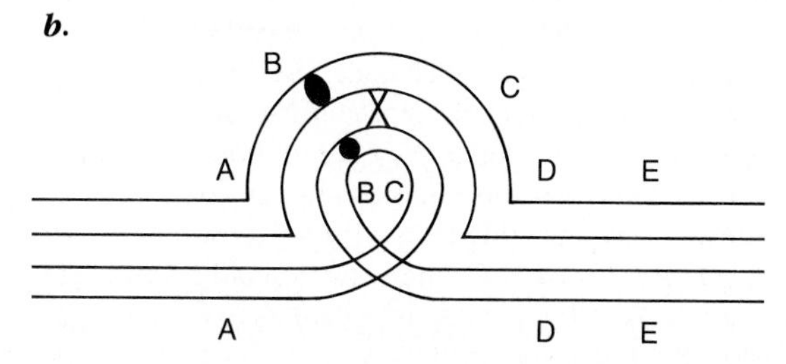


***c.***

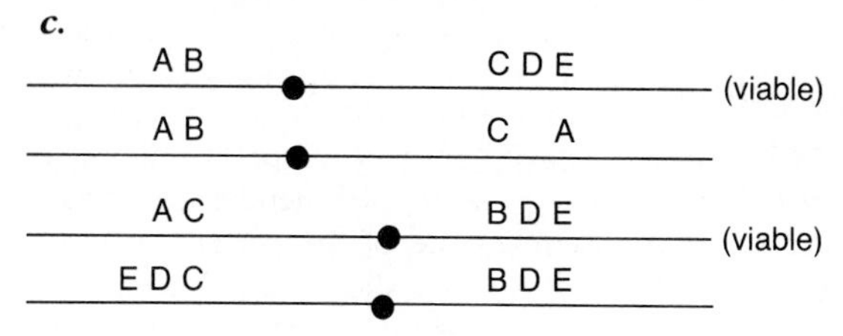


***12.10*** ***a.***

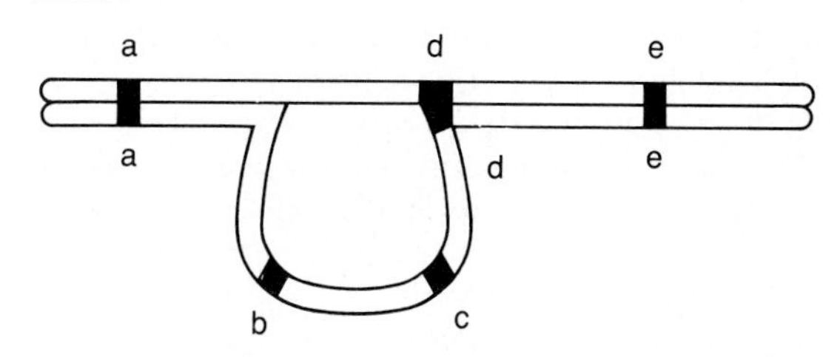


***b.***

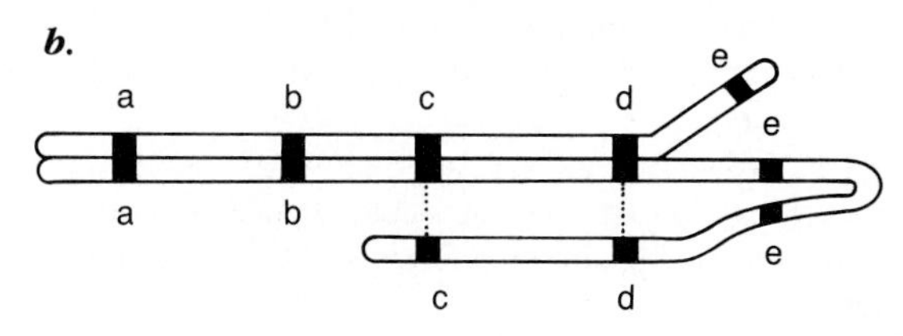


***c.***

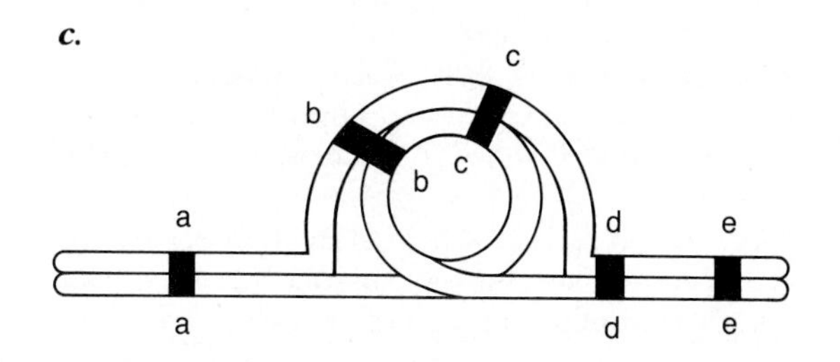


***d.***

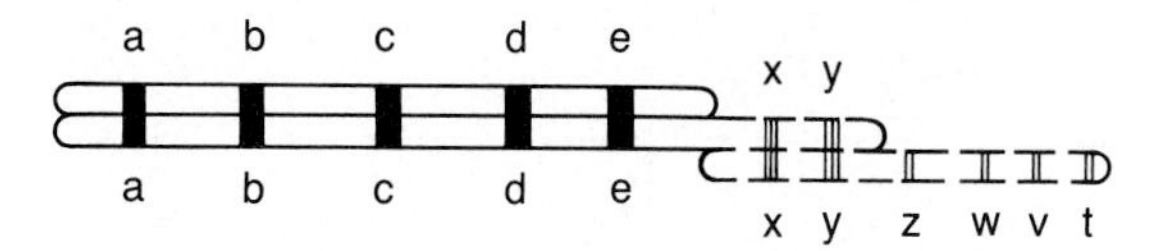


***e.***

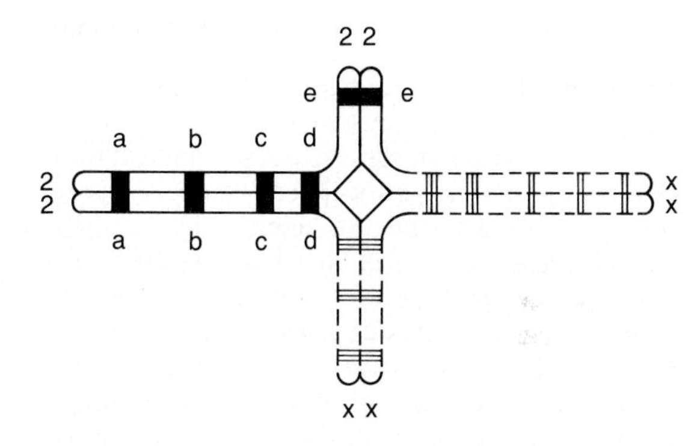


***12.11*** ***a.*** *Tk,* chromosome 17; *Ldh,* chromosome 11; *Pgk,* X chromosome; *Ahh,* chromosome 2. ***b.*** no.

***12.12*** ***a.*** $\alpha$, chromosome 7; $\beta$, chromosome 20; $\gamma$, chromosome 5; $\delta$, chromosome 3; $\epsilon$, chromosome 17; $\zeta$, chromosome 5; $\eta$, chromosome 17; $\theta$, chromosome 21. ***b.*** Yes; genes for $\gamma$ and $\zeta$ on chromosome 5, and $\epsilon$ and $\eta$ on chromosome 17.

***12.13*** ***a.*** Melting of duplex rDNA in alkaline CsCl gradient or by heating to 100°C, and collection of H- and L-strand fractions by centrifugation.

***b.*** H-strand from one species with L-strand from the other species and vice versa; require complementary strands to obtain duplex hybrid molecules.

***c.*** H-strand from one species with 18S and 28S rRNA of the other, L-strand from one species with 18S and 28S rRNA of the other, and vice versa; complementary strands are required to obtain duplex hybrid molecules, but the coding rDNA strand is unknown; hybridization would indicate complementarity (and homology) between coding rDNA of one species and the transcript of the coding strand of the other species, just as DNA-DNA hybrid molecules indicate complementarity of DNA strands. ***d.*** Pairing would indicate homology between coding regions, and unpaired strands would indicate nonhomologous spacers, which have divergent base sequences. (The illustration for this problem is shown at the top of the following page.)

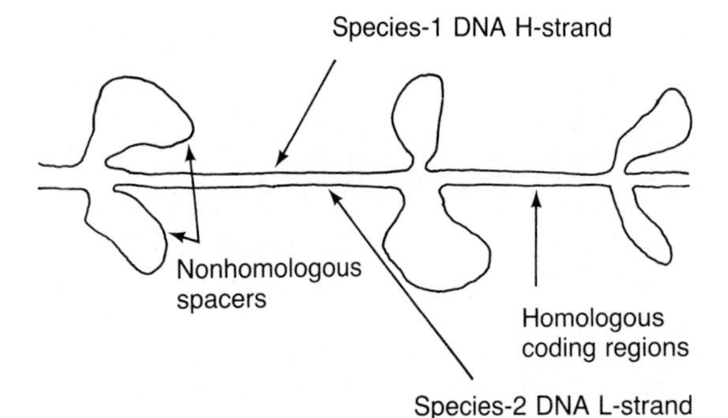


***12.14*** ***a.*** The major rRNA gene was simpler to isolate because its high G-C content causes rDNA to sediment in a different part of the CsCl gradient from bulk nuclear DNA; it is present in repeated copies and can be detected because of amplification of these repeated genes or as repetitious DNA; purified rRNA was available for DNA-RNA molecular and *in situ* hybridizations, as well as heteroduplex mapping by R loops. ***b.*** Specificity of DNA-RNA molecular hybridization between genes (DNA) and gene products (RNA). ***c.*** Various methods can be used, for example, heteroduplex analysis showing the *Xenopus* DNA region present; gel electrophoresis revealing the rDNA within the bacterial DNA material; observation of rDNA gene product (*Xenopus* rRNA) in the clones carrying toad genes.
***12.15*** ***a.*** Two in each parent. ***b.*** Two on one chromosome and none on the other, in each parent. ***c.*** One parent would have two genes on one chromosome 16 and none on the other, and the other parent would have one gene on each chromosome 16. ***d.*** (c).
***12.16*** ***a.*** Turned on transcription of the $\gamma$-globin genes. ***b.*** Hb F ($\alpha_2\gamma_2$) would provide normal hemoglobin functions that cannot be carried out by Hb S; there are no $\beta$ globin chains in Hb F and the mutation would be ineffective in Hb F synthesis.
***12.17*** ***a.*** Two bands: one band of 7.0- or 7.6-kbp fragments + one band of 13-kbp fragments. ***b.*** Gel electrophoresis of *Hpa* I restriction fragments obtained from fetal cellular DNA in amniotic fluid extracted during amniocentesis. ***c.*** 13-kbp band only, no 7.0-kbp or 7.6-kbp band in the gel; indicates $\beta^S\beta^S$.
***12.18*** ***a.*** Cell or individual with more than two complete sets of chromosomes. ***b.*** Cell or individual with at least one more or one less than the diploid number of chromosomes. ***c.*** Cell or individual with three copies of a chromosome. ***d.*** Alteration in the linear ordering of the DNA or chromosome sequence by deletion, duplication, inversion, or translocation. ***e.*** 180° reorientation of an excised and reinserted sequence that includes the centromere of the chromosome. ***f.*** Product of fusion between somatic (body) cells in culture, often involving cells from different species. ***g.*** A selective growth medium containing hypoxanthine (purine precursor), aminopterin (drug that blocks major pathways of DNA synthesis), and thymidine (pyrimidine precursor), in which only $Hprt^+/Tk^+$ cells can multiply since they may make DNA by a "salvage" pathway. ***h.*** Genes associated with the same chromosome as determined by concordance of chromosome and phenotype presence or absence in a culture. ***i.*** Inherited hemoglobin disorders characterized by quantitative changes in the amounts of $\alpha$ and $\beta$ globins produced.

## Chapter 13

***13.1*** ***a.*** (1) 4 *str-s*:0 *str-r*, 2 $mt^+$:2 $mt^-$; (2) 0 *str-s*:4 *str-r*, 2 $mt^+$:2 $mt^-$. ***b.*** *str*, non-Mendelian since all the progeny resemble one of the parents in each cross; *mt*, Mendelian since alleles segregate 2:2 at meiosis. ***c.*** (1) 2 *str-s* $mt^+$ and 2 *str-s* $mt^-$ per tetrad; (2) 2 *str-s* $mt^+$ and 2 *str-s* $mt^-$ per tetrad. ***d.*** (1) 2 *str-r* $mt^+$ and 2 *str-r* $mt^-$ per tetrad; (2) 2 *str-r* $mt^+$ and 2 *str-r* $mt^-$ per tetrad.
***13.2*** ***a.*** 4 $arg^+$:4 $arg^-$, 4 $pro^+$:4 $pro^-$, all large or all small. ***b.*** Mendelian inheritance involving two alleles of one gene for Arg and two alleles of one gene for Pro; non-Mendelian inheritance through the female parent for colony size.
***13.3*** Mendelian inheritance for one pair of alleles, with grande dominant over petite ($F_1$ diploid has grande phenotype and $+/p$ genotype), yielding 2:2 tetrad ratio.
***13.4*** Half will be grande and half will be petite, due to segregation of nuclear alleles, since grande × neutral petite produces grande only, in 4:0 tetrad ratio.
***13.5*** The orange phenotype is probably the result of an extranuclear mutation.
***13.6*** ***a.*** *Rf/rf* individuals, which are normal.
***b.*** Progeny would segregate 1 normal:1 male-sterile, according to *Rf/rf* and *rf/rf* genotypes in male-sterile cytoplasm.
***13.7*** In the cross A ♀ × B ♂, an X chromosome from the B parent is necessary for survival of a hybrid with cytoplasm from parent A. Since male offspring do not

receive their X chromosome from the B (♂) parent, they do not survive and all offspring are females. In the reciprocal cross, an X chromosome from species A has a generally lethal effect on hybrids with cytoplasm from species B.

***13.8*** See Section 13.1.

***13.9*** See appropriate figures for diagrams. Electron micrographs for each replication mechanism would show:

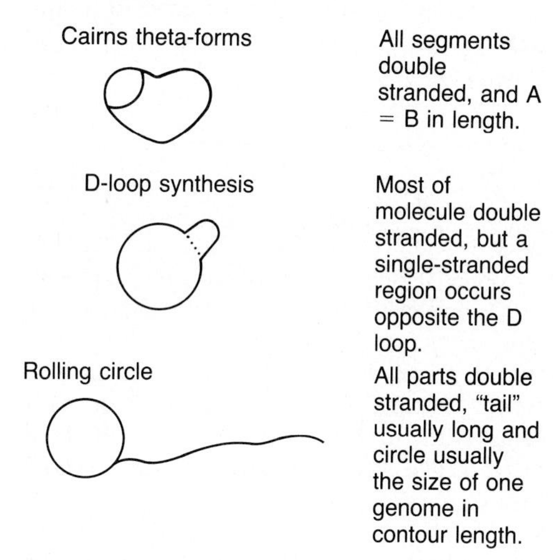


***13.10*** ***a.*** Add poly(U) or other appropriate messenger in a protein-synthesizing system and observe for polypeptide chain synthesis. ***b.*** Using poly(U) *in vitro,* see that either subunit fraction alone cannot sponsor polypeptide chain synthesis, but that monosomes made of unequal-sized subunits can sponsor such synthesis. ***c.*** Chloramphenicol or erythromycin will inhibit mitochondrial ribosome functions, but not cytoplasmic ribosome activity. ***d.*** Mitochondrial ribosomes are small in size (55S–60S), sensitive to chloramphenicol but not to cycloheximide, have no 5S (except for plants) or 5.8S rRNA molecules present; have 12–13S and 16–17S small and large subunit rRNAs, respectively; compared with cytoplasmic 80S ribosomes that are sensitive to cycloheximide, contain 28S, 5.8S, and 5S rRNAs in the 60S subunit and 18S rRNA in the small subunit.

***13.11*** Conduct molecular hybridizations using organelle rRNA with organelle DNA (forms hybrids) and nuclear DNA (no hybrids); and do a control using cytoplasmic rRNA with nuclear DNA (hybrids) and organelle DNA (no hybrids). When organelle rRNA is hybridized with each DNA single-stranded fraction, hybrids forming in only one fraction indicate the template DNA strand, or hybrids with both DNA single-stranded fractions indicate two template strands in duplex DNA.

***13.12*** ***a.*** $\alpha$, $\beta$, $\delta$ in mtDNA and $\gamma$, $\epsilon$, $\zeta$ in nuclear DNA. ***b.*** First two hours: $\gamma$, $\epsilon$, $\zeta$ with $^3$H; next two hours: $\alpha$, $\beta$, $\delta$ with $^{14}$C.

***13.13*** Grow cells in media supplemented with chloramphenicol, cycloheximide, neither drug, and both drugs. Upon electrophoresis, you should see the chloroplast-encoded unit present in cycloheximide-media and the nuclear-encoded enzyme subunit in chloramphenicol-supplemented media. Neither subunit forms in media with both drugs, and the whole enzyme is synthesized in drug-free media; the last two systems serve as experimental controls.

***13.14*** Genes for $tRNA^{Lys}$, $tRNA^{Val}$, and $tRNA^{Leu}$ are encoded in the H strand; gene for $tRNA^{Pro}$ is encoded in the L strand.

***13.15*** ***a.*** At least six. ***b.*** A and B are missing in strain 3; A–F are missing in strain 2. ***c.*** At least seven. ***d.*** Yes; no difference in base sequence of exons, whose continuous reading frame in mature mRNA would be translated into the same amino acid sequence.

***13.16*** ***a.*** UGA = stop in cytoplasm but = Trp in mitochondria. ***b.*** AUA = Met in HeLa mitochondria but = Ile in the other two; CUU = Thr in yeast mitochondria but = Leu in the other two.

***13.17*** (1) killer, (2) sensitive, (3) killer, (4) sensitive, (5) killer (unstable), (6) sensitive.

***13.18*** ***a.*** 1 × 2: one would be *KK* killer, the other would be *KK* sensitive; 1 × 6: one would be *Kk* killer, the other would be *Kk* sensitive; 2 × 6: each would be *Kk*-sensitive. ***b.*** 1 ×2: each would be *KK* killer; 1 ×6: each would be *Kk* killer; 2 × 6: each would be *Kk* sensitive. ***c.*** Each ex-conjugant produces descendants like itself by asexual fission. ***d.*** Each ex-conjugant produces descendants like itself by asexual fission.

***13.19*** The original parent was heterozygous *D/d.*

***13.20*** ***a.*** A pattern of non-Mendelian inheritance in which alleles are transmitted by only one of the parents to the progeny. ***b.*** A cell or individual with two or more genetically different nuclei in a common cytoplasm. ***c.*** Separation of alleles during mitosis, leading to segregation of parental alleles in mitotic rather than meiotic progeny cells. ***d.*** Semiconservative replication of DNA duplex in which the L template strand replicates first, producing a loop of the original complementary H sequence opposite the L-strand origin of replication. ***e.*** DNA sequences in a genome that consist of a contiguous array of amino acid-specifying codons flanked by initiation and termination codons, but whose polypeptide translation product is unknown. ***f.*** A proposed mechanism for cleavage of a polygenic RNA transcript at the site of a tRNA terminus, through recognition signals for cleavage enzymes at looped secondary structures of the transcribed tRNA segments, thus liberating tRNA and mRNA sequences from the original transcript molecule. ***g.*** An organism that lives inside another living organism, with mutually beneficial effects to the partners. ***h.*** False inheritance pattern that results from transmission of bacteria, viruses, or other infectious agents from parent to offspring; the genes responsible for phenotypes are those of the infectious agent and not of the organism expressing the acquired phenotype. ***i.*** Influence of maternal substances on the phenotype of a developing organism, which may produce a false pattern of inheritance from earlier generations.

## Chapter 14

***14.1*** In virus chromosomes, genes active in the same part of the life cycle are often clustered together and transcribed together. The genome is ordered sequentially and so is transcription (all rightward or all leftward).
***14.2*** The heads must be defective in this T4 mutant because it makes normal tails.
***14.3*** Group B makes tails and tail fibers, group D makes heads and tail fibers, group E makes heads and tails, and group F makes tail fibers.
***14.4*** Complementation map is:

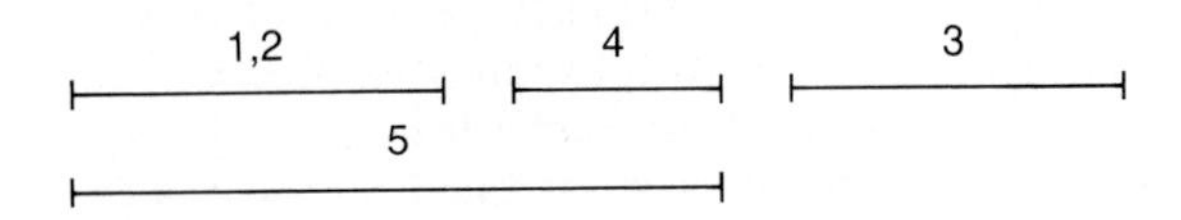


Three unique gene loci are defined: locus 1-2, 4, and 3.
***14.5*** The transformer ($t$) gene probably acts first since normal male sex organs are produced in XX, homozygous males. If the $z$ gene acted first, defective female structures would be produced and the $t$ gene would then direct formation of defective male structures from these defective female structures.
***14.6*** Vermilion eye disks cannot synthesize formylkynurenin, but they do develop pigment when supplied with this substance in the body of a cinnabar host ($v^+$). Cinnabar flies require hydroxykynurenin in order to make pigment, but cinnabar eye disks do not receive this substance in vermilion hosts since vermilion flies themselves are blocked in its production (at an earlier developmental stage). These results define the sequence.
***14.7*** ***a.*** *cdc* 28 → *cdc* 4 → *cdc* 7. ***b.*** They are independent of each other.
***14.8*** Gene *A* products are needed at all times during phage maturation; the involvement of gene *A* in head protein assembly is independent of genes *B* and *C*. Genes *B* and *C* act sequentially (dependently) in head assembly, as follows:

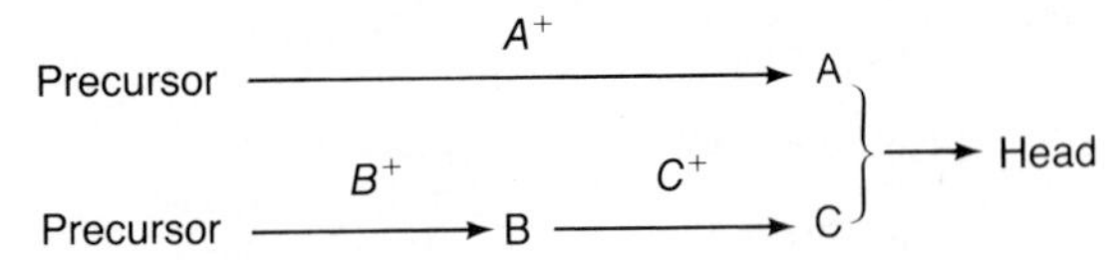


***14.9*** Grow each mutant at 25°C during the early stage of the cycle and switch to 42° at the later stage. Of the two, only the mutant defective in DNA replication (an early function) will form plaques. In the reciprocal experiment (harvest phage growing at 42° in early stage and switch to 25° for later stage), only those mutants defective in tail formation (a late function) will form plaques.
***14.10***

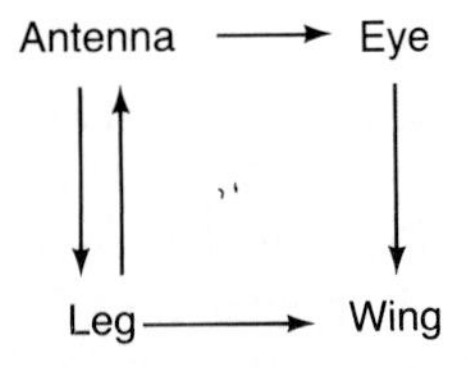


***14.11*** The viral DNA had probably been integrated into mouse chromosomal DNA, existing there in proviral form. The integration of viral DNA may have been involved in tumor induction since normal cell DNA lacked viral DNA sequences, whereas tumor cell DNA had such sequences.
***14.12***

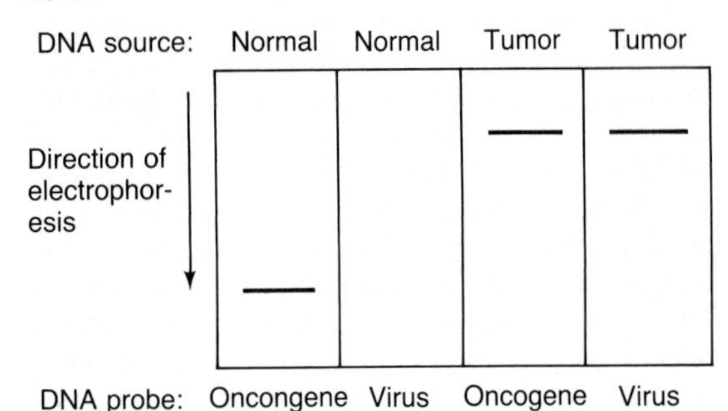

***14.13*** ***a.*** The development of form and function in the organism. ***b.*** System of development in which required information to build a structure is part of the component molecules of that structure. ***c.*** *In vitro* complementation that depends on interactions directly between gene products. ***d.*** The time at which a mutation acts to arrest cellular development at a morphogenetic landmark, or characteristic terminal phenotype expressed by the mutant. ***e.*** A nucleus that can provide all the genetic information needed for development of the individual. ***f.*** An expanded swollen segment of the chromosome, engaged in active transcription of the gene(s) present in that segment; usually observed in polytene chromosomes. ***g.*** Establishment of a particular developmental program in an organism or some part of it. ***h.*** The process of producing form and function specified by a determined genetic program in the organism. ***i.*** An individual in which a normal structure or part of a structure develops in an abnormal location. ***j.*** A gene that can initiate and maintain the tumorous state in the organism. ***k.*** Response in cell cultures characterized by uncontrolled growth, disordered distribution of cultured cells to form heaps rather than confluent single layers on solid substratum and growth in semisoft substratum; analogous to tumor development in the organism. ***l.*** Potentially oncogenic RNA viruses of animal species that make their DNA from RNA using enzyme reverse transcriptase.

## Chapter 15

***15.1*** ***a.*** Total $C^B$ alleles is $622 + 2(554) + 108 = 1838$, and total alleles at this X-linked locus in the sample is 2058. Frequency of $C^B$ is $1838/2058 = 0.893$ and frequency of $C^Y$ is $1 - 0.893 = 0.107$.
***b.*** Expected ♀♀ $= (0.893)^2 + 2(0.893 \times 0.107) + (0.107)^2$; observed ♀♀ $= 0.82 + 0.16 + 0.02$ [fits Hardy-Weinberg predictions]. ***c.*** Expected ♂♂ $= 0.893 + 0.107$, observed ♂♂ $= 0.88 + 0.12$ [fits Hardy-Weinberg predictions].
***15.2*** ***a.*** $p(v^+) = 85/100 = 0.85$, and $q(v) = 1 - 0.85 = 0.15$. ***b.*** $q^2 = (0.15)^2 = 0.0225$, $0.0225 \times 100 = 2.25\%$.
***15.3*** ***a.*** $L^M = 0.737$, $L^N = 0.263$. ***b.*** $2pq = 2(0.7 \times 0.3) = 0.42$; $0.42 \times 1000 = 420$.
***15.4*** $q^2 = 1/10{,}000$, $q = 1/100$, $p = 1 - 0.01 = 0.99$; heterozygote frequency is $2pq = 2(0.99 \times 0.01) = 0.0198$, probability of heterozygote × heterozygote mating is $(0.0198)^2 = 0.0004$, and probability of recessive child is $1/4$ (0.0004), or $0.0001 = 0.01\%$ (1 in 10,000).
***15.5*** $q^2$(black, *bb*) $= 0.4131 + 0.3969 = 0.81$, and $q(b) = 0.9$; $q^2$ (ebony, *ee*) $= 0.0931 + 0.3969 = 0.49$, and $q(e) = 0.7$.
***15.6*** ***a.*** $1/100$. ***b.*** $1/100 \times 1/100 = 1/10{,}000$.
***15.7*** ***a.*** $480/500 = 0.96$. ***b.*** All but $(0.04)^2$ females would be homozygous and heterozygous normal, that is, 99.84% would be phenotypically normal.
***15.8*** ***a.*** Only $TT \times tt$ marriages between tasters and nontasters can produce taster children exclusively (all *Tt*). Since $q^2(tt) = 0.3$, $q(t) = 0.55$ and $p(T) = 0.45$; $p^2(TT) = 0.20$, $2pq(Tt) = 2(0.45 \times 0.55) = 0.495$. Marriages between tasters and nontasters include $TT \times tt$ $(0.20 \times 0.30 = 0.06)$ and $Tt \times tt$ $(0.495 \times 0.30 = 0.148)$; and $0.06/(0.06 + 0.148) = 0.39$, or 39% of all marriages between tasters and nontasters have no chance of producing a nontaster child.
***b.*** $(0.3)^2 = 0.09 = 9\%$.
***15.9*** $(0.6)^3 = 0.216$, or 21.6%.
***15.10*** $p(I^A) = 0.2$, $q(I^B) = 0.1$, $r(i^O) = 0.7$.
***15.11*** Dominance and recessiveness cannot be determined from genotypic frequencies in populations, as Hardy and Weinberg discussed. Breeding tests must be conducted.
***15.12*** ***a.*** $q^2 = 6/2400 = 0.0025$; $q = \sqrt{0.0025} = 0.05$.
***b.*** $2pq = 2(1 - 0.05)(0.05) = 0.095$.
***c.*** $p^2 = (0.95)^2 = 0.9025$. ***d.*** $12/4800$ gametes carried the *co* mutant allele, so the mutation rate of $co^+$ to *co* is 0.0025. In fact, the mutation rate equals the birth rate for recessive lethals since $\hat{q}^2 = u/s$ and for lethals, $s = 1$ and $\hat{q}^2 = u/1$.
***15.13*** ***a.*** $q^2 = 1/25 = 4\%$ are *y/y*, so 96% have white fat; $0.96 \times 1000 = 960$ sheep have normal fat color.
***b.*** $2pq = 2(0.8)(0.2) = 0.32$; $.32 \times 1000 = 320$ sheep are heterozygous. ***c.*** Recessives continue to be produced in heterozygote × heterozygote matings, and the recessive allele is protected against selection in the heterozygote. ***d.*** The allele cannot be eliminated, since $\hat{q}^2 = u/s$; recessive alleles would be maintained in the gene pool by recurrent mutation.

## Chapter 16

***16.1*** ***a.*** $\hat{q} = s_1/(s_1 + s_2) = 0.1/(0.1 + 0.7) = 0.12$; $\hat{p}(Hb^A) = 1.00 - 0.12 = 0.88$.
***b.*** $(0.12)^2 \times 100 = 1.44\%$.
***16.2*** See Fig. 16.1 and Section 16.2.

***16.3***

| characteristic | *D. colorata* | *D. willistoni* |
|---|---|---|
| recombination index | higher | lower |
| frequency of inversions | lower | higher |
| frequency of supergenes | lower | higher |
| habitat stability | higher | lower |

*D. willistoni* is the more highly adapted of the two species.

***16.4*** Crossing over will not take place between genetically differentiated X and Y sex chromosomes, and their separate supergene systems are maintained generation after generation.

***16.5***

| relative fitness | genotype *YY* | *Yy* | *yy* | |
|---|---|---|---|---|
| *W* | 0.6 | 0.6 | 1.0 | $p(Y) = 0.2$ |
| *s* | 0.4 | 0.4 | 0 | $q(y) = 0.8$ |

For dominant alleles with $s > 0$, the proportion of the dominant allele lost in each generation from the population is $sp^2 + pqs$ (see Table 15.13). In this case, $p_0(Y) = 0.2$, and $p_1(Y) = p - (sp^2 + pqs) = 0.2 - [(0.4)(0.2)^2 + (0.2)(0.8)(0.4)] = 0.2 - 0.08 = 0.12$. In 10 generations of predation, the frequency of *Y* would be $p_{10}(Y) = p_0/[1 + 10(p_0)] = 0.2/[1 + 10(0.2)] = 0.067$, as discussed in Section 15.6, and the frequency of *Y* after *n* generations would be $p_n = p_0/[1 + n(p_0)]$.

***16.6*** ***a.*** The effect is achieved through directional selection of polygenes affecting leaf number. ***b.*** The original mean is not attained because some alleles governing "low leaf number" have been eliminated from the population during the 15 years of selection leading from 32 to 58 leaves per plant.

***16.7*** ***a.***

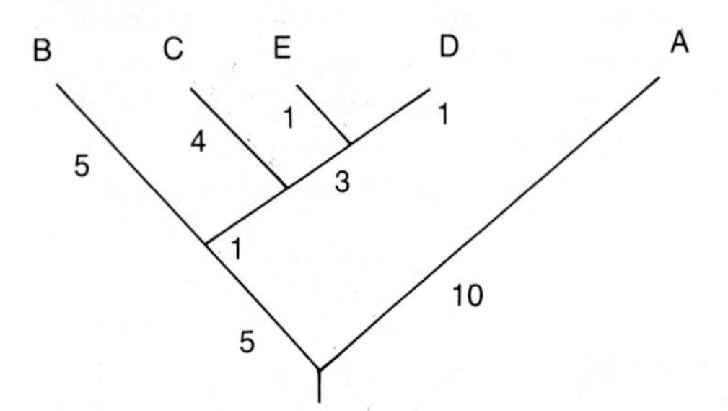


***b.*** Species A is probably most ancient since it differs from every other species by the greatest number of nucleotides.

***16.8*** Genotype *e/e* was completely lethal, and allele *e* was reduced in frequency by selection against *e/e* according to the formula:

$$p_n = \frac{p_0}{1 + np_0} = \frac{0.5}{1 + 8(0.5)} = \frac{0.5}{5.0} = 0.1$$

***16.9*** ***a.*** Interbreed members of the three populations and determine whether they produce hybrids (same species) or do not produce hybrids (different species). ***b.*** If hybrids are not produced by the three populations of the single species, one or more isolating mechanisms must be preventing interbreeding. They may reproduce at different seasons, show different sexual behaviors, occupy different habitats on the island, and so forth, thereby producing no hybrids even though they are genetically close according to experimental breeding tests.

***16.10*** ***a.***

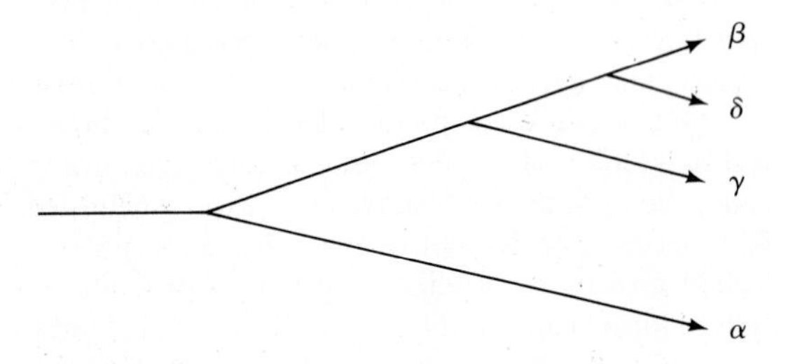


***b.*** The loci occupy different chromosomal sites presumably because of translocations during genome evolution.

***16.11*** ***a.*** Five very different genes specified the five very different proteins. ***b.*** Different and largely nonhomologous base sequences for the five genes would be expected since different codons would occupy the different reading frames leading to different amino acid compositions. ***c.*** The horse *A* gene and pig *X* gene arose from a common ancestral gene and diverged in base sequence but not in gene organization.

***16.12***

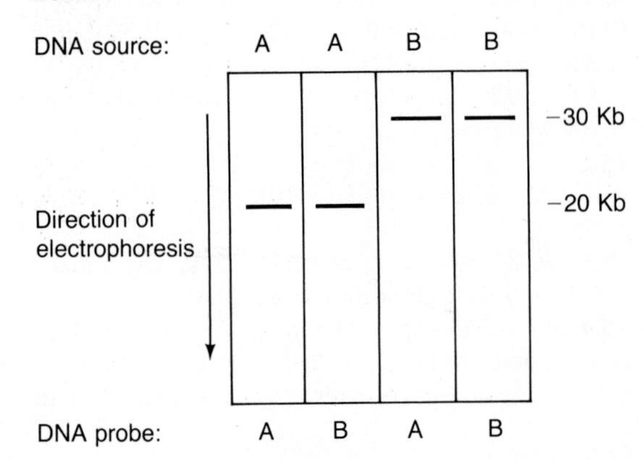

***16.13*** ***a.*** Duplication of one of the three original exons and retention of the duplicate plus the original three exons would lead to a gene with four exons, two of which were highly homologous. The origin by chance of two highly homologous coding segments would be less probable than their origin by duplication and substantial retention of homology. ***b.*** Unequal crossing over, as follows:

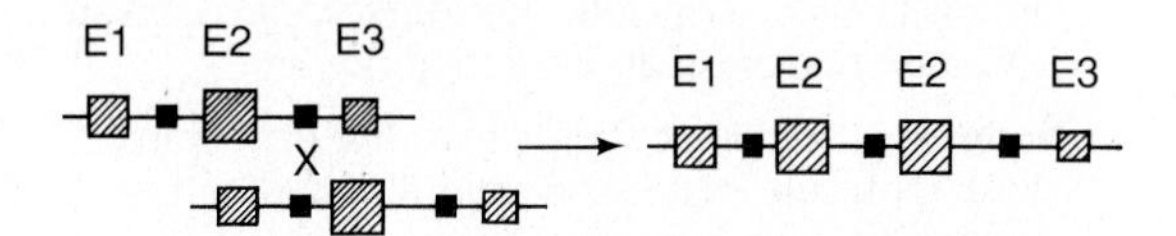


***16.14*** ***a.*** The third gene may have been lost by an unequal crossover event in the pig genome if no selective disadvantage was involved, or a duplication of a keratin gene may have occurred in the horse genome but not in the pig genome. Divergence of duplicate genes might obscure their common origin in the distant past. ***b.*** Pseudogenes have grossly distorted exon-intron organization when compared with their functional homologues and may even have lost all the introns originally present in the interrupted gene sequences.

***16.15*** ***a.*** The relative fitness of a genotype in relation to its environment. ***b.*** An evolutionary strategy in which changes in the gene pool lead to population shifts in the direction of greatest adaptedness in relation to changing environmental conditions. ***c.*** Molecular differences in a gene product due to allelic differences in the coding sequence of the gene. ***d.*** The relative adaptedness of individuals and populations to their environment. ***e.*** The ability of individuals or populations to adapt to changing environments. ***f.*** A measure of genotypic diversity in populations, derived from chiasma frequency/number of chromosome pairs. ***g.*** The inability for fruitful gene exchange between species. ***h.*** A compact structural region in a polypeptide or protein, corresponding often to an exon in the encoding gene sequence. ***i.*** An exchange of homologous chromosome or DNA segments in which the products of recombination are unequal, thereby leading to deletion in one strand and duplication in the other.

# Glossary

**acentric:** designating a chromosome without a centromere

**acrocentric:** designating a chromosome whose centromere is located near one end, making one chromosome arm much longer than the other.

**adaptability:** the capacity to undergo modification so as to function better in different environments, thus improving the chances for leaving descendants; also called flexibility

**adaptation:** the process by which organisms and populations become better suited to function in a given environment, thus improving their chances for leaving descendants

**adaptedness:** the state of being best fit to function in a given, specific environment; fitness

**adenine:** a purine base found in nucleic acids

**allele:** one of the alternative forms of a gene

**allopolyploid:** a polyploid organism that originates from the combination of two or more genetically distinct sets of chromosomes

**Ames test:** an assay desiged by Bruce Ames to identify mutagenic chemicals, utilizing histidine-requiring strains of *Salmonella typhimurium* and observing the frequency of histidine-independent reverse mutants in the cultures; an assay that identifies potentially carcinogenic agents, since these often are also mutagenic agents

**aminoacyl-tRNA synthetase:** one of a group of enzymes that catalyze the activation of an amino acid to the aminoacyl form and the bonding of the aminoacyl residue to a specific transfer RNA (tRNA) carrier

**amniocentesis:** a procedure for removal of a sample of amniotic fluid and fetal materials from a pregnant woman, inserting a hypodermic syringe needle through the woman's abdominal wall and penetrating into the amniotic sac in which the fetus is located

**anaphase:** a stage of mitosis or meiosis when sister chromatids or homologous chromosomes separate and move toward opposite poles of the cell

**aneuploidy:** having one or more chromosomes in excess of or less than the usual number

**antibody:** a protein secreted by plasma cells that interacts with a specific invading antigen in an immune response; an immunoglobulin

**anticodon:** the triplet of nucleotides in a transfer RNA molecule that associates by complementary base pairing with a specific triplet (codon) in the messenger RNA molecule during its translation at the ribosome

**antigen:** a foreign substance that stimulates the body to produce specific neutralizing antibodies in an immune response

**arithmetic mean:** an average; the number found by dividing the sum of a series by the number of items in the series

**ascospore:** a spore produced by meiosis and contained within an ascus sac in ascomycetous fungi, such as yeast and *Neurospora*

**ascus:** a sac that contains a tetrad of ascospores produced by meiosis, or contains eight ascospores when each of the four meiotic products has undergone a subsequent mitosis

**asexual reproduction:** reproduction without the sexual process; vegetative propagation

**A site:** the region of the ribosome that accepts an incoming aminoacyl-tRNA unit during translation of a polypeptide from genetic instructions

**$\alpha$ thalassemia:** one of a group of human blood disorders associated with variations in the amounts of hemoglobin made, generally due to the presence of fewer than the normal number of four $\alpha$-globin genes on chromosome pair 16

**attenuation:** a control mechanism that regulates gene expression through termination of transcription of biosynthetic operons in bacteria

**attenuator:** a stem-and-loop secondary structure of messenger RNA that prevents transcription by RNA polymerase of structural genes in bacterial operons

**autopolyploid:** a polyploid organism that contains more than two sets of the same chromosomes

**autoradiography:** a method for localizing radioactive atoms in biological materials by exposing of a photographic emulsion laid over these materials and observing silver grains or darkened film where radioactive decays have occurred

**autosomal inheritance:** a pattern of transmission of alleles on autosomal chromosomes, usually characterized by production of identical progeny in reciprocal crosses

**autosome:** a chromosome other than a sex chromosome

**auxotroph:** a mutant microorganism that can grow on minimal medium only when supplemented by growth factors not required by wild-type (prototroph) strains

**avirulent:** lack of virulence; the incapacity to initiate infection

**backcross:** the cross of an $F_1$ heterozygote with an individual that is genotypically identical to one of the two parental types

**bacteriophage:** a virus whose host is a bacterial cell; phage

**balancing natural selection:** a selection strategy that acts to maintain a high frequency of a deleterious allele through selective advantage of the heterozygote over its homozygous genotypic alternatives

**Barr body:** any condensed or inactivated X chromosome in the interphase nucleus; sex chromatin

**base analogue:** a purine or pyrimidine base other than a standard nucleic acid base of similar molecular structure

**base substitution:** a type of gene mutation in which the base or base-pair at one site in the wild-type DNA sequence is different in the mutant DNA sequence

**binomial expansion:** $(a + b)^n = 1$, where $a$ and $b$ are the probabilities of occurrence of two alternative outcomes in $n$ number of events or individuals, and the successive terms of the expansion are preceded by successive coefficients, as in $(a + b)^3 = 1\,a^3 + 3\,a^2b + 3\,ab^2 + 1\,b^3$

**biochemical mutation:** a modified gene that alters the phenotype as the result of a metabolic defect in the organism

**bivalent:** a pair of synapsed homologous chromosomes seen in prophase I of meiosis

**breakage and reunion:** a mechanism of crossing over between linked genes involving breaks and rejoinings of DNA strands catalyzed by enzymes in DNA repair reactions, and giving rise to allelic recombinations of the linked genes

**$\beta$ thalassemia:** one of a group of human blood disorders associated with variations in the amounts of hemoglobin made, generally due to deletions in the $\beta$-globin gene cluster located on chromosome pair 11

**bulk nuclear DNA:** the major DNA component of the nucleus, distinct from satellite and other minor DNA components in the same nucleus

**cancer:** a class of diseases characterized by uncontrolled growth, invasiveness, and spread to distant sites (metastasis) in the body

**capping:** addition of a residue of 7-methylguanosine to the 5′ end of a pre-messenger RNA molecule during its posttransciptional processing to mature messenger RNA

**carcinogen:** a physical or chemical agent that induces cancer development, or carcinogenesis

**carcinogenesis:** the process or phenomenon of cancer induction in an organism

**catabolite repression:** the failure to synthesize inducible enzymes when glucose is present in addition to the inducing metabolite; the glucose effect

**C-banding:** a method for staining chromosomes differentially to show heterochromatic regions as stained bands and nonheterochromatic regions as unstained or weakly stained

**cDNA:** copied DNA; DNA synthesized from a template strand of nucleic acid

**cell cycle:** the sequence of events in dividing cells in which an interphase consisting of $G_1$, $S$, and $G_2$ periods separates one mitosis from the mitosis of a successive cell division cycle

**centric fusion:** whole-arm fusions of chromosomes, producing one larger chromosome with a single centromere region from two individual chromosomes each with its own centromere region

**centromere:** the chromosome structure to which the spindle fibers are attached and which is required for directed chromosome movement to the poles at anaphase

**chiasma:** a site of exchange between two chromatids of a bivalent, visible as a crossover figure in diplonema, diakinesis, and metaphase of Meiosis I

**chi-square ($\chi^2$) test:** a statistical procedure used to determine how closely a set of values obtained experimentally fits a given theoretical expectation

**chromatid:** one half of a replicated chromosome, joined to its sister at the centromere region of the whole chromosome

**chromatin:** the deoxyribonucleoprotein material of the chromosomes

**chromatin fiber:** the continuous deoxyribonucleoprotein molecule of the chromosome

**chromosomal mutation:** an alteration in the genome due to rearrangement, loss, or duplication of more than one base or base-pair

**chromosome:** the gene-containing structure in the eukaryotic nucleus; sometimes used to describe the genetically encoded nucleic acid molecule in prokaryotes and viruses

**cis arrangement:** coupling arrangement of allelic or nonallelic genes in which mutant loci are on one chromosome and wild-type loci are on the homologous chromosome in the individual ($ab/++$), in contrast with trans or repulsion ($a+/+b$)

**cis-trans test:** a complementation test of functional allelism in which growth occurs if the mutant sites on the same strand are in different genes, growth fails to occur if the mutant sites on the same strand are located in the same gene, but growth occurs in either case when the mutants are in the control trans arrangement

**cistron:** the segment of a DNA molecule that specifies a particular polypeptide chain or RNA gene product, sometimes equated with a gene in bacterial or viral genomes

**ClB method:** a technique used to detect X-linked mutations in *Drosophila melanogaster,* and so named because one X chromosome in the female parent is marked with a *C*rossover suppressor, *l*ethal recessive allele, and *B*ar eye allele

**clone:** a group of cells or individuals that are genetically identical, having descended from a single cell by mitotic divisions

**cloned DNA:** a collection of identical DNA molecules produced by replication of a single molecule in a suitable host or system

**coding sequence:** *see* **exon**

**codominant:** designating alleles of a gene for which the alternatives are fully expressed in the heterozygote, as with the development of the AB blood group phenotype in $I^AI^B$ heterozygotes

**codon:** a triplet of nucleotides that specifies an amino acid or stop signal in translation of a polypeptide from encoded genetic instructions copied into mRNA

**coefficient of coincidence:** an experimental value equal to the observed number of double crossovers divided by the expected number of double crossovers

**colinearity:** the spatial correspondence between the order of codons in DNA and the order of amino acids in the polypeptide translated from the DNA blueprint

**colony:** a cluster of single cells derived from a single ancestral cell and growing on a solid medium or surface

**complementary base pairing:** specific hydrogen bonding between a particular purine and a particular pyrimidine component in duplex nucleic acid molecules; for example, guanine pairs with cytosine and adenine pairs with thymine or uracil

**complementation:** the mutual interaction between different genetically defective components that together can produce the wild-type phenotype; mutants that produce wild-type phenotypes in complementation tests can be ordered in complementation maps; different defective units can compensate or complement each other's defects and display wild-type characteristics

**concordance:** the expression (or lack of expression) of a given trait in both members of a pair; agreement

**conditional mutant:** a mutant whose viability or phenotypic expression is dependent on environmental or growth conditions, as in temperature-sensitive mutants which grow only within a particular range of temperature

**conidia:** asexual haploid spores that are produced by various fungi and can germinate and form new hyphae when incubated on a suitable medium

**conjugation:** a process involving physical contact between participating cells, thus permitting gene exchange. The process involves reciprocal exchange in sexual reproduction in unicellular organisms, but unidirectional transfer of DNA in *E. coli* strains that carry the fertility factor F.

**consensus sequence:** a sequence of nucleotides found in different genes and believed to act as a recognition site for identical events in the different cases, as the nucleotide segment that may signal the site for excision of introns in pre-mRNA transcripts of interrupted (mosaic) genes in eukaryotes

**conservative replication:** the process of synthesis of new duplex DNA molecules in which the entire duplex acts as a template for synthesis of an identical duplex DNA

**constant (C) region:** a segment of the heavy- and light-chain polypeptides of an immunoglobulin molecule, existing in limited variety among the diversity of other segments of immunoglobulin molecules

**constitutive:** constant or unchanging; for example, a constitutive enzyme which is synthesized at a constant rate and is not subject to regulation, or constitutive heterochromatin, which is permanently condensed nuclear chromatin

**continuous variation:** variation for some characteristic that exhibits a spectrum of values from one extreme to the other in a continuous spread or gradation

**copied DNA:** *see* **cDNA**

**copy choice:** an explanation of genetic recombination based on the proposition that newly made strands switch back and forth along the template parental strands during DNA replication, giving rise to recombinant molecules consisting only of new DNA

**corepressor:** a metabolite that binds with repressor protein and blocks transcription of messenger RNA, which specifies a repressible enzyme polypeptide

**$C_0t$ plot:** a graphical representation of the reassociation kinetics for a given sample of single-stranded DNA made by plotting the fraction of molecules remaining single-stranded against the product of DNA concentration and time ($C_0t$ value)

**cotransduction:** the simultaneous transduction of two or more genes due to the presence of these genes in the same segment of transduced DNA, indicating close linkage of the genes

**coupling:** the arrangement of allelic or nonallelic mutant loci in one chromosome and their wild-type alternatives in its homologous chromosome ($ab/++$); cis arrangement, in contrast with trans, or repulsion ($a\ +/+\ b$)

**covalent bond:** interaction between atoms with shared electron shells

**crossing over:** exchange of homologous chromosome or DNA segments, leading to recombination of linked genes

**culture:** a population of microorganisms or the dissociated component cells of a tissue growing in nutrient medium

**cyclic AMP:** adenosine monophosphate with the phosphate group bonded internally to form a cyclic molecule; a mononucleotide involved in positive control of transcription; cAMP

**cytogenetics:** the study of biological systems using the combined methods of cytology (microscopy) and genetics (breeding analysis or molecular analysis)

**cytoplasm:** the protoplasmic contents of a cell, exclusive of the nucleus

**cytoplasmic inheritance;** *see* **extranuclear inheritance**

**cytosine:** a pyrimidine base found in nucleic acids

**degrees of freedom *(df)*:** the number of items of data that are free to vary independently, equal to $n - 1$ for $n$ items, since the $n$th item is fixed or determined by the other values

**deletion:** loss of a segment of a chromosome or DNA molecule, including the loss of a single base in a sequence; deficiency

**deletion mapping:** use of overlapping deletions to localize the site of a gene or a DNA sequence on a linkage map or a gene map

**denaturation:** weakening and disruption of secondary or tertiary structure or both structures of proteins and nucleic acids, leading to loss of molecule function; *see also* **melting**

**deoxyribonucleic acid (DNA):** the genetic material, which is comprised of the nitrogenous bases, adenine, guanine, cytosine, and thymine, covalently bonded to a repeating chain of deoxyribose and phosphate residues; usually found in double helix configuration, or duplex molecules

**determination:** the establishment of a particular developmental program in an organism or in some part of it, and the fulfillment of the program irrespective of any subsequent situation

**detrimental mutation:** an inherited change that causes reduced viability or lower life expectancy of an individual

**development:** an orderly sequence of progressive changes resulting in a more complex biological system; *see also* **morphogenesis**

**developmental genetics:** the study of mutations that produce developmental abnormalities, in order to gain understanding of how normal genes control growth, form, function, and other biological expressions in individuals

**diakinesis:** the end of prophase I in meiosis

**differentiation:** the complex of changes involved in the progressive diversification of cellular structure and function in the organism; *see also* **morphogenesis**

**dihybrid cross:** interbreeding between individuals that differ in two specified genes

**diploid:** a cell or individual that has two copies of each chromosome (not always including the sex chromosomes) in its genome ($2n$)

**diplonema:** the penultimate substage of prophase I of meiosis, during which the paired homologous chromosomes open out (separate) everywhere except at the chiasmata

**directional natural selection:** *see* **selection strategies**

**discontinuous variation:** variation that falls into two or more nonoverlapping classes

**discordance:** the expression of a given trait in only one member of a pair

**disjunction:** the separation of homologous chromosomes or sister chromatids at anaphase of mitosis or meiosis and the movement of each member of a pair toward an opposite pole

**diversifying natural selection:** *see* **selection strategies**

**diversity (D) region (diversity segment):** a sequence of amino acids situated between the joining and constant regions of heavy chains in immunoglobulin

**dizygotic twins:** two genetically dissimilar individuals produced simultaneously by two separate ova fertilized by separate sperm; fraternal twins; *see also* **monozygotic twins**

**D-loop synthesis:** a mode of DNA replication in mitochondria and chloroplasts that is characterized by a displacement loop of heavy-strand DNA opposite the origin of replication of the light strand, which begins synthesis long before its complementary partner strand in the duplex molecule

**DNA:** *see* **deoxyribonucleic acid**

**DNA-dependent RNA polymrase:** *see* **RNA polymerase**

**DNA ligase:** the enzyme that catalyzes covalent bonding between free ends of the sugar-phosphate "backbone" of DNA strands during DNA replication and DNA repair

**DNA polymerase:** an enzyme that catalyzes the synthesis of DNA polymers from deoxyribonucleoside triphosphate precursors, using single-stranded DNA as template; DNA polymerase I is a gap-filling polymerase with endonuclease capability, and DNA polymerase III is the main replicating enzyme in DNA synthesis

**DNA repair synthesis:** replacement with new segments of nucleotides in damaged or recombinant DNA molecules, catalyzed by DNA polymerase and DNA ligase

**DNA-RNA hybrid:** a duplex nucleic acid comprised of complementary strands of DNA and RNA: *see* **molecular hybridization**

**domain:** a compact structural region of a polypeptide, often associated with a particular function of the molecule

**dominant:** an allele or phenotype that is expressed in the homozygous or heterozygous state, in contrast with the recessive allele or phenotype, which is expressed only in the homozygous state (and only when not masked by the dominant)

**donor:** a cell or individual that contributes DNA or other material to a recipient by some method of transfer

**dosage effect:** variations in phenotypic expression of a trait in an individual with fewer or more than the normal number of identical genes governing the trait

**double crossover:** the exchange that results from two occurrences of breakage and reunion within a bivalent, the two crossovers involving two, three, or all four of the chromatids

**Down syndrome:** a human clinical condition usually associated with an extra chromosome 21 in the nucleus; trisomy-21

**duplication:** an extra copy of a nucleotide sequence, gene, part of a chromosome, or whole chromosome in an individual or cell

**dyad:** a replicated chromosome consisting of two chromatids in a meiotic cell

**ecdysone:** a hormone derived from cholesterol, produced by the prothoracic gland of insects, and required for molting and pupation

**electrophoresis:** the movement of charged molecules in solution in an electrical field. The solution is generally held in a porous supporting medium such as cellulose nitrate or a gel made of polyacrylamide, agar, or starch.

**endonuclease:** an enzyme that breaks one or both strands of a nucleic acid molecule by disrupting internal phosphodiester bonds of the sugar-phosphate "backbone"

**endosymbiont:** an organism that lives within another organism in a mutually beneficial, or symbiotic, relationship

**endosymbiont theory:** the proposal stating that mitochondria and chloroplasts in eukaryotic cells are evolutionary descendants of ancient free-living prokaryotic organisms

**enrichment method:** a method for isolating a desired type of cell or organism by providing conditions that enhance the desired type and diminish the other type(s) in the culture

**enzyme induction:** *see* **inducible enzyme**

**enzyme repression:** *see* **repressible enzyme**

**episome:** a genetic element that may behave as an autonomous replicating unit in the host, independent of the host genome, or as an integrated component of the host chromosome, replicating in synchrony with it

**epistasis:** the nonreciprocal interaction of nonallelic genes, in which the expression of one gene or its allele (the epistatic gene) masks the expression of another gene or its allele (the hypostatic gene)

**equilibrium density gradient centrifugation:** a method used to separate macromolecules and cellular components according to differences that cause them to come to rest in regions of the gradient with solute densities corresponding to their own buoyant densities in the solute. The gradation of densities is produced by centrifugation at very high speeds.

**equilibrium population:** a population in which selection pressure and mutation pressure have interacted long enough to bring the allelic frequencies into equilibrium (no net change)

**euchromatin:** the noncondensed, active chromosomes or chromosome regions of the interphase nucleus; *see also* **heterochromatin**

**eukaryote:** a cellular organism with a well-defined nucleus enclosed within a nuclear membrane and usually having one or more other membranous structures in the cytoplasm; any cellular organism that is not a prokaryote

**euploid:** a diploid or polyploid cell or individual whose chromosome number is an exact multiple of the basic number of the species ($n$)

**evolutionary tree:** a branching depiction of related systems that have diverged to different degrees from a common ancestral system, through accumulated mutational changes in successive generations or time intervals

**excision repair:** a light-independent process of enzymatic repair of ultraviolet light-damaged DNA through removal of damaged segments containing TT dimers and replacement of the dimers by newly synthesized segments complementary to the opposite undamaged strand of the duplex DNA

**exon:** a coding sequence within an interrupted, or mosaic, gene in eukaryotes and their viruses, which alternates with noncoding intervening sequences, or introns, of the gene sequence

**exonuclease:** an enzyme that digests DNA, beginning at the exposed or free ends of the strands

**expressivity:** the degree of phenotypic expression of a gene in relation to the expected phenotype

**extract complementation test:** an *in vitro* test that examines complementation which is due to interactions between gene products directly and not between cells or individuals

**extranuclear inheritance:** a pattern of allelic transmission in crosses which indicates that the genetic information is not in chromosomes of the nucleus, and which usually is evident from the non-Mendelian ratio of alternative alleles or phenotypes in the progeny

**$F_1$ generation:** first filial generation; the offspring resulting from the first experimental crossing of individuals of the parental or P generation

**$F_2$ generation:** second filial generation; the progeny produced by interbreeding or self-fertilization of $F_1$ individuals

**factorial:** a product of factors obtained by multiplying together all the positive integers from one to a specified number, or the reverse. The factorial symbol is an exclamation point (!) following the specified term in the series; for example, factorial four is written 4! and means $1 \times 2 \times 3 \times 4 = 24$, or $4 \times 3 \times 2 \times 1 = 24$.

**family method:** a method of genetic analysis of inheritance patterns based upon the pedigree or history of one or more families in which the particular trait has been observed

**fate map:** a map of an embryo in an early stage of development that indicates the various regions whose prospective significance has been established by marking methods

**fertility factor (F):** an episome that determines the donor or recipient status of a bacterial cell, with $F^+$ acting as donor of DNA and $F^-$ acting as recipient of DNA that is transferred during conjugation by the donor cell. In Hfr strains the F factor is physically integrated into the chromosome, in $F^+$ strains the F factor is independent of the bacterial chromosomes, in $F^-$ strains the F factor is lacking. The F′ (F prime) factor carries a genetically recognizable fragment of bacterial chromosome that may lead to production of merodiploid recipient cells.

**fine structure mapping:** the high resolution analysis of intragenic recombination, even to the level of single nucleotides in the gene sequence

**first-division segregation:** the separation of a pair of alleles into different nuclei during the first meiotic division, due to the lack of crossing over between the gene and centromere of the homologous chromosomes

**fitness:** the relative ability of an organism to survive and transmit its genes to the next generation; adaptedness

**flexibility:** *see* **adaptability**

**fluctuation test:** a statistical analysis first used by Luria and Delbrück to distinguish between mutations induced in direct response to some environmental condition (Lamarckian) versus selection of pre-existing spontaneous mutations in that environment (Darwinian)

**formylmethionyl-tRNA:** fMet-tRNA; the initiator aminoacyl-tRNA in polypeptide translation at the ribosome in prokaryotes and in mitochondria and chloroplasts

**forward mutation:** a genetic change from the wild-type to a mutant gene sequence; *see also* **reverse mutation**

**frameshift mutation:** an alteration in the DNA sequence due to addition or deletion of one or more nucleotides, causing a change in the translational reading frame and the synthesis of a modified, often nonfunctional, polypeptide

**frequency-dependent selection:** *see* **selection strategies**

**gamete:** a reproductive cell that fuses with another reproductive cell, such as eggs with sperm, to produce the new individual of the next generation

**gametophyte:** the haploid phase or individual of plants, which develops from a meiotic spore and which, in turn, produces gametes by mitosis

**G-banding:** a method of staining chromosomes with Giemsa stain to reveal patterns of deeply stained bands separated by lightly stained regions which permits the identification of each chromosome in the genome as a unique entity

**gel electrophoresis:** *see* **electrophoresis**

**gene:** a unit of inheritance, which occupies a specific site, or locus, on the chromosome and specifies a polypeptide or RNA in its encoded DNA sequence, and which can mutate to various allelic forms

**gene amplification:** differential replication of some genes, producing many copies, at the same time that other genes in the same cell do not replicate

**gene conversion:** a phenomenon in which one or more pairs of alleles segregate in some ratio other than the expected 1:1 in meiotic products, while neighboring alleles segregate in the expected 1:1 ratio, presumably as the consequence of mismatch repair of DNA in recombinant strands

**gene expression:** the readout of encoded genetic instructions in DNA during transcription and translation which leads to phenotypic development in the cell or individual

**gene flow:** the spread of new genes within a population as the result of interbreeding with immigrants

**gene pool:** the total genetic information possessed by a population of sexually reproducing organisms

**generalized transduction:** the unidirectional transfer of genes from one bacterium to another bacterium through the agency of a generalized transducing phage, which packages bacterial genes instead of its own genes during its formation in the host cell

**genetic code:** the base triplets of DNA and RNA, which carry the genetic information for protein synthesis, specifying 20 amino acids and start and stop codons for translation at the ribosome

**genetic drift:** the random fluctuations of gene frequencies due to chance rather than selection, particularly in small populations

**genetic engineering:** the production of recombinant DNA and of host strains carrying recombinant DNA, usually derived from different species and capable of synthesizing proteins specified by one or more genes from another species; for example, *E. coli* may synthesize human insulin that is encoded by the human insulin gene in a recombinant DNA molecule which is resident in the *E. coli* host

**genetic load:** the average number of lethal equivalents per individual in a population

**genetic map:** the linear arrangement of mutable sites or of base sequences on a chromosome or DNA molecule as deduced from genetic recombination frequencies or from molecular analysis

**genetic polymorphism:** the long-term occurrence in a population of two or more genotypes or molecular forms of a gene in frequencies that cannot be accounted for by recurrent mutation

**genetics:** the science of heredity

**genome:** all the genes present in the chromosome or set of chromosomes in a cell or an individual

**genomic library:** a collection of cloned DNA, which is derived from restriction fragments of the chromosomes and includes all or part of the genetic material of the species

**genotype:** the genetic constitution of an organism as distinguished from its expressed features (phenotype)

**glucose effect:** *see* **catabolite repression**

**goodness of fit:** a statistic indicating that observed experimental values correspond with theoretical expected values and support a stated hypothesis, as in the chi-square test

**grandfather method:** a means by which the cis or trans arrangement of linked genes in an individual can be deduced from the arrangement of the genes in the chromosomes of the individual's parent; used in the family method of linkage mapping of human chromosomes

**guanine:** a purine base found in nucleic acids

**gyrase:** a colloquial term for topoisomerase II, the DNA replication enzyme that sponsors supercoiling of duplex DNA through nicking and closing reactions

**haploid:** a cell or individual having one copy of each chromosome in the genome

**Hardy-Weinberg law:** the law stating that gene frequencies and genotypic frequencies will remain constant from generation to generation in an infinitely large, interbreeding population for which mating is at random and there is no mutation, selection, or migration; the gene frequencies for one pair of alleles ($A$ and $a$) are $p$ and $q$, respectively, and at equilibrium the frequencies of the genotypic classes are $p^2$ ($AA$), $2pq$ ($Aa$), and $q^2$ ($aa$).

**HAT medium:** a selective growth medium, which contains the nucleic acid precursors hypoxanthine and thymidine and the drug aminopterin, and which permits growth of somatic cell hybrids that have the alleles to produce the enzymes hypoxanthine-guanine phosphoribosyl transferase (HPRT) and thymidine kinase (TK), which catalyze nucleic acid synthesis by a pathway not blocked by aminopterin

**heavy chain of immunoglobulins:** the heavy polypeptide chain consisting of variable, joining, diversity, and constant regions, each specified by a different gene

**heavy isotope:** a stable atom that contains more neutrons than the common isotope of an element, and thus is heavier; for example, $^{15}N$ versus $^{14}N$

**heavy strand:** the nucleic acid chain in duplex DNA which has a higher buoyant density in CsCl than its complementary light strand; often the sense or coding strand of genetically informational DNA

**HeLa cells:** an aneuploid strain of human epithelial-like cells carried in cell culture since 1952, and originally obtained from a tissue specimen from a carcinoma of the cervix in a patient named *He*nrietta *La*cks

**helicase:** one of the enzymes that is involved in DNA replication, acting ahead of the replication fork and breaking hydrogen bonds between complementary strands of the duplex through an energy-generating hydrolysis of ATP

**helix:** a curve on the surface of a cylinder that cuts all the elements of the solid at a constant angle

**hemizygous:** a state of having one copy of a chromosome or gene, usually with reference to diploid cells or individuals that have two copies of the other chromosomes or genes: for example, males are hemizygous for the X chromosome and X-linked genes, and females may be homozygous or heterozygous since they have two copies of the X chromosome and of X-linked genes

**heritability:** a measure of the degree to which a phenotypic expression is genetically influenced and can be modified by selection

**heterochromatin:** the chromatin or chromosomes in the interphase nucleus that are either condensed at all times (constitutive) or, in some cells, at some times (facultative); *see also* **repetitive DNA, X inactivation**

**heterodulex:** a duplex nucleic acid molecule that is comprised of two strands of DNA, or of RNA, or of one of each, that have been obtained from different sources but are held together in a duplex by complementary base pairing

**heterogametic:** referring to an individual or class of individuals that produce two or more kinds of gametes with reference to chromosome constitution; for example, human males may produce X-bearing and Y-bearing sperm

**heterokaryon:** a cell or individual having nuclei derived from genetically different sources, particularly characteristic of fungi

**heterosis:** the greater vigor of growth, survival, and fertility of hybrids, usually from highly inbred parent lines, and usually associated with increased heterozygosity; hybrid vigor

**heterozygosity:** the condition of having one or more pairs of dissimilar alleles

**heterozygote:** a diploid or polyploid individual having dissimilar alleles at one or more loci and therefore producing segregating classes of progeny

**heterozygous:** the characterization of a genotype, cell, or individual having one or more pairs of dissimilar alleles

**Hfr strain:** a strain of *E. coli* that shows *h*igh *f*requencies of *r*ecombination, and carries the fertility factor F integrated into the bacterial chromosome

**highly repetive DNA:** DNA consisting of millions of repeated short nucleotide sequences, usually occurring in large clusters in a eukaryotic genome

**histone:** a major protein of the chromosome, characterized by a high content of the basic amino acids arginine and lysine; *see also* **nucleosome**

**Hogness box:** a term applied to the TATA box in a eukaryotic genome, which is found in the promoter sequence of the gene and is involved in binding RNA polymerase in the initiation of transcription of messenger RNA; also called the Hogness-Goldberg box

**homeotic mutant:** an individual, usually *Drosophila*, having a transdetermined organ in place of the normal organ in the normal location in the body; for example, a leg in place of a wing in the normal wing location; *see also* **transdetermination**

**homogametic:** referring to an individual or class of individuals that produce a single kind of gamete with reference to chromosome constitution; for example, human females produce only X-bearing eggs

**homologous:** having the same or similar gene content, usually with reference to pairs of chromosomes, to parts of chromosomes, or to amino acid and DNA sequences and compositions in related species descended from a common ancestor

**homozygote:** a diploid or polyploid individual having one or more pairs of identical alleles and therefore breeding true to type in successive generations; *see also* **pure line**

**homozygous:** referring to a genotype, cell, or individual having one or more pairs of identical alleles

**HPFH disorder:** hereditary persistance of fetal hemoglobin; an inherited human condition characterized by synthesis of fetal hemoglobin, Hb F, ($\alpha_2\gamma_2$) instead of HbA ($\alpha_2\beta_2$) after birth, and usually found in individuals with deletions of the $\delta\beta$-globin genes but with intact $\gamma$-globin gene sequences in chromosome 11

**hybrid:** an individual resulting from a cross between genetically dissimilar parents

**hybrid vigor:** *see* **heterosis**

**hydrogen bond:** a weak chemical interaction between a hydrogen atom covalently bonded to an oxygen or nitrogen atom and a neighboring electronegative atom of oxygen or nitrogen

**hydrophobic interaction:** a weak interacting force between water-repelling, usually nonpolar, residues or molecules, such as between stacked aromatic bases in duplex DNA or between fatty acid residues of phospholipids in membranes

**hyphae:** the filaments comprising the body, or mycelium, of a fungus

**hypoxanthine-guanine phosphoribosyl transferase:** HPRT; the enzyme that catalyzes purine synthesis from hypoxanthine precursors of DNA and is encoded in the X-linked *Hprt* locus of the human X chromosomes; *see also* **HAT medium**

**imaginal disk:** a larval structure in certain insects, such as *Drosophila,* that gives rise to a predetermined adult organ during pupation, which leads to the adult insect

**immunoglobulin:** the antibody molecule, which is comprised of two heavy polypeptide chains and two light polypeptide chains and is made in B lymphocytes of the immune system; *see also* **antibody, antigen, heavy chain, light chain**

**inbreeding:** the crossing of closely related individuals, usually leading to increased homozygosity in a population or family

**inbreeding coefficient (*F*):** the proportion of loci at which a population is homozygous as a consequence of lost heterozygosity due to inbreeding; first proposed by Sewall Wright

**incomplete dominance:** a pattern of inheritance in which the heterozygous offspring are phenotypically different from both homozygous parents by virtue of partial expression of two different alleles of a gene in the heterozygote; for example, pink heterozygotes produced in crosses between red and white parents; *see also* **codominant, dominant**

**independent assortment:** the random distribution to the gametes of alleles of genes located on different chromosomes, as described in Mendel's Second Law of Inheritance

**inducible enzyme:** one of a class of enzymes that is synthesized only in the presence of its substrate (inducer) and is usually regulated by operator-repressor control of messenger RNA transcription in bacterial operons

**inheritance:** the transmission of genetically determined characteristics from generation to generation in a family

**initiation codon:** the base triplet AUG in messenger RNA to which the initiating fMet-tRNA or Met-tRNA binds in the initiation of polypeptide translation at the ribosome; it specifies the first amino acid at the *N*-terminus of the polypeptide chain that is made from encoded genetic instructions

**initiation complex:** the aggregate of messenger RNA, small ribosomal subunit, and initiator aminoacyl-tRNA that marks the start of polypeptide translation from encoded genetic instructions

**insertion sequence (IS):** a transposable genetic element, which is less than 2000 bases long, contains only genes involved in its insertion into DNA, and characteristically has inverted repeat base sequences at its termini

***in situ* hybridization:** a cytological method used to localize the chromosomal DNA regions that are complementary to specific nucleic acid molecules; for example, DNA that specifies ribosomal RNA is localized by hybridizing radioactive rRNA to the slide preparation and observing the complementary DNA regions indicated by silver grains in the overlying photographic emulsion of an autoradiograph of the chromosomes

**interference:** the interaction between crossover events such that the occurrence of one exchange between homologous chromosomes reduces (positive interference) or increases (negative interference) the likelihood of another exchange in the same vicinity

**interphase:** the interval characterizing the eukaryotic nucleus when it is not undergoing mitosis or meiosis; the $G_1$, $S$, and $G_2$ periods of the cell cycle

**interrupted gene:** the hereditary unit in a eukaryote or its viruses organized as a mosaic of coding segments (exons) and noncoding intervening sequences (introns)

**interrupted mating experiment:** a genetic experimental design in which the manner and extent of gene transfer between conjugating *E. coli* cells is studied by withdrawing samples of these cells at various times and subjecting them to strong shearing forces in an electric blender before plating the cells to observe their phenotypes as indications of transferred genes

**intervening sequence:** a transcribed noncoding segment of an interrupted or mosaic gene that occurs in eukaryotes and their viruses and alternates with coding segments (exons), but is excised from pre-mRNA during processing leading to mature mRNA; also called an intron

**intron:** *see* **intervening sequence**

**inversion:** a structural rearrangement of part of a chromosome such that genes within that part end up in inverse order from their original arrangement, having been rotated 180°; a paracentric inversion does not include the centromere, and a pericentric inversion includes the centromere within the inverted region

**isolating mechanism:** a genetic, chromosomal, physiological, behavioral, ecological, or geographic difference that prevents successful interbreeding between two or more groups of organisms; *see* **prezygotic isolating mechanism, postzygotic isolating mechanism**

**isotope:** an alternative form of an element characterized by a difference in the number of neutrons (atomic weight) but not in the number of protons (atomic number) of the atomic nucleus; *see also* **heavy isotope, radioactive isotope**

**joining (J) segment:** $J_H$ of the heavy-chain and $J_L$ of the light-chain polypeptides of the immunoglobulin molecule; it joins with the variable segment in light-chain assembly and with the variable and diversity segments in heavy-chain assembly before the V-J or V-J-D aggregate combines with the constant (C) segment of the two kinds of polypeptide chains to complete the sequence

**kappa:** the endosymbiotic bacterium in certain strains of *Paramecium tetraurelia* that releases a toxin which kills sensitive paramecia. Cells with kappa are called killers, and the maintenance of kappa requires the dominant nuclear allele *K*.

**karyotype:** a distribution of photographed chromosomes from a cell or an individual showing chromosomes arranged in pairs in order of centromere location and decreasing size

**kinetochore:** centromere

**Klinefelter syndrome:** a clinical condition of human males who have one or more extra X chromosomes

**lambda (λ):** a temperate phage of *E. coli*

**leader:** a base sequence that is located between the promoter and the structural genes in a biosynthetic operon in bacteria and is involved in attenuation control of transcription; the leader sequence is transcribed but usually is not translated

**leptonema:** the first substage of prophase I of meiosis

**lethal mutation:** a mutation that results in death of the individual before the age of reproduction

**ligase:** *see* **DNA ligase**

**light chain of immunoglobulins:** the light polypeptide chain which consists of variable, joining, and constant regions, each specified by a different gene, and which occurs as an identical pair of light chains together with an identical pair of heavy chains in an immunoglobulin molecule

**light strand:** the nucleic acid chain in duplex DNA that has a lower buoyant density in CsCl than its complementary heavy strand; often the noncoding strand of genetically informational DNA

**linkage:** the greater association in inheritance of genes that are located on the same chromosome and are inherited together more often than is expected from independent assortment

**linkage map:** a chromosome map showing the relative locations of the known genes on the chromosome or chromosomes of a given species

**linked genes:** genes that are located on the same chromosome or nucleic acid molecule

**locus:** the particular location that a gene occupies on a chromosome

**lymphocyte:** a type of white blood cell that functions in immune response systems, and of either the antibody-secreting B cell type or the T cell type of the cell-mediated immune response system

**Lyonization:** *see* **X inactivation**

**lysogenic bacterium:** a bacterium that carries a temperate phage in the prophage state and is therefore not subject to lysis by the noninfectious form of the virus

**map unit:** a number that corresponds to a recombination frequency of 1%

**maternal effect:** the development of phenotypic differences among individuals of identical genotype, due to either an effect of some substance that is synthesized in the egg cytoplasm from maternally specified genes or an effect produced by a virus or other agent transmitted from mother to offspring, as during nursing

**mature mRNA:** the colinear copy of DNA-encoded information specifying the amino acid sequence in a protein; spliced and processed pre-mRNA in eukaryotes

**Maxam-Gilbert base sequencing method:** a set of procedures in which four equal aliquots of a preparation of single-stranded DNA fragments are exposed to chemical agents that cause breaks at A, T, G, or C residues, and from which the exact sequence of bases in the original fragment can be read from the bands that represent the broken fragments separated by gel electrophresis for each of the four aliquots

**mean:** *see* **arithmetic mean**

**median:** the middle value in a set of data arranged in order of size

**medium:** the nutritive substance provided for the growth of cells or organisms in the laboratory

**meiocytes:** reproductive cells capable of undergoing meiosis, such as oocytes, spermatocytes, and sporocytes

**meiosis:** the reduction division of the nucleus in a sexual organism, which produces daughter nuclei having half the number of chromosomes as the original meiocyte nucleus; the process consists of two successive divisions of the nucleus and therefore results in a quartet or tetrad of gametes, spores, or other reproductive cell types

**melting:** the dissociation of duplex nucleic acids to form single strands upon disruption of hydrogen bonds, which hold the strands together; *see also* **denaturation**

**Mendel's First Law of Inheritance:** the law which states that there is a segregation of alleles during sexual reproduction (meiosis)

**Mendel's Second Law of Inheritance:** the law which states that members of different pairs of alleles undergo independent assortment during sexual reproduction (if the genes are in different chromosomes)

**merodiploid:** a partially diploid bacterium whose second copy of one or more particular genes may be introduced by fertility factor F′ during conjugation or by transduction; also called merozygote

**messenger RNA (mRNA):** the complementary copy of the coding strand of DNA, which is made during transcription and provides the encoded information for polypeptide synthesis at the ribosome; *see also* **pre-mRNA, transcription**

**metacentric:** designating a chromosome that has a centrally placed centromere and therefore has two arms of equal length

**metaphase:** the stage of mitosis or meiosis when chromosomes are aligned along the equatorial plane of the spindle

**metastasis:** the spread of malignant cancerous cells from the original site to another part of the body

**7-methylguanosine:** an unusual purine derivative that binds in its nucleoside triphosophate form to the 5′ terminus of pre-messenger RNA and serves to "cap" that terminus; $m^7G$

**micrometer ($\mu m$)**: a unit of length ($1 \times 10^{-6}$ meters) that is equivalent to 3000 bases or basepairs in a nucleic acid molecule and to $2 \times 10^6$ molecular weight of nucleic acid

**middle-repetitive DNA:** DNA that consists of hundreds or thousands of repeated nucleotide sequences which encode rRNA, tRNA, and certain proteins in eukaryotes

**minimal medium:** a medium providing only those substances essential for growth and reproduction of wild-type cells or organisms; *see also* **auxotroph**

**mismatch repair:** the incorporation of incorrect bases during DNA repair synthesis of recombinant duplex DNA in meiotic cells that have experienced crossing over; proposed to be the process responsible for the phenomenon of gene conversion

**misreading:** the incorrect translation of one or more codons in messenger RNA, producing polypeptides with amino acid substitutions at the misread sites

**missense mutation:** a mutation in which a codon is altered to one that specifies a different amino acid in the polypeptide translation

**mitosis:** the division of the nucleus that produces two daughter nuclei identical to the original parental nucleus; somatic or vegetative nuclear division

**mitotic segregation:** postmeiotic segregation; segregation of members of pairs of alleles during mitotic divisions in successive generations; a characteristic of extranuclear inheritance

**molecular biology:** a branch of modern biology concerned with explaining biological phenomena in molecular terms, and often utilizing techniques of physical chemistry to investigate genetic and other biological problems

**molecular hybridization:** the formation of DNA-DNA, DNA-RNA, and RNA-RNA duplex molecules by hydrogen bonding of complementary single-stranded molecules or parts of molecules under suitable experimental conditions; a test for complementarity

**molecular polymorphism:** *see* **genetic polymorphism**

**monohybrid cross:** interbreeding of individuals differing with respect to one specified pair of alleles of a gene

**monomer:** the basic unit of a larger functional molecule, particle, or cellular entity

**mononucleotide:** a molecule comprised of one nitrogenous base, one pentose sugar, and one phosphate residue; the monomeric unit of nucleic acids

**monosome:** a single ribosome, in contrast with an aggregate of ribosomes, or polysome

**monosomy:** the condition in which one chromosome of a pair is missing, as in monosomic females with Turner syndrome who lack one of the two X chromosomes normally present

**monozygotic twins:** two genetically identical individuals produced from a single fertilized egg that separated into two zygotic or embryonic entities during early development; identical twins; *see also* **dizygotic twins**

**morphogenesis:** the developmental processes leading to the characteristic mature form of an organism or part of an organism; *see also* **development, differentiation**

**morphological mutation:** an inherited change that produces an alteration in a structural aspect of the phenotype or some other visible change in the appearance of an individual

**multiple alleles:** a set of three or more alternative forms of a single gene

**multiple genes:** three or more genes that govern the same quantitative phenotypic trait through additive effects on phenotypic expression; also called polygenes

**mutagen:** a physical or chemical agent that raises the rate of mutation of a gene significantly above its spontaneous mutation rate; such an agent is mutagenic

**mutagenesis:** the process of mutation induction

**mutation:** a process by which a gene undergoes a change in its base composition, base sequence, or organization; a modified gene resulting from mutation

**mutation frequency:** the number of mutations in populations of individuals

**mutation pressure:** the continued production of an allele by mutation

**mutation rate:** the number of mutations per gamete or per cell per generation

**mycelium:** the vegetative portion of a fungus composed of a network of filaments called hyphae

**natural selection:** the differential reproduction of genetically diverse types in populations which results in types of greater fitness leaving more progeny in successive generations than will other genetic types of less fitness, with consequent changes in gene frequencies; Darwinism; *see also* **selection strategies**

**negative control:** one of a class of regulation mechanisms by which transcription is turned off through blocking of the operator site adjacent to one or more structural genes, or through attenuation, in contrast to positive controls which turn on transcription by binding of molecules to regulatory sites

**neutral evolution:** non-Darwinian evolution; the preservation of mutations with no obvious selective advantage (neutral mutations) by chance rather than selection pressure

**neutral petite:** a respiratory deficient yeast mutant totally lacking mitochondrial DNA

**N-formylmethionine (fMet):** a modified methionine molecule or residue that has a formyl group substituted for a hydrogen in its amino radical; the initiating amino acid of protein synthesis in prokaryotes and organelles

**nondisjunction:** the faulty separation of homologous chromosomes or sister chromatids during nuclear division, producing cells or individuals with an aneuploid number of chromosomes

**nonhistone protein:** a chromosomal protein of acidic or neutral character, in contrast to basic histone proteins of the chromosome

**nonparental ditype (NPD) tetrad:** a set of meiotic products having two recombinant genotypes and none with parental genotypes

**nonpermissive:** referring to a condition either of growth or of the host that does not allow growth or phenotypic expression of a conditional mutant

**nonsense mutation:** a mutation in which a codon specifying an amino acid is altered to a stop codon, causing premature termination of translation in the mutant

**normal distribution:** a probability distribution in statistics, graphically depicted by a bell-shaped curve whose dimensions are determined by the standard deviation ($\sigma$), such that the larger the standard deviation, the broader the curve

**normalizing natural selection:** *see* **selection strategies**

**nuclear transplantation:** the injection of a diploid somatic nucleus into an enucleated egg in order to determine the developmental potentialities of the implanted nucleus in guiding development of the organism; *see also* **totipotent**

**nucleic acids:** a class of biologically important organic compounds existing in single- or double-stranded polymeric forms called deoxyribonucleic acid (DNA) and ribonucleic acid (RNA), and comprised of four nitrogenous bases (A, C, and G in DNA and RNA) and T (DNA) or U (RNA), a pentose sugar (deoxyribose in DNA, ribose in RNA), and phosphate residues; nucleic acids are the principal components of the genetic apparatus in viruses, prokaryotes,and eukaryotes

**nucleolar chromatin:** the portion of the nucleolar-organizing chromosome that encodes ribosomal RNAs and is attached to the nucleolus

**nucleolar-organizing (NO-) chromosome:** one or more chromosomes in the genome that contains various genes, including genes for ribosomal

RNA, and that has the capacity to generate a nucleolus at the site of the ribosomal RNA genes (rDNA)

**nucleolar-organizing region (NOR):** the specific portion of the nucleolar-organizing chromosome where ribosomal RNA genes are situated and where the nucleolus is produced

**nucleolus:** a discrete structure in the nucleus that is associated with ribosomal RNA synthesis

**nucleoprotein:** a conjugated molecule comprised of nucleic acid and protein; the principal constituent of the chromatin fiber

**nucleoside:** a nucleic acid monomer consisting of a nitrogenous purine or pyrimidine base covalently bonded to a pentose sugar

**nucleosome:** the repeating nucleoprotein unit of chromatin structure, consisting of approximately 140 base-pairs of DNA wound around an octamer of histone proteins, and connected to neighbor nucleosomes by about 60 base-pairs of DNA to which histone H1 is bonded

**nucleotide:** a nucleic acid monomer consisting of a nitrogenous purine or pyrimidine base, a pentose sugar, and a phosphate residue; a nucleoside phosphate

**nucleus:** the membrane-bounded spheroidal structure containing the chromosomes in eukaryotic cells

**Okazaki fragment:** a newly synthesized single-stranded DNA fragment, about 2000 nucleotides long, that is covalently joined to similar fragments during DNA replication to lengthen the growing DNA chain along each complementary template strand of duplex DNA in the mode of discontinuous synthesis

**oncogene:** a gene that can initiate and maintain the tumorous state in the organism, and is found in oncogenic viruses and in normal cells in equivalent forms (v- and c-*onc* genes)

**oncogenesis:** tumor or cancer induction; synonymous with carcinogenesis and tumorigenesis

**oncogenic virus:** one of a potentially carcinogenic class of RNA viruses of animal species which copies DNA from an RNA genome using the specific enzyme reverse transcriptase

**operator:** a regulatory region of the chromosome that is capable of interacting with a specific repressor, thereby controlling transcription of structural genes in the same operon

**operon:** a gene cluster consisting of a promoter, an operator, and one or more structural genes that function coordinately in gene expression

**origin of replication:** a particular segment of a DNA strand that is the site of initiation of DNA replication in one direction away from the origin or in both directions away from the origin

**overlapping deletion method:** a genetic mapping procedure by which point mutations can be located in a gene or chromosome through progeny analysis in crosses between the mutant and one or more deletion mutants whose lesions overlap one another in a reference map of the region

**overlapping deletions:** regions of missing DNA sequences in the chromosome that are partly the same in various mutant strains of an organism, and are utilized in gene mapping studies to deduce the location and the order of mutant sites

**overlapping genes:** different genes that share a common base sequence, which is translated in two or three overlapping reading frames and thereby can specify different gene products

**pachynema:** the substage of prophase I of meiosis when synapsis of homologous chromosomes is completed and the synaptonemal complex is fully formed

**paracentric inversion:** *see* **inversion**

**parasexual cycle:** a reproductive asexual cycle that mimics the sexual cycle in having fusions between somatic cells and random loss of individual chromosomes, which is analogous to allelic segregation

**parental ditype (PD) tetrad:** a set of meiotic products that is genotypically like the two parents but has no recombinants present

**pedigree:** a family history of inheritance of one or more traits, usually depicted by branching lines connecting individuals in a series of consecutive generations

**penetrance:** the proportion of individuals of a specified genotype that show the expected phenotype under a defined set of environmental conditions. If all individuals carrying a dominant allele express the expected phenotype, the gene is said to show *complete penetrance;* if only some show the expected phenotype the gene is said to be *incompletely penetrant.*

**penicillin enrichment method:** a method for the isolation of auxotrophic mutants in bacteria, in which nongrowing auxotrophic mutants remain viable in the presence of penicillin in the medium and can be isolated for further study

**peptide:** a compound formed from two or more amino acids

**peptide bond:** the universal link between amino acids in proteins, formed when the amino group of one monomer joins with the carboxyl group of an adjacent monomer amino acid in a dehydration reaction

**peptidyl transferase:** the ribosomal enzyme that catalyzes peptide bond linking between the existing peptidyl-tRNA at the P site and the newest aminoacyl-tRNA bound to the A site of the ribosome, making the growing peptidyl chain longer by one aminoacyl residue

**pericentric inversion:** *see* **inversion**

**permissive:** referring to a condition either of growth or of the host that allows growth or phenotypic expression of a conditional mutant

**petites:** mutant strains of yeast characterized by slow growth and small colonies as the result of respiratory deficiency; vegetative petites are extranuclear mutants with one or more altered mitochrondrial genes and segregational petites are nuclear gene mutants

**P generation:** the immediate parents of the $F_1$ generation in a genetic cross

**phage:** a virus whose host is a bacterium; bacteriophage

**phage lysate:** the contents of lysed infected bacteria containing infective phage particles

**phenotype:** the observable properties of a cell or organism, which are produced by the genotype in conjunction with the environment

**3′,5′-phosphodiester bridge:** the diester link between a phosphate residue and the 3′ carbon of one pentose above it and the 5′ carbon of the pentose below it; the repeating linkage of the sugar-phosphate "backbone" of a nucleic acid molecule

**photoreactivation:** reversal in visible light of DNA damage induced by exposure to ultraviolet radiation, through the catalytic action of light-dependent photoreactivating enzyme(s), which excise pyrimidine dimers

**phylogenetic tree:** *see* **evolutionary tree**

**plaque:** a clear area on an otherwise opaque growth of bacteria where bacteria have been lysed by a virulent virus and its clonal descendants

**plasmid:** an extranuclear genome that may influence the cellular phenotype, and may exist integrated within the host chromosome or free and autonomously replicating in the host cytoplasm; *see also* **episome**

**pleiotropy:** the phenomenon of a single gene being responsible for a number of distinct and seemingly unrelated phenotypic effects

**point mutation:** a mutation caused by the substitution of one nucleotide by another

**poly(A) tail:** the polyadenylated 3′ terminus of eukaryotic pre-mRNA and mature mRNA, consisting of a sequence of adenosine monophosphate residues of varying number

**polycistronic transcript:** a messenger RNA molecule specifying the sequence of two or more proteins or ribosomal RNAs transcribed from adjacent genes in an operon

**polygenes:** *see* **multiple genes**

**polymer:** an association of monomer units in a large molecule

**polymerase:** one of a class of enzymes catalyzing the synthesis of DNA or RNA from nucleoside triphosphate monomeric precursors; *see also* **DNA polymerase, reverse transcriptase, RNA polymerase**

**polymorphism:** *see* **genetic polymorphism**

**polypeptide:** a polymer of amino acids linked together by peptide bonds in a long unbranched chain; *see also* **protein**

**polyploid:** an individual or cell having more than two sets of chromosomes

**polysome:** an aggregate of ribosomes bound to a messenger RNA molecule and actively engaged in translation of a polypeptide from encoded genetic instructions

**polytene chromosome:** a multi-stranded chromosome whose many replicated DNA molecules remain together in a single functional unit

**poly(U):** an artificial RNA molecule consisting entirely of uracil residues bonded to a conventional sugar-phosphate "backbone"

**population genetics:** the analysis of the genetic structure of natural populations through the application of Mendelian principles, using quantitative models and methods to determine the factors and forces that shape biological evolutionary patterns

**position effect:** a change in the expression of a gene when its position is changed with respect to neighboring genes, as by structural aberration or transposition

**positive control:** one of a class of regulation mechanisms by which transcription is turned on by binding of a particular molecule or molecules to the regulatory sites (promoter, operator) adjacent to one or more structural genes; *see also* **negative control**

**postmeiotic segregation:** *see* **mitotic segregation**

**posttranscriptional control:** a class of mechanisms that regulate gene expression through processing or modification of the messenger RNA transcripts of genes

**postreplication repair:** the light-independent repair of DNA damaged by exposure to ultraviolet irradiation, involving filling in "daughter-strand gaps" by polymerase-directed synthesis and ligation of the new segment to the main DNA strand, without excision of pyrimidine dimers formed as the primary lesion of UV irradiation in parental DNA strands from which the defective daughter strands are copied

**postzygotic isolating mechanism:** any one of a class of isolating mechanisms that prevents successful interbreeding between related populations as a consequence of hybrid sterility or inviability

**pre-messenger RNA (pre-mRNA):** the immediate transcript of the gene, which is processed in eukaryotes by capping the 5′ terminus, polyadenylating the 3′ terminus, and excising introns and splicing exons to make the mature mRNA transcript that is translated at the ribosome

**prezygotic isolating mechanism:** any one of a class of isolating mechanisms that prevents successful interbreeding between related populations as a consequence of the failure to mate or of successful fertilization to produce a zygote

**Pribnow box:** a term applied in prokaryotic genomes to the TATA box that is part of the promoter sequence of the gene and is involved in binding RNA polymerase in the initiation of transcription of messenger RNA

**primary structure:** the sequence of amino acids in a polypeptide chain or of nucleotides in a nucleic acid, as specified by the base sequence of the gene

**probability:** the ratio of a specified event to total events; chance of occurrence

**prokaryote:** a cellular organism lacking a membrane-bounded nucleus, typically one of the bacteria and blue-green algae groups

**promoter:** a specific nucleotide sequence to which RNA polymerase binds in the initiation of transcription and, therefore, a regulatory genetic component necessary for expression of structural genes

**prophage:** in lysogenic bacteria, the nucleic acid component that carries genetic information necessary for the production of a particular type of phage and confers hereditary properties on the host; usually integrated within the host chromosome and existing in the noninfective state

**prophase:** the first stage of mitosis or meiosis, occurring after DNA replication and before chromosome alignment on the equatorial plane of the spindle

**protein:** one of a class of biologically important organic compounds consisting of one or more polypeptide chains of amino acids

**protein kinase:** an enzyme that phosphorylates proteins by transfer of the phosphoryl group from ATP to one or more kinds of amino acid residues

**prototroph selection:** a method of isolating wild-type recombinants from mixtures of growing auxotrophic strains with complementary genetic deficiencies, by plating such mixtures on minimal media where only wild-type (prototrophic) colonies can develop

**provirus:** the equivalent in eukaryotic cells to a prophage in bacterial cells; *see also* **prophage**

**pseudogene:** a nonfunctional duplicate or derivative of a functional gene

**pseudowild:** a mutant having the wild-type phenotype as a consequence of a second mutation that suppresses the mutant effect of a first mutation

**P site:** the region of the ribosomal large subunit to which peptidyl-tRNA is bound during translation of a polypeptide from genetic instructions

**puff:** an expanded chromosomal region undergoing active transcription, usually observed in giant polytene chromosomes

**punctuation processing of RNA:** a proposed mechanism of processing polycistronic or polygenic transcripts of mammalian mitochondrial genomes through cleavage of the transcript at sites of tRNA termini, involving recognition signals for cleavage enzymes at looped secondary structures of the transcribed tRNA segments, thus liberating tRNA and mRNA sequences from the original transcript molecule

**pure line:** a strain of an organism that is homozygous because of continued inbreeding

**purine:** the parent compound of the nitrogenous bases adenine and guanine of nucleic acids

**pyrimidine:** the parent compound of the nitrogenous bases cytosine, thymine, and uracil of nucleic acids

**pyrimidine dimer:** the compound formed by ultraviolet irradiation of DNA whereby two thymine residues, or two cytosine residues, or one thymine and one cytosine residue at adjacent positions in the nucleic acid strand become covalently joined by carbon-carbon bonding and produce a bulge at the site of the dimer

**Q-banding:** a method of staining chromosomes using a fluorescent dye so that stained banded regions are differentiated from unstained areas between bands

**qualitative variation:** *see* **discontinuous variation**

**quantitative inheritance:** inheritance of a measurable character that depends on the cumulative action of many genes, each of which produces a small additive effect; *see also* **multiple genes**

**quantitative variation:** *see* **continuous variation**

**quaternary structure:** specific assemblages in a protein of two or more polypeptide chains that have different properties in the protein molecule than as individual chains; proteins comprised only of one polypeptide chain do not develop quaternary structure

**radioactive isotope:** an unstable form of an element having a different number of neutrons but having the same number of protons as the more commonly occurring isotope, which emits ionizing radiations as it stabilizes itself; for example, $^{3}H$ versus $^{1}H$, and $^{32}P$ versus $^{31}P$

**random genetic drift:** *see* **genetic drift**

***r* determinant:** the portion of an R plasmid that includes genes for antibiotic resistance and, like other transposons, has typical insertion sequences (IS) at both termini

**rDNA:** *see* **ribosomal DNA**

**reassociation:** reannealing or renaturation of single-stranded nucleic acid molecules through complementary base pairing to form duplex molecules

**reassortment:** the occurrence of progeny having new combinations of alleles other than those in the parents, due to random segregation of member of pairs of unlinked alleles

**recessive:** an allele or phenotype that is expressed in the homozygous state but is masked in the presence of its dominant allele; an individual homozygous for a specified allele that affects the phenotype only in the absence of its dominant allele

**recipient:** the cell or individual that receives genes or other material from a donor

**reciprocal crosses:** crosses of the forms, A ♀ × B ♂ and B ♀ × A ♂; crosses that are complementary to each other

**reciprocal translocation:** *see* **translocation**

**recombinant:** the new cell or individual arising from recombination; DNA molecules or genotypes or phenotypes having genetic information from two or more sources

**recombinant DNA:** DNA molecules spliced from DNA derived from more than one source

**recombinant DNA technology:** the methods employed to amplify recombinant DNA in a suitable host in order to obtain large amounts of the DNA or its products

**recombination:** the occurrence of progeny with combinations of alleles other than those in the parents, due to crossing over between linked genes

**recombination frequency:** the value found by dividing the number of recombinants by the total number of progeny and used as a guide in assessing the relative distance separating loci on a genetic map

**recombination index:** an estimate of genotypic diversity based on the chiasma or crossover frequency divided by the number of chromosome pairs in the individual or species, with higher values indicating greater diversity potential

**recombination intermediate:** a molecular configuration of duplex DNA believed to indicate a stage in crossing over leading to recombinant molecules or genotypes

**regulation of gene expression:** the modulation through control mechanisms of transcription and translation leading to phenotype development

**regulatory gene:** a gene whose primary function is control over expression of other genes by modulating the synthesis of the products of these genes

**renaturation kinetics:** the pattern of reassociation for a given sample of single-stranded DNA to produce duplexes, which indicates genome size and complexity; *see also* $C_0t$ **plot**

**repetitive DNA:** DNA consisting of repeated nucleotide sequences of variable lengths, occurring in hundreds, thousands, or millions of repeated sequences; sometimes referred to as satellite DNA

**replica plating:** imprinting of cells on (replica) plates containing different factors, by transfer from a master plate of cells adhering to velvet covering a cylindrical holder

**replication:** a duplication process requiring copying from a template

**replication fork:** the Y-shaped region of duplex DNA marking the site where replication is in progress

**replicon:** a unit of replication that has a unique origin sequence

**repressible enzyme:** one of a class of enzymes whose synthesis is decreased or stopped in the presence of an increased concentration of specific metabolites, through transcriptional control

**repressor:** the protein that is specified by a repressor gene and regulates bacterial operon transcription through binding to the operator site or leaving the site open for RNA polymerase to catalyze transcription of adjacent structural genes

**repressor gene:** a regulatory gene that encodes a repressor protein which modulates transcription of bacterial structural genes in particular operons

**repulsion:** *see* **trans arrangement**

**resistance transfer factor (RTF):** the portion of R plasmids that mediates transfer of the plasmid from cell to cell, the other portion of the R plasmid being *r* determinant(s)

**restriction enzyme:** one of a class of endonucleases that cuts both strands of duplex DNA at sites of particular base-pair sequences

**restriction fragments:** the cut pieces of duplex DNA arising from restriction enzyme action

**restriction map:** a physical map of the gene or genome derived from the ordering of restriction fragments produced by restriction enzymes

**retrovirus:** one of a class of RNA viruses of animal species, which copies DNA from RNA genomes using the specific enzyme reverse transcriptase, and which is often oncogenic

**reverse mutation:** a genetic change that restores the original phenotype of an individual by restoring the original codon(s) of the gene sequence; *see also* **pseudowild**

**reverse transcriptase:** an RNA-dependent DNA polymerase that guides the synthesis of complementary DNA from RNA templates, which is a reversal of the usual situation involving synthesis of RNA complements along DNA templates in transcription (hence the name of the enzyme)

**ribonucleic acid (RNA):** any of a type of nucleic acid characterized by the component sugar (ribose) and by the occurrence of uracil rather than thymine as the fourth kind of base; the three major classes of RNA are messenger RNA, ribosomal RNA, and transfer RNA, and another class is that of RNA genomes in certain viruses

**ribosomal DNA (rDNA):** the repeated genes specifying ribosomal RNA, usually found clustered in the nucleolar-organizing region of nucleolar-organizing chromosomes in eukaryotic species

**ribosomal RNA (rRNA):** the class of ribonucleic acid molecules that are present in both prokaryotes and eukaryotes in the small and large subunits of the ribosome, which functions in translation

**ribosomal subunit:** one of two classes of structures comprising the ribosome, one class being small subunits and the other being large subunits; each ribosome is made from one small and one large subunit

**ribosome:** a complex structure that is comprised of RNAs and proteins and is the site of translation in the cytoplasm, mitochondria, and chloroplasts when aggregated as polysomes

**R-loop mapping:** a method of determining nucleic acid complementarity from electron micrographs of slightly denatured duplex DNA hybridized with single-stranded RNA molecules, in which duplex DNA-RNA hybrid regions are seen at the looped-out portions of DNA produced by denaturation

**RNA:** *see* **ribonucleic acid**

**RNA polymerase:** one of a class of enzymes that catalyzes the formation of RNA molecules from DNA templates during transcription

**RNA primer:** a short sequence of ribonucleotides that is located at the 5′ end of a segment of replicating DNA at the replication fork and is required for DNA polymerase III to begin guiding DNA strand synthesis; the priming segment is excised by DNA polymerase I afterward, and this enzyme fills in the gap with deoxyribonucleotides which are then ligated to the completed portions of the DNA strand

**roentgen (r):** the quantity of ionizing radiation that liberates 2.083 × $10^9$ ion pairs in a cubic centimeter of air or approximately two ion pairs per cubic micrometer of a substance such as nucleic acid or protein

**rolling circle mechanism:** a process of replication of duplex DNA in which multiple lengths of DNA sequences are synthesized and are then cut by endonucleases, after which the linear molecules are circularized through complementary base pairing of their redundant termini ("sticky" ends)

**R plasmid:** an autonomously replicating duplex DNA molecule that is comprised of a resistance transfer factor (RTF) and one or more *r* determinants and that can be transferred from one bacterial cell to another and can thus influence cellular phenotypic expression

**rRNA:** *see* **ribosomal RNA**

***S* period:** the interval during a cell cycle when DNA replication occurs

**S value:** a quantitative measure expressed in Svedberg units of the rate of sedimentation of a given substance in a centrifugal field, based upon the sedimentation coefficient *s; s* of 1 × $10^{-13}$ sec is defined as one Svedberg unit (S)

**salivary gland chromosomes:** giant polytene chromosomes found in the interphase nuclei of the salivary gland cells in dipteran larvae, such as *Drosophila* and *Chironomus*

**satellite DNA:** a population of DNA molecules that sediments in a CsC1 density gradient separately from the main band of nuclear DNA of a cell-free preparation, and usually is repetitive nuclear DNA or organellar DNA

**secondary structure:** the local structure of a polypeptide or nucleic acid chain arising from chemical bonding between nearby residues to produce forms such as stem-and-loop or cloverleaf RNAs and α-helix configurations in polypeptides

**second-division segregation:** the separation of a pair of alleles into different nuclei during the second meiotic division, due to crossing over between the gene and centromere of the homologous chromosomes

**selection:** the process that determines the relative proportions of different genotypes in the propagation of a population

**selection coefficient *(s)*:** a measure of the disadvantage of a given genotype in a population

**selection pressure:** the effectiveness of natural selection in altering the genetic composition of a population over successive generations

**selection strategies:** different ways in which natural selection influences the genetic composition, or gene pool, of a population over successive generations; examples include **normalizing natural selection,** which acts to reduce or eliminate disadvantageous alleles or genotypes with lowered fitness; **directional natural selection,** which leads to an increase in the proportion of a favored genotype and a reduction in the frequency of less favored genotypes in relation to changing environmental conditions; **diversifying** or **disruptive natural selection,** which leads to greater diversity in the gene pool, since two or more classes of genotypes are greatly favored over other classes of genotypes in a heterogeneous environment; and **frequency-dependent natural selection,** which acts by decreasing the frequency of the more common phenotypes and increasing the frequency of the rare phenotypes

**selective disadvantage:** a less favored state of one or more genotypes in a population, which can be quantitated by the selection coefficient *s*

**self-assembly:** the organization of structure in the absence of a parent template or structure, occurring only by interactions between assembling components

**semiconservative replication:** the usual mode of DNA synthesis in which each parental template strand of duplex DNA guides the synthesis of a new complementary partner strand, producing two duplexes each comprised of one parental and one newly synthesized strand

**sense strand:** the coding strand of duplex DNA

**Sex Balance theory:** a proposal stating that sex is determined at fertilization by the ratio of X chromosome number to number of autosome sets, where a ratio of 1 or greater leads to females and a ratio of 0.5 or less leads to males

**sex chromatin:** *see* **Barr body**

**sex chromosome:** any chromosome involved in sex determination, such as the X and Y chromosomes

**sex determination:** the mechanism in a given species by which the developmental program is established specifying the particular gonads (structures producing the gametes) in an individual

**sexduction:** the process of merodiploid production through the mediation of the modified fertility factor F′, which carries a fragment of bacterial genes into another bacterial cell during conjugation in *E. coli*

**sex-influenced inheritance:** a pattern of allelic transmission in which different phenotypes are expressed in males and females of the same genotypes such that one allele is dominant in one sex but is recessive in the other sex

**sex-limited inheritance:** a pattern of allelic transmission in which one of the phenotypes is expressed in only one sex

**sex-linked inheritance:** a pattern of transmission involving genes on a sex chromosome, such as X-linked inheritance of genes located on the X chromosome

**sexual reproduction:** reproduction involving the fusion of gametes that are produced by meiosis or from cells arising by meiosis

**sigma ($\sigma$) factor:** a polypeptide subunit of the RNA polymerase in *E. coli* that serves to recognize specific binding sites in the promoter sequence for the initiation of transcription

**sigma virus:** a virus that confers $CO_2$ sensitivity upon *Drosophila melanogaster*

**single-burst experiment:** a procedure for recovering phage progeny from individual host cells upon lysis following infection

**single-strand binding (SSB) protein:** a noncatalytic protein that binds cooperatively to single-stranded DNA regions, leading to unwinding of the double helix at the replication fork and to tautness of the opened single-stranded regions

**sister chromatid exchange (SCE):** a result of crossing over between sister chromatids of a replicated somatic chromosome, visible as a harlequin pattern of reciprocal heavily and lightly stained chromatid segments

**somatic cell genetics:** genetic analysis of the genome using somatic (body) cells grown in culture, involving cell fusions and random loss of particular chromosomes from the hybrid cells

**somatic cell hybrid:** the product of fusion between two somatic cells in culture

**somatic cell hybridization:** the process of fusion between somatic cells to produce somatic cell hybrids for genetic analysis

**somatic mutation:** a mutation occurring in any body cell not destined to become a gamete or equivalent reproductive cell

**specialized transduction:** a process of unidirectional gene transfer in bacteria in which a specialized transducing phage carries some of its own genes and some of its host's genes into the recipient cell. The host's genes are ones that are adjacent to the site of integration of the prophage form of the temperate virus

**speciation:** the evolutionary development of new species from an originally unitary population or parental species

**species:** one or more interbreeding populations whose members can exchange genes with one another but not with members of populations belonging to another species

**spindle:** an aggregate of microtubules, seen during nuclear division, that functions in the alignment and movement of chromosomes at metaphase and anaphase

**splicing:** joining together of component parts derived from separated sources or locations

**spontaneous mutation:** a naturally occurring mutation

**spore:** a single-celled reproductive unit that can give rise to a new individual directly upon germination

**sporophyte:** the diploid phase or individual of plants that develops from a zygote and, in turn, produces sexual spores by meiosis

**standard deviation ($\sigma$):** a measure of the variability in a population of items—the square root of the variance ($\sigma^2$)

**standard error:** a measure of variation of a population of mean values—the standard deviation of a population mean divided by the square root of one less than the number of items in the population

**statistics:** the discipline concerned with the collection, analysis, and presentation of data, and dependent upon the application of probability theory

**structural aberration:** a physical modification of the chromosome, including deletion, duplication, inversion, and translocation, all of which involve breakage and subsequent rejoining of broken chromosome parts to produce a new arrangement

**structural gene:** the sequence of DNA encoding a polypeptide, rRNA, or tRNA

**submetacentric:** referring to a chromosome whose centromere is submedian in location, making one arm somewhat longer than the other

**supercoiled DNA:** coiling of the helix-coil conformation of an extended double helix of DNA; a coil superimposed on a coiled duplex

**supergene:** a chromosomal segment protected from crossing over and so transmitted from generation to generation in the aggregate as if it were a single gene

**suppressive petite:** a respiratory deficient yeast extranuclear mutant that produces up to 99% all-petite tetrads in crosses with respiratory sufficient wild-type strains

**suppressor mutation:** a mutation at a genetic locus, separate from the site of a primary mutation, that reverses the effect of the primary mutation and produces the wild-type phenotype in the double mutant; *see also* **pseudowild**

**synapsis:** the specific pairing of homologous chromosomes or chromosome regions, typically occurring during zygonema of prophase I of meiosis but also evident in certain somatic cells, such as dipteran larval cells having paired polytene chromosomes

**synaptonemal complex:** a structural component that is situated between a pair of synapsed meiotic chromosomes during pachynema in prophase I of meiosis and holds the chromosomes in register during crossing over

**synteny:** the assignment of genes to the same chromosome in somatic cell hybrids, as determined by concordance of phenotype and particular chromosomes occurring in a clone; such genes may or may not be linked in the normal genome

**$T_m$:** midpoint melting temperature; the temperature at the midpoint of transition of duplex DNA molecules to single strands during melting (denaturation)

**TATA box:** a sequence in the promoter of six or seven nucleotides consisting of adenine- and thymine-containing residues which are a recognition site for the binding of RNA polymerase in the initiation of transcription

**telocentric:** referring to a chromosome whose centromere is located at one end, so that the chromosome has only one arm

**telophase:** the final stage of mitosis or meiosis, when nuclear reorganization takes place

**temperate phage:** an avirulent bacterial virus that infects but rarely causes lysis of the host cell, and often exists in the host in prophage form

**temperature-sensitive mutation:** a conditional mutation that is manifested only at certain temperatures

**template:** the macromolecular mold or blueprint for the synthesis of a complementary molecule

**termination codon:** a punctuation codon that signals the end of a translation; a stop codon

**terminator:** the terminal sequence which is located at the 5′ end of the coding strand of DNA and is not transcribed into pre-messenger RNA

**tertiary structure:** the three-dimensional folding of a polypeptide or nucleic acid chain brought about by interactions among side-groups or other residues at some distance from one another in the primary structure of the molecule

**testcross:** a cross between a heterozygote and an individual homozygous for the recessive alleles of the gene(s) in question, yielding progeny in which each genotype class is represented by a different phenotypic class

**tetrad analysis:** a genetic analysis of recombination and independent assortment of members of pairs of alleles by the study of tetrads, each of which is a set of four products of a single meiotic cell

**tetratype (T) tetrad:** a set of meiotic products that includes both parental genotypes and a pair of reciprocal recombinant genotypes

**thalassemia:** a group of human blood disorders characterized by alterations in the amounts produced of one or more kinds of hemoglobin; *see* **$\alpha$ thalassemia, $\beta$ thalassemia**

**three-point testcross:** a cross between individuals that are heterozygous or homozygous recessive for three pairs of alleles of linked genes, designed to determine the order and distance of the three loci for a genetic map of the chromosome or linkage group

**thymidine:** the deoxyribonucleoside of thymine

**thymidine kinase (TK):** an enzyme catalyzing the phosphorylation of thymidine monophosphate or thymidine diphosphate to thymidine diphosphate and thymidine triphosphate, respectively

**thymine:** a pyrimidine base found in DNA

**thymine dimer:** *see* **pyrimidine dimer**

**topoisomerases:** DNA replication enzymes that catalyze reactions involving nicking and closing of duplex DNA; topoisomerase I is an unwinding enzyme that relaxes supercoils in duplex DNA, and topoisomerase II (or, gyrase) induces relaxed DNA to undergo supercoiling in a reaction that requires ATP hydrolysis

**totipotency:** the capacity of a nucleus to underwrite a complete program of development of an organism

**trailer:** the nucleotide sequence near the 3′ end of the messenger RNA, just downstream of the encoded region, which is usually not translated at the ribosome

**trans arrangement:** repulsion arrangement of allelic or nonallelic genes in which both mutant and wild-type loci are on each homologous chromosome in the individual ($a+/+b$), in contrast with cis or coupling ($a\,b/+\,+$)

**transcription:** a process by which the nucleotide sequence of the coding strand of DNA is copied into a single-stranded complementary molecule of RNA, under the catalytic direction of an RNA polymerase

**transcriptional control:** a mechanism by which regulation of gene expression is accomplished by modulating transcription (on, off, increase, decrease)

**transdetermination:** a change in the determined state of a larval imaginal disk in *Drosophila* such that another organ develops instead of the organ originally determined for the adult insect

**transduction:** the unidirectional transfer of genes from one bacterial cell to another through the agency of a virus, which acts as a vector for transfer and is called a transducing phage; *see also* **generalized transduction, specialized transduction**

**transfer RNA (tRNA):** the RNA molecule that carries an amino acid to a specific codon in messenger RNA during polypeptide translation at the ribosome

**transformant:** a cell or individual that is genetically altered upon receiving DNA from a donor during transformation

**transformation:** 1. the heritable modification of the properties of a recipient cell by transfer of DNA from a donor cell; 2. the modification in properties of growth and movement of normal cells in culture after the introduction of a potentially oncogenic virus, presumably analogous to tumor induction in an organism

**transforming response:** the alteration in growth and movement of normal cells in culture after the introduction of a presumably oncogenic virus, a state believed to be analogous to tumor induction in an organism

**transition:** a mutation caused by the substitution in DNA or RNA of one purine by the other or one pyrimidine by the other

**translation:** the process by which amino acids are joined into a polypeptide chain at the ribosome, under the direction of a messenger RNA transcript of encoded DNA

**transcriptional control:** a mechanism by which regulation of gene expression is accomplished by modulating transcription (on, off, increase, decrease)

**translocase (G factor):** the ribosomal enzyme responsible for the translocation of the lengthened peptidyl-tRNA from the A site to the P site of the ribosome during translation

**translocation:** a structural aberration of chromosomes, involving an interchange between nonhomologous chromosomes (reciprocal translocation) or the addition of all or part of one chromosome to a nonhomologous chromosome (simple translocation)

**transposable genetic element:** a unique DNA sequence that can move from one site to another in a genome through insertions and excisions and influence gene expression; a movable genetic element

**transposon (Tn):** a transposable genetic element which is more than 2000 base pairs long and includes genes for insertion into a chromosome as well as genes unrelated to insertion; *see also* **insertion sequence (IS)**

**transversion:** a mutation caused by the substitution in DNA or RNA of a purine for a pyrimidine, or vice versa

**trihybrid cross:** interbreeding between individuals that differ in three different pairs of alleles

**trisomy:** the condition in which one chromosome is present in three copies instead of the usual number, producing a trisomic cell or individual

**trisomy-21:** *see* **Down syndrome**

**tritium:** $^3H$, a radioactive isotope of hydrogen

**tRNA:** *see* **transfer RNA**

**tumor:** an uncontrolled growth in the organism

**tumorigenic:** responsible for tumor induction in an organism

**tumorigenesis:** the induction of a tumor in an organism; *see also* **carcinogenesis**

**Turner syndrome:** a clinical condition of human females who have only one X chromosome (*45,X*)

**ultraviolet radiation (UV):** that part of the invisible electromagnetic spectrum of radiation with wavelengths between 100 and 400 nanometers; a physical mutagenic agent

**unequal crossing over:** an exchange between chromosomes or DNA strands that produces one product with missing genes or sequences and the other copy with duplicated genes or sequences

**uniparental inheritance:** a pattern of nonchromosomal inheritance in which the progeny receive nonchromosomal genes from only one of the two parents during sexual reproduction

**unique copy sequence:** a DNA sequence that occurs once or a few times in a genome, in contrast with repetitive DNA, which occurs in many copies per genome

**uracil:** a pyrimidine base found in RNA

**uridine:** the nucleoside of uracil

**variable region** (variable segment): the amino-terminal sequence of amino acids in the light and heavy chains of an immunoglobulin molecule, and the portion of the antibody which combines with an antigen in an immune response

**variance (V):** the mean of the squared deviations from the population mean ($\sigma^2$), which is obtained from all values in a population expressed as plus and minus deviations from the population mean

**vector:** an agent that transfers an item from one host to another

**virulence:** the capacity to initiate infection

**virulent phage:** a phage that causes lysis of the host bacterium; a lytic phage, in contrast with a temperate phage

**virus:** an acellular organism that requires a living host for its reproductive cycle

**wild type:** the most common allele or phenotype, or one arbitrarily designated as "normal"

**wobble hypothesis:** a hypothesis developed by F.H.C. Crick to explain how a tRNA may recognize two or more codons in mRNA; the proposition that the 5′ base in the anticodon triplet of tRNA has less restraint in pairing with the 3′ base of the mRNA codon and can thus pair with more than one base in certain instances because of this "wobble" or property of reduced restraint

**X chromosome:** the sex chromosome found in two copies in the homogametic sex and in one copy in the heterogametic sex

**X inactivation:** a phenomenon in which all the X chromosomes but one in a nucleus become genetically inactive by virtue of condensation to the heterochromatic state, and the one remaining active X chromosome is responsible for expression of X-linked genes in the cell or individual; also referred to as Lyonization in recognition of the hypothesis having been proposed by Mary Lyon

**X-linked inheritance:** a pattern of allelic transmission involving genes on the X chromosome, usually evident from the production of nonidentical progeny in reciprocal crosses, in contrast with identical reciprocal progeny in autosomal inheritance

**X-ray diffraction:** a crystallographic method by which diffraction patterns produced by x-ray scattering from crystals of a substance are used to determine the three-dimensional structure of the atoms or molecules in the crystal

**X-rays:** that part of the invisible electromagnetic spectrum of radiation with wavelengths shorter than UV but longer than gamma radiation, which are high-energy ionizing radiations; a physical mutagenic agent

**Y chromosome:** the sex chromosome found only in the heterogametic sex; in mammals, the male-determining sex chromosome

**zygonema:** the substage of prophase I of meiosis when homologous chromosomes begin to synapse

**zygote:** the product of the fusion of gametes in sexual reproduction, such as the fertilized egg produced by fusion of egg and sperm

# INDEX

*Pictorial information indicated by italics.*